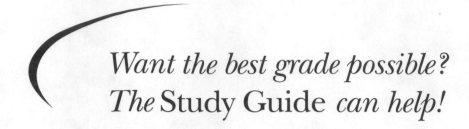

Want the best grade possible?
The Study Guide can help!

COMPLETE STUDY GUIDE
(0-669-34081-2)

The *Complete Study Guide* contains many useful features to help you succeed in your biology course. For every chapter in your text, the *Guide* provides supplemental exercises that reinforce what you've learned and provide the extra practice that can help you get better grades.

- Section overviews highlight the most important topics in each chapter.
- Key Terms help you master the vocabulary of biology.
- Exercises for Chapter Mastery review the key concepts and content of each chapter.
- Quick Recall questions allow you to check yourself, so you'll know before the test which topics you need to spend the most time studying.
- Recent magazine articles for each major section of your text keep you up-to-date on current research and exciting discoveries in biology.

The *Guide* doesn't cost much, but it will give you valuable help in preparing for quizzes and exams. And better preparation can result in better grades.

Look for the Complete Study Guide *in your bookstore. If you don't find it, check with your bookstore manager or call D. C. Heath toll free at 1-800-334-3284 to place an order. As a student customer, you will be charged the retail price of the item plus $2.00 shipping and handling, plus state tax where applicable (tell the operator you are placing a #1-PREFER order).*

BIOLOGY

Discovering Life

BIOLOGY

Discovering Life

SECOND EDITION

Joseph S. Levine
Boston College

Kenneth R. Miller
Brown University

D. C. HEATH AND COMPANY
Lexington, Massachusetts   Toronto

Address editorial correspondence to:
D. C. Heath and Company
125 Spring Street
Lexington, MA 02173

Dedication
To Carol and Bob, Marion and Ray, our parents, who gave us our lives, the best of our values, our love of learning, and our appreciation for the importance of education in a complex world.

Acquisitions Editor: Elizabeth Coolidge-Stolz
Editorial Director: Kent Porter Hamann
Developmental Editor: Barbara Withington Meglis
Production Editor: Kathleen A. Deselle
Designer: Cornelia Boynton
Photo Researcher: Sue McDermott
Art Editor: Gary Crespo
Production Coordinators: Lisa Merrill, Dick Tonachel
Artists: Patrice Rossi, Mikki Senkarik, Marlene DenHouter, Arleen Frasca, Lynette Cook, Marcia Smith, Charles Boyter, Andrew Robinson, Michael Woods, Illustrious Inc., Sanderson Associates, Maryland CartoGraphics, Inc., TecDocPub, Inc., Marsha Goldberg, Peggy Jefferson
Cover: The cover features a close-up of an adult whooping crane (© Perry Conway) and a wider view of wild cranes on their wintering grounds in Aransas, Texas (© Tom Bean).

International Standard Book Number: 0–669–33494–4

Library of Congress Catalog Number: 93–71201

10 9 8 7 6 5 4 3 2 1

This book is printed
on recycled paper

Action and Theory: Discovering Life

ACTION

With an adult height of 4–5 feet, the whooping crane is North America's tallest native bird. The cranes, with white bodies, black wing tips, and red-capped heads on long, elegant necks, are beautiful. The whooping crane is also one of the continent's most famous endangered species, with fewer than thirty birds remaining before people rallied to save it. Even so, after more than fifty years of intensive scientific effort on the cranes' behalf, it is still unclear whether whoopers will survive in the wild.

In the early 1800s, John James Audubon observed the arrival of a family of 20–30 birds at their southern wintering ground and painted an adult male for *The Birds of America*, which he published in 1827. At that time, the birds' breeding grounds probably ranged from the Hudson Bay westward across Canada, to as far south as the current states of Nebraska and Iowa. Their wintering

grounds lay along the east coast, from New Jersey to Florida, on the Gulf of Mexico, and as far southwest as central Mexico.

By 1900, the conversion of western North America from wilderness to farmland and ranches had destroyed much of the cranes' wetland habitat. In the following years, scientists observed the whoopers disappearing from their wintering grounds until only two populations were left, one migratory group on the Texas Gulf coast and one nonmigratory flock in Louisiana. In 1936, a scientist with the U.S. Bureau of Biological Survey (the forerunner of the U.S. Fish and Wildlife Service) recommended that the Texas habitat be purchased as a wildlife refuge. Thus, the U.S. government became active in whooping crane preservation efforts with its establishment of the Aransas National Wildlife Refuge on December 31, 1937.

During the winter of 1941, scientists counted only 22 whooping cranes, 6 of which were in Louisiana, where they died out by 1949. Biologists from the government and the National Audubon Society worked hard to preserve the Texas habitat, to try to find the northern breeding grounds of the Aransas flock, and to learn more about whoopers' natural history. In June 1954, an observer in a Canadian Forest Service helicopter returning from a fire in Wood Buffalo Park, Northwest Territories, spotted a pair of adult cranes and one juvenile bird. The last remaining breeding ground of the whooping crane had been found.

Both the Canadian and U.S. governments increased efforts to protect the Wood Buffalo and Aransas sites while scientists learned more about the birds' behavior from monitoring both refuge sites and tracking birds visually on the 2500-mile migration (see map on p. vi). In 1965, when scientists believed field work only was insufficient to increase the cranes' population significantly, they used knowledge of the birds' breeding behavior to try a captive breeding program.

Biologists had learned that most whooping cranes lay two eggs, but only one juvenile bird survives to migrate south. The scientists, with government support, decided to remove one egg from some two-egg nests, transport the eggs to the Patuxent Research Center in Maryland (see map), and try to raise birds there. By the mid-1970s, Patuxent-raised whooping cranes were laying their own eggs through selective artificial insemination.

Because the Patuxent program proved successful, scientists decided to use similar techniques to establish a second whooping crane population in the wild. In 1975, biologists took eggs to Gray's Lake National Wildlife Refuge in Idaho, where they switched them with sandhill crane eggs. The unknowing foster parents raised both sets of chicks and led them to wintering grounds in New Mexico (see map). As the surviving whooping cranes matured, they tended to stay together. It is unknown if the whoopers recognized that they were different from their sandhill neighbors or not, but none of them has passed the first step toward a self-sustaining population by mating successfully. This group now numbers 9 whoopers.

After reviewing the success in raising whooping cranes in captivity at Patuxent and the apparent failure of the techniques used in the wild in Idaho, biologists developed a new plan for establishing a nonmigratory whooping crane population in Kissimmee Prairie, Florida (see map). In the winter of 1992, biologists took 14 juvenile whoopers, 6 hatched at Patuxent and 8 hatched at the International Crane Foundation in Wisconsin (see map), and released them in Kissimmee Prairie. At first, scientists kept the birds in a protected environment. Over the course of the winter, as the cranes adjusted to their surroundings and to each other, the biologists allowed them to range more freely through the reserve, forcing the young cranes to become increasingly independent.

As we write this in the spring of 1993, biologists are preparing to establish a third captive flock. Some juvenile cranes hatched at Patuxent and some from the International Crane Foundation are being sent to Calgary, Alberta (see map), to become the foundation of a Canadian captive breeding group. In the future, biologists may try to establish a new migratory population on breeding grounds in Canada.

There are now approximately 270 adult whooping cranes and 50 chicks in the world, including both wild and captive birds. They are descended from the one breeding population left in the wild, which breeds at Wood Buffalo Park in Canada and winters at Aransas, Texas. (The marshes of Aransas are shown on the cover.) If the restoration programs are successful in the long term, there is hope of establishing 3–4 independently reproducing whooping crane populations living free in North America.

THEORY

The story of the whooping cranes is interesting for its own sake, but we believe it also points out many of the themes vital to understanding the topics of introductory biology:

1. The process of scientific investigation includes both observation and experimentation. The techniques used by a group of biologists change over time as they learn from previous work and take advantage of new technologies. In the first monitoring of crane migration, for example, spotters along the suspected migration route counted birds as they passed through. Current monitoring can be done by radio tracking of banded birds.

2. Scientists are an integral part of society. Biologists were able to capture public attention and government support in the 1930s because the public had become reawakened to the environment through the photos of Ansel Adams and the essays of Aldo Leopold, to name only two figures of the era. It is still important to look at cultural values when studying what scientific research receives government funding, what work is done unsupported, and what, if any, work is banned.

3. The first biologists were limited to observing nature and asking questions based on their observations. By the nineteenth century, scientists could use newly developed tools and could benefit from a new mental

Map showing locations: Wood Buffalo, NWT; Hudson Bay; Calgary, AB; Gray's Lake, ID; Baraboo, WI; Patuxent, MD; Bosque del Apache, NM; Mississippi; Rio Grande; Atlantic Ocean; Kissimmee Prairie, FL; Aransas, TX

openness to pose many questions, which they addressed by both observations and planned experiments. The quantum leaps in biological understanding of the twentieth century arose from increasingly complex experiments that were drawn from this continuing tradition of observing, questioning, and experimenting.

4. Modern biology rests on two related themes: genetics and evolution. Genetics examines the makeup of individuals while evolution studies the flow of inheritance over many generations. Biologists are concerned that today's whooping cranes are descended from a small number of related birds, which is why they want to establish several independently reproducing populations. In the short term, if disease or other natural disaster destroys one population, even though the others are closely tied genetically, geographic separation may save them. In the long term, each population will change genetically, and the random change that marks each will be distinct from the change occurring in the other populations. The overall increase in genetic variation for the species will give the whooping crane population a better chance to adapt to diverse evolutionary pressures.

5. In order to have the best understanding of the genetic makeup of an individual or the variation within a species, we need to take advantage of recent work to view biology in the light of biochemistry. Chemistry may seem more abstract than biology, but all organisms are composed of a wondrous array of simple and complex molecules. Biologists can assess individual genetic profiles and measure overall variation for an endangered species by analyzing DNA, the complex molecules that carry inheritance from parent to offspring through the generations.

6. As our understanding of biology progresses, we see that the fields within biology are increasingly interdependent. For instance, laboratory techniques have had a profound impact on field biology. DNA fingerprinting technology is now used by scientists starting captive breeding programs for other species. They use this sensitive technique to establish matches between the least-related prospective parents, optimizing the genetic variation in offspring. Many lab researchers continue to look to biological diversity for molecules with specific behaviors. Taxol, an extremely promising anticancer drug, was originally isolated from the Pacific yew tree. (See the story of Hawaiian geese captive breeding in Chapter 22 and discussions of taxol in Chapters 7 and 41.)

Acknowledgments

We thank Nell Baldacchino and George Gee of the U.S. Fish and Wildlife Service, Patuxent Wildlife Research Center, for graciousness in our telephone conversations about current work with whooping cranes and for reading "Action and Theory" to ensure accuracy in the information about whoopers.

Bibliography

Audubon, John James. *The Birds of America: From Original Drawings*. Foreword and captions by William Vogt. New York: Macmillan, 1961.

McNulty, Faith. *The Whooping Crane: The Bird that Defies Extinction*. New York: E. P. Dutton, 1966.

Sanders, Scott, ed. *Audubon Reader: The Best Writings of John James Audubon*. Bloomington, IN: Indiana University Press, 1986.

U.S. Government and the International Crane Foundation. Various pamphlets, fact sheets, and other informational materials made available through the courtesy of the Aransas National Wildlife Refuge staff.

Further Readings

Bergman, Charles. *Wild Echoes: Encounters with the Most Endangered Animals in North America*. New York: McGraw-Hill, 1990. An English professor and nature author, Bergman movingly writes of his experiences with the biologists who work with endangered species, as well as his encounters with the animals themselves.

Doughty, Robin. *Return of the Whooping Crane*. Austin, TX: University of Texas Press, 1989. This comprehensive book on whoopers is recommended by the staff at Aransas National Wildlife Refuge.

Discovering Life: Organizing a Textbook

Each instructor balances the time available for lecture with his or her course goals in order to develop a curriculum. We realize most instructors will not teach all major topics, so we organized the seven sections of our complete text both to be logical if the entire progression is followed and to be sufficiently flexible for instructors to teach the sections they wish in the order that makes most sense for them. The themes discussed in Action and Theory: Discovering Life led to the following organization:

PART 1 Introduction to Biology

The opening two chapters explain how the processes of scientific inquiry work and how scientists have an impact on, and receive an impact from, their society. We believe that if students do not clearly understand this procedural and historical background, the rest of the book will not be as accessible or as meaningful as it should be.

PART 2 Organisms and Ecology

The first biological material to be presented was also the first to be studied historically. Although we teach ecology from the modern perspective, students see its historical development from observation, through experimentation, to current theory and research, making it easier for them to understand Part 3, which introduces the key themes of evolution and genetics by looking chronologically at the work of Darwin and Mendel.

PART 3 Evolution and Mendelian Genetics

By presenting Darwin and Mendel in their historical context, students see from what background each worked, what contributions to scientific knowledge each made, and where each came to obstacles in understanding he could not overcome. We have found that this intellectual context gives students a greater appreciation for molecular biology: they understand why they know more than either Darwin or Mendel and how our current questions about cellular structure and function developed over time.

PART 4 Molecules of Life

These chapters present basic chemistry, biochemistry, and molecular biology, as well as the experimental foundations for future molecular understanding of genetics, evolution, and artificial evolution (genetic engineering).

PART 5 Diversity of Life

Part 5 starts with biological classification and geologic history and then builds a context for understanding biodiversity through a discussion of the probable origins of life on Earth, evolutionary changes over time, and an examination of current species. We use the molecular perspective established in Part 4 to show how recent laboratory advances are used in the ongoing field work of cataloging and protecting species.

PARTS 6 AND 7 Plant Systems and Animal Systems

We use both the macroscopic (ecological and evolutionary) and microscopic (cellular and molecular) perspectives to study the functioning of plants and animals. In Part 6, we look at how plants have adapted to the needs to reproduce, develop to maturity, and survive in their environments. In Part 7, we take a comparative view of how different animals have responded to the same challenges, placing most of our attention on the mammals, especially humans.

What Students and Professors Should Know About this Book

We wrote this book to be read and enjoyed. We use a number of terms because we need precise language to describe the phenomena of biology. However, we have used only those terms we feel you need to learn at this level. In other places, we have used nontechnical explanations to teach concepts. We also have tried to emphasize the *process* and *development* of biological knowledge in addition to the knowledge itself. Whether or not you study biology after this course, it is vital to understand that science is dynamic. Not only do the "facts" change over time, but our understanding of them changes as well. Learning how to learn and relearn is crucial for anyone interested in biology at the casual or professional level.

The seven parts of *Biology: Discovering Life*, Second Edition, open with two pages that give a context for the section's topics and lay out its organization. If you read the opening pages before you read the chapters, you will understand how the chapters fit together. Each chapter should be read thoughtfully, not just memorized. We recommend you read each chapter and then use the Study Focus at the end to organize your review. If you feel more comfortable, you can read the Study Focus first. Chapters can be reviewed in detail by checking your recollections against the chapter headings, which are also part of the complete table of contents in the front of the book. If the meaning of a heading seems unclear, you should reread that part of the chapter.

Finally, we hope that through your course, you will share the excitement, vitality, and urgency of biology that bring biologists to the field and keep them there. Please let us know your comments on the textbook so we can improve it for future students.

Acknowledgments

If the appearance of the second edition of a book can be taken as a vote of confidence, then we have many people to thank for their votes. First and foremost, we are grateful to the instructors who chose the first edition of *Biology: Discovering Life* for their course. We take your

selection of this book as a high personal trust placed by one colleague in another, and we have done our best to earn your trust again in the second edition. We are also grateful to the thousands of students who have used the book, whose comments and suggestions have been invaluable in writing this new edition. On a personal level, we are especially grateful to Steve, Jody, Lauren, Tracy, and the other members of our families who have endured our single-minded fixation with this project for more months than we can remember.

Like all scientists, we owe enormous intellectual debts to those who have trained us, worked with us, shared their research and their ideas with us, and to those who have learned from us, our students. We are especially grateful to our scientific friends and colleagues who have contributed photographs, diagrams, and ideas to the new edition, and we trust that you will find your kindnesses reflected in a textbook that captures the excitement of biological research in the 1990s.

If writing a textbook sometimes seems like a solitary art, publishing a textbook is not. We will always be grateful to the extraordinary College Division at D. C. Heath. Our special thanks go to Kent Porter Hamann, whose confidence, support, and encouragement made *Biology: Discovering Life* a reality. We are grateful for the help, professionalism, friendship, and patience of our indispensable editors, Elizabeth Coolidge-Stolz and Barbara Withington Meglis. We extend our thanks and appreciation to the hard-working and capable Heath staff, including Kathleen Deselle, Production Editor; Cornelia Boynton, Designer; Sue McDermott, Photo Researcher; Gary Crespo, Art Editor; Lisa Merrill and Dick Tonachel, Production Coordinators; and Jim Porter Hamann, Marketing Manager. It is exhilarating and challenging to work with the best, and that is exactly the standard you have set for us. We thank you for the simple pleasure of working together on this project.

J. S. L.
K. R. M.

Reviewer Acknowledgments

Second Edition reviewers:

James Anderson, *DeAnza College*
Robert C. Anderson, *Idaho State University*
Paul E. Barney, Jr., *Penn State Erie*
William Barstow, *University of Georgia*
Gerald Bessey, *L.A. Valley College*
John S. Campbell, *Northwest College*
Linda Fleet Chapman, *University of Missouri*
Patricia B. Cox, *University of Tennessee*
Wynn Cudmore, *Chemeketa Community College*
Alfred Diboll, *Macon College*
William D. Elliott, *Hagerstown Junior College*
Thomas Emmel, *University of Florida*
Paul W. Gabrielson, *William Jewell College*
Marcel H. Gregoire, *Merrimack College*
John P. Harley, *Eastern Kentucky University*
Michael Hawkes, *University of British Columbia*
Anne Morris Hooke, *Miami University*
George Hudock, *Indiana University*
Thomas Hutto, *West Virginia State College*
Bill Jacobson, *Blue Mountain Community College*
Ronald Jenkins, *Samford University*
Vida C. Kenk, *San Jose State University*
Robert Kitchin, *University of Wyoming*
Cheryl Knox, *St. John's University*
Deborah M. Langsam, *University of North Carolina–Charlotte*
Patricia Parsley Lapennas, *St. Bonaventure University*
Gary Lipton, *Trenton State College*
James E. Mickle, *North Carolina State University*
Richard J. Montgomery, *Hagerstown Junior College*
Joseph Moore, *California State University–Northridge*
James W. Morrow, *Community College of Allegheny County*
Norman C. Negus, *University of Utah*
Robert Neill, *University of Texas at Arlington*
Sandra J. Newell, *Indiana University of Pennsylvania*
Gary Ogden, *Moorpark College*
Patricia Pagni, *Knoxville College*
Debra Pearce, *Northern Kentucky University*
Finn Pond, *Whitworth College*
Michael Renfroe, *James Madison University*
Quentin Reuer, *University of Alaska*
John Rushin, *Missouri Western State College*
Richard R. Ryno, *College of Marin*
Barbara W. Smigel, *Community College of Southern Nevada*
Bruce N. Smith, *Brigham Young University*
Timothy A. Stabler, *Indiana University–Northwest*
Judy Sullivan, *Antelope Valley College*
Donald L. Terpening, *Ulster County Community College*
Karen Trevors, *University of Waterloo*
Robin W. Tyser, *University of Wisconsin*
Joanne Westin, *Case Western Reserve University*

First Edition reviewers:

Joseph Allamong, *Ball State University*
Margaret Anglin, *St. Louis Community College at Meramec*
Lois Bailey, *Southeastern Community College*
Frank Baron, *Duquesne University*
May R. Berenbaum, *University of Illinois at Urbana–Champaign*

Joseph S. Bettencourt, *Marist College*
P. K. Bhattacharya, *Indiana University Northwest*
David L. Bishop, *Moorpark College*
Donald Bissing, *Southern Illinois University at Carbondale*
Robert V. Blystone, *Trinity University*
Richard Boohar, *University of Nebraska–Lincoln*
Maynard C. Bowers, *Northern Michigan University*
Jean A. Bowles, *Metro State College*
J. H. Brown, *University of Houston*
Peter Burn, *Suffolk University*
Thomas E. Byrne, *Roane State Community College*
William Cain, *University of Delaware*
Thomas R. Campbell, *Los Angeles Pierce College*
Joseph Chinnici, *Virginia Commonwealth University*
Norm Christensen, *Duke University*
David Cotter, *Georgia College*
Joe R. Cowles, *University of Houston*
Alfred H. Crawford, *Southern Florida Community College*
Judy Daniels, *Eastern Michigan University*
Manuel E. Daniels, Jr., *Tallahassee Community College*
Deborah S. Dempsey, *Northern Kentucky University*
Raymond Dillon, *University of South Dakota*
Linda Dion, *University of Delaware*
Nathan Dubowsky, *Westchester Community College*
Frank Duroy, *Essex County College*
M. Duvall, *Smithsonian Institution*
David Eberiel, *University of Massachusetts at Lowell*
H. W. Elmore, *Marshall University*
Susan Ernst, *Tufts University*
Robert C. Evans, *Rutgers University*
Darrel Falk, *Point Loma Nazarene College*
Carl D. Finstad, *University of Wisconsin–River Falls*
Jim Fowler, *State Fair Community College*
David J. Fox, *University of Tennessee, Knoxville*
Bernard L. Frye, *University of Texas at Arlington*
Ric Garcia, *Clemson University*
Michael S. Gaines, *University of Kansas*
Thomas C. Gray, *University of Kentucky*
Elizabeth C. Hager, *Trenton State College*
Thomas Earl Hanson, *Temple University*
Robert E. Herrington, *Georgia Southwestern College*
Carl Hoegler, *Marymount College*
Alfred J. Hopwood, *Saint Cloud State University*
Daniel Hornbach, *Macalester College*
Patricia J. Humphrey, *Ohio University*
Robert N. Hurst, *Purdue University*
Mary Keim, *Seminole Community College*
Donald L. Kimmel, Jr., *Davidson College*
Valerie M. Kish, *Hobart and William Smith Colleges*
F. M. Knapp, *Stetson University*
Helen G. Koritz, *College of Mount Saint Vincent*
James Lampky, *Central Michigan University*
John H. Langdon, *University of Indianapolis*
Ron W. Leavitt, *Brigham Young University*
Michael Lee, *Joliet Junior College*
James Luken, *Northern Kentucky University*
Joseph Marshall, *West Virginia University*
John Matsui, *University of California at Berkeley*
John Mattox, *Northeast Missouri Community College*
Fred McCorkle, *Central Michigan University*
Dorothy Minkoff, *Trenton State College*
David J. Morafka, *California State University, Dominquez Hills*

Robert L. Neill, *University of Texas at Arlington*
Mary Nossek, *Ohio University*
Thaddeus Osmolski, *University of Massachusetts at Lowell*
Joel Ostroff, *Brevard Community College*
Helen Oujesky, *University of Texas at San Antonio*
Gail R. Patt, *Boston University*
Peter Pedersen, *Cuesta College*
Kathryn Stanley Podwall, *Nassau Community College*
David M. Prescott, *University of Colorado*
Jeffrey Pudney, *Harvard Medical School*
Ralph Reiner, *College of the Redwoods*
Donald Reinhardt, *Georgia State University*
Robert Rinehart, *San Diego State University*
Robert Romans, *Bowling Green State University*
David Rose, *Trenton State College*
Peter E. Russel, *Chaffey Community College*
A. G. Scarbrough, *Towson State University*
Allen B. Schlesinger, *Creighton University*
Robert Schodorf, *Lake Michigan College*
Erik Scully, *Towson State University*
Richard W. Search, *Thomas College*
Margaret Simpson, *Sweet Briar College*
Susan Singer, *Carlton College*
Jerry W. Smith, *St. Petersburg Junior College*
Marshall Smith, *Los Angeles Mission College*
Beatrice L. Snow, *Suffolk University*
Gilbert D. Starks, *Central Michigan University*
Howard J. Stein, *Grand Valley State University*
Philip Stein, *State University of New York at New Paltz*
Robert W. Sterner, *University of Texas at Arlington*
Eric Strauss, *Tufts University*
Stephen Subtelny, *Rice University*
Gerald Summers, *University of Missouri–Columbia*
Marshall Sundberg, *Louisiana State University*
Daryl Sweeney, *University of Illinois at Urbana–Champaign*
Dan Tallman, *Northern State University*
William J. Thieman, *Ventura College*
Kathy S. Thompson, *Louisiana State University*
Bruce Tomlinson, *State University of New York at Fredonia*
Michael Treshow, *University of Utah*
Marenes Tripp, *University of Delaware*
Kent M. Van De Graaff, *Brigham Young University*
Pauline E. Washington, *Forest Park Community College*
Jean Werth, *William Paterson College*
George J. Wilder, *Cleveland State University*
Garrison Wilkes, *University of Massachusetts, Boston*
Wayne Wofford, *Union University*
Paul Wright, *Western Carolina University*
Tommy Elmer Wynn, *North Carolina State University*
Gerald W. Zimmerman, *University of Indianapolis*

Art Reviewers:

H. W. Elmore, *Marshall University*
Michael C. Kennedy, *Hahnemann University*
Richard E. Morel, *Strong House Incorporated*

Brief Contents*

*NOTE: Parts 1–4 (Chapters 1–22) are covered in the *Core Concepts* paperback version of the text. Part 5 (Chapters 23–29) is covered as Chapters 1–7 of the *Diversity of Life* paperback.

Contents

PART 3 Evolution and Mendelian Genetics

8 Darwin's Dilemma: The Birth of Evolutionary Theory 148

9 Continuity of Life: Cellular Reproduction 174

10 Genetics: The Science of Inheritance 189

11   Human Genetics   217

12   Darwinian Theory Evolves   241

13   Evolution of Species   259

PART 4 Molecules of Life

PART 6 Plant Systems

PART 7 Animal Systems

40   Reproduction   805

41   Embryology and Development   828

42   Immunity and Disease   865

Appendix

Teaching and Learning Package

Lecture Support

Instructor's Edition for complete text (34075–8)

Instructor's Edition for Core Concepts (34080–4)

Instructor's Guide for complete text by Frank Romano (34076–6)

Biology: Core Concepts (34077–4)

Biology: Diversity of Life (34078–2)

Transparencies, Slides, and Transparency Notes for complete text (34087–1)

Transparencies, Slides, and Transparency Notes for Core Concepts (34079–0)

Correlation Guide for BioSci II Videodisc for complete text (35687–5)

Test Item File for complete text by Paul E. Barney, Jr. (34086–3)

Computerized Test Item File for complete text, in IBM and Macintosh formats

Laboratory Support (customized packages available)

Investigations in Biology, Second Edition, by Richard J. Montgomery and William D. Elliott (34084–7)

Instructor's Guide for Investigations in Biology, Second Edition, by Richard J. Montgomery and William D. Elliott (34085–5)

Ecology Simulations for Use in the Introductory Biology Laboratory by Neil A. Miller (27697–9)

Process-Oriented Laboratory Activities for Introductory Biology by Kerry Kilburn and Thomas Hutto (available through D. C. Heath)

Study Support

Study Guide for complete text by George Karleskint, Jr. (34081–2)

Study Guide for Core Concepts by George Karleskint, Jr. (34083–9)

Between us, we've taught introductory biology for more than 30 years. We believe that a book's opening chapters should enable students to appreciate the procedural and historical background of biology. In Chapter 1, we explain the general process of scientific inquiry from initial observations through hypothesis formation and testing to interpretation of experimental results. We include two case studies so students can evaluate the distinctions between a poorly-designed and a well-designed experiment.

---

imental treatment or control factors previously thought extraneous. Many circuits around this loop are often necessary before the final hypothesis is supported.

OBSERVATIONAL SCIENCE

This classic sequence of hypothesis, experiment, interpretation, and reevaluation, as central as it is in scientific thought, is not the only way that science proceeds. For much of its history biology depended on observation; biologists watched, sketched, dissected, and classified living things into groups. Even today, observational science combines with experimental science in important ways.

For example, if we could learn about the sun only by visiting it and directly conducting chemical analyses, we would know nothing about it. But, in fact, we know a great deal about the chemical composition of the solar surface. How is this possible? Gases in the sun are heated to such high temperatures that they radiate light into space. Here on Earth, we can analyze that light and compare it with light given off by gases that we heat in laboratory apparatus. Thus, by comparing experimental results with observations, we can infer a great deal about the nature of an object beyond our direct reach. So powerful is this technique that the element helium was detected and named on the sun before it was found on Earth.

Historical Science

In a similar vein, one might conclude that we cannot determine anything about the history of the earth. After all, we can't do experiments involving things that happened 100 million years ago. In one sense that is true; we can't visit the past. But we can observe evidence of past events, compare them with events that occur today, and draw useful conclusions though inference.

Figure 1.10, for example, shows one of hundreds of similar objects recovered from a deep-sea drilling core taken in the Gulf of Mexico. That core sample brought to the surface layer upon layer of sediment that had accumulated over millions of years. Those sediments contained many fossils—mineralized remains of animals that died and were buried. The particular object pictured here was buried roughly 170 million years ago.

As the figure caption explains, researchers used observation and comparison to determine that this fossil represents an ancient relative of modern-day organisms called sea urchins. This fossil thus enables us to conclude that sea urchins existed 170 million years ago and that they differed from living sea urchins in certain ways. It also

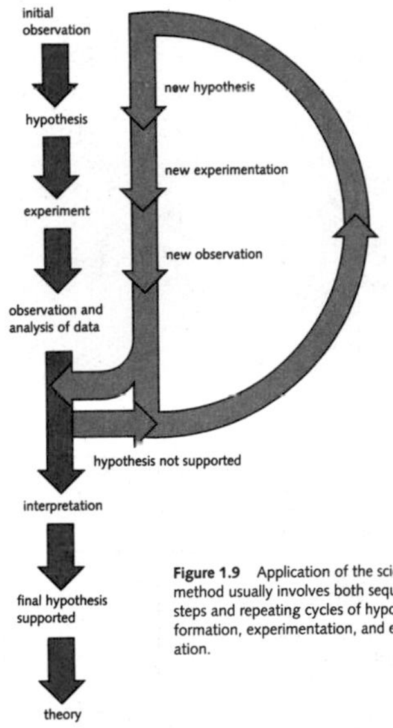

Figure 1.9 Application of the scientific method usually involves both sequential steps and repeating cycles of hypothesis formation, experimentation, and evaluation.

provides one of many pieces of information showing that life has changed over time. So although we can't travel backwards through time, we can use scientific methods to make strong inferences about life on the ancient earth.

Scientific Theories

Sometimes an hypothesis will continue to expand in complexity and to grow in the power and the accuracy of its predictions until it becomes worthy of the name **theory**. Charles Darwin's original observations and hypotheses about change in living organisms over time grew and expanded for several years before he presented them to the world as the *theory of evolution by natural selection*.

CHAPTER 1   *Understanding Life: A Crucial Responsibility*   **11**

Chapter 2 looks at the interactions among science, scientists, and society. The chapter opens with a review of social responses to epidemics, including an examination of society's reaction to Jenner's discovery of immunity to smallpox after exposure to the less virulent cowpox. The chapter concludes by looking at the connections between ecology and economics, another area where science has an impact on society and is influenced by it.

2
Science and Society

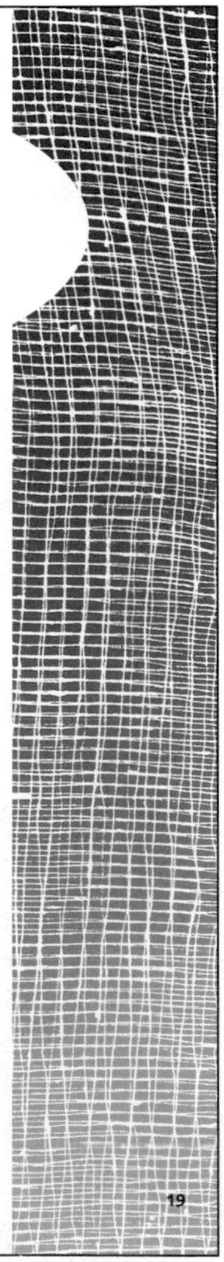

When Mark O'Donnell of Holbrook, Massachusetts, was a boy, he and his friends played in "the pastures," a field near their home that offered acres to romp in and fascinating things to play with. Among the kids' favorite toys were abandoned barrels filled with a green, jelly-like substance they called "moon glob." Mark and his best friend spent hours rolling in those barrels and tossing handfuls of moon glob at each other.

At age 27 Mark was diagnosed with a rare cancer of the adrenal glands. He died a year later. Mark's best friend contracted Hodgkin's disease, a cancer of the immune system. Within a 3-year period, 10 other people in that small town—most of whom had played in the pastures—were diagnosed with cancer. On one street not far from Mark's house, a woman died of breast cancer in nearly every other house.

For a time, no one connected these apparently separate tragedies. Then an evening news program rated a chemical plant near the pastures as among the most toxic places in the nation. Within minutes Mark's mother Joanne got a phone call. "That's what got Mark," her friend said.

Joanne O'Donnell wasn't trained as a biologist and hadn't given environmental issues much thought. But those tragic events drew her into a tangle of scientific and legal issues. What is cancer? What causes cancer and how? What happens to such compounds as dioxin, arsenic, DDT, and chlordane in the environment? What kinds of studies and what sort of data are needed to test and either confirm or disprove the hypothesis that toxic compounds in the environment can increase cancer risk (Fig. 2.1)?

What Joanne learned about these scientific issues and their links with law and politics transformed her life. She campaigned to close the plant—which, in addition to dumping toxic compounds onto the ground, had poured them into a brook that fed the local reservoir. Her efforts facilitated impressive legal battles. The chemical plant was closed, the Environmental Protection Agency placed the pastures on its Superfund cleanup list, and the company alleged to have polluted the reservoir was sued for negligence.

But were those real victories? The corporation declared bankruptcy, and its top officials maneuvered legally to protect their personal assets. Of course, legal decisions couldn't bring Mark back. Medical science

19

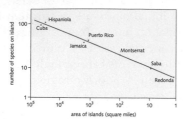

Figure 6.15 Large islands support a greater number of species than smaller islands, all other things being equal. This graph shows the diversity of reptiles and amphibians on seven similar, adjacent Caribbean islands of various sizes. In accordance with the area effect, a reduction in island size of 90 percent correlates with a 50 percent loss of species. [Adapted from Wilson (1992), Harvard University Press.]

The Distance Effect

Another phenomenon of island life noticed by MacArthur and Wilson is known as the **distance effect:** islands close to a large continent are likely to have more species than equal-sized islands located farther away. Once again, part of the explanation for this effect is obvious; the closer an island is to a rich source of potential colonizing species, the more of those species are likely to make the trip. A more complete explanation, however, is both more interesting and more complex.

Area, Distance, and Diversity: A Dynamic Equilibrium

As Wilson notes in his description of succession on Rakata, just because a species arrives on an island and survives for a time doesn't mean that it will persist. On the contrary, island species constantly arrive, survive, disappear, and reappear over time. On Rakata today, Wilson notes, "the community of species remains in a highly fluid state. . . . The reticulated python, recorded as recently as 1933, was not present in 1984–85. . . . Owls and flycatchers arrived after 1919 . . . while several old residents such as the bulbul and gray-backed shrike disappeared." Why did these changes occur?

Rakata's living community is too large and too complex to track with precision. So to study how species colonize and persist on islands, Wilson and colleague Daniel Simberloff performed an experiment that examined species diversity on more manageably sized islands over time. The study sites they chose were four isolated groups of mangrove trees called islets, each about 15 m across (Fig. 6.16a). After the researchers identified these study sites came the hard part—time-consuming biological surveys. "Crawling over each of the islets from mud bottom to the tops of the trees," Wilson writes, "examining every millimeter of leaf and bark surface, probing every crack and seam, photographing and collecting, we made as complete a list as possible of the insect and other arthropod species on the four i

Their survey results
The islet closest to a la
most species (43); the isl
est number (25), with th
tween (Fig. 6.16b).

Then the researcher
fumigated with a short-l
insects but harmless to n
tent. In this way, they vi
sects on the islets. But w
back and, within a year, r
diversity they held at the

diverse its physical habitats are likely to be. Thus, large islands would be expected to offer more distinct kinds of habitats that can support more species.

Other explanations are more subtle and depend on understanding the importance of population biology as discussed in the previous chapter. From this perspective, the larger an island is, the more space—of any kind—it has. The more space available on an island, the larger the populations of a given species it can support. Why is that important? Because when facing the ups and downs of growth and decline, larger populations tend to last longer than smaller ones. If your luck changes for a while, Wilson explains in his recent book, *The Diversity of Life,* "you are less likely to go completely broke if you are rich at the start."

114 PART 2 *Organisms and Ecology*

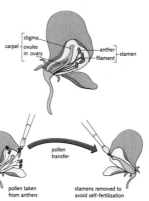

Figure 10.3 Although pea plants are normally self-fertilizing, Mendel carried out crossbreeding by brushing the pollen of one plant on the stigma of another.

ductive cells in the very same flower. To cross one plant with another, Mendel opened the pea flowers and removed the male pollen-producing anthers. He then dusted the female portions of the flower with pollen from another plant (Fig. 10.3). In this way, he could carry out controlled pollination from one plant to another—a process known as *crossing.* Because each plant produced many seeds, Mendel could obtain reliable statistics on the offspring of each cross.

Mendel's First Experiments

In one experiment, Mendel fertilized the flowers of a strain that produced purple flowers with pollen from plants that produced white flowers. He also reversed the process by using pollen from purple-flowered plants to fertilize the white-flowered strain. One of his first observations was that it did not matter which plant contributed the pollen and which received it.

The original plants involved in the cross are known as the **parental,** or P, generation. The seeds produced from this first cross Mendel called the **first filial,** or F_1, generation (*filial* comes from the Latin root for "son"). All of these seeds produced plants that bore purple flowers, so

Figure 10.4 Mendel's first experiments involved the crossing of parental (P generation) plants with white flowers and plants with purple flowers. The seeds produced by this cross (F_1 generation) grew into plants which produced purple flowers. When these plants were allowed to cross naturally, both white and purple flowers were produced in the F_2 generation.

it seemed that the white-flower trait had disappeared. Mendel took the seeds from this cross, planted them, and allowed the resulting plants to self-fertilize. This $F_1 \times F_1$ cross produced the **second filial,** or F_2, generation. Mendel collected the seeds and planted 929 of them. To Mendel's surprise, the white flowers reappeared in some of these plants; 224 plants bore white flowers and 705 had purple flowers, a ratio of purple to white flowers of 3.1 to 1 (Fig. 10.4).

In the previous chapter, we saw how genetic information is encoded and expressed. We explored the way in which the base sequences of DNA transfer information to direct the synthesis of RNA and protein, and we examined some of the cellular machinery that regulates gene expression. As we have done in other fields of science, it's only natural that we would seek to use that basic knowledge to alter living things for our own purposes. In the past 15 years, that's exactly what has happened in molecular biology laboratories throughout the world.

Of course, we have been manipulating the genes of organisms *indirectly* for thousands of years—through experiments in plant and animal breeding. The commonsense approach of breeding only those individuals that best suit human needs has produced many of our most important crops and farm animals. The plant or animal breeder, however, has always been limited by the diversity of genetic material that existed within a species. In other words, if farmers wanted to breed a sheep with brown wool, they had to wait for a brown individual to appear, breed that individual with their existing flock, and work gradually over several generations towards the desired characteristic.

Genetic engineering is something different: it is the direct introduction of *new* genes, producing *new* characteristics, into an organism. This process is much more rapid than traditional breeding methods, and it allows researchers to combine genes from different sources, producing organisms that serve specific purposes. Now, for example, it is theoretically possible to find the genetic instructions that produce brown hair in goats—or even humans—and to transfer those instructions into sheep. This emerging power has earth-shaking implications for the future of our species and our relationships with other organisms. Because the technologies of genetic engineering are recent developments, our first task is to understand how these technologies work and what they are capable of doing. We will then highlight some of the exciting—and potentially troubling—issues that this technology raises.

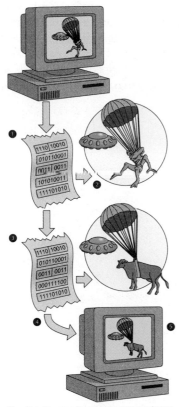

Figure 21.1 Reprogramming a computer program to change a game is a complex process. At least five steps, as illustrated here, would be necessary to make a small change in the appearance of the game. We can compare this simple job to the more complex task of modifying the genetic program of a living organism.

Editing a Complex C...

Living organisms are c...
ask how it would be p...
without having a comp...
systems. That's a fair q...
means of an analogy.

Video games are...
much, much simpler...
might compare the lin...
grams for these gam...
coded in an organism...
ine a computer game...
aliens that parachute...

suppose, for reasons that are not necessarily obvious, that we'd like to change the game so that *cows* parachute from the spaceship instead of aliens. What would we have to do?

First, armed with a general understanding of how a computer game works, we'd have to get a copy of the program for the game, written in the computer code that the machine interprets to produce the game. Somewhere in that code is the information that produces the image of an alien on the video screen. Our plan might work in five steps:

1. Extract the coded program.
2. Find the lines in the program that produce the images of aliens.
3. Change those lines so that they now produce images of cows.
4. Reinsert the altered program code into the computer.
5. Run the modified program to see if we have made a successful change.

As anyone who's ever written a computer program knows, every step of this project will be tricky. If we understand very little about how the program works, steps 2 and 3 will be particularly difficult. If the program is long, we might have to spend a great deal of time looking for the small portion that produces the images of aliens. Once we did find it, it might not be obvious how to change those lines to produce cows, and that might force us to take a shortcut. For example, by borrowing lines of program code from another video game that already has pictures of cows, we could put cows in our game without knowing exactly how the image of a cow is produced. Eventually, with luck and persistence, our modified game might be ready to play.

Changing a Genetic Code

In a certain sense, the information coded in DNA is similar to the program that runs our video game. As we have already seen, the elements "written" in DNA code, the genes, help to determine the characteristics of an organism. Remember what we had to do to modify the computer game: remove, modify, and reinsert the program code. To change one of the genes in a living organism, we could follow a scheme not very different from the way in which we modified our computer program (Fig. 21.2):

1. Extract the DNA code.
2. Find the genetic information for the characteristic we wish to change.
3. Insert new or modified genetic information into the RNA.
4. Reinsert the modified DNA into a living cell.
5. Test the modified cell to see if the new gene is expressed.

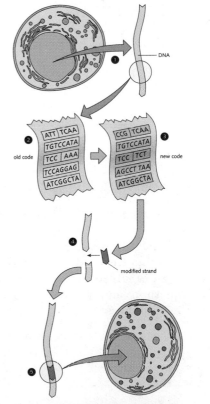

Figure 21.2 Similar to the reprogramming of a computer game, we can visualize genetic engineering as a multistep process involving the isolation, analysis, modification, and reinsertion of a DNA sequence, followed by the expression of that sequence in a transformed cell.

Students find difficult conceptual material much easier to understand if the topic is introduced with the use of an everyday analogy. In this case, we discuss and illustrate the steps necessary to alter a video game program and then show how the same general steps are involved in genetic engineering.

No text is complete unless students can both understand its concepts and see the relevance to their lives. Through *Theory in Action* boxes, we try to maximize the personal impact of science on the individual student. In telling the story of Nettie Stevens and the discovery of the *Y* chromosome, students share the work of an actual, extraordinary woman biologist.

THEORY IN ACTION

Nettie Stevens and the Y Chromosome

The history of genetics, like that of any science, is filled with wrong turns, mistakes, and errors in judgment that time and the experimental method have corrected. The identification of the *X* chromosome is a case in point. The first suggestion that sex could be a chromosomal characteristic was made in 1901 by Clarence E. McClung, who suggested that a so-called *accessory chromosome* in insects determined maleness. The term had been coined 10 years earlier by an investigator named H. Henking. Henking noticed this chromosome in studies on the cells of *male* insects. Unlike the other chromosomes, it did not have a paired chromosome, so he called it an accessory. He correctly noted that it divided in only one of the two meiotic divisions, which showed that it was present in only two of the four sperm cells. This unusual behavior made many biologists doubt that it even *was* a chromosome, and the designation *X* reflects that doubt. Because McClung found the unpaired accessory chromosome only in male cells, he concluded that the *X* chromosome determined maleness.

We can look back on the work of Henking and McClung with the benefit

of hindsight, but only because a more patient observer followed them—one who noticed an even smaller structure in these insect cells, a structure they had missed. That scientist was Nettie M. Stevens. Nettie Stevens was born in rural Vermont, studied in Massachusetts at a state teachers' college, and then taught school for nearly 10 years in order to save enough money to attend Stanford University to study science. From Stanford she went to Bryn Mawr College in Pennsylvania in 1900 and began a series of studies focused on the way in which the sex of an animal is determined. While many other investigators believed that sex was *not* genetically determined, Stevens was convinced that McClung was on the right track.

Stevens made a series of very careful observations with *Tenebrio molitor*, the common mealworm. She found that female cells contained 20 chromosomes, whereas male cells contained 19 large chromosomes and 1 small one. She correctly concluded that this small difference was the result of chromosomal sex determination! Nettie Stevens had discovered the *Y* chromosome. The determinant of maleness was not the *X*,

for females had two of those, but the small *Y* chromosome, which was found only in males. Stevens's discovery of the *Y* chromosome was a pivotal event in the development of genetics, because it made possible for the first time a correct explanation of sex-linked inheritance.

---

cells, however, have both an *X* and a *Y* chromosome (and no tetrad forms). During meiosis, each gamete receives either an *X* or a *Y* chromosome. Therefore, about half of the sperm cells produced by a male fly contain the *X* chromosome and about half contain the *Y*. Because all eggs contain an *X* chromosome, the sex of the zygote is determined by the sex chromosome carried by the sperm. If the sperm cell donates an *X* chromosome, the zygote will be female; if it donates a *Y* chromosome, the zygote will be male.

The *X-Y* mechanism, which is also how sex is determined in humans, produces a 1:1 ratio of males to fe-

males (Fig. 10.21).
on one of the sex c
covered by the Ame
inheritance can be s

Sex Linkage

Thomas Hunt Morg
his first work at Col
Morgan pioneered
melanogaster, in gene

Current Controversies boxes draw on recent research to pull students into the process of science. The featured stories, such as the one here about a recently discovered cellular organelle, show students how research both answers questions and raises new problems that need to be addressed.

CURRENT CONTROVERSIES

Discovering a New Organelle

With all the years that cells have been studied by thousands of scientists throughout the world, is it possible that most cells contain a structure that has never been noticed? If you found something completely new, would you have the confidence to trust your own observations, or would you assume that you'd made some silly mistake and ignore your own results?

In 1986 Nancy Kedersha and Leonard Rome of UCLA had the chance to answer all these questions for themselves. They were purifying a certain class of vesicles normally associated with a basket-shaped protein called *clathrin*. Examining her samples in the electron microscope, Kedersha saw some peculiar egg-shaped structures unlike *clathrin* or anything else she ex-

pected to be present in her preparation. Their delicate, arching internal structure so reminded her of the arches in cathedral vaults that she called them "vaults."

At first, Kedersha and Rome thought that the vaults might be rearrangements of *clathrin* or one of the other proteins known to be in vesicle preparations. Other experiments clearly showed, however, that the vaults were composed of five hitherto unknown proteins and a small (140 base) RNA. Perhaps the vaults were a peculiarity of the cells she was studying? Not so. They found them in cells from cows, chickens, frogs, and even from amoebae. Genes for vault proteins and RNA are found throughout the living world. They must be doing something important.

Detailed studies with the electron microscope have shown how the parts of the vault fold together to form the complete, basket-like structure, but to date no one has been able to discover exactly what vaults do. Kedersha's personal view is that the vault is involved in the transport of material between nucleus and cytoplasm. The vault's combination of RNA and protein suggests such a role, and she's pointed out that the vault is almost exactly the right size to fit into a nuclear pore.

She may be right, of course, but the vault story points out something that biologists in all fields have known for a long time. There are new discoveries to be made even in territory that we think is familiar.

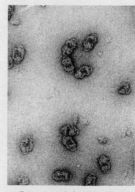
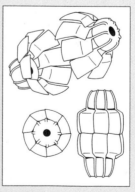

LEFT: Electron micrograph of isolated "vaults." Each vault measures 36 x 65 nm.
RIGHT: Artist's conception of the structure of a single RNA–protein vault.

Diversity boxes in Part 5 combine drawings and photographs to organize anatomical and life-cycle information into one understandable package. Each box has a phylogenetic-tree icon to help students locate each illustrated group within the appropriate kingdom.

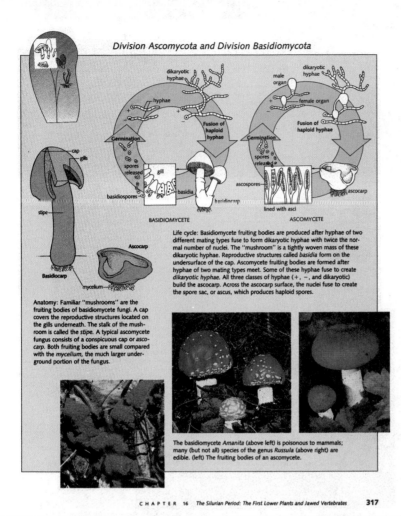

Division Ascomycota and Division Basidiomycota

BASIDIOMYCETE

ASCOMYCETE

Life cycle: Basidiomycete fruiting bodies are produced after hyphae of two different mating types fuse to form dikaryotic hyphae with twice the normal number of nuclei. The "mushroom" is a tightly woven mass of these dikaryotic hyphae. Reproductive structures called *basidia* form on the undersurface of the cap. Ascomycete fruiting bodies are formed after hyphae of two mating types meet. Some of these hyphae fuse to create *dikaryotic hyphae*. All three classes of hyphae (+, −, and dikaryotic) build the ascocarp. Across the ascocarp surface, the nuclei fuse to create the spore sac, or ascus, which produces haploid spores.

Anatomy: Familiar "mushrooms" are the fruiting bodies of basidiomycete fungi. A cap covers the reproductive structures located on the gills underneath. The stalk of the mushroom is called the *stipe*. A typical ascomycete fungus consists of a conspicuous cap or *ascocarp*. Both fruiting bodies are small compared with the *mycelium*, the much larger underground portion of the fungus.

The basidiomycete *Amanita* (above left) is poisonous to mammals; many (but not all) species of the genus *Russula* (above right) are edible. (left) The fruiting bodies of an ascomycete.

In-text icons help students place individual topics, such as organelle organization, within a larger discussion, such as eukaryotic cell structure. Arrow icons, located in the margin, show students where the same topic will be discussed again in another context. This arrow icon directs students to the chapter where the interrelated functions of organelles will be examined.

J. S. L.
K. R. M

...article into the cell, ...od fuses with a lyso- ...lysosomal enzymes ...wn into smaller com- ...l and are easily re- ...omes of white blood ...destroy the bacteria ...rposes as well. Lyso- ...ocess of destroying a ...the rupture of many ...tion of the cell that ...*autolysis* (*auto* means ...hy should the lyso- ...anelles help the cell dispose of damaged or defective organelles, and they also play a creative role in helping destroy cells in strategic locations where a developing organism needs to shape the pattern of growing tissues.

form to another. Every activity that we associate with life requires a source of energy, and in most organisms these are the organelles that provide that energy.

Chloroplasts trap the radiant energy of sunlight and use that energy to produce energy-rich compounds like carbohydrates. In turn, these carbohydrates serve as a convenient form of chemical energy that can be stored by the cell until needed or may serve as food for another organism (Fig. 16.17). In Chapter 19 we will examine the details of how chloroplasts use solar energy to drive the synthesis of these compounds.

Mitochondria (singular, *mitochondrion*) carry out the next step in processing cellular energy. Before the energy in carbohydrates and other food molecules can be used by the cell, it must be converted to a convenient form. By far the most convenient form of chemical energy in the living cell is a nucleotide known as **ATP (adenosine triphosphate).** The task of the mitochondrion is to use the chemical energy available in food molecules to produce ATP (Fig. 16.18). That ATP is then made available to the rest of the cell. In Chapter 18 this process will be explored in depth.

The ATP molecules produced by mitochondria may be thought of as little chemical "batteries" that the cell can use to provide instant energy for just about any purpose. And we do mean *any* purpose. ATP provides the energy for a firefly's twinkle on a summer night, the whip-like movement of flagella, and even the separation of chromosomes during mitosis. Not surprisingly, cells use lots of energy, and therefore they have to produce lots of ATP.

ENERGY-PRODUCING ORGANELLES

Mitochondria and Chloroplasts

Two other organelles are absolutely essential to any discussion of the organization of living cells: **mitochondria** and **chloroplasts.** Mitochondria and chloroplasts are organelles that change energy from one

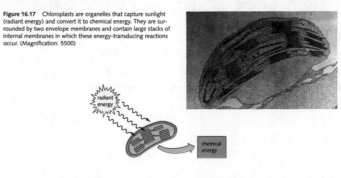

Figure 16.17 Chloroplasts are organelles that capture sunlight (radiant energy) and convert it to chemical energy. They are surrounded by two envelope membranes and contain large stacks of internal membranes in which these energy-transducing reactions occur. (Magnification: 5500)

radiant energy

chemical energy

PART 1

Introduction to Biology

This photograph of a sunrise in the Great Smoky Mountains, between Tennessee and North Carolina, juxtaposes plants, earth, sky, and sun: all elements that humans must understand to safeguard our future.

*M*onumental questions about humankind's place in nature and within the greater cosmos have puzzled and inspired human beings for millennia. In our time, we seek to answer many of those questions through the disciplines we call the sciences: biology, chemistry, physics, astronomy, and others. But what, exactly, is science? What, if anything, sets the thought processes of scientists apart from those of persons in other fields of intellectual inquiry, such as philosophy or religion? And what bonds to the race of humans, if any, are so strong that they mark scientists indelibly as the products of their society?

In Part 1, we seek to explore these primary questions about the nature of science as a human endeavor, and of biology in particular: what is science, how do scientists approach intellectual questions, and how do scientists and their resultant works interact with society?

In Chapter 1, we place science within the context of other intellectual approaches to the world by explaining science as an empirical "Way of Knowing" as opposed to numerous belief-based systems, including philosophy and religion. Once readers understand what science is, and as importantly, what it is not, the nature of scientific methodology is explained in general terms through the introduction of the processes of observation, hypothesis formation, hypothesis testing through experimentation, and re-evaluation of an hypothesis. The chapter concludes by examining two actual experiments so students can see what constitutes both good and poor experimental design.

Chapter 2 looks at the interaction of science and society through two examples: social responses to epidemics of the past and present and the social and scientific dialogue (and occasionally argument) on the interaction of ecological knowledge and economic planning. Again, actual examples are used wherever possible so readers get accustomed to seeing how scientific inquiry

The question of all questions for humanity, the problem which lies beyond all others and is more interesting than any of them is that of the determination of man's place in Nature and his relation to the Cosmos. Whence our race came, what sort of limits are set to our power over Nature and to Nature's power over us, to what goal are we striving, . . . [these] . . . are the problems which present themselves afresh, with undiminished interest, to every human being born on earth.
— T. H. Huxley, 1863

works in the real world, and they can also consider how they might respond to questions in the future that involve science and society.

1

Understanding Life:
A Crucial Responsibility

*W*e are living in the midst of a revolution in the relationship between humanity and the rest of the living world. Never before in history has biological knowledge accumulated and changed our understanding of life with such dizzying speed. Never has information gathered by biologists offered such extraordinary and far-reaching opportunities to improve the human condition. And never has ignorance of biological concepts been so potentially dangerous to individuals and society.

These statements might seem, at first glance, like exaggerations. They are not. The practical applications of modern biology span the entire spectrum of human concerns and cover the entire range of life's phenomena, from the microscopic world of molecules, genes, and cells to the blanket of life that covers our planet. That wide-ranging relevance is what makes biology so exciting and so important.

On the level of life's tiniest particles, molecular biologists are learning to read the genetic code that controls life's most fundamental activities. In doing so, researchers are struggling to comprehend how genes direct certain cells to form bones in our legs and other cells to conduct thoughts through our brains. They are also seeking to understand how genes and cells change with age, how they fight (or contribute to) disease, and how they evolve from generation to generation. The early fruits of this research are already revolutionizing the way we grow our food, and ongoing efforts in molecular medicine will soon transform the way we are conceived, the way we are born, the way we fight degenerative and infectious diseases, and the way we die.

On a scale so vast that it encompasses our entire planet, ecologists are using information gathered from airplanes and satellites to recognize and evaluate human influence on global environments and are learning to anticipate and moderate those impacts. We now know that certain human actions are destroying Earth's protective ozone layer, while others are threatening supplies of clean air, water, and food. We also know that activities such as the destruction of forests and the burning of coal and oil are increasing the amount of certain gases (such as carbon dioxide, CO_2) in the atmosphere. An ongoing debate is now raging at the interface between ecology, geochemistry, and climatology about whether or not CO_2 levels will continue to increase and, if so, whether or not that increase will cause a global-warming trend.

The importance of these issues obliges all of us—scientists and nonscientists alike—to learn enough about the science behind them to make informed judgments on perplexing questions. Do we know enough to create new forms of life in the laboratory safely? Have we enough ethical and legal safeguards in place to allow alteration of human genes in efforts to diagnose and cure inherited diseases? How can we weigh moral concerns about experimentation on fetal tissues against the potentially lifesaving benefits of those procedures? What should (and should not) be done to evaluate and deal with the effects of human activity on the global environment?

None of these questions have simple answers. All of them demand the attention of scientifically well-informed citizens, able to appreciate the power and responsibility conferred upon us by discoveries in the world of life. Yet precisely *because* biology is tied so closely to so many facets of human life, our science can be more controversial, in more different ways, than it has ever been before. Almost any active and important field in biology—from the study of evolution, to genetic engineering, to evaluations of human impact on local and global environments—directly affects the beliefs, economic interests, or political convictions of some segment of the population.

Perhaps in the past there was a time when biology was a quiet, isolated discipline, remote from the concerns of society. That certainly is not the case today. The biologically related challenges we face range from protecting endangered species and fragile environments to deciding whether or not to use genetic engineering in efforts to design new organisms or cure inherited disease. Like it or not, biology finds itself at the center of some of the most serious and controversial issues of our time (Fig. 1.1).

No book should presume to provide answers to these controversial questions. We do hope, however, to point out such questions where they exist and to provide you with the basic scientific understanding you need to recognize important issues, participate in democratic debate, and make your own decisions. As first steps in that direction, this chapter and the next introduce you to three basic themes: the nature of life, the nature of science, and the relationship between science and society.

LIFE: A GLOBAL PHENOMENON

From space, Earth is a dazzling sphere of blue, green, and white against a stark background (Fig. 1.2). In striking contrast to our desolate planetary neighbors, Earth is warm, moist, inviting—and *alive*. From space we can actually see that life is a global phenomenon, involving the entire **biosphere**—the planetwide network of physical environments and living organisms.

That biosphere contains living things that span an enormous range of sizes and operate on time scales ranging from fractions of a second to millions of years. Yet all living things share certain traits: they take in and process materials from their surroundings, they grow and reproduce, and they respond to events around them. Over

Figure 1.1 The relationship between science and society affects all our lives. The enormous Names Project Quilt, each square of which is assembled in memory of a casualty of AIDS, is now so large that it cannot be displayed in its entirety even on the huge mall between the U.S. Capitol building and the Washington Monument. Scientific research uncovered an enormous amount of information about AIDS with incredible speed during the early 1980s. Yet more than a decade later, many Americans are still tragically misinformed about this disease. Vital information about transmission and prevention often has proved difficult to disseminate, primarily for political, rather than scientific, reasons.

Figure 1.2 A NASA photograph showing the entire planet Earth—a scene astronauts have described as both beautiful and spiritually uplifting.

long periods of time, living things change, or *evolve*, in ways that produce new varieties. A salmon, an orchid plant, a monkey, a mosquito, and a bacterium are all alive by these criteria; as assemblages of active, interdependent living parts, they are called *organisms*.

Organisms, in turn, are composed of *cells*; the basic units of living things, and the smallest units of any organism considered to be alive in the full sense of the word. Cells are amazingly diverse in size, shape, and function. The human body alone contains at least 85 completely different cell types, and elsewhere, cells exist in almost infinite variety (Fig. 1.3). Yet all cells share the ability to grow, reproduce, respond to their surroundings, and adapt to change. A great many organisms consist of nothing more than a single living cell.

Ignoring some of life's complexity for the moment, we can say that a "typical" animal cell is about 20 microns (μm) [2.0×10^{-5} meters (m)] in diameter (Fig. 1.4). That cell is surrounded by a *cell membrane* which separates it from its surroundings, and it contains many tiny structures called *organelles* that function much like miniature organs. It also has a prominent, central structure, about 6 μm in diameter, called a *nucleus*, after a Latin word meaning "kernel," because it looks like a seed within a fruit.

If we were to take a handful of such cells and dissolve them in detergent, we would release a clear, sticky substance that can be twisted like caramel around a glass rod. This is *deoxyribonucleic acid,* or *DNA*, the molecule of

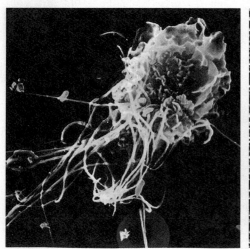
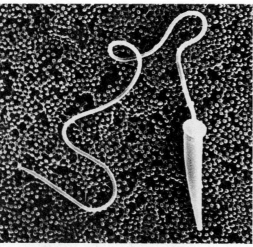
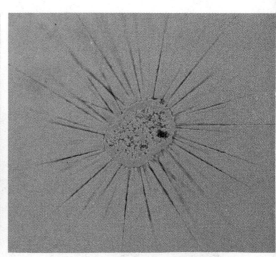

Figure 1.3 Living cells display a diversity of structures and adaptations. LEFT: Macrophage: a key player in the defense against infection and disease, as we will see in Chapter 42. CENTER: Sperm: a starfish sperm on the egg surface. RIGHT: Protozoan: *Actinophrys sol*, one of the many single-celled organisms.

heredity. Each DNA molecule is a double string of billions of atoms coiled up tightly into an elegant spiral chain (Fig. 1.5). Inscribed on that chain is the genetic code—a linear sequence of four simple molecules that carries the information necessary to build and operate a fly, an oak tree, or a human being. That code, conceived along with life itself nearly 3.5 billion years ago, has been evolving ever since, has made possible the diversity of life on Earth, and has only recently begun to reveal its secrets to human observers.

From DNA to Biosphere: Spaceship Earth? Or a Living Planet?

Making the connection among all the levels of life between DNA and the biosphere is difficult. As we concentrate on cells, tissues, organisms, and ecosystems in the chapters to come, it will be easy to lose track of the connections between the molecular "intelligence" that directs life's processes and the planet that houses us all. So while we are still focused on these extremes of the living spectrum, let us consider briefly two views of the connections between life and our planet.

Years ago, Buckminster Fuller coined the term *spaceship Earth* to suggest that our planet is a giant spaceship on whose life-support systems all organisms depend. Expanding Fuller's concept into a model of planetary function, many scientists believe Earth operates as a sort of giant, orbiting geochemical machine whose workings fortuitously sustain life but whose processes are not controlled by life. In this view, living organisms are simply "passengers" on a planet whose environment they may affect but do not control.

Other scientists postulate that physical and chemical conditions on Earth are shaped by life's processes. These scientists do not deny that living organisms adapt over time to changing conditions around them. They simply argue that as life on Earth has evolved, it has brought about major changes in global conditions that would not have occurred otherwise. In a sense, these researchers believe that DNA molecules contain information that directs not only the lives of individual organisms but also transformations that encompass our entire planet. As far-fetched as this idea might seem to you now, you will see in the next part that even single-celled organisms (in sufficient numbers) can profoundly affect the earth and its atmosphere.

Some researchers go even further, maintaining that the biosphere is not just a collection of plants and animals but a global "superorganism" whose components have evolved in ways that maintain the delicate balance necessary for planetary life. This view has been termed the **Gaia hypothesis** after the earth goddess of the ancient

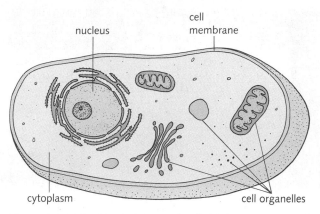

Figure 1.4 The "typical" animal cell exhibits structures common to most multicellular animals.

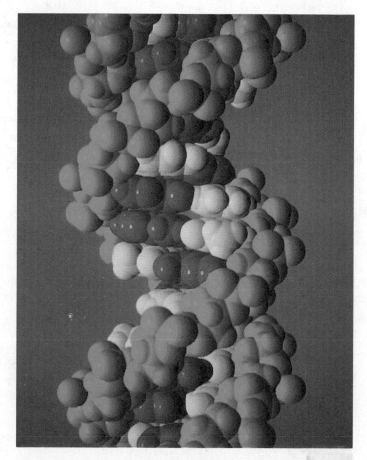

Figure 1.5 A computer graphic reconstruction shows the beauty and the complexity of the extraordinary molecule called DNA.

The Gaia Hypothesis

James Lovelock, a British biochemist, and his American colleague Lynn Margulis do not believe that long-term constancy in Earth's temperature and the life-sustaining levels of oxygen in the atmosphere have arisen incidentally.

Life's influence on planetary conditions struck Lovelock as so powerful and so precisely regulated that he proposed the *Gaia hypothesis*. According to Lovelock, "the physical and chemical condition of the earth's surface, of the atmosphere, and of the oceans has been and is actively made fit and comfortable by the presence of life itself."

His hypothesis rests on evidence that living organisms *interact* with, and powerfully affect, Earth's atmosphere and geochemical cycles. Lovelock goes further, however; he proposes that all life on Earth has evolved into a global superorganism—Gaia—whose parts monitor and manipulate carbon dioxide concentration, oxygen levels, and other environmental parameters. He argues that only this active monitoring and correction keep global conditions within the narrow margins essential for life.

In this view, Gaia's atmosphere and oceans act like a global circulatory system, carrying compounds across the globe and dumping them where necessary. Plants and animals living and dead

process and store such critical compounds as oxygen and carbon dioxide, releasing them as necessary to control Earth's temperature and atmospheric composition.

Homeostasis is the term physiologists use to describe an organism's maintenance of stable internal conditions in the face of a changing external environment. Lovelock was bold enough to propose that this global superorganism has actively maintained planetary homeostasis over the 3.5 billion years it has been alive.

The Gaia hypothesis is controversial, but it is extremely useful as an alternative point of view. It reminds us that though we tend to think of Earth as stable, we as living organisms have altered it in many ways. What will come of those changes no one knows. In the meantime, Lovelock's ideas can help us view the evolution of life and the physical evolution of Earth not as two separate series of events but as a single, tightly integrated process.

Greeks (see Current Controversies, The Gaia Hypothesis, above).

The Gaia hypothesis is highly controversial—so much so, in fact, that some detractors dismiss it as "unscientific." That assertion brings us to a fundamental issue that we need to address before we can go any further: Just what *is* science, anyway?

THE NATURE OF SCIENCE

Ever since our species became truly human, we have pondered the origin of living things and the forces of nature. At first, people explained these phenomena with tales of nature gods and goddesses, magic, and witchcraft. Over time, human explanations of the world evolved and converged into two lines of thought—science and religion—that represent parallel efforts to put human life and the natural world in perspective. Although science and religion are often portrayed as conflicting, they need not contradict each other, for their aspirations and emphases are as different as night and day.

Science, Myth, and Religion

Typically, formal religions are organized around dogmas, usually based on divine revelation, that are often not open to interpretation and revision. Orthodox religion, especially, is often based on trust in inviolable scriptures and depends on unquestioning belief. In most Western societies today, most religious leaders see religion's primary purpose as providing the faithful with moral guidance, spiritual meaning for existence, and rituals to help them through high and low points in their lives. Inasmuch as human judgments on personal conduct, morals, and ethics often coincide, the world's great religions often find common ground in such matters, as interfaith councils affirm.

When it comes to explaining the natural world, though, religions differ widely. The Chewa tribe of Malawi credits a lonely chameleon with giving birth to all other forms of animal life. The Hindu Manava Sastra describes how God dispelled the darkness with an egg as brilliant as the sun, from which Brahma, the creator of all thinking beings, emerged. And according to James Ussher, archbishop of Armagh and a distinguished Irish scholar of the 1600s, the earth was created at 9 A.M. on Sunday the 23rd of October in the year 4004 B.C., and Adam and Eve were evicted from Eden 18 days later. The accuracy of these stories cannot be tested in any practical sense, but each has been accepted within its own culture and time, and each serves its purpose within its faith.

Science as a Way of Knowing

Science, on the other hand, deals exclusively with the natural world and utilizes a completely different way of explaining natural events. The main concerns of modern science have been summarized by Francisco Ayala, a leading biologist, as follows:

1. Science tries to collect and organize information about the world in a systematic way. While engaged in this process, scientists look for recurring patterns and relationships among events and processes.
2. Science attempts to explain observed events and processes in terms of natural phenomena that can themselves be observed or demonstrated in a scientific fashion.
3. The explanations of events proposed by scientists—called *hypotheses*—must be *testable*. This means that all scientific hypotheses are constantly subject to verification or disproof. Any scientific explanation may be altered or rejected if it no longer fits all the available evidence.
4. Scientists continually attempt to construct powerful hypotheses that explain a large number of observed events.

Most important, science is neither dogma nor an assemblage of immutable "facts." Science is, rather, a "way of knowing"—an ongoing process of observing the world, forming ideas about how that world operates, conducting tests of those ideas, and revising conclusions. Science requires both the ability to place events in perspective across time and distance and the talent to combine the unpredictable power of the human mind with the precision of meticulous measurement.

Importantly, when scientists attempt to explain "why" something happens, that explanation concerns only the actions of natural forces that are themselves understandable in scientific terms. Although biology endeavors to explain what life is, how life began, and how humans evolved, it cannot address issues such as why life exists, what the meaning of life is, how humans should behave, or what our purpose in the world should be.

It is also vital to understand that "facts" and principles aren't science; they are the *products* of scientific investigation. That's a good thing, because scientific "facts" change often, and scientific principles are always open to skepticism and question based on further scientific tests. (Remember, the "fact" that the earth is flat was once widely accepted!) You should therefore never be content simply to *believe* any "scientific fact" you are told; aim instead to *understand the processes* by which researchers reach conclusions. Science and scientists should *never* be approached with the sort of faith better reserved for religion, because controversy, questioning, and change are the heart of the scientific process.

Strictly scientific concerns are thus completely separate and distinct from (rather than antagonistic to) the goals of most religions. Many leading scientists throughout history have been devoutly religious, and they have had no problem believing that a god who controls the universe through immutable, physical laws is just as awe inspiring and worthy of worship as one who governs by intervening through the use of supernatural power.

Science as a Mirror of Society

Science is a quintessentially human endeavor, and as such, it is subject to both the advantages and the pitfalls of the human condition. On the positive side, science benefits from majestic leaps of intuition that are the hallmark of human thought. From the revolutionary ideas of Copernicus, who first understood the sun to be the center of the solar system, to the discovery of the structure of DNA, which illuminated the mechanism of heredity, science has depended on the power of human creativity.

But precisely *because* scientific thought is abstract and creative, it is inevitably colored by the mind that creates it. Scientists are subject to emotion, opinion, pride, and prejudice, just like everyone else. Data gathered by honest scientists in quantitative experiments are seldom open to question. The important things, however, are not what *numbers* scientists gather but what *kinds* of experiments they choose to do in the first place, what *interpretations* they draw from their data, and what *applications* their conclusions endorse. At each step of this process prejudice can cloud thinking.

The latter half of the nineteenth century offers many examples of prejudice in science, for it was a time of great social upheaval in Western civilization. In the midst of intellectual and philosophical ferment in Europe, several white, male researchers employed the scientific method—consciously or unconsciously—to justify their own prejudices and reinforce their privileged positions in the fragile status quo.

One branch of contemporary "science" called craniometry, for example, claimed to evaluate intelligence and general mental superiority by examining sizes and shapes of human skulls and brains (Fig. 1.6). One group of craniometers led by Paul Broca, a respected professor of surgery at the Faculty of Medicine in Paris, tried to prove that women and members of all nonwhite races were mentally inferior to Caucasian men. Today we dismiss these discredited assertions as groundless pseudoscience. In their day, however, the "scientific" authority of Broca's school lent these highly prejudiced statements a credibility not merited by the flimsy data on which they were ostensibly based. And it does not take much imagination to extrapolate such "findings" to exclusionary racist policies regarding education, employment, and equal protection under the law.

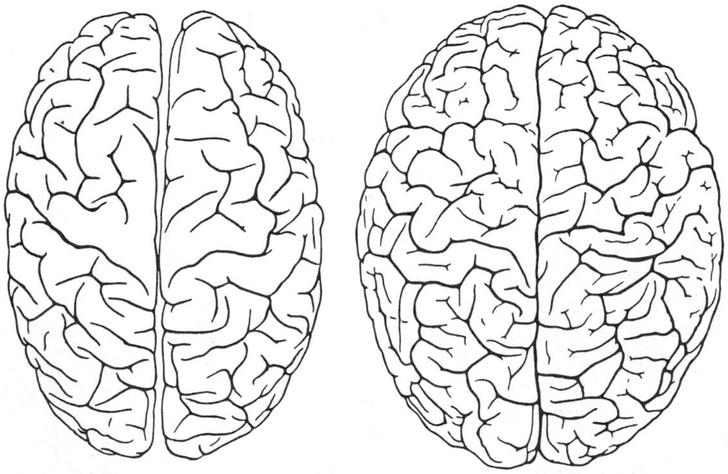

This is not to say that all (or even many) scientists either have an ax to grind or are sloppy thinkers. But the influence of preexisting ideas on scientific thought is often subtle and unpredictable. The best researchers know that, depending on what they *expect* to find, they may collect different data or interpret sets of data in different ways. Consequently, as you will see shortly, many of the best experiments are designed to eliminate (as much as possible) the subjective effects of unavoidable researcher bias.

The important point here is that prejudices, societal attitudes, and researchers' expectations affect science today as surely as they did in Broca's time. But because we are part of the society that creates those attitudes, we may not be able to spot their effects on scientific judgment as readily as we can recognize mistakes of the past. Oddly, it was Broca himself who once pointed out that "the least questioned assumptions are often the most questionable."

For that reason, it is dangerous to place blind faith in the judgment of any single "expert" on scientific findings with ethical, legal, or social implications. In such matters science improperly interpreted, accepted with unquestioning faith, or applied incautiously can be dangerous. In a democracy such as ours the public *must* understand science sufficiently to weigh various arguments and make reasoned judgments at election time.

Figure 1.6 The renowned mathematician K. F. Gauss's brain (RIGHT) weighed 1492 grams, only slightly larger than the average brain (LEFT). This dispelled a number of theories regarding brain size and intelligence. Gauss's brain was, however, more convoluted than the average brain.

AT THE HEART OF SCIENCE: SCIENTIFIC METHODOLOGY

The power of science resides in its ability to organize individual observations into a logical system that attempts to explain past events and predict the outcome of current phenomena. To accomplish that goal, science proceeds along one or another series of ordered steps that are often called the **scientific method.** In reality, science depends on much more than such "cookbook" methodology, and there are actually several scientific methods. Still, examining the idealized form of the scientific method provides vital insights into the nature of science.

Observation

Every scientific investigation begins with *observations*. In some cases these observations may be the predominant component of the scientific process (see Observational Science, p. 11). Sometimes observations are made over long periods, as in the case of Charles Darwin, who spent years traveling around the world and noting natural phenomena. Other observations may occur in an instant, as did Alexander Fleming's discovery of penicillin. Fleming noticed one day that a culture dish of bacteria was contaminated with a fungus. He also noticed that there were no bacteria growing around the fungal colony. He immediately realized that the fungus might be producing a substance that killed bacteria (Fig. 1.7).

Of course, exciting scientific discoveries are not generated by people who simply note things they see but, rather, by people who combine observation with luck, creativity, and insight. (Fleming, for example, could have said, "Darn! Another contaminated culture!" and thrown the fateful dish away.) A good researcher can, as the philosopher Schopenhauer pointed out, "think something that nobody has thought yet, while looking at something that everybody sees." Nobel laureate Albert Szent-Gyorgyi, who (among other accomplishments) first isolated and described vitamin C, put it another way. "If you know in advance what you are going to do," Szent-Gyorgyi admonished, "then it is not research at all; then it is only a kind of honorable occupation."

Figure 1.7 Alexander Fleming working in his laboratory.

Forming an Hypothesis: Inductive Reasoning

If a series of observations is reproducible, scientists invoke **inductive reasoning** to generate an **hypothesis.** In reasoning inductively, scientists try to work from *particular* events to formulate *general* principles. In other words, they try to use unifying concepts to link events that were previously not seen as related to one another. Thus, they try to establish a concept that completes the sentence "All these things happen as they do because ————."

A simple hypothesis may propose a cause-and-effect relationship to explain a series of observations. In other words, it may suggest that event A caused event B. Most hypotheses, however, try to make general predictions that extend beyond the limited number of events originally observed. (Event A caused event B because of such and such a general phenomenon.) A more complex hypothesis may propose a *model* of the way a complex system operates. Darwin's early hypotheses about evolutionary change created such a model to explain the changes he had observed to occur in plants and animals over time.

Experimental Design: Deductive Reasoning

Once an hypothesis has been created, **deductive reasoning** can be employed to make *specific* new predictions from the *general* statements of the hypothesis and from other general principles assumed or believed to be true. Often, deductive reasoning can be summarized in an "if–then" format: if event A caused event B because of such and such a phenomenon, then event C should cause event D.

These predictions are then tested by performing **controlled experiments.** In most controlled experiments, scientists perform at least two sets of parallel trials. In one set of trials they vary a single factor, called the **experimental variable.** This is known as the *experimental* set. The other set of trials, known as the **control,** is subjected to the same conditions *except* the changes in the experimental variable. Whenever possible, scientists design such experiments to provide *quantitative data*—results that can be described in numerical form and can thus be subjected to statistical analysis. Designing experiments in this manner

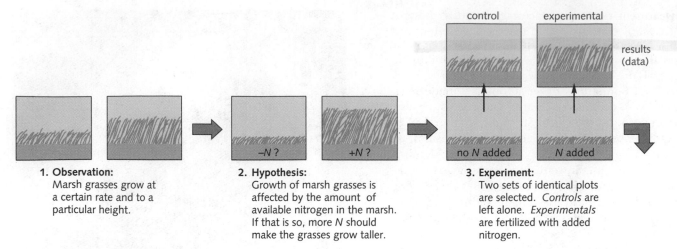

1. Observation:
Marsh grasses grow at a certain rate and to a particular height.

2. Hypothesis:
Growth of marsh grasses is affected by the amount of available nitrogen in the marsh. If that is so, more *N* should make the grasses grow taller.

3. Experiment:
Two sets of identical plots are selected. *Controls* are left alone. *Experimentals* are fertilized with added nitrogen.

Figure 1.8 A diagrammatic representation of a well-designed experiment that examined the effects of nitrogen on the growth of salt marsh grasses.

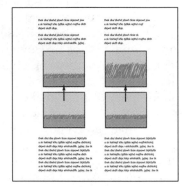

4. Observation:
Fertilized plots grow taller and more rapidly.

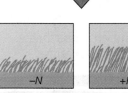

5. Interpretation and conclusion:
Growth of marsh grasses is affected by *N*.

helps ensure that any effect observed in the experiment is due to the factor or factors under investigation.

While Ivan Valiela was studying the ecology of a New England salt marsh, for example, his observations led him to hypothesize that the growth of marsh grass was limited by the availability of nitrogen, a nutrient necessary for plant growth (Fig. 1.8). A straightforward prediction based on this hypothesis is that marsh grasses will grow faster or larger (or both) when more nitrogen is available.

To test this hypothesis, Valiela chose several marsh plots that were as similar to one another as possible in plant density, soil type, freshwater input, and height above average tide level. In a series of *experimental* plots he added nitrogen fertilizer, thus manipulating the experimental variable that was the subject of his hypothesis. Other plots, selected as *controls*, were exposed to all the same environmental conditions as the experimental plots *except* for the addition of nitrogen.

Throughout the growing season Valiela sampled the plots to measure growth rates, chemical composition of leaves, decay rates of dead leaves, and so on. By statistically comparing the results of these tests, he determined (among other things) that several important marsh grasses grew taller and larger than controls when they were given additional nitrogen. The experiments were checked and the results replicated. The hypothesis was supported.

Note that a single isolated experiment is never sufficient; scientists must be able to *reproduce* or *replicate* their observations to take them seriously. Valiela used several control and experimental plots, rather than a single pair. The outcomes on all these plots together were then evalu-

ated by rigorous statistical tests to ascertain whether his observations could be accounted for by chance alone.

Reevaluation

Often, hypotheses are neither supported nor discredited by one set of experiments. Rather, new data indicate that researchers have basically the right idea but were wrong about a few particulars. The process then reenters the loop diagrammed in Fig. 1.9. The original hypothesis is reevaluated and revised. New predictions are made, and new experiments are designed that either alter the experimental treatment or control factors previously thought extraneous. Many circuits around this loop are often necessary before the final hypothesis is supported.

OBSERVATIONAL SCIENCE

This classic sequence of hypothesis, experiment, interpretation, and reevaluation, as central as it is in scientific thought, is not the only way that science proceeds. For much of its history biology depended on observation; biologists watched, sketched, dissected, and classified living things into groups. Even today, observational science combines with experimental science in important ways.

For example, if we could learn about the sun only by visiting it and directly conducting chemical analyses, we would know nothing about it. But, in fact, we know a great deal about the chemical composition of the solar surface. How is this possible? Gases in the sun are heated to such high temperatures that they radiate light into space. Here on Earth, we can analyze that light and compare it with light given off by gases that we heat in laboratory apparatus. Thus, by comparing experimental results with observations, we can infer a great deal about the nature of an object beyond our direct reach. So powerful is this technique that the element helium was detected and named on the sun before it was found on Earth.

Historical Science

In a similar vein, one might conclude that we cannot determine anything about the history of the earth. After all, we can't do experiments involving things that happened 100 million years ago. In one sense that is true; we can't visit the past. But we can observe evidence of past events, compare them with events that occur today, and draw useful conclusions though inference.

Figure 1.10, for example, shows one of hundreds of similar objects recovered from a deep-sea drilling core taken in the Gulf of Mexico. That core sample brought to the surface layer upon layer of sediment that had accumulated over millions of years. Those sediments contained many fossils—mineralized remains of animals that died and were buried. The particular object pictured here was buried roughly 170 million years ago.

As the figure caption explains, researchers used observation and comparison to determine that this fossil represents an ancient relative of modern-day organisms called sea urchins. This fossil thus enables us to conclude that sea urchins existed 170 million years ago and that they differed from living sea urchins in certain ways. It also

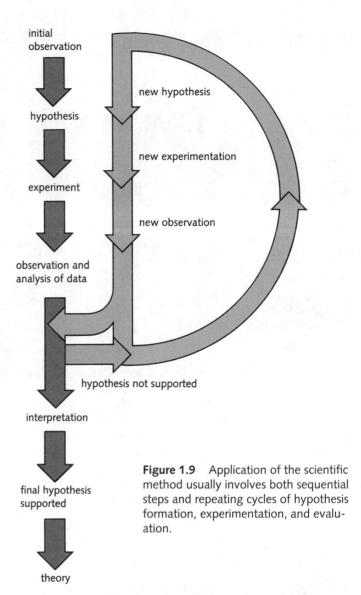

Figure 1.9 Application of the scientific method usually involves both sequential steps and repeating cycles of hypothesis formation, experimentation, and evaluation.

provides one of many pieces of information showing that life has changed over time. So although we can't travel backwards through time, we can use scientific methods to make strong inferences about life on the ancient earth.

Scientific Theories

Sometimes an hypothesis will continue to expand in complexity and to grow in the power and the accuracy of its predictions until it becomes worthy of the name **theory.** Charles Darwin's original observations and hypotheses about change in living organisms over time grew and expanded for several years before he presented them to the world as the *theory of evolution by natural selection.*

Figure 1.10 This fossil is roughly 4 cm across, has fivefold symmetry, and had a tough calcium carbonate skeleton. What was it? Are there similar organisms alive today? There certainly are: they are called sea urchins (see Chapter 28). The pockmarks on the surface of the fossil are places where spines would have attached, just as they do in present-day urchins. Despite many other general similarities, however, this fossil is not *precisely* identical to any living sea urchin. Thus it represents an urchin species from the Jurassic period (roughly 170 million years ago) that has become extinct.

Note that the word *theory* has a very specific meaning in science. When you say, "I have a theory," you probably mean something like "I have a hunch." But when scientists talk about gravitational theory or evolutionary theory, they refer to a long-established body of observations and experimental tests.

SCIENTIFIC METHODS IN THE PUBLIC EYE: CASE STUDIES

Applying the scientific method to noncontroversial events in nature can be straightforward. But today scientists are often called upon to pass judgments on matters of great emotional or financial concern—to place the label *scientifically proven* on products ranging from toothpaste to potentially lifesaving new medications. People with cancer demand to know which treatments will save their lives. Governments, corporations, and individuals want specific answers to environmental questions: How much waste treatment is enough? Will another coastal housing development destroy local shellfish beds?

Designing and evaluating experiments dealing with such important questions is very difficult. In many cases no single trial can satisfy everyone. As examples of how important experimental design and interpretation are, consider the following two case studies.

Medication for Bacterial Infections in Aquarium Fishes

Hypothesis   Brand X antibiotic is effective in treating certain diseases of ornamental aquarium fishes.

Background   Fishes in home aquaria get sick, so hobbyists want medications to cure them. The term *scientifically proven* carries weight with consumers, so companies scramble to put it on product labels. There is no equivalent of the Food and Drug Administration (FDA) to set standards for "proving" the value of aquarium medications, and *Consumer Reports* has not yet evaluated these products. As a result, pet supply manufacturers are left on their own to test, evaluate, and advertise products. With no standard, experiments may or may not be competently designed.

Experimental design   The manufacturer of Brand X antibiotic hired independent academic researchers to test drug action on four bacterial fish diseases. Thirty-six aquarium fishes were chosen and observed to make certain that they were healthy and vigorous. Then four types of bacteria (labeled here B1, B2, B3, and B4) were each inoculated into nine fishes. Disease appeared in all injected individuals after 3 days. No treatment was instituted until the eighth day to ensure that several animals in each group would be seriously ill. Medication was then administered to all individuals at recommended doses, and the results were followed for 5 days.

Results   Results were presented in two forms. Survival of animals is shown in Table 1.1. An additional part of the report contains verbal descriptions of visible changes in the degree of infection of survivors. Those verbal reports generally indicate lessening of disease symptoms; in two cases animals recovered completely by the end of the experiment. The experiment is summarized in a different manner in Fig. 1.11.

Conclusions according to company spokesperson   Before treatment, fishes were dying at a rate of approximately two per day. After treatment began, only two died in 5 days. Other surviving fishes were infected but showed significant improvements during treatment. Therefore, Brand X is of significant value in treating these diseases in these fishes.

Possible alternative conclusions   Because of the way this experiment was designed, the data it yields cannot exclude the following alternative interpretations:

1. Fishes that survived 7 days of illness might have been strong enough to recover on their own without treatment. Fishes, like humans, have immune systems—defenses against infection—that can successfully battle bacteria. Fish immune systems, however, usually take about a week to mobilize fully. Thus the infections could have killed physically or genetically weaker individuals before their immune systems sprang into action. Stronger individuals, meanwhile, might have survived with or without antibiotic treatment.
2. It is possible that other procedures associated with medication (such as water changes) reduced mortality independently of drug action.

Discussion   The manufacturer's interpretation of the results follows the logic "The fishes had disease; they were treated; they stopped dying of the disease; therefore, the medication works." But this sort of "before-and-after" interpretation is not valid in the absence of proper controls. To confirm the efficacy of Brand X, the experiment should have involved twice as many fishes divided into groups as shown in Fig. 1.12. One set (the experimentals) would have been treated just as these were. Another set (the controls) would have been treated exactly the same way until the eighth day. At that point, instead of adding medication to the controls, the researchers would have added what is called a *placebo:* a look-alike substance that contained no antibiotic. Then, by *comparing* survivorship and health of controls and experimentals, the researchers

Table 1.1   *Survival of Fishes After 7-Day Infections and 5-Day Treatment*

	Initially	7 Days After Infection	After 5 Days of Treatment
B1	9	3	3
B2	9	3	3
B3	9	5	3
B4	9	6	6
Total	36	17	15

Figure 1.11   A procedural outline of the fish drug experiment, as conducted, summarizes the investigators' reasoning. Do you see any problems with this experimental design?

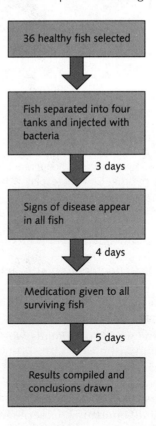

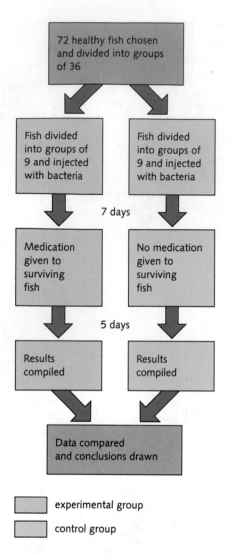

Figure 1.12 A properly controlled experiment to test the efficacy of an aquarium drug. Can you explain why this design is preferable to the one shown in Fig. 1.11?

could have made a judgment about whether the medication affected the course of disease.

Furthermore, the verbal descriptions of surviving fishes should have been made by trained observers *who did not know which animals received medication and which did not.* In this manner, the subjective process of describing the animals' condition could not have been biased by expectations that medicated fish would be better off. Finally, those descriptions should have been evaluated by the head researchers, who (for the same reason) *should not have known which descriptions applied to experimentals and controls until after final evaluation was complete.* This would have constituted a **double-blind study,** because it contained two sets of "blind" evaluations that could not have been influenced by anticipated or hoped-for results.

Because of these flaws, this study would never be accepted for publication by a scientific journal whose articles are reviewed by other scientists. Unfortunately, this sort of flawed logic is accepted by many nonscientists. Experiments not unlike this one are the source of "scientific evidence" used in marketing many products, such as the once widely touted but worthless "cancer drug" laetrile. Unfortunate individuals in the grip of a serious disease for which there is no accepted cure hear that some people were sick, that they took the medication, and that they later recovered. Although that sequence of events may have occurred, it proves nothing about the drug's ability to cure disease.

Clinical Trial of Azidothymidine (AZT) in the Treatment of AIDS

Hypothesis   The drug azidothymidine (AZT) is effective in the treatment of acquired immune deficiency syndrome (AIDS).

Background   AIDS is caused by a virus that cripples the body's ability to fight disease. (We will cover AIDS in detail in Chapter 42.) Most people infected with this virus die of *opportunistic infections,* diseases that the body normally fights off but that ravage people with AIDS. Once afflicted with full-blown AIDS (as defined at the time), patients have no hope of survival without treatment. AZT first showed promise by interfering with the reproduction of the causative virus in test-tube experiments. Preliminary trials in seriously ill patients were not conclusive.

Experimental design   This experiment, sponsored by the manufacturer of AZT, was a double-blind, randomized, placebo-controlled study. Obviously, researchers did not inject healthy individuals with the virus; persons already infected by the virus were asked to volunteer. A computer-generated code randomly assigned each subject to receive either AZT or a placebo. Neither the subjects nor the researchers evaluating their condition knew which patients received AZT. The placebo itself was designed to look and taste just like AZT. Subjects were asked not to take any other medications without the permission of the researchers. No other treatment to prevent or control opportunistic infections was permitted to either experimental or control subjects during the course of the study. Data were rigorously analyzed by advanced statistical methods to determine whether differences between experimentals and controls could have been due to chance alone (Fig. 1.13).

Ethical considerations weighed heavily in experimental design. The study was approved by committees at each of 12 participating medical centers across the country. An additional independent board was established to review the study data on a regular basis. It was this group's responsibility to protect the subjects if it appeared that AZT was causing unacceptable toxic side effects. This board also kept watch for any sign that the drug was clearly benefiting those who received it. Every subject signed a consent form certifying that he or she understood the risks and benefits of the procedure. (Imagine the courage of these participants; they knew that half of them were voluntarily forgoing treatment that might save or prolong their lives. They did so in order to help produce evidence about whether or not the drug worked.)

Results Because many of the data from this experiment deal with complex phenomena that we will not cover until later, we will present only the most basic results here. Figure 1.14(a) and (b) show the number of patients who developed opportunistic infections during the treatment period. Figure 1.14(a) shows results among those people with full-blown AIDS; Figure 1.14(b) shows similar information for subjects with the less severe illness then called AIDS-related complex (ARC). Note that in both groups subjects receiving AZT had a lower rate of opportunistic infections than did those receiving the placebo. Table 1.2 shows that AZT recipients also had significantly higher chances of surviving the 24-week treatment period. Additional data indicated that AZT caused side effects of varying severity in different subjects. Side effects ranged from muscle aches, insomnia, and nausea to serious anemia and the loss of certain classes of white blood cells.

Conclusions This study indicated that AZT produced the best short-term results of any medication available for AIDS treatment at the time it was conducted. Researchers therefore recommended the use of AZT in the treatment of AIDS and AIDS-related complex (ARC) in patients who could tolerate its side effects.

Discussion This study was terminated earlier than planned because emerging data showed that AZT produced positive results. When the study was terminated, placebo recipients were offered the opportunity to receive AZT. Additional information summarized 3 months after the study credited AZT with a fourfold to sixfold reduction in deaths in the study population over the entire period.

The data from this study, along with other data not discussed here, strongly influenced the continued use of AZT in treating AIDS. (Today certain other drugs are used in conjunction with or instead of AZT for individuals who cannot tolerate AZT's side effects.) These data were also used to support the contention that AZT treatment might be useful for individuals infected with HIV but not yet showing visible symptoms of the disease. Can you see how these data hint at that possibility? What sort

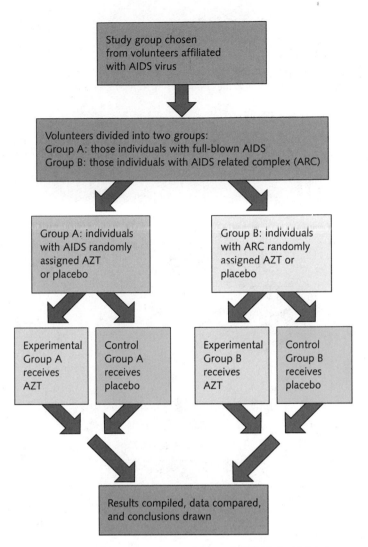

Figure 1.13 Schematic representation of AZT clinical trial design. Because this experiment dealt with human subjects, researchers had no control over the selection of individuals infected with HIV. Beyond that point, however, patients were diagnosed, divided into two medical categories (AIDS and ARC), and further subdivided into experimentals and controls. Double-blind procedures ensured that neither experimenters' expectations nor those of the patients would affect data gathering or interpretation.

Figure 1.14 Patients with AIDS who developed opportunistic infections during the treatment period. **(a)** Treatment results among patients with full-blown AIDS. **(b)** Treatment results among patients with AIDS-related complex (ARC). Adapted from the paper by Fischl et al. referenced at the end of the chapter.

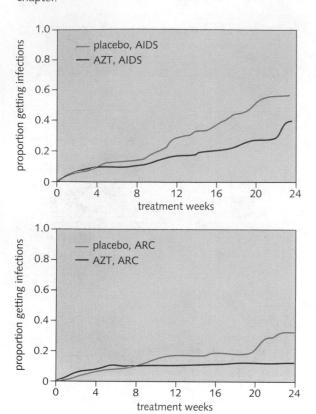

Table 1.2 *Projected Probability of 24-Week Survival*

Experimental Group	Survival Rate
AIDS	
AZT	96%
Placebo	76%
AIDS-related complex (ARC)	
AZT	100%
Placebo	81%

NOTE: This table was adapted and simplified from Table 1 in the paper by Fischl et al. referenced at the end of the chapter. Those interested in the statistical analyses accompanying these data are encouraged to read the paper in its original form.

of experiment might you design to test that hypothesis? Can you see the problems inherent in actually carrying out such a study?

Science in the Real World: Between a Rock and a Hard Place

As you can see, designing experiments is far from simple. The first example we examined was poorly designed and could neither support nor discredit the hypothesis it was intended to test. The second, though extraordinarily difficult to conduct, was well designed and—despite shortcomings associated with human experiments—still stands, years later, as a landmark in its field.

Yet the AZT study is a "landmark" in an unsettling way as well. Because the double-blind, placebo-controlled trial was ended soon after signs that the drug was working, the trial was neither large enough (in numbers of participants) nor long enough (in light of AIDS's time course of a decade or longer) to produce statistically solid evidence that AZT prolongs life. It clearly kept people healthier in the short term, but whether or not it would actually keep them alive for significantly longer periods of time is another question, and one that cannot be answered from the data presented.

Since that first trial, medical researchers have been plagued by a dilemma, articulately posed in *Science* magazine: "Should (researchers) insist on scientific accuracy, designing long-term trials (or enormous shorter ones) and withholding unproven drugs from patients on the grounds that untested therapies could do more harm than good? Or should they heed compassion and release drugs as soon as they show any hint of effectiveness, running the risk that—in the absence of carefully controlled trials—it may never be possible to tell which drugs are actually more effective?"

In fact, the efficacy of AZT—like many other issues surrounding the AIDS epidemic—is still the subject of considerable debate. Since this study, there have been other studies of the usefulness of AZT. The data from this ongoing research, including a 1993 British study, may significantly revise our understanding of the value of AZT. Physicians' descriptions of the disease are also constantly changing; the term *ARC*, for example, is no longer widely used. These redefinitions come from the data of scientific inquiry; researchers have realized that there is a broad spectrum of symptoms ranging from relatively minor opportunistic infections to full-blown AIDS, where the immune system barely functions at all.

Problems in designing definitive tests for hypotheses and the uncertainties generated by our constantly changing scientific view of the world are not issues that concern AIDS drug research alone. Precisely the same difficulties

(handwritten top margin) cell (4)

apply to evaluation of drugs and other therapies aimed at treating other lethal diseases such as cancer and heart disease. As a consequence, certain well-informed groups of patients are pressuring researchers to devise new, scientifically acceptable alternatives to traditional experimental formats. Real-world experimental design problems aren't restricted to clinical research; slightly different issues in experimental design, interpretation, and policy decisions plague those trying to understand pollution and climate change.

These scientific difficulties—and the need for society at large to make policy decisions despite uncertainty—underscore the excitement and challenge of understanding biological research today. The science of life has never been more important to *your* life than it is today. To ensure that you don't become a pawn in a game whose rules you don't understand and, in a positive sense, to fulfill your responsibility as an informed participant in our democracy, you must become familiar with important issues in science and with the nature of scientific information itself.

SUMMARY

Life encompasses a great range of phenomena, from interactions among molecules to events that affect our entire planet. All living things grow, reproduce, take in and process materials from their surroundings, and respond to their environment. At the root of all life is the biological information stored in molecules of DNA.

Throughout history, humans have sought to explain the natural world. The development of science has been one response to that attempt to explain. Scientific knowledge is assembled from direct experience with nature, depends on experimentation, and is never final. Scientific methodology depends heavily on investigations involving an hypothesis, design and execution of experiments to test that hypothesis, and interpretation of results. This process is complex and highly subjective and can be influenced by human limitations such as cultural bias.

STUDY FOCUS

After studying this chapter, you should be able to:

- Describe some of the ways in which life affects Earth.
- Understand the essential attributes of scientific thought.
- Describe at least two styles of scientific investigation.
- Analyze and criticize experimental results and conclusions.
- Appreciate the influence of cultural values on science.

TERMS AND CONCEPTS

biosphere   3
Gaia hypothesis   5
inductive reasoning   9
hypothesis   7, 9
deductive reasoning   9
controlled experiments   9
experimental variable   9
control   9
theory   11
double-blind study   14

(handwritten) organism (4)
cell membrane ⎫
organelles ⎬ fig 1.4
nucleus ⎭
DNA (4)
genetic code (5)
homeostasis (6)
Science (7)
scientific method (8)

REVIEW

Objective Questions (Answers in Appendix)

1. The Gaia hypothesis states that
   (a) life processes are regulated solely by planetary conditions.
   (b) over millions of years global conditions have fluctuated greatly.
   (c) global conditions are regulated by life processes.
   (d) living organisms have no influence on global conditions.

2. What does a scientist do with an hypothesis?
   (a) controls it
   (b) changes it to a theory
   (c) refers it to a world authority
   (d) tests it

3. The following steps are integral to scientific methodology. Which is the last step?
   (a) Make observations.
   (b) Conduct experiments.
   (c) Generalize from test results.
   (d) Collect and organize test results.

4. What happens when an hypothesis is neither supported nor discredited by one set of experiments?
   (a) The results are published.
   (b) The steps involved in the scientific method are repeated.
   (c) The original observations and hypotheses become a theory.
   (d) More variables are added to the experiment.

Discussion Questions   *(handwritten)* Fig 1.9 p.11

5. Describe the steps in the standard experimental protocol often labeled "the scientific method." How does that ideal differ from real scientific experimentation?

6. What are some current issues that involve both biology and *(handwritten)* p.14 questions of ethics or of public policy? Can science alone offer a definitive solution to any of these issues?

7. Can science ever provide the absolute truth about anything in nature? Why or why not? Give an example of something that was once accepted as a "fact" in any branch of science but has since been disproven.

8. Have you or any of your friends or family members been afflicted with cancer, heart disease, AIDS, or other potentially lethal diseases with no cure? What did that experience do for your understanding of the relationship between research and medicine? What, in your opinion, is the best approach to evaluating the efficacy of drugs aimed at such devastating illness?

READINGS

Cohen, Jon. "Searching for markers on the AIDS trail." *Science* 258 (1992): 388–390.

Lovelock, James. *The Ages of Gaia.* New York: Norton, 1988.

Fischl, Margaret, et al. "The efficacy of azidothymidine in the treatment of patients with AIDS and AIDS-related complex—A double-blind, placebo-controlled trial." *New England Journal of Medicine* 317 (1987):185–191.

Richman, Douglas D., et al. "The toxicity of azidothymidine in the treatment of patients with AIDS and AIDS-related complex—A double-blind, placebo-controlled trial." *New England Journal of Medicine* 317 (1987): 192–197.

Thompson, William I., ed. *Gaia: A Way of Knowing.* Great Barrington, MA: Lindisfarne Press, 1987.

Futuyma, D. J. *Science on Trial: The Case for Evolution.* New York: Pantheon Books, 1983.

Radner, Daisie, and Michael Radner. *Science and Unreason.* Belmont, CA: Wadsworth, 1982.

Gould, Stephen J. *The Mismeasure of Man.* New York: Norton, 1981. A masterfully written, enlightening, and often infuriating exposé of abuses of science in society over the past century.

Kuhn, Thomas S. *The Structure of Scientific Revolutions.* Chicago: University of Chicago Press, 1970.

2

Science and Society

When Mark O'Donnell of Holbrook, Massachusetts, was a boy, he and his friends played in "the pastures," a field near their home that offered acres to romp in and fascinating things to play with. Among the kids' favorite toys were abandoned barrels filled with a green, jelly-like substance they called "moon glob." Mark and his best friend spent hours rolling in those barrels and tossing handfuls of moon glob at each other.

At age 27 Mark was diagnosed with a rare cancer of the adrenal glands. He died a year later. Mark's best friend contracted Hodgkin's disease, a cancer of the immune system. Within a 3-year period, 10 other people in that small town—most of whom had played in the pastures—were diagnosed with cancer. On one street not far from Mark's house, a woman died of breast cancer in nearly every other house.

For a time, no one connected these apparently separate tragedies. Then an evening news program rated a chemical plant near the pastures as among the most toxic places in the nation. Within minutes Mark's mother Joanne got a phone call. "That's what got Mark," her friend said.

Joanne O'Donnell wasn't trained as a biologist and hadn't given environmental issues much thought. But those tragic events drew her into a tangle of scientific and legal issues. What is cancer? What causes cancer and how? What happens to such compounds as dioxin, arsenic, DDT, and chlordane in the environment? What kinds of studies and what sort of data are needed to test and either confirm or disprove the hypothesis that toxic compounds in the environment can increase cancer risk (Fig. 2.1)?

What Joanne learned about these scientific issues and their links with law and politics transformed her life. She campaigned to close the plant—which, in addition to dumping toxic compounds onto the ground, had poured them into a brook that fed the local reservoir. Her efforts facilitated impressive legal battles. The chemical plant was closed, the Environmental Protection Agency placed the pastures on its Superfund cleanup list, and the company alleged to have polluted the reservoir was sued for negligence.

But were those real victories? The corporation declared bankruptcy, and its top officials maneuvered legally to protect their personal assets. Of course, legal decisions couldn't bring Mark back. Medical science

Figure 2.1 Nearly all industries generate waste products. How dangerous are these wastes? Can they be recycled? If not, can they be disposed of safely? The waste shown here is low-level radioactive material dumped into a trench near Hanford, Washington.

may have helped some affected residents of Holbrook to live more or less normal lives, but not without great emotional hardship and enormous medical bills.

This story is but one example of the impact of science on society and of the ways scientific issues can suddenly change from abstract concepts—about which you may think you don't much care—to matters that directly affect your life. This story also makes the general point that, in the long run, we will almost always be better off if we can understand, predict, and avoid environmental problems in advance, rather than trying to correct and seek redress for errors after the fact. Joanne O'Donnell has had a major positive impact on the quality of her local environment—but that success doesn't alter the fact that her son is dead. If regulatory agencies had paid proper attention *in advance* to the threat of toxic wastes in Holbrook, the O'Donnells and their neighbors could have been spared a great deal of pain, suffering, and loss. They could also have saved large sums of money spent on legal fees and even larger sums that are still being paid out for medical bills and environmental cleanup.

The residents of Holbrook have learned a lot about the effects of science on society. Some of them have also acquired insights into the influence society has on both the way science is done and the effects scientific findings have on the way we live. They have learned, for example, that if members of the news media leap onto one side of a scientific debate or the other, or if the public is "primed" by news events to receive an issue, inconclusive scientific data can quickly alter public opinion. They also learned

that an unusually large number of local cancer cases are not direct proof that something is amiss in the environment. Why? Because "cancer clusters," as they are sometimes called, can occur by chance. Consequently, medical histories must be taken and statistical studies must be performed to determine the likelihood that environmental carcinogens are to blame.

Society has other effects on the process of science as well. Public attitudes affect the sorts of questions scientists are interested in, the specific questions they ask as they conduct their investigations, the way they evaluate their experimental data, and the way those data are translated into public policy. Two issues from current headlines demonstrate this relationship quite clearly: the history of research and public health actions taken around the AIDS epidemic, and both national and international response to global environmental issues.

SOCIETY, DISEASE, AND MEDICINE: IN TIME OF PLAGUE

Plague. The word has an ominous ring to it—and for good reason. Diseases ranging from yellow fever, bubonic plague, and smallpox, to syphilis, leprosy, and tuberculosis have devastated humans since prehistoric times. One plague wiped out nearly a quarter of Europe's population in the fourteenth century. A single epidemic of influenza in 1918 killed more than 20 million people. And as recently as 1970, another epidemic killed 10,000 people in the Indian State of Bihar alone.

In every culture people have tried to understand disease, and for most of history religion offered the only explanation. In Asia, where smallpox was a fact of life for generations, parents knew that if sick children recovered, they would be safe from future attacks. They therefore prayed to a smallpox goddess to spare their sons and daughters. In Europe diseases were often considered God's punishment for sin or for "deviant" behaviors of the poor, minority groups, or foreigners. Epidemics were nurtured as much by ignorance and fear as by crowding and poor sanitation. The primary reactions to early outbreaks of bubonic plague in Europe, for example, were panic, fear, and denial, as people sought desperately for someone to blame. Many unfortunate scapegoats were tortured and executed rather than treated (Fig. 2.2).

Why did people react this way? Because by blaming disease on individuals accused of being "different," the majority accomplished two goals. By persecuting the victims, they created for themselves the illusion of power over a situation in which they were otherwise helpless. They also created a psychological defense against fear by assuring themselves that only "that other sort of person" would get ill.

By the eighteenth century, physicians began to suspect that certain diseases were contagious, although they still didn't understand what made people sick. The late nineteenth century saw a fundamental breakthrough: it was recognized that microorganisms could make people sick and that proper treatment could cure many ailments. The twentieth century saw the rise of modern medicine and with it the assumption that most infectious diseases could be cured (or at least controlled) by scientific means.

Ignorance, fear, and superstition gave way to confidence in medicine's power to cure and prevent infection. Response to disease became straightforward: identify the problem, determine how the disease is spread, find a way to relieve symptoms while searching for a cure, and educate the public on how to avoid infection. Importantly, public health education has proven repeatedly to be as important as, or even more important than, medical advances per se in curbing epidemics. In other words, telling people how to avoid disease is more effective (and less costly) than attempting to cure the sick.

By the time you were born, a generation of Americans had grown up in a society unaccustomed to epidemics. (Vaccinations and public health measures had ended the last epidemic of polio in this country during the early 1950s.) By the last quarter of the twentieth century it seemed that any new disease would be faced with the same calm scientific resolve and focus on public education that triumphed over killers of the past. That self-assurance was destroyed early in the 1980s by a disease called AIDS.

Figure 2.2 In Europe diseases were often seen not only as divine actions but also as specific punishments for wrongdoing against the laws of God and man. As a result, public response to infection was often less than enlightened. This scene shows plague sufferers in the Italian city of Milan being tortured and executed.

AIDS: Society's Reaction

Acquired Immune Deficiency Syndrome, or AIDS (described in detail in Chapter 42), perplexed the medical community when it surfaced in the early 1980s. The first cases presented a baffling collection of ailments, including a rare form of cancer and fatal infections from microorganisms never previously known to cause disease in healthy people. The first clear message about the syndrome was its final effect: a crippled immune system that left the body defenseless against infections. In the United States (although not in Africa and other countries) AIDS first appeared among members of the gay community, in which it spread rapidly. Soon, cases appeared among intravenous (IV) drug users and their sexual partners. The challenge of this new disease brought out both the best and the worst in the relationship between science and society.

On the positive side, an important part of the scientific community responded admirably and courageously to the challenge. Given the fact that AIDS is caused by a virus then new to science, the speed with which researchers accomplished their ground-breaking detective work is astonishing. And as devastating as the epidemic has turned out to be around the world, we are extraordinarily lucky that it did not begin to spread 20 years earlier. Why? Because the basic scientific understanding of molecular biology that proved essential to identifying this new virus simply didn't exist until the late 1970s.

As it happened, research proceeded on several fronts. First, epidemiologists conducted studies among gay and IV-drug-using AIDS patients. By the fall of 1981 they suggested that AIDS was a contagious disease spread through blood (by sharing contaminated needles) and through sexual intercourse. By 1982 the first cases of AIDS were reported among infants born to infected mothers and among hemophiliacs and surgical patients who had received transfusions of blood or blood products (Fig. 2.3). Physicians closest to the data became convinced that AIDS was caused by an infectious agent contained in semen and blood. They soon determined that this agent (which was still unidentified) could spread from one person to another by only three means: through unprotected sexual intercourse, through injection (or transfusion) of contaminated blood, and from infected mother to unborn child. It wasn't long before the human immunodeficiency virus (HIV) was isolated and identified.

In recent years, new drugs have been developed and pushed through testing procedures to help people with AIDS. One researcher attempting to develop a vaccine showed his boldness and dedication by injecting himself with a test vaccine. What's more, the international effort against AIDS has led to an explosion of work on the immune system, the target of HIV infection. This research

Figure 2.3 No one in this family fit the media's image of a "high risk group" for AIDS in the early 1980s. But by the time this picture was taken, the father —a hemophiliac— had acquired AIDS from a transfusion and had unwittingly passed it on to his wife. She, in turn, passed it to her newborn son during pregnancy or through breast feeding. The young girl—though she lived in close contact with three infected family members—was not infected.

will ultimately benefit all medical science by leading to a better understanding of the body's natural defenses of disease, which we will explore later.

Making a Bad Situation Worse

Unfortunately, serious problems arose in translating this spectacular burst of scientific and medical knowledge into public health education programs. Large portions of American society—including some physicians, researchers, journalists, and members of the public—could not face this lethal and (thus far) incurable disease in a rational manner. Instead, the fear, ignorance, and denial of old fueled a return to prejudice and superstition. In the United States minority groups (gay men and IV drug users) and outsiders (Haitians) were singled out for blame. In Africa and Asia AIDS was called "the European disease." The Japanese government delayed efforts at widespread public preventive education until late in 1992 based, in part, on a popular belief that the Japanese had superior immune responses. These reactions served the same psychological purposes as they did during the Middle Ages: by insisting that the disease is caused by and restricted to "others," people vent their frustrations and thus feel safer.

But such reactions seriously hindered our ability to control the spread of AIDS until thousands of people had died and hundreds of thousands had been infected. Some people in the medical community refused to treat people with AIDS altogether. Governments around the world were slow to react to the epidemic, ultimately missing the chance to stop its spread at an early stage by failing to support both AIDS research and effective public health measures. To cite just two examples, nearly a year after cases of transfusion-related AIDS were reported and confirmed, U.S. blood banks denied that there was enough risk to merit the expense of screening blood and blood products from donors at risk of HIV infection. Shortly thereafter, French physicians supervising blood products given to hemophiliacs decided against taking steps to eliminate HIV contamination of these materials. The result in both countries was the infection of enormous numbers of hemophiliacs and transfusion recipients, hundreds of deaths, and needless suffering.

For the most part, scientific efforts against AIDS have followed a classic pattern: study the disease, identify its cause, determine how it is spread from person to person, and take steps to prevent that spread—all the while racing for treatments aimed at prevention and cure. The great irony of AIDS is that we have known for more than a decade the principal means by which the virus is spread: through contaminated blood and by unprotected sexual contact. One would like to think that, armed with this scientific knowledge, society would have responded quickly and compassionately to check the spread of this lethal virus. Sadly, fear, denial, and prejudice have repeatedly stood between science and effective public health action, and AIDS has continued to spread throughout the world.

Current estimates suggest that up to a million and a half Americans—an increasing number of whom are women, children, sexually active college students, and others who do not fit into the categories originally identified as "high risk"—are infected with HIV, although only a fraction of those individuals are currently ill. Between 30 and 110 million people are likely to be infected worldwide by the year 2000, according to recent WHO estimates. Statistically, it is possible that one or more members of your biology class may be infected.

What is the general lesson regarding science and society here? That even the most useful scientific knowledge can benefit society only to the extent that society accepts that knowledge and acts on it responsibly.

ECOLOGY AND ECONOMICS: TENDING OUR HOUSES

Francisco (Chico) Mendez was born into modest circumstances and spent most of his life deep in an Amazon jungle town called Xapuri. There he earned a living as a *seringueiro,* someone who taps rubber trees for their sap. But soon after he became the leader of the local rubber tappers' union, Mendez found himself embroiled in a national dispute that made global headlines.

Seringueiros harvest products of the rain forest for a living (Fig. 2.4). By virtue of both temperament and profession, they prefer to see most of the jungle left standing. They correctly surmise that their region will produce more valuable goods for a longer period of time if the rain forest remains intact, allowing them to harvest its biological wealth as renewable resources. That philosophy puts them squarely at odds with wealthy and powerful ranchers, who prefer to destroy the rain forest to plant grasses for cattle. For reasons we will discuss in the next several chapters, Amazonian rain forest soils cannot support such grasses for long, so within years after clearing such areas are reduced to wastelands.

Campaigning to save the rain forest, Mendez found himself first labeled an environmentalist, then lauded as a self-styled ecologist. His work received international acclaim, culminating in an award from the United Nations. Then in December 1988 Mendez was shot and killed outside his jungle home by the son of a cattle rancher. Like the O'Donnells, he had been caught in a fire storm that, though steeped in science, was driven by money, power, and intrigue. That's not surprising, because ecology and economics (and, therefore, politics) have affected one another since ancient times.

An Historical Perspective on Ecology and Economics

The words *ecology* and *economics* are both derived from the Greek word *oikos,* which means "house." Economics, nicknamed "the dismal science," is concerned primarily with human household management, whereas ecology deals with the study of nature's "houses" and their inhabitants. For a while, the disciplines ran in parallel, but by the eighteenth century their philosophies had diverged.

Figure 2.4 Rubber tappers—who settle in the rain forest and exploit its resources in a potentially sustainable way—are often caught in the center of economic and ecological controversy. On one side are indigenous forest peoples who resent any intruders. On the other side are developers, cattle ranchers, and an increasing number of subsistence farmers whose actions cause wholesale forest destruction.

The split began in 1749, when Swedish botanist Carl von Linné (Linnaeus) wrote "The Oeconomy of Nature," an essay in which he argued that God created "nature's economy" solely to serve the human economy. Adam Smith, the founder of economics and a disciple of Linnaeus, took his mentor's views to heart and championed humanity's birthright to dominate and exploit nature.

Most later economists adopted a similarly human-centered view of our interactions with nature. Over time, economists developed economic models that analyze many aspects of human activity accurately. But until recently, traditional economic theory dealt only with strictly *human* wealth. It assigned no economic value to *biological* wealth—such as living species and ecosystems—and broader *environmental* wealth—such renewable resources as clean air and water. For that reason, Robert Repetto of the World Resources Institute in Washington, D.C., has argued in *Scientific American* that traditional economic models "misrepresent the policy choices nations face. ...A country can cut down its forests, erode its soils, pollute its aquifers, and hunt its wildlife and fisheries to extinction, but its measured income is not affected as these assets disappear. Impoverishment is taken for progress."

In contrast, biologists and natural philosophers have long appreciated the economic and aesthetic value of environmental assets and have realized the consequences of unrestricted growth and development. The proponents of this view best known to the public were not ecological researchers but gentleman–naturalist–authors such as Henry David Thoreau. Thoreau, who built a cabin on a pond in Massachusetts that he immortalized in his most famous book, *Walden*, embraced the natural world as home to *all* life. He counseled society to recognize, and learn to live within, environmental limitations.

During the mid-twentieth century it became apparent that human activities were disturbing local ecosystems. In 1962 biologist Rachel Carson published *Silent Spring*, a disturbing documentary of the destructive power of chemical pesticides then widely used in agriculture. *Silent Spring* inaugurated the environmental movement by warning of ecological apocalypse—a spring without the sounds of birds and insects.

The scientific environmental movement, often at odds with economists, grew during the 1960s and early 1970s as ecological researchers strengthened the scientific underpinnings of environmentalists' philosophical appeals. Researchers monitoring interactions between human activities and the global environment detected what they believed was a human-caused increase in concentrations of atmospheric carbon dioxide. That increase, they warned, might lead to a rise in global temperatures dubbed "global warming." (That hypothesis is presented in more detail in Chapter 4.) Growing concern over environmental matters was reflected in several important

Rubber Tappers, Ranchers, and Rainforest Crunch®

It is morning in Xapuri, Brazil. Children have left for school, and most men are heading through the rain forest on their way to tap rubber trees. Other than American design T-shirts scattered through the crowd, the scene could be set in the present, a decade ago, or even further in the past.

But one not-so-visible element is new. Many women (who previously had no way to add to family income) are heading off to work, too, at a community nut-processing plant where they roast, sort, crack, and pack Brazil nuts. Many of those nuts were collected by their husbands within the Chico Mendez Extractive Reserve in a manner that does not disturb the ecological balance of the forest.

Those nuts are sold to Cultural Survival, a nonprofit group that promotes the well-being of indigenous peoples by selling nuts to manufacturers of such products as Rainforest Crunch® candy and Rainforest Crisp® cereal. Income generated by those sales enables Brazilian workers to purchase commercial goods for their families and provides an economic alternative for the community at times when the price for natural rubber falls.

The 4000-square-mile reserve, named in Chico Mendez's memory, was set aside by the Brazilian government to encourage the extraction of forest products. That extraction can be sustained over long periods, in contrast to large-scale cattle ranching, which fails after a few years. The project works because of a unique application of ecological and economic principles in determining the way that nuts are collected and sold. Working together, the World Wildlife Fund, the Inter-Ameri-

A rubber tapper cracks the hull of a Brazil nut with his machete.

can Foundation, the Ford Foundation, the Brazilian rubber tappers' union, and local conservation groups have determined the best way to harvest forest products and have helped the community set up processing facilities. In addition, Cultural Survival pays a fair price for the nuts—much more than traditional intermediaries have paid in the past.

Experiments elsewhere—involving local farmers in Ecuador, Bolivia, El Salvador, and the Philippines—are aiming for similar success with small-scale pro-

ducers of sugar, coffee, cocoa, and cashews. The goal of all these efforts is to make the economic changes necessary to encourage habitat conservation among indigenous people.

The success of the arrangement in Xapuri thus far is summarized by an elderly resident. "Everything we need is here," he told a reporter from the World Wildlife Fund, while standing in the midst of the forest. "We want to make sure that our children also have a future here."

Figure 2.5 Public demonstrations and other actions (not always as colorful or as high-profile as this one) have been—and will continue to be—vital in encouraging both federal and state environmental legislation and changes in the attitudes and practices of major corporations.

pieces of legislation designed to protect the environment and in regular demonstrations (Fig. 2.5).

Then during the summer of 1988 much of the United States sweltered under a heat wave, while agricultural areas from California to Georgia were scorched by drought that cut grain harvests by 31 percent. Scientists detected holes in the earth's protective ozone layer. Raw sewage, garbage, and discarded syringes washed up onto public beaches around New York and Boston. During one weekend coastal waters in Alaska, Rhode Island, California, and Delaware were being fouled by oil from separate tanker accidents (Fig. 2.6). One of those spills, caused by navigational errors aboard the tanker Exxon *Valdez*, befouled more than 1355 miles of previously pristine Alaskan coastline.

Scientists rushed to point out that the American heat wave and drought could *not* be cited as evidence that anything resembling global warming had begun. But somehow, all these disturbances together changed mass media attitudes and the public mood. *Time* magazine replaced its "Man or Woman of the Year" cover story with "Endangered Earth: Planet of the Year" (Fig. 2.7), and *National Geographic* ran a similar global issue. *Money* magazine ran an article about the *Valdez* spill entitled "Tanker from Hell" that described not the environmental effects of the spill but its long-term effect on business. "Exxon's oil spill," the article warned, "may usher in an age when companies—and their shareholders—will be penalized for a devil-may-care approach to the environment."

Soon thereafter, the fall of the iron curtain exposed

Figure 2.6 Coastal waters and beaches suffer incalculable harm from oil spills such as this one near Port Angeles, Washington.

Figure 2.7 This cover illustration is emblematic of the mass media's "rediscovery" of global ecological issues.

staggering problems in former eastern bloc countries and made it clear that capitalism held no monopoly on pollution. Vaclav Havel, then president of Czechoslovakia, declared in the *New York Times* that "I live in a country that suffers from serious environmental problems and is one of the greatest polluters in Europe. ...These are the consequences of Marxist ideology—the consequences of the arrogance of modern man, who believes he understands everything and knows everything, who names himself master of nature and the world."

Environment and Development

A United Nations conference on environment and development, nicknamed "The Earth Summit," was held in Rio de Janeiro in June 1992. That conference produced two important documents, the Rio Declaration on Environment and Development, and Agenda 21. The declaration enumerated 27 principles concerning the rights, responsibilities, and relationships of developing and developed nations as they form a "new global partnership" aimed at improving the human condition while protecting Earth's life-support systems. Agenda 21 outlined 115 program

areas addressing problems, challenges, and opportunities in ecological issues you will learn about in the next several chapters. The agenda also recognized data suggesting several points: that the global environment is deteriorating; that biological and environmental assets, the earth's primary source of natural "capital," are in grave danger; that the ozone layer protecting the earth's surface from harmful solar radiation is being damaged; and that increasing emissions of carbon dioxide may cause global warming.

Still, environmental legislation and applied ecological research remain controversial and politically charged issues. Whenever an industrialized nation suffers a recession, as the United States did in the mid-1970s and at both ends of the 1980s, it is convenient to some to claim that environmental protection has hurt the economy. Such politicians and business leaders sometimes create a false dichotomy between "jobs" and "the environment," often blaming environmental legislation for job losses and other problems actually caused mainly by long-term economic trends, changes in the national economy, and patterns of international trade. One old bumper sticker popular in New England read "Hungry? Cold? Eat an environmentalist!" A newer version in the Pacific Northwest proclaims "We need jobs, not owls!"

Our point here is to emphasize strongly that in many cases there are far greater long-term costs associated with *failing* to protect the environment—costs that are reflected in the loss of fresh water, viable farmland, valuable forests, and fishing stocks. Successful long-term management of global resources is possible only if both economic *and* environmental concerns are considered together in formulating local, national, and international policy. According to *The Global Partnership*—the action plan drawn up at the Earth Summit—humans in all societies must be willing to make changes in attitudes and lifestyles if our children and their children are to inhabit the earth responsibly.

Some changes will necessitate major international economic innovations. There is still no consensus, for example, on how best to encourage changes in energy use and manufacturing aimed at controlling certain widespread pollutants. What would work best: strict legal emission controls and disciplinary fines, economically positive cleanup incentives, or consumption taxes? We aren't yet sure, although many ecologists and numerous environmentalists point out that we do not have an infinite amount of time to solve the problem.

Other adjustments needed will be relatively minor, requiring only the commitment and determination to set ourselves on a sensible course. There are many creative ways to combine basic ecological understanding, sophisticated environmental analysis, economic principles, and the will of local people to produce ecologically, socially, and politically sound agricultural alternatives to current practices (see Theory in Action, Rubber Tappers, Ranchers, and Rainforest Crunch®).

To understand why any changes are necessary, we must understand the rules that govern the global phenomenon we call life. Furthering that vital understanding is a major goal of this book.

SUMMARY

Science has positive and negative effects on society. The attitudes of society, in turn, affect the way in which science is conducted and the uses to which it is put. The similarities and differences between society's response to the AIDS epidemic and to global environmental problems emphasize the reality that science does not operate in a vacuum. In an ever-more crowded and technological world, both the process of science and its findings increasingly have profound economic, political, and social repercussions. All of us must be willing and able to recognize those repercussions and to use fruits of science in the most logical way possible.

STUDY FOCUS

After studying this chapter, you should be able to:

- Give examples of historical and contemporary interactions between science and society.

- Describe obstacles that societal attitudes have placed in the path of the dissemination of scientific information aimed at fighting the spread of AIDS.

- Relate some of the efforts that environmentalists have made to save the Amazon rain forests.

- Discuss why economists and ecologists must work together to address issues involving human interactions with local and global environments.

REVIEW

Objective Questions (Answers in Appendix)

1. A modern publication that inspired public awareness of environmental concerns was
   (a) Adam Smith's *Wealth of Nations*.
   (b) Rachel Carlson's *Silent Spring*.
   (c) Carl von Linné's "Oeconomy of Nature."
   (d) William Fitzgerald's *Exxon Valdez*.

2. AIDS is *not* transmitted by
   (a) contaminated food and water.
   (b) blood.
   (c) sexual intercourse.
   (d) birth.

Discussion Questions

3. Have there ever been either bona fide environmental problems or overblown "poison scares" in your hometown? If so, how did you, your family, and your friends and neighbors respond to news coverage? What sorts of experts were consulted? Was the debate about what action to take "scientific" in the technical sense of the word?

4. Did you know, *before* reading this chapter's coverage of AIDS, how HIV is and is not transmitted? Do you know what sorts of behavior do and do not place individuals at risk? Where did you get that information? Was it available to you as soon as you felt you needed it? Why or why not? Check pages 22 and 23 and see if your understanding of the situation is supported by current data. Compare your answers with those of friends and classmates.

5. Whether or not the Gaia hypothesis discussed in Chapter 1 is "true" in an absolute sense, its thesis has value in directing studies of the global environment. How?

READINGS

The Global Partnership for Environment and Development: A Guide to Agenda 21. UNCED, United Nations Publications, Document #92-1-100481-0, New York and Geneva, 1992.

Repetto, Robert. "Accounting for environmental assets." *Scientific American* 266 (June 1992): 94–100. A well-informed, concise effort to incorporate ecological thinking into economic models.

Burnham, J. C. *How Superstition Won and Science Lost: Popularizing Science and Health in the United States.* New Brunswick, NJ: Rutgers University Press, 1987. A well-documented and perceptive explanation of at least some reasons for the decline in science education in this country.

Shilts, R. *And The Band Played On: Politics, People, and the AIDS Epidemic, A First-Person Account of AIDS in America.* New York: St. Martin's Press, 1987. The title tells it all. This is a no-holds-barred, controversial, often shocking account of the interplay of science and society during the first years of the AIDS epidemic.

Berman, M. "The cybernetic dream of the twenty-first century." *Journal of Humanistic Psychology* 26 (1986): 24–51.

Freidman, S. M., S. Dunwoody, and C. L. Rogers. *Scientists and Journalists: Reporting Science as News.* New York: American Association for the Advancement of Science, originally published for AAAS by The Free Press, a division of Macmillan, 1986. A dispassionate look at the opportunities, challenges, and problems faced by those who try to report on scientific developments in mass media.

Evernden, N. *The Natural Alien: Humankind and the Environment.* Toronto: University of Toronto Press, 1985.

Berman, M. *The Reenchantment of the World.* Ithaca, NY: Cornell University Press, 1981.

Fritsch, A. J. *Environmental Ethics: Choices for Concerned Citizens.* Garden City, NY: Anchor Books, Doubleday, 1980.

Worster, Donald. *Nature's Economy: The Roots of Ecology.* Garden City, NY: Doubleday, 1979. A fascinating and engaging history of ecological and economic thought from the eighteenth century to the present.

McNeil, William H. *Plagues and Peoples.* New York: Doubleday, 1977. An easy-to-read, engrossing history of the effects of disease on human societies around the world and throughout the ages.

PART 2

Organisms and Ecology

In a marsh of the Pacific Northwest, a lifeless snail and the decaying leaf upon which it rests symbolize the cycle of birth, growth, death, and decay that rules the natural world. As these organisms decompose, nutrients return to the environment, making possible the growth of generations of organisms to come.

*t*he biosphere is in many ways like an extraordinarily complicated chess game in which each species is a single piece on the board. Unlike chess, however, both the biosphere and its pieces are dynamic; they interact with each other constantly and powerfully. The earliest biologists observed these interactions and asked themselves the first formal questions about the rules of nature. The modern branch of biology that examines these interactions and attempts to predict the response of living systems to environmental change is called ecology.

Understanding the interactions among various organisms and their environments is more important today than ever before. Why? Human activity is altering the global biosphere, destroying habitats and species at record rates. Our release of certain chemicals into the air is destroying the atmospheric layer of ozone that protects us from damaging ultraviolet radiation from the sun, and rates of some radiation-associated skin cancers have risen in areas where ozone depletion has been most pronounced. An increase of carbon dioxide in the atmosphere is associated with the possibility of global warming, which would have powerful effects on all living things.

For most of history, decisions on industrial and agricultural development were based almost entirely on social, political, and economic considerations. Today, those criteria alone are clearly insufficient. Biologists have learned that some basic principles govern the interactions of all organisms, including humans, with the environment. If we work within those principles, we flourish; if we ignore them, we do so at our peril and at the peril of all other species, as well.

In order to understand the social background against which ecological concepts are evaluated for significance and guidance, we first introduce readers to the basic principles noted previously. Chapter 3 introduces the physical and biological factors that define an ecosystem and explains the types of terrestrial and aquatic biomes found on Earth. Chapter 4 discusses two types of "currency," nutrients and energy, that must cycle through ecosystems for them to be self-sustaining and discusses how those cycles actually work in nature. Chapter 5 applies those concepts to study the limitations within which any population of a species will inter-

The chess board is the world, the pieces are the phenomena of nature, the rules of the game are what we call the rules of nature. The player on the other side is hidden from us. We know that his play is always fair, just, and patient. But we also know, to our cost, that he never overlooks a mistake, or makes the smallest allowance for ignorance.

— *T. H. Huxley*, A Liberal Education

act with its environment, while Chapter 6 discusses the rules within which communities of several species interact with each other and the environment. Finally, Chapter 7 serves as a capstone by discussing several specific environmental problems, their probable causes, and their possible solutions. The ongoing nature of scientific inquiry in ecology is represented by the conclusion of the chapter, which presents four case studies: one of a problem without a currently feasible solution, one of a problem defined and solved, and two where definition of the problem has begun, but any possible answer is still in the future.

3

Environmental and Biological Diversity

*V*isitors to tropical rain forests often assume that their soil is the richest on Earth—and with good reason. The forests support hundreds of tree species, of all heights and shapes, whose branches are festooned with orchids, ferns, and mosses. So dense is the leafy canopy that little light reaches ground level, and so dense are the forest walls that the still air deep inside is saturated with moisture (Fig. 3.1a). Big animals are scarce, but smaller ones are everywhere, including countless insects, birds, reptiles, amphibians, and small mammals.

This living tapestry has remarkable ability to repair itself. Now and then, thunderstorms sweep down, hurling lightning bolts and bearing gusts of winds that break branches, topple trees, and leave gaps in the canopy. Beneath those gaps, more sunlight strikes the ground, temperatures rise, and humidity drops. Yet those areas are quickly colonized by pioneering plant species that need sun and high temperatures. Although these plants don't thrive in mature forest, their seeds germinate quickly in "damaged" areas. As they grow, the pioneers change conditions around them, shading the ground and raising humidity. Ultimately, those changes allow regrowth of plants accustomed to mature forest conditions. Within a century, the gap in the canopy is closed by a new generation of trees. The forest has healed itself.

In contrast, America's Great Plains don't speak to the untutored eye of biological wealth. Their grasses and wildflowers are seasonal; leaves grow in spring and summer and seeds ripen as foliage dries. Before the great buffalo hunts, vast herds of bison roamed here (Fig. 3.1b). In those days, too, droughts paved the way for prairie fires ignited by lightning in tinder-dry grass. Sweeping across the plains, those conflagrations seemed to leave little in their wake besides charred roots. But grassland, too, "knows" how to heal itself. Some species, in fact, need the stimulus of occasional fires to do well.

These two ecological systems—one constantly wet and cloaked in trees up to 60 m (roughly 180 ft) tall, the other seasonally dry and composed of plants barely 2 m (about 6 ft) tall—both exhibit remarkable ability to recover from natural disturbances. Yet they exhibit dramatically different responses to human intervention.

Grassland often can be converted to farmland quite successfully. Vast tracts of former grassland are blanketed with corn and wheat (our

(a)

(b)

(c)

(d)

Figure 3.1 Varied climates and soil types support different ecosystems and determine those systems' suitability for farming. **(a)** "The land," wrote Charles Darwin about Amazon rain forests, "is one great, wild, untidy luxuriant hothouse, made by nature for herself." **(b)** America's Great Plains didn't *look* as lush as rain forests but were productive enough to support vast bison herds. **(c)** The deep, rich, fertile soil of this Illinois field dependably produces abundant crops. **(d)** This patch of former rain forest near Manaus, Brazil, can be farmed for a few years, but will then degenerate into wasteland.

"amber waves of grain") that produce rich yields after year (Fig. 3.1c). Many areas have been farmed for over a century, and can remain farmland for many more years if protected against soil erosion.

But farmers who clear rain forests to raise cattle fail miserably. Clear-cut areas sustain crops of grass for a year or two but deteriorate to wasteland within 5 years. The soil—stripped of vegetation that held and protected it—becomes so hard that it can break a plow and so depleted of nutrients that native vegetation cannot return (Fig. 3.1d).

Why do attempts to cultivate one ecosystem pay off handsomely, while efforts in another system fail miserably? Why do clear-cut rain forests turn into barren wasteland, while abandoned fields in the Midwest can return (after many years) to a semblance of their past condition? Is there any way to predict which human interactions with nature will succeed and which will turn sour?

Answers to these questions are complex yet fascinating, because they require understanding how plants and animal species "make a living" in the biological sense and how they interact with each other and their physical surroundings. To begin answering them, this chapter explains how and why Earth's environments—which range from wet tropics to frigid deserts, and from mountaintops to the ocean floor—differ from one another and affect the organisms living in them.

ECOSYSTEMS AND BIOLOGICAL DIVERSITY

Ecology is the branch of biology that attempts to explain how and why organisms interact with each other and their environments as they do. The science of ecology itself should not be confused with *environmentalism*, a social movement which, at its best, brings the findings of ecologists to bear on environmental issues of importance.

Among the largest units of concern to both ecologists and environmentalists are **ecosystems**: sizable interacting systems composed both of living organisms and their physical environment (*oikos* means "house"; *systema* means "that which is put together"). But before we study ecosystems and their components, it is important to recognize three vital concepts that are integral to the study of ecology: biological diversity, interconnectedness within and between ecosystems, and evolutionary adaptation.

Biological Diversity

The term **biological diversity** (or biodiversity) refers to variety and variability in living organisms and the ecological systems in which they live. As Harvard biologist Edward O. Wilson points out, "Biological diversity is the key to the maintenance of the world as we know it." Over the next several chapters, you will see how and why this is true.

In the meantime, we can say that biodiversity exists on several levels:

- **Ecosystem diversity is the variety of ecosystems in a region.** An area with patches of farmland, forests, ponds, marshes, and grasslands has higher diversity than a landscape devoted exclusively to farming. Ecosystem diversity varies a great deal from region to region; in some areas, grassland or rain forest cover vast tracts of land, while in others, natural patchiness is higher. Ecosystem diversity will be discussed extensively in Part 2.

- **Species diversity refers to the myriad kinds of organisms alive today.** To date, 1.7 million species of plants and animals have been identified, but experts estimate that our planet may harbor up to 40 million species. Species diversity will be introduced here in Part 2 and explained in detail in Part 3 on evolutionary theory.

- **Genetic diversity consists of the heritable variability among individual members of a single species.** Genetic

diversity is only mentioned in Part 2, but not because it isn't important. As you will discover when studying evolutionary theory and molecular biology, genetic diversity can be essential to species' survival.

Interconnectedness Within and Between Ecosystems

Organisms constantly interact with each other in positive and negative ways. The nature of those interactions, as you will see in Chapters 5 and 6, affects the structure and function of ecosystems. In a real sense, organisms that live in an area are as much a part of each other's environment as the physical features around them.

Organisms interact with their physical environment. These interactions are a two-way street: physical conditions are powerful factors in shaping ecosystems, and organisms can have a significant impact on their surroundings.

Ecosystems, although among the largest functional units of living systems, are not independent entities. As mentioned in Chapter 1, Earth's atmosphere and oceans flow over and between ecosystems, carrying water and nutrients from one place to another. Migratory animals cross the boundaries of ecosystems in their travels to feeding and breeding grounds. There are many other connections among global ecosystems, as you will see throughout Part 2.

Evolutionary Adaptation

Over time, evolutionary change has generated biodiversity and equipped organisms in ways that enable them to deal with physical environmental conditions and interactions with one another. Evolution, therefore, is background to ecology; its mechanisms will be described and explained in detail in Part 3.

ENVIRONMENTS AND THE DISTRIBUTION OF LIFE

The term **environment**, which we've used informally so far, refers to the sum of all the conditions surrounding an organism. Note that the word **habitat** is sometimes used interchangeably with *environment*. Often, however, habitat refers to a particular place or to a specific collection of organisms. In a manner of speaking, the habitat of an organism is its biological "street address."

Environmental influences are often divided into two subgroups: those of the physical, or *abiotic*, environment and those of the biological, or *biotic*, environment. The physical environment includes all the conditions created by the nonliving components of the organism's surroundings: sunlight, heat, moisture, the speed of wind and water currents, the size of sand or sediment grains, and so on. The **biological environment** consists of the living organisms in the habitat—that is, other species that may serve as food, parasites, predators, or competitors.

In practice, biotic and abiotic environmental factors are closely intertwined. Earthworms, for example, live in organic soil and moldy leaves in woods and forests. That decaying vegetation absorbs rainwater and stays moist longer than clean sand would under similar conditions. The plant material decays because it houses fungi, bacteria, and protozoans. Soil in a forest is held in place by plant roots, shaded from direct sun, and protected from strong winds by the leafy forest canopy. Thus the environment experienced by the worm (and most other organisms) is composed of both physical and biotic factors.

Abiotic Environmental Factors

The range of physical conditions found in Earth's environments has been a major factor in the generation and maintenance of biodiversity. To assemble a global picture of how physical environments affect life, we will first consider the physical factors themselves, and then we examine how those factors differ around the planet.

Oxygen   The concentration of oxygen is important in many habitats. Most multicellular organisms are **aerobes** that require oxygen for respiration. But many bacteria are **anaerobes** that thrive only in the absence of oxygen. Too much oxygen in the surroundings of anaerobic organisms can be just as fatal to them as a lack of oxygen is to aerobes.

In aquatic environments, and beneath the surface of wet soil or mud, oxygen concentrations can fall too low to support aerobic life. Under these conditions, anaerobic organisms thrive. Air-breathing animals that spend time underwater, as well as aquatic animals that are periodically exposed to air, must conserve oxygen when it is not readily available (Fig. 3.2). We will learn more about how organisms obtain oxygen and deliver it to their tissues in Chapter 36.

36⟩

Sunlight   Photosynthesis, the vital reaction that plants use to harness solar energy, is powered by sunlight (Fig. 3.3). As you will see in the next chapter, photosynthesis ultimately supplies energy to nearly all life on Earth. For that reason, light dim enough to limit photosynthesis and curtail plant growth can limit the productivity of an ecosystem. Sunlight is also essential for vision, which many animals rely on for catching food, spotting predators, and communicating with each other.

Figure 3.2   TOP: This diving seal's ability to hold its breath and regulate its body's demand for oxygen allows it to feed underwater for extended periods of time. BOTTOM: These mussels growing on a tidal flat in an estuary can "breathe" only underwater. When exposed to air at low tide, they must alter their metabolism dramatically to survive.

Figure 3.3 These plants use their large, thin leaves to catch sunlight in the shade beneath the canopy of tropical rain forests.

Light intensity is especially important in marine environments, because water absorbs much more light than the atmosphere does, particularly in the shortwave (ultraviolet) and longwave (infrared) regions of the spectrum. This means that the farther beneath the surface a marine organism lives, the less light reaches it and the more restricted the distribution of wavelengths in that light is (Fig. 3.4a). In the open sea and on tropical coral reefs, light much below 10 m (33 ft) is a vibrant, turquoise blue (Fig. 3.4b). Closer to shore in the temperate zone, it has a more yellow–green hue, because dissolved compounds carried into the sea by rivers and streams absorb blue light.

These dramatic changes in light intensity and quality have profound effects on marine plants. Why? Because photosynthesis best utilizes light in the violet and red regions of the spectrum—precisely those colors removed by seawater. Thus despite the biochemical adaptations of marine plants that you will learn about in the next chapter, photosynthesis can only occur fairly close to the ocean's surface, where light is sufficiently strong. The fairly shallow region near the water's surface in which sunlight is strong enough to power photosynthesis is called the *photic zone*, and it can be as shallow as 30 m in the turbid North Atlantic or as deep as 200 m in the crystalline tropical Pacific. Below the photic zone are vast volumes of water in which darkness prevents plants from growing.

On land, by contrast, too much sunlight can be as problematical as too little. Trees and cacti that live in full sunlight must protect delicate tissues from overheating, desiccation, and damaging ultraviolet radiation. Organisms accustomed to protective shade have as much difficulty tolerating brilliant sunshine as open-field organisms have surviving in dense shade.

Figure 3.4 **(a)** As sunlight passes through clear tropical seawater, most of the longwave (red) light and much of the shortwave (ultraviolet) light is absorbed. The light that reaches deep water, therefore, is restricted primarily to the blue region of the spectrum. **(b)** This photograph, taken roughly 30 ft beneath the Red Sea, shows how blue the light in clear oceans becomes below the surface.

(b)

(a)
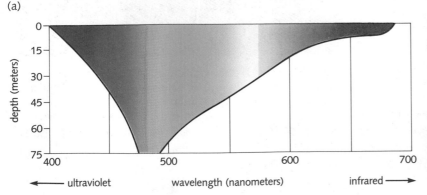

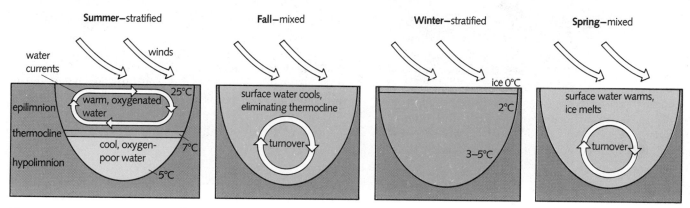

Figure 3.5 Water currents and temperature stratification in a deep lake. In summer, surface waters—the *epilimnion*—are warm and saturated with oxygen but may be depleted of nutrients for plant growth. Bottom waters—the *hypolimnion*—may be dark, cold, and nearly anaerobic. Low light levels make plant growth difficult, while sinking fecal material and decomposing plants and animals encourage the growth of bacteria that simultaneously use up oxygen and return nutrients to the water. In fall, as air cools (and winds pick up), surface waters cool until the thermocline nearly vanishes, and winds mix the lake from top to bottom. This *turnover* redistributes oxygen and nutrients. In winter, ice coats the surface, and mild stratification occurs, this time with the coldest water on top. In spring, ice melts and turnover occurs once again. Some lakes turn over only in the spring, while others mix in both spring and fall.

Temperature Most organisms survive only within a specific, limited range of temperatures. If their body temperatures rise above or fall below that range, critical chemical reactions in their tissues get "out of sync" with one another, and metabolic chaos results. The effects of either cold or heat on body functions can be equally deadly.

The temperatures of soil, air, and water all affect both local and global environments. But certain peculiarities of water's response to temperature change are especially important in structuring aquatic ecosystems. For when water is cooled below 4°C, its density begins to decrease. That's why water in its solid phase (ice) floats on its liquid phase, unlike metals or rocks.

In addition, stationary water masses of two different temperatures don't mix easily with each other because they have different densities. As a result, temperature differences between surface water and deeper water can create two or more distinct layers, a phenomenon known as **thermal stratification.** When thermal stratification occurs, upper and lower water masses are separated by a **thermocline,** or zone of temperature change (Fig. 3.5), across which little mixing takes place. As a result of such stratification, other environmental variables—such as oxygen or nutrients—can vary independently in separate water masses. Thus shallow waters of lakes and oceans can offer dramatically different environmental conditions than deeper portions of the same body of water.

Stratification can occur in air masses, too. But those events are fairly transient. Their duration is usually measured in terms of hours or days, rather than in the weeks or months of analogous events in aquatic ecosystems.

The importance of soil Terrestrial ecosystems are strongly affected by their soil. The basic inorganic components of soil are partially weathered rocks and minerals of varying size and composition. Sand is composed of large grains often made of quartz. Clays are formed from much finer grains. These inorganic materials combine with varying amounts of organic matter to form productive soils.

The physical and biological characteristics of a region's soil determine not only which plants and animals can live there but also the use to which humans can put an area (Fig. 3.6). In the grasslands that once covered the midwestern United States, for example, roots of native plants penetrated 1 to 3 m (3 to 10 ft) into the ground, forming dense sod that held moisture and prevented erosion. Because both leaves and roots of grasses die back each year, organic matter built up over time and created deep, rich soil. Across New England and the Mid-Atlantic states, an equally fortuitous combination of rainfall, soil structure, and temperature favors lush temperate forests. The dense roots of these plant communities hold soil in place, and the annual fall of leaves contributes organic materials that decompose to form **humus,** a spongy bed of organic matter that holds and delivers both water and nutrients. The deep, highly productive, organically rich soils of both these regions thus make them well suited to the demands of long-term, intensive agriculture.

In many tropical rain forests, on the other hand, dense clay subsoils, heavy rainfall, and high temperature create shallow soils that do not hold on to nutrients for very long. Dead organic material breaks down rapidly, so the humus layer is very thin and restricted to the soil surface. Soluble nutrients that aren't bound up in living tissue or in the thin humus layer are quickly washed away by heavy rains. When stripped of forest cover, such soils support agriculture for only a few years before they become exhausted, compacted, and useless.

Desert soils hold little organic matter but, because of low rainfall, do not lose those nutrients they do contain. With irrigation, some desert areas (such as California's Imperial Valley) can become productive farmland. But as you will see in Chapter 7, agriculture in deserts can be a very tricky business.

Water and dissolved salts   All organisms must maintain a precise balance of water, dissolved salts, and organic molecules in body fluids to keep their cells alive. For this reason, both the availability of water and the concentration of dissolved salts are important environmental variables.

Figure 3.6   The nature of a region's soil interacts with both averages and extremes of temperature and rainfall to determine the density and types of plants—and therefore, indirectly, the number and kinds of animals—that can survive.

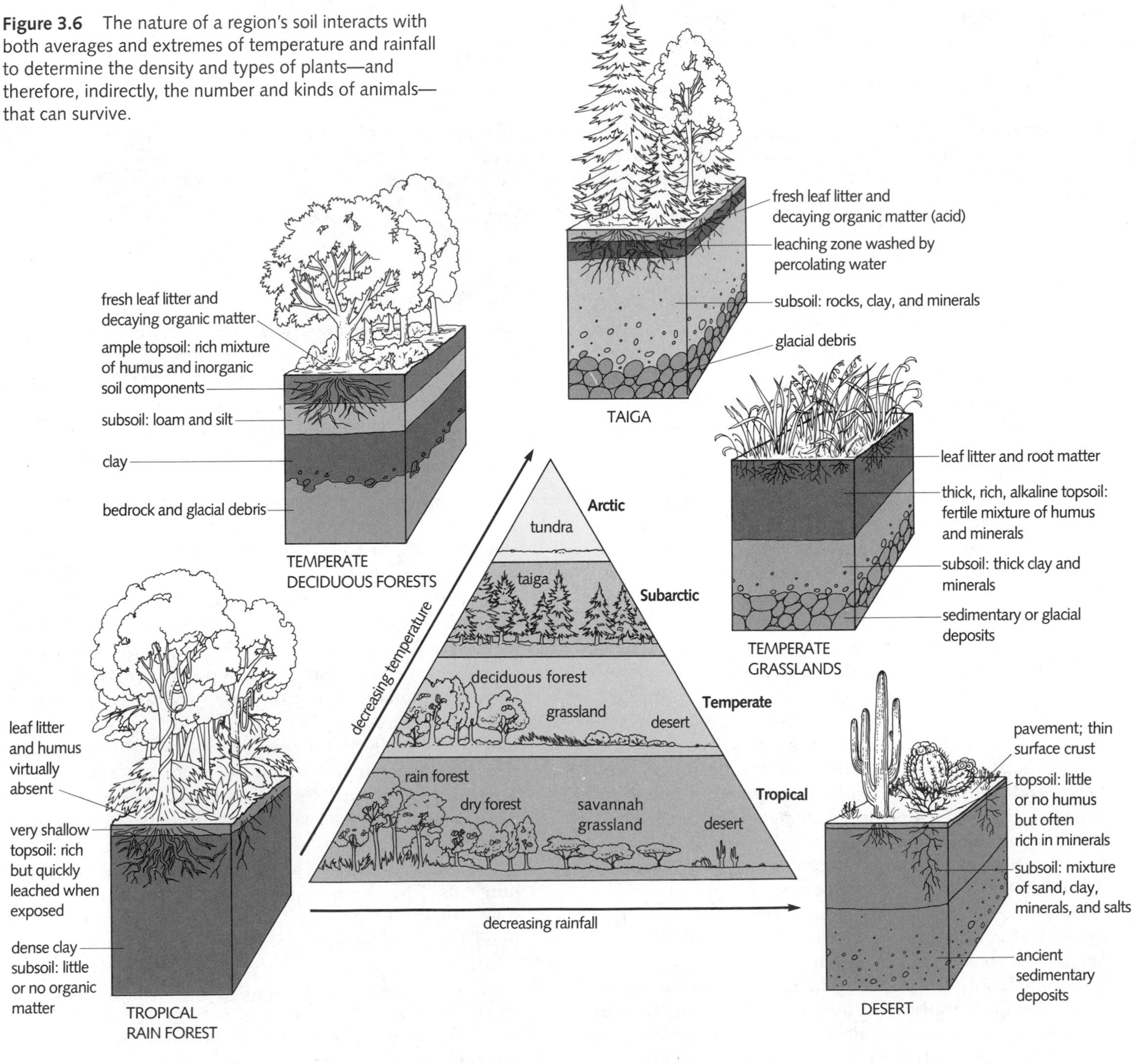

TAIGA

fresh leaf litter and decaying organic matter (acid)

leaching zone washed by percolating water

subsoil: rocks, clay, and minerals

glacial debris

fresh leaf litter and decaying organic matter

ample topsoil: rich mixture of humus and inorganic soil components

subsoil: loam and silt

clay

bedrock and glacial debris

TEMPERATE DECIDUOUS FORESTS

leaf litter and root matter

thick, rich, alkaline topsoil: fertile mixture of humus and minerals

subsoil: thick clay and minerals

sedimentary or glacial deposits

TEMPERATE GRASSLANDS

decreasing temperature

Arctic

tundra

taiga

Subarctic

deciduous forest

grassland

desert

Temperate

rain forest

dry forest

savannah grassland

desert

Tropical

decreasing rainfall

leaf litter and humus virtually absent

very shallow topsoil: rich but quickly leached when exposed

dense clay subsoil: little or no organic matter

TROPICAL RAIN FOREST

pavement; thin surface crust

topsoil: little or no humus but often rich in minerals

subsoil: mixture of sand, clay, minerals, and salts

ancient sedimentary deposits

DESERT

Many plants and animals cannot survive in deserts, for example, because they cannot acquire and store water under such dry conditions. But an equally large number of organisms cannot live in swamps and marshes because they cannot survive where there is too much water (and hence too little oxygen) in the soil.

Salinity, the concentration of dissolved inorganic salts, affects organisms' abilities to control their water balance and therefore plays a critical role in structuring both aquatic and terrestrial communities. Very few freshwater organisms can survive immersion in seawater for very long, and few marine plants and animals can tolerate fresh water. The ability of crop plants to tolerate salty soil is of great importance to world agriculture. In dry regions, especially in places where fields have been irrigated for a long time, salts can accumulate in soil and make it difficult for plant roots to absorb water.

Metabolic wastes

All organisms produce metabolic wastes. As you will see in more detail in the next chapter, animals exhale carbon dioxide and excrete nitrogen. Plants release oxygen by day, give off carbon dioxide by night, and discard leaves and stems seasonally. In most cases, these waste products are broken down or carried away. But under certain circumstances, wastes can accumulate. In mud, for example, noxious gases such as hydrogen sulfide (which causes the odor of rotten eggs) and methane (commonly called marsh gas) may build up, making the immediate environment unsuitable for many organisms. (There are, however, certain bacteria that actually thrive in these conditions.)

Nutrients

The presence or absence of essential nutrients—and their relative abundance—profoundly affects where organisms can and cannot grow. Where nutrients are present in the correct amounts, photosynthetic organisms can utilize them to construct complex organic molecules needed by animals. Where nutrients are deficient or absent, plant growth may be stunted or blocked. Soils formed from volcanic rock, for example, often contain far more magnesium, calcium, nitrogen, and phosphorus than soil formed from sandstone or shale, and may therefore support different plant communities. In the sea, the distribution of nutrients by ocean currents is a major factor in determining the nature and density of life. Often, as you will soon see, nutrients become concentrated in water masses below the photic zone, where photosynthesis doesn't occur.

Global Climate and Life

The earth's major climate zones are formed by unequal heating of the planet's atmosphere, land masses, and oceans, and the rotation of the earth on its axis. Now that

you have seen the importance of local environmental factors, you can understand how global climate patterns affect the large-scale distribution of life on the earth by creating variations in temperature, rainfall, and humidity in different regions of the globe.

Climate versus weather The **climate** consists of the *means* or *averages* of temperature, rainfall, hours of sunlight, wind speed, and so on. Climate data, such as average annual temperature and rainfall, enable biologists to determine whether averages for particular environmental variables fall within an organism's optimal range. On the basis of such averages, certain areas are said to be "wet" or "dry," "temperate," "subtropical," or "tropical."

Weather comprises the day-by-day variations and extremes in those same factors. Weather data describe the full range of environmental conditions that test organisms' tolerance limits and therefore determine where those organisms can survive (Fig. 3.7).

Figure 3.7   Plant hardiness maps developed by the U.S. Department of Agriculture help gardeners determine which plants can survive in their areas. The map divides the country into 10 zones based on *average minimum temperatures*: the *coldest* it gets during a typical winter. The map uses *average minimum* rather than *average* temperature because it doesn't matter to a tree in your backyard that the usual temperature is 4°C (25°F) if it *actually* gets down to −23°C (−10°F)! Many factors, such as altitude, snow cover, exposure, humidity, and soil type, create variations even within these zones.

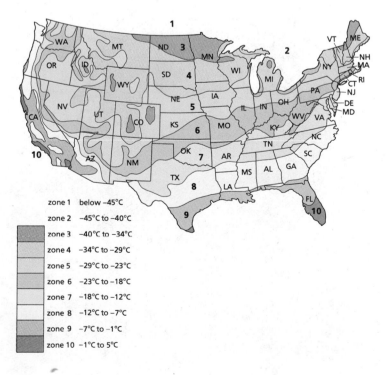

zone 1	below −45°C
zone 2	−45°C to −40°C
zone 3	−40°C to −34°C
zone 4	−34°C to −29°C
zone 5	−29°C to −23°C
zone 6	−23°C to −18°C
zone 7	−18°C to −12°C
zone 8	−12°C to −7°C
zone 9	−7°C to −1°C
zone 10	−1°C to 5°C

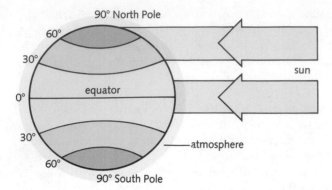

Figure 3.8 Unequal heating of the earth's surface caused by two imaginary beams of solar energy from the sun. The same amount of energy is spread over a larger area at the poles than at the equator, so each square kilometer near the poles receives only a fraction of the energy received by the same area at the equator. Furthermore, the energy reaching the equator travels straight through the earth's atmosphere, while that hitting the poles travels at an angle, passing through more atmospheric gases, water vapor, clouds, and suspended dust that remove energy.

Figure 3.9 This schematic view of the globe shows the symmetrical patterns of atmospheric circulation in each hemisphere and the surface winds they cause.

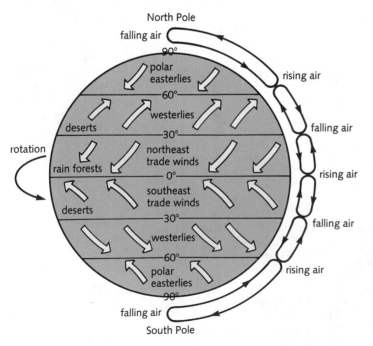

Solar energy and temperature Every region on Earth receives the same *total number of hours* of daylight and darkness each year, but regions near the equator receive far more solar energy than regions near the poles (Fig. 3.8). The unequal heating of land masses, oceans, and atmosphere that results is further increased because polar regions radiate more heat back into space than do equatorial regions.

This differential heating sets Earth's climatic machinery in action. Why? Because both air and water become less dense as they warm and denser as they cool (until water reaches 4°C). As a result, warm air and water masses near the equator rise, while cooler air and water near the poles tend to sink. Because Earth is so large, however, the behavior of atmosphere and oceans is complicated.

Global air movement Figure 3.9 shows that as equatorial air rises and moves toward the poles, it loses heat, cools, and becomes more dense. By the time it reaches 30° North and South latitudes, it has cooled sufficiently to sink towards the ground. In the meantime, cold air that sinks at the poles moves southward near the earth's surface, picking up heat from the ground as it flows. By the time this polar air reaches 60° North and South latitudes, therefore, it becomes warm enough to rise.

Whenever air rises or sinks, nearby air moves in sideways to replace it. As a result, air near the earth's surface is constantly moving from areas of sinking air to areas of rising air. This horizontal air movement forms prevailing winds such as the *trade winds* that blow steadily on either side of the equator. Furthermore, because Earth spins on its axis from west to east, air currents traveling north or south are deflected sideways. In the northern hemisphere, the deflection is to the right; in the southern hemisphere, the deflection is to the left. This gives rise to the circular air movements seen on weather maps.

Winds and oceanic surface currents These prevailing winds act steadily on the ocean's surface, pushing water from east to west near the equator, from west to east further north or south, and from east to west again close to the poles. This influence, combined with the shapes and locations of the continents and certain other factors, generates surface currents that govern marine ecosystems the way air currents shape life on land (Fig. 3.10). Several important ocean currents have profound effects on both terrestrial and marine life over vast regions of the globe (see Theory in Action, El Niño).

Winds, mountains, and rainfall Whenever humid air rises and cools, the water vapor it carries condenses and falls as rain, leaving the air drier than it was before. For this reason, whenever warm, moist air at sea level is pushed up over a mountain range, it drops most of the

El Niño

In 1983 world climate seemed to be going haywire. Floods devastated Ecuador and Peru, and unusually heavy rains drenched California. Fishes and seabirds died by the thousands off the coast of South America, while seals, whales, and dolphins took off for parts unknown. Record drought fueled brushfires in Australia and devastated farmlands throughout Indonesia.

These seemingly unconnected events, spanning half the globe, were in fact caused by a single set of interrelated changes in global wind and ocean current patterns. Scientists studying the phenomenon soon adopted the name Peruvian farmers had been using to describe it for years—El Niño, or "The Child," after its regular occurrence around Christmas. Normally, El Niño appears as a short-term, local phenomenon: a slight warming of the surface waters off the coast of South America that Peruvians associate with a spell of bad fishing. But during 1982–1983, El Niño was much stronger than usual, and the warming of the eastern Pacific was combined with a change in air pressures over the western Pacific called the southern oscillation. The worldwide effects of this joint phenomenon sparked an international investigation into its causes.

The story of El Niño–southern oscillation (ENSO) is complex. For most of the year, steady winds blow from east to west across the equatorial Pacific. These easterly trades, as they are called, pull surface water away from the west coast of South America and bring nutrient-rich deeper water to the surface. Described in the text, this upwelling supports a rich and productive marine food web. Half a world away in the western Pacific, the same easterly winds interact with local winds to control the seasonal monsoon rains.

When ENSO begins, trade winds slack off and then reverse. The upwelling of deep water off South America stops and is replaced by a southward flow of warm water from the equator. Deprived of the nutrient-rich deep water, phytoplankton growth stops, and the coastal food web off Peru collapses. Related changes in ocean temperatures and wind patterns caused drought in the west and heavy rains in the east.

Such phenomena seem to occur about every 4 or 5 years, but their timing and intensity vary. The most recent El Niño began in late 1991. During 1992 it contributed to North America's mild winter, the severe drought in southeast Africa, and the heavy rains and snow that relieved the drought in California, Nevada, and Oregon. The warming trend, which continued into 1993 in the central Pacific, might (or might not) indicate an actual climate change there.

Exactly what sets El Niño in motion is not known, and it is not clear what causes the effect to be stronger in some years than in others. Ocean surface temperatures, unexplained changes in wind speed and direction, habitat alteration on land, and even undersea earthquakes and major volcanic eruptions may be involved. Though their causes remain a puzzle, ENSOs are clear reminders of the interconnectedness of the earth's ecosystems.

These false-color satellite images illustrate the differences in sea surface temperatures in the tropical Pacific during January 1984, a normal winter (LEFT), and 1983, during an El Niño year (RIGHT). Note the "finger" of colder water stretching westward from the Pacific coast of South America in the normal pattern and the solid band of warm water in 1983.

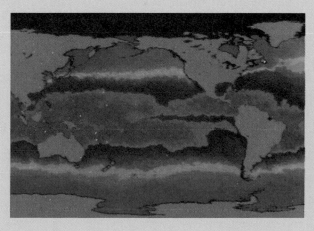

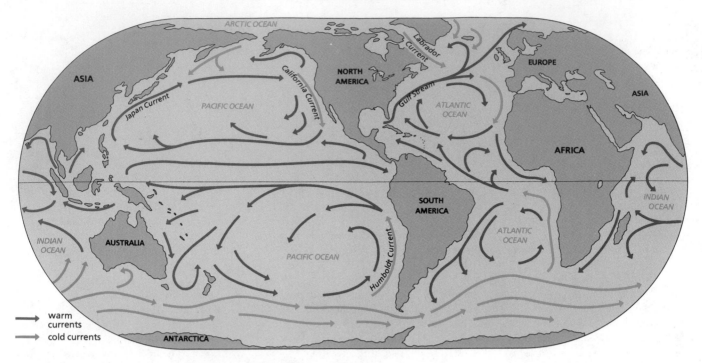

Figure 3.10 Major oceanic surface currents. Note the warm Gulf Stream circulating northward along our Atlantic coast and the cold California current running southward along the Pacific coast. The Gulf Stream transports somewhere around 30 billion gallons every second—more than 65 times the amount of water carried by all the rivers of the world combined.

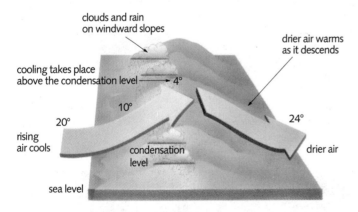

Figure 3.11 The effect of mountain ranges on wind and rainfall.

water it carries on the windward side of those mountains. As this drier air sinks downwind of the mountains, it warms and dries even more. Therefore, wherever mountain ranges abut the sea, heavy rains drench the windward slopes and desert conditions prevail in the lee. This phenomenon is called a "rain shadow" (Fig. 3.11 and Fig. 3.12).

Lakes and oceans: Climate moderators   It takes substantially more energy to raise the temperature of water than it does to raise the temperature of air, soil, or rock by the same number of degrees. Thus sunlight warms land and air much faster than bodies of water. Conversely, land and air cool more rapidly than water.

Because oceans and large lakes warm more slowly than continents in spring and cool more slowly in the fall, they help keep summers cooler and winters warmer over nearby land areas. On sunny summer days the land warms rapidly, heating air immediately above it. That warm air rises, pulling cool air in off the water to create refreshing sea breezes. In winter, on the other hand, oceans retain more heat than land, so winds off the water are warmer than those that blow from the interior of a large land mass.

Figure 3.12 World precipitation patterns and their effects on organisms. Areas near the equator receive the most rainfall, creating wet tropical forests. North and south of the equator, at 30°N and 30°S latitudes on the west sides of the continents, are the world's largest deserts. Farther north and south—in northern California and western Europe—more moderate temperatures and seasonal rainfall encourage extensive animal and plant life. Still farther northward in Oregon and Washington, and southward into southern Chile, heavy rainfall becomes common once again, encouraging the growth of great temperate rain forests.

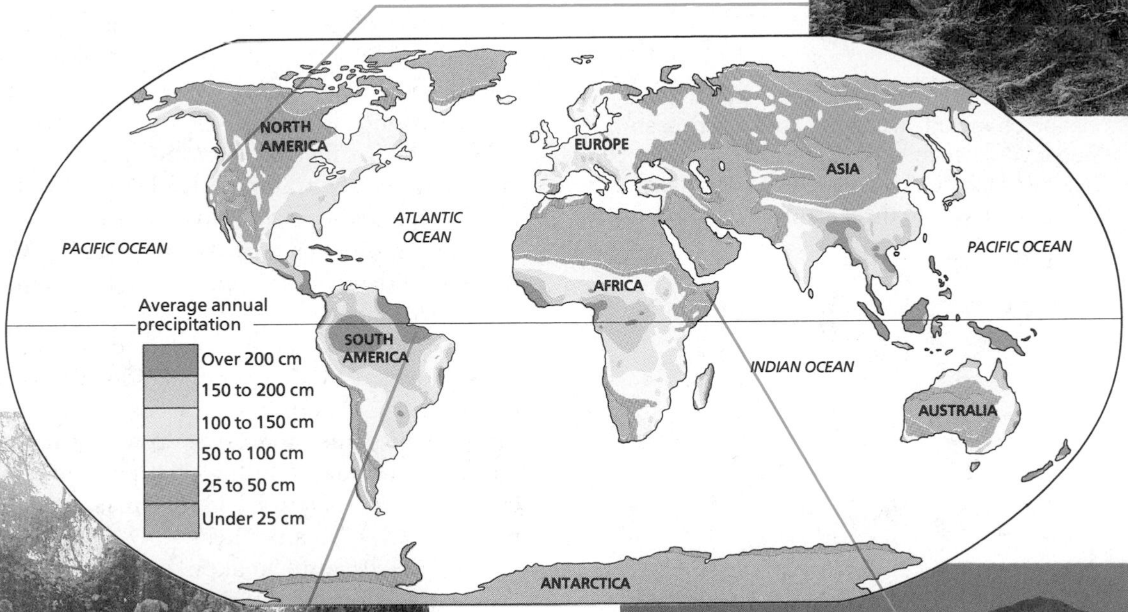

Average annual precipitation

- Over 200 cm
- 150 to 200 cm
- 100 to 150 cm
- 50 to 100 cm
- 25 to 50 cm
- Under 25 cm

NORTH AMERICA
EUROPE
ASIA
PACIFIC OCEAN
ATLANTIC OCEAN
AFRICA
SOUTH AMERICA
PACIFIC OCEAN
INDIAN OCEAN
AUSTRALIA
ANTARCTICA

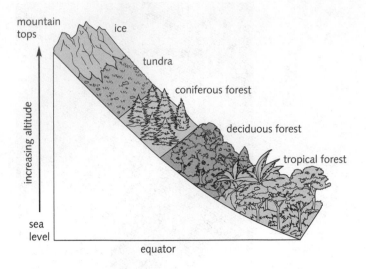

Labels on figure:
mountain tops — ice
tundra
coniferous forest
deciduous forest
tropical forest
increasing altitude
sea level
equator

Figure 3.13 Because temperature decreases with increasing altitude, ecological zonation on mountains mirrors the pattern seen when changing latitude. We encounter the same ecological zones when we increase in altitude at the equator as we would encounter by increasing latitude from the equator to the poles.

The influence of elevation on environmental variables
Temperature, rainfall, humidity, and winds vary not only with latitude but also with elevation. At every latitude, locations at higher altitudes experience conditions similar to those at higher latitudes. For this reason, the zonation of life from river valleys to mountaintops at the equator, for example, mirrors the zonation of life from the tropics to the poles (Fig. 3.13).

The Impact of Environmental Variables on Organisms

What determines how organisms respond to environmental variables? As mentioned earlier, most organisms must maintain internal conditions within certain limits. Interposed between the "raw" environmental conditions that surround an organism and its internal milieu is a balancing act called *homeostasis*—a series of physiological and behavioral strategies to keep internal conditions within limits the organism can tolerate (*homeo* means "similar"; *sta* means "stand or remain"). (One biochemical definition of death, incidentally, is equilibrium with the environment.)

One illustration of the costs and benefits of homeostasis is provided by **thermoregulation,** the control of body temperature. Mammals, for example, generate heat within their tissues and use several techniques to conserve that heat and maintain high body temperatures. Such organisms are called **endotherms** (*endo* means "internal"; *therm* means "heat"). Other animals, such as turtles, lizards, and snakes, bask in the sun to warm themselves up during the day and shelter in underground burrows to conserve heat during cold nights. Because these animals obtain most of their body heat from their surroundings, they are called **ectotherms** (*ecto* means "external").

We will discuss thermoregulation in more detail in Chapter 38. For the time being, realize that temperature regulation under extreme conditions is energetically expensive. Ectotherms trying to stay active in the cold, for example, must spend lots of time basking. During winters that are both cold and dark, ectotherms either enter a dormant state (as many amphibians and reptiles do in the temperate zone during winter), migrate to warmer climes (as monarch butterflies do), or die. Endotherms, on the other hand, may be able to remain active during cold weather—if they can find enough food to stoke their metabolic furnaces. In some cases they too, must resort to inactive winter periods, as temperate zone bears and squirrels do.

Law of tolerance Those conditions of temperature, moisture, sunlight, oxygen concentration, or nutrient availability under which an organism grows and reproduces most vigorously comprise the organism's optimum range for that particular variable. Unless other factors are unfavorable, the organism should survive and reproduce within this optimum range. When conditions fall outside the optimum range, the organism's chances for survival are described by the ecological *law of tolerance*, which states that the growth of a species is limited by that environmental factor for which the organism has the narrowest range of tolerance.

Of course, in reality, organisms' tolerance of extremes of one environmental variable is not absolute and depends on other biotic and abiotic factors. During hot weather, for example, animals that sweat require more water, and both endo- and ectotherms must often shelter in the shade (Fig. 3.14). Thus certain organisms can tolerate more heat if they have sufficient water, and others can live in warmer areas if trees or caves offer sufficient respite from the heat.

Figure 3.15 shows the response of a typical organism to a gradient of some environmental variable. Above or below the optimum range, too much or too little of that factor forces the organism to work harder to maintain internal conditions within acceptable limits. These regions along the scale are referred to as *zones of stress*. Under low stress, organisms usually survive. But the greater the environmental stress on an organism, the more work it has to do to stay alive, and the less energy it has left over for growth and reproduction. Under conditions of higher stress, therefore, individual organisms may survive but lack sufficient energy to reproduce. Under these conditions, *individuals* may subsist, but *populations* cannot maintain themselves. Still higher and lower on the scale are ranges beyond which organisms cannot survive; these are *zones of intolerance*.

Given a range of environmental conditions and the opportunity to behave as they choose, animals actively

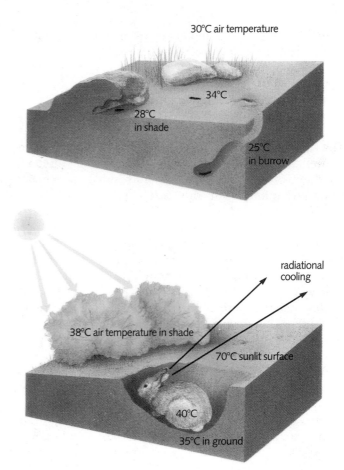

Figure 3.14 TOP: Body temperatures of an ectothermic desert insect in different locations at noon in the desert. The body temperatures of this ectothermic animal match those of its microhabitat. BOTTOM: Many endothermic animals take advantage of microclimate differences as well. Here, a jackrabbit shelters by digging a hole shaded by vegetation to help keep cool in a hot environment.

Figure 3.15 This graph shows the response of a typical organism to a range of a single environmental variable such as sunlight, temperature, or oxygen concentration. At the center of the optimum zone, the organisms are potentially most abundant. They become rarer in zones of physiological stress and are absent from zones of intolerance.

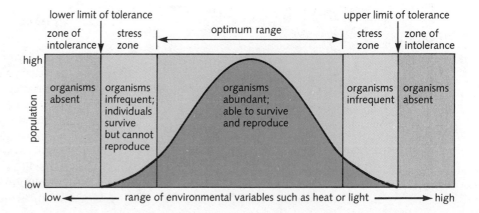

select conditions that suit them from those available. If animals are forced to occupy unfavorable habitats, they either die or fail to reproduce. Similarly, plants grow and flower best under certain conditions. When seeds are spread over a large area, only those that encounter favorable conditions survive.

The importance of microclimate It is important to recognize that when *we* as humans observe environmental conditions, we perceive most easily situations encountered by animals our own size. Yet organisms much smaller than ourselves can be affected by conditions that vary over very small distances, creating **microenvironments** or *microhabitats* not always obvious to human observers. Each microhabitat can have its own temperature, humidity, and wind speed—conditions referred to as **microclimate.**

Microclimates are important to germinating seeds and small plants, small insects, or bacteria, because what matters most to these organisms are the amounts of light, nutrients, temperature, and moisture available within a few millimeters of their location (Fig. 3.16). Conditions even a meter away are of less consequence. Medium-sized organisms also respond to small-scale variations in environmental conditions. Within a single, easily recognizable habitat such as a lake, for example, there may be a wide range of microhabitats whose differing environmental

conditions govern the distributions of organisms (see Theory in Action, Where Are the Bass?, p. 50).

Biotic Environmental Factors

Thus far, we have acted as though the physical factors of an organism's environment depend exclusively on nonliving aspects of its surroundings. But organisms can profoundly modify the conditions that they and their neighbors experience. Some of these biotic influences occur on a grand scale, others only over short distances.

Vegetation and climate Substantial evidence exists that vegetation can influence climate over large areas. Plants (especially dense forests) affect the amount of solar radiation a region absorbs or reflects, the amount of water retained in the soil after each rain, the speed at which that water is returned to the air, and the rate at which soil builds up or is eroded.

In Amazonian rain forests, for example, much of the water that falls as rain isn't brought to the region from elsewhere and doesn't leave after it falls. Up to three-fourths of local rainfall is returned to the atmosphere by either evaporation from leaf surfaces or transpiration (loss of water by leaf tissues) from living foliage. Thus much of the water in the area is continuously recycled. When large tracts of forest are cleared, this critical cycle is altered. When too much forest is destroyed, the cycle is broken and local rainfall decreases dramatically.

Biotic contributions to microclimate A large organism such as a tree can provide a wide range of microclimates in what would otherwise be a monotonous environment. As Fig. 3.17 shows, the presence of trees creates a host of microenvironments, ranging from leaf litter and rotting logs to a host of ecological "apartments" beneath bark, within leaves, and on twigs and branches.

BIOMES

By now, you have seen how local environmental variables affect organisms and how global wind and ocean current patterns combine with geological features such as mountain ranges to create variations in those conditions around the globe. Now we can (rather subjectively) divide the globe into regions characterized by different combinations of environmental conditions. Ecologists call these major life zones **biomes.** Each biome is home to certain communities of plants and animals adapted to local temperatures and adapted to both the quantity and the timing

Figure 3.16 Air and soil temperatures near the surface of a sand dune during typical rainy and sunny days. Note the dramatic temperature differences across a vertical distance of only a few centimeters. Seeds of microorganisms only a few centimeters below the surface would encounter dramatically different conditions from those sitting directly on top of the sand.

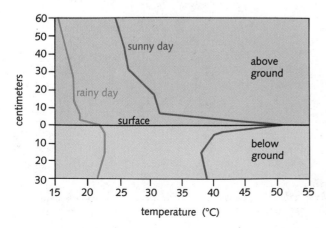

Figure 3.17 Microclimates within a temperate woodland. These are but a few of the organisms sheltering in various microclimates in and around a single deciduous tree. CLOCKWISE FROM TOP: A moth and its larvae in the uppermost branches, a bird parasite crawling through the down feathers of its host, a bark beetle drilling cavities beneath the bark, nitrifying bacteria beneath the soil surface, an earthworm in the leaf mold within the topsoil, termites within the decaying log, fungi in the process of decomposing moist wood, a salamander beneath a rotting log, and a gall wasp.

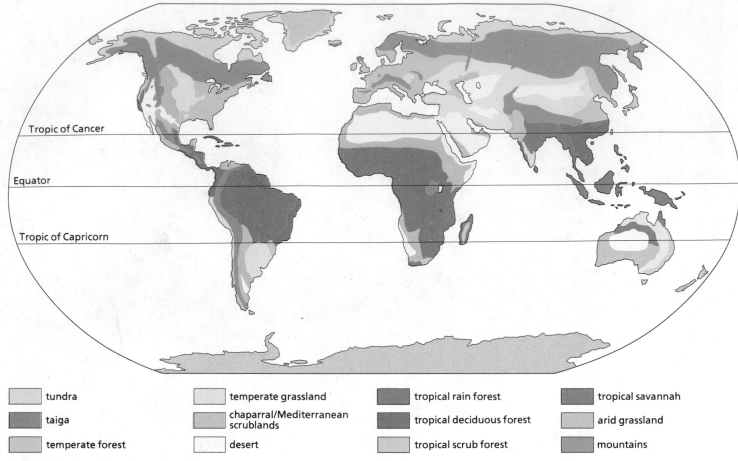

Tropic of Cancer

Equator

Tropic of Capricorn

	tundra		temperate grassland		tropical rain forest		tropical savannah
	taiga		chaparral/Mediterranean scrublands		tropical deciduous forest		arid grassland
	temperate forest		desert		tropical scrub forest		mountains

Figure 3.18 Major biomes of the world, identified by the ecosystem that existed in each before human intervention. Note that on this scale, nearly the entire eastern United States is classified as a single biome entitled "temperate forest."

of precipitation (Fig. 3.18). Though physical conditions are similar throughout a single biome, local variations within each biome can encourage the growth of different living communities.

Terrestrial Biomes

Tundra Near the North and South poles, winter is a way of life; average monthly temperatures rarely exceed 10°C (50°F). The top few feet of soil thaw during the 6–8-week summer, but the subsoil stays permanently frozen in a state called **permafrost.**

This is the **tundra,** an environment with predictable but harsh conditions. What little precipitation does fall near either pole comes as snow that remains frozen and therefore unavailable to plants for most of the year. This makes winter on the tundra as inhospitable as summer in the desert. When summer arrives, continuous daylight

warms the soil surface, but that warmth does not penetrate far. Any water that melts or runs onto the tundra from higher ground is trapped at the surface by the permafrost below, forming bogs and ponds.

Microclimates are important on the tundra, where most plants and animals live just at, just above, or just below ground level. During a typical midsummer day, the permafrost retreats to a depth of 35 cm below the surface, the temperature of the soil surface is about 38°C (100°F), air temperature just above the ground hovers around 15.5°C (60°F), air temperature a meter above the ground drops to 4.7°C (40.5°F), and air temperature above that level may be below freezing.

The dominant plants are sphagnum mosses and lichens, because summers are too short and too cool to support the growth of any trees except a few dwarf willows. The few species present make the most of short summers, reproducing quickly and in large numbers (Fig. 3.19). As soon as surface ice thaws, flies and mosquitoes

Figure 3.19 Plants of the tundra, predominantly mosses and quick-growing annuals, make the most of the region's brief summers to grow and reproduce.

Figure 3.20 The taiga supports a diverse and complex community, including evergreens, deciduous trees, and a wide variety of animals.

form vast swarms, while herbs and grasses grow and flower. Flocks of migratory birds feed on abundant seeds and insects and breed before heading south in the fall. Because no small area produces enough vegetation to support large herds of plant eaters for very long, reindeer range far and wide, and caribou migrate south to places where food is more plentiful. Smaller plant eaters such as arctic hares and lemmings burrow and hibernate for much of the winter. Meat eaters include arctic foxes, snowy owls, arctic wolves, weasels, and (near the coast) polar bears.

Taiga South of the tundra lies the evergreen forest called **taiga** (Fig. 3.20), which stretches in broad bands across the continents. The taiga's midwinter climate resembles that of the tundra, but its summers are longer and warmer. Especially in the southern-most fringe of taiga, the summer growing season is still short and temperature fluctuates wildly, but soil thaws completely, and water is liquid for longer periods.

Evergreen, cone-bearing trees such as spruce, fir, pine, and larch are important here, although deciduous trees such as birch, aspen, and willow appear. Evergreens predominate for several reasons. Their tough, needle-like leaves retain water when the soil is frozen. Those leaves also remain on the stems through the winter, ready to carry on photosynthesis as soon as temperatures moderate in the spring. The reproductive habits of evergreens are also suited to short growing seasons. Pine cones do not release mature seeds until nearly a year after pollination, giving the seeds much longer than a single growing season to mature.

A host of plant-eating insects live here, along with rodents, porcupines, rabbits and hares, moose, elk, and deer. Carnivores range from weasel, mink, and polecat to lynx, wolves, and bears. Birds are everywhere in summer; seed-eating thrushes, finches, buntings, grosbeaks, nuthatches, and jays are joined by such insect-eating migrants as sandpipers, oystercatchers, and ducks that

Where Are the Bass?

Many animals sense light, temperature, and oxygen and respond by selecting microhabitats whose conditions suit their needs. The striped bass, or "striper," a favorite of sports and commercial fishers, is one animal whose behavior directs it to select water masses of certain specific temperatures.

As ectotherms, bass cannot control their body temperatures independently of water temperatures around them. Yet their metabolic processes function best at specific temperatures that vary with the age of the fish. Adult striped bass normally live in coastal ocean waters and migrate up rivers to spawn in the early spring. Eggs hatch in the frigid water of early spring, juveniles grow as streams warm in summer, and adults return to the cooler sea. Over time, bass physiology has adapted to these different temperature regimes. Newly hatched fish survive best near 20°C (68°F), young fish grow fastest between 24°C and 26°C (75°F and 79°F), and adults grow best at around 22°C (72°F). Lab and field experiments have shown that bass of every age sense the temperature best suited to their needs and seek out water near that temperature.

The natural tendency of bass to seek out a specific temperature at each stage in their life cycle serves the fish well in nature but can frustrate biologists' attempts to cultivate these fishes in certain environments. After researchers found that stripers can survive in fresh water, for example, they stocked them in many rivers and reservoirs around the country. The fish thrive in some deep, cool reservoirs but not in others with different characteristics. In Tennessee's Cherokee Reservoir, for example, juveniles did well, but scores of adults died mysteriously every summer.

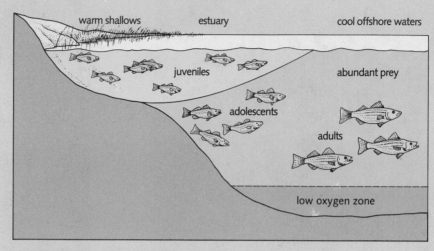

Separation of striped bass by age and thermal preference in a healthy estuary.

Autopsies revealed that those adults were covered with body sores caused by disease and had empty stomachs—despite the fact that abundant prey was available near the surface.

What had happened to the healthy juvenile bass? The figure to the right shows how behavior in search of preferred temperature provides a probable explanation. Adult bass need water cooler than 25°C (77°F) and an oxygen concentration of at least 2 milligrams per liter. In spring, conditions in the Cherokee Reservoir suit them just fine. But problems develop during the hot Tennessee summer, when decomposing organic matter depletes oxygen in bottom water and sunlight heats surface water. In early summer, adults are squeezed into a shrinking "comfort zone" sandwiched between surface water that is too warm and bottom water that carries too little dissolved oxygen. Then as juveniles mature and change

preferences to cooler water in late summer, they crowd the acceptable microhabitats. Adults overflow into water warmer than they can cope with, and they die.

Understanding the role of temperature in bass biology may help fishery managers create conditions more acceptable to bass in such habitats. Fishery personnel could, for example, provide cool water for adult bass in artificial lakes by controlling the release of water from dams. They might also increase the oxygen content of bottom water by cutting down on the inflow of organic material. Elsewhere, in the polluted river systems of New York harbor and Chesapeake Bay, striped bass stocks have been declining drastically. Understanding the relationship between bass behavior, low oxygen concentrations in bottom water, and the effects of toxic pollutants may some day help save these fish as well.

Bass behavior in Cherokee Reservoir. ◊

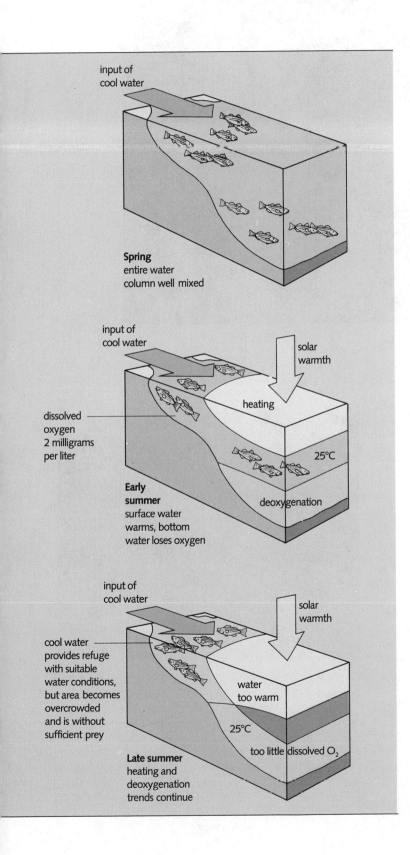

input of
cool water

Spring
entire water
column well mixed

input of
cool water

solar
warmth

heating

dissolved
oxygen
2 milligrams
per liter

25°C

**Early
summer**
surface water
warms, bottom
water loses oxygen

deoxygenation

input of
cool water

solar
warmth

cool water
provides refuge
with suitable
water conditions,
but area becomes
overcrowded
and is without
sufficient prey

water
too warm

25°C

too little dissolved O₂

Late summer
heating and
deoxygenation
trends continue

frequent the taiga's bogs and streams. Eagles, falcons, and buzzards are the most common birds of prey.

Temperate forest   South of the taiga, milder winters and longer growing seasons support deciduous forests. Although precipitation falls throughout the year, much of it falls during spring and summer. Often considered a single biome, the **temperate forest** actually comprises at least three different communities (Fig. 3.21): the temperate deciduous, the moist–temperate coniferous, and the broad-leaved evergreen forests. In all these forests, fallen leaves produce a deep, rich humus layer which serves as a dependable reservoir of water and nutrients.

Temperate forests provide enough buds, leaves, seeds, and fruits to support a wide variety of animal life. Insect diversity and abundance are both high. In spring, summer, and fall, temperate forests provide food for scores of bird species. Some birds live there year-round; others migrate back and forth between temperate breeding grounds and tropical or subtropical wintering grounds.

Mammals, once present in great numbers and diversity, included wolves, coyotes, foxes, wildcats, martens, and lynx. Those predators helped control populations of elk, moose, and deer, as you will learn in Chapter 5. But farming and hunting displaced or killed large predators. Deer and moose are now managed by wildlife authorities, who control their numbers by regulating hunting. Recently, under the protection of law, wolves, coyotes, and other native predators have been making a comeback.

Desert   Annual rainfall is less than 25.5 cm in a **desert.** Mean temperature records from deserts are nearly meaningless because both daily and seasonal fluctuations can be enormous. African deserts, such as the Kalahari, sometimes experience temperature fluctuations of up to 38°C (from just below freezing to 99°F) in the space of only 24 hours.

Deserts are extraordinarily diverse (Fig. 3.22). Some, like the Sinai Peninsula between Africa and the Middle East, are nearly barren year-round. Others, like North America's Sonoran Desert and Israel's Negev, appear dead for most of the year but harbor drought-tolerant plants that burst into life after seasonal rains. Many of these are annuals that sprout, grow, mature, flower, and die within weeks of the first rain.

Desert plants called *xerophytes* (*xeric* means "dry"; *phytos* means "plant") gather and conserve water in different ways. The root systems of desert plants may spread horizontally, like those of the saguaro cactus, or plunge 50 m into the ground, like those of the mesquite. To conserve water, some plants, such as cacti, have reduced their leaves to spines and carry on photosynthesis in thick, waxy stems. Plants that bear leaves often lose them during

Figure 3.21 LEFT: Temperate deciduous ecosystems such as this oak–hickory forest once covered most of eastern North America, Europe, and China. Light reaches the ground in these forests, encouraging the growth of shrubs and herbs. RIGHT: Moist–temperate coniferous forests, such as this stand of Douglas fir, may also be dominated by conifers such as redwoods and spruces. These areas receive both ample rainfall and lots of fog.

Figure 3.22 Cacti, native to the Americas, are premier examples of plants adapted to life in arid environments.

Figure 3.23 Large animals, such as camels, that cannot burrow for shelter cope with desert life through physiological adaptations that permit them to survive the loss of far more of their body water than most other vertebrates. Contrary to myth, the hump does not store water; it stores energy in the form of fat. By concentrating fat in one place, this mammal allows metabolic heat to escape more readily from the rest of its body. In addition, camels and their kin allow their body temperatures to fluctuate more when water is in short supply.

Figure 3.24 In their pristine state, temperate and tropical grasslands—such as the Serengeti seen here—can support vast herds of several herbivores.

long dry spells, whereas others such as yuccas rely on thick, fleshy leaves to store water.

Desert animals exhibit remarkable adaptations to heat and cold (Fig. 3.23). Numerous insects and other small animals escape the extremes of desert life by burrowing to create hospitable microclimates beneath the sand. Desert reptiles, small mammals, spiders, and scorpions generally remain inactive below ground during the day and emerge to feed only at night.

Grassland Both temperate and tropical regions have large areas where rain is more abundant than in deserts but is either too sparse or too seasonally concentrated to support forest growth. These areas, which have played crucial roles in human history, are the **grasslands** or **savannahs.** In the United States they are called prairies; in South America they are known as pampas; in South Africa they are referred to as veldt; in central Asia they are called steppes (Fig. 3.24).

The grasses that dominate this biome are among the world's most efficient plants at converting sunlight and inorganic nutrients into living tissue. They have widely spreading, fibrous root systems that hold soil firmly and capture water from violent storms. Grasses grow with extraordinary speed when conditions are right, and they enter dormancy during dry seasons. During dry times, the aboveground portions of grass plants die, leaving only underground stems alive.

Grasslands of one sort or another directly or indirectly provide food for most of the world's human population. Corn, wheat, rye, oats, and barley are just a few of the grasses American farmers grow; and only three grass species—maize (corn), wheat, and rice—support more than half the world's human population. Grasslands can be farmed for generations if treated properly, because, as we discussed earlier, their soils are deep and fertile. Even here, however, careless exploitation can turn formerly fertile areas into deserts or dust bowls (Fig. 3.25).

Tropical rain forest The world's **tropical rain forests** harbor more species than any other terrestrial environment. A single 25-acre plot in a rain forest in Borneo is home to more than 700 species of trees alone, although each species is represented by few individuals. Many rain forest

Figure 3.25 Careless replacement of native grasses with less hardy agricultural crops created the great American dust bowl of the 1930s. Much of Texas and parts of Oklahoma, once fertile prairies, became wastelands and are healthy once again only because of extensive rehabilitative work.

organisms are long-lived and slower to reproduce than related forms elsewhere.

Trees in a tropical rain forest create a three-dimensional canopy with up to five distinct layers (Fig. 3.26). On these trees grow climbers, or lianas, which may be as thin as a clothesline or as thick as tree trunks. In clumps on the branches grow epiphytes (*epi* means "on top of"; *phyte* means "plant"), including ferns, mosses, orchids, and bromeliads. Most obtain nutrients from decaying leaves that accumulate around their roots.

Live vegetation at ground level is scanty because light there is dim. Seeds and fruits are scarce on the ground as well, because high rainfall, high temperatures, and high humidity cause organic matter to decay rapidly. (That's why the humus layer and topsoil are so thin.) As a result, ground-dwelling animals (except ants and termites) are rare. Most reptiles, birds, mammals, and amphibians live in the canopy, where flowers, seeds, and insects are plentiful. The canopy is home to scores of fruit eaters ranging from birds such as parrots, toucans, and hornbills to mammals such as bats, apes, and monkeys. As described earlier, soils beneath many rain forests are easily leached by abundant rainfall, and their thin humus layer makes them—and rivers flowing through them—vulnerable to disturbance.

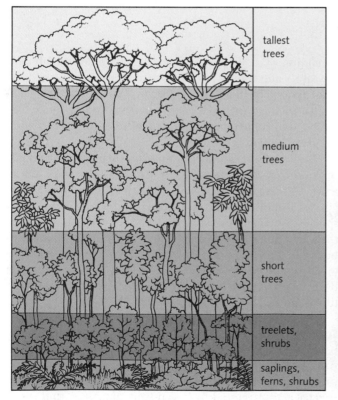

tallest trees

medium trees

short trees

treelets, shrubs

saplings, ferns, shrubs

Figure 3.26 LEFT: The multilayered canopy of the tropical rain forest creates a complex, three-dimensional habitat that houses thousands of animal and plant species. RIGHT: The rain forest canopy is home to animals, such as this slow loris, that are normally seen by humans only when trees are cut down.

Aquatic Ecosystems

Because many plants and animals float in water without expending energy, aquatic environments (particularly those in the sea) can be more three-dimensional than terrestrial habitats. Tropical rain forests, for example, range in height from 30 to 60 m (90 to 180 ft). The photic zone in marine habitats, on the other hand, is regularly deeper than that, and animal life in the sea ranges down to a depth of several kilometers.

Life under water is often controlled by different environmental factors than life on land. As noted earlier, light availability restricts photosynthesis to the fairly shallow photic zone near the surface, and thermal stratification can effectively separate water masses within a single body of water. For reasons we will discuss in the next chapter, nutrients tend to accumulate in deep water far below the photic zone. This separation of sunlight and nutrients limits the productivity of many bodies of water, either seasonally or year-round. In certain places, particular combinations of winds and currents pull large quantities of nutrient-laden deep water up to the surface. This phenomenon, known as *upwelling*, encourages highly productive plant and animal growth.

Both freshwater and saltwater ecosystems are often based wholly or partially on floating or swimming plants and animals, collectively called *plankton*, or "drifters." The plant members of this assemblage are called **phytoplankton** (*phyto* means "plant"; *phytoplankton* means "drifting plants"), while the animals are called **zooplankton** (*zoo* means "animals"; *zooplankton* means "drifting animals"). Because many aquatic animals are bizarre by terrestrial standards and not well known to the general public, we will defer discussion of them to Part 5.

Freshwater ecosystems   *Lakes* and *ponds* are open bodies of fresh water with aerobic surface layers. Clear lakes with low nutrient content and relatively little phytoplankton growth are called **oligotrophic** (Fig. 3.27a). Lakes with heavy phytoplankton and attached plant growth and high nutrient concentrations are said to be **eutrophic.** Most large lakes in nature are oligotrophic, because rivers flowing into them and groundwater seeping through the soil around them carry relatively few nutrients.

Bogs are small, deep, stagnant bodies of water in which mosses such as *Sphagnum* grow densely (Fig. 3.27b). *Sphagnum* mosses acidify the bog waters they populate. Few aquatic animals can tolerate this acidity, which also makes it difficult for plants to take up the nutrients (particularly nitrogen) they need for growth. Some bog plants have evolved unusual ways to obtain necessary nutrients: carnivorous plants trap and digest insects, while nitrogen-fixing bacteria pull nitrogen from the atmosphere.

(a)

(b)

(c)

Figure 3.27 (a) Oregon's Crater Lake is an extreme example of an oligotrophic lake, low in nutrients, with limited productivity and clear blue water. **(b)** This quaking bog in New Jersey's Pine Barrens is filled with aquatic vegetation and floating mats of living sphagnum moss. Living on the sphagnum mat are a wide variety of bog plants including azaleas, several types of conifers, and a number of insectivorous plants. **(c)** A freshwater marsh in Pennsylvania's Pocono mountains displays water lilies, several types of submerged and partially emergent pondweeds, and several species of aquatic grasses, rushes, and sedges along the shore.

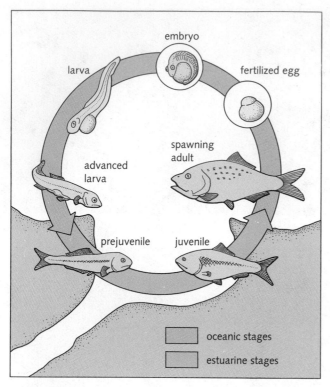

Figure 3.28 Estuaries serve as both nursery grounds (as shown here) and as spawning grounds by many fishes and invertebrates that spend their adult lives elsewhere.

Swamps and *marshes* often look similar (Fig. 3.27c). The distinction between them is that water moves (however slowly) through marshes, whereas it just sits in swamps. In the northern United States, freshwater swamps and marshes shelter a variety of tall grasses, although some are dominated by cedar trees. In the South, especially in Louisiana and Mississippi, they may contain either grasses and bog plants or cypress trees.

Estuaries An **estuary** is a partially enclosed arm of the sea that is fed by freshwater runoff from the land. Estuaries experience extremely variable conditions, from full-strength saltwater on the ocean side to completely fresh water upstream, with a brackish zone of intermediate salinity in between. This salinity gradient often shifts dramatically as tides rise and fall and as seasonal rains come and go. Estuaries in the temperate zone experience great extremes in temperature; mud flats exposed at low tide can bake on a sunny summer afternoon or freeze solid on a frigid winter night. Relatively few species can handle these fluctuations in environmental conditions, but those species able to survive thrive in enormous numbers.

The mixing of fresh water and saltwater makes estuaries highly productive. Why? Because fresh water usually contains nitrogen and silica but may lack phosphorus. Surface seawater often carries phosphorus but tends to be deficient in nitrogen and silica. Each type of water can thus support limited plant growth, because each lacks one or more essential nutrients. When combined and mixed in shallow estuaries with ample sunlight, their nutrients complement each other and encourage rapid, dense growth of plants. Many fishes and shellfish either use estuaries as spawning grounds or prey upon species that do (Fig. 3.28). Commercially valuable pink shrimp, spiny lobsters, blue crabs, bluefish, spotted sea trout, and groupers all enter estuaries to feed on abundant small prey.

Estuaries, among the most productive and vital of all aquatic ecosystems, are in grave danger around the world. Many of the world's great cities were founded on or near estuaries, so pollution and development have eliminated or poisoned many of them. Coastal development for both housing and resort purposes has also taken its toll. Because both temperate and tropical estuaries are critical as spawning grounds and nurseries, healthy marshes must be preserved at regular intervals along the shore to ensure the health of vital commercial fisheries offshore.

Salt marshes and mangrove swamps Many temperate estuaries contain **salt marshes** (Fig. 3.29), which are both created and dominated by one or more species of the marsh grass *Spartina*. *Spartina's* tightly woven roots stabilize sediments, providing stability and protection from the ravages of hurricanes.

In the tropics and subtropics, *Spartina* gives way to salt-tolerant shrubs and trees called **mangroves.** Mangrove swamps fringe tropical islands from southern Florida to Indonesia and straddle the deltas of tropical rivers such as the Mekong, Amazon, Congo, and Ganges. The peculiar root systems of red and black mangroves, the keys to success in difficult environments, are also their Achilles' heels. Increases in suspended sediments, long-lasting increases in water level, or oil spills can damage those roots enough to harm or even kill the entire forest (Fig. 3.30).

Salt marshes and mangrove swamps both support large populations of small fishes, burrowing shrimps and crabs, clams and mussels, and resident and migratory birds.

Between the tides: Life on rocky shores Wherever hard surfaces meet the sea, a thriving community lives in the space bounded by high and low tides. These communities differ between temperate and tropical shores but usually include barnacles, mussels, snails, shrimp, crabs, and starfish that survive exposure to air for varying lengths of time. Several species of sea urchins, anemones, and fishes survive in tidal pools—pockets of water retained by depressions in the rocks after the tide falls.

The plants and animals of all intertidal areas are arranged in distinct horizontal bands, a pattern called vertical zonation (Fig. 3.31). As you will learn in Chapter 6, the upper edge of many species' ranges is determined

Figure 3.29 This salt marsh at Monomoy National Wildlife Refuge in Massachusetts serves as feeding ground, nursery, and breeding ground for many species of fishes, shellfish, and migratory birds.

Figure 3.30 Mangroves create estuarine forests whose uppermost branches can reach 30 m in height. Their roots survive in thick organic mud that is often anaerobic. The arching roots of these red mangroves are covered with pores and filled with air channels that conduct oxygen to buried roots.

Figure 3.31 This photograph shows the distinct bands of life formed by the various attached animals and plants of the rocky intertidal zone.

Figure 3.32 In the Pacific, the tall, narrow fronds of *Macrocystis* create open, roomy forests.

by the physical stresses (largely desiccation and overheating) caused by living high up on the shore. The lower edge of many species' ranges, on the other hand, is often determined by biological factors such as competition for food or predation.

Kelp forests Giant, cold-water algae, termed **kelp,** grow along the Pacific coasts of North America, Japan, Siberia, and South America, and off the Atlantic coasts of Canada, New England, South Africa, Great Britain, and Scandinavia. These remarkable plants can grow up to 50 cm per day and, when fully grown, can extend 20 to 40 m between the rocky bottom and the surface of the water (Fig. 3.32). Kelp forests truly deserve the name *forests,* because of their appearance and their extremely high productivity, and because they are complex, three-dimensional habitats. Kelp forest denizens include such commercially important species as abalone (large, flattened, snail-like animals) in the Pacific and American lobsters, *Homarus americanus,* along the Atlantic coast.

Hunting and fishing have upset the ecological balance in many kelp forests. As you will learn in Chapter 6, both the Pacific sea otter (Fig. 3.33) and the American lobster perform vital roles in maintaining the ecological stability of their habitats.

Coral reefs Both productive and ecologically diverse, **coral reefs** support abundant life in nutrient-poor tropical seas. Like the trees of the tropical rain forest and giant kelp, hard corals—which build for themselves intricate branching skeletons that accrete over time—create elaborate three-dimensional habitats in which other organisms live. With nearly constant, hospitable environmental conditions, coral reefs in the western Pacific shelter more species that exhibit more complex ecological interconnections among them than most other habitats on Earth.

Many reef animals are active only by day and seek shelter by night (Fig. 3.34). Others wander only after dark and hide from sunrise to sunset. These animals alternate occupancy in coral caves so regularly that one could describe the reef as a collection of boardinghouses where the same rooms are occupied day and night by different guests.

Reefs around the world are under serious pressure from human activities. Coastal development has blanketed some reefs with debris, smothered others in silt, and literally buried others in asphalt. Although reefs are productive enough to permit reasonable harvesting of their inhabitants, overfishing for food, overcollection of animals for the home aquarium trade, and removal of coral for jewelry and decoration have caused serious damage around the world. Conservation efforts parallel to those aimed at the better-known rain forests are currently underway.

The open sea In the open sea, all higher life depends on phytoplankton and zooplankton. The planktonic world is a dynamic, multilayered ecosystem that covers the oceans, reaching down into the depths for hundreds, or even thousands, of feet (Fig. 3.35).

The surface waters of tropical seas offer intense sunlight but carry few nutrients and are therefore not very productive. In areas along the coasts of major continents, freshwater runoff provides enough nutrients for more aquatic growth. And in upwelling areas such as Georges Bank in the North Atlantic and the Pacific coast of Peru, dense populations of phytoplankton support high concentrations of zooplankton. Such areas are known for their highly productive and commercially important fisheries.

A diverse group of animals—from small shrimp and fishes to whales, and a wide range of fishes in between—either hunt zooplankton individually or strain them from the water. Coastal waters host fishes familiar to anglers and restaurateurs—including bluefish, flounder, halibut, cod, and sea bass. These fishes feed either on small fishes or on other marine animals such as worms, clams, and mussels. Ranging out into deeper water are magnificent fast-swimming fishes such as tuna, marlin, and swordfish, not to mention the spectacular great white shark and a

Figure 3.33 Sea otters are carnivores that play vital roles in structuring their own habitat by controlling populations of herbivorous abalone and sea urchins.

Figure 3.34 LEFT: By day, huge schools of scarlet *Anthias* roam the waters of the coral reef, as do many other diurnal species. RIGHT: The change in the reef at sunset is striking, as these and many other daytime fishes seek shelter in coral nooks and crannies. At the same time, nocturnal animals, hidden throughout the day, emerge to prowl the darkening waters.

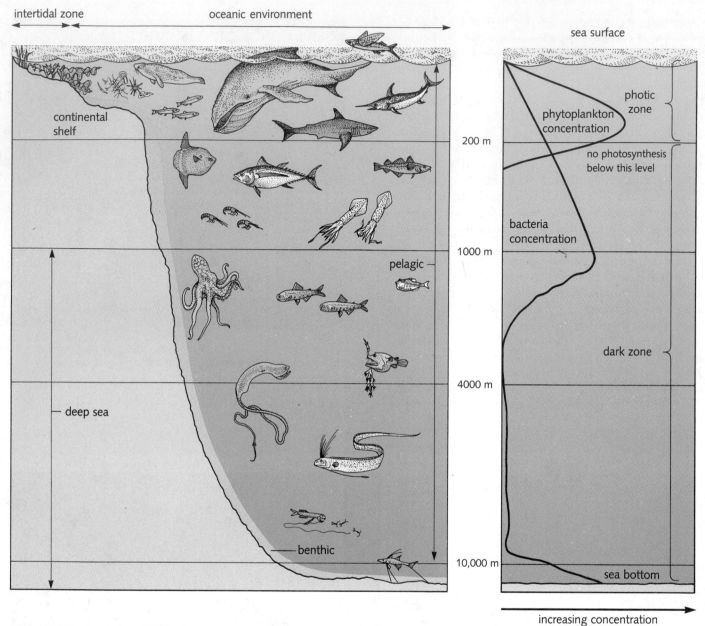

Figure 3.35 Provinces of life in the ocean. LEFT: Organisms that live on the bottom are called *benthic,* while those that live in open water are called *pelagic.* Animals and plants that spend their time floating in the water column are called *plankton* or "drifters." Plant plankton are called *phytoplankton;* animal plankton are called *zooplankton.* Animals that actively swim through the water are called *nekton.* RIGHT: Concentration of organisms is highest in the photic zone and at the sea bottom.

wide range of marine mammals. These animals generally feed on small- to intermediate-sized fishes and squid. The deep sea contains a menagerie of fantasy creatures with enormous mouths, huge goggle-eyes, and even organs that light up in the dark.

Ecosystem Structure and Function

This simple description of biomes and habitats raises a host of questions about the organisms that inhabit them. How do they make their livings? What factors control how many of them can survive in a given area? How do interactions among organisms either help or hinder them? We will attempt to answer these questions in order in the next several chapters.

SUMMARY

Biological diversity refers to the variety of ecosystems and organisms in the biosphere. A substantial portion of that diversity is important for the stability and resilience of the earth's living systems.

An organism's ability to survive in a given area is determined both by the environmental conditions it encounters and by its ability to cope with those conditions. Both animals and plants have certain limits within which they can grow and reproduce. Environmental conditions encountered by organisms fall into two broad categories: (1) Physical, or abiotic, conditions—the means and extremes of sunlight, temperature, rainfall, and humidity—are determined by global meteorological phenomena, such as wind and ocean currents, and by local geographic features, such as the presence of a lake or mountain range. (2) Biological, or biotic, conditions are determined by the presence of other organisms that can provide food or shelter or may act as predators. Organisms affect local conditions around them, creating microhabitats with distinct microclimates.

Because living systems are adapted to prevailing environmental conditions, changes in those conditions—from either natural or human-caused events—can dramatically reshape many ecosystems.

Ecologists divide the world into major habitat areas called biomes, characterized by similar physical environmental characteristics. Each biome may house several different living communities.

Certain ecosystems, because of both their physical and their biological characteristics, can withstand considerable pressure from human use, while others cannot. Grasslands, when properly treated, can support intensive, large-scale farming of corn, wheat, and other plants for many years; tropical rain forests cannot.

STUDY FOCUS

After studying this chapter, you should be able to:

- Define *biological diversity*, and explain its importance to natural systems.
- Describe the several major classes of terrestrial and aquatic environments, and list some of their resident species.
- Explain how environments influence plant and animal communities.
- Define *environment*, describe environmental variables, and explain their importance to plants and animals.
- Illustrate the role of organisms themselves in shaping their environments.
- Explain the importance of global climate patterns.
- Show how environmental changes can affect plant and animal growth.

TERMS AND CONCEPTS

ecosystems *34*	microenvironments *46*
biological diversity *34*	microclimate *46*
environment *34*	biomes *46*
habitat *34*	tundra *48*
physical environment *35*	taiga *49*
biological environment *35*	temperate forest *51*
aerobes *35*	desert *51*
anaerobes *35*	grasslands/savannahs *53*
thermal stratification *37*	tropical rain forests *53*
thermocline *37*	oligotrophic *55*
humus *37*	eutrophic *55*
climate *39*	estuary *56*
weather *39*	salt marshes *56*
thermoregulation *44*	mangroves *57*
endotherms *44*	kelp *58*
ectotherms *44*	coral reefs *58*

REVIEW

Objective Questions (Answers in Appendix)

1. Which of the following variables does not determine the distribution of plants and animals?
   ✓(a) the size of the habitat
   (b) incoming solar radiation
   (c) salinity of the waters
   (d) available nutrients

2. The biome that can have enormous fluctuations of daily and seasonal temperatures is the
   (a) temperature deciduous forest.
   (b) tundra.
   (c) savannah.
   (d) desert.

3. In tropical rain forests
   (a) there is intense competition for available sunlight.
   (b) diversity of species is limited.
   (c) the soil is fertile.
   (d) the bulk of mineral nutrients is in the soil.

4. The difference between eutrophic and oligotrophic lakes is that the oligotrophic lake
   (a) has little phytoplankton growth.
   (b) has decaying plants on the bottom.
   (c) contains many nutrients
   (d) tends to be shrinking in size.

5. What biome would be best suited for cropland if cleared?
   (a) tropical rain forest
   (b) taiga
   (c) temperate forest
   (d) tundra

6. In a tropical rain forest, the majority of the nutrients are present in
   (a) plants.
   (b) animals.
   (c) decaying humus, seeds, and fruits at ground level.
   (d) soil.

7. In marine ecosystems, the critical limiting factor for plants below the photic zone is the
   (a) intensity of solar radiation.
   (b) availability of nutrients.
   (c) salinity of the water.
   (d) amount of fertilizer runoff from adjacent farmlands.

Discussion Questions

8. What does the term *environment* mean? What conditions determine the effects a given environment has on organisms living in it?

9. What is a habitat? Why is it important to consider microhabitats when looking for the effects of environmental conditions on small organisms?

10. How do global air circulation and ocean currents determine climate and weather where you live?

11. Using examples from natural ecosystems, how would you support or refute the idea that harsh physical conditions support lower species diversity than do more benign areas?

12. Why does a hiker, climbing from the foot of a tropical mountain to its summit, encounter ecological zones similar to those found across great distances from the equator to the poles?

13. Why are tropical rain forests so interesting to scientists and so potentially valuable to human society?

14. Why do upwelling areas produce more useful food for humans and other large animals than the open sea?

15. What is an estuary? Why is life there so abundant?

16. What is biological diversity? Why is it so important? Why is the world's biological diversity threatened?

READINGS

See integrated list of readings for Part 2 after Chapter 7.

4

Energy and Nutrients in Ecosystems

On Africa's Serengeti plain, grasses 2 m tall ripple in the wind that carries seasonal rains. Growing with astonishing speed when water is available, these plants spend the dry season in dormancy (Fig. 4.1). Wandering across these plains are herds of wildebeests, zebras, gazelles, and many other plant eaters. Those animals are preyed upon by carnivores such as lions and cheetahs, and their carcasses are devoured by scavengers ranging from hyenas to carrion-eating birds.

In the lowland rain forests of Ceylon, trees form an emerald canopy to catch sunlight more than 100 m (300 ft) from the ground. Blessed by abundant rainfall and warmth, many of these trees constantly add new leaves and shed old ones throughout the year. Each species flowers and sets seeds encased in fruits that are as different from one another as they are from such familiar tree crops as apples and oranges. Those leaves and fruits are relished by countless insects, scores of monkeys, rainbow-colored birds, and even strong-jawed fishes waiting in the rivers below.

Off the California coast, a 30-m (90-ft) forest of kelp sways gently in the frigid water, adding new tissue to its blades at the remarkable rate of 50 cm per day. Abalones and urchins graze at its base, seeking out and devouring immature plants wherever they find them. Abraded by waves, kelp blades fray at their edges, are chewed into tiny pieces by minute shrimp, and decompose beneath a blanket of bacteria and fungi. Ultimately, products of that decomposition enable an entire menagerie of aquatic creatures to make a living without moving—by filtering food from the water around them while glued in place to kelp stems or rocks beneath them.

And in the stormy waters of Georges Bank off the New England coast, floating, single-celled algae grow and divide as they drift with the currents. Were we to follow their fate, we would see a chain of large animals eating smaller animals before being devoured themselves by something larger still. Near the surface, that chain begins with tiny shrimp-like creatures and may end with bluefish, striped bass, sharks, or humans. On the ocean floor the chain is buried in clams, worms, mussels, and other bottom dwellers that in turn fall prey to halibut, cod, and lobster.

Figure 4.1 Africa's Serengeti plain.

ENERGY AND NUTRIENTS: LIFE'S VITAL CURRENCIES

In every ecosystem, plants, animals, fungi, and bacteria are linked together into complex relationships based on the most fundamental needs of all living things: *energy* to power the processes of life and *nutrients* for use as building blocks to assemble living tissue. In a sense, energy and nutrients serve as two indispensable "currencies" in nature's economy. Without a steady supply of both in one form or another, living systems simply cannot function.

Borrowing terms from economists, ecologists often divide living things into two large, diverse groups based on how they obtain their biological "cash supply":

1. **Producers,** including nearly all plants, harness energy from the sun to assemble simple, inorganic nutrients into the complex molecules that comprise their cells. For this reason, they are called **autotrophs** (*auto* means "self"; *troph* means "food or feeding"; *autotroph* means "self-feeding"). Such organisms are also called **primary producers** because they are the first producers of complex compounds (Fig. 4.2).

2. **Consumers,** including all animals, most fungi, and many bacteria, depend directly or indirectly on primary producers for energy and nutrients. For this reason, they (or, more correctly, *we*) are called **heterotrophs** (*hetero* means "other"; *troph* means "feeding"; heterotroph means "other feeding"). Heterotrophs can *store* energy and certain nutrients in many ways, such as by laying down deposits of fat beneath the skin. But

we must *obtain* both energy and nutrients by eating either primary producers or other heterotrophs.

Understanding human economies requires an appreciation of how currency and commodities move through economic systems. The same is true of nature's economy. Nature's "currencies" (energy and nutrients) accumulate and move through ecosystems in the form of various "commodities" (living tissue). As you will soon see, "cash flow" varies enormously between different ecosystems, as do the types of "commodities" available. Despite that variation, certain basic biological and chemical rules regulate all possible transactions. Because those rules govern the total productivity of ecosystems, the types of organisms that live in them, and their interactions, we examine first the nature and movement of energy in living systems, and then we follow nutrients as they move through in tandem.

Life's Power Supply

In some forms, energy is easy to measure and understand. The heat energy of a roaring bonfire is obvious. But how can energy be stored in chemical form? What forms of energy are useful to life? And how is energy channeled through living systems? Answering these questions requires taking a brief look at a few basic chemical concepts relevant to ecology.

Figure 4.2 This freshwater alga is an aquatic primary producer.

Energy can be stored chemically in the form of a *chemical bond*, a linkage that binds atoms together to form molecules, the basic units of chemical compounds. Energy is required to form a chemical bond between two atoms. In a sense, we can think of that energy as being stored in the bond, to be released whenever and however the bond is broken.

One fact about chemical bonds and energy is vitally important for biology: **some bonds contain more energy than others.** We can see why that is so important from a simple example. Fig. 4.3 shows a photovoltaic cell that converts energy from sunlight into electric current. That current is applied to two metal rods (electrodes) in a beaker containing water, which is composed of hydrogen and oxygen. The energy in the current breaks apart the bonds that link hydrogen to oxygen in water molecules. Oxygen gas (O_2) bubbles up along the positive electrode and hydrogen gas (H_2) along the negative electrode.

What has actually happened here is that one set of chemical bonds (between hydrogen and oxygen) were broken and new bonds (H—H and O—O) were formed. In this case, the energy required to make the new bonds is *greater* than the energy released by breaking the old bonds. For that reason, this reaction requires energy to run. The electric current from the photovoltaic cell provides that energy.

But the energy poured into this reaction isn't lost. The hydrogen and oxygen can be gathered into separate vessels, mixed together as shown, and ignited with a spark. Once ignited, the mixture burns brightly, as H_2 and O_2 combine to form water. Note that this is the reverse of the reaction that produced the two gases to begin

with. When the reaction proceeds in this direction, the energy contained in the H—O bonds is less than the energy in the O—O bonds and H—H bonds. Consequently, a tremendous amount of energy is given off in the form of heat. (This reaction powers many booster rockets used in the U.S. space program.) The energy given off as heat in this rapid burning was present all along, stored as chemical energy in bonds between atoms of hydrogen gas and oxygen gas.

This ability of chemical bonds to store energy makes life possible, as you will see shortly. The rules that govern the storage and release of energy are contained in a branch of science known as thermodynamics. The first and second laws of thermodynamics, which govern the relationship between matter and energy in chemical reactions, will be covered in Chapter 17.

Autotrophs and Energy

Living systems trap and store energy in chemical form, too, by synthesizing (building up) a variety of energy-rich compounds. Most primary producers do that through **photosynthesis,** a process in which the green pigment *chlorophyll* harnesses solar energy to convert carbon dioxide and water into organic compounds known as **carbohydrates** (*carbo* means "containing carbon"; *hydrate* means "containing hydrogen and oxygen").

The simplest and most biologically important carbohydrates are *sugars,* such as glucose, sucrose, and fructose. All sugars are composed of individual units known as simple sugars or *monosaccharides,* each of which contains from

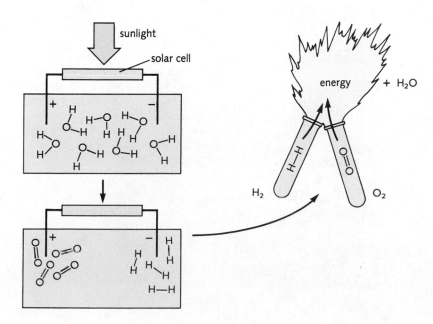

Figure 4.3   In this diagrammatic representation of simple electrolysis, solar energy produces an electric current that breaks the bonds between the hydrogen and oxygen molecules in water and forms new bonds between pairs of oxygen atoms (in oxygen molecules) and pairs of hydrogen atoms (in hydrogen molecules).

three to seven carbons. One of the most important simple sugars made by plants is *glucose* (Fig. 4.4). The energy that plants pour into the formation of these glucose molecules can later be released by breaking them back down to carbon dioxide and water in the presence of oxygen. That is what happens in living cells; both animals and plants power their vital functions by burning glucose as fuel in the process known as cellular respiration (see Chapter 18).

Glucose is also important because it provides a convenient way for organisms to store energy. Living cells contain a great deal of water, and sugars are readily soluble in water. This means that sugar can be held in solution anywhere in the living cell and kept available as an energy source.

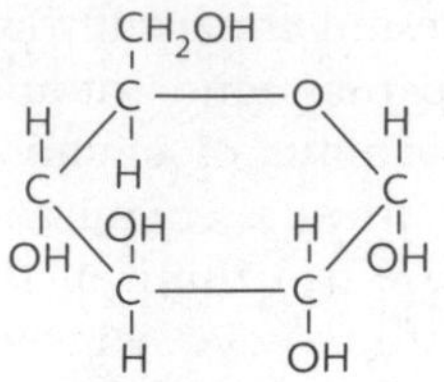

Sugars and edible carbohydrates   Just about every molecule found in one organism can be used as food by some other organism. (That's one of the things that can make life dangerous.) But there are no molecules more important as a food source to living things than carbohydrates.

Monosaccharides are important not only because they store energy well but also because they are chemically versatile. Simple sugar molecules can be joined together into complex carbohydrates such as *starch* and *cellulose*. Figure 4.5, for example, shows a chemical reaction in which two glucose molecules are joined by a chemical bond. The precise nature of that bond is very important. As the figure shows, the oxygen attached to the carbon atom can be in either of two positions when the bond is made. And although the two bonds—named α and β linkages—do not look very different in the diagram, there is a world of difference between them in living systems.

Why are the differences between those bonds important? Before cells can use the energy stored in complex carbohydrates, those large molecules must be broken down—or digested—into their component sugars. As we

Figure 4.4   The structure of glucose, an important carbohydrate molecule involved in storage, transfer, and release of energy in biological systems.

will learn in more detail in Chapter 37, many steps in digestion depend on specific *digestive enzymes* that break down large molecules, such as complex carbohydrates, into small molecules, such as glucose. But not all enzymes can break down all kinds of bonds, so the difference between α and β linkages can make the difference between one carbohydrate that can be digested and another that cannot.

Plants, for example, store much of their energy in the form of a carbohydrate called starch, in which the simple sugars are held together by α linkages. Many other organisms (including humans) produce digestive enzymes that can break α linkages very easily. Thus starch is an excellent source of food; in fact, starches found in potatoes, corn, rice, wheat, and barley are the major food sources for our species.

Cellulose   Many plants, however, use another carbohydrate for an entirely different purpose. By linking thousands of glucose molecules together with β linkages, plants form cellulose, which they use to build the tough cell walls that support and protect their cells. In larger plants, cellulose fibers form the basis of an even stronger material: wood.

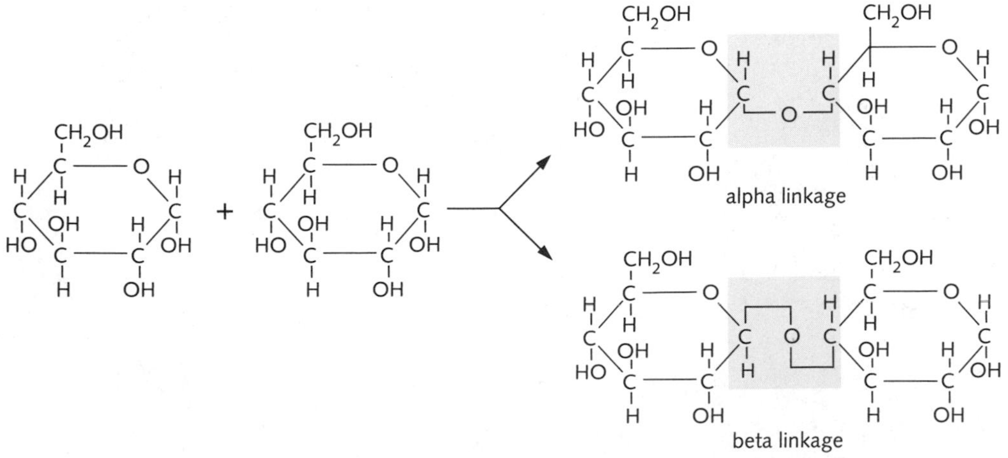

Figure 4.5   Two simple sugars may be joined by a chemical bond. Two possible bonds, the α linkage and the β linkage, are shown. While most organisms can break (and therefore digest) the α linkage, very few can break the β linkage.

Cellulose is very similar to starch in chemical composition, but its β linkages make it very difficult to digest. Why? Because the β linkages in cellulose can be broken only by a single enzyme called cellulase, which most animals cannot manufacture. Only a handful of snails, some clams, and a few species of microorganisms produce cellulase and so can digest cellulose. Actually, we can be thankful that the cellulose bond is difficult to break. Large plants such as trees could not survive long in a world where every bug, fungus, and mammal could use them as easy meals! Our digestive tracts have gotten used to a certain amount of indigestible material; cellulose is a major component of what is commonly called "roughage" in foods.

Heterotrophs: Obtaining Energy Captured by Plants

If all living organisms were autotrophs, understanding energy would be easy; chemical energy hoarded in sugars would be stored by organisms that make them and then used as needed. But heterotrophs obtain energy through one of three main strategies: eating plants (herbivory), eating animals (carnivory), and decomposing the remains of other organisms and their waste products.

Herbivores: The diverse plant eaters The **herbivores** (*herb* means "plant"), such as rabbits and deer, eat plants that have stored energy in the forms of sugars, starches, and other complex carbohydrates. Because they are the first organisms to consume energy and carbon "fixed" by primary producers, herbivores are also called **primary consumers.** The diversity of plants has spawned an equally diverse set of herbivores, with a wide range of feeding strategies (Fig. 4.6).

Leaf eaters The most familiar plant eaters, such as cattle and sheep, are *grazers* who harvest the leaves of grasses. But leaves, though plentiful and easily obtained, contain large quantities of cellulose which, as you've just learned, is nearly indigestible. To extract energy from that tough food, grazing animals depend on strategies that involve both feeding behavior and the design of their digestive organs.

First, grazers chew leaves into a pulp with powerful grinding molars. When this pulp is swallowed, it enters the grazer's long and complex digestive tract. There, cellulose is broken apart with the aid of beneficial microorganisms (Fig. 4.7). Even with microbial assistance, however, grazers can extract relatively little energy from each mouthful of food and must spend most of their waking hours eating. The fascinating story of how this system evolved will be described in Chapter 22.

Figure 4.6 Herbivore diversity. TOP: A giraffe, which feeds on the leaves of tall trees. CENTER: A sea urchin uses a five-part jaw to rasp encrusting marine algae off hard surfaces. BOTTOM: The toucan is a colorful herbivore that feeds on berries and fruits.

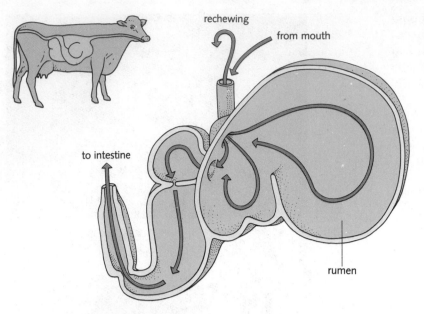

rechewing

from mouth

to intestine

rumen

Figure 4.7 Cows have four stomach chambers. In the rumen, the first and largest chamber, bacteria and other microorganisms produce the enzyme cellulase, which combines with the cow's digestive enzymes to attack leaf pulp. Cows and some other grazers regurgitate the pulp back into their mouths, chew it a second time, grind it into an even finer paste, and add more digestive enzymes from saliva before swallowing it again. This is what cows are doing when they "chew their cud." After the food has been processed again in the rumen, it passes through the other stomach chambers and into the intestines.

Seed, fruit, and berry pickers Other herbivorous animals, including many fishes, birds, and mammals, feed on seeds and fruits. Some, such as many finches, specialize in eating seeds of particular sizes and types. Others, such as South American toucans, use long, sharp bills to pluck small berries or to chop large fruits into bite-sized pieces (see Fig. 4.6). Humans also depend heavily on plant seeds and fruits. Most of the world's human population lives on the seeds of a single plant family, the grasses, which includes rice, corn, wheat, oats, and barley.

Table 4.1 *Comparison of Birds' Intestines Showing Longer Length with Poorer Diet*

	Quail	Wood Pigeon	Starling
Type of Diet			
Poor diet	Artificial	Brassica	Plant
Rich diet	Artificial	Grain	Animal
Intestine Length (cm)			
Poor diet	51	220	33
Rich diet	46	157	27
Poor/rich	1.1	1.4	1.2
Cecum Length (cm)			
Poor diet	17.0	—	0.8
Rich diet	14.5	—	0.6
Poor/rich	1.2	—	1.3

Source: Sibly, 1981.

Carnivores: The meat eaters **Carnivores,** such as cats and hyenas, eat other animals that have stored energy in the form of fats, oils, and proteins. Animals that eat herbivores are sometimes called **secondary consumers.** Carnivores that eat other carnivores are called **tertiary consumers.**

Many meat eaters expend more energy than grazers in actually obtaining their food; stalking and running down a zebra or subduing a wildebeest is no easy matter. Other carnivores, such as chameleons, lie in wait to ambush their prey; these animals may spend much of their time waiting for unsuspecting meals to wander by (Fig. 4.8).

The ease with which meat is digested is reflected in several aspects of carnivore anatomy and behavior. Even tame carnivores, such as pet cats, hardly bother to chew. Their razor-sharp teeth quickly chop food into pieces small enough to be swallowed. Because chunks of meat yield energy so readily, carnivore intestines are shorter on average than those of herbivores. (There is more than circumstantial evidence linking intestine length to food quality. Table 4.1 records experimental evidence that birds fed on poorer diets grew longer guts in as short a time as 25 days!) Many meat eaters feed only occasionally, and some—such as the big cats and boa constrictors—spend much of their time sleeping or relaxing.

Omnivores, such as humans and several other primates, eat both plants and animals. Some omnivores, including certain fishes of the Amazon river system, vary their diets according to the seasonal availability of preferred foods. Others eat both meat and vegetable matter year round. As you might expect, the teeth and digestive systems of these animals often (though not always) combine characteristics of both carnivores and herbivores.

Decomposers: Organisms of decay *Saprophytes,* such as many fungi and bacteria, obtain energy by breaking down, or *decomposing,* complex molecules in tissues of plants and animals that have died (Fig. 4.9). These organisms of decay are essential to the health of all ecosystems, for they play indispensable roles in the cycling of important nutrients. Without these nearly invisible, yet vital, heterotrophs, bodies of dead animals would litter the landscape and nutrients those organisms accumulated would be lost to the ecosystem.

Other feeding techniques Many heterotroph feeding techniques don't fit neatly into familiar categories. Some herbivores and carnivores, for example, are **liquid feeders** that feed only on liquids extracted from their prey (Fig. 4.10). Aphids and other sucking insects pierce plant stems to suck out the sugary sap. Mosquitoes and ticks attack animals in the same way, using syringe-like mouthparts to drink the blood of birds and mammals.

Another group is **scavengers** or **waste feeders.** Egypt's scarab beetles feed on the dung of other animals. Still other animals feed largely on **detritus,** a combination of bits of decaying organisms and dung containing partially digested food. On land, detritus mixes with soil particles and is eaten by earthworms and soil-dwelling insects. Worms swallow large amounts of soil and digest detritus particles as they pass through their guts. Many marine animals, from corals to sea cucumbers, either filter floating detritus from the water or pick it up off the bottom after it settles.

Aquatic environments offer several food sources not found on land. In addition to carrying floating detritus, phytoplankton, and zooplankton, water carries dissolved organic molecules. To exploit these foods, many marine

Figure 4.9 Fungi are vital decomposers that recycle nutrients locked up in the complex molecules of dead animals and plants.

Figure 4.8 Carnivore diversity. TOP: Sparrow hawks use aerial agility and sharp talons and beaks to prey upon small birds. CENTER: Chameleons snare insects, using long, sticky-tipped tongues that shoot out with lightning speed. BOTTOM: Lions hunt in groups and generally share their prey, usually large herbivores like this zebra.

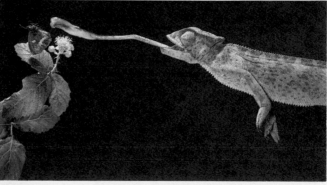

Figure 4.10 Spiders paralyze prey trapped in their webs, inject digestive enzymes into the prey's body, and suck out the partially digested tissues as they liquefy.

species have become *filter feeders,* with mouthparts, gills, and limbs that filter food from the water around them (Fig. 4.11).

Solar Energy and Life

Because all heterotrophs depend on organic molecules they cannot make themselves, their lives are structured around the need to obtain those molecules. Herbivores must live within reach of growing plants. Carnivores survive only where there is prey, and saprophytes only where there are dead plants or animals. Thus *any conditions that control the growth of primary producers in an ecosystem indirectly control the lives of all other organisms as well.*

Although many factors affect plant growth both on land and in the sea, marine plant life is limited by the fact that sunlight penetrates only a short distance from the surface, while nutrients tend to accumulate in deep water where photosynthesis cannot occur. This phenomenon has profound effects on both life in the sea and human interactions with that life, because most of the earth's surface is underwater.

As you learned in the previous chapter, light intensity decreases markedly with depth in water. This phenomenon has important effects on marine plants. Chlorophyll —the premier light-absorbing pigment in photosynthesis—most efficiently harnesses light in the regions of the spectrum that are removed by seawater. In response to this situation, some algae have evolved a different kind of chlorophyll that captures energy from a different part of the spectrum than that of terrestrial plants.

Even this "aquatic" chlorophyll, however, can't carry on photosynthesis deeper than a few dozen meters be-

Figure 4.11 Filter feeders, such as these marine worms, stretch out feather-like gills covered with tiny hairs and sticky mucus. Food particles are caught in the mucus and funneled by the beating hairs down into the worm's mouth. Variations of this technique are used by clams, mussels, and scores of other species.

Figure 4.12 Accessory pigments—which absorb light of wavelengths not absorbed efficiently by chlorophyll and make its energy available for photosynthesis—endow many algae with striking colors.

neath the surface. So marine algae use other compounds, called *accessory pigments,* to collect more energy by absorbing light of wavelengths that chlorophyll does not absorb (Fig. 4.12). This extra energy is passed on to the machinery of photosynthesis, enabling plants or algae to survive in deeper water than they could otherwise.

Still, as you saw in Fig. 3.4 on page 36, the sunlit photic zone forms little more than a thin skin across the surface of the world's oceans. That's one reason why estuaries and the various marine ecosystems along the shallow margins of continents are vital in the grand scheme of marine ecosystems: those are the only places where marine primary producers receive adequate amounts of both sunlight and the nutrients they require.

FOOD CHAINS AND FOOD WEBS

Obviously, if some animals eat plants, other animals eat one another, and organisms of decay eat everybody sooner

or later, all forms of life in an ecosystem are linked together by feeding relationships. Examining ecosystems briefly as we did in the chapter introduction, looking at just a few species at a time, we can imagine that a primary producer, an herbivore, and a carnivore form a simple **food chain.** For example, in a temperate woodland, a grass-eating mouse becomes prey for a predatory owl:

and in tropical Africa, impala feed on range grasses and are fed upon by the big cats:

But if we observe all the animals and plants in most ecosystems, we find that feeding relationships are rarely that simple. Usually, they weave numerous organisms into large, complex networks called **food webs,** in which many animals eat several different kinds of food. Fig. 4.13, for

Figure 4.13 Feeding relationships among selected organisms of the African savannah, still enormously simplified. This food web gives a better idea than the simple food chain in the text of the complexity of feeding relationships in most environments. Arrows are color-coded to highlight simpler food chains within the web. In order to simplify this diagram and emphasize the complexity of even the most familiar connections among animals, we have left out the vital role of decomposers, as discussed in the text shortly.

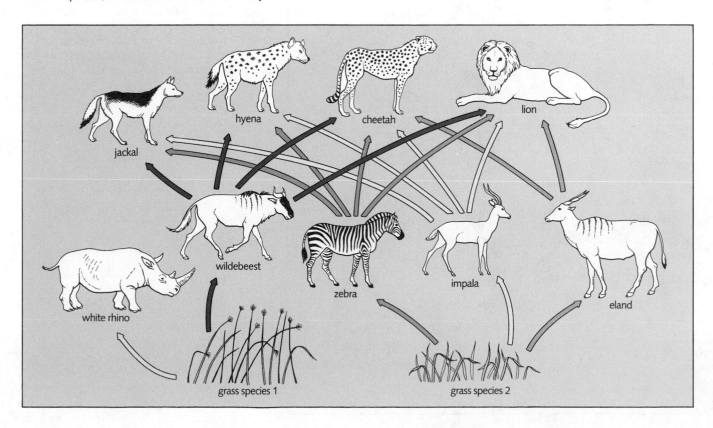

example, shows a somewhat more realistic (though still highly simplified and incomplete) representation of a food web on an African savannah. Note that different primary consumers choose different plants as food, and that not all herbivores are equally palatable (or available) to carnivores.

Many estuarine and marine food webs differ from this terrestrial example because detritus-feeding animals play a central role. In salt marshes and mangrove swamps, detritus—along with phytoplankton and zooplankton—supports a wide range of filter-feeding animals (Fig. 4.14). Detritus also plays a major role in kelp forest ecosystems. Kelp blades are constantly eroding as they grow, producing a steady stream of detritus that supports a food web much like that described for salt marshes and mangrove swamps.

Importantly, food webs are rarely contained within a single ecosystem; nearly all have branches that interconnect with food webs of adjacent areas. Mangrove and sea grass systems export both dissolved organic matter and detritus, which serve as food for filter-feeding organisms on nearby coral reefs. Predators from reefs stalk small fishes in their mangrove swamp nursery grounds. Osprey,

herons, and pelicans eat in one habitat, deposit their nutrient-bearing feces in another, and raise their young in a third. Back and forth, animals and currents carry energy and nutrients among these systems, binding them together.

Energy Flow Through Ecosystems

Understanding the movement of energy, or *energy flow*, through living systems is essential to understanding how ecosystems work. Beginning with the capture of sunlight by primary producers, energy flows through food chains and food webs in a steady, *one-way stream*. As it flows, energy is alternately stored and used to power the life processes of animals through which it moves, much as the potential energy of a river stored behind a dam can be used to drive a mill wheel. The actual use of energy in living systems, of course, is more complex than that simple analogy, because several important processes take place at each step along life's energy chain.

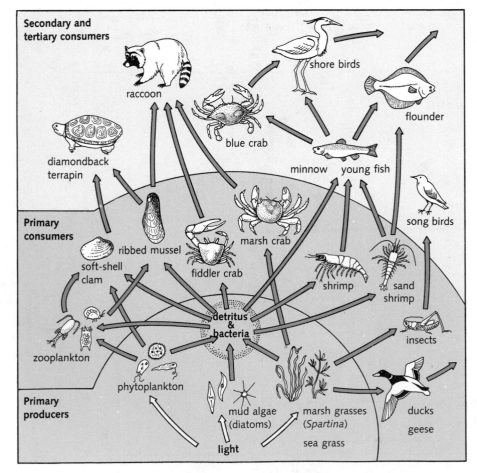

Figure 4.14 This simplified schematic of an estuarine food web shows the central position of detritus in these ecosystems. Scores of small fish, zooplankton, and fiddler crabs ingest detritus particles, digesting decomposers growing on them. The nitrogen these animals release back into the water is quickly reabsorbed and recycled by primary producers.

Trophic levels Energy captured by producers and consumers is temporarily stored until one organism eats another. Each of these storage steps in a food chain is called a **trophic level.** Primary producers are the first trophic level, herbivores occupy the second, carnivores that eat herbivores form the third trophic level, and so on.

Although theoretically a productive ecosystem could support many trophic levels, in reality there are practical limitations. Every time one organism eats another, only a small fraction of the energy present in the lower trophic level is stored in the next higher level. Finding, catching, and ingesting food uses up energy, as does digesting food. And very few animals come close to extracting all the energy in their diet. In addition, life processes such as cellular respiration convert a great deal of chemical energy into heat, and much of that heat (nearly all of it in some organisms) is lost to the environment (Fig. 4.15).

Estimates of energy loss in living systems vary widely. One widely quoted estimate, based on studies in aquatic ecosystems, is called the **rule of 10** or the *10 percent rule.* It

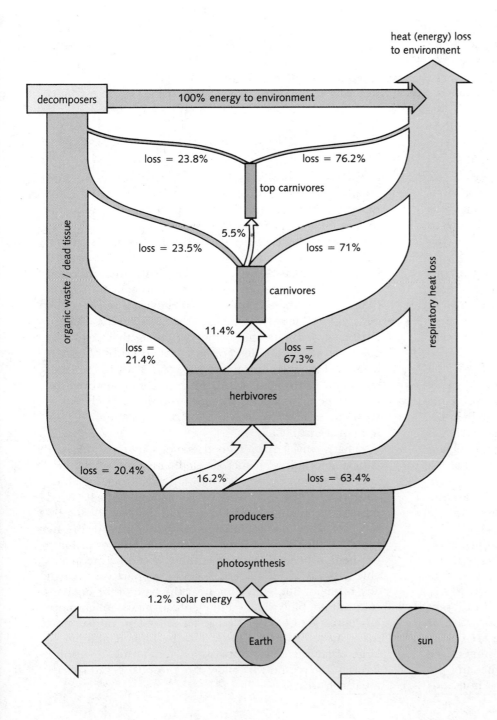

Figure 4.15 Energy flow through a typical ecosystem. Note the major losses of energy to the environment through respiration (heat) and the loss through dead tissue at each step. Note also that this diagram shows dramatically how the entire system depends on the capture of energy by primary producers.

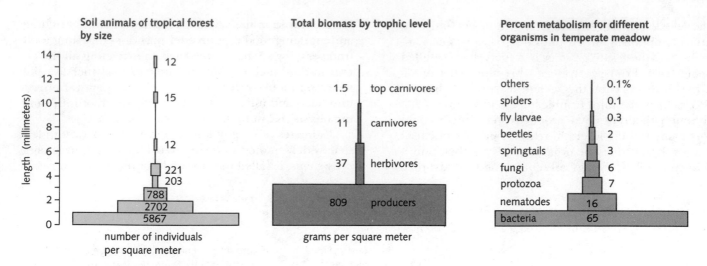

Figure 4.16 Ecological pyramids and pyramids of numbers, biomass, and energy. LEFT: Numbers of animals of different sizes in a tropical forest. CENTER: Dry biomass of organisms at Silver Springs, Florida. RIGHT: Percentage of total ecosystem metabolism for different groups of organisms in a temperate meadow.

states that, on average, only about 10 percent of the energy fixed by plants is ultimately passed on to herbivores. Only 10 percent of the energy herbivores accumulate is passed on to the carnivores that eat them. And only 10 percent of *that* energy is successfully transferred to carnivores on the third trophic level.

This generalization paints a horribly inefficient picture of energy transactions in nature—and one that does not seem to hold in many natural systems. More recent studies have shown that the actual efficiency of energy transfer in food webs varies from a minimal 0.05 percent to as high as 20 percent.

Ecological pyramids Although the actual efficiency of energy transfer varies widely, the general principle demonstrated by the rule of 10 is important. Why? Because the less energy available at any trophic level, the less living tissue that trophic level can support. Inefficient energy chains thus contribute to the creation of what are called **ecological pyramids,** in which each level is substantially smaller than the level below it (Fig. 4.16). In an ecological pyramid, each trophic level contains only a fraction as much living tissue as the layer beneath it.

Often, the restrictions of energy transfer in living systems also create different sorts of pyramids. One of these, a *pyramid of numbers,* shows that each trophic level contains fewer individual organisms than the level below it. Yet another type of pyramid ranks organisms of different types according to their relative metabolic importance. As is usually the case in biological systems, there are exceptions to general "rules." In some cases, consumers are much smaller than the organisms they consume, and the pyramid of numbers is turned upside down. Thousands

of insects may graze on a single tree, for example, and countless mosquitoes can feed off a few deer or humans. But even in situations where the pyramid of numbers is reversed, the pyramid of biomass still applies.

These ecological rules have dramatic ecological consequences, as we can demonstrate if we accept, for simplicity, the mathematics of the 10 percent rule. In that case, if a human chooses to eat only red meat to gain a kilogram of weight, that person must eat 10 kilograms of beef. The cow responsible for producing that 10 kilograms of flesh must have originally eaten at least 10 times that weight, or 100 kilograms, of fodder. For us to obtain 1 unit of energy from beef thus requires the storage of at least 100 units of energy in cattle feed. (And this calculation ignores the energy spent to process, store, transport, and sell food.) We will consider the implications of this situation in the human arena in Chapter 7.

That's why it makes good ecological sense that the largest land animals—such as elephants—are vegetarians and that the largest marine animals—such as blue whales—are plankton feeders. Such large animals need prodigious quantities of food to build their own living tissue. Were they to feed on large carnivores, already on the fourth or fifth trophic levels, the total amount of primary production needed to support them would be inconceivably large. A given ecosystem could support very few, if any, of them. But by feeding near the base of the ecological pyramid, blue whales and elephants make much more efficient use of energy, so a given ecosystem can support many more of them. Conversely, the limited number of large organisms that can be supported on higher trophic levels is one reason why many rare or endangered species .are carnivores.

THE CYCLING OF NUTRIENTS

In addition to obtaining energy from food, organisms must obtain the nutrients necessary to construct body tissues. For proper growth, most organisms require roughly 17 chemical elements or *essential nutrients*. Six of these—carbon, hydrogen, oxygen, nitrogen, phosphorus, and potassium—are required in large amounts. Calcium, magnesium, sulfur, iron, and manganese are required in lesser quantities. Still smaller amounts of sodium, boron, molybdenum, copper, zinc, and chlorine are needed. Other elements, including vanadium, cobalt, iodine, selenium, silicon, fluorine, and barium, are required in very small quantities (*trace* amounts) by some organisms. Several members of these last two groups, though essential in minute amounts, may be toxic in high doses.

Interestingly, though both energy and nutrients are passed from one organism to another within the same complex molecules, the paths that energy and nutrients take through ecosystems are quite different. Energy *flows through* ecosystems, arriving steadily from the sun, passing from one trophic level to the next, and dissipating in the environment along the way. But the earth receives no continuous supply of chemical elements from space; they are neither produced nor used up but are *passed around* from one organism to another in closed loops called *nutrient cycles*. Because most of these cycles involve the passage of elements through both living organisms and geological features of the globe, they are often referred to as **biogeochemical cycles.**

Nutrient cycling and recycling have been going on since life began, and atoms circulating today have seen many previous incarnations. Atoms of carbon that reside in your body for the time being may once have been part of rocks on the ocean floor, of single-celled algae in the central Pacific, or of the tail of a long-extinct dinosaur.

The Hydrological Cycle

The cycling of water through clouds, rain, snow, and rivers is called the **hydrological cycle.** This is the most familiar and visible biogeochemical cycle, and its processes (evaporation, condensation, precipitation, and runoff back to the sea) are fairly well understood (Fig. 4.17).

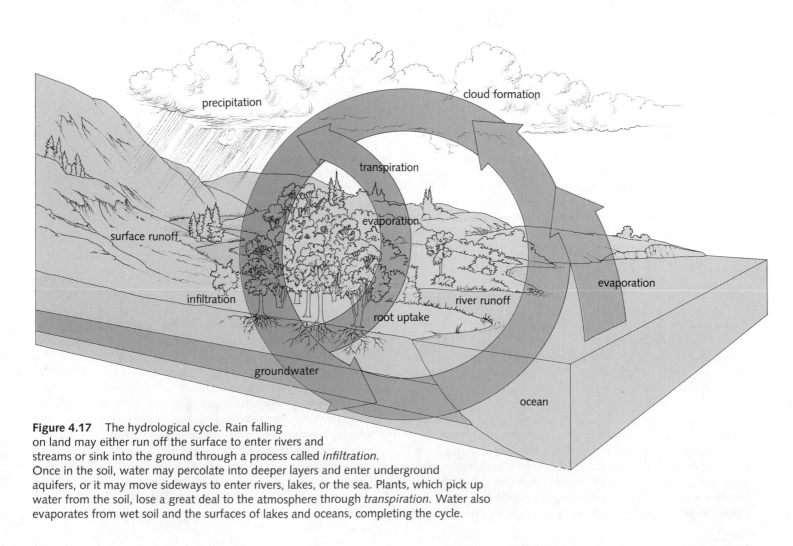

Figure 4.17 The hydrological cycle. Rain falling on land may either run off the surface to enter rivers and streams or sink into the ground through a process called *infiltration*. Once in the soil, water may percolate into deeper layers and enter underground aquifers, or it may move sideways to enter rivers, lakes, or the sea. Plants, which pick up water from the soil, lose a great deal to the atmosphere through *transpiration*. Water also evaporates from wet soil and the surfaces of lakes and oceans, completing the cycle.

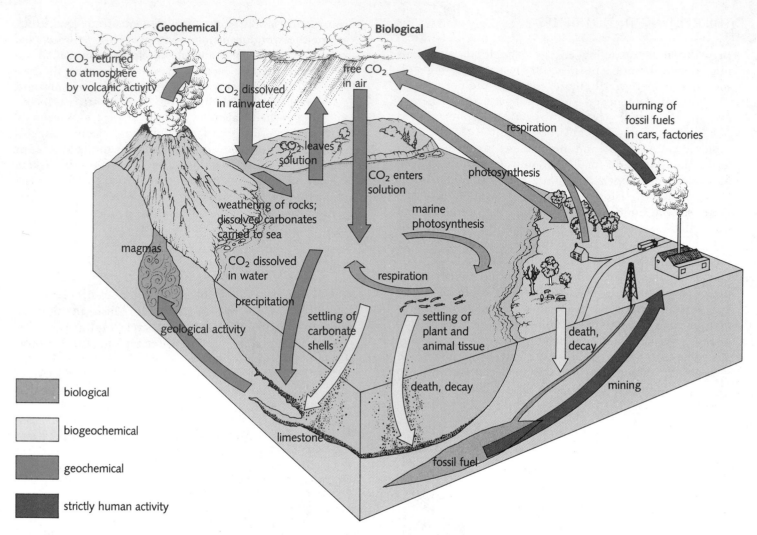

Geochemical

Biological

CO$_2$ returned to atmosphere by volcanic activity

CO$_2$ dissolved in rainwater

free CO$_2$ in air

burning of fossil fuels in cars, factories

respiration

CO$_2$ leaves solution

CO$_2$ enters solution

photosynthesis

weathering of rocks; dissolved carbonates carried to sea

marine photosynthesis

magmas

CO$_2$ dissolved in water

respiration

precipitation

geological activity

settling of carbonate shells

settling of plant and animal tissue

death, decay

death, decay

mining

limestone

fossil fuel

biological

biogeochemical

geochemical

strictly human activity

Figure 4.18 The global carbon cycle. Atmospheric carbon dioxide is incorporated by green plants and marine algae and is both passed through food webs and returned to the atmosphere as CO$_2$ by respiration. Carbon in organic matter, including wood and fossil fuels, can also be returned to the atmosphere by burning. Cutting and burning of forests and combustion of fossil fuels return large amounts of previously fixed CO$_2$ into the atmosphere. Most carbon exchange between the biosphere and the atmosphere takes place at the ocean surface, where phytoplankton incorporate carbon into their tiny shells. The shells sink to the ocean floor, where many are eventually pressed together into carbonate rocks.

Global air movements often carry moisture for hundreds or even thousands of kilometers before it condenses and falls as rain. It is tempting, therefore, to view the hydrological cycle as a strictly physical phenomenon that *affects* life but is itself little affected *by* life. It is now clear, however, that water often cycles quite locally, and that living organisms can strongly affect rainfall patterns. As mentioned in the previous chapter, trees in tropical rain forests return lots of water to the atmosphere from their leaves. Much of that moisture feeds the heavy local rainstorms that keep the forest so well watered. Thus removing large areas of rain forest interrupts this cycle and can cause major, long-lasting changes in local climate.

Human activity also affects the hydrological cycle by redistributing surface and underground water supplies. Humans use enormous quantities of fresh water for irrigating farm crops, mining mineral resources, producing steel, and converting shale and coal to liquid fuels. Much of the world's current water supply is pumped from underground reservoirs called *aquifers*. Some aquifers are continuously replenished by percolation of rainfall through the ground. Other aquifers, especially in desert and prairie areas, were established thousands of years ago when those regions received much more rainfall than they do today. Water in that type of aquifer is called *fossil water*, because it has been trapped underground for long periods of time. Much of the water currently used to irrigate farmland is fossil water that is not being replenished as quickly as it is being consumed. This situation requires careful attention, as you will see in Chapter 7.

The Carbon Cycle

Carbon is an element of many faces. Bonded to hydrogen and oxygen, it is the backbone of the organic molecules on which life is based. In fact, almost all organisms consist of at least 49 percent carbon by dry weight. Carbon is also an important component of carbonate rocks. Finally, carbon in atmospheric carbon dioxide helps regulate Earth's temperature.

Not surprisingly, the movement of carbon through the biosphere is of paramount importance to life on Earth. Scientists have understood the basics of the **carbon cycle** for some time, but many questions still remain about exactly how much carbon travels along which part of the pathway at any particular time. In trying to evaluate the importance of human activity on carbon dioxide levels in the atmosphere, for example, different researchers tend to stress either geochemical or more biogeochemical pathways (Fig. 4.18).

The **geochemical view** argues that the most important pathways of the carbon cycle involve strictly physical processes. Atmospheric carbon dioxide dissolves in rainwater, forming a weak acid that erodes rocks composed of calcium, silicon, and oxygen. The resulting mixture of elements travels through rivers to the oceans, where the carbon dioxide may combine with calcium and magnesium to form insoluble compounds called carbonates. These carbonates precipitate out of solution and accumulate on the ocean bottom, where they harden into sedimentary rocks called limestone and dolomite. In certain places, geological activity forces those rocks underneath the continents, sometimes so deeply that intense heat drives the carbon dioxide out in gaseous form. When volcanoes erupt, this underground carbon dioxide is reinjected into the atmosphere.

According to this view, changes in atmospheric carbon dioxide content result exclusively from changes in this cycle that are caused by variations in planetary temperature and rates of volcanic activity. Although those who hold this view do not deny the participation of living organisms in the carbon cycle, they believe that geologic processes are far more important.

The **biogeochemical view** is based on evidence that organic compounds—both in the bodies of living organisms and in the remains of ancient organisms (such as coal, oil, and natural gas)—are significant carbon storage sites. Adherents of this view point out that most carbonate rocks created today are formed biologically, rather than geochemically. Myriads of single-celled marine organisms incorporate dissolved carbon dioxide into their shells. When these organisms die, their shells fall to the ocean bottom and accumulate into vast deposits, such as the famous white cliffs of Dover.

As this process removes carbon dioxide from the sea, more of the gas in the atmosphere enters the oceans to replace it. At the same time, terrestrial plants incorporate carbon dioxide from the air into living tissue. Much of this organic carbon is continually recycled, but a great deal was buried in the form of vast organic deposits that, over time, became coal and oil. According to this hypothesis, changes in growth rates of plants and other organisms, together with changes in the amount of organic compounds buried beneath the sea and in the ground, can significantly affect the amount of atmospheric carbon dioxide.

Which view is correct? As is often the case in biology, it seems to many ecologists that carbon dioxide levels are affected by *both* geochemical and biological activities. It is even possible that the relative importance of the two routes may have changed recently as a result of human population growth and industrialization. How? It took millions of years for substantial amounts of organic carbon to be stored and buried in coal and oil, but human activity is releasing those carbon stores at a rate many times faster than that (Fig. 4.19). Can we afford to delay attempts to reduce carbon dioxide emissions while more studies are performed to see which view is "correct" (or more correct)? That question is the center of an ongoing

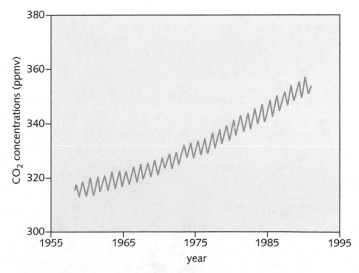

Figure 4.19 Researchers began recording atmospheric carbon dioxide in 1957 at Mauna Loa, Hawaii. Since that time, CO_2 concentration has increased 12%, from 315.83 parts per million per dry volume of air (ppmv) to 353.95 ppmv. Annual fluctuations are caused by seasonal fixation of CO_2 by green plants in the northern hemisphere. (Data from Oak Ridge National Laboratory.)

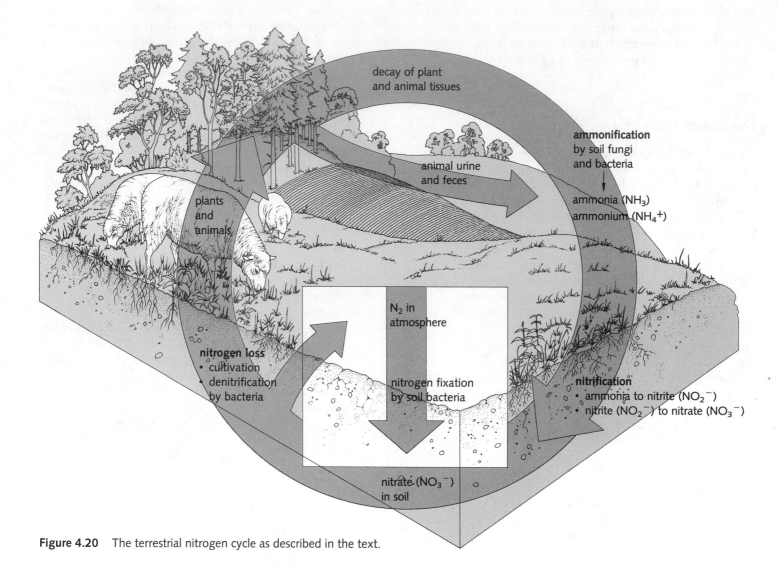

Figure 4.20 The terrestrial nitrogen cycle as described in the text.

controversy involving ecologists, environmentalists, climatologists, geologists, and business leaders (see Current Controversies, The Greenhouse Effect, p. 82).

The Nitrogen Cycle

From primary production to consumers Primary producers absorb nitrogen in a simple form, usually either ammonia (NH_3), ammonium (NH_4^+) or nitrate (NO_3^-). Plants use energy from photosynthesis to concentrate nitrogen in their tissues and then use more solar energy to fuel the assembly of nitrogen, hydrogen, carbon, and other elements into proteins and amino acids, the complex molecules used as building blocks in living systems.

From organisms to the environment When animals eat plants or other animals, they digest proteins, breaking them down into nitrogen-containing amino acids. Some of these amino acids are reassembled into the "personal proteins" of the animal that ate them. Others are broken down to liberate energy they contain, releasing nitrogen in the form of ammonia (NH_3).

Because ammonia is toxic in low concentrations, it must be eliminated from body fluids or changed to a less poisonous form. Most aquatic animals eliminate ammonia continuously into the water around them. Terrestrial animals often convert ammonia into *urea* and concentrate it in urine before eliminating it. Other strategies for eliminating waste are described in Chapter 38.

Microorganisms in the environment   Nitrogenous wastes don't last long in the environment, because bacteria go right to work on them. One group, the *Nitrosomonas* bacteria, combine ammonia with oxygen and convert it into *nitrite* (NO_2^-). The nitrite may then be converted into *nitrate* (NO_3^-) by bacteria called *Nitrobacter.* These two processes together are called **nitrification.**

Although they rarely appear in simple food chains, decomposers are critical to the **nitrogen cycle,** because nitrogen locked up in dead animals and plants is useless to primary producers. Decomposers disassemble the organic molecules in animal and plant carcasses and release their component elements in simpler form. This process is called **ammonification,** because bacteria and fungi usually release nitrogen in the form of ammonia.

Completing the cycle   When nitrate, nitrite, and ammonia are released into the soil of a healthy forest or into water filled with growing algae, they may be quickly reabsorbed by primary producers (Fig. 4.20). Many terrestrial plants absorb nitrate efficiently, which is one reason why most fertilizers contain a source of nitrogen in nitrate form. Some terrestrial plants and many algae, however, pick up ammonia more readily than nitrate.

In the open sea, dead marine organisms and their nutrient-rich solid wastes sink rapidly out of the *photic zone.* Large quantities of nutrients, therefore, end up in slow-moving currents near the ocean floor, far out of reach of photosynthetic primary producers. In certain places around the world, winds and ocean currents pull large quantities of this nutrient-laden water back up from the ocean floor into the photic zone, a phenomenon known as *upwelling* (Fig. 4.21).

In areas of steady upwelling, phytoplankton thrive and grow at rapid rates, providing the basis for a vigorous and productive food web that may include anchovies, herring, lobster, cod, hake, bluefish, tuna, and many other commercially important kinds of seafood. The productivity of upwelling areas is staggering; up to 50 percent of the worldwide fish catch comes from upwelling areas that comprise a mere tenth of 1 percent of the total ocean surface.

Earth's nitrogen reserve: The atmosphere   While aquatic and terrestrial primary producers scramble for nitrogen, an enormous reservoir of gaseous nitrogen floats out of reach in the atmosphere. Unfortunately, the paired atoms in nitrogen gas are held together by a powerful chemical bond that cannot be broken by most organisms. In fact, only a few bacteria can break that bond and incorporate atmospheric nitrogen into living tissue, a process known as **nitrogen fixation.** The most familiar of these are **nitrogen-fixing bacteria** that live on the roots of such terrestrial plants as peas, soybeans, and other members of the

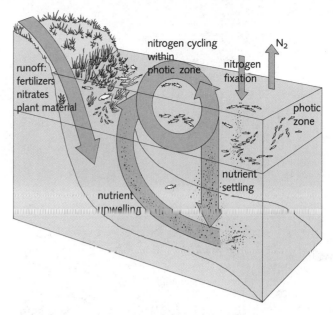

Figure 4.21   The oceanic nitrogen cycle. Note the recycling of nitrogen within the photic zone, the input from land runoff, the steady sedimentation of organic matter out of the photic zone into the depths, and the periodic return of those nutrients to the photic zone in areas of upwelling.

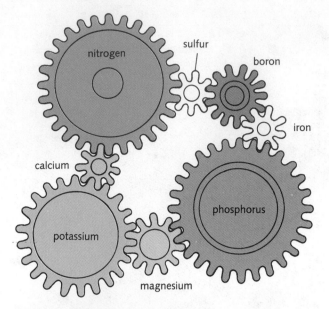

Figure 4.22 This schematic representation of interlocking nutrient cycles demonstrates metaphorically how the movement of each nutrient through ecosystems is dependent upon the movements of all others. If even a minor nutrient is present in limiting quantities, it can slow down the entire assemblage.

legume family. The most important nitrogen fixers in the sea are photosynthetic cyanobacteria, often called blue–green algae.

Denitrifying bacteria carry out precisely the reverse reaction; they convert organic nitrogen back into nitrogen gas and release it to the atmosphere. Ironically, the constant addition of nitrogen fertilizers to farm soil can encourage the growth of denitrifying soil bacteria that pump significant amounts of nitrogen out of the soil and back into the atmosphere!

Nutrient limitation   Major and minor nutrients all revolve through their own complex biogeochemical cycles. Of course, as long as atoms are bound together in a single organic compound, they must travel together through the ecosystem. For example, protein molecules contain carbon, nitrogen, phosphorus, sulfur, iron, magnesium, and other elements. Yet in order for an ecosystem to function, each cycle must revolve at its proper speed. The whole assembly of cycles works like a complicated set of interlocking cogwheels (Fig. 4.22). The steady, one-way flow of energy through the system provides the power to drive the gears through their perpetual revolutions, each at its own specific rate.

The **law of limiting factors** states that the growth of an organism will be limited if any essential growth factor is present in insufficient quantity relative to the other factors. In deserts, for example, most plant and animal growth is limited by lack of water. In most ocean surface waters, nutrients are limiting. Below the photic zone—where the largest volume of ocean water falls—the limiting factor is insufficient light.

The productivity of an entire ecosystem can thus be curtailed by a single nutrient in short supply, a phenomenon called *nutrient limitation*. In the cogwheel analogy, the speed of the entire assembly is controlled by the speed of the slowest wheel. When any single nutrient cycle is interfered with—when any wheel sticks—the whole system must slow down or stop.

Nutrients, Productivity, and Ecosystem Structure

Nutrient limitation in the abstract is neither a "good" phenomenon nor a "bad" one; it is simply a fact of life and an important part of the dynamic balance of nature. In some situations, nutrient limitation keeps ecosystems functioning in ways we find attractive and desirable, while in other cases it curtails the productivity of ecosystems we would like to exploit more efficiently.

In ponds and lakes, for example, nitrogen, iron, magnesium, and several other nutrients are often present in quantities that could support much higher rates of plant growth than usually occur. The primary producers in

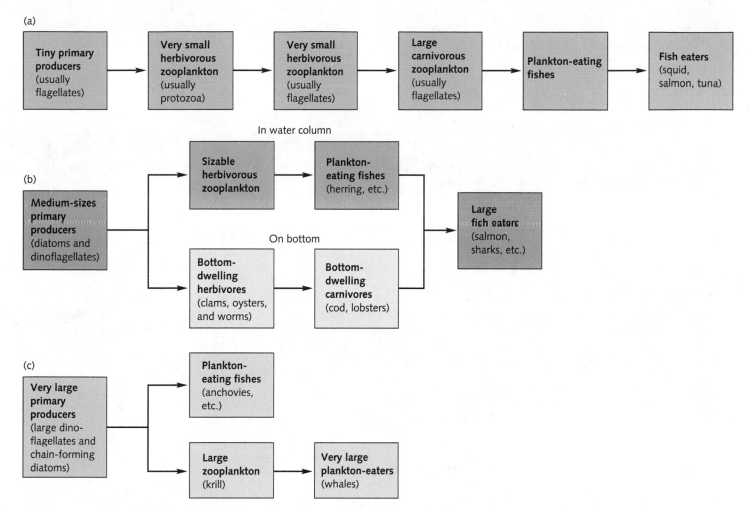

Figure 4.23 The availability of nutrients in various marine environments contributes to different food chain structures. (NOTE: decomposers and parasites omitted for clarity.) In the open sea where nutrients are scarce **(a)**, tiny primary producers form the basis of a food chain with as many as five or six trophic levels. In an ordinary coastal environment **(b)**, larger primary producers may be consumed by either free-swimming or bottom-dwelling consumers, forming a two- or three-step food chain. In upwelling areas **(c)**, very large primary producers directly support sizable fishes and, with the addition of a single extra step, can support animals as large as whales.

these ecosystems—and hence the system as a whole—are held back by a lack of phosphorus. If phosphorus is suddenly added in large amounts, the phosphorus "wheel" becomes unstuck, and a prodigious growth of algae called an algal bloom can quickly cover ponds and rivers, choking out life beneath it.

On land and in the sea, on the other hand, nitrogen, rather than phosphorus, is commonly the limiting nutrient. For this reason, farmers use extra nitrogen—either in natural forms such as "green manure" (plants and other organic matter added to the soil) or in manufactured chemical fertilizers—to increase crop growth. Similarly, the addition of nitrogen-rich sewage and agricultural runoff to coastal waters can greatly increase the

productivity of marine systems. But massive discharges of untreated sewage can seriously upset the local ecological balance, and in some cases they foster blooms of undesirable—even poisonous—algae.

In many ecosystems—particularly, though not exclusively, aquatic ones—nutrient availability can affect not only the rate of primary production but also the structure of the food web itself (Fig. 4.23). As mentioned in the previous chapter, for example, surface waters of tropical seas offer intense sunlight but carry very few nutrients. Only tiny phytoplankton, little larger than 20 μm can thrive there. Those tiny primary producers are suitable food only for very small zooplankton, which, in turn, are eaten by slightly larger zooplankton. Thus a system based

The Greenhouse Effect

Earth's nearest planetary neighbors have wildly varying climates. On Venus, daytime temperatures hover near 480°C (over 900°F), hot enough to melt lead. On Mars, temperatures vary more than 150°C (270°F) between noon and midnight, plummeting to an icy −128°C on the planet's dark side. On Earth, by comparison, average surface temperature is 15°C (59°F), and extreme temperatures—from summer in tropical deserts to midwinter in Antarctica—cover only a fraction of the range on Mars. What's more, Earth's av-

erage temperature has remained nearly constant since life evolved, never dropping below 5°C (41°F) or rising above 25°C (77°F). That stability is even more remarkable because the sun is now 30 percent hotter than it was 3.5 billion years ago!

To what do we owe our planet's hospitable and stable temperature? To a combination of atmospheric carbon dioxide, water vapor, and other gases that admit sunlight but retard the escape of heat. Because these gases act much like the glass in a greenhouse,

this phenomenon is called the **greenhouse effect,** and the gases involved are called *greenhouse gases* (see figure). Greenhouse gases are vital in balancing incoming solar radiation and escaping heat. When their concentration increases, heat is retained and the planet gets warmer. If, on the other hand, greenhouse gas concentrations fall, Earth's temperature drops.

The stability of Earth's temperature over time seems to be explained by changes in one particular greenhouse gas: carbon dioxide. During Earth's

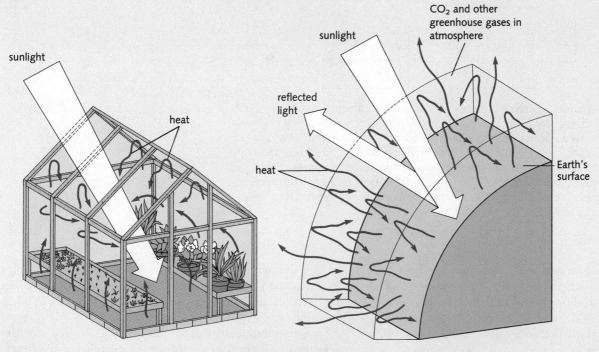

TOP LEFT: When sunlight enters a greenhouse, much of its energy is absorbed by objects inside, and some is reradiated as heat. Because greenhouse glass doesn't let infrared energy pass, much of that heat is trapped inside, warming the greenhouse. TOP RIGHT: Greenhouse gases in the atmosphere allow sunlight to enter and strike Earth's surface. Much of the solar energy reradiated as heat is absorbed by greenhouse gases and reradiated again in all directions. About half escapes into space, but the other half is directed back at Earth's surface, warming the planet.

early years, our atmosphere had close to 1000 times as much carbon dioxide as it does today. The greenhouse effect of all that carbon dioxide kept Earth warm, even though the sun was much dimmer. But over millions of years, as the sun grew warmer, much of that free carbon dioxide dissolved in the oceans, formed carbonate rocks, and left the atmosphere. After life evolved, living organisms incorporated carbon dioxide into their bodies. Atmospheric carbon dioxide concentrations dropped to the present level, allowing more and more heat to escape from Earth's surface. Somehow, the decrease in the greenhouse effect paralleled the rise in the sun's output closely enough that planetary temperatures stayed nearly constant.

That relationship is clear. But controversy surrounds the effect of human activity on the concentration of carbon dioxide in the atmosphere and the effect of that change on global climate. The cutting of global forests, particularly (though not exclusively) in tropical countries where many trees are burned either as waste or for fuel, destroys living tissue that fixes atmospheric CO_2 and returns the carbon they contain to the atmosphere. At the same time, burning of fossil fuels—coal, oil, and natural gas—is releasing organic carbon that was stored in the biosphere over millions of years. Together, these activities seem to be raising atmospheric carbon dioxide levels slowly but steadily. Is this human-caused addition of carbon dioxide sufficient to set off a global warming trend? If so, how will global climate respond? Thus far, no one knows, and it will be at least two decades before we can say with certainty. We will discuss this question in more detail in Chapter 7.

on tiny phytoplankton creates a food chain with many levels between primary producers and sizable carnivores. As we saw earlier, food chains containing many trophic levels can support very little living tissue at their upper end with a given weight of primary producers. Accordingly, open-ocean food chains are very inefficient at producing large organisms.

In areas along the coasts of major continents, however, freshwater runoff provides enough nutrients for larger phytoplankton, including *dinoflagellates* and *diatoms*. These larger primary producers can be either eaten directly by good-sized zooplankton or filtered from the water by clams, mussels, worms, and other bottom-dwelling filter feeders. Here, higher primary productivity and a shorter food chain combine to support a greater mass of high-level consumers.

Upwelling areas, with their still higher availability of nutrients, support abundant growth of large, chain-forming diatoms which are eaten directly, not only by sizable zooplankton such as krill, but also by fishes such as anchovies. Krill and anchovies, in turn, are large enough to serve as acceptable food for seabirds, seals, whales, and even humans. Thus, upwelling areas combine higher productivity and still shorter food chains to support the largest amount of large animal tissue of any planktonic system—up to 36,000 times the final productivity of the open sea.

In the next two chapters we will see how the availability of energy combines with the storage and cycling of nutrients to control the growth of animal and plant populations, the structures of natural ecosystems, and the response of natural communities to human activities.

SUMMARY

Autotrophs, also called primary producers, harness the energy of sunlight through photosynthesis and create organic compounds from simple inorganic nutrients. Many autotrophs store energy as carbohydrates. Heterotrophs obtain both energy and nutrients by eating other organisms and digesting their organic molecules. Heterotrophs obtain energy through strategies including herbivory, carnivory, and decomposing the decaying tissues of plants and animals.

In every ecosystem, feeding relationships link plants and animals into food chains and food webs. Energy *flows* through food chains and food webs in a one-way stream from lower to higher trophic levels. Nutrients, such as carbon and nitrogen, are passed around from primary producers to consumers, to decomposers, and back again, as they *cycle* continuously between organisms and the abiotic environment.

Limiting factors—the lack of any single requirement for growth—can constrain productivity in living systems. Any conditions that control the growth of primary producers in an ecosystem indirectly control the lives of all other organisms as well.

After studying this chapter, you should be able to:

- Explain how energy flows through collections of plants and animals.

- Describe how the principles of energy flow affect the structure of ecosystems.

- Demonstrate how nutrients cycle, both within and between ecosystems.

- Show how nutrients and energy connect organisms and ecosystems to each other.

- Explain the law of limiting factors.

- Give an example of how nutrient availability can affect the structure of a food web.

TERMS AND CONCEPTS

autotrophs *64*	food chain *71*
primary producers *64*	food webs *71*
photosynthesis *65*	trophic level *73*
carbohydrates *65*	rule of 10 *73*
herbivores *67*	ecological pyramids *74*
primary consumers *67*	hydrological cycle *75*
carnivores *68*	carbon cycle *77*
secondary consumers *68*	nitrogen cycle *79*
tertiary consumers *68*	law of limiting factors *80*

REVIEW

Objective Questions (Answers in Appendix)

1. The energy for most communities comes from
   (a) the sun.
   (b) decomposition of organisms.
   (c) the oceans and lakes.
   (d) secondary consumers.

2. Which of the following statements about ecosystems is false?
   (a) All the energy entering the ecosystem is passed on to the decomposers.
   (b) The primary producers control the lives of herbivores and carnivores.
   (c) The efficiency of energy transfer in food webs varies widely.

(d) Ecosystems must start with the capture of energy by autotrophs.

3. Organisms that synthesize their own food are called
   (a) microorganisms.   (c) autotrophs.
   (b) heterotrophs.   (d) secondary consumers.

4. Which of the following nutrients is often a limiting factor in ponds and lakes?
   (a) nitrogen   (c) phosphorus
   (b) magnesium   (d) iron

5. Which process causes a loss of nitrogen from the soil because of bacterial activity?
   (a) upwelling   (c) ammonification
   (b) nitrification   (d) denitrification

Discussion Questions

6. Many aquatic organisms use feeding techniques that are not used by terrestrial creatures. Give two examples, and explain how these feeding techniques help aquatic organisms obtain nutrients.

7. What is the photic zone? How do the existence of the photic zone and the movements of nitrogen in the sea combine to cause nutrient limitation of phytoplankton in the open sea?

8. What are the three steps in the nitrification process? What organisms perform these transfers of nutrients, and why are they important?

9. Why is it difficult for grazing animals to extract energy from their feed? What physical and behavioral characteristics of a grazing animal increase the efficiency of its digestion?

10. What is the greenhouse effect? Why are the concentrations of greenhouse gases important in determining conditions for life on Earth?

11. The burning of fossil fuels seems to be causing an increase in the amount of carbon dioxide in the atmosphere. What could happen to global climate as a result?

12. What are the differences between the ways in which energy and nutrient elements move through living systems? What is energy flow? Why is it possible for energy to flow through ecosystems, whereas nutrients must cycle?

13. Assume that a particular human lives entirely on the meat of a bird that, in turn, eats only herbivorous insects. If the human gains 1 kg on this diet, how many kilograms of plant material has the human eaten indirectly?

READINGS

See integrated list of readings for Part 2 after Chapter 7.

5

Population Ecology

across the eastern United States, caterpillars of native moths and butterflies go about their business, consuming roughly as many leaves as they always have. Their numbers do rise and fall a bit; some species do better in cool, wet years, while others prosper in warm, dry ones. Still, their populations are fairly stable over time. Yet only decades after a few European gypsy moths were accidentally released in Massachusetts, that species periodically multiplies out of control, invades woodlands by the millions, and defoliates thousands of acres of forests (Fig. 5.1). During peak years, these pests reach incredible densities—literally covering cars and driveways with their bodies and feces.

Rabbits, likewise, are hardly environmental threats where they are normally found. Yet in England, where conquering Normans introduced rabbits more than 800 years ago, and in Australia, where a farmer released 12 pairs in 1859, rabbits are pernicious pests whose populations fluctuate wildly and whose actions threaten native species and agriculture alike. In Australia, periodic plagues of grass-gobbling rabbits threaten not only to drive certain kangaroos and other native herbivores to extinction but also to outcompete sheep for pasture. And in Britain, where an estimated 100 million rabbits roamed free in 1953, they inflict nearly $2.25 million in damages on food crops every year. There are still parts of Britain today in which crops can flourish only when surrounded by rabbitproof enclosures.

DYNAMICS OF POPULATIONS

Gypsy moths and rabbits are examples of apparently harmless organisms that turned into plagues when introduced to new habitats. But why? Gypsy moths don't multiply out of control in their native Europe—only in North America where they are foreigners. Rabbits don't lay waste to the countryside in their native habitats either—they cause trouble in Great Britain and Australia, where they had never lived before.

Most plant and animal populations don't fluctuate wildly. Instead, they are kept under control by interactions with each other and with their environments. As a first step in understanding why that is so, we

Figure 5.1 Gypsy moth caterpillars, an introduced exotic species, have multiplied out of control. These pests regularly defoliate thousands of acres of forests.

need to examine the factors that influence the growth of natural populations.

Populations: Functional Units of Ecology

To ecologists and geneticists, the word **population** has a specific meaning: a group of individuals of a single species that interact and interbreed with each other.

Having made that point, we should emphasize that populations can be of very different sizes and cover different amounts of territory. Because individuals within a population must be able to breed with one another, single populations may cover large areas or be restricted to small ones, depending on a species' ability to find and mate with one another.

The entire North Atlantic, for example, might contain only a single population of wide-ranging animals such as open-water marine fishes or whales. Yet a narrow strip of land may form an impenetrable barrier for aquatic animals living in two adjacent ponds. In that situation, a barrier just a few yards wide can divide certain organisms into two distinct (if tiny) populations. Terrestrial species can be divided into separate populations by rivers, oceans, canyons, mountain ranges, or even smaller geographic features.

The area covered by plant populations can be determined by factors that relate directly to the ecological needs of the plants themselves or to animals that pollinate their flowers or disperse their seeds. Because many plants depend upon insect pollinators, the distance over which their flowers exchange genetic material may be controlled by the movements of those insects. Additionally, because many plants rely on birds or mammals to spread their seeds, plants' "movements" may be tied to the activities of seed- or fruit-eating animals.

Predicting Population Growth

One major goal of population biology is to predict and explain increases and decreases in populations over time. Under very simple conditions, very general predictions are fairly easy to make. It seems intuitively obvious, for example, that if a population is free of predators, unencumbered by disease, and well fed, it will grow in size. But that simple prediction isn't very useful without more specific information. How fast will the population grow? How large will it get?

To answer those questions, we must deal with both ecological concepts and numbers. Why? Because numbers enable us to translate abstract concepts into concrete, practical predictions about how populations change over time. In order to show clearly how the concepts and numerical predictions of population biology relate to one another, our discussion will switch back and forth between verbal descriptions of population growth and the quantitative language of population biology.

Exponential Growth

Before we try to predict what happens to a real-life population in a complicated environment, it is helpful to think first about what happens in an idealized group of organisms, protected from predators and disease and provided with abundant food. For the time being, we'll also assume that the organisms we're discussing are neither leaving the population (emigrating) nor coming in from somewhere else (immigrating).

The most basic information we can gather about such a population is its *size,* the number of *births* that occur during a given period of time, and the number of *deaths* that occur during the same period. Those three statistics enable us to predict the increase in the size of the population over time.

In the symbolic language of population biology, the increase in a population is designated by the letter *I.* Now, in our hypothetical group of organisms, any increase in population size occurs when more births than deaths occur. We can summarize that statement with a simple equation:

$$I = \text{births} - \text{deaths}$$

It is often more convenient, however, to consider the **birth rate** and **death rate** for a population. These rates are simply the number of births or deaths over a certain period of time *divided by the size of the population.* Once again, the symbolic language of population biology represents the birth rate by *b,* the death rate by *d,* and the size of the population by *N.* We can therefore describe the birth and death rates as follows:

$$\text{birth rate} = \frac{\text{number of births}}{\text{population size}} \qquad \text{death rate} = \frac{\text{number of deaths}}{\text{population size}}$$

$$b = \frac{\text{births}}{N} \qquad d = \frac{\text{deaths}}{N}$$

What does this mean in real terms? Let's say we have a town of 1000 people, in which 46 babies are born each year. The annual birth rate for that town is 46/1000, or 0.046. If there were 500 babies born in that town over the same period of time, the annual birth rate would be 500/1000, or 0.5. If during that same period, 36 people in the town were to die, its annual death rate would be 36/1000, or 0.036.

But why is it useful to get these birth and death rates? Let's say that we are civic planners, and we want to know how fast the town's population is growing. A population biologist would say that we want to know the *rate of population growth*—a value usually represented by *r.* Given the information we already have, we can calculate *r* by using the following relationship:

$$\text{rate of population growth} = \text{birth rate} - \text{death rate}$$

or symbolically,

$$r = b - d$$

Now in the particular town we've just considered, we can plug in our imaginary birth and death rates to obtain a real numerical value for *r* as follows:

$$r = 0.046 - 0.036$$
$$= 0.010$$

What does that mean? We have just determined that this particular population of 1000 people is growing at the rate of 1 percent, or 10 people, each year. But why have we bothered to go to all this trouble? After all, we could have obtained this piece of information much more easily by subtracting the number of deaths in the town each year (36) from the number of births (46). Why calculate the rate of population growth at all?

It turns out that *r* comes in very handy if we want to know more than just the rate of population growth in one particular year. Let's say, for example, that we want to know how much the population of that town will increase over more than a single year—two years, perhaps, or even five or ten years. To determine that kind of long-term growth, we simply multiply the rate of population growth (*r*) by the size of the population, which we designate by *N.* We can therefore calculate the expected increase in the population (*I*) as follows:

$$\begin{array}{c} \text{increase in} \\ \text{population} \end{array} = \begin{array}{c} \text{rate of} \\ \text{population growth} \end{array} \times \text{population size}$$

or again in symbolic form,

$$I = rN$$

But we're not quite finished. This equation tells us how many *more* people there will be at the end of the growth period in question. To know the *total* number of people in the town at that time, we have to add that increase (*I*) to the size of the population we started with (*N*). If we represent the size of the population at time 0 by the symbol N_0, and the size of the final population as simply *N,* we could summarize this relationship as

$$\begin{array}{c} \text{size of} \\ \text{final} \\ \text{population} \\ (N) \end{array} = \begin{array}{c} \text{initial} \\ \text{population} \\ \text{size} \\ (N_0) \end{array} + \left(\begin{array}{c} \text{rate of} \\ \text{population} \\ \text{growth} \\ (r) \end{array} \times \begin{array}{c} \text{initial} \\ \text{population} \\ \text{size} \\ (N_0) \end{array} \right)$$

or symbolically,

$$N = N_0 + rN_0$$

We could also rewrite that same equation like this:

$$N = N_0(1 + r)$$

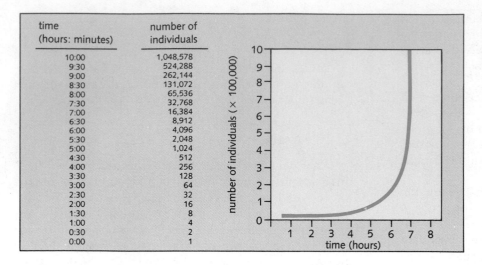

Figure 5.2 If provided with an ideal environment, continually supplied with food, and protected from the buildup of waste products, bacteria—like many other organisms—grow exponentially. The larger the population becomes, the faster it increases in size.

This relationship enables us to calculate how much our hypothetical population will grow over *several* time periods. To do that, we represent time n in years or generations, whichever we've been considering thus far. Then we place n as an exponent around the $(1 + r)$ term, and our equation becomes

$$N = N_0(1 + r)^n$$

Now we can see how useful this equation is. Suppose a particular population currently contains 250 individuals. Suppose we also know that this population increases in size by 6 percent per year, which translates into an annual growth rate of 0.06. If we want to know how big that population will be in 10 years, we simply plug those values into the equation, as follows:

$$N = N_0(1 + r)^n$$
$$N = 250 \, (1 + 0.06)^{10}$$
$$= 250 \, (1.06)^{10}$$
$$= 448 \text{ animals}$$

Admittedly, this equation assumes that the rate of growth (r) does not change over time—an assumption that rarely holds for real populations in the real world. It also assumes (as we did at the beginning of this discussion) that animals are neither entering nor leaving the population. But what if these assumptions do hold true for a limited period of time?

To give you a hint, we can note that our equation describes what we call **exponential growth.** If this equation really did predict the growth of populations over time, what could we expect?

Let us suppose that we have a single imaginary bacterium able to divide to form 2 new cells every 30 minutes. This means that as long as local conditions remain favorable, the size of this population doubles every 30

minutes. At the end of the first half hour, there are 2 individuals; at the end of the first hour, 4 individuals; and so on. If we graph the growth of this population over time, we get a curve that rises slowly at first and then accelerates rapidly as the population continues to double—from 4 to 8 to 16 to 32, 64, 128, and on and on and on (Fig. 5.2). Note that the larger the population is, the faster it grows. Looking at the graph, you can see why exponential growth curves are sometimes also called **J-shaped** growth curves.

Can you see what will happen if our imaginary bacterial colony continues in this unchecked exponential growth? After 10 hours, there will be 1,048,576 bacteria. After 20 hours, there will be 1,099,511,600,000 of them. And within four days (given the size of a typical bacterium), the mass of this single bacterial population would exceed the mass of the entire earth! Clearly, natural populations cannot grow exponentially for very long.

It is true that organisms do grow exponentially under certain circumstances and for brief periods of time. In fact, that's precisely what happens when species such as rabbits and gypsy moths are introduced by humans to favorable new environments. (It is also, as you will see shortly, what has been happening to human populations over the last few hundred years.) But exponential growth rarely lasts very long in nature. Why does exponential growth stop when it does, and what happens next? Much of ecological theory attempts to answer these questions.

Logistic growth The simplifying assumptions that enabled us to derive our growth equation created a totally unrealistic prediction. Why? Because we assumed that the rate of population growth remained constant over time.

But growth rates in the real world *do* change over time. If a laboratory population of bacteria is grown with

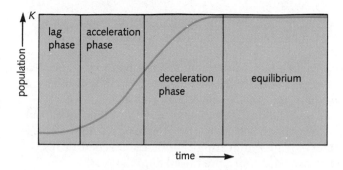

Figure 5.3 The S-shaped, or logistic growth, curve has four parts. In the *lag phase,* the population is established and begins to grow. In the *acceleration* phase, births greatly exceed deaths, and the population grows exponentially. During the *deceleration* phase, the rate of population growth slows down. Ultimately, the rate continues to decrease until the population reaches a constant, maximum size.

a limited amount of space and nutrients, for example, its initial growth does, in fact, resemble the graph in Fig. 5.2. But sooner or later, growth begins to slow down.

Typically, growth in such situations can be graphed with what is called an **S-shaped,** or **logistic growth,** curve (Fig. 5.3). At the top of such a curve, population growth is zero; although new individuals are continually being born, others are dying at the same rate. As a result, $(b - d)$ equals zero, so I and r equal zero. The total number of individuals therefore remains constant, and the population exists in a *steady state,* or **equilibrium.**

So where do we go with our predictive equations now? Although by this point you may be disillusioned with a simple equation that looked good at first, there's no need to start from scratch. All we need to do is adjust our predictions to account for a simple fact: as populations grow, their rate of growth changes. Why does the rate change? Perhaps all those organisms use up food or oxygen. Or perhaps their waste products accumulate to toxic levels. Whatever the reason, population growth slows down sooner or later.

Intrinsic Rate of Growth

Before growth slows, during that brief period when ideal conditions prevail, the population grows at a constant rate. Ecologists call this the *intrinsic growth rate* of a population, because theoretically only *intrinsic* factors—how fast the organism can reproduce, for example—affect it. The intrinsic growth rate is usually represented by the symbol r_0.

Thus the *maximum* expected increase in the population (I) is simply the intrinsic growth rate times the population size:

$$I = r_0 N$$

A population would grow at this rate only if it found unlimited resources, and only if its increasing numbers had no effect on its birth and death rates.

The Concept of Carrying Capacity

In real-life situations, however, resources *are* limited, and large populations *do* encounter crowding problems. As a result, growth ultimately slows down. That realization leads us to one of the most important concepts in ecology: populations cannot grow at exponential rates indefinitely because natural environments can only support a finite number of organisms.

Let's return for a minute to consider the difference between the exponential growth curve in Fig. 5.2 and the logistic growth curve in Fig. 5.3. Instead of continuing to increase indefinitely, the population in Fig. 5.3 ultimately reaches a point at which it contains the largest number of individuals that its environment can sustain for extended periods of time. As mentioned earlier, we call that point equilibrium.

Each environment has a finite limit to the number of individuals of a given species it can support. This limit is called the **carrying capacity** for that species. Carrying capacity is often represented by K, the maximum number of individuals of a particular species able to survive in a particular environment.

Using K, we can modify our equation to represent more realistic environmental limits to population growth. How? By factoring in the relationship between carrying capacity and population size in a way that simulates what happens as the population size (N) approaches the maximum number of individuals able to survive (K). We can do so as follows:

$$I = r_0 N \left(\frac{K - N}{K} \right)$$

In this equation, as before,

I = rate of increase in the population over time
r_0 = intrinsic rate of population growth
N = size of the population
K = carrying capacity of the environment

Now consider what this equation means for populations of different sizes (see Table 5.1).

At the beginning, when the population is very small and far below the carrying capacity, N is much less than K. In that situation, the term $(K - N)/K$ is close to 1, so

Table 5.1 *Effect of Carrying Capacity on Population Growth*

N	$(K - N)/K$	I	
1	999/1000	$0.999r_0N$	
10	990/1000	$0.99r_0N$	
100	900/1000	$0.9r_0N$	
500	500/1000	$0.5r_0N$	Growth
900	100/1000	$0.1r_0N$	slows
990	10/1000	$0.01r_0N$	
999	1/1000	$0.001r_0N$	
1000	0/1000	0.0	Growth stops: equilibrium
1010	−10/1000	$-0.01r_0N$	Population
1100	−100/1000	$-0.1r_0N$	declines

K for this population is 1000.

it has little effect on population growth. Under these conditions, this equation makes almost the same predictions as its exponential predecessor.

As the population grows, it approaches the carrying capacity. In other words, N approaches K. As that happens, the term $(K - N)$ becomes smaller and smaller, and the term $(K - N)/K$ multiplies r_0N by a smaller and smaller fraction. What effect does that have? This steadily decreasing fraction in the equation reduces the effective rate of population growth, thereby slowing down the increase in population size (Fig. 5.4).

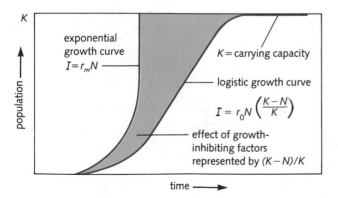

Figure 5.4 This graph represents the relationship between the theoretical potential of a species to grow exponentially and the effect of environmental limiting factors on actual population growth. The shaded area between the two curves represents the growth-inhibiting effects of density-dependent regulating factors represented by the term $(K - N)/K$.

When the population size reaches the carrying capacity, N equals K, so the term $(K - N)/K = 0$. Thus I becomes zero as well. In plain English, population growth stops altogether. At this point, when birth rate and death rate precisely balance each other, the population reaches a point of *zero population growth.*

If population size ever exceeds the carrying capacity, in other words, if N is ever greater than K, $(K - N)/K$ becomes *negative*, indicating that deaths exceed births. Under these conditions, the population decreases.

Factors determining carrying capacity   The logistic growth equation elegantly describes the limitations placed by carrying capacity on populations' tendency to grow exponentially. But what factors in a real environment determine K for a particular species? And how do those factors alter birth rates and death rates to slow population growth?

Nearly any of the ecological limiting factors described in Chapters 3 and 4 can limit population size. Insufficient sunlight or a deficiency in any essential nutrient can limit plant growth and reproduction. A shortage of suitable prey can limit animal populations. Too much or too little water, heat, or humidity can affect both plants and animals.

Note that although we have treated K as a constant, in reality the carrying capacity of a specific environment for a particular species is often anything *but* constant. Physical conditions (weather) change from season to season, and biological conditions (interactions with predators, prey, and parasites) also vary. And because K is determined by these factors, it often changes too.

Population Growth in Nature

The formula for logistic growth accurately predicts growth in certain laboratory populations and for certain species under specific conditions in nature. The graphs in Fig. 5.5 (a) and (b) show laboratory and field populations whose growth is reasonably well described by the logistic growth curve.

Other natural populations, however, may exhibit any of several different growth patterns. Figure 5.5(c), for example, shows the annual fluctuations in populations of marine phytoplankton and zooplankton in temperate coastal waters. These cyclical changes are typical of interactions between herbivores and plants and between predators and prey, as we will see in the next chapter.

Figure 5.5(d), on the other hand, represents a wild "boom and bust" growth curve. In this situation, organisms grow exponentially without limit and far overshoot the carrying capacity of the environment. Such populations usually crash violently, grow again, and crash again, instead of reaching a steady state. Often this situation results because each time these populations grow out of control, they seriously degrade their environment.

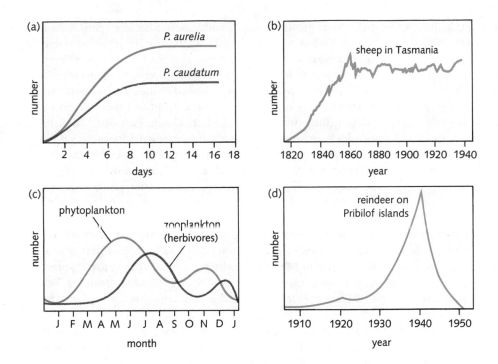

Figure 5.5 *Patterns of population growth.* **(a)** Logistic growth of two species of the single-celled *Paramecium* in a laboratory culture. Notice that *P. aurelia* has a higher *K* under these conditions than *P. caudatum.* **(b)** Near-logistic growth of sheep when first introduced to the island of Tasmania. **(c)** Seasonal fluctuations in marine phytoplankton and zooplankton in temperate coastal waters. **(d)** "Boom and bust" growth curve of reindeer introduced to one of Alaska's Pribilof Islands. The 26 original animals grew exponentially and overgrazed their food supply. The population then plummeted to 18 in less than a decade.

POPULATION-REGULATING FACTORS

Ecologists have attempted to divide determinants of carrying capacity into two main categories. The first type exerts stronger effects when population density is high than when it is low. Because their influence varies with population size, these are called **density-dependent factors. Density-independent factors,** on the other hand, influence population growth to the same extent whatever the population density.

Density-Independent Population Regulation

Sudden, extreme, or unpredictable changes in environmental conditions (such as droughts, floods, cold spells, or heat waves) periodically wipe out large numbers of organisms. Such events may decimate certain populations in an area without any regard to their size or density. A sudden frost in Florida, for example, is likely to kill a certain proportion of the orange trees it hits, regardless of how many trees are growing in any particular orchard or how many orchards there are in an area. Two areas of the United States have recently seen dramatic demonstrations of density-independent factors: the forest fire that burned large areas of Yellowstone Park at the beginning of the decade, and the violent volcanic explosion of Mount Saint Helens.

Even temperature changes, however, rarely have effects that are *completely* unrelated to population density. A hard frost, for example, may kill only those insects, lizards, or birds that are left out in the cold after other members of their species have occupied all available shelters. Thus, in a sense, this sort of random event can "enforce" a particular carrying capacity on a population. The higher the population density, the higher the proportion of "homeless" individuals—and the more organisms are killed by the cold.

Density-Dependent Population Regulation

Many of the most important factors contributing to carrying capacity act in a manner that varies proportionately to population density. In other words, the larger the population, the larger the effect on further population growth.

Competition, the classic example of a density-dependent factor, occurs when organisms of the same or different species require common resources—such as sunlight, food, water, or space—that are present in limited supply. Alternatively, if resources are not actually in short supply, competition can occur when the animals that utilize those resources harm one another in pursuit of their needs. To many ecologists, competition is the single most important density-dependent population-regulating factor. It is also a potent process that directs the course of evolution. Because competition is such an important and complex phenomenon, and because it often involves organisms interacting in groups within living communities, we will consider it in detail in the next chapter.

Predation and parasitism are two other important influences on population growth that act largely through increasing mortality rates. Both predators and parasites can selectively attack individuals of specific sizes or ages. In addition, predators often find prey more easily, and parasites spread from one individual to another more rapidly in dense populations. Once again, these population-regulating factors are complex, important phenomena that play integral roles in structuring living communities; so they, too, will be treated in detail in the next chapter.

Emigration, the mass exodus of individuals from a population, is a response to population pressure found among some rodents and many insects. For example, when locusts in one area reach a certain critical density, they gather and migrate in search of food, forming vast swarms that can literally darken the sky.

Immigration, the addition to an area of individuals born elsewhere, can sometimes maintain a population in an area where local conditions do not permit individuals to reproduce. Populations on small islands near large continents are sometimes maintained in this fashion by a steady trickle of wanderers from the mainland.

Cannibalism, the eating of members of one's own species, increases in some animal species under the stress of crowded conditions. Either male or female rodents, for example, may eat their own babies when population densities become too high.

Metabolic wastes released by animals or plants may accumulate to toxic levels in dense populations. Wastes may affect the survival of the species that creates them, or they may be more toxic to other species, thus affecting the outcome of competition between species. Yeasts grown in culture, for example, produce alcohol as a waste product. The toxicity of this alcohol can act either to stem the population growth of the strain that produces it or to suppress the viability of competing strains of yeast, or both.

In the final analysis, natural populations are regulated by the combined action of *all* potential population-regulating factors, most of which act most of the time in a density-dependent manner.

LIFE HISTORY STRATEGIES

Organisms vary enormously in what are called their **life history strategies**—the speed with which they grow, the age at which they mature, the number of offspring they bear, and the number of times during their life that they reproduce. Plants that we commonly call "annuals," for example, grow rapidly from seed, mature in a single growing season, bloom their heads off, and die after producing prodigious quantities of seeds. Other plants, called biennials, take two years to mature, but then they put all

their energy into seed production and die. Still other plants, called perennials, take anywhere from two to more than a dozen years to mature, but they reproduce every year after maturity and may live for a century or more.

Animals have a similar range of strategies. Salmon, for example, may take between three and five years to mature (depending on species) and then sacrifice their lives to produce prodigious quantities of offspring in a single frenzied bout of reproduction. Humans, by contrast, take roughly 12–14 years to mature, are physically able to reproduce (without adverse effects on either mother or offspring) for 30 years or so, and live for several decades afterwards.

Each of these strategies has advantages and disadvantages. Under certain circumstances, organisms that reproduce early in life and produce many offspring are favored in the long term. That is not always the case, however, because reproduction "costs" organisms energy they might otherwise put into growth, protection, and maintenance. Producing lots of offspring may leave parents weakened in the face of competition or harsh environmental conditions (Fig. 5.6). This is an evolutionarily acceptable trade-off if and only if significant numbers of those offspring can themselves survive to reproduce.

The "best" approach for any particular species thus depends on many factors, including its size, position in the food web, and exposure to disease and predation. In addition, different environments make different demands on individual organisms and populations. In some environments, such as tropical rain forests and coral reefs, physical conditions are stable and predictable throughout the year. In others, such as mountain highlands and shifting sand dunes, physical conditions may be widely variable, less predictable, or continually disturbed. Environments that are either extremely stable and predictable or extremely variable and unpredictable favor different life history strategies.

Unpredictable Environments

In physically difficult or unpredictable habitats, environmental upsets often kill organisms in large numbers, opening up spaces to newcomers. It is usually advantageous for species living in such environments to reproduce as quickly as possible, because *adults don't survive for long* and because their *offspring can survive for brief periods.* Many bacteria, protozoans, and plants that we call "weeds" have evolved reproductive strategies geared to unstable environments: they typically reproduce at a very young age, produce large numbers of offspring, devote little or no care to their brood, and may die after reproducing.

Such species are often called *opportunistic species:* their reproductive strategy allows them to invade new habitats quickly and grow maximally during brief periods of environmental benevolence. These species are often the first to invade new or disturbed environments, such as small patches of rain forest damaged by storms or patches of grassland denuded by fires.

Predictable Environments

In stable, predictable environments, competition, predation, and parasitism are often the major sources of mortality. Under these conditions, organisms gain little by stretching their resources to produce large numbers of offspring. Why? Because there isn't room in the environment for lots of new individuals. Food, nesting sites, or other factors limit the size of the adult population and make it difficult for newcomers to get established. For this reason, most offspring—particularly small or weak ones—don't survive.

Organisms in predictable environments, therefore, usually mature and reproduce later in life. Although they may reproduce repeatedly once they mature, they produce fewer offspring than opportunistic species. They also spend more time and/or energy nurturing and protecting their young. In this way, they can produce larger, stronger, and better-developed offspring that are more likely to survive in a highly competitive environment. Such organisms (including many tropical mammals and birds) are often called *equilibrium species,* because they make up a large proportion of mature, stable ecosystems.

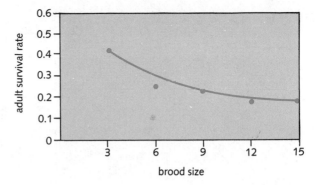

Figure 5.6 Survival of female songbirds in the year after breeding. Survivorship is plotted against brood sizes manipulated experimentally. Females who raised large broods had lower survival rates than females raising smaller numbers of young.

Table 5.2 compares opportunistic and equilibrium species on several dimensions.

r- and *K-*strategists It is instructive to compare organisms that experience conditions at opposite ends of this "predictable versus unpredictable" continuum. When we do so, we can identify two archetypical life history strategies.

Organisms whose populations are regularly decimated tend to reproduce early and prolifically; because this

Table 5.2 *Comparison of Opportunistic and Equilibrium Species*

	Opportunistic	Equilibrium
Environmental conditions	Variable and/or unpredictable; uncertain	Fairly constant and/or predictable; more certain
Survivorship*	Often type III	Usually types I and II
Population size	Variable in time; nonequilibrium; periodic ecological vacuums allow recolonization	Fairly constant in time; equilibrium at or near carrying capacity of the environment; stable, ecologically saturated communities
Intraspecific/ interspecific competition	Variable, often low	Usually intense
Typical characteristics	1. Rapid growth 2. High *r* 3. Early reproduction 4. Small body size	1. Slower growth 2. Greater competitive ability 3. Delayed reproduction 4. Larger body size
Length of life	Short, usually less than 1 year	Longer, usually more than 1 year

Source: Adapted from Pianka, "On *r* − 1 and *K*-selection," *American Naturalist* 104 (1970): 592–597.
*Note: Survivorship is classified into three types. Refer to Fig. 5.8 for a complete explanation.

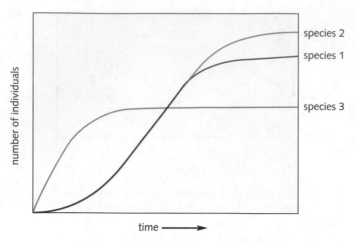

Figure 5.7 Characteristics of *r*- and *K*-strategists compared graphically for three imaginary species. Species 1 exhibits a standard logistic growth curve, growing exponentially until it reaches the carrying capacity of its environment. Species 2, more of a *K*-strategist, has the same growth rate as species 1 but tolerates crowding better, and so it reaches equilibrium at a larger population size. Species 3, more of an *r*-strategist, has a higher growth rate than either of the other species and a lower *K*.

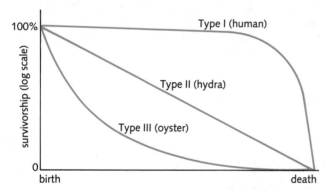

Figure 5.8 Three survivorship curves. The human curve, type I, shows that humans who survive their early years usually live to a ripe old age. Hydra exhibit a type II curve; these animals have a fairly equal chance of dying throughout life. Many organisms, such as fishes and oysters, exhibit a type III curve. Type III organisms produce tremendous numbers of offspring, most of which die at a very early age. Once these organisms reach a certain age, however, they are likely to live a long time.

strategy maximizes r_0, their intrinsic rate of reproduction, they are called **r-strategists.** Species whose populations are more often in equilibrium, on the other hand, are often called **K-strategists** because they seem to have maximized their carrying capacity, or *K.* Most organisms, however, are neither pure *r*-strategists or *K*-strategists but lie somewhere along a continuum between these extremes (Fig. 5.7).

DEMOGRAPHY: THE STUDY OF POPULATIONS

So far, we have paid attention only to the total number of individuals present in a population. We have also treated populations as though they were very, very simple, with each individual equally capable of producing offspring indefinitely. In reality, of course, that is not the case. As we've just noted, members of our own species cannot reproduce until puberty, and females ultimately lose the ability to reproduce after menopause. Therefore, a group of older humans introduced into an environment has a rather different potential for population growth than a group of younger ones.

How can we handle these complexities? To fully understand populations and predict their growth, ecologists must take a group of individuals, examine their ages, and determine both their reproductive abilities and their chance of survival at each age. This type of study is called **demography** (*demos* means "the people"; *graphos* means "measurement").

Demography is a crucial component of the biology of all species. Because extremely accurate and detailed population data are available for humans, and because knowledge of human population growth is vital to understanding global ecology, we will use data on humans as we discuss demography. All the principles that emerge, however, are just as valid for frogs, flies, and oak trees as they are for *Homo sapiens.*

Population Age Structure

The **age distribution** of a population refers to the relative number of individuals of each age, from newborns to the oldest survivors present.

Mortality and survivorship Individuals of different ages have different chances of living for the same length of time into the future. For example, a 90-year-old person is less likely to survive another 20 years than a teenager is. The chances that individuals of a particular age will live or die in a finite amount of time are expressed in **life tables** (Table 5.3). Life tables reflect two complementary sets of data: the percentage of the population that can expect to live to a given age, or its **survivorship,** and the population's age-related death rate, or its **mortality.**

Different species have different patterns of mortality and survivorship. Demographers recognize three classic patterns, although there are many intermediates (Fig. 5.8). The human data shown in Table 5.3 translate into what is called a type I survivorship curve.

Fertility There is another reason why age structure influences a population's ability to grow. Individuals of dif-

Table 5.3 *Life Tables*

	MALES					FEMALES					
Age	Population	Deaths	Probability of Dying Within Five Years	Survivorship per Population of 100,000 at Birth	Life Expectancy	Population	Deaths	Births	Probability of Dying Within Five Years	Survivorship per Population of 100,000 at Birth	Life Expectancy
0	414700	8706	0.020609	100000	68.646	393600	6274	0	0.015716	100000	74.831
1	1716199	1489	0.003465	97939	69.088	1631199	1193	0	0.002923	98428	75.024
5	1980499	868	0.002191	97600	65.323	1882499	572	0	0.001520	98141	71.239
10	1715999	676	0.001977	97386	60.461	1633099	430	222	0.001319	97992	66.344
15	1734799	1544	0.004445	97193	55.575	1681799	624	81853	0.001854	97862	61.428
20	1857199	1711	0.004598	96761	50.812	1838899	739	295946	0.002011	97681	56.538
25	1538699	1346	0.004370	96316	46.035	1500799	793	240807	0.002651	97484	51.646
30	1504099	1553	0.005159	95896	41.226	1426099	1076	125316	0.003776	97226	46.777
35	1508799	2446	0.008080	95401	36.426	1443299	1738	58083	0.006009	96859	41.944
40	1540799	4652	0.014997	94630	31.700	1522099	3152	15904	0.010309	96277	37.181
45	1626399	8274	0.025168	93211	27.141	1644899	5568	1140	0.016797	95284	32.540
50	1411599	13022	0.045200	90865	22.771	1488799	7847	1	0.026037	93684	28.050
55	1458299	23367	0.077259	86758	18.720	1585299	12600	0	0.039025	91245	23.728
60	1292799	34784	0.126664	80055	15.063	1483799	19195	0	0.062862	87684	19.583
65	991000	43709	0.200187	69915	11.863	1280499	27232	0	0.101552	82172	15.717
70	631100	43425	0.295219	55919	9.183	1033400	37935	0	0.169360	73827	12.191
75	399200	42010	0.415613	39411	6.970	753600	47723	0	0.275469	61324	9.138
80	206600	32674	0.558305	23031	5.182	464300	49948	0	0.424249	44431	6.632
85	101100	27036	1.000000	10173	3.739	275100	58912	0	1.000000	25581	4.670

SOURCE: Data from N. Keyfitz and Wilhelm Flieger, *Population: Facts and Methods of Demography* (San Francisco: 1971), W. H. Freeman, pp.154–157.
NOTE: Data collected from populations in England and Wales, 1968.

ferent ages have different **fertility rates;** they do not produce offspring at the same rate. At any given time, some are too young to reproduce, others too old. Even those capable of reproduction differ in *fertility* according to age.

The peak biological reproductive years for human females, for example, fall between the ages of 15 and 44. But ethnic traditions and trends in different societies interact with biological potential to influence actual human fertility rates very strongly. Consider the total fertility rate in the United States: from a post–World War II baby boom peak of 3.8 births per woman, our birth rate dropped to 1.8 in the late 1970s and 1980s—slightly below the level of 2.0 required for zero population growth (Fig. 5.9). More recently, the U.S. fertility rate has risen slightly. It is now just above 2.0. These shifts reflect social changes rather than biological or medical changes.

Age structure and population growth When demographers know the age structure of a population, they can predict its future by combining information on the number of females, their ages, and the fertility rate for each age class. Once they have all this information, they can describe the population as belonging to one of four basic types of population age structures:

- Stable
- Slowly expanding
- Rapidly growing
- Gradually declining

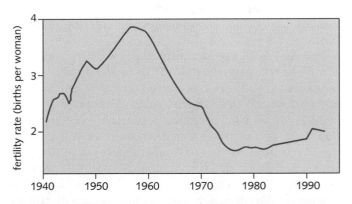

Figure 5.9 This graph of the U.S. fertility rate shows fluctuations influenced by economic and social trends. U.S. fertility rose dramatically from a rate of just above 2 births per woman in 1940 to a post-war baby boom high, then dropped to a low of 1.8 during the 1980s to produce the so-called "baby bust," and then rose to a high of 2.1 in 1990, producing what demographers call a "baby boomlet." According to recent data from the Population Reference Bureau, the rate has dropped slightly to a 1993 level of barely above 2.0.

Rapidly declining populations can be recognized, too, but they don't last very long in nature.

Figure 5.10 illustrates three of these situations in the human populations of different countries. Sweden's age

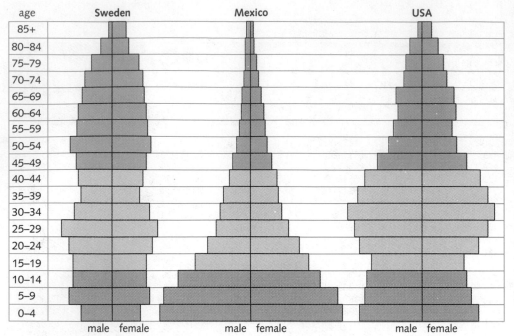

age	Sweden	Mexico	USA
85+			
80–84			
75–79			
70–74			
65–69			
60–64			
55–59			
50–54			
45–49			
40–44			
35–39			
30–34			
25–29			
20–24			
15–19			
10–14			
5–9			
0–4			
	male female	male female	male female

Figure 5.10 Age structure diagrams of human populations; males are represented on the left-hand side of each figure and females on the right. In Sweden, a country with zero population growth, many individuals are far beyond primary reproductive age. This population structure reflects a low, stable birth rate. In Mexico, a country with rapid population growth, a large percentage of the population will soon reach prime reproductive age. This structure reflects a high birth rate that will continue into the future. The United States has a fairly stable population, as indicated by the most recent available data based on the 1990 census. Note the post–World War II baby boom bulge in 20–45 year-olds, the relatively small number of 10–20 year olds, and the slightly larger number of very young children.

structure reflects a low, stable birth rate and an even distribution of individuals across all age classes. This sort of population usually remains stable because the same number of females are constantly entering and leaving their peak reproductive years. Mexico, by contrast, exhibits a distribution that indicates very rapid growth, because each year more and more females are entering their reproductive period.

The American age structure has been interpreted by various experts as growing slowly, as heading toward zero population growth, and as declining slowly. During the early part of the past decade, some demographers predicted that the United States would soon achieve zero population growth. Since that time, however, our population has continued to grow, and experts now disagree about whether or not it will stabilize over the next 50 years. Some factors contributing to the current apparent increase in population growth include a recent rise in childbearing by maturing "baby boomers," high birth rates among some recent immigrant groups, as well as continuing immigration itself.

HUMAN POPULATION GROWTH

Many biologists are worried about global human demographic data, because our species' growth to date describes a classic J-shaped curve that cannot continue indefinitely (Fig. 5.11). It is to be hoped that human population growth will slow down, taking on the S shape of a population in equilibrium with the carrying capacity of

the global environment. If it doesn't, our species may be faced with the sort of "boom and bust" growth characteristic in which unpleasant circumstances limit growth for us.

One of the first to recognize the problems of unchecked human population growth was the eighteenth-century English economist Thomas Malthus. Malthus observed that human populations of his day were growing exponentially, but that society's ability to increase food production was growing much more slowly. If exponential human population growth continued, Malthus reasoned, sooner or later we would run out of food and space. Malthus believed that runaway human population growth could be checked only by food shortages, plagues, and wars. These pessimistic observations are called the *Malthusian doctrine*.

To Malthus's gloomy predictions, modern ecologists have added questions about the effects of the earth's human population and pollution on the biosphere. There is spirited debate among ecologists, human demographers, and social scientists about what will happen to the world's human population in the next few decades (see Current Controversies, Human Population Growth).

CAN WE PREDICT (AND CONTROL) OUR OWN IMPACT?

Humans are affecting environments around the world in many ways. In order to predict the effects of our own population on the environment, ecologists must understand

Human Population Growth

A farmer has a water lily in his pond that grows exponentially, doubling in size every day. If nothing interferes with it, the lily can cover the water's surface in 80 days, suffocating all other life in the pond. But the farmer is busy, and he decides not to bother with the water lily until it covers half the pond.

That may sound reasonable, but think about it. The lily won't cover half the pond until the twenty-ninth day. The farmer will then have only 24 hours to save his pond.

This parable aptly illustrates the hidden danger of exponential growth, and it should explain why many biologists are worried about human population. World population growth now averages about 1.7 percent per year, a rate that causes our population to double about every 41 years.

Why is Earth's human population growing so quickly? To answer that question, we must consider the phenomenon called *demographic transition*—a drop first in death rates and then in birth rates, often considered the most crucial event in the history of human populations.

Remember that *r*, the rate of population growth, is equal to the birth rate minus the death rate. The human death rate begins to fall first as countries begin to develop, because improvements in health care, nutrition, and sanitation help people of all ages live longer. Some time later on in the development process, birth rates, too, begin to fall. Researchers often attribute this decline in birth rates to several factors associated with changes in socioeconomic conditions:

1. As urbanization proceeds, overcrowding and migration from farms to cities make it hard for families to house many children. Also, extra hands aren't as useful in the city as they would have been on a farm.

2. Increased educational and vocational opportunities offer new social and economic alternatives for women in societies where they once spent most of their lives producing and raising children.

3. The availability of birth control technology enables couples to choose how many children they will have.

Whatever the reasons for this decline, when *b* and *d* are both small, $(b - d)$ is small or zero, so population growth stabilizes. The United States, Europe, and Japan completed the demographic transition between 1850 and 1950. In those countries today, population growth averages between 0.5 and 1 percent per year.

More than two-thirds of the world's population, however, lives in the "Third World" countries of Central and South America, Asia, and Africa. There, death rates have dropped as they did in the West—a triumph of humanitarian efforts that everyone applauds. But birth rates in those countries have thus far declined only slightly. As a consequence, annual growth rates average between 1.2 percent (in the South Pacific) and a staggering 4.1 percent in parts of Africa.

Few question the harsh realities of life in the Third World today. Disagreements arise immediately, however, when discussion shifts to the importance of population growth in development. For this reason, we will return to this question in Chapter 7.

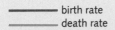

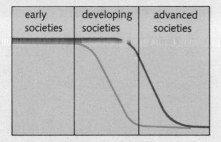

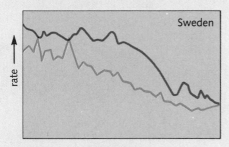

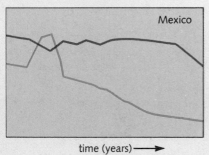

time (years) ⟶

Birth and death rates in societies at three stages: "early," "developing," and "advanced." TOP: In early societies, high birth rates are matched by high death rates, so population growth is slow. CENTER: In societies that complete the demographic transition, both birth rates and death rates fall, as they did in Sweden during the nineteenth and early twentieth centuries. In such societies, population growth is also slow. BOTTOM: In Mexico, a country in the midst of the transition, death rates have fallen, while birth rates have just begun to drop. This population is growing rapidly.

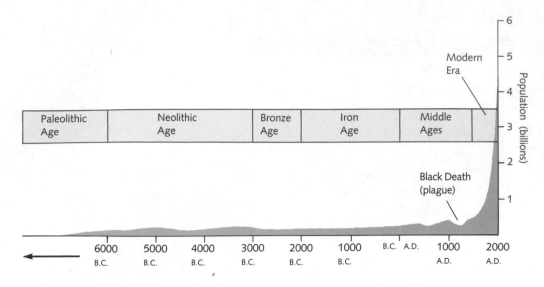

Figure 5.11 Growth curve for the global population of *Homo sapiens*. After a long lag period—lasting until the industrial revolution—human populations entered exponential growth that has been maintained to date.

the population biology of natural populations. We now appreciate the significance of competition, predator–prey interactions, and host–parasite interactions in controlling natural populations; but in most cases our knowledge works only "backwards." We can explain why things go awry *after* the fact, but we can seldom predict outcomes in advance. Our knowledge has helped us reverse the ill effects of tampering with some natural ecosystems (Fig. 5.12), but it has proved useless in other situations. To improve that record, we must learn a good deal more about population growth and ecological interactions than we know today.

Figure 5.12 Correcting an ecological error. LEFT: The prickly pear cactus, introduced into Australia from Latin America, grew wildly and uncontrollably and covered vast areas with tangled, spiny growth. RIGHT: After careful research, *Cactoblastis,* a ravenous, cactus-eating moth, was released in Australia. This insect predator, adapted to feed on prickly pear species in ways that Australian insects were not, soon helped bring the cactus under control.

SUMMARY

Populations are collections of individuals, belonging to a single species, that regularly interbreed with one another. Under ideal conditions, most populations grow exponentially, increasing in size more rapidly as they get larger. In nature, one or more environmental factors normally place a ceiling on population size, either by decreasing the birth rate, increasing the death rate, or both. Natural environments cannot support infinitely large populations of any species; instead, each environment is said to have a finite carrying capacity for various types of organisms.

Environments vary over time in their ability to support organisms. Some environments are stable and predictable, while others experience hard frosts, droughts, or unpredictable environmental fluctuations of other kinds. Over time, organisms have evolved different life history strategies adapted to predictable and unpredictable environments. Species that inhabit stable, predictable environments often exhibit logistic growth and have relatively low growth rates and high carrying capacity; they are said to be *K*-strategists. Organisms that experience major environmental fluctuations often switch between exponential growth and calamitous declines.

Demography—the study of individuals' ages, their reproductive abilities, and their chance of survival at each age—is important in predicting the future growth of a population. When population-regulating factors in the environment operate in a density-dependent fashion, population growth follows an S-shaped or logistic growth pattern, leveling off as the population reaches the environmental carrying capacity, or *K*. The most important density-dependent population-regulating factors are competition, predation, and parasitism.

STUDY FOCUS

After studying this chapter, you should be able to:

- Explain how and why populations grow, and describe the basic equations used to predict population growth.

- Discuss the factors that control populations in nature.

- Explain the basic phenomena important to demography, and contrast the demographic character of developed and developing nations.

TERMS AND CONCEPTS

REVIEW

Objective Questions (Answers in Appendix)

1. The birth rate minus the death rate determines a population's
   (a) reproductive rate.
   (b) intrinsic rate of growth.
   (c) logistic growth.
   (d) carrying capacity.

2. When a species overshoots its carrying capacity in the environment,
   (a) the population will plummet.
   (b) population growth will not be affected.
   (c) the population growth will reach a steady state.
   (d) none of the above apply.

3. A J-shaped curve represents a population that
   (a) has a limited supply of food.
   (b) has reached the carrying capacity of its environment.
   (c) is growing exponentially.
   (d) was growing exponentially and then crashed.

4. Which of the following represents a density-independent factor regulating population growth?
   (a) competition for food and raw materials
   (b) activities of predators
   (c) parasites
   (d) climate

5. Which of the following descriptions represents *K*-strategists?
   (a) large number of young produced
   (b) little or no care devoted to young
   (c) large investment of energy in reproduction
   (d) advantageous to reproduce as quickly as possible

Discussion Questions

6. What is a population? How do organisms' physical abilities and environmental conditions combine either to limit populations to small areas or enable them to cover large regions?

7. What is exponential growth? Under what conditions in nature is exponential growth likely to occur? What would happen if exponential growth continued unchecked?

8. What are the four stages of growth in a population that follows a logistic growth curve?

9. What is the difference between density-dependent and density-independent population-regulating factors? Which are most likely to result in a stable equilibrium? Why?

10. Compare and contrast opportunistic and equilibrium organisms, and give examples of each.

READINGS

See integrated list of readings for Part 2 after Chapter 7.

6

Community Ecology

One otherwise ordinary morning, the South Pacific island of Krakatau disappeared in a series of cataclysmic volcanic eruptions. The blasts were so intense that the crew of a ship 84 km to the northeast feared the apocalypse was upon them. "A heavy shower of ashes came . . . ," the first officer wrote, "the air being so thick it was difficult to breathe . . . all hands expecting to be suffocated; the terrible noises from the volcano, the sky filled with forked lightning . . . the howling of the wind . . . formed one of the most awful scenes imaginable. . . ."

The island was obliterated—except for a small chunk of land at its southern tip. That remnant, called Rakata, was smothered with a thick layer of volcanic rock and ash heated to several hundred degrees Celsius. Needless to say, the tropical forest that had covered Krakatau before that fateful day in 1883 was destroyed.

Nine months later, scientists found no signs of life—save for a spider spinning a web amidst the debris. But a few at a time, organisms swam, drifted, floated, or flew from adjacent islands. A year after the explosion, 2 grass species took root. Over the next two years, 15 more grass and shrub species appeared. Fourteen years later, 49 plant species called Rakata home, and lizards, birds, bats, and scores of insect species were arriving. In 1919, Rakata sported patches of forest surrounded by grass; ten years later, forest was strangling the remaining grassy patches, and nearly 300 plant species had taken hold. Over the years, species appeared and disappeared. In fits and starts, an ecosystem "bootstrapped itself" to the point that today, just over a century after the eruption, it resembles in many respects any other tropical forest (Fig. 6.1). But it doesn't yet contain certain tree species found deep in old-growth rain forests on Java and Sumatra. That may take another century or so.

At about the same time that Krakatau blew its top, a much more gradual and more subtle series of events caused dramatic changes in the marine ecosystems of North America's Pacific coast. There, sea otters—active and endearing marine animals—were hunted nearly to extinction for their highly valued pelts. That was unfortunate in and of itself. But the decline in otter populations had startling and unexpected effects on the kelp forests those animals called home.

Figure 6.1 The volcanic island of Krakatau (Krakatoa), once covered with verdant rain forests, was virtually destroyed (and effectively sterilized) by a violent eruption in 1883. Since that time, its tropical flora and fauna have slowly but steadily reestablished themselves, thanks to colonizers from nearby Java and Sumatra.

As we saw in Chapter 3, otters are carnivores who feed on globe-shaped, spiny sea urchins that live in kelp forests' lower depths. Why is that predatory connection so important? Because urchins are voracious herbivores, who especially favor the rope-like organs that anchor kelp plants to the bottom. When otters—who had been the urchins' major predators—disappeared, urchin populations exploded. Wave after wave of them bulldozed through kelp forests, nibbling on and weakening the plants' anchors. Slowly but steadily, storm waves uprooted the kelp and tossed them on shore. Once adult kelp were gone, the insatiable urchin hordes devoured any infant kelp that settled on the bare rocks.

The results were devastating. Kelp plants, as discussed in Chapters 3 and 4, provide both three-dimensional structure and a supply of energy and organic nutrients to their habitat. After the kelp vanished, scores of once common species—ranging from sponges to crabs, shrimp, squid, fishes, and even grey whales—died or moved away in search of greener pastures. Large stretches of what had been highly productive and densely populated habitats degenerated into wastelands.

Happily, the story didn't end there. Over time, conservationists successfully "transplanted" otters from small populations that had survived the slaughter in the remote Aleutian Islands and the vicinity around Big Sur in California. The newly seeded otter populations were assiduously protected from hunting, and slowly they began to grow. As otters staged a comeback, urchin populations dropped. And as urchin populations declined, kelp were freed from their intense grazing pressure. Young kelp plants slowly reappeared and, in some spots, now approach their former luxuriance.

ECOLOGICAL COMMUNITIES

The regrowth of Rakata's forest and the ups and downs of the otter–urchin–kelp system are classic examples of relationships among populations that inhabit the same environment and interact with each other. Ecologists have a special name for these coherent assemblages of organisms; they call them **communities.** Interactions within ecological communities take many forms and have wide-ranging effects: typically, they prevent some species from persisting, enable other species to survive, and control the population densities of those species that do endure.

This chapter will focus on basic principles of community ecology that relate to three related questions: What determines a species' success or failure in an ecosystem? What determines which organisms coexist in an ecosystem? And finally, how do ecosystems respond to disturbances of different types?

The Concept of Niche: How an Organism Makes a Living

Before we can appreciate events at the community level, we must have a clear idea of the role individual species play in their environment. In describing that role, ecolo-

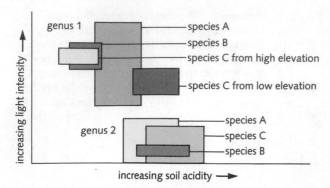

Figure 6.2 The response of several moss species to two physical factors: light and soil acidity. Each species prefers a different light intensity and different range of pH values. Because of these preferences, these species inhabit different microhabitats within the same forest. For example, genus 1 species A tends to live in a less acidic and more intensely lighted habitat than genus 2 species C.

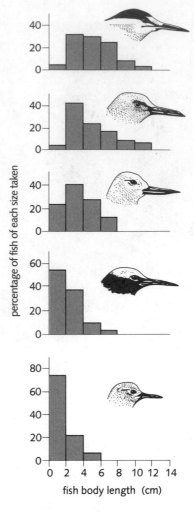

Figure 6.3 LEFT: Beak length and food choice in seabirds. Each of these tern species from Christmas Island in the Pacific has a different-sized bill. Each chooses fish of various sizes in different proportions, as shown by the bar graphs with each species. BELOW: Ecologists translate these bar graphs into curves called resource utilization curves. Resource curves make it easy to compare the feeding habits of different species.

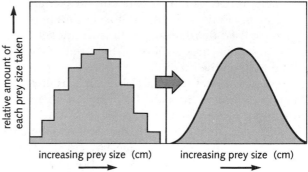

gists often invoke the concept of **niche.** The word *niche* refers to the functional relationship of an organism to its physical and biological environment. At first glance, niche may remind you of habitat, but it shouldn't. An organism's habitat, as you learned earlier, can be viewed as its "address"—the physical location where it lives. An organism's niche, in contrast, is more like its "profession"—the various things it does to "make a living." A full description of an organism's niche includes three important sets of parameters:

1. The range of physical factors in which the organism can survive and reproduce: temperature, humidity, salinity, pH, grain size of the soil, and other such variables

2. The biological factors with which the organism interacts: predators, prey, parasites, organisms that provide shelter, and those that compete for the same limiting resources

3. The organism's behavior: when, where, and upon what it feeds, its social organization, and its behavioral interactions with other organisms

We can't provide a picture or a complete graphical representation of a niche, because pictures and graphs cannot represent so many different factors at once. We can, however, look at two or three aspects of a niche at a time as we try to determine which factors are most important in the ecology of a particular species.

Physical aspects of the niche One view of niches concentrates on the physical factors that determine where particular organisms can live. Figure 6.2, for example,

shows how several different moss species divide up available habitats according to the strength of incoming light and the acidity or alkalinity of the soil.

Biological aspects of the niche Another perspective on the niche focuses on ways that organisms share habitats by dividing up available resources. For example, similar species that live in the same area often prefer slightly dif-

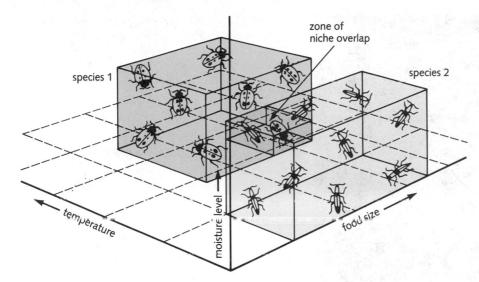

Figure 6.4 Niches visualized as three-dimensional spaces created by environmental conditions and food characteristics. Shown here are partial representations of adjacent niches occupied by two imaginary insect species, each of which has slightly different preferences for temperature, moisture, and food size. Each insect has part of its niche space to itself and shares a portion with the other species. Where overlap occurs, competition will take place.

ferent types of food, obtain that food in different places, or hunt at different times of day.

To show those preferences clearly, researchers create what are called *resource utilization curves*. Figure 6.3, for example, uses such curves to show how five species of seabirds, all of which live in the same place, eat fish of slightly different sizes.

Note that it is possible to use resource curves to describe many types of interactions between organisms and selected aspects of their surroundings: the part of the habitat in which they feed, the time of day at which they forage, the temperature at which they are active, and many more. If, starting with a single resource utilization curve, we could add to our graph all these other interactions, we could create a multidimensional picture of an organism's niche (Fig. 6.4).

Fundamental versus realized niches   The broadest of all possible niches an organism theoretically can occupy is called its **fundamental niche.** But both positive and negative interactions with other organisms often restrict a species to a smaller niche, called the **realized niche.** In order to understand why realized niches are invariably smaller than fundamental niches, we will consider in more detail certain interactions among species that we introduced in the previous chapter as population-regulating factors.

Competition

Competition, a potent force in controlling populations of many organisms, is also a key to understanding how pairs or groups of species interact in natural communities.

Plants and animals may compete for a variety of important resources. Plants in crowded forests or aquatic communities stretch their leaves out over one another in a battle to intercept the most sunlight, and they use aggressive root systems to compete for inorganic nutrients and water. Animal populations compete for water, edible plants, or suitable prey. Many animal species, from sea gulls to lions, also need various kinds and amounts of space in which to feed, breed, and/or hide from predators (Fig. 6.5).

Ecologists recognize two main classes of competition. In **intraspecific competition,** members of the same species compete among themselves (*intra* means "inside or within"). In **interspecific competition** (*inter* means "between"), members of two or more species compete with each other.

When population densities rise, any ecological commodity in short supply can function as a limiting resource; shortages curtail the growth of all populations present, while most severely stressing the least successful competitors. Why does competition make things difficult? Because competing for resources takes both time and energy, individuals who actively compete with one another have less energy left for growth and reproduction than individuals with no competitors. The stress caused by competition can decrease a population's birth rate, increase its death rate, or both—profoundly affecting its ability to grow or even to survive.

Competitive exclusion   In 1934, Soviet ecologist G. F. Gause performed a classic series of experiments involving two species of single-celled *Paramecium*. Gause grew each species in culture under two sets of conditions: first alone and then together with the other species. Although both populations grew well and survived when grown alone,

Figure 6.5   Sea gulls and many other birds stake out territories in which to build nests and raise young. Nesting space is in short supply on many breeding grounds, and gulls fight fiercely to defend their turf. Pairs that do not win suitable pieces of ground simply cannot breed.

when they were grown together, one species drove the other species to extinction (Fig. 6.6).

These and similar experiments led ecologist G. Hardin to propose **the principle of competitive exclusion.** This principle states that two species cannot live together in the same place at the same time if they are both limited by one or more of the same limiting resources. Under those conditions, the principle states, one species always drives the other to extinction locally. If you think for a moment about this concept, you will see that it is directly relevant to the concept of realized niche. Why? Because it asserts that no two organisms can occupy the same niche in the same place at the same time.

Imagine, for example, that two species living in the same area compete with one another for a certain type of resource—say seeds in a certain range of sizes—as shown by the resource utilization curves in Fig. 6.7. The zone of overlap between those two curves represents the degree of overlap between one particular aspect of their niches and, hence, the extent of interspecific competition over that particular resource.

This sort of competitive overlap, as you will learn in Part 3 on evolutionary theory, can be a potent force in shaping the evolution of these species over time. But for now, we'll concentrate on the short-term ecological consequences for the populations in question.

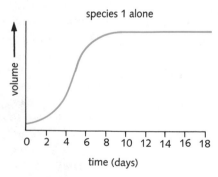
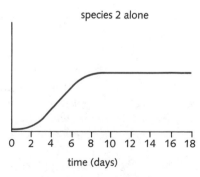
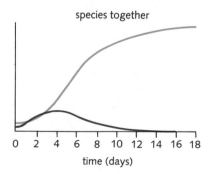

Figure 6.6   Competition in the laboratory. The two species Gause studied have similar requirements. When grown in culture alone, both exhibit logistic growth, reach the carrying capacity of their container, and persist indefinitely. When grown *together*, however, species 1 drives species 2 to extinction.

Because competition affects birth and death rates of competing populations, it can exclude the less efficient competitor from any habitats the two species share. Suppose, for example, that species 1 in Fig. 6.7 is a better competitor for the contested resource than species 2. Depending on precisely which resource is in short supply and how the contest is decided, that competitive advantage may mean that species 1 has sharper eyes, has a stronger beak, or defends its feeding territory more aggressively. By whatever strategy it comes out on top, if species 1 is an abler competitor, it may exclude species 2 from the portion of its fundamental niche represented by the zone of competitive overlap.

Competition in the real world The model of competitive exclusion that we've just discussed is reassuringly straightforward. It is also too simple to be taken literally outside carefully controlled laboratory experiments. In the real world, the outcome of competition between two species can be strongly affected by the response of the competing species to other factors in the environment.

In Gause's classic experiments, for example, it turned out that the successful competitor won *not* because it was better at grabbing available food, but because it was more resistant to chemical waste products that accumulated in the densely populated culture medium. The competitive outcome could be *completely reversed* if the culture medium was changed frequently enough to limit the buildup of those waste products. In other words, keep the "environment" free of metabolic wastes, and *Paramecium* species 2 in Fig. 6.6 successfully outcompetes species 1 instead of being driven to extinction.

These and other laboratory experiments make it clear that any change in the physical environment has the potential to alter the outcome of competition. That observation, in turn, led ecologists to question whether differences in physical conditions might cause one competitor to win out in one microhabitat and another to prevail somewhere else.

The first clear indication that competition in nature works this way was provided by a study of two barnacle species conducted by J. H. Connell from the University of California. Biologists had known for years that plants and animals on rocky seashores are arranged in distinct, horizontal layers, as mentioned in Chapter 3. These layers form a predictable pattern stretching from a perpetually wet zone below the lowest low tide to a splash zone above the highest high tide that is rarely completely underwater but is regularly washed by waves and spray (Fig. 6.8).

But *why* are the organisms arranged as they are? Working on the rocky coast of Scotland, Connell noticed that two species of barnacles, *Balanus* and *Chthamalus,* were found in adjacent, but separate, layers. Adult *Chthamalus* lived only on the highest rocks, just above normal high

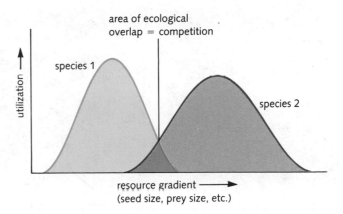

Figure 6.7   The resource utilization curves of two similar species whose uses of a common resource overlap. The extent of competition between the species is indicated by the shaded area on the graph.

tides. Below a specific height relative to the tides, adult *Chthamalus* virtually disappeared and were replaced by dense communities of *Balanus.* Connell wanted to find out why the barnacles divided up their potential habitats in this way. Because adult barnacles are small, spend their adult lives glued to the rocks, and live at high densities, Connell could perform several experiments that would have been difficult or impossible to do with animals that are rare or that move around a great deal.

Connell's experiments had several components. As controls, he carefully watched several rocks over time in order to monitor barnacles' natural growth and survival at various heights. He did not otherwise tamper with these sample areas. At the same time, he did manipulate conditions experienced by barnacles in several experimental areas. Some rocks he transplanted to different heights. Others he scraped clean of adult barnacles to see which species would settle and grow. Still others he picked clean of one barnacle species but not the other.

One of the first things Connell noticed was that if he kept his experimental rocks completely free of *Balanus, Chthamalus* larvae settled, and adults survived, much *lower* down on the rocks than they were normally found. He also found that *Balanus* larvae settled *higher* up on the rocks than adults of that species normally lived—but adults did not survive there. The realized niches of both species seemed significantly smaller than the areas in which their larvae settled. But why?

Connell found the answer by studying the growth of barnacles at different heights. Low down on the rocks, in places not often exposed to drying air and hot sun, *Balanus* grew much more vigorously than *Chthamalus.* This rapid growth made *Balanus* a superior competitor for a vital limiting resource: space on crowded rocks. In the lower

Figure 6.8 This photograph shows the sharp line dividing several species that inhabit rocky, temperate seacoasts. The experiments of J. H. Connell sought to explain the factors behind this sort of rigid zonation.

part of the intertidal zone, *Balanus*'s rapid and aggressive growth enabled it either to overgrow its slower-growing competitors (thereby starving them), or to undercut them—literally prying them off the rocks. Thus, competition for space with *Balanus* excluded *Chthamalus* from the lower part of its potential range.

Higher up on the rocks, however, *Chthamalus* did better because it had higher tolerance for the physical stresses of heat and drying out. In other words, above a certain height in the intertidal zone, *Balanus* was more seriously weakened by physical stress than *Chthamalus*. In that region, therefore, *Balanus*'s ability to survive was impaired, and *Chthamalus* "won out" by default. Thus, differential effects of physical environmental stress on the two species shifted the competitive balance, enabling *Chthamalus* to survive on higher rocks than *Balanus* could.

The combined effects of physical stress and competition on *Balanus* and *Chthamalus* thus divide the rocky intertidal zone between these species (Fig. 6.9). The upper range of each species is determined largely by physical stress. The lower range of *Chthamalus* is controlled by interspecific competition with *Balanus*. Other physical and biological interactions combine to control *Balanus* populations at the lower edge of its range. Data like Connell's indicate that interspecific competition often works together with physical environmental factors (such as heat, desiccation, stress, cold, or exposure) to regulate plant

and animal populations and shape the communities in which they live.

Predation

Predation is another biological interaction that shapes natural communities, because herbivores and carnivores can profoundly shape plant and animal populations. In simple laboratory environments, predators may completely eliminate their prey during the course of an experiment. In nature, that usually doesn't happen. Instead, predators can usually only harvest their prey until prey populations fall to a certain level. Beyond that point, the physical complexity of natural environments—caves, holes, cavities in logs, tunnels or crevices in rocks or coral reefs, and so forth—offers hiding places for prey, and predators have a tough time locating and catching them.

Despite those limitations, predators can play instrumental roles in regulating what would otherwise be uncontrolled exponential growth of their prey. Around the turn of the century, for example, Isle Royale in Lake Superior was colonized by wandering moose. At that time, the island supported no large predators. The moose population grew exponentially, exceeded the island's carrying capacity, and severely overgrazed the vegetation. The result was a dramatic decline, or crash, in the moose population in 1930. With the moose population reduced, the vegetation recovered. But the moose population rebounded in turn, only to crash once more in the 1940s. This sort of "boom and bust" growth cycle, as you may recall from the previous chapter, is a pattern commonly observed when population size is not adequately controlled in a density-dependent manner.

On Isle Royale the situation changed, however, when a few timber wolves crossed winter ice to the island in 1948. The wolf population grew to about 24 individuals, and the moose population decreased to just under 1000. Since then the wolves and moose have reached a dynamic equilibrium in which both populations remain at relatively constant levels. Today, the situation seems to be somewhat less stable, although the reasons are not completely clear.

Predator–prey oscillations Of course, predator–prey interaction is a two-way street; prey availability affects predator populations, too. Sometimes, predators and prey enter a predictable series of **oscillations**—regular population fluctuations in which both species alternately rise and fall in numbers. One classic example is the oscillation of lynx and snowshoe hare populations, as graphed in Fig. 6.10.

It is easy to interpret the data simplistically; one can assume that available prey are the major limiting factor for lynx and that predation is the major factor limiting hare population size. But these data are obtained by observation and not experiment; they lack the controls and experimental manipulation necessary to prove the point and exclude competing hypotheses.

Recently, ecologists have tried to determine whether factors other than predator–prey interactions could be responsible for these oscillations. Observations on Anticosti Island in the Gulf of St. Lawrence showed that hare populations oscillate even in the absence of lynx or other large predators. These new data suggest that early interpretations focused on the wrong predator–prey interaction. How? More work is needed to fully understand the interactions between hares and the plants they eat; it is possible hares may grow exponentially until they overgraze their food source and then crash. The presence of predators might (or might not) lessen the severity of the peaks and crashes.

Some of the same arguments may apply to the wolves and moose on Isle Royale. In that situation, it seems reasonable to assume that the wolf population is limited by food and the moose population by predation. Some ecologists disagree, however, arguing that the moose population is primarily controlled even today by food availability, disease, and factors other than predation.

In any case, it is clear that in most ecological interactions in nature, populations and distribution patterns are controlled by several interacting factors—including food

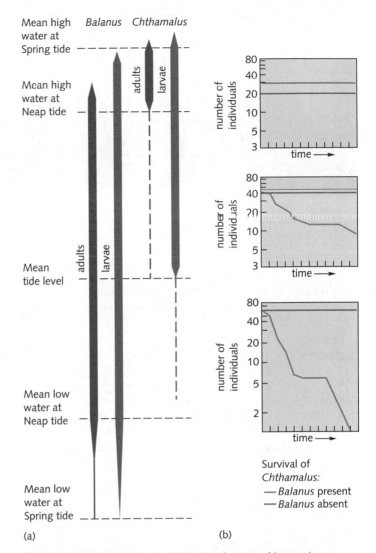

Figure 6.9 Competition and the distribution of barnacles on rocky shores as studied by J. H. Connell. [Adapted from Connell (1961), Ecological Society of America, by permission.] **(a)** Connell discovered that the larvae of both barnacle species settled over a broader range than that in which adults of both species survived. Why? **(b)** In one series of experiments, Connell compared survival of *Chthamalus* at different heights under two sets of conditions: when *Balanus* was present and when *Balanus* was artificially removed from the study site. As these graphs show, removal of *Balanus* made no difference high up on the rocks. Further down, however, *Chthamalus* survived well when growing by itself, but suffered significant mortality when growing with *Balanus*. These experiments, along with others, enabled Connell to determine that competition for space limited *Chthamalus* to the upper reaches of the intertidal zone (see text).

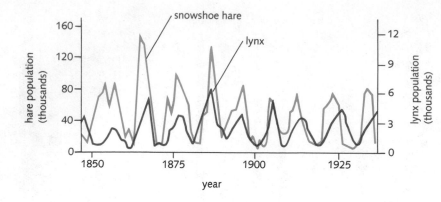

Figure 6.10 Lynx and hare population fluctuations. These data, provided by records of pelts sold by fur trappers, have been used for years to tell a simple story. Whenever hare populations increase, lynx populations can increase as well. At some point, the predators devour the hares faster than the prey can reproduce, and the hare population declines rapidly. As food becomes scarce, the lynx population plummets, releasing hares from predation pressure, and the cycle begins again. For more information on the real complexity of the situation, see text.

limitation, weather conditions, predation, and parasitism—rather than by any single phenomenon alone.

Parasitism

Parasites are organisms that live on or in other organisms, taking nutrients from their hosts without killing them—at least, not right away. In a sense, parasites act like "predators from within," and so they can also act as density-dependent population regulators. Parasites can profoundly affect the structure and function of ecological communities. In laboratory settings, as well as on farms and in zoos, parasites may completely eliminate their hosts. In nature, however—as we saw in the case of predation—this rarely happens. Instead, parasites and hosts usually strike some sort of balance in which both organisms survive.

One classic example of a host–parasite relationship reaching such an equilibrium brings us back to the previous chapter's story about rabbits in Australia. Rabbits, you'll recall, were intentionally introduced to that island continent as a source of meat and game. But unencumbered by their natural population-limiting factors, their populations grew exponentially and devastated the countryside.

In a desperate effort to control the rabbits, biologists introduced the viral disease myxomatosis. At first, nearly all strains of the virus were lethal, and most rabbits were highly susceptible. The disease spread like wildfire, killing off most of the rabbits. But the most lethal strains of the virus disappeared with the rabbits they killed, so less virulent strains became more common. And as the most susceptible rabbits perished in large numbers, the survivors developed resistance to the disease. After several years, virus and rabbits reached an equilibrium that allowed host and parasite to coexist.

Predation, Parasitism, and Perturbations to Community Structure

Predators and parasites, independently and together, may affect ecological communities in more complex and far-reaching ways than the simple examples given above might suggest. Recall from Chapters 4 and 5 that food webs and nutrient cycles involve all the inhabitants of an ecosystem. For that reason, factors that maintain or perturb populations of a single important species in a community can indirectly affect many more species.

A case in point is the relationship mentioned at the beginning of this chapter between sea otters, urchins, and kelp. Otters are sizable animals who are heavy feeders. They weigh, on average, around 23 kg, and in order to maintain body temperature in cold water, they consume between 20 and 30 percent of their body weight each day. When otter populations are high, therefore, their favored foods are hunted intensely.

As described earlier, removal of otters from substantial areas of the Pacific coast led to massive increases in local sea urchin populations, which in turn caused the virtual destruction of entire kelp forest habitats. These changes affected not only otters, kelp, and urchins but also scores of other organisms that called the kelp forest home. Individual species that have this kind of central role in structuring the surrounding community are often called **keystone predators;** remove them and—just like an arch whose keystone is removed—the entire ecosystem collapses. Parasites, through their direct effects on predators, prey, or primary producers, can also have profound effects on community structure. One example of a completely unexpected influence of disease on a seminatural community in a national park is discussed in the Theory in Action, Competition, Predation, and the Complexity of Natural Ecosystems.

Competition, Predation, and the Complexity of Natural Ecosystems

The complex relationship between predation, competition, and environmental conditions in nature is illustrated by a recent study in Africa's Etosha National Park. There, the excavation of gravel pits during road construction, a seemingly harmless change in the park ecosystem, had unexpected results: wildebeest and zebra populations crashed; cheetah, brown hyena, and eland populations declined, and lion, spotted hyena, and springbok populations increased! Biologist Hugh Berry pieced together the following explanation.

Stagnant rainwater in the gravel pits, which were used as watering holes by many animals, became living reservoirs for deadly anthrax bacteria. Wildebeest and zebra became sick in large numbers, and lions and spotted hyenas (immune to anthrax) easily captured these usually elusive prey.

As lion and spotted hyena populations increased, they captured larger numbers of eland, which are immune to anthrax. Concurrently, lions and spotted hyenas began to outcompete cheetahs and brown hyenas, the next largest carnivores. As cheetah and brown hyena numbers dropped and the number of carcasses continued to rise, another carnivore–scavenger, the black-backed jackal, increased in numbers.

The declines in wildebeest and zebra populations freed the springbok from competition for food. Cheetahs (now in decline) had been the springbok's major predator. With fewer adversaries and fewer competitors for food, the springbok population grew rapidly.

This story demonstrates several important principles about predation and competition as population-regulating factors. First, a variety of ecologically similar organisms, such as Africa's many grazing animals, may coexist in a stable fashion only because predators remove enough to prevent any one species from forcing others to extinction. Anything that shifts the balance of predators in such a system can also change the relative numbers of their prey, resulting in the extinction of one or more competitors.

Second, when a change in the environment aids one or two of several competing species, increased numbers of those species can overpower their competitors.

Third, when one or more of a group of competing species is hit hard by an environmental change, its competitors benefit from the end of the competitive struggle.

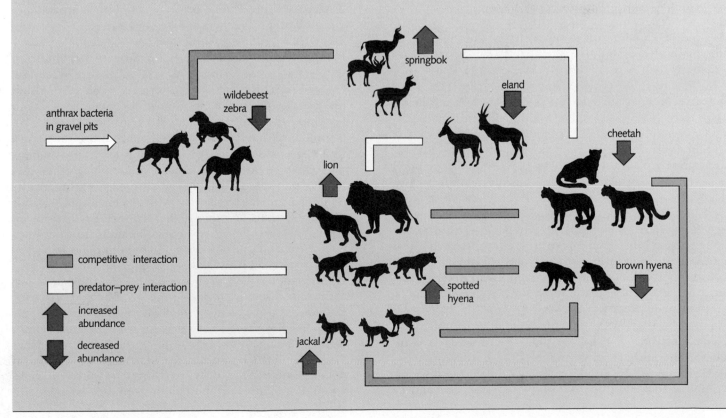

Figure 6.11 Commensal relationships in nature. LEFT: Commensal anemone shrimp on its host. RIGHT: Orchids clasp trees with their roots and gather nutrients from decaying leaves and rainwater. The orchids obtain support and a place in the sun from their host tree, which is neither helped nor harmed.

Positive Interactions Between Organisms

It would be easy for you to suppose that the structure of most natural communities is controlled by "negative" interactions such as predation and competition. But nature is not exclusively Machiavellian; in many cases, the presence of one organism *helps* rather than hinders another—and hence makes it possible for both to exist in a particular community.

Situations in which two or more species form close relationships are called **symbioses** (*sym* means "together"; *bios* means "life"). Parasitism, because it involves a close relationship between two species, is technically considered a form of symbiosis. The types of symbiosis we are referring to here, however, concern positive associations among organisms. Many organisms involved in positive symbioses depend utterly on their partner and cannot exist without them. For that reason, any environmental factors that affect the distribution of one member of a symbiotic pair also affect the other.

Commensalism A common type of symbiosis is **commensalism,** in which one member of the association benefits while the other is neither helped nor harmed. Commensal relationships are abundant on coral reefs and in

tropical rain forests (Fig. 6.11). Some of them are well understood, whereas others are still mysteries.

Mutualism In **mutualism,** both partners benefit from the symbiotic association. As you will learn in several chapters to come, mutualisms are common throughout the biosphere. Human intestines, for example, contain beneficial bacteria that aid in the digestion of food. We provide the bacteria with a warm, moist environment and with food, and they in turn synthesize certain vitamins our bodies could neither obtain nor produce otherwise. (That's one reason why taking a broad-spectrum antibiotic for an infection can cause problems such as gas or diarrhea; it kills or inhibits the actions of beneficial intestinal bacteria.) And, of course, animals from termites to cows rely on gut symbionts for help in digesting cellulose in wood and plant leaves (Chapter 4).

Some of the most dramatic (and ecologically most pivotal) mutualisms are found in the sea. The coral reef, for example, is a major ecosystem whose very existence depends on a symbiotic relationship involving hard corals. In fact, calling these organisms "corals" and classifying them as animals is almost misleading from an ecological point of view. Why? Because living coral tissue is composed of a tightly knit mutualistic symbiosis between animals related to sea anemones and single-celled algae called *zooxanthellae*. This symbiosis, which enables both partners to grow more efficiently than either could alone, is based on a steady exchange of nutrients between corals and their algal guests.

The corals generate carbon dioxide and ammonia as waste products. Though toxic to the corals themselves, these wastes are essential nutrients for the zooxanthellae. (Because the algae grow within coral tissues, they can absorb those compounds with minimal effort. (The nutrients don't get diluted in seawater, so the plants don't have to work to concentrate them again.) In return, the algae facilitate coral growth, both by providing their hosts with certain complex products of photosynthesis, and by somehow helping the corals lay down their calcium carbonate skeletons. This combination of primary and secondary producers in a single unit thus enables both algae and animals to function more efficiently than either could alone. As a result, hard corals can construct and maintain a highly productive, three-dimensional habitat in seawater that contains relatively few nutrients.

Many other inhabitants of the reef also engage in mutualistic symbioses. The best-known are clown fishes that live within the stinging tentacles of sea anemones (Fig. 6.12). The anemone offers the fish protection from some of its enemies. The clown fish, in turn, defends its living home against other fish species that like to eat the anemone's tentacles.

Figure 6.12   A mated pair of symbiotic clown fish in the tentacles of their anemone host. The clown fish's body is covered with mucus that protects it from the anemone's sting.

Succession

Long ago, ecologists noticed that natural communities in a given area often change slowly but steadily over time. They gave the name **succession** to this gradual process of environmental change and species replacement. Formally, succession is defined as a series of changes in the species composition of an ecological community over time. In many (though not all) cases, succession seems to be an orderly and predictable process that ends with the establishment of a final, stable ecological association called a **climax community.**

Primary succession   When succession occurs for the first time in newly formed habitats—such as lava flows, deep bogs or lakes, or sand dunes—it is called **primary succession.** In one sense, therefore, the story of Rakata is a description of primary succession.

Most early arrivals in newly formed habitats tend to be what we described in Chapter 5 as opportunistic species or *r*-strategists—they tolerate harsh physical conditions and aren't too picky about how they make a living. When such species appear during the first stages of succession, they grow rapidly, mature quickly, and reproduce in large numbers.

As these colonizers grow, however, they change conditions around them. As grasses and hardy herbs grow on a newly formed coastal sand dune, for example, they hold the sand down, shade the surface, and add organic matter in the form of dead leaves and roots. These changes affect the stability of the sand, its temperature on sunny days, the amount of moisture the soil holds, and the availability of nutrients within it. Similarly, as water-loving mosses and other plants at the edges of a bog grow and die, they build up layers of peat that gradually fill in the bog, change its pH, and alter its nutrient content.

These biologically caused environmental changes often allow other organisms, less tolerant of the original harsh conditions, to survive. Sometimes, new arrivals ultimately outcompete the pioneers that paved the way for them. In dune habitats, shrubs and pine trees gradually move in, shading and replacing pioneering grasses and herbs. As these taller plants grow, they change the environment further, encouraging the growth of still other plants that ultimately replace them. Pine woods give way to oak, and in some places, oaks are replaced by beech–maple forest. In bogs, azaleas and tamarack trees follow mosses, to be replaced in turn by birches and spruce (Fig. 6.13). (Of course, as the plant life of an area changes, its animal life changes, too.)

Secondary succession   The process of **secondary succession** occurs in places where either human activity or natural events destroy or disturb ecosystems. Secondary

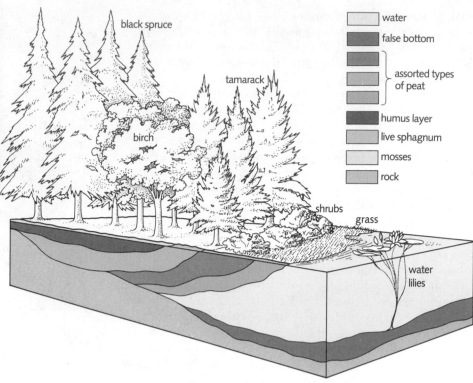

water
false bottom
assorted types of peat
humus layer
live sphagnum
mosses
rock

black spruce
tamarack
birch
shrubs
grass
water lilies

Figure 6.13 LEFT: This typical northern peat bog shows a number of plants ranging from floating sphagnum moss and small shrubs, to spruce and other trees in the background. RIGHT: This bog in cross section shows a view of primary succession "frozen in time." The open bog is to the right, the forest to the left. Over many years, the open area will slowly fill with peat and vegetation, and the floating mosses and shrubs will encroach on the open water. As the bog fills in, larger shrubs and trees will follow.

Figure 6.14 When storms damage the canopy of tropical rain forests, as the great hurricane of 1989 did to this patch of rain forest in Puerto Rico, secondary succession begins almost immediately. In this photograph, taken 6 months after the storm, sun-loving, opportunistic plant species have already begun to grow. Over time, succession will restore a mature rain forest to this area.

succession regularly occurs in the United States, for example, when a plowed field is abandoned. As the field lies fallow, it is colonized first by annual and perennial grasses and small herbs. Over the years, its flora progresses through small shrubs and trees to larger trees and, ultimately, mature forest.

Secondary succession often occurs without human interference—thanks to fires, severe storms, outbreaks of disease, or even volcanic eruptions. These events, as described in Chapter 3, can destroy mature forest trees, creating openings in the canopy that allow sunlight to reach the normally shady forest floor (Fig. 6.14).

Along rocky coastlines, storm-driven waves can scour the shore of its normal cover of algae or mussels, exposing bare rock.

In some types of secondary succession, disturbance knocks the system back to a relatively pristine state. In such cases, secondary succession proceeds by the same mechanisms as primary succession; early colonizers pave the way for later arrivals. (One could argue that because Rakata is a remnant of Krakatau, what happened there was really secondary succession, but quibbling over semantics isn't our point here.) In other cases, such as the recolonization of old fields or the regeneration of a rain forest canopy after storm damage, the seeds of both early and late colonizers may already be present and ready to grow. In these situations, the first plants to appear may simply be those that grow fastest. Their early advantage is ultimately lost, however, as slower-growing but taller shrubs and trees overgrow and shade the field.

Climax communities: Stability and replacement For years, many ecologists felt that succession in any particular area always proceeded in a predictable fashion towards a specific climax community that remained stable over long periods of time. Succession in one region, for example, would always lead to pine woods; in another area, the climax community would invariably be a beech–maple forest.

More recently, however, it has become clear that chance plays a major role in many forms of succession. Different organisms reproduce at different times of year, and the seeds or eggs of one colonizing species may win out over those of another just because they happened to arrive in a disturbed area first. Whether a storm-scoured rocky shore ends up covered with barnacles or mussels, for example, may depend on the season of the storm.

Furthermore, many so-called climax communities are disturbed so often that they can't really be called stable. We've mentioned several such situations here in Part 2; lightning sporadically sets forest and prairie fires; storms knock patches out of rain forests; periodic hurricanes cause local damage to parts of coral reefs. Each of these disturbances decimates some species and encourages the growth of others. Fairly common, localized disturbances of this type create such a patchwork of early, middle, and late successionary stages that some ecologists have begun to question whether the concept of climax has any real validity.

Finally, although succession is driven by biological events, it is profoundly shaped by the environmental conditions described in Chapters 3 and 4. You may have noticed, for example, that we've presented two dramatically different stories about how tropical rain forests respond to disturbance in different places. And you may well have wondered why succession on Rakata proceed-ed so smoothly, while clear-cut Amazon forests don't regenerate.

The reason lies partly in the specific identity of the ecological communities involved, but largely in the physical and climatic differences between the parts of the Indonesian archipelago and the Amazon basin. In central Amazonia, soils are nutrient-poor and shallow, and weather patterns are strongly influenced by the surrounding terrestrial environments. Removal of the living forest in that region therefore interrupts both nutrient and water cycles, stressing the ecosystem so severely that it cannot recover. On Rakata, by contrast, the eruption that destroyed previous life left behind incredibly nutrient rich soil. (Volcanoes in and around Java are known for their chemically basic ash that is rich in nitrogen, phosphorus, calcium, and magnesium.) In addition, Rakata is but a speck of land surrounded by ocean; the loss of its patch of rain forest had little effect on abundant monsoon rains that inundate the region from December through April.

ISLAND COMMUNITIES: SPECIFIC STORIES WITH GENERAL LESSONS

As you will see in Part 3, islands are favorite subjects for ecologists and evolutionary biologists—not just because they happen to be nice places to visit, but because they are so much smaller and more geologically youthful than major continents. As a result, islands support smaller, simpler living communities, in which interactions and trends can be easier to spot. We will talk about the importance of islands to evolutionary theory in subsequent chapters, so here we will deal with only one lesson learned from islands—the relationship between habitat size and species diversity.

The Area Effect

Back in 1963, widely traveled researchers Robert MacArthur and E. O. Wilson noticed that larger islands supported more species than smaller ones. That relationship between island size and species diversity, often called the **area effect,** is shown for several Caribbean islands in Fig. 6.15. As a rough estimate, the number of species an island can hold doubles for every tenfold increase in island area. Since the area effect was first proposed, it has been shown to hold true on virtually all island groups studied, from the compact British Isles to the thousands of islands in the far-flung Indonesian archipelago around Rakata.

Some explanations for the area effect are intuitively obvious. For a start, the larger an island is, the more

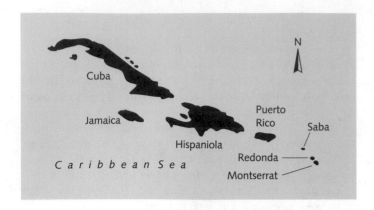

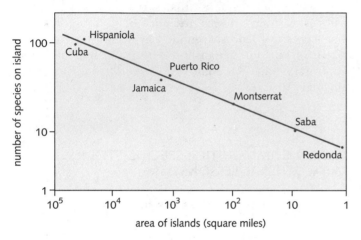

Figure 6.15 Large islands support a greater number of species than smaller islands, all other things being equal. This graph shows the diversity of reptiles and amphibians on seven similar, adjacent Caribbean islands of various sizes. In accordance with the area effect, a reduction in island size of 90 percent correlates with a 50 percent loss of species. [Adapted from Wilson (1992), Harvard University Press.]

diverse its physical habitats are likely to be. Thus, large islands would be expected to offer more distinct kinds of habitats that can support more species.

Other explanations are more subtle and depend on understanding the importance of population biology as discussed in the previous chapter. From this perspective, the larger an island is, the more space—of any kind—it has. The more space available on an island, the larger the populations of a given species it can support. Why is that important? Because when facing the ups and downs of growth and decline, larger populations tend to last longer than smaller ones. If your luck changes for a while, Wilson explains in his recent book, *The Diversity of Life*, "you are less likely to go completely broke if you are rich at the start."

The Distance Effect

Another phenomenon of island life noticed by Mac-Arthur and Wilson is known as the **distance effect:** islands close to a large continent are likely to have more species than equal-sized islands located farther away. Once again, part of the explanation for this effect is obvious; the closer an island is to a rich source of potential colonizing species, the more of those species are likely to make the trip. A more complete explanation, however, is both more interesting and more complex.

Area, Distance, and Diversity: A Dynamic Equilibrium

As Wilson notes in his description of succession on Rakata, just because a species arrives on an island and survives for a time doesn't mean that it will persist. On the contrary, island species constantly arrive, survive, disappear, and reappear over time. On Rakata today, Wilson notes, "the community of species remains in a highly fluid state. . . . The reticulated python, recorded as recently as 1933, was not present in 1984–85. . . . Owls and flycatchers arrived after 1919 . . . while several old residents such as the bulbul and gray-backed shrike disappeared." Why did these changes occur?

Rakata's living community is too large and too complex to track with precision. So to study how species colonize and persist on islands, Wilson and colleague Daniel Simberloff performed an experiment that examined species diversity on more manageably sized islands over time. The study sites they chose were four isolated groups of mangrove trees called islets, each about 15 m across (Fig. 6.16a). After the researchers identified these study sites came the hard part—time-consuming biological surveys. "Crawling over each of the islets from mud bottom to the tops of the trees," Wilson writes, "examining every millimeter of leaf and bark surface, probing every crack and seam, photographing and collecting, we made as complete a list as possible of the insect and other arthropod species on the four islets."

Their survey results supported the distance effect. The islet closest to a large mangrove island carried the most species (43); the islet farther away carried the smallest number (25), with the two in the middle falling in between (Fig. 6.16b).

Then the researchers covered each islet with a tent, fumigated with a short-lived chemical that was lethal to insects but harmless to mangroves, and then removed the tent. In this way, they virtually eliminated all resident insects on the islets. But within days, insects started moving back and, within a year, reached virtually the same species diversity they held at the beginning of the experiment.

(a)

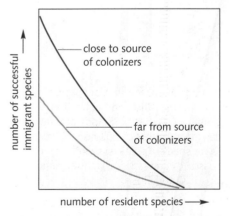

(b)

Figure 6.16 **(a)** In a landmark experiment aimed at investigating the distance effect on islands, E. O. Wilson and Daniel Simberloff chose a series of mangrove islets not unlike this one. Because the islets were manageably sized, researchers could first survey their insect faunas, fumigate to remove those faunas, and then resurvey the islets repeatedly to watch as species repopulated and reached equilibrium. **(b)** When Wilson and Simberloff surveyed the islets (shown here schematically), their results supported the distance effect; islets closer to a sizable island contained more species than those farther away.

The most interesting discovery, however, was yet to come. Simberloff and Wilson discovered that once their islands' insect populations reached equilibrium, the *number* of species on each island stayed constant over two years. That seemed reasonably straightforward. But the *identity* of those species varied from time to time. "The equilibrium was . . . dynamic," Wilson recalls, "with many of the arthropod species colonizing a given islet, vanishing after a month or two, then making a second appearance or else giving way to one or two similar species. The fauna tracked through time was kaleidoscopic, with the total number more or less balanced but the composition constantly changing, like travellers in an air terminal."

This, then, is a more realistic picture of island communities: species arrive from large continents at a rate affected strongly by distance and by the number of species that are already there (Fig. 6.17). Species present on an island tend to persist for varying lengths of time, depending on the size of the island and the number of species it already holds (Fig. 6.18). The total number of species we expect to find on an island at any one time, therefore, is

Figure 6.17 This graph shows that survival of immigrant species on an island is affected by the number of species already present and by distance from a source of colonists. The more species an island already has, the lower the probability is that new species will be able to gain a foothold. Islands closer to a source of new species, however, have consistently higher rates of successful immigration than similar islands farther away.

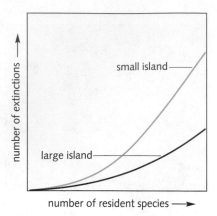

Figure 6.18 This graph shows that the rate of extinction on an island is affected both by the number of species present and by the island's size. The more species an island contains, the more likely it is that competitive interactions will accelerate extinction. Large islands are expected to have lower extinction rates for a given number of species than small islands.

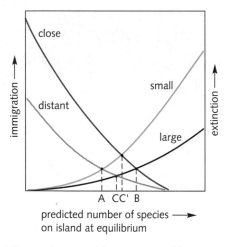

Figure 6.19 Predicting an equilibrium number of species on islands. By combining graphs of immigration and extinction, we can predict the relative number of species on islands of different sizes and different distances from sources of colonists. Small, distant islands will support fewer species (point A) than large, near ones (point B). Small, close islands and large, distant ones fall somewhere in between (points C and C').

the result of a balance between the rate of successful immigration and the rate of extinction (Fig. 6.19).

If this discussion sounds somewhat abstract and theoretical to you right now, hold your judgment until the next chapter. For although this work on island communities was originally done as "pure" or "basic" ecological research, it is turning out to have both unexpected and important applications in the rapidly growing field of conservation biology.

SUMMARY

Populations of organisms that inhabit the same environment and interact with each other are called communities. Interactions within communities typically control the success or failure of species in an ecosystem. An organism's ability to survive in a given area is determined by physical factors, biological factors, and the organism's behavior. These factors constitute the organism's niche.

Competition, both intraspecific and interspecific, is a potent force in controlling populations. Predation is another biological interaction that shapes communities, and predators along with parasites can affect the structure of communities in far-reaching ways. Positive interactions within communities include such symbiotic relationships as commensalism, in which one member benefits and the other is neither helped nor harmed, and mutualism, in which both partners benefit from the association.

Succession is the gradual process of environmental change and species replacement. Primary succession occurs in newly formed habitats; secondary succession occurs in places where either human activity or natural events destroy or disturb ecosystems.

Island communities are valuable subjects for studying the relationships between habitat size and species diversity. Both the area effect and the distance effect influence the diversity of species on an island community.

STUDY FOCUS

After studying this chapter, you should be able to:

- Explain competition, predation, and symbiosis.
- Trace some of the ecological effects of both positive and negative interactions among organisms.
- Understand how ecosystems change over time.
- Discuss the importance of islands in ecological studies.

TERMS AND CONCEPTS

REVIEW

Objective Questions (Answers in Appendix)

1. Which of the following describes primary succession?
    (a) natural reforestation of a burned-out forest
    (b) may go more rapidly because favorable conditions are already established
    (c) mosses growing on newly hardened lava rock
    (d) plants growing are less tolerant of harsh conditions

2. Which of the following describes secondary succession?
    (a) farmers abandon a field and move to the city
    (b) organisms that can tolerate harsh conditions
    (c) organisms growing on a lake formed by a recent earthquake
    (d) mosses growing on newly hardened lava rock

3. A climax community
    (a) is self-sustaining.
    (b) never changes.
    (c) is one in which growth proceeds in a predictable pattern.
    (d) is not likely to be disturbed by localized climatic changes.

Discussion Questions

4. What is a niche? What are some ways organisms divide their habitats into niches?

5. How can interactions between predator and prey result in density-dependent control of both populations?

6. How can interactions among organisms amplify apparently minor environmental changes into major shifts in population size?

7. What is succession? How does it exemplify biologically caused, local environmental change?

8. What interacting factors determine how many species an island is likely to support?

9. Give an example of how a change in the population of a keystone predator can have major consequences for an ecosystem.

READINGS

See integrated list of readings for Part 2 after Chapter 7.

7

Humanity and the Biosphere: A Dynamic Balance

*f*ar out in the Pacific Ocean, west of South America, lies Easter Island. Settled more than 1500 years ago, this tiny island supported a culture sophisticated enough to produce the haunting stone monuments that draw visitors from around the world (Fig. 7.1). Today, Easter Island is virtually uninhabited, its imposing culture is dead, its monuments abandoned, its civilization lost forever.

What happened to the Easter Islanders? Apparently, while building a civilization on a small island with limited resources, they made serious environmental errors. Clearing the island of forests, they eliminated the wood needed to build fishing canoes. Exposed to rain and wind, once-fertile topsoil washed into the sea and crops failed. Unable either to fish or to farm, islanders starved. Their culture might have collapsed in a storm of warfare, cannibalism, and slavery.

There is a lesson to be learned from this small island. As you've seen in earlier chapters, all organisms affect their physical and biological environments by consuming resources, producing wastes, and interacting with predators, prey, parasites, and symbiotic partners. You have seen examples of populations that grow quickly under good conditions, exhaust their resources, disrupt their environments, and suffer the consequences.

Could humans on larger "islands" (such as continents) suffer similar ecological misfortune? Surprisingly, yes. That has happened in places as diverse as the American Southwest, Central America, and the area once called "the fertile crescent" in the Middle East. Could similar difficulties affect the entire globe? Efforts to answer that question constantly stir a simmering brew of ecology, economics, and politics—and raise many more questions in the process.

ISLAND EARTH

Viewed from space, Earth is not a universe with endless horizons but a finite island of life in the desolate void of space. It is a huge island, and one endowed with considerable resources. But it is an island, nonetheless, and one that is teeming with *Homo sapiens.*

Thus far, we have triumphed. We have transformed Earth, converting forests and grasslands into fields and cities. Agricultural, technological, and medical advancements allow us to live longer and support more offspring than at any other time in history. Yet success brings environmental challenges. Despite crowding, human populations in many parts of the world are still growing exponentially. The activities of all those people affect the air we breathe, the water we drink, the soil we grow crops in, and the global climate. Collectively, human actions threaten the existence of millions of other species with whom we share the earth and with whose destinies our future is intertwined.

How has this happened? Our species, like every other, shapes its environment and is affected by that environmental change. In that respect, humans are no different from tropical leaf-cutter ants—which began cultivating mushrooms in underground gardens long before humans ever planted a seed (Fig. 7.2)—or rabbits—whose rates of reproduction are legendary. The same biological principles that shape all ecological interactions also govern *Homo sapiens*.

But in several crucial ways, our species *is* different. We have transformed local and global environments to a degree unmatched by any other species, and our intelligence enables us to recognize that fact. We can learn from past successes and failures and plan for the future. Unlike leaf-cutter ants, we can choose to grow different crops or decide to grow them in different ways if we find that certain agricultural practices are no longer viable. And unlike rabbits, we can influence the growth of our own population by voluntarily controlling birth rates, without waiting for disease or food shortages to force us to do so.

These uniquely human abilities to shape our future for the better are precisely what makes learning about human environmental issues so valuable. Successful application of these abilities to global environmental issues requires recognizing three important themes that recur throughout this chapter: the finite nature of Earth and its resources, the significance of exponential human population growth, and the practical, ethical, and aesthetic aspects of conservation.

HUMAN IMPACT ON THE BIOSPHERE

For just under 2 million years, small human populations manipulated their surroundings to provide life's necessities. As those populations grew, some of them succeeded in preserving their local environments (see Case Study: The Goddess and the Computer, p. 134), and in other cases they failed. Until recently, these environmental

Figure 7.1 These abandoned giant statues on Easter Island bear mute witness to the collapse of the ecosystem that once supported a sophisticated culture.

Figure 7.2 Tropical leaf-cutter ants may have been the natural world's first "farmers." They cut leaves from selected plants, carry them to special chambers within their nest, chew them into a pulp, and then use them to feed a fungus colony that they harvest for food. Though remarkably complex, this behavior is innate. Human farmers have much more flexibility in their agricultural pursuits—and should use that flexibility wisely.

Table 7.1 *Projected Population Size at Stabilization*

Country	Population in 1986 (million)	Annual Rate of Population Growth	Population at Stabilization (million)	Change from 1986
Slow-Growth Countries				
China	1,050	1.0%	1,571	+ 50%
Former Soviet Union	280	0.9	377	+ 35
United States	241	0.7	289	+ 20
Japan	121	0.7	128	+ 6
United Kingdom	56	0.2	59	+ 5
Former West Germany	61	− 0.2	52	− 15
Rapid-Growth Countries				
Kenya	20	4.2	111	+455
Nigeria	105	3.0	532	+406
Ethiopia	42	2.1	204	+386
Iran	47	2.9	166	+253
Pakistan	102	2.8	330	+223
Bangladesh	104	2.7	310	+198
Egypt	46	2.6	126	+174
Mexico	82	2.6	199	+143
Turkey	48	2.5	109	+127
Indonesia	168	2.1	368	+119
India	785	2.3	1,700	+116
Brazil	143	2.3	298	+108

SOURCE: World Bank, *World Development Report 1985*. New York: Oxford University Press, 1985.
NOTE: These data summarize population sizes and growth rates for selected countries and project the ultimate size of those countries when growth stabilizes. If these projections are correct, Nigeria will ultimately house more people than all of Africa did in 1986, and Kenya and Ethiopia will quintuple in size.

successes and failures were regional stories that left the biosphere at large unaffected. But today, human activity is affecting the physical and biological life-support systems of the entire planet.

Many Western environmentalists suggest that our assault on global ecosystems has been fueled by a Judeo-Christian worldview that encourages human domination of nature (see Chapter 2) and a scientific paradigm that believes technology will always be able to correct any problems that arise. But *all* dominant cultures—democratic, autocratic, and totalitarian; capitalist, socialist, and communist; Judeo-Christian, Muslim, Hindu, and Buddhist—have caused their share of environmental calamities. According to several philosophers, the problem can be traced to value systems within all these cultures which ignore humanity's roots in the natural world and overlook our dependence on local and global environments.

But this is all abstract talk. What tangible aspects of human activity produce our environmental problems? Over the past half century, in particular, four interrelated factors began to have multiplicative effects.

Human population growth, which continues in most parts of the world, is a driving force behind our species' growing ecological impact. More people need more food, consume more energy, and produce more wastes. As you learned in Chapter 5 (pp 94–96 and Current Controversies, Human Population Growth, p 97), global human population has grown exponentially over the past 2000 years, soaring from around 130 million to more than 5.4 billion persons today.

By some estimates, humans will number 8.5 billion by the year 2025 and possibly as many as 10 to 15 billion by the time growth levels off sometime during the middle of the twenty-first century. These numbers are so large that they are difficult to fathom; it is clear, however, that this incredible increase in human biomass is placing—and will continue to place—increasing demands on the planet and its resources (see Table 7.1 and Fig. 7.3).

Many ecologists argue persuasively that human population growth is the single most important factor behind current and future ecological problems. Why? Remember the models of population growth and carrying capacity from Chapter 5. Application of those models to the human situation suggests that each continent—and "island earth" as a whole—has some finite K for humans that places ultimate limits on human population size. As we approach that limit, many argue, the world will become more crowded, more polluted, less stable ecologically, and more vulnerable to disruption. But what is the world's K for humanity? That critical question lies at the heart of bitter and ongoing disputes between many ecologists, economists, and politicians today. Following the logic presented at the beginning of this chapter—that avoiding problems is easier than correcting them—many who follow ecologically informed reasoning recommend strong national and international efforts to control population growth.

Industrialization, the transition from an agricultural society to one based largely on manufacturing and technology, is a potent force for change. Among its benefits, industrialization provides an economic base to support many people at high standards of living. However, it also magnifies the environmental impact of human populations by enabling large-scale transformation of the landscape and by producing compounds that can be dangerous if handled improperly.

Economic growth, an increase in the total of goods and services produced nationally, is expected by all governments to rise fairly steadily. To keep the economy growing in this manner, individuals in countries such as the United States are encouraged to consume more resources. That encouragement continues despite the fact that we in the West already consume a disproportionate share of energy and other raw materials; a single baby born in the United States can have the equivalent ecological impact of more than 200 babies born in the Third World. That high per capita consumption keeps the ecological impact of countries such as ours growing significantly, even though our population growth is slower than that of "less developed" nations.

Over the past century, global economic production, as measured by an indicator called the gross domestic product, or GDP, has grown *15-fold*. Until recently, that growth was achieved by consuming resources largely without regard to future supply and by disposing waste products with relatively little regard to adverse environmental effects. That approach kept costs down and profit margins high—but only in the short term. Today, pollution is global in scale and cost, and past errors are catching up with us—as this chapter will show. A new breed of economic analysts now emphasize that resource depletion and pollution control must be included in figuring both the *real* costs of production and the *real* growth in GDP (Fig. 7.4).

Agriculture may be the only activity absolutely essential to the survival of humans in large groups. Its fundamental importance to human civilization cannot be

Figure 7.3 Many analysts urge developed countries to support family-planning programs in the Third World in order to head off ecological impoverishment that may hinder economic stability. Meanwhile, those nations address the problem in their own style. China's government, faced with severe population pressure, instituted the world's strictest birth control measures, encouraging families to limit themselves to a single child and prohibiting large families. In many areas, women are coerced (psychologically) into sterilization or abortion after bearing their first child. These harsh measures have slowed China's growth rate to 1.0 percent and helped agricultural production catch up with national food needs.

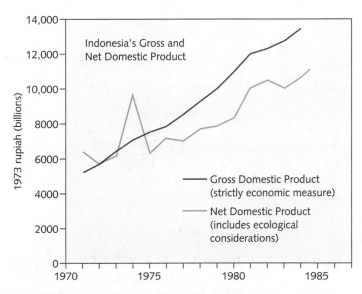

Figure 7.4 When economic growth is calculated according to such strictly exclusively economic indicators as consumption, savings, and investments, the figures are often misleading. The top line in this graph shows GDP for Indonesia as calculated in this manner. When degradation of natural resources—such as loss of soil fertility and deforestation—is taken into account, the result (bottom line) shows that a more realistic *net* domestic product is lagging behind.

Figure 7.5 Mercury released from a chemical plant in Minamata Bay, Japan, accumulated in the marine food chain and produced tragic debilitation and death in over 100 local residents who consumed large quantities of contaminated fish. Once the problem was recognized, residents warned, and mercury sources eliminated, "Minamata disease" disappeared.

understated, and the need to devote land to agriculture shapes our most significant interactions with the biosphere. In the 1950s, the need to produce more food from less farmland drove the development of high-productivity farming techniques that led to a "**Green Revolution,**" which transformed agriculture throughout the world and produced an unprecedented increase in world agricultural output. Not surprisingly, the Green Revolution was hailed as an unqualified success, and its techniques were implemented around the world. Its very success, however, has increased stresses on farmland to the point where penalties for misuse of resources are greater than ever. In addition, certain Green-Revolution techniques have caused some of our planet's most serious environmental problems, as you will see shortly.

CURRENT ENVIRONMENTAL PROBLEMS

Together, these four factors—population growth, economic growth, increased industrialization, and changes in agricultural practices—have created specific stresses on physical and biological environments. We will survey the most important of them and examine how researchers are learning to evaluate environmental stresses and mitigate their negative effects.

Water Pollution

No resource is more basic to human civilization than water; most great human cities are located near abundant sources of water that once were fresh and pure. Unfortunately, this vital resource is one of the first to be threatened by human activity.

Industrial pollution　Water may be polluted by industrial and household wastes. Certain industrial and agricultural chemicals, such as the insecticide **DDT,** break down slowly and are thus very long-lived in nature. They are also highly toxic to animals and humans. DDT and several other compounds, including **PCBs** (polychlorinated biphenyls), persist in the environment and accumulate in the food chain. Heavy metals such as cadmium, lead, mercury, and zinc can also be concentrated in the food chain and pose serious threats to human health. Even at low concentrations, lead causes neurological problems in young children, and mercury can cause serious brain damage in children and adults (Fig. 7.5).

It is hard to remove many such compounds from the environment. One promising line of research involves the discovery of bacteria that can break down PCBs and similar compounds, much as the microorganisms involved in nutrient cycles process animal wastes. Laboratory and field tests with these organisms show promise, but no commercially viable applications have been developed yet.

Surface pollution The rains that sustain life also wash pollutants from streets, fields, and farms into freshwater supplies. Fertilizer and animal waste from farms and ranches are major sources of nitrate pollution; and roads contribute oil, gasoline, and salt (from the salting of roads in winter).

Groundwater contamination Pollution of underground aquifers is frighteningly common; in the mid-1980s it was discovered that many towns across the nation were pumping water from contaminated wells (Fig. 7.6). For example, **TCE** (trichloroethylene), a suspected cancer-causing chemical, was discovered in 39 wells that supplied water to 13 cities in California's San Gabriel Valley. TCE contamination also forced the closing of municipal wells in several Massachusetts towns.

Residential sewage Humans, like all animals, produce wastes containing nitrogen and phosphorus that are normally processed by bacteria involved in nutrient cycles. On a small scale, these wastes can be absorbed into natural systems and can actually increase local primary productivity. But in large towns and cities, human wastes combine with other materials flushed into drains and sewers to produce *sewage* that carries bacteria and viruses into water supplies and can overload aquatic ecosystems with excess nutrients. Sewage-contaminated water is a major health hazard in Third World countries, where it can spread cholera and other potentially fatal forms of diarrhea. Furthermore, marine filter feeders (such as clams and mussels) concentrate bacteria and viruses in their guts, forcing the closing of thousands of acres of clam and mussel beds harvested for human consumption each year.

The damage caused by sewage is unfortunate, because readily available technology can solve many sewage-related problems. *Primary treatment* employs settling tanks and screens to filter out solid wastes. *Secondary treatment* uses huge bacterial cultures to degrade organic compounds, releasing nutrients in forms that algae and plants can utilize and generating enough heat to kill many dangerous microorganisms. *Tertiary treatment* employs algae or plants to "scrub" nutrients from wastewater; it releases a discharge pure enough to return to the environment. Regulations requiring tertiary sewage treatment can cause immediate improvement in local water quality (Fig. 7.7) and provide a clear demonstration that pollution problems can be solved—if sufficient funds are made available to construct and operate the necessary facilities.

Air Pollution

Our atmosphere is a dynamic mix of gases that interacts with plants, animals, soil microorganisms, and oceans.

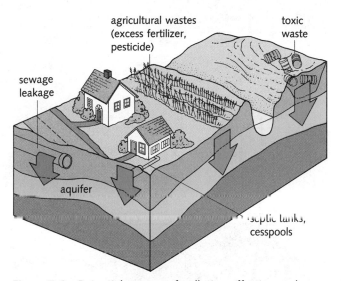

Figure 7.6 Potential sources of pollution affecting underground aquifers. Groundwater contaminants range from inorganic fertilizer salts, such as chloride and nitrate, to complex organic compounds and radioactive isotopes. Particularly insidious is the groundwater pollution that occurs when industrial chemicals such as TCE leak out of chemical waste dumps.

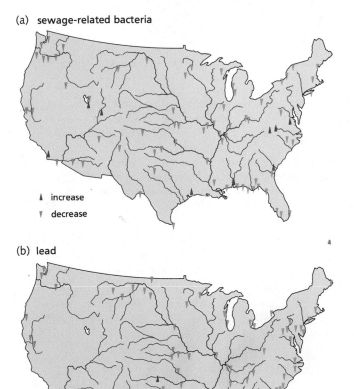

Figure 7.7 Trends in water contaminants at sampling stations from 1974 to 1981. Upward-pointing triangles indicate increase; downward-pointing triangles indicate decrease. Map **(a)**: sewage-related bacteria; map **(b)**: lead.

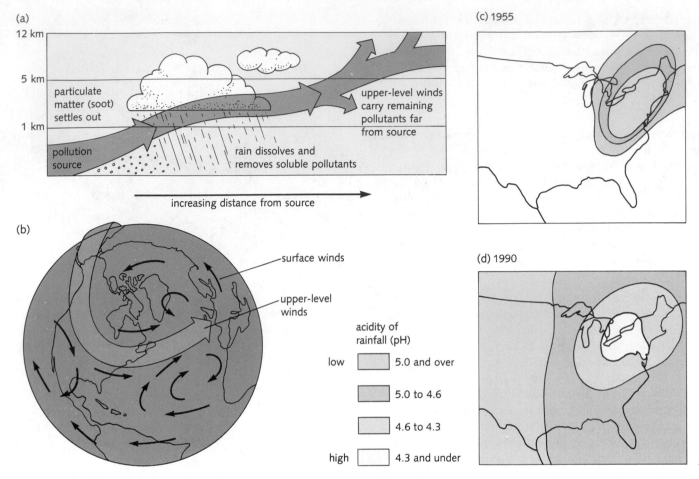

(a)

12 km

5 km

particulate matter (soot) settles out

1 km

pollution source

rain dissolves and removes soluble pollutants

upper-level winds carry remaining pollutants far from source

increasing distance from source

(b)

surface winds

upper-level winds

acidity of rainfall (pH)

low ☐ 5.0 and over

☐ 5.0 to 4.6

☐ 4.6 to 4.3

high ☐ 4.3 and under

(c) 1955

(d) 1990

Figure 7.8 Airborne pollutants can cross international boundaries. **(a)** While particulate pollutants generally settle close to their source, gases such as sulfuric and nitric oxides and chlorofluorocarbons often travel great distances. **(b)** Some regions are more seriously affected by acid rain because of wind and rainfall patterns and because of soil composition factors. Pollutants released in the Midwest, for example, are carried by winds in the upper atmosphere to Canada and the northeastern United States. Maps **(c)** and **(d)** show the increase in severity of acid rain since the 1950s. (Data collected by the National Atmospheric Deposition Program/National Trends Network.)

Human activity is changing the atmosphere in several ways faster than it has ever changed in the history of life on Earth. Although most forms of air pollution are linked to more than one source, we can divide them up into several categories for discussion purposes.

Smog   *Smog* is a catchall term used to describe a collection of airborne dust, smoke, particles produced by internal combustion engines, and a variety of gases. One particularly noxious component of certain types of smog is ozone. Ozone is vital where it naturally occurs in the upper atmosphere (as you will see shortly), but at ground level it poses serious health hazards to both animals and plants.

Acid rain   The burning of fossil fuels releases sulfur dioxide and nitrous oxides into the atmosphere, where winds carry them far from their sources (Fig. 7.8). When these compounds dissolve in fog or raindrops, they produce **acid rain** that contains sulfuric acid and nitric acid. Acid rain, along with other airborne pollutants, can damage the leaves of plants, can harm roots by releasing aluminum and other metals from some soils, and can interfere with bacterial decay in topsoil and humus. Acidification of surface water kills aquatic organisms from algae to fishes (Fig. 7.9). A number of lakes and streams in New England, eastern Canada, Scandinavia, and central Europe are classified as "dead" because of acid rain, and 14,000 Canadian lakes show serious damage. Nearly half

Figure 7.9 LEFT: Healthy lake trout from unpolluted waters. RIGHT: A lake trout suffering from the effects of acid rain (pH 5.0 in the lake).

of central Europe's forests already show damage so serious that Germans coined a word to describe it—*Waldsterben,* or "forest death."

Changes in the ozone layer Earth's upper atmosphere is surrounded by a layer of **ozone (O_3)** produced when oxygen molecules are split by ultraviolet radiation in sunlight. Ozone absorbs ultraviolet light, preventing much of that potentially dangerous radiation from reaching the earth's surface.

Unfortunately, a range of chemicals including chlorofluorocarbons (CFCs) like Freon (used in refrigerators and air conditioners) combine with ozone and break it down prematurely. Research flights over the Antarctic have shown that ozone levels are dropping at an alarming rate (Fig. 7.10). Recent evidence suggests that ozone depletion is not limited to the polar regions but is already spreading to the skies over densely populated temperate zones.

For every 1 percent loss of stratospheric ozone, 2 percent more ultraviolet light reaches Earth's surface. Ultraviolet light is such an important cause of skin cancer and cataracts in humans that the Argentine Health Ministry has advised residents of Patagonia (located beneath the expanding Antarctic ozone "hole") to avoid exposure to sunlight during September and October. And in southern Australia, children are legally required to wear wide-brimmed hats and neck scarves on their way to and from school.

A treaty to limit and slowly roll back the use of ozone-destroying chemicals in 43 nations was signed in 1987. Though the treaty failed to limit CFC use as strictly as some environmental analysts think necessary, it was a landmark step in international legal action on environ-

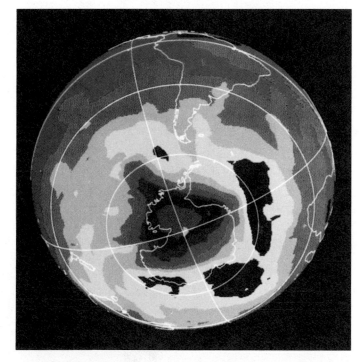

Figure 7.10 This false color image based on satellite data shows an unusually large and persistent ozone "hole" over Antarctica as it appeared on November 30, 1992—more than two weeks later than it was expected to break up. Such holes were restricted to north and south polar regions when first discovered but have begun spreading over populated areas in both hemispheres. Decreased ozone protection from powerful ultraviolet radiation means increased risk of skin cancer, cataracts, and immune system damage, as well as crop damage and decrease in productivity of marine algae.

mental issues. In 1992, the nations that had signed the 1987 treaty agreed that the scientific evidence for ozone depletion was so clear that the 1987 guidelines for phasing out these compounds were accelerated.

Carbon dioxide and global warming   Ever since the Industrial Revolution, humans have been conducting a global experiment on the role of carbon dioxide (CO_2) in regulating climate. For much of that time we were unaware of the experiment, but today there is no doubt what is taking place. Since 1900, the carbon dioxide content of Earth's atmosphere has risen from 280 parts per million to 350 parts per million. If the trend continues, by the middle of the twenty-first century human activity will have *doubled* atmospheric CO_2 in less than 250 years.

Why the rapid increase? Recall from Chapter 4 that atmospheric CO_2 is part of a complex global carbon cycle that involves both biological and geochemical pathways (see Fig. 4.18). Apparently, the burning of fossil fuels such as oil and coal and the clearing of global forests are returning carbon dioxide to the atmosphere faster than plants can remove it.

But why should this trend have concerned enough scientists to place the issue on the agenda at the 1992 United Nations Earth Summit in Rio de Janeiro? After all, CO_2 at such levels is not toxic to animals, and CO_2 is required by plants to carry out photosynthesis.

Recall from Chapter 4 that CO_2 in the atmosphere is a major factor in regulating our global "greenhouse" because it retards the loss of heat into space. With much less CO_2 in the atmosphere, Earth would be cooler than it is at present; with a great deal more of the gas around, evidence suggests that Earth could become much warmer. If that hypothesis is correct, increases in atmospheric CO_2

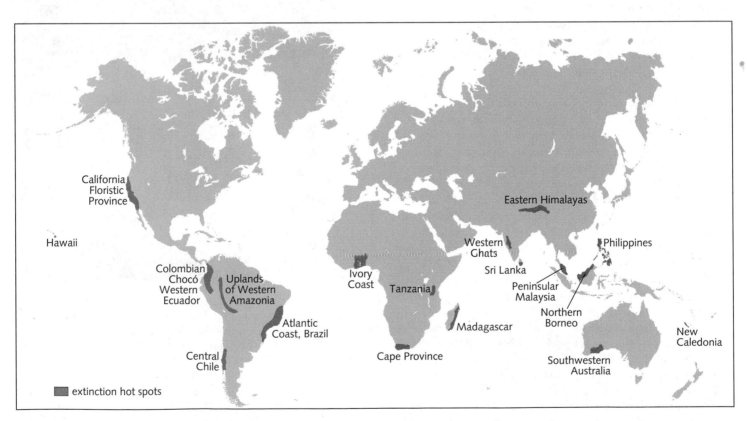

Figure 7.11   The threat of extinction is not spread evenly across all habitats but is concentrated in what some ecologists call "extinction hot spots"—places where many species live and human activity is altering natural environments rapidly. This map, based on studies that are still in progress, does not include certain types of forests, lakes, rivers, or coral reefs. Note that American hot spots cover much of our West Coast and the entire Hawaiian archipelago. [Adapted from Wilson (1992), Harvard University Press.]

could cause temperature and rainfall patterns to change throughout the world. Some agricultural areas could become too dry or too hot to support the crops grown there today. Sea levels could rise by several meters, flooding coastal cities from Miami to Bangladesh. Global warming could also seal the fate of numerous endangered species now restricted to isolated patches of protected natural habitat surrounded by fields, cities, and suburbs.

But both the global carbon cycle and the atmosphere's effects on climate are extremely complex, so predictions about the greenhouse effect are hampered by scientific uncertainty. Most climatologists agree that atmospheric CO_2 levels will double by the middle of the next century, and that this doubling may increase average global temperatures from as little as 1°C at the equator to as much as 5°C at the poles. Yet other scientists argue that a slight rise in global temperature will produce more clouds that could *prevent* excessive warming by reflecting heat out of the atmosphere.

At this time, uncertainties about the magnitude of global warming and the best way to combat it have prevented the kinds of international treaties now in place to protect the ozone layer. As our understanding of atmospheric chemistry increases, so will the chances for action to deal effectively with this issue.

Biological Damage: Habitat Destruction and Loss of Diversity

Human activity is now a major cause of extinction of species on this planet. It is difficult to state with any real certainty what current global extinction rates are, because many of the species that are disappearing haven't even been cataloged yet! By studying the recent loss of species in certain threatened ecosystems such as tropical rain forests, some researchers have estimated that by the year 2000, nearly 1 million species will have probably vanished forever, and that somewhere between one-third and one-half of the world's species will have disappeared within 500 years (Fig. 7.11). How does this square with great extinctions of the past? Quite simply, no extinction of this magnitude has occurred for 65 million years.

Some experts project that if **deforestation** continues at current rates, the world's last fully functioning rain forest ecosystems may be destroyed by the beginning of the next century (Fig. 7.12). Unfortunately, deforestation is also occurring in the United States. In the Pacific Northwest, logging of old-growth forests is a major bone of contention between environmentalists and lumber companies. Why should we care? As a practical matter, we should care because inhabitants of both tropical and temperate forests may be useful to us.

In their struggle for existence, many organisms have evolved chemicals that protect them from enemies.

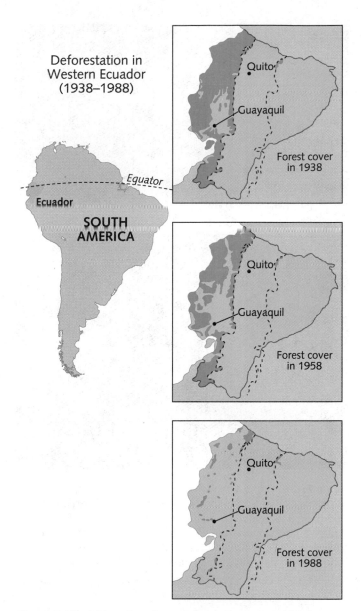

Figure 7.12 This series of maps shows the extent of deforestation in a single South American country over a 50-year period. Note that more than 90 percent of that nation's forests have already been destroyed. Note also that the studies in island biogeography discussed in the previous chapter suggest that this reduction in area has either eliminated or doomed more than half of those forests' species. [Adapted from Wilson (1992), Harvard University Press.]

Figure 7.13 The rosy periwinkle, a plant native to Madagascar, contains two compounds now used in chemotherapy for Hodgkin's disease and several other forms of cancer, including acute lymphocytic leukemia. This plant was on the brink of extinction when its therapeutic properties were discovered. Worldwide sales of these drugs now top $100 million each year.

Humans have long used such compounds in food, medicine, and industry. In fact, more than half the drugs in use today were discovered in wild plant species (Fig. 7.13 and Table 7.2). One of the most interesting new drugs with potential anticancer activity, for example, is taxol, an anticancer drug extracted from the Pacific yew tree, *Taxus brevifolia.* This yew, native to old-growth forests of the Pacific Northwest, was considered a "trash tree" by the lumber industry until the experimental value of taxol was established.

No one knows how many more species like this one contain similar treasures; as every unstudied specimen disappears forever from the face of the earth, its potential applications for human welfare vanish also. How large must pieces of natural habitat be to preserve their flora and fauna? In many cases, we simply don't know, but some researchers are already performing experiments designed to find out (see Case Study: Islands of Life, p. 138).

Agriculture: Sustaining What Sustains Us

Large-scale agriculture as practiced in much of the world today was born during the 1950s, when researchers launched the Green Revolution and dramatically increased world agricultural output. New land was brought under cultivation, new strains of high-yielding crops were bred and distributed, and chemical-intensive techniques were introduced. The short-term *economic* benefits were immediate and dramatic. Yields per harvest increased, and more crops were grown each year. Between 1950 and 1970, wheat production in Mexico increased tenfold, while China and India became self-sufficient in food production for the first time in recent history. In many regions, harvests continue to increase, and prices for grain, rice, and other commodities are stable or falling.

The Green Revolution's central principles are simple: Plow large fields, plant high-yielding crops, irrigate to minimize dependence on rainfall, maintain soil fertility with chemical fertilizers, and control pests with insecticides. The high yields produced by this technology use less land and reduce food cost, but they put greater stresses on the long-term productivity of the land and often have serious environmental repercussions.

One of the most important challenges facing agricultural science today is to make certain that agricultural techniques can be sustained with minimal ecological damage over the long term. As large-scale agriculture is currently practiced today, it seriously impacts several resources upon which modern farming depends, including biological diversity, soil, water, and energy.

Biological diversity and agriculture At the core of Green-Revolution agriculture is a technique called **monoculture**—the planting of large areas with a genetically uni-

Table 7.2 *Some of the Drugs in Common Use Derived from Plants and Fungi*

Drug	Plant Source	Use
Atropine	Belladonna (*Atropa belladonna*)	Blocks transmission of nerve impulses in part of nervous system
Bromelain	Pineapple (*Ananas comosus*)	Controls tissue inflammation
Caffeine	Tea (*Camellia sinensis*)	Stimulant, central nervous system
Camphor	Camphor tree (*Cinnamomium camphora*)	Increases blood flow to local areas of the skin
Cocaine	Coca (*Erythroxylon coca*)	Local anesthetic
Codeine	Opium poppy (*Papaver somniferum*)	Pain relief
Colchicine	Autumn crocus (*Colchicum autumnale*)	Anticancer agent
Digitoxin	Common foxglove (*Digitalis purpurea*)	Cardiac stimulant
Diosgenin	Wild yams (*Dioscorea* species)	Source of female contraceptive
L–Dopa	Velvet bean (*Mucuna deeringiana*)	Parkinson's disease suppressant
Ergonovine	Smut-of-rye or ergot (*Claviceps purpurea*)	Control of hemorrhaging and migraine headaches
Glaziovine	*Ocotea glaziovii*	Antidepressant
Gossypol	Cotton (*Gossypium* species)	Male contraceptive
Indicine *N*–oxide	*Heliotropium indicum*	Anticancer (leukemias)
Menthol	Mint (*Menta* species)	Increases blood flow to local areas of the skin
Monocrotaline	*Crotalaria sessiliflora*	Anticancer (topical)
Morphine	Opium poppy (*Papaver somniferum*)	Pain relief
Papain	Papaya (*Carica papaya*)	Dissolves excess protein and mucus
Penicillin	Penicillium fungi (especially *Penicillium chrysogenum*)	General antibiotic
Pilocarpine	*Pilocarpus* species	Treats glaucoma and dry mouth
Quinine	Yellow cinchona (*Cinchona ledgeriana*)	Antimalarial
Reserpine	Indian snakeroot (*Rauvolfia serpentina*)	Reduces high blood pressure
Scopolamine	Thornapple (*Datura metel*)	Sedative
Strychnine	Nux vomica (*Strychnos nuxvomica*)	Stimulant, central nervous system
Taxol	Pacific yew (*Taxus brevifolia*)	Anticancer (especially ovarian cancer)
Thymol	Common thyme (*Thymus vulgaris*)	Cures fungal infection
D–Tubocurarine	*Chondrodendron* and *Strychnos* species	Active component of curare; surgical muscle relaxant
Vinblastine, vincristine	Rosy periwinkle (*Catharanthus roseus*)	Anticancer

SOURCE: E. O. Wilson, *The Diversity of Life*. Cambridge, MA: Harvard University Press, 1992.

Figure 7.14 New disease strains find uniform fields easy pickings. The corn blight that struck the United States in 1970 proved resistant to every defense modern agriculture could muster. Once a field was infected, nothing could eradicate the disease, and roughly 15 percent of the crop was destroyed. Future blights could be worse.

In response, crop geneticists introduce new, disease-resistant strains every 10 to 15 years. But disease-resistant genes must be found in wild strains and bred into hybrid stock or transferred by genetic engineering. For that reason, wild relatives of domestic crops, though seldom grown for food today, are invaluable to agriculture's future. Unfortunately, habitat destruction makes these wild populations harder to find every year, and farmers in remote parts of many countries who once grew crops from their own genetically diverse seed stock are switching to higher-yielding hybrid seeds. The situation threatens wild relatives of corn in Mexico and Central America, rice in Southeast Asia, cereals in Turkey, and barley in Ethiopia.

Chemical fertilizers Fertilizers play an essential role in the Green-Revolution approach, because planting the same crops year after year exhausts even the richest soils. While fertilizers are not toxic in moderate doses, excessive, long-term use can cause problems. Fertilizers inhibit the growth of nitrogen-fixing organisms in the soil and stimulate organisms that export oxygen from the soil to the atmosphere. Excess fertilizer is often washed into streams and lakes, where it can cause algal blooms and other problems associated with eutrophication. In addition, certain major constituents of fertilizers, such as nitrates, can make both surface water and groundwater unfit for human consumption.

Pesticides Chemical compounds aimed at controlling pests and diseases are essential to Green-Revolution farming techniques. When farmers set out large populations of identical plants, any insects (or bacteria or fungi) which feed on those plants discover a resource bonanza. As a consequence, pest and pathogen populations grow rapidly and can cause ruinous damage. In addition, insect populations regularly exposed to pesticides invariably develop resistance to them. Thus farmers must escalate their offense with a stream of new and ever-more-potent chemical weapons that pose a risk to humans and wildlife.

Unfortunately, if pesticides persist in the environment, they can be picked up by plants and passed along through food chains. Even if they have no ill effects on wildlife in the concentrations at which they are applied, such toxins can accumulate in the fatty tissues of animals on higher trophic levels (Fig. 7.15). This process, called **biological magnification,** is not restricted to pesticides (it also occurs with industrial waste products such as PCBs), but it was first noticed because of damage done to wildlife by DDT.

Irrigation The process of irrigation, the provision of water when rainfall is insufficient, seems innocuous—and it can be, under the right circumstances. But when large-

form, single-species crop chosen for high yields and rapid maturity. Yet despite the fact that most farming involves the cultivation of just a few species in this manner, the future of agriculture depends on one of the earth's most valuable natural resources—biological diversity.

Modern crops have been derived from wild species through selective breeding for improved yield, strength, and ease of cultivation. Their extraordinary success, however, carries hidden dangers. Why? Because a handful of top-ranking, highly inbred strains are *so* productive that they are planted to the exclusion of almost everything else. Almost 70 percent of American corn, for example, is derived from no more than five hybrid lines. As a result, crop plants covering hundreds of square miles are nearly identical genetically, so new diseases can spread rapidly (Fig. 7.14).

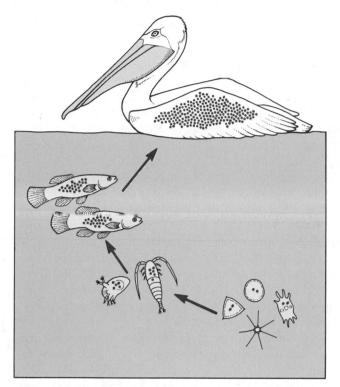

Figure 7.15 Concentration of toxic compounds through biological magnification. Pesticides such as DDT and heavy metals such as lead and mercury are neither eliminated nor broken down by animals that eat them. Though present in the environment in small amounts, these pollutants are concentrated in food webs, accumulating in the fatty tissues of animals on higher trophic levels. Though they rarely kill organisms outright, some compounds damage reproductive organs and cause sterility, while others cause cancer.

Figure 7.16 Although Costa Rica is known and widely respected in Central America for its ecological preservation programs, deforestation has occurred in several places throughout the country. Here, in what was once a mountain forest in Monteverde, soil erosion (lower left of photo) has begun to take its toll.

scale farming is extended into arid regions, irrigation places heavy demands on local water supplies, and dissolved salts in irrigation water applied to desert soil can cause serious difficulties (see Case Study, Death of an Inland Sea, p. 132).

Soil erosion Erosion, the removal of soil by wind and water, is more or less balanced in natural ecosystems by the creation of new soil and humus. But intensive monoculture often leaves large tracts of soil surface exposed to wind and water between crops, causing serious and permanent increases in erosion. Loss of soil causes steady deterioration of croplands covering nearly 35 percent of Earth's land surface. Fields on the high plains of the American Midwest lose an average of 20.9 tons of topsoil per acre per year to erosion. In the mountains of

Guatemala, where crowding pushes farmers into marginal areas, some fields lose 770 tons of soil per acre each year. Heavy fertilization cannot compensate for this loss of nutrient-rich topsoil even when affected areas are relatively few in number and small in size, as they are in Costa Rica (Fig. 7.16).

Energy use The consumption of fossil fuels such as coal and oil increases dramatically when Green-Revolution techniques are introduced. Rice production in the United States, for example, requires 10 times as much energy as it does in the Philippines, and American wheat farmers require 1000 times as much energy as Indian farmers to produce a unit of grain. Careful management of energy resources is especially important to modern farming.

CASE STUDIES IN HUMAN ECOLOGY

The foregoing ecological problems arise from human activity in the biosphere. Yet as you learned earlier, ecosystems are complex enough that it is difficult to predict what the effect of any given human activity will be. The four case studies that follow examine specific situations that illustrate the range and diversity of ecological issues and some of the ways that an understanding of basic biology can be used to solve them.

DEATH OF AN INLAND SEA

The town of Muynak, located in northwestern Uzbekistan in the former Soviet Union, was once a busy fishing port that supplied the Soviet Union with 3 percent of its total fish catch. Muynak was located on the Aral Sea, a productive body of saltwater only slightly smaller than Lake Superior that supported 24 native fish species (Fig. 7.17a).

To Soviet economic planners of the 1930s, the fresh water that flowed through the central Asian desert into the Aral was an irresistible target. Government planners decided that millions of gallons of river water should be diverted from the Amu Darya and Syr Darya rivers in order to irrigate millions of acres of arid land. True to their hopes, the desert bloomed; before long 90 percent of the cotton produced by the Soviet Union (along with substantial quantities of rice and melons) was grown on this irrigated land. But something went terribly wrong.

The first signs of trouble surfaced in the 1960s, when the water level of the Aral Sea began to drop. In the years since, there have been very few good fishing days in Muynak. Waves that once lapped at Muynak's piers are now nearly 25 miles away—separated from the town by parched, salt-encrusted desert (Fig. 7.17b). Since 1960, the sea has lost a volume of water greater than 1.5 times the total content of Lake Erie and has shriveled to barely 50 percent of its former size. The water that remains is so salty that the local fish have died.

What happened? So much water was diverted for irrigation that not enough was left to replace the water lost by evaporation, and the sea began to shrink—and to die.

The transformation of the Aral has had other ecological effects as well. The sea is no longer large enough to

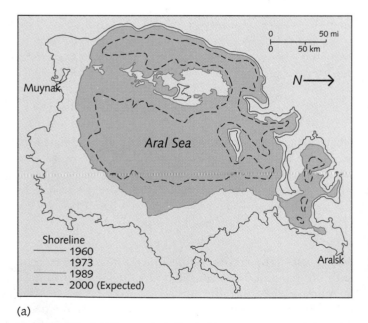

(a)

(b)

Figure 7.17 **(a)** This map of the Aral Sea compares its former borders with its current size. The Aral Sea once covered 67,000 square kilometers (26,00 square miles). **(b)** When the Aral Sea began to shrink, fishers attempted to keep ahead of falling water levels by digging a navigation canal. Ultimately, they gave up; these boats, stranded in what was once part of that canal, are now miles from both the former port and what is left of the Aral.

moderate local climate, so local winters have become colder and summers have become hotter. In addition, hundreds of square kilometers of sandy salt flats have spawned a new meteorological phenomenon: salt storms. Each year, strong winds carry 45 million tons of salt and sand from the former seabed, stinging inhabitants' eyes, clogging the carburetors of their cars, adding salt to farmland, and possibly contributing to a significant increase of respiratory and eye diseases and throat cancer.

To make matters worse, many areas that "bloomed" under irrigation are no longer suitable for farming. The water used for irrigation (like virtually all surface and groundwater) contains dissolved mineral salts. In addition, desert soils, because they are not leached by rainwater, have a higher salt content generally than the soil of grasslands or forests. Normally that salt concentration is low enough to be harmless. But irrigation in arid regions dissolves salts from soil and brings them to the surface by capillary action. That salt, combined with salt left behind as irrigation water evaporates, and produces extensive deposits at the surface. The result is **salinization** that prevents many crops from growing. In addition, local drinking water is now contaminated with salt, pesticides, and other chemicals used on the fields.

Discussion

The specifics of the Aral Sea story are unique, and the magnitude of the disaster is yet unequaled. The general points to be learned from this case, however, are universal.

First, the Soviet government undertook this project for sound economic reasons: it saw cotton as an essential source of hard currency. The error was not in short-term economic goals but in ignoring long-term ecological considerations. This thinking is often typical in the genesis of major environmental problems and is part of what makes them so difficult to rectify.

Second, the problem is far more difficult to *solve* than it would have been to *avoid*. The only way to restore the Aral now is to curtail irrigation. But this means taking huge chunks of farmland out of cultivation. What then would happen to the people now farming those 18 million acres? Devising an ecologically, economically, and politically acceptable compromise will be a formidable task.

Related Issues Elsewhere

The lessons of the Aral are not confined to central Asia. Only about one-seventh of American farmland depends on irrigation, but that land provides a disproportionately large share of our annual harvest. Farmers across much of the Midwest depend on aquifers composed mainly of fossil water (Fig. 7.18; also see Chapter 4, p. 76). Until recently, this water was pumped from the ground much faster than it was being replenished; aquifer levels had dropped between 23 and 40 percent by 1980.

Facing predictions that wells would soon run dry, farmers responded. Some now use advanced irrigation techniques to conserve water (see the solutions that follow), and others have planted some cropland in native grasses to allow the aquifer to recover. Happily, the water table in many places is dropping only one-third as rapidly as it was in the 1970s. Still, the long-term outlook for irrigation in the area remains uncertain. Similar problems in a different setting face Californian farmers who, like their Uzbeki counterparts, depend on government-sponsored river diversions for irrigation in the arid yet fertile Imperial Valley.

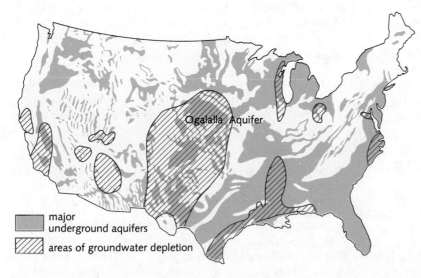

Figure 7.18 This map shows the distribution of the 134,000-square-mile Ogalalla Aquifer, a major fossil aquifer that varies in thickness from a few meters to more than 200 m.

Ogalalla Aquifer

major underground aquifers

areas of groundwater depletion

Solutions

Because agriculture is essential, we must try to improve the efficiency and ecological sensitivity of water use in agriculture and avoid the sorts of conditions that are destroying the Aral. In southern California and in much of the Middle East, researchers and farmers are developing water-conserving irrigation techniques. One such approach is *drip irrigation,* which uses pipes to provide water slowly, steadily, and directly to plant roots.

In other cases, agricultural researchers are rethinking policies that encourage production of water-hungry crops, such as cotton and rice, on arid desert land. Researchers are currently testing a host of little-known "alternative" crops that thrive without irrigation. One example is Kallar grass, a plant used in Pakistan for animal fodder, that can be grown in full-strength seawater. Another is the Tamarugo tree of Chile's Atacama Desert, which is accustomed to growing in places where rain may not fall for seven years at a stretch, and yet is still able to produce leaves that can be used as animal feed. These experiments in progress are prime examples of the need for biological diversity discussed earlier.

Solving water problems will require courage, creativity, and access to plant biodiversity. But above all, it will require intelligence and foresight to prevent the overuse of water resources that led to the Aral tragedy.

CASE STUDY

THE GODDESS AND THE COMPUTER

On the Indonesian island of Bali, a high priest invokes the blessing of Dewi Danu, the water goddess, as he assigns water allotments, irrigation schedules, and planting dates to farmers. This conjunction of religion and agriculture is taken for granted among the Balinese. Nature gods and natural rhythms of growth, maturation, and decay shape Balinese rituals, their social life, and the ways they make their living. That's why all obey the priest's instructions; to plant at the wrong time or to take too much water from communal irrigation ditches would be disrespectful to Dewi Danu.

The priests, who have worked with farmers for centuries, have devised a system that coaxes abundant crops from ancient terraces that were first cultivated at least 1000 years ago (Fig. 7.19a). Amazingly, productivity has been maintained over all that time without any need for either chemical fertilizers or pesticides. The system works well because the priests use their authority to encourage a style of farming that exploits the ecological attributes of traditional rice strains and several "companion species" which thrive in rice paddies.

The logic and organization of this system are the subject of ongoing studies by anthropologist Steven Lansing and biologist James Kremer, both of the University of Southern California. Their research to date has revealed that the system works as follows. Traditional strains of rice are cultivated in flooded paddies, where plants mature in about 210 days. In the water around the rice, nitrogen-fixing bacteria supply a constant trickle of nitrogen—the main limiting nutrient in the island's otherwise rich volcanic soil. Frogs, ducks, and freshwater eels live in the paddies, where they feed on insect pests and leave nutrient-rich droppings behind. All three of these animals are harvested by farmers, so they provide protein in the Balinese diet (Fig. 7.19b).

At harvest time, only the rice-bearing tassels are removed; the remainder of the rice plants are plowed into the soil. Paddies are then planted with a dry-ground crop whose roots aerate the soil and reach deeply to retrieve nutrients flushed down by flooding. This crop rotation, practiced in a coordinated manner across the island, conserves water and preserves soil nutrients and texture.

In addition, priests control the timing of fallow and crop rotation periods in a manner which ensures that sizable areas are taken out of rice cultivation on a rotating schedule (Fig. 7.19c). The result is an agricultural patchwork that discourages rice pests and diseases, because most rice pests and diseases can survive *only* on rice plants, and travel relatively slowly. Because of this crop-rotation system, as soon as pests begin to reach destructive levels in any one spot, their only source of food is removed, and their populations crash.

Some years ago, however, in efforts to increase yields, Green-Revolution agriculturists introduced a dwarf, high-yielding rice strain that grows fast enough to permit two or three successive crops each year. Soon, farmers began to double- and triple-crop that new "miracle rice," planting a new batch of rice as soon as possible after previous crops were harvested. True to Green-Revolution form, rice yields rose dramatically for a few years. But then, several unexpected things happened.

- The fast-growing rice needed more nitrogen, and needed it more quickly than the nitrogen-fixing bacteria could provide. For the first time, farmers had to buy and apply chemical fertilizers to their fields.

- With genetically identical rice growing everywhere all the time, pest populations skyrocketed. Ducks and other natural predators couldn't keep up with them, so farmers had to buy and apply pesticides.

- Although the agricultural chemicals controlled pests, they also either killed eels and frogs or accumulated in their flesh, making them unfit for human consumption. With pesticides in the paddies, farmers could no longer allow ducks intended for food to graze there.

(a)

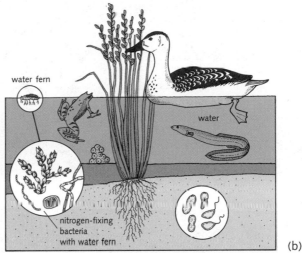

(b)

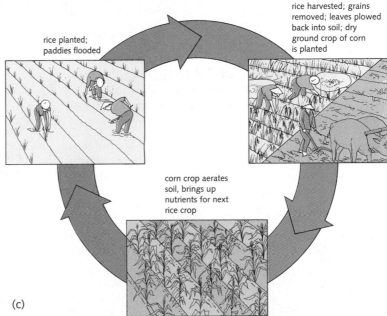

rice planted; paddies flooded

rice harvested; grains removed; leaves plowed back into soil; dry ground crop of corn is planted

corn crop aerates soil, brings up nutrients for next rice crop

(c)

Figure 7.19 **(a)** In the traditional Balinese rice paddy ecosystem, rice plants grow in flooded paddies, maturing in about 210 days. **(b)** Nitrogen-fixing bacteria growing in association with water ferns supply fertilizer. Frogs, ducks, and freshwater eels in the paddy provide protein in the Balinese diet, feed on harmful insects, and leave nutrient-rich droppings. **(c)** In a typical rice paddy, a community of plants and animals form a self-perpetuating system that thrives under Balinese management practices.

- Even with steady application of chemical fertilizers, rice yields declined after a few years. Paddy soil hardened and became difficult to cultivate.

Weighing yields against the added cost of chemicals, and factoring in Balinese preference for more flavorful native rice, many farmers returned to their homegrown system. But Bali's priests and farmers are resourceful, reject simplistic all-or-nothing thinking, and are impressed by modern agricultural science. The high priest mentioned earlier is currently working with American anthropologists and ecologists, helping them to construct a model of his irrigation system on portable computers. Although the thought of a high priest manipulating a Mac-

intosh between ritual ceremonies may seem incongruous to some, it bodes well for the future. With luck, the combination of two complementary views of agriculture and ecology will produce higher, more consistent yields than either method alone.

Discussion

The Balinese situation is unique in certain ways; the island is blessed with ample rainfall, fertile soil, and a cohesive culture. Still, this agricultural system has remained productive because it incorporates a functional understanding of ecological principles.

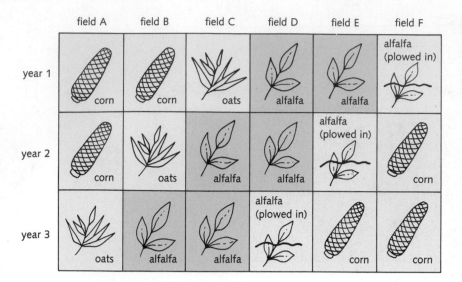

	field A	field B	field C	field D	field E	field F
year 1	corn	corn	oats	alfalfa	alfalfa	alfalfa (plowed in)
year 2	corn	oats	alfalfa	alfalfa	alfalfa (plowed in)	corn
year 3	oats	alfalfa	alfalfa	alfalfa (plowed in)	corn	corn

Figure 7.20 In the United States, intensive, repetitive monoculture is the rule on large farms, but it does not need to be. *Crop rotation* involving nitrogen-fixing plants such as alfalfa improves soil fertility by adding both organic content and nitrogen.

The Balinese system is a prime example of **sustainable agriculture** because it has the following characteristics:

1. *Stability:* It preserves a functional ecological balance in agricultural systems by managing natural resources in a renewable fashion and avoiding overexploitation of living organisms.

2. *Resilience:* Because the system not only is stable but also operates with a reasonable margin of error, it can survive unavoidable natural perturbations such as floods, droughts, and heat waves.

3. *Efficiency and productivity:* The system offers farmers a satisfactory return on investment of time and effort. It also provides not only the means for survival but also sufficient profit to improve the local economy.

4. *Appropriate technology:* The system relies on equipment and practices that are suited both to the local ecosystem and to the abilities, training, and economic system of the people using it.

Sustainable agricultural systems have been developed in numerous cultures. In the rush of excitement generated by the short-term success of the Green Revolution, such local wisdom was often ignored. But times are changing. The American and Indonesian researchers working in Bali, for example, are now involved in an international collaborative effort to recognize such systems where they exist elsewhere and to study the best ways to incorporate their successes into future agricultural development.

Related Issues Elsewhere

With world populations at record levels, the issue of sustainability is one of the great keys to the future health of the planet. Bali is a success story with respect to sustainable agriculture, but sustainability is an important issue with respect to other activities as well, including fishing, hunting, and timber harvesting.

Solutions

Fortunately, farmers interested in moving towards sustainable agriculture don't have to move to Bali. In many cases, they don't even have to change the crops they farm.

More than 25 years ago, for example, Nebraska farmer Delmar Akerlund stopped using chemical fertilizers and pesticides on his family's farm and cut back as much as possible on irrigation. Herbicide and pesticide suppliers foretold certain doom. Friends in the farming community laughed or shook their heads. Yet Akerlund's farm is thriving; when he takes his organically grown corn, soybeans, wheat, alfalfa, or beef to market, he can (within reason) name his price.

Akerlund is internationally known, both because of his business acumen and because of his pioneer stance in the organic-farming movement. Yet other "mainstream" American farmers are discovering the economic viability of minimizing or eliminating the use of pesticides and chemical fertilizers. Here in the United States, these techniques often take farming a step or two away from dangerous monoculture practices (Fig. 7.20). In tropical countries, innumerable variations on rice and wheat monoculture are possible because there are so many unused or underexploited food and forage crops (Table 7.3). In nearly all situations, an approach called "integrated pest management" combines the best features of organic and chemical-dependent farming techniques; it relies on organic methods as much as possible while allowing for occasional use of pesticides if things get out of hand.

Table 7.3 *New or Underutilized Potential Food Crops*

Species	Location	Use
Arracacha (*Arracacia xanthorrhiza*)	Andes	Carrot-like tubers with delicate flavor
Amaranths (3 species of *Amaranthus*)	Tropical and Andean America	Grain and leafy vegetable; livestock feed; rapid growth, drought-resistant
Buffalo gourd (*Curcurbita foetidissima*)	Deserts of Mexico and southwestern U.S.	Edible tubers, source of edible oil; rapid growth in arid land unusable for conventional crops
Buriti palm (*Mauritia flexuosa*)	Amazon lowlands	"Tree of life" to Amerindians; vitamin-rich fruit; pith as source for bread; palm heart from shoots
Guanabana (*Annona muricata*)	Tropical America	Fruit with delicious flavor; eaten raw or in soft drinks, yogurt, and ice cream
Lulo (*Solanum quitoense*)	Colombia, Ecuador	Fruit prized for soft drinks
Maca (*Lepidium meyenii*)	High Andes	Cold-resistant root vegetable resembling radish; with distinctive flavor; near extinction
Spirulina (*Spirulina platensis*)	Lake Tchad, Africa	Cyanobacterium-producing vegetable supplement; very nutritious; rapid growth in saline waters; cultivated in U.S. for health food industry
Tree tomato (*Cyphomandra betacea*)	South America	Elongated fruit with sweet taste
Ullucu (*Ullucus tuberosus*)	High Andes	Potato-like tubers, leafy part a nutritious vegetable; adapted to cold climates
Uvilla (*Pouroma cecropiaefolia*)	Western Amazon	Fruit eaten raw or made into wine; fast-growing and robust
Wax gourd (*Benincasa hispida*)	Tropical Asia	Melon-like flesh used as vegetable, soup base, and dessert; rapid growth, several crops each year

SOURCE: E. O. Wilson, *The Diversity of Life.* Cambridge, MA: Harvard University Press, 1992.

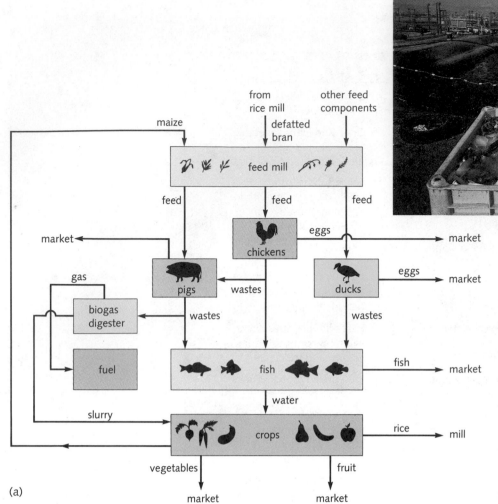

Figure 7.21 **(a)** This schematic of a Thai energy–aquaculture–agriculture facility shows how resource conservation and nutrient recycling can support high-yielding food production units. Virtually all plant and animal "wastes" are recycled in some way, creating a productive system that runs with minimal inputs of chemical fertilizers. **(b)** This Israeli experimental aquaculture facility uses an expanded understanding of marine species' ecological requirements in order to culture commercially valuable fishes (in this case, Sea Bream) in saltwater. Such projects have great potential to relieve pressure on limited freshwater supplies in arid areas.

Aquaculture, the farming of aquatic animals and plants, offers other opportunities for sustainable food production. Some modern aquaculture facilities operate on the same sort of monoculture model that dominates terrestrial agriculture. But researchers are rediscovering techniques known in Egypt and Asia for millennia. A single agriculture–aquaculture farm in Thailand, for example, can support acres of crops, thousands of ducks, chickens, and pigs, and over a million fishes (Fig. 7.21a). The secret is intensive internal recycling; crops are fed to animals whose wastes fertilize pond water that is used to grow algae and other fish food. Water from fish ponds is used to irrigate and fertilize crops, and other wastes are either recycled into fertilizer or digested by bacteria to produce methane gas used as fuel. And in the arid Middle East, researchers in Israel and Egypt are working together on pilot projects for culturing saltwater fishes (Fig. 7.21b), thus relieving pressure on freshwater supplies.

ISLANDS OF LIFE

It takes more than good intentions to solve serious environmental issues. Sometimes we simply lack answers to basic biological questions that would help conservation efforts. If, for example, we want to save chunks of tropical rain forest to preserve endangered species, how large must those chunks be? For that matter, how much old-growth temperate rain forest should we leave standing in our own Pacific Northwest? Good questions.

Imagine that you hold a high government office and are able to set aside enough of one particular habitat to preserve certain endangered species—*if you can prove that you know how much habitat each species needs to survive.* Would you have sufficient data to facilitate such a policy decision? With few exceptions, the answer is no. But ecologists are now conducting experiments to fill in the missing information.

One such project, begun by Thomas Lovejoy in the late 1970s, is based on theoretical studies of island biogeography discussed in Chapter 6. Lovejoy's project, arguably the largest (and possibly the longest-running) biological experiment in history, seeks to determine the minimum size a rain forest preserve must be to sustain viable populations of native plants and animals. The basic idea is brilliant in its simplicity: Lovejoy arranged to cut a large stand of rain forest into "islands" of various sizes and is now observing the fate of resident species living in those different-sized patches.

The experiment was made possible by a Brazilian law requiring that landowners who wanted to clear their property for farming or development could do so only if they left 50 percent of the native rain forest in its natural state. With the help of the World Wildlife Fund, Lovejoy convinced several landowners to distribute that 50 percent in a novel way; under his direction, they cut the forest around square patches which range in size from 1 hectare (10,000 m^2, or roughly 2.47 acres) to 1000 hectares (Fig. 7.22). The experiment has been in progress for more than a decade and is still producing vast quantities of data. Some of the more striking findings were summarized by E. O. Wilson in *The Diversity of Life:*

> The diversity of the smaller "islands" is decreasing the most rapidly, as expected. The extinction of species has been accelerated by the unexpectedly deep penetration of daytime winds, which dry out the forest from the edge in and kill deep-forest trees and shrubs for distances inward of 100 meters or more. Many plant and animal species have disappeared from the smaller plots, but a few are increasing their numbers. The reasons for the changes are sometimes obvious, but just as often baffling. Army ant colonies, which require more than 10 hectares to maintain their worker force, quickly disappeared from the 1- and 10-hectare plots. With them went five species of ant birds that make their living by following ant swarms and feeding on the insects driven forward by the 10-meter-wide raiding front. Shade-loving butterflies of the deep forest declined rapidly from the wind-drying effect, but other species specialized to live around forest edges and second growth flourished. Large, metallic green and blue euglosine bees, which are premier pollinators of orchids and other plants, were hard hit in plots of up to 100 hectares. Saki monkeys, which eat fruit, dropped out of the 10-hectare plots. But red howler monkeys, which are leaf-eaters and hence able to harvest more food, stayed on. Larger ground-dwelling mammals, including margay cats, jaguars, pumas, pacas, and peccaries, simply walked away from the smaller plots and out of the fauna altogether.

Figure 7.22 This photograph shows one of the forest patches left undisturbed in a clear area as part of an experiment in rain forest island biogeography.

By the late 1980s, second-order effects could be seen spreading through the food web. With the peccaries gone, there were no wallows in which temporary forest pools could form. Without the pools, three species of *Phyllomedusa* frogs failed to breed and disappeared. As the mammal and bird populations declined, dung and carrion beetles became scarcer. Scarab beetles that feed on these materials dropped in number of species and individuals. The average size of the surviving beetles became smaller.

As the study continues, researchers are spotting a few third-order interactions and speculating about others. Some long-term effects involve improbable cause-and-effect relationships. Why, for example, should a decline in populations of dung and carrion beetles make a difference to other animals? Because those beetles, which eat and bury both dead animals and their feces, help remove several kinds of parasites from circulation. In the beetles' absence, carcasses and dung persist on the ground longer and can spread disease. (Remember Theory in Action from Chapter 6, p. 109, about the effect of stagnant water on anthrax and the mammalian food web.)

Discussion

This study, worthwhile from the standpoint of basic research in island biogeography alone, will provide valuable information on the preservation of biodiversity. Even from Wilson's qualitative report, it is clear that different species require different amounts of habitat and for different reasons. It is natural to suspect that certain animals might need large patches of undisturbed forest to survive—because of their foraging behavior, perhaps, or their need for breeding space. But notice that certain deep-forest trees and shrubs can't survive in tiny forest "islands" either, because they are unaccustomed to fluctuations in temperature and humidity.

Species that require significant amounts of habitat in good condition are sometimes used as **indicator species** to monitor the health of an ecosystem. Once such species are identified, they are watched in much the same way as old-time coal miners used to watch canaries they took into coal mines to warn of lethal (but odorless) gas in the shafts. If indicator species thrive, the ecosystem is probably in reasonably good shape; if those species falter, they offer an early warning that other species may follow suit. The case of the spotted owl in the old-growth temperate rain forests of the Pacific Northwest is one example of such a situation. These owls, because they fall under the protection of the Endangered Species Act, have been singled out as the focus of current legal and political battles. But more importantly, because of their requirement for large tracts of healthy old-growth forest, they are representatives of the health of their entire habitat.

CASE STUDY

TO SAVE A CORAL KINGDOM

The Florida Keys are a chain of low limestone islands strung out along the eastern coast of Florida (Fig. 7.23). Stretching from Key Biscayne near Miami southwesterly to Key West, they dangle from the southern tip of Florida like the whisker on a catfish's chin. The shallow, turquoise waters that bathe these islands shelter an underwater chain of coral reefs—the only example of that rich, colorful, and fascinating habitat attached to the mainland United States.

These outposts of a tropical ecosystem are located close to the northern limit of their range, where water temperatures can get colder in winter than some species can tolerate. As a result, Florida's reefs contain fewer species and are not as physically complex as those farther south in the Caribbean. Reefs in the southern keys are the most diverse and best developed, while those in the north near Biscayne Bay have lower coral diversity and more restricted coral growth. Yet all these reefs are special; many of them are protected by inclusion in national parks and sanctuaries.

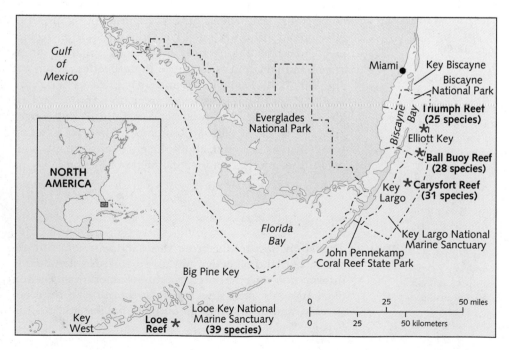

Figure 7.23 The Florida Keys and coral reefs under study. This map shows the relative locations of the Florida Keys, Florida Bay, Everglades National Park, and Biscayne Bay. Note that the diversity of coral species is highest in the southern part of the chain and lowest in the north.

But despite their legally protected status, Florida's reefs are in trouble. James Porter and Ouida Meier of the University of Georgia have found that the total coverage of live corals and the number of coral species are declining precipitously. Some reefs lost as much as 29 percent of their species richness during the study period, and five out of six reefs surveyed lost up to 45 percent of their live coral cover between 1984 and 1989 (Fig. 7.24). Porter estimates that if the current rate of loss continues, some of the keys' reefs will be bare of living corals by the year 2000.

Paradoxically, the study sites hardest hit were those in the south around Looe Key, historically home to the best-developed and fastest-growing reefs in the chain. Even more surprising, the only station that gained in live coral cover during the study was the northernmost study site near Biscayne Bay. If something as obvious as local coastal pollution were behind the reefs' trouble, that station would be expected to suffer the most—because it is closest to metropolitan Miami.

What could be behind this alarming trend? For a time, Porter and his colleagues were stymied. Then they made what may be a breakthrough by combining their efforts with those of other researchers in the region and standing back from the reefs in two ways: moving physically back from the study sites to examine conditions over all of southern Florida and moving back in time to observe environmental changes in the region over several decades. Together, the researchers reconstructed the following local environmental history.

Florida Bay was once a highly productive, brackish-water estuary, with salinity usually running around 18 parts per thousand (ppt). (Normal seawater usually runs between 32 and 36 ppt.) Today, however, that same body of water is now more like a bitter lake, with salinities regularly exceeding 42 ppt and occasionally reaching as high as 70 ppt.

The results of that change in physical environment have been far-reaching. Apparently as a result of high salinity, nearly 90 percent of the underwater sea grass meadows that once carpeted the bay have died. With the death and subsequent decay of all that sea grass, nutrient levels in the bay have risen, the water is cloudier, and the amount of dissolved oxygen in the water has fallen. In the process, commercial fisheries for lobster, shrimp, and stone crabs—together worth more than $250 million each year—have collapsed.

Why have conditions in Florida Bay changed so much? And how might these changes affect the reefs along the bay's southeastern edge? Porter and his colleagues have combined their observations with those of researchers in the Everglades to create what they call the "Florida Bay water hypothesis"—a tentative answer to these questions. Keeping in mind that this remains an unproved hypothesis, the reasoning is as follows:

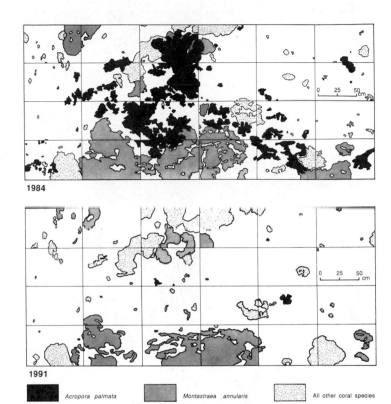

Figure 7.24 These schematic representations of live coral cover were taken from precisely the same locations on Looe Key reef in 1984 (TOP) and 1991 (BOTTOM). Note the extensive loss of live coral during the study period in this hard-hit, southern location.

Historically, much of the rain that fell over central and southern Florida collected in Lake Okeechobee, a land-locked lake with no obvious outlet. But all that water had to go somewhere. Overflowing from Okeechobee's southern shore, it moved in a giant, shallow, sheet over and through porous limestone rock in an arc through the great wetlands called the Everglades (Fig. 7.25). Ending up in Florida Bay, it diluted the seawater and created a productive estuary.

But beginning several decades ago, the Army Corps of Engineers, seeking to "control" surface water, constructed an elaborate network of levees, dikes, and canals. Their efforts at that time were aimed at controlling seasonal floods on low-lying wetlands, making more land available for agriculture and development. Today, the northern half of what was once the Everglades is thoroughly drained by that system; water that once flowed through it is now diverted eastward, where much of it empties into Biscayne Bay. As a consequence, so little water reaches the southern Everglades that much of it is seriously stressed by dehydration, and very little water ever reaches Florida Bay.

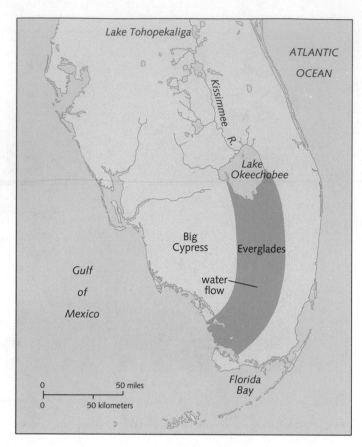

Figure 7.25 This map shows the former path of water overflow from Lake Okeechobee, through the Everglades, and out into Florida Bay and the nearby waters of the Gulf of Mexico.

According to Porter's hypothesis, the diversion of all that fresh water caused the following chain of events in Florida Bay:

- Salinity rose.

- Sea grasses died *en masse.*

- Dead sea grass began to decompose, and as a consequence, nutrients were released from decaying organic matter; sea grasses no longer removed nutrients as they once did.

- Nutrient levels rose and stayed high.

- Water turbidity increased.

- Dissolved oxygen concentrations fell markedly.

Why should any of this affect the reefs? Wind and current patterns in the Gulf of Mexico and Caribbean, Porter explains, cause water to "pile up" in Florida Bay and spill out through the keys. Before the diversion of fresh water on land, Florida Bay water was less saline (and therefore less dense) than normal seawater; so as it

flowed out of the bay, it "floated" near the surface and therefore passed above the reefs (Fig. 7.26). Now, however, that water is more saline and therefore more dense than seawater. As it passes out of the bay, it sinks to the bottom and flows out directly across the corals. That high-salinity, oxygen-poor, turbid, nutrient-rich water stresses corals, directly killing some of them and severely weakening others.

If this hypothesis is true, reversing the diversion of Okeechobee's water could remedy the situation. Surprisingly, much of the diverted water is not actually used; it is simply dumped into Biscayne Bay. According to several analyses, most of what *is* used can be replaced with water from other sources. There are some areas of southern Florida where alternative methods of flood control might be needed during the rainy season, but that doesn't seem to be an insurmountable problem. Thus a great deal of water *could* be redirected to save both the reefs and the Everglades without enormously upsetting either agriculture or urban life.

Discussion: Testing the Hypothesis

If Porter's hypothesis is correct, saving the coral reefs (and resurrecting Florida Bay fisheries) will require extensive changes in Florida's water system, at an estimated cost of at least $1 billion. Before the state makes those changes, more concrete proof of the hypothesis is certainly necessary. In particular, at least two major questions must be answered.

Question 1 Is water from Florida Bay actually moving as the hypothesis proposes? Thus far, water with characteristics similar to those of Florida Bay water has been found on the seaward margin of the Atlantic continental shelf, but did it definitely come from the bay?

Test Oceanographic studies using floating buoys or drogues, as well as satellite mapping of water flow, could produce the hydrological equivalent of a "smoking gun" by actually following Florida Bay water across the reefs.

Question 2 Water with high nutrient content, low oxygen, and high turbidity has been detected on the reef. But is it definitely coming from Florida Bay? Could it instead be coming from the sewage systems of residential and resort communities scattered throughout the lower keys?

Test Sampling stations could perform a series of analyses to more fully characterize water flowing over the reef. If elevated nutrients are found in water with *higher* than normal salinity, that water is most likely coming from Florida Bay. If, on the other hand, elevated nutrient levels are associated with *fresh water*, or with water of *lower*-than-normal salinity, Florida Bay may not be the problem. Instead, the lethally altered water may be coming from sewage

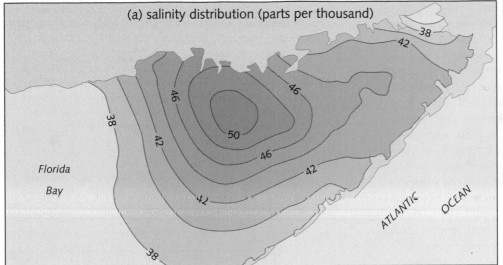

Figure 7.26 **(a)** Typical, present-day salinity distribution in Florida Bay. Values are given in parts per thousand (ppt). Note that this body of water was once a brackish estuary with salinities as low as 18 ppt, but it is now significantly saltier than standard seawater. **(b)** In the past, water from Florida Bay, warmer and less saline than water on the Atlantic side of the keys, floated above the submerged reefs. **(c)** Today, Florida Bay water is denser than seawater because of its high salinity. As a result, according to the Florida Bay water hypothesis, it sinks during its passage through the keys, engulfing the reefs in water that is too warm, too salty, too high in nutrients, and too turbid for most coral species.

systems on the keys. In that case, changes to those sewage systems may be necessary to save the reefs.

The final answer is not in yet, but both of these studies are under way. Right now, one can only hope that in the long term the Florida Keys will serve as an example of how basic science can be spurred by the need to protect a unique system, and how science and public policy can rise to the challenge.

TOWARDS AN ENVIRONMENTAL ETHIC

It would be naïve to think that biological science alone can solve these sorts of problems. Philosophers suggest that modern environmental dilemmas reflect fundamental flaws in our ethical and moral relationship with the biosphere. We fail to include in our definition of humanity the recognition that we are a powerful natural force in the world and that we have an intimate symbiotic relationship with the rest of the biosphere. These biologically erroneous attitudes, coupled with economic imperatives, too often lead us to sacrifice the eternal for the expedient.

It is unlikely that continued environmental folly will force our species into extinction, and it would be virtually impossible for us to extinguish all life on Earth. It is probable, however, that continued disregard for environmental management could seriously and permanently diminish the *quality* of life our descendants lead. Why? Because the much-touted "conflict" between "jobs" and "the environment" is not only an artificial dichotomy, but a fundamentally erroneous representation of the relationship between economics and ecology described in Chapter 2. In reality, the only way to ensure long-term economic prosperity is to guide human development along paths that utilize ecological understanding to guide sustainable development.

In November 1992, more than 1500 scientists, including 101 Nobel laureates, issued an urgent letter addressed to 160 world leaders. "The greatest peril," the letter warned, "is to become trapped in spirals of environmental decline, poverty and unrest leading to social, economic, and environmental collapse. . . . No more than one or a few decades remain before the chance to avert the threats we now confront will be lost and the prospects for humanity immeasurably diminished." These and other scientists urged governments to make the preservation and restoration of the global environment a central organizing principle for post–Cold War civilization.

Thankfully, a long overdue dialogue between economic and environmental interests has begun and shows promise of avoiding past mistakes. Back in the 1970s and 1980s, for example, "development projects" that encouraged Brazilian farmers to clear rain forests and grow cash crops for export were encouraged by funding from the World Bank. Brazilian environmentalists, however, argued forcibly that properly managed and harvested rain forests could produce up to 500 pounds per acre per year of tropical fruits, nuts, game, and fishes. If that same rain forest is carved into pastureland for cattle, however, it produces only 45 pounds of meat per acre for the few years before the soil becomes worthless. Following that reasoning, in 1986 the World Bank began to require projects it funds to maintain or improve the world's resource base.

In the meantime, ecologists are coming up with development projects of their own that encourage local people to participate in projects that benefit them economically and ecologically in both the short and the long term (see Chapter 2, Theory in Action, Rubber Tappers and Rainforest Crunch®). And in Bolivia, an innovative "debt-for-nature" swap involving Conservation International, Citicorp Bank, and a private conservation foundation allowed the government to trade $650,000 of its foreign debt in return for guaranteed protection of a tract of virgin Amazon forest.

Our best hope for developing a sustainable global ecosystem lies with approaches that incorporate economic and political reality, the resourcefulness of the human mind, and respect for the world around us. Many of the world's great religions hold views supportive of what some have called environmentalism of the spirit. There is challenge, inspiration, and warning in the conclusion of a letter written by a Native American chief to the "Great Chief in Washington" over a century ago:

> One thing we know which the white man may one day discover. Our God is the same God. . . . [and]

. . . This earth is precious to him. And to harm the earth is to heap contempt on its creator. . . . Continue to contaminate your bed, and you will one day suffocate in your own waste. . . .

When the buffalo are all slaughtered, the wild horses all tamed, the secret corners of the forest heavy with the scent of many men, and the view of the ripe hills blotted by the talking wires, where is the thicket? Gone. Where is the eagle? Gone. And what is it to say good-bye to the swift and the hunt? . . . [It is] the end of living and the beginning of survival.

—"This Earth is Sacred" *Letter from Chief Sealth (Seattle) to President Franklin Pierce, 1885*

SUMMARY

During the mid-twentieth century, the public became more aware of the environmental effects of human activities on global ecosystems. Still, environmental damage persists. Water pollution from industrial and agricultural chemicals and residential sewage endanger both surface water and underground aquifers; atmospheric pollution may raise the earth's temperature through increased levels of carbon dioxide and cause widespread damage to streams and forests through acid rain. Pollution and habitat destruction result in the extinction of plant and animal species and the loss of genetic diversity. Faced with these ecological dilemmas, ecologists and economists are searching for a new style of interacting with the global environment—a system of sustainable development that provides for human needs in ways that do not adversely affect the biosphere.

Population growth, flawed economic systems, and shortsighted government policies contribute to the problems of hunger and poverty throughout the world. Our best hope for developing a sustainable global ecosystem lies with approaches that incorporate economic and political reality, the resourcefulness and creativity of the human mind, sound knowledge, and a healthy respect for the living world around us.

STUDY FOCUS

After studying this chapter, you should be able to:

- Explain why human activity is affecting the biosphere more powerfully today than ever before.

- Give examples of ecologically imprudent human actions and their consequences.

- Explain the concept of sustainability.

- Cite successes and failures in "agricultural ecology."

- Give an example of research aimed at answering an important question in conservation biology.

TERMS AND CONCEPTS

REVIEW

Objective Questions   (Answers in Appendix)

1. The least common type of sewage treatment used is
   (a) primary treatment.
   (b) secondary treatment.
   (c) tertiary treatment.
   (d) a combination of primary and secondary treatment.

2. Ozone
   (a) is a highly poisonous gas that combines with oxygen in the lungs, causing breathing difficulties.
   (b) is a gas that absorbs ultraviolet light and prevents it from reaching the earth's surface.
   (c) is one of several gases that combines with sulfur and oxygen to produce acid rain.
   (d) combines with carbon in the atmosphere to produce the air pollutant carbon monoxide.

3. Nonrenewable energy sources do not include
   (a) oil.   (c) hydroelectric power.
   (b) natural gas.   (d) coal.

4. Secondary sewage treatment involves the
   (a) biological breakdown of organic material by bacteria.
   (b) settling out of solid waste.
   (c) chemical removal of toxic compounds.
   (d) heating of toxic compounds using blast furnaces to decompose them.

5. In which organism would DDT most likely be in the heaviest concentration once it had been introduced into the ecosystem?
   (a) grasshopper   (b) toad   (c) snake   (d) seal

Discussion Questions

6. What economic arguments are used to justify industrial and municipal pollution in your area? How are they similar to or different from arguments put forward in developing tropical countries? Are these arguments convincing in the long term? Why or why not?

7. What common practices in modern agriculture can be detrimental to farmland over the long term? Why are they in such widespread use? What measures can be taken to moderate the effects of large-scale agriculture on the environment?

8. What is happening to the concentration of carbon dioxide (CO_2) in the atmosphere? Why? Why do some scientists think that change is important?

9. Why is damage to the ozone layer of both local and global significance?

10. When a large midwestern industrial facility spews sulfur dioxide and nitrous oxides into the atmosphere, what are the environmental effects, and where are these effects felt?

11. Have you personally noticed any differences in species diversity between small and large patches of native habitat in your area? If so, what are the similarities and differences? Choose several key plant and animal species common (or once common) in your area. Using the information in Part 2, propose experiments and observations to determine each species' habitat needs.

READINGS

(Integrated list for Part 2)

Brown, Lester, et al. *State of the World: A Worldwatch Institute Report on Progress Toward a Sustainable Society.* New York: Norton, annually since 1984. Published each year, these volumes offer a unique combination of ecological and economic analysis of progress towards a sustainable global society. (Available through the Worldwatch Institute in Washington, D.C.)

Monastersky, R. "Once bashful El Niño now refuses to go." *Science News* 143:4 (January 23, 1993): 53.

Gore, Al. *Earth in the Balance.* New York: Houghton Mifflin, 1992.

Wilson, E. O. *The Diversity of Life.* Cambridge, MA: Harvard University Press, 1992.

Fowler, Cary, and Pat Mooney. *Shattering: Food, Politics, and the Loss of Genetic Diversity.* Tucson, AZ: University of Arizona Press, 1990.

Kaufman, Les, and Kenneth Mallory, eds. *The Last Extinction.* Cambridge, MA: MIT Press, 1986.

Coutant, Charles C. "Thermal niches of striped bass." *Scientific American* (1986): 98.

Meyers, Norman. *The Primary Source: Tropical Forests and Our Future,* New York: Norton, 1984.

Meyers, Norman, ed. *Gaia: An Atlas of Earth Management.* Garden City, NY: Anchor Books, Doubleday, 1984.

Fritsch, Albert J. *Environmental Ethics: Choices for Concerned Citizens.* Garden City, NY: Anchor Books, Doubleday, 1980.

Teal, John, and Mildred Teal. *Life and Death of a Salt Marsh.* New York: Ballantine, 1969. A timeless classic that combines American history with estuarine ecology.

Connell, J. H. "The influence of interspecific competition and other factors on the distribution of the barnacle *Chthamalus stellatus.*" *Ecology* 42 (1961): 710–723

PART 3

Evolution and Mendelian Genetics

Pea flowers such as these were the plants with which Gregor Mendel first demonstrated the inheritable variation within species that we know makes agricultural breeding and evolutionary change possible.

m odern biology rests on the themes of genetics and evolution. Genetics looks at the inheritable makeup of individuals and how that inheritance is passed from parent to offspring. Evolution, as Darwin noted in a passage from his *On the Origin of Species,* looks at changes in organisms that occur over very long periods of time and result in the creation of new species.

We approach these themes from an historical perspective by taking a chronological look at the work of Darwin and his contemporary, Mendel. By studying the social and scientific backgrounds from which each man began work, how each worked from his observations and/or experiments to produce hypotheses, and where each came to intellectual obstacles that he could not surmount, we can see both the brilliance of these two thinkers and the limitations of the nineteenth century that delayed until the twentieth century the corroboration and enlargement of their work into the powerful, recognized theories we will study in the next section, Part 4, Molecules of Life.

Chapter 8, Darwin's Dilemma, examines the turbulent social and scientific world into which Darwin was born, follows Darwin on his voyages as naturalist on the HMS *Beagle*, explains how he drew the basic premises of evolutionary theory from his observations, and concludes by explaining how knowledge from contemporary geologists and paleontologists corroborated the fact that evolution had occurred and that Darwin's basic explanatory premises were correct. Chapter 9 turns from the grand vision of Darwin's work to the microscopic view of cells and their behavior during growth and cell division, mitosis. Chapter 10 looks directly at Mendel, his experiments and data, and his resulting genetic principles. This chapter also explains meiosis, the cellular process

responsible for inheritance from parent to offspring, and correlates Mendel's principles to chromosomal behavior during meiosis. The predictive power of genetic principles and how it can be explored through problems and problem solving are also covered. Chapter 11 moves from Mendel's general principles to our modern understanding of the human genome, including the route of inheritance and clinical presenta-

> There is grandeur in this view of life, with its several powers, having been originally breathed by the Creator into several forms or one; and that, whilst this planet has gone cycling on according to the fixed law of gravity, from so simple a beginning endless forms most beautiful and wonderful have been and are being evolved.
> — Charles Darwin,
> On the Origin of Species

tion of some well-known and little-known genetic diseases. Chapter 12 returns to evolution. This chapter places Darwin's principle of natural selection in the context of the principles of inheritance defined by Mendel, explaining several forms of natural selection and showing the basis for the mathematical expression of population genetics, the Hardy–Weinberg law. Last, in Chapter 13, we take an integrated (Darwinian—Mendelian) look at mechanisms that can cause the creation of new species over time, thereby pulling together the major conceptual threads of evolution and genetics.

8

Darwin's Dilemma: The Birth of Evolutionary Theory

*t*he word **evolution** literally means "unrolling or unfolding," though in common usage it simply means "change." A theory of evolution is nothing more (or less) than a theory of biological change over time. Yet *On the Origin of Species,* Darwin's first collection of facts supporting a theory of biological evolution, has been described as "the book that shook the world." Because our species is among the organisms that have evolved, facts and theories about evolution have profound practical and philosophical implications for humanity. It should come as no surprise, therefore, that evolutionary thought has guided, inspired, frightened, and infuriated scientists, philosophers, religious authorities, and laypeople from the nineteenth century to the present. A century after Darwin published *On the Origin of Species,* philosopher J. Collins asserted that "there are no living sciences, human attitudes, or institutional powers that remain unaffected by the ideas . . . catalytically released by Darwin's work." To understand why Darwin's work was so important, we must first place his theory within the context of eighteenth- and nineteenth-century Western scientific philosophy.

WESTERN WORLD VIEW BEFORE DARWIN

From the time of the Greeks, most Western philosophers viewed the material world as rigid, static, and innately flawed. To Plato and Aristotle, both living organisms and inanimate objects were inferior mimics of perfect models called **ideal types.** Ideal types, found only in the transcendent world of ideas, were perfect and unchanging. But their imperfect copies on Earth were full of flaws that human naturalists saw as variations among members of plant and animal species.

Later, Western philosophers combined many Platonic and Aristotelian ideas with Christian thought to create a world view that encompassed religion, science, and society. Two beliefs in particular constrained the natural sciences. First, theologians believed that because God was perfect, all of His work had to be perfect. Second, philosophers asserted that perfection *necessarily* implied stability; things that were divine and perfect should not change.

For those reasons, orthodox Christian philosophy taught that after God created the first ideal types, species were fixed for all time. Imperfections appeared in living things because the material world, unlike the spiritual world, is corrupt and imperfect. But because the original Creation was complete and perfect, no organisms had appeared or disappeared, and the "type" for each species did not change.

Furthermore, each living species had a permanent place in the divine order of things called *the Great Chain of Being*. Derived from Aristotle's *Scala Naturae*, the Great Chain of Being stretched from nonliving matter, through lower forms of life, to humans at the top of the earthly chain. Most important, nearly all European scientists shared the belief that because humans were created in God's image, our species was unique and essentially different from all the other forms of life. Together, these beliefs formed the original doctrine of creationism.

Stability in Biology and Society

The idea of the unchanging type permeated biology into the nineteenth century. Biologists sought to look beyond the visible flaws of earthly organisms—to ignore individual variation—and to reveal God's master plan by studying the ideal types those organisms represented. When a new species was discovered, a specimen thought to resemble its ideal type most closely was deposited in a museum collection, where it represented its species as the "type" specimen. It was this philosophy that guided Swedish naturalist Carolus Linnaeus (1707–1778) when he established the system still used today for naming and classifying organisms (Chapter 23).

Through the eighteenth century, this view of a divinely ordered and stable world governed not only the natural world but social systems as well. Just as the human race had dominion over creation, kings ruled over humanity by "divine right." The rigid system of upper and lower social classes was thus seen as an extension of the immutable world order represented by the Great Chain of Being. Talk of change was immoral; such change was unthinkable, whether in the human social order or the natural world.

DARWIN'S TIME: A WORLD IN FLUX

Charles Darwin (1809–1892) was born and raised in a privileged family within a society growing uncomfortable

Figure 8.1 Charles Darwin was born into a wealthy family immersed in the social change of nineteenth-century Europe. Both his maternal grandfather, Josiah Wedgewood, and his paternal grandfather, Erasmus Darwin, rose from poverty to become industrial magnates. These self-made men didn't fit into the static, hereditary social structure; they and their peers wanted society to change and wanted to affirm upward mobility based on individual achievement.

with the rigid status quo (Fig. 8.1). Social and technological change was sweeping through Europe. The merchant class created by the Industrial Revolution was not satisfied with the hereditary social structure and struggled to change it, while philosophers searched for a new world view that could accommodate the emerging, competitive social order. At the same time, some of the greatest scientists Europe ever produced were making major contributions to the emerging scientific view of the world (Fig. 8.2).

By the time Darwin was born, nearly every branch of science except biology had challenged the established view of a static, divinely ordered world. Astronomers argued that the earth was not the center of the universe, as religious dogma had maintained. Newton provided mathematical explanations both for the previously mysterious orbits of planets and for the movements of objects on Earth.

Then, during the eighteenth and early nineteenth centuries, global explorers made discoveries that defied

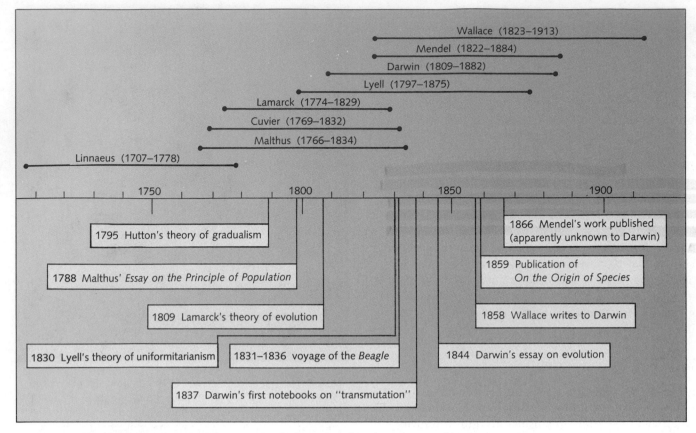

Figure 8.2 Darwin's life and work in the context of other scientific achievements of his era.

traditional biological perspectives. They found that Asia, Africa, and the New World harbored hundreds of exotic plant and animal species unknown in Europe and that many of these animals and plants lived only in certain parts of the world. If all living things had been recently created in the same place at the same time and had later been released from Noah's ark, what could explain these distribution patterns? Biology and geology were ripe for their own revolution.

Fossils and Catastrophism

Fossils presented another problem for biologists. As more and more preserved remains of plants and animals were found, it became impossible for biologists to deny that fossils represented extinct organisms (Fig. 8.3). Where did these fossils come from? And why did the animals they represented die out?

The French anatomist Cuvier was so overwhelmed by fossil diversity that he suggested there had been not one but six separate creations! Other scientists suggested that

there had been several successive creations followed by floods or other catastrophes, a doctrine that became known as **catastrophism.** Note that catastrophism did not really challenge the dominant philosophy; although catastrophists suggested that the earth was far older than scriptures implied, they still assumed that the world and its inhabitants were stable and had been specially created. Every now and then, the state of the world changed, but those changes were rare events that occurred only by divine decree.

GEOLOGISTS CHALLENGE THE STATIC EARTH

Geologist James Hutton produced the first scientific challenge to this concept of a static world in his revolutionary *Theory of the Earth,* published in 1788. Hutton proposed that the same geological principles in action today—weathering and erosion, deposition of sediment, and volcanism—had shaped the world over time (Fig. 8.4). In

Figure 8.3 This fossil sampler shows examples of the sort of fossils unearthed before and during Darwin's life. The diversity of these fossils and their abundance posed serious problems for creationists of the Victorian era.

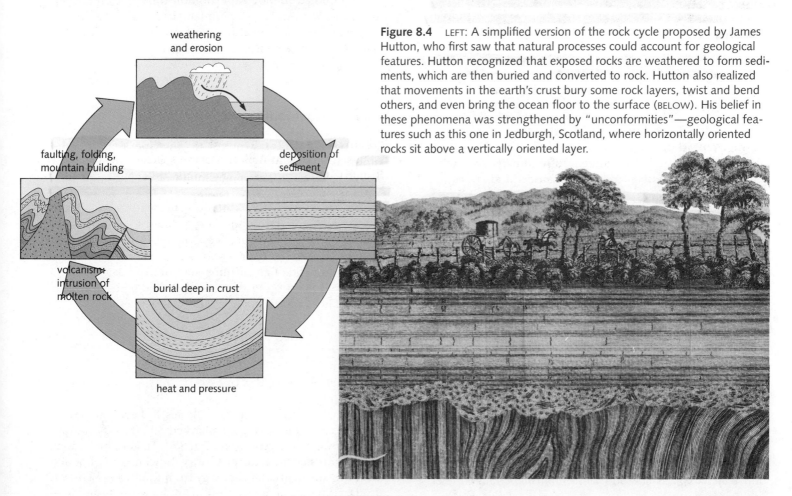

Figure 8.4 LEFT: A simplified version of the rock cycle proposed by James Hutton, who first saw that natural processes could account for geological features. Hutton recognized that exposed rocks are weathered to form sediments, which are then buried and converted to rock. Hutton also realized that movements in the earth's crust bury some rock layers, twist and bend others, and even bring the ocean floor to the surface (BELOW). His belief in these phenomena was strengthened by "unconformities"—geological features such as this one in Jedburgh, Scotland, where horizontally oriented rocks sit above a vertically oriented layer.

weathering
and erosion

deposition of
sediment

faulting, folding,
mountain building

burial deep in crust

volcanism:
intrusion of
molten rock

heat and pressure

order for these processes to have occurred, of course, the earth had to be *very* old. Hutton was the first to introduce to both geology and biology the critical concept of *deep time*. According to Hutton, "The result . . . of our present enquiry is, that we find no vestige of a beginning—no prospect of an end." That statement emphasizes Hutton's realization that the history of Earth's geological features is measured not in hundreds or thousands of years but in hundreds or thousands of *millions* of years.

The man best known for championing geological change was Charles Lyell, whose *Principles of Geology* was published in 1830. Lyell's book expanded on Hutton's work, putting forth three principles that constitute the theory of **uniformitarianism.** (The first two principles must apply not only to geology but to all scientific inquiry.)

1. *Natural laws are constant in space and time.* As contemporary Harvard biologist Stephen Jay Gould notes, this is a statement of method that any scientist must make to analyze the past. For "if the past is capricious, if God violates natural law at will, then science cannot unravel history."

2. *Scientists should attempt to explain events of the past through the same sort of natural processes that we can observe directly today.*

3. *Most geological change occurs slowly and gradually, not through sudden, catastrophic events.* This rule of uniformity of rate was an extremely important influence on Darwin's thinking.

By 1830 most scientists believed that the earth was old (though they didn't agree on *how* old) and that the structure of the earth had changed substantially over time (though they argued over the causes and rates of change). By current geological estimates, the earth is 4.5 *billion* years old and has changed in many ways since it first formed, as you will see. Biology, of all the sciences, was the slowest to accept theories of change.

BIOLOGICAL SCIENCE AND CHANGE

Mid-nineteenth-century intellectuals were intensely interested in theories of change in nature. Darwin's grandfather, Erasmus Darwin, discussed the origin and evolution of life as early as 1794. And in 1857, two years before Darwin published *On the Origin of Species,* the English philosopher Herbert Spencer argued that life *must* have evolved, because change was universal in all other domains. But though people suggested that organisms *could* evolve, no one could explain *how* and *why* organisms changed over time.

Lamarck's Theory of Evolution

The first to propose a mechanism for evolutionary change was the French scientist Jean Baptiste de Lamarck. Lamarck's theory of **evolution through inheritance of acquired characteristics** was published in 1809, the year Darwin was born. Though he has often been maligned, Lamarck was an innovative theoretician who combined evolutionary and ecological thinking. Lamarck, who believed in divine creation and the Great Chain of Being, actually conceived of the chain as more like an escalator; he felt that each species was created with a God-given drive toward perfection that, combined with environmental factors, impelled it along a relentless journey up the chain.

Lamarck made two basic assumptions that we know now to be incorrect. First, his theory was **teleological;** he believed that evolution had a goal, or directed purpose, and that species changed over time because they "wanted" to "better" themselves. Second, he believed that characteristics acquired during the life of an organism could be passed on to its offspring—a belief we now know to be generally false. To Lamarck's lasting credit, however, he championed biological evolution during a time when such ideas were not at all popular. He was also the first to propose a truly *scientific* theory of change, as well as the first to recognize that evolution involved interaction between organisms and their environments.

Darwin's Contribution

When Darwin and his contemporary, Alfred Russel Wallace (see Theory in Action, Darwin's Delay and Wallace's Insight), independently suggested a different scientific explanation for evolution, the last of the pieces of the puzzle finally fell into place. Both the scientific community and the general public were primed to accept a biological theory of change. The publication of Darwin's book in 1859 was not just a scientific event but also a public sensation. The first printing sold out the day it was released, and many other printings followed before the year ended. The world hasn't been the same since.

DARWIN ON THE *BEAGLE*

Just after Christmas in 1831, Charles Robert Darwin, an educated gentleman of 22, embarked on HMS *Beagle* for a global voyage of exploration (Fig. 8.5). Darwin had always been interested in natural history, but he had obtained his university degree in theology from Cambridge University. Darwin was, in fact, a pious man; he wrote home mid-

Darwin's Delay and Wallace's Insight

When Darwin returned from his voyage on the *Beagle* in 1836, his theories were still in the formative stages. But though he had completed most of his important work by 1844, he chose not to publish it. Instead he put it aside, instructed his wife to publish the work in the event of his death, and turned for more than a decade to a study of worms and barnacles.

Several of Darwin's colleagues repeatedly urged him to publish his ideas before someone else beat him to it. Admonished, Darwin started assembling his thoughts, chapter by chapter, in 1856. Meanwhile, half a world away, an English naturalist named Alfred Russel Wallace was traveling through Malaysia. Wallace worked extensively in Southeast Asia and was impressed, as Darwin had been, with the diversity of life in the tropics. Wallace had also read the essay on population growth

written by mathematician Thomas Malthus decades earlier (see Chapter 5). Then, while ill with malaria, Wallace awoke in a fever with an inspiration. Writing with the same fever that had inspired him, Wallace outlined a theory of evolution nearly identical to Darwin's and mailed it to Darwin asking for his opinions.

Darwin was downcast as he conveyed the paper to Lyell for public presentation. All his work was for nought; Wallace would receive credit for publishing his ideas first. But Lyell presented both Wallace's paper and excerpts from Darwin's earlier, though still unpublished, essay. Darwin worked furiously to publish *On the Origin of Species*, and everyone (including Wallace) agreed that Darwin's exhaustive research and documentation of evolutionary phenomena entitled him to the lion's share of the credit.

Alfred Russel Wallace

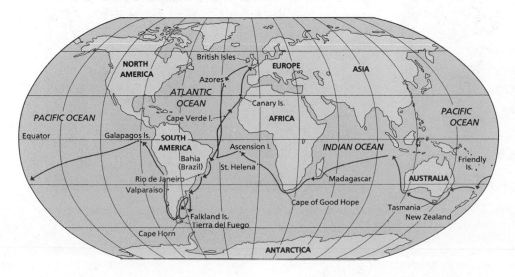

Figure 8.5 The voyage of the *Beagle*, 1831–1836.

Figure 8.6 Darwin explored as far inland as time allowed. In Argentina, he discovered exotic fossil animals, including giant ground sloths, peculiar horses, and this bizarre, extinct relative of the armadillo called Glyptodon. As numerous as Darwin found living organisms to be, he was soon convinced that they were far outnumbered by forms no longer living.

way through his voyage that he could see himself spending the rest of his days as a country preacher. But during the *Beagle*'s voyage, Darwin's intellect, his naturalist's eye, and some extraordinary luck led him far from the commonly accepted religious views of nature.

Geological Observations

Darwin carried with him a gift from his mentor Professor J. S. Henslow—the first volume of Lyell's *Principles of Geology*. Darwin was so impressed by Lyell's work that he had subsequent volumes delivered to him en route.

With Lyell's ideas in mind, Darwin had the good fortune to witness the forces of geology in action. In January of 1835, he saw the eruption of a volcano in Chile and later learned that another volcano 480 miles away had blown its top the same night. Just over a month later, he experienced an earthquake that lifted beds of marine mussels "still adhering to the rocks, ten feet above the high water mark." Still later, he observed beds of fossil mussels a thousand feet and more above sea level. Whereas others were simply awed by these phenomena, Darwin realized that he had glimpsed the geological processes that were gradually building the Andes Mountains. His belief in geological uniformitarianism was established.

Figure 8.7 These photographs show two remarkably similar yet unrelated animals—a marsupial mouse (LEFT) from Australia and a common wood mouse (RIGHT). Occupying similar ecological niches, they exhibit strikingly similar forms and behaviors.

Figure 8.8 Fitness for animals includes the design of legs, wings, and claws. Fitness for plants includes the features of leaves, stems, roots, and flowers. ABOVE: This mole has no use for vision but obtains a great deal of information through a highly developed sense of smell. Its limbs and digits are modified into efficient tools for digging tunnels. RIGHT: Many large tropical plants, such as this banyan tree (a species of fig), develop extensive systems of prop roots that both gather nutrients and serve as extra trunks to support the spreading canopy.

Biological Diversity

Darwin was staggered by the variety of animals and plants he encountered during his voyage. Everywhere he looked, he saw dozens of new and oddly shaped trees, hundreds of exotically colored flowers and birds, and beetles and other insects literally beyond easy counting. But Darwin quickly realized that the diversity of living organisms was only part of the mystery of life, for he found an even greater number of fossil species. In Argentina, he discovered fossil armadillos, giant ground sloths, peculiar horses, and creatures that reminded him of the hippopotamus.

As numerous as living organisms were, Darwin was soon convinced that they were vastly outnumbered by extinct forms (Fig. 8.6). That extraordinary diversity of fossil species convinced Darwin that both extinction and the appearance of new species were real phenomena that had to be explained. (Biologists today estimate, in fact, that more than 99.9 percent of species that have lived on the earth are now extinct. Because current estimates place the number of living species as somewhere between 2 and 30 million, over 2 billion species must have come and gone since life began.)

Darwin also discovered that both flora and fauna differed markedly from continent to continent and on opposite sides of natural barriers such as mountains, deserts, and large rivers. The Argentinean pampas, for example, supported very different animals from the superficially similar grasslands of Australia. And Darwin noticed that although plants and animals in ecologically similar but geographically separate areas differed from each other, they often possessed similar structures and behaviors (Fig. 8.7).

Fitness

Throughout his journey, Darwin marveled at the "perfection of structure" that made it possible for organisms to do whatever they needed to do to stay alive and produce offspring. Darwin called this perfection of structure **fitness,** by which he meant the combination of all traits, physical and behavioral, that help organisms survive and reproduce in their environment (Fig. 8.8). As he observed, Darwin also began to question what process had "fit" these organisms to their physical environments and to each other.

Figure 8.9 Map of the Galapagos Islands drawn by the officers of the *Beagle* in 1835.

The Galapagos Islands

Off the west coast of South America, the *Beagle* visited a cluster of tiny, rocky islands called the Galapagos after the Spanish word for "tortoise" (Fig. 8.9). Darwin noted that these islands were inhabited by a surprising number of bizarre and often beautiful plant and animal species. He surmised correctly that many of these species were *endemic*, which means they are found nowhere else (Fig. 8.10). One group of Galapagos birds has since been named Darwin's finches in his honor.

These islands and their inhabitants made lasting impressions on Darwin, and his notes and collections served him well in later years. In his first *Journal of Researches*, published in 1837, he noted that the Galapagos Archipelago was like "a little world within itself" that was "very remarkable" in its organic productions. But not until the second edition of his *Journal*, published in 1845, did Darwin begin to recognize the full significance of what he had seen there. In one of his most famous passages, he wrote:

> Considering the small size of these islands, we feel the more astonished at the number of aboriginal beings, and their confined range. Seeing every height crowned with its crater, and the boundaries of most lava-streams still distinct, we are led to believe that within a period geologically recent the unbroken ocean was here spread out. Hence both in space and time, we seem to be brought somewhat near to that great fact—that mystery of mysteries—the first appearance of new beings on this earth.

But it took Darwin nearly a quarter of a century after his Galapagos sojourn to publish his solution to that great mystery. On his return to England, he discussed his discoveries with prominent botanists and zoologists. Later, in 1837, he started his first notebook on "transmutation" (changing to a different form). There, and in other notebooks to follow, Darwin documented his growing belief in evolution and his search for a mechanism that could explain it. Then in October of 1838, he read "for amusement" a 40-year-old essay that crystallized his thinking, the *Essays on the Principle of Population* by mathematician Thomas Malthus (see Chapter 5).

Malthus, you will recall, believed that without some external control, human population growth would invariably outstrip our ability to feed ourselves. Darwin realized that if this were true for humans, who usually had fewer than ten children, it was doubly true for plants and animals that produced hundreds or even thousands of offspring. Darwin calculated, for example, that if all the descendants of even a slowly reproducing pair of elephants survived, that pair would have 19 million descendants in 750 years.

Figure 8.10 ABOVE: Four species of Galapagos finches, each with head and beak sizes suited to different diets. So different in appearance and habits were some of these birds that Darwin erroneously placed them in separate subfamilies. RIGHT: Darwin found that each of the Galapagos Islands had its own peculiar type of giant tortoise. These tortoises are from the islands of Pinta (TOP), Hood (CENTER), and Isabella (BOTTOM). Although Darwin learned of these differing forms during his visit, he did not recognize their evolutionary significance until much later.

DARWIN'S THEORY OF EVOLUTION BY NATURAL SELECTION

After a great deal of work, Darwin finally devised a scientific explanation for why populations of plants and animals did not continually grow out of control. That explanation allowed him to interpret and explain the change he knew must be occurring in the natural world.

Variation

Darwin began by abandoning the idea of species as perfect and unchanging. Instead of viewing differences among members of a species as imperfections or deviations from an ideal type, Darwin realized that these **variations** were a basic fact of life (Fig. 8.11).

From plant and animal breeders, Darwin learned that seed from a single ear of corn produces both plants with ears that are larger than average and plants with ears that are smaller. Among crops of garden flowers, flocks of pigeons, and packs of dogs, some individuals always differ from the norm in color and size. Breeders knew that much of this variation could be passed on from parent to offspring—in other words, it was **heritable.**

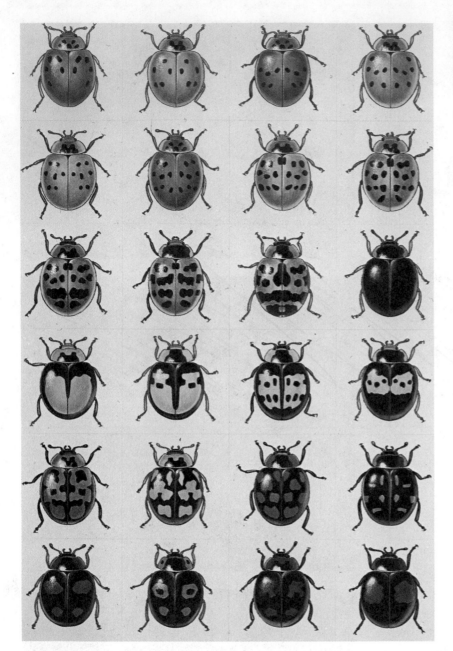

Figure 8.11 The variation in color patterns found in "ladybug" beetles across their range. We now know that in addition to such visible variation, organisms also exhibit "invisible" variation in body chemistry, physiology, and behavior.

From his field observations, Darwin knew that wild species varied in nearly every characteristic he could observe and measure. Of course, he had no idea how variations arose or how they were inherited. (That understanding had to await the later emergence of modern genetics.) But Darwin did realize that heritable variation—in both domestic and wild organisms—is *random, purposeless, and in no way subject to control.* Farmers cannot *cause* any particular kind of variation to appear among their crop plants, any more than wild animals can will their necks to grow longer or their fur to grow darker.

Artificial Selection

Darwin built his argument brilliantly. His first chapter, "Variation Under Domestication," details how farmers take advantage of random variation in crops and livestock. In a process Darwin called **artificial selection** (intuitive selective breeding), farmers choose the most desirable cows, sheep, or tomato and corn plants for breeding. Over several generations, this selective process produces individuals that differ markedly from their forebears (Fig. 8.12). In Darwin's words, "The key is man's

power of accumulative selection: nature gives him successive variations; man adds them up in certain directions useful to him."

Natural Selection

Darwin's next insight was to see in nature an analogue to the farmer as selective agent. Here again, Darwin took a giant philosophical step. Rather than postulating a mysterious or supernatural force behind change, as his predecessors had done, Darwin searched for a direct material force, or scientific mechanism, to drive evolution. Darwin called that material force **natural selection** and described it as a process that favors the survival and reproduction of those organisms exhibiting variations best suited to their environment.

> As many more individuals of each species are born than can possibly survive, and as, consequently, there is a frequently recurring struggle for existence, it follows that any being, if it vary however slightly in any manner profitable to itself under the complex and sometimes varying conditions of life, will have a better chance of surviving, and thus be *naturally selected*. From the strong principle of inheritance, any selected variety will tend to propagate its new and modified form.

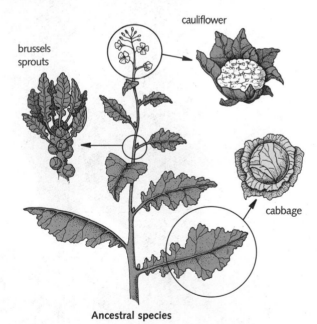

Figure 8.12 TOP: Darwin was a pigeon fancier, and he knew that the fancy breeds popular in his day were descended from the rock dove, a species resembling common city pigeons. All these breeds were created through artificial selection. BOTTOM: Cabbage, cauliflower, and brussels sprouts are all descendants of a single ancestral species, shown at the center here. The modern vegetables acquired their forms through generations of artificial selection favoring individuals with larger leaves, denser heads of flower buds, or large leaf buds.

But though Darwin compared natural and artificial selection, he emphasized that natural selection operates without the foresight and purpose of the farmer. Darwinian evolution is **nonteleological,** which means that the process of natural selection operates without any ultimate goal of perfection. Natural selection operates only in the here and now for each organism, selecting those variants that are most fit to survive and reproduce *under local environmental conditions.*

Note that the forces of natural selection and farmers practicing artificial selection share a major handicap: neither has the ability to *cause* desirable variations. Particular variants either arise, or do not arise, by chance. Only after such variations arise can they be favored by either artificial or natural selection.

Adaptation

According to Darwin, organisms are "selected," or molded, over many generations by natural selection to become better suited or "fitted" to their environment in a process called **adaptation.** Thus it is through the process of adaptation that organisms acquire fitness. The word *adaptation* is also used to describe any characteristic of an organism that increases its fitness.

Adaptation is a complex process that involves all parts of an organism's anatomy, physiology, and behavior. Woodpeckers, for example, have evolved a suite of adaptations, all of which work together to enable these birds to feed on insects that live in the bark of trees (Fig. 8.13).

SUMMARY OF NATURAL SELECTION AS PRESENTED BY DARWIN

The evolutionary process as Darwin envisioned it can be summarized as follows:

1. Organisms alive today were not specifically created as we see them but have descended from species that lived before them. This concept of **common descent** links plants or animals together into groups descended from ancestors they share.

2. More organisms are produced than can possibly survive, most die before reaching sexual maturity, and many that do survive fail to reproduce. Individual organisms are constantly struggling against each other, and often against hostile environmental conditions, for the necessities of life.

3. The physical characteristics of individual members of each species vary a great deal, and much of this variation can be inherited.

4. Some variants in each generation are better suited to life in their environment—that is, are better adapted—than others.

5. Better-adapted individuals are more likely than others to survive and reproduce; hence the phrase "survival of the fittest."

feet with toes
that can grip bark

stiffened tail feathers
provide support

Figure 8.13 Woodpeckers are superb examples of evolutionary adaptation of a basic bird body plan that suits a specific niche—in this case, eating insects that live in and beneath the bark of trees. Woodpecker adaptations include a powerful, chisel-tipped beak; strong neck muscles for hammering; a sturdy skull with extra padding that protects the brain from impact; a flexible, protrusile tongue that snags insects in crevices; feet adapted for grasping onto tree trunks; and stiff tail feathers that support the bird's body against the tree.

6. Over long periods of time, natural selection can both produce changes in existing species and create new species from preexisting ones.

Scientific and Philosophical Significance

Evolutionary theory has profound practical and philosophical repercussions that make it essential for *all* educated people to understand the essentials of Darwinian thought.

Philosophical ramifications Darwin knew that accepting his theory required believing in *philosophical materialism,* the conviction that matter is the stuff of all existence and that all mental and spiritual phenomena are its by-products. Darwinian evolution was not only purposeless but also heartless—a process in which the rigors of nature ruthlessly eliminate the unfit.

Suddenly, humanity was reduced to just one more species in a world that cared nothing for us. The great human mind was no more than a mass of evolving neurons. Worst of all, there was no divine plan to guide us. These realizations troubled Darwin deeply, for in his day, materialism was even more outrageous than evolution (Fig. 8.14). Some scholars speculate that fear of being branded a heretic for his materialism contributed to Darwin's 21-year delay in publishing his theory. The same antimaterialistic reasoning also drives much modern-day opposition to evolutionary thought.

Yet as pointed out by contemporary evolutionary scholar Douglas Futuyma, seldom do the detractors of the Darwinian world view take note of its positive implications. In Darwin's world we are not helpless prisoners of a static world order but, rather, masters of our own fate in a universe where human action can change the future. And from a strictly scientific point of view, rejecting biological evolution is no different from rejecting other natural phenomena such as electricity and gravity.

Darwin remained to the end a devout, if somewhat unorthodox, Christian. "I see no good reason why the views given in this volume should shock the religious feelings of anyone," he wrote. Like religious scientists of many faiths today, he found no less wonder in a god that directed the laws of nature than in one that circumvented them.

Darwin and politics Political theorists have always had a field day with Darwin's materialistic world view, although different individuals have interpreted and extended its message in diametrically opposite directions. Karl Marx and Friedrich Engels, for example, saw in evolution both justification for the overthrow of the aristocratic order and proof of the inevitability of the class struggle. Yet

Figure 8.14 "I see no good reason why the views given in this volume should shock the religious feelings of anyone," Darwin wrote earnestly. But contemporary clergy quickly condemned his implying that humans had descended from ape-like ancestors, and cartoonists responded much as they would today.

Henry Ford, America's preeminent capitalist, found in Darwinism the perfect rationale for the free-enterprise system.

Herbert Spencer championed the twisted logic of *social Darwinism,* which had nothing to do with Darwin himself. According to social Darwinism, society should operate according to the same principle of "survival of the fittest" as nature does. Although social Darwinism itself is now a discredited philosophy, its underlying principles have had considerable impact in modern history. For example, the ideology that led Adolph Hitler into power in Germany was largely founded on the ideas of social Darwinists.

Evolution and other sciences Evolutionary biology has exerted powerful influences on anthropology, the social sciences, and psychology. By drawing attention to evolutionary trends and similarities in the behaviors of animals and humans, Darwin laid the groundwork for modern studies of comparative psychology and animal behavior. Both Freud's theories of the id and sexual development and Jung's notions of the collective unconscious were influenced by evolutionary thought.

Biological significance Last, but more certainly not least, evolutionary theory provides the current scientific underpinning of all the biological sciences and medicine. Because evolution from shared ancestors has shaped organs and physiological processes that operate in similar ways from worms to humans, we learn effectively about our own bodies and cells by studying bacteria, dogs, and

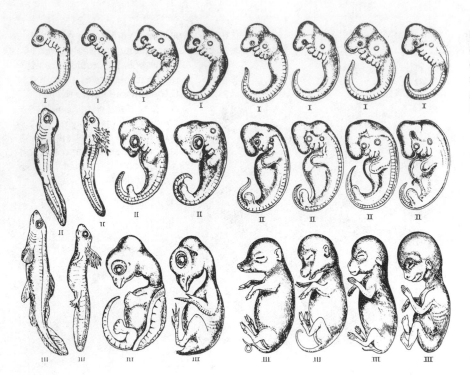

Figure 8.15 The early embryos of humans and other vertebrates look so similar that it takes an expert to tell them apart. During the earliest stages of development, all these embryos have gill pouches and a tail—remnants of structures needed by our aquatic ancestors.

monkeys. Because ecological assemblages of plants and animals—herbivores and plants, parasites and hosts, predators and prey—have evolved together over time, many features of their anatomy and physiology are intimately interconnected in ways that we ignore to our peril and can understand to our advantage.

DATA SUPPORTING THE FACT OF EVOLUTIONARY CHANGE

Darwin's success at proving the *fact* that life had changed over time was admirable; evolution as a fact of natural history was widely accepted by the time of his death in 1882. The honored position that Darwin's work earned him in the world of ideas is evidenced by his burial next to Sir Isaac Newton in Westminster Abbey.

Why was evolution accepted so rapidly? Part of the reason was that laypeople and scientists who had already accepted social change were ready for a theory of change in nature. But much of the credit goes to Darwin himself for his impressive skills in gathering data, both during his voyage of discovery and from observations in his own backyard. Darwin was able, for example, to point out how well his concept of evolution by common descent explained a variety of puzzling biological phenomena that had been documented thoroughly over the years. More recent discoveries enable us to understand many of those phenomena better than Darwin could, but the basis of our understanding is still rooted in his original observations.

Similarities in Anatomy and Development

Darwin and his contemporaries knew that early embryos of many animals look nearly identical and that the earliest stages of development in "lower" animals seem to be repeated in the early development of "higher" animals such as ourselves (Fig. 8.15). Darwin realized that the similar developmental paths followed by animal embryos make sense if all of us evolved long ago from common ancestors through a lengthy series of evolutionary changes.

These striking embryological similarities led some of Darwin's contemporaries (though apparently not Darwin himself) to believe that the embryological development of an individual repeats its species's evolutionary history.

Why, then, should the embryos of related organisms retain similar features when adults of their species look quite different? The cells and tissues of the earliest embryological stages of any organism are like the bottom levels in a house of cards. The final form of the organism is built upon them, and even a small change in their characteristics can result in disaster later. It would hardly be adaptive for a bird to grow a longer beak, for example, if it lost its tongue in the process.

The earliest stages of the embryo's life, therefore, are essentially "locked in," whereas cells and tissues that are produced later can change more freely without harming the organism. As species with common ancestors evolve over time, divergent sets of successful evolutionary changes accumulate as development proceeds, but early embryos stick more closely to their original appearance.

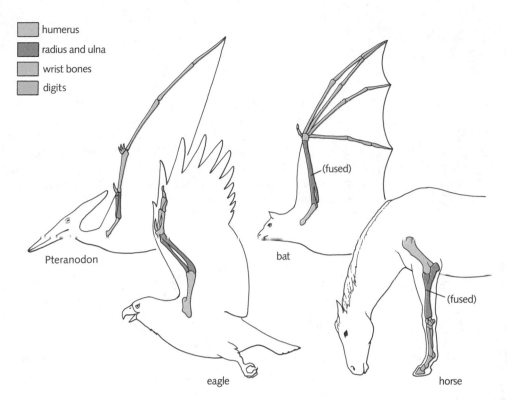

humerus
radius and ulna
wrist bones
digits

Figure 8.16 Diverse vertebrates, all descended from the same four-limbed ancestors, exhibit forelimb structures built upon modifications of the same bones.

Pteranodon

(fused)

bat

eagle

(fused)

horse

Homologous structures and adaptive radiation Darwin knew of many remarkable similarities among body parts of radically different animals. "What can be more curious," Darwin wrote, "than that the hand of man, formed for grasping, . . . the leg of a horse, . . . the paddle of the porpoise, and the wing of the bat should all be constructed on the same pattern, and should include similar bones in the same relative positions?" Figure 8.16 shows how the components of several such limbs are all derived from the same structures in developing embryos. These structures are said to be **homologous** (*homo* means "same") (Fig. 8.17).

Figure 8.17 Homologous flight structures. The wings of bats and birds, though different in certain particulars, are derived from homologous limbs.

Cuvier and other creationist-catastrophists had found it difficult to explain such obviously homologous structures. Why, they were forced to ask, would the creator have chosen to patch together such different animals out of the same mix-and-match bag of parts instead of using unique structures best suited to each?

Darwin's evolutionary theory, on the other hand, provided a satisfying explanation for homology. Sometimes a population of a species or populations of several related species may develop a new, advantageous adaptation or migrate to a new environment that offers many ecological opportunities. Under these circumstances, Darwin hypothesized, organisms would rapidly evolve new adaptations that enabled them to occupy those different niches. We call this evolutionary "spreading out," or divergence of organisms from a common heritage, an **adaptive radiation.**

During an adaptive radiation, evolution shapes new body parts for flying, digging, hunting, or hopping, not by continually reinventing the wheel, but by modifying existing structures to serve new functions. Bats, for example, did not evolve from mouse-like ancestors by sprouting a new set of wings and losing their front legs. Instead, the digits of their forelimbs elongated and developed a membrane of skin stretched between them.

Analogous structures and convergent evolution   There is another kind of similarity in nature: the resemblance between a bird's wing and a butterfly's wing and among a whale's flukes, a fish's fins, and a penguin's webbed feet (Figs. 8.18 and 8.19). Such structures, which look and function similarly but are built of embryologically *un*related parts, are called **analogous.**

Evolutionary theory explains analogous structures as a result of **convergent evolution,** which occurs when different organisms face similar environmental demands (such as flying and swimming) but start out with quite different "raw materials" for evolution to work with. Natural selection then molds these divergent structures along similar lines. Note that when two groups of organisms undergo

Figure 8.18   The wings of butterflies, though they serve the same function as the homologous wings of bats and birds, are strictly analogous.

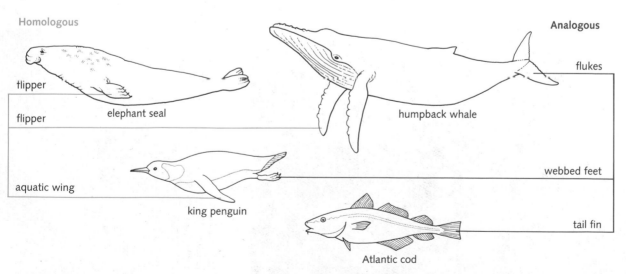

Figure 8.19   Paddle-like structures have developed from different parts of various swimming animals. The tail flukes of whales, the tail fins of fishes, and the webbed feet of water birds are analogous structures. The flipper-like wings of penguins and the front flippers of seals, on the other hand, are homologous.

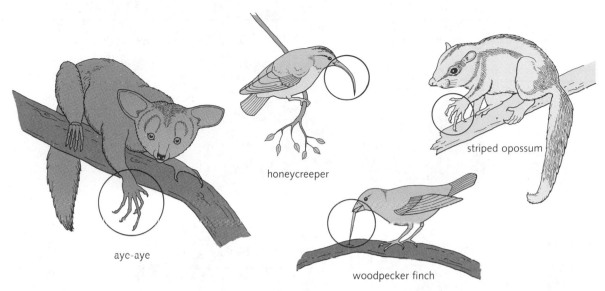

honeycreeper

striped opossum

aye-aye

woodpecker finch

Figure 8.20 Different animals have adapted to the "woodpecker" niche in habitats lacking woodpeckers. The New Guinea striped opossum and the Madagascar aye-aye both have one elongated finger for digging through bark and spearing insects. The Akiapolaau, one of the Hawaiian honeycreepers, uses its lower beak to dig and its upper beak for impaling prey. The Galapagos woodpecker finch holds a cactus spine in its beak to accomplish the same feat.

adaptive radiations in separate, but ecologically similar, environments, convergent evolution often produces remarkably similar-looking—though unrelated—organisms (Figs. 8.20 and 8.21).

Vestigial structures   Darwin found further evidence for evolution in remnant structures that either seem to serve no function at all or have, at best, functions that are *vestigial,* or very much reduced. These are commonly called **vestigial structures.** He recognized that perfection of form was acceptable evidence of evolution if one believed in evolution at the outset. But because creationists argued that perfection in nature reflected the inspired engineering of the divine architect, the mere existence of perfect creatures would not shake their views.

Useless (or nearly useless) structures, on the other hand, made no sense from the creationist viewpoint. Why would the creator have burdened humans not only with the troublesome appendix but also with the useless remains of an embryonic tail that never completely disappears? These sorts of structures make sense only as evolutionary remnants of organs that served an important

Figure 8.21   Deserts in the New World have spawned our familiar cacti (family Cactaceae) (LEFT), while similar conditions in Africa have given rise to remarkably cactus-like plants in a completely different family (the Euphorbiaceae), which includes this relative of the familiar "crown of thorns" (RIGHT).

Relative Time Span		Era	Period	Epoch	Began (millions of years ago)	Length (millions of years)
Cenozoic		Cenozoic	Quaternary	Recent	0.01	
Mesozoic				Pleistocene	1.5	
Paleozoic			Tertiary	Pliocene	12	10
				Miocene	25	13
				Oligocene	34	9
				Eocene	56	22
				Paleocene	63	7
		Mesozoic	Cretaceous		135	70
			Jurassic		180	45
			Triassic		225	45
		Paleozoic	Permian		280	55
	Carboniferous		Pennsylvanian		310	30
			Mississippian		350	40
			Devonian		400	50
Precambrian			Silurian		430	30
			Ordovician		500	70
			Cambrian		570–600	70–100
		Precambrian	Proterozoic		2500	2000
			Archaeozoic		4600	2000

Figure 8.22 LEFT: The Grand Canyon, where the Colorado River has eroded through sedimentary rock formed over millions of years. ABOVE: By observing such slices of Earth's history, geologists assembled the geological time scale.

function to the organism's ancestors. Such remnants, Darwin argued, are proof that today's species were not created in their present form but evolved through time.

Darwin and Fossils: Possibilities and Problems

Darwin knew that the fossilized remains of extinct organisms provide critical evidence that life has evolved over time. But he had trouble reconciling the fossil record as he knew it, in the late nineteenth century, with certain aspects of his theory.

First, Darwin worried about the incompleteness of the fossil record, for he believed that biological evolution, like Lyell's geological evolution, proceeded slowly and gradually. "Nature," Darwin asserted, "does not take leaps." Thus Darwin felt it should be possible to find a long line of fossils documenting gradual transitions between one species and another. In Darwin's time those examples were not forthcoming.

Second, primitive members of most major groups of organisms appeared abruptly at the beginning of the geological period known as the Cambrian (Fig. 8.22). Of the simplest, earlier life-forms there was no record. This "Cambrian explosion," as it was called, looked to many of Darwin's contemporaries like the sudden population of the earth by a Creator, rather than the outcome of gradual evolution.

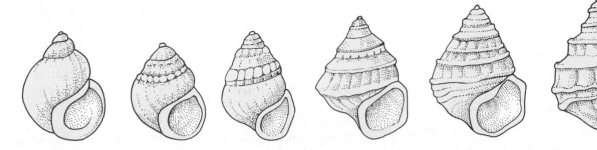

Figure 8.23 These snails illustrate a smooth series of intermediate fossils from the oldest known ancestors to present-day forms. The transitions along the way are gradual, often making it difficult to divide the lineage into distinct species.

Furthermore, nineteenth-century scientists could not prove precisely *how* old any fossils were. No one doubted that ancient sedimentary rocks had formed as sediments accumulated in the sea and that younger rocks had formed on top of them. Thus, as long as fossil-bearing rocks under study had not been disturbed, fossils found in lower layers were clearly older than those found in upper layers. But although these straightforward techniques could arrange fossil series in chronological order, they could not assign an absolute age to any specimen. Thus, in an absolute sense, they could neither prove fossils' antiquity nor provide evidence on rates of evolutionary change.

Darwin spent two chapters in *On the Origin of Species* wrestling with these problems. He pointed out that soft-bodied organisms and nonwoody plants do not fossilize unless they are buried in the right types of sediments before they decompose or are torn to shreds by predators. He also realized that even hard body parts, such as bones and shells, could be destroyed or scattered before fossilization occurred. And he noted that many habitats—such as deserts and mountain ranges—don't offer proper conditions for fossilization. In conclusion, he argued that the fossil record was "extremely imperfect" and allowed that anyone who did not accept that imperfection could "rightly reject my whole theory."

Paleontology and Evolution Today

As biologists applied new insights and new techniques to evolutionary biology and paleontology, they learned that Darwin was overly pessimistic. Although the fossil record is undeniably patchy, thousands of fossil finds have filled in many of the gaps in the evolutionary record over which Darwin agonized. (In addition, we will see in Chapter 13 that evolution may not always proceed at the slow, steady pace on which Darwin insisted.)

We now know of many examples of transitional forms in evolutionary sequences, and even in cases where some pieces are missing, striking intermediate creatures docu-

ment the transition between one group and another (Fig. 8.23). Paleontologists have recently uncovered single-celled organisms that date back much further than the Cambrian explosion (Chapter 25). And several theories involving the physiology and evolution of the first modern cells help explain why this explosive increase in new species occurred when it did.

Furthermore, we can now date fossils reliably by examining their ratios of various **radioisotopes,** unstable atoms that break down to form other atoms. Solar radiation, for example, generates radioactive carbon (carbon-14 or ^{14}C) in the atmosphere. Plants take in that isotope along with normal carbon (^{12}C) as they photosynthesize, and animals pick it up from the plants they eat. Once an organism dies, the ^{14}C in its tissues decays steadily to ^{14}N (nitrogen-14). Thus the older a fossil is, the smaller its ratio of ^{14}C to normal carbon (^{12}C).

Carbon-14 has a half-life of roughly 5700 years, which means that half the molecules of ^{14}C that were present at the organism's death decay to ^{14}N over that length of time. A fossil sample that has a ratio of ^{14}C to ^{12}C that is one-half of the ratio of ^{14}C to ^{12}C in a living specimen will be about 5700 years old. Because ^{14}C has a relatively short half-life, other radioisotopes are used to date really ancient samples. One long-term method measures the decay of radioactive potassium (^{40}K) to argon (^{40}Ar) over a half-life of about 1.3 billion years. And the clock provided by the decay of uranium into lead can be read as far back as 713 million years (Table 8.1).

Table 8.1 *Radioactive Isotopes and Half-Lives*

Radioactive Isotope	Half-Life (years)	Useful Range (years)
Carbon-14	5730	0–60,000
Potassium-40	1.25 billion	> 100,000
Uranium-235	713 million	> 100 million
Thorium-232	14 billion	> 200 million

Coevolution: Support from Plant and Animal Relationships

Some of the most fascinating pieces of living evidence for evolutionary change are the products of a phenomenon called **coevolution.** When we say that two species have co-evolved, we mean that they have evolved in a tightly knit, reciprocal fashion, each evolutionary change in one occurring as a response to a change or changes in the other. Coevolution has occurred innumerable times among animals, among plants, and between plants and animals. Because these relationships are seen most clearly in the context of evolutionary history, you will more fully appreciate the significance of coevolution during our discussion of the history of life on Earth in Chapter 25. It is worthwhile, however, to examine selected dramatic examples here.

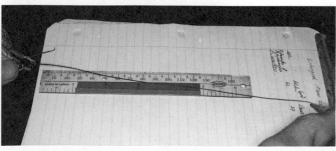

Figure 8.24 TOP: This orchid, *Angraecum orchidglade* (a hybrid between *A. sesquipedale* and *A. eburneum*), carries a long spur with nectar at its base. Darwin, aware that this unusual nectary must have evolved with a specific pollinator able to reach it, predicted the existence of a moth with a long proboscis (BOTTOM). The moth was discovered years later.

Plant–pollinator coevolution Darwin, a master at recognizing the significance of common phenomena in nature, observed that most flowering plants require pollen from another plant to fertilize their flowers, and that many insects spend hours collecting nectar, pollen, or both for food. Darwin correctly hypothesized that flowers and pollinating insects have evolved together in ways generally beneficial to all parties.

After studying the many intricate and unusual flower parts of many orchids, for example, Darwin predicted—although he did not directly observe—one of the most extreme examples of flower–pollinator coevolution. When sent a sample from Madagascar, Darwin recognized an orchid (*Angraecum orchidglade*) whose flower sports a foot-long spur with nectar at its base. He correctly predicted the existence of a moth with a proboscis long enough to sip that nectar (Fig. 8.24).

Plant–herbivore coevolution Over time, many plants have evolved tough leaves, armored spines, indigestible compounds such as tannins, and potent poisons such as alkaloids. (Pyrethrum, one of the most powerful insecticides known, was first extracted from a species of *Chrysanthemum.*) Sometimes these toxic compounds enable plants to live free from insect predators. But sooner or later, one insect species or another evolves resistance to a particular herbivore deterrent, often in the form of an enzyme that inactivates it. From then on, plant and insect are locked in a coevolutionary "arms race."

Fighting the deterrents of several plants is apparently not feasible for an insect. For this reason, it is not unusual to find that closely related species of insects have become specialized to feed only on closely related plant species. The larvae of butterflies in the genus *Heliconius*, for example, feed on either passion flower vines of the genus *Passiflora* or on related species in the same plant family. These plants all contain similar poisonous compounds, and these larvae are among the few insects that can ingest these compounds without harming themselves. In response, several *Passiflora* species (Fig. 8.25) have evolved stiff, hooked hairs that trap, injure, and kill caterpillars.

Symbiosis: Life Together

As we noted in Chapter 5, many organisms engage in symbiotic relationships that have powerful effects on their ability to survive and reproduce. Each type of symbiosis—parasitism, commensalism, and mutualism—provides examples of animal and plant species that have evolved intricate and often essential relationships with one another.

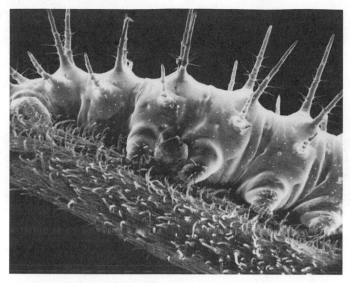

Figure 8.25 This scanning electron micrograph of a caterpillar impaled on the spines of a *Passiflora* leaf shows a striking example of a plant that "bites back" at its predators. Although the caterpillar has evolved resistance to toxins present in the leaves, it must still contend with these physical defenses.

Parasitism Technically, *parasitism* is a form of symbiosis in which one partner benefits from the association while the other partner is harmed. Although parasites can cause serious illness or death, no evolutionarily successful parasite could be rapidly lethal to all its hosts or it would soon be out of business! As we saw in the case of rabbits and myxomatosis in Chapter 6, parasite virulence and host resistance evolve together in a continually shifting balance.

Commensalism In a form of symbiosis called *commensalism*, one of the participants benefits but the other is neither helped nor harmed. Orchids and Spanish moss, for example, may look like parasites but they are not; they attach harmlessly to the branches of host trees (Fig. 8.26). They obtain their nutrients from rainwater and decaying leaves that collect around their roots, not from the host plant.

Mutualism In *mutualism,* both members of the association benefit. In fact, the participants in many mutualistic relationships cannot survive without their partners. We have already encountered such examples of mutualistic symbioses as wood-digesting microorganisms in the guts of herbivores, clownfish in the tentacles of sea anemones, and lichens. In fact, one hypothesis (discussed in detail in Chapter 24) proposes that the very cells of all higher organisms are symbiotic associations between two simpler cell types.

Mimicry

Mimicry, in which two or more species resemble each other closely, demonstrates intimate evolutionary relationships between species but may or may not fall precisely in the strictly defined category of coevolution. In **Batesian mimicry** one organism evolves in ways that enable it to resemble another organism that has a defense against a common predator. In **Müllerian mimicry** two species with similar defense mechanisms evolve in such a way that they ultimately come to resemble each other. Note that no individual (or species) consciously "learns" to "trick" another; the evolution of mimicry is a long, random evolutionary process guided by natural selection.

Figure 8.26 This orchid of the genus *Cattleya* depends on its host tree for support in a simple commensal relationship. The orchid obtains nutrients from decomposing organic matter and leaf wash that run down the trunk over its spongy, adhesive roots.

Figure 8.27 RIGHT: Poisons, resistance, and mimicry in butterflies. Monarch caterpillars pick up toxic compounds from the milkweed plants they eat. The monarchs not only tolerate those compounds but also store them in their tissues and retain them after metamorphosing into adult butterflies. Any predator unfortunate enough to swallow a monarch containing those compounds becomes violently sick to its stomach. BELOW LEFT: The similar-looking queen butterfly also tastes bad to birds and so is classified as a Müllerian mimic. BELOW RIGHT: The viceroy butterfly contains no such poison but gains protection by mimicking the monarch; viceroys are therefore considered Batesian mimics.

The best-known group of North American animal mimics includes monarch, viceroy, and queen butterflies (Fig. 8.27). At the center of the group is the conspicuously colored monarch, whose noxious taste protects both itself and its tasty mimics (Fig. 8.28). Mimicry also occurs in plants, where the selective advantage it confers may enhance either pollination or protection from predators (Chapter 31).

One example of coevolutionary mimicry occurs in the plant–herbivore "arms race" between *Passiflora* and *Heliconius*. Several *Passiflora* species have evolved a remarkable defense that specifically exploits the egg-laying behavior of their insect enemies. The larvae of *Heliconius* butterflies—which compete for food on small patches of *Passiflora*—are often aggressive and cannibalistic. For that reason, larvae emerging on leaves that already host young caterpillars are less likely to survive than those emerging

on vacant leaves. Research has shown that female butterflies tend not to lay eggs on leaves that already have eggs on them. In a remarkable defensive adaptation, several *Passiflora* species have evolved structures that mimic the shape and color of the *Heliconius* eggs, ostensibly "fooling" females into avoiding them!

Support from Comparative Biochemistry

Just as anatomists have found structural homologies among related animals, biochemists have found chemical homologies among related species. Living organisms share many biologically important chemical compounds, and the more closely related two species are, the more closely their chemical constituents resemble each other

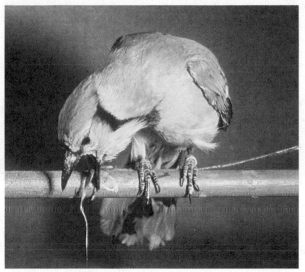

Figure 8.28 Once a bird has been made sick by eating a monarch, it avoids anything that resembles the distasteful species.

 (Fig. 8.29). The degree of resemblance among these compounds is often shocking. For example, anatomical evidence convinced biologists long ago that humans and chimpanzees are close relatives. But it is fascinating to discover that we share more than 98 percent of our genetic code with chimps! We will discuss chemical homologies in more detail in Chapter 22.

DARWINIAN THEORY EVOLVES

Despite his ability to convince the world that life has evolved (and continues to evolve), Darwin was initially less successful in promoting his theory of natural selection, the process he credited with directing evolutionary change. But though natural selection as Darwin envisioned it has never been universally accepted, most of its basic principles have been reinforced and extended as biological knowledge has grown over the last century. Part of the strength of Darwinian theory lies in its ability to absorb information and evolve itself as our knowledge of the world evolves.

One major weakness of the theory in Darwin's time was that neither Darwin nor his colleagues understood how heritable traits are controlled and passed from generation to generation. In Darwin's day, it was believed that parents' traits blended in their offspring, somewhat like mixing paints: blend a strong color and a weak one, and you get a mixture of intermediate hue. Everyone knew, for example, that crossing large animals with small ones produced offspring of intermediate size and that crossing red flowers with white ones often yielded pink progeny. The idea made sense, but no one knew why things worked out that way.

Darwin realized that this view of inheritance could spell trouble for natural selection. When an organism appeared with a new, favorable characteristic, it would have to mate with an organism which probably did not have the new characteristic to pass that adaptation to the next generation. But if all its characteristics were "blended" with those of typical members of the species, they would be blended toward the species' average and would diminish in strength generation after generation. Any new trait, no matter how useful, would disappear so quickly that evolutionary change would be impossible.

This problem concerned Darwin. In later editions of *On the Origin of Species*, he suggested that the environment might somehow cause variation and that characteristics acquired in this way might be inherited. Without a better understanding of inheritance, Darwin was up against a formidable obstacle.

The Contribution of Genetics

Ironically, even as Darwin was grappling with this problem, ground-breaking studies on the nature of inheritance were being conducted by an Austrian priest named Gregor Mendel. But Mendel's elegant work, which set the stage for a radically new way of thinking about heredity, was ignored by the scientific community for over 30 years, so Darwin may never have learned about it.

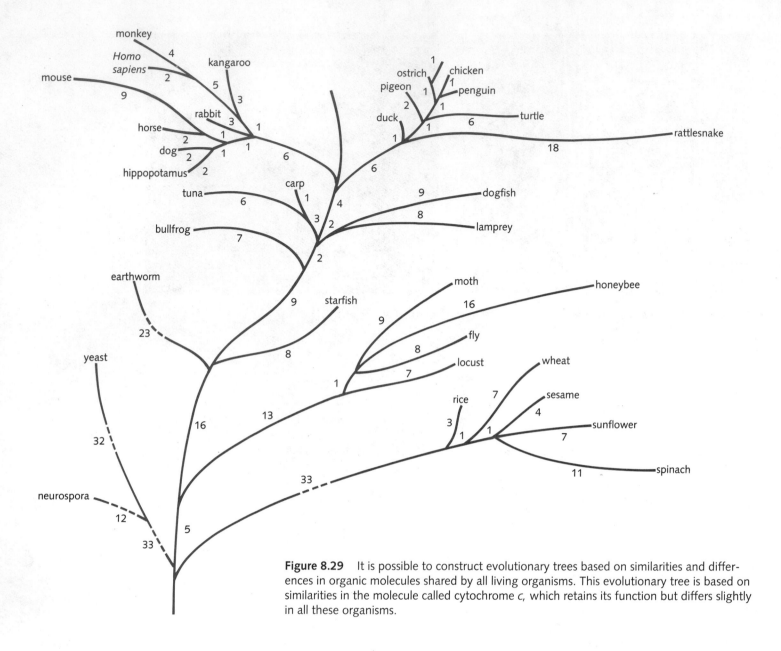

Figure 8.29 It is possible to construct evolutionary trees based on similarities and differences in organic molecules shared by all living organisms. This evolutionary tree is based on similarities in the molecule called cytochrome *c*, which retains its function but differs slightly in all these organisms.

Then, around the beginning of the twentieth century, several independent researchers obtained results similar to Mendel's, and Mendelian genetics was born. The maturation of genetics as a discipline since that time has revolutionized evolutionary theory. Because genetics is fundamental to modern evolutionary theory, we will spend the next three chapters examining the principles of Mendelian genetics. We will then explore the ways in which genetics, ecology, and population biology are woven together in the exciting and ever-changing world of evolutionary biology today.

SUMMARY

Darwinian evolutionary theory has profoundly influenced scientists, philosophers, religious leaders, and laypeople since the day *On the Origin of Species* was published. Today evolutionary theory provides the basis for understanding all the biological sciences and medicine, and over the years it has affected the development of the social sciences, anthropology, psychology, and politics.

Though Darwin was not the first to propose that life has evolved, he provided the first plausible scientific mechanism by which evolution could occur. His thinking was influenced both by the accumulating evidence for gradual geological change and by his travels and reading. Darwin's theory of evolution by natural selection explains evolutionary change on the basis of the differential survival and reproduction of those individual organisms best adapted to their local environments. The variation in living species that makes such change possible is random, purposeless, and not subject to control.

Many current lines of evidence prove that evolution has occurred. Fossils document extinct transitional forms between living groups of organisms. Similarities among diverse living organisms—ranging from the structure of bones and muscles to similarities among proteins and genetic material—prove that modern organisms share common ancestors. And the phenomena of coevolution, symbiosis, and mimicry demonstrate intimate, long-term evolutionary relationships between and among species.

STUDY FOCUS

After studying this chapter, you should be able to:

- Outline the nineteenth-century scientific world view, and explain the developments in science and society that made the world view indefensible.

- Place Charles Darwin and his ideas in the context of nineteenth-century society.

- Describe the events that led Darwin to formulate his theory.

- Explain classical Darwinian evolutionary theory.

- Appreciate the importance of evolutionary theory in the biological and social sciences.

TERMS AND CONCEPTS

evolution *148*	common descent *160*
catastrophism *150*	homologous *163*
uniformitarianism *152*	adaptive radiation *164*
teleological *152*	analogous *164*
fitness *155*	convergent evolution *164*
variations *157*	vestigial structures *165*
heritable *157*	radioisotopes *167*
artificial selection *158*	coevolution *168*
natural selection *159*	Batesian mimicry *169*
adaptation *160*	Müllerian mimicry *169*

REVIEW

Objective Questions (Answers in Appendix)

1. Evolution can be described as
   (a) a continuing process.
   (b) a catastrophic event in the past.
   (c) static.
   (d) the attaining of an ideal type.

2. During Darwin's time, inheritance was considered to result from
   (a) the passing of genes from generation to generation.
   (b) the passing of chromosomes from generation to generation.
   (c) dominant maternal traits.
   (d) the blending of parents' traits.

3. Of the greatest importance in evolution is the number of individuals who, under local environmental conditions,
   (a) are born and survive.
   (b) survive and reproduce.
   (c) undergo coevolution.
   (d) undergo adaptive radiation.

4. Which of the following is *not* an example of coevolution?
   (a) flowers and their pollinators
   (b) predators become more effective hunters, and their prey evolve a better means of escape
   (c) plants and specialized plant eaters
   (d) plants that can be eaten by a variety of insect species

5. A major weakness in Darwin's theories was that
   (a) there was no explanation of how characteristics are transmitted from parents to offspring.
   (b) the concept of survival of the fittest was omitted.
   (c) the concept of natural selection was omitted.
   (d) he did not explain how the theory of special creation supported his observations on the Galapagos Islands.

Discussion Questions

6. How do natural selection and artificial selection differ? Can pressure from either natural or artificial selection cause specific variations in organisms?

7. What was the Platonic view of variation in nature, and how did Darwin's view differ from it?

8. How did the Malthusian doctrine inspire Darwin?

9. What changes in the scientific world view were necessary before Darwin could begin to think about biological evolution?

10. What three principles constituted Lyell's theory of uniformitarianism?

11. Darwin's theory dispensed with the teleological thinking of Lamarck and his colleagues. Using descriptions of variation and natural selection, explain why Darwinian evolution can have no ultimate purpose.

12. Why were Darwin's ideas both welcomed and feared by nineteenth-century Europeans?

13. Give three examples of homologous structures in familiar animals. Give three examples of analogous structures.

14. Why did Darwin so enthusiastically advance vestigial structures as proof of his theory?

READING

"Alfred Russel Wallace's malarial 'fit.' *Natural History* (July 1991): 53.

9
Continuity of Life: Cellular Reproduction

Death, be not proud, though some have called thee
Mighty and dreadful, for thou art not so;
For, those whom thou think'st thou dost overthrow,
Die not, poor Death, nor canst thou kill me.

—John Donne (1633)

*O*f all things that are certain in life, death is the surest. Every living thing will ultimately die, ourselves included. Yet despite this fact, life itself seems almost immortal. For about 3 billion years, the thread of life on this planet has been spun into a wide variety of organisms, each of which is linked to the whole by common ancestry as well as by the common properties of life itself.

Although the triumph of death over the biological individual is inevitable, death, even as John Donne noted, does not triumph over life. Why not? The ability of cells and organisms to reproduce, to pass life from one generation to the next, has ensured that life itself outlasts the death of any one individual.

For simple organisms, death is *not* inevitable. Think about that for a moment. If a single-celled organism divides to produce two identical cells, which is parent and which is offspring? The answer is that for two *identical* cells, that question has no meaning. In a sense, each of the two cells has the same identity as the one that gave rise to them. We could trace each cell backward through time in an unbroken series of cell divisions that reach as far into the past as we care to go. As long as such cells are able to divide, they do not age in the sense that we do. One might almost think of them as immortal.

The ability to outlast death via reproduction is a property of the individual cell. Ultimately, life *does* triumph over death. It does this not by magic but by the simple act of growth and cell division.

THE DIVERSITY OF CELLS

The invention of the microscope in the seventeenth century led directly to the discovery that living organisms are composed of cells. Although all cells share certain similarities, in many ways the most striking thing about the cells that make up a living organism is their diversity (Fig. 9.1). It is commonly estimated that the human body is made up of at least 85 completely different types of cells, each distinguishable from the other—and that is just for our species alone. When we examine the larger world, we find that cells exist in an almost infinite variety of sizes, shapes, colors, and structures. It is still tempting to generalize, of course, and try to describe a typical cell. Such a cell would be about 20 μm (2.0×10^{-5} m) in diameter; it would have a prominent, centrally placed structure about 6 μm in diameter that we call a **nucleus** (after a Latin word meaning "kernel," because it looks like a seed in the center of a fruit); it would be surrounded by a cell membrane that separates it from its surroundings; and it would contain a

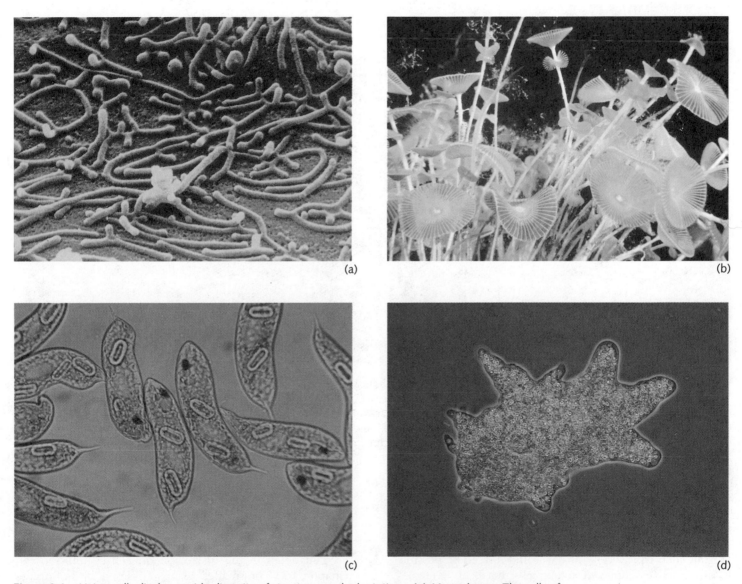

(a)   (b)   (c)   (d)

Figure 9.1 Living cells display a wide diversity of structures and adaptations. **(a)** *Mycoplasma:* The cells of this small and simple prokaryote grow, often undetected, in a variety of environments (scanning electron micrograph). (Magnification factor: 18,000) **(b)** *Acetabularia:* The mushroom-shaped cells of this green alga are among the largest in nature. Many species are 1–3 cm in length, like these specimens, photographed beneath the waters of the Mediterranean. **(c)** *Euglena:* These motile cells swim rapidly, propelled by two whiplike flagella. (Magnification factor: 400) **(d)** *Amoeba:* These large and flexible cells move by streaming their cytoplasm into false feet, or pseudopodia. (Magnification factor: approx. 60)

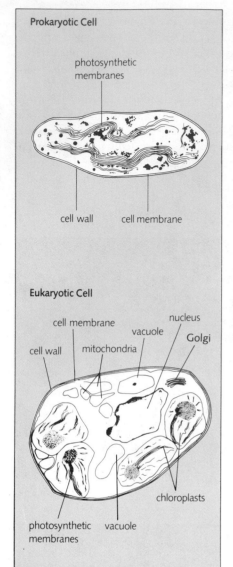

Prokaryotic Cell

photosynthetic membranes

cell wall   cell membrane

Eukaryotic Cell

cell membrane   nucleus
cell wall   vacuole   Golgi
mitochondria

chloroplasts

photosynthetic   vacuole
membranes

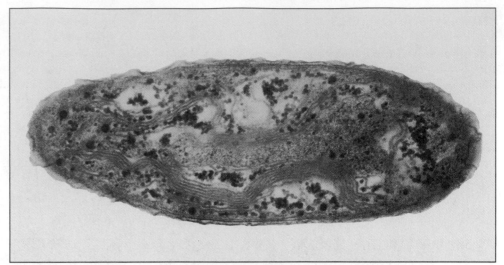

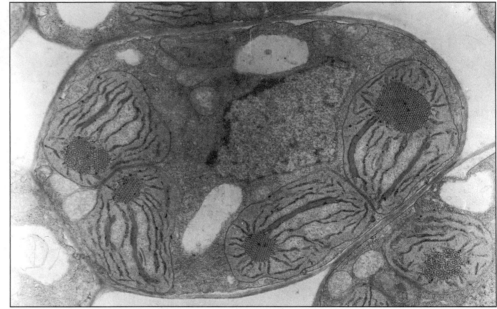

Figure 9.2   These cell diagrams and electron micrographs of two photosynthetic organisms provide a basis for comparing the structures of eukaryotic and prokaryotic cells. Eukaryotic cells possess nuclei, while prokaryotic cells do not. Each cell type possesses different photosynthetic membranes. In the eukaryotic cell, these membranes are enclosed in an organelle called the chloroplast. In the prokaryotic cells, these membranes are free in the cytoplasm. (Magnification factors: TOP, approx. 16,500; BOTTOM, approx. 2000)

number of tiny structures that might be thought of as miniature organs: **organelles.** We might be bold enough to draw such typical cells, and it would indeed be possible to find a few examples of actual cells that bear a remarkable resemblance to our drawings (Fig. 9.2).

Generalizations are always useful, but we must not allow them to keep us from appreciating the enormous number of cases in which the generalization does not apply. Let's consider, for example, our generalization about cell *size*. Cells are small and that's why the use of the mi-

croscope, which expanded our range of vision, preceded the development of the cell theory. (Microscopes are covered in detail in the discussion on cell organization in Chapter 16.) The smallest living cells are a kind of organism known as *mycoplasma*, a form of bacteria. They are less than 0.2 μm across and can be observed only in an electron microscope. But many cells are large enough to be seen with the unaided eye. In oceans throughout the world, tiny plant-like organisms grow on the surfaces of submerged rocks. A particularly beautiful one is *Acetabu-*

laria mediterranea, which grows in the Mediterranean. Individual *Acetabularia* measure up to 2 or 3 cm, yet each one is a single cell. Even larger cells are found elsewhere in nature. The nerve cells carrying the impulses that control the muscles in your lower limbs, for example, can be as long as 1.5 m.

Types of Cells

Living cells can be divided into two distinctly different classes. Cells with nuclei are commonly known as **eukaryotic** (*eu* means "true"; *karyon* means "kernel" or "nucleus") cells, whereas living cells that lack nuclei are called **prokaryotic** (*pro* means "before") cells. Typical prokaryotic cells are bacteria such as *Escherichia coli* and blue-green algae such as *Oscillatoria* (Fig. 9.3).

Although prokaryotic cells may contain some internal membranes, as a rule they are less complicated internally than eukaryotic cells. Eukaryotic cells generally contain a large number of specialized, membrane-enclosed organelles that enable the cell to carry out a variety of activities in different compartments. In other words, eukaryotic cells seem to display a greater degree of internal specialization. The fundamental distinction between prokaryotes and eukaryotes, of course, remains the presence of absence of a nucleus.

We will use some terms that are related to the earlier terminology, however. In eukaryotic cells we might think of the cell as divided into two large sections: the nucleus and the cytoplasm. The word **cytoplasm** ("cell fluid") is widely used by biologists to refer to the major *compartment* in most living cells. In this way, we can quickly describe molecules found outside the nucleus as "cytoplasmic" and can speak of specific organelles as being found in the cytoplasm (as opposed to the nucleus). In prokaryotic cells, where there is no nuclear compartment to describe, we can still speak of the cytoplasm as the large compartment bounded by the cell membrane.

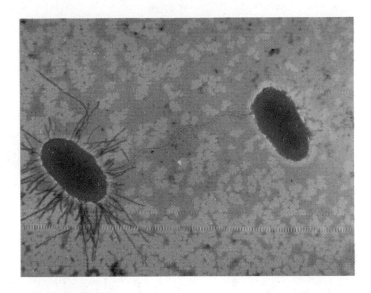

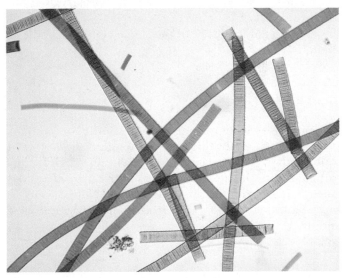

Figure 9.3 *Escherichia coli* (TOP) and *Oscillatoria* (BOTTOM) are prokaryotes, cells that do not contain nuclei. The *E. coli* are engaged in conjugation, a process that transfers genetic information from one cell to another.

GROWTH AND DEVELOPMENT: THE PROBLEM OF REPRODUCING A CELL

Growth is one of the principal characteristics of life; in fact, it is one of the attributes by which we recognize an organism as being alive. Because living things are made up of cells, growth is a process that occurs first at the level of the cell. Within limits, cells may increase in size, but no process of growth can be sustained for very long without involving **cellular reproduction**—the formation of new cells.

The basic patterns of cellular reproduction are deceptively simple: A cell passes through a phase in which it increases in size; then it divides and two new "daughter"

cells are formed. In many respects, each of these daughter cells is identical to the single cell that produced it. Therefore, as we examine the process of cellular reproduction, we should expect to find two things: (1) a process of *duplication*, so that each daughter cell will possess a complete set of cellular structures, including the information-carrying structures, and (2) a process of *separation*, so that cell division carefully and precisely parcels structure and information between the two daughter cells. In this chapter we will examine these two processes, which are at the heart of life's most essential mystery, the formation of new cells.

Cellular Growth

As a cell increases in size, its needs increase as well. It requires larger amounts of nutrients and other materials from its environment, and it produces larger amounts of waste products. These increasing demands place an effective limit on how big a cell can be. The mathematics of size and shape dictate that when a cell doubles in diameter, its surface area increases by a factor of 4. Its volume, however, increases by a factor of 8. The simplest solution for the problem of increasing cell size is the process of cell division, the formation of two smaller cells by the splitting of a preexisting one.

In single-celled organisms, such as the bacterium *Escherichia coli,* cell division is often the means by which members of the species *reproduce,* because the production of two new cells means the production of two completely new individuals (Fig. 9.4). Except in special cases, the process of cell division produces two cells, each of which has all the characteristics of the single cell that produced them.

The Cellular Life Cycle

The process of cell growth cannot be continuous and unbroken. A cell is not filled with a nondescript molecular soup. Rather, each cell contains within it the necessary information, in biochemical terms, to synthesize the molecules it needs for life. In order for cell division to be successful, this essential information must be duplicated and then carefully divided between the two daughter cells.

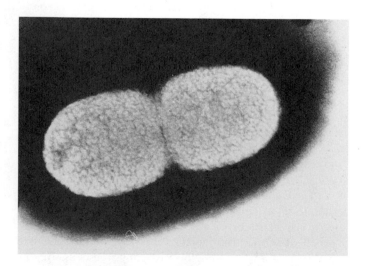

Figure 9.4 "Binary fission" aptly describes the process of cell division in most prokaryotes. When it has completely copied its genetic material in the form of DNA, the cell splits in two.

The most important form of biochemical information is carried from one generation to the next in the molecule known as **DNA** (deoxyribonucleic acid). DNA molecules contain instructions for synthesizing proteins, and this enables them to carry the information needed to build nearly all of the important molecules of the cell. Therefore, one of the critical points in the time between one cell division and the next is the time when the duplication of DNA molecules occurs (Fig. 9.5).

When essential information has been duplicated, the process of cell division must then ensure that each daughter cell receives a complete set of that vital information. Mechanisms must exist to distribute DNA molecules between the two daughter cells and, finally, to break what had been a single cell into two cells independent of each other. The cycle of changes involved in cellular reproduction occurs repeatedly as living cells progress from one generation to the next.

CELL DIVISION IN PROKARYOTES

Prokaryotic cells—cells without nuclei—are able to divide in a relatively simple way. After the cell has grown to the point where division is possible and has duplicated its DNA molecule(s), a small fissure begins to grow inward from the cell membrane. Prokaryotic DNA molecules are attached at several points to the cell membrane, and these DNA molecules can be separated at their membrane attachment points, as shown in Fig. 9.6. As the DNA molecules are gradually separated, the cell membrane and cell wall continue to grow inward until the point at which the two cells will divide becomes apparent.

This process resembles a simple splitting of one cell to form two, and it is often called **binary fission** (*fission* means "splitting"). Bacteria placed in rich nutrient broth can carry out binary fission as often as once every 20 minutes, which gives them the capacity to increase in numbers at a fantastic rate.

CELL DIVISION IN EUKARYOTES

The process of cell division in eukaryotes is known as **mitosis.** Mitosis differs somewhat from one cell type to the next, but its general features are the same in most eukaryotic cells. Eukaryotic cells have an additional feature that makes cell division more complex: The majority of their DNA molecules are collected in structures known as **chromosomes,** which are found within the nucleus. Chromosomes (*chromo* means "colored"; *soma* means "body") contain the genetic information of the cell.

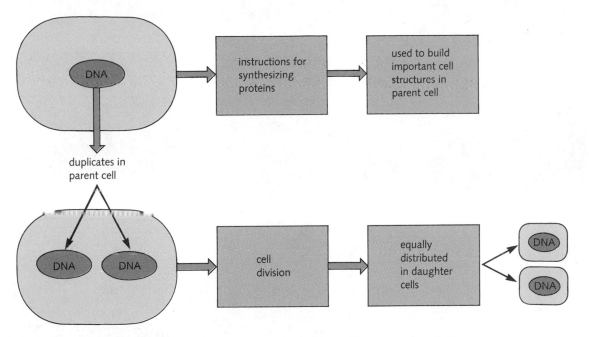

Figure 9.5 Cellular information in the form of DNA is duplicated between one cell division and the next. DNA carries a number of forms of information, including the instructions for building proteins.

Chromosomes were originally discovered in eukaryotic cells because of their distinctive appearance during mitosis. Chromosomes are composed of a coiled web of protein and DNA, which is dispersed within the nucleus during most of a cell's life cycle. During cell division, however, each chromosome condenses into a distinct, visible structure as the cell enters mitosis (Fig. 9.7).

The Eukaryotic Cell Cycle

The life cycle of a eukaryotic cell can be divided into two general phases: mitosis (the process of cell division) and **interphase** (the time between cell divisions). Because many cells proceed quickly from one mitosis to the next, we can visualize some of the repeating events as laid out graphically in a circular pattern that is known as the **cell cycle** (Fig. 9.8).

The term *interphase* is misleading because it gives the impression that little is happening in the time between cell divisions. However, this is the period when a cell is rapidly growing, increasing in size, developing new structures, and synthesizing new molecules. In order to prepare for the next round of cell division, each rapidly growing cell must have a period of time in which it duplicates its DNA. This is known as the *S phase* of the cell cycle, where S stands for the synthesis of DNA.

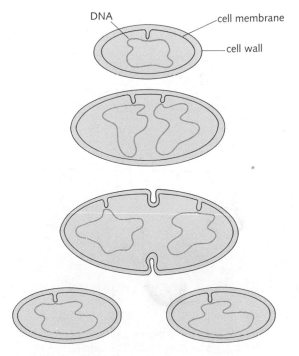

Figure 9.6 In many prokaryotes, attachments to the cell membrane ensure that each cell receives a complete DNA molecule.

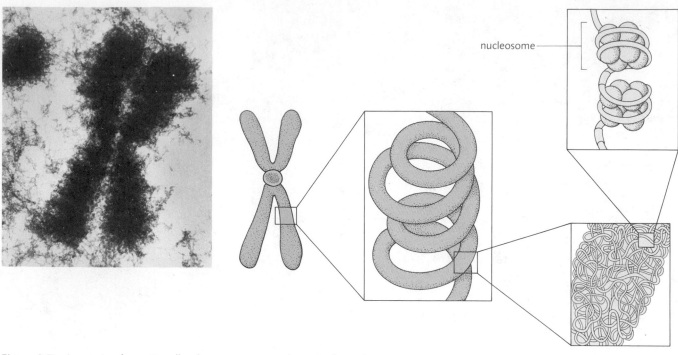

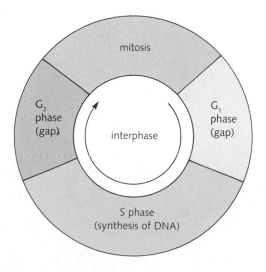

nucleosome

tightly coiled
DNA and protein

Figure 9.7 In most eukaryotic cells, chromosomes condense to form distinct structures during mitosis. The electron micrograph shows a single mitotic chromosome. Protein and DNA make up the structure of a chromosome. DNA is wound around protein cores and folded into a complex pattern.

Figure 9.8 The eukaryotic cell cycle. DNA synthesis occurs during the S phase, which is separated from mitosis by two "gaps," G_1 and G_2.

The fact that DNA is synthesized at a particular point in the cell cycle enables us to break interphase down a bit further. The G_1 *phase* is the "gap" between mitosis and the beginning of DNA synthesis, and the G_2 *phase* is the time between the S phase and the beginning of mitosis.

Scientists studying the cell cycle have found that certain events take place during each phase of the cycle. The synthesis of some proteins, for example, is confined to G_1. Other molecules are made only in G_2, just prior to mitosis, whereas DNA-associated proteins are often synthesized preferentially during the S phase.

The lengths of the various phases of the cell cycle are extremely variable. In rapidly growing yeast cells, the entire cycle takes as little as 2 hours. Human skin cells may take 24 hours to complete the cycle, and some cells may spend days, weeks, or even years in just one phase of the cell cycle. In an important sense, the G_1 phase of the cycle is the "decision" period for each cell. When a cell completes the G_1 phase and enters the S phase, it has made the "decision" to divide—there is no point in beginning DNA replication unless mitosis will take place. This is why G_1 is the phase in which nongrowing cells find themselves locked. For example, sugar-conducting cells of a woody plant may spend decades in the G_1 phase of the cycle.

Cellular changes during the cycle Cells go through a series of dynamic changes during the cell cycle, some of which are reflected in changes at the cell surface. Although there are slight differences in cellular appearance during the G_1, S, and G_2 phases, the most profound changes occur as a cell enters mitosis. Cells grown in culture "round up" as they begin to enter mitosis, withdraw from the supportive substrate upon which they have been growing, and mobilize their resources for the dramatic changes of mitosis. Scanning electron microscopic images of cells in various phases of the cycle confirm this (Fig. 9.9).

Mitosis: The Beginning of Division

Like life, mitosis is a continuing process. And just as there are no clear-cut distinctions between adolescence and adulthood, two phases of human life, there are few obvious points at which the process of cell division can be divided into stages. Nonetheless, since the end of the nineteenth century, scientists have broken mitosis into four distinct stages or phases to make studying the process a little easier (Fig. 9.10). The four stages are as follows:

- **Prophase.** Preparation for cell division. Chromosomes condense, and the nuclear envelope disappears.

- **Metaphase.** Chromosomes align along a central plane.

- **Anaphase.** Chromosomes separate and move.

- **Telophase.** Mitosis ends and the two daughter cells separate.

Figure 9.9 These cells in culture change shape as they pass through the phases of the cell cycle. The cell in the center of the field has rounded up and is beginning mitosis. The two cells to its right are in late G_2 and have begun to separate from the substrate.

Figure 9.10 The four stages of mitosis.

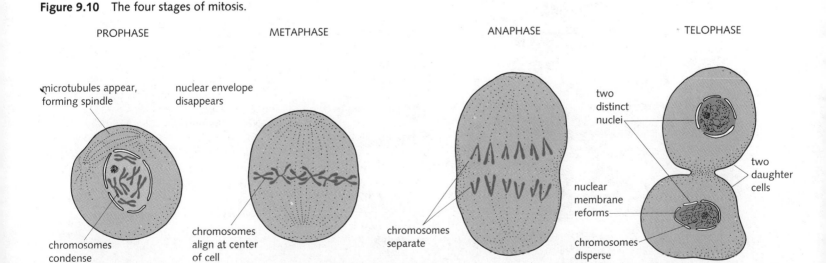

PROPHASE — microtubules appear, forming spindle — chromosomes condense

METAPHASE — nuclear envelope disappears — chromosomes align at center of cell

ANAPHASE — chromosomes separate

TELOPHASE — two distinct nuclei — nuclear membrane reforms — chromosomes disperse — two daughter cells

Figure 9.10 A schematic representation of the four stages of mitosis.

Prophase After interphase (Fig. 9.11), the cell begins the **prophase** stage of mitosis. In animal cells, the first hint that mitosis is about to take place is the duplication of two small structures known as **centrioles** (Fig. 9.12). The centrioles are found just outside the nuclear envelope, and after duplication, they slowly move to either side of the nucleus. Gradually, the **mitotic spindle,** a structure of fine filaments, develops near each of the centrioles and envelops the nucleus. The mitotic spindle is made up of tiny structures known as **microtubules.**

Another important prophase event is the rapid disintegration of the nuclear envelope (see Theory in Action, Where Does the Nuclear Envelope Go?), which can take

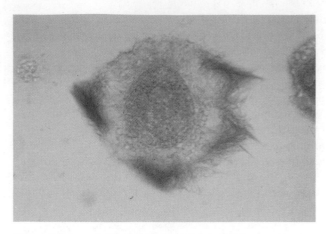

Figure 9.11 A typical cell late in G$_2$ contains a nucleus with dispersed chromatin, as shown in this light micrograph of an interphase cell from the endosperm seed tissue of a plant from the genus *Haemanthus*, commonly known as a blood lily.

THEORY IN ACTION

Where Does the Nuclear Envelope Go?

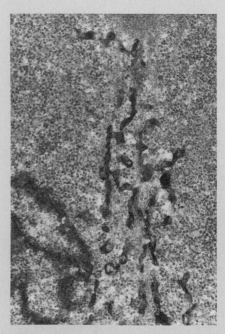

These small vesicles, seen in an electron micrograph of a plant cell in prophase, are the remnants of the nuclear membranes.

Near the end of the nineteenth century, biologists studying cell division began to wonder about one of the very first events of prophase, the apparent dissolution of the nuclear envelope. Even under the highest-power light microscope, the envelope, which is composed of two distinct membranes, seems simply to disappear. Just as magically, two nuclear envelopes re-form at the conclusion of telophase. Where do the nuclear membranes go during mitosis? Until very recently, this was just one of many unsolved mysteries of cell division.

Peter Hepler, a scientist at the University of Massachusetts, seems to have found the answer to the riddle of the disappearing membrane. By using the electron microscope to study cells just after the disappearance of the membranes, he has identified a large number of small vesicles, bounded by membranes, that appear late in prophase of mitosis. Late in anaphase, these vesicles disappear. It seems that the nuclear membranes don't dissolve after all. Instead, the material that makes up these membranes becomes rearranged into a series of smaller structures. At the end of mitosis, these smaller structures are brought back together to assemble the nuclear membranes of each daughter cell. At the level of the light microscope—which is how scientists studied mitosis and we still usually look at mitosis—the membranes seem to disappear, but the fact is that they simply change into a form that keeps them out of the way during cell division.

place in as lirttle as 15 seconds in animal cells. Also in prophase, dense, thread-like chromosomes begin to appear (Figs. 9.12, 9.13, and 9.14). Gradually, these chromosomes untangle themselves and gather near the center of the cell. With a good microscope, it is possible to see that each chromosome consists of two identical strands called **chromatids** (Fig. 9.13, *bottom*) joined in a region that biologists have named the **centromere** ("central zone"). The point at which each chromatid is attached to the mitotic spindle is known as the **kinetochore.**

Metaphase When the centromeres of each chromosome are aligned in a single plane at the center of the

Figure 9.12 In most animal cells, one of the first clues that mitosis is about to begin is the replication of centrioles. RIGHT: Following their replication, a pair of centrioles moves to either side of the nucleus.

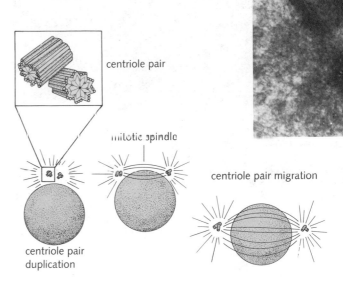

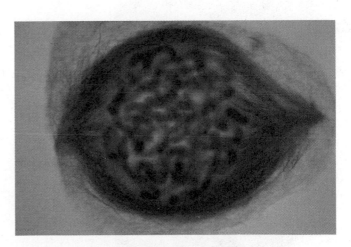

centriole pair

mitotic spindle

centriole pair migration

centriole pair duplication

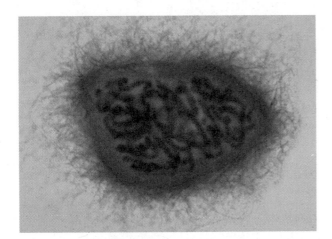

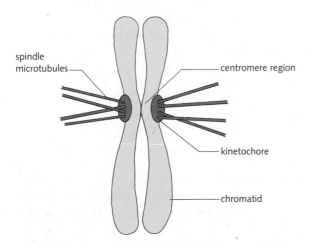

spindle microtubules

centromere region

kinetochore

chromatid

Figure 9.13 TOP: Early prophase, shown in a cell of the plant *Haemanthus*, is marked by the appearance of individual chromosomes and a change in the structure of the cytoskeleton that is associated, in animal cells, with the separation of centrioles. BOTTOM: During prophase, each chromosome becomes visible as a pair of chromatids. The chromatids are joined to each other at the centromere, and will be joined to the mitotic spindle at the kinetochore after the breakdown of the nuclear envelope.

Figure 9.14 The end of prophase and the beginning of metaphase is marked by the complete disintegration of the nuclear envelope. Fully condensed chromosomes are now visible.

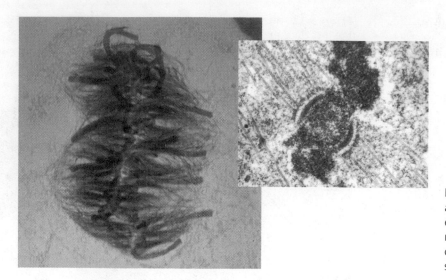

Figure 9.15 At metaphase, the chromosomes are aligned along the central plane of the cell. INSET: This electron micrograph of the centromere region of an animal cell shows both kinetochores. (The detailed structure of the kinetochore in a plant, like the *Haemanthus* cell shown at left, is somewhat different.)

cell, **metaphase** begins (Fig. 9.15). At metaphase, the entire set of chromosomes can be seen across the center of the cell. The mitotic spindle extends from one "pole" of the cell to the other (Fig. 9.15, *inset*).

As we have seen in animal cells, the poles of the spindle are oriented around tiny structures in the cytoplasm known as centrioles. Plant cells lack centrioles, but the poles of their mitotic spindles have a similar shape. The spindle is a dynamic structure. Tension between spindle fibers extending to each pole seems to be responsible for moving chromosomes toward the center of the cell. In what can be described as a gentle tug-of-war, each side pulls with roughly the same pressure, moving the chromosomes toward the center of the plant cell.

Metaphase, often the briefest stage of mitosis, lasts only until the chromosomes begin to separate (just a few minutes in many cells). But metaphase is also the stage at which many chromosomes can be observed most clearly, and the fact that each chromosome is composed of two identical chromatids is often apparent at metaphase.

Anaphase The most dramatic phase of mitosis is **anaphase.** Almost as if in response to a signal, attachments that have held the paired chromatids at the centromeres break, and the single *chromatids* (now called *chromosomes* in their own right) move in opposite directions toward the regions where two new nuclei will be organized. Each chromosome comes apart in this fashion, contributing one identical chromatid to each of the new cells. Anaphase concludes when each set of chromosomes has arrived at its pole (Fig. 9.16).

A great deal of scientific effort has gone into studying the mechanism of anaphase movement. It is clear that microtubules play a principal role in moving the chromosomes to opposite poles of the cell, but the actual nature of the force-generating mechanism is still not clear.

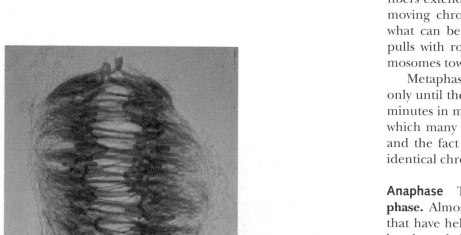

Figure 9.16 At anaphase, the chromatids separate and begin to move to the poles of the spindle.

Telophase When chromosome movement is complete, the two separated groups of chromosomes enter **telophase,** during which they start to organize into two cell nuclei. The condensed chromosomes begin to disperse, the nuclear envelope reforms, the mitotic spindle disappears, and the formation of two complete nuclei has been accomplished. Mitosis is complete when the two nuclear envelopes form once more (Fig. 9.17).

Cytokinesis

In most cells, the cytoplasm of the cell divides and produces two distinct cells, a process known as **cytokinesis** (*cyto* means "cell"; *kinesis* means "movement"). In animal cells, cytokinesis begins with the formation of a **cleavage furrow,** a progressive constriction of the cytoplasm between the two nuclei (Fig. 9.18). Contractile proteins in the cytoplasm gradually pull the furrow inward until the cell is broken in two. In many plant cells, small vesicles and vacuoles gather to form a structure known as the *phragmoplast.* The vesicles of the phragmoplast gradually fuse to form a **cell plate** that separates the cytoplasm into two distinct cells (Fig. 9.19).

Control of Cell Division

No large organism could survive without strict controls over the rates of cell division. Some cells, such as those in the nervous and muscular systems of individuals, rarely divide after the organism's embryonic stage. Others, such as those lining the digestive system, divide regularly to replace cells lost to wear and tear. Still others maintain low rates of cell division until they are stimulated. When part of the liver is destroyed by surgery or disease, healthy cells in the remaining portions divide rapidly until the liver returns to its normal size. The rate of cell division then slows down to its usual level. How is the rate of cell division controlled?

Contact inhibition One of the key factors in regulating cell division is the physical contact of neighboring cells. The importance of cell contact can be illustrated by a simple laboratory experiment. Mammalian cells in a culture dish grow until they form a single layer, as shown in Fig. 9.20. Cell division stops when the layer is only one cell thick. However, if cells are removed from the center of the dish, the cells surrounding the empty area enter the cell cycle and divide rapidly until that empty space is filled. Cell division stops when cells on both sides of the empty space make contact with each other. This contact-dependent regulation of cell growth and division is known as **contact inhibition.** The cellular signals that produce contact inhibition seem to be activated by the bind-

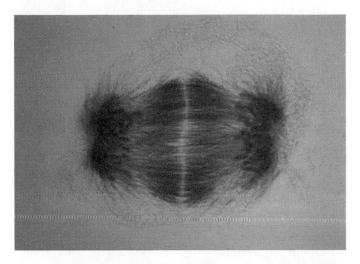

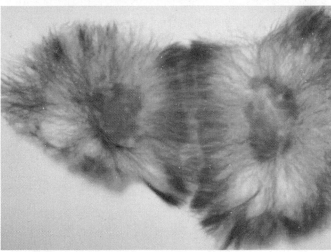

Figure 9.17 TOP: Telophase begins when anaphase movement is complete. Two new nuclei begin to organize around the two sets of chromosomes. BOTTOM: Late telophase marks the beginning of cytokinesis, the division of the cytoplasm.

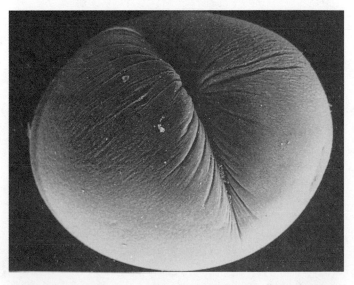

Figure 9.18 Cytokinesis in this dividing animal cell is a dramatic event, as the cell membrane forms a large cleavage furrow to complete cell division. (Magnification factor: 80)

ing of protein molecules on the surface of one cell to receptor molecules on its neighbor. If these molecules are disrupted, another round of cell division takes place.

External controls of cell division The simple experiment outlined in Fig. 9.20 can be compared to the healing of a small wound in the skin. Cells on the edges of the wound are stimulated to divide, and rapid growth fills the wound area. When the healing is complete, cell division slows down and returns to its normal rate. Biologists have discovered that chemical factors known as **growth factors** or **mitogens** ("*mito*sis-*gen*erator*s*") are responsible for such rapid changes in the rate of cell division. A host of different growth factors exist; some affect skin cells involved in the healing of wounds, some regulate the growth and development of nerve cells, and others control the rate of cell division in blood-forming cells. Still other factors, which are less well understood, inhibit cell division and

which are less well understood, inhibit cell division and slow down the rate of cell growth.

Cancer: When control is lost The importance of maintaining control over the rate of cell division cannot be overemphasized. When that control fails, the result may be an uncontrolled growth of abnormal cells that forms a mass known as a **tumor.** Tumors may damage the body by absorbing nutrients and interfering with the functions of vital organs. If tumor cells are able to spread throughout the body, they are considered cancerous. Normal body cells may become **cancer cells** when they lose the ability to respond to the normal controls that regulate cell division. Cancer cells do not grow or divide more rapidly than normal cells. The problem is that cancer cells do not know when to stop. Cancers are leading causes of death in developed countries, and we will discuss them in detail in Chapter 41.

41

Figure 9.19 LEFT: A diagrammatic view of phragmoplast and cell plate formation in plant cells. RIGHT: The phragmoplast forms from hundreds of small vesicles. These vesicles fuse to form a cell plate, thereby completing the process of cytokinesis.

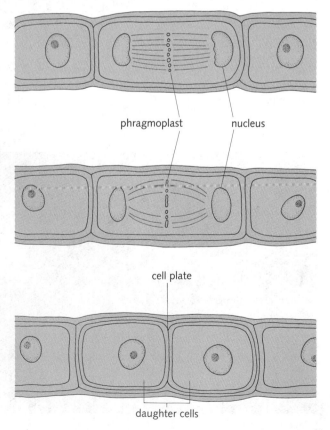

phragmoplast   nucleus

cell plate

daughter cells

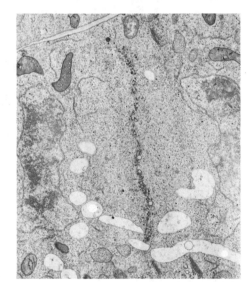

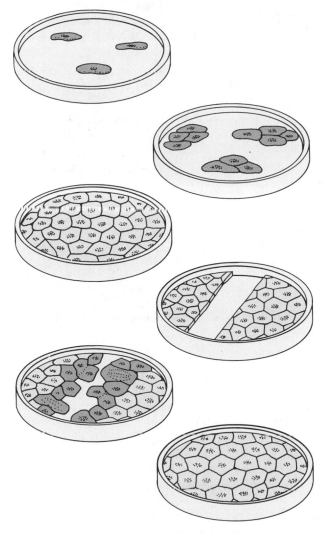

Figure 9.20 When cells are grown in culture, they often stop dividing when they make contact with each other. This is known as contact inhibition.

SUMMARY

Cells are the basic units that make up living organisms. Cells can be classified into two types: eukaryotic cells, which contain nuclei, and prokaryotic cells, which do not. Cells display an enormous range of diversity in both structure and function.

Most cells have the capacity to meet the need for growth by dividing to produce new cells. However, division cannot serve as a means of cellular reproduction unless two conditions are met: First, a cell must make a duplicate copy of the information-carrying molecules that its offspring will need. Second, it must efficiently divide that information between the "daughter" cells. After cellular information has been duplicated, the information carrying DNA molecules move to the two ends of the cell, and the cell gradually splits in two.

In eukaryotic cells, cell division is accomplished by mitosis. Eukaryotic cells duplicate their information-carrying DNA molecules during interphase, the time between cell divisions. The cell cycle—the sequence of events during the lifetime of a growing cell—has four principal phases: mitosis (M), DNA synthesis (S), the period or "gap" between the end of mitosis and the S phase (G_1), and the gap between the S phase and the beginning of mitosis (G_2). The timing of events during the cell cycle is critical to the proper control of cellular growth and reproduction.

Mitosis also has four distinct phases: prophase, when the nuclear envelope disappears and individual chromosomes become visible; metaphase, when the mitotic spindle arranges chromosomes along the central plane of the cell; anaphase, when the chromatids separate from each other and move toward the poles of the cell; and telophase, when the chromosomes reorganize into two clusters and two distinct nuclei begin to form. Following these four phases is cytokinesis, the process by which the cytoplasm of the cell separates into two, completing the process of cell division.

The rate of cell division in a multicellular organism is closely regulated. Cell division can be stimulated by growth factors, which may be released when rapid cell division is necessary to heal a wound or complete a process of development. If control over cell division is lost, cells may proliferate wildly, resulting in a disease known as cancer.

STUDY FOCUS

After studying this chapter, you should be able to:

- Explain what must be accomplished to produce two new cells from a parent cell.
- Describe the simple process of cell division in prokaryotes.
- Outline the phases of mitosis, and discuss the events that occur within the cell during mitosis.
- Describe the phases of the cell cycle.
- Describe how the rate of cell division is controlled.

TERMS AND CONCEPTS

nucleus *175*	metaphase *184*
binary fission *178*	anaphase *184*
mitosis *178*	telophase *185*
chromosomes *178*	cytokinesis *185*
interphase *179*	cleavage furrow *185*
prophase *182*	cell plate *185*
centrioles *182*	contact inhibition *185*
mitotic spindle *182*	mitogens *186*
chromatids *183*	tumor *186*
centromere *183*	cancer cells *186*

REVIEW

Objective Questions (Answers in Appendix)

1. DNA is replicated
   (a) during interphase.
   (b) immediately before telophase.
   (c) immediately after telophase.
   (d) as part of cytokinesis.

2. During the division of cytoplasm, a cell plate appears in
   (a) plant cells.
   (b) animal cells.
   (c) both plant and animal cells.
   (d) neither plant nor animal cells.

3. The mitotic spindle
   (a) holds the cells together.
   (b) helps in the movement of proteins from the cytoplasm to the nucleus.
   (c) links the cell plate to the nuclear membrane.
   (d) is a network of fibers to which the chromosomes are attached.

4. In prophase the number of chromatids per chromosome is
   (a) one.   (c) three.
   (b) two.   (d) four.

5. In anaphase the number of chromatids per chromosome is
   (a) one.   (c) three.
   (b) two.   (d) four.

6. The nuclear envelope disintegrates and the contents of the nucleus are released into the cytoplasm
   (a) at the end of the S phase.   (c) during metaphase.
   (b) during prophase.   (d) at the beginning of telophase.

Discussion Questions

7. What are some of the differences between cell division in eukaryotes and in prokaryotes?

8. What are the four phases of the cell cycle and the four phases of mitosis? Sketch a unified cell cycle that displays each of these stages.

9. Distinguish between mitosis and cytokinesis. Why is the process of cell division sometimes not completed at the end of telophase?

10. How does the process of cytokinesis differ between plants and animals? What difference between a typical plant cell and a typical animal cell does this distinction point out?

11. In many adult animals, cells in certain organs reach a stage where they no longer enter mitosis. These cells are often said to be "arrested" at a certain stage of the cell cycle. Which stage do you think this is likely to be, and why?

READINGS

Gallagher, G. L. "Evolutions: The mitotic spindle." *Journal of NIH Research* 2 (March 1990): 103–104. A marvelous two-page summary of the structure of mitosis in animal cells, with a very clear summary diagram.

Pickett-Heaps, J., D. Tippit, and K. R. Porter. "Rethinking mitosis." *Cell* 29 (1982): 729–744. This article describes the mitotic structure and offers some intelligent speculation on the evolution of mitosis and meiosis. Difficult in parts, but thought provoking.

Sloboda, R. D. "The role of microtubules in cell structure and cell division." *American Scientist* 68 (1980): 290–298. Microtubules are the major structural components of the mitotic spindle. This article explores some theories of how they behave during cell division.

Mazia, D. "The cell cycle." *Scientific American* 230 (January 1974): 53–64. A classic article on the cell cycle by one of the pioneers of cell biology.

10

Genetics: The Science of Inheritance

or thousands of years people have selectively bred plants and animals. The idea of selective breeding is simple. One chooses a few organisms with desirable characteristics to be the parents of a new generation. When that generation is grown, the hardiest plants, the fastest racehorses, or the most obedient dogs are selected for the next round of breeding. It's simple in practice and simple in theory. Offspring tend to resemble their parents, so by patiently choosing parents with the proper characteristics, the breeder can accentuate those characteristics.

Is inheritance really that simple? For hundreds of years, people regarded inheritance as a *blending*—the characteristics of both parents were thought to blend to produce offspring. In most cases, the blending explanation seems to make sense. Most of us look a little like our mothers and a little like our fathers. A cross between a large dog and a small dog, usually produces medium-sized dogs. Everything seems to make sense. Well, almost everything.

INHERITANCE: A PROBLEM

Many people enjoy the common parakeet (*Melopsittacus undulatus*), an attractive bird native to Australia. Because you have a preference for birds with green wings, you select a matched pair of such birds and place them in a breeding cage at home. The birds get along, conditions are right, and before long they are the parents of five little parakeets. But something is wrong. The new generation is not a simple blend of the two green parents. In fact, some of the birds don't look anything like their parents at all.

Two of the new birds are green, one is blue, one is bright yellow, and one is completely white (Fig. 10.1)! Where did the new colors come from? And why don't all of the new birds look like their parents? More to the point, what should you do now? You might rush back to the pet store, complaining that the green parakeets do not "breed

Figure 10.1 These five parakeets are the offspring of two green-colored parents.

Figure 10.2 Gregor Mendel (1822–1884).

true." Or you could wonder whether another parakeet had slipped into the cage while you were away.

However, there is something else that you could do. Puzzled by the results of this cross, you might *experiment*. You could conduct more breedings and try to learn the rules by which parakeets inherit their color. You would need more cages, some extra time, and a lot of birdseed. But if you had done this 150 years ago with patience and care and insight, you just might have founded **genetics,** the science of inheritance. As luck would have it, someone else got there first.

MENDEL AND THE BIRTH OF GENETICS

As we saw in Chapter 8, Darwin realized that the blending theory of inheritance presented special problems for evolution. Specifically, he wondered whether favorable characteristics might be "blended away" before natural selection had a chance to increase their frequency in a population. Ironically, and unknown to Darwin, this problem was being solved just as *On the Origin of Species* was being published. This ground-breaking work, performed by an Austrian priest named Gregor Mendel, set the stage for a radically new way of thinking about heredity. But because Mendel's work did not receive widespread recognition until the early twentieth century, Darwin probably was unaware of it.

Gregor Mendel was born in 1822 in the town of Heinzendorf. A bright student, he entered the Augustinian monastery at Brno, which, like his birthplace, is now part of the Czech Republic (Fig. 10.2). Mendel became a priest when he was 25. He had a special interest in natural science and took an exam to qualify for a science teaching certificate. Although he failed that exam, his superiors at the monastery thought enough of the young priest to send him to the University of Vienna for two years to study science and mathematics. When he returned to the monastery he was given two assignments: to teach physics and biology in a local high school and to supervise the monastery's gardens. We do not know how well he fulfilled the first assignment. But we know a great deal about his work in the garden.

Mendel took a special interest in the garden's peas. The other gardeners helped Mendel isolate several strains of plants with distinct characteristics: one always produced tall plants, another only short plants; one always produced purple flowers, another only white flowers. Mendel selected plants that bred true—that is, each variety produced seeds that grew into plants identical to the parent.

Mendel knew that peas are self-fertilizing. Because of the structure of the pea flower, the pollen that produces the male reproductive cells usually fertilizes female repro-

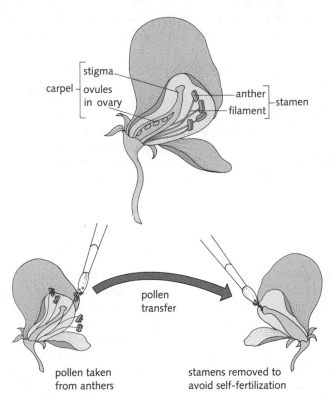

Figure 10.3 Although pea plants are normally self-fertilizing, Mendel carried out crossbreeding by brushing the pollen of one plant on the stigma of another.

ductive cells in the very same flower. To cross one plant with another, Mendel opened the pea flowers and removed the male pollen-producing anthers. He then dusted the female portions of the flower with pollen from another plant (Fig. 10.3). In this way, he could carry out controlled pollination from one plant to another—a process known as *crossing*. Because each plant produced many seeds, Mendel could obtain reliable statistics on the offspring of each cross.

Mendel's First Experiments

In one experiment, Mendel fertilized the flowers of a strain that produced purple flowers with pollen from plants that produced white flowers. He also reversed the process by using pollen from purple-flowered plants to fertilize the white-flowered strain. One of his first observations was that it did not matter which plant contributed the pollen and which received it.

The original plants involved in the cross are known as the **parental,** or P, generation. The seeds produced from this first cross Mendel called the **first filial,** or F_1, generation (*filial* comes from the Latin root for "son"). All of these seeds produced plants that bore purple flowers, so

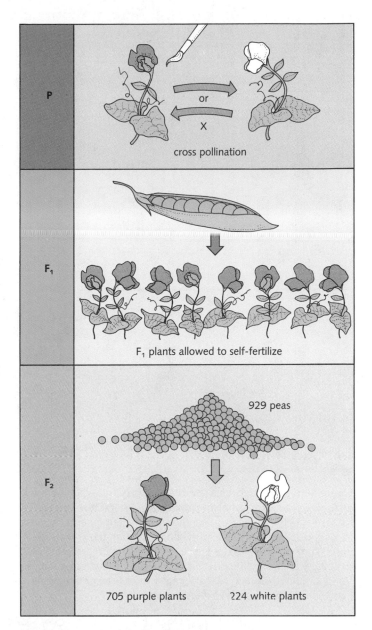

Figure 10.4 Mendel's first experiments involved the crossing of parental (P generation) plants with white flowers and plants with purple flowers. The seeds produced by this cross (F_1 generation) grew into plants which produced purple flowers. When these plants were allowed to cross naturally, both white and purple flowers were produced in the F_2 generation.

it seemed that the white-flower trait had disappeared. Mendel took the seeds from this cross, planted them, and allowed the resulting plants to self-fertilize. This $F_1 \times F_1$ cross produced the **second filial,** or F_2, generation. Mendel collected the seeds and planted 929 of them. To Mendel's surprise, the white flowers reappeared in some of these plants; 224 plants bore white flowers and 705 had purple flowers, a ratio of purple to white flowers of 3.1 to 1 (Fig. 10.4).

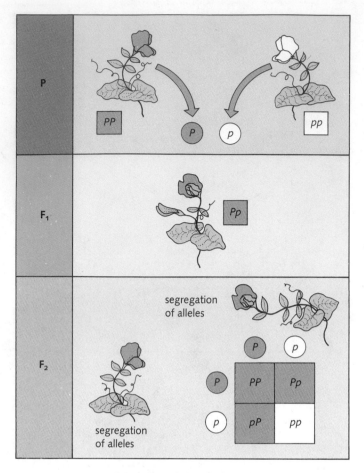

segregation of alleles

segregation of alleles

Figure 10.5 Mendel explained the results of his experiment by assuming that each pea plant carried two "characters," or alleles, for flower color. In the F$_1$ generation, each plant inherits an allele for purple (P) and an allele for white (p). When the F$_2$ generation is formed, these alleles are segregated as gametes are produced. By analyzing the possible combinations of these gametes using a Punnett square, we can predict the allele combinations that the F$_2$ generation will inherit.

A Theory of Particulate Inheritance

At first, these results must have seemed every bit as confusing as the blue and yellow parakeets mentioned in the opening pages of this chapter. But Mendel was able to explain them. He proposed a system of **particulate inheritance** in which heritable characteristics were controlled by individual "units." Mendel assumed (correctly, as it turned out) that each plant had two such units for each trait. Mendel called each unit a "Merkmal," the German word for "character." Today we call these units **genes,** and we know that the cells of the pea plant carry two genes for most characteristics. If we represent the units with symbols, we might use P for Mendel's purple-flower character and p for the white-flower character.

P and p are known as **alleles,** which are alternative forms of a single gene, in this case the gene for flower color. In Mendel's first cross, the original, true-breeding purple-flowered parent would have been PP, and the original white-flowered parent would have been pp.

One of Mendel's most extraordinary insights was the realization that when organisms produce their reproductive cells, or **gametes,** each gamete carries only one allele for each gene. Therefore, the two alleles for each gene are **segregated** from each other when gametes are formed. In Mendel's experiment, for example, the F$_1$ generation received an allele for purple flowers from one parent and an allele for white flowers from the other parent. In this way, each parent makes an individual genetic contribution to its offspring.

The Punnett square   We can visualize the alleles involved in each cross by making a diagram known as a **Punnett square** (Fig. 10.5). On one side of the square we write down all the *gametes* that can be formed by one parent, and on the other side all the gametes that can be formed by the other parent. Then we use the blocks within the square to represent all the possible *combinations* of alleles that may occur in the offspring. When two plants in the F$_1$ generation are crossed, as shown, the result is that, on average, ¼ of the F$_2$ offspring have PP alleles for flower color, ¼ are pp, and ½ are Pp. The Punnett square is a powerful predictive tool. In this case, the square predicts that ¾ of the offspring from the F$_1$ × F$_1$ cross will be either PP or Pp.

Dominance

One of the remarkable results of the crosses shown in Figs. 10.4 and 10.5 is the fact that plants with contrasting alleles (P and p) produce purple flowers. The purple allele of the flower-color gene is **dominant** over the white allele. (That is why purple is represented by the capital letter P.) Mendel carried out crosses with six other characteristics, including plant size, seed color, and seed shape (Fig. 10.6). In each case, one characteristic was "dominant" in the F$_1$ generation. The characteristic that seemed to disappear in the F$_1$ generation and to reappear in the F$_2$ generation he called a **recessive** trait. Mendel explained this phenomenon by proposing that whenever a dominant allele and a recessive allele were found in the same organism, the dominant allele alone controlled the appearance of the plant. Only when two recessive alleles occurred together did the recessive characteristic, or "character," emerge.

Now we can look at the results of the crosses in Fig. 10.4 and appreciate the value of Mendel's data. Because both *PP* and *Pp* plants produce purple flowers, the Punnett square predicts a 3:1 ratio of purple to white flowers in the F$_2$ generation. Mendel's 3.1:1 ratio was a close match.

The same 3:1 ratio held for crosses involving the six other traits. For example, when Mendel crossed plants producing round seeds (alleles: *RR*) with those producing wrinkled seeds (alleles: *rr*), all the F$_1$ plants produced round seeds, showing that the *R* allele is dominant over the *r* allele. When the *Rr* F$_1$ plants were crossed with themselves, the F$_2$ generation included 5474 plants that bore round seeds and 1850 plants with wrinkled seeds, a ratio of 2.96:1. Obviously, Mendel's patience in counting seeds knew no bounds!

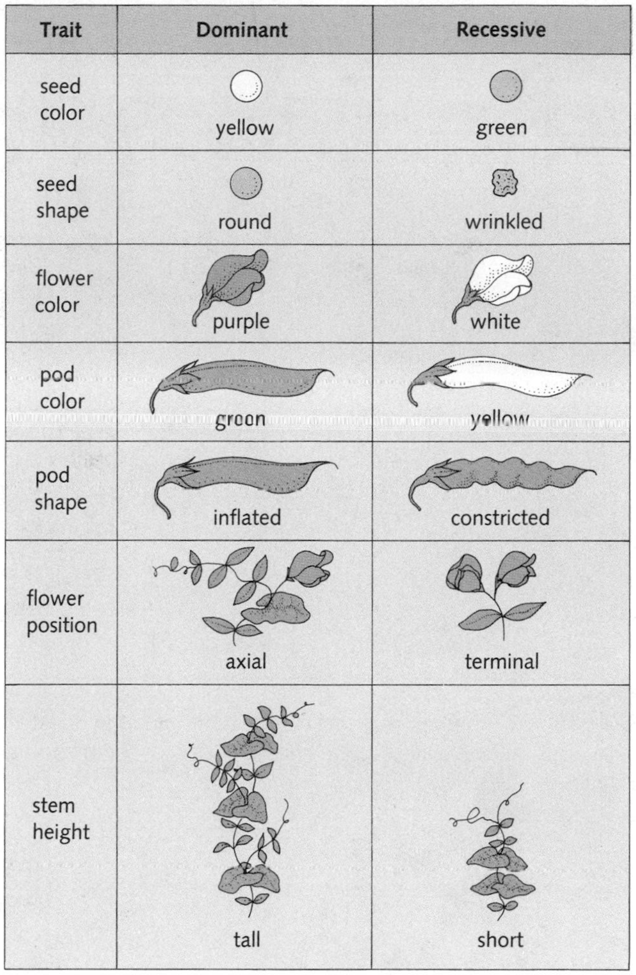

Mendel's First Principles

It is hard for us today to realize just how revolutionary Mendel's theory of particulate inheritance was. With only the evidence of his garden pea crosses, he established the first two of three important principles on which the science of genetics is founded:

- The characteristics of an organism are determined by individual units of heredity called genes. Each adult organism has two alleles for each gene, one from each parent. These alleles are *segregated* (separated) from each other when reproductive cells are formed. This is known as the **principle of segregation.**

- In an organism with contrasting alleles for the same gene, one allele may be *dominant* over another (as round is dominant over wrinkled for seed shape in the garden pea). This is known as the **principle of dominance.**

Although Darwin could not take Mendel's principles into account, it is clear they solved the problems presented by the previously held blending theory of inheritance. When a new and beneficial allele appears in a population, the characteristic that allele produces won't disappear by repeated "blending." Instead, the allele is preserved from one generation to the next. Even a recessive characteristic can appear in a later generation through the changing gene combinations generated by sexual reproduction.

Genotype and Phenotype

A second major contribution of Mendel's experiments was the idea that every organism has a *genetic* makeup called its **genotype.** The actual characteristics an organism exhibits are called its **phenotype.** The genotype is in-

Trait	Dominant	Recessive
seed color	yellow	green
seed shape	round	wrinkled
flower color	purple	white
pod color	green	yellow
pod shape	inflated	constricted
flower position	axial	terminal
stem height	tall	short

Figure 10.6 The seven contrasting traits investigated by Mendel.

herited, whereas the phenotype is produced under the influences of the environment and the genotype. Plants may have the same phenotype and different genotypes. For example, about ⅔ of the 705 purple-flowered plants of the F$_2$ generation in Fig. 10.4 have the *Pp* genotype, and ⅓ have the *PP* genotype.

Plants with two identical alleles for the same gene (*PP* or *pp*) are said to be **homozygous** for that gene. Plants with two contrasting alleles for a gene (*Pp*) are **heterozygous** for that gene. How can we determine which of the 705 purple-flowered plants are homozygous for the purple allele of the flower-color gene and which are heterozygous? Mendel developed a simple technique known as a **test cross** that enabled him to determine the genotype of

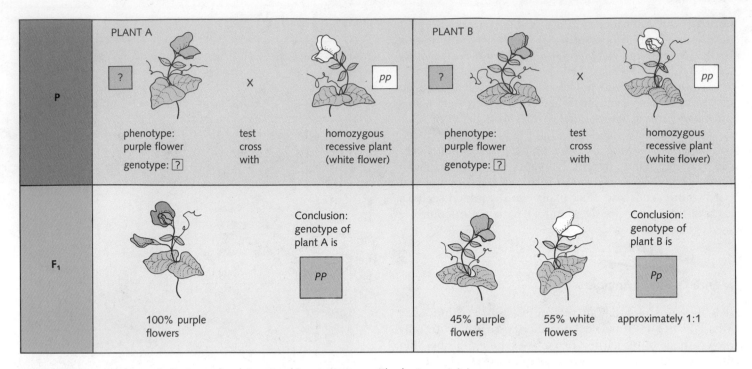

Figure 10.7 Genotypes of plants can be determined in a test cross with plants containing homozygous recessive alleles. The characteristics of the offspring of a test cross reveal the genotypes of the parents.

any plant. The plant in question is crossed with another plant that is homozygous for the recessive version of the gene in question. The phenotypes of the offspring from the test cross then reveal the genotypes of the parents.

Figure 10.7 shows how this works in the case of plants with purple flowers. One test cross produces only plants with purple flowers, proving that the plant used for that cross had the *PP* genotype. The second test cross produces 9 plants with purple flowers and 11 with white flowers (roughly a 1:1 ratio), proving that the original genotype was *Pp*. The test cross is a powerful technique that can be used to determine genotype in animals as well as plants. Can you imagine a test cross that might be useful in the case of our parakeets? We'll come back to this problem later.

Independent Assortment

Knowing that Mendel had seven different traits available for study, each of which had two contrasting alleles, what do you think his next experiment might have been? If you guessed that he might have tried to follow the alleles

for two different genes at the same time, then you guessed right.

Mendel wondered whether the alleles for different genes would segregate in the same pattern as alleles of a single gene did. He carried out a series of experiments like those shown in Fig. 10.8. A parental line of plants with yellow-colored, round seeds was crossed with a line that produced green, wrinkled seeds. The seeds in the F_1 generation were all round and yellow, showing that those two alleles are dominant over the alleles for wrinkled and for green seeds.

If the alleles for the seed coat and seed color genes were able to assort independently during gamete formation, then plants grown from the F_1 generation seeds could produce four different kinds of gametes:

$$RrYy$$

$$RY \quad Ry \quad rY \quad ry$$

But these four types of gametes could be produced only if the alleles for the two genes were free to segregate and to assort independently of each other. **Independent assortment** means that the allele a gamete receives for the seed coat (**R** or **r**) has no effect on which allele it receives for the seed color gene (**Y** or **y**). The Punnett square for this cross shows that the 16 possible offspring of an F₁ cross display four different phenotypes and that these plants are present in a 9:3:3:1 ratio. Mendel's breeding experiments produced 556 seeds whose phenotypes matched these ratios almost exactly. These results show that the *R* and *Y* genes assort independently.

- Each of the seven genes that Mendel investigated obeys the third of Mendel's three important principles, the concept of independent assortment.

It's interesting to note that Mendel confirmed the genotypes of each of his F₂ plants by conducting test crosses. A round, yellow pea from the F₂ generation might have any of four possible genotypes: *RRYY, RRYy, RrYY,* or *RrYy*. By growing a plant from each pea and then crossing that plant with a double-recessive (genotype: *rryy*) plant, Mendel determined the genotypes of the offspring (Fig. 10.9).

Problem Solving with Mendelian Genetics

We are now in a position to think about parakeets. Feather color in these birds is controlled by two genes. The "B" gene controls black and blue color in the feathers. The dominant allele (*B*) produces blue and black pigmentation, whereas the recessive allele (*b*) does not produce any color. The "C" gene controls yellow color. The dominant allele (*C*) produces yellow feathers, whereas the recessive allele (*c*) produces no color.

Can we use this information to explain the surprising results of our attempts to breed the two green parakeets we introduced at the beginning of this chapter? As shown in Fig. 10.10, we could begin by assuming that our mating pair had the genotype *BbCc*. Such birds would contain a combination of blue and yellow pigments that would appear green. By using a Punnett square to analyze the gametes that such birds would produce, we can predict the outcome of a cross between our two parents. As you can see, we would expect about ¹⁄₁₆ of the offspring to be white (genotype: *bbcc*), so the single white bird in our cross can be explained as a result of Mendelian genetics. We would also expect green, yellow, and blue birds in the offspring, and that prediction came true. Just for fun, you might plan a way to test our explanation of the parental genotypes with a test cross. One of the five parakeets that resulted from crossing the two parental green parakeets is the ideal bird to use for such a test. Can you identify it? (See Appendix for the answer.)

Genetics and the Cell

Mendel's three important principles, the principles of *dominance, segregation,* and *independent assortment,* place

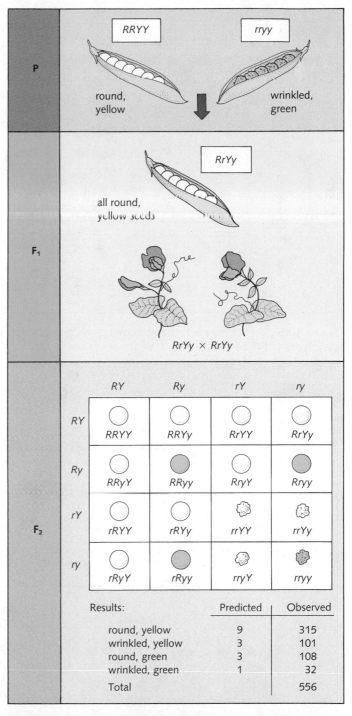

Figure 10.8 A two-factor cross carried out by Mendel. The dominant characteristics of the parents appear in the F₁ generation. When the F1 plants are crossed among themselves, independent assortment of the alleles for both characteristics produces four distinct phenotypes in the F₂ generation with a 9:3:3:1 ratio.

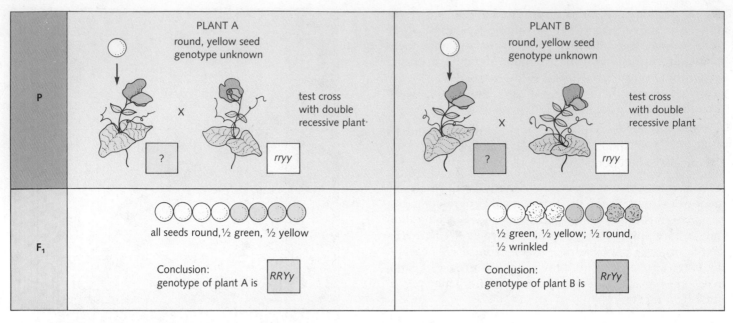

Figure 10.9 Test crosses with plants homozygous recessive for two characteristics reveal the genotypes of the parent plants.

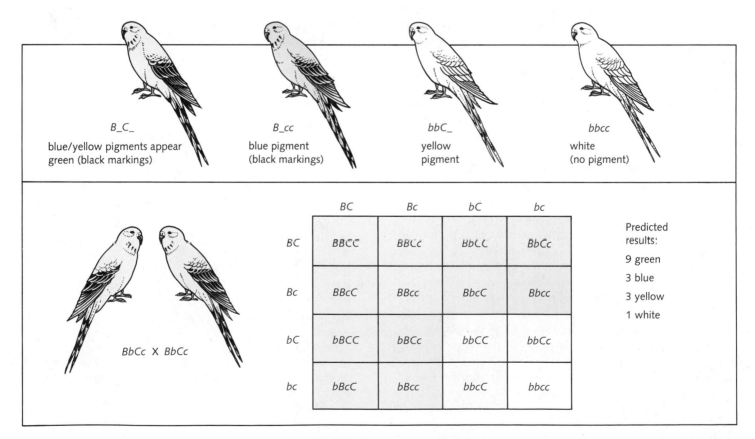

Figure 10.10 A genetic explanation using a Punnett square of how two green-colored parakeets could produce offspring with four different color patterns. Each of the green parents is heterozygous for the two genes that control feather color. The Punnett square shows how birds with four different colorations may be produced by the cross.

certain restrictions on the way genes and their alleles behave. For example, the fact that individuals inherit one allele for each gene from each parent implies that the cells of an adult organism contain two sets of alleles. Similarly, because segregation requires that the two alleles for each gene be separated when gametes are formed, there must be a mechanism that accomplishes this separation. What an opportunity! To Mendel, the "gene" was a purely hypothetical factor that controlled heredity. But if genes were real, then we should be able to find physical structures within the cell that behave in accordance with Mendel's three principles.

Unfortunately, no one was ready to seize upon Mendel's work when it was published. It was, after all, just one of hundreds of papers on plant breeding experiments. Mendel's work was generally ignored for more than 30 years after it was published. By the turn of the century, though, other scientists began to take an interest. Mendel's work with peas was confirmed by two other scientists (Erich von Tschermak and Carl Correns), and similar results were found for more than a dozen different plants and for many animals as well. There was no doubt that an important principle had been discovered.

However, as striking as Mendel's work was, it was just one of a number of competing theories of inheritance, and it was destined to remain so as long as his units of inheritance were strictly hypothetical. The ability of theory to explain events is powerful evidence, but biology is never content with theoretical units. It was necessary to actually *find* the units Mendel had postulated. Finding those units was the first step toward making genetics a modern experimental science.

MEIOSIS: FORMING A NEW ORGANISM

As we saw in Chapter 9, during mitosis each chromosome of the cell appears as a pair of *chromatids*. As mitosis proceeds, the chromatids of each chromosome separate and two daughter cells are formed, each containing the same number of chromosomes as the original cell. To many biologists, the fact that chromosomes were duplicated just prior to mitosis and then carefully separated into the two daughter cells suggested that chromosomes were important structures. Some even suggested that they contained hereditary information.

The critical evidence, however, came from studies of what happened to chromosomes during the formation of reproductive cells, or *gametes*. In sexually reproducing organisms, a new individual is formed by the fusion of male and female gametes to form a single cell known as a **zygote.** The zygote then goes through a process of development and growth to become an independent organism.

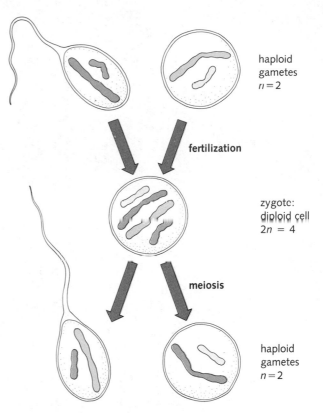

Figure 10.11 A diploid organism contains two complete sets of chromosomes, one from each parent. The contributions of each parent are illustrated in the formation of a fertilized zygote in this figure. When the diploid organism reproduces, it too will produce haploid gametes.

Gametes contain only *half* the number of chromosomes found in the cells of the body. The sperm and egg cells of cats, for example, contain 19 chromosomes each. A fertilized zygote, about to develop into a kitten, contains a total of 38 (19 + 19) chromosomes. How does this work? Does the sperm contribute chromosomes A, B, C, and D while the egg contains E, F, G, and H? No. In fact, with one exception that we will encounter shortly, the two sets of chromosomes are very similar to each other. A better way to describe it would be to say that the egg contributes A, B, C, and D while the sperm donates A', B', C', and D'. For each chromosome contributed by one parent, a corresponding chromosome, or **homologous chromosome,** is contributed by the other parent. Each cell of an adult cat contains two sets of homologous chromosomes; one set is derived from the male parent and one from the female.

Biologists use the term **diploid** to describe a cell that contains two sets of homologous chromosomes. The *somatic* (body) cells of peas, cats, and humans are diploid. Their gametes, however, which contain only a single set of chromosomes, are **haploid** (Fig. 10.11). We use the

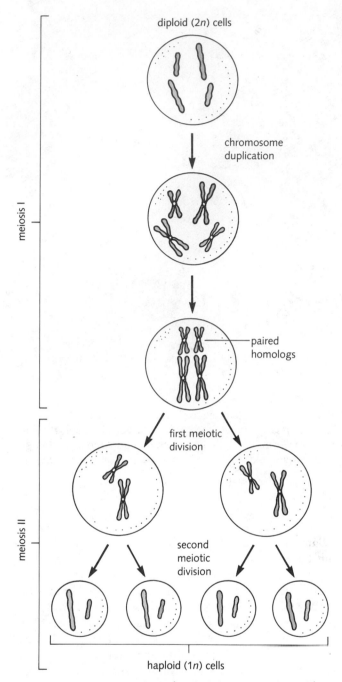

diploid (2n) cells

chromosome duplication

meiosis I

paired homologs

first meiotic division

meiosis II

second meiotic division

haploid (1n) cells

Figure 10.12   An overview of meiosis in an organism with a diploid chromosome number of 4.

symbol n to represent a single set of chromosomes. Diploid cells, therefore, are $2n$ and haploid cells are n. (The term *haploid* does not mean that such cells have *half* a set of chromosomes. In fact, haploid cells have a full, single set. A better term for n might be *monoploid*, but *haploid* and *diploid* are now firmly entrenched in the scientific vocabulary.)

Reduction Division

Haploid gamete cells are formed by a special process that is often called *reduction division* because it reduces the chromosome number of cells that are developing into sperm or eggs. Reduction division is commonly known as **meiosis,** and it occurs in all organisms that reproduce sexually. Meiosis generally takes place over the course of two rounds of cell division. It may superficially resemble mitosis, but there are big differences in how chromosomes behave in the two processes.

Meiosis is a two-stage process that produces four haploid cells from a single diploid cell. Homologous chromosomes are segregated from each other in the **first meiotic division.** The **second meiotic division** is not preceded by an S phase of the cell cycle, so DNA and chromosome duplication do not occur between the two divisions. This results in a reduction of chromosome number. Figure 10.12 illustrates the features of meiosis.

One of the first organisms in which meiosis was intensively studied was the ordinary fruit fly, *Drosophila melanogaster* (*Drosophila* means "dew lover" and *melanogaster* means "dark belly"). Diploid *Drosophila* cells have eight chromosomes ($2n = 8$). Four of these chromosomes were originally provided by the fly's mother and four by its father. We shall use a female *Drosophila* to trace the details of meiosis (Fig. 10.13).

The first meiotic division   Prior to meiosis, each of the eight chromosomes is duplicated, so we now have eight chromatids. This duplication is no different from that which occurs before an ordinary cell division. However, as the process continues, something interesting happens: the chromosomes *pair* to form bundles known as **tetrads** (the same structure is sometimes referred to as a *bivalent*) composed of four chromatids each. The pairing of homologous chromosomes, also known as **synapsis,** produces an appearance at metaphase that is very different from a mitotic metaphase. As shown in Fig. 10.13, a *Drosophila* cell at metaphase I of meiosis has four tetrads.

Which chromosomes pair to form the tetrads? The *homologous* chromosomes derived from both parents. Chromosome A from the mother pairs with A′ from the father, B with B′, and so on. At metaphase, all four tetrads are aligned across the center of the cell.

Anaphase begins as the tetrads draw apart and the homologous chromosomes are separated. Each daughter cell receives one chromosome from each pair. The alignment of each chromosome pair is random with respect to the others. In other words, the fact that the maternal copy of one chromosome pair moves to a particular cell does not affect the way the maternal copy of a different chromosome pair goes. The chromosomes exhibit *independent assortment!*

Genes That Cheat at Meiosis

As we have seen, during meiosis the homologous chromosomes of an organism pair to form tetrads. The alignment of chromosomes during metaphase is random, which gives each cell roughly a 50–50 chance of getting either the maternal or the paternal homologue when the first meiotic division occurs. Each gamete, therefore, has a gene composition decided by the chance mechanisms of meiosis. All of this is true—*assuming* that the gene-sorting mechanism plays fair. But what would happen if a gene "cheated" during meiosis?

A gene could cheat by making certain that it was *always* included in reproductive cells, ensuring that it, and not its homologue, was passed along to succeeding generations. How could it do this? James F. Crow and his associates at the University of Wisconsin have discovered exactly such a system in *Drosophila*.

The genetic system is known as **segregation distorter,** or **sd** for short. The genes that produce *sd* carry on a kind of genetic warfare during meiosis—one that ensures that chromosomes carry-

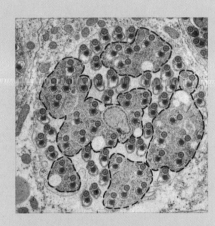

ing the *sd* genes, and *only* the *sd* genes, are passed on to the next generation. When *sd* is present in the heterozygous form, it actually sabotages the homologous chromosome, the one that doesn't have *sd*. The sabotage occurs at the beginning of meiosis, and the results are evident a few days later when sperm cells begin to develop. The accompanying photo shows the effect of the sabotage. *Drosophila* sperm are normally produced in packets of 64 cells each. The electron micrograph shows *sd*

"cheating" in action: 32 of the cells have begun normal development, but 32 are failing to develop. Which 32? The ones that lack *sd*. Somehow, the *sd* gene has destroyed every cell containing its competition in the next generation. It actually breaks the rules of meiosis: it cheats!

How does the rest of the genetic system respond to this outlandish behavior? The surprising answer may be that the entire system is organized to keep the cheaters under control. Several biologists have suggested that genetic recombination (crossing over) is actually a defense mechanism against *sd* and other cheating genes. By reshuffling the combinations of genes on chromosomes, recombination occasionally separates cheating genes from the closely linked marker genes that usually protect the chromosome containing *sd* from its own deadly effects. This prevents *sd* and other cheaters from spreading throughout the population and gives "honest" genes a chance to compete fairly.

The first meiotic division (meiosis I) produces a random shuffling of the genetic deck: each cell has an equal chance of getting either the maternal or the paternal copy of any individual chromosome. As you know, Mendel's principle of *segregation* states that the maternal and paternal alleles of a gene are segregated during gamete formation. The pairing of homologous chromosomes during the first meiotic division makes that possible. As we observe anaphase of the first meiotic division, we are watching the microscopic events that underlie the segregation of alleles. Mendel never worked with a microscope, but microscopic observation confirms that chromosomes obey the principles of independent assortment and segregation.

The second meiotic division Each of the two cells produced by the first meiotic division now enters a second meiotic division (meiosis II). The cell cycle that occurs between the first and second meiotic divisions is exceptional, because the chromosomes are *not* duplicated during it. Each of our two *Drosophila* cells received four chromosomes (with two chromatids each) from the first division, and these two cells now enter another round of division with exactly the same number of chromatids. In metaphase, four chromosomes are visible in each cell, and anaphase separates these four chromosomes into two new cells, each of which now receives four single chromatids. The four cells formed by the second meiotic division are true haploid cells, each containing four distinct

Figure 10.13 Meiosis in *Drosophila*.

maternal chromosomes

paternal chromosomes

INTERPHASE
Cell prior to chromosome duplication.

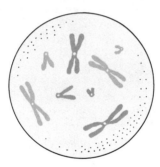

PROPHASE I
Each chromosome is composed of two identical chromatids attached at a centromere.

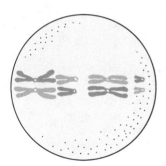

METAPHASE I
Tetrads of homologous chromosomes align on the metaphase plate.

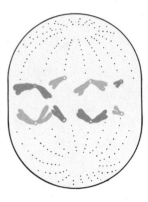

ANAPHASE I
Tetrads separate and homologous chromosomes are separated.

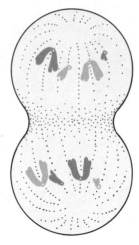

TELOPHASE I
Cytokinesis occurs, forming two daughter cells.

chromosomes ($n = 4$). Although the cellular details of meiosis differ in male and female animals (a sperm cell is produced quite differently from an egg cell), the nuclear details of meiosis are similar in both sexes and in plants as well as animals.

Meiosis and the Rise of Mendelian Genetics

An American graduate student, Walter Sutton, and the German biologist Theodor Boveri were the first to appreciate the fact that meiotic chromosomes behave *exactly* as one would expect for Mendel's genes. In 1902 they were bold enough to say that Mendel's genes "are on chromo-

somes." They proposed a model in which genes were located on chromosomes like beads on a string. And they were right. Mendelian genetics was rapidly accepted because studies on chromosome behavior during cell division and meiosis showed that the Mendelian model was essential to account for the inheritance of genetic units located on chromosomes and passed from one generation to the next by means of haploid reproductive cells.

Mendelian genetics was also accepted because of its ability to *predict* the results of experimental crosses. In the remainder of this chapter, we will investigate some aspects of Mendelian genetics and examine how the laws of genetics can be used to solve problems in inheritance—and even to predict genetic disorders.

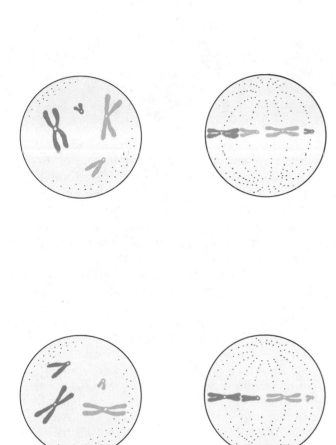

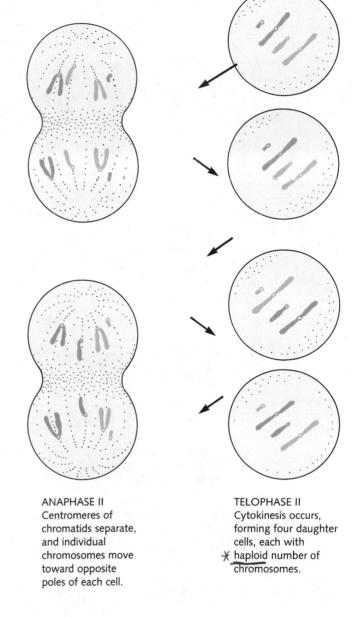

PROPHASE II
Each daughter cell
has one chromosome
from each pair;
chromosomes migrate
toward metaphase plate.

METAPHASE II
Chromosomes align
on metaphase plate.

ANAPHASE II
Centromeres of
chromatids separate,
and individual
chromosomes move
toward opposite
poles of each cell.

TELOPHASE II
Cytokinesis occurs,
forming four daughter
cells, each with
haploid number of
chromosomes.

Gene Linkage and Crossing Over

Although one of Mendel's important principles was the law of independent assortment, geneticists soon discovered that not all genes assorted independently. Some were inherited in groups, as though they were *linked* together. The explanation for *gene linkage* turned out to be a simple one. Genes that are located on the same chromosome are physically linked because whole chromosomes separated together during meiosis.

One of the first examples of linked genes appeared in an experiment done in 1905 by William Bateson and R. C. Punnett (yes, the inventor of the Punnett square) in Cambridge, England. Their experiments were conducted on peas, and they had developed two lines of plants whose genetic composition they had determined. One line contained an allele for purple flowers (P) that was dominant over its allele for red flowers (p); it also contained an allele to produce long pollen grains (L) that was dominant over round pollen grains (l). Their first cross was a mating of plants that were homozygous for both alleles:

$$PPLL \quad \times \quad ppll$$

(purple, long)   (red, round)

The cross produced a crop of heterozygotes (*PpLl*) as its F₁ generation. All of these plants produced purple flowers and long pollen grains. When the F₁ plants were crossed among themselves, an F₂ generation was produced containing 381 plants with the following phenotypes:

284	purple, long	21	red, long
21	purple, round	55	red, round

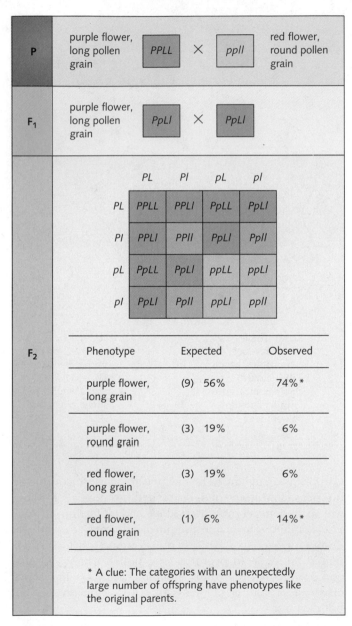

Figure 10.14 An experimental cross of peas showing evidence of gene linkage. If the two genes were unlinked, independent assortment would produce a 9:3:3:1 ratio in the F₂ offspring. Instead, the F₂ offspring have an unexpectedly large number displaying parental phenotypes.

This does not match the 9:3:3:1 ratio we would have expected for such a cross. In fact, the expected and observed numbers of offspring are quite different (Fig. 10.14).

Phenotype	Number Observed	Percent Observed	Percent Expected
Purple, long	284	74%	56% (9)
Purple, round	21	6%	19% (3)
Red, long	21	6%	19% (3)
Red, round	55	14%	6% (1)

Why should there be *more* of the purple, long plants, and more of the red, round plants than expected? And why *fewer* than expected of the purple, round and of the red, long? Bateson and Punnett did not have an immediate explanation. Neither did scientists wrestling with data on other systems which, like these, suggested that Mendel's ratios did not correspond to reality. They entertained the notion that Mendel had been wrong. Can you think of an explanation for these results?

The most striking thing about the data from this cross is that the phenotypes present in larger-than-expected amounts were **"parental"** phenotypes, in the sense that they resembled the original *P* generation parents. How could this have happened? Suppose that the two genes under study had been "linked" together in some way? Those genes would then not assort independently, and they would be inherited "together."

As shown in Fig. 10.15, the gametes produced by the parental plants must contain the linked *P* and *L* alleles or the linked *p* and *l* alleles. The F₁ organisms, therefore, contain a pair of alleles linked in this fashion. What will happen when the F₁ plants produce their gametes? If the genes are unlinked, we would expect four different kinds of haploid gametes to be formed in equal proportions:

PL   Pl   pL   pl

However, if the genes are linked, then only two combinations of alleles are possible:

PL   and   *pl*

If only two kinds of gametes can be produced, then the possibilities for offspring are severely limited (see Fig. 10.15). But the simple explanation that these genes might be linked to each other by virtue of being located on the same chromosome was not sufficient. The real data conform to *neither* model:

Phenotype	Unlinked	Real Data	Linked
Purple, long	56%	74%	75%
Purple, round	19%	6%	—
Red, long	19%	6%	—
Red, round	6%	14%	25%

Why would these genes behave as though they were partly linked, but not completely? We now know that the answer can be found by watching how chromosomes behave dur-

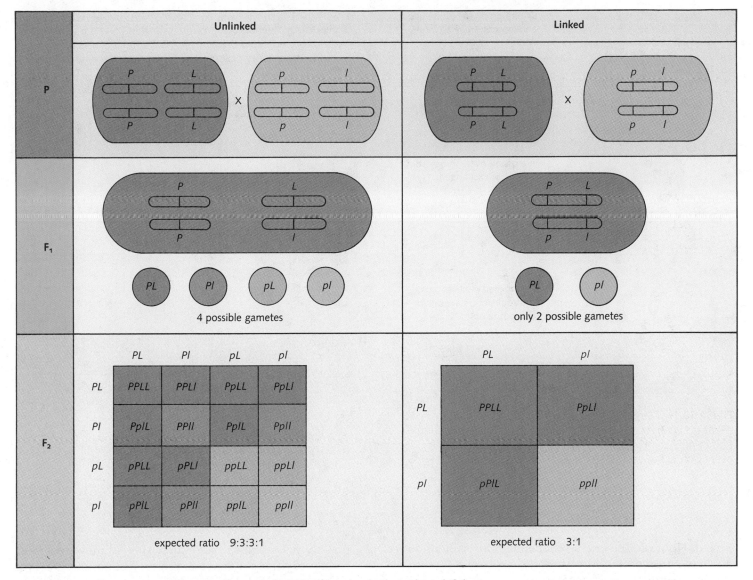

Figure 10.15 The experimental results shown in Fig. 10.14 can be partly explained if the two genes in question are linked.

ing meiosis. During meiosis, the homologous chromosomes pair to form tetrads.

While the homologues are paired, microscopists have observed exchanges of material between one chromosome and its homologue (Fig. 10.16). These exchanges allow alleles to cross over from one chromosome to its homologue.

Exchanges like these can occur anywhere along a pair of homologues in synapsis, and the process is sometimes known as **crossing over.** Because the existing genetic material is *recombined* as a result of crossing over, the process is also called **genetic recombination.** Now we have a mechanism that can be used to explain the results from our genetic cross.

If two genes are located on the same chromosome, then they are **linked** and will be inherited together *unless* a cross-over event occurs between them. As shown in Fig. 10.16, if a cross-over event occurs between two linked genes, their alleles will be separated onto different chromosomes. Thus linkage can be variable, just as Bateson and Punnett's breeding experiment with peas suggested. The reason why the actual data are intermediate between what we might expect for linked and for unlinked genes is the occurrence of occasional cross-over events that separate the alleles onto separate chromosomes.

Crossing over requires us to modify the law of independent assortment. As we have just seen, genes on the same chromosome do not assort independently; they are

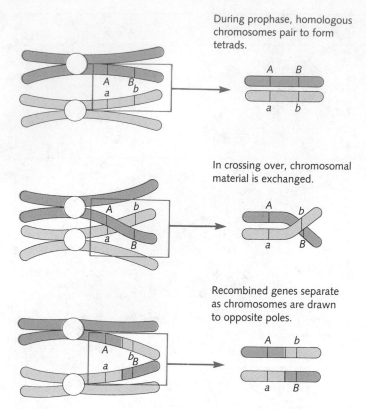

During prophase, homologous chromosomes pair to form tetrads.

In crossing over, chromosomal material is exchanged.

Recombined genes separate as chromosomes are drawn to opposite poles.

Figure 10.16 Linked alleles can still be separated if a cross-over event occurs between them. Crossing over results in the formation of new combinations of linked alleles. Here a cross-over event occurs between alleles *A* and *B* during meiosis.

Figure 10.17 Because cross-over events occur at random, the alleles for two genes located close together (*A* and *B*) are less likely to be separated by a cross-over than the alleles for two genes that are some distance apart (*C* and *A*, for example).

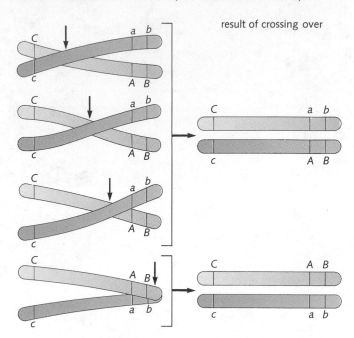

result of crossing over

linked. However, the law of independent assortment does apply to chromosomes themselves, so we may still refer to the independent assortment of genes on different chromosomes. These results, therefore, do not invalidate the Mendelian model for inheritance but, rather, modify it by adding another dimension to it.

GENETIC MAPS

In 1911 Alfred Sturtevant, while still an undergraduate at Columbia University, noticed that although some genes were tightly linked, others were not. The tightness (or the looseness) of the coupling was reproducible from one experiment to the next. It occurred to Sturtevant that the degree of linkage could be explained in a very straightforward way. If the genes were arranged on the chromosomes and if cross-over events occurred at random, then the closer together two genes were, the less likely they were to be separated by cross-over events. Random cross-over events are relatively unlikely to separate two genes

that are right next to each other, whereas they are almost certain to separate genes that are at opposite ends of a chromosome (Fig. 10.17).

Geneticists quickly realized that the frequency of cross-over events between two genes could be used to measure how far apart two genes were on a chromosome. In other words, they could construct a **genetic map** showing the location of genes on a chromosome (Fig. 10.18). Gene mapping is generally done in a number of individual steps:

- Test crosses between a homozygous recessive and an unknown genotype are used to determine the genotype of an organism.

- If two genes fail to display independent assortment in test crosses, it is likely that they are located on the same chromosome.

- And finally, if several genes are tightly linked (meaning that they are rarely separated by cross-over events), we can construct a genetic map of the chromosome by measuring the frequency of cross-over events between pairs of genes.

Sturtevant, incidentally, later wrote that he constructed the first genetic map (involving five genes on a *Drosophila* chromosome) in a single night in 1911. He wrote that this project took place "to the neglect of my undergraduate homework," an example that you may or may not wish to emulate.

Giant Chromosomes in Tiny Organisms

Although gene mapping has shown us how genes are arranged on the chromosome, in most cases the chromosome itself is so tiny that very little of its own structure can be seen. Luckily there are some exceptions. The nuclei of a few cell types contain giant chromosomes that are as much as one hundred times larger than usual.

In the formation of giant chromosomes, many copies of a single chromosome are produced without mitosis, and hundreds of chromosome strands lie side by side (Fig. 10.19). Because of the way they are formed, these chromosomes are called **polytene,** which means "many-stranded." Once polytene chromosomes have formed in a cell, the cell never divides again. The cells containing them, therefore, represent a final stage of development. By a great stroke of luck, *Drosophila*—the very same organism that had been so useful in studies of genetics—was discovered to have polytene chromosomes in its salivary glands.

A comparison of polytene chromosomes with normal metaphase chromosomes illustrates why these structures have been so useful in cellular genetics. The tiny individual bands enable observers to recognize different regions of the chromosome, and that has given geneticists an important opportunity. Each band marks the location of an individual gene on the chromosome, and genetic mapping techniques have made it possible to identify a great many of the bands with particular genes.

Regions of genetic activity can also be seen on the chromosomes. When a gene becomes active, the band corresponding to it sometimes swells to form a distinct **chromosome puff** (a puff appears in Fig. 10.19). The presence of puffs in certain genes enables geneticists to determine which genes are active at particular stages of development. For more than half a century, these giant chromosomes have given biologists an unparalleled opportunity: the chance to "see" the positions of genes on chromosomes and to "watch" genes in the process of being turned on and off.

THE INHERITANCE OF SEX

Mendel's simple principles account for many types of inheritance, but the reality of nature is of course more complex. One of the first areas to require modification

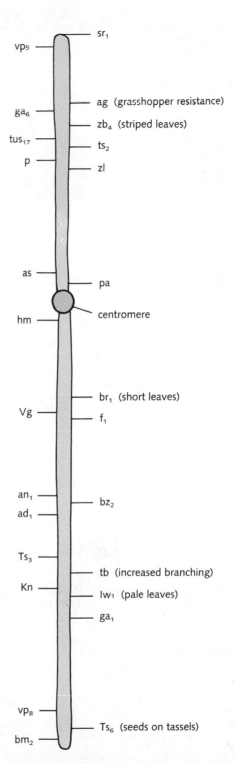

Figure 10.18 A gene linkage map from chromosome number 1 of corn (*Zea mays*). The locations of genes on this map were determined by the frequency of cross-over events between them.

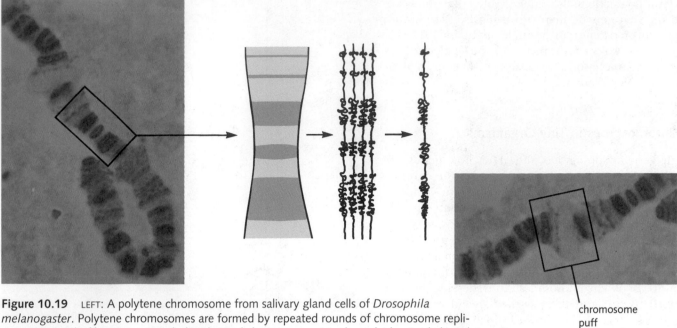

Figure 10.19 LEFT: A polytene chromosome from salivary gland cells of *Drosophila melanogaster*. Polytene chromosomes are formed by repeated rounds of chromosome replication. Instead of being separated, the identical chromosome strands are laid out side by side to produce a polytene chromosome that contains hundreds or thousands of copies of the single chromosome. RIGHT: A chromosome puff.

chromosome puff

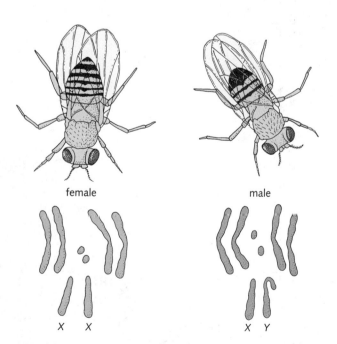

female   male

X   X   X   Y

Figure 10.20 Chromosomes of male and female *Drosophila melanogaster*.

involved differences in genetics between the sexes. These differences have their origin in the fact that in most organisms, the sexual identity of an individual is determined by special chromosomes known as the *sex chromosomes*. When we examine each of the eight chromosomes found in a diploid female *Drosophila* cell, for example, we can easily group them into four pairs of homologous chromosomes. In a cell taken from a male fly, however, the story is a little different. We can arrange six of the chromosomes in pairs identical to those found in the female cell, but two dissimilar chromosomes are left over (Fig. 10.20). One of these unmatched chromosomes is identical to the last pair in the female cell, but one is unlike any other chromosome. This is the exception (which we referred to earlier) to the rule that nearly all chromosomes occur as one member of a homologous pair. We call these two sex-related chromosomes *X* and *Y*. Between them, they determine the sex of an organism.

Inheritance of the Sex Chromosomes

The *X* and *Y* chromosomes are passed from one cell generation to the next like any other chromosomes, except during the process of meiosis. Female cells have two *X* chromosomes. These constitute a homologous pair that in the sex cells produces a tetrad along with the other chromosome pairs during the first meiotic division. Male

Nettie Stevens and the Y Chromosome

The history of genetics, like that of any science, is filled with wrong turns, mistakes, and errors in judgment that time and the experimental method have corrected. The identification of the *X* chromosome is a case in point. The first suggestion that sex could be a chromosomal characteristic was made in 1901 by Clarence E. McClung, who suggested that a so-called *accessory chromosome* in insects determined maleness. The term had been coined 10 years earlier by an investigator named H. Henking. Henking noticed this chromosome in studies on the cells of *male* insects. Unlike the other chromosomes, it did not have a paired chromosome, so he called it an accessory. He correctly noted that it divided in only one of the two meiotic divisions, which showed that it was present in only two of the four sperm cells. This unusual behavior made many biologists doubt that it even *was* a chromosome, and the designation *X* reflects that doubt. Because McClung found the unpaired accessory chromosome only in male cells, he concluded that the *X* chromosome determined maleness.

We can look back on the work of Henking and McClung with the benefit of hindsight, but only because a more patient observer followed them—one who noticed an even smaller structure in these insect cells, a structure they had missed. That scientist was Nettie M. Stevens. Nettie Stevens was born in rural Vermont, studied in Massachusetts at a state teachers' college, and then taught school for nearly 10 years in order to save enough money to attend Stanford University to study science. From Stanford she went to Bryn Mawr College in Pennsylvania in 1900 and began a series of studies focused on the way in which the sex of an animal is determined. While many other investigators believed that sex was *not* genetically determined, Stevens was convinced that McClung was on the right track.

Stevens made a series of very careful observations with *Tenebrio molitor*, the common mealworm. She found that female cells contained 20 chromosomes, whereas male cells contained 19 large chromosomes and 1 small one. She correctly concluded that this small difference was the result of chromosomal sex determination! Nettie Stevens had discovered the *Y* chromosome. The determinant of maleness was not the *X*,

for females had two of those, but the small *Y* chromosome, which was found only in males. Stevens's discovery of the *Y* chromosome was a pivotal event in the development of genetics, because it made possible for the first time a correct explanation of sex-linked inheritance.

---

cells, however, have both an *X* and a *Y* chromosome (and no tetrad forms). During meiosis, each gamete receives either an *X* or a *Y* chromosome. Therefore, about half of the sperm cells produced by a male fly contain the *X* chromosome and about half contain the *Y*. Because all eggs contain an *X* chromosome, the sex of the zygote is determined by the sex chromosome carried by the sperm. If the sperm cell donates an *X* chromosome, the zygote will be female; if it donates a *Y* chromosome, the zygote will be male.

The *X-Y* mechanism, which is also how sex is determined in humans, produces a 1:1 ratio of males to females (Fig. 10.21). But what happens to a gene located on one of the sex chromosomes? The answer, as first discovered by the American geneticist T. H. Morgan, is that inheritance can be *sex-linked*.

Sex Linkage

Thomas Hunt Morgan was a professor of biology who did his first work at Columbia University beginning in 1904. Morgan pioneered the use of the fruit fly, *Drosophila melanogaster*, in genetics research. These organisms offer

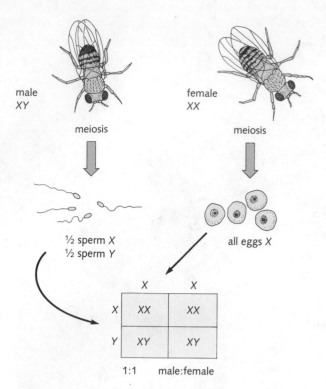

½ sperm X
½ sperm Y

all eggs X

	X	X
X	XX	XX
Y	XY	XY

1:1   male:female

Figure 10.21 A Punnett square shows why male and female *Drosophila* are produced in roughly equal numbers.

several advantages over Mendel's peas. Fruit flies can be grown in small bottles on a laboratory shelf, they will breed at any time of the year, and above all, the generation time for the fly is only 10–14 days. Using these flies, Morgan and his co-workers were able to show that the basic laws of genetics Mendel had discovered in peas governed inheritance in flies as well.

A few experiments, however, yielded baffling results. Morgan and his students were constantly on the lookout for interesting traits to study, and they examined first those traits that could be observed quickly and easily even in a tiny fly. Consequently, one of the first characteristics they studied was the color of its eyes.

Drosophila found in the wild (wild-type flies, as geneticists call them) have reddish-colored eyes, a color produced by a screening pigment that is located in the cells surrounding the light-sensitive cells and prevents excessive scattering of light in the eye. As Morgan examined his flies, he found a single male fly with white eyes. He isolated this individual and decided that he would try to determine, by mating this fly with "normal" females, whether the white eyes were a heritable trait. In his first cross, he produced an F_1 generation made up entirely of red-eyed flies (Fig. 10.22).

It looked as though the allele for red eye color was dominant over the allele for white eye color. The best way to test this idea was to cross members of the F_1 generation with each other. Morgan carried out an $F_1 \times F_1$ cross and got the 3:1 ratio of red to white eyes in the F_2 generation, just as one would expect if the allele for red eyes were dominant over the allele for white eyes. However, he noticed something puzzling about the offspring. *All of the white-eyed flies were male!* In other words, none of the female flies showed white eyes, and about half of the males did. Why were the white eyes concentrated in the male flies?

Drawing on the observations of Nettie Stevens (see Theory in Action, Nettie Stevens and the *Y* Chromosome, p. 207) and Edmund Wilson, Morgan produced a simple explanation for his results. *Drosophila* females had four chromosome pairs, whereas males had three pairs and a nonmatching *XY* pair. Assuming that the females were *XX*, Morgan suggested that the gene for eye color was located on the *X* chromosome. If we use the symbol *W* for the red-eye allele and *w* for the white-eye allele, his original white-eyed male must have been X_wY. The wild-type females were X_wX_w. No white-eyed flies appeared in the first filial (F_1) generation, because all flies in that generation, whether male or female, had an *X* chromosome with the *W* (red-eye) allele. In the second filial (F_2) generation, only one combination of gametes could produce a white-eyed fly: a sperm cell carrying the *Y* chromosome and an egg cell carrying the X_w chromosome. Although about half the females in the F_2 generation *carried* the *w* allele, the fact that they also carried a dominant *W* allele meant that they had red eyes (Fig. 10.23).

Morgan was the first to show that sex-linked inheritance could be explained if the gene in question were located on the *X* chromosome. The same pattern is seen for other genes located on the *X* chromosome. It is important to appreciate why a gene found on the *X* chromosome produces sex-linked inheritance. When a recessive allele is located on one of the two *X*s in a female, it may or may not be expressed, depending on which allele is present on the other *X* chromosome. But in a male, *any* allele on the single *X* chromosome is expressed. The tip-off that a particular gene is sex-linked is the preferential expression of a recessive allele in males.

Other Patterns of Sex Determination

Humans and fruit flies happen to follow one pattern of sex determination, but a great many organisms follow different patterns. In some insects there is no *Y* chromosome. Males in such species are *XO* and females are *XX*. (The letter *O* represents the absence of a chromosome.) In birds and moths, the males possess a pair of identical sex chromosomes referred to as *ZZ*, whereas the females are *ZW* or

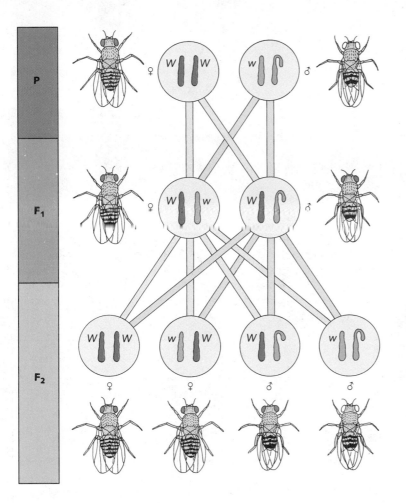

Figure 10.22 The results of two generations of crosses involving eye color in *Drosophila*. The original breeding couple are a white-eyed male and a red-eyed female. In the F₁ generation all flies have red eyes, but in the F₂ generation ½ of the male flies are white-eyed. (Magnification factor: 20)

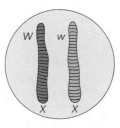

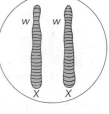

ZO. (The letters *Z* and *W* are used to prevent confusion with the *XY* system we have become familiar with.)

In bees and ants, sex determination occurs by means of a system called **haplo-diploidy.** In haplo-diploidy there are no sex chromosomes as such. The males develop from unfertilized eggs and are *haploid,* whereas the females develop from fertilized eggs and are *diploid.*

females, being *XX,* must have two copies of the white-eye allele (*ww*) to display the white-eye phenotype

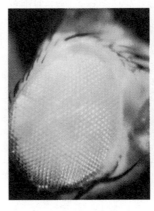

female

heterozygous = red eyes

homozygous recessive = white eyes

OTHER FORMS OF GENETIC VARIATION

Thus far we have considered only traits in which one allele is recessive and one allele is dominant. Such situations are known as **simple dominance** or **complete dominance.** However, not all genes behave quite so neatly. Flower color in one strain of snapdragons, for example, is controlled by a system in which alleles exist for red flowers (*R*) and white flowers (*r*). When both alleles for a particular trait are identical, we say that an organism is homozygous for that trait. In snapdragons, the homozygous organisms have either red flowers (*RR*) or white

males, carrying only one *X* chromosome, will display the white-eye phenotype with just one *w* allele

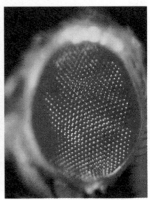

male

red eyes

white eyes

Figure 10.23 Sex-linked inheritance explains the appearance of the white eye color in half of the F₂ males in Fig. 10.22.

Using Genetics to Solve Problems

Part of the power of genetics as a scientific tool stems from the ability it confers on us to analyze and solve problems and, often, to predict the results of crosses between two organisms. Let's look at one example of a genetics problem.

Two *Drosophila* flies, each of which seemed to be phenotypically wild type, were crossed. Their offspring (the F_1 generation) were then collected, and each one was test-crossed with a fly displaying purple eye color, a recessive trait. Of these crosses (the F_2 generation), half produced flies with wild-type eye color, and the other half produced 50 percent flies with wild-type eyes and 50 percent flies with purple eyes. *Determine the genotype of the original parents.*

By reasoning backwards from the last cross, we can solve the problem. Purple is a recessive trait, so each of the purple-eyed flies must have been genotype *ww*. Because half of the F_1 flies that were mated to the purple flies produced offspring with 50 percent purple eyes, these F_1 flies must have had a copy of the *w* allele. Their own eye color was wild type, so they must also have had a copy of the wild-type allele, making them genotype *Ww*. The other half of the F_1 generation did not produce any purple-eyed flies in their F_2 offspring. This means that their genotype must have been *WW*. What would the genotypes of the two original fly parents have to have been to produce an F_1 generation that was half *WW* and half *Ww*? The answer is that they must have been *WW* and *Ww*.

It is also possible to use genetics to predict future events. Remember that Mendel discovered that tall is dominant over short in garden peas. Two tall peas are crossed and 100 seeds are collected. Two seeds are planted and grown under identical conditions. One produces a tall plant and the other a short plant. On the basis of these results, *determine*

how many of the remaining 98 plants can be expected to grow into tall plants.

Remember that we were not told the genotypes of the original parents—only their phenotypes (they were both tall). However, because they were both tall, we know that each had at least one of the dominant alleles for tallness (*T*). The single seed of the first filial generation that grew into a short plant had to be genotype *tt*, because the *t* allele is recessive. The only way in which a short plant could have been produced was for each of the parents to have had at least one *t* allele. Therefore, the two parents were both genotype *Tt*. Simple genetic analysis tells us that 25 percent of the seeds in a cross between two such plants will grow to display the short phenotype. Because 75 percent of 98 = 73.5, 73 or 74 of the remaining seedlings can be expected to be tall.

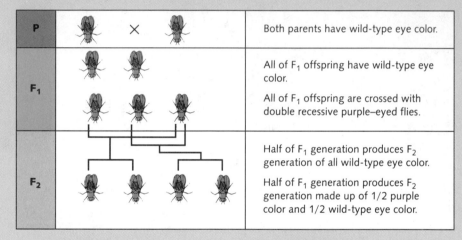

P	Both parents have wild-type eye color.
F_1	All of F_1 offspring have wild-type eye color. All of F_1 offspring are crossed with double recessive purple–eyed flies.
F_2	Half of F_1 generation produces F_2 generation of all wild-type eye color. Half of F_1 generation produces F_2 generation made up of 1/2 purple color and 1/2 wild-type eye color.

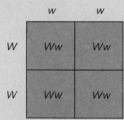

F_1 flies that produced all wild-type flies when mated with purple flies must have been genotype *WW*.

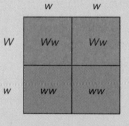

F_1 flies that produced 1/2 purple flies and 1/2 wild-type flies when crossed with purple flies must have been genotype *Ww*.

To produce these two types of F_1 flies the parents must have been genotype *WW* and *Ww*.

flowers (*rr*). The heterozygous organism, of course, has different alleles for the same trait. Which allele is dominant in the heterozygote?

Incomplete Dominance

In snapdragons *neither* allele is dominant. Instead, the flowers produced by *Rr* plants are pink, a blend of red and white. Such a situation is called **incomplete dominance,** because neither allele is clearly dominant. In incomplete dominance, the characteristics of the heterozygote are not the same as those of either homozygous organism. In some respects, one might consider this a form of "blending" inheritance, in which the offspring look a little bit like one parent and a little bit like the other. However, incomplete dominance still follows the basic ratios of Mendelian inheritance. When an F$_1$ generation of snapdragons is crossed with itself, the ratio of flower color in the F$_2$ offspring is 1:2:1 (red:pink:white) (Fig. 10.24). This ratio is identical to the 3:1 ratio that would be obtained for simple dominance in such a cross *except for one thing:* the heterozygote has different characteristics from the homozygous dominant (it is pink rather than red).

Snapdragon flower color is not the only example of incomplete dominance. It is observed in a host of different genes, including a gene that controls feather structure in chickens. One allele causes disrupted feathers (known as frizzle). A homozygous chicken with the frizzle allele (*ff*) shows "extreme frizzle" in its feathers. The heterozygous cross of such a chicken with a normal chicken (*FF*) displays "mild frizzle" as a result of its *Ff* genetic composition.

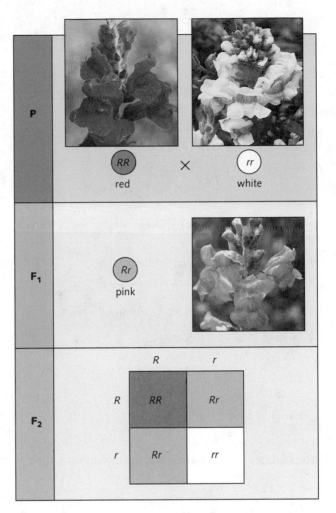

Figure 10.24 A cross showing incomplete dominance in red, white, and pink snapdragons. The pink color results from incomplete dominance of the red allele over the white allele.

Codominance

When each of two different alleles contributes to the phenotype, the alleles are said to be **codominant.** In such situations, it is not possible to say that one allele is dominant over the other. One of the best-known examples of codominance occurs in humans, where it determines the ABO blood group. The gene locus for this blood group system has three possible alleles: I^A, I^B, and i. The I^A and I^B alleles determine sugars present on the surfaces of blood cells. An individual with only the I^A allele has A-type sugars and is said to be blood type A. An individual with only the I^B allele has B-type sugars and is blood type B. I^A/I^B heterozygotes have *both* A-type and B-type sugars and are blood type AB. Because both the I^A and I^B alleles contribute to the phenotype, these alleles are said to be codominant. Individuals homozygous for the i allele (i/i) are blood type O.

Multiple Alleles

Although many of the genes we have examined may have two different alleles, not all genes do. In fact, for many genetically determined characteristics, three or four or even ten different varieties of the same gene are found in a population. The seven characteristics investigated by Mendel were all examples of two-allele systems. Only two types of each gene were present in his breeding population. However, Mendel was either very lucky or very clever in his choice of genes for study. Multiple alleles are not at all uncommon.

One of the most familiar examples of a multiple allele system is found in the rabbit (Fig. 10.25). The coat color of rabbits is controlled by a series of four alleles for the same gene:

C full color
c^{ch} chinchilla

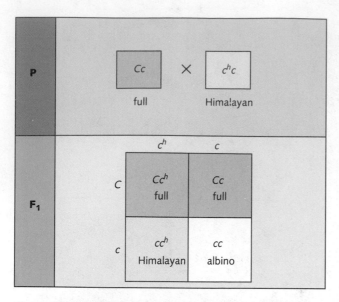

full $C\_$

chinchilla $c^{ch}c^{h}$ or $c^{ch}c$

Himalayan $c^{h}c$

albino cc

Figure 10.25 An experimental cross involving the multiple allele coat color system in rabbits. (CLOCKWISE FROM LEFT: full color, chinchilla, white, and Himalayan.)

c^{h} Himalayan
c albino (white)

These four alleles show a pattern of simple dominance:

$$C > c^{ch} > c^{h} > c$$

This means that the heterozygote cc^{ch} would have a chinchilla coat, Cc would have a full-color coat, and $c^{h}c^{ch}$ would also have a chinchilla coat. There are many other examples of genes with multiple alleles (including the ABO blood-group gene in humans), as we will see in the next chapter.

Traits Controlled by Gene Interaction

The apparent simplicity of the principles that govern the inheritance and expression of some traits should not be taken to suggest that all characters are controlled as simply as the color of a flower or the fur of a rabbit. In reality, many phenotypic differences that seem clear-cut are controlled by gene interaction.

An interesting one is a two-gene system that seems to control coat color in Labrador retrievers (Fig. 10.26). These dogs commonly come in both black and yellow varieties, and they occasionally exhibit a rich brown color called chocolate. The genetics of coat color in these dogs has been investigated at breeding colonies where several generations of controlled matings have been carried out. It turns out that their coat color is controlled by genes

found at two **loci** (singular: *locus*)—that is, at two different positions in the genetic system.

The two loci are known as the B locus and the E locus. Each locus has two alleles. At the E locus the double-recessive genotype e/e produces yellow fur regardless of the situation at the B locus. But if the E locus contains the dominant E allele (E/E or E/e), then control shifts to the B locus. Two alleles are possible at the B locus: a dominant B allele, which produces black coat color, and a recessive b allele, which produces chocolate color if it is in the homozygous form (b/b). A system like this, in which the genotype at one locus controls the expression of genes at another, is an example of **epistasis.** The E gene is said to be epistatic to the B gene, which literally means that it "stands above" the B locus.

In Labrador retrievers an interesting twist enables us to determine the condition of genes at the other locus. Look closely at Fig. 10.26. Both dogs are yellow, which means that they are genotype e/e at one locus. However, they have different pigmentation around the nose and lips. Breeding experiments show that dogs with black pigmentation around the nose and lips are genotype B/B or B/b in the B locus and that dogs with pale noses and lips are genotype b/b. Therefore, coat color in Labrador retrievers is governed by an epistatic two-locus system, but one in which it is sometimes possible to determine the status of both loci by looking very closely at the phenotype.

A question: The coat color system in parakeets is also a two-locus system, as we have seen. Can you determine whether this coat color system is epistatic? (See Appendix for the answer.)

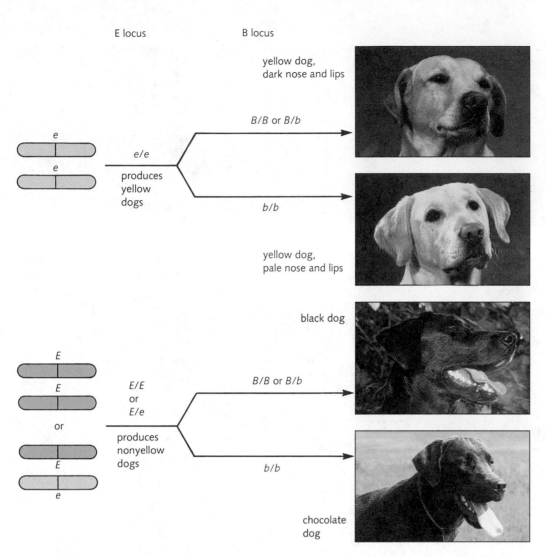

Figure 10.26 Coat color in Labrador retrievers is controlled by four alleles found at two different loci. Note how the alleles present at the E locus affect the expression of the alleles at the B locus.
RIGHT: Labrador retrievers showing the four possible coat and skin color combinations.

E locus

B locus

yellow dog, dark nose and lips

e/e produces yellow dogs

B/B or *B/b*

b/b

yellow dog, pale nose and lips

black dog

E/E or *E/e* produces nonyellow dogs

B/B or *B/b*

b/b

chocolate dog

Polygenic Systems

Systems wherein a single trait is controlled by more than two genes are said to be **polygenic** ("many genes"). Many polygenic systems, particularly those that control such traits as shape and form, are so complex that we are years away from understanding them fully. However, we do know the details of a few simple systems.

The reddish color of wheat kernels is controlled by a three-locus system. There are two contrasting alleles at each gene locus, a red allele (R_n) and a white allele (r_n). (The n designates a particular locus in the three-locus system, as shown in Fig. 10.27.) The red alleles show incomplete dominance over the white alleles, and each locus makes a contribution to the color of the kernel. Therefore, the overall kernel color is determined by the total number of red and white alleles. Kernels with six red alleles are bright red, those with six white alleles are white, and those with intermediate numbers are somewhere between red and white.

Figure 10.27 illustrates the expected result of crossing a white-kernel plant with a red-kernel plant. The F_1 kernels would be medium-red in color. When the F_1 plants are crossed, 64 possible allele combinations result. Only one of these would produce a white kernel, and only one would produce a bright red kernel. The remaining 62 kernels would be intermediate between those two colors. It's easy to see how the kernels themselves would give the impression that the white and red colors of the original kernels had *blended* to produce the kernels of the F_2 generation.

Polygenic traits sometimes give the impression that **continuous variation**—a full range of phenotypes between two extremes—exists in a population. The traits that Mendel investigated involved **discontinuous variation;** all phenotypes fell into a few well-separated categories. But you would be hard-pressed to sort each of the F_2 kernels into seven categories of pigmentation. As the number of gene loci involved in a polygenic trait increases, the range of variation becomes almost completely continuous.

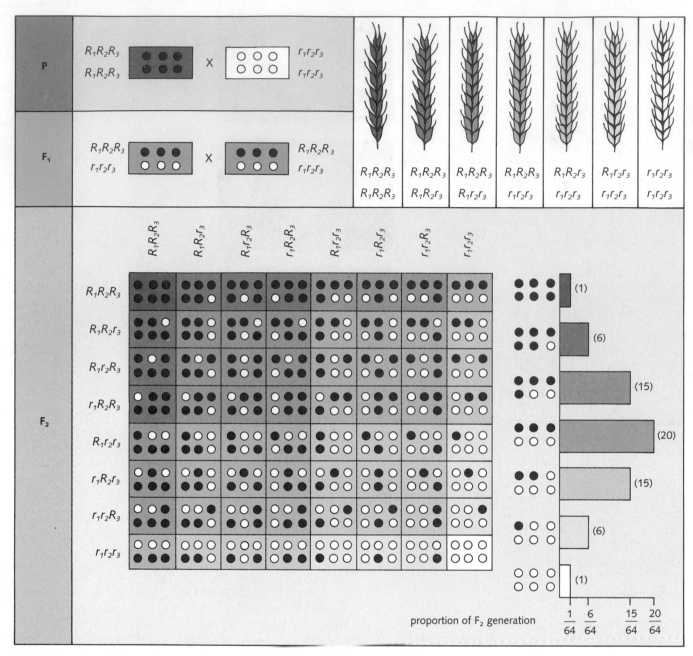

Figure 10.27 Many complex traits are controlled by polygenic systems. Kernel color in wheat is determined by three loci, which produce a range of phenotypes from white to dark red.

Figure 10.28 The expression of some genes is affected by the environment. The expression of the "curly-wing" allele in fruit flies is dependent on the temperature at which the flies develop.

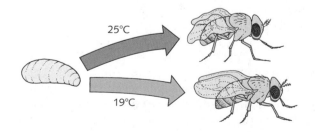

Figure 10.29 The expression of coat color in Siamese cats varies with temperature. Black pigment is produced only in those areas of the skin which are lowest in temperature, such as the ears and tail. This variation produces the typical Siamese markings.

Environmental Effects on Gene Expression

Earlier in the chapter, when we made the distinction between genotype and phenotype, we were careful to say that the phenotype of an organism develops under the influence of its genotype *and* its environment. Some of the effects of the environment are obvious. As Mendel learned, the height of pea plants is genetically determined. One of his seven traits was determined by a height gene, for which there were two alleles, tall and short. The actual size of each plant, however, was influenced by environmental factors, including the amount of water, sunlight, and nutrients that each plant received. The effect of environment means that even a group of plants with identical genotypes displays variability in phenotype.

Curly is a dominant allele that is found in *Drosophila* and influences the shape of the wing. The effect of this allele is controlled by the temperature of the environment. If the flies develop at 19°C, most wings are normal. If they develop at 25°C, however, the wings are curly (Fig. 10.28).

Temperature also influences gene expression in the Siamese cat (Fig. 10.29). Siamese cats are pale in color, with gradually darkening fur near the tips of the tail, feet, ears, and nose. This coloration is produced only in cats that are homozygous for a particular recessive allele that produces the black pigment melanin. Unlike the normal melanin allele, the recessive allele produces melanin only at temperatures just below the usual body temperature. Black pigment, therefore, is produced only in the cooler regions of the skin, creating the characteristic Siamese markings.

SUMMARY

The modern science of genetics traces its origins to the work of Gregor Mendel, who determined that a set of characters, or genes, was passed from one generation to the next by reproductive cells, or gametes. Mendel's three important principles of dominance, segregation, and independent assortment describe the way the genotype is determined. The genotype and the organism's environment determine its phenotype, the sum total of all its expressed characteristics.

Although Mendel was aware of their existence, other biologists had discovered that cell nuclei contained chromosomes, visible during mitosis. Cells that contain one maternal and one paternal set of chromosomes are said to have the diploid number $(2n)$ of chromosomes. During meiosis, the chromosome number is reduced from diploid to haploid (n).

Genes are located in distinct positions on chromosomes. Genes located on the same chromosome are said to be "linked," since they are inherited together. This linkage is not absolute, however, and linked genes may be separated by cross-over events which take place during meiosis. Because random cross-over events are more likely to separate distant genes than nearby ones, the frequency of crossing over between genes provides geneticists with the information they need to construct a map of genes on a chromosome.

The existence of giant polytene chromosomes has enabled geneticists to visualize chromosome regions that are associated with specific genes, and even to watch genes in the process of being activated.

In many organisms, sex determination occurs by means of sex chromosomes such as the *XY* system, in which individuals of genotype *XX* are female and those of genotype *XY* are male. Genes carried on either sex chromosome display a sex-linked pattern of inheritance in which the phenotypes they determine are preferentially expressed in one sex or the other. Living organisms exhibit a wide range of genetic systems involved in the control of different characteristics. These include incomplete dominance, codominance, multiple alleles for a single gene, interactions between different genes, and polygenic control of a single characteristic.

STUDY FOCUS

After studying this chapter, you should be able to:

- Explain how the concept of a gene originated and how the science of genetics developed.
- Apply basic Mendelian genetics to real-life situations in the breeding of animals and plants.
- Describe the nature of the hereditary connection between one generation and the next.
- Give evidence that genes are real structures that occupy definite positions on chromosomes.
- Explain how two green parakeets could produce blue, yellow, and white offspring, as described in the opening pages of this chapter.

TERMS AND CONCEPTS

genes *192*

alleles *192*

dominant *192*

recessive *192*

principle of
   segregation *193*

principle of
   dominance *193*

genotype *193*

phenotype *193*

homozygous *193*

heterozygous *193*

test cross *193*

independent
   assortment *195*

zygote *198*

homologous
   chromosome *198*

diploid *198*

haploid *198*

meiosis *198*

synapsis *199*

genetic map *204*

incomplete
   dominance *209*

Discussion Questions

7. How did Mendel's discoveries help to support Darwin's theory of evolutionary change?

8. In what respect does the behavior of *chromosomes* during mitosis and meiosis resemble the behavior of *genes* during the same two processes?

9. Why is a two-factor cross necessary to demonstrate the principle of independent assortment?

10. The existence of chromosomes requires an important modification of the principle of independent assortment. What is it? Why is the process of crossing over an exception?

11. Is a 50:50 balance of males and females maintained in a population wherein sex determination is based on the number of *X* chromosomes present: *XX* for females and *XO* for males? How about a population wherein *ZZs* are male and *ZWs* are females?

REVIEW

Objective Questions (Answers in Appendix)

1. An organism of unknown genotype of gene *P* was test-crossed with a double-recessive organism in order to determine the unknown genotype. The results of the cross conclusively showed that the genotype of the unknown organism was *Pp*. The phenotypic ratio in that F_1 generation must have been
   (a) 9:3:3:1. (c) 2:2:1.
   (b) 3:1. (d) 1:1.

2. The outward appearance of an organism is called its
   (a) phenotype. (c) allele.
   (b) genotype. (d) chromosome.

3. The genetic makeup of an organism is called its
   (a) phenotype. (c) allele.
   (b) genotype. (d) chromosome.

4. The principle of independent assortment assumes that
   (a) gametes combine in an ordered fashion.
   (b) genes segregate freely during mitosis.
   (c) genes segregate freely during gamete formation.
   (d) genes are linked together in the chromosome.

5. For a diploid species, fertilization of the gametes will
   (a) reduce the diploid number by one-fourth.
   (b) reduce the diploid number by one-half.
   (c) restore the diploid number.
   (d) result in a haploid number in the zygote.

6. A gamete produced by a sexually reproducing organism contains 8 chromosomes. The diploid number of this organism is
   (a) 2. (c) 8.
   (b) 4. (d) 16.

READINGS

Bull, J. J., I. J. Molineux, and J. H. Warren. "Selfish genes." *Science* 256 (1992): 65. *Sd* (segregation distorter) is not the only gene that "cheats." This brief report describes a gene called "Medea," a gene in flour beetles that kills any offspring that do not carry it.

Fincham, J. R. S. "Mendel—Now down to the molecular level." *Nature* 343 (1990): 208–209. More than 100 years after Mendel we know a great deal more about each of the seven genes he studied. This article gives the latest information on these seven famous genes.

Crow, J. F. "Genes that violate Mendel's rules." *Scientific American* 240 (February 1979): 134–146. A clear and thoughtful discussion of the *sd* (segregation distorter) and other genes that "cheat" at meiosis.

Brush, S. G. "Nettie M. Stevens and the discovery of sex determination by chromosomes." *ISIS* 69 (1978): 162–172. A superb sketch of the life and scientific achievements of the discoverer of sex chromosomes.

John, B. "Myths and mechanisms of meiosis." *Chromosoma* 54 (1976): 295–325. Some aspects of meiosis can be confusing. This clearly written article explores meiosis in a way that focuses on the essential aspects of the process and its relationship to Mendelian genetics.

Stern, C., and E. R. Sherwood, eds. *The Origin of Genetics: A Mendel Source Book.* New York: W. H. Freeman, 1966. An interesting collection of papers and essays on the origin of the science of genetics. It includes an English translation of Mendel's original papers and also papers by T. H. Morgan and other pioneers of genetics.

11

Human Genetics

"**k**now thyself." The philosopher's first commandment is an important one, for it stresses the strengths and limitations of human inquiry. More to the point, the desire to know about ourselves is one of the things that motivates many of us to study biology in the first place. That is particularly true of the science of genetics. Therefore, it is appropriate that we spend a good deal of time reviewing some of what we have learned about ourselves as genetic organisms.

Except for the fact that we find our own species interesting, *Homo sapiens* is a most inappropriate organism for the study of genetics. It is large and complex and cannot be maintained in the laboratory. Its generation time is very long—about 20 years in most societies. It usually produces only a single offspring after mating and most individuals produce no more than three or four offspring in a lifetime. Finally, we cannot perform experimental crosses, as we can for other species, so we must depend on the organisms to tell us their own life histories. Knowing that any sensible person would study fruit flies, roundworms, or bacteria, we plunge ahead anyway.

THE HUMAN GENETIC SYSTEM

Humans are multicellular organisms that reproduce sexually. An analysis of human genetics begins with individual cells. A diploid human cell contains 46 chromosomes, and nearly all of its genetic content, or *genome*, is contained on these chromosomes. Two of these are the **sex chromosomes** (*XX* in females and *XY* in males), and the other 44 are referred to as **autosomal chromosomes.** The autosomes consist of 22 pairs of homologous chromosomes, with one member of each pair inherited from each parent. In order to describe human chromosomal composition quickly, we may write 46 *XX* for a normal female and 46 *XY* for a normal male. This shorthand indicates the *total* number of chromosomes and the nature of the sex chromosomes.

We can get a good look at human chromosomes via a technique known as **karyotyping** (*karyon* means "nucleus"). Karyotyping is a standard procedure in genetics and medicine and is used to diagnose

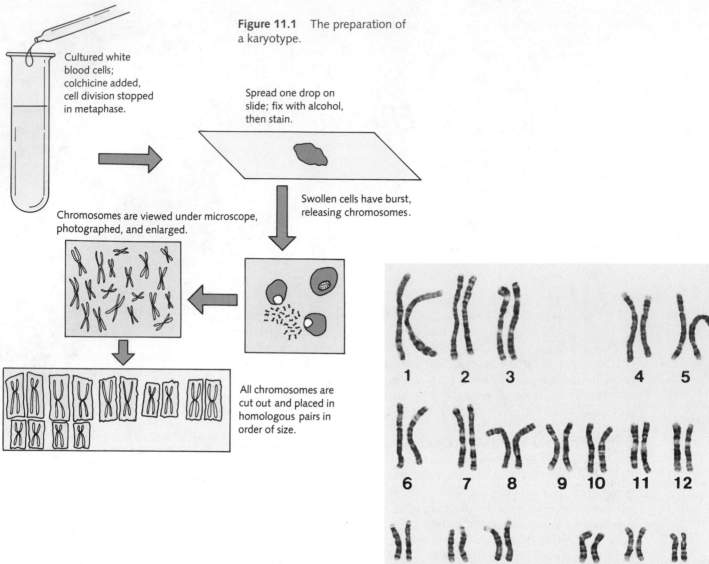

Figure 11.1 The preparation of a karyotype.

Cultured white blood cells; colchicine added, cell division stopped in metaphase.

Spread one drop on slide; fix with alcohol, then stain.

Swollen cells have burst, releasing chromosomes.

Chromosomes are viewed under microscope, photographed, and enlarged.

All chromosomes are cut out and placed in homologous pairs in order of size.

Figure 11.2 Karyotype of a human male.

chromosomal abnormalities. A few white blood cells are removed from a blood sample and grown in culture (Fig. 11.1). Then *colchicine,* a chemical that causes the microtubules making up the mitotic spindle to disassemble, is added to the cells. This enables the cells to enter mitosis but prevents the chromosome separation that normally occurs during anaphase. The cells are now locked into metaphase, and each chromosome is visible as a pair of identical chromatids joined by a single centromere. The cells are then fixed on a slide and stained to make it easier to visualize the chromosomes released from cells that have burst.

Photomicrographs of these metaphase chromosomes can be analyzed in a very simple way. We take a pair of scissors and cut out each of the chromosomes from the photograph. When we sort through the loose pictures, we find a match for each chromosome in our pile, and they can be arranged in order of size (Fig. 11.2). A male karyotype contains two chromosomes for which no match can be made: the X and the Y.

Genetics with Human Subjects: The Pedigree

In order to apply the principles of Mendelian genetics to human beings, we must first identify an inherited character that is controlled by a single gene. This is not easy, for it is often difficult to determine which characters are directly inherited and which are related to environmental influences. To make things more difficult, we cannot carry out test crosses with humans as we do with plants or fruit flies. Instead, scientists must rely on family records of births, marriages, and deaths.

These records are often displayed as a **pedigree,** a chart on which we can trace the genetic relationships of individuals. Figure 11.3 shows a Norwegian family and its pedigree, tracing the inheritance of *wooly hair,* an unusual trait in which the hair is tightly kinked and brittle, causing it to break off before it becomes very long. The pedigree summarizes the relationships among the eight family members in the photo. As you can see, the mother and three of her six children have wooly hair. Can you guess what sort of gene produces the wooly hair trait? (A hint: The trait did not appear in the father's family.)

Finding Genes: Abnormalities and Disorders

The wooly hair trait is a good example of how we learn about the existence of a human gene. It is only because the wooly hair trait is *abnormal*—uncommon in the population—that we can recognize it as a distinct characteristic. In that sense, it is a **genetic abnormality,** a characteristic very different from the norm, or average.

Some genetic abnormalities produce medical problems. A **genetic disease** is a significant disorder caused by a gene or group of genes. The alleles that cause such diseases may be either dominant or recessive. A great many genetic diseases have been described in the past two centuries, and our acquaintance with these diseases makes up a large part of our knowledge of the human genome. This is not due to any inclination on the part of geneticists to be fascinated with disease; rather, it is because the presence of a genetic disorder highlights the "missing" function that the gene normally carries out.

HUMAN GENES

More than 4000 human genes and their alleles have been described, and many of these have been mapped to specific chromosomes. **Sex-linked genes** are carried on either the *X* or the *Y* chromosome. Genes located on any of the other 44 chromosomes, or *autosomes,* are known as **autosomal genes.**

Autosomal Recessive Inheritance

At least 1000 known human traits are produced by recessive autosomal alleles. Such alleles do not affect the phenotype unless an individual is homozygous for them.

Albinism The genetic disorder known as **albinism** is caused by a recessive allele found on an autosomal chromosome and present at low frequencies in all human population groups (Fig. 11.4). Because the allele is recessive, individuals who are heterozygous for the trait express their normal skin color, so the presence of the allele is "hidden" by the dominance of the normal allele. Albinos are unable to synthesize **melanin,** the pigment molecule responsible for most human skin coloring. This makes their skin and eyes extremely sensitive to light, and they must avoid exposure to bright sunlight.

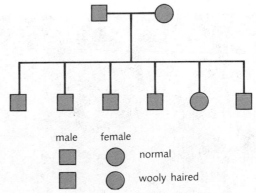

Figure 11.3 LEFT: A Norwegian family, some of whose members have the wooly hair trait. A pedigree chart of the family (ABOVE) shows the passage of this trait from one generation to the next.

What happens when an albino and a normally pigmented person have children? We can represent the genotype of the normal individual as *A/A* and that of the albino as *a/a*. A simple genetic analysis reveals that their children will carry the genotype *A/a*. But these children will not be albinos; they will be normally pigmented. Now what happens when two people who carry the genotype *A/a* have children? We can analyze this with a simple Punnett square in which the possible gamete combinations are enumerated (see Fig. 11.4).

Our result is the classic 3:1 ratio for a heterozygote cross. What this means is that the probability that a child of this couple will display the normal phenotype for this gene is three in four, or 75 percent. However, chances are two out of three (66 percent) that a normally pigmented child will carry a copy of the defective allele. The passage of an inherited disorder from one generation to the next can be predicted by Mendelian genetics. Because it can be used to help counsel individuals who carry a genetic disorder by determining the odds of their having healthy children, the science of genetics is a practical medical tool (see Theory in Action, Genetic Counseling: Knowing the Odds).

Tay–Sachs disease   In the 1880s two physicians, Warren Tay of Great Britain and Bernard Sachs of the United States, described a fatal disease that took the lives of many of their young patients. After about six months of normal development, babies with this disease developed a reddish patch on the retina of their eyes. Gradually they became blind and deaf, suffered convulsions, and usually died by the age of 3 or 4. Both physicians noted that the disease occurred in Jewish families of eastern European ancestry.

Today it is recognized that **Tay–Sachs disease** is caused by a recessive allele. Therefore, each parent of a Tay–Sachs child carries a single copy of the Tay–Sachs allele. The Tay–Sachs allele is a defective version of a normal allele that helps cells in the nervous system dispose of fatty molecules known as *gangliosides*. A child who is homozygous for the defective allele cannot dispose of these compounds. They gradually build up within the cells of the nervous system, leading to mental degeneration and death. There is a simple test that can detect heterozygotes who carry the Tay–Sachs allele, warning them that they are at risk of giving birth to children with the disease.

Cystic fibrosis   The most common fatal genetic disease in the United States is **cystic fibrosis,** which affects 1 child in 2500 and is most common among Americans of European ancestry. Victims of the disease suffer from digestive disorders and produce a thick, heavy mucus that clogs their lungs. With medical intervention, breathing passages can be kept clear, and some victims survive until adulthood. Many, however, die as children. The cystic fi-

Figure 11.4   BOTTOM: An albino child with his family. RIGHT: The inheritance of the recessive albinism allele.

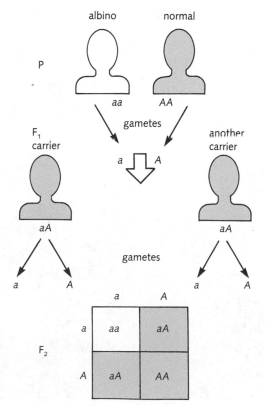

Genetic Counseling: Knowing the Odds

One of the most interesting aspects of genetics is its predictive power. Genetics can be used to predict the future course of events. It is possible, for example, to examine the genetic system by which human sex is determined and predict that male and female babies will continue to be born in roughly a 1:1 ratio. The statistical power of genetics is useful in animal and plant breeding and even in predicting the possibilities of genetic diseases in human offspring.

At the very heart of genetics is the principle that the assortment of individual genes is a matter of chance. **Genetic counseling** is a discipline which makes use of the principles of genetics to advise potential parents of the risk that they will have children who may suffer from a genetic disease. Parents who have supported one child in his or her struggle against a genetic disease must know the odds of facing the same battle with another child.

Consider a healthy couple whose first child was a boy with hemophilia. If they have a second child, what are the chances that it will suffer from the same disease? This is a question you should be able to answer yourself. One of the mother's two X chromosomes carries the hemophilia allele. If their second child is a girl, she cannot suffer from hemophilia, because one of her two X chromosomes will be the normal X chromosome from her father. If their second child is a boy, he will have a 50:50 chance of inheriting his mother's X chromosome with the hemophilia allele. Once they know the odds, a couple can make an informed decision.

As another example, say a healthy couple gives birth to a boy who develops Duchenne muscular dystrophy (p. 230). Can you calculate the odds of their having a second child suffering from the disease? If your father were to develop Huntington's disease (below), what are the chances that you would develop it sooner or later?

---

brosis allele is a recessive gene carried on chromosome number 7. From 3 to 4 percent of the U.S. population carries the cystic fibrosis allele in heterozygous form. The molecular basis of cystic fibrosis will be discussed in detail in Chapter 21.

Other autosomal recessives   A host of other genetic diseases are produced by alleles that, like Tay–Sachs, are defective versions of normal alleles. These include **galactosemia,** a disorder that makes it impossible for babies to digest lactose (milk sugar), and **sickle-cell anemia,** a sometimes fatal disorder of the oxygen-carrying proteins of the blood. We will examine the molecular basis of sickle-cell anemia in detail in Chapter 22.

Autosomal Dominant Inheritance

Autosomal dominant alleles affect the phenotype even when they are present in the heterozygous condition with a normal allele. Therefore, children showing an autosomal dominant phenotype may have inherited the allele from either parent.

Darwin tubercle   A small thickening of cartilage near the upper rim of the ear is called the **Darwin tubercle** (Fig. 11.5). This trait is caused by an autosomal dominant allele. Therefore, if you have it, one of your parents is likely to have it as well.

Achondroplasia   This disorder affects the conversion of cartilage to bone, which is a normal part of the growth process in young people. Individuals with a defective allele for this gene are unable to make the conversion rapidly enough, and this slows down the rate at which their long bones grow, producing dwarfism. Growth and development in other tissues of the body and development of the nervous system are unaffected by the allele.

Huntington's disease   This rare disease takes its name from the physician (George Huntington) who first showed that the disease has a genetic basis. Most individuals with **Huntington's disease** have no symptoms until their late thirties or forties, when they begin to lose some control over their voluntary muscles. Early symptoms include muscle twitches and convulsions. Later the nervous system itself begins to degenerate, and the patient usually dies within 12 years after symptoms first appear.

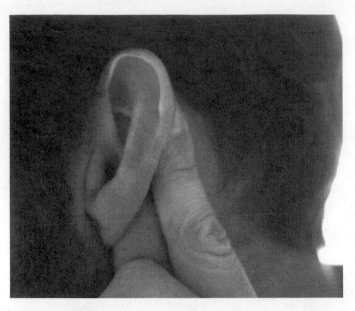

Figure 11.5 This thickening of cartilage near the upper rim of the ear is a Darwin tubercle, caused by an autosomal dominant allele.

Huntington's disease is controlled by a dominant allele carried on chromosome number 4. Therefore, individuals who are *either* homozygous or heterozygous for the gene suffer from Huntington's disease. Because this disease does not manifest itself until later in life, many individuals had to decide whether or not to have children *before* they knew whether they carried the allele. As we will see in Chapter 22, all of this has now changed. Molecular biologists have now devised a test for the presence of the Huntington's allele, though the test itself has raised new issues and new problems.

Polydactyly   A dominant allele produces **polydactyly,** a condition that results in extra fingers and toes (Fig. 11.6). This allele is found throughout the world, although it is especially common in the Ukraine, a former republic of the Soviet Union. Polydactyly is variable in its influence on phenotype, sometimes producing extra fingers, sometimes extra toes, and sometimes both. Polydactyly also displays **partial penetrance,** meaning that sometimes the allele has no phenotypic effect, and an individual carrying it has the normal number of fingers and toes.

Multiple Alleles

Many human genes display more than two alleles, and this can complicate inheritance considerably. One important gene that falls into this category is the blood group gene that is involved in blood *transfusion* (the replacement of lost blood with blood from a donor). When the first recorded human blood transfusions were carried out in 1818 by James Blendell, the results were mixed. His patients were women who were in danger of bleeding to death after childbirth. Although he was able to save the lives of three patients by transfusion, he lost four. These patients' reactions to the transfused blood were fatal.

A physician named Carl Landsteiner discovered that human blood could be classified into four different groups that made it possible to predict whether a transfusion would be successful. The four groups, often referred to as blood "types," are determined by molecules known as *antigens* that occur at the surface of the red blood cells. An **antigen** is a molecule that the body's immune system recognizes. Violent reactions against transfused blood occur when the new blood brings with it an antigen that the body recognizes as foreign. Success in a transfusion, therefore, depends on making sure that the blood being used does not contain such antigens.

There are four blood groups in the system discovered by Landsteiner. The **A** blood group contains a molecule on the surface of red cells known as the *A antigen*, **B** contains the *B antigen*, **AB** contains both *A and B antigens*, and **O** contains *neither antigen*. As previously described in Chapter 10, these four different phenotypes are deter-

Figure 11.6 Polydactyly, the presence of extra fingers or toes, is caused by a dominant allele.

Table 11.1 *Blood Group Genotypes and Corresponding Phenotypes*

Genotype	Phenotype
$I^A I^A$	A
$I^B I^B$	B
$I^A I^B$	AB
ii	O
$I^A i$	A
$I^B i$	B

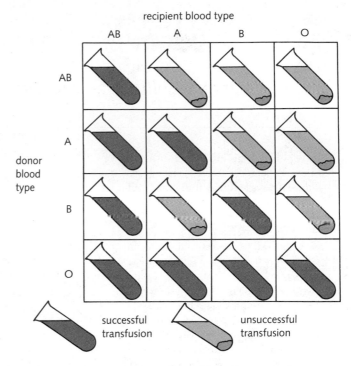

Figure 11.7 Compatibility of blood transfusions in the ABO blood groups.

mined by the alleles found at a single gene locus, the **immunoglobulin** (*I*) locus, or gene. This gene has three different alleles (I^A, I^B, and *i*) which determine the antigens (Table 11.1).

Blood type O, lacking both A and B antigens, is sometimes known as the "universal donor," because it was thought that any of the ABO blood types could safely receive O blood (Fig. 11.7). Similarly, blood type AB is sometimes known as "universal recipient," because it was thought that an individual with this blood type could safely receive any blood type in transfusion. In actual practices, physicians insist on an exact match of blood types for transfusion. As we noted in Chapter 10, the ABO blood group system is an example of **codominance;** that is, neither of two alleles is clearly dominant and both affect the phenotype. The alleles I^A and I^B are *codominant;* the combination produces blood type AB, a phenotype different from that produced by either I^A or I^B alone, and the allele *i* is recessive to both.

It is worth noting that the ABO blood grouping is not the only example of human blood typing. There is another important blood antigen known as the **Rh factor,** named for the Rhesus monkey, in which it was discovered *before* it was found in humans. Individuals are classified as Rh-negative when it is absent. In medical work it is important to match both blood groups, if possible, so blood is generally classified according to both sets of antigens: O+ for type-O blood, Rh positive; B- for type-B blood, Rh negative; and so on.

The Rh blood group is particularly important during pregnancy. When an Rh-negative mother carries an Rh-positive fetus, a bit of the fetus's blood may leak into her circulation during birth and "sensitize" her immune system to Rh-positive blood. Generally this does not cause a problem with a first child. But if a second pregnancy with Rh-positive blood occurs, medical attention is required to prevent her immune system from attacking the blood of

the "foreign" Rh+ fetus, causing damage to both mother and fetus.

Fortunately, there is a routine medical treatment to lessen Rh compatibility problems. A substance known as *rhogam* can be added to the mother's blood shortly after her Rh+ baby is born. This substance is a form of **antibody,** a protein that binds directly to the Rh antigen. The binding of antibody to antigen prevents the mother's own immune system from being exposed to the Rh antigen, and strong reactions that might develop against the next child are avoided. We will discuss this effect and the immune system more thoroughly in Chapter 42.

Polygenic Traits or "You've Got Your Uncle's Nose!"

Size and shape   We can speak with precision about the inheritance of well-defined traits such as blood type and Darwin's tubercle, but what about the other characteristics that common experience suggests we inherit from our parents: a tendency to be short or tall, the shape of our ears, a tendency to be athletic, and our very faces, which often are similar enough to enable total strangers to recognize brothers and sisters? Do genetic systems operate in these cases, too?

Human Intelligence: Are There Genes for IQ?

Every person is an individual with unique capacities, abilities, strengths, and weaknesses. Just as individuals differ in physical characteristics, they differ in mental characteristics too. In 1903 Alfred Binet, a French social scientist, introduced a test to measure intelligence—the *IQ test*. The premise of the test was simple. Just as all people have a physical age, Binet reasoned, they have a mental age. He also assumed that an *intelligence quotient (IQ)* could be computed by dividing the mental age by the physical age and multiplying by 100. A child of 10 with a mental age of 9 would have an IQ of 90, for example, whereas a child of the same age with a mental age of 11 would score 110 on the test.

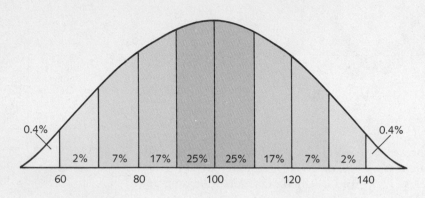

The distribution of IQ scores in an idealized human population.

Intelligence quotient scores differ widely among the human population. A typical distribution is shown above. However, the claim that IQ tests measure intelligence has been challenged by many investigators, some of whom have charged that the tests are flawed by hidden racial and ethnic bias. We will leave those questions unanswered and instead will ask a question that is a bit more biological. Are individual differences in IQ scores the result of genetic differences between individuals or the result of differences in the individual environments that children experience in their formative years? In short, are there genes for IQ?

The answer to this question is emotionally charged for a number of reasons. One of them is finding that average IQ scores of American blacks are about 15 points lower than average IQ scores of American whites. Some

scholars, including Arthur Jensen and William Shockley (the latter is one of the inventors of the transistor), have cited this as proof that differences in average academic performance between white and black children are biological and cannot be eliminated by improved schooling. There is substantial evidence that the ability to score well on an IQ test may indeed be inherited to some degree. Identical twins, whether they are reared together or apart, are more likely to have similar IQ scores than nonidentical siblings reared together. Unrelated children who are adopted into the same home, by contrast, do not show any strong similarity of scores.

However, the situation is not as simple as these statistics may make it appear. Studies of children adopted from orphanages have shown that adoption itself may raise the adopted child's IQ

score by as much as 10 points, and there is a very high correlation of IQ level with social and economic status. Individual IQ is now known to be variable, which is to say that it can be changed by study and a positive learning environment, and it may be dramatically affected by self-image (what a child thinks of himself or herself). Therefore, many other social scientists have cautioned that the prevalence of social and economic inequality is not entirely the *result* of differences in IQ scores between different racial groups but rather acts as one of the *causes* of such differences. At this point, there is little doubt that intelligence is shaped by both genetic and environmental factors, but there is little scientific support for the notion that biological differences will undermine the positive effects of improved schooling for any children.

In most respects, it is clear that these characteristics are genetically determined. Nothing illustrates this more clearly than so-called identical twins (Fig. 11.8). Such twins are properly called **monozygotic**, because they have developed from a single cell that divided at an early stage to form a pair of genetically identical embryos. The resemblance between such twins, even those who have been raised apart, is uncanny. It is clear that complex traits such as the appearance of the face and the shape of the hands and feet are inherited. However, it is also apparent that the genetics of the systems that determine *morphology* (size and shape) is extremely complicated.

Fingerprints   S. B. Holt carried out a detailed analysis of a simple morphological system in humans, the shape of our fingerprints (Fig. 11.9). The patterns of fingerprints are genetically controlled, and the fingerprints themselves are formed very early in embryonic development. Using a system that law enforcement officers developed to analyze fingerprint patterns, Holt tested the fingerprints of members of various families for similarity. He found that identical twins, who are genetically identical, had roughly 95 percent of their fingerprint features in common. This is consistent with the hypothesis that genetics determines most (although not 100 percent) of fingerprint characteristics. Parent and child had about 48 percent similarity, which is consistent with the fact that parent and child share an average of 50 percent of their genes. Siblings also exhibit roughly a 50 percent similarity and can be expected to have about 50 percent of their genes in common. These figures, Holt emphasizes, do not mean that 50 percent of the *fingerprints* of siblings are identical. In fact, generally *none* of their fingerprints are identical. Their individual fingerprints merely show about 50 percent similarity in the patterns of lines and ridges that occur on corresponding fingers (Table 11.2).

Figure 11.8   Identical twins.

Figure 11.9   Analysis of fingerprint patterns indicates that their major features are inherited as polygenic traits.

Table 11.2   *Correlation Between Relatives for Total Dermal Ridge Count**

Relationship	Observed Correlation	Expected	Comments
Parent–child	0.48	0.50	Indicates 50% of genes in common
Father–mother	0.05	0.00	Indicates no relationship
Sibling–sibling	0.50	0.50	Indicates 50% of genes in common
Identical twins	0.95	1.00	Indicates 100% of genes in common
Fraternal twins	0.49	0.50	Indicates 50% of genes in common

Source: S. B. Holt. "Quantitative genetics of fingerprint patterns." *British Medical Bulletin* 17 (1961): 247–250.
*The "total dermal ridge count" is a way of quantifying one of the key features of the fingerprint pattern. It is the number of ridges crossed by a line drawn between the center of the pattern and the "triradial point," a place where three groups of ridges meet to produce a small triangle.

In genetic terms, this means that a fingerprint cannot be the product of a single gene in the same way that ABO blood type is. Instead, such traits are **polygenic:** They are specified by more than one gene. The similarity of fingerprints among members of the same family does not mean that the police will ever confuse a father's fingerprints with his son's. Instead, it reflects the fact that shared individual genes produce similar features that make up each fingerprint, generating many regions of similarity rather than identical patterns.

Skin color   Our species is remarkably diverse. We inhabit every corner of the globe, have successfully invaded the extreme environments on the planet, and display an extraordinary range of genetic diversity. One measure of this diversity is human skin color. The coloration of human skin is caused by a dark pigment called *melanin* that is found in a number of cells, including specialized cells near the surface of the skin known as *melanocytes* (Fig. 11.10). More than 30 different shades of human skin color have been described, and they follow a general adaptive trend: native populations with darker skin color occur in areas closest to the equator, where sunlight is most direct, and populations with lighter skin color occur in regions toward the north that receive less sunlight. There are many exceptions to that rule, however, and these have prevented us from developing a complete explanation for the origin of differences in skin color.

Human skin color is a *polygenic* trait. If a single gene with two alleles governed color, humans would come in three colors at most: dark-skinned, light-skinned, and an intermediate skin tone if the alleles displayed incomplete dominance. But the many different shades of human skin color tell us differently (Fig. 11.11).

Calculations based on the degree of variation in human skin color suggest that at least four different genes, each with several alleles, govern skin pigmentation. Such complexity is just one of many reasons why the traditional concept of a limited number of well-defined human races is not tenable. There are no clear categories into which the observed human phenotypes can be neatly fit.

The same is quite probably true for such subtle traits as facial appearance. Each of these characters is polygenic—the result of interactions among many genes—so it would not be appropriate to expect Mendelian ratios in the inheritance of a specific earlobe shape or profile. The science of genetics, then, offers us no help in deciding which side of the family the baby most resembles. This important task still must be left in the capable hands of grandparents.

SEX-LINKED HUMAN INHERITANCE

Genes located on the *X* and *Y* chromosomes are inherited in a **sex-linked** pattern. As we have already seen, fe-

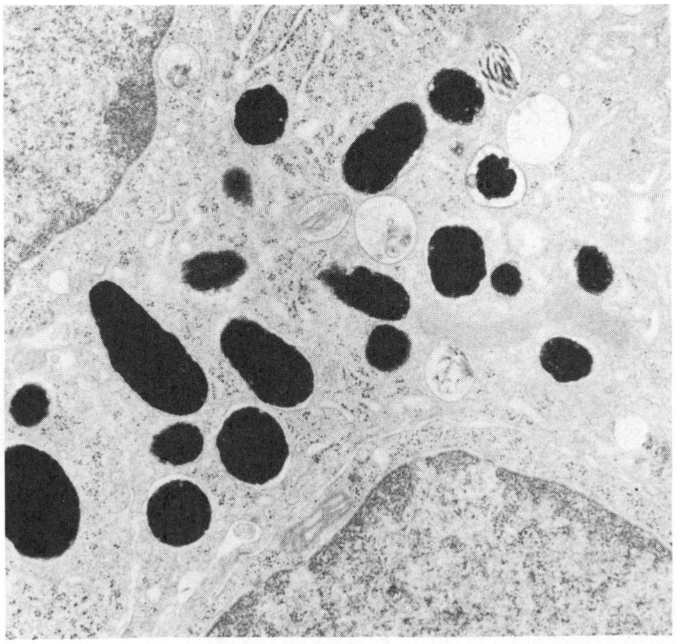

Figure 11.10   A human melanocyte, containing the dark, melanin-rich granules that produce skin color.

Figure 11.11   Human skin color is highly variable, suggesting that it is a polygenic trait.

males are diploid with respect to the sex chromosomes (they have two Xs, whereas males are functionally haploid for both the X and the Y).

The human X chromosome contains a large number of important genes. The complexity of the X chromosome is particularly striking when the large number of X-linked genes is compared to the miniscule number traced to the Y chromosome (Fig. 11.12). Because males have a single copy of the X chromosome, they are particularly susceptible to X-linked gene defects. The two X chromosomes of females, however, mean that they must have two copies of the recessive allele before being affected by them. Therefore, a gene defect on one of the X chromosomes of a female is no more likely to be expressed than a gene defect on any other chromosome, except in cases of Lyonization (as we will discuss on p. 234). But because males have only a single X chromosome, *any* defective gene on that chromosome is expressed.

The mathematics of sex-linked disorders are revealing. If 1 percent of the X chromosomes in a population contain a particular recessive gene defect, that genetic disorder is expressed in 1 percent of the males but in only 0.01 percent of the females (because the chances of *both* X chromosomes having the defect are 0.01 × 0.01 = 0.0001). Thus 1 in 100 males, but only 1 in 10,000 females, will express the disorder.

Sex-Linked Disorders

Hemophilia   "Bleeders' disease," or **hemophilia,** results from a defect in one of the genes required for the normal clotting of blood. The two most common forms of hemophilia are both X-linked recessive disorders. When a normal allele for one of the genes is not present, a **clotting factor,** one of the proteins needed for normal blood clotting, is missing (Fig. 11.13). Even a small wound can be serious for a hemophiliac because it is so hard to stop the bleeding completely. Rough physical activity often has to

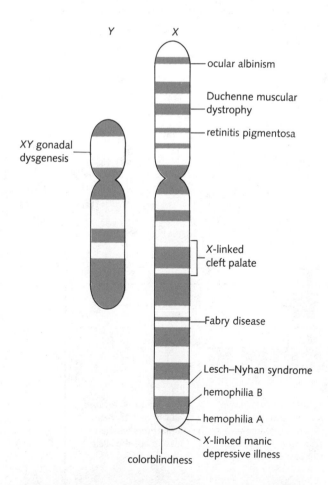

Figure 11.12   Genetic maps of the human X and Y chromosomes. The markers indicate the positions of genes that have been mapped to the chromosomes by conventional genetic techniques.

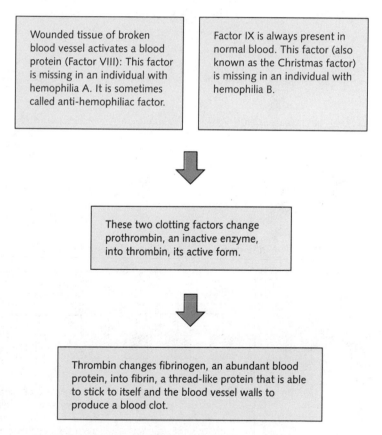

Figure 11.13   Hemophilia is an inherited disorder that affects the protein factors needed for normal blood clotting.

be avoided because of the possibility that a bump or bruise will cause continuing internal bleeding. Although the disease cannot be cured, it can be successfully treated with regular administration of clotting factors prepared from the blood of healthy donors.

As is true of the victims of other human sex-linked disorders, the overwhelming majority of hemophiliacs are male. For a female to suffer from the disease, *both* of her X chromosomes must carry the recessive allele for hemophilia. This means that she must be the daughter of a man who suffered from hemophilia himself and of a woman who carried at least one allele for the disease. Because the hemophilia allele is relatively rare, this seldom happens (Fig. 11.14). However, the presence of the gene on the X chromosome of the male has some interesting consequences that apply to other sex-linked traits as well.

For example, although the allele for the disease is expressed in males, it cannot be passed from father to son. Every boy carries the Y chromosome he got from his father, but his X chromosome is derived from his mother. Therefore, fathers can pass the allele only to their daughters.

Furthermore, although women rarely suffer from sex-linked diseases, they are the only parents capable of pass-

ing such a disease on to their sons. Because every boy receives his X chromosome from his mother, the birth of a son with a sex-linked disease to a normal woman indicates that such a woman is a carrier for the sex-linked genetic disease. A healthy woman who gives birth to a hemophiliac son, for example, must have the hemophilia allele present on one of her X chromosomes. Because her other X chromosome must contain the normal allele, the odds of her having a second son with the disease are 50:50 (see Fig. 11.14).

Hemophilia has played an important role in world history. Queen Victoria (1819–1901) of Great Britain had nine children during her reign. One son, Leopold, suffered from hemophilia, and at least two daughters, Alice and Beatrice, carried a single copy of the hemophilia allele (Fig. 11.15). Alice passed the allele along to her daughter Alexandra, who married Nicholas II, the tsar of Russia. Their only son, Alexis, suffered from hemophilia. Alexis's hemophilia, and his parents' desperation in seeking spiritual and medical treatment, preoccupied the tsar to the extent that it limited his ability to deal with issues of state. What happened during this time? The Russian Revolution deposed the tsar, and he and his immediate family were executed.

Figure 11.14 The inheritance of hemophilia in human families.

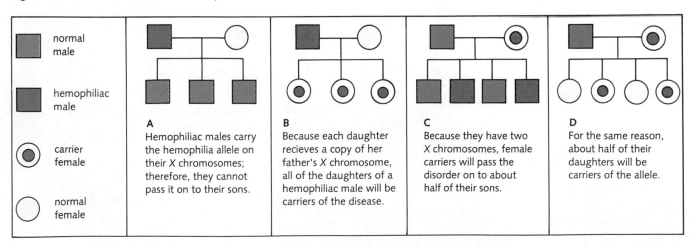

normal male				
hemophiliac male				
carrier female	**A** Hemophiliac males carry the hemophilia allele on their X chromosomes; therefore, they cannot pass it on to their sons.	**B** Because each daughter recieves a copy of her father's X chromosome, all of the daughters of a hemophiliac male will be carriers of the disease.	**C** Because they have two X chromosomes, female carriers will pass the disorder on to about half of their sons.	**D** For the same reason, about half of their daughters will be carriers of the allele.
normal female				

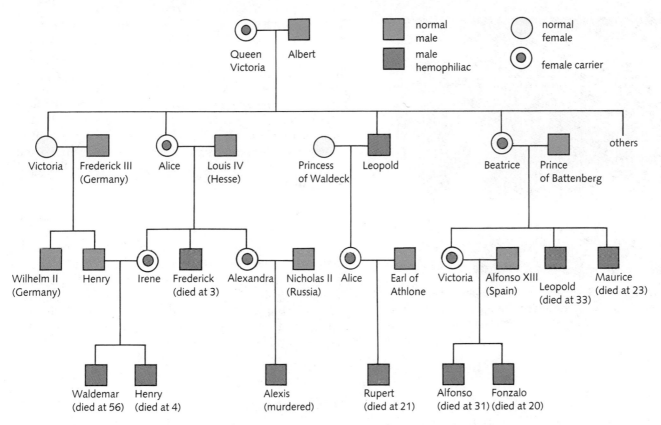

Figure 11.15 The inheritance of hemophilia in a famous human family, that of Queen Victoria of Great Britain. Only a few of her children are shown, to simplify the diagram. There was no known incidence of this disorder in the royal family before Victoria, so that it is possible that a mutation in her germ line was responsible for the disease in her offspring.

Colorblindness One of the most common *X*-linked recessive traits is **colorblindness.** About 10 percent of the male population of the United States suffers from one form or another of colorblindness, which is caused by a defective allele for one of three *X*-linked genes. In the most common form of colorblindness, individuals are unable to distinguish pale shades of red from green.

Check the color test pattern reproduced in Fig. 11.16. The chances are good that one of your male friends (assuming that you have at least 10 male friends) will be unable to read the pattern correctly. (There are several types of colorblindness. Total colorblindness is a very rare disorder caused by an autosomal recessive allele and is known as *achromatopsia*.) When a colorblind male has children, is the allele for colorblindness passed on to his sons or his daughters?

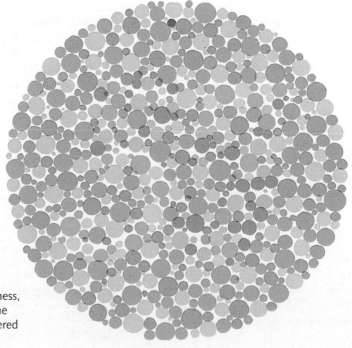

Figure 11.16 This pattern tests for red–green colorblindness, although difficulties in maintaining precise colors during the printing process make it less accurate than a test administered by a physician.

Figure 11.17 A young girl with muscular dystrophy. Females suffering from muscular dystrophy are very rare, because their fathers must have the disorder and their mothers must be carriers.

Duchenne muscular dystrophy One boy in 3000 in the United States begins to experience a sudden weakness of skeletal muscles between the ages of 3 and 6. **Muscular dystrophy,** as this disorder is known, produces crippling paralysis and eventually leads to death, often before the early twenties. The most common form is **Duchenne muscular dystrophy,** caused by a recessive X-linked allele. The normal version of this allele is responsible for a protein found in small amounts in muscle cells. There is no specific treatment or cure for the disease, but it is the object of intensive medical research (Fig. 11.17).

Hypophosphatemia Not all sex-linked disorders are recessive. Several are caused by dominant alleles carried on the X chromosome. One such disorder, **hypophosphatemia,** causes a severe deficiency of phosphates in the blood. Because this disease is a dominant X-linked disorder, males with the disorder transmit it to *all* of their daughters but to *none* of their sons. Conversely, females with the disorder transmit it to about *half* of their children, regardless of the children's sex.

Lesch–Nyhan syndrome A rare and devastating X-linked genetic disease, **Lesch–Nyhan syndrome** was first recognized in 1965. Boys suffering from this disease have chemical imbalances in the body resulting from the absence of a critical enzyme. Some symptoms of this disease can be successfully treated with diet and medication. However, in some unknown way Lesch–Nyhan affects the nervous system, producing jerky, uncoordinated movements, violent behavior, and self-mutilation including biting of the hands and arms. Victims of the disease rarely live through childhood.

Retinitis pigmentosa Tristan de Cunha is a small group of islands in the Atlantic Ocean between Africa and South America. No more than 300 people live on the islands. For years they have known a strange form of progressive blindness that affects mostly young men. At first, victims of this disease must twist their heads to see clearly. Growing patches of pigmented tissue cover portions of their retinas, the light-sensitive tissue at the back of the eye. As these pigmented tissues merge, the victims lose their sight completely. **Retinitis pigmentosa,** as this condition is called, is a rare X-linked recessive disorder. The few women who have suffered from this disease were daughters of men who also had the disease. These islands were first settled in 1814 by 15 British colonists, one of whom may have carried this allele, explaining why this tiny population has the highest incidence of retinitis pigmentosa in the world.

CHROMOSOMAL INHERITANCE

Human reproduction begins with the cellular events that produce sperm and egg (Fig. 11.18). Meiosis, tetrad formation, crossing over, and the segregation of homologues occur just as they do in other species. When meiosis is completed, the male and female gametes contain 23 chromosomes each: 22 autosomes and 1 sex chromosome. When **fertilization** occurs, the haploid male and female gametes fuse with each other to form a diploid **zygote** with a full complement of 46 chromosomes.

Disjunction Abnormalities

Normally, each chromosome pair separates during the first meiotic division. However, in some cases the separating mechanism fails, and a pair of homologous chromosomes ends up in one of the two cells produced by that division. The failure of a chromosome pair to separate properly is known as **nondisjunction** (literally, "not coming apart").

When nondisjunction involves the sex chromosomes, zygotes may be formed with abnormal combinations of sex chromosomes (Fig. 11.19). The X chromosome contains a great many genes that are indispensable to human development. Zygotes that contain only a Y chromosome (they are abbreviated *OY*) fail to develop. However, cells containing other unusual combinations of sex chromosomes can and do develop to maturity.

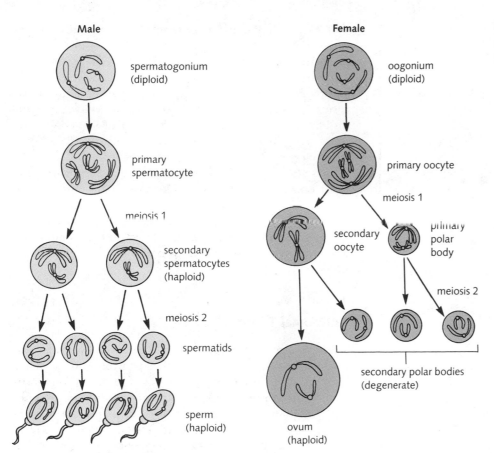

Male

spermatogonium (diploid)

primary spermatocyte

meiosis 1

secondary spermatocytes (haploid)

meiosis 2

spermatids

sperm (haploid)

Female

oogonium (diploid)

primary oocyte

meiosis 1

secondary oocyte

primary polar body

meiosis 2

secondary polar bodies (degenerate)

ovum (haploid)

Figure 11.18 A simplified version of mitosis and meiosis. Human gametes have 23 chromosomes.

Turner syndrome People with only a single *X* chromosome suffer from **Turner syndrome.** Genetically, they are 45 *XO* (44 autosomes and 1 *X* chromosome; the *O* is included to draw our attention to the missing sex chromosome). Even though these individuals are females, their ovaries do not develop fully, they are infertile, and they do not reach sexual maturity. Individuals with Turner syndrome tend to be shorter than average and to have a number of characteristic features, including a slight webbing of the skin at the back of the neck, enlarged feet when they are babies, slight abnormalities in the aorta (the main blood vessel leading from the heart), and a slightly lowered hairline. Turner syndrome has been thought to involve a mild degree of mental retardation, although this may not be the case. In fact, many individuals who are afflicted with the syndrome are able to lead normal lives.

Klinefelter syndrome Individuals who are genetically 47 *XXY* suffer from **Klinefelter syndrome,** named after the physician who first recognized and described it. These individuals are sterile males. They tend not to develop the sex-related body structure of most males (wide shoulders

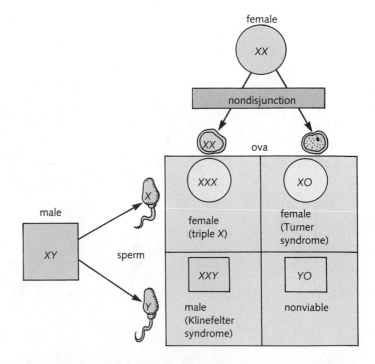

female

XX

nondisjunction

ova

XX

male

XY

sperm

X

Y

XXX
female (triple *X*)

XO
female (Turner syndrome)

XXY
male (Klinefelter syndrome)

YO
nonviable

Figure 11.19 When nondisjunction, a failure of chromosomes to separate during meiosis, occurs in the sex chromosomes, the zygote may contain an abnormal number of sex chromosomes.

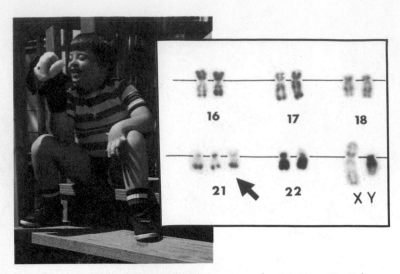

Figure 11.20 LEFT: A child with Down syndrome. RIGHT: Partial karyotype of a child with Down syndrome. Note the presence of an extra copy of chromosome 21.

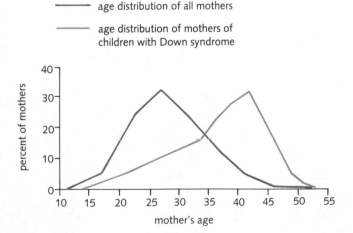

Figure 11.21 Down syndrome shows a strong correlation with maternal age.

and narrow hips), and they tend to be unusually tall. At puberty about half of them develop breast tissue that is female in appearance. Males with Klinefelter syndrome tend to be mentally retarded, although the severity varies greatly from one individual to the next. There are more severe variants of Klinefelter syndrome in which individuals have a genetic makeup of 48 *XXXY* or 49 *XXXXY*. In these cases, mental retardation is pronounced.

Other sex chromosome abnormalities There are also rarer cases of people whose chromosomal compositions are 47 *XXX* and 47 *XYY*. As you might expect, the former are females and the latter males. Although there have been some controversies in each case, no clear pattern of abnormalities is associated with either of these conditions.

Autosomal Trisomy

We have already seen how nondisjunction can cause abnormalities in the number of sex chromosomes. What happens when nondisjunction occurs in one of the autosomal chromosomes? One of the most common genetic abnormalities is known as **trisomy 21,** in which a child is born with *three* copies of chromosome number 21. It might seem that having an extra copy of a particular chromosome would cause no great harm, but this is not the case. Trisomy 21 is also known as **Down syndrome** and is associated with mental retardation, reduced resistance to infection, and a lowered life expectancy. The degree of mental retardation associated with Down syndrome varies from one individual to the next, and Down children are often warm, loving individuals who are capable of developing strong relationships with those around them (Fig. 11.20; see Theory in Action, Prenatal Genetics).

Why is an extra copy of chromosome 21 so serious? The real answer may be that trisomy 21 is the *least* serious of the many possible trisomies. In fact, there are medical reports of at least two other trisomies (chromosome number 18 and 13), each of which results in a more serious set of abnormalities than Down syndrome. Many other trisomies are so serious that fetuses affected by them do not develop to term. Ironically, we may see so many children with trisomy 21 (about 15 per 10,000 births) precisely because the effects of Down syndrome are so *mild*.

The incidence of Down syndrome increases quite dramatically in mothers over the age of 35 (Fig. 11.21). Mothers aged 20 to 30 have about 1 chance in 1000 of having a child with Down syndrome. Mothers over 40 years of age have better than 1 chance in 100. Why should maternal age make such a difference? On the day that a girl is born, every egg cell that she will release during her lifetime is already in prophase of the first meiotic division. Each egg cell remains in the first meiotic prophase

Prenatal Genetics

Until relatively recently, the only way that a woman could find out whether the baby she was carrying was suffering from a genetic disease was to wait for the child to be born. But it is now possible to routinely diagnose a number of genetic disorders while a child is still developing within the uterus. *Amniocentesis* is a medical procedure in which a needle is carefully inserted into the uterus to withdraw a small amount of the amniotic fluid that surrounds the developing fetus. This fluid contains cells that were released from the developing embryo and are genetically identical to it. The cells can be grown in culture and examined by a number of methods in the lab (see the accompanying figure).

In recent years a new technique, *chorionic villi biopsy*, has become common. A thin tube is inserted through the vagina into the tissue surrounding the placenta, and cells from the chorion (a tissue surrounding the fetus) are removed. These cells are derived from the developing fetus, and they can be examined directly for abnormalities. This method has the advantage that the cells collected can be examined immediately. With amniocentesis it often takes three or four weeks to grow enough cells for analysis. Chorionic villi biopsy can also be performed earlier in a pregnancy than amniocentesis. Unfortunately, there may be an increased risk of birth defects in children born after chorionic villi sampling owing to the early stage of development at the time of the procedure, although this has not yet been proven.

The number of defects that can be detected by these techniques is growing larger every year. A few disorders, such as Down syndrome and Turner syndrome, are easily discovered in karyotypes of the fetal cells. Karyotypes can also inform prospective parents whether their child is a girl or a boy. Chemical tests on the fetal cells can detect a host of other genetic disorders. Some of these are diseases that result in abnormally low levels of specific carbohydrates, lipids, or proteins in the developing embryo. Tay–Sachs disease and sickle-cell anemia can be detected by such tests. The pathways by which these molecules are produced are genetically determined, and more than 250 such diseases can now be detected before birth, some in time for effective treatment of the fetus if the prospective parents decide to continue the pregnancy.

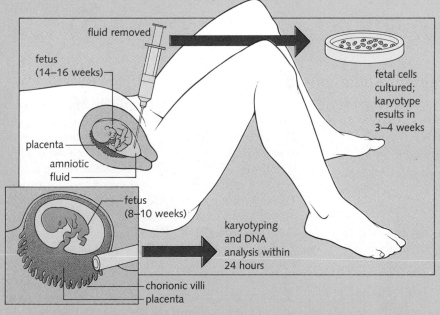

Amniocentesis and chorionic villi biopsy are two methods used to obtain cell samples from a developing fetus.

What Turns Off the X?

When Mary Lyon first proposed her theory of X-chromosome inactivation in 1961, geneticists immediately began to wonder *how* a whole chromosome could be turned off while leaving the homologous X chromosome fully active. For more than 30 years, it remained a puzzle. During those 30 years, however, researchers discovered that a small handful of genes (six so far) remain active on the inactivated X chromosome. In 1991, Carolyn Brown, Huntington Willard, and their co-workers at Stanford discovered that one of these six genes is active *only* on the inactivated X chromosome and is not expressed on the "active" X chromosome.

At first, this was a puzzling finding. But Brown and Willard believe that this gene may hold the key to X-chromosome inactivation. It could very well produce a regulatory RNA that keeps the other genes on its own chromosome turned off, even while corresponding genes on the active X chromosome are being expressed. This would explain why this gene is active only on the *inactivated X* chromosome. Since it is only expressed in cells that contain an inactivated X chromosome, it may also explain the abnormalities associated with Turner and Klinefelter syndromes (p. 231). As a result of finding this remarkable gene, in a couple of years we may have to rewrite the book on X-chromosome inactivation.

until just before it is released. This long period of inactivity may increase the chance that a nondisjunction error will occur when the egg cell finally completes meiosis.

X-Chromosome Inactivation

Females have two X chromosomes and males have one. In 1961 Mary Lyon, a British geneticist, proposed that only one X chromosome is active in a female cell. Her model, called **Lyonization,** suggests that one of the two X chromosomes is inactivated during the process of embryonic development. The other X chromosome remains active and genes on that chromosome are expressed. Because this occurs early in development, females are actually "mosaics," composed of small patches of tissue in which alternating X chromosomes are active.

The most obvious examples of female mosaicism are calico cats. Many domestic cats have dark patches of fur on a white background. In males, the dark patches may be either yellow or black, depending on which allele is found on the single X chromosome. However, when a female is heterozygous for coat color, the patches of skin in which one X chromosome is active produce orange fur, and those patches in which the other X is active produce black fur. The result is the striking three-color calico (Fig. 11.22). In females, the patches of fur on a calico cat occur because of random inactivation of the X chromosome in the cells that produce coat color. Calico cats are almost always female, and the few males that have been reported have been sterile, implying that they may be *XXY.* (On the theory that a male, if found, would be so rare as to be very valuable, calico cats are sometimes called "money cats.")

Lyon's ideas were also supported by the discovery of a dark-staining structure called a **Barr body** (Fig. 11.22) within the nucleus of female cells. A Barr body is actually a single condensed, inactivated X chromosome. The active X chromosome remains dispersed in the nucleus and is not visible during interphase.

Lyonization occurs in humans, too. Patches of color-insensitive cells have been demonstrated on the retinas of women who carry one copy of a colorblindness allele, although enough of the retina responds normally to light to produce overall normal color vision. *Anhidriotic dysplasia,* a disorder that prevents the formation of sweat

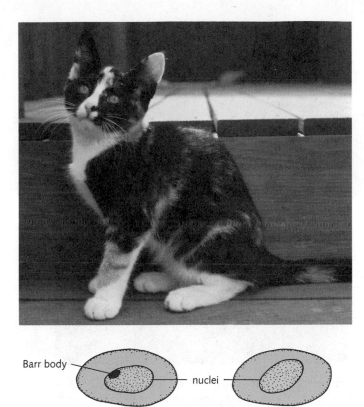

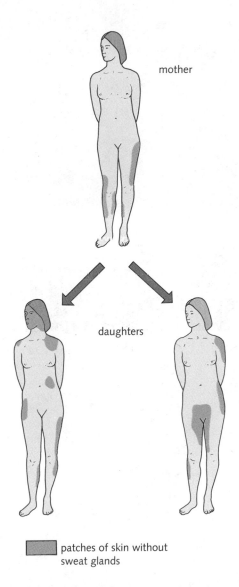

Figure 11.22 TOP: The three-color coat of a calico cat is the result of X-chromosome inactivation. BOTTOM: The inactivation of an X chromosome produces a dense structure in the nucleus known as a Barr body. Barr bodies are formed in cells with two or more X chromosomes, including normal females (XX) and males with Klinefelter syndrome (XXY).

Figure 11.23 Mosaic women showing patches of skin without sweat glands.

glands, also shows a mosaic pattern of development. Males with the allele have no sweat glands. Heterozygous females have patches of skin with normal sweat glands and patches of skin without sweat glands (Fig. 11.23).

Chromosome Deletions and Translocations

For quite some time, biologists have studied chromosomes as though the genes they contain were locked in place. However, it has recently been revealed that this is not necessarily so and that changes in chromosome organization and structure are common. The several well-known kinds of changes include **deletions** (part of a chromosome is missing), **inversions** (part of a chromosome is inverted with respect to the rest of the chromosome), and **translocations** (a portion of one chromosome is broken off and attached to another chromosome).

We can use these events to map gene locations on human chromosomes. Figure 11.24 shows a human chromosome *deletion* that occurred in a young boy with crippling birth defects. Researchers were able to grow a few of the boy's cells in culture and to determine that a part of chromosome number 2 had been deleted (Fig. 11.24). Chemical studies of the cells confirmed that the gene for *acid phosphatase*, an important enzyme, had been lost as a result of the deletion. Therefore, the gene must have been on that part of chromosome number 2.

Translocations have made it possible to determine the portion of chromosome 21 that is responsible for Down syndrome. Researchers discovered a small fraction of

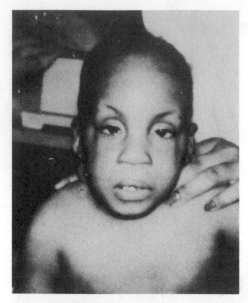

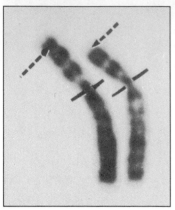

Figure 11.24 TOP: A young boy with crippling birth defects caused by a chromosomal deletion. The deletion took place in the terminal portion of chromosome 2. LEFT: The dotted arrows point to the position of the deletion.

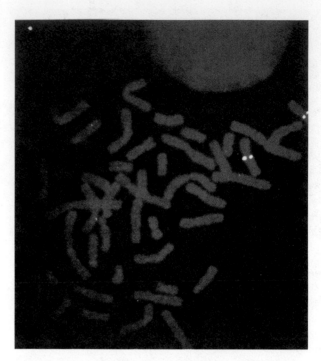

Figure 11.25 A fluorescent probe for the human muscle glycogen phosphorylase gene reveals the location of this gene on the long arm of chromosome 11.

Down syndrome patients whose karyotypes were 46 *XY* instead of the usual 47 *XY*. Soon it was discovered that three patients had a *translocation* in which a tiny portion of chromosome 21 had become attached to one of the other autosomal chromosomes. Because they also had two normal copies of chromosome 21, they actually had *three* copies of the portion involved in the translocation. Although the translocated fragments were different sizes in different patients, each contained one critical region of chromosome 21. Researchers are now trying to analyze the genes in this segment of the chromosome to understand the cause of the disorder.

MAPPING HUMAN GENES

We saw in Chapter 10 how experimental crosses in other species can be used to determine recombination frequencies between two genes located on the same chromosome, and how recombination frequencies can then be used to produce a genetic map. Classical genetics can use the same techniques to map human genes, but there are serious limitations. You may remember that our ability to calculate recombination frequencies depends on having large numbers of offspring available to generate reliable statistics. This is usually not possible with humans. It also is not possible, for obvious reasons, to generate stocks of organisms that are homozygous recessive and use them in experimental crosses with heterozygous organisms. *Homo sapiens* is just not a very good organism for studies of genetics. As recently as 1968, 68 genes had been mapped to the *X* chromosome, but not a single autosomal (non–sex-linked) gene had been mapped to any of the other 22 chromosomes.

As we will see in Chapter 21, however, new genetic tools have been produced by molecular biology, and these tools have surpassed classical genetics to supply a wealth of information about the organization of the human genome. It is now possible to locate genes to particular chromosomes and then to fix their approximate location on that chromosome with remarkable speed. Some of these tools are complex biochemical techniques, and some are as simple as new fluorescent markers that mark the location of a gene so that it can be detected with the light microscope (Fig. 11.25).

From Mom or from Dad? Gene Imprinting

We have seen that the genes located on the human *X* chromosome are inherited in a sex-linked pattern. According to classical genetics, genes located on the autosomal chromosomes should not be inherited in a sex-linked pattern. In other words, it shouldn't matter whether you inherited an autosomal gene from Mom or from Dad. In most cases, that's absolutely true. But recently there have been some surprises.

One of the most striking came from the study of two inherited human disorders, both of which produce mental retardation. Patients with **Angelman syndrome** show extreme obesity and are short of stature, whereas individuals with **Prader–Willi syndrome** are thin, of normal height, and tend toward hyperactivity. Researchers first discovered that both disorders could be traced to a mutation on chromosome 15. Further work revealed that both syndromes were caused by the same gene. The only difference was that individuals with Prader–Willi syndrome inherited the gene from their mothers, while those with Angelman inherited it from their fathers.

If the gene is the same in both cases, why should it make any difference which parent it comes from? Good question. These results made some investigators recall experiments done in the 1980s in which the origin of chromosomes had a critical effect on development. By transplanting chromosomes with fine pipettes, they had constructed embryos that contained the full diploid set of chromosomes, obtained either from a female mouse or from a male mouse. Without exception, the embryos failed to develop. When a mixture of male- and female-derived chromosomes were used, the embryos developed normally.

What is happening in these cases is not completely clear, but current theories suggest that certain genes are chemically "imprinted" by their presence in a male or female parent (see figure). When these genes are passed along to the next generation, they retain that **genetic imprint** in a way that affects their activity, so that the same gene inherited from a female parent does not always have the same effect as an identical gene inherited from a male parent. Imprinting may account for the differences between Angelman and Prader–Willi syndromes, and it may also explain why some chromosomes from *both* parents are necessary for development of a mouse embryo. Researchers now suspect that a score of other human disorders may be affected by genetic imprinting. Whatever the mechanism of imprinting may be, it turns out that it *does* matter whether you got a gene from Mom or from Dad.

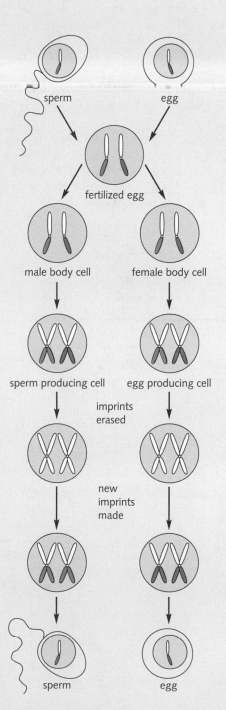

A model for genetic imprinting of chromosomes. The passage of a gene through the male parent attaches one imprint to a chromosome (blue). When the same gene passes through a female parent, it is imprinted differently (red), so that the same genes may have different effects in the offspring, depending upon which parent was its source.

The Human Genome Project

An international effort, known as the **Human Genome Project,** is now underway to map each of the 24 human chromosomes (22 autosomes plus *X* and *Y*). The ultimate goal of this project, still far in the future, is to determine the complete molecular code for every human gene (see Theory in Action, Mapping the Territory, Chapter 21). A much more modest goal, a set of genetic linkage markers for each chromosome, was reached in 1992. These markers are like surveyor's points that will make it easier to find the genes located between them. They have already allowed geneticists to construct maps of each chromosome showing the general locations of scores of important genes on each human chromosome (Fig. 11.26).

Rapid progress on human gene mapping is to be expected in the years ahead, and it may not be too long before we can say, with complete honesty, that we know just as much about the genetics of *Homo sapiens* as we do of *Drosophila melanogaster*.

Figure 11.26 Markers for many very important genes on each human chromosome have already been identified.

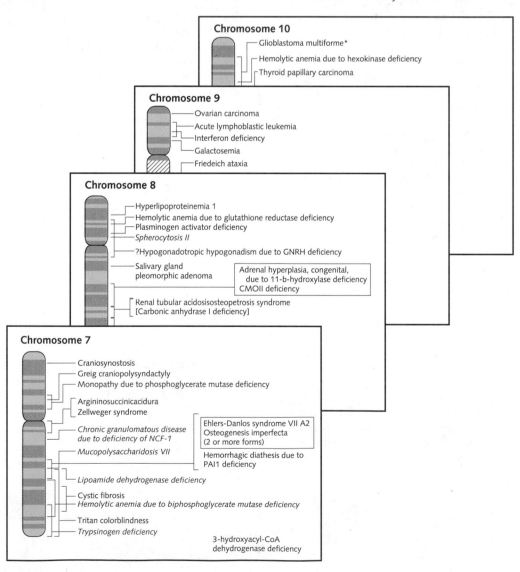

SUMMARY

Humans are not ideal organisms for classical genetics studies. Nonetheless, because of our obvious interest in human genes, considerable scientific efforts have been made over time to develop human genetic maps and to detect human genetic defects. Human cells contain 46 chromosomes, 44 autosomal chromosomes, and 2 sex chromosomes. More than 4000 human genes have been described, and many of these have been localized to particular chromosomes. Recessive human alleles include those for albinism, Tay–Sachs disease, cystic fibrosis, and sickle-cell anemia. Dominant human alleles include those that cause achondroplasia, Huntington's disease, and polydactyly. Many human traits are produced by genes with multiple alleles, and others, like skin color and fingerprint patterns, are polygenic.

A number of human disorders are sex-linked, and the most common sex-linked conditions are caused by genes on the X chromosome, including hemophilia, colorblindness, and muscular dystrophy. Nondisjunction, the failure of chromosomes to separate properly during meiosis, results in a number of chromosomal disorders, including Turner syndrome (45 *XO*) and Klinefelter syndrome (47 *XXY*). Down syndrome results from a nondisjunction event.

The positions of genes on human chromosomes have been mapped by a number of techniques, including cell fusion in culture and the analysis of chromosome breaks and translocations.

STUDY FOCUS

After studying this chapter, you should be able to:

- Show how human inheritance can be analyzed by techniques used in genetics.

- Describe special problems, diseases, and disorders caused by defective genes.

- Explain the techniques used in chromosome mapping.

- Discuss the heritability of eye color, facial appearance, and other traits.

TERMS AND CONCEPTS

sex chromosomes *217*
autosomal
    chromosomes *217*
karyotyping *217*
pedigree *219*
albinism *219*
Tay–Sachs disease *220*
Huntington's disease *222*
antigen *223*
codominance *223*
Rh factor *223*
antibody *223*
hemophilia *227*

nondisjunction *230*
Turner syndrome *231*
Klinefelter syndrome *231*
trisomy 21, or Down
    syndrome *232*
Lyonization *234*
Barr body *234*
deletions *235*
inversions *235*
translocations *235*
genetic imprint *237*
Human Genome
    Project *238*

REVIEW

Objective Questions (Answers in Appendix)

1. A karyotype is used to
    (a) determine the stage of mitosis.
    (b) determine the number of chromosomes.
    (c) substitute a normal allele for a defective allele.
    (d) determine the stage of meiosis.

2. A chromosome map is used to determine
    (a) the order of genes on a chromosome.
    (b) the type of gene on a chromosome.
    (c) if Lyonization has occurred.
    (d) if nondisjunction has occurred.

3. Which genetic disorder is caused by a recessive allele?
    (a) Tay–Sachs disease
    (b) Darwin tubercle
    (c) Huntington's disease
    (d) polydactyly

4. An example of polygenic inheritance is
    (a) blood type.
    (b) Darwin tubercle.
    (c) fingerprints.
    (d) albinism.

5. A woman who is heterozygous for colorblindness marries a normal male. What is the probability that their child will be colorblind?
    (a) 25 percent
    (b) 50 percent
    (c) 75 percent
    (d) 100 percent

Discussion Questions

6. What are some of the reasons why the fruit fly (*Drosophila melanogaster*) is a better organism than the human (*Homo sapiens*) as a subject in classical studies of genetics?

7. At which stage of meiosis are tetrads formed in human gamete formation? At which stage does crossing over occur?

8. Carriers of Huntington's disease are generally unaware until late adulthood that they carry the gene. Why?

9. What evidence suggests that fingerprints are determined by a polygenic system, rather than by a system of one or two genes with multiple alleles?

10. If a healthy couple were to give birth to a child with cystic fibrosis, what odds would they face if they decided to have a second child?

11. What information would you give the same couple if their first child were afflicted with Down syndrome instead of cystic fibrosis? How would the ages of the two parents be significant?

READINGS

"Genomes." *Science* 258 (October 2, 1992): 1–188. An issue devoted largely to news and scientific reprints from the Human Genome Project.

Davies, K. "The essence of inactivity." *Nature* 349 (1991): 15. This brief paper describes experimental reports on the mechanisms of *X*-chromosome inactivation (discussed in Theory in Action, What Turns Off the *X*?, p. 234). The same issue of *Nature* contains two research papers on the genes that may be responsible for *X*-chromosome inactivation.

McKusick, V. A. "Human genetic disorders." *Journal of NIH Research* 3 (1991): 143–168. Would you like to find out on which chromosome a particular genetic disorder is located? This is the place to look. This brief reference article lists more than 700 disorders and their chromosomal locations. It also contains a map of each human chromosome.

Culliton, B. J. "Mapping terra incognita (humani corporis)." *Science* 250 (1990): 210–212. A lively and readable progress report on the Human Genome Project.

Sapienza, C. "Parental imprinting of genes." *Scientific American* 263 (October 1990): 52–60. A very readable summary of current theories regarding parental imprinting of genes.

Patterson, D. "The causes of Down syndrome." *Scientific American* 257 (August 1987): 52–61. Only recently have the actual causes of Down syndrome become clear. This article explores the mechanisms that produce this important human disorder.

Lewin, R. "Cultural diversity tied to genetic differences." *Science* 212 (1981): 908–910. A short but provocative article reviewing the book *Genes, Mind, and Culture* by E. O. Wilson and C. Lumsden, which explores the possibility that there is a genetic basis for culture.

12

Darwinian Theory Evolves

*d*uring the early twentieth century, geneticists explained the nature of heritability and demonstrated several sources of variation among individuals in populations. Then, as fields related to Mendelian genetics matured, mathematicians began to build on the simple models of gene segregation and recombination represented by Punnett squares. Soon they began modeling the more complicated behavior of genes in plant and animal populations, creating the field of *population genetics* that transformed evolutionary theory.

The 1930s and 1940s were filled with intense debate and creative thinking among evolutionary biologists. During that time, researchers combined the essence of Darwin's original theory with insights afforded by new developments in Mendelian and population genetics, paleontology, and natural history. The result was a body of theory known as the modern evolutionary synthesis—the core of the biological sciences today.

Because the tenets of the modern synthesis are so important, we will spend most of the next two chapters discussing them. It is important to remember, however, that evolutionary theory, like all scientific theory and the organisms whose nature it seeks to explain, is always subject to change. The birth of molecular biology and recent advances in population genetics that began in the 1960s and 1970s (fueled in part by the computer revolution) have initiated another transformation in evolutionary theory that is still in progress today. We will mention some of the more fundamental revelations of molecular evolution in these chapters, but will reserve detailed treatment for Chapter 22.

22

SPECIES AND FITNESS: GENETIC DEFINITIONS

Because genetics is now the keystone of evolutionary biology, we will begin our discussion of evolutionary theory by redefining several important Darwinian concepts in genetic terms.

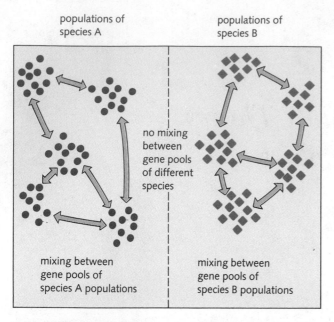

Figure 12.1 Alleles of genes mix among interbreeding populations of a species, but because of reproductive isolation, they do not mix between species.

Figure 12.2 Evolutionary fitness is ultimately determined, not by any particular characteristics of its physiology, but by the number of copies of its alleles an organism successfully passes on to the next generation.

The Species

Biologists have historically recognized and classified species strictly by physical characteristics, or what we can now call phenotype (see Chapter 10). But genetics and population biology suggest that species can and should be defined in terms of genes and the reproductive behaviors that affect the movements of alleles of genes.

From a genetic perspective, a **species** is a group of natural populations that can (at least potentially) interbreed among themselves to produce fertile offspring and are reproductively isolated from other such groups. Because mating within and between populations of a species means that members mix alleles among their offspring, a species is also a genetic unit whose members share a group of alleles of genes called a **gene pool.**

This genetic definition is important. It means that when a genetic change occurs in some individuals of a species, that change can spread through the gene pool of that species but not to other gene pools (Fig. 12.1). Thus each species evolves as a separate unit. You will understand the importance of this genetic definition of species more clearly after we discuss, in the next chapter, the way new species originate.

Evolutionary Fitness and Adaptation

Genetics also provides a concise definition of evolutionary or Darwinian fitness: An individual's **evolutionary fitness** is defined by the size of its probable genetic contribution to the next generation. In other words, an organism that produces many offspring, carrying many copies of its alleles, has high fitness (Fig. 12.2).

An **evolutionary adaptation** is defined as any genetically controlled characteristic that increases an organism's fitness. Larger muscles that are built up through exercise (and thus are an acquired characteristic) are not considered an evolutionary adaptation. An allele that causes an individual's muscles to grow faster in response to exercise, on the other hand, could be considered an adaptation—*if* it increased an individual's fitness.

EVOLUTION AND NATURAL SELECTION

Defined in genetic terms, **evolution** is any change over successive generations in the relative frequencies of different alleles in the gene pool of a population. We can see the logic in that definition and examine the process of evolution by natural selection by looking at any heritable, adaptive characteristic in a population of organisms.

Figure 12.3 Military recruits arranged in groups by height. Notice that this grouping produces a distribution much like those represented in the bar graphs and histograms you have seen for characteristics in animals and plants.

Observable Variation in Organisms

Every characteristic of organisms in a population has a frequency distribution (Fig. 12.3). That is, if you look at any heritable trait in a large enough population, you will see that it has some average value and some degree of variation or deviation from that average value. Much of this variation represents the constant reshuffling of alleles that occurs each time an organism reproduces sexually, but it also includes the much rarer occurrence of new mutations in existing alleles. To the extent that this phenotypic variation reflects differences in genotype, it can be affected by evolutionary change. And to the extent that this variation affects the fitness of organisms in the population, it can provide the raw material on which natural selection operates.

Remember that genetic variation is random. It does not occur because an organism *needs* or *wants* to evolve. (That's close to what Lamarck thought.) There is no way for an organism to cause variation in its alleles, and there is no way to prevent it. Similarly, specific environmental conditions do not give rise to specific sorts of mutations or other variations in genotype that are useful to organisms. In other words, alleles do not "know" precisely how to mutate or recombine in order to benefit their organism.

The occurrence of a genotype that endows an organism with resistance to an environmental poison, for example, does not occur *because* the poison is present in the environment. In other words, the poison in the environment neither produces new alleles that confer resistance to poison nor directs the reorganization of existing alleles to confer resistance. Instead, the existence of the poison simply increases the fitness of those individuals that *already* carry resistance alleles or allele combinations. This was demonstrated by a series of elegant experiments ex-amining the resistance of bacteria to penicillin (Fig. 12.4) and was later shown to hold true for the development of resistance to DDT in *Drosophila*.

Note that humans can increase the *speed* at which new variants appear in a population by exposing organisms to radiation or chemicals that cause mutations to occur more often. And as you will learn in Chapter 21, genetic engineers can move certain genes and groups of genes from one organism to another, dramatically increasing the power of artificial selection. But we still have no idea how to create new genes "from scratch."

Natural Selection: Effects on Phenotype

Natural selection, however, never touches genes themselves. Rather, it operates only through the effects that variations in genotype have on organisms' phenotypes. Selection can influence the distribution of phenotypic characters in a population in three basic ways: *stabilizing selection,* which tends to maintain the genetic status quo, *directional selection,* which tends to shift the mean of the frequency distribution in one direction or another, and *disruptive selection,* which tends to split a population into two divergent forms.

We will examine these three categories of natural selection in two ways. We will first propose a series of hypotheses about the effects of different environmental conditions on a bird population modeled after Darwin's finches. We will then examine data from controlled observations and experiments that test our hypotheses on other organisms.

The hypothetical population Suppose a population of seed-eating birds exhibits substantial variation in beak

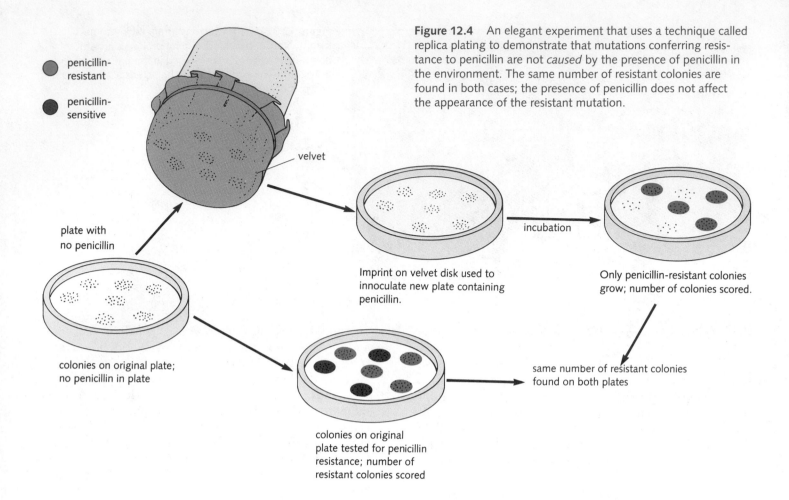

Figure 12.4 An elegant experiment that uses a technique called replica plating to demonstrate that mutations conferring resistance to penicillin are not *caused* by the presence of penicillin in the environment. The same number of resistant colonies are found in both cases; the presence of penicillin does not affect the appearance of the resistant mutation.

penicillin-resistant

penicillin-sensitive

velvet

plate with no penicillin

Imprint on velvet disk used to innoculate new plate containing penicillin.

incubation

Only penicillin-resistant colonies grow; number of colonies scored.

colonies on original plate; no penicillin in plate

same number of resistant colonies found on both plates

colonies on original plate tested for penicillin resistance; number of resistant colonies scored

size, a genetically controlled characteristic. Suppose also that this variation in beak size affects the birds' ability to feed on seeds of different types. Birds with small beaks can feed most effectively on smaller seeds, whereas birds with bigger, stronger beaks, can handle larger, thicker, harder seeds more easily. The frequency distribution of beak size in our hypothetical population at the outset is shown in Fig. 12.5.

Stabilizing selection If the sizes of seeds available to these birds remain unchanged, natural selection will favor individuals with beaks suited to average-sized seeds. **Stabilizing selection** will tend to eliminate individuals with beaks very much larger or much smaller than average, because those individuals will be able to handle most efficiently only some of the available seeds and so will have lower fitness (Fig. 12.5a). Because sexual reproduction tends to maintain the original degree of genotypic variation in each generation, the frequency distribution of beak size will not necessarily narrow, but it is not likely to broaden either. Stabilizing selection discourages evolution away from the average condition.

Directional selection If, however, the population of small-seeded plants declines for some reason, or if the population of large-seeded plants increases, the relative fitness of birds with the smallest beaks will fall (Fig. 12.5b). After a few generations, this type of **directional selection** will cause the average beak size in the population to increase. The resulting bird population will be better adapted to feeding on larger seeds.

Disruptive selection Now suppose instead that the supply of intermediate-sized seeds suddenly decreases. (In nature this could happen either because plants producing those seeds suddenly decline in numbers or because a new bird species appears that competes with the original birds and eats large numbers of midsized seeds.) If this were to happen, the relative fitness of individuals near the middle of the frequency distribution would fall relative to that of those individuals at either extreme (Fig. 12.5c). The resulting **disruptive selection** would tend to split the population into two groups, one with larger beaks for large seeds and one with smaller beaks for small seeds.

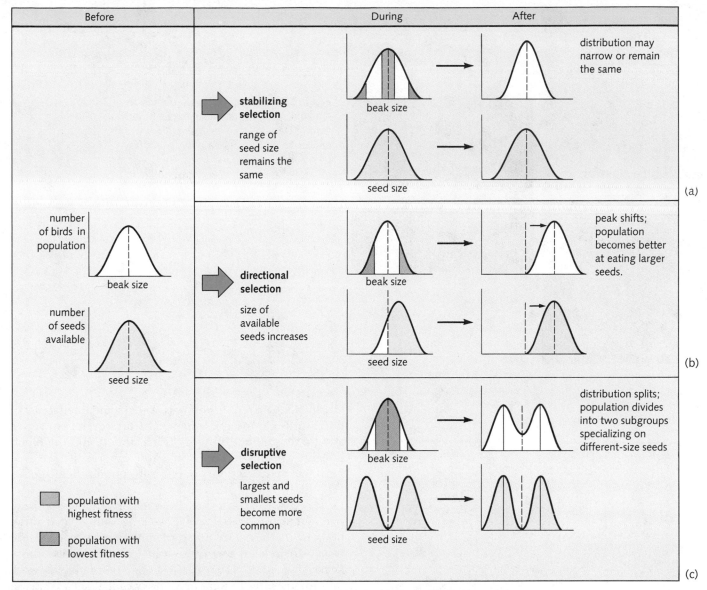

Before

number of birds in population

beak size

number of seeds available

seed size

population with highest fitness

population with lowest fitness

During

stabilizing selection

range of seed size remains the same

directional selection

size of available seeds increases

disruptive selection

largest and smallest seeds become more common

beak size

seed size

beak size

seed size

beak size

seed size

After

distribution may narrow or remain the same

peak shifts; population becomes better at eating larger seeds.

distribution splits; population divides into two subgroups specializing on different-size seeds

(a)

(b)

(c)

Figure 12.5 Distribution of beak length in a hypothetical population of birds and the effect of changes in seed availability on that distribution. In each case, the distribution of beak length in the population changes over time, because some individuals in the population (blue) have lower fitness and others (orange) higher fitness under the new conditions. **(a)** Stabilizing selection: Average-sized seeds become more common, and the birds become more specialized with around the same (average) beak length. **(b)** Directional selection: Larger seeds become more common, and the bird population evolves larger beaks. **(c)** Disruptive selection: Average-sized seeds become less common, and larger and smaller seeds become more common. In response, the bird population splits into two subgroups specializing in eating the larger and smaller seeds, respectively.

VARIATION AND SELECTION: A CASE STUDY

The preceding discussion makes straightforward predictions from evolutionary theory, but it is entirely hypothetical. Can we test any of its predictions on a real population in nature? Thanks to a long series of observations and ex-periments on the fascinating birds from the Galapagos Islands called Darwin's finches, we can. These 14 species of birds are so ecologically diverse that Darwin first classified them as members of several separate subfamilies, including finches, orioles, and warblers. It was not until much later, after showing his specimens to ornithologists in England, did he realize that they were, in fact,

Figure 12.6 These photos show the remarkable diversity of beak shapes found among Darwin's finches. TOP LEFT: A woodpecker finch, *Cactospiza pallida.* BOTTOM LEFT: A medium ground finch, *Geospiza fortis.* RIGHT: A sharp-beaked ground finch, *Geospiza difficilis.*

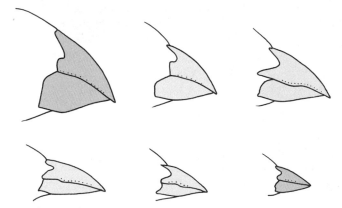

Figure 12.7 Here you can see clearly the variations in beak size among species that characterize Darwin's finches. CLOCKWISE FROM TOP LEFT: *magnirostris, fortis, conirostrus, fuliginosa, difficilis,* and *scandens.* [From Grant (1986) after Abbott et al. (1977).]

all finches. His confusion was understandable; some of these birds eat insects, others eat fruit, several eat seeds, and there is even a "vampire finch" that drinks the blood of larger seabirds (Fig. 12.6). You can readily see that beak size and shape vary extensively among these species (Fig. 12.7).

It is easy to accept that these major variations in beak size and shape *between* species—along with related variations in the size and shape of muscles that control the beak—are related to the ways in which birds use those beaks in feeding (Fig. 12.8). Long, pointed bills work well for probing flowers or woody plant tissues. Massive, deep beaks are well suited to crushing large seeds. Curved beaks with extremely sharp tips are good for biting and grasping small insects. Thus we can hypothesize (if we agree with the premise of natural selection in advance) that evolutionary pressure to avoid competition has somehow "pushed" these birds into different ecological niches.

But the charge of evolutionary biologists is to *test* such hypotheses rigorously, not to accept them on faith. And to test an hypothesis about the way natural selection has acted on finch populations over time, we must make observations at a finer level of detail. We must look for variation *within* species and search for evidence that natural selection can, in fact, cause differential survival based on that variation.

Because these birds, quite apart from their place in the history of evolutionary thought, are biologically fasci-

Platyspiza	Geospiza	Pinaroloxias	Certhidea
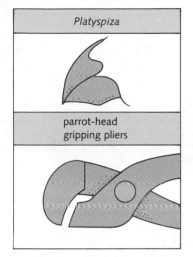	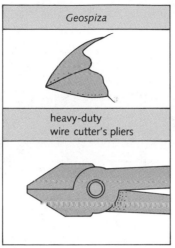	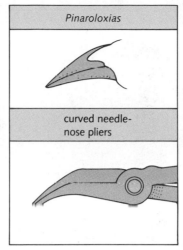	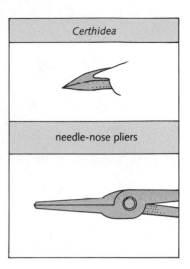
parrot-head gripping pliers	heavy-duty wire cutter's pliers	curved needle-nose pliers	needle-nose pliers

Figure 12.8 A relatively simplistic—yet revealing—comparison between selected finch beak types and certain human tools that apply force in different ways for different ends. [From Grant (1986) after Bowman (1963).]

nating, they have been studied extensively over the last century. The results of those studies fill several books, and we cannot do them all justice here. In this chapter, to give an example of modern evolutionary biology in action, we will concentrate on experiments examining variation and directional selection in seed-eating finch species. These studies, summarized, organized, and significantly expanded by Peter and Rosemary Grant of Princeton University, are among the most elegant examinations ever performed into the nature of natural selection. In the next chapter, we'll expand our observations to consider how this group of species arose.

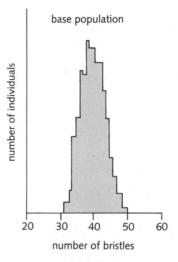

Figure 12.9 This graph, showing the frequency distribution of *Drosophila* individuals with various numbers of abdominal bristles, demonstrates the classic pattern of variation found in many heritable characteristics of organisms. The same distribution occurs in final beak sizes, as you will see in Chapter 13.

Variation in Beak Size: Raw Material

Measurements of both living birds and preserved specimens in museums make it clear that there is substantial variation in beak size among individuals of each of several species of seed-eating Galapagos finches. This sort of variation, evident in graphs such as those shown in Fig. 12.9, agrees with the first premise in the hypothetical case outlined previously.

But are these heritable differences in beak size of any adaptive significance to the birds? In other words, is the difference in fitness caused by these variations sufficient to cause differential survival of these birds in their natural environment? Early workers in the Galapagos guessed "yes," but they had no quantitative data to support their

hypothesis. The Grants and other workers since that time have provided solid evidence that beak size is, in fact, of real importance.

Beak size and seed-handling ability By observing birds feeding in their natural habitats, several researchers found that individuals with different beak sizes fed on different-sized seeds. Birds seen feeding on the smallest and softest seeds had smaller beaks than those seen feeding on larger, harder seeds that are tougher to crack.

This relationship is true both within and between species; small-beaked species eat a higher percentage of smaller seeds than large-beaked species, and smaller-beaked individuals within each species prefer smaller seeds on average than their larger-beaked fellows. These

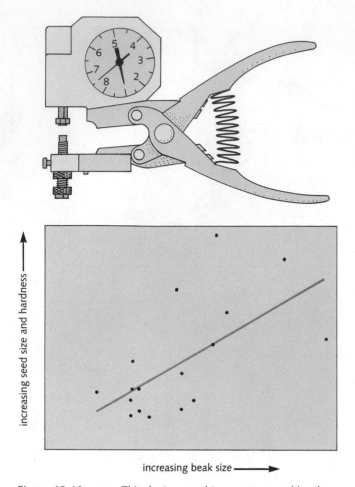

Figure 12.10 TOP: This device, used to measure seed hardness, shows the ingenuity often required in the design of field experiments in evolutionary biology. To test the hypothesis that beak size affects the birds' ability to handle large, hard seeds, the experimenters had to do more than guess at seed toughness; this device provided replicable, quantitative measurements of seed hardness. BOTTOM: This figure shows the positive correlation between beak size in ground finches and an index that represents the relative size and hardness of seeds in their diets. [From Abbott et al. (1977).]

data were bolstered by studies showing that larger, tougher seeds were more readily accepted by larger-beaked birds (Fig. 12.10). Smaller-beaked birds either would not or could not utilize these seeds.

The importance of food limitation It has become clear that in most years, food is a limiting resource for these finches. The Galapagos Islands have extremely seasonal climate; rains, during which plants grow and set seed are followed by dry months during which many plants enter dormancy. The supply of seeds thus rises and falls markedly (Fig. 12.11a). Finch populations also fluctuate

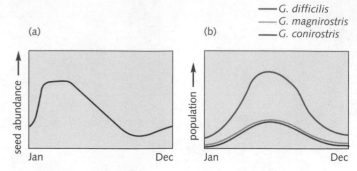

Figure 12.11 Relationship between seed availability and finch populations in the highly seasonal Galapagos climate. **(a)** Annual fluctuations in seed abundance. **(b)** Annual fluctuations in the numbers of three species of ground finches. The rise in bird population represents the onset of the breeding season. The subsequent fall occurs as individuals starve when food is scarce. [From Grant and Grant (1980).]

seasonally, rising as new individuals hatch and falling as many of them fail to survive food shortages during the dry season (Fig.12.11b).

Such seasonal fluctuations, it turns out, are part of a larger cycle during which wetter years alternate with drier ones. These long-term cycles, too, affect finch population size. During one exceptionally wet period during 1982–1983, plants grew more abundantly, and finches bred far more prolifically than usual. And during a prolonged drought, documented by the Grants from mid-1976 through the end of 1977, seed production fell markedly and finch populations plummeted from around 1200 individuals to no more than 180 (Fig. 12.12a,b).

Response to Drought: Directional Selection

Interestingly, the birds that survived the seed shortage of 1977 were *not* simply a representative sample of the birds present before the drought began. Population surveys taken before, during, and after the drought showed clearly that large-beaked birds survived the famine in greater numbers than small-beaked members of the same species.

Why might this have been the case? Both small and large seeds were available at the beginning of the drought, offering food for both large- and small-beaked birds. But during the drought, as the seed supply ran low, the supply of small seeds was exhausted. By early 1977 the birds had to depend almost exclusively on large, hard-to-crack varieties. Large-beaked birds, which could exploit this food, survived to reproduce later on; small-beaked individuals starved. The result, a classic demonstration of directional selection in action, was an increase in the average beak size of surviving birds (Fig. 12.13a,b).

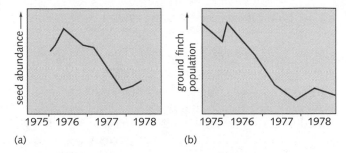

(a)   (b)

Figure 12.12   During a particularly severe drought in 1977, both seed abundance **(a)** and ground finch populations **(b)** dropped steadily. [From Boag and Grant (1981).]

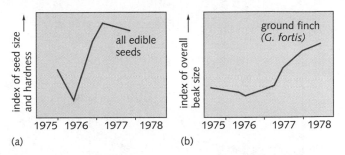

(a)   (b)

Figure 12.13   **(a)** During the first several months of the drought, small, easy-to-crack seeds were consumed in large numbers. As a result, the average size and hardness of remaining seeds rose steadily. **(b)** Forced to feed on large, hard seeds, the population of *G. fortis* was subjected to intense natural selection for large beaks. This graph shows a steady increase in an index related to both beak size and body size; the line on the graph represents the average for the remaining living birds at each point in time. [From Boag and Grant (1981).]

OTHER EXPERIMENTAL STUDIES OF NATURAL SELECTION

In addition to ambitious field studies of finches and other organisms in nature, many laboratory studies of both plant and animal populations have sought to test the effects of natural selection or to search for evidence of selection in action. Here we will present one such study examining each of the modes of natural selection we have described.

Effects of Stabilizing Selection

We can find evidence of stabilizing selection acting on our own species by looking at the relationship between infant survival and birth weight. The weight of newborn human infants varies widely, ranging from less than a pound in cases of extremely premature births to more than 10 pounds (Fig. 12.14). The survival rate for those infants also varies widely; both very small infants and unusually large ones have much higher early mortality rates than babies averaging between 6 and 8.5 pounds at birth. You can see from the figure that the actual average birth weight and the theoretical optimal birth rate are nearly identical—evidence for the operation of stabilizing selection on our species over time.

Effects of Directional Selection

The effects of directional selection have been demonstrated in the laboratory through artificial selection in a number of organisms. One variable character in laboratory populations of *Drosophila*, for example, is the number of bristles the flies carry on their abdomens. It is a simple matter to count these bristles and to select for breeding

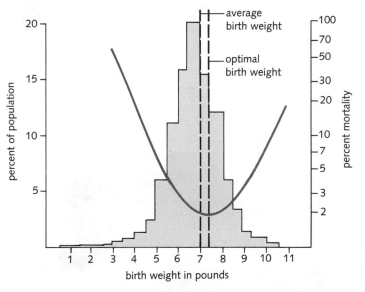

Figure 12.14   Stabilizing selection and human birth weight. The curve on this graph shows that human infants have the best chance of surviving the trials of birth if they weigh between 7 and 8 pounds; at higher and lower weights, mortality is higher. The histogram shows that average birth weight matches that of the optimal weight quite closely. [From Cavalli-Sforza and Bodmer (1971).]

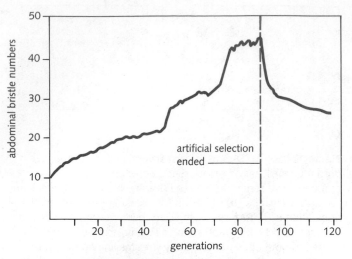

Figure 12.15 This graph shows the response of a laboratory *Drosophila* population to artificial selection for increased numbers of abdominal bristles. In generations 1 through 90, only those flies with the highest numbers of bristles were allowed to mate. Note that when this artificial selection ended after 90 generations, the average bristle number began to decline. [From Yoo (1980).]

from a "wild-type" population those individuals with higher or lower number of bristles. Figure 12.15 shows the steady increase in bristle numbers in a laboratory fly population selected for high numbers of abdominal bristles.

Effects of Disruptive Selection

Disruptive selection has also been demonstrated experimentally through artificial selection with *Drosophila,* as shown in Fig. 12.16. In these experiments, a population was subjected to selection for both higher and lower bristle numbers. The result, after 35 generations, was two subpopulations with mean bristle numbers significantly higher than normal and lower than normal, respectively. Note that Fig. 12.16 lends support to the claim that even strong directional or disruptive selection—like stabilizing selection—does not necessarily decrease variation in the population.

THE HARDY–WEINBERG LAW

Once biologists defined evolution as change in the frequencies of alleles in populations, they could examine the circumstances in which such genetic change occurs. In 1908 two independent investigators, George H. Hardy of Cambridge University and Wilhelm Weinberg, a German physician, realized that although the segregation and recombination of alleles during mating provide genetic variability among offspring, sexual reproduction by itself would not change relative allele frequencies. They went on to describe a restricted set of conditions under which the relative frequencies of alleles in a population remain the same indefinitely.

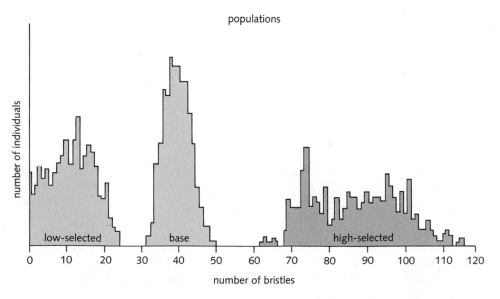

Figure 12.16 The effect of disruptive selection on laboratory population of *Drosophila.* Beginning with a base population (CENTER), artificial selection favored flies with the highest and lowest numbers of abdominal bristles. The resulting populations (LEFT and RIGHT) were obtained in 34–35 generations. Note that the pressure of selection did not decrease the amount of variation in the resulting populations. [From Yoo (1980).]

When Evolution Will Occur and When It Will Not

According to what is now called the **Hardy–Weinberg law,** the relative allele frequencies in a population remain constant from generation to generation—in other words, evolution does not occur—*as long as:*

1. Chance events do not affect the genetic frequencies in the population.

2. Mutations do not occur, or, if they do occur, they balance each other out.

3. All genotypes have equal reproductive success, in other words, there is no natural selection operating on the population.

4. There is no net flow of alleles into or out of the gene pool. This means that there is no net migration of organisms into or out of the population.

5. All mating in the population is completely at random. That is, no genotype in the population can have a preference for mating with any other particular genotype.

If and only if all these conditions are met, the gene pool of a population remains in a state of genetic equilibrium. But only rarely do all these conditions exist in nature for all of an organism's alleles at the same time. Chance *can* affect genetic frequencies in small populations. Mutations *do* occur, and it is impossible for them to be completely balanced by opposite mutations. Organisms are *constantly* entering and leaving most populations, though it is only the net genetic change that counts. And mating in natural populations is *rarely* completely random; certain phenotypes (and hence certain genotypes) are almost always preferred over others as mating partners.

Why is the Hardy–Weinberg law important if its conditions are seldom met? First, the law provides equations that predict the frequencies of different genotypes in a stable gene pool. If the relative proportions of genotypes in a population match those predicted by the equations that express the Hardy–Weinberg law, that population's gene pool is stable and the particular locus in question is not under strong selective pressure. If the genotype ratios are different from those the equations predict, the gene pool is under pressure, from natural selection or some other factor, to change.

Second, the Hardy–Weinberg law clearly defines the set of conditions under which evolution does *not* occur. Biologists who look at natural populations find that these conditions rarely all prevail at once for prolonged periods of time. We therefore know that many natural populations do not remain in genetic equilibrium for long, and we have additional evidence that change is a fact of life on earth.

GENETIC DRIFT

As important as natural selection is under certain circumstances, both field and theoretical studies in population genetics have shown that it is not the source of all change in gene frequencies in natural populations. It is possible for a mutation to become common in a population entirely by accident through a phenomenon called **genetic drift.** Even a harmful mutation can become established, and can remain in a gene pool, as long as a single copy of it is not lethal.

Operating by chance, genetic drift works most strongly in either small populations or populations of dramatically rising and falling size. When a nonlethal mutation occurs in such a population, it is possible for the organism carrying that mutation to have more offspring than its neighbors *not* because it is better adapted, but just by chance.

Also, a sudden change in the environment of that population may just happen to wipe out many of the individuals that do not carry that mutation. When either of these situations arise, a new mutation can appear in most members of the population after only a few generations (Fig. 12.17). In large, stable populations, this is much less likely to happen by accident.

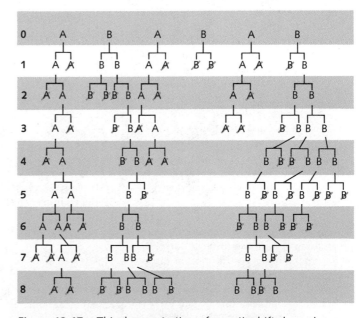

Figure 12.17 This demonstration of genetic drift shows how evolution can occur by chance. The population begins with equal numbers of *A* and *B* genotypes. Each type produces identical offspring. In each generation, half of the population dies *at random*, regardless of genotype. The relative numbers of *A* and *B* genotypes vary randomly until *A* becomes extinct, not because it is poorly adapted, but by chance alone.

The Hardy–Weinberg Equilibrium

Imagine that you are a geneticist studying a trait controlled by two alleles, *A* and *a*, that follow the rules of simple dominance at a single locus. When you survey a natural population of organisms for this trait, you discover that only 4 percent of the population exhibits the phenotype representing genotype *aa*. Fully 96 percent of the population is either *AA* or *Aa* and thus exhibits the dominant phenotype. Not surprisingly, you assume that over many generations, the dominant allele will somehow displace the recessive allele, not because it confers any adaptive advantage, but just because of its greater apparent frequency in the population. Your assumption, however, would be wrong, for reasons demonstrated by the Hardy–Weinberg principle.

The Hardy–Weinberg principle, simplified for a discussion of a single locus with only two possible alleles, *A* and *a*, can be demonstrated as follows:

Let us represent the frequency of the *A* allele as *p* and the frequency of the *a* allele as *q*. Because all "places" at this locus in the population must be occupied by one allele or the other, the sum total of all occurrences of *p* and *q* must always equal 100 percent of those loci. Stated in another way, $(p + q) = 1$.

In any cross involving these alleles, there are three possible genotypes: *AA*, *Aa*, and *aa*.

As you can see from the accompanying diagram, one generation of random mating among the three genotypes creates a binomial distribution of those genotypes in the next generation. This occurs because segregation of these alleles produces gametes that carry the alleles in the same relative frequencies at which those alleles occur in the population. Thus the relative frequency of *A*-carrying eggs is *p*, and the relative

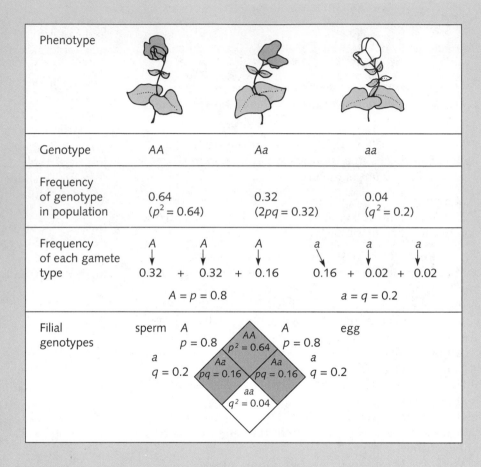

frequency of *a*-carrying eggs is *q*. The same is true for sperm. Accordingly, the three types of resulting zygotes are produced in the same relative numbers as the individuals in the Punnett squares you met earlier:

$$(p + q)^2 = p^2 + 2pq + q^2$$

Note that because the sum of all genotypes in the population must also equal 100 percent, $(p^2 + 2pq + q^2) = 1$.

If 4 percent of the population exhibit the *aa* genotype, $q^2 = 0.04$ and *q* therefore is 0.2. If $q = 0.2$, then $p = 0.8$. In each generation, therefore, the rela-

tive frequency of *AA* homozygotes will be p^2, or 0.64; the number of *aa* homozygotes will be q^2, or 0.04; and the number of *Aa* heterozygotes will be $2pq$, or 0.32. As long as the necessary conditions continue to be met, neither the frequency of the genotypes nor the frequencies of the alleles (*p* and *q*) will ever change from generation to generation.

You can demonstrate this stability over time to yourself by modeling this situation through several generations. To do that, begin with an imaginary population of 1000 organisms in which the first generation consists of 640 *AA* individuals (p^2), 320 *Aa* individuals ($2pq$), and 40 *aa* individuals (q^2).

You can also see how population geneticists and evolutionary biologists actually use this seemingly abstract relationship by trying your hand at the following problems.

1. In a particular human population, roughly 1 in 5000 individuals is afflicted with a recessive genetic disorder that causes mental degeneration and death, but usually not until after age 55. What percentage of this population would you expect to be heterozygous for the gene?

Hint: Assume that this population meets all the criteria needed to satisfy the Hardy–Weinberg principle. Remember that those afflicted with the disease are homozygous recessives (*aa*) and that the total number of these individuals is q^2.

2. A researcher collects individuals from a natural population of butterflies that displays the three phenotypes indicated below. Is this particular locus at or near Hardy–Weinberg equilibrium?

Hint: Set up a diagram for this population similar to the figure included in the box, but use as your numbers the numbers of individuals of the three genotypes instead of frequencies. Calculate p and q using the binomial equation ($p^2 + 2pq + q^2$) to predict the percentage of individuals of each genotype in the population.

Phenotype	Genotype	Number of Individuals
Heavy white spotting	(*AA*)	1469
Moderate spotting	(*AA'*)	138
Little spotting	(*A'A'*)	5
	Total	1612

The Founder Effect

If a small number of individuals migrate off to found a new population, they may carry alleles in different relative frequencies from the main population just by chance. This phenomenon, known as the **founder effect,** is an extreme case of genetic drift. Figure 12.18 offers an illustration of how the founder effect might work.

The founder effect can be especially important in places such as the Hawaiian or Galapagos islands, where populations of mainland species may be established by a handful of stranded wanderers. The tiny gene pool of those individuals may not be at all representative of the gene pool of the main population from which they came. Founding populations, therefore, can start out with peculiar collections of alleles.

There are several well-documented cases of the founder effect among small groups of humans whose practices isolate their gene pool from that of the surrounding community. One example is the Amish community of Pennsylvania, an orthodox religious group that

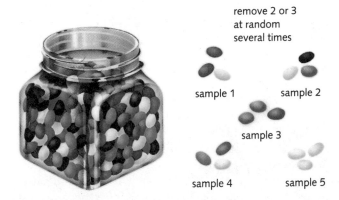

Figure 12.18   A schematic demonstration of the founder effect. Imagine a jar of 1000 jellybeans, containing equal numbers of five colors. If you poured out half of the jellybeans, you would probably find roughly equal numbers of all five colors in your sample. But if you picked out only two or three beans at random, your sample would be missing at least two of the colors present in the main population, and you could conceivably end up with three yellow ones. The same thing can happen with the alleles in a small founding population.

was founded by a very small number of immigrants and allows no marriage to outsiders. Among the founders of the American Amish community was the family of a Mr. and Mrs. Samuel King. The Kings happened to carry a recessive allele that causes dwarfism, the growth of extra fingers, and heart defects in babies homozygous for it. Those who are afflicted with this condition usually die at a very early age. In the outside world, the frequency of this allele is less than 1 in 1000, but the Kings and their descendants had larger families than the others in their Amish community. Thus, through a combination of the founder effect and genetic drift, this allele occurs in the Amish population today at the extraordinarily high frequency of about 1 in 14.

PLEIOTROPY AND HETEROZYGOUS ADVANTAGE

One unexpected finding of modern genetics is that a single allele often has *several* important effects on the phenotype of an organism. This phenomenon is called **pleiotropy**, which means "many effects." The same allele can have several very different, simultaneous effects, some of which may be beneficial while the others are harmful. The selective advantage of one trait controlled by such an allele may be powerful enough to make the allele common in a population, even though its ill effects would lower the organisms' fitness if they occurred alone.

    One example of pleiotropy in human populations is an allele that produces an unusual form of hemoglobin, called Hemoglobin-S, most commonly, but not exclusively, in blacks of African descent. In people homozygous for the Hemoglobin-S allele (those who have inherited copies from both parents), red blood cells become distorted into rigid, sickle-shaped forms. These cells lose some of their oxygen-carrying ability and get stuck in tiny blood vessels throughout the body. This causes a number of serious health problems (Fig. 12.19). These homozygotes become very ill and often die before puberty. Even the red cells of heterozygotes, who inherit only a single copy of the allele from one parent, may "sickle" under certain conditions.

Why, then, has this "sickle-cell" allele never been eliminated by natural selection? In addition to its negative effects on homozygotes, this allele has two additional, unexpected, beneficial effects on phenotype. Individuals who are heterozygous for the Hemoglobin-S allele are endowed with increased resistance to malaria and with slightly enhanced fertility. For this reason, in areas of western Africa where malaria is a common and

often lethal disease, the allele significantly increases the fitness of individuals who carry a single copy of it.

Such a situation, in which an individual with one copy of a particular allele has an advantage over an individual with either two copies or none at all, is called *heterozygote advantage*. Note that in malaria-free environments (such as the United States) this advantage disappears. Here, as expected, the fitness of the heterozygote is slightly lower than that of individuals who carry no Hemoglobin-S allele, so the frequency of that allele has decreased significantly.

MODERN STUDIES OF GENETIC VARIATION

Note that all our models of evolutionary change assume that sufficient genetic variation exists in nature to provide raw material on which natural selection can operate. In order to understand why some species evolve rapidly while others stagnate or become extinct, evolutionary biologists are trying to test that assumption in a variety of species. To do so, they must answer two related questions: How many different alleles of each gene are typically present in a population? And how much do members of a population differ from one another in their genetic makeup? In other words, what is the degree of genetic **polymorphism** in the population? (By definition, a population is polymorphic for a given trait whenever more than one allele of a given gene is present. Mendel's peas, for example, were polymorphic for both seed color and seed shape.)

Darwin observed many heritable differences among the organisms he studied. But a lot of genetic variation cannot be discerned simply by looking at an organism's physical characteristics. Molecular biology has provided the tools that evolutionary biologists needed to study this previously undetectable genetic variation, and with those tools they have uncovered more polymorphism than anyone ever thought existed. In the fruit fly *Drosophila*, for example, average individual flies carried more than one allele for as many as 12 percent of their genes, a level of polymorphism that turns out to be about average for insects. Plants, on the whole, have a polymorphism rate of about 18 percent, while average vertebrates exhibit about 7 percent.

In many species, therefore, there is plenty of genetic variation for natural selection to work on. But what maintains all those different alleles in the population? Are all of them adaptive? Is polymorphism itself adaptive? In at least some cases the answer seems to be "yes." We have seen, in the case of sickle-cell anemia, how heterozygote

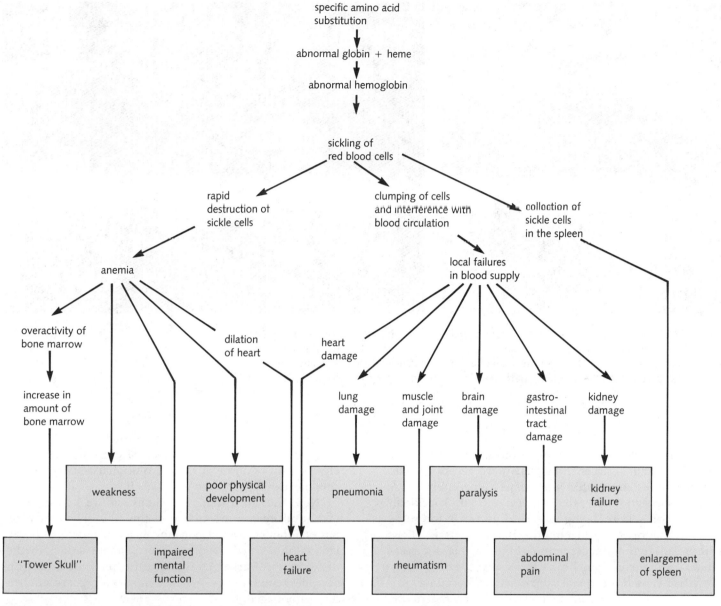

Figure 12.19 Pleiotropic effects of the allele for sickle-cell anemia (Hemoglobin-S). [From Raff and Kaufman (1983).]

advantage can maintain polymorphism, even when one of the alleles is lethal in the homozygous condition.

The *neutral theory* suggests that in other cases of polymorphism, natural selection neither favors nor opposes the spread of alternative alleles. In such cases, adherents of the neutral theory argue, polymorphism may be maintained entirely by the action of genetic drift.

But environments in nature change regularly. Under different environmental conditions, an originally neutral trait—such as a different flower color or the ability to digest a certain chemical compound—might become advantageous. The existence of neutral variation in a population, therefore, can prove useful in the long term.

Evolution Due to Human Activity: British Moths

Human activity in the biosphere has caused evolution to occur in many organisms over very brief periods of time, demonstrating dramatically how the existence of previously neutral variability in a population can prove useful in changing environments. In one such example, an ecological side effect of the Industrial Revolution in Britain around Darwin's time caused several British moth species to change color.

Before Britain's industrialization, most of these moths were lightly mottled and blended superbly with the tree

Figure 12.20 The adaptive significance of industrial melanism in *Biston betularia*. LEFT: On normally colored tree trunks, the light-colored form is nearly invisible, whereas the dark form can be seen clearly. RIGHT: On tree trunks darkened by soot, the light form is clearly seen— and hence more vulnerable to bird predators—whereas the dark form is well camouflaged.

trunks on which they rested, motionless, for much of the day (Fig. 12.20). Darker individuals were always present, but they never constituted more than a small percentage of the population. By 1880, however, many decidedly darker moths were being observed in such industrial areas as Manchester and London. Before the end of the century, more than 98 percent of the individuals of *Biston betularia* caught in Manchester were of the darker form. What had caused this remarkable shift in coloration?

E. B. Ford of Oxford University proposed that the moth populations had shifted toward darker color because of natural selection. The moths' major predators, Ford reasoned, were birds that hunted by sight. In pre-industrial England, the light-colored moths were so well camouflaged as they clung to normally colored tree trunks that those birds had trouble locating them. Industrial burning of coal, however, noticeably blackened the bark of trees around major cities. Ford proposed that darkened bark of trees had conferred a selective advantage on dark-colored moths and that natural selection had caused the black form to take over the population. The phenomenon thus described was called *industrial melanism* after the pigment, melanin, that darkened the moths' wings.

Later experimental studies by H. B. D. Kettlewell confirmed that moth-eating birds hunt primarily by sight and that those moths that most resemble their backgrounds

in color have the highest chance of escaping predation. Here was an example of rapid evolutionary change occurring practically in Darwin's backyard!

Note, however, that the patchy nature of industrial and rural environments across the English countryside has maintained overall polymorphism in the moth population, rather than favoring a permanent switch to the darker form. Darker moths have higher fitness only on soot-blackened trees; elsewhere, lighter forms are still better camouflaged.

Evolution Today: Out of Control

Not all examples of human-caused evolution are quite so innocuous. Efforts to eradicate weeds from cultivated areas and to eliminate disease and parasites from both humans and domestic plants and animals have imposed new selective pressures on wild plant and animal populations. The hand weeding of cultivated fields, for example, creates selective pressures in favor of weeds that resemble crop plants. Over successive generations, many weeds converge in appearance with crops, making later weeding much more difficult.

Additionally, humans now use antibiotics to control bacterial infections, pesticides to kill insects on farms and around human dwellings, and herbicides to kill weeds.

All three of these uses have important positive effects on human health and well-being. But our regular use of those compounds has made them part of the environment of insects and microorganisms. As it turns out, many natural populations of insects and bacteria contain enough genetic variation for a few of them to be resistant to any given toxic compound. By killing susceptible individuals, these insecticides dramatically increase the relative fitness of resistant variants.

Already, DDT and several other insecticides no longer kill many of the insect species they controlled a few years ago. Even more worrisome is the fact that many bacteria and other parasitic microorganisms are rapidly evolving resistance to antibiotics and other drugs. In the tropics, for example, the organism that causes malaria has evolved resistance to the drug that once effectively controlled the disease.

Note, however, that resistance to poisons is rarely a "free ride" for either insects or other organisms, because the selective trade-offs imposed by pleiotropy often maintain polymorphism either within or between populations of a species. Some populations of Norway rats, for example, have evolved resistance to the rat poison *warfarin*. Where the poison is in widespread use, homozygotes for the allele that confers resistance are common. But that allele also lowers rats' ability to synthesize vitamin K, a compound essential in allowing blood to clot, and they bleed more easily. For that reason, in places where warfarin is not used, individuals homozygous for this allele are at as much as a 54 percent selective *dis*advantage compared to "wild-type" rats, and the allele is far less common. The same sort of phenomenon has been demonstrated for the alleles that confer resistance to DDT and to dieldrin in mosquitoes.

SUMMARY

Evolutionary biology today, which has absorbed significant contributions from Mendelian genetics, population genetics, and ecology, deals both with the fact of evolutionary change and with theories about how and why evolution takes place.

The gradual accumulation of evolutionary change within populations can often be explained by the differential action of selective pressures on different phenotypes that represent different genotypes. Recent advances in molecular biology have revealed unexpected riches of genetic variation in natural populations. Some of these variations are sufficient to provide the raw material necessary for evolution to operate; others may be neutral, or selectively meaningless.

Natural selection is not necessarily responsible for all evolutionary change. Additional mechanisms, including genetic drift, operate more according to chance than in terms of selective advantage.

STUDY FOCUS

After studying this chapter, you should be able to:

- Define *adaptation*, *selection*, and *fitness* in genetic terms.
- Explain how Mendel's principles of genetics affected evolutionary biology.
- Understand the significance of the Hardy–Weinberg principle in population genetics.
- Describe the way natural selection can affect heritable characteristics in a population.
- Appreciate that natural selection is not necessarily responsible for all evolutionary change.

TERMS AND CONCEPTS

species *242*	disruptive selection *244*
gene pool *242*	Hardy–Weinberg law *251*
evolutionary fitness *242*	genetic drift *251*
evolutionary adaptation *242*	founder effect *253*
evolution *242*	pleiotropy *254*
stabilizing selection *244*	polymorphism *254*
directional selection *244*	

REVIEW

Objective Questions (Answers in Appendix)

1. The critical source of variability within a population is

    (a) sexual reproduction.
    (b) recombination of alleles.
    (c) environmental changes.
    (d) mutations.

2. When the mean of the frequency distribution of a trait in a population shifts in one direction or another, this describes

    (a) directional selection.   (c) pleiotropy.
    (b) disruptive selection.   (d) stabilizing selection.

3. For a mutation to be important in the evolution of a species, it must be found in

    (a) the reproductive cells.
    (b) all of the body cells.
    (c) random body cells.
    (d) those cells that keep an organism healthy.

4. Evolutionary fitness is defined as

    (a) producing just enough offspring to be supported by the existing food supplies.
    (b) being physically fit for the age and type of species.

(c) adapting to changes in the environment.

(d) contributing to the gene pool of the next generation.

5. A type of natural selection in which two extreme pheno-types take hold in a population is known as

    (a) divergent selection.

    (b) environmental selection.

    (c) disruptive selection.

    (d) stabilizing selection.

Discussion Questions

6. How did genetic theory provide biologists with a new way to define species?

7. What is genetic drift? How can genetic drift cause changes in allele frequencies in a small population, even in the absence of natural selection?

8. What is the founder effect? How did it cause unusual allele frequencies among the American Amish?

9. How does pleiotropy help to maintain the existence of the sickle-cell allele in Africa? How does it oppose the spread of the allele for warfarin resistance in Norway rats?

10. What do antibiotic resistance in bacteria and pesticide resistance in insects have in common? Why are both cause for concern?

READINGS

Grant, P. R. *Ecology and Evolution of Darwin's Finches*. Princeton, N. J.: Princeton University Press, 1986.

Raff, R. A., and T. C. Kaufman. *Embryos, Genes, and Evolution: The Developmental Genetic Basis of Evolutionary Change*. New York: Macmillan, 1983.

Boag, P. T., and P. R. Grant. "Intense natural selection in a population of Darwin's finches *(Geospizinae)* in the Galapagos finches." *Science* 214 (1981): 82–85.

Grant, P. R., and B. R. Grant. "The breeding and feeding characteristics of Darwin's finches on Isla Genovesa, Galapagos." *Ecological Monographs* 50 (1980): 381–410.

Yoo, B. H. "Long-term selection for a quantitative character in large replicate populations of *Drosphilia melanogaster*. I. Response to selection." *Genetic Research* 35 (1980): 19–31.

Abbott, I., L. K. Abbott, and P. R. Grant. "Comparative ecology of Galapagos ground finches (*Geospiza* Gould): Evaluation of the importance of floristic diversity of interspecific competition." *Ecological Monographs* 47 (1977): 151–184.

Cavalli-Sforza, L. C., and W. F. Bodmer. *The Genetics of Human Populations*. San Francisco: Freeman, 1971.

Bowman, R. I. "Evolutionary patterns in Darwin's finches." *Zoology* 58 (1963): 1–302.

13

Evolution of Species

*i*f you ever read *On the Origin of Species,* you will notice a rather paradoxical omission; despite its title, the book doesn't really say much about the origin of species! Throughout the book, Darwin discussed his ideas about how natural selection, operating on heritable variations within species, produces steady, gradual change in *existing* species. Today biologists often use the term *microevolution* to describe these sorts of small changes, such as shifts in beak size in finches and the darkening of wing color in moths.

But though Darwin wrote elegantly about the way microevolutionary changes enable species to adapt to changing environments, he wrote very little about the multiplication of one species into two or more—the process we call **speciation.** The accumulation of genotypic and phenotypic changes large enough to create new species, genera, and higher taxonomic categories (Chapter 23) is often called *macroevolution.* In this chapter we will discuss several hypotheses about the conditions under which new species are formed, and we will examine the rates at which existing species may evolve.

THE ORIGIN OF SPECIES

In Chapter 12 we defined a species as a population of physically similar, interbreeding organisms that are *reproductively isolated* from other such groups. Thus, in genetic terms, speciation occurs when two or more populations diverge in their heritable characteristics in such a way and to such an extent that breeding between them no longer takes place and, with a few exceptions, is no longer possible.

What sorts of mechanisms could cause reproductive isolation? Organisms can be prevented from breeding with one another by any of several **reproductive isolation mechanisms**—characteristics that prevent them from interbreeding. Depending on whether these mechanisms have their effect before or after mating, they are separated into two main categories: *prezygotic* and *postzygotic isolating mechanisms.*

Figure 13.1 Behavioral isolating mechanisms. Courtship displays of birds, such as these great frigate birds, are species-specific and greatly reduce the likelihood of accidental interspecific mating.

Prezygotic Isolating Mechanisms

Prezygotic isolating mechanisms prevent eggs and sperm from ever coming together to form fertilized eggs (*pre* means "before"; *zygote* means "fertilized egg"). This type of isolation can occur in several ways.

Mechanical isolation In some species of animals and plants, males and females have reproductive organs that have coevolved for so long that they are uniquely suited to one another. Among insect species, for example, male reproductive organs vary enormously in size and shape, preventing the effective transfer of sperm to females of other species. Such species are said to exhibit **mechanical isolation** from one another.

Similarly, plant–pollinator coevolution in certain groups of plants ensures that interspecific pollination never happens in nature. In many cases, a particular plant species is pollinated by a single insect species that is adapted both physically and behaviorally to deal with that plant's unique flower characteristics. Such specific pollinators either do not or cannot pollinate the flowers of other species. Numerous orchids are reproductively isolated in nature in this manner; although flowers can be hand-pollinated to produce fertile interspecific hybrids in cultivation, such hybrids are rarely (if ever) seen in the wild.

Behavioral isolation Animals of many species engage in elaborate courtship rituals before mating (Fig. 13.1). During such rituals, males may display brightly colored body parts, serenade females with species-specific songs, or release unique chemicals as odor signals. Differences in particular aspects of these rituals enable the female member of the pair to be certain that her prospective mate belongs to her own species. If courtship begins between two closely related yet different species, the male soon gives the wrong signals and the female refuses to mate with him. Such species illustrate **behavioral isolation,** which we will discuss in more detail in Chapter 46. 46

Behavior not directly involved in mating can also lead to reproductive isolation. For example, the closely related lions and tigers once lived in broadly overlapping ranges throughout India, but their behavioral ecology kept them separated. Lions are highly social creatures that prefer to hunt in open grasslands. Tigers, on the other hand, are solitary and choose forest over open space. Thus, although these animals can interbreed and produce fertile hybrids in zoos, there is no record of hybrids ever having occurred in nature.

Temporal isolation If two species breed at different times of the year (or even different times of day) in nature, they do not exchange gametes even if they could interbreed in captivity. This is called **temporal** (time-related) **isolation.** Many species of potentially interfertile plants, for example, flower at different seasons, so even if they can be pollinated by the same insects, their flowers never open at the same time. Closely related animals, too, may be reproductively isolated from each other because they breed at different seasons or at different times of day.

Gamete incompatibility Even if pollen is passed from one flower to another, or if sperm from one animal reaches the reproductive tract of another, successful fertilization may not occur. In some cases, the female reproductive tract is inhospitable to sperm from different species; in other cases, egg and sperm simply do not fuse. These are cases of **gamete incompatibility.**

Postzygotic Isolating Mechanisms

Even if mating and fertilization occur between members of different species, the zygotes that result may not give rise to functioning individuals. Genetic and physiological characteristics that keep hybrid zygotes from developing properly, or from becoming established in nature if they do survive past birth, are called **postzygotic isolating mechanisms** (*post* means "after").

Often, genetic differences between species are so great that the mismatched sets of genes cannot function together to direct the development of the embryo. This is a situation known as **hybrid inviability.**

In other cases, interspecifically fertilized embryos do develop into viable organisms. But though they are able to survive as individuals, these hybrid plants and animals cannot reproduce. In some cases, they are simply too weak or stunted to reproduce. In other cases, such as the mules produced by crossing horses and donkeys, hybrids are healthy, vigorous animals but are sterile. Because of this **hybrid infertility,** a self-sustaining mule population could never develop; mules must always be produced by repeating the original cross.

Ecological isolation Sometimes two species that can produce viable and fertile hybrids under laboratory conditions do not do so in nature because they are adapted to habitats that may be adjacent to one another but are ecologically quite different. For example, *Quercus lobata* and *Q. dumosa,* two species of oak trees found in the western United States, respectively inhabit fertile valley grasslands and drier, less fertile chaparral habitats on steeper slopes (Fig. 13.2). These species experience **ecological isolation,** a phenomenon that can act as both a prezygotic and postzygotic isolating mechanism.

Adaptations to distinct habitat types such as these can discourage hybridization in two ways. First, the different ecological conditions the plants prefer may minimize the number of places where they occur close enough to one another to exchange gametes. Second, when viable hybrids do occur, they are poorly suited to either parent's habitat (and thus fail to compete successfully for resources there) and have no intermediate environment in which they can thrive and reproduce.

Figure 13.2 *Quercus dumosa* (TOP) and *Q. lobata* (BOTTOM) in their respective habitats.

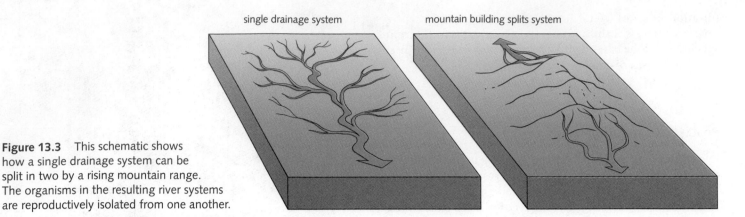

Figure 13.3 This schematic shows how a single drainage system can be split in two by a rising mountain range. The organisms in the resulting river systems are reproductively isolated from one another.

single drainage system

mountain building splits system

MECHANISMS OF SPECIATION: HOW REPRODUCTIVE ISOLATION DEVELOPS

Because reproductive isolation is a critical part of the definition of a species, any explanation of how populations of one species evolve into two or more species must explain how those populations become reproductively isolated from one another. There are several mechanisms that describe how reproductive isolating mechanisms can arise, and each explains certain cases of speciation that have been documented in nature.

Allopatric Speciation

Allopatric speciation (*allos* means "different"; *patria* means "native land") occurs when two or more populations are separated from one another by a geographic barrier that may be as large as a mountain range or as small as a narrow stream. Any barrier, large or small, can isolate populations of a species if that barrier creates conditions in which individuals of that species cannot survive or across which they will not travel. Once populations are isolated in this way, their members cannot interbreed, and gene flow between them is blocked. This sort of separation may occur in several ways, some of which depend on geological events and others of which result from the habits of the organisms themselves.

Wide-ranging populations may be separated from one another by events in the physical world around them. The major continents of the world have united, drifted apart, and collided again, while global climate has warmed and cooled during and between ice ages. Although major geological or climatic changes and long-distance migrations are not necessarily the most common causes of allopatric speciation, they have produced some of the most spectacular adaptive radiations known.

Less drastic geological change can also isolate populations. Changes in continental topography may cause major river systems to change course, separating once-contiguous populations of freshwater organisms (Fig.

13.3). Over time, those streams may carve valleys or create new lakes that can also harbor isolated aquatic populations. And those very same valleys can isolate populations of certain (though not all) terrestrial organisms living on either side of them. A chasm as wide and as deep as the Grand Canyon, for example, is a formidable barrier to small mammals such as squirrels, although it is trivially easy for large birds to traverse.

On an even smaller scale, patches of one sort of habitat are often separated from one another by bands or patches of other habitat types. Mountains in the tropics, for example, may be capped with snow-covered alpine habitats while valleys at their feet shelter tropical rain forests. Not surprisingly, high-altitude populations of plants and animals are often unable to survive in tropical valleys, and organisms adapted to wet tropical forests may find cool mountain ranges and alpine deserts equally inhospitable. Thus these patches of similar habitat are home to local populations of organisms that only rarely interbreed with neighboring populations.

Occasional individuals, however, may cross such barriers. When they do, they may find new homes in previously vacant patches of suitable habitat on the other side of the ecologically hostile zone. These infrequent colonizers can establish populations that are fairly isolated from one another even if they are separated by relatively short distances. In other cases, highly mobile animals and seeds traverse major geographic barriers to establish populations in truly remote locations.

Effects of separation Geographic isolation of populations is often the first step in the evolutionary processes that eventually give rise to other isolating mechanisms. Why should this be the case? Once two populations are physically separated, gene flow between them stops. From then on, all factors that affect gene frequencies, such as genetic drift and natural selection, can operate on those populations independently. Especially when separated populations experience different environmental conditions, natural selection may favor changes in gene frequency between them over time.

Worlds Apart: Island Mountains and Inland Seas

When most people hear the word *island,* they envision islands of land in bodies of water. But in biological terms, an island is any habitat that is isolated in some way from other, similar habitats. Islands of all sorts are living laboratories for the study of evolution in isolation.

Towering 9000 ft above the steamy rain forests of Venezuela, the craggy, fog-shrouded plateau called Neblina is the image of a world lost in time. In fact Neblina, like other Venezuelan table-topped mountains, is one surviving piece of a giant plateau more than 1.5 billion years old. Over time, the soft sediments that made up most of the plateau eroded away, leaving isolated outcrops of harder rock, like Neblina, behind.

Environmental conditions at the mountain's cool, misty summit differ dramatically from those in the surrounding lowland jungle. The jungle, therefore, isolates Neblina's plants and animals from those of the other scattered plateaus as effectively as any ocean could. Over time, this ecological isolation has spawned scores of endemic insects and amphibians, peculiar giant earthworms, unusual carnivorous pitcher plants, and a whole genus of plants called *Neblinaria* whose members grow nowhere else.

Allopatric speciation does not necessarily require separation by major geological features. In East Africa's Lake Malawi, for example, hundreds of species of cichlid fishes have evolved from a few common ancestors within 30,000 to 40,000 years—a mere instant in geological time. Biologists who study these species believe that allopatric speciation has occurred repeatedly in at least one group, the Mbuna, or rock dwellers.

The shallow regions of Lake Malawi are dotted with rock piles separated by long stretches of sandy bottom. Why should that be important? Because the Mbuna really are rock dwellers; they hover above those rock piles and rarely swim out over the sand flats. Each rock pile thus serves as a tiny "island" whose isolated population has minimal genetic contact with other populations. Given the right combinations of chance migrations, the founder effect, genetic drift, and natural selection, these fish populations could have evolved as Darwin's finches did in the Galapagos archipelago.

This fish, discovered on Neblina, is believed to be a kind of catfish.

The English oak, *Quercus robur,* for example, easily hybridizes with either of the two American oak species mentioned earlier if they are planted together in a botanical garden. Under natural conditions, however, the Atlantic Ocean and North America separate them (Fig. 13.4). Though their ancestors may have been identical, these isolated populations have evolved independently. And although they have reached a point where they *look* sufficiently different to have convinced morphologists to classify them as different species, they have not yet developed intrinsic reproductive isolating mechanisms.

Sooner or later, however, separated populations such as these may diverge in such traits as color, habitat preference, breeding behavior, or breeding (or flowering) season that act as reproductive isolating mechanisms. Once that happens, the populations can no longer interbreed, even if the physical barrier between them is removed or if they jointly colonize a common area. Those populations have, in effect, become separate species.

There are numerous examples of allopatric animal populations that seem to be diverging into separate species. Often these populations form what is called a

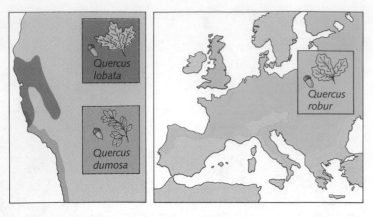

Figure 13.4 Distribution maps of three oak species, one native to Europe, the other two native to the West Coast of the United States.

"ring species" in which a series of slightly dissimilar populations are arranged in the shape of a ring. Adjacent populations within the ring—which can and do interbreed with their immediate neighbors on either side—are frequently assigned the status of subspecies. But typically, populations on either end of the ring do not interbreed—and are classified as distinct species (Fig. 13.5).

Many of the most spectacular cases of allopatric speciation have occurred on various kinds of islands: terrestrial islands in the sea, aquatic "islands" in lakes and streams, and mountainous islands in tropical lowlands (see Theory in Action, Worlds Apart: Island Mountains and Inland Seas, p. 263). There, in seclusion from the rest of the world, a few stranded travelers have evolved into **endemic species**—organisms found nowhere else on earth. Two of the best-studied cases of allopatric speciation in island faunas involve the honeycreepers of the Hawaiian archipelago and Darwin's finches (Fig. 13.6). Both cases are explained by similar logic; we will examine in detail here a plausible scenario for the finches' adaptive radiation in the Galapagos.

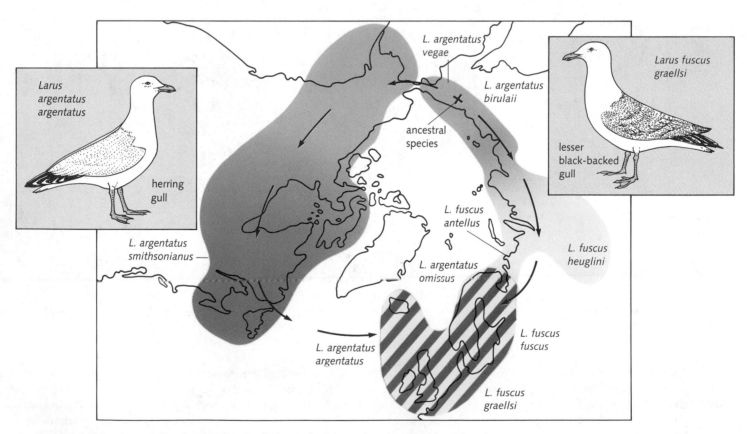

Figure 13.5 This series of closely related gull species rings the North Pole today. It appears that the ancestral species first evolved in Siberia and then spread around the northern continents in both directions. Genetic divergence among those populations has resulted in differences among adjacent populations. At opposite ends of this "ring," those differences are enough to have produced two populations that interbreed so rarely that they are considered distinct species. [From Darwin, *Illustrated Origin of Species* (1979).]

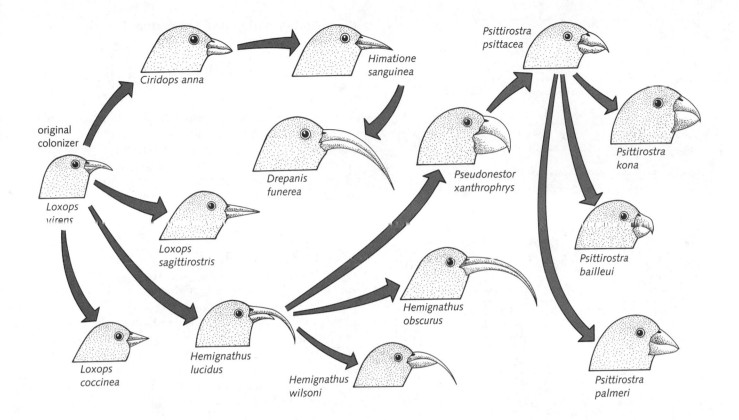

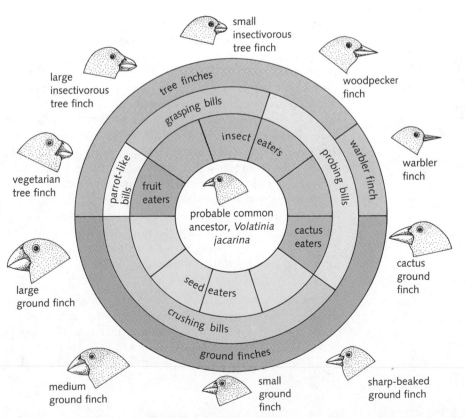

Figure 13.6 TOP: Hawaiian honeycreepers form a remarkable adaptive radiation. Founded by a small number of ancestral birds that wandered far from land, these species evolved to fill the many vacant niches for birds on the islands. BOTTOM: In the process, two groups, specialized to feed on seeds and insects, respectively, converged in beak size and shape with several ecologically similar Galapagos finches.

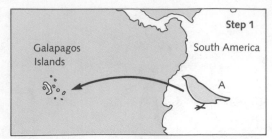

Small mainland population (A) is blown to archipelago. Individuals colonize adjacent islands.

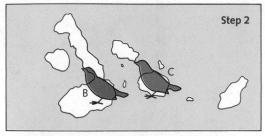

Two populations on different islands are isolated by distance. As they adapt to their different environments, they become behaviorally isolated from one another, and become distinct species, (B) and (C).

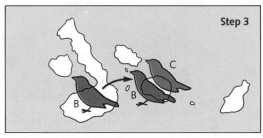

Following reproductive isolation, a small population of species (B) migrates to the island of species (C). The two species enter into ecological competition with each other.

Directional selection driven by competition leads to further change in (B). This population becomes reproductively isolated from both species (C) and its parent population (B), and becomes a new species (D).

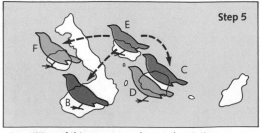

Repetition of this process and several variations of colonization, divergence, and competition can create many species on an archipelago.

A Case Study in Allopatric Speciation: Galapagos Finches

The Galapagos Islands arose, devoid of life, in the Pacific Ocean between 3 and 5 million years ago. More than 1000 km west of Ecuador, these islands have never been connected either to the mainland or to each other. Somehow, more than a dozen of the finch species we met in the last chapter have evolved on this archipelago from a small founding population that strayed from the mainland (Fig. 13.7). Why and how could that single population have given rise to so many distinctively different kinds of birds? Let us reconstruct a plausible scenario.

Colonization   Assume that a small finch population (it could even have been a single impregnated female) was blown to the archipelago in a storm (Fig. 13.7, *step 1*). Once a breeding population was established on the island, wandering individuals could occasionally colonize adjacent islands. But finches seldom stray far from their nesting sites and do not like flying over open water. Because the islands are separated by enough water to make island hopping uncomfortable, their finch populations are isolated not only from the mainland but also from each other.

Genetic divergence in allopatry   Ecological conditions across the archipelago vary significantly, so the various islands offer different kinds and quantities of food. Under those conditions, small, reproductively isolated bird populations would be expected to evolve different feeding characteristics—a supposition confirmed by studies published in the early 1980s. Species on different islands have evolved beak sizes and feeding behaviors that enable them to exploit efficiently the seeds, fruits, and insects present on their particular islands. Knowing for certain whether that divergence was originally or primarily driven by natural selection or whether it occurred through genetic drift is not critical for us here. Recall from Chapter 12, however, that natural selection, particularly during droughts, has been shown to cause measurable changes in heritable characteristics related to feeding.

Reproductive isolation   At some point, the gene pools of these isolated populations diverged from one another enough so that their members would no longer interbreed, even if they lived in the same place. At that point, the divergent island forms could legitimately be called different species (*step 2*).

Figure 13.7   Hypothetical mechanism for the evolution of Galapagos finch species through repeated episodes of allopatric speciation.

How might these populations have been reproducing if isolated from one another? Recent studies have shown that, except in rare cases, finches reliably choose mates of their own species by using a combination of physical cues (such as beak size) and behavioral cues (such as song patterns learned from their parents). Thus, as the isolated island populations evolved different-sized beaks adapted to feeding on different foods, and as they diverged (possibly at random) in their song patterns, they became increasingly behaviorally isolated from each other.

Competition and further divergence   After behavioral reproductive isolation occurred, birds from one island that occasionally colonized adjacent islands were still reproductively isolated from the local population (*step 3*). Under these circumstances, the two species might have entered into competition with one another for food. Such competition could favor the survival of those individuals in each species that were most different in feeding habits from their competitors (*step 4*). This kind of divergent evolution should push the competitors ecologically farther apart from one another.

This part of the hypothesis sounds satisfying, but are there any quantitative data to back it up? Indeed there are, because several finch species occur on several islands, and because each island has its own assortment of species. It is thus possible to compare, for example, beak sizes between two populations of the same species of seed-eating finch, one of which occurs with a similar competing species and the other of which has no competitors. In several such cases, beak measurements show clearly that these birds differ more when the species occur together than when they occur separately (Fig. 13.8). It is thus reasonable to assume that competition has, in fact, caused directional selection.

Further speciation   Theoretically, divergent evolution, either driven by this sort of competition or allowed by genetic drift, could cause sufficient changes in beak size to lead to the formation of yet another species (*step 5*). Repeated over time and over the entire archipelago, this process could have produced the 13 living species and the several extinct forms that we know once existed in the Galapagos Islands.

Parapatric Speciation: Life on the Fringe

In **parapatric speciation** (*para* means "next to"; hence *parapatric* means "next to the native land"), a small population on the fringe of a larger population diverges, often despite the absence of a major geographic boundary and despite the existence of some gene flow between it and the main group.

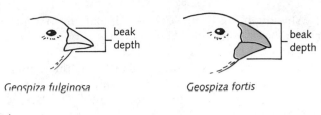

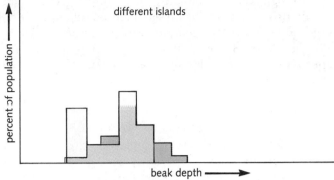

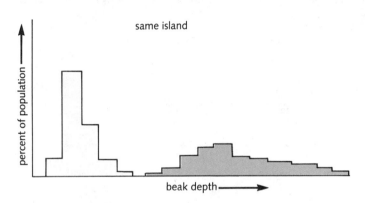

Figure 13.8   Quantitative evidence for the effects of interspecific competition on beak sizes in Galapagos ground finches. When populations of *Geospiza fortis* and *G. fuliginosa* occur on separate islands, their range of beak sizes is fairly similar. When populations of those same species occur together on the same island, however, their average beak sizes shift markedly in a manner that reduces the overlap (and hence the competition) between them. [From Futuyma (1986) after Lack (1947).]

Why might this occur? Fringe populations of a species are regularly subjected to different and often more stressful conditions, and they may be exposed to different food and predator species than populations in the center of the species' range. Under these circumstances, and if interbreeding with individuals from other populations is rare enough, the combination of natural selection and genetic drift might cause these fringe populations to diverge sufficiently to become distinct species. Because both natural selection and genetic drift operate more rapidly in small populations than in large ones, some biologists believe that small populations on the fringe of large populations' ranges are the primary sites of macroevolutionary changes.

Researchers disagree about whether or not several variant populations of organisms recorded in nature represent parapatric speciation in progress. In several industrial areas, for example, heavy metals that are normally toxic to plants have entered the soil. In some locations, populations of grass species have slowly evolved tolerance for those toxins and are thus able to grow in patches of habitat where tainted soil excludes other members of their species. Some of those populations appear to be diverging from wild-type populations around them in both flowering time and gamete compatibility. This partial reproductive isolation may be an example of parapatric speciation in progress. The pair of American oak species mentioned earlier may represent a similar situation, in which nearly complete reproductive isolation has evolved.

Sympatric Speciation: Alone in a Crowd

Under certain conditions, it is possible for new species to arise in the midst of their parent populations. Such speciation, which occurs "together in the native land," is called **sympatric speciation.**

There is convincing evidence that sympatric speciation has occurred repeatedly, at least among plants. It is not uncommon for individual plants to appear with double the normal amount of genetic material, either through errors in cell division in plant buds or through errors in the production of pollen and eggs. These tetraploid individuals can perpetuate themselves only by breeding with one another, for they are instantaneously reproductively isolated from other members of what had been their species. (Although such genetically doubled individuals may be able to cross with their parents and produce viable offspring, those offspring are usually sterile and are thus an evolutionary dead end.)

Many horticulturally and agriculturally important plant species, including several strains of wheat, have more than the normal amount of genetic material and may well have arisen sympatrically through this sort of genetic event. Some botanists estimate that more than half of today's flowering plant species may have arisen as the result of chromosome doubling.

In addition, several laboratory experiments have shown that strong selection can lead to reproductive isolation within controlled populations. For example, disruptive selection for high and low bristle numbers in *Drosophila* (Chapter 12) unexpectedly created subpopulations of few-bristled and many-bristled flies that preferred to mate with others of their own kind. Similarly, one researcher working with corn planted two genetically marked strains together, but planted, in each generation, only seed from plants that carried the smallest number of hybrid kernels. In just a few generations, the rate of fertile crosses between the strains dropped from 40 percent to less than 5 percent (Fig. 13.9).

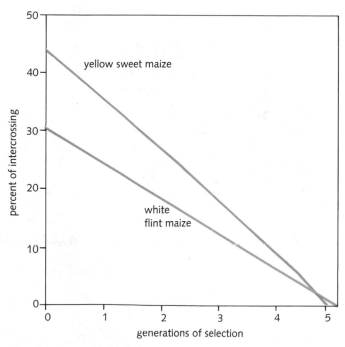

Figure 13.9 When two genetically marked strains of corn are planted together, cross fertilization invariably occurs. After several generations of artificial selection against hybrids (selecting for plants less likely to interbreed) the amount of intercrossing drops markedly, as shown here. Experiments such as this one support the contention that, under certain conditions, reproductive isolation can arise within sympatric populations. [From Futuyma (1986).]

Why Shouldn't Species Interbreed?

Why is reproductive isolation so important? Why, in other words, is it evolutionarily worthwhile for species to invest time and energy in complex, species-specific breeding behaviors and other prezygotic isolating mechanisms to prevent hybridization? There is no simple general answer to these questions, but in certain cases, at least, interspecific hybridization can result in lowered fitness.

In southern Germany, for example, the ranges of two species of European house mice (*Mus musculus* and *M. domesticus*) overlap in a region about 20 miles wide. That limited area harbors mice rarely seen elsewhere; they are hybrids between the two dominant species. While studying these animals,

Richard Sage of the University of California noticed that the hybrid mice had more fleas than either parent species. Further investigation showed that the hybrids also had many more parasitic worms and more different kinds of worm parasites than either parent. Although these infestations are not fatal, they can damage the host's internal organs and increase its susceptibility to other diseases.

There is evidence that resistance to parasites has a genetic basis and can therefore be affected by natural selection. The two mouse species in question seem to have specific alleles that control resistance to the bacterium *Salmonella*. Furthermore, resistance to parasitic worms in these species acts like a

simple, dominant trait; it can be passed from generation to generation in the sorts of ratios Mendel found in his peas.

It is not clear precisely how interspecific hybridization decreases resistance to parasites in these species. But whatever the genetic cause, its effect is to lower the fitness of the hybrids. For that reason, natural selection would favor any isolating mechanism that reduces the likelihood of a mismatch.

Models of sympatric speciation in natural populations, however, are highly controversial. Although several theories have been proposed to explain how reproductive isolation could arise gradually within a natural population of animals, none has yet been conclusively proved. And although some biologists believe that sympatric speciation has been clearly demonstrated in the case of several insects, others insist that those cases are actually examples of parapatric speciation.

CURRENT DEBATE ON EVOLUTIONARY THEORY

Recent advances in genetics, molecular biology, and paleontology have called into question certain assumptions about evolutionary change. Is natural selection the *only* force behind genetic change? Are all evolutionary changes adaptive? Does evolution really have to progress

slowly and gradually? Spirited debate on these issues continues today.

In an honest attempt to interpret these debates for the public, the popular science press has featured articles implying that the very existence of evolutionary change is in serious doubt. Nothing could be further from the truth. Certain aspects of Darwinian theory *have* come under scrutiny. But as Stephen Jay Gould eloquently stated, "Facts don't disappear while scientists debate theories.... Einstein's theory of gravitation replaced Newton's, but apples did not suspend themselves in mid-air pending the outcome." Similarly, the fact of evolutionary change over time remains, regardless of whether scientists agree on the mechanisms of *how* evolution proceeds.

Yet these debates are informative, both for what they tell us about evolution and for what they have to say about the process of science. To give you some insight into contemporary debates on evolutionary theory, we will recall two basic pillars of Darwinian thought: the nature of species and the pace of evolutionary change.

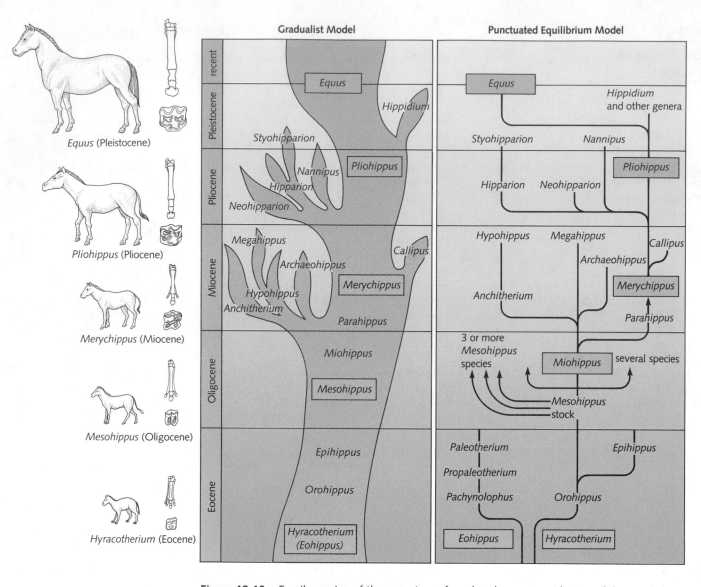

Figure 13.10 Fossil remains of the ancestors of modern horses provide one of the most nearly complete records of evolutionary change for any lineage of mammals. LEFT: Evolution and speciation in equine lineages. Several trends are obvious: increase in size, loss of side toes to form a single, central hoof, and changes in tooth structure. This figure illustrates the older idea of evolution in animal lineages—that a single, dominant species gradually changes in response to natural selection. But where in this constantly changing series of ancestors does one species end and another begin? The definitions are necessarily arbitrary.

RIGHT: The evolution of the modern horse as reconstructed according to the modern, expanded fossil record and arranged to more closely reflect the ideas of punctuated equilibrium theory. This theory relies on the same fossil history as the gradualist model, with two critical differences. First, we now know of scores of fossil horses, only a few of which are shown here. These species represent a highly branched evolutionary "bush" rather than a tree with a straight central trunk. Modern horses are merely one twig off a side branch of this bush that happened to survive. Second, this view attributes most evolutionary change to brief periods of speciation, rather than to long periods of gradual evolution.

The Darwinian View of Species

Recall that Darwin abandoned the idea of species as fixed entities because in his day "fixed" meant forever, and immutable ideal types left no opportunity for change over time. In what may have been overreaction, Darwin went to the other extreme, essentially denying species any real existence over time. As paleontologist Niles Eldredge wrote, Darwin came to view species as nothing more than "momentary collections of similar and interbreeding organisms that looked somewhat different in the not-too-distant past, and are destined to become modified in the not-too-distant geological future."

Species and Gradual Change

Following in Darwin's footsteps, the current mainstream of evolutionary thought stresses the importance of slow, continuous transformation of one species into another. This view, known as **gradualism,** holds that natural selection, genetic drift, and other processes lead to the accumulation of small (microevolutionary) changes within species. Large-scale (macroevolutionary) events result from the accumulation of these smaller changes over time.

This view, though easily explained, causes some problems with definitions of species over time. Anyone who studies mammals, for example, can give you a precise definition of the species we call the modern horse. But that clearly defined entity loses its identifiable characteristics when we try to follow it back over geological time. Remember, too, that Darwin's often unsuccessful search for the intermediate fossils this sort of change should produce led him to criticize the fossil record as grossly incomplete.

Punctuated Equilibrium

In 1972, Niles Eldredge and Stephen Jay Gould challenged the doctrine of gradualism, adjusting their view of evolutionary change to accommodate their greater respect for the physical record in the rocks. They say the fossil record shows that most species remain "virtually unchanged" from the time they appear to the time they disappear. They also argue that new species, when they do arise, appear more quickly than Darwinian gradualism allows.

Gould and Eldredge propose that evolution proceeds in a manner they call **punctuated equilibrium.** According to this theory, very little microevolutionary change occurs over a species' lifetime, because each species is a stable, relatively unchanging entity. This stability, or "equilibrium," is periodically broken, or "punctuated," by brief periods of change during which new species arise from preexisting ones. The new species evolve for a time but then, like their predecessors, remain intact for the rest of their geological lives, perhaps as a result of stabilizing selection (Fig. 13.10).

There are some extreme examples of species that have remained in equilibrium since they first appeared in the fossil record, despite the fact that, around them, other species have evolved considerably while countless more have suffered extinction. A few living species, such as the coelacanth, closely resemble fossils over 250 million years old. And the living tadpole shrimp, *Triops cancriformis,* is so close in morphology to a well-preserved series of 180-million-year-old fossils that both living and fossil animals are placed in the same species. Steven Stanley, another participant in the punctuated equilibrium debate, suggests that this small shrimp is the oldest known living animal species (Fig. 13.11).

Figure 13.11 Visible characteristics of this small shrimp species have remained unchanged since it first appeared in the fossil record over 180 million years ago. [From Kaestner (1970).]

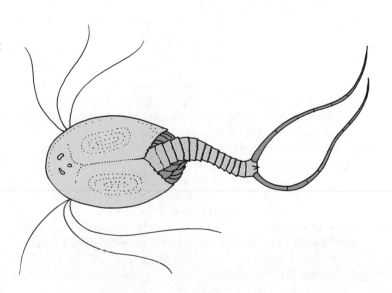

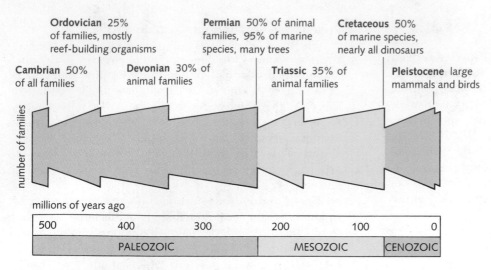

Figure 13.12 Periodic mass extinctions, as you will learn in later chapters, have been important episodes in the evolution of life on Earth.

Several mechanisms have been proposed to explain punctuated equilibrium. Some researchers point out that changes in the gene pool of large populations tend to occur relatively slowly. In small populations, on the other hand, such changes can accumulate much faster. For this reason, parapatric speciation events could occur rapidly enough to produce changes during periods very much shorter than the lifetime of the more stable, parent species. In addition, sympatric speciation (in plants) through polyploidy can produce new species literally in an instant. And the colonization of new habitats (such as the Galapagos Islands) by either animals or plants can result in rapid adaptive radiations.

There have also been several times in the earth's history when large numbers of species have disappeared during what are called **mass extinction events** (Fig. 13.12). During such episodes, global ecosystems have been disrupted and many previously filled ecological niches have been opened up. The best-known mass extinction occurred at the end of the Cretaceous period, eliminating the ruling dinosaurs and paving the way for the earliest mammals (Chapter 25). Although such mass extinctions are not necessary for rapid speciation to occur, it is clear that they have created several major "punctuation marks" in the history of life on Earth.

But more significant than these views on rates of change is the new vitality this theory infuses into the concept of species. For in this view of evolution, species are not constantly shifting biological will-o'-the-wisps but distinct individuals—from their birth at speciation to their death at extinction. That critical change in perspective opens up an entirely new series of questions about the

ways in which natural selection operates. If punctuated equilibrium continues to gain acceptance and comes to dominate evolutionary thinking, it may set the stage for yet another period in the evolution of evolutionary thought.

SUMMARY

Speciation—often called macroevolution—is the process by which populations of a single species diverge evolutionarily to produce more species. The first step in speciation is usually the creation of at least a partial barrier to gene flow between two populations of an existing species. Speciation is complete when the two populations are reproductively isolated by any of several reproductive isolating mechanisms: mechanical, behavioral, temporal, or ecological isolation; gamete incompatibility; hybrid inviability; or hybrid infertility.

Allopatric speciation, the most widely accepted mechanism for speciation among animals, presumes that an initial barrier to gene flow occurs because of a physical barrier between populations. This isolation may be caused by major geographic features or through the colonization of patches of habitat separated by environments hostile to the organism in question. Parapatric speciation involves the isolation of small populations at the fringes of a species' range. Subsequent genetic divergence in both cases occurs because of both natural selection and genetic drift.

Sympatric speciation, the probable mechanism for nearly half the speciation events among flowering plants, occurs principally as a result of genetic changes that reproductively isolate certain individuals from surrounding members of their parent population.

Classical Darwinian theory holds that speciation occurs gradually, as small changes accumulate over long periods of time. The theory of punctuated equilibrium, on the other hand, holds that most species remain relatively unchanged for most of their existence and that speciation occurs during relatively brief periods.

STUDY FOCUS

After studying this chapter, you should be able to:

- Discuss modern theories on the evolution of new species.
- Define the term *reproductive isolation*, and explain how populations of a species may be reproductively isolated from one another.
- Explain the theories of speciation.
- Outline the modern evolutionary theory of punctuated equilibrium.

TERMS AND CONCEPTS

speciation *259*
reproductive isolating
   mechanisms *259*
prezygotic isolating
   mechanisms *260*
mechanical isolation *260*
behavioral isolation *260*
temporal isolation *260*
gamete incompatibility *261*
postzygotic isolating
   mechanisms *261*

hybrid inviability *261*
hybrid infertility *261*
ecological isolation *261*
allopatric speciation *262*
endemic species *264*
parapatric speciation *267*
sympatric speciation *268*
gradualism *271*
punctuated
   equilibrium *271*
mass extinction events *272*

REVIEW

Objective Questions (Answers in Appendix)

1. A postzygotic barrier prevents
   (a) the union of sperm and egg.
   (b) the development of the embryo.
   (c) fertilization.
   (d) difficulties during meiosis.

2. The theory of punctuated equilibrium
   (a) can account for sudden appearance and disappearance of species.
   (b) explains that evolution occurs during periods of slow, gradual change.
   (c) assumes that speciation is an ongoing process.
   (d) cannot be substantiated using fossil records.

3. Speciation cannot occur unless there is _____ isolation.
   (a) prezygotic   (c) divergent
   (b) genetic   (d) convergent

4. Sympatric speciation
   (a) most likely involves sudden genetic changes.
   (b) most likely involves gradual genetic changes.
   (c) is dependent on allopatric speciation.
   (d) requires geographical barriers.

5. Hybrids, if they live, are
   (a) always sterile.   (c) usually sterile.
   (b) never sterile.   (d) unusually fertile.

Discussion Questions

6. Why is reproductive isolation so important in the process of speciation?

7. What are two ways in which the behaviors of animals can result in reproductive isolation?

8. What is the difference between hybrid inviability and hybrid infertility?

9. How does the punctuated equilibrium theory challenge Darwin's view of the fossil record?

10. How can geographical or ecological isolation eventually lead to the evolution of isolating mechanisms that keep new species separate even when the external barriers are removed?

READINGS

Grant, P. R. "Natural selection and Darwin's finches." *Scientific American* (October 1991): 82–87.

Futuyma, D. J. *Evolutionary Biology.* 2d ed. Sunderland, MA: Sinauer Associates, 1986.

Sage, R. D., et al. A letter in *Nature* (November 6, 1986): 324.

Darwin, C. *Illustrated Origin of Species.* Abridged and introduced by Richard E. Leakey. New York: Hill and Wang, 1979.

Kaestner, A. *Invertebrate Zoology.* Volume III. Translated by H. Levi and C. R. Levi. New York: Wiley, 1970.

Lack, D. *Darwin's Finches.* New York: Cambridge University Press, 1947.

PART 4

Molecules of Life

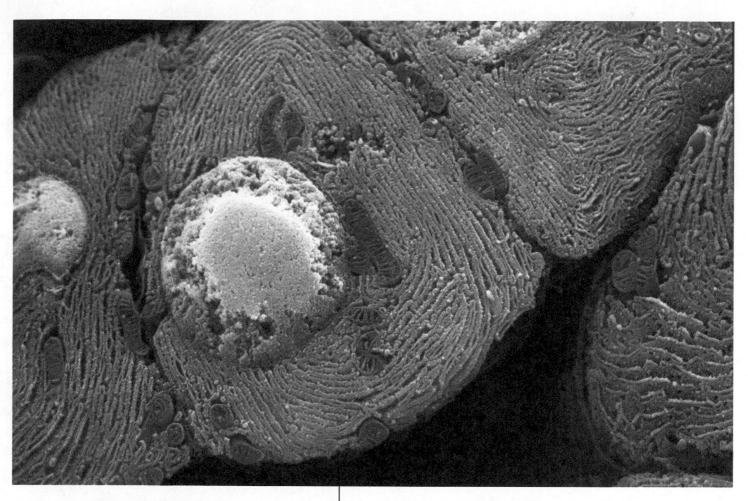

A scanning electron micrograph reveals the internal structure of human cells.

*h*ermann J. Muller, a pioneer in genetics, was probably kidding when he wrote about physically "grinding" and "cooking" genes in the laboratory. To many of his colleagues, the possibility that geneticists might be able to manipulate genes chemically must have seemed impossibly optimistic. It was a wild idea.

Despite the scientific advances of the early twentieth century, there was no apparent way to link the physical world with the living one. Chemists had discovered that elements found in living organisms were similar to those found in nonliving matter. A few chemists had even succeeded in producing molecules in the test tube that previously had only been found in living cells—a sure sign that there was no fundamental chemical barrier between life and nonlife. Nonetheless, an explanation of the material basis of life seemed almost beyond reach. The most ordinary properties of living things, including the abilities to grow and reproduce, could not be explained in chemical terms. This inability led many to conclude that there was something mystical about living things. The idea was particularly attractive in studies of inheritance, where it was argued that no mere molecule could carry genetic information from one generation to the next.

It was Muller himself who helped to convince biologists that it was possible to think of the gene as a molecule. Muller irradiated fruit flies with X-rays and discovered that the number of mutations found in the flies was directly proportional to the X-ray dosage. Because X-rays cause atomic changes, Muller guessed that the gene must have a molecular basis.

As we now know, Muller was correct. Little by little, biologists have done what he imagined: unravel the physical and chemical basis for inheritance, to develop a new understanding of the material nature of living things.

Chapters 14 and 15 present the basic chemistry and biological chemistry that underlie the functioning of cells. Chapter 16 explains how cells are constructed and organized from macromolecular components. Chapter 17 looks at cellular energy needs and at how evolutionarily successful cells have improved the efficiency of chemical reactions involving energy through the use of enzymes. Chapter 18 discusses how the compartmentalized organization of cells and

. . . we cannot categorically deny that perhaps we may be able to grind genes in a mortar and cook them in a beaker after all. Must we geneticists become bacteriologists, physical chemists and physiologists simultaneously with being zoologists and botanists? Let us hope so.
— H. J. Muller, 1922

the use of enzymes in chemical pathways enables cells to obtain needed energy from fuels such as sugars. Chapter 19 looks at the opposite process by examining how plant cells produce sugar through another set of chemical reactions that harness the energy of sunlight. The last three chapters bring readers up to date on genetics and evolution by discussing the molecular nature of each. Chapter 20 explains the molecular nature of genes, while Chapter 21 explores the techniques that "grind" and "cook" genes: a process called genetic engineering. Finally Chapter 22 shows how these techniques have been used to improve our molecular understanding of evolution as it has occurred in nature.

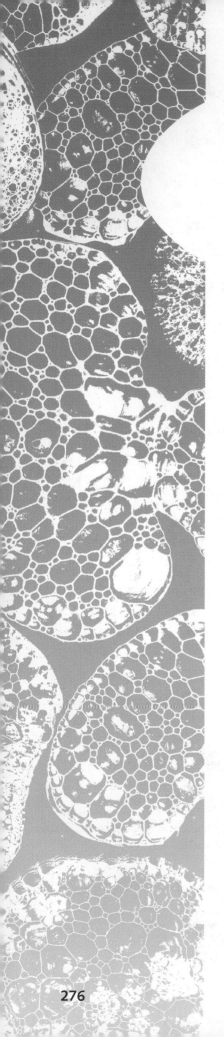

14

The Molecules of Life

On a cold winter night, the snowy landscape is quiet, even desolate, and sound carries easily through the barren woods. Most trees have lost their leaves, the birds have flown south, and insects have all but disappeared. At times it may seem that life itself has vanished in the cold weather. Despite the cold, we all know that things will change in the spring. The snow will melt, and the songbirds will return. We know that living things are subject to the seasons. But *why* should seasonal changes affect living things so profoundly? Why should the cold of winter cause life to all but disappear from the forest? And why should so many forms of life reappear with the warm winds of spring? The answers to these questions hold the key to one of the most fundamental principles of biology: *life is chemical.*

The changes we see in the forest during the winter are a reflection of the fact that living organisms, plants, and animals contain water, and when water becomes solid, living organisms must adapt to the change. Living things and nonliving things are made of the very same types of atoms, and living matter is composed of molecules that are subject to the same laws of physics and chemistry that govern nonliving matter. Life really *is* chemical. In this chapter, we will begin to investigate the material organization of life. We will look closely at some of the ways in which the "ordinary" atoms of nonliving things have been put together to produce the "extra-ordinary" phenomenon called *life* (Fig. 14.1).

LIFE IS CHEMICAL

Just as buildings are made from bricks, steel, glass, and wood, living things are made from chemical compounds. If it's fair to say that the first task of an architect is to understand steel and stone and glass, then it's also fair to say that the first task of a biologist is to understand the chemistry upon which life is built. It goes without saying that naturalists could not appreciate the cycles of carbon, nitrogen, and oxygen that exist on the earth until those elements were discovered, and that cell biologists could not understand the nature of cellular membranes until the basic chemistry of *lipids,* which form the backbones of these membranes, was understood.

Figure 14.1 This medieval engraving portrays the laughable chaos of an alchemist's shop. Alchemists took some of the first steps along the road that led to an understanding of the chemical nature of living things.

Let's reconsider a bit of the biology that we have just covered. We have seen how the laws of genetics provide a systematic basis for passing distinct traits from one generation to the next. These laws also form the basis for the genetic change that makes evolution possible. One of the ultimate goals of biology is to understand complex processes, including genetics, in chemical terms. By understanding the details of how ordinary matter can be organized into the components of a living organism, we can approach a much deeper understanding of life itself. What we call modern biology has developed because biologists have sought chemical explanations for biological events.

Atoms and Molecules

The basic unit of chemical structure is the **atom.** Atoms themselves are composed of three different kinds of **subatomic particles: electrons, protons,** and **neutrons.** Each of these subatomic particles has its own set of physical characteristics. Electrons have very little mass and are negatively charged. Protons and neutrons are much heavier (1836 and 1839 times the mass of an electron, respectively). The proton carries a positive charge and the neutron is neutral.

Protons and neutrons form the **nucleus,** where most of an atom's mass is concentrated. The simplest atom, hydrogen, has a single proton for its nucleus. All other atoms have at least one neutron in their nuclei.

At one time, physicists thought that the negatively charged electrons moved around the nucleus in circular orbits, much like the planets of our solar system move around the sun. Sometimes the hydrogen atom is represented in this way (Fig. 14.2a).

More complicated atoms—helium, for example— have more than one electron surrounding their nuclei and have several protons and neutrons clustered together to form their atomic nuclei (Fig. 14.2b).

It's now clear that this is an oversimplification. Electrons don't move in circular orbits. Instead, they follow patterns that are determined by the amount of energy they possess and actually represent the *probability* of finding electrons in particular positions. These patterns are known as **orbitals.**

The orbitals occupied by electrons are not randomly arranged around the atomic nucleus. They are organized into a distinct series of energy levels, or shells, that are limited in terms of how many electrons they contain. Electrons at the lowest energy levels are found in orbitals closest to the nucleus.

The lowest energy level can hold no more than 2 electrons. The second shell can hold as many as 8 electrons, and the third shell can hold 18. Tracking the locations of electrons in these shells can be a tricky business. As Fig. 14.2(c) shows, the probability distribution for the first shell, which holds 2 electrons, is a simple sphere.

But that's where simplicity ends. There are 4 possible orbitals in the second shell, each of which can hold up to 2 electrons, for a total of 8. These 4 orbitals (Fig. 14.3) include a spherical *s* orbital and 3 dumbbell-shaped *p* orbitals. Superimposed, the 4 orbitals of the second shell make for a complex picture. The third shell, which has 9 different orbitals, is even more complex.

Atoms are electrically neutral, meaning that their charges balance out. The number of positively charged

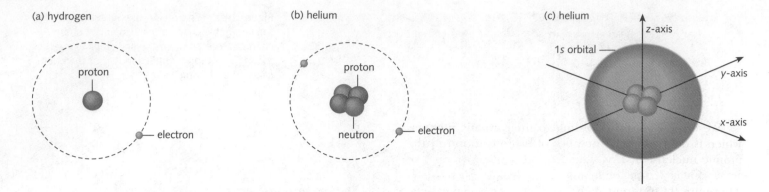

(a) hydrogen

proton

electron

(b) helium

proton

neutron · electron

(c) helium

z-axis

1s orbital

y-axis

x-axis

Figure 14.2 **(a)** This drawing portrays hydrogen, the simplest atom, which consists of a single proton and a single electron. **(b)** Larger atoms, such as the helium atom shown here, have nuclei that contain protons and neutrons. **(c)** Any attempt to draw an atom must be inexact, because the precise locations of electrons cannot be fixed at any one time. This drawing of helium shows a "probability shell" for the orbital containing helium's two electrons. The spherical envelope shows the space in which the electrons are found 90 percent of the time.

(a)

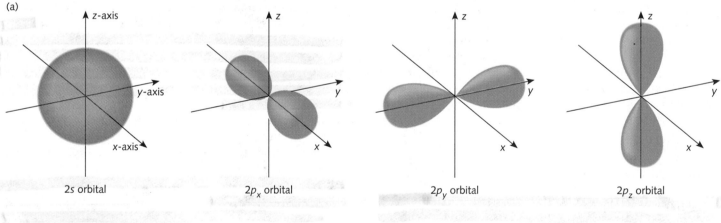

z-axis

y-axis

x-axis

2s orbital

z

y

x

$2p_x$ orbital

z

y

x

$2p_y$ orbital

z

y

x

$2p_z$ orbital

(b)

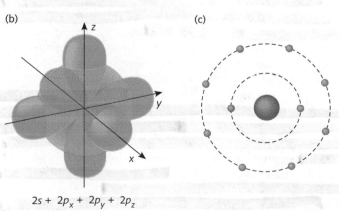

z

y

x

$2s + 2p_x + 2p_y + 2p_z$

(c)

Figure 14.3 Orbitals in the second energy level. **(a)** There are four orbitals in the second shell, each of which may be occupied by up to two electrons. **(b)** A combined drawing of the three p orbitals and the single s orbital shows how complex the distribution of electrons in the second shell is. **(c)** Too much to think about? Leave it for the chemists and remember that the first shell holds up to two electrons and the second shell can hold as many as eight electrons.

protons in the nucleus is equal to the number of negatively charged electrons found in the orbitals. Equality in charge doesn't mean equality in mass, however. The protons and neutrons found in the nucleus are far more massive than the electrons that surround it—more than 99 percent of the mass of an atom is found within its nucleus.

We classify atoms according to their **atomic number,** which is defined as the number of protons found in the atomic nucleus. By this standard, the atomic number of hydrogen is 1, that of helium 2, and that of carbon 6. There are 92 different naturally occurring **chemical elements,** and each of these chemical elements is composed of a different type of atom with its own atomic number. The most important elements in living things (hydrogen, oxygen, nitrogen, and carbon) are very common at the surface of the earth; each is found in the gases of the atmosphere. The heaviest naturally occurring element is uranium, which has an atomic number of 92. Heavier elements with larger atomic numbers can be created experi-

mentally by bombarding other elements with high-energy radiation.

Only about a third of the naturally occurring chemical elements are found in living organisms. Table 14.1 lists the 26 elements that are most commonly found in living things and also indicates their relative abundance in the earth's crust.

Chemical and Physical Properties

The **physical properties** of a material are the characteristics we can see or measure, and they include the color, density, physical state (solid, liquid, or gas), and melting and boiling points of a substance. Each of the 92 chemical elements has different physical properties. Some, such as gold, zinc, and sodium, are shiny solids. Some are gases, such as hydrogen, chlorine, and oxygen; and some, including bromine and mercury, are liquids at room temperature.

Table 14.1 *Elements Important to Life*

Symbol	Element	Atomic Number	Percentage of Earth's Crust by Weight	Percentage of Human Body by Weight
H	Hydrogen	1	0.14	9.5
B	Boron	5	Trace	Trace
C	Carbon	6	0.03	18.5
N	Nitrogen	7	Trace	3.3
O	Oxygen	8	46.6	65.0
F	Fluorine	9	0.07	Trace
Na	Sodium	11	2.8	0.2
Mg	Magnesium	12	2.1	0.1
Si	Silicon	14	27.7	Trace
P	Phosphorus	15	0.07	1.0
S	Sulfur	16	0.03	0.3
Cl	Chlorine	17	0.01	0.2
K	Potassium	19	2.6	0.4
Ca	Calcium	20	3.6	1.5
V	Vanadium	23	0.01	Trace
Cr	Chromium	24	0.01	Trace
Mn	Manganese	25	0.1	Trace
Fe	Iron	26	5.0	Trace
Co	Cobalt	27	Trace	Trace
Ni	Nickel	28	Trace	Trace
Cu	Copper	29	0.01	Trace
Zn	Zinc	30	Trace	Trace
Se	Selenium	34	Trace	Trace
Mo	Molybdenum	42	Trace	Trace
Sn	Tin	50	Trace	Trace
I	Iodine	53	Trace	Trace

Although the physical properties of the different elements can vary considerably, their **chemical properties**—the ways in which atoms combine with other atoms to form chemical compounds—show certain important similarities among groups of elements. Fluorine and chlorine, both highly reactive gases, are examples of elements with such similarities. These similarities were noticed in the last century by Mendeleev, a Russian chemist. Mendeleev thought that it would be useful to group the elements in a table according to their chemical properties and atomic weight. Today the product of his idea is a chart known as the **periodic table** (see Appendix). Mendeleev's efforts were successful because the chemical properties of an element are determined mainly by the number of electrons in its outer atomic orbital. Two different elements have similar chemical properties if they have the same number of electrons in their outermost orbitals.

Figure 14.4 shows how the periodic table predicts chemical similarities. As the number of electrons in an atom increases, orbitals are occupied by electrons in a regular pattern: The orbital closest to the nucleus can hold only two electrons; the second shell, which has four orbitals, can hold as many as eight; and the third shell may also hold eight. The reasons for similarities between fluorine (F) and chlorine (Cl) become clear when we ex-amine the arrangements of electrons in their atoms: each needs but a single electron to complete its outermost orbital.

Atomic Numbers and Atomic Weights

We have defined the atomic number of an element as the number of protons found in its nucleus. We can also speak of **atomic weight** (more properly, **atomic mass**) as the sum of the weights of the elementary particles in a single atom. The unit of measurement that is used for this purpose is the *atomic mass unit* (amu), which is defined as $1/12$ the atomic mass of the carbon atom. Therefore, ordinary carbon has an atomic mass of 12.0, and hydrogen's atomic mass is 1.0078. For our purposes the proton and neutron have masses that are nearly equal to 1.0, but the electron is so small that its mass ($1/1839$ the mass of the neutron) is often neglected.

Atomic weights are important because they reflect the relationship between numbers of atoms (atoms are difficult to count!) and mass (which we can measure easily with a scale or balance). If we know that the atomic mass of oxygen is 16, we can place 2 grams of hydrogen and 16 grams of oxygen in a reaction vessel and be certain that we have exactly 2 atoms of hydrogen for every atom of oxygen.

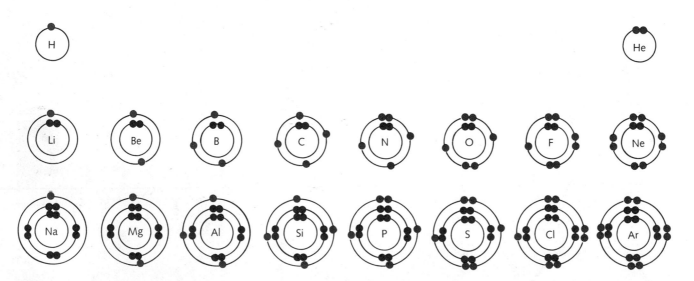

Figure 14.4 Orbital diagrams of the 18 lightest elements show the pattern in which electrons are arranged. The first orbital is complete with just two electrons, while the second may hold up to eight. The arrangement of elements shown here, which is the basis of the periodic table, places elements that have the same number of electrons in their outer shells into the same column. Because chemical properties are determined by the number of electrons in the outermost shell, elements in the same column should have similar chemical properties. Note that helium, neon, and argon have completed outer electron shells, which make them chemically inert, generally unable to form chemical bonds with other atoms.

Isotopes

The identity of a particular atom is determined by the number of protons in its atomic nucleus. If the nucleus contains 6 protons, the atom is carbon; 7 protons, nitrogen; 8 protons, oxygen. The number of neutrons, however, can vary. For example, the most common form of carbon has a nucleus of 6 protons and 6 neutrons: a total of 12 elementary particles. Therefore, we refer to it as carbon-12 (in atomic shorthand: ^{12}C). However, there are other forms of carbon as well. One of the most important, ^{14}C, has 8 neutrons in its nucleus. Carbon-14 is chemically identical to carbon-12, despite the fact that it has a different atomic nucleus, because it has the same number of electrons. Both ^{12}C and ^{14}C are **isotopes** of carbon.

Atomic isotopes are nearly identical in chemical behavior because they have the same configuration of outer electron shells. It is the arrangement of electrons that determines chemical properties. However, isotopes differ from each other in one and sometimes two respects:

- Because their nuclei have different numbers of neutrons, they have different atomic weights. For example, ^{12}C has an atomic weight of 12 (actually 12.0113), whereas ^{14}C has an atomic weight of 14 (actually 14.1023). Therefore, molecules that contain ^{14}C are a bit heavier than ones that contain only ^{12}C.

- Some chemical isotopes have unstable nuclei, which means that they may break down, or decay, releasing atomic *radiation*. Such isotopes are said to be *radioactive*. For example, ^{14}C emits **beta particles** (fast-moving electrons), a form of radiation, and changes to ^{14}N after the beta particle is released.

These two differences mean that isotopes can be used as *tracer molecules*—they can be used to "label" a molecule and identify compounds that contain the isotope. This enables investigators to trace the movements of specific molecules through cells and tissues (Fig. 14.5).

We describe the rate at which a radioactive atom decays by stating its **half-life.** The half-life for ^{14}C is 5700 years, which means that if we were to take 1000 atoms of ^{14}C and wait for 5700 years, roughly *half* of these atoms would have released one beta particle each and changed into ^{14}N by that time. In another 5700 years, half of the 500 remaining particles (or about 250) would do the same, and at the end of another 5700 years only about 125 ($\frac{1}{8}$) of the original ^{14}C atoms would be left.

Not all chemical isotopes are radioactive. An important isotope of nitrogen is ^{15}N, which has a single neutron more than ^{14}N, the more common form of nitrogen. Neither ^{15}N nor ^{14}N is radioactive, and the only difference between them is in atomic weight.

Figure 14.5 **(a)** Isotopes are forms of an element with different atomic masses. Carbon-14, a radioactive isotope, has two more neutrons in its atomic nucleus than carbon-12, as shown. **(b)** Radioactive isotopes can be used to locate compounds in living organisms. The presence of ^{125}I, a radioactive isotope of iodine, reveals the outline of the thyroid gland in this scan. The thyroid concentrates iodine atoms for use in synthesizing the thyroid hormone, thyroxine.

(a)

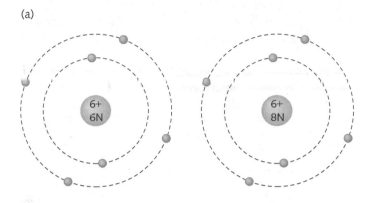

(b)

Isotopes are particularly useful in biology. In order to follow a certain molecule through a chemical reaction, all that is necessary is to find an appropriate isotope of one of its chemical elements. The labeled molecule can be recognized either by virtue of its mass or by its emission of radioactivity.

CHEMICAL COMPOUNDS AND CHEMICAL BONDS

Atoms may be combined in fixed ratios to form **compounds.** Water, for example, is a chemical compound composed of oxygen and hydrogen atoms. Because there are two atoms of hydrogen for each oxygen, we write the chemical formula of water as H_2O. The chemical and physical properties of a compound are different from those of the atoms from which they are formed. Oxygen and hydrogen are highly reactive gases at room temperature, whereas water is a stable liquid.

Just as the atom is the basic unit of an element, the **molecule** is the basic unit of a compound. A single molecule of sodium chloride is composed of one atom of chlorine and one of sodium. We can even consider each molecule to have a *molecular weight* in the same manner that single atoms have atomic weight. How does this work?

We can estimate the molecular weight of a compound simply by adding up the atomic weights of its individual atoms. In the case of NaCl, the molecular weight is found as follows:

$$
\begin{array}{ll}
22.98 & \text{(the atomic weight of Na)} \\
+\,35.45 & \text{(the atomic weight of Cl)} \\
\hline
58.43 & \text{(molecular weight of NaCl)}
\end{array}
$$

Molecules that contain 10, 15, or 20 atoms have much larger molecular weights, and the very large **macromolecules** found in living cells may have weights of a million or more.

The atoms in a compound are linked together by **chemical bonds.** Although there are several types of chemical bonds, we will begin our study of biological chemistry by looking at two of the most important ones: ionic bonds and covalent bonds.

Ionic Bonds

The atoms we have looked at so far have all been chemically *neutral.* They contain equal numbers of electrons and protons, so that their positive and negative charges balance. However, most atoms are able to gain or lose electrons in their outer shell. When this happens, the balance of electrons and protons is upset, and the atoms become electrically charged. An atom or molecule that becomes charged as the result of gaining or losing an electron is called an **ion.** Under the right conditions, an electron may transfer from one atom to another. The two atoms involved in such a transfer may then form an **ionic bond.**

Sodium (Na), a shiny metal, and chlorine (Cl), a greenish gas, are two very different and very dangerous materials. Sodium reacts violently when it comes in contact with water, and chlorine is a deadly poison that has been used as a weapon of war. Sodium has one electron in its outermost shell; chlorine has seven. These two atoms can undergo a **chemical reaction** in which sodium loses its outermost electron and chlorine gains an electron. This transfer allows each atom to fill its outermost shell with the maximum complement of eight electrons (Fig. 14.6). Chemically, this is the most stable form in which the electrons can be arranged, and reactions in which an outermost electron shell can be filled are very common. Because the electron has a negative (–) charge, the loss of an electron makes sodium positively charged (we symbolize it Na^+), and the gain of an electron makes chlorine negatively charged (Cl^-). Both Na^+ and Cl^- are ions; Na^+ is a **cation,** or positively charged ion; and Cl^- is an **anion,** or negatively charged ion.

Electrostatic forces exist between charged chemical groups: Opposite charges attract, and like charges repel. Because the sodium and chlorine ions are oppositely

Figure 14.6 Sodium and chlorine can participate in the formation of an ionic bond. The transfer of an electron from sodium to chlorine completes the outermost shell of each ion.

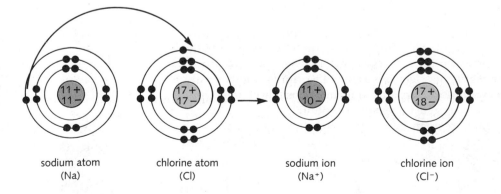

sodium atom (Na)   chlorine atom (Cl)   sodium ion (Na^+)   chlorine ion (Cl^-)

charged, they are attracted to each other by electrostatic forces, and an ionic bond exists between them. Earlier we said that the chemical properties of an atom depend on the number of electrons in its outermost shell. That principle applies to ions as well. Therefore, the chemical properties of the two ions are completely different from what they were before they gained or lost electrons. The name for the compound NaCl is sodium chloride. Despite the noxious nature of each of the elements that go into it, sodium chloride (ordinary table salt) is an essential nutrient in the human diet.

Covalent Bonds

Although many compounds are formed by ionic bonds, another kind of chemical bond is equally important. Hydrogen has a single electron in its outermost orbit and carbon has four. When hydrogen and carbon combine to form methane (CH_4), they actually *share* electrons, so that carbon seems to have eight electrons in its orbital and each of the four hydrogen atoms seems to have two. This sharing of electrons produces four *combined* orbitals for the compound different from that formed by either atom alone. The atomic bonds that exist between carbon and hydrogen are said to be **covalent bonds,** because electrons are shared in the combined orbits of the atoms (see Fig. 14.7).

Each covalent bond results from the sharing of a pair of electrons. When a single pair of electrons is shared, the single covalent bond is represented as a single line drawn between the symbols for the atoms. However, it is also possible for two or even three pairs of electrons to be shared. In carbon dioxide, for example, each oxygen atom shares four electrons with the carbon atom, and we represent the compound with a symbol that shows the existence of a **double covalent bond** between the atoms. Hydrogen cyanide contains a **triple covalent bond** (Fig. 14.8).

The Differences Between Covalent and Ionic Bonds

Although both types of chemical bonds are important, there are clear differences between them. Covalent bonds are formed between two distinct atoms and involve the actual *sharing* of electrons (Fig. 14.8). Ionic bonds involve the *exchange* of electrons to form oppositely charged ions. In a salt crystal, for example, sodium and chloride ions are in a regular lattice that has the appearance of a schoolyard jungle gym (Fig. 14.9). Each sodium is surrounded by six chlorines. Each chlorine is surrounded by six sodiums. Each atom in the lattice is equally attracted to all six surrounding atoms, but the sodium and chlorine

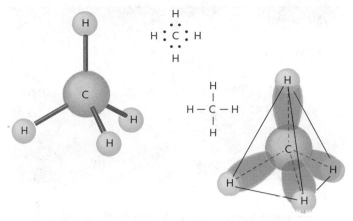

Figure 14.7 Formation of a covalent bond can be represented in many different ways. Here the four covalent bonds in methane (CH_4) are shown in a ball-and-stick model (LEFT), as a structure showing paired electrons (CENTER TOP), as a molecular diagram drawn in a flat plane (CENTER BOTTOM), or as overlapping electron orbitals (RIGHT).

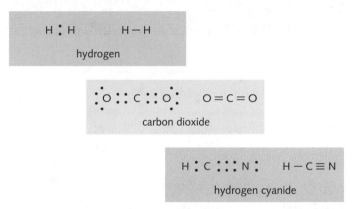

Figure 14.8 A single covalent bond links hydrogen atoms together. Two double covalent bonds are present in carbon dioxide (CO_2). Hydrogen cyanide (HCN) contains a triple bond linking carbon and nitrogen.

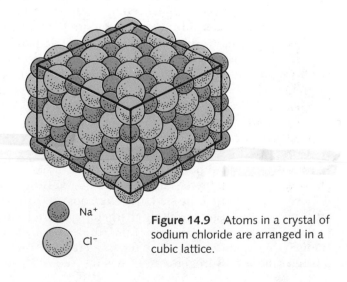

Figure 14.9 Atoms in a crystal of sodium chloride are arranged in a cubic lattice.

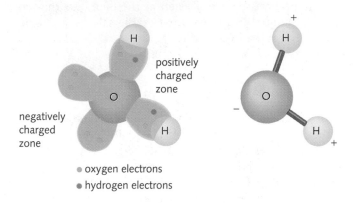

Figure 14.10 LEFT: From the deep blue of the Pacific Ocean to the delicate clouds that shade the surface, this view of the Hawaiian Islands from space is dominated by water. BELOW: Water is a polar molecule. The angle between the two hydrogen atoms is 105°, so that they help to produce a positively charged zone on one side of the molecule. The distribution of electrons forms a negatively charged zone near the oxygen.

positively charged zone

negatively charged zone

● oxygen electrons
● hydrogen electrons

atoms are not distinctly bonded to any of their six neighbors. That is not true for atoms held together by covalent bonds. Even in a crystal of water (ice!) each hydrogen is covalently bonded to *one* and only one oxygen atom.

WATER: SOMETHING SPECIAL

From outer space, the single most striking feature of planet Earth is its color, the deep blue of its oceans and the striking white puffs of clouds (Fig. 14.10). Clouds and oceans, of course, are composed of water. Considering how water dominates the surface of our planet, it is not surprising that the single most common molecule in living organisms is **water.** Living cells are more than 50 percent water by weight, and some are as much as 98 percent water. Without water, the kind of life that we know on earth would not be possible. Water *seems* to be a simple molecule. Two atoms of hydrogen are held by single covalent bonds to a single atom of oxygen. But that simplicity is more apparent than real.

Chemists say that the water molecule is **polar,** which means that different parts of the molecule have different electrical charges. Although the entire water molecule is electrically neutral, the electrons in the shared orbits are not evenly distributed. Instead, the electrons are strongly attracted to the large, positively charged oxygen nucleus, as shown in Fig. 14.10. Because of this, the electron cloud is much denser in the region of the oxygen nucleus than it is near the two hydrogen nuclei. When we consider this

(along with the fact that the two hydrogen atoms are arranged to one side of the oxygen), it is obvious that the two ends of a water molecule have slightly different charges. In other words, water molecules have two positive poles and one negative pole.

The polarity of water has a number of important consequences. The polar ends of the molecule have a strong attraction for each other, and they also have a strong attraction for other charged molecules. The polar water molecule forms weak bonds between the hydrogen and oxygen atoms of different molecules, as shown in Fig. 14.11. This interaction, which is called a **hydrogen bond,** is not nearly as strong as a covalent bond, but it does cause water molecules to be attracted to each other. Other molecules can form hydrogen bonds, and in the next chapter we will encounter several examples of the importance of these bonds in biological systems.

When other molecules are placed in water, their ability to dissolve is determined by how strongly they are attracted to water molecules. The most **soluble** molecules are those that can interact strongly with polar water molecules. Ions dissolve very well, so that NaCl, which actually dissociates in solution to Na^+ and Cl^-, dissolves very quickly in water (Fig. 14.12). Larger compounds (such as sugars), which are uncharged, can still dissolve well if they too are polar because of uneven electron distributions. In fact, the ability to dissolve an enormous range of substances is precisely what makes the water molecule so special. No other solvent can keep such large amounts of so many chemicals in solution, a property that is at the

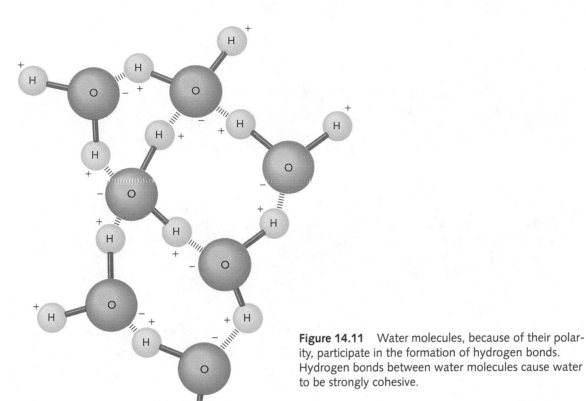

Figure 14.11 Water molecules, because of their polarity, participate in the formation of hydrogen bonds. Hydrogen bonds between water molecules cause water to be strongly cohesive.

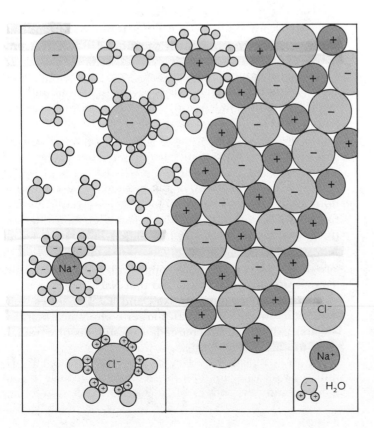

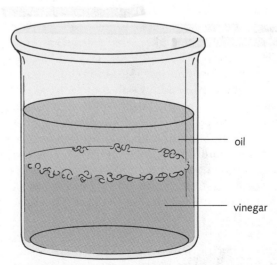

oil

vinegar

Figure 14.12 LEFT: Many substances, such as NaCl, interact with the polar water molecules and dissolve very quickly. These materials that interact with the water molecule are said to be hydrophilic ("water-loving"). TOP: Hydrophobic ("water-fearing") molecules are those that cannot interact with the polar water molecule.

Making Chemical Solutions: The Mole

Because the molecular weights of compounds are different, it is often difficult to calculate just how much of a chemical should be added to a solution. For example, suppose we wanted to have equal numbers of sodium chloride and glucose molecules in a solution. If we added 100 grams of salt, how much glucose would we have to add? The molecular weight of salt is 58.5, and that of glucose is 180. So we would have to add 100 grams x 58.5/180, or . . . well, you get the picture: It would be complicated. But there's a simpler way to do it.

Scientists generally make up chemical solutions in terms of moles. A *mole* is a *gram molecular weight* (the molecular weight in grams) of a compound. One mole of salt would be 58.5 grams; 1 mole of glucose, 180 grams. If we were to add 1 mole of each to a container, we could be certain (within the limits of our ability to measure) that there were equal numbers of each compound in the mixture. When describing solutions, we take this a step further. We might, for example, make up a 1 *molar* solution of NaCl (designated 1.0 M

NaCl), which would contain exactly 1 mole of NaCl per liter of solution. A 0.1 molar solution would have 0.1 moles/liter, and so forth. By mixing solutions with their contents described in moles, a worker can be sure of the relative amount of each compound in a particular solution, even if the numbers of individual molecules can't be counted.

A flask containing 1.0 M NaCl, 0.5 M KCl, and 0.1 M $MgCl_2$ has exactly two sodium (Na) atoms for every potassium (K) and five potassiums for every magnesium (Mg).

heart of life itself. Water is sometimes called the "universal solvent."

Hydrophilic ("water-loving") or *polar* molecules are those that interact strongly with water. Of course, many substances do not dissolve in water, and that is important for the existence of life as well. **Hydrophobic** ("water-fearing") or *nonpolar* molecules do not interact with water and are *insoluble* in it, like salad oil in an oil-and-vinegar dressing (Fig. 14.12).

Hexane, a molecule commonly found in oil and gasoline, has a molecular structure in which the electrons are evenly distributed; therefore, the hexane molecule has no charged regions to interact with water. Hexane molecules do not dissolve well in water and tend to cluster together, excluded from the bonds that link water molecules to each other. The interactions between hydrophilic and hydrophobic molecules are important in the formation of a number of biological structures, including cellular membranes.

Acids, Bases, and Buffers

Besides its polarity, water has another chemical property that is important in living things. The water molecule can reversibly dissociate to form two different ions. The process goes like this:

$$HOH \rightarrow H^+ + OH^-$$

The H^+ ion is called the **hydrogen ion;** it is merely a *proton* and is often referred to as such. The OH^- group is called the **hydroxide ion.**

The chemical properties of these ions require us to pay special attention to their concentrations. A solution that contains excess hydrogen ions is said to be **acidic;** a solution in which hydroxide ions predominate is **basic.** Lemon juice, vinegar, and battery acid are examples of solutions that contain an excess of hydrogen ions and are acidic. Ammonia, hair remover, and oven cleaner are examples of bases. To determine the degree to which a solution is acidic or basic, we can calculate a quantity called **pH,** which tells us at a glance just how acidic or basic a solution is.

The pH is determined by the hydrogen ion concentration. If the hydrogen ion concentration is 10^{-9} molar, then pH is 9; if it is 10^{-2} molar (10 million times higher than 10^{-9} molar), the pH is 2. An *acidic* solution has a pH lower than 7; a *basic* solution has a pH higher than 7. A solution of pH 7 is *neutral* because it has equal concentrations of H^+ and OH^- ions.

An **acid** is a compound that, when dissolved in water, releases hydrogen ions into solution, lowering the pH. A **base** is a compound that raises the pH by removing hydrogen ions from solution.

As you can see, the pH scale is logarithmic: each pH unit represents an increase or decrease in hydrogen ion concentration by a power of 10 (Fig. 14.13). A solution at

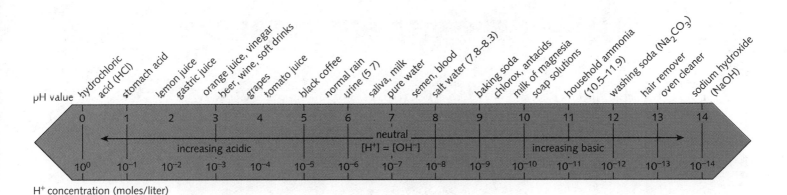

Figure 14.13 The pH scale measures the concentration of hydrogen ions in a solution. Strong acids produce low pHs, while strong bases produce high pHs.

pH 5 has 100 times as many hydrogen ions per liter as a solution at pH 7, and therefore it is 100 times as acidic.

The dissociation of water is of enormous significance. A solution is called acidic simply on the basis of H^+ ions, ignoring other ions such as K^+ or Na^+. We are free to do this because the hydrogen ion, as we've emphasized, is nothing more than a single proton, lacking any electrons. In water, this powerful positive charge is surrounded by an envelope of water molecules. The hydrogen ion is strongly attracted to negatively charged ions and is able to attract electrons from a great many larger molecules. Therefore, the concentration of hydrogen ions in a solution is more important than the concentration of any other ion, and there is good reason for inventing a special terminology to keep track of it. Knowing whether a solution is acidic or basic tells us what the concentration of protons is, and it enables us to predict to some extent what kinds of reactions will occur in solution.

The fact that pH is directly determined by the hydrogen ion concentration might lead us to believe that cellular pH fluctuates whenever large numbers of hydrogen ions are released into solution. To some extent that is true. However, a number of compounds found in living systems tend to stabilize pH at certain levels. Such pH-stabilizing compounds are known as **buffers.** Buffers absorb hydrogen ions as the pH of a solution declines or release hydrogen ions as the pH rises, thereby stabilizing pH. Acids or bases can be added to a strongly buffered solution without dramatically altering pH. Carbonic acid and bicarbonate ion buffer pH changes in the blood. If the pH of blood begins to rise, carbonic acid tends to dissociate and release hydrogen and *bicarbonate* ions:

$$H_2CO_3 \rightarrow H^+ + HCO_3^-$$

The release of hydrogen ions tends to lower the pH, "buffering" the tendency to increase pH. Similarly, if the pH of a solution begins to fall, the additional hydrogen ions can be "soaked up" by the bicarbonate ions:

$$H^+ + HCO_3^- \rightarrow H_2CO_3$$

Buffers such as the carbonic acid/bicarbonate ion pair make it possible for living systems to resist rapid changes in pH and to maintain a constant internal environment.

Other Properties of Water

Water has a number of other special properties that affect living systems. It is relatively transparent, making it possible for light to penetrate through some living tissues as well as deep into lakes, streams, and oceans. Water expands slightly as it freezes, making solid water (ice) somewhat less dense than liquid water at most temperatures. This causes ice to form at the surfaces of ponds and lakes, insulating many forms of life below the ice against the harsh extremes of cold weather. Water is strongly **cohesive,** meaning that water molecules are attracted to each other, principally due to hydrogen bonding. Water's cohesiveness makes it possible for water molecules to be drawn into small openings or tubes and then to rise against the force of gravity, a phenomenon known as **capillary action** (Fig. 14.14).

Hydrogen bonds produce such strong attractions between water molecules at an air–water interface that water displays a measurable **surface tension.** This surface attraction is so great that small animals are able to walk directly on the surface of water, and many (such as water striders) spend their lives skimming the boundary between air and water (Fig. 14.14). Finally, water has a high heat capacity. A substantial amount of heat energy is required to change the temperature of a body of water by just a few

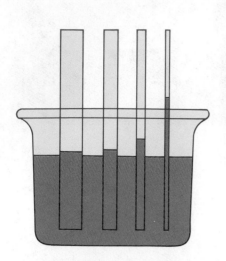

Figure 14.14 Capillary action. LEFT: The cohesive properties of the water molecule enable water to be drawn upwards into small openings, provided that the walls of the capillary tube also interact with water. RIGHT: The surface tension of water enables small creatures, such as this water strider, to live safely on its surface.

degrees. Even more energy is required for water to move through a phase transition: from solid to liquid or from liquid to gas. These properties enable water to act as a "heat buffer," producing stability in the face of rapid changes in air temperature.

THE MOLECULES OF LIVING THINGS

"Organic" Chemistry

As chemistry developed in the nineteenth century, scientists were able to determine the general patterns in which the chemical elements combine to form compounds. They developed a science of chemistry that led directly to our modern understanding of the nature of matter. Not surprisingly, some chemists began to explore the applicability of this knowledge to living things as well as to the materials under study in their labs. At first, such efforts were unsuccessful and led to confusion. Some chemists suggested that living organisms obeyed a different kind of chemistry from nonliving material. In fact the term *organic chemistry* was originally coined to distinguish the chemistry of a living organism from the *inorganic chemistry* that chemists had been investigating in the lab.

To scientists today, however, **organic chemistry** is the chemistry of carbon compounds. This change of emphasis reflects the fact that carbon compounds dominate the chemistry of living things. Nineteenth-century scientists experienced difficulties in studying the molecules found in living organisms because they did not appreciate the extreme complexity of carbon-based molecules. What's so special about carbon? As we already know, carbon's atomic number is 6, and it has four electrons in its outer-

most shell. This means that carbon can form a total of four covalent bonds with other atoms. The key to carbon's importance is the fact that it is able to form stable bonds with other carbon atoms. This allows it to form long chains that may contain single, double, and even triple carbon-to-carbon covalent bonds. The other elements that have four electrons in their outer shells (including silicon and germanium) can also form bonds to themselves, but none of them comes close to matching the stability of a C–C bond. We might say that carbon is the world's best electron sharer. This key property allows carbon to form the backbone of large and complex molecules—the very molecules that make life possible (Fig. 14.15).

Figure 14.15 Carbon's ability to bond to itself in chain-like fashion enables it to form a wide variety of molecular structures with other atoms. Note that more complex molecules, like benzene, can be represented in an abbreviated style in which hydrogen atoms are not drawn out and a carbon is understood to exist at each vertex of the diagram.

The Octet Rule (and Why It Doesn't Always Work)

As you've seen, in many chemical reactions atoms of the lighter elements seem to gain or lose electrons in order to complete their outermost shell. Because eight electrons are needed to complete a second-level shell, this tendency is sometimes called the "octet rule," and it's a handy predictor of the compounds that can be formed by many elements. For example, the octet rule would lead us to predict that nitrogen could form a stable compound with three hydrogen atoms:

Nitrogen has five electrons in its second shell, and hydrogen has one in its first shell. The combination of three hydrogens with one nitrogen fits the octet rule perfectly. It allows nitrogen to fill its outermost second-level shell with eight electrons, and it allows three hydrogens to fill their first-level shells with two electrons, completing each of them. This compound, NH_3, is called *ammonia*.

The formation of water (H_2O) can be analyzed in the same way:

For most of the elements that are commonly found in living organisms, the octet rule is a handy predictor of the number of chemical bonds that an atom can form. But there's a problem.

Phosphorus has five electrons in its outermost, third-level shell (see Fig. 14.4). Therefore, like nitrogen, it should form three chemical bonds to complete the octet, right?

Well, sometimes it does. When combined with chlorine, it forms PCl_3, fitting the octet rule perfectly. But it's also possible to prepare a different compound, PCl_5, in which phosphorus forms five bonds. How is this possible?

Remember that the third electron shell can actually hold a maximum of eighteen electrons in nine different orbitals. The first four orbitals are at a slightly lower energy level than the other five, and therefore a compound

with eight electrons in its third shell is relatively stable. That's why the octet rule usually works, even in the third shell. However, only a slight increase in energy is necessary for the five electrons in phosphorus to rearrange themselves so that one electron occupies each of the first five orbitals. When that happens, phosphorus has five positions available to share electrons, making it possible to form five covalent bonds with other compounds.

Therefore, when you see a compound like phosphoric acid (H_3PO_4), in which phosphorus forms five bonds, don't be surprised.

$$
\begin{array}{c}
H \\
| \\
O \\
| \\
H-O-P=O \\
| \\
O \\
| \\
H
\end{array}
$$

It's just phosphorus's way of telling us that even in chemistry, some rules are made to be broken.

Building a Carbon Compound

It's possible to think of carbon atoms as a flexible "skeleton" that can be used to construct just about any type of molecule. To see how this works, start with a simple carbon chain:

$$
\begin{array}{c}
\quad | \quad | \quad | \quad | \\
-C-C-C-C- \\
\quad | \quad | \quad | \quad |
\end{array}
$$

The first variable is size. The chain can be short—just one or two carbons—or it can be long—up to thousands of

carbons. The second crucial point of variation comes from the chemical groups that may be attached to the chain.

For example, let's suppose that we wanted to construct an oily compound that would not be soluble in water. We could attach hydrogen atoms to an eight-member carbon chain to produce this compound:

$$
\begin{array}{c}
H \; H \; H \; H \; H \; H \; H \; H \\
| \; | \; | \; | \; | \; | \; | \; | \\
H-C-C-C-C-C-C-C-C-H \\
| \; | \; | \; | \; | \; | \; | \; | \\
H \; H \; H \; H \; H \; H \; H \; H
\end{array}
$$

Table 14.2 *Seven of the Most Common Biochemical Functional Groups*

Group	Found in	Behavior	Example
Hydroxyl R—OH	Alcohols, sugars	Polar O—H Interacts strongly with water	Ethanol
Methyl R—C—H	Hydrocarbons, oils	Nonpolar C—H Incapable of interacting well with water Larger molecules are insoluble	Ethane
Aldehyde R—C—O	Aldehydes	Polar C=O Interacts with water Highly reactive chemically	Acetylaldehyde
Keto R C=O R	Ketones	Polar C=O Interacts with water Somewhat reactive chemically	Acetone
Carboxyl R—C=O	Organic acids	—COOH dissociates to —COO$^-$ Acts as an acid Interacts strongly with water	Acetic acid
Amino R—N—H	Organic bases	Takes up proton to form —NH$_3^+$ Acts as a base Interacts strongly with water	Methyl amine
Phosphate R—O—P—OH	Organic phosphates	Can release 2 protons: $H_2PO_4 \rightarrow PO_4^{2-}$ Powerful organic acid Interacts strongly with water Can be linked to form di- and triphosphates	Glycerol phosphate

Because electrons are evenly distributed around each C—H bond, this compound, called **octane,** is nonpolar and doesn't interact with water. Octane, which is one of the compounds found in common gasoline, is oily and doesn't dissolve well in water.

If we wanted to make an entirely different molecule, perhaps an acid that interacted strongly with water, we could use carbon for this task, too. We could take two carbons and attach hydrogens to one and a group of oxygens and hydrogen to the other:

$$H-\overset{\overset{\displaystyle H}{|}}{\underset{\underset{\displaystyle H}{|}}{C}}-\overset{\overset{\displaystyle O-H}{|}}{C}=O \;\leftrightarrow\; H-\overset{\overset{\displaystyle H}{|}}{\underset{\underset{\displaystyle H}{|}}{C}}-\overset{\overset{\displaystyle O^-}{|}}{C}=O \;+\; H^+$$

In water at neutral pH, this compound releases a proton into solution. The compound is **acetic acid,** and it helps to produce the sour taste in vinegar. After the proton is released, the rest of the molecule has a negative charge, which ensures that water will interact with it, making acetic acid highly soluble.

Functional Groups

Functional groups have similar chemical functions and properties in a variety of different compounds. The acidic part of the molecule in our drawing is the four-atom group including the right-hand carbon (—COOH). This is known as a **carboxyl group.** The carboxyl group appears so frequently in nature that we refer to it as a functional group—in this case, an organic acid. Any carbon compound containing a carboxyl group has the potential to act as an acid, releasing a proton into solution. We can summarize how a carboxyl group acts by a diagram showing how it releases a proton:

$$R-\overset{\overset{\displaystyle OH}{|}}{C}=O \;\rightarrow\; R-\overset{\overset{\displaystyle O^-}{|}}{C}=O \;+\; H^+$$

In this diagram, we've simply written **"R"** to stand for the *rest* of the molecule, whatever it happens to be.

Are there organic bases, too? Absolutely.

If a nitrogen with two hydrogens(—NH$_2$) is attached to a carbon compound, it will act as a *base,* removing a proton from solution:

$$R-\overset{\overset{\displaystyle H}{|}}{N}-H \;+\; H^+ \;\rightarrow\; R-\overset{\overset{\displaystyle H}{|}}{\underset{\underset{\displaystyle H}{|}}{N^{\pm}}}-H$$

The **amino group,** as –NH$_2$ is called, is another kind of functional group. When the amino group is present in a compound, one can predict that the amino portion will act as a base and will carry a positive charge after acquiring a proton, tending to make it soluble in water.

Table 14.2 summarizes the seven most common and most important functional groups in biological chemistry. Why should we go to the trouble to name and classify these groups? In a word, simplicity. No matter how complex a molecule may be, if we can recognize a few functional groups within that molecule, we can go a long way to predicting and understanding how it will behave in living systems.

SUMMARY

The basic unit of matter is the atom. Atoms are composed of subatomic particles: protons, electrons, and neutrons. The chemical and physical properties of an atom are determined by the number of subatomic particles in its nucleus and its electron orbitals. Atoms may be joined by chemical bonds to form compounds. The chemical and physical properties of compounds are often very different from the atoms that form them. Chemical bonds may be ionic or covalent, depending on whether electrons are transferred from one atom to another or shared between them.

The most abundant molecule in living things is water. Water has a number of special properties, including an asymmetric arrangement of electrons around the three atomic nuclei that make up the compound. This produces a polarity in the molecule that makes it capable of dissolving other molecules, one of water's most important characteristics. Water molecules may dissociate to produce protons (H$^+$) and hydroxide ions (OH$^-$), and the amounts of these ions in a particular solution may be represented on a system known as the pH scale.

The properties of larger molecules can be analyzed in terms of the chemical groups that make them up. Biological molecules contain a number of distinct chemical groups whose basic properties are similar from one molecule to the next. These groups make it possible to describe the chemical properties of complex molecules.

After studying this chapter, you should be able to:

- Appreciate that chemistry and chemical processes underlie biological systems.

- Describe chemical systems in general terms.

- Explain why biological research increasingly involves chemical techniques and chemistry as a level of analysis.

- Recognize some of the most important biological molecules and chemical groups that are crucial to the study of living things.

TERMS AND CONCEPTS

subatomic particles *277*	anion *282*
orbitals *277*	covalent bonds *283*
atomic number *279*	polar *284*
isotopes *281*	hydrogen bond *284*
beta particles *281*	hydrophilic *286*
half-life *281*	hydrophobic *286*
compounds *282*	pH *286*
molecule *282*	buffers *287*
macromolecules *282*	capillary action *287*
ionic bond *282*	surface tension *287*
cation *282*	

REVIEW

Objective Questions (Answers in Appendix)

1. Atoms of different elements react differently from one another because of
   (a) the number of neutrons in their nuclei.
   (b) the number of electrons in their outer orbitals.
   (c) the number of neutrons in their outer orbitals.
   (d) their atomic weight.

2. Isotopes of an element differ in all the following *except*
   (a) the number of neutrons in their nuclei.
   (b) atomic weight.
   (c) stability of the nuclei.
   (d) chemical behavior.

3. Electrons that are transferred from one atom to another form
   (a) covalent bonds.   (c) hydrogen bonds.
   (b) ionic bonds.   (d) beta bonds.

4. Electrons that are shared between elements form
   (a) covalent bonds.   (c) hydrogen bonds.
   (b) ionic bonds.   (d) cooperative bonds.

5. What type of bond is found between molecules of water?
   (a) covalent   (c) ionic
   (b) hydrogen   (d) nonpolar

6. H^+ and OH^- are
   (a) atoms.   (c) ions.
   (b) compounds.   (d) molecules.

Discussion Questions

7. What is the difference between atomic number and atomic mass? Why does adding additional neutrons to an atom not affect its chemical properties, although adding protons does?

8. What are the three most important types of chemical bonds, and how are they different from each other?

9. What is the origin of the term *organic chemistry,* and how does its original meaning differ from its contemporary one?

10. Why is water such an important compound to living things?

11. Which solution has a higher concentration of H^+ ions, one at pH 3 or one at pH 6?

READINGS

Zumdahl, S. S. *Introductory Chemistry.* 2d ed. Lexington, MA: D. C. Heath, 1993. An excellent introductory chemistry textbook.

Mertz, W. "The essential trace elements." *Science* 213 (1981): 1332–1338. A brief review of the biological roles of elements needed by living organisms in small amounts.

Frieden, E. "The chemical elements of life." *Scientific American* 227 (October 1972): 52–60. A description of the key elements found in living systems.

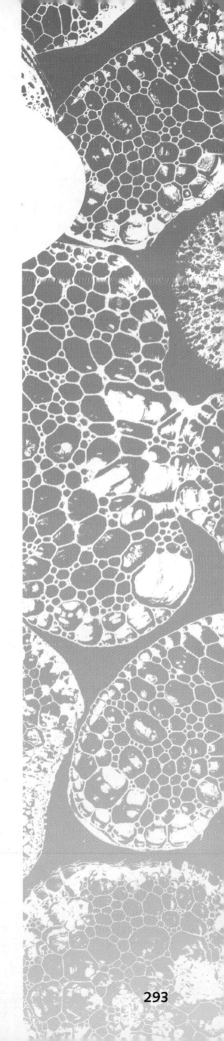

15

Macromolecules

*f*or years, chemists were baffled by the sheer complexity of living matter. Although living organisms are composed of atoms that are common on Earth's crust, many of the compounds found in living things are unlike anything in the nonliving world. The complexity of life, the chemists began to understand, was due to the way in which individual elements were joined together in living things.

One of the first discoveries was that many of the important molecules in living things are very, very large. Unprepared to deal with molecular weights larger than a thousand, chemists suddenly found themselves confronted with molecules that were more than a hundred times as large. Quite properly, they called these compounds *macromolecules*. The name is well deserved.

How can we begin to explore the properties of these gigantic compounds? The answer, of course, is that we'll start at the beginning, by seeking to understand how these molecules are constructed. Like a complex machine, each of them is put together by assembling smaller parts. An engineer knows that the key to understanding such a machine is a thorough knowledge of how its parts work. That's exactly what we seek in this chapter—an understanding of the biochemical parts of living things.

There are four principal classes of macromolecules found in living organisms:

    lipids
    carbohydrates
    nucleic acids
    proteins

To study these chemical machines, we will see how each of these molecules is put together, one at a time. To do that, we will start with one of the simplest things in the world—a drop of oil in a bowl of water.

BUILDING A MACROMOLECULE

Everyone has heard the expression "Oil and water don't mix." It's true, of course, and the reason is that oils are mostly hydrocarbon, which means that they are composed of long chains in which carbon atoms are surrounded by hydrogens:

$$H-\overset{\displaystyle H}{\underset{\displaystyle H}{C}}-\overset{\displaystyle H}{\underset{\displaystyle H}{C}}-\overset{\displaystyle H}{\underset{\displaystyle H}{C}}-\overset{\displaystyle H}{\underset{\displaystyle H}{C}}-\overset{\displaystyle H}{\underset{\displaystyle H}{C}}-\overset{\displaystyle H}{\underset{\displaystyle H}{C}}-\overset{\displaystyle H}{\underset{\displaystyle H}{C}}-\overset{\displaystyle H}{\underset{\displaystyle H}{C}}\cdots\cdots$$

In Chapter 14 we saw that chemical groups like this are nonpolar. They don't have charge differences that enable them to form hydrogen bonds to water. In the midst of millions of water molecules, all strongly attracted to each other, the oil molecules are gradually pushed together into droplets that separate from the water (see Fig. 14.12 on page 285).

Let's suppose that we had to design a molecule that would dissolve oil droplets. To do this, we'd have to find a way to make the oil molecules capable of interacting with water. The ideal way to do this would be to construct a molecule that could interact with *both* oil and water. An obvious way would be to produce a molecule that was hydrophobic enough to interact with the oil on one end and hydrophilic enough to interact with water on the other.

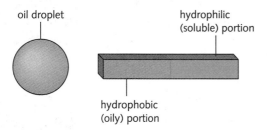

oil droplet

hydrophilic (soluble) portion

hydrophobic (oily) portion

Is it possible to actually make such a molecule? Using our knowledge of functional groups, we'd first want to construct a hydrophobic carbon skeleton that could interact with the oil. To do this, what could be better than a hydrocarbon chain? A long one—perhaps 17 carbons.

$$H-\overset{\displaystyle H}{\underset{\displaystyle H}{C}}-\overset{\displaystyle H}{\underset{\displaystyle H}{C}}-\overset{\displaystyle H}{\underset{\displaystyle H}{C}}-\overset{\displaystyle H}{\underset{\displaystyle H}{C}}-\overset{\displaystyle H}{\underset{\displaystyle H}{C}}-\overset{\displaystyle H}{\underset{\displaystyle H}{C}}-\overset{\displaystyle H}{\underset{\displaystyle H}{C}}-\overset{\displaystyle H}{\underset{\displaystyle}{C}}\cdots\cdots-\overset{\displaystyle H}{\underset{\displaystyle H}{C}}-$$

Our next step would be to link the hydrocarbon chain to a functional group that is strongly charged and that would be attracted to the polar water molecule. What better candidate than the strongly polar carboxyl group?

$$H-\overset{\displaystyle H}{\underset{\displaystyle H}{C}}-\overset{\displaystyle H}{\underset{\displaystyle H}{C}}-\overset{\displaystyle H}{\underset{\displaystyle H}{C}}-\overset{\displaystyle H}{\underset{\displaystyle H}{C}}-\overset{\displaystyle H}{\underset{\displaystyle H}{C}}-\overset{\displaystyle H}{\underset{\displaystyle H}{C}}-\overset{\displaystyle H}{\underset{\displaystyle H}{C}}-\overset{\displaystyle H}{\underset{\displaystyle}{C}}\cdots\cdots-\overset{\displaystyle H}{\underset{\displaystyle H}{C}}-\overset{\displaystyle O^-}{C}=O$$

What we've just done is to design a molecule with two very different parts, one that interacts with oil and one that interacts with water. Add a few drops of this substance to a glass of oil and water, stir the mixture, and guess what happens. The oil droplet disappears. The molecule works just as we had hoped, dispersing the large droplet into microscopic oil molecules surrounded by our bifunctional molecule.

The 18-carbon molecule we've just constructed, by the way, is called **stearic acid.** It's a member of a group of compounds called **fatty acids.** The sodium salt of stearic acid has an even more familiar name. We call it "soap."

We all know why soaps are important. Fatty acids are notable for two reasons. First, they illustrate how a molecule with different functional groups can carry out two functions: interacting with both water and oil. Second, fatty acids are the building blocks of one of the most important classes of macromolecules: **lipids.**

LIPIDS

Lipids are biological compounds that are waxy or oily and that dissolve in organic solvents such as acetone, ether, or alcohol. That is not a very precise chemical definition, and frankly, it's not intended to be. The lipids are a very diverse group that includes the compounds most of us know as "fats." Lipids dissolve in organic solvents because a major part of most lipid molecules is hydrocarbon, like the first 17 carbons we used to "construct" stearic acid.

Many lipids are formed from fatty acids, like stearic acid. A fatty acid consists of a long hydrocarbon chain with a carboxyl group attached to one end.

There are many different kinds of fatty acids, and they differ from each other in two fundamental ways: the number of carbon atoms they contain and the number of carbon-to-carbon double bonds. As you can see from Fig. 15.1, stearic acid doesn't have any carbon-to-carbon double bonds. But oleic acid has a single double bond near the middle of its hydrocarbon chain. Other fatty acids may have two or even three such double bonds.

These double bonds are involved in an interesting bit of jargon that you are probably familiar with. Because the hydrocarbon chain of stearic acid has the maximum number of hydrogens per carbon, it is said to be **saturated.** But when a double bond is inserted, there are two fewer hydrogens in the molecule: therefore, a fatty acid with a double bond is said to be **unsaturated.** If the fatty acid has two or three double bonds, we say it is **polyunsaturated,** a term that food companies often use to describe certain kinds of cooking oil and margarine. Unsaturation affects how the body is able to use food containing the fatty acids and also determines whether the lipid is liquid or solid.

Figure 15.1 Fatty acids are the principal building blocks of many lipids. Fatty acids consist of long hydrocarbon chains with a carboxyl group at one end of the molecule. Palmitic (16 carbons), stearic (18 carbons), and oleic acid (18 carbons with one double bond) are typical fatty acids.

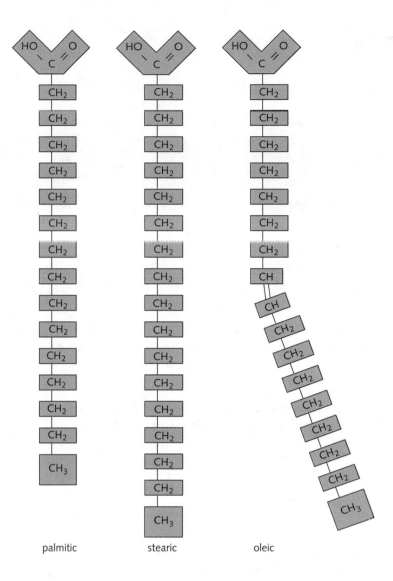

palmitic   stearic   oleic

Saturated fats are often solid at room temperature, whereas unsaturated fats are generally liquid.

Lipids are used for a variety of purposes. The hydrocarbon regions of lipids are loaded with chemical energy, almost like fuel oil or gasoline, making lipids especially suitable for long-term energy storage. Lipids are also used for chemical signaling and for constructing cellular membranes.

Triglycerides and Neutral Lipids

Fatty acids are the building blocks of many lipids. One example is a compound known as a **triglyceride.** A triglyceride is constructed by attaching three fatty acids to a single glycerol molecule (Fig. 15.2). **Glycerol,** which has three hydroxyl (—OH) groups, is highly soluble in water. A chemical reaction between these hydroxyls and the carboxyl groups of three fatty acids produces a single large molecule. Triglycerides are a very common form of lipid found in fat cells and are used for food storage in both plants and animals. Triglycerides are called neutral lipids because they are relatively nonpolar and are uncharged.

Polar Lipids

Many other kinds of lipids can be assembled on this general pattern, using glycerol as a kind of "backbone." When one of the groups that is attached to glycerol is strongly polar, a polar lipid is formed. Fatty acids, as we've seen, have both polar and nonpolar regions. Polar lipids take this concept a step further: their nonpolar and their polar regions are *both* larger than those of individual fatty acids.

One of the most important of the polar lipids is a class of compounds known as **phospholipids.** Phospholipids are constructed much like triglycerides: a glycerol molecule is covalently linked first to one fatty acid and then to

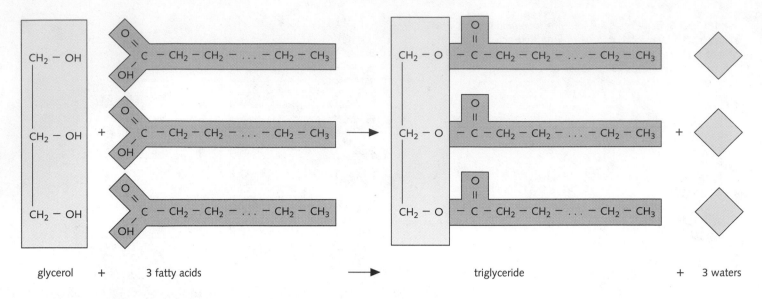

glycerol   +   3 fatty acids $\longrightarrow$ triglyceride   +   3 waters

Figure 15.2 Triglycerides are simple lipids that form from three fatty acids and glycerol. Condensation reactions form covalent bonds between each fatty acid and one of the three carbon atoms of glycerol.

a second. But the third hydroxyl group on glycerol is used for something a little different: Here a phosphate group is attached to the molecule, and yet another group can then be attached to the phosphate. If that final group is a choline group, a lipid called phosphatidyl choline is formed (Fig. 15.3a). That complicated name simply indicates that this is a phospholipid molecule that includes a choline group.

Phospholipids are particularly interesting because the opposite ends of each molecule have quite different properties. The hydrocarbon chains derived from their fatty acids are quite oily and do not interact well with water. But the opposite end of the molecule is strongly polar and carries a positive charge, a negative charge, or both. This end interacts very strongly with water and would be able to dissolve easily were it not for the attachment to the opposite, nonpolar end. We often draw simple diagrams that represent the structure of polar lipids like phospholipids and that can be used to emphasize the point that one end of the molecule is hydrophilic and the other end is hydrophobic (Fig. 15.3b). This property of polar lipids is the basis for one of their most important functions in living systems: they can self-associate in water to form lipid bilayers. Lipid bilayers are the structural backbones of biological membranes.

Steroids

Besides polar lipids and triglycerides, both of which are based on glycerol, there are many other types of lipids, and some of them are constructed quite differently. One type is a class of molecules known as **steroids** (Fig. 15.4).

Steroids are based on an intricate ring structure that is best exemplified by one of the most common steroids, **cholesterol.** Steroids are important molecules that play a wide variety of roles in living systems. Some of them aid in the assembly of cell membranes, and other steroids function as critical **hormones**—chemical messengers that regulate the activity of cells throughout the body.

CARBOHYDRATES

The compounds that most of us call sugars are known by the chemical name **carbohydrates.** Carbohydrates have at least three carbon atoms and follow the general chemical formula $(CH_2O)_n$. Just as we did with lipids, we'll start by looking at the chemical properties of a single small molecule.

Glucose is a six-carbon sugar with the chemical formula $C_6H_{12}O_6$. To construct the compound, we might use a six-carbon chain:

$$-\overset{|}{\underset{|}{C}}-\overset{|}{\underset{|}{C}}-\overset{|}{\underset{|}{C}}-\overset{|}{\underset{|}{C}}-\overset{|}{\underset{|}{C}}-\overset{|}{\underset{|}{C}}-$$

We could attach a number of hydrogens to the carbons and then five hydroxyls, which should make the compound very soluble in water:

$$H-\overset{H}{\underset{O}{\underset{|}{C}}}-\overset{H}{\underset{O}{\underset{|}{C}}}-\overset{H}{\underset{O}{\underset{|}{C}}}-\overset{\overset{H}{O}}{\underset{H}{\underset{|}{C}}}-\overset{H}{\underset{O}{\underset{|}{C}}}-\overset{H}{\underset{|}{C}}-$$

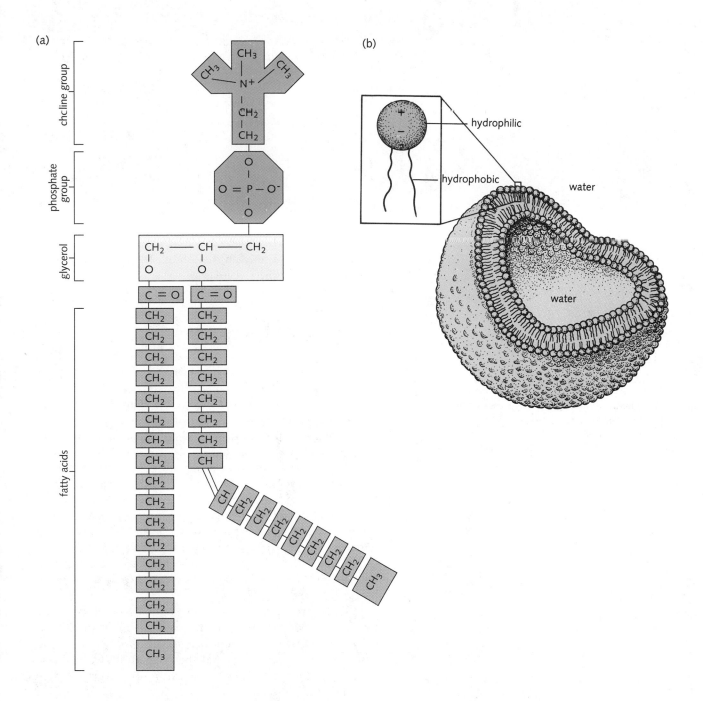

Figure 15.3 **(a)** Phospholipids, like this phosphatidyl choline molecule, are polar lipids. The positive and negative charges at the top of the molecule make this portion, the "head" group, strongly hydrophilic. By contrast, the hydrocarbon "tails" are strongly hydrophobic. **(b)** Polar lipids can associate to form structures like this liposome, a sphere bounded by a lipid bilayer. Note how the strongly hydrophilic head groups of the liposome make contact with water, and how the hydrophobic lipid tails are shielded from water.

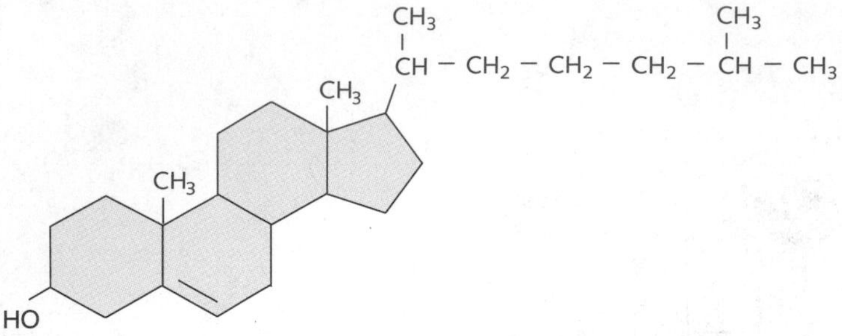

Figure 15.4 Cholesterol, a typical steroid lipid. Steroids have a basic four-ring structure like cholesterol but differ in the chemical groups attached to the rings.

And then we could finish the job by attaching a very reactive aldehyde group to one end:

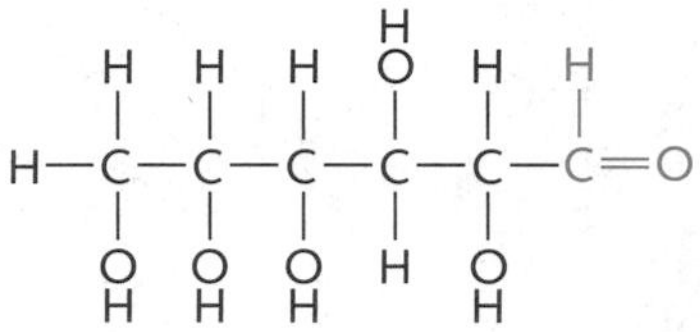

Can we predict the chemical behavior of glucose by looking at its functional groups (Fig. 15.5a)? As we've indicated, the fact that every one of the carbons is attached to a polar group should make the glucose *very* soluble in water. What about that aldehyde group on the first carbon? Aldehydes can react with a number of other chemical groups, including hydroxyls. Could it react with one of the hydroxyls on the very same molecule? Yes, indeed.

In fact, in solution the glucose molecule does not actually exist in a linear form. Instead, the aldehyde group attacks the hydroxyl group of the number 5 carbon to form a covalent bond linking the two carbons by an oxygen atom. The new bond causes glucose to form a closed, six-membered ring (Fig. 15.5b).

There are actually two forms of glucose, α-glucose and β-glucose, which differ in the position of the hydrogen atom attached to the number 1 carbon when the ring closes (Fig. 15.5c). As long as they remain individual molecules, α-glucose and β-glucose can interconvert (change from one form to another) in solution.

Figure 15.5 Glucose can exist in several forms. Although we sometimes draw it as a linear molecule **(a)**, it is generally in a closed-ring form. In solution, the ring can open spontaneously and close with very little loss of energy. **(b)** Note that the position of the hydroxyl group on the number 1 carbon determines whether α- or β-glucose **(c)** is formed when the ring closes.

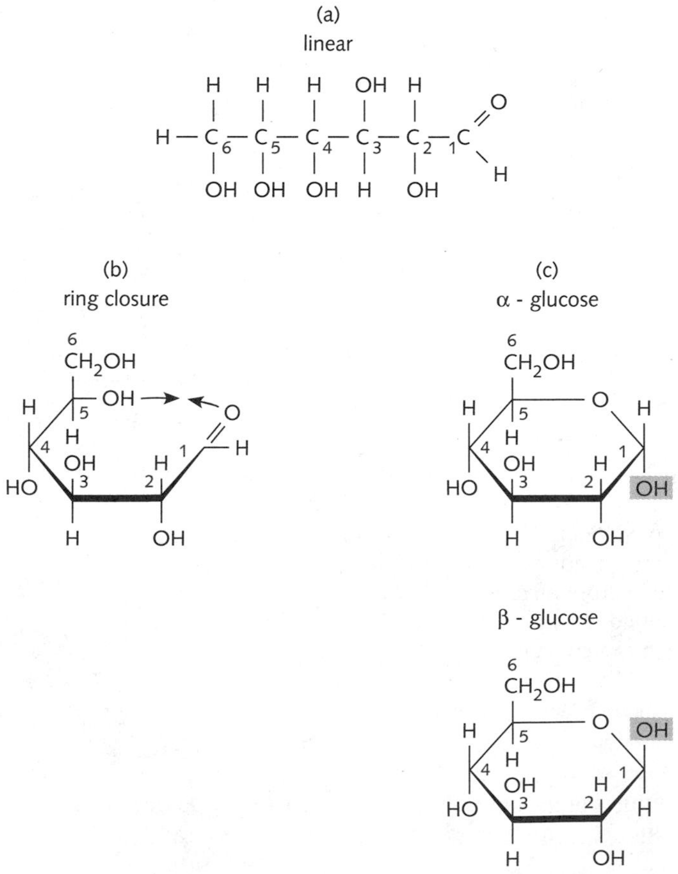

(a)
linear

(b)
ring closure

(c)
α - glucose

β - glucose

Simple Sugars

Individual sugar molecules, like glucose, are known as **simple sugars.** A simple sugar with three carbons is known as a triose, and those with five, six, and seven carbons are called pentose, hexose, and heptose sugars, respectively. Simple sugars are also known as **monosaccharides** (*mono* means "one"; *saccharide* means "sugar"), because each molecule consists of a single sugar unit. Glucose is a hexose monosaccharide, a six-carbon simple sugar.

Although all sugars are chemically similar, consisting of a carbon chain, a series of hydroxyls, and a reactive chemi-

Figure 15.6 Carbohydrates are a diverse group of molecules that fit the general chemical formula of $(CH_2O)_n$, where $n \geqslant 3$. They include five-carbon (pentose) sugars like ribose, six-carbon (hexose) sugars such as glucose, and its structural isomers fructose and galactose.

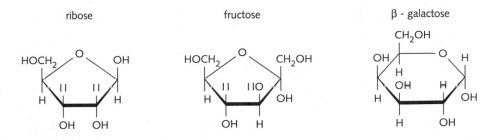

ribose   fructose   β - galactose

Figure 15.7 The formation of β-maltose, a disaccharide. Maltose is formed in a reaction between α- and β-glucose, which releases a water molecule. Sucrose, ordinary table sugar, is a disaccharide composed of α-glucose and β-fructose. Lactose, another disaccharide, is composed of β-galactose and β-glucose.

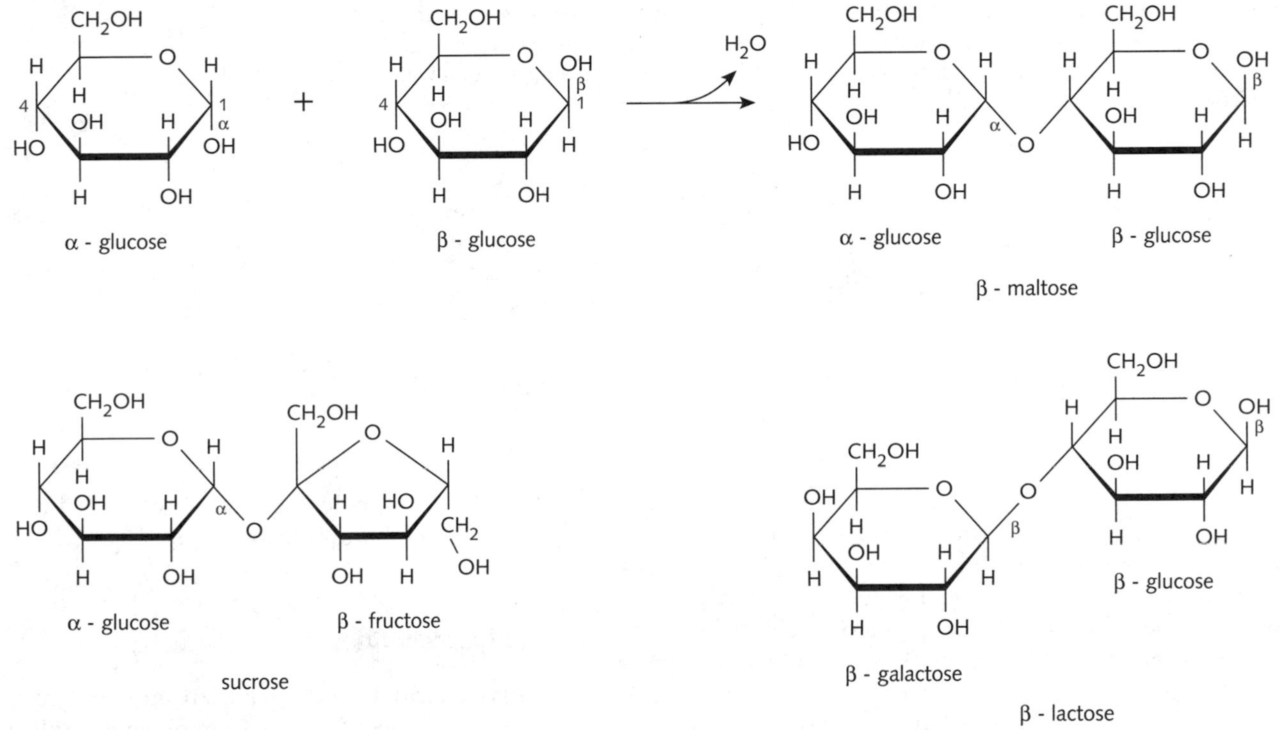

α - glucose   β - glucose   α - glucose   β - glucose

β - maltose

α - glucose   β - fructose

sucrose

β - galactose

β - glucose

β - lactose

cal group, there's plenty of flexibility in that basic design. Hexose sugars, for example, differ from each other in terms of the orientation of their hydroxyl groups and the kind of reactive group. Fructose, for example, is a hexose sugar with the same chemical formula as glucose, $C_6H_{12}O_6$. However, unlike glucose, it has a keto group on its number 2 carbon and forms a 5-membered ring (Fig. 15.6).

Disaccharides

You may have noticed that glucose, with a molecular weight of about 180, doesn't exactly qualify as a macro-

molecule. You might even be tempted to ask why it occupies a place in this chapter. The reason is that monosaccharides are used as building blocks for much larger molecules. These complex carbohydrates are formed by linking simple sugars together. When two sugars are joined by a covalent bond, the compound is known as a **disaccharide** (*di* means "two").

Some of the compounds we think of as sugar in our everyday experience are in fact disaccharides. Ordinary table sugar, **sucrose,** is a disaccharide formed by linking glucose and fructose. Other important disaccharides include *maltose* and *lactose* (Fig. 15.7).

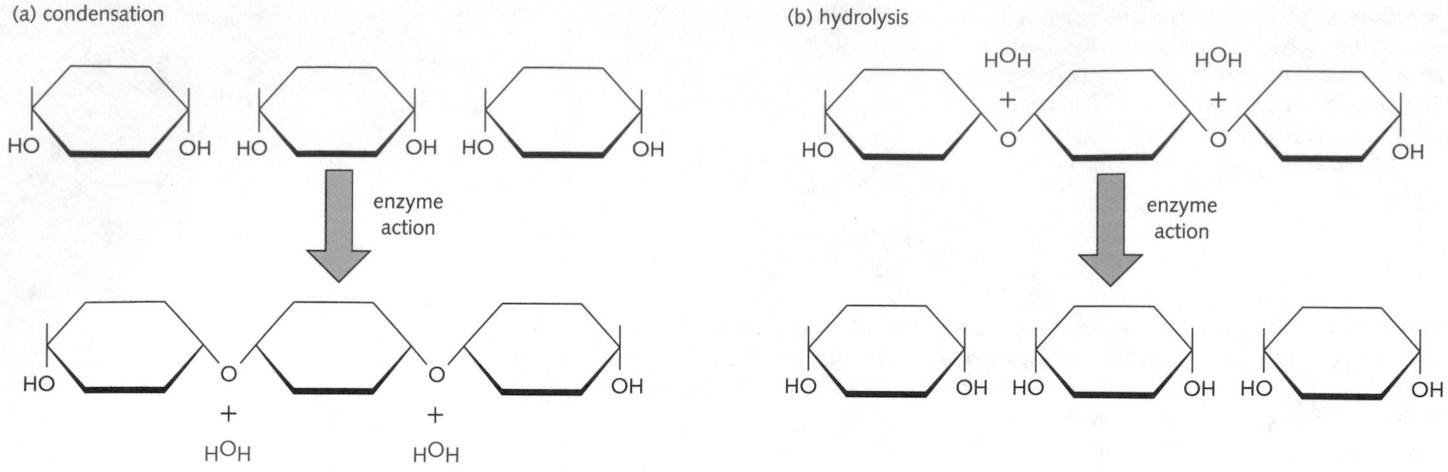

(a) condensation

(b) hydrolysis

enzyme action

enzyme action

Figure 15.8 **(a)** Formation of a trisaccharide by covalently bonding three monosaccharides releases two molecules of water in a condensation reaction. **(b)** These two covalent bonds may be broken by subsequent enzyme action in which hydrolysis, the addition of a water molecule, occurs at each linkage.

Maltose is produced by linking two glucose molecules. A covalent bond links the number 1 carbon of one glucose to the number 4 carbon of the other. The fact that this bond involves the number 1 carbon has an important consequence. Although α- and β-glucose can interconvert as long as they are individual molecules, once any sugar is linked to another through its number 1 carbon, the orientation of the hydroxyl on that number 1 carbon is locked into either the α or the β position. In maltose, the hydrogen atom on glucose's number 1 carbon is in the α position, so the link between the two sugars is referred to as an α-1,4 bond.

As you can see, the formation of the bond is accompanied by the loss of a hydrogen atom (H) from one sugar and the loss of a hydroxyl group (—OH) from the other (see Fig. 15.7 on the previous page). Because these two ions condense to form a water molecule, this kind of reaction is frequently known as a **condensation reaction** (Fig. 15.8a). Condensation reactions release water molecules.

Condensation reactions are common in the formation of biological polymers. The reverse reaction, in which polymers are broken down to monomers, consumes a water molecule and is known as **hydrolysis** ("water split-

ting"); water molecule atoms are divided between the subunits as the bond is broken (Fig. 15.8b).

Polysaccharides

Disaccharides are larger than monosaccharides, but they still don't qualify as macromolecules. Still, the ease with which two sugars can be joined should prepare us for the fact that it's possible to link sugars into very long chains. These chains may contain thousands of individual sugars, and these complex carbohydrates are known as **polysaccharides** (*poly* means "many").

An enormous variety of compounds can be formed by linking simple sugars (Fig. 15.9). **Starch** consists of long chains of glucose molecules connected by α-1,4 linkages. Plants store much of their excess carbohydrate in the form of starch. **Glycogen** is a polysaccharide similar to starch, except for the presence of a few α-1,6 linkages that allow the chain to branch. Glycogen is used in animal cells to store carbohydrate in a form that can be broken down quickly to use as a source of energy.

Although nearly all organisms can break α linkages, very few are able to break β-1,4 linkages. As we saw in

Chapter 3, carbohydrates that include the β-1,4 linkage are highly resistant to digestion by animals. Such carbohydrates include *chitin* and *cellulose.* **Cellulose** is the major molecule found in plant cell walls and in wood. Cellulose is identical to starch, except that its linkages are in the β-1,4 form. **Chitin** is a very important polysaccharide that consists of a sugar called *N*-acetylglucosamine joined by β-1,4 linkages. Chitin is tough and resilient, and it forms the external skeletons of insects and crustaceans.

These carbohydrate chains are almost unlimited in size; they may contain more than a million linkages. Starch granules can be found in a raw potato, clusters of glycogen in a liver cell, cellulose in a bristlecone pine, and chitin constituting the bulk of a crab's exoskeleton. They are carbohydrates, every one.

starch

glycogen

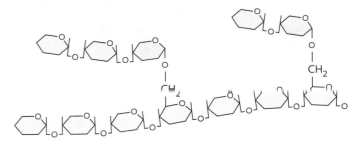

POLYMERIZATION

Don't be tempted to think that polysaccharides are the only macromolecules that are produced by hooking smaller molecules together to form chains. In fact, this "construction technique" is so common that a special name exists for it: **polymerization.**

Polymerization is the process of producing large molecules by joining smaller ones. The individual molecules (like simple sugars) are known as **monomers.** When two monomers are hooked together, a **dimer** is formed; three form a **trimer,** four a **tetramer,** and a large number is said to form a **polymer.** We've just seen that complex carbohydrates are polymers of simple sugars. The remaining two classes of macromolecules, nucleic acids and proteins, are polymers, too.

cellulose

chitin

Figure 15.9 The chemical structures of major polysaccharides.

NUCLEIC ACIDS

Nucleic acids are macromolecules that perform a task that was barely suspected when they were first discovered in the nineteenth century: they carry genetic information. We will not explore this function here (skip ahead to Chapter 20 if you're curious), but we will examine how nucleic acids are assembled.

The name *nucleic acid* reflects the fact that these molecules were first discovered in the cell nucleus and have a mildly acidic character. Nucleic acids are polymers of monomers known as **nucleotides.** The structure of a single nucleotide is shown in Fig. 15.10. As you can see, a nucleotide consists of three main parts: a pentose (five-carbon) sugar, a phosphate group, and a nitrogen-containing base. There are five different bases commonly found in nucleic acids (we'll examine these in detail in Chapter 20).

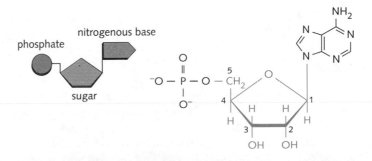

Figure 15.10 A nucleotide is formed from a five-carbon sugar, a nitrogenous base, and a phosphate group.

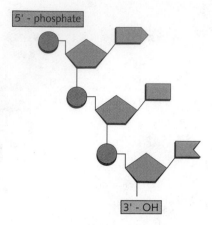

Figure 15.11 Covalent bonds between the sugar and phosphate groups of adjacent nucleotides form nucleic acids. Note that the two ends of a polynucleotide chain are chemically different: the bottom of the chain ends with a hydroxyl (—OH) on the number 3 carbon, and the top ends with a phosphate attached to the number five carbon.

The sugar can be either ribose or deoxyribose (so named because it's missing an oxygen on the number 2 carbon).

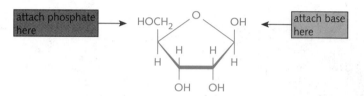

Like other sugars, these five-carbon sugars are very soluble in water, and they are able to form chemical linkages to other compounds. The two other parts of a nucleotide, the phosphate group and the nitrogenous base, are attached to the sugar at its number 1 and number 5 carbons, respectively:

The phosphate group is acidic, and once it has released a proton into solution, it carries a negative charge. The phosphate group also plays a key role in polymerization. As shown in Fig. 15.11, nucleotide polymers are formed by covalent bonds that join the phosphate group of one nucleotide and the number 3 carbon of another. When two nucleotides are linked by such a bond, a **dinucleotide** is formed. Long chains produced by many such linkages are **polynucleotides.**

When ribose sugars are used in a polynucleotide chain, the molecule is known as **ribonucleic acid (RNA).** If deoxyribose sugars are used, the polynucleotide is **deoxyribonucleic acid (DNA).** Polynucleotide chains may exceed 1 million nucleotides in length.

We should make two final points about nucleic acids. First, it may occur to you from Fig. 15.11 that the polynucleotide chain is really just a sugar–phosphate chain. The nitrogen bases stick out on the side and seem to play no role in the polymerization process. That impression is *superficially* correct. The bases don't play a role in linking nucleotides in the chain. However, remember that the biological role of nucleic acids is to carry information. That's where the different bases of DNA and RNA come into play. In effect, the sugar–phosphate chain is the paper upon which a message is written, but the bases are the written words themselves.

Second, it's very important to note that polynucleotide chains always have a particular orientation. One end of the chain always contains a free phosphate group attached to the number 5 carbon of the sugar, and the other end has a free —OH group on the number 3 carbon. Therefore, we can call the two ends of the chain the 5′ and 3′ ends, respectively. Why is this important? Remember that nucleic acids carry information. Quite simply, the 3′ and 5′ ends tell the cell in which direction the message is to be read.

PROTEINS

Proteins are probably the most diverse class of macromolecules. Proteins are formed by the polymerization of monomers known as **amino acids.** There are structural proteins like *keratin,* which forms hair, skin, and fingernails; proteins that transmit chemical messages, such as *insulin;* proteins that carry oxygen in the blood, such as *hemoglobin;* and proteins that catalyze chemical reactions, such as *amylase,* which is found in saliva and breaks starch down into simple sugars. Proteins are involved in every activity in the cell. They pump molecules across cell membranes, form a cytoskeleton that supports cell movement, synthesize other macromolecules, and help to regulate gene expression and cell recognition.

Amino Acids

More than 40 kinds of amino acids are found in living organisms, but only about 20 different types are so common that they are found in all cells.

Amino acids are simple chemical compounds that contain three principal functional groups: an amino group ($—NH_2$), a carboxyl group ($—COOH$), and a third chemical group that differs from one amino acid to the next. This third group is known simply as an **R group.**

amino acid $NH_2—\overset{\overset{\textstyle R}{|}}{\underset{\underset{\textstyle H}{|}}{C}}—COOH$

The simplest amino acid is *glycine,* in which the R group is merely a hydrogen atom. The structures of the most common amino acids and their R groups are shown in Table 15.1 on p. 304. Because all amino acids have an amino (base) group and a carboxyl (acid) group, amino acids have some properties of both acids and bases. You'd expect any small molecule with both an acidic and a basic functional group to be very soluble, and that's true for amino acids.

Polypeptides

Amino acids can be joined one to another by forming a covalent bond that links the amino group of one to the carboxyl group of another. This condensation reaction removes a molecule of water (Fig. 15.12). Two amino acids

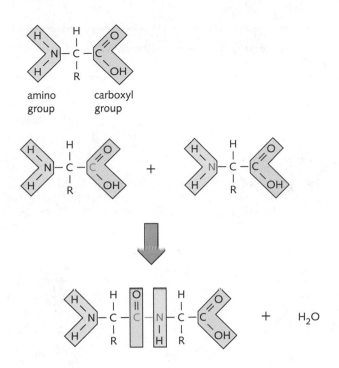

Figure 15.12   The formation of a peptide bond. The bond is formed between the carbon of a carboxyl group and the nitrogen of an amino group. A molecule of water is split off when the bond is formed.

Table 15.1 *Amino Acids and the Characteristics of Their R Groups**

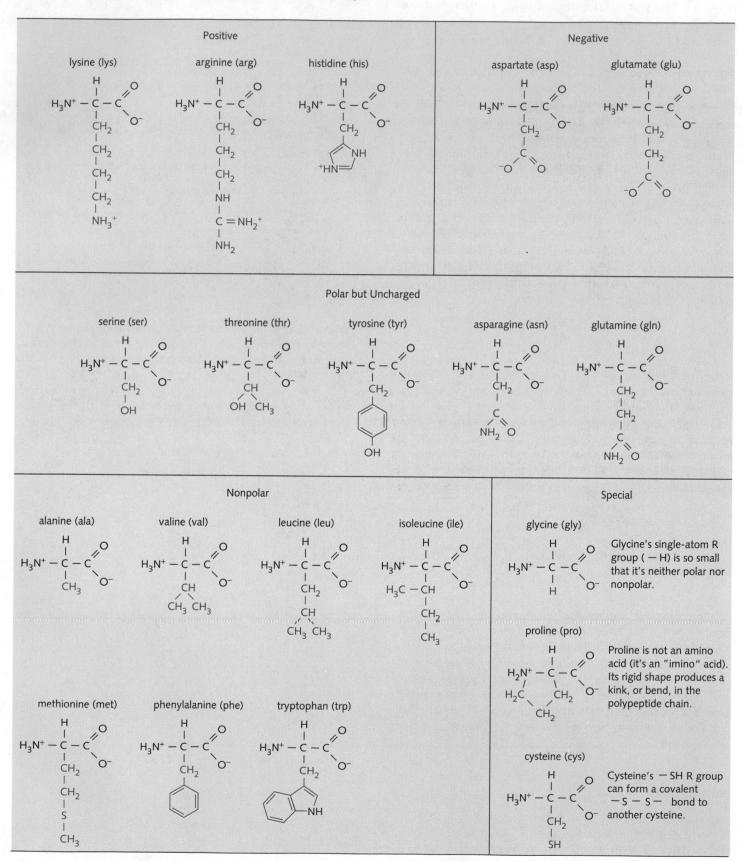

*The 20 most common amino acids as found in their dissolved state in living organisms. They are grouped according to the chemistry of their R groups.

joined in this way are a **dipeptide,** three a **tripeptide;** and long chains are known as **polypeptides.** Polypeptides as long as 1000 amino acids are common.

As shown in Fig. 15.13, one end of a polypeptide chain has a "free" amino group and one end has a free carboxyl group. We call the carboxyl end the **C-terminus** and the amino end the **N-terminus.** Polypeptide chains have a direction because each end is chemically different.

Why *polypeptide* and not *polyaminoacid?* The word *peptide* comes from the Greek word *pepticos,* which means "cooked," as in cooked food. Cooked? Though that seems like a stretch, the first substances that could break down proteins were identified from digestive fluids, where, naturally enough, they break down proteins in food. The fragments left over when these substances broke down proteins were called peptides long before the smallest of them were identified as amino acids. The name *peptide* has remained part of the scientific vocabulary ever since. The covalent bond that links one amino acid to another is known as a **peptide bond.**

PROTEIN STRUCTURE

The Possibilities of a Polypeptide

At first glance, polypeptides look a little dull. After all, they're nothing but chains of amino acids. If we were to examine nothing but the carbon–nitrogen backbone of a polypeptide, this view would be confirmed (Fig. 15.13).

There's a wild card lurking in this deceptive simplicity, however: the R groups. Each amino acid has a different R group or side group. Some of them are small hydrocarbon chains (as in leucine and isoleucine). Some have complicated ring structures (as do phenylalanine and tryptophan). Some side groups are actually acids (glutamic acid) or bases (arginine) themselves. One amino acid has a sulfhydryl side group (—SH) that can form covalent bonds on its own. Such diversity means that the carbon–nitrogen chain is just a backdrop for the principal chemical action, which is carried out by the R groups.

Think about what this means for a polypeptide. The molecule could be acidic or basic, charged or uncharged. It might be neutral at one end and basic at the other; it could contain large bulky R groups or small ones; and these R groups might be able to form covalent bonds to tie different polypeptide strands together. The amino acids represent a chemical construction set with which it is possible to build a vast number of different molecules with completely different chemical properties.

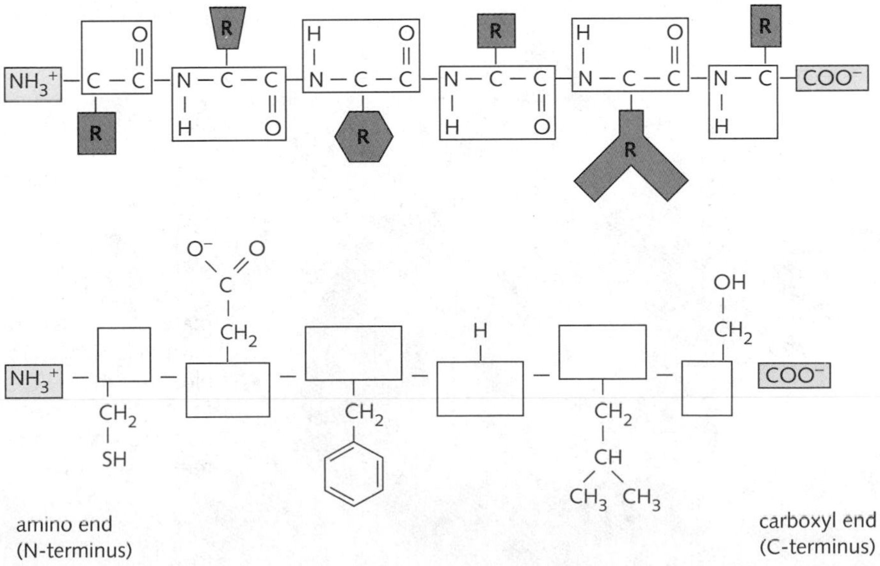

Figure 15.13 When actual side groups are shown instead of the representative "R," we see how the differences in side-group size and charge will affect the behavior of the polypeptide as a whole. Note that we show the amino- and carboxyl-termini as uncharged groups. In nature, because polypeptides and amino acids are in their dissolved state, the termini are actually charged, as shown for amino acids in Table 15.1.

amino end
(N-terminus)

carboxyl end
(C-terminus)

(a)

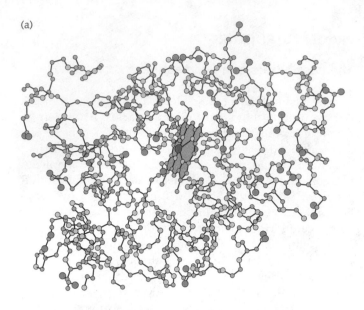

(b) primary

valine
leucine
serine
glutamate
glycine
glutamate
tryptophan
glutamine
leucine
valine
leucine
histidine
valine
tryptophan
alanine
lysine

(c) secondary

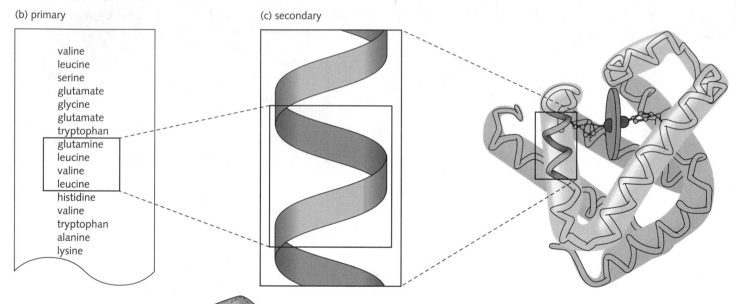

(d) tertiary

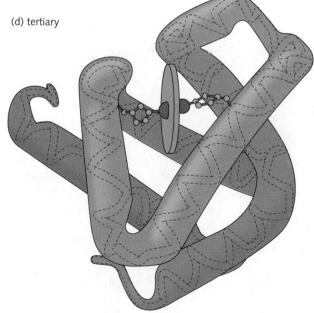

Figure 15.14 **(a)** A detailed drawing of the principal atoms in myoglobin, an oxygen-binding protein with 153 amino acids. **(b)** The primary structure of myoglobin is its amino acid sequence, part of which is shown here. **(c)** A schematic drawing of the carbon–nitrogen backbone of myoglobin is easier to understand. Note the many coiled (helical) regions in the polypeptide. Secondary structure (RIGHT) is the localized pattern in which the polypeptide chain is folded or coiled. Approximately 78 percent of the myoglobin polypeptide chain is folded into a pattern known as the α-helix, although other folding patterns are visible, too. **(d)** The tertiary structure is the three-dimensional pattern formed by the complete folding pattern of the polypeptide chain. Note how the small region of helical secondary structure fits into the larger tertiary structure of myoglobin.

Folding a Polypeptide

The three-dimensional nature of a macromolecule is always important, but for proteins it is so crucial that biochemists have made a special field of it. Although polypeptides are initially made as linear chains (primary structure), they don't stay that way for long. They quickly fold into complex shapes, or "conformations" (secondary and tertiary), that are maintained by attractive forces between the amino acids in the polypeptide chain.

Figure 15.14(a) shows the three-dimensional conformation of myoglobin, an oxygen-binding protein found in muscles. At first glance, the pattern seems almost too complex to describe. Indeed, a two-dimensional drawing of the protein gives us very little understanding of its actual shape.

Primary structure Is there a way to analyze this twisted pattern? To begin with, we could list the amino acid sequence of the polypeptide chain, starting with the amino terminus and going all the way to the carboxyl terminus, a total of 153 amino acids. This sequence is known as the **primary structure** of a protein (Fig. 15.14b).

Secondary structure Can we go a little further? Let's draw the protein in a slightly different way, showing the carbon–nitrogen backbone of the polypeptide chain as a ribbon-like band (Fig. 15.14c). In this simplified drawing, an interesting pattern emerges. It turns out that many regions of the polypeptide chain are twisted into regular patterns that resemble coiled springs, although other portions are folded quite differently. The short-range folding of the polypeptide chain is known as the **secondary structure,** and the coils visible in Fig. 15.14(c) and (d) are just one form of short-range folding.

As much as 78 percent of the myoglobin polypeptide chain is folded in the coil pattern, which is more properly known as a **helix.** Clearly, there has to be a reason for this, and Fig. 15.15(a) shows what the reason is. Hydrogen bonds between —C=O and —N—H groups hold the polypeptide chain in this helical shape. Several different helical folding patterns are known in nature, but this one, called the **α-helix,** is by far the most common. The α-helix was first described in 1951 by Linus Pauling, the great American chemist, who realized that the presence of so many hydrogen bonds would give the helix exceptional stability.

As common as it is, the α-helix is not the only form of secondary structure. Another folding pattern, the **β-sheet,** is a type of secondary structure that is very common in fibrous proteins, such as those that make up bird feathers and silk. In the β-sheet configuration, hydrogen bonds link regions of the polypeptide chain that run in parallel or antiparallel directions (Fig. 15.15b). The β-sheet configuration is strong, stable, and flexible.

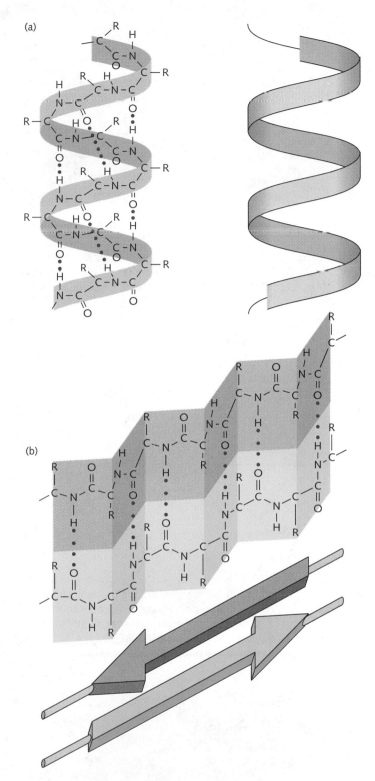

Figure 15.15 **(a)** The α-helix (LEFT) and its diagrammatic representation. The helix is stabilized by hydrogen bonds (dotted lines). Note that the R groups point out to the sides of the helix. **(b)** The β-sheet and its diagrammatic representation. The flattened configuration of the β-sheet is stabilized by hydrogen bonds between adjacent chains (dotted lines). The β-sheets may be formed in either the antiparallel direction (as shown, where two chains run in parallel, opposite directions) or the parallel direction (where both chains run the same way).

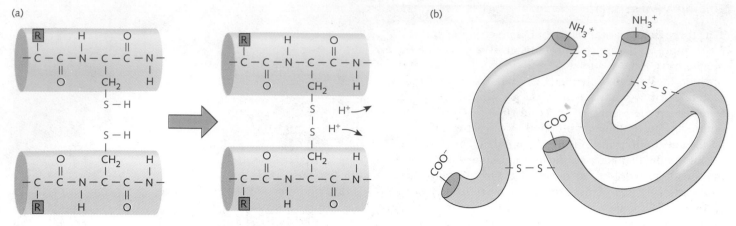

Figure 15.16 **(a)** Formation of a disulfide bond between the sulfhydryl (—SH) groups of cysteine amino acids. **(b)** These covalent bonds may link adjacent polypeptide chains, or two regions of the same chain, as shown in this diagram of human insulin.

Figure 15.17 A model of the oxygen-binding protein, hemoglobin. The quarternary structure of a protein describes the arrangement of its subunits, each of which contains a single oxygen-binding heme group. Heme is a prosthetic group, a non–amino acid portion of the protein.

prosthetic group

There are other kinds of secondary structure as well, and we have mentioned the α-helix and β-sheet configurations only because they are so common. In fact, many proteins have large regions of chain folding that do not follow a definite pattern and are completely unique. Some biochemists describe such regions as folded in an "open-chain" pattern, meaning that their secondary structure fits no special pattern.

Tertiary structure Even if we were to discover that the secondary structure of a polypeptide was 50 percent α-helix and 50 percent β-sheet, we would still know very little about the actual three-dimensional structure of the polypeptide. To learn this, we would have to study the complete folding of the polypeptide in space. This large-scale folding pattern is called **tertiary structure** (Fig. 15.14d on p. 306). Tertiary structure is stabilized by a number of bonds that can be formed between the R groups of amino acids. These bonds include hydrogen bonds, ionic and electrostatic interactions, weak interactions between uncharged side groups, and covalent bonds.

The covalent bonds are particularly interesting. They form between two cysteines, amino acids that have the —SH (sulfhydryl) side group. Under the right chemical conditions, a covalent bond can form between two cysteines (Fig. 15.16). These bonds may link two regions of the polypeptide chain that are far apart in the primary sequence, or they may join two different polypeptide chains. This bond, sometimes called a **disulfide bond,** is very important in holding together the shapes of large and complicated proteins.

Disulfide bonds are the basis of some very practical chemistry. A hair permanent, a series of washes and rinses used to produce curls in normally straight hair, works by

Zippers, Fingers, Barrels, and Hairpins

The structural possibilities for proteins are so great, one might be tempted to think that no two proteins would ever have similar shapes. As it turns out, that's just not true. In fact, many different proteins are built along common themes, or "motifs." When a certain pattern appears time after time in different proteins, biochemists cannot resist the temptation to give it a colorful name. As a result, students of protein structure must master the intricacies of the "leucine zipper," the "zinc finger," the "α,β-barrel," and the "β-hairpin." The figure shows some of these common themes.

As interesting as these shapes are, their existence leads to an obvious question: *Why should proteins with different functions have such similar parts?* Although the final answer is not yet in, it may turn out to be deceptively simple. Think of two machines with vastly different functions, such as a pocket watch and a bicycle. As different as the

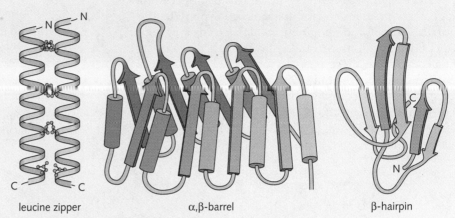

leucine zipper   α,β-barrel   β-hairpin

Three examples of common protein structure "motifs."

watch and the bike are, some of their parts, such as their circular gears, are nearly identical in structure (although not in size). Both a bicycle and a watch must transfer rotary motion between two wheels of different sizes, and a gear is the perfect way to do the job. There-

fore, they both have a few similar parts.

In proteins, the story is probably the same. Different proteins may have different tasks to perform; but like the pocket watch and the bicycle, at least a few of those tasks may be solved by the use of similar biochemical parts.

altering protein tertiary structure. First, the hair is rinsed with a solution containing a chemical agent that breaks the disulfide bonds in keratin, the principal protein found in hair. Then the hair is rolled into the desired shape, and finally, it is treated with another chemical that re-forms disulfide bonds. The new bonds help to make the wave permanent—all because of protein chemistry.

Quaternary structure The fourth level of protein structure, **quaternary structure,** describes the way a protein is assembled when it consists of more than one polypeptide chain. As an example, we might take hemoglobin, the oxygen-carrying protein found in human blood, which is composed of four polypeptides. Its quaternary structure is the way in which these four polypeptide subunits fit together (Fig. 15.17).

The quaternary structure of a protein is determined by the same kinds of forces that control tertiary structure: hydrogen bonds, ionic attractions, disulfide bonds, and weak interactions between amino acid R groups. All of

these interactions influence the ultimate shape of a protein, and shape determines which chemical groups are exposed on the protein's outer surface.

Prosthetic groups Although the tremendous chemical diversity of the amino acids would seem to give a protein the chemical potential to do just about anything, there are some tasks that are best performed by another type of molecule. Good examples are hemoglobin and myoglobin, both of which bind oxygen. In each case, oxygen (O_2) molecules bind not to the polypeptide but to a chemical group known as *heme,* which itself contains an iron atom in its center (see Fig. 15.17 as well as Fig. 15.14).

Heme is not an amino acid. It is almost a "foreigner" within the folded structure of the polypeptide chain. Heme is called a **prosthetic group,** a colorful name which suggests that it is almost like a prosthesis, an artificial limb. Prosthetic groups are non–amino acid portions of proteins, and many proteins contain them.

Polypeptides and Proteins

If polymers of amino acids are called *polypeptides,* then what's a *protein?* Isn't it just another name for polypeptide? Not exactly. Polypeptides are polymers of amino acids, but **proteins** are functional units containing one or more polypeptides. Hemoglobin (as shown in Fig. 15.17) is a perfect example of why we make this distinction. Although some proteins consist of a single polypeptide, this is not the case for hemoglobin, which consists of four polypeptides and four prosthetic groups, all properly assembled. None of these parts is considered a protein in and of itself, because none of them is a functional unit. It's only when the complete hemoglobin molecule is assembled that we are justified in calling it a *protein.*

STUDY FOCUS

After studying this chapter, you should be able to:

- Describe the way the presence of different functional groups in a lipid molecule gives the molecule water-insoluble parts.

- Explain the principle of polymerization and its importance in the synthesis of natural and artificial macromolecules.

- Describe the four major classes of biological macromolecules and the nature of the chemical bonds that are involved in each of them.

- Illustrate the richness and chemical flexibility of protein structure, and describe the levels of protein structure.

- Present a number of examples of the importance of certain macromolecules in key cellular structures.

SUMMARY

The chemistry of living things is dominated by macromolecules, large compounds that contain a variety of functional groups. The four major classes of biological macromolecules are lipids, carbohydrates, nucleic acids, and proteins. Lipids, assembled from fatty acids and other components, are oily macromolecules that dissolve in organic solvents. They include polar and neutral lipids and also the steroids. Lipids play key roles in cell organization and communication, forming the structural basis for cell membranes and frequently being used as chemical messengers.

Carbohydrates are based on simple sugars like ribose and glucose. These simple sugars may be joined by covalent bonds to form disaccharides, trisaccharides, and complex polysaccharides such as starch, glycogen, and cellulose. The production of complex compounds by joining individual units, a process known as polymerization, is the principal way in which macromolecules are built.

Nucleic acids are macromolecules produced by the polymerization of nucleotides. Nucleotides are themselves composed of a five-carbon sugar, a nitrogenous base, and a phosphate group. The two principal forms of nucleic acids are DNA and RNA. These molecules serve as the carriers of biological information.

Proteins are formed by the polymerization of amino acids. Amino acids themselves are joined together by covalent bonds, known as peptide bonds, to form chains called polypeptides. The folding of polypeptide chains to produce the complete structure of a protein can be described on four different structural levels. Primary structure is the amino acid sequence of a polypeptide. Secondary structure is the localized folding of the polypeptide chain, as typified by the α-helix and β-sheet patterns. Tertiary structure is the complete three-dimensional folding of such a chain. And quarternary structure describes the spatial arrangement of polypeptides in a protein that contains more than one polypeptide subunit. Proteins may also contain prosthetic groups, components like metal atoms or heme groups, that are not derived from amino acids.

TERMS AND CONCEPTS

fatty acids *294*	nucleotides *301*
lipids *294*	ribonucleic acid (RNA) *302*
saturated *294*	deoxyribonucleic acid
polyunsaturated *294*	(DNA) *302*
phospholipids *295*	amino acids *303*
steroids *296*	polypeptides *305*
carbohydrates *296*	peptide bond *305*
monosaccharides *298*	α-helix *307*
condensation reaction *300*	β-sheet *307*
hydrolysis *300*	disulfide bond *308*
polysaccharides *300*	prosthetic group *309*
polymerization *301*	proteins *310*
nucleic acids *301*	

REVIEW

Objective Questions (Answers in Appendix)

1. A saturated fat has _____ than an unsaturated fat of the same size.
   (a) more fatty acid molecules
   (b) more hydrogens per carbon
   (c) shorter hydrocarbon chains
   (d) fewer carbon atoms

2. A lipid that exhibits a ring structure is known as a
   (a) fatty acid.   (c) triglyceride.
   (b) glycerol.   (d) steroid.

3. The process of hydrolysis involves
   (a) evaporation of excess water upon heating the polymer.
   (b) breaking down a long chain of carbohydrates by adding water.
   (c) linking small chains of carbohydrates by removing water.
   (d) removal of hydrogen ions from the long chain of carbohydrates by heating.

4. Which of the following is a polysaccharide?
   (a) cellulose
   (c) glucose
   (b) hemoglobin
   (d) keratin

5. Which of the following is an amino group?
   (a) —COOH
   (c) H—
   (b) OH
   (d) NH$_2$

6. The major linkage between amino acids is the
   (a) hydrogen bond.
   (c) peptide bond.
   (b) nucleotide bond.
   (d) polymer bond.

7. The primary structure of a protein consists of
   (a) a linear sequence of amino acids.
   (b) a β-sheet configuration.
   (c) an α-helix pattern of amino acids.
   (d) more than one polypeptide chain with hydrogen bonds between each.

8. A quaternary protein structure has
   (a) four different kinds of peptide bonds.
   (b) more than one polypeptide chain.
   (c) four infoldings of the polypeptide chain.
   (d) four types of amino acids.

Discussion Questions

9. What are the four principal kinds of biological macromolecules? Give an example of each.

10. How is polymerization associated with the formation of each of the four main biological polymers?

11. How do the four levels of structure in proteins differ from one another?

12. What important characteristic of lipids makes it possible for them to self-assemble into larger structures?

13. Polynucleotides and polypeptides are linear polymers with certain similarities, including the fact that each polymer has chemically distinct ends (5′ and 3′ in the case of polynucleotides and carboxyl and amino in the case of polypeptides), and each contains a series of side groups (R groups and nitrogenous bases). Despite these similarities, they display remarkable chemical differences. Focusing on the side groups, what are the greatest chemical differences between polypeptides and polynucleotides?

READINGS

Brandon, C., and J. Tooze. *Introduction to Protein Structure.* New York: Garland Publishing, 1991. An extremely well-illustrated book that summarizes a wealth of knowledge on protein architecture and function.

Richards, F. M. "The protein-folding problem." *Scientific American* 264 (January 1991): 54–60. Predicting protein folding is *not* an exact science, at least not yet. This lively article explains recent progress in trying to solve this important problem.

Stryer, L. *Molecular Design of Life.* New York: W. H. Freeman, 1989. A readable and beautifully illustrated book based on Stryer's excellent biochemistry text. Intended for general audiences, it is a very good reference on proteins and nucleic acids.

Sharon, N. "Carbohydrates." *Scientific American* 243 (November 1980): 90–116. A straightforward description of the structure of carbohydrates and the biological roles they play.

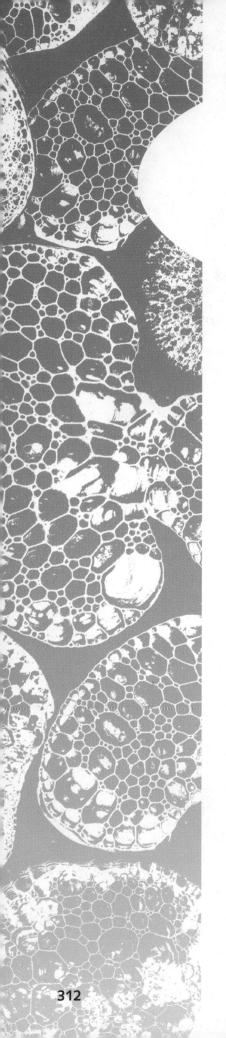

16

Cell Organization

S ome animals know the world through sound. They sense vibrations so precisely that they can fly through the night sky or glide through dark waters, turning and darting through the smallest openings. Others know a chemical world of tastes and scents. They can follow a chemical clue for miles to find food or water or to locate a mate. Humans have senses of sound and smell, of course, but the structure of our brains reveals that we humans *see* the world around us. We are visual creatures, and we know the world through our eyes. The dominance of the visual sense is reflected in our brains, where high priority is given to the processing of visual information, and also in our culture, where images and written words are perhaps the most powerful means of communication.

The limits of the visual sense are obvious, too. Some of the most interesting things in the world are just too small to be seen. Or at least they *were* too small. Less than four centuries ago, humans first extended the powers of sight with a new invention, a device that has opened new worlds to human eyes—the microscope.

The invention of the microscope was important to all of the sciences, of course, but to none more than to biology. History tells us that science and technology move hand in hand, advances in one fueling new achievements in the other. Never was the dependence of science upon technology more obvious than in the case of the microscope. By using it, biologists were able to see, for the very first time, the basic unit of life itself. The microscope made possible the discovery of the living cell.

THE MICROSCOPE: EXTENDING THE SENSES

The basic principles of a microscope are simple and easy to describe. A thin piece of glass, ground down into a lens, will focus parallel light rays to a single point, called the *focal point,* on the opposite side of the lens (Fig. 16.1). Figure 16.1 also shows what happens when the same lens is used to focus light from an object that is not at the focal point: An *image* of the object is formed at some distance from the lens. Depending on the

LS
N
LL

placement of the original object and the power of the lens, this image can be magnified or reduced in size from that of the object. The ability to change the apparent sizes of objects with lenses means that magnified images of objects can be made that reveal detail too fine to be seen without assistance. A lens enables us to see things that are either very small or very far away.

The Invention of the Microscope

History does not tell us precisely who invented the microscope, although certain names do stand out. Anton van Leeuwenhoek (1632–1723) was a young man in Holland when he was apprenticed to a dry goods store by his family. During his apprenticeship, he learned to use magnifying lenses to examine the weave of fine cloth, a standard tool of the trade, but van Leeuwenhoek was interested in other things. As far as we know, he never went very far in the dry goods trade. Instead, history records him as one of the first to use the microscope for biology.

Van Leeuwenhoek's microscope was a simple affair. A single lens mounted in a metal plate enabled him to observe a world of tiny animals and fantastic plants. They inhabited the very water his neighbors swam in and drank, and they were more numerous than the birds of the air.

Others added to his observations, including an Englishman named Robert Hooke (1635–1703). While describing a thin slice of cork, Hooke wrote that the material was like a "honeycomb" and that the air within the spongy cork "is perfectly enclosed within little boxes or cells distinct from one another" (Fig. 16.2). The substructures seen by van Leeuwenhoek and Hooke are called **cells.** The first people to claim that all living things were composed of cells were Matthias Schleiden (1838) for plants and Theodore Schwann (1839) for animals. Today we often consider Schleiden and Schwann the authors of the **cell theory,** a premise that all living organisms are composed of individual, self-reproducing, living structures known as cells.

The Light Microscope

A typical **light microscope** contains two main lenses: an *objective lens,* which forms an image of the specimen within the barrel of the microscope, and a *viewing lens* (or *ocular*), which produces an image for an observer (Fig. 16.3). The enlargement of small regions in a sample enables us to see fine detail. When we use more powerful lenses, the image is bigger still and the detail finer and finer—but only up to a point. The wavelength of light places a limit on the detail that a microscope can produce.

Under the very best conditions, the *resolution limit* of a light microscope is about 0.2 μm. This means that the light microscope enables us to see about 500 times

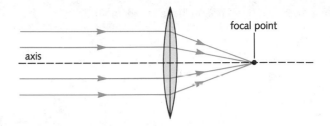

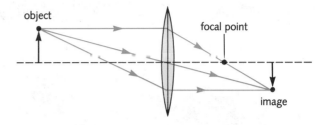

Figure 16.1   An ideal converging lens refracts parallel rays of light so that they converge at a single focal point.

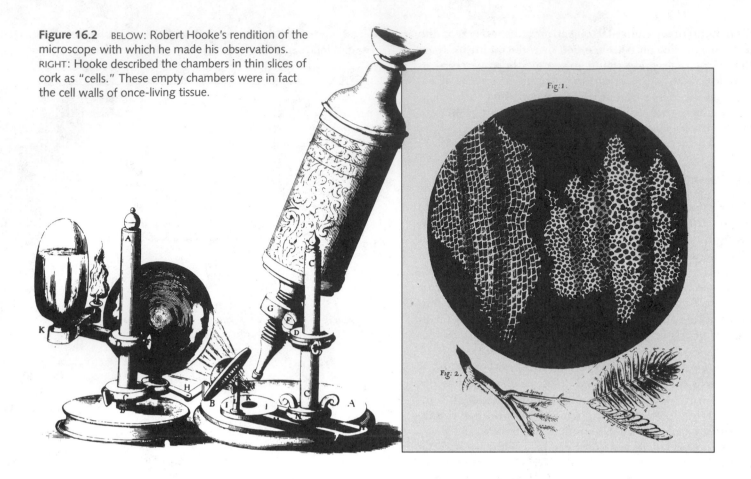

Figure 16.2 BELOW: Robert Hooke's rendition of the microscope with which he made his observations. RIGHT: Hooke described the chambers in thin slices of cork as "cells." These empty chambers were in fact the cell walls of once-living tissue.

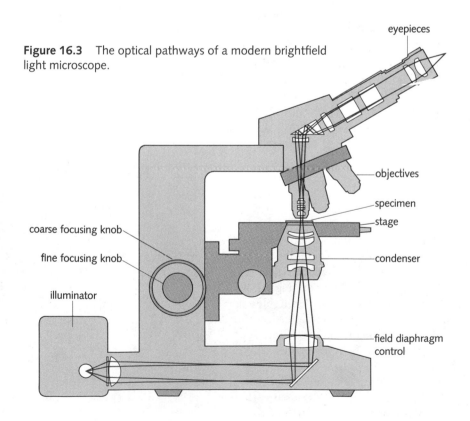

Figure 16.3 The optical pathways of a modern brightfield light microscope.

eyepieces

objectives

specimen

stage

condenser

field diaphragm control

coarse focusing knob

fine focusing knob

illuminator

the detail that we can see without it (0.1 mm = 500 × 0.2 μm). Because living cells are much larger than this resolution limit (a typical cell is about 30 μm in diameter), cells and some of their internal structures are quite visible. However, we are not able to see the smallest structures within the cell, nor can we view viruses, tiny infectious particles, or individual molecules, all of which are below the resolution limit of the instrument.

The Electron Microscope

In the 1920s scientists discovered that beams of electrons can be focused by magnetic fields in the same way that glass lenses focus light (Fig. 16.4). This made it possible to construct an **electron microscope.** Because the wavelength of an electron beam is much shorter than the wavelength of light, electron microscopes have resolution limits in the vicinity of 0.2 nm, almost 1000 times smaller than the best light microscopes.

However, this resolution has come at a price. Electrons are scattered by the molecules in air. Therefore, any material to be observed in an electron microscope must be placed in a vacuum. Samples are first treated with a chemical preservative, or fixative, to preserve their structure. Then they are dehydrated (all of their water removed) so that there is no water to boil in the vacuum of the microscope. Finally, biologists must slice most samples into thin slices (usually about 50 nm thick) to enable

electrons to penetrate them. Despite these limitations, biologists have used electron microscopes to gain a great deal of knowledge about the organization of living things.

In a **transmission electron microscope,** the image is formed on a screen at the bottom of the microscope column. When large numbers of electrons strike the screen, it glows, and the user can "see" an image of the sample.

In the **scanning electron microscope,** a narrow beam of electrons is focused on one spot at the surface of a sample. When electrons strike the surface of the sample, their energy dislodges other electrons in the sample. These "secondary" electrons can be measured by electron

Figure 16.4 TOP: A modern scanning electron microscope. LEFT: In transmission electron microscopes, a series of electromagnetic lenses focuses a beam of electrons on a thin sample inserted into the specimen chamber. Electrons that pass through the specimen are focused into an image by another series of lenses. RIGHT: In scanning electron microscopes, a narrow beam of electrons is scanned across the surface of a specimen in a two-dimensional pattern. Electrons or other radiation that this beam produces at the surface of the sample are collected by a series of detectors and then shown on a television-like display, which scans in steps with the electron beam. Samples observed in each type of electron microscope must be placed within the vacuum of the microscope column.

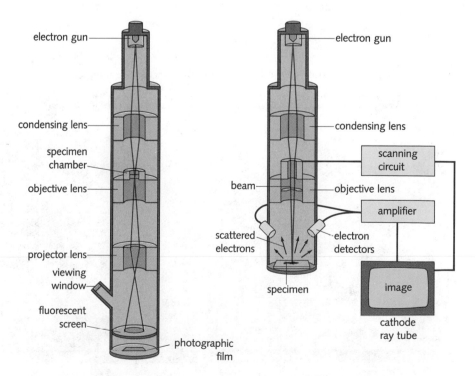

detectors. As the electron beam is scanned across the surface of the sample, continuous measurements are made of the secondary electrons striking the detector, and these measurements are displayed on a video screen.

The scanning electron microscope enables us to get a three-dimensional look at the surfaces of objects with great accuracy. The scanning electron microscope, however, suffers from some of the same limitations as the transmission electron microscope, including the fact that its specimens must be placed in a vacuum chamber.

Figure 16.5 shows different views of a single cell type as it is seen through the light microscope, the scanning electron microscope, and the transmission electron microscope.

THE CELL: A DETAILED LOOK

There is, of course, no such thing as a "typical" eukaryotic cell, any more than there is a "typical" college student. For better or for worse, however, it's sometimes necessary to pretend that there is (in both cases), because it makes study and generalization much easier. Figure 16.6 shows the basic structure of two such typical cells, at least one of which might be found in a typical college student. Figures 16.7 and 16.8 on the following pages show animal and plant cells in greater detail.

How do we analyze the complexity of the living cell? Our strategy is one that cell biologists have been using for several decades. One at a time, we'll look at the distinctive structures within the cell, study their composition and function, and then see how they relate to each other.

Many structures within the cell are known as **organelles** (literally, "little organs"). Organelles are membrane-bounded structures that perform a series of specialized tasks. In a way, organelles represent one of the major themes of the eukaryotic cell: important jobs are carried out in specialized compartments of the cell. The organelles in most eukaryotic cells can be broken down into three categories according to their function, as shown in Table 16.1.

Table 16.1 *The Function of Organelles in Eukaryotic Cells*

Function	Organelle
Genetic information and control: Genetic information in the form of DNA is stored, copied, and expressed.	Nucleus
Synthesis, transport, recycling: Proteins, carbohydrates, and lipids are synthesized and moved through the cell.	Endoplasmic reticulum Golgi apparatus Secretory vesicles, vacuoles Lysosomes
Energy: Usable energy is obtained from food or from sunlight.	Mitochondria Chloroplasts

Figure 16.5 Three different views of a single cell type produced in the (LEFT) light microscope, (CENTER) transmission electron microscope, and (RIGHT) scanning electron microscope.

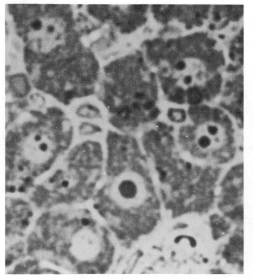 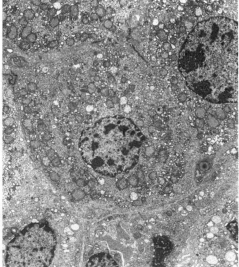 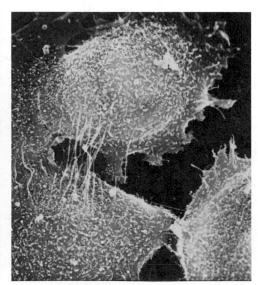

Figure 16.6   Diagrams showing the basic organization of an animal cell (RIGHT) and a plant cell (BELOW). Although these models do not do justice to the tremendous diversity of cellular structure in animals and plants, they do show some of the common features of many cell types.

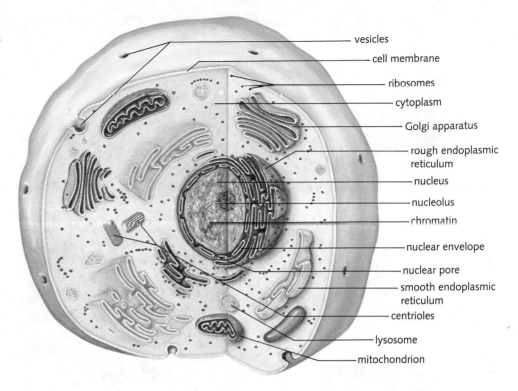

vesicles
cell membrane
ribosomes
cytoplasm
Golgi apparatus
rough endoplasmic reticulum
nucleus
nucleolus
chromatin
nuclear envelope
nuclear pore
smooth endoplasmic reticulum
centrioles
lysosome
mitochondrion

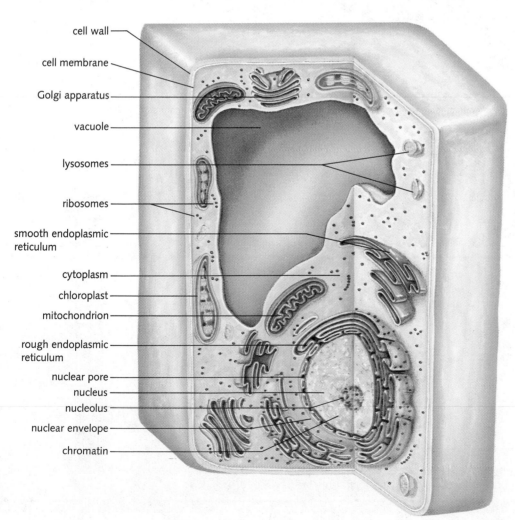

cell wall
cell membrane
Golgi apparatus
vacuole
lysosomes
ribosomes
smooth endoplasmic reticulum
cytoplasm
chloroplast
mitochondrion
rough endoplasmic reticulum
nuclear pore
nucleus
nucleolus
nuclear envelope
chromatin

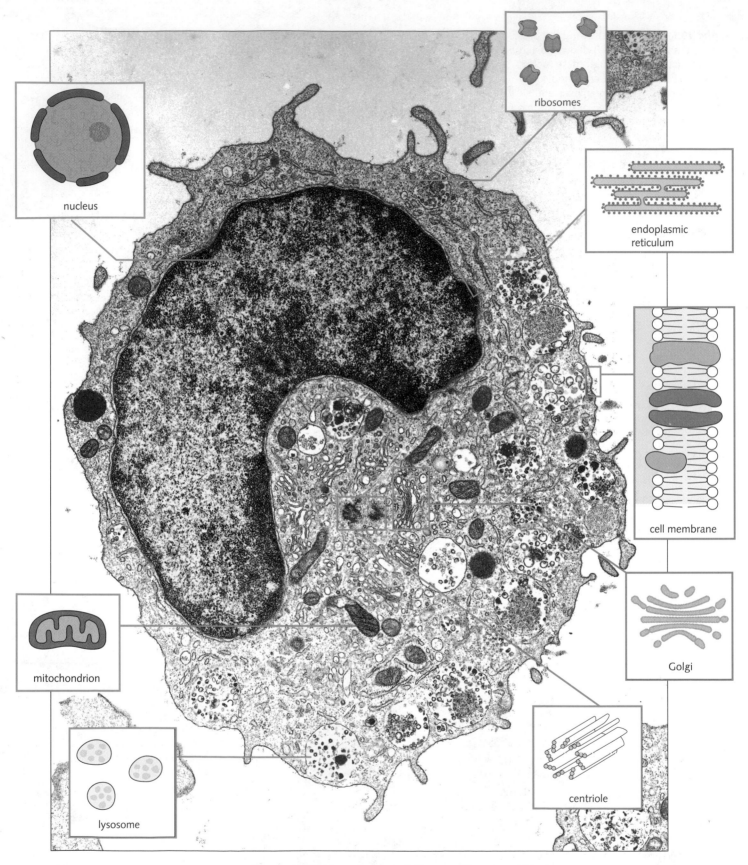

Figure 16.7 This electron micrograph shows a fibroblast cell from rat connective tissue. (Magnification factor: 18,000)

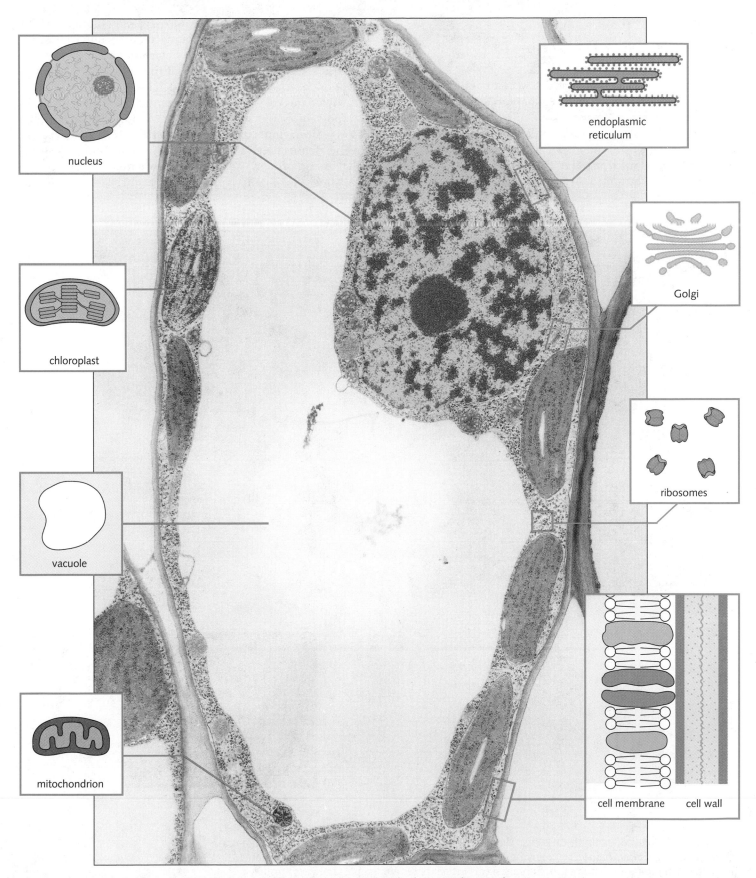

nucleus

endoplasmic reticulum

chloroplast

Golgi

vacuole

ribosomes

mitochondrion

cell membrane   cell wall

Figure 16.8   This electron micrograph shows a cell from a bean seedling leaf. (Magnification factor: 12,000)

The Nucleus: The First Organelle

The nucleus is the main repository of genetic information in the cell. As we saw in Chapter 9, the chromosomes, on which genes are located, are found in the nucleus. The nucleus is surrounded by a **nuclear envelope** consisting of two membranes that are interrupted in several places by **nuclear pores** (Fig. 16.9). These pores allow material to pass into and out of the nucleus without passing directly through a biological membrane. The nucleus is a busy place, controlling many cellular activities, responding to changes in the environment, and synthesizing RNA. During the S phase of the cell cycle, DNA replication takes place within the nucleus.

The internal structure of the nucleus is difficult to decipher in many cases; the chromosomes that are so obvious during mitosis become dispersed during interphase. The dense material in the nucleus, consisting mostly of DNA and protein, is known as **chromatin.** Between cell divisions, chromatin is a dispersed, thread-like material.

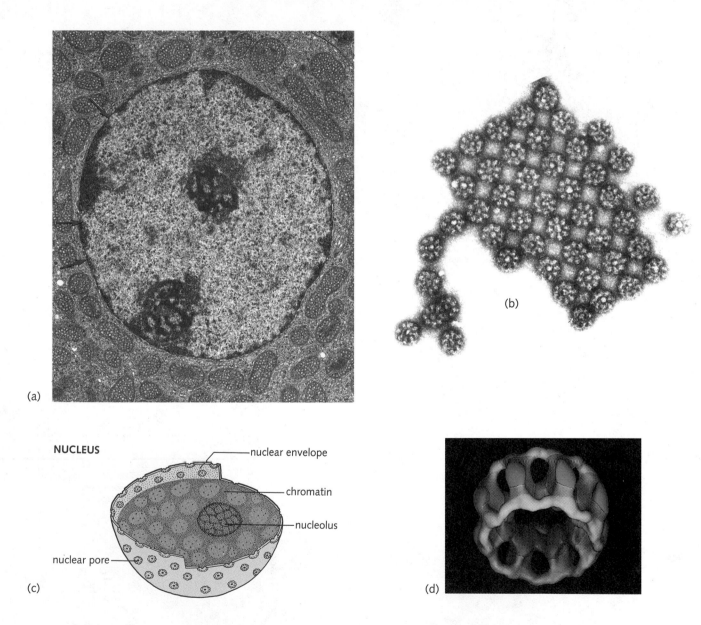

Figure 16.9 **(a)** This nucleus, from the rat adrenal gland, shows two dense nucleolar regions. Chromatin is dispersed within the nucleus. The nucleus is surrounded by two envelope membranes interrupted by a series of nuclear pores (arrows). **(b)** A cluster of nuclear pore complexes, isolated by gentle disruption of the nuclear membranes. **(c)** Nuclear pores are the passageways through which material enters and exits the nucleus. **(d)** A three-dimensional reconstruction of the basic structure of a nuclear pore, calculated from electron micrograph images like those in **(b)**.

Figure 16.10 Ribosomes are particles in the cytoplasm that are sites of protein synthesis. **(a)** Transmission electron micrograph of ribosomes. **(b)** Model of a eukaryotic ribosome, showing the large and small subunits. Each of the subunits is composed of RNA and protein.

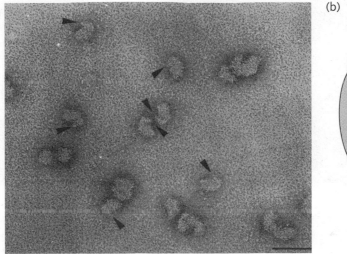

(a)

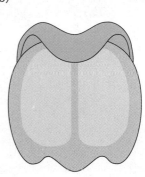

(b)

In mitotic prophase, chromatin condenses to form distinct, individual chromosomes that are visible in the microscope.

The Ribosome

Life is a dynamic process. To keep up with it, cells must produce new macromolecules at a rapid rate. Proteins are produced by small particles in the cytoplasm known as **ribosomes,** which are found in the cytoplasm of both eukaryotic and prokaryotic cells (Fig. 16.10).

Ribosomes are themselves made of protein and RNA, and their role in protein synthesis will be discussed in detail in Chapter 20. Most cells contain many copies of the genes that produce the RNA portion of ribosomes, and these genes are often clustered together in a dark region of the nucleus called the **nucleolus** (see Fig. 16.9).

Many of the ribosomes found in the cell are "free" ribosomes, meaning they are not attached to any other organelle. The proteins that are synthesized on free ribosomes are released directly into the cytoplasm.

The Endoplasmic Reticulum

In many cells, a majority of the ribosomes are *membrane-bound;* they are directly attached to internal cellular membranes. The membranes to which ribosomes are most often attached are known as **endoplasmic reticulum (ER)** (Fig. 16.11). Portions of the endoplasmic reticulum that

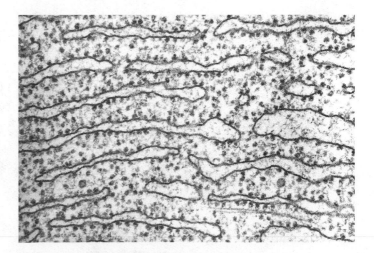

Figure 16.11 This section of rough endoplasmic reticulum shows the close association of some ribosomes with internal cell membranes. Ribosomes on the rough ER produce proteins for export as well as membrane proteins. The newly synthesized proteins produced by these membrane-bound ribosomes are inserted directly into the rough ER.

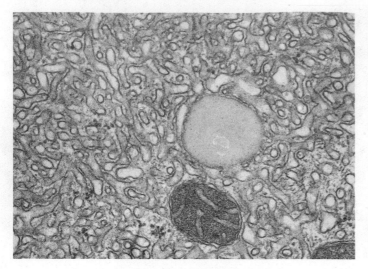

are covered with ribosomes are known as **rough endoplasmic reticulum,** because the scientists who discovered the association thought that the presence of ribosomes made the membrane look "rough," like sandpaper (see Fig. 16.10). Regions of the ER that do not have ribosomes attached are known as **smooth endoplasmic reticulum** (Fig. 16.12).

Why are ribosomes attached to the rough endoplasmic reticulum? Figure 16.13 shows the reason. These ribosomes insert their newly synthesized polypeptides di-

Figure 16.12 Smooth endoplasmic reticulum is involved in a variety of biochemical pathways. Proteins involved in lipid biosynthesis as well as drug detoxification are contained in smooth ER.

Figure 16.13 The movement of secretory proteins through the cytoplasm of a eukaryotic cell. Newly synthesized proteins are inserted into the rough ER and then passed to the Golgi apparatus, where a variety of chemical modifications may take place, including the attachment of sugars to produce glycoproteins. Secretory vesicles that bud off from the Golgi lead to the cell membrane, where their contents are released by exocytosis.

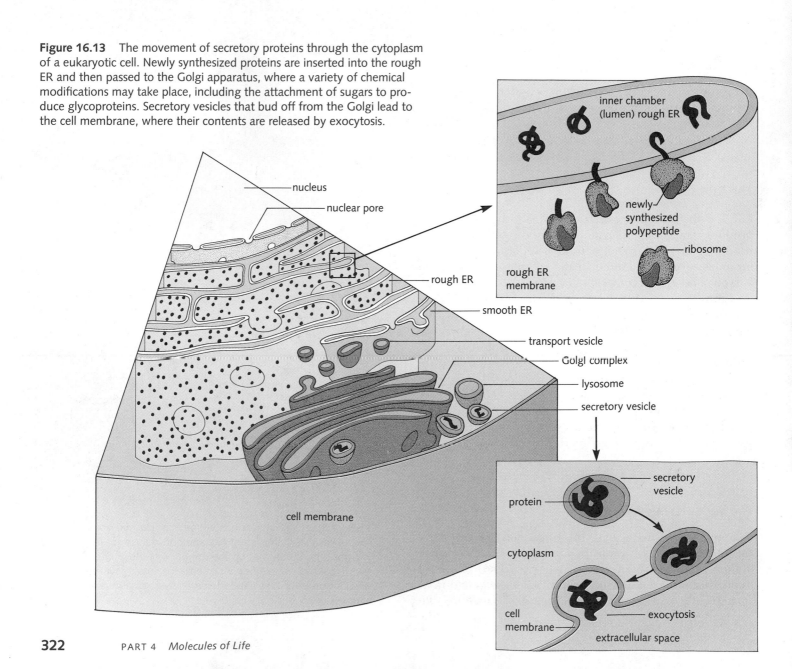

rectly through the ER membrane. The proteins synthesized in this way may be released inside the ER, or they may remain in the ER membrane. Two types of proteins are synthesized in this way: membrane proteins and proteins that will be released ("secreted") from the cell. Recent work has shown that the first 15 or 20 amino acids of a newly synthesized polypeptide may serve as a "signal sequence" that helps to direct a growing polypeptide into the endoplasmic reticulum.

Why should proteins destined to leave the cell be inserted into the rough ER? Because this organelle is the first stop on a pathway that leads newly synthesized proteins through a series of organelles to the cell surface.

The smooth ER is not directly involved in protein synthesis. In many cases, the smooth ER seems to contain a series of enzymes that are responsible for carrying out other types of biochemical jobs that vary from cell to cell. In some, the enzymes responsible for steroid synthesis are found within them. In the liver, smooth ER often contains the enzymes responsible for drug detoxification.

The Golgi Apparatus

The **Golgi apparatus** was discovered by Camillo Golgi, an Italian microscopist. The Golgi apparatus generally appears as a stack of flattened vesicles that is closely associated with the rough ER. Small vesicles bud off from the rough ER and fuse with one side of the Golgi, and other vesicles leave from the other side (Fig. 16.14).

A great many proteins are not finished when the last amino acid is put in place in the rough ER. These proteins require a bit more tinkering ("modification") before they are ready for action. The Golgi is the site at which these chemical modifications take place. Depending on the particular protein, sugar molecules may be attached, the polypeptide chain may be cut in a strategic location, or prosthetic groups such as the heme in hemoglobin may be attached—to name just a few possibilities. In cells that are very active in protein synthesis, the Golgi may be very large and well developed.

Secretory Vesicles

When the synthesis of a protein is complete, that protein can be used by the cell. However, many proteins are released, or secreted, from cells to the exterior. Such proteins pass from the Golgi into **secretory vesicles** (Fig. 16.15). In the pancreas, for example, the outer rim of the cell is filled with vesicles that are packed with protein products from the Golgi. One of the jobs of the pancreas is to produce enzymes that aid in the digestion of food. These secretory vesicles contain enzymes that were first

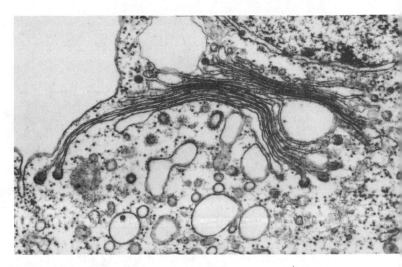

Figure 16.14   The Golgi apparatus, processing site for many secretory proteins, often appears in thin sections as a stack of flattened vesicles.

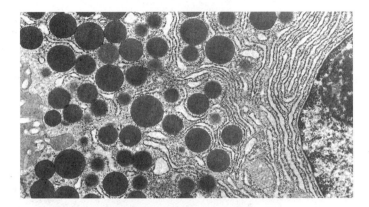

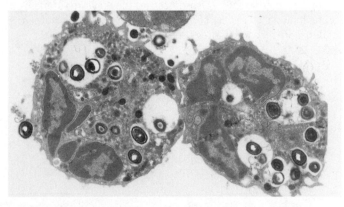

Figure 16.15   TOP: These secretory vesicles store zymogen, a collection of the principal secretory products of the exocrine pancreas. The secretory proteins will be released from the cell by exocytosis. (Magnification factor: 28,000) BOTTOM: Endocytosis, which brings material into a cell, is important in a variety of cellular activities. These white blood cells are engaged in the phagocytosis ("cell eating") of *Streptococcus* bacteria. (Magnification factor: 8500)

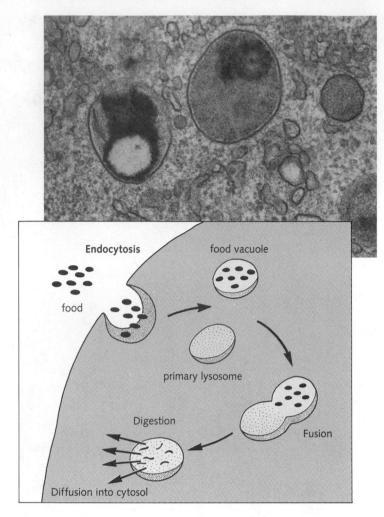

synthesized on the rough ER, were modified in the Golgi, and were then passed along to the vesicles, where they are stored until they are released from the cell.

The contents of a secretory vesicle are released from the cell in a process known as **exocytosis** (*exo* means "outside"), in which the vesicle membrane fuses with the cell membrane (see Fig. 16.13). Many cells release their secretory contents in response to specific signals, and the link between stimulus and exocytosis is an area of much research interest.

Endocytosis (*endo* means "inside"), the opposite of exocytosis, in which material is brought into the cell enclosed in vesicles, is also an important research topic. Many cell membranes contain protein *receptors* that bind to specific molecules. When large numbers of these molecules bind to the surface receptors, the cell membrane turns inward, forming an endocytic vesicle that can then be delivered to a destination within the cell. Liver cells contain receptors for cholesterol-containing lipoproteins, and these enable the liver to remove excess cholesterol from the blood by endocytosis. When very large particles are brought into a cell, the process is known as **phagocytosis** (see Fig. 16.15, bottom).

Figure 16.16 TOP: Lysosomes are organelles that contain degradative enzymes enclosed within a limiting membrane. They enable cells to degrade and digest endocytosed material. BOTTOM: Newly endocytosed material is brought into a pathway that involves fusion with lysosomes and the recycling of membrane material to the cell surface. Small molecules and nutrient material are recovered after lysosomal digestion is complete.

Vacuoles

Although most cellular organelles are much more complicated, a few organelles consist of little more than a single large membranous sac. Such organelles are often known as **vacuoles.** Many types of cells contain vacuoles, which may enclose everything from food particles to waste products. In plant cells, however, vacuoles are much more prominent, and they may occupy as much as 90 percent of a cell's total volume (see Fig. 16.8 on p. 319). The fluid within these vacuoles serves as a storage reservoir containing water, salts, and sugars. In many plants, certain cells contain large amounts of water-soluble pigments in their vacuoles; these pigments give leaves and flowers their characteristic brilliant colors.

Lysosomes

The small membrane-bounded organelles called **lysosomes** are filled with enzymes (Fig. 16.16). Lysosomes are produced by the Golgi apparatus, and the synthetic organelles of the cell direct a flow of very specialized proteins into them. Lysosomes are filled with *lytic* enzymes—enzymes capable of breaking down macromolecules such as proteins, carbohydrates, and nucleic acids into their smaller building blocks. Lysosomes are used for intracellular digestion and destruction.

When endocytosis brings a food particle into the cell, the vesicle containing the ingested food fuses with a lysosome (Fig. 16.16). The action of lysosomal enzymes quickly breaks the food molecules down into smaller compounds that can be used by the cell and are easily removed from the lysosome. The lysosomes of white blood cells are loaded with enzymes that destroy the bacteria they trap and engulf.

Lysosomes seem to have other purposes as well. Lysosomes are occasionally seen in the process of destroying a cell's own organelles, and sometimes the rupture of many lysosomes actually causes the destruction of the cell that contains them, a process known as *autolysis* (*auto* means "self"; *lysis* means "destruction"). Why should the lysosome turn on its own cell? These organelles help the cell dispose of damaged or defective organelles, and they also play a creative role in helping destroy cells in strategic locations where a developing organism needs to shape the pattern of growing tissues.

ENERGY-PRODUCING ORGANELLES

Mitochondria and Chloroplasts

Two other organelles are absolutely essential to any discussion of the organization of living cells: **mitochondria** and **chloroplasts.** Mitochondria and chloroplasts are organelles that change energy from one form to another. Every activity that we associate with life requires a source of energy, and in most organisms these are the organelles that provide that energy.

Chloroplasts trap the radiant energy of sunlight and use that energy to produce energy-rich compounds like carbohydrates. In turn, these carbohydrates serve as a convenient form of chemical energy that can be stored by the cell until needed or may serve as food for another organism (Fig. 16.17). In Chapter 19 we will examine the details of how chloroplasts use solar energy to drive the synthesis of these compounds.

Mitochondria (singular, *mitochondrion*) carry out the next step in processing cellular energy. Before the energy in carbohydrates and other food molecules can be used by the cell, it must be converted to a convenient form. By far the most convenient form of chemical energy in the living cell is a nucleotide known as **ATP (adenosine triphosphate).** The task of the mitochondrion is to use the chemical energy available in food molecules to produce ATP (Fig. 16.18). That ATP is then made available to the rest of the cell. In Chapter 18 this process will be explored in depth.

The ATP molecules produced by mitochondria may be thought of as little chemical "batteries" that the cell can use to provide instant energy for just about any purpose. And we do mean *any* purpose. ATP provides the energy for a firefly's twinkle on a summer night, the whip-like movement of flagella, and even the separation of chromosomes during mitosis. Not surprisingly, cells use lots of energy, and therefore they have to produce lots of ATP.

Figure 16.17 Chloroplasts are organelles that capture sunlight (radiant energy) and convert it to chemical energy. They are surrounded by two envelope membranes and contain large stacks of internal membranes in which these energy-transducing reactions occur. (Magnification: 5500)

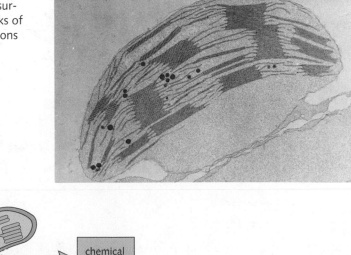

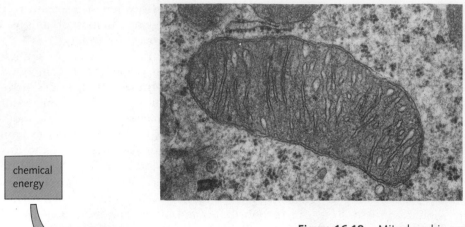

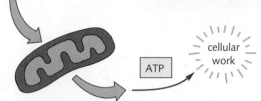

Figure 16.18 Mitochondria are the key energy-releasing organelles of eukaryotic cells. A deeply folded inner membrane contains the components involved in the synthesis of ATP (adenosine triphosphate). ATP is then used to power a wide range of cellular activities. (Magnification: 10,000)

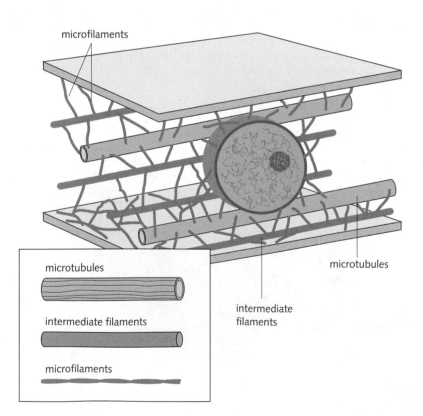

Figure 16.19 Cell shape and cell movement are the principal functions of the cytoskeleton. There are three major classes of cytoskeletal structures: microtubules, intermediate filaments, and microfilaments.

New mitochondria and chloroplasts seem to arise only by the division of preexisting mitochondria and chloroplasts. These organelles also contain their own DNA molecules and their own small genetic systems. These facts have led many biologists to suggest that the persistence of DNA in mitochondria and chloroplasts may reflect the evolutionary history of each organelle and even of the eukaryotic cell itself—a theory we will explore further in Chapter 19.

THE CYTOSKELETON

From red blood cells to amoebae to nerve cells, one of the most striking attributes of individual cells is their ability to maintain a characteristic shape. Earlier in the history of biology, the ability of different cell types to maintain different shapes seemed so mysterious that biochemists ignored the problem of cell shape altogether in order to concentrate on cellular chemistry. That is no longer the case. A host of discoveries about the structure of the cytoplasm have led to the development of an important subtopic within cell biology: the study of the cytoskeleton.

The term **cytoskeleton** really means "cellular support," and the word itself reflects one of the basic discoveries of the past decade and a half: the eukaryotic cell has a skeleton-like substructure (Fig. 16.19). When we think

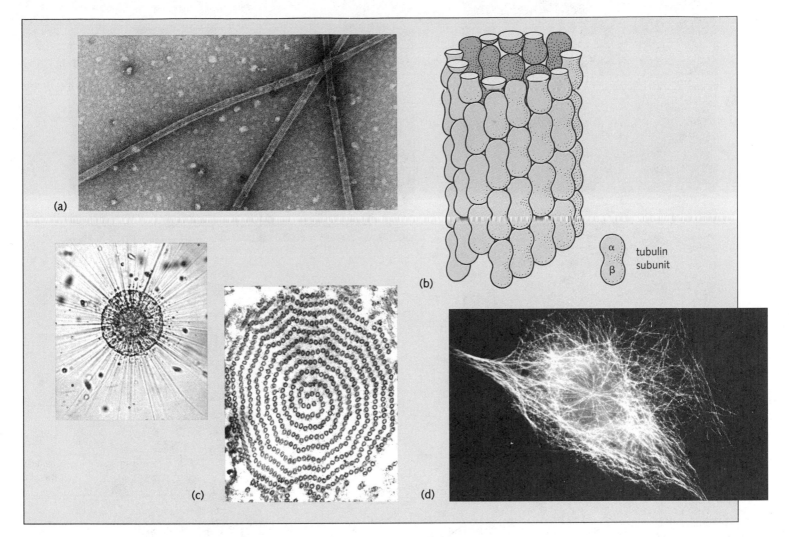

Figure 16.20 **(a)** Microtubules appear in electron micrographs as hollow tubes roughly 25 nm in diameter. **(b)** Microtubules are composed of a helical array of dimeric subunits. Each subunit contains one molecule of α-tubulin and one of β-tubulin. **(c)** Microtubules are often involved in the maintenance of cellular shape. *Echinosphaerium,* a protist, is surrounded by a stunning series of bristling spikes. These spikes are projections of the cytoplasm, each supported by a double spiral of microtubules. **(d)** Microtubules help to maintain the shape of many cell types, as fluorescently labeled antibodies to tubulin illustrate in this tissue culture cell.

about the skeleton of an animal, we imagine a structure composed of many parts that helps to support the organism and that also aids in movement, because muscles are attached to it. The cytoskeleton is similar in both respects. As shown in Fig. 16.19, the cytoskeleton is associated with three major classes of filaments: microtubules, intermediate filaments, and microfilaments.

Microtubules

The first cytoskeletal structure to be appreciated (because it is the largest and most prominent) was the **micro-**

tubule. Microtubules are hollow, tube-like structures with a diameter of about 25 nm, and they can easily be spotted in electron micrographs (Fig. 16.20a). Microtubules are composed of two proteins, α-**tubulin** and β-**tubulin,** which form the helical wall of the microtubule itself (Fig. 16.20b). Microtubules can *self-assemble* from soluble tubulin under the right conditions, and changes in cellular shape are sometimes associated with the sudden assembly or disassembly of microtubules.

Microtubules, which are often found in small groups or bundles within the cytoplasm (Fig. 16.20c shows *Echinosphaerium* as an extreme example), are sometimes used to provide support for the cell surface (Fig. 16.20d). They play a critical role in mitosis and are often associated with

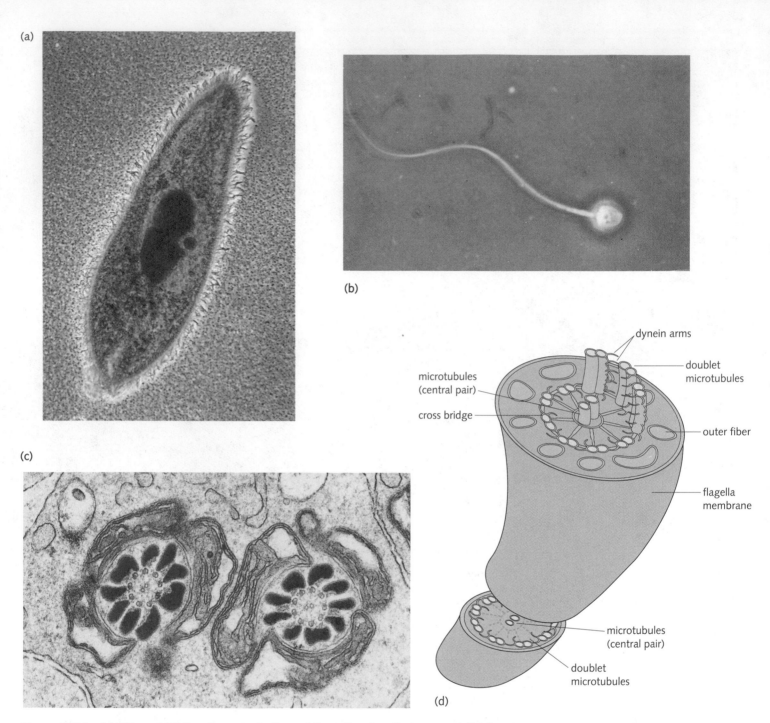

Figure 16.21 **(a)** Cilia and **(b)** flagella are projections of the cell surface that are specialized for movement. The cilia of the *Paramecium* beat in a regular pattern, which propels the cell through water. The whip-like motions of the sperm's single flagellum enable it to swim quickly. **(c)** These electron micrographs of sperm cells show some of the key structures in both cilia and flagella. Each organelle contains an axoneme, a barrel-shaped arrangement of microtubules with a central pair and nine doublet microtubules arranged in a cylinder. Inter-actions between tubules in the axoneme generate force in cilia and flagella. Sperm flagella often contain additional "outer fibers," as shown in this micrograph of developing rat sperm. **(d)** Force in cilia and flagella is produced by means of connections between neighboring mi-crotubules. The cross bridges between adjacent doublets are made of a protein known as *dynein,* which uses ATP to provide the chemical energy to bend the axoneme and produce movement.

cellular movements. Tubulin molecules very similar to those in microtubules are also used to build motile structures, such as cilia, in the cell.

Cilia and **flagella** are specialized structures that protrude from the cell surface and are used to produce motion (Fig. 16.21a,b). [The distinction between cilia and flagella is somewhat arbitrary. The term *cilium* (meaning "hair") is used for structures less than 20 μm in length, and *flagellum* ("whip") for structures 20–100 μm in length.] The internal structure of cilia and flagella in eukaryotes consists of nine microtubule-like doublets that surround a central pair of microtubules (a "9 + 2" arrangement). The individual tubules are held together by a series of cross bridges made up of a protein called *dynein* (Fig. 16.21c,d). These cross bridges use energy from ATP to generate force, causing the whole structure to whip to one side or the other. Both cilia and flagella are attached to small "rootlets" within the cytoplasm and terminate in structures known as **basal bodies** (Fig. 16.22).

Centrioles are structures that are remarkably similar to basal bodies in appearance. Cross sections show that they are composed of nine groups of three tubule-like structures (a "9 + 0" arrangement). A typical animal cell contains two centrioles, which remain closely associated throughout most of the cell cycle (Fig. 16.23). Then, just before mitosis, the centrioles are replicated and two "daughter" centrioles are formed. The pairs of centrioles then move to opposite sides of the nucleus as mitosis begins; here they serve as the poles of the mitotic spindle. Surprisingly, plant cells produce perfectly functional mitotic spindles despite the fact that they do not contain centrioles.

Microfilaments and Intermediate Filaments

Two other types of filaments are also components of the cytoskeleton. **Microfilaments** are fibers about 6 nm in diameter that are made up of a protein known as *actin*. Like the microtubule, the microfilament is capable of spontaneous assembly and disassembly. Microfilaments are found in nearly all cells and, like microtubules, help to stabilize cell shape. Actin is also one of the major proteins found in muscle cells, and microfilaments are associated with a number of different forms of cell movement. The streaming cytoplasm of the amoeba, for example, is produced by the actions of actin filaments and other proteins in the cytoplasm of the cell (Fig. 16.24).

The third and final class of cytoskeletal proteins is made up of the **intermediate filaments.** These filaments are intermediate in size between microtubules and microfilaments: about 10 nm in diameter. Unlike the other

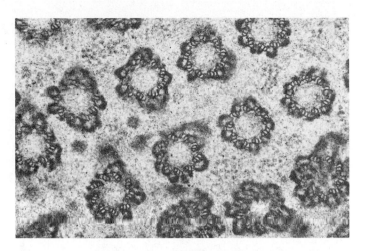

Figure 16.22 These basal bodies, from the lining of the oviduct, are the cytoplasmic anchors to which cilia are attached. Nine groups of three microtubule-like structures are arranged to form this barrel-like structure. Flagella have similar basal bodies.

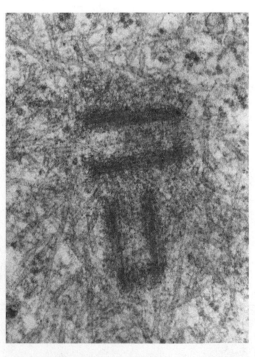

Figure 16.23 These paired centrioles are surrounded by microtubules.

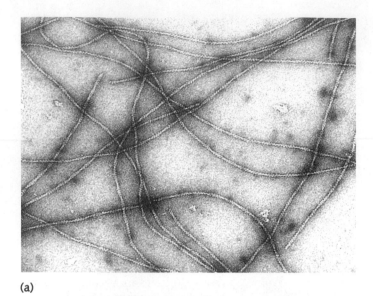

(a)

actin molecules

(b)

actin filament

Region of actin
depolymerization
(converts rigid ectoplasm
to more fluid endoplasm)

Region of actin
polymerization
and cross-linking
(pulls endoplasm
forward; converts
to more rigid
ectoplasm)

(c)

Figure 16.24 **(a)** Microfilaments, which are roughly 6 nm in diameter, are cytoskeletal components made up of actin. Actin filaments are associated with cell movement and with changes in cell shape. **(b)** Microfilaments are composed of helical arrays of actin proteins. **(c)** The polymerization and depolymerization of actin filaments are associated with changes in cytoplasmic structure, including the movement of cells such as the amoeba.

Figure 16.25 The bundle of intermediate filaments in the center of the micrograph is another class of skeletal proteins. They are approximately 10 nm in diameter and are involved with the maintenance of cell shape.

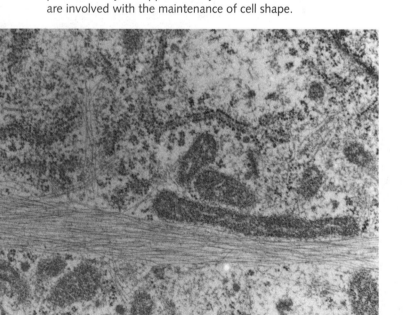

cytoskeletal components, which are made up of actin and tubulin proteins, intermediate filaments are composed of a range of related proteins that vary from one cell type to another. The proteins do seem to have similar properties, however, and scientists are now beginning to make progress in understanding how the intermediate filaments affect cell shape and structure (Fig. 16.25).

It is possible to visualize the different cytoskeletal proteins within a single cell by introducing fluorescently labeled compounds that bind to each of them. When this is done, the impression that emerges is that the cytoskeleton is a complex, interconnected structure. We do not yet understand all the relationships between the different protein types the cytoskeleton comprises, and we know only a little about how the cell is able to regulate changes in shape by acting on the cytoskeleton. The answers to these questions will be very important for biology, because such mechanisms help determine the way an organism grows and develops.

Figure 16.26 Lipids dispersed in water may aggregate to form a number of different structures, including spherical or cylindrical *micelles,* or sheet-like *bilayers.* In each structure, the polar head groups are in direct contact with water, while the hydrocarbon chains are shielded from water.

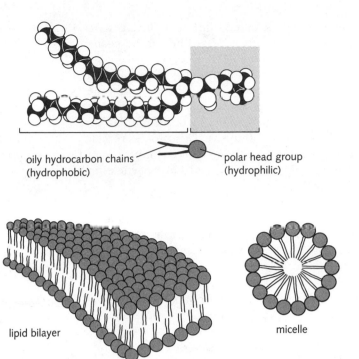

oily hydrocarbon chains
(hydrophobic)

polar head group
(hydrophilic)

lipid bilayer

micelle

BIOLOGICAL MEMBRANES

It should be clear by now that some of the most important structures in the cell are the thin membranes that divide the cell into its many compartments. In fact, the electron microscopic image of the cell, whether plant or animal, is dominated by membranes (see Figs. 16.7 and 16.8 on pp. 318–319). First among these, of course, is the **cell membrane** that separates the cell from its environment.

The Basic Structure of a Membrane

The structural backbone of a biological membrane is a **lipid bilayer.** We have already seen how the basic chemical structure of many lipid molecules enables them to form bilayers in which individual molecules are arranged with the most polar portions of the molecules facing the water that surrounds the bilayer (Fig. 16.26). Lipid bilayers can form

spontaneously in mixtures of lipid and water, and they are relatively stable. This is because the oily, or *hydrophobic* ("water-fearing"), hydrocarbon chains are gathered together in the middle part of the bilayer, whereas the polar, or *hydrophilic* ("water-loving"), parts of the lipid molecules are exposed to the water at the surface of the bilayer.

Membrane Proteins

For years many scientists believed that the proteins associated with biological membranes were only found at the surfaces of the lipid bilayer. But more recent work has shown that proteins can actually span the lipid bilayer and make contact with both surfaces of the membrane.

New techniques used in electron microscopy have made it possible to visualize membrane proteins within the lipid bilayer. In a procedure known as freeze-etching, cell or tissue samples are rapidly frozen and then placed in a special vacuum chamber where they can be split in half at low temperatures. A metal film, or replica, is then cast on the fractured surface, and that film can be examined in the electron microscope (Fig. 16.27).

When the frozen material fractures near a biological membrane, the fracture tends to split the membrane open between the two halves of the lipid bilayer (because the hydrophobic bonds that hold the tails of lipids together are weak at low temperatures). This process reveals the inner region of the membrane, and complexes of transmembrane proteins are visible as distinct particles in the replica (Fig. 16.27d).

Membrane proteins may be associated with a variety of other molecules. In some cases, there are direct connections between filaments of the cytoskeleton and membrane proteins (Fig. 16.28), allowing each to influence the other. At the outer surface of the cell, *carbohydrate* molecules may be attached to proteins, forming glycoproteins, and they may also be associated with lipids to form glycolipids. Protein and glycoprotein molecules exposed at cell surfaces can function as *receptors* for chemical messages, as *markers* that enable cells to identify each other, and as *control points* that regulate cell attachment and cell growth.

All of this may make biological membranes seem like mosaics or conglomerates, which is not far from the truth. In fact, the most widely accepted theory of membrane organization today is known as the **fluid mosaic model.** The name is appropriate. The membrane is a mosaic of many different components, and biological membranes tend to be fluid structures. The "fluidity" of biological membranes reflects the fact that membrane components are free, to some extent, to move about within the plane of the membrane. Both lipid and protein components of the membrane are able to move laterally in the plane of the membrane, and this fluidity is important for many membrane functions.

FUNCTIONS OF BIOLOGICAL MEMBRANES

Membranes as Barriers

The first and foremost task of the biological membrane is to serve as a barrier. Without such a barrier, the components of the cell would diffuse away, and any molecule in the environment would be free to enter. The fluid mosaic plan of organization makes membranes ideal barriers.

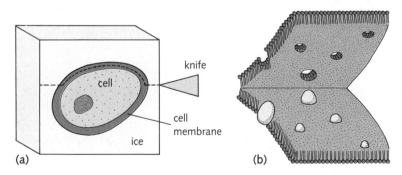

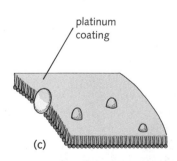

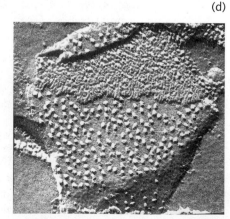

Figure 16.27 The visualization of a cell membrane structure by *freeze-etching*. **(a)** Frozen cells or tissues are placed in a special vacuum chamber where they can be fractured at low temperature. Usually, fracturing is done by forcing a sharp knife through the frozen sample. **(b)** The hydrophobic forces that hold membranes together are weak at low temperatures. Therefore, each membrane has a preexisting plane of weakness that may split open as the sample is fractured, revealing the internal details of membrane organization. **(c)** A *replica* is then made of the fractured membrane. A heavy metal, usually platinum, is shadowed on the fractured surface to produce the replica. The metal highlights the topology of the fracture face. The metal replica is then stabilized by covering it with a layer of carbon, the frozen sample is thawed, and the tissue is dissolved in bleach or acid. The metal replica, a nearly exact copy of the fractured surface, may then be examined in a transmission electron microscope. **(d)** Biological membranes (from a chloroplast) prepared by the freeze–fracture technique. The particles visible within the membrane are internal structures that have been revealed as the membrane split open during the fracture process. (Magnification: 165,000)

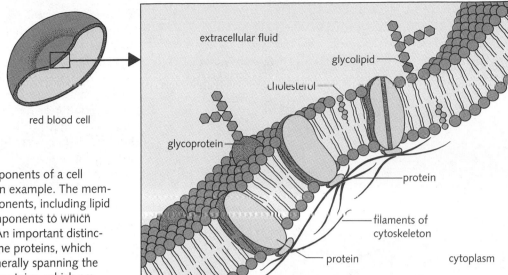

Figure 16.28 Some of the major components of a cell membrane, with the red blood cell as an example. The membrane is a *fluid mosaic* of several components, including lipid and protein, as well as cytoskeletal components to which membrane proteins may be attached. An important distinction is made between integral membrane proteins, which are associated with the lipid bilayer generally spanning the membrane, and peripheral membrane proteins, which are attached to the membrane at its surface, as shown.

Biological membranes can best be described as *selectively permeable,* which means that some things can pass through and some can't. Lipid bilayers, as we have already seen, are held together by the hydrophobic forces that exist between the oily hydrocarbon tails of lipid molecules. Charged molecules (ions such as K⁺ and Na⁺), which are strongly attracted to water, are not attracted by the hydrocarbon region in the middle of a lipid bilayer, so they are not able to pass through bilayers very easily. Neither are larger, electrically neutral molecules such as glucose, which cannot squeeze between the tightly packed lipid molecules (Fig. 16.29).

The molecules that can pass through lipid bilayers most easily are water itself and nonpolar (oily) molecules such as ethanol and propanol, which can "dissolve" in the bilayer as they move through. The fact that membranes are selectively permeable leads to some properties that have important biological consequences. We will examine those properties in the next few pages. But first, we must understand the basic physical process of diffusion.

Diffusion

The molecules in living things are in constant motion. Under a light microscope, the movement of water molecules can be seen to cause very small objects to move about in short, sudden jumps—a phenomenon known as

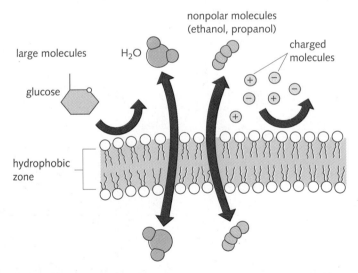

Figure 16.29 Biological membranes are *selectively permeable.* While water and most nonpolar molecules can readily cross such membranes, other molecules cannot. Most biological membranes effectively exclude larger molecules and charged molecules unless the membranes contain specific transport channels that allow such molecules to cross.

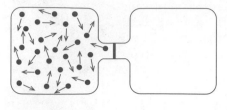

A barrier separates two compartments with different concentrations of a substance.

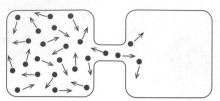

When the barrier is opened, there is a movement of molecules from the region of higher concentration to the region of lower concentration—diffusion.

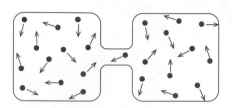

After a while, a state of equilibrium is reached in which the concentrations of material in each compartment are identical.

Figure 16.30 Diffusion is the movement of molecules between two compartments that have different concentrations of solutes. The removal of the barrier in this example allows the random movement of individual molecules to produce an equilibrium in which the concentration of the solute is uniform throughout.

Brownian movement. This molecular motion is an effect of temperature—the higher the temperature, the greater the rate of movement—and it is random, or undirected.

The random motion of molecules in a gas or liquid means that no molecule within a liquid or gas can remain stationary for very long. Consider what happens when we open a bottle of perfume in one corner of a room. The molecules that produce the scent are moving within the liquid perfume (usually a solution of alcohol). As they approach the surface of the liquid, many of the scent molecules pass directly into the air, and the layer of air near the open bottle gradually fills with such molecules. If the movement of air molecules were to stop, the scent would go no farther. However, the air molecules continue to move, and their motion bounces the molecules carrying the scent in every direction. Some of them cross back into the liquid and return to solution. But many of them move farther into the air, away from the bottle. Little by little, the scent spreads from the bottle until it fills the room. This process is known as *diffusion.*

Strictly speaking, **diffusion** is the process by which a substance moves from an area of higher concentration to an area of lower concentration, and it commonly occurs in liquids and gases (although in some special cases it can also take place in solids). The gradual movement of scent from our opened bottle of perfume, the dispersal of color from a droplet of dye placed in water, and even the spreading of a puff of smoke in a quiet room are examples of the process of diffusion.

Diffusion occurs because of a *concentration gradient:* a difference in the concentration of a substance between two areas. Diffusion depends only on the random thermal motion of molecules. When two compartments with different concentrations of compound X are prepared, and the barrier between them is opened, diffusion causes a net movement of molecules from one compartment to another until a point is reached at which the concentrations of X in the two compartments are equal. At that point, individual molecules still continue to move from one compartment to the other (molecular motion is random!), but there is no *net* movement. A *diffusion equilibrium* has been reached between the two compartments (Fig. 16.30).

Diffusion Across a Biological Membrane

We can modify the example of a molecule diffusing from one compartment to another and use it to consider one of the most important kinds of molecular movement in biological systems: diffusion across a biological membrane. Because most biological membranes are selectively permeable, at least a few molecules can cross them. The way in which a particular molecule crosses a biological membrane falls into one of three general categories: simple diffusion, facilitated diffusion, and active transport.

Simple diffusion   As we have just seen, **simple diffusion** causes the movement of molecules from a region of higher concentration to a region of lower concentration. Diffusion can take place across a membrane if two conditions are met: First, the molecule is present on one side of the membrane at a higher concentration than it is on the other side (a concentration gradient). Second, the molecule can actually pass between the molecules that make up the membrane. This second condition requires that the molecule be relatively *small* (big ones can't slip between the molecules of the membrane itself) and *nonpolar* (so that it can pass through the hydrophobic region at the interior of the membrane).

Facilitated diffusion   A special case of diffusion in which molecules cross membranes by passing through pore-like transport molecules that *facilitate* their passage is called **facilitated diffusion.** Facilitated diffusion is similar to simple diffusion in that a concentration gradient is required to drive it. But the important difference between this process and simple diffusion is the existence of the pores or carriers that are part of the membrane itself and that enable facilitated diffusion to occur at a faster rate than simple diffusion. Many cells have transport proteins built into their membranes that allow glucose to pass through. These proteins allow glucose to enter the cell as much as 100 times faster than simple diffusion. A diagram suggesting that such proteins may function like pores is shown in Fig. 16.31 (in fact, it's not yet known if they are actually shaped like pores).

Active transport   The use of energy to move molecules across a membrane is known as **active transport.** The energy is derived from active transport molecules, which are sometimes referred to as pumps, because they use energy to move material. The most common source of this energy for active transport is ATP, and many membrane proteins use the energy available in ATP to pump molecules across a membrane. Because these molecules can use

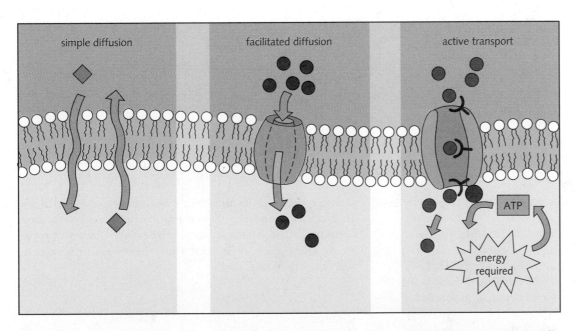

Figure 16.31   Three ways in which a substance may pass through a biological membrane are simple diffusion, facilitated diffusion, and active transport. Facilitated diffusion across biological membranes is made possible by special pore-like carrier molecules. Such carriers are membrane proteins that form channels across the membrane. These channels can be quite specific, allowing some molecules to pass through while excluding others. Active transport involves the use of energy to pump specific substances across a cellular membrane. Many such pumps use the energy of ATP to move material from one side of the membrane to the other. Unlike facilitated diffusion, active transport can move material in a direction *opposite* a concentration gradient.

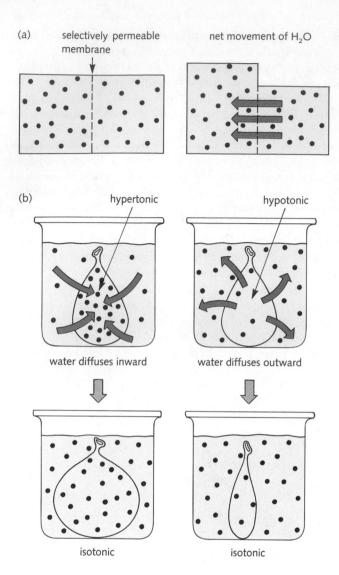

(a)
selectively permeable membrane

net movement of H₂O

(b)
hypertonic

hypotonic

water diffuses inward

water diffuses outward

isotonic

isotonic

Figure 16.32 **(a)** Osmosis is the movement of water across a selectively permeable membrane. When two compartments contain different concentrations of a solute that cannot pass through a membrane, a net movement of water will occur in the direction of the compartment with a higher solute concentration. The concentration of water in this compartment increases as the result of osmosis. **(b)** The effects of osmosis can be seen in a membrane-like sac placed in two different solutions. If the solution within the sac is more concentrated (hypertonic) than the surrounding solution, osmosis will cause a net movement of water into the sac, and it expands. If the solution within the sac is less concentrated (hypotonic), osmosis will cause a net movement of water out of the sac, and it will shrink. When the concentration of material in the sac is the same as the surrounding solution, net movement of water stops, and the concentration of material in the sac is said to be isotonic with respect to the surrounding solution.

energy, they are capable of moving material across membranes *against* a concentration gradient (Fig. 16.31). Working against a concentration gradient is like trying to move perfume particles from the air back into the bottle, a task that would demand patience and energy. This is a key difference between active transport and facilitated diffusion, which depends on a concentration gradient for movement of material.

Many cells contain an active transport molecule known as sodium–potassium ATPase. This ATPase is a membrane protein that can pump both sodium and potassium across the cell membrane, using ATP as its source of chemical energy.

The pump works in two directions at once, pumping sodium ions out of the cell and potassium ions into it. Because of the actions of this active transport protein, the cytoplasm of a typical cell contains much more potassium and much less sodium than the fluid surrounding it. Al-

though cell membranes are capable of transporting a number of ions, the concentration gradients involving sodium and potassium are particularly important in nerve and muscle cells, which use these ions to help conduct electrical impulses.

Osmosis

A special case of diffusion involves the movement of water molecules across a membrane. **Osmosis** is the movement of water across a selectively permeable membrane as a result of a concentration gradient (Fig. 16.32a). Although water is a polar molecule, it passes through biological membranes quite easily. Why water is able to do this is still something of a mystery and is under active investigation.

If we separate two compartments with a selectively permeable membrane and fill each with pure water, water molecules pass through the membrane in each direction, but there is no net movement because the system is in equilibrium—the solutions on both sides of the membrane are identical. But if we dissolve another molecule—sucrose, for example—in one of the compartments, the situation changes. Sucrose cannot cross the selectively permeable membrane. However, the presence of sucrose in one compartment can be thought of as actually lowering the concentration of water in that compartment. An unequal concentration means that diffusion may occur. In this case, water diffuses from the compartment of higher concentration (pure water) to the compartment where its concentration is lower (sucrose solution).

Because cells contain high concentrations of dissolved material surrounded by cell membranes, osmosis can cause water to move into or out of the cell. We can get a good idea of what effect this might have on a cell by examining a system in which a small balloon made from a selectively permeable membrane has been suspended in a beaker. If the concentration of dissolved material within the balloon is the same as that of the fluid in the beaker, there is no net movement of water (equilibrium). However, if the concentration of dissolved material within the balloon is greater than that of the surrounding fluid, water begins to diffuse into the balloon. The net movement of water as a result of concentration differences gradually causes the balloon to swell (Fig. 16.32b).

The inward movement of water in this case occurs because the concentration of dissolved material in the balloon is *greater* than the concentration outside the balloon. We refer to the solution with a greater concentration of dissolved material as **hypertonic** (*hyper* means "greater"; *tonic* means "strength") to the other solution. The solution with the lower concentration is described as **hypotonic** (*hypo* means "lesser"). The net movement of water across a selectively permeable membrane is always from the hypotonic solution toward the hypertonic one. When equilibrium is reached, the net movement of water stops, and the two solutions are said to be **isotonic** (*iso* means "the same") with respect to each other.

As we have just seen, a selectively permeable balloon containing a hypertonic solution swelled when osmosis caused water to move into the balloon. The direct cause, of course, is that water moved into the balloon, following the concentration gradient and increasing the balloon's volume. What would happen if the balloon were placed in a wire meshwork so that it could not expand? Although osmosis could not force water to move into the restricted volume of the balloon, the tendency of water molecules to move into the balloon would still generate a force, causing pressure to build up. The force generated by osmosis is known as **osmotic pressure.**

Effects of Osmosis on a Living Cell

The cell membrane is a selectively permeable membrane, and water passes across it easily, although most salts, sugars, and complex molecules do not. Therefore, a single cell suspended in pure water is subjected to a severe osmotic pressure. If the pressure is not counteracted, the cell will swell and burst (Fig. 16.33).

Living organisms have evolved three different strategies to contend with osmotic pressure. The first, and most direct, is to solve the problem by changing the environment of the cell. In a large organism, individual cell membranes rarely come into contact with the environment. The organism produces a "microenvironment" around each cell, and that microenvironment includes an isotonic solution. The isotonic environment ensures that cells are not subjected to osmotic pressure, and there is little net movement of water into or out of most cells.

Figure 16.33 Osmosis can cause dramatic changes in the shape of an unprotected cell. These three micrographs show the appearance of red blood cells in hypotonic (LEFT), isotonic (CENTER), and hypertonic (RIGHT) solutions.

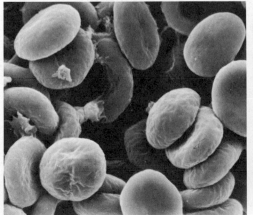
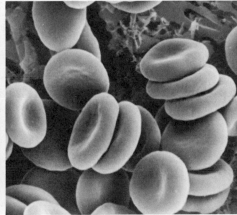
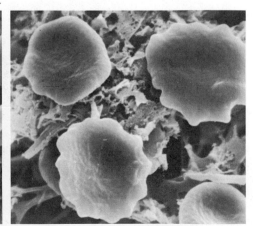

Discovering a New Organelle

With all the years that cells have been studied by thousands of scientists throughout the world, is it possible that most cells contain a structure that has never been noticed? If you found something completely new, would you have the confidence to trust your own observations, or would you assume that you'd made some silly mistake and ignore your own results?

In 1986 Nancy Kedersha and Leonard Rome of UCLA had the chance to answer all these questions for themselves. They were purifying a certain class of vesicles normally associated with a basket-shaped protein called *clathrin*. Examining her samples in the electron microscope, Kedersha saw some peculiar egg-shaped structures unlike *clathrin* or anything else she ex-pected to be present in her preparation. Their delicate, arching internal structure so reminded her of the arches in cathedral vaults that she called them "vaults."

At first, Kedersha and Rome thought that the vaults might be rearrangements of *clathrin* or one of the other proteins known to be in vesicle preparations. Other experiments clearly showed, however, that the vaults were composed of five hitherto unknown proteins and a small (140 base) RNA. Perhaps the vaults were a peculiarity of the cells she was studying? Not so. They found them in cells from cows, chickens, frogs, and even from amoebae. Genes for vault proteins and RNA are found throughout the living world. They must be doing something important.

Detailed studies with the electron microscope have shown how the parts of the vault fold together to form the complete, basket-like structure, but to date no one has been able to discover exactly what vaults do. Kedersha's personal view is that the vault is involved in the transport of material between nucleus and cytoplasm. The vault's combination of RNA and protein suggests such a role, and she's pointed out that the vault is almost exactly the right size to fit into a nuclear pore.

She may be right, of course, but the vault story points out something that biologists in all fields have known for a long time. There are new discoveries to be made even in territory that we think is familiar.

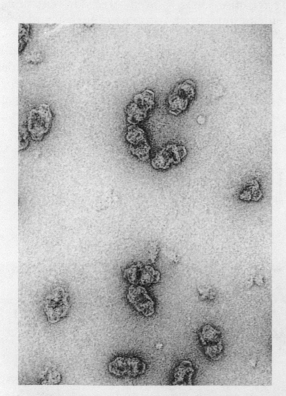

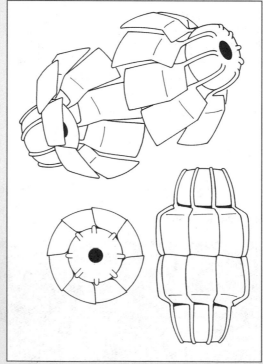

LEFT: Electron micrograph of isolated "vaults." Each vault measures 36 x 65 nm.
RIGHT: Artist's conception of the structure of a single RNA–protein vault.

A second strategy is used by a number of single-celled organisms that live in water and cannot alter their immediate environment. These cells contain special organelles known as *contractile vacuoles* that help them cope with osmosis. The details of contractile vacuole function are not completely known. In a general way, however, we do understand that contractile vacuoles help collect the water that osmosis has brought into the cell and periodically release it to the exterior. In a sense, the vacuole works like a bilge pump in a boat, pumping out water as soon as it leaks in. The forces of osmosis are kept in check and the cell maintains its normal shape (Fig. 16.34).

For many cells, however, there is an even simpler solution, one that is a lot like our balloon in a cage. Many bacteria, algae, and plants produce tough, rigid **cell walls** that surround their cell membranes (Fig. 16.35). The wall is porous enough to allow food molecules to diffuse into, and waste products to diffuse away from, the cell. But it is also tough enough to withstand tremendous pressure. That is important, because the forces of osmosis, which tend to force water into the hypertonic cytoplasm, create a substantial osmotic pressure. The cell wall counteracts that pressure and prevents the cell from expanding.

Figure 16.34 Many cells deal with osmotic pressure by eliminating the excess water through contractile vacuoles. *Paramecium*, a protist, has well-developed contractile vacuoles that take up water from the surrounding cytoplasm, pass it through canals to a central vacuole, and then expel the accumulated water through a pore in the cell surface.

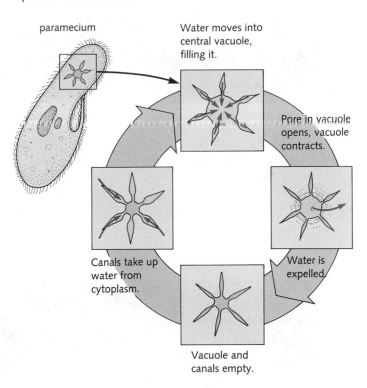

paramecium

Water moves into central vacuole, filling it.

Pore in vacuole opens, vacuole contracts.

Water is expelled.

Vacuole and canals empty.

Canals take up water from cytoplasm.

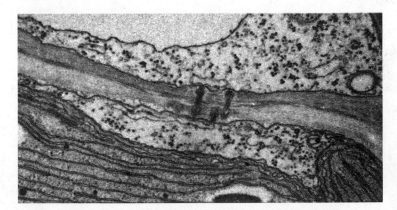

Figure 16.35 Osmotic pressure can be controlled by cell walls. These structures are formed outside the cell membrane and provide a semirigid casing that prevents excessive cell expansion due to osmotic pressure. This micrograph illustrates the structure of cell walls in a corn seedling. In many plants, cytoplasmic connections known as plasmodesmata (center of micrograph) provide direct contact between adjacent cells despite the presence of the walls. (Magnification: 8500)

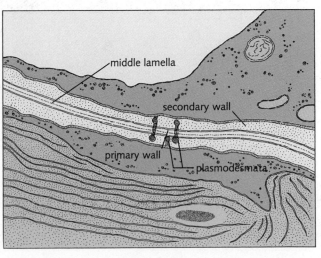

middle lamella

secondary wall

primary wall

plasmodesmata

Seeing It All

Just because the light microscope has been around for a couple of hundred years doesn't mean that light microscopists have run out of ways to make the instrument more useful than ever. The figure shows one of the key directions of light microscopy in the immediate future. By combining carefully designed fluorescent marker compounds with high-resolution imaging and computer-controlled video processing, Lans Taylor and his associates at Carnegie-Mellon University produced this informative micrograph of wound healing in cultured mouse cells.

DNA has been stained with a blue fluorescent label, actin glows green, yellow shows vesicles formed by endocytosis, and a red marker has been taken up by mitochondria. By labeling four components simultaneously, researchers can get a great deal of information about the dynamics of cell movements and internal activities.

The cell near the center of the micrograph, for example, is migrating upward to fill a "wound" in the culture dish. The green margin of the cell is filled with actin—which plays an important role in cell movement—but it has few mitochondria or endocytic vesicles.

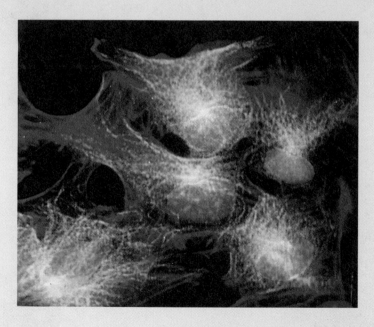

Multiple-label fluorescent image of mouse cells.

SUMMARY

Microscopes, which magnify images produced by light, electrons, or other forms of radiation, make it possible to discover and to study the cell. Eukaryotic cells contain a wide variety of complex organelles. The primary organelle in any eukaryotic cell is the nucleus, which contains nearly all of the cell's genetic information in the form of DNA. The nucleus is bounded by a nuclear envelope, consisting of two membranes interrupted in several places by nuclear pores that allow material to pass between nucleus and cytoplasm.

Proteins are synthesized on ribosomes, which may be free in the cytoplasm or bound to the membranes of the rough endoplasmic reticulum. Proteins destined for release from the cell are generally synthesized on the rough endoplasmic reticulum and are passed to the Golgi apparatus, where many proteins are modified. Finally, the proteins are released from secretory granules to the cell exterior. Mitochondria and chloroplasts, which are self-replicating and contain their own DNA, are involved with energy transformations in the cell.

Eukaryotic cells also contain a cytoskeleton, which functions in supporting cellular shape and in producing cell movement. The cytoskeleton contains three general types of structures: microtubules, intermediate filaments, and actin microfilaments. Each of these structures is assembled from smaller subunits, and two of them, microtubules and microfilaments, can produce cell movement when they interact with other filaments and cross-bridging proteins.

The many compartments of the cell are separated by biological membranes. Membranes possess a structure based on the lipid bilayer but also contain proteins and carbohydrates. Membranes can serve as barriers; however, membranes are permeable to some molecules and are able to regulate the passage of these molecules. Material may pass across a membrane by simple diffusion, through special carrier molecules, or by means of an energy-requiring active transport system. Osmosis is a special form of diffusion in which water moves across membranes. The rate and direction of osmosis is determined by the concentration of solute on either side of a membrane.

STUDY FOCUS

After studying this chapter, you should be able to:

- Explain how light and electron microscopes form images, and distinguish between transmission and scanning electron microscopes.

- Describe the major organelles of the eukaryotic cell.

- Cite some of the most recent information relating to cell structure and function, including the targeting of proteins to organelles and the molecular basis of cytoplasmic shape and structure.

- Explain the structural organization of biological membranes, including the roles played by lipids and proteins in the organization of these membranes.

- Distinguish between active transport and facilitated diffusion.

- Explain why animals in fresh water face osmotic problems that are different from those faced by animals that live in saltwater.

TERMS AND CONCEPTS

cell theory *313*	cytoskeleton *326*
transmission electron	microtubule *327*
microscope *315*	α-tubulin *327*
scanning electron	β-tubulin *327*
microscope *315*	cilia *329*
organelles 316	flagella *329*
nuclear envelope *320*	basal bodies *329*
nuclear pores *320*	centrioles *329*
ribosomes *321*	microfilaments *329*
nucleolus *321*	intermediate filaments *329*
rough endoplasmic	lipid bilayer *331*
reticulum *321*	fluid mosaic model *332*
smooth endoplasmic	diffusion *334*
reticulum *322*	facilitated diffusion *335*
Golgi apparatus *323*	active transport *335*
secretory vesicles *323*	osmosis *336*
vacuoles *324*	hypertonic *337*
lysosomes *324*	hypotonic *337*
mitochondria *325*	isotonic *337*
chloroplasts *325*	cell walls *339*
ATP *325*	

REVIEW

Objective Questions (Answers in Appendix)

1. The resolution of the light microscope is principally limited by
   (a) the presence of air around the sample.
   (b) the wavelength of light.
   (c) the wavelength of electrons.
   (d) imperfections in lens design.

2. Which of the following is involved in protein synthesis?
   (a) lysosomes.   (c) secretory vesicles.
   (b) ribosomes.   (d) vacuoles.

3. The Golgi apparatus is directly involved in
   (a) polypeptide synthesis.   (d) chemical modification
   (b) DNA replication.   of proteins.
   (c) ribosome production.

4. Molecules can most easily pass through membranes if they are
   (a) water molecules only.
   (b) nonpolar and of a small size.
   (c) polar and of a small size.
   (d) charged molecules and of a small size.

5. Facilitated diffusion is an example of
   (a) osmosis.   (c) active transport.
   (b) passive transport. (d) pinocytosis.

6. When a free ribosome completes the synthesis of a protein, the protein is generally
   (a) passed to secretory vesicles.
   (b) released inside the ER.
   (c) inserted into the ER membrane.
   (d) released directly into the cytoplasm.

7. The forces of osmosis dictate that if a cell is placed in a solution that is hypertonic to its cytoplasm, the net movement of water will be
   (a) equally balanced in both directions across the cell membrane.
   (b) into the cytoplasm from the surrounding solution.
   (c) into the surrounding solution from the cytoplasm.
   (d) into the cell nucleus.

Discussion Questions

8. Describe the major organelles of a plant cell. How are these different from the typical organelles found in an animal cell?

9. The amino acid sequences of a large number of integral membrane proteins have now been analyzed in detail. Where such proteins span the lipid bilayer, they tend to have amino acids with side chains (R groups) that are nonpolar. Explain why this observation makes sense.

10. As you have seen, ribosomes active in protein synthesis are either free or membrane-bound. Liver cells synthesize a protein called albumin that is released from the liver into the bloodstream. Would you expect albumin to be synthesized on free or membrane-bound ribosomes? Why? How about tubulin?

11. Cytoskeletal proteins like actin and tubulin are found in nearly all cells in the body. However, many tissues express genes that code for slightly different versions of these proteins. Develop a hypothesis about why these proteins might be tissue-specific.

12. A number of human genetic disorders result from improper targeting of lysosomal enzymes. Speculate on why the inability to make fully functional lysosomes might have serious consequences for an organism.

READINGS

Lienhard, G. E., J. W. Slot, D. E. James, and M. M. Mueckler. "How cells absorb glucose." *Scientific American* 266 (January 1992): 82–92. An up-to-date discussion of how one of the most important metabolic molecules is brought across the cell membrane.

Taylor, D. L., M. Nederlof, F. Lanni, and A. S. Waggoner. "The new vision of light microscopy." *American Scientist* 80 (1992): 322–330. An extraordinary description of how new technologies in video microscopy are revolutionizing the study of the cell.

Rome, L., N. Kedersha, and D. Chugani. "Unlocking vaults: Organelles in search of a function." *Trends in Cell Biology* 1 (1991): 47–50. A short summary of these "organelles in search of a function."

Fawcett, D. *The Cell.* Philadelphia: Saunders Publishing, 1981. A wonderful and thorough atlas of cell structures as seen through the electron microscope.

Gunning, B. E. S., and M. W. Steer. *Ultrastructure and the Biology of Plant Cells.* London: Arnold Publishers, 1975. A superbly illustrated account of the main structural features of plant cells.

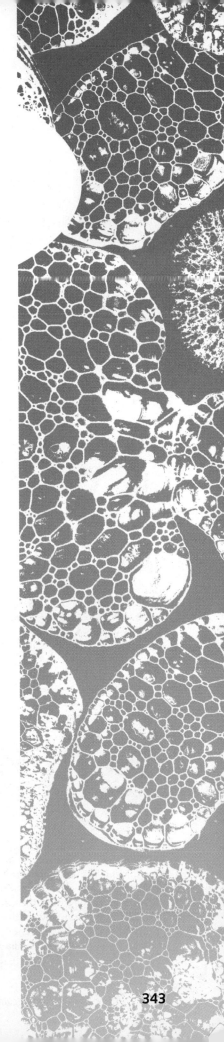

17

Chemical Reactions and Energy

*i*t's a warm summer morning at the beach. The sun has just risen and it casts long shadows across the mounds of sand and grass. Waves rise gently up the shoreline and breezes carry gulls over the shallow waters. Birds swoop down to poke at passing fish and drifting seaweed, while small clams, exposed by the falling tidal waters, dig into the sand for protection.

Like every other part of the living world, the organisms at the beach might be described in terms of matter: proteins, carbohydrates, and other chemicals, simple and complex. Yet such a description would be incomplete. The beach is full of movement and sound. Organisms search for food, they react to their environment, and they change.

To deal with the real world, we must consider more than the materials of life. We must consider how matter moves and changes from one form to another and, even more important, the cause of those transformations, a factor known as energy.

ENERGY AND LIFE

A physicist defines **energy** as *the capacity to do work.* In everyday usage, work is the expenditure of energy to move something: to lift a bale of hay or throw a ball, each of which requires energy.

Energy can exist in several forms and can be converted from one form to another. **Kinetic energy** is the energy of a moving object, whether a baseball, planet, or a molecule of carbon dioxide. The kinetic energy of that baseball can be changed into other forms. When a fast-moving baseball strikes a glass window, its kinetic energy is reduced as the baseball is slowed down by the impact. Some of the original kinetic energy is converted into *sound*, some into energy used to break the bonds that hold the glass together, and some into *heat* (the glass is slightly hotter as a result of the impact). Energy can also exist in a stored form known as **potential energy.** The kinetic energy used to compress the spring in a child's dart gun becomes potential energy. When the spring is released, that potential energy is converted back into kinetic energy (Fig. 17.1).

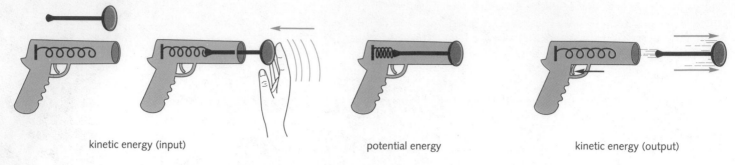

kinetic energy (input)   potential energy   kinetic energy (output)

Figure 17.1 Stored energy is potential energy. When the toy dart is pushed into the barrel, energy is stored in the compression of the spring. The coiled spring contains potential energy. When the trigger on the toy gun is pulled, the spring is released and the potential energy is transformed back into kinetic energy.

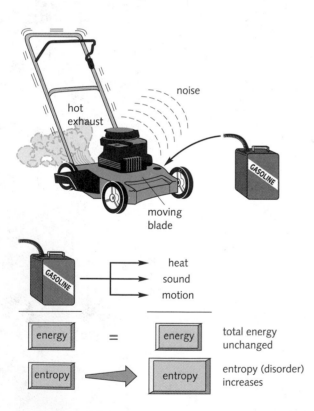

Figure 17.2 Chemical energy released from burning gasoline in a lawn mower is converted into a variety of forms, including noise, motion, and heat. The first law of thermodynamics (conservation of energy) tells us that energy will not be lost in the conversion processes. However, the second law of thermodynamics dictates that entropy will increase, converting the concentrated chemical energy into less available forms.

Chemical energy is a special form of potential energy. Chemical energy is not contained in the compression of a spring but in the chemical bonds that hold molecules together. When bonds are broken, energy is consumed. When new bonds are formed, energy is released. If the chemical energy in a liter of gasoline is suddenly released, its rapid conversion into kinetic energy, heat, sound, and light produces a powerful explosion.

As we saw much earlier in the text, nearly all of the energy available on the earth comes from the sun. Most of this solar energy is immediately transformed into other forms of energy. Sunlight warms the planet, and the uneven distribution of solar energy causes powerful currents in the atmosphere and the seas. A small but significant amount of solar energy is captured by living things and converted into chemical energy. It's not an exaggeration to say that the conversion of sunlight into chemical energy is what makes life possible on Earth.

Thermodynamics

The energy changes that occur in the physical world are studied in a branch of science known as **thermodynamics.** The **first law of thermodynamics** is also known as the principle of **conservation of energy.** It states that "energy can be neither created nor destroyed." In other words, *the total amount of energy available in the universe does not change.* Einstein showed that matter and energy are interchangeable, and the first law is sometimes written in a way that points this out. The first law tells us that we must account for all the matter and energy involved in a chemical reaction.

The laws of thermodynamics apply to all systems, even to devices as ordinary as a gasoline-powered lawn mower.

Chemical energy is available in the gasoline tank. The expansion of gasoline and air as they burn in the cylinders of the engine provides the mechanical force to do work—spinning a rotary blade at high speeds and sending pieces of grass scattering out the cutting chute. The first law tells us that *none* of the energy from the gasoline is lost. That does not mean that *all* of the energy from the gasoline is transferred to the blades of grass that fly out of the mower. Lots of chemical energy is converted into the heat in the hot exhaust, the heat of the engine, and the noise of the mower (Fig. 17.2). All the energy is still there—it's just been converted to many different forms.

The lawn mower can also introduce us to the **second law of thermodynamics,** which says that *once energy has been used to do work, it becomes less available to do additional work.* The chemical energy from the gasoline has not been lost; it has been converted into forms much less available once the grass has been mowed. It is not possible to collect the noise and exhaust heat released during the pumping process, combine them with the kinetic energy of the clippings, and reconcentrate them to produce more fuel. The energy has become dispersed. It is less available. This is the change that is at the heart of the second law.

Chemists use the word **entropy** in connection with the second law of thermodynamics. Entropy is a measure of disorder. When the energy in a system becomes dispersed, we say that the system's entropy (disorder) has increased. The second law says that *in the universe as a whole and in any isolated system, the entropy associated with any chemical or physical change always increases.* Energy becomes more dispersed, less available, and more disorderly, and entropy increases.

Although the second law of thermodynamics seems to say that things must always run "downhill," as energy becomes increasingly dispersed and the entropy of the universe increases, it is important to realize that our statements of the second law have been phrased carefully to apply to closed systems: the universe as a whole is a closed system, but the earth is not. It is an open system with a strong and steady input of energy every day from the sun in the form of light.

What does this say about the second law and biology? Without an input of energy, living systems would be unable to maintain the organization and molecular order characteristic of life. Life depends on energy, and the first two laws of thermodynamics explain why.

ENERGY AND CHEMICAL REACTIONS

Chemical reactions involve the making and breaking of chemical bonds, and chemical bonds involve energy. If you've ever used a gas kitchen stove, you've seen the hot

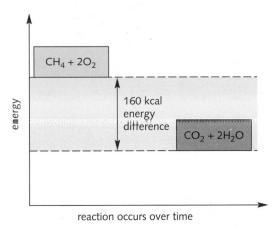

Figure 17.3 When 1 mole of methane (about 16 grams) is burned, 160 kcal of energy are released. The sources of this energy are the rearrangements of chemical bonds between carbon, hydrogen, and oxygen.

blue flame of a natural gas such as methane burning in the presence of oxygen. In chemical terms, this is a straightforward reaction:

$$CH_4 + 2O_2 \rightarrow CO_2 + 2H_2O$$

Why does the reaction produce enough heat to power a stove? The answer can be found in an important difference in the amount of energy in the chemical bonds of the products and reactants. As you may know, the heat energy produced by a chemical reaction is measured in units called **calories.** A calorie is the *amount of heat required to raise the temperature of* 1 cm^3 *of water by* 1° *C.* A **kilocalorie** (abbreviated **kcal**) is equal to 1000 calories. As we will see in Chapter 37, the energy found in food can also be expressed in terms of kilocalories (1 food "calorie" equals 1 kilocalorie of heat energy).

When 1 mole of methane is burned in oxygen, 632 kcal are required to break the chemical bonds. Remember that a *mole* is equal to the molecular weight of a compound in grams, and for methane, that's about 16 grams. The 632 kcal needed to break the chemical bonds in methane is a substantial amount of energy. However, 792 kcal of energy are released when new bonds in the products of the reaction (CO_2 and H_2O) are formed, exceeding the amount required to break the initial bonds by 160 kcal (Fig. 17.3). The 160 kcal are released as heat when methane burns. Just 1 gram of methane has more than enough chemical energy to bring 100 milliliters of water

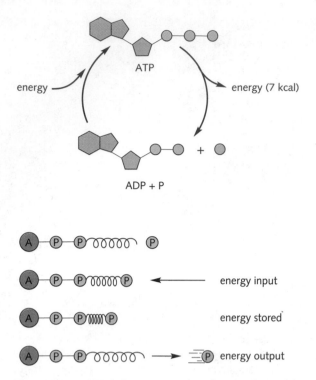

Figure 17.4 ATP (adenosine triphosphate) is formed by the process of *phosphorylation,* in which a third phosphate group is added to ADP (adenosine diphosphate). Energy is required to make ATP, and energy is released when ATP is converted back into ADP and phosphate. One way to visualize the role that ATP plays as a carrier of chemical energy is to compare the attachment of the third phosphate group to the compression of a spring.

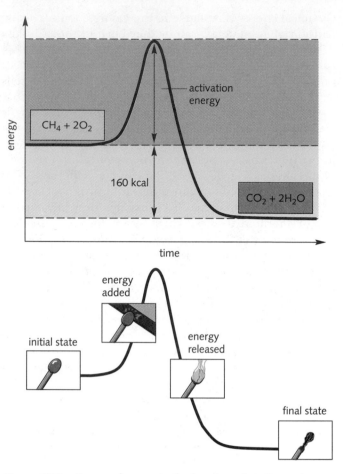

Figure 17.5 Energy changes in the burning of methane. In order to participate in a chemical reaction, the reactants must first be raised to a certain energy level. The energy required to reach this level is known as *activation energy*. When an ordinary match is struck, the potential energy barrier is overcome from the heat produced by friction, thus igniting the match and releasing energy.

to the boiling point. That's why methane gas is such a powerful and efficient fuel.

Chemical reactions, like the burning of methane, that release energy are known as **exergonic reactions.** Reactions that require energy are termed **endergonic reactions.** Living organisms carry out both types of reactions, and exergonic reactions are often used to provide the energy necessary for endergonic reactions to take place.

Chemical Energy and ATP

You may wonder how the energy *produced* in one reaction could be transferred to *provide* energy for a second reaction. This is a trick that living organisms carry out in a number of ways, depending upon their particular needs. However, the most common way of transferring chemical energy is by means of the compound called **adenosine triphosphate (ATP).** ATP is actually a nucleotide, one of the building blocks of nucleic acids. But it is also used to transfer energy within the cell.

ATP is produced by adding two phosphate groups to **adenosine monophosphate (AMP),** or by adding one phosphate group to **adenosine diphosphate (ADP).** This is an energy-requiring (endergonic) reaction. Energy must be "invested" in the production of ATP (Fig. 17.4). Breaking the bonds that attach either the second or third phosphate groups to ATP releases about 7000 calories (7 kcal) of energy per mole of ATP. Of course, 7 kcal is much less than the 160 kcal released by burning methane, but that's just the point. ATP is a carrier of small, convenient packets of chemical energy—just what the cell needs.

Frequently, these bonds are described as "high-energy" phosphate bonds. Actually, the bonds that attach the second and third phosphates to ATP release relatively small amounts of energy. The principal distinctions of these chemical bonds are that they are easily made and broken and that small amounts of energy are consumed or released in each. This enables the cell to use the energy released by exergonic reactions to produce ATP and then to use that ATP to provide the energy necessary for endergonic reactions. As we continue our study of biology, we will see many examples of biological processes that involve ATP as a carrier of chemical energy.

ACTIVATION ENERGY

Just because a chemical reaction produces energy does not mean that the reaction will begin spontaneously. Mixing methane and oxygen is not enough to start a flame. Something else is necessary. We all know from common experience that we must *ignite* combustible material to start a fire. A small amount of energy is needed to get the reaction under way. Why is that energy necessary? Remember that 632 kcal/mole were needed to break the chemical bonds in methane and oxygen before the products of the reaction were formed to release even more energy.

The energy required to start a chemical reaction is known as **activation energy.** A common way to supply activation energy is by heating the reactants. Heating increases the frequency of collisions between the molecules involved in the reaction, and it may weaken or break chemical bonds that must be opened for the reaction to begin.

An energy level diagram, like Fig. 17.5, shows how activation energy affects a chemical reaction. In this example, the reactants are at a higher energy level than the products, so energy will be released if the reaction takes place. But activation energy is a barrier that prevents the reaction from taking place unless enough energy is supplied to lift the reactants over that barrier. In some cases, supplying that energy can be as simple as striking a match. In others, the task can be much more difficult.

Catalysts

The fact that many chemical reactions require substantial activation energy has important consequences for the world around us. To begin with, it lends some stability to things. If every molecule that was capable of participating in an energy-yielding reaction were to go ahead and enter such a reaction, most of the molecules around us would break down very quickly. Our clothes, most of our homes, our books and papers, and our CDs and tapes would quickly react with oxygen and burn if the requirement for activation energy were lifted.

Nonetheless, for a living organism the need for activation energy could be troublesome. It would seem, for example, that a cell would have to heat food molecules to the point of burning in order to release the energy they contain. Fortunately, that's not the case. Living cells have discovered ways to lower activation energies, ways that chemists did not suspect until the past hundred years.

A **catalyst** is a chemical compound that lowers the activation energy of a chemical reaction (Fig. 17.6). Catalysts are extremely important in industry and are used in everything from the curing of plastic to the refining of petroleum. In automobiles, catalytic converters remove unburned fuel from the exhaust gases. In the converter, platinum and palladium catalysts lower the activation energy of unburned fuel enough to break it down into products that will not harm the environment.

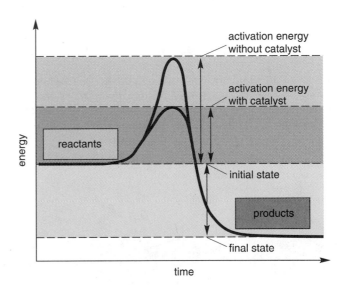

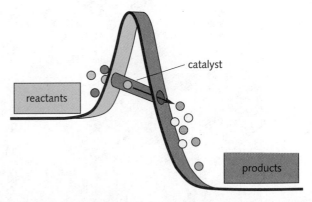

Figure 17.6 Catalysts act by lowering the activation energy of a chemical reaction. We can visualize a catalyst as a tunnel through the activation energy barrier.

Cells have a similar need for the right kind of catalyst. Cells need something like platinum and palladium that will lower the activation energy of important reactions so that they can take place at ordinary temperatures. These biological catalysts do indeed exist, and they are known as **enzymes.**

ENZYMES

Enzymes were discovered in the nineteenth century when Eduard and Hans Büchner showed that an extract prepared from broken yeast cells could perform the reactions of fermentation. (History lives on in the vocabulary of science: the word *enzyme* comes from *zyme,* the Greek word for "yeast.") As the molecules in the extract were analyzed, it was discovered that the catalytic portion of the extract was made up of proteins.

Enzymes are biological catalysts, and they get involved directly in the reactions they catalyze. The reactants of an enzyme-catalyzed reaction are known as **substrates.** The substrate or substrates bind directly to the enzyme and are converted into product:

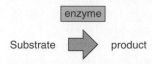

The binding of substrate to enzyme is extremely specific. Many enzymes will bind one, and only one, group of substrate molecules. Enzymes are able to distinguish these substrates from other molecules that are nearly identical. The binding itself takes place in a region called the **active site** of the enzyme. The active site is generally shaped to fit the substrate and contains chemical groups that bind the substrate tightly (Fig. 17.7). Biochemists, impressed with the right fit between substrate and active site, originally compared it to a lock and key, and that's not a bad analogy.

Sometimes, enzymes change shape as they bind substrate, producing an even tighter fit (Fig. 17.8). This is

Figure 17.7 Enzymes are chemical catalysts that bind substrates to a region known as the active site. This diagram shows one example of a substrate, a uracil base in a strand of RNA, bound to an enzyme, ribonuclease. Chemical groups in the active site stabilize the binding of the substrate, as shown.

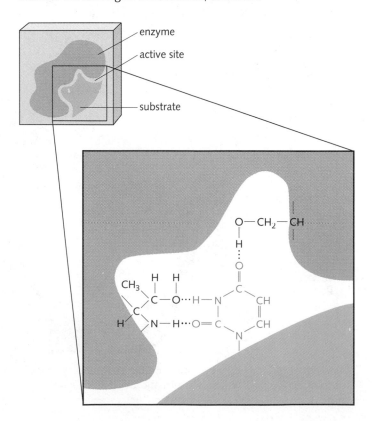

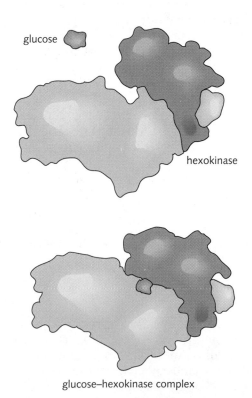

Figure 17.8 In some cases, the binding of substrate is made more effective by a change in enzyme shape, a phenomenon known as induced fit. Note how the two major regions of hexokinase fold around glucose when it binds to the active site of the enzyme.

known as **induced fit.** It's called that because the presence of the substrate *induces* a change in shape that enables the enzyme to *fit* the substrate better. This is essentially what happens when you shake hands—you change the shape of your hand when contact is made to ensure a better fit.

This tight binding produces an **enzyme–substrate complex** (Fig. 17.9), and that's where the real work of catalysis, the lowering of activation energy, takes place. Substrate is converted into product at the active site, and the product is then released. This opens the active site to the binding of more substrate, and the process can begin all over again.

How quickly can all this happen? Urease is an enzyme that catalyzes a reaction between urea and water to produce carbon dioxide and ammonia. At room temperature, a single molecule of the enzyme can catalyze 10,000 reactions in a single second!

Until very recently (see Current Controversies, Ribozymes, p. 354), all known enzymes were proteins. The tremendous chemical flexibility of proteins has provided evolution with the material to develop enzymes that will catalyze almost any reaction.

Cofactors

Many enzyme-catalyzed reactions require more than just an enzyme and its substrates. Some enzymes require a particular ion or even a small molecule. These molecules are called **cofactors.** If a cofactor is an organic molecule, it is called a **coenzyme.** Cofactors are not considered substrates to a reaction because they are not used up and converted to product. They may be used over and over again. This means that a cell needs very little of a particular cofactor.

Many of the *vitamins* that we need in our daily diet are actually coenzymes that human cells are unable to synthesize themselves. Many of the B-type vitamins, including vitamin B_2 (riboflavin), pantothenate, and niacin, are required as coenzymes for the chemical reactions that release energy within the cell. The ill effects caused by deficiencies in these vitamins are the direct results of having a particular set of reactions slowed down by the absence of a coenzyme.

Enzymes and the Laws of Chemistry

As remarkable as they are, even enzymes cannot violate the laws of thermodynamics. Therefore, although enzymes can lower activation energy and dramatically increase the *rate* of a chemical reaction, they do not affect its energy requirements (Fig. 17.9). A reaction in which the products are at a higher energy level than the reactants still requires a source of energy to power the reac-

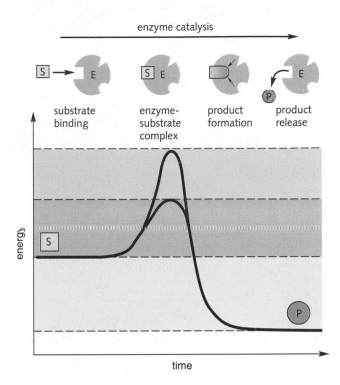

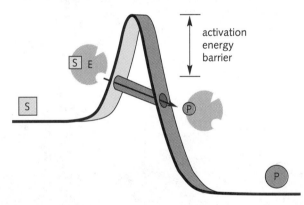

Figure 17.9 Enzyme catalysis is a multistep process. The first step involves the binding of substrate to enzyme, forming an enzyme–substrate complex. The substrate is then converted to product. In the final step, the product is released and the enzyme is ready to bind another substrate. Enzymes increase the rate of chemical reactions by lowering activation energy.

tion. An enzyme may speed up a chemical reaction, but it cannot change its direction.

Like any other compound, an enzyme is affected by its chemical environment. Remember that the binding of enzyme to substrate depends on a close three-dimensional fit between the active site of the enzyme and the substrate. Therefore, any factor that changes the shape of an enzyme has the potential to change its chemical characteristics.

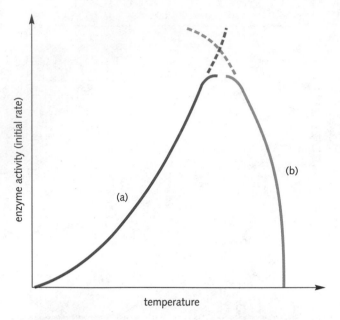

Figure 17.10 Enzyme-catalyzed reactions are temperature-sensitive. Up to a point, increasing temperature speeds up such reactions (curve a). However, because most enzymes are proteins, when the temperature reaches a point where it affects protein structure, the shape of the active site changes and reaction rates slow down (curve b). Not all enzymes have the same optimum temperature.

Figure 17.10 shows how temperature affects the rate of a typical enzyme-catalyzed reaction. Initially, increasing temperature speeds up the reaction rate, as it does with most chemical reactions. However, as temperatures exceed 40° and 50°C, something else happens. Increasing molecular motion begins to disrupt protein structure around the active site, and the enzyme begins to lose its ability to catalyze the reaction. Very high temperatures may permanently inactivate, or **denature,** an enzyme.

Enzymes are also sensitive to pH (Fig. 17.11), which can affect the structure of protein and substrate alike. It's interesting to note that many enzymes do their optimum work at different pH values, and there's a reason for that. Pepsin, which breaks down proteins in food, works best at low pH, typical of the acidic stomach fluids into which pepsin is released. By contrast, the starch-digesting amylase found in saliva has an optimum close to that of saliva itself, pH 7.0.

ENZYME REGULATION

Enzymes can be affected by molecules other than their substrates. For example, molecules that are similar in shape to a substrate molecule may bind to the active site and prevent the enzyme from making contact with its substrate. When this occurs, the enzyme is said to be *inhibited.* Molecules that block the active site in this way are known as **competitive inhibitors,** because they *compete* with the substrate for the active site (Fig. 17.12).

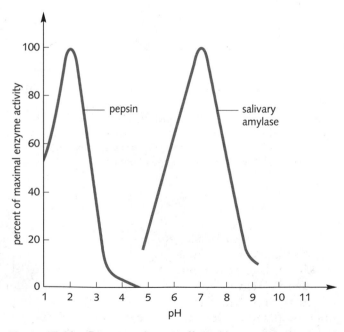

Figure 17.11 Enzyme activity is affected by pH. Enzymes generally display a pH optimum that is appropriate for their location in the body or the environment.

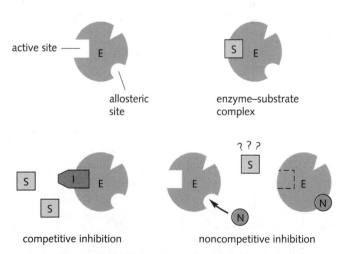

Figure 17.12 Enzyme inhibition. *Competitive inhibitors* (I) resemble the substrate and compete with substrate for binding to an enzyme's active site. *Noncompetitive inhibitors* (N) bind outside the active site and alter enzyme activity by causing changes in the enzyme shape that affect the active site. These shape changes are known as *allosteric effects.* The binding site where allosteric changes can be produced is known as an *allosteric site.*

Enzymes can also be affected by inhibitors that bind outside the active site. These molecules are often called **noncompetitive inhibitors,** because they inhibit the reaction but *do not compete* with substrate for the active site (Fig. 17.12).

How can a molecule that binds outside the active site affect the activity of an enzyme? Noncompetitive inhibitors cause a change in the shape—the **conformation**—of an enzyme after they bind to it. This change in shape sometimes closes the entrance to the active site or makes the site disappear altogether. In its new conformation, the enzyme is no longer active. The inactivation will usually last as long as the inhibitor remains bound to the enzyme.

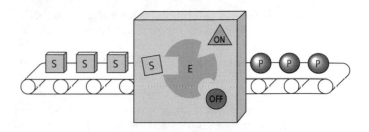

Figure 17.13 An enzyme is a bit like a machine that takes in substrate and converts it to product. Allosteric sites may be thought of as *on* and *off* switches. (This analogy has its limits, of course. One is that most machines use an outside source of energy to carry out their jobs. Enzymes *do not* provide energy for chemical reactions.)

Allosteric Effects and Feedback Pathways

Strictly speaking, any molecule that binds to an enzyme outside its active site and affects its activity is said to be **allosteric** (*allo–steric* means "other–shape"). Noncompetitive inhibitors are just one type of allosteric compound. **Allosteric activators** also bind outside the active site, but they change enzyme structure in a way that *increases* an enzyme's reaction rate.

A great many enzymes are sensitive to allosteric effects, which means that molecules other than their substrates can either increase or decrease their rate of activity. The possibilities that open up are remarkable. Just for a moment, think of an enzyme as a machine (Fig. 17.13). Raw material (substrate) goes in one side and finished product comes out the other side. No machine is useful unless it can be regulated. In an important sense, allosteric sites provide the *on* and *off* switches that allow the machine to be slowed down or speeded up, as necessary.

Let's further suppose, also just for a moment, that we want to automate a factory full of such machines. We'll place the machines on an assembly line. Substrate goes into one end, and four steps (four different enzymes) are necessary to produce finished product at the other end. We'd like our line to be automatic, making product when it's needed and shutting down when more product is not needed. How can we do this?

Figure 17.14 shows one way. We can set up the production line so that the amount of final product controls the *on/off* switch on the first machine. If product piles up, the first machine is switched off. When product becomes scarce, the first machine is switched on and the line begins to work. When you think about it, the final product acts as an "allosteric inhibitor" to the first machine. A factory specialist would say that this production line is

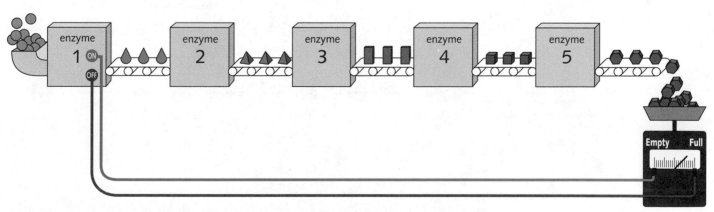

Figure 17.14 The synthesis of complex product may require the cooperation of many machines (enzymes). In this self-regulating system, a scale at the end of the production line weighs the amount of accumulated product. When the bin is nearly empty, the green *on* switch starts the production line. When the bin is nearly full, the red *off* switch stops it. This system of feedback regulation ensures that the line will not run unless more product is needed.

governed by **feedback inhibition,** since the amount of final product "feeds back" to the first machine to "inhibit" further production.

Believe it or not, some enzymes in the cell use just such systems to regulate their own activity.

Many enzymes are involved in chemical pathways in which a sequence of reactions synthesizes a molecule step by step from a starting compound. The level of the pathway's end product is often kept constant by a **feedback system,** in which the end product *inhibits* the activity of the first enzyme in the pathway by binding to an allosteric site. This allows the pathway to be self-regulating. When the level of the end product is high, noncompetitive inhibition shuts the pathway off. When the level of the end product drops, the number of enzyme molecules that are inhibited also drops; and the pathway starts up again, replenishing the supply. The synthesis of *isoleucine,* an amino acid, is regulated in exactly this way (Fig. 17.15). Isoleucine is synthesized in a five-step pathway from another amino acid, *threonine.* The first enzyme in the pathway, *threonine deaminase,* contains an allosteric site that binds the end product, isoleucine. This self-regulating pathway, in which feedback inhibition from the end product controls the first reaction, enables the cell to regulate the level of isoleucine needed for protein synthesis.

Epilogue: What's in the Bag?

When biochemists began to investigate the properties of enzymes, they came to appreciate the powers of these extraordinary molecules. The color of a flower, the materials we can eat for food, the shape of an eggshell, and the strength of a muscular contraction are all determined by enzymes. It is not incorrect to say that the process of life itself is under the control of enzymes. However, we are left with an interesting question. If enzymes can do so many things, where does the information for *their* synthesis come from? As we will see shortly, attempts to find an answer to that question have led to the most important biological discovery of the twentieth century: the chemical nature of the gene.

As the field of biochemistry developed in the 1930s and 1940s, many scientists joked that the cell could be thought of as nothing more than a "bag of enzymes." Even at the time, scientists realized that this was not quite true, but the sweeping generalization was their way of emphasizing the importance of understanding the chemical pathways that enzymes control. In the next chapter, we will look at two of the most critical of those pathways—those that produce and those that transfer cellular energy.

Figure 17.15 Feedback inhibition in a biochemical pathway. Isoleucine is an amino acid synthesized in a multistep pathway catalyzed by a series of enzymes. Isoleucine, the final product of the pathway, acts as a noncompetitive inhibitor to threonine deaminase, the first enzyme in the pathway. This feedback inhibition allows the pathway to be self-regulating.

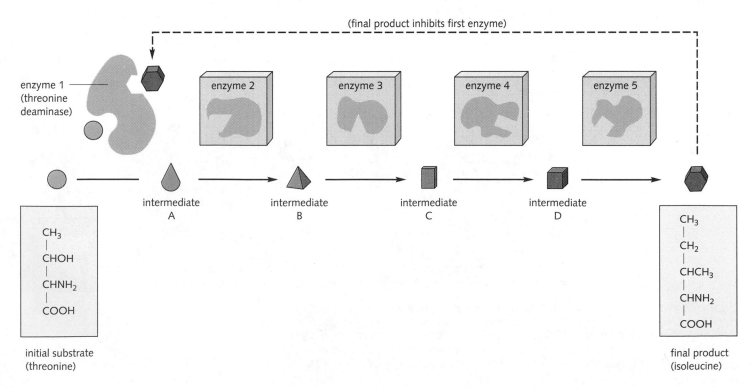

The Active Site: A Peek Inside

How does an enzyme actually work?

The short answer is that the active site of an enzyme provides a chemical environment that speeds up a reaction. The details of how an enzyme does this are different in every enzyme.

One of the best-studied active sites is found in a class of protein-cutting enzymes called serine proteases. The figure (part a) shows what a serine protease does. It catalyzes the cutting (hydrolysis) of peptide bonds in a polypeptide chain. Here's how it works:

- On either side of the enzyme's active site are a serine and a histidine side group (part b of the figure).
- When a polypeptide substrate enters the active site, the SER and HIS side groups attack different sides of the peptide bond. The SER actually forms a covalent bond with the substrate (part c of the figure).
- The HIS side group donates a proton to the substrate, causing the peptide bond to break and releasing the left-hand fragment of the peptide (part d of the figure).
- Then a water molecule enters the active site, splitting the SER link and releasing the second fragment, which returns the active site to its original configuration (part e of the figure).

The way a serine protease works is typical of the chemical "tricks" used by enzymes to catalyze chemical reactions. Some R groups of amino acids in the active site contain chemical groups that hold the substrates in the best possible positions for the reaction to occur. Other R groups donate or remove a proton to the substrate at just the right moment to assist the reaction and convert substrate to product. Some active sites actually push or pull the substrate to make or break particular chemical bonds.

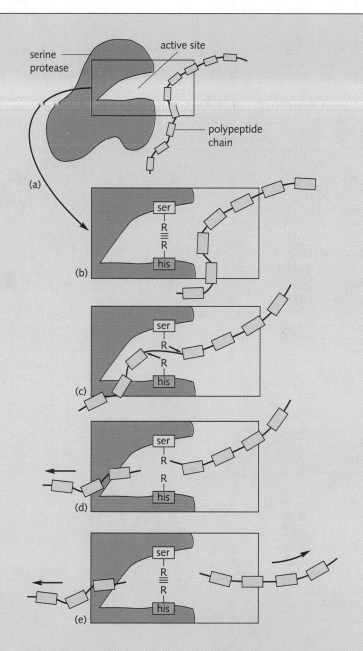

(a) Serine proteases are enzymes that cut polypeptide chains. (b) The active site of such a protease contains a serine and a histidine on opposite sides, held close to each other by hydrogen bonds. (c) Binding of substrate to active site. Hydrogen bonds hold the chain in place. Side groups of serine and histidine attack one of the peptide bonds. (d) The peptide bond is broken and half of the polypeptide is released. The other half is covalently bound to serine. (e) The serine bond is broken, and the remaining polypeptide is released. The active site returns to its original structure.

Ribozymes

Anyone who studies a biochemical reaction usually expects to find an enzyme at the heart of it. Enzymes are often purified first by trying to produce a cell extract that shows the greatest activity in catalyzing a particular reaction. In that fraction, the investigator expects to find the enzyme. Imagine the surprise, therefore, of Thomas Cech, Paula Grabowski, and Arthur Zaug at the University of Colorado a few years ago. They were studying the processing of one particular form of ribonucleic acid (RNA) in *Tetrahymena*, a protozoan.

This RNA molecule is synthesized as a large precursor and "processed" to its final form: An intervening sequence, or "intron," of 413 nucleotides is cut out of the precursor molecule. Once the intron is removed, the two ends of the RNA are spliced back together to form the mature RNA. When these researchers maximized the rate of the cutting and splicing reactions, they expected to find a series of proteins that catalyzed the reactions. Instead, they found that the solution contained only the RNA that was being cut and spliced. There was no protein at all in the fractions that had the highest enzymatic activity!

Concerned that they had made a mistake, they set the results aside and continued to study the cutting and splicing reactions. Gradually, they returned to their original results, confirmed them, and confirmed them again. Only one conclusion was possible: *the RNA molecule was acting as an enzyme that catalyzed its own splicing.* The enzyme for the reaction was RNA itself. It became clear that the reaction proceeded in three steps. First, a guanosine base attacks and breaks a linkage in the precursor. Next, the intervening sequence is removed and the two ends of the mature RNA are spliced together. Finally, a 15-base fragment

breaks off the intron, and the remaining portion of the intervening sequence joins to itself to form a small circular RNA. Strictly speaking, the whole RNA molecule cannot be considered a true enzyme because it is altered by the reaction. But further studies have shown that the portion of the RNA molecule that is removed from the RNA itself can act as a true enzyme, catalyzing nucleotide sequences.

We now know that many chemical reactions are catalyzed by RNA. To many scientists, the knowledge that RNA can catalyze chemical reactions suggests something else: the possibility that the evolution of life may have begun with small RNA molecules capable of controlling chemical reactions and even making copies of themselves.

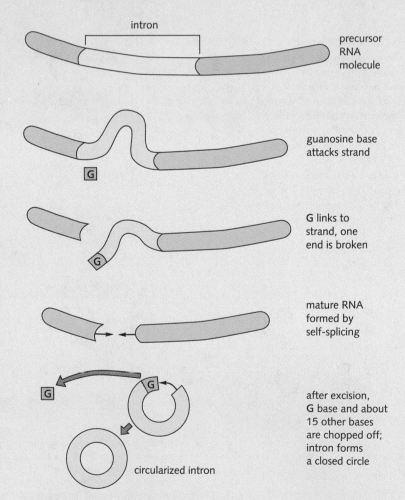

An outline of the self-splicing RNA molecule discovered by Cech, Zaug, and Grabowski.

Energy, the ability to do work, is fundamentally important to living things. The vast majority of energy available to living things on the earth is derived from sunlight. Energy transformations are governed by the first two laws of thermodynamics, which state that the total amount of energy in the universe is not changed by chemical reactions, and that once energy has been used to do work, it becomes less available to do additional work. One of the biological consequences of these laws is that organisms require a continual supply of energy to remain alive.

Energy is involved in chemical reactions, where it may be either absorbed or released during the formation and breaking of chemical bonds. Biological systems make use of the ATP molecule as the common currency of energy storage and transfer, storing small amounts of energy by attaching a third phosphate group to ADP to produce ATP, and then using the energy available in ATP to power energy-requiring chemical reactions.

Chemical reactions, even those that release energy, cannot begin until the reactants surpass an energy level known as the activation energy. This activation energy can be supplied by heating the reactants, but the requirement for activation energy can also be met via catalysts or enzymes, which are biological catalysts. Enzymes promote chemical reactions by lowering activation energies, and this can dramatically increase the rates of many chemical reactions. Enzymes act by binding substrate molecules to an active site, and they may interact directly with their substrates to lower activation energy. Enzymes may also require additional organic molecules, known as cofactors, to catalyze reactions.

Enzyme activity may be affected by inhibitors, which slow down reaction rates, or by activators, which speed up rates. The activities of enzymes directly control the chemical pathways and many of the characteristics of living organisms.

STUDY FOCUS

After studying this chapter, you should be able to:

- Explain the relevance of chemical energy to biology.
- Explain the concept of a chemical reaction, and analyze what happens when a reaction takes place.
- Explain the role of ATP in storing and transferring chemical energy.
- Describe catalysis and activation energy.
- Illustrate how enzymes can catalyze biological reactions.
- Give an example of feedback inhibition, and explain how it helps to regulate the amounts of some molecules found in cells.

TERMS AND CONCEPTS

kinetic energy *343*
potential energy *343*
chemical energy *344*
first law of
  thermodynamics *344*
second law of
  thermodynamics *345*
entropy *345*
ATP *346*
activation energy *347*

catalyst *347*
enzymes *348*
substrates *348*
active site *348*
cofactors *349*
coenzyme *349*
competitive inhibitors *350*
noncompetitive
  inhibitors *350*
allosteric *351*

REVIEW

Objective Questions (Answers in Appendix)

1. Which of these statements is not true about enzymes?
   (a) Most enzymes can catalyze a variety of reactions.
   (b) An enzyme lowers the activation energy of the chemical reaction.
   (c) An enzyme temporarily binds substrates at an active site.
   (d) An enzyme is usually a protein.

2. Entropy is defined as
   (a) disorder.
   (b) conversion from one state of matter to another.
   (c) coupling.
   (d) a state of equilibrium.

3. When an energy change takes place, the entropy of the universe increases. This statement best describes the
   (a) first law of thermodynamics.
   (b) second law of thermodynamics.
   (c) law of activation energy.
   (d) third law of thermodynamics.

4. Activation energy can be defined as the
   (a) amount of randomness in a system.
   (b) total amount of energy used in biological reactions.
   (c) energy input needed to start a reaction.
   (d) conversion of chemical to heat energy.

5. By lowering the activation energy of a reaction, enzymes
   (a) become competitive inhibitors.
   (b) compete with substrate for the active site.
   (c) follow the principle of mass action.
   (d) increase the rate at which chemical reactions occur.

6. A noncompetitive inhibitor
   (a) does not affect reaction rates as much as competitive inhibitors.
   (b) has a close chemical resemblance to substrate.
   (c) has a close chemical resemblance to product.
   (d) binds outside the active site of an enzyme.

Discussion Questions

7. What happens during a chemical reaction? Are changes in energy always associated with chemical reactions?

8. What do investigators in thermodynamics study? Describe the two principal laws of thermodynamics.

9. What is entropy? How is entropy associated with chemical reactions?

10. What is the role of ATP in energy storage and transformation?

11. Explain the role of activation energy in a chemical reaction. How can the activation energy be lowered?

12. What is an enzyme? What role does an enzyme play in a chemical reaction? How are enzymes controlled and regulated?

13. Is it possible to have two enzymes catalyzing a reaction, one making the reaction go forward while the other makes it go in the reverse direction?

14. A chemical reaction in which compound A is converted to compound B is being studied in the lab. The reaction is endergonic. Will it be possible to find an enzyme that can make the reaction exergonic? Why or why not?

READINGS

Gibbons, A. "Molecular scissors: RNA enzymes go commercial." *Science* 252 (1991): 521. Interested in the possibilities of RNA-based enzymes? This article explores the commercial possibilities of these novel enzymes.

Kraut, J. "How do enzymes work?" *Science* 242 (1988): 533–540. A detailed discussion of the principles of active site structure and catalysis.

Karplus, M., and J. A. McCammon. "The dynamics of proteins." *Scientific American* 238 (March 1986): 42–51. Enzyme reaction mechanisms often include changes in shape initiated by binding of substrate. This well-written article explains how proteins change shape.

18
Cells and Energy

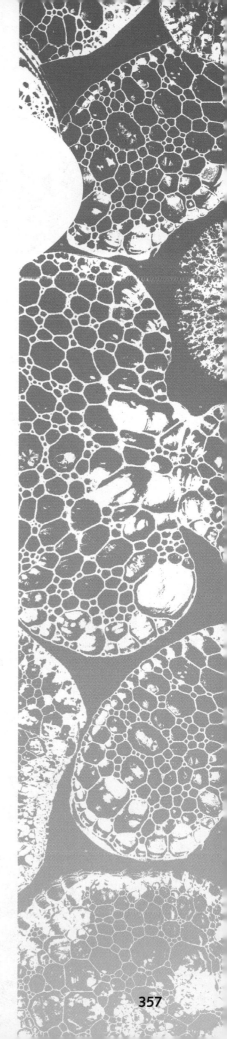

i t's surprising to think how many aspects of daily life depend directly on sources of energy. When a car runs out of gas, it comes to a sputtering halt. Without fuel, a warm house is at the mercy of the elements. Without electricity, computers, lights, appliances, and mass media such as radio and television grind to a halt. Nearly every thread in the fabric of modern society depends on energy.

Living things depend on energy, too. Nearly everything that happens within a living cell—from cell division, to DNA replication, to the synthesis of biological membranes—requires energy. We've already seen that autotrophic organisms solve their energy needs by harnessing sunlight directly and that heterotrophic organisms must find a source of food to obtain chemical energy. On a global scale, the flow of energy from sunlight through autotrophs to heterotrophs affects the flow of carbon, nitrogen, and oxygen throughout the biosphere.

We can see the effects of global energy flow all around us, from winds and storms to the gentle force with which a seedling breaks through the earth. But the critical events involving energy and living organisms, including the *transformation* of one kind of energy into another kind, are often invisible. How do photosynthetic organisms capture the energy of sunlight and transform it into chemical energy? How do organisms release chemical energy from the food they eat or produce? How do they change that energy into a form that can be used for movement and growth? In this chapter, we will study the ways in which energy transformation occurs at the cellular level and examine in detail how cells handle the flow of energy.

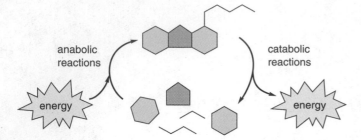

Figure 18.1 Anabolic reactions and catabolic reactions exist within the chemical activity of the cell's metabolism, in which the demands for growth and energy production are balanced.

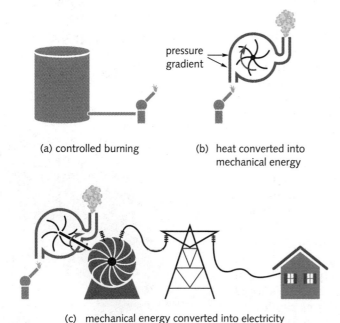

(a) controlled burning

(b) heat converted into mechanical energy

(c) mechanical energy converted into electricity

Figure 18.2 Converting the energy in fuel oil to electricity is a multistep process. **(a)** The oil is burned slowly, under controlled conditions, so that the release of heat energy is manageable. **(b)** The expanding gases are produced in a turbine or boiler, so that the pressure they produce can be converted into mechanical energy. **(c)** Mechanical energy is converted into electricity by a generator. Electrical energy is then distributed over a wire network.

METABOLISM

In the preceding chapters, we studied a number of biochemical pathways in which macromolecules were synthesized from simpler chemicals. Proteins are assembled from amino acids, and polysaccharides are built from individual sugars. Chemical pathways in which larger molecules are assembled from smaller molecules are said to be **anabolic.** Anabolic pathways generally require an input of energy.

In most energy-producing pathways, larger molecules are broken down into smaller ones. Such pathways are **catabolic.** In every cell, anabolic and catabolic pathways operate at the same time, and the relationship between them is complex (Fig. 18.1).

We refer to the sum of all chemical and physical reactions associated with life as **metabolism.** In a way, the metabolism of a cell is nothing more than its internal biochemical "economy." The economy of a cell is every bit as varied and complex as the economy of a nation, and studying the metabolism of a cell is just as challenging. In this chapter, we will focus on some of the catabolic pathways that provide the energy that makes life possible.

Energy Conversion

All living cells depend on chemical energy from compounds that we call food—principally, carbohydrates, lipids, and proteins. Cells face the challenge of converting that energy to a more convenient form—most commonly, adenosine triphosphate (ATP). The production of ATP using food energy is one of the most important tasks carried out by any living cell. In this chapter, we will investigate that process in detail, asking, quite simply, how the cell does it.

One way to begin our study is to look at a similar yet nonbiological problem: taking a single source of chemical energy (fuel oil) and converting it to a convenient form (electricity) that can be used by all of a small city's residents. As we all know, this process is carried out at a power plant in which oil is burned and converted to electrical energy distributed throughout the city on power lines. But there's more to the process, including a host of details that will be useful when we study energy conversion in the cell.

In our power plant example (as with a living cell), an uncontrolled burning of fuel with oxygen would release plenty of heat energy; however, it might not transform that energy into a useful form. Uncontrolled burning, therefore, has to be avoided.

The production of electricity from fuel oil is a multistep process:

1. The plant burns the oil gradually, so as to conserve energy and avoid the damage (explosion and fire) that

would be caused by releasing all the energy at once (Fig. 18.2a).

2. There's no practical way to convert oil directly into electricity. Therefore, when the oil is burned, the first product is heat. Engineers are careful not to allow that heat to escape directly up the smokestack, where its energy would be lost. Instead, they design devices to capture the energy. These devices have one thing in common: they trap expanding gases against a barrier so that pressure builds up. The difference in pressure across a barrier, a pressure "gradient," forces the barrier to move, producing mechanical energy in the form of a sliding piston or spinning turbine (Fig. 18.2b).

3. An electrical generator, by forcing some wires to move through magnetic fields, converts that mechanical energy to electricity. Then, a system of wires distributes electricity throughout the city, bringing it into the devices that require it: TVs, radios, lights, toasters, and refrigerators (Fig. 18.2c).

At first, this multistep process might seem complex. Why not just truck the oil to all households and let them use it right there? However, after a few minutes of trying to make a computer work on fuel oil, you, too, might come to the conclusion that the complexity of power generation makes it simpler to convert chemical energy to a more convenient form, electricity. Much simpler.

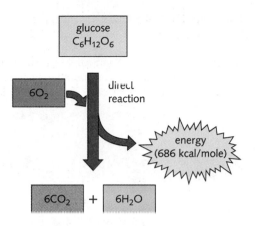

Figure 18.3 A direct reaction with oxygen releases 686 kcal of energy per mole of glucose (180 grams). Living cells, which cannot harness this enormous amount of energy in one step, must release the energy gradually.

The Energy in Glucose

The cell is not that different from a municipal power plant. Cells burn more than one fuel, of course, but we can concentrate on one food molecule that is representative of many others—the simple sugar, glucose. Just like fuel oil, glucose is a source of chemical energy. When glucose is burned in oxygen, it releases 686 kcal of energy per mole (180 grams). That's more than enough heat to bring a gallon of water (6.86 liters) to the boiling point. In the power plant, that heat is used to create pressure and drive a generator; but in the cell, there are no pistons, cylinders, camshafts, and wheels. To make matters more complex, the cell could not tolerate the release of so much direct heat. There has to be another way to capture the energy of glucose.

How can a cell capture the energy of a chemical reaction without producing massive amounts of heat? To answer that question, first think about *where* the heat comes from when glucose is burned. The chemical reaction looks like this:

$$C_6H_{12}O_6 + 6O_2 \rightarrow 6CO_2 + 6H_2O$$

If glucose is burned directly, then nearly all of the 686 kcal of energy show up as heat. But where does all that energy come from? Since the energy does not come from

the atomic nuclei, which are unaffected by the chemical reaction, it must come from the *electrons*. Rearranging the covalent orbitals of the electrons in the two compounds that entered the reaction lowers their energy levels and releases that excess energy as heat. Remember: *The energy comes from the electrons.* All cells have to do is to trap those high-energy electrons and find a way to use them to synthesize ATP, instead of producing heat.

Trapping the Energy from Glucose

To accomplish this task, the chemical pathways in the cell must meet *three* requirements. When glucose and oxygen enter a direct reaction, all of the available energy is released right away and lost as heat (Fig. 18.3). This uncontrolled loss of energy has to be avoided at all costs, even though ultimately we will produce the very same products, CO_2 and H_2O. Therefore, the *first* requirement is to break down that six-carbon sugar *gradually,* one carbon at a time. When the breakdown is complete, we should have nothing left but carbon dioxide, as shown in Fig. 18.4.

What does this mean to us in chemical terms? Remember that glucose ($C_6H_{12}O_6$) contains 12 hydrogen

Figure 18.4 The chemical pathways that capture the energy available in glucose must meet three requirements. This figure illustrates the first one: the six-carbon glucose molecule must be broken down to carbon dioxide gradually, so that a little bit of the available energy can be captured in each stage of a multistep pathway. (The six carbon atoms of glucose are shown as six black dots in these diagrams.)

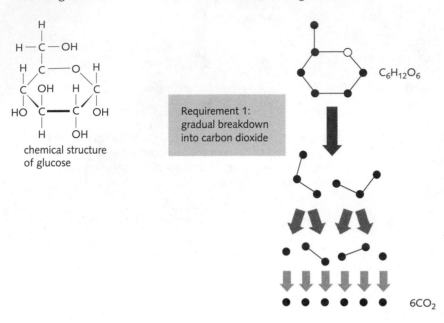

chemical structure
of glucose

Requirement 1:
gradual breakdown
into carbon dioxide

$C_6H_{12}O_6$

$6CO_2$

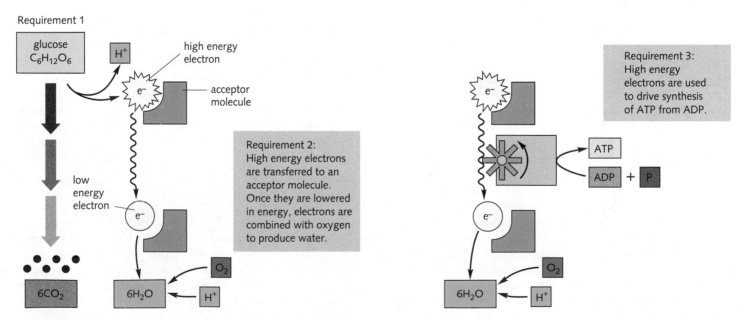

Figure 18.5 LEFT: The second requirement for energy capture is that the high-energy electrons from glucose *not* be allowed to react directly with oxygen. Instead, these electrons should be passed to acceptor molecules so that they can gradually be lowered in energy. RIGHT: The third requirement for energy capture deals with how glucose's high-energy electrons are used. They should be linked to a system that uses their energy to synthesize ATP from ADP and phosphate. The ATP produced in this way can then be used to drive other cellular processes.

Figure 18.6 Three major biochemical pathways are associated with the transformation of chemical energy from glucose. As shown, they are known as glycolysis, the Krebs cycle, and electron transport.

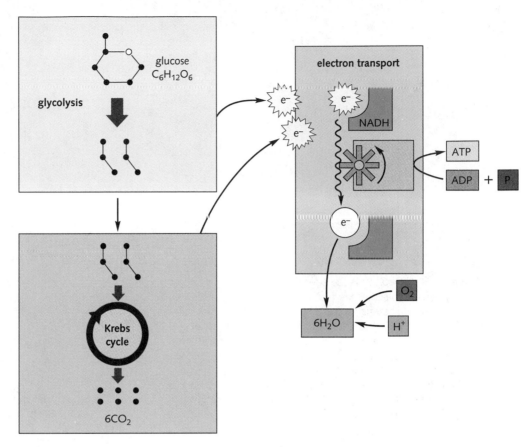

atoms. Little by little, those hydrogen atoms have to be taken away from the carbons until only carbon and oxygen are left, producing low-energy carbon dioxide.

Most of the energy liberated in the reaction of glucose with oxygen comes from the formation of new bonds between oxygen and hydrogen. Therefore, those hydrogens (more properly, their high-energy electrons) have to be handled very, very carefully. The electrons cannot be allowed to react with oxygen while they are at high energy levels, because the energy would then be lost as heat. So the *second* requirement is that the electrons must be taken very carefully from glucose and lowered in energy. Only then may they be allowed to react with oxygen to produce water. To do this, we need a carrier molecule, an electron acceptor, to accept and hold the high-energy electrons removed from glucose, as shown in Fig. 18.5.

Finally, just how does one "lower" an electron's energy level? This can be done by using the high-energy electrons to do work. Somehow, these high-energy electrons have to be harnessed to synthesize ATP. Therefore, our *third* requirement is that the electron acceptors must pass these electrons to a "machine" that can use their energy to drive the synthesis of ATP, as shown in Fig. 18.5. How could that happen? What's the cellular equivalent of an "electrical generator"? Those are questions we'll answer later in the chapter. What we can say for sure right now is that such a machine must exist, and it must run on electron energy.

The remarkable part of this story is that all of these processes work together to satisfy the ATP needs of the cell. With this in mind, we have some idea of where we're going, and now we will examine the detailed pathways that will get us there.

Figure 18.6 shows the three principal biochemical pathways that carry out the jobs outlined in Fig. 18.5. As you can see, the carbon skeleton of glucose is broken down in two sequential pathways known as glycolysis and the Krebs cycle. As we examine these pathways, keep Figs. 18.5 and 18.6 in mind. However complex the chemistry, the goal of the process is the same—to extract high-energy electrons and use their energy to drive the synthesis of ATP.

GLYCOLYSIS: THE FIRST PATHWAY

The first chemical pathway involved in the breakdown of glucose is known as **glycolysis,** which literally means "sugar-breaking." The name is well chosen, for the nine-step pathway does indeed break the six-carbon sugar molecule in two. The pathway itself is easy to summarize:

- One glucose (six-carbon) molecule enters the pathway.
- Two pyruvate (three-carbon) molecules are produced.

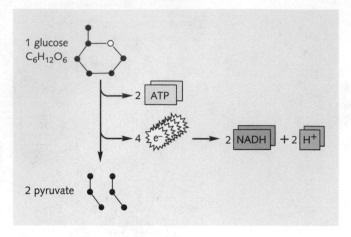

Figure 18.7 Glycolysis is the energy-yielding pathway by which glucose molecules are first broken down. The pathway consists of distinct steps. In the first four steps, two molecules of ATP are required, and therefore, there is a net input of cellular energy. In the later reactions, however, a total of four molecules of ATP are synthesized from ADP, providing the cell with a net gain of two ATPs per glucose. In the fifth stage of the pathway, two pairs of high-energy electrons are passed to the electron carrier, NADH. Pyruvate, the final product of glycolysis, may then be processed further by other chemical pathways.

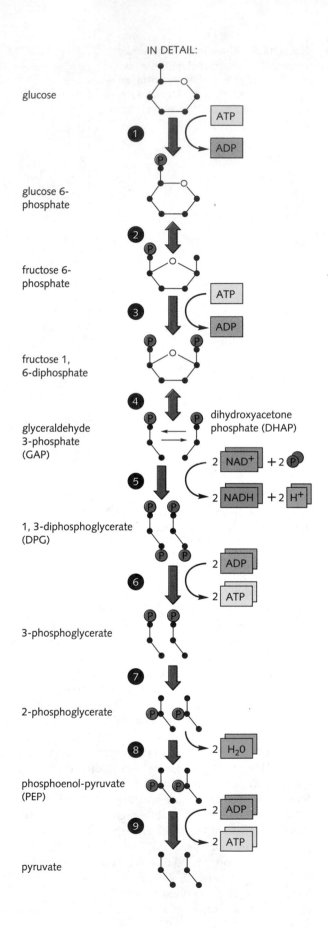

- Two molecules of ATP are produced from ADP.
- Two pairs of high-energy electrons are passed to an electron carrier.

As you can see, relatively little energy is released in the glycolytic pathway. In the presence of oxygen, glycolysis is only the first of two pathways that ultimately lead to the production of carbon dioxide and capture high-energy electrons (Fig. 18.6). However, as we will find later in the chapter, the glycolytic pathway can be the cell's *only* way to make ATP in the absence of oxygen.

Investing ATP to Make ATP

The enzyme catalyzing the first reaction of the pathway binds two very different substrates: a molecule of glucose and a molecule of ATP. Once they are in the active site of the enzyme, a phosphate group is transferred from ATP to glucose, producing a new molecule called **glucose-6-phosphate** (the 6 indicates that the phosphate group is attached to the number 6 carbon) and releasing ADP (Fig. 18.7). At first, this reaction seems to be working backwards. After all, the point of the pathway is to *produce* ATP, not to consume it! However, we can think of the ATP molecule used in this reaction as a kind of investment. The cellular energy represented by the ATP is necessary to "prime the pump" for the flow of energy that will follow.

In step 2 of the pathway, glucose-6-phosphate is rearranged a bit to form **fructose-6-phosphate** in a reaction that requires very little energy. In step 3, a second phosphate group is added to the sugar molecule, using another molecule of ATP. At this point, we have used up two ATP molecules and produced a molecule of **fructose-1,6-diphosphate.** The cell has yet to gain energy, and one might begin to doubt that these chemical reactions have any point to them at all. In the next few steps of the pathway, however, the investment of ATP begins to pay off significantly.

Making ATP: Substrate Phosphorylation

In step 4, glycolysis lives up to its name. The six-carbon sugar molecule is split into two three-carbon compounds, each of which contains a single phosphate group. In the reactions that follow, two important events take place. These three-carbon compounds pass through five chemical steps, each catalyzed by a different specific enzyme. When the pathway is complete, we are left with two molecules of **pyruvate.** Each pyruvate contains three carbon atoms, so all six carbons of the original glucose are still present. However, two very important events take place in these last few reactions that produce pyruvate.

In two of these reactions, steps 6 and 9, a molecule of ATP is formed, and in each case the phosphate group is transferred directly from one of the substrate molecules in the reaction to ADP. This transfer of a phosphate converts ADP to ATP. Biochemists describe these reactions as **substrate phosphorylation,** because the phosphate groups come directly from the substrates of the reaction. *Phosphorylation* is a general term that simply means the attachment of a high-energy phosphate group to another molecule. Because each of the three-carbon compounds yields a pair of ATPs in these reactions, we actually obtain four molecules of ATP for every molecule of glucose that enters the pathway. Since we invested two molecules to get things going, we actually have a *net gain of two molecules of ATP per glucose in glycolysis* (Fig. 18.7).

Accepting Electrons: Oxidation and Reduction

In addition to the ATP molecules that are produced in glycolysis, in step 5 two pairs of high-energy electrons are transferred to an electron carrier, a molecule known as **nicotinamide adenine dinucleotide** (NAD⁺). As Fig. 18.8 shows, NAD⁺ is capable of accepting a pair of high-energy

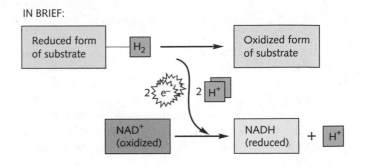

IN DETAIL:

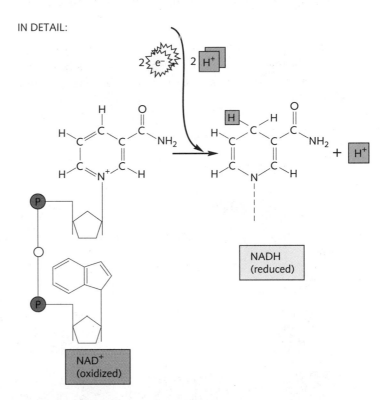

Figure 18.8 Redox (reduction/oxidation) reactions are those that involve a transfer of electrons. NAD⁺ (nicotinamide adenine dinucleotide) is an acceptor of high-energy electrons produced during the breakdown of glucose. The reduced form of the acceptor is NADH.

electrons and a proton (H^+), changing it to a form abbreviated as **NADH.**

Transfers of electrons from one molecule to another are so common that chemists use a special terminology to describe them. The compound that donates electrons is said to have become **oxidized** in the reaction, and the compound that accepts electrons is said to be **reduced.** NAD^+ is the oxidized form of the acceptor. When it accepts a pair of electrons, it is converted to the reduced form, NADH.

What does oxygen have to do with electron transfer? Oxygen is one of the world's best electron acceptors, or "oxidizing agents." Remember that when the double bond that holds molecular oxygen (O_2) together is broken, each oxygen atom can accept as many as four high-energy electrons. That's why so many carbon compounds will *burn* in the presence of oxygen.

It's important to note that when electrons are transferred from one compound to another, *both* oxidation and reduction reactions take place. The compound losing electrons is oxidized, and the compound receiving electrons is reduced. In fact, oxidation and reduction reactions occur together so commonly that they are often referred to as **redox** reactions ("**red**uction–**ox**idation").

Energy Release in Glycolysis

As we've seen, glycolysis produces only two net molecules of ATP per glucose. Although the cell can make good use of those two molecules of ATP, they don't amount to much when we compare them to the total energy available in glucose. The synthesis of two molecules of ATP from ADP accounts for only about 14 kcal of energy per mole, while glucose itself, if completely oxidized to H_2O and CO_2, would release 686 kcal per mole. The ATP molecules formed directly from glycolysis therefore represent only about 2 percent of the total energy available in glucose itself (Fig. 18.9). Where is the rest of that energy?

The answer can be found not so much in what glycolysis has done but in what it has not done. The molecules left over when glycolysis is finished, two pyruvates and two NADHs, contain a great deal of chemical energy that is still untapped. To release the remaining energy, the cell must find an oxidizing agent to accept the high-energy electrons still present in these compounds. The agent that is used is the most powerful oxidizer of them all: oxygen itself.

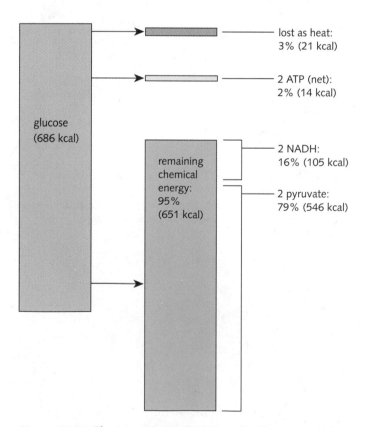

THE KREBS CYCLE: THE NEXT STEP IN ENERGY RELEASE

If oxygen is available to the cell, then a chemical pathway called the **Krebs cycle** is the next step in breaking down the products of glycolysis. This cycle is known by several other names, including *citric acid cycle* and *tricarboxylic acid cycle*, but *Krebs* is the most common usage, after its discoverer, Hans Krebs.

Releasing Energy in the Presence of Oxygen

The Krebs cycle begins with pyruvate, which is where glycolysis left off. It completes the work of breaking down the carbon skeleton from glucose, and by the time it is finished, the carbon atoms from glucose have been released as carbon dioxide.

In brief, for each glucose (six-carbon) molecule that entered glycolysis:

Figure 18.9 The two ATPs made during glycolysis represent only about 2 percent of the chemical energy available in glucose.

- Two pyruvate (three-carbon) molecules enter the Krebs pathway.
- Two molecules of an ATP-like compound are produced (GTP).
- Ten pairs of high-energy electrons are passed to electron carriers.
- Six carbon dioxide (one-carbon) molecules are released.

As the Krebs cycle begins, pyruvate is oxidized in a reaction in which one of the products, a two-carbon compound called acetate, is attached to a much larger molecule known as **coenzyme A.** The compound that results is called *acetyl coenzyme A (acetyl CoA).* One carbon atom from pyruvate is released in the form of carbon dioxide (CO_2), and a pair of electrons is used to reduce NAD^+ to NADH.

The two carbon atoms from acetyl coenzyme A enter the Krebs cycle by means of a reaction in which they are combined with a four-carbon compound, *oxaloacetate.* The molecule that results is a six-carbon compound called citrate. Figure 18.10 shows the details of the Krebs cycle, including the 10 intermediate compounds in the cycle. We can take a closer look at the Krebs cycle by examining it from the perspective of the intermediate compounds in the cycle.

A Closer Look at the Krebs Cycle

Carbon At two points in the cycle, a *decarboxylation* reaction occurs in which one carbon atom is removed and released in the form of carbon dioxide. Because of these reactions, by the time the cycle returns to the starting point, only four carbons are left, so that with the addition of two more carbons from acetyl coenzyme A, citrate (six carbons) is regenerated. Of the original six-carbon glucose molecule, two carbon atoms are released after glycolysis during the formation of acetyl CoA, and four are released during the Krebs cycle. Nearly all the carbon dioxide that we exhale in the process of breathing comes from the CO_2 released in these two pathways.

ATP One of the Krebs cycle reactions is a substrate phosphorylation that produces a molecule of GTP (guanosine triphosphate) by adding a phosphate group to GDP (guanosine diphosphate). In terms of energy, GTP is equivalent to ATP, so each turn of the Krebs cycle produces the energy equivalent of one molecule of ATP.

Oxidation/reduction Four of the Krebs cycle reactions warrant special attention, for they involve the transfer of high-energy electrons to an electron acceptor. In three cases, the acceptor is NAD^+, the same carrier molecule we saw earlier in glycolysis. One of the redox reactions involves FAD (flavine adenine dinucleotide), a carrier that, like NAD^+, can accept a pair of high-energy electrons, reducing it to $FADH_2$. In each turn of the cycle, four pairs of high-energy electrons are removed, producing three NADHs and one $FADH_2$. Because two pyruvates are produced from each glucose, a single glucose molecule produces six NADHs and two $FADH_2$s in the Krebs cycle (Fig. 18.10).

It is important to note that the Krebs cycle is, in a sense, the end of the line for the carbon atoms of the glucose molecule. Sooner or later, every carbon atom of a glucose molecule used by the cycle is released to the environment in the form of carbon dioxide. The cycle is *not* the end of energy production, however. As the cycle disassembles carbohydrate molecules, one step at a time, the energy from those molecules is captured in the high-energy electrons used to reduce NAD^+ and FAD. In a final series of pathways, the energy from those electrons is used to produce ATP.

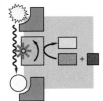

ELECTRON TRANSPORT: ENERGY FROM OXIDATION AND REDUCTION

So far we have seen a direct gain of only four molecules of ATP per molecule of glucose: two from glycolysis and two from the Krebs cycle. However, a careful accounting of the pathways so far shows that the energy in a single molecule of glucose has been used to produce 12 reduced electron carriers: 10 NADHs (2 from glycolysis, 2 from the entry point to the Krebs cycle, and 6 from the cycle itself) and 2 $FADH_2$s.

You might also have noticed that although we refer to the Krebs cycle as an **aerobic** (oxygen-requiring) pathway, so far we haven't seen any oxygen involved in any of its reactions. The need for oxygen doesn't become apparent until we consider what happens to these reduced electron carriers. *Oxygen serves as the final acceptor of electrons.*

The Electron Transport Chain

The cell must deal with the high-energy electrons that have been passed to NADH and $FADH_2$. Ultimately, they will be passed to oxygen to produce water. If this were done directly from NADH, all the available energy would be wasted as heat. Therefore, the cell must have a way to capture some of that energy in the form of ATP. The question is *how.*

We can get our first clue by following the high-energy electrons that were passed to NADH. These electrons are transferred to another electron carrier known as FMN

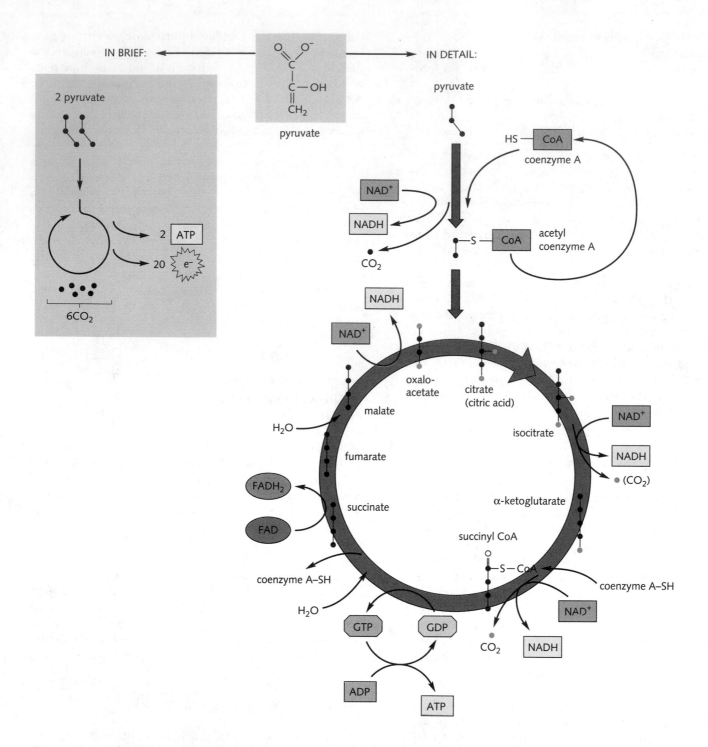

Figure 18.10 The Krebs cycle. Pyruvate, the end product of glycolysis, enters the cycle in a reaction which produces citric acid (for this reason, the pathway is sometimes known as the citric acid cycle). As carbon atoms from pyruvate are passed through the eight intermediates of the Krebs cycle, two carbon atoms are removed with each turn. Each turn of the cycle also produces a molecule of GTP, reduces a molecule of FAD to $FADH_2$, and reduces three molecules of NAD^+ to NADH. At two points in the cycle, a molecule of CO_2 is released. The overall chemical effect of the cycle, therefore, is to pass high-energy electrons from pyruvate to a series of electron carriers.

Figure 18.11 The electron transport chain is a series of electron carriers which pass electrons from the Krebs cycle to oxygen, the final electron acceptor. Electrons from the Krebs cycle may enter the pathway from two sources (NADH and $FADH_2$). The energy levels of the electrons drop a bit at each step of the chain. The abbreviations for electron carriers are FMN (flavine mononucleotide), Q [coenzyme Q (ubiquinone)], and cyt b, c_1, and so on [various cytochromes (electron transport proteins with heme prosthetic groups)].

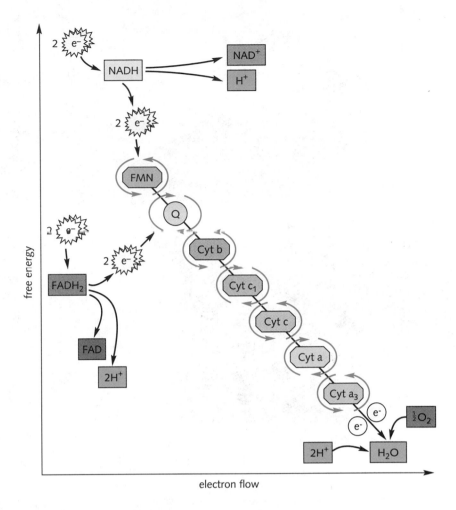

(*flavine mononucleotide*). The transfer of electrons reduces FMN, while NADH (having lost electrons) is oxidized back to its starting form of NAD^+.

Things don't stop there. Electrons are passed from FMN to another electron carrier, then another, then another. This sequence of electron carriers is known, appropriately enough, as the **electron transport chain** (Fig. 18.11). When electrons reach the end of the chain, they are passed to oxygen, the final electron acceptor. Protons are supplied from solution, and water is produced.

At every step of the chain, the energy level of the electrons is reduced a bit. When they are finally passed to oxygen, very little energy is left to be converted to heat. This multistep electron transport chain puzzled scientists for many years. It seemed logical to assume that the high-energy electrons were being used to do the work of ATP synthesis, but no one was able to find any evidence that the synthesis of ATP or any other high-energy compound occurred during any of the chain's many steps.

Chemiosmosis:
The Machine That Makes ATP

In the early 1960s, while other scientists were puzzling over this problem, a British scientist named Peter Mitchell believed that important clues could be found by studying *where* electron transport takes place. In a eukaryotic cell, glycolysis takes place in the cytoplasm. The pyruvate produced by glycolysis enters the mitochondrion (Fig. 18.12), and the reactions of the Krebs cycle take place within the mitochondrial matrix. The surprising thing was that the electron transport chain was membrane bound. All of the carriers of the chain were part of the inner mitochondrial membrane (Fig. 18.13). In prokaryotes, which do not have mitochondria, the electron transport chain is nonetheless part of the cell membrane.

Why should the electron transport chain be membrane bound? Mitchell and many others believed that the membrane itself must play a pivotal role in the process of ATP synthesis.

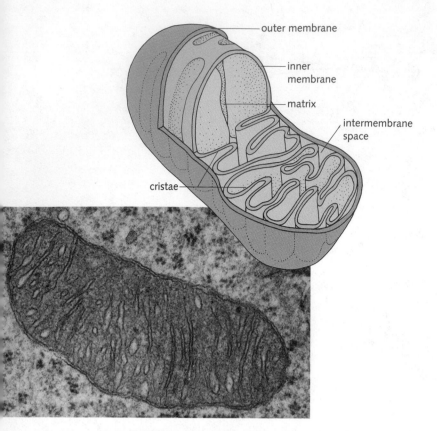

What would a membrane have to do with the synthesis of ATP? Remember the problems solved by engineers in the construction of an oil-fired power plant. Heat in such a plant is not converted directly to electricity. Rather, hot gases are released on one side of a barrier (a piston or turbine), where their expansion produces mechanical energy. That mechanical energy is then converted to electricity. Could something akin to this be taking place inside the mitochondrion?

Mitchell may have reasoned like this:

- High-energy electrons are passed from the Krebs cycle to NADH inside the mitochondrion.

- These electrons are then passed to the inner mitochondrial membrane.

- Membranes serve as barriers, especially to the flow of charged molecules. Is the membrane perhaps the equivalent of the blade of a turbine, converting energy from one form to another?

Figure 18.12   The mitochondrion. These energy-transducing organelles are separated from the cytoplasm by an outer membrane. The inner membrane, which contains the electron transport system, is deeply folded in many cells forming structures known as *cristae,* which increase its surface area.

Figure 18.13   The compartmentalization of metabolic pathways. In prokaryotic cells, glycolysis and the Krebs cycle take place in the cytoplasm, and the electron transport chain is part of the cell membrane. In eukaryotic cells, glycolysis takes place in the cytoplasm. The end product of glycolysis, pyruvate, enters the mitochondrion to begin the Krebs cycle. The electron transport chain is located in the inner mitochondrial membrane.

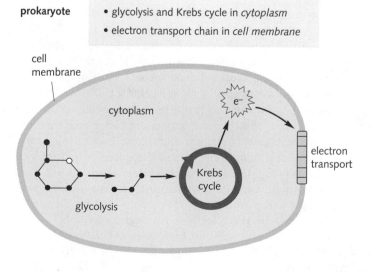

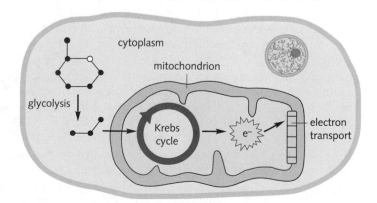

Figure 18.14 shows the general outline of Mitchell's thinking. High-energy electrons are passed to the electron transport chain. As they pass from carrier to carrier, their energy is used to pump protons (H^+ ions) across the membrane. After the proton pumping has gone on for a while, it produces a gradient of charge across the membrane: the outside is positively charged compared with the inside. In chemical terms, this charge gradient is every bit as powerful as the pressure gradient across the blades of a power plant turbine.

What could the proton gradient be used for? It could provide more than enough energy to attach a phosphate group to ADP. It could drive ATP synthesis. Mitchell understood that this chemical gradient would produce a transmembrane pressure very much like osmotic pressure. Appropriately, Mitchell called his idea, which combined *chemical* gradients with *osmotic* forces, **chemiosmosis.**

Experimental Support

At first, there was little support for Mitchell's "chemiosmotic hypothesis." Soon, however, experiments carried out with isolated, inner mitochondrial membranes produced surprising results that fit his ideas. One such experiment was carried out on vesicles made from inner mitochondrial membranes. The electron transport chain remained attached to such vesicles, and when oxygen and NADH were added, the chain functioned just as it did within the cell. However, something unexpected happened as electrons began to flow: the pH of the medium containing the vesicles changed dramatically.

As you know, pH is a measure of the hydrogen ion content of a solution. The pH changes confirmed that the inner membrane did pump protons during electron transport, just as predicted by the chemiosmotic theory.

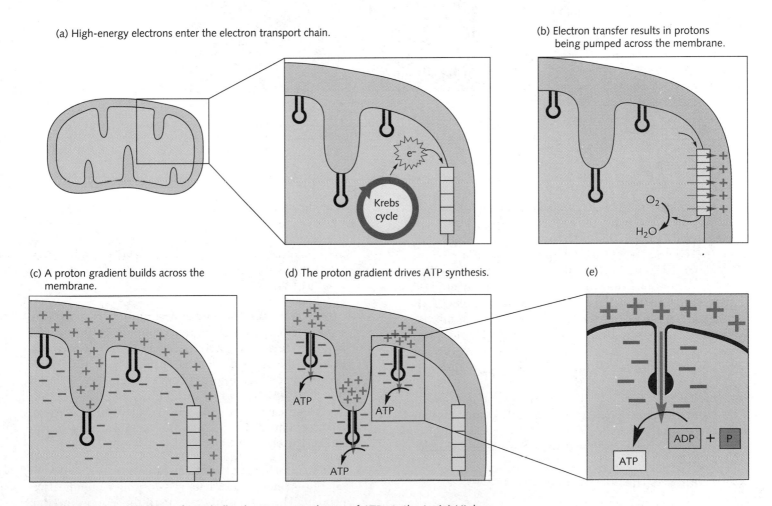

(a) High-energy electrons enter the electron transport chain.

(b) Electron transfer results in protons being pumped across the membrane.

(c) A proton gradient builds across the membrane.

(d) The proton gradient drives ATP synthesis.

(e)

Figure 18.14 A summary of Mitchell's chemiosmotic theory of ATP synthesis. **(a)** High-energy electrons are passed to the membrane-bound electron transport chain. **(b)** During electron transport, protons are pumped across the inner mitochondrial membrane. **(c)** Continued electron transport results in a proton gradient across the inner membrane. **(d)** The proton gradient provides the energy to drive the synthesis of ATP. **(e)** ATP-synthetase enzymes produce ATP, using the proton gradient.

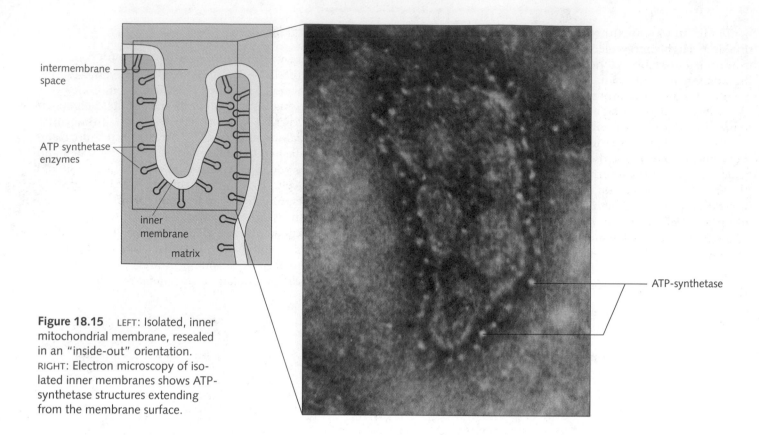

intermembrane
space

ATP synthetase
enzymes

inner
membrane

matrix

ATP-synthetase

Figure 18.15 LEFT: Isolated, inner mitochondrial membrane, resealed in an "inside-out" orientation. RIGHT: Electron microscopy of isolated inner membranes shows ATP-synthetase structures extending from the membrane surface.

The isolated membranes produced other interesting results. When these membranes were studied with the electron microscope (Fig. 18.15), scientists noticed that there were large structures protruding from one surface. A number of experiments showed that these particles were the actual enzymes that synthesized ATP from ADP. This complex membrane-bound enzyme is called **ATP-synthetase.** At the present time, we know that this enzyme is capable of synthesizing ATP (from ADP and phosphate) only if the enzyme is present in a membrane across which there is a strong pH gradient, as shown in Fig. 18.14. Quite frankly, we do not yet understand exactly how the enzyme harnesses the energy of the gradient to attach the third phosphate group to ADP. Understanding the mechanism by which ATP-synthetase works would be a great scientific prize, and many scientists are seeking it.

ATP Synthesis: The Mechanism

As a result of Mitchell's work (he was recognized with the Nobel Prize in 1982), the basic mechanisms by which electron transport results in ATP synthesis are now clear:

1. Carriers in the electron transport chain are built into the inner mitochondrial membrane. When electrons flow through them, part of the energy from the electrons is used to pump protons across the membrane. Because the membrane itself is otherwise impermeable to protons, once they have been pumped from one side to the other, they cannot "leak" back easily (Fig. 18.16a).

2. As electron transport continues, more and more protons are pumped across the membrane into the space between the inner and outer mitochondrial membranes (the intermembrane space)—so many that a noticeable change in pH occurs. A gradient of protons has built up across the membrane. Because protons are electrically charged, this gradient produces an electrochemical field across the membrane, as shown in Fig. 18.16b.

3. The electrochemical gradient can serve as a source of energy—almost like a battery—to power chemical reactions. The ATP-synthetase uses the energy in that gradient to drive the synthesis of ATP from ADP. The energy to attach the third phosphate group comes directly from the electrochemical gradient across the membrane and the flow of protons back across the membrane through the ATP synthetase enzymes.

The beauty of Mitchell's theory is its simplicity. Electron transport produces a gradient, and the gradient drives the synthesis of ATP. Because there is no direct link between electron transport and ATP synthesis, we often say that the coupling between these two processes is *indirect*. The complexes within the membrane that are responsible for electron transport do not have to be near those responsible for ATP synthesis.

For each pair of high-energy electrons transferred to the electron transport chain from NADH, enough protons can be moved across the membrane to make a gradient large enough to drive the synthesis of three molecules of ATP. The electrons derived from $FADH_2$ are somewhat lower in energy and are able to drive the synthesis of only two molecules of ATP.

Cellular Respiration

The word *respiration* is often used as a synonym for *breathing*. Organs such as gills and lungs that help to bring oxygen to the various tissues are parts of the *respiratory system*. However, we also refer to the oxidation of glucose and of other molecules within cells as "respiration." At the level of the cell, the major need for oxygen is found at the end of the electron transport chain: oxygen is needed to accept electrons. If the flow of oxygen is stopped, the elec-

tron transport chain stops working and the synthesis of ATP cannot continue. (This is why suffocation causes death so quickly. Cells need a constant supply of ATP, and without oxygen, ATP synthesis cannot keep pace with the energy demands of the cell.) Thus, **cellular respiration** is nothing more than the sum total of the pathways we have examined: glycolysis, the Krebs cycle, and the electron transport pathway taken together.

Summing Up the Oxidation of Glucose

Figure 18.17 gives an accounting of the total amount of chemical energy that is trapped in the form of ATP when the oxidation of a molecule of glucose is complete. As you can see, 36 molecules of ATP can be synthesized from each molecule of glucose that enters the pathways. How much energy is captured in the form of ATP? Taking the figure of 7 kcal/mole for the production of ATP from ADP, we find that 36 x 7 = 252 kcal are captured for each mole of glucose. Because the total chemical energy available in a mole of glucose is 686 kcal, this represents an efficiency of 252/686 = 37 percent. Most of the remaining 63 percent is lost in the form of heat.

You might notice from Fig. 18.17 that a small correction is applied for the NADH molecules synthesized during glycolysis. This correction is necessary because these

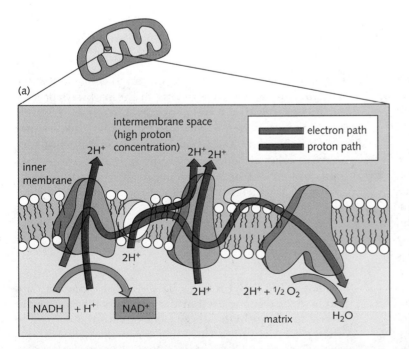

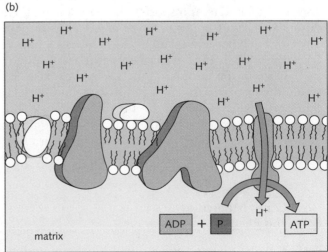

Figure 18.16   **(a)** The electron transport chain. The passage of electrons from one carrier to another results in a movement of protons across the membrane, from the matrix to the intermembrane space. **(b)** ATP synthesis. A proton gradient is produced across the inner mitochondrial membrane by electron transport. The energy in this gradient is sufficient to allow a membrane-bound enzyme, the ATP-synthetase, to produce ATP from ADP and phosphate.

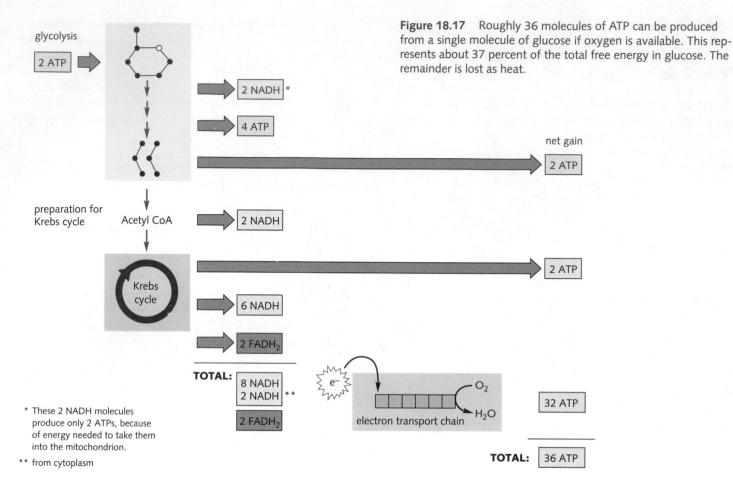

Figure 18.17 Roughly 36 molecules of ATP can be produced from a single molecule of glucose if oxygen is available. This represents about 37 percent of the total free energy in glucose. The remainder is lost as heat.

molecules are produced in the *cytoplasm,* and a bit of energy is required to transport them into the mitochondrion. This result points out one of the most important aspects of cellular energy use: the chemical pathways are *compartmentalized.* In a typical eukaryotic cell, glycolysis takes place in the cytoplasm, the Krebs cycle takes place within the matrix of the mitochondrion, and the electron transport chain is bound to the inner mitochondrial membrane. In prokaryotes, the situation is a bit different. Prokaryotes don't have mitochondria, and their electron transport chains are built into their cell membranes.

How Other Molecules Enter the Respiratory Pathways

Earlier in the chapter, we justified all the attention we have paid to glucose by emphasizing the fact that other food molecules are broken down to release energy in much the same way. Figure 18.18 shows how this is done. Proteins are broken down into individual amino acids,

and other enzymes then convert them to compounds that can enter the Krebs cycle or one of the pathways preceding it. Carbohydrates are generally broken into simple sugars and converted to glucose. Lipids are broken down to fatty acids and glycerol; these enter the mitochondria, where special enzymes cut them up, two carbons at a time, and attach them to coenzyme A to produce acetyl coenzyme A, which can enter the Krebs cycle.

FERMENTATION

Oxygen is the living world's best electron acceptor, and that is why oxygen is used in the pathways that follow glycolysis: to accept high-energy electrons from glucose and other food molecules. Does that mean that living organisms are restricted to environments where they can find oxygen?

You will remember that a small amount of energy (two ATPs per molecule of glucose) can be trapped just from

Figure 18.18 Glycolysis and the Krebs cycle are the basic energy-releasing pathways of the cell. Proteins, fat, and carbohydrates may enter the pathways at a variety of different points, enabling cells to draw energy from virtually any food source.

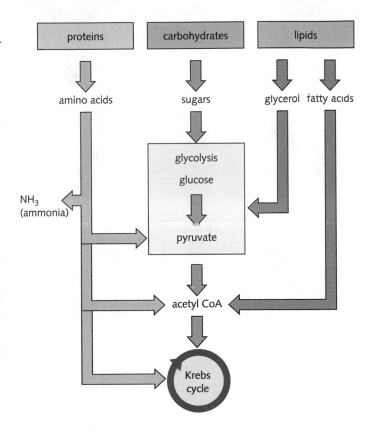

glycolysis, a pathway that doesn't require oxygen. Therefore, if an organism is able to get a large supply of food material, or if it has a low metabolic rate, it can use glycolysis to produce enough ATP to survive. This process is known as **fermentation.**

However, an organism that begins to ferment its food has a serious problem to overcome. The glycolytic pathway transfers electrons from glucose to an electron acceptor, NAD$^+$, which is reduced to form NADH. Without a constant supply of NAD$^+$, the process would quickly stop. Therefore, in order to keep glycolysis moving, an organism that is fermenting its food must be able to *oxidize* NADH to NAD$^+$. Living cells have come up with a number of different ways to do this.

Alcohol Fermenters

A small number of organisms—the best known are the **yeasts**—oxidize their NADH in a pathway that reduces *pyruvate*, the end product of the glycolytic pathway, and

converts it to **ethanol** (also known as ethyl alcohol, or grain alcohol). One of the carbon atoms of pyruvate is removed in the process and is released in the form of *carbon dioxide*. **Alcoholic fermentation** is the source of alcoholic beverages, and the bubbles in beer and sparkling wine come from the carbon dioxide released during the fermentation process (Fig. 18.19).

The same fermentation process takes place when bread is made. Tiny carbon dioxide bubbles released by yeast cells cause the dough to rise, and the alcohol evaporates during baking.

Lactate Fermenters

Many other organisms are capable of carrying out fermentation in another chemical pathway that reduces pyruvate. The final product of this pathway is **lactate,** or lactic acid; and unlike alcoholic fermentation, lactate fermentation gives off no carbon dioxide. Humans are lactate fermenters to a limited extent, and during very brief

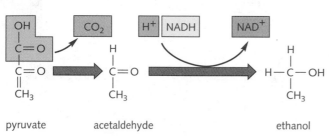

pyruvate   acetaldehyde   ethanol

Figure 18.19 Alcoholic fermentation. LEFT: The NAD⁺ used up in glycolysis is regenerated in a two-step pathway which converts pyruvate to ethanol, releasing carbon dioxide gas in the process. This pathway enables yeasts to obtain energy from glucose in the absence of oxygen. The by-products of this pathway, alcohol and carbon dioxide, help to produce alcoholic beverages. RIGHT: The fermentation process in a winery rapidly converts grape mash into wine, releasing large amounts of carbon dioxide in the process, which must be carefully vented off from the tanks. It is critically important to keep oxygen out of the vats during fermentation. Can you explain why?

periods without oxygen, many of the cells in our bodies can generate ATP in this way. The cells best-adapted to it, however, are muscle cells, which often need very large supplies of ATP for rapid bursts of activity (Fig. 18.20).

By-products of Fermentation

The use of fermentation as a sole source of energy has a number of drawbacks. For one thing, it is much less efficient: the two ATPs formed from each glucose represent only a fraction of the available energy (14/686 = 2 percent). For another, the end products (lactate or ethanol) can produce problems for the organism. Alcohol is toxic to most cells, yeast included, and if fermentation occurs in a closed system, then the concentration of alcohol rises until the yeast cells are killed. A similar problem occurs with lactate. During bursts of rapid exercise, lactate accumulates in muscle tissue. The buildup of lactate causes a painful burning sensation, with which every athlete is familiar.

Fortunately, there are chemical pathways that can eliminate lactate, although first it must be taken to the liver by the circulatory system. These pathways require oxygen, and they are responsible for the out-of-breath sensation, known as *oxygen debt,* that may last for several minutes after only 10 or 20 seconds of intense exertion. In Chapter 36 we will examine this effect more closely.

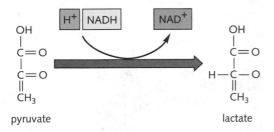

pyruvate   lactate

Figure 18.20 Lactate fermentation. The NAD⁺ used up in glycolysis can be regenerated in a reaction that produces lactate. This process is widely used by animals whenever muscles and other active tissues require more ATP than can be released from oxidative pathways. Olympic runners, beginning the 100-m dash, will use up all of the oxygen reserves in their large muscles in a matter of 2–3 seconds. For the rest of the sprint, their muscles will produce ATP anaerobically, flooding the bloodstream with lactate. Ten seconds later, this biochemical oxygen "debt" will cause them to pant and breathe heavily for a minute or more as that lactate is removed from the bloodstream.

Trouble in the Power Plant

Nearly every cell in the human body depends upon mitochondria for ATP production. What would happen if something went wrong with the cell's "power plant"? For many years, that was nothing more than a hypothetical question. That's not true any more.

In 1871 a physician named Leber described a number of patients who had been outwardly healthy until young adulthood. Then seemingly without warning, they began to lose sight, first in one eye and then sometimes in the other. Leber's examinations of the patients convinced him that the eye itself was still healthy in the blinded patients. But the eye depends upon the optic nerve to pass information to the brain. Cells in the optic nerves of Leber's patients had degenerated, leading to blindness. The physician noticed something else about the disorder—it ran in families. Since his reports, this condition has been known as Leber's hereditary optic neuropathy.

But there's something peculiar about Leber's neuropathy. It's inherited, sure enough, but it quickly became clear that it is only passed from mother to child. There were no examples of a male patient passing the disorder to his children. This pattern of **maternal inheritance** led some investi-

gators to suggest that Leber's neuropathy might result from an infectious organism that invaded embryos still in the uterus. An Australian investigation ruled this out.

In 1989 Douglas Wallace and Gurparlash Singh at Emory University in Atlanta hit upon the real source of the disease: a tiny mutation in one of the genes in the human mitochondrion. Mitochondria and chloroplasts (as noted in Chapter 16) have their own DNA molecules. These organelle–chromosomes are very small—the human mitochondrion contains just 36 genes. In Leber's neuropathy patients,

a single defective gene produces a change in one of the components of the electron transport chain in the inner mitochondrial membrane. How serious is the change? The chain still functions, but its slightly reduced efficiency lowers the rate at which mitochondria can produce ATP. The relationship of this change to blindness isn't precisely clear, but Wallace and Singh note that the optic nerve has an unusually high demand for ATP. They theorize that the mitochondria of Leber's neuropathy patients eventually fail to meet that demand, and the cells in the nerve begin to break down.

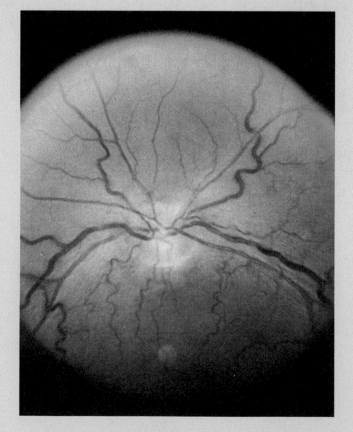

Blood vessels in the eye of a Leber's hereditary optic neuropathy patient are twisted and distorted in shape. This is one of the first indications that mitochondria in the optic nerve are failing to keep up with the demand for ATP.

Their research neatly explained another facet of the disorder. Your nuclear genes are inherited from both parents, but nearly all your mitochondria came from your mother's egg cell, and that's why Leber's neuropathy, like any mitochondrial gene defect, shows maternal inheritance.

Incidentally, it turns out that this is not an isolated case. Three other human genetic disorders have already been mapped to the mitochondrial genome, and as many as 10 others are possible candidates that may yet be traced to defects in these cellular power plants.

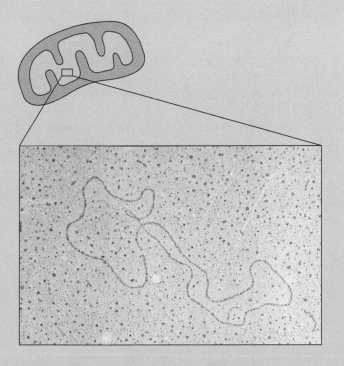

Mitochondrial DNA. The 36 human mitochondrial genes are located on a circular DNA molecule located in the matrix of the organelle.

SUMMARY

Energy is central to the activities of living organisms, and the transformation of energy from one form to another is one of the key processes that takes place within the cell. Metabolism, the sum total of the biochemical and physical transformations that take place in living organisms, includes both anabolic reactions (in which molecules are built up from smaller ones) and catabolic reactions (in which larger molecules are broken down).

The oxidation of glucose is a key catabolic pathway that serves as a model for how energy is extracted from organic molecules. Glucose molecules are processed through a series of pathways, beginning with glycolysis, in which the glucose molecule is split into two three-carbon compounds. Glycolysis results in a small net gain in energy for the cell (two molecules of ATP per molecule of glucose) and produces pyruvate, a three-carbon molecule that enters the Krebs cycle if oxygen is present.

In eukaryotic cells, the Krebs cycle takes place within the mitochondrion. Reactions within the Krebs cycle reduce electron carrier molecules, which pass electrons along to a series of electron transport complexes associated with the inner mitochondrial membrane. Electrons move through this transport chain to oxygen, the final electron acceptor. As they do so, they cause a movement of protons across the inner mitochondrial membrane, producing a chemiosmotic gradient that is used to provide the energy for ATP synthesis, a process known as oxidative phosphorylation.

In the absence of oxygen, cells can capture a small amount of the energy available in glucose by a process known as fermentation, in which electron carriers for glycolysis are regenerated. Yeasts and related organisms carry out alcoholic fermentation, producing carbon dioxide and ethyl alcohol as chemical byproducts of the process. Most animals produce lactate, which is released into the bloodstream.

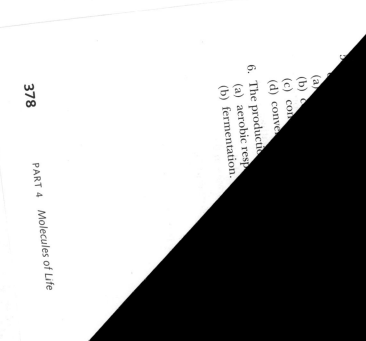

6. The product
   (a) aerobic resp
   (b) fermentation.

(a)
(b)
(c) co
(d) conve

TERMS AND CONCEPTS

anabolic *358*
catabolic *358*
metabolism *358*
glycolysis *361*
substrate
  phosphorylation *363*
NADH *364*
oxidatized *364*
reduced *364*

Krebs cycle *364*
coenzyme A *365*
aerobic *365*
electron transport
  chain *367*
chemiosmosis *369*
ATP-synthetase *370*
cellular respiration *371*
fermentation *373*

REVIEW

Objective Questions (Answers in Appendix)

1. A substance that loses electrons in a chemical reaction is said to be
   - (a) an enzyme.
   - (c) oxidized.
   - (b) a catalyst.
   - (d) reduced.

2. Among the products of glycolysis are
   - (a) 38 molecules of ATP.
   - (b) 38 molecules of ADP.
   - (c) one six-carbon molecule.
   - (d) two molecules of pyruvate.

3. Glycolysis takes place in which compartment of the eukaryotic cell?
   - (a) nucleus
   - (c) plasma membrane
   - (b) cytoplasm
   - (d) mitochondrion

4. Two different electron acceptors are used in the Krebs cycle. They are
   - (a) coenzyme A and coenzyme Q.
   - (b) ATP and GTP.
   - (c) NAD^+ and FAD.
   - (d) pyruvate and acetyl coenzyme A.

5. In the presence of oxygen, the pyruvates formed at the end of glycolysis will be
   - oxidized in the Krebs cycle.
   - converted to lactic acid.
   - converted to glucose.
   - converted to ethyl alcohol.

6. ...on of ethanol from glucose occurs as a result of
   - ...iration.
   - (c) phosphorylation.
   - (d) electron transport.

Discussion Questions

7. Describe the differences between anabolic and catabolic pathways. Is the synthesis of RNA by RNA polymerase an example of a catabolic or an anabolic pathway?

8. Oxidation and reduction reactions always occur together. When hydrogen gas burns in the presence of oxygen, which molecules are reduced? Which are oxidized? Is the product of the reaction an example of oxidation or reduction?

9. Describe the basic elements of Mitchell's chemiosmotic theory. If a preparation of inner mitochondrial membranes actively synthesizing ATP was treated with a substance that made the membranes permeable ("leaky") to protons, would this increase or decrease the rate of ATP formation?

10. You have read that the amount of energy released from aerobic respiration (36 ATPs per glucose) is much greater than the amount released from anaerobic respiration, or fermentation (2 ATPs per glucose). Use this fact to explain the following observation, which Louis Pasteur made on the making of wine. When yeast cells are added to a sealed vat of grape juice, they use up the sugar in the juice at a very slow rate—until the dissolved oxygen in the juice is gone. When the last trace of oxygen disappears, however, the rate of sugar consumption by the yeast increases dramatically. How would you explain this phenomenon, which is known as the Pasteur effect?

READINGS

Palca, J. "The other human genome." *Science* 249 (1990): 1104–1105. A concise summary of recent work on human mitochondrial disorders.

Dratch, P. "Mitochondria DNA defects: disease from outside the nuclear family." *Journal of NIH Research* 1 (1989): 73–76. A more detailed summary of mitochondrial disorders, with excellent photos and research references.

Hinkle, P. C., and R. E. McCarthy. "How cells make ATP." *Scientific American* 238 (March 1978): 104–123. One of the best summaries ever on how chemiosmosis drives the synthesis of ATP.

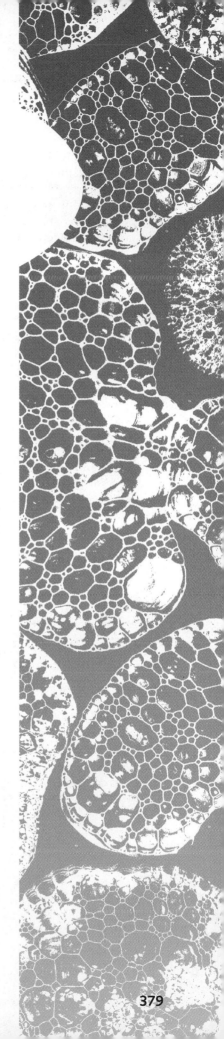

19
Photosynthesis

When humans first imagined space travel, their thoughts were outward, toward the moon, the planets, and the stars. As a species, we have long dreamed of leaving our home planet to explore. Given that dream, it is ironic that so many space travelers have said that the most remarkable sight they have seen from space is Earth itself. Its gentle blur of ocean, cloud, and land stands in marked contrast to the harshness of the moon and our planetary neighbors. It seems almost obvious that this great blue planet harbors life.

What makes Earth so hospitable to life? We can find one of the answers in a biochemical pathway. The green plants and microorganisms that dominate the earth's surface have mastered the process of **photosynthesis,** the capture of solar energy and its transformation into energy-rich compounds that can be used by living organisms. Photosynthesis not only provides nearly all of the energy used by living organisms on the earth, but it has also completely changed the atmosphere of the planet by releasing enormous amounts of oxygen.

In the last chapter, we saw how cells are able to release the energy stored in glucose and other energy-rich compounds. In this chapter, we will see how that chemical energy was originally trapped. The ultimate source of energy for life is the sun, and photosynthetic organisms are the crucial links on which other forms of life depend. Without them, this planet would be a very different place.

METABOLISM STEP BY STEP

In Chapter 18 we explored the pathways that synthesize ATP, using the energy available in compounds like glucose. As we saw, the covalent bonds in glucose are broken, liberating high-energy electrons; those electrons can be used to drive the synthesis of ATP. That ATP can then be used directly by energy-requiring reactions throughout the cell. But where did that glucose come from in the first place? What pathway produced it, and what was the source of that pathway's energy?

Imagine Building a Pathway

One way to answer those questions might be to analyze the events of cellular respiration and see what it would take to run them backwards, to make them anabolic rather than catabolic.

Figure 19.1 illustrates, in a minimalist way, what takes place in respiration. Bonds are broken between the carbon atoms in glucose to release carbon dioxide, and high-energy electrons are produced as a result. Much of the energy in these electrons is used to produce ATP, and the electrons, now reduced in energy, react with oxygen and hydrogen ions to form water.

Running the reaction backwards would mean starting with carbon dioxide and water and synthesizing glucose. Since breaking the bonds in glucose gives us carbon dioxide and high-energy electrons, then presumably, if we put carbon dioxide and those high-energy electrons back together, we would be able to make glucose, right? Well, al-

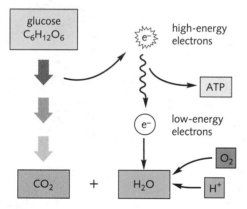

Figure 19.1 A diagrammatic summary of respiration.

Figure 19.2 Four components would be needed to run respiration backward: carbon dioxide, water, ATP, and high-energy electrons.

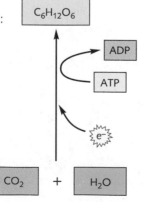

most right. Some of the energy of glucose is lost as heat when the high-energy electrons are removed.

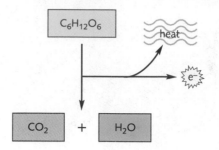

Therefore, we've got to supply some extra energy to run the reaction backwards. Naturally, we would assume the cell would provide it in the form of chemical energy, ATP.

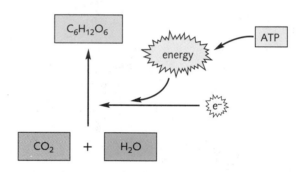

Figure 19.2 shows the basic requirements for the synthesis of glucose from carbon dioxide and water. Biochemically, the scheme is perfectly reasonable. There is no reason why we couldn't take carbon dioxide, water, ATP, and high-energy electrons and use them to make glucose. In fact, as we've drawn things out, our plan needs just one thing to work—a source of energy to produce ATP and the high-energy electrons.

The logical solution, of course, would be to make use of the most abundant form of energy on our planet—sunlight. If the energy from the sun could be used to produce ATP and high-energy electrons, it would be possible to make sugar from water and carbon dioxide. It even suggests a name for the process, which would use light (*photo*) to *synthesize* carbohydrate—*photosynthesis*.

Trapping Sunlight

To trap enough energy to drive photosynthesis, we would have to start with a **pigment,** a light-absorbing molecule that could capture sunlight; but the energy absorbed by most pigments is quickly lost as heat. Therefore, our ideal pigment must be located in a structure that quickly captures that energy in the form of high-energy electrons (Fig. 19.3).

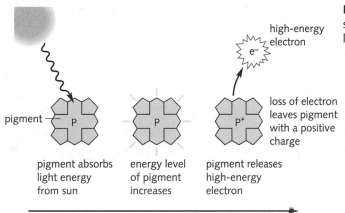

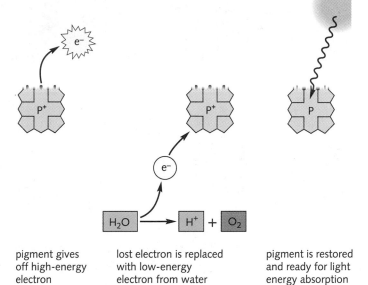

Figure 19.3 Light absorption by a pigment (P). Energy from sunlight is transferred to a high-energy electron (e⁻), which leaves the pigment oxidized (P⁺) when the electron is lost.

The idea seems sound, but it poses a potential problem. That high-energy electron came from the pigment molecule itself. If our pigment keeps on losing electrons every time it absorbs light, it will change chemically after losing only a couple of electrons. Therefore, we need a system that replaces the lost high-energy electrons with replacement, low-energy ones. Where could we find a source of low-energy electrons? Since low-energy electrons are most easily extracted from hydrogen, why not use the most abundant source of hydrogen atoms on the planet—water?

As Fig. 19.4 shows, water serves as a ready source of low-energy electrons; therefore, our pigment could function over and over again, using the sun's energy to release high-energy electrons. We need to make one more high-energy compound, ATP. We can't be sure where the ATP will come from, but since we've already built a system to capture solar energy, it's sensible to use that system to make ATP too (Fig. 19.5).

Now we've made all the high-energy compounds necessary to make glucose from carbon dioxide and water; therefore, there's no need for solar energy in the rest of the process. We could call everything we've imagined so far the "light-dependent reactions," because they are driven by solar energy.

Building Carbohydrate

The next step would be to use the products of the light-dependent reactions (ATP and high-energy electrons) to produce carbohydrate. The atmosphere contains more than enough carbon dioxide (CO_2) to serve as a source of carbon, and we could simply take those carbon atoms and our high-energy compounds and assemble a molecule of glucose.

Figure 19.4 Replacing the lost electron. Hydrogen atoms in water can be used to replace the missing electron, preparing the pigment to absorb more light energy. The removal of electrons from water releases oxygen.

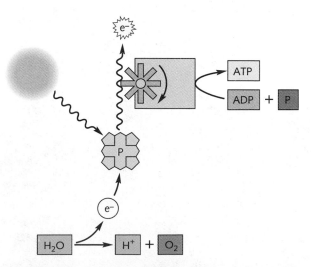

Figure 19.5 Using high-energy electrons to make ATP. We can imagine that machinery exists, driven by high-energy electrons, to synthesize ATP.

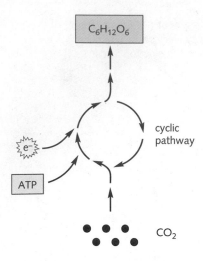

Figure 19.6 Borrowing from the Krebs cycle, we can guess that sugars are synthesized from CO_2 in a complex, multistep pathway driven by ATP and high-energy electrons.

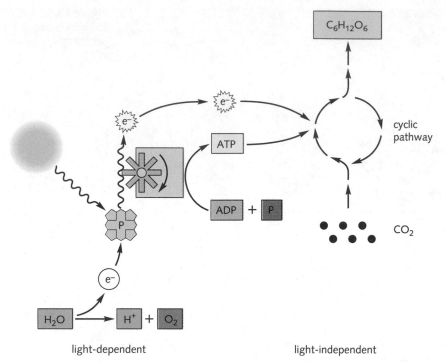

light-dependent   light-independent

Figure 19.7 Photosynthesis—light-dependent and light-independent reactions. Note the passage of ATP and high-energy electrons from the light-dependent to the light-independent reactions.

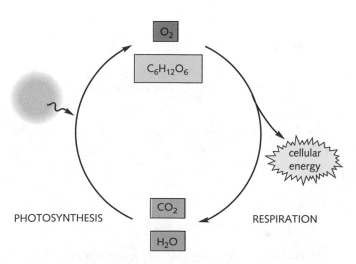

Figure 19.8 The ultimate source of nearly all energy used by living organisms is the sun. Photosynthesis and respiration exist in a global balance.

However, if we learned anything from our studies of glycolysis and the Krebs cycle in the last chapter, it was that living cells do not take glucose apart in a single reaction. More than 16 different steps are required. Since we have so many pairs of high-energy electrons to add, we should not expect to add them all at once, either, but rather in a complex, multistep pathway, maybe even one with a cycle (Fig. 19.6).

We could name these complex pathways the "light-independent reactions," since this part of photosynthesis doesn't require sunlight directly. Figure 19.7 illustrates how our hypothetical light-dependent and light-independent reactions would fit together. Perhaps we should say, "how they *do* fit together," because this is almost exactly how photosynthesis works:

1. Trap the light.
2. Produce ATP and high-energy electrons.
3. Build the carbohydrate.

Our hypothetical exercise has already shown us the basic organization of the anabolic side of the great global carbon cycle. Now we will look at photosynthesis itself for the real story.

Figure 19.9 The chloroplast. These energy-capturing organelles are surrounded by two enveloping membranes. The chloroplast contains photosynthetic membranes, or thylakoids, which are stacked in several places to form *grana*. The "stroma" region of the chloroplast contains soluble enzymes, ribosomes, and several dark lipid droplets.

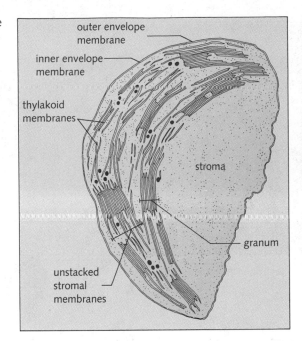
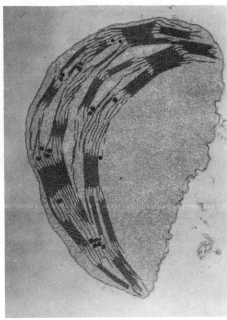

PHOTOSYNTHESIS

The global importance of photosynthesis cannot be overstated. Life on Earth depends on the great store of chemical energy that is built up by photosynthesis, and the atmosphere itself contains the gases that link these processes. Life requires energy, and on Earth that energy is supplied by photosynthesis (Fig. 19.8).

The Chloroplast

In eukaryotic cells, the process of photosynthesis takes place in organelles called **chloroplasts.** A chloroplast is surrounded by a pair of envelope membranes and contains internal membranes known as **thylakoids.** The thylakoids are sac-like membranes that are stacked in some parts of the chloroplast to form **grana** (singular, *granum*) (Fig. 19.9).

As we've seen, photosynthesis consists of **light-dependent reactions** and **light-independent reactions.** This much we anticipated in our hypothetical discussion at the beginning of this chapter.

We did not anticipate, however, that these reactions might take place in different places in the chloroplast. As it turns out, the light-dependent reactions are associated with the photosynthetic membranes (also called thylakoids) of the chloroplast. The light-independent reactions take place in the **stroma,** the semifluid matrix within the chloroplast (Fig. 19.10). This separation parallels the situation in mitochondria, where the Krebs cycle takes place in the soluble mitochondrial matrix, and electron transport is associated with the inner membrane.

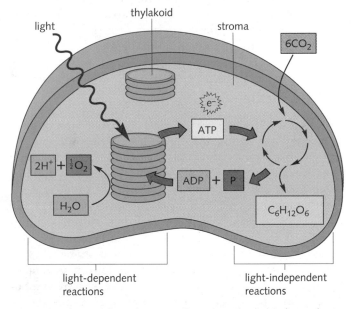

Figure 19.10 As this schematic illustrates, the light-dependent reactions, which take place in the thylakoid membranes, use the energy of sunlight to produce ATP and high-energy electrons. Oxygen is released during the light reaction as electrons are removed from water. The light-independent reactions, which take place in the stroma, use ATP and high-energy electrons to produce carbohydrates from carbon dioxide.

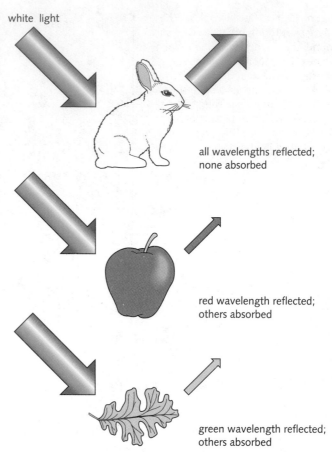

white light

all wavelengths reflected;
none absorbed

red wavelength reflected;
others absorbed

green wavelength reflected;
others absorbed

We will explore photosynthesis by starting at the beginning—the interaction of light with the photosynthetic membranes. These membranes contain a molecule that is absolutely essential to photosynthesis, a pigment known as **chlorophyll.** Since chlorophyll traps solar energy, it gives us the perfect place to start—the interaction between sunlight and energy-trapping pigment.

Sunlight: Starting Out

Light is just one small part of the larger spectrum of **electromagnetic radiation** that includes X-rays, microwaves, and gamma rays. Our eyes are capable of detecting only what we call visible light, at wavelengths between 400 nm (violet) and 730 nm (deep red). Ordinary sunlight is actually a mixture of different wavelengths (Fig. 19.11).

Electromagnetic radiation has a dual nature, exhibiting the characteristics of both waves and particles. The energy carried in electromagnetic radiation behaves as though it were bundled into particles known as **quanta** (singular: *quantum*). Quanta of light are known as **photons.** When photons strike an object, they are either **reflected** or **absorbed.** A surface that absorbs all wavelengths of visible light appears black, whereas one that reflects all wavelengths appears white. A colored object absorbs some wavelengths while reflecting others (Fig. 19.11). When light is absorbed by an object, the energy carried in the photons is converted into other forms, especially heat. As we know, in photosynthesis this energy is trapped and converted to chemical energy.

Chlorophyll: Capturing Solar Energy

Chlorophyll is a pigment, and like most pigments, it can absorb only part of the visible spectrum of light. Chloro-

Figure 19.11 LEFT: Refraction of sunlight by a prism. "White" light is actually a mixture of light of many different wavelengths. RIGHT: Reflection. The color of any object, including a leaf, is determined by which wavelengths of light are absorbed and which are reflected.

phyll absorbs red and blue light very well, but it absorbs only slightly in the middle region of the spectrum, where the predominant color is green (Fig. 19.12a). Chlorophyll-containing plants appear green to us because that green light is reflected, not absorbed.

Chlorophyll comes in a number of slightly different forms. The most important is chlorophyll a, but there are other chlorophylls as well as several forms that are found only in prokaryotic cells. Chlorophyll molecules consist of a long hydrocarbon tail attached to a complex cyclic structure called a **porphyrin ring** (Fig. 19.12b). The ring has a magnesium atom in its center, and the atoms that make up the ring are linked by an alternating series of single and double bonds. These alternating bonds enable the chlorophyll molecule to absorb a tremendous amount of light. Just a few milligrams of chlorophyll in a liter of water will absorb nearly all of the red and blue light passing through it.

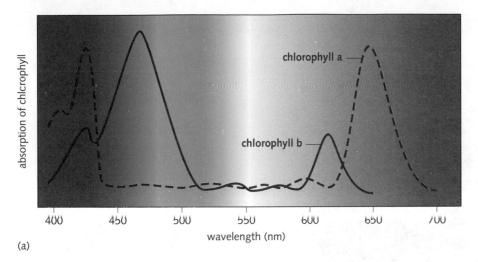

(a)

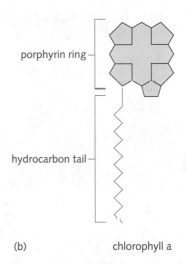

porphyrin ring

hydrocarbon tail

(b)　　chlorophyll a

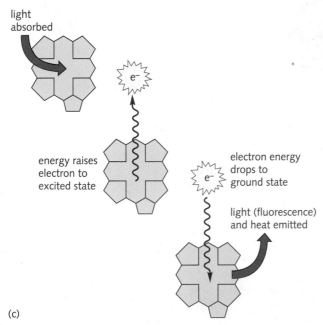

light absorbed

energy raises electron to excited state

electron energy drops to ground state

light (fluorescence) and heat emitted

(c)

Figure 19.12 **(a)** Absorption spectra of chlorophylls a and b. These pigments absorb very strongly in the red and blue regions of the spectrum. The poor absorption of these pigments in the middle of the visible spectrum gives chlorophyll its green color. **(b)** The chlorophyll molecule. The porphyrin ring structure encloses a magnesium atom, and attached to the ring is a long hydrocarbon chain known as the phytol tail.
(c) Chlorophyll is extremely efficient at absorbing light energy. When a photon strikes chlorophyll, one of the electrons of the pigment is raised to a higher energy level. **(d)** When this excited electron returns to the ground state, its energy is lost as heat or fluorescence, except in the chloroplast, where this energy is not lost but is captured in photosynthesis.

(d)

When light is absorbed by a chlorophyll molecule, energy is trapped in the electrons of the porphyrin ring. The electrons are raised to an excited state (Fig. 19.12c). If a chlorophyll molecule is alone when this happens, the electron usually returns to the ground state by giving off a quantum of light in the form of fluorescence (Fig. 19.12d). But if an acceptor molecule is nearby, the electron can be passed from chlorophyll to the acceptor: a classic redox reaction. In one sense, the ability of chlorophyll to produce high-energy electrons (by absorbing light) and then to lose them to another molecule in a redox reaction is the inner secret of photosynthesis—this is where high-energy electrons are generated.

Figure 19.13 The beautiful colors of autumn leaves come from β-carotene and other accessory pigments.

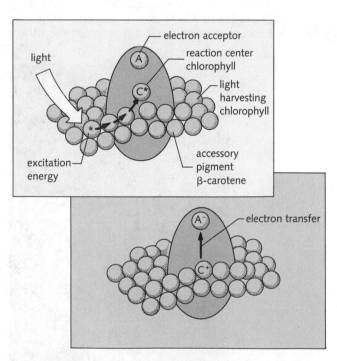

Figure 19.14 Photosynthesis begins with the absorption of light energy by pigment molecules surrounding the reaction center. This excitation energy is passed to the reaction center, which results in the transfer of an excited electron from the central chlorophyll to an electron acceptor.

The Photosynthetic Reaction Center

The heart of the light-dependent reactions is the photosynthetic reaction center in the chloroplast.

What happens in the reaction center is easy to describe. Chlorophyll absorbs light and releases high-energy electrons, and then those electrons are passed to an electron acceptor.

To gather sunlight more efficiently, the reaction center is surrounded by light-harvesting proteins that contain chlorophyll, as well as "accessory pigments" that absorb light in regions of the spectrum where chlorophyll doesn't absorb light as well. In higher plants one of the most common accessory pigments is β-**carotene,** which gives some leaves their characteristic reddish orange color in the autumn (Fig. 19.13).

The reaction center and the light-harvesting proteins are found within the photosynthetic membranes of the chloroplast. When a photon is absorbed by a pigment molecule, the excitation energy from that photon is passed from molecule to molecule until it arrives at the reaction center chlorophylls, as shown in Fig. 19.14. Within the reaction center, the pair of chlorophyll mole-

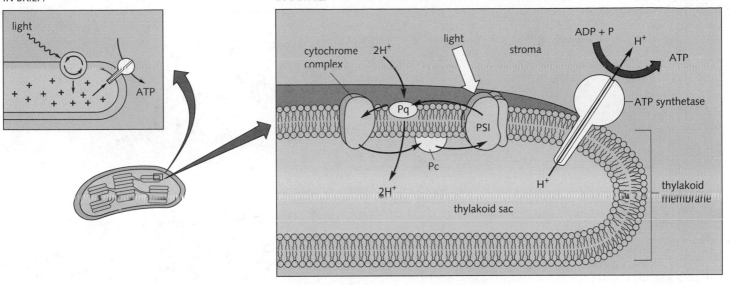

IN BRIEF:

IN DETAIL:

Figure 19.15 Cyclic electron flow. The photosystem I reaction center is part of the photosynthetic membrane. High-energy electrons produced in the reaction center are passed to an electron transport chain that includes plastoquinone (Pq), plastocyanin (Pc), and a cytochrome complex, and finally delivers the electrons back to the photosystem I reaction center. Protons are pumped across the membrane during the process, producing a powerful electrochemical gradient that drives the synthesis of ATP by an ATP-synthetase enzyme on the membrane surface.

cules now lose their excited electrons to an electron acceptor. This is the first step in **photosynthetic electron transport,** and it begins in the reaction center itself.

Cyclic Electron Flow and ATP

Electron transport in photosynthesis is only slightly more complicated than it is in respiration. In plants and many algae, there are two different kinds of photosynthetic reaction centers that absorb light of slightly different wavelengths. They are called **photosystem I** and **photosystem II.** There are two different patterns of electron transport that involve these reaction centers. **Cyclic electron transport,** which involves only photosystem I, is the simplest, so we will look at that pattern first (Fig. 19.15).

When the photosystem I reaction center (PSI) absorbs light, it produces a steady flow of high-energy electrons. These electrons are passed to a series of carrier molecules that are very similar to the electron transport chain of the inner mitochondrial membrane. [These carriers include lipids such as *plastoquinone* (Pq) and proteins such as the *cytochrome complex* and *plastocyanin* (Pc).]

Where do these electrons go? In one sense, they go nowhere. In cyclic electron transport, high-energy electrons from the photosystem I reaction center pass through a series of carrier molecules and then are returned to the very same photosystem I reaction center. At first glance, nothing seems to be accomplished by the cyclic pathway.

But look closely at Fig. 19.15. As the high-energy electrons move from one carrier to another, their energy is used to pump protons across the photosynthetic membrane. This should sound familiar. An almost identical mechanism is used in mitochondria to create a proton gradient and drive the synthesis of ATP by chemiosmosis.

The cyclic pathway produces a powerful gradient of protons across the photosynthetic membrane. Just like the inner mitochondrial membrane, the photosynthetic membrane contains an ATP-synthetase enzyme, an enzyme that performs *exactly* the same function as the mitochondrial ATP-synthetase, producing ATP by using the energy of the proton gradient. Therefore, cyclic electron transport *does* accomplish something: it drives the synthesis of ATP.

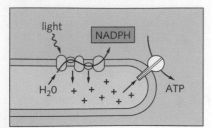

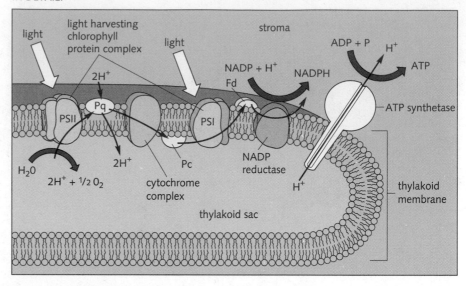

Figure 19.16 Noncyclic electron flow. High-energy electrons produced in the photosystem I and photosystem II reaction centers pass through a series of electron carriers, including plastoquinone (Pq), plastocyanin (Pc), and ferredoxin (Fd), ultimately reducing NADP⁺, a soluble electron carrier, to produce NADPH. The source of electrons for this chain is water, and a "water-splitting" apparatus associated with photosystem II releases oxygen when electrons are removed from water. The buildup of protons within the thylakoid membrane, from water splitting and electron transport, produces a proton gradient, which drives the synthesis of ATP.

Cyclic electron flow may have been the first form of photosynthetic electron transport to evolve, and it is still the predominant form in photosynthetic bacteria.

Noncyclic Electron Flow

Cyclic electron transport, although it does make ATP, does not produce the stable form of high-energy electrons that would be needed to synthesize carbohydrate. That job is accomplished by another pattern of electron flow known as **noncyclic electron transport.** This pattern involves both photosystem II and photosystem I (Fig. 19.16).

When the photosystem II reaction center (PSII) absorbs light, it produces a steady flow of high-energy electrons that are passed through an electron transport chain to photosystem I. Photosystem I absorbs more light energy and passes high-energy electrons along to other carriers. At the end of the chain, high-energy electrons are passed to a soluble electron carrier called *nicotinamide adenine dinucleotide phosphate* (NADP⁺), reducing it to a form abbreviated as **NADPH.** NADP⁺ is almost identical to the NAD⁺ in mitochondria, and it accepts electrons in exactly the same way.

As shown in Fig. 19.16, the noncyclic electron transport chain accomplishes two things:

1. It boosts electrons up to an energy level high enough to produce NADPH.

2. It pumps protons across the membrane, producing an electrochemical gradient that drives the synthesis of still more ATP.

By producing a steady supply of ATP and NADPH, the two patterns of electron transport ensure that everything is in place to begin the synthesis of carbohydrate. There is only one small problem posed by noncyclic electron flow—replacing the electrons lost in the reduction of $NADP^+$ to NADPH.

Getting Electrons: Oxygen Evolution

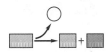

In noncyclic electron flow, photosystem II becomes oxidized every time it passes a high-energy electron to the electron transport chain. As we saw earlier in our hypothetical reaction, unless these lost electrons are promptly replaced, the pigment will change chemically after losing only a couple of electrons. In this case, the reaction center chlorophyll molecules in photosystem II will be permanently altered. Fortunately, the photosynthetic membrane has a means of replacing the electrons that chlorophyll has lost. The membrane contains a special set of enzymes that can remove electrons from water. When this happens, the water molecule is split, a process called **photolysis:**

$$H_2O \rightarrow 2H^+ + 1/2O_2 + 2e^-$$

As you can see, photolysis provides a source of electrons (e^-) that can reduce photosystem II. Photolysis also produces oxygen (O_2) and frees a pair of protons (H^+) into the thylakoid sac. This oxygen is released from the plant into the atmosphere. The process of photosynthetic oxygen evolution is the major source of oxygen on the planet, and it all comes from the need to supply electrons to photosystem II.

It's worth noting that there is no absolute requirement that water be used as the source of electrons, and some photosynthetic organisms make use of different compounds. Some photosynthetic bacteria, for example, use hydrogen sulfide as a source of electrons:

$$H_2S \rightarrow 2H^+ + S + 2e^-$$

These bacteria live near abundant sources of hydrogen sulfide (underwater sulfur springs and volcanic regions are typical habitats), and they produce rich deposits of pure sulfur as they grow.

Summary of the Light-Dependent Reactions

The light-dependent reactions of photosynthesis, as we have seen, take place within the thylakoid membranes. Photons are absorbed by light-harvesting pigments, and the energy they pass to photosynthetic reaction centers elevates electrons to higher energy levels. These high-energy electrons pass through a series of carrier molecules, eventually ending up at the soluble electron carrier, $NADP^+$. Water-splitting enzymes remove electrons from water to replace those that are used up in this pathway. As a result of water splitting and electron flow, protons move into the thylakoid sac, producing a strong electrochemical gradient. The energy from this gradient drives the synthesis of ATP. The products of the light-dependent reactions are NADPH and ATP, both of which are used in the light-independent reactions.

THE LIGHT-INDEPENDENT REACTIONS OF PHOTOSYNTHESIS

The ATP and NADPH produced in the light-dependent reactions of photosynthesis provide the chloroplast with a light-driven source of chemical energy. These molecules make possible the next steps in the photosynthetic pathway, the light-independent reactions. We can remind ourselves of where we are going by looking at the overall, balanced equation of photosynthesis:

$$6CO_2 + 12H_2O \rightarrow C_6H_{12}O_6 + 6O_2 + 6H_2O$$

ATP and NADPH supply the energy that will be used to produce carbohydrate from carbon dioxide and water.

The light-independent reactions take place in the stroma, the soluble portion of the chloroplast outside the thylakoid membranes. These reactions require ATP and NADPH from the light-dependent reactions, CO_2 from the atmosphere, and a complex series of enzymes and cofactors found in the chloroplast stroma.

Biochemists began to explore the pathways by which photosynthetic organisms produce carbohydrate by following the path that the carbon atoms of carbon dioxide took after they entered the cell. As radioisotopes became available at about the time of World War II, radioactive carbon-14 (^{14}C) was used in an effort to find out which compounds were formed directly from radioactive CO_2. The first scientists who tried this approach, however, were badly disappointed. Even with a brief pulse of labeled CO_2, nearly every kind of carbohydrate found in the cell

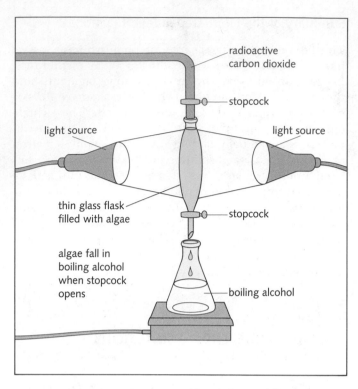

Figure 19.17 The apparatus used by Calvin and his associates to determine the pathways of carbon fixation in photosynthesis.

was labeled with ^{14}C. The "fixation" of carbon dioxide, as it was called, must be a very fast reaction.

The speed of photosynthesis demanded drastic experiments. In 1946 Melvin Calvin, a young scientist at the University of California at Berkeley, began a series of experiments in which a new approach was tried. Calvin and his co-workers exposed a flat flask of algae to radioactive CO_2 and a few seconds of light, and then they opened a valve at the bottom of the flask, allowing the algae to drop into hot alcohol. The cells were instantly killed, and the chemical products of photosynthesis were preserved so that they could be analyzed (Fig. 19.17). After many years these scientists were able to put together a complete chemical scheme for how the process of CO_2 fixation takes place.

The Calvin–Benson Cycle

Calvin and his associates discovered that CO_2 from the atmosphere is taken up by plants and enters a biochemical

pathway with nearly two dozen enzymes. The very first step of that pathway is the key reaction, however, because it establishes the first chemical link between CO_2 and an organic molecule, "fixing" the carbon atom. The CO_2 is linked to a five-carbon sugar known as *ribulose bisphosphate* (*RuBP*), briefly forming an unstable six-carbon molecule. This compound quickly breaks apart to form two molecules of *phosphoglycerate* (*PGA*):

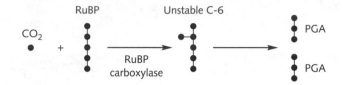

PGA was the first organic molecule to be labeled with radioactive carbon in Calvin's experiments. *Ribulose bisphosphate carboxylase* (*RuBP carboxylase*), the enzyme that catalyzes this reaction, works very slowly. Nonetheless, the light-independent reactions cannot proceed without the steady supply of new carbon atoms that this reaction provides. Plants respond by producing enormous quantities of the enzyme. In some chloroplasts, as much as 50 percent of the total protein is RuBP carboxylase, making this enzyme the single most abundant protein in the world—eloquent testimony to the importance of carbon fixation for life on Earth.

The two PGA molecules are part of a cycle that produces the carbohydrate used by the plant and regenerates RuBP so that carbon fixation can continue. This pathway is known as the **Calvin–Benson cycle,** after its discoverers, or the **C3 cycle,** because the first stable compounds formed from CO_2 are *three-carbon* compounds.

The phosphoglycerate molecules formed in the first steps of the cycle are used to produce *fructose-6-phosphate*, a six-carbon sugar. Fructose is formed in a four-step pathway that uses the energy of ATP and NADPH:

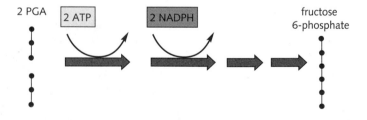

However, if all PGA molecules were converted into fructose, the plant cell would quickly use up the pool of RuBP molecules needed to fix CO_2. In fact, PGA molecules pass through the Calvin–Benson cycle to produce more RuBP. This pathway requires the energy of ATP and involves a series of steps with five-carbon, six-carbon, and seven-carbon intermediates. It eventually results in the formation of enough RuBP to keep carbon fixation "in business."

IN BRIEF:

IN DETAIL:

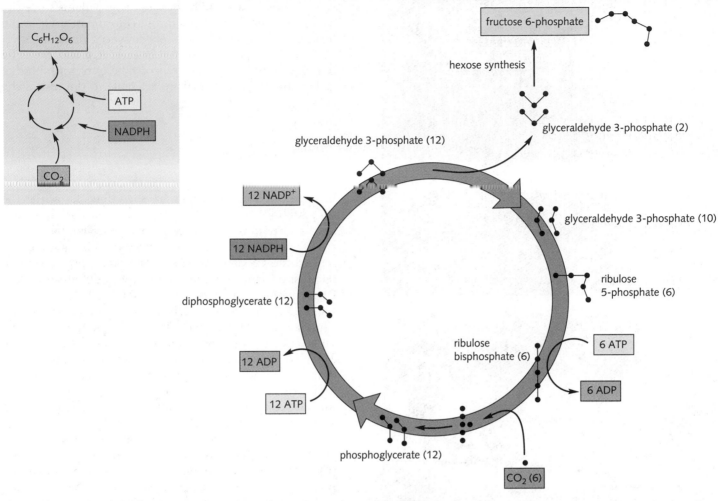

Figure 19.18 The Calvin–Benson cycle. In this pathway the products of the light reaction, ATP and NADPH, provide the chemical energy needed to fix CO_2 to produce carbohydrate. A series of soluble enzymes in the chloroplast stroma catalyzes the reactions of the cycle. For every six CO_2 molecules entering the cycle, a single six-carbon sugar is produced. Fructose, the immediate product of the cycle, can be converted into complex carbohydrates, lipids, and other organic compounds.

Figure 19.18 is a summary of the light-independent steps of photosynthesis. In order to visualize the synthesis of a six-carbon fructose, we must follow the fates of six CO_2 molecules as they are fixed by RuBP carboxylase.

As shown in Fig. 19.18, 6 carbon atoms enter the Calvin cycle from 6 molecules of CO_2. A total of 18 ATPs and 12 NADPHs are required to regenerate RuBP and produce a single six-carbon fructose molecule. Can we compare the energy required to synthesize a six-carbon sugar with the energy obtained by breaking it down? Here's a quick calculation:

- The energy of NADPH is roughly equivalent to that of NADH, and as we saw in Chapter 18, 3 ATPs

can be synthesized from each NADH in oxidative phosphorylation.

- Therefore 12 NADPHs are more than equivalent to 36 ATPs.

- Summing it up, we have 18 + 36 = 54 ATPs, compared with the 36 ATPs produced by aerobic respiration.

Therefore, the energy required to synthesize a sugar molecule is at least 50 percent greater than the energy obtained by breaking it down. And even this simple calculation understates the energy of synthesis, because 1 NADPH contains more available energy than 3 ATPs.

Figure 19.19 Sugarcane (TOP) and corn (BOTTOM) are important commercial C4 plants. The special adaptations of C4 plants make them more efficient than C3 plants where light intensity is high and water is limited.

The carbohydrate that is produced in photosynthesis can be used in a number of ways. Some plants store the products of photosynthesis in the chloroplast itself in the form of starch granules. Others convert glucose to lipids and oils, which are stored within the organelle, too. The products of photosynthesis may also be transported out of the organelle so that they are available to the rest of the cell.

C4 Photosynthesis

The light-independent reactions that were first described by Calvin and his associates are sometimes known as the **C3 pathway** of carbon fixation because carbon dioxide is first incorporated into a compound that contains three carbon atoms. Not long after Calvin and his associates described the C3 pathway, Hatch and Slack, two Australian scientists, discovered that several types of plants use a different pathway in which carbon dioxide is first incorporated into *oxaloacetate*, which contains four carbon atoms. The pathway Hatch and Slack described has become known as the **C4 pathway.** The C4 pathway turns out to be quite common among desert grasses and other plants that are well adapted for growth under conditions of low moisture and high light intensity. Common C4 plants include corn, sugarcane, and Bermuda crabgrass (Fig. 19.19).

One of the problems faced by plants growing in dry climates is the need to conserve water. The leaves of such plants tend to close the openings through which water vapor might be lost to the air, but in so doing, they also make it difficult for carbon dioxide to diffuse into the leaf. Because photosynthesis is rapid in bright sunlight, oxygen is released from the splitting of water, while the small amount of carbon dioxide within the leaf is quickly reduced by carbon fixation. This leads to a very high O_2-to-CO_2 ratio, and that leads to a problem called **photorespiration.**

RuBP carboxylase is the enzyme that normally catalyzes the first reaction of the C3 cycle:

$$CO_2 + RuBP \rightarrow 2PGA$$

However, when the O_2-to-CO_2 ratio is extremely high, oxygen occupies the active site of the enzyme, and the following reaction takes place:

$$O_2 + RuBP \rightarrow PGA + phosphoglycolate$$

Eventually, phosphoglycolate is broken down elsewhere in the plant cell and CO_2 is released instead of being used to form carbohydrate. Because it releases CO_2, photorespiration limits the efficiency of photosynthesis

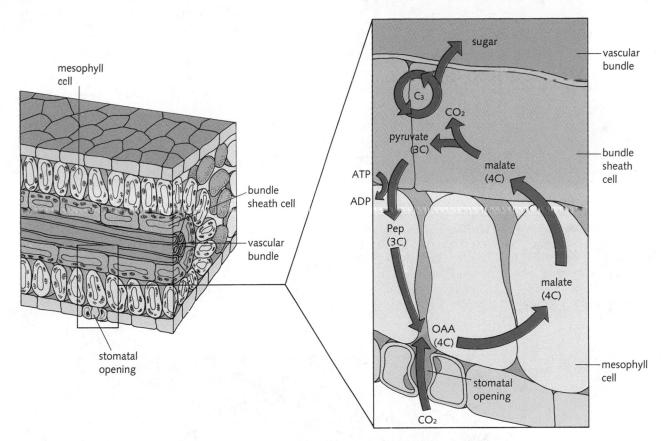

Figure 19.20 C4 plants are so named because they fix CO_2 to produce oxaloacetate (OAA), a four-carbon compound. This reaction is the first in a cycle in which malate transfers newly fixed carbon to the bundle sheath tissue of the leaf, where the Calvin–Benson cycle takes place. In a sense, the mesophyll tissue pumps carbon into the bundle sheath by means of these reactions. Although the C4 cycle requires more chemical energy than the Calvin–Benson cycle alone, it can take place at much lower CO_2 concentrations, increasing efficiency at high light intensities.

and substantially reduces the productivity of crops grown at high light intensity. Although photorespiration is a common problem among C3 plants, it is almost nonexistent among C4 plants.

Figure 19.20 shows why this is the case. The C4 plants carry out the first stages of carbon fixation in leaf tissues known as the mesophyll, passing reduced carbon in the form of malate (a C4 compound) or oxaloacetate into adjacent tissue known as the *bundle sheath*. Within the bundle sheath, CO_2 is released and used in the ordinary C3

cycle. The C4 pathway within the mesophyll cells uses extra energy in the form of ATP, but it keeps the concentration of CO_2 in the bundle sheath high enough to ensure that photorespiration does not occur. And this, in turn, increases the efficiency of C4 plants in bright, arid environments. Many plant breeders and geneticists hope to breed some aspects of C4 photosynthesis into useful C3 plants, such as soybeans, in an effort to obtain higher crop yields.

An Alternative for Desert Plants

A few plants solve the problems posed by dry climates in a quite different way. These plants admit air into their leaves at night, when temperatures are cooler and they are less likely to lose moisture. In the darkness, carbon dioxide diffuses into the leaf and is combined with existing molecules to produce organic acids, "trapping" the carbon within the leaves. This allows these plants to accumulate large amounts of carbon at night. During the daytime, when the leaves are tightly sealed to prevent the loss of water, these organic acids release carbon dioxide *within* the leaf, allowing it to be used in the Calvin–Benson cycle for carbohydrate production.

The plants that use this form of carbon fixation include members of the family *Crassulaceae,* and therefore the process is called **crassulacean acid metabolism (CAM).** CAM plants include many desert cacti and also the fleshy "ice plants" that are frequently planted near freeways along the West Coast to retard brush fires.

These desert cacti, known as "ice plants," are CAM plants, capturing CO_2 at night to conserve water.

MITOCHONDRIA AND CHLOROPLASTS: REMARKABLY SIMILAR ORGANELLES

Despite their obvious differences, there are striking similarities between mitochondria and chloroplasts. Each organelle is surrounded by a pair of membranes, each one contains a cyclic chemical pathway (the Krebs cycle and the Calvin–Benson cycle), each contains a membrane-bounded electron transport system, and each synthesizes ATP in response to an electrochemical gradient established by the process of electron transport. Interestingly, there are other similarities as well.

Mitochondria and chloroplasts contain their own *genes*. They each contain multiple copies of circular chromosomes, which are composed of DNA. These organelles depend on the genes located on these chromosomes, and some very careful experiments have shown that they cannot survive without them. What sorts of genes do these organelles contain? They vary from one organism to another. Most mitochondria and chloroplasts contain genes for some of the electron transport components found in the inner mitochondrial membrane and the photosynthetic membrane. These organelle chromosomes also contain genes that are necessary for chromosome replication and gene expression.

Does this mean that chloroplasts and mitochondria are genetically independent of the rest of the cell? Actually, no. The majority of the proteins found within these organelles are synthesized in the cytoplasm of the cell and are under the control of nuclear genes. These proteins are then "imported" into the organelles by a mechanism that is just beginning to be investigated. The chromosomes within these organelles may be remnants of complete genomes from a time when the organelles were actually independent, free-living prokaryotic cells.

There is no fossil record to tell us exactly when the first mitochondrion or chloroplast appeared. However, scientists have tried to understand how these organelles might have evolved by looking at organisms that are alive today. One suggestion, which has been promoted by Lynn Margulis of the University of Massachusetts (see Chapter 24), is that these organelles may have evolved from independent, free-living prokaryotic cells that took up residence in the cytoplasm of early eukaryotes. If this suggestion were correct, it would be easy to understand why each organelle maintains its own genetic system.

More recently, however, other evidence has emerged to suggest that the story may not be so simple. Molecular biologists have found that some mitochondrial genes closely resemble genes found in the eukaryotic nucleus. Recent work has also shown that some genes from mitochondria are almost identical to genes found in chloroplasts, leading to speculation that these genes may have been moved from one organelle to another. Findings like this have made it more difficult to construct an evolutionary history for mitochondria and chloroplasts, but only time and more research will tell for sure. In the meantime, it is safe to say that these remarkable organelles may have a few more surprises in store for us.

SUMMARY

In many respects, photosynthesis is the reverse of respiration. In respiration, sugar molecules are broken down to carbon dioxide and water, and energy is released for cellular activities. In photosynthesis, the energy from sunlight is used to produce sugar from carbon dioxide and water. Photosynthesis consists of a series of light-dependent reactions, which trap the energy of sunlight in chemical form, and a series of light-independent reactions, which involve the fixation of carbon dioxide to form sugar.

The light-dependent reactions begin when electrons in chlorophyll, the primary pigment molecule of green plants, are raised to higher energy levels by the absorption of light energy. These excited electrons are then passed along an electron transport pathway found in the thylakoid membranes of the chloroplast. Electron transport in these membranes produces a chemiosmotic gradient that drives ATP synthesis and also reduces a soluble electron carrier. The light-independent reactions use ATP and the reduced electron carriers to capture atmospheric carbon dioxide and reduce it to form sugar, which then becomes the primary chemical energy source of the plant. This process occurs by means of a chemical pathway known as the Calvin–Benson cycle. Some plants in arid environments use modifications of this pathway to help conserve water. Two examples are the C4 carbon fixation pathway used by desert grasses, and crassulacean acid metabolism (CAM), used by many cacti and their relatives.

Chloroplasts and mitochondria, which carry out photosynthesis and cellular respiration, have many features in common. Both possess their own genetic material, and both may be related to ancestral prokaryotic organisms that took up residence within the eukaryotic cell.

STUDY FOCUS

After studying this chapter, you should be able to:

- Explain how energy-rich food molecules are made by the process of photosynthesis, which uses the sun as its energy source.

- Describe how photosynthesis takes place within the structure of the chloroplast.

- Distinguish between the light-dependent and light-independent reactions, between cyclic and noncyclic electron flow, and between C3 and C4 pathways.

- Explain how some plants are adapted to carry out photosynthesis in hot and dry environments.
- Cite similarities between respiration and photosynthesis and between chloroplasts and mitochondria.

TERMS AND CONCEPTS

chloroplasts *383*
thylakoids *383*
light-dependent
  reactions *383*
light-independent
  reactions *383*
stroma *383*
chlorophyll *384*
β-carotene *386*
photosynthetic electron
  transport *387*
photosystem I *387*

photosystem II *387*
cyclic electron
  transport *387*
noncyclic electron
  transport *388*
photolysis *389*
Calvin–Benson cycle *390*
C3 pathway *392*
C4 pathway *392*
photorespiration *392*
crassulacean acid
  metabolism (CAM) *394*

REVIEW

Objective Questions (Answers in Appendix)

1. In a hot, dry environment, a plant undergoing C4 photosynthesis will do better than a plant that can undergo C3 photosynthesis only because
   (a) C4 plants can take in oxygen more efficiently than C3 plants.
   (b) C4 plants can store carbon dioxide more efficiently.
   (c) C3 plants have a lower rate of energy use.
   (d) C4 plants have a higher rate of photorespiration.

2. Chlorophyll is located in the
   (a) thylakoids.   (c) cell membrane.
   (b) mitochondria.   (d) cell nucleus.

3. The part of the chloroplast that takes part in the light-independent reaction is the
   (a) thylakoids.   (c) stroma.
   (b) grana.   (d) photosynthetic membranes.

4. The electrons that pass to the NADP$^+$ during the noncyclic part of the photosynthetic electron flow come from
   (a) carbon dioxide.   (c) sunlight.
   (b) water.   (d) oxygen.

5. In a C3 plant, the carbon atom that first enters the light-independent reactions as carbon dioxide can first be found in which molecule?
   (a) NADPH   (c) RuBP
   (b) PGA   (d) protons

Discussion Questions

6. It is sometimes pointed out that the direction of proton movement in the photosynthetic membrane, or thylakoid, is reversed with respect to the inner mitochondrial membrane. Draw a diagram to illustrate what is meant by this.

7. Compare and contrast cyclic and noncyclic electron flow. Which one results in the reduction of a soluble electron carrier?

8. You are about to install lights in a greenhouse. However, ordinary "white" lights are not available, and you must pick from a selection of colored lights. They include blue, green, and red lamps. To drive photosynthesis most efficiently, which colors would you choose, and which would you avoid?

9. Normally, the thylakoid membrane is not permeable to protons. What would happen to photosynthesis if a compound were added to plant cells to make the thylakoid membrane permeable to protons? Be specific. Would the rate of oxygen production, for example, speed up or slow down? How about the rate of carbon dioxide consumption?

10. Many C4 plants are grasses found in arid, desert-like regions. Among the most important C4 crop plants is corn. Corn has a remarkable ability to grow rapidly during long, hot summer days with little rainfall. Which aspects of C4 metabolism contribute to this ability?

READINGS

Govindjee, and W. J. Coleman. "How plants make oxygen." *Scientific American* 262 (February 1990): 50–61. The machinery that "splits" water to release oxygen is only partly understood, and this article shows some of the fascinating elements of that machinery that are already apparent.

Youvan, D., and B. Marrs. "Molecular mechanisms of photosynthesis." *Scientific American* 256 (June 1987): 42–50. A detailed examination of what happens inside a photosynthetic reaction center.

Moore, P. "The varied ways plants trap the sun." *New Scientist* 12 (1981): 394–397. A clear exposition of C4 photosynthesis.

Miller, K. R. "The photosynthetic membrane." *Scientific American* 241 (October 1979): 102–113. This article describes the basic structural organization of the photosynthetic membranes of higher plants.

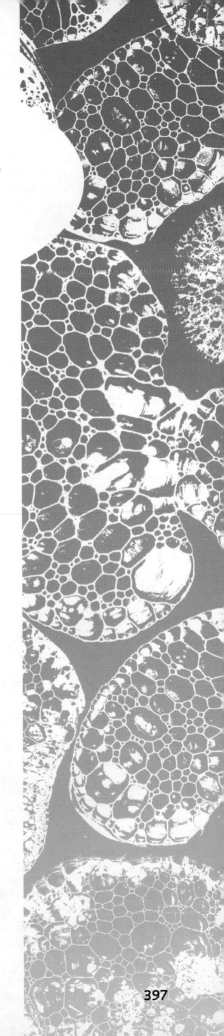

20
Molecules and Genes

*l*ife requires *information*. A living cell contains thousands of different proteins, each with a specific sequence of amino acids and a specific three-dimensional structure. The cell contains a bewildering variety of membranes, organelles, and filaments. Within the cell an interlocking series of biochemical pathways carries out the chemical reactions that make life possible. The organization of the cell requires a small mountain of information, and that information must be passed on from one generation to the next. But how can a cell carry information?

In our everyday experience, we are all familiar with systems that carry information: the alphabet, the dots and dashes of Morse code, and the binary code that represents information on a computer. For a molecule inside a living cell to carry information, it must have a structure as variable as these more familiar information systems. Such a molecule might be structured like a series of symbols strung out to produce a coded message. The cell would have to contain a decoding system to put that information to useful work, and it would need something else as well: a system to *copy* the code so that a complete copy could be passed to each daughter cell when cell division takes place. If living cells contain a molecular code, the molecule that carries it must be capable of **replication**—of making an exact copy of itself.

In this chapter we will consider how a living cell might carry information, and we will retrace the pathway of one of the greatest discoveries of twentieth-century science—the molecular nature of the gene.

PROLOGUE: WHAT IS LIFE?

As the first half of the twentieth century was drawing to a close in the midst of a tragic world war, a number of scientists thought that the time was right to explain life in chemical terms. One of these was Erwin Schrödinger, a famous physicist who had fled Germany at the beginning of the war.

In 1944 he gave a series of lectures that were rewritten into a book entitled *What Is Life?* The logic of Schrödinger's approach was powerful and simple. Genetics, he reasoned, held the key to the nature of life. The genes that are passed from one generation to the next do much more than simply determine the colors of flowers and the shapes of wings. They contain the information that makes life possible. If we understood the nature of the gene, Schrödinger argued, we would be much closer to understanding life.

Because of his training in the physical sciences, Schrödinger believed that genes should be made of ordinary matter, like everything else in the physical universe. Yet the behavior of living matter was *so* different from that of nonliving matter that he believed it had to be organized along principles that (in 1944) remained to be discovered.

How is the *information* that living things require passed from generation to generation? What is the molecular nature of the gene? Schrödinger suggested that the basis of cellular information was something he called an "aperiodic crystal." This confusing term was widely misunderstood, but Schrödinger meant it to imply a molecule in which individual subunits, each of which might be different from its neighbors, held fixed positions (Fig. 20.1). Such a molecule could contain information coded at the molecular level and would be able to pass that information from one generation to the next.

Although many of Schrödinger's speculations were just plain wrong, his insights on the molecular nature of information were right on the mark.

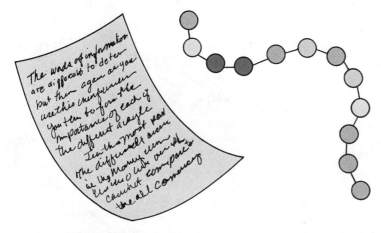

Figure 20.1 Schrödinger's concept of aperiodic crystals can be likened to a page of written text, capable of carrying information. In molecular terms, the closest analogy to a coded message might be a polymer constructed from a sequence of variable subunits.

What Is Life? is an important book in the history of biology. First, it convinced a number of talented scientists to leave physics and begin to work on biological problems. Second, it pointed with astonishing clarity to two ideas that were crucial in understanding the molecular nature of the gene: (1) Life is so very different from nonlife that new principles will have to be developed to account for its properties. (2) These new principles will be based on the ordinary laws of physics and chemistry that are familiar to us. What are these "new principles?" That's the subject of this chapter.

THE CHEMICAL NATURE OF THE GENE

We have seen how studies of inheritance led to the idea that the characteristics of an organism are controlled by individual elements called *genes*. We've also explored some aspects of biological chemistry and examined the kinds of molecules found in living cells. If we accept the idea that genes are specific molecules, the first question we must answer is an obvious one: *which molecule?*

Coding Capacity: Proteins Seem to Be the Best Bet

Could protein be the genetic material? Eukaryotic chromosomes contain very little carbohydrate; they are about 30 percent nucleic acid and from *60 to 75 percent* protein. Most scientists in the first half of the twentieth century, Erwin Schrödinger included, agreed that proteins were most likely to be the molecules that carry information. To begin with, protein molecules are composed of 20 amino acids, but nucleic acids are composed of just 4 nucleotides. Therefore, proteins should be better coding molecules (an alphabet with 20 letters would be far more expressive than one with just 4). Proteins also differ a great deal more from one species to another than nucleic acids do. Finally, although scientists at the time were familiar with the ability of protein enzymes to control and regulate chemical reactions, there was no evidence that nucleic acids were able to *do* anything of interest. Proteins looked like a very good bet to be the genetic material.

The Transforming Principle

Well before Schrödinger's famous lectures, a few people were working on projects that eventually were to identify the molecules responsible for inheritance. Frederick Griffith was a British scientist studying the way in which certain types of bacteria cause pneumonia, a serious and occasionally fatal lung disease.

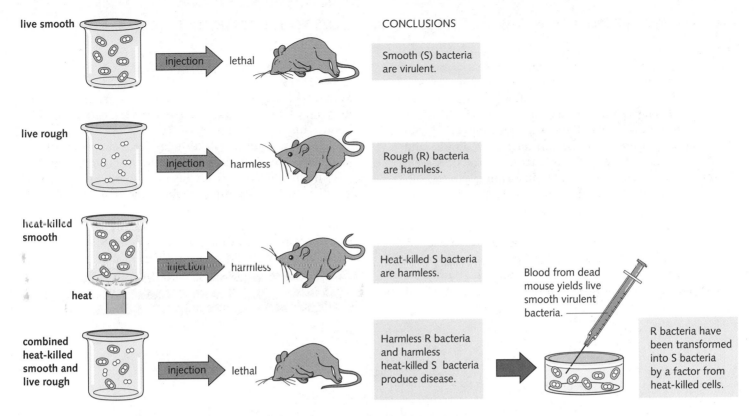

live smooth → injection → lethal → Smooth (S) bacteria are virulent.

live rough → injection → harmless → Rough (R) bacteria are harmless.

heat-killed smooth / heat → injection → harmless → Heat-killed S bacteria are harmless.

combined heat-killed smooth and live rough → injection → lethal → Harmless R bacteria and harmless heat-killed S bacteria produce disease.

Blood from dead mouse yields live smooth virulent bacteria.

R bacteria have been transformed into S bacteria by a factor from heat-killed cells.

CONCLUSIONS

Figure 20.2 A diagrammatic summary of Griffith's experiments on transformation. Something present in a heat-killed extract of virulent pneumonia bacteria was able to transform a nonvirulent strain of bacteria so that they caused pneumonia. Griffith recovered live, virulent bacteria from the infected mice.

In 1928 Griffith had in his laboratory two different types—*strains*—of pneumonia bacteria. Both strains grew very well in his lab, but only one of them actually caused pneumonia in mice. Griffith noticed that he could distinguish one strain from the other simply by its physical appearance on a culture dish. The **virulent** (disease-causing) bacteria produced a shiny coating around themselves that made their colonies look smooth. The **nonvirulent** (harmless) strains made no such coating and instead grew into colonies with rough, jagged edges.

Griffith first tested the hypothesis that the coating contained a disease-causing poison. To do this, he killed a culture of virulent cells and injected the dead cells (coatings and all) into the mice. The mice were not harmed by the injections. Griffith concluded that his suspicion about a poison was incorrect, since the dead cells did not poison the mice. Next, he injected both live nonvirulent and killed virulent cells into a mouse (knowing that neither one of these injections by itself made any of the mice sick). To Griffith's surprise, when the two types of cells were injected at the same time, the mice developed pneumonia (Fig. 20.2).

Today it is hard to appreciate just how startling this result was. The live nonvirulent bacteria never made the mice sick, and neither did the killed virulent ones. Why should the *combination* of these two have an effect that was different from what was seen when the two were administered *separately*?

To confuse matters further, Griffith recovered live bacteria from the animals that had developed the disease. Were these bacteria the same nonvirulent ones he had injected? He grew the bacteria on plates to find out. Now they formed the smooth colonies that were characteristic of the virulent strain. *Griffith's extract had transformed one kind of bacterium into another!* Griffith called the process he had discovered **transformation.**

Griffith speculated that when the live and killed bacteria were mixed together, a "factor" was transferred from the killed cells into the live ones. This factor changed the characteristics of the live cells in a permanent way, so that they henceforth acted like the virulent ones. What had actually happened? The molecule of inheritance had been transferred, and it had changed the characteristics of the cell that received it.

The Transforming Principle Is DNA

In 1942 Oswald Avery, a scientist at the Rockefeller Institute in New York City, devised a simple research project: he would determine the chemical nature of the material responsible for the transformation effect in Griffith's experiments. Using Griffith's transformation system, Avery and his co-workers treated the transforming extract in ways that destroyed proteins, carbohydrates, lipids, and ribonucleic acid (RNA). Transformation still occurred. However, when they treated the transforming extract with an enzyme that destroyed the other nucleic acid, deoxyribonucleic acid (DNA), transformation did not occur. Proteins did not carry genetic information. DNA did.

Avery's results were treated with some skepticism by his colleagues, if only because his answer seemed to make so little sense. The scientific community had not suspected that nucleic acids were capable of playing such an important role.

CLUES TO THE STRUCTURE OF DNA

Base Composition

We have already seen that nucleic acids are polymers of nucleotides. Long chains, **polynucleotides,** can be built by linking nucleotides together (Fig. 20.3). By the late 1940s, scientists understood the general chemistry of the polynucleotide. Still, there was nothing that might distinguish the nucleic acid from any other biological polymer. Was there something unusual about the structure of DNA that would enable it to carry information?

Some of the first clues to the structure of DNA came from experiments in which scientists determined the *composition* of DNA obtained from different organisms. DNA is a nucleic acid containing four nucleotide bases: **adenine, cytosine, guanine,** and **thymine.** Each of these bases is represented by a single letter: **A, C, G,** and **T.** Erwin

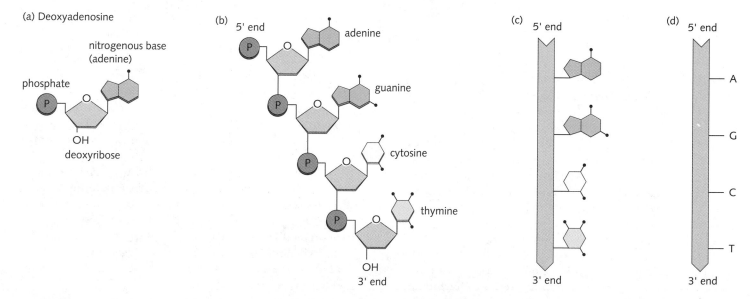

Figure 20.3 The basic structure of nucleic acids. **(a)** A nucleotide consists of a 5-carbon sugar (either ribose or deoxyribose), a phosphate group attached to the number 5 carbon, and a nitrogenous base attached to the number 1 carbon. **(b)** A polynucleotide. Note that the sugar–phosphate chain has two chemically different ends: the 5' end (TOP) has a phosphate group attached to the number 5 carbon, and the 3' end (BOTTOM) has an —OH group attached to the number 3 carbon. The bases extend out to the side of the sugar–phosphate chain. The polynucleotide chain can be drawn in a simplified fashion that emphasizes the bases, as shown in **(c)** and **(d)**.

Chargaff, a biochemist, was the first to notice a pattern in the relative percentages of the four bases. This pattern is apparent in Table 20.1.

Do you see the same pattern that Chargaff did? The percentages of the four bases are not static. They vary over a wide range, but the relative proportions of guanine (G) and cytosine (C) are always nearly equal; the same is true for the proportions of adenine (A) and thymine (T). In more symbolic form, we might express this observation as follows:

$$[A] = [T]$$

$$[C] = [G]$$

This observation became known as **Chargaff's rule.** Chargaff himself had no explanation for the rule, but it did suggest something about the structure of DNA. Because there were always equal amounts of adenine and thymine, for example, the molecule had to be organized in a way that would account for the equivalence of the two bases. Therefore, to our basic knowledge of the chemical structure of a polynucleotide we can add that bit of information about the ratios of the various bases.

The X-Ray Pattern

If a molecule can be crystallized, X-ray diffraction may produce a scattering pattern from the crystal that yields information about the molecule's internal structure. The repeating pattern of atomic bonds that exists within a crystal scatters a beam of X-rays in a regular pattern. When researchers realized that DNA might be an interesting and important molecule, a number of X-ray crystallographers began to work on its structure. In most cases, however, the formation of good crystals from DNA turned out to be both a practical and a theoretical problem. Crystals were difficult to produce, and the patterns from successful crystals were difficult to interpret.

A few scientists, however, believed that DNA was such an interesting molecule that X-ray diffraction should be attempted even if perfect crystals couldn't be formed. One such person was Rosalind Franklin, a young scientist working with Maurice Wilkins, a crystallographer in London. Franklin drew a thick suspension of the fiber-like DNA molecules up into a glass capillary tube and used this DNA sample to scatter X-rays. In the tube, she hoped, the thick suspension of DNA molecules would be forced to line up so that the molecules were parallel to the tube. Like spaghetti drawn through a straw, the molecules were all arranged in the same direction—not perfect enough to give a true crystal-like pattern, but good enough to yield vital clues about the structure of the DNA molecule.

Table 20.1 *Base Composition (Percent of Total DNA Bases)*

Source of DNA	A	T	G	C
E. coli	26.0	23.9	24.9	25.2
Streptococcus p.	29.8	31.6	20.5	18.0
Yeast	31.3	32.9	18.7	17.1
Herring	27.8	27.5	22.2	22.6
Human	30.9	29.4	19.9	19.8

SOURCE: Chargaff, E., and J. Davidson, eds. *The Nucleic Acids.* New York: Academic Press, 1955.

Interpreting the X-Ray Pattern

One of the X-ray patterns produced by Franklin's DNA samples is shown in Fig. 20.4. The pattern contains two critical clues to the structure of the DNA molecule (graphically summarized in the figure).

Clue 1: The two large dark patches at the top and bottom of the figure showed that some structure in the molecule was arranged at a right angle to the long axis of DNA and repeated at a distance of 3.4 Å (see Appendix, The Metric System). In other words, something in the molecule was arranged like the rungs of a ladder.

Clue 2: The X-like mark in the center of the pattern showed that something in the molecule was arranged in a zigzag fashion at a spacing of about 20 Å. If DNA had a twisting, helical configuration, the X-ray pattern it produced from the side would indeed look like that "X." The X-ray pattern suggested that the molecule was a *helix* and that its diameter was about 20 Å.

The scientific challenge remaining was to put all of these clues together to determine the structure of DNA.

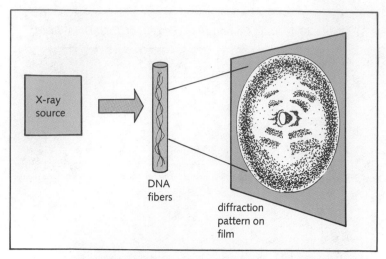

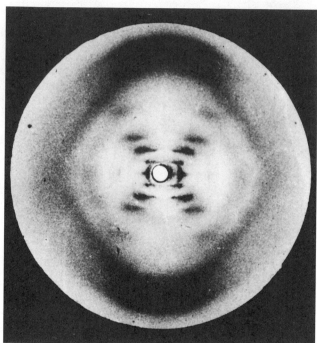

Figure 20.4 LEFT: DNA fibers were taken up in a thin tube so that most of them were oriented in the same direction. An X-ray diffraction pattern was then recorded on film. BOTTOM: X-ray diffraction pattern of DNA in the "B" form, as taken by Rosalind Franklin in 1952. Franklin's X-ray pattern contained two important clues to the structure of DNA. The large spots on the top and bottom of the pattern indicate that there is a regular spacing of 3.4 Å along the length of the fiber. The "X"-shaped pattern in the center indicates that there is a zigzag feature in the molecule, which might be consistent with a helix.

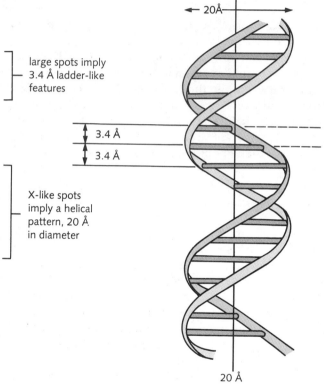

large spots imply 3.4 Å ladder-like features

3.4 Å

3.4 Å

X-like spots imply a helical pattern, 20 Å in diameter

20 Å

20 Å

In 1953 two young scientists working at Cambridge University in England learned about Franklin's remarkable X-ray pattern. The scientists, James Watson and Francis Crick, had been working with molecular models, twisting little bits of wood and paper into various shapes in an effort to determine how nucleotides could form a structure that would do all that DNA was able to do. According to their own accounts of that discovery, one look at Franklin's X-ray pattern was the last bit of evidence they required, and the problem was solved.

In his book *The Double Helix,* James Watson wrote, "The instant I saw the picture my mouth fell open and my pulse began to race. . . . The black cross of reflections which dominated the picture could arise only from a helical structure."

Watson and Crick had to account for several things. An ideal model for the structure of DNA would:

1. Explain Chargaff's rule.

2. Be able to carry coded information.

3. Be capable of being replicated.

4. Fit the chemistry known for polynucleotides.

5. Agree with the X-ray pattern's three predictions:

    (a) The molecule is helical.

    (b) Some structure within the molecule is stacked at a spacing of 3.4 Å like the rungs of a ladder.

    (c) The width of the molecule is about 20 Å.

Because the X-ray data were not detailed enough to determine a structure directly, Watson and Crick hoped to find an arrangement of nucleotide subunits that would be consistent with each piece of the puzzle, including the X-ray pattern.

The Double-Helix Model

In early 1953 Watson and Crick believed that they had a sensible structure for the molecule: the **double-helix** model. They published their ideas in a brief paper that appeared in April of that year. The details of their model were surprisingly simple. Watson and Crick realized they could account for the 3.4-Å spacing in the X-ray pattern if they arranged the nitrogenous bases (adenine, guanine, cytosine, and thymine) so that they were "stacked" on top of each other.

The 20-Å width of the molecule could be accounted for if they placed two *antiparallel* strands side by side in opposite directions and arranged the bases facing each other.

The helical twist that was evident in the diffraction patterns could be accounted for as well. All they had to do was twist the molecule so that the two strands twisted about each other.

Initially, however, there were two problems with the model. First, what kinds of forces might hold the two strands together? Second, how could one solve the problems posed by the sizes of the nitrogenous bases? Two of the bases, adenine and guanine, belong to a chemical group known as the **purines.** They have *two* carbon–nitrogen rings in their basic structures. The other two, thymine and cytosine, are **pyrimidines.** They have a single ring, meaning that they are quite a bit smaller than the purines. This would cause a problem in the model. If two pyrimidines were paired, the two strands would have to be much closer than when two purines were paired, making the model "lumpy" (Fig. 20.5).

Chargaff's rule showed how a "lumpy" helix could be avoided. If a purine was always paired with a pyrimidine, the helix wouldn't be lumpy. But was it possible for bonds to form between purines and pyrimidines to hold the two strands together? Watson and Crick remembered how hydrogen bonds (weak, noncovalent interactions) seemed to stabilize the structure of the α-helix in proteins. To their delight, when James Watson drew a sketch of the bases, they could find perfect places for hydrogen bonds to form between A and T and between G and C (Fig. 20.6).

The specific hydrogen bonding between the bases is known as **base pairing.** Watson, by the way, had never done very well in chemistry, and his sketch showed it. He didn't notice that a third hydrogen bond (highlighted in Fig. 20.6) existed between G and C. Watson and Crick's insight solved one of the critical problems regarding the

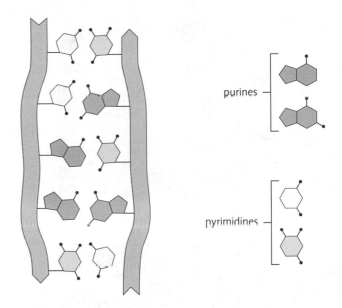

Figure 20.5 There were many false steps along the road to solving Franklin's X-ray pattern. For example, placing the nitrogenous bases inside the sugar–phosphate chains would give a fiber of the correct average width; but the different sizes of the bases, if there were no restrictions on how they were placed, would produce a "lumpy" fiber.

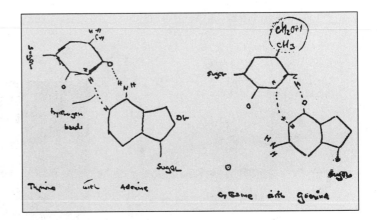

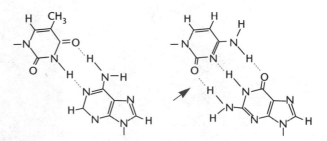

Figure 20.6 TOP: The problem of placing nitrogenous bases was solved by James Watson, who drew sketches showing how hydrogen bonds might pair thymine with adenine and guanine with cytosine. BOTTOM: The modern representation of A–T and G–C base pairing shows *three* hydrogen bonds in the G–C pair, something that was missing from Watson's first sketch.

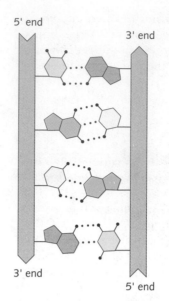

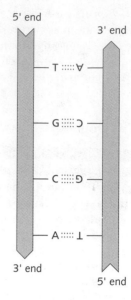

Figure 20.7 LEFT: James Watson and Francis Crick shortly after the publication of their double-helix model for the structure of DNA. RIGHT: The two strands of a DNA double helix are held together by hydrogen bonds between the nitrogenous bases. The strands run in antiparallel directions, forming a helix with a diameter of 20 Å.

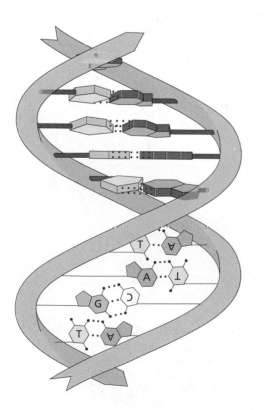

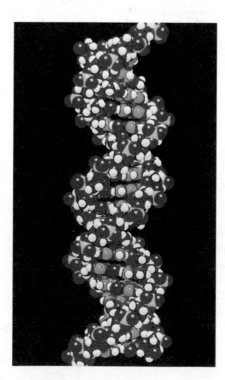

Figure 20.8 LEFT: A schematic model of the DNA double helix. RIGHT: A computer-generated space-filling model of DNA.

biological role of DNA. Prior to 1953, no one had been able to come up with a reasonable scheme for how a molecule might be *replicated*. But the structure of DNA itself contained an obvious answer to the riddle.

Each strand is a *complementary* "copy" (not an *identical* copy) of the other, which means that each contains the "information" required to reproduce the other strand. All that is required to copy the molecule is to separate the two strands and then to form a new strand complementary to each original one by using the base-pairing rules suggested by Watson and Crick. James Watson and Francis Crick (Fig. 20.7) had put together a three-dimensional model that accounts for the biological properties of DNA (Fig. 20.8; see Theory in Action, The Prize).

DNA REPLICATION

The DNA molecule, as we see, suggests a method for its own replication. Watson and Crick pointed this out in a paper published shortly after the announcement of the double-helix structure.

Figure 20.9 illustrates how the double helix is unwound to enable each strand to serve as the *template* for the synthesis of a new strand. The rules of complementary base pairing, linking C with G and A with T, help to control the process and ensure that each newly synthesized strand has the appropriate base sequence.

The replication process is **semiconservative,** meaning that the two original strands of the helix are separated

THEORY IN ACTION

The Prize

The X-ray pattern that provided the critical clues for Watson and Crick was made by Rosalind Franklin. Her paper, which contained the pattern, was actually published in the very same issue of *Nature* that contained the double-helix paper.

Watson and Crick have never denied that Franklin's pattern contained the important new information that made their model possible. But some writers have questioned whether their actions, interpreting the data after an advance look, were ethical. One of Franklin's biographers has argued that Watson and Crick deprived Franklin of the credit she deserved for making one of the fundamental scientific discoveries of the twentieth century. In their defense, one might point out that the double-helix model was simply the solution to clues provided by the work of Griffith, Avery, Chargaff, Schrödinger, and

many others in addition to Franklin. Isaac Newton wrote that if he had seen farther than others, "it is because I have stood on the shoulders of giants." Ultimately, Franklin, Crick, and Watson, like Isaac Newton, stood on those shoulders, too.

Unfortunately, Rosalind Franklin lived long enough to see only the first hint of how important her work had been. She died of cancer in 1957, just as DNA was beginning to become a central focus in biological research. A few years later, the Nobel Prize committee decided that the time was right to award the prize to the group that had developed the double-helix model. The Nobel Prize is never given posthumously, and it is never divided into more than three parts. It was awarded to Watson, Crick, and Franklin's associate, Maurice Wilkins.

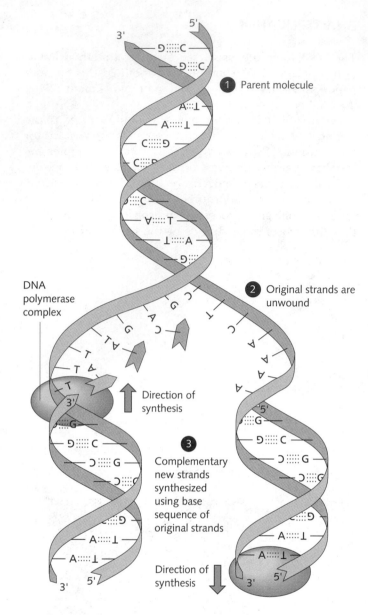

1. Parent molecule

2. Original strands are unwound

DNA polymerase complex

Direction of synthesis

3. Complementary new strands synthesized using base sequence of original strands

Direction of synthesis

Figure 20.9 Because each strand of the DNA molecule is complementary to the other strand, each may serve as a *template* against which a new strand may be constructed. DNA replicates in semiconservative fashion. Each strand of the helix serves as a template against which a new strand is assembled, following base-pairing rules. DNA replication is carried out by DNA polymerase, a complex enzyme.

and that at the end of replication, each strand is paired with one of the newly synthesized strands. In this way, the replication process produces two identical DNA molecules, each composed of one "old" strand and one "new" strand.

DNA Polymerase

It was soon discovered that DNA does not replicate in isolation but, rather, requires a number of special enzymes to unwind the double helix and to synthesize new DNA strands. The most important enzyme in this process is known as **DNA polymerase.** The action of DNA polymerase is illustrated in Fig. 20.9. DNA polymerase contains a binding site for attachment to the DNA strand and also a binding site for nucleotides. The nucleotides that attach to the enzyme are *triphosphates:* like ATP, they have three phosphate groups attached to them. DNA polymerase binds the correct nucleotide to the growing DNA strand and uses the energy from splitting off two of the three phosphates to form a covalent bond linking the nucleotide to the growing chain.

As DNA polymerase moves along the chain to attach the next molecule, part of the enzyme "proofreads" the work it has just done by checking the nucleotide pair to ensure that the proper base pairing has taken place. If an incorrect nucleotide has been inserted, this portion of the enzyme swiftly removes the nucleotide from the chain, and the molecule starts work over again.

This intricate process of base pairing, covalent bond formation, and proofreading is at the heart of DNA replication in all organisms. In tiny bacteria, a few DNA polymerase molecules may work to replicate the single DNA molecule that contains all of the cell's genetic information. In larger organisms, thousands of DNA polymerase molecules may work at scattered sites throughout many chromosomes to complete DNA replication in time for cell division to begin.

THE FLOW OF INFORMATION

DNA as a Coding Molecule

Although the double-helix model makes it clear that the sequence of bases in DNA can be copied, it does not explain how DNA can encode information that is important to the cell or how DNA actually works on a day-to-day basis. What, exactly, does DNA do? A quick answer is that *DNA directs the synthesis of proteins.* Why proteins? On a biochemical level, the reason for emphasizing proteins is very simple. The formation of most other biological molecules, and therefore most of the cell, is catalyzed by

Stutters, Stumbles, and RNA

As shown in Fig. 20.9, DNA replication looks pretty simple. The DNA double helix separates, and DNA polymerase makes a complementary copy of each strand, right?

If only life were that simple. For openers, DNA polymerase can actually go to work only after RNA polymerase makes a small piece of RNA to serve as a "primer" for the new strand. In addition, DNA replication requires a small mob of proteins and cofactors that do things such as unwind the DNA, smooth out kinks and twists, and keep single strands separated. However, the most fundamental complication is something that should be apparent if you look closely at Fig. 20.9. DNA polymerase works only in one direction, producing a new strand of DNA by adding bases onto its 3' end. That means, of course, that it must work in a different direction on each strand of the double helix. And there's the rub.

When replication begins, the two strands of the double helix separate and DNA polymerase complexes go to work on each of the two strands:

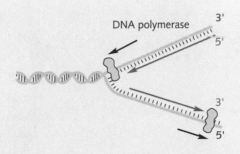

DNA polymerase

For the DNA complex working on the upper strand, things are simple. The double helix continues to separate and it continues to replicate. However, the one on the lower strand is at the end of the line. As the helix opens up, a new complex must form upstream of the old one and begin to make another piece of DNA.

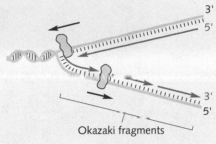

Okazaki fragments

Before long, the new DNA on the lower strand consists of a series of pieces known as **Okazaki fragments,** named after their Japanese discoverer. What happens to the fragments? An enzyme called **DNA ligase** seals the pieces together, completing replication. DNA polymerase is a remarkable enzyme, but like any superstar, it doesn't work alone.

---

proteins. Enzymes, nearly all of which are proteins, are responsible for the synthesis of nearly all other macromolecules. If DNA codes for proteins, then the proteins are capable of making everything else.

These discoveries and others laid the foundation for what is generally called **molecular biology,** the study of biology at the level of the molecule. Once the role of DNA was understood, "molecular biology" began to take on a special meaning as it focused on the structure and function of nucleic acids.

Nucleic Acid Templates

A **template** is a mold or model by which something is constructed. A cookie cutter acts as a template, allowing the baker to cut consistent cookie shapes from rolled out dough. A stencil sheet serves as a template for producing lettering. Nucleic acids can act as templates, too, and we

have already seen one example of this: DNA replication. The source of this ability is the *base-pairing* mechanism proposed by Watson and Crick. The double-stranded DNA molecule is held together by a series of relatively weak hydrogen bonds. The weakness of these bonds is an important feature of the DNA molecule, because it enables the two strands to be separated when the molecule is replicated.

A single nucleic acid strand contains a sequence of bases that are ready to form hydrogen bonds with any other molecule that will fit. In molecular terms, that strand forms a surface against which new molecules can be placed and to which they will bind if the molecules "fit." This means that one nucleic acid molecule can serve as a template against which another can be assembled. It also means that *information* in the form of a sequence of bases can be copied from one molecule to another by using the existing molecule as a template.

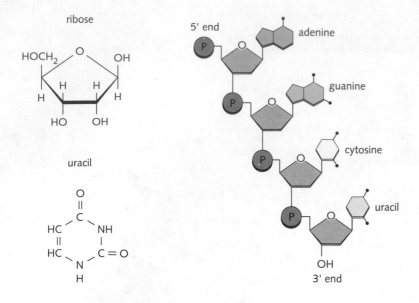

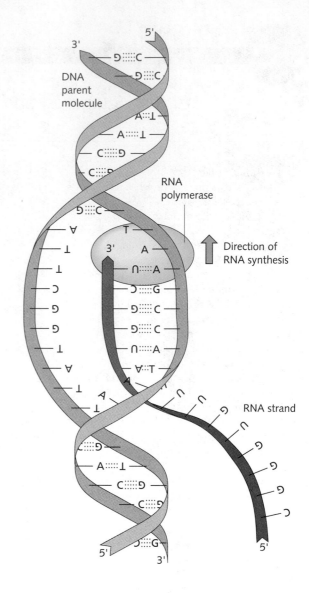

Figure 20.10 The structure of RNA. There are two important differences between RNA and DNA: RNA contains *ribose* (in place of *deoxyribose*) and employs the nitrogenous base *uracil* (in place of *thymine*). RIGHT: Under the same base-pairing rules as DNA, a complementary RNA strand may be made by the enzyme RNA polymerase.

The Role of RNA

We have said that the sequence of bases in DNA directs the synthesis of proteins. *How* does DNA do this? To begin with, DNA does not directly determine the amino acid sequence of a protein. Instead, the cell makes a complementary copy of one strand of a DNA molecule by synthesizing a single-stranded molecule of **ribonucleic acid (RNA).** RNA differs from DNA in two important respects (Fig. 20.10):

1. It uses a different sugar in its backbone (ribose instead of deoxyribose).
2. The four bases found in RNA are adenine, cytosine, guanine, and **uracil.**

Thymine, which is present in DNA, is replaced in RNA by uracil. (Scientists are divided about *why* RNA and DNA should differ in one of their bases. One theory is that the very slight difference between the shapes of the two bases makes it easier for enzymes to distinguish RNA from DNA.)

The process of RNA synthesis is known as **transcription.** An enzyme known as **RNA polymerase** synthesizes RNA as a complement to one strand of a double-stranded DNA molecule (see Fig. 20.10). What does RNA do? The simple answer is that many RNA molecules are like working copies of a master blueprint. Like the plans on a blueprint, the sequence of bases in RNA serves as a set of directions for the construction of a protein. The detailed answer is a little bit more complicated.

PROTEIN SYNTHESIS

We can think of the instructions for protein synthesis as a "message" that is written in RNA, a set of specifications to the cellular machinery that builds proteins. The language of that message is known as the **genetic code**.

Twenty different amino acids are commonly used to construct the proteins in living cells. How can a code that consists of only 4 different bases in an mRNA strand specify 20 different components? If each base stood for a single amino acid, only 4 different amino acids could be specified. If *pairs* of bases were used to stand for amino acids, only 16 different amino acids could be represented (4 × 4 = 16 different combinations). But if 3 bases were used for each amino acid, we would have more than enough coding capacity: 4 × 4 × 4 = 64 different combinations. Remarkably, this little bit of numerical reasoning is borne out in the lab.

The genetic code is a *triplet* system in which combinations of three bases are "read" by the protein-synthesizing machinery, and one amino acid is brought into position according to each *three-base triplet,* or **codon** (Fig. 20.11). In the early 1960s biologists began to unravel the genetic code. One codon at a time, they learned what each three-base combination "stood for." Today we can represent these data in a simple chart that shows the precise amino acid specified by each codon (Fig. 20.12). You will note that there are three codons that don't stand for *any* amino acid. They are **stop codons,** and their occurrence signifies the end of a polypeptide chain.

Most amino acids can be specified by more than one codon, and this gives the genetic code a great deal of redundancy. For example, there are six different codons for the amino acid *arginine* and four for *proline.*

Decoding the Message

Because the RNA molecule containing instructions for protein synthesis can be thought of as a "message," it is called **messenger RNA (mRNA).** What kind of "machine" does the cell need to follow the instructions carried in mRNA? First, the machine would have to find the proper end of the message (that's the 5' end) and the right place to start. Then the machine would have to "read" the message, one codon at a time. Reading is more complicated than one might think. For example, when the codon CGG appeared, the machine would have to recognize that "CGG" specifies the amino acid "arginine." Then it would have to find an arginine, attach it to the polypeptide chain by means of a peptide bond, and then move on to the next codon. When the machine encountered a "stop" codon, it would release the finished polypeptide.

Translating a 3-letter code in English

T h e d o g s a w h e r r u n

The | dog | saw | her | run

Step 1: Divide message into 3-letter words.

Step 2: Decoding system (human mind) interprets each word.

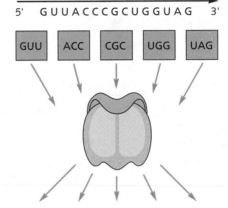

Translating a 3-letter code in RNA

5' GUUACCCGCUGGUAG 3'

GUU | ACC | CGC | UGG | UAG

Step 1: Divide message into 3-letter codons.

Step 2: Decoding system (ribosome and transfer RNA) interprets each codon as an amino acid.

valine threonine arginine tryptophan STOP

Figure 20.11 A message written in English can be decoded into three-letter groups that make up a five-word sentence. Key steps in the process are (1) to divide the message into three-letter words and (2) to decode the message using a system that assigns meaning to each word (the human mind). A message written in RNA can be decoded by a similar two-step process: (1) the message is divided into three-letter codons, and (2) a decoding system interprets each codon as an amino acid or a "stop" instruction.

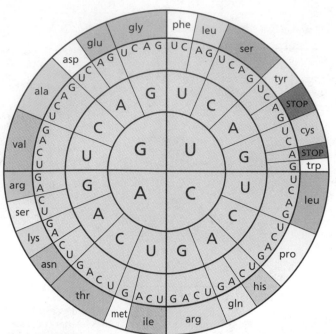

Figure 20.12

A convenient representation of the genetic code. The innermost circle of this chart represents the first base in a codon, the next circle the second base, and the third circle the third and final base. The outermost circle lists the amino acid specified by the codon. For example, to translate the codon AUG, start at the middle of the circle. Because the first base is an *A,* move a pointer to the *A* in the first circle. From there move the pointer to the *U* in the second circle, and then to the *G* in the third circle. That *G* borders *MET* in the outer circle, indicating that the codon AUG specifies methionine. Note that three codes (UAA, UAG, and UGA) are "stop" codons, meaning that they specify the termination of a polypeptide chain.

The minimum requirements for such a machine are shown in Fig. 20.13. It would have to hold on to an existing polypeptide chain, recognize the next codon in mRNA, find the correct amino acid called for by that codon, attach the amino acid to the chain by forming a new peptide bond, and then move ahead three bases to read the next codon.

Is there a single cellular machine that can do all these things? Not exactly. The job is divided up into several specialized parts. Holding the existing chain, moving along the message, and forming the peptide bond are carried out by structures called ribosomes. Recognizing the codon and bringing the appropriate amino acid into position are carried out by a specialized class of RNA known as transfer RNA. To fully understand protein synthesis, we have to look closely at these two parts of the cellular machinery.

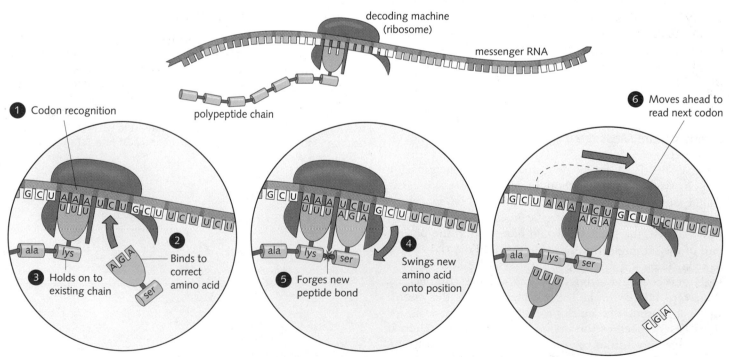

Figure 20.13 A *hypothetical* machine designed to decode an RNA message. The machine would have to have (1) a binding site for codon recognition, (2) a binding site for the correct amino acid for the codon, (3) a holding site attached to the existing polypeptide chain, (4) a mechanism to place the new amino acid in position and (5) forge a new peptide bond, and (6) a translocation system to move the machine along to the next codon.

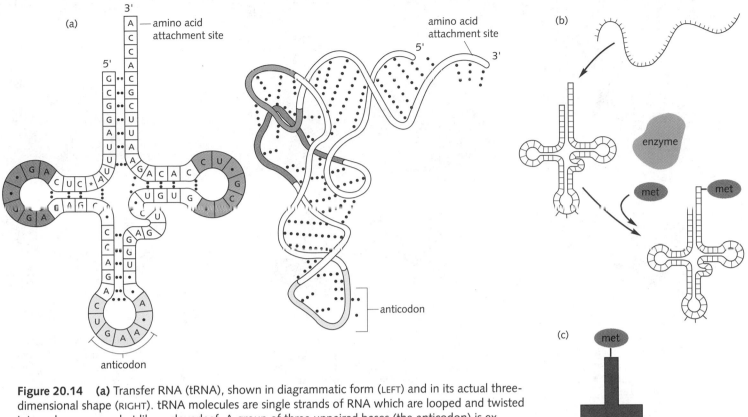

Figure 20.14 **(a)** Transfer RNA (tRNA), shown in diagrammatic form (LEFT) and in its actual three-dimensional shape (RIGHT). tRNA molecules are single strands of RNA which are looped and twisted into a shape somewhat like a cloverleaf. A group of three unpaired bases (the anticodon) is exposed at one end of the molecule. An amino acid is covalently joined at the attachment site, as noted. **(b)** tRNAs are synthesized as single strands that fold into their characteristic three-dimensional shape, and then are joined to the appropriate amino acid by a special enzyme. **(c)** A simplified representation of a tRNA molecule with an attached amino acid (met).

Transfer RNA

The job of **transfer RNA (tRNA)** is quite literally to "read" the codons in mRNA and ensure that the proper amino acids are brought into position. Each tRNA molecule (Fig. 20.14) contains an **anticodon**, a string of three bases that is complementary to a codon. At the other end, a special enzyme has attached the proper amino acid. Therefore, the tRNA molecules can line up the proper amino acids simply by base-pairing their anticodons to the appropriate codons. We can think of each tRNA as an adaptor that uses the base-pairing mechanism to bring the proper amino acid into position to incorporate into a growing polypeptide.

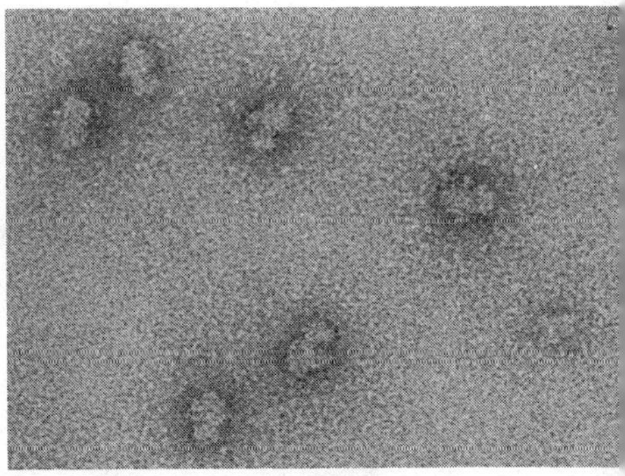

Figure 20.15 TOP: Electron micrograph of ribosomes. BOTTOM: Ribosomes consist of large and small subunits, each of which contains RNA and proteins.

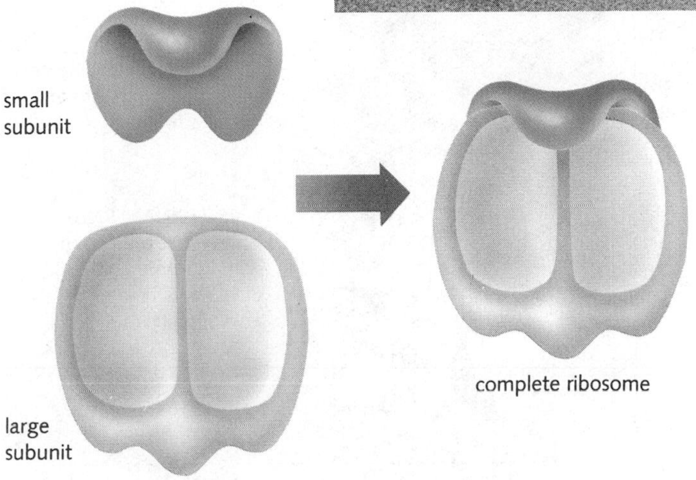

small subunit

large subunit

complete ribosome

The Ribosome

Protein synthesis also requires the presence of a **ribosome** (Fig. 20.15). *Ribosomes* are structures about 250 Å in diameter—large enough to be seen in the electron microscope. Each ribosome is composed of two subunits (one large, one small), and each of those subunits is composed of a large number of ribosomal proteins and from one to

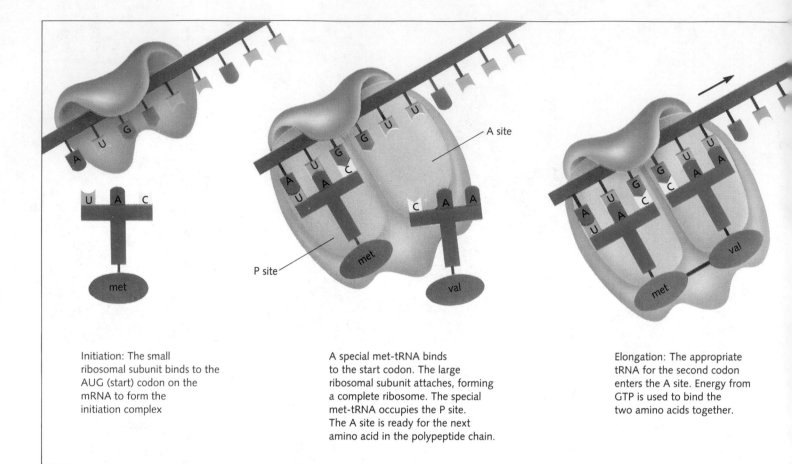

Initiation: The small ribosomal subunit binds to the AUG (start) codon on the mRNA to form the initiation complex

A special met-tRNA binds to the start codon. The large ribosomal subunit attaches, forming a complete ribosome. The special met-tRNA occupies the P site. The A site is ready for the next amino acid in the polypeptide chain.

Elongation: The appropriate tRNA for the second codon enters the A site. Energy from GTP is used to bind the two amino acids together.

Figure 20.16 Protein synthesis takes place on ribosomes as a multistep process. The two essential processes are shown "in brief" (BOTTOM RIGHT) to emphasize the different roles of tRNA and ribosomes in protein synthesis.

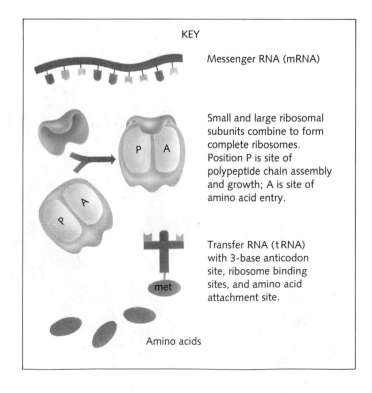

KEY

Messenger RNA (mRNA)

Small and large ribosomal subunits combine to form complete ribosomes. Position P is site of polypeptide chain assembly and growth; A is site of amino acid entry.

Transfer RNA (tRNA) with 3-base anticodon site, ribosome binding sites, and amino acid attachment site.

Amino acids

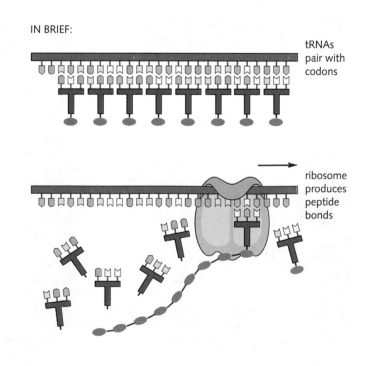

IN BRIEF:

tRNAs pair with codons

ribosome produces peptide bonds

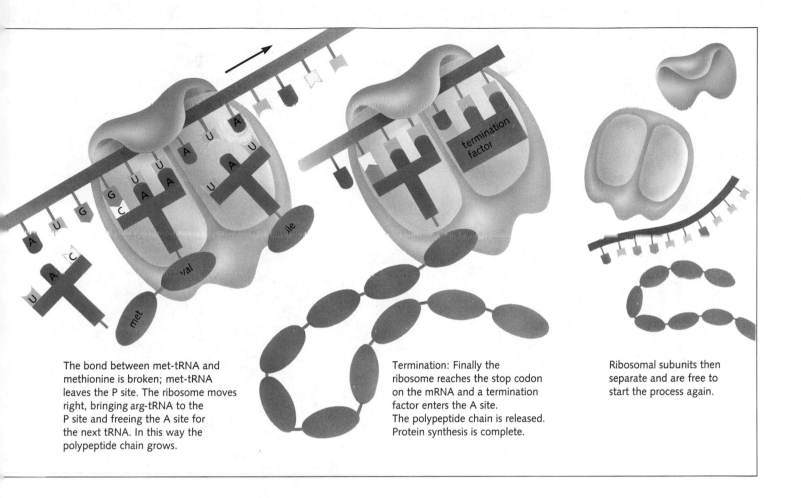

The bond between met-tRNA and methionine is broken; met-tRNA leaves the P site. The ribosome moves right, bringing arg-tRNA to the P site and freeing the A site for the next tRNA. In this way the polypeptide chain grows.

Termination: Finally the ribosome reaches the stop codon on the mRNA and a termination factor enters the A site.
The polypeptide chain is released. Protein synthesis is complete.

Ribosomal subunits then separate and are free to start the process again.

three RNA molecules. The RNA that makes up the ribosome is called the **ribosomal RNA (rRNA).** The function of rRNA is not completely clear, although there are now some indications that it plays an important role in producing the polypeptide chain (see Current Controversies, The Soul of a Small Machine, p. 415).

The process of protein synthesis is known as **translation.** Protein synthesis occurs in a number of distinct steps, which are listed below and diagramed in Fig. 20.16.

A messenger RNA molecule is released into the cytoplasm. Near one end of the mRNA molecule, there is a codon with the base sequence *AUG*. In nearly all mRNA molecules, the *start signal* is *AUG,* which codes for the amino acid *methionine.*

The *small-ribosomal subunit* (along with several soluble proteins called *initiation factors*) recognizes the end of the mRNA and binds to the AUG codon to form an *initiation complex.* A special version of met-tRNA (tRNA with the amino acid methionine attached to it) enters the small subunit and binds to the start codon, and then the *large subunit* attaches, forming a complete ribosome.

As far as we can tell, the first tRNA molecule occupies a special site inside the ribosome called the *P site.* ("P" stands for polypeptide, because it is at this site that the growing *polypeptide* chain is later found.) When the right tRNA for the second codon enters the ribosome at the *A site* (because the *amino acids* enter there), we have the conditions necessary to begin building a polypeptide.

Now part of the ribosome acts like an enzyme, forming a peptide bond between methionine and the second amino acid—in this case, valine. The energy to form the bond comes from the hydrolysis of GTP (guanidine triphosphate, a molecule very similar to ATP). The ribosome briefly contains a dipeptide in the A site.

Next, the tRNA that has lost its amino acid leaves the P site, and the ribosome moves relative to the mRNA so that the second tRNA containing the dipeptide now occupies the P site. The ribosome is now ready for a third tRNA to enter at the A site, allowing the growth of the polypeptide chain to continue.

Finally, after 50, 60, or maybe 300 amino acids have been added, the mRNA reaches the point of its stop

Figure 20.17 Summary of protein synthesis.

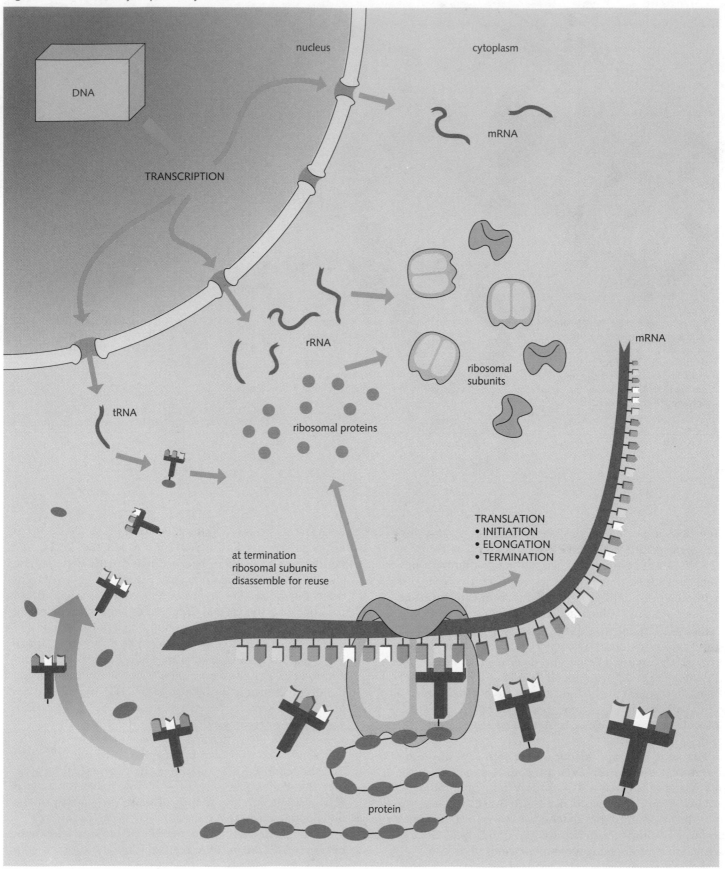

The Soul of a Small Machine

The machine that produces peptide bonds is the ribosome. Like any complex machine, the ribosome must perform a wide variety of functions: it must contain binding sites for tRNAs, for the growing polypeptide chain, and for mRNA; and deep inside, the ribosome must contain the enzymatic machinery for joining each new amino acid to the growing polypeptide chain.

What kind of enzyme produces the peptide bond? A eukaryotic ribosome contains more than 80 ribosomal proteins. Therefore, it was logical to assume that the enzymatic activity was found in a cluster of these proteins. It now looks as though that assumption was not correct.

In 1992, Harry Noller and his coworkers (University of California at Santa Cruz) carried out a series of experiments in which every protein was

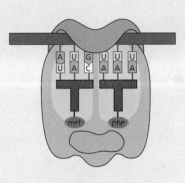

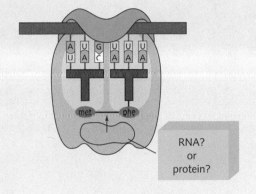

carefully extracted from a preparation of ribosomes. Incredibly, the ribosomal RNA (rRNA) that remained still retained the ability to catalyze peptide bond formation. The logical conclusion is that peptide bond formation is catalyzed by an RNA enzyme, a ribozyme, that is part of rRNA.

The classic view of ribosomes was that rRNA was just a scaffold upon which the really interesting molecules (the proteins) were arranged. That view is now completely changed, and molecular biologists are one key step closer to understanding this marvelous little machine.

---

codon. As we have seen, there is no tRNA or amino acid corresponding to the stop sequences. Instead, when the stop codon takes up a position within the A site, special molecules known as *termination factors* enter the A site and cause the release of the polypeptide chain. Synthesis of the protein is complete.

At termination, not only is the chain released, but also the ribosome actually falls apart into large and small subunits. These are released into the cytoplasm and are free to start the process all over again.

The role of the ribosome is to organize the process of protein synthesis: from the recognition of the start codon to the formation of every peptide bond to the termination of the chain (Fig. 20.17). The ribosome is really a small factory with a host of different enzymatic functions directed towards the synthesis of a finished product, the protein (see Theory in Action, Seeing Genes at Work, p. 421).

The flow of information from DNA through RNA to protein can be summarized in a simple diagram:

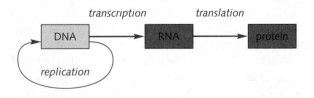

Among molecular biologists, this pattern of information flow has become known as the "central dogma." The word *dogma* is used almost as a joke, because science must never be dogmatic. As we will see shortly, the central dogma is not universally correct.

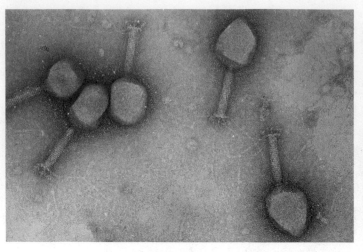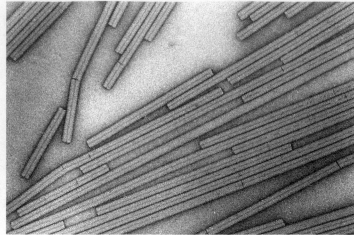

Figure 20.18 Electron micrographs of (LEFT) bacteriophage T4 and (RIGHT) tobacco mosaic virus.

THE VIRUSES: A LESSON IN INFORMATION TRANSFER

Our short history of the birth of molecular biology has been, out of necessity, incomplete. Until now we have left out one branch of biology that was very important in the identification of DNA as the molecule of heredity: the study of viruses. The viruses have taught us some important lessons about the role of DNA in living cells.

The word *virus* means "poison" in Latin, and for quite a few years the term was applied to any organism that could cause disease. Viruses as we understand them today were discovered in the early 1900s by the Russian biologist Dimitri Iwanowski.

Iwanowski was interested in a disease that had ruined thousands of acres of tobacco plants. The leaves of the infected plants were covered with large yellow spots, forming a pattern that the farmers called a "mosaic." Searching for an infectious organism, Iwanowski passed juice from infected plants through a filter so fine that no cell could pass through. Nonetheless, the filtered juice *still* passed on mosaic infection to other plants! He concluded that an infectious organism so small that it could pass through the finest filter must be causing the disease.

Today we call such tiny, subcellular particles *viruses,* and the organism that Iwanowski encountered is called the *tobacco mosaic virus (TMV).* **Viruses** are subcellular particles that are composed of nucleic acid and protein and that can infect a living cell. Viruses exist that can infect virtually any type of cell: animal or plant, large or small,

prokaryote or eukaryote. The electron microscope has made viruses "visible" to us, and these small particles often have exotic and beautiful structures (Fig. 20.18).

Bacteria Eaters

One important class of viruses attacks bacteria. They are known as **bacteriophage,** which means "*bacteria eaters.*" We will take a close look at the life cycle of just one bacteriophage, a virus with the scientific name *T2,* and we will glance briefly at bacteriophage T4.

Bacteriophage T2 contains a single long DNA molecule enclosed in a protein coat (Fig. 20.19). Left to itself, a single virus particle is almost *inert;* it does not grow or eat or metabolize or reproduce. But when a T2 particle makes contact with the right kind of bacterium, all of that changes. First, the tail fibers of the protein capsule make contact with the cell wall of the bacterium. The virus changes its shape a bit when it contacts the cell wall, and the genetic material of the virus is injected into the bacterium.

In 1952, Alfred Hershey and Martha Chase realized that this system could be used to determine the chemical nature of the infectious viral genetic material. They prepared particles of bacteriophage T2 that contained two radioactive labels: their DNA was made radioactive with ^{32}P, and their proteins with ^{35}S. The scientists mixed these labeled viruses with bacteria and waited a few minutes for the virus particles to attach to the bacteria and begin the process of infection. After infection, they analyzed the pattern of radioactivity in the bacteria. Hershey and Chase found that most of the ^{35}S label was still with the virus particles, whereas the ^{32}P label had been inserted into the bacteria. The virus particles that eventually burst from the infected bacteria also contained the ^{32}P label in their DNA (Fig. 20.20).

This simple experiment showed that the DNA carried the genetic material necessary to direct the production of virus particles during an infection and furnished powerful proof that the basic molecule of inheritance was DNA.

Once inside a cell, the viral DNA goes to work and, just like a rebel army, begins to take over the cell. To the molecules of the host cell, the viral genes seem no different from the cell's own. A number of viral mRNAs are

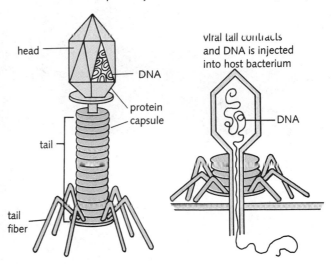

Figure 20.19 Bacteriophage T2. The DNA core of the virus is surrounded by an intricate protein coat (LEFT) that protects the nucleic acid and helps to inject it into a bacterial cell (RIGHT).

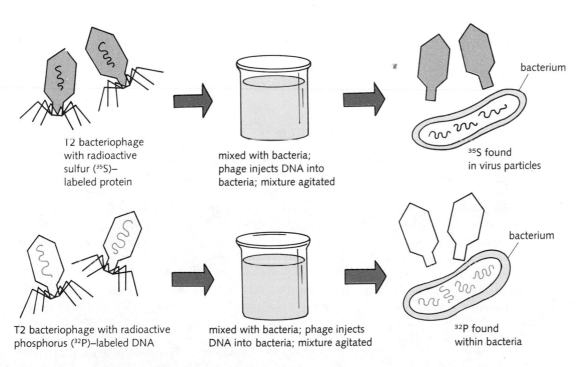

Figure 20.20 The Hershey–Chase experiments. In order to determine whether DNA or protein carried viral genetic information, Alfred Hershey and Martha Chase labeled viral DNA with ^{32}P and viral protein with ^{35}S. By shearing virus particles off the surface of bacteria, they discovered that ^{32}P (DNA) entered the infected cell.

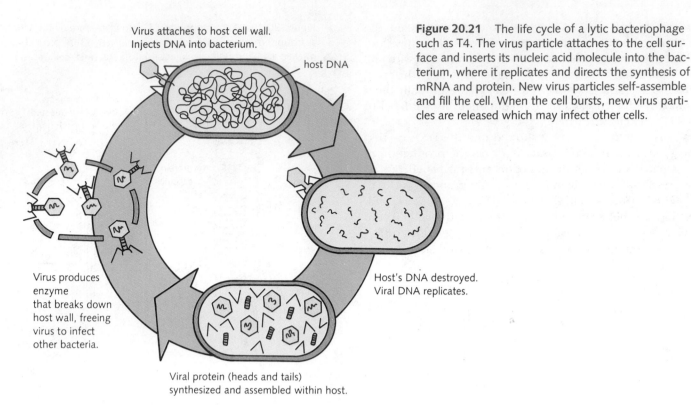

Virus attaches to host cell wall.
Injects DNA into bacterium.

host DNA

Host's DNA destroyed.
Viral DNA replicates.

Virus produces
enzyme
that breaks down
host wall, freeing
virus to infect
other bacteria.

Viral protein (heads and tails)
synthesized and assembled within host.

Figure 20.21 The life cycle of a lytic bacteriophage such as T4. The virus particle attaches to the cell surface and inserts its nucleic acid molecule into the bacterium, where it replicates and directs the synthesis of mRNA and protein. New virus particles self-assemble and fill the cell. When the cell bursts, new virus particles are released which may infect other cells.

made and translated into proteins, which begin to shut down the bacterium's own genetic machinery. Then the viral DNA begins to replicate, again using the infected cell's own enzymes, and before long the cell is filled with hundreds of copies of viral DNA.

In the late stages of a T4 infection, genes on the viral DNA molecules coding for the proteins that make up the coat of the virus are activated. As the cell manufactures and accumulates these proteins, complete virus particles begin to assemble inside the cell, until the cell is filled with them. Finally, the cell bursts, releasing hundreds of virus particles into the surrounding medium. The whole process takes only about 40 minutes, and all of it is the work of a single DNA molecule from the virus (Fig. 20.21).

The power of a single DNA molecule to be capable of destroying a healthy cell tells us quite a bit about the role of DNA as the central molecule in the process of heredity. The viral DNA, after all, contains the genetic information of the virus—information that is capable of shutting down the host cell, replicating its own DNA, and producing new copies of the virus particle, coat proteins and all, that are ready to infect other cells.

Other Viruses

Not all viruses work in the same way as bacteriophage. A great many viruses do not destroy the cells they infect but, rather, grow inside them and "bud" new virus particles off from the surface of the cell. A few others insert their DNA into the DNA of the host cell, and for a time the viral genes are actually carried in the genome of the infected cell. In addition, not all viruses contain DNA. An important group of viruses contain RNA as their genetic information. At first, this puzzled investigators, and many biologists wondered whether these RNA viruses worked by a completely different genetic mechanism. In a sense, they do.

In 1972, biologist Howard Temin suggested that there might be an enzyme, which he termed **reverse transcriptase,** that could synthesize DNA from an RNA template. Temin thought that some RNA viruses might first make a DNA transcript that could then function as an ordinary piece of DNA. Only a few years passed before he was proved right and the enzyme that he postulated was discovered. These RNA-containing viruses are now called **retroviruses,** because they seem to work in a *retro* grade (backward) fashion; that is, they make a DNA molecule

Figure 20.22 Electron micrograph of an *E. coli* chromosome. The great length of the single, circular DNA molecule in this bacterium is apparent and is attached to fragments of the cell membrane.

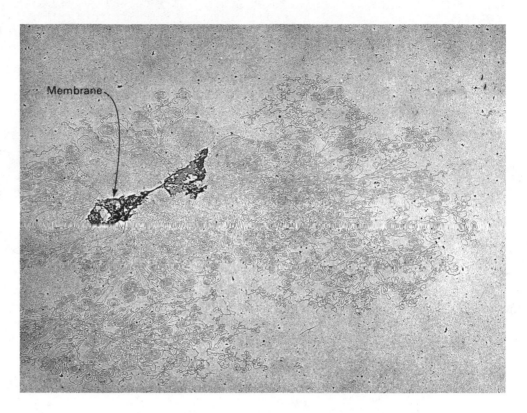

Membrane

from RNA. The DNA molecules that are synthesized by retroviruses are sometimes used to produce messenger RNAs that help the virus replicate and produce new virus particles. Sometimes these DNA molecules are integrated into the chromosomes of the infected cells. Many of the viruses that cause tumors are retroviruses, as is the AIDS virus. In Chapters 41 and 42 we will examine the biology of retroviruses and their involvement in cancer and AIDS, respectively.

CHROMOSOMES AND DNA

One question central to the study of cell biology is how DNA molecules are organized within a cell. The answer to this question depends on whether the organism involved is a prokaryote or a eukaryote.

Prokaryotic Chromosomes

The DNA molecules in prokaryotes are free in the cytoplasm. Originally, the term *chromosome* was applied only to eukaryotes, because only eukaryotic chromosomes are large enough to be visualized in the light microscope. More recently, however, the term has also been applied to the DNA molecules found in prokaryotic cells. If a prokaryotic cell is broken open, it is possible to visualize the chromosome with the aid of an electron microscope (Fig. 20.22). The chromosomes of most prokaryotes are single, circular DNA molecules. In each cell, there is generally only a single chromosome that contains the organism's complete genome.

There's an important exception to this rule, however. A great many prokaryotes contain additional "minichromosomes," or **plasmids.** A cell may contain as many as 30 plasmids, which are generally identical copies of the same short DNA segment. Plasmids contain special DNA sequences that ensure that they are replicated along with the rest of the cell's DNA. Occasionally, plasmids can insert themselves into the cell's chromosome.

Eukaryotic Chromosomes

As we have already seen, eukaryotic cells contain more than a single chromosome, and those chromosomes are enclosed within a special cellular organelle known as the *nucleus.*

Figure 20.23 **(a)** An electron micrograph of extracted chromatin reveals the *nucleosome* structures which are associated with DNA in most eukaryotic cells. Nucleosomes appear as "beads on a string" and are composed of histone proteins and DNA. **(b)** Diagram of the structure of a nucleosome, which contains two copies of each of four different histone proteins.

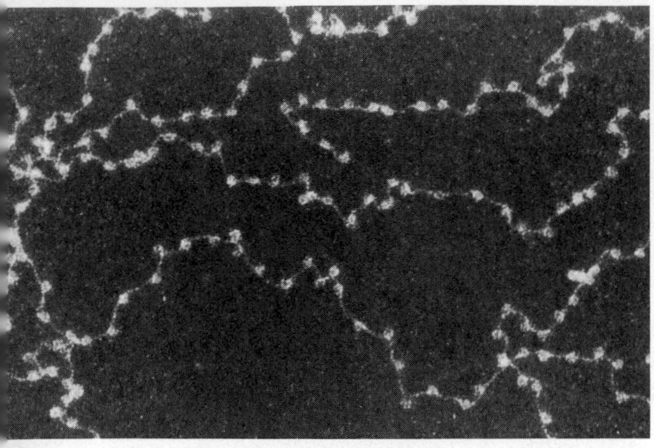

(a)

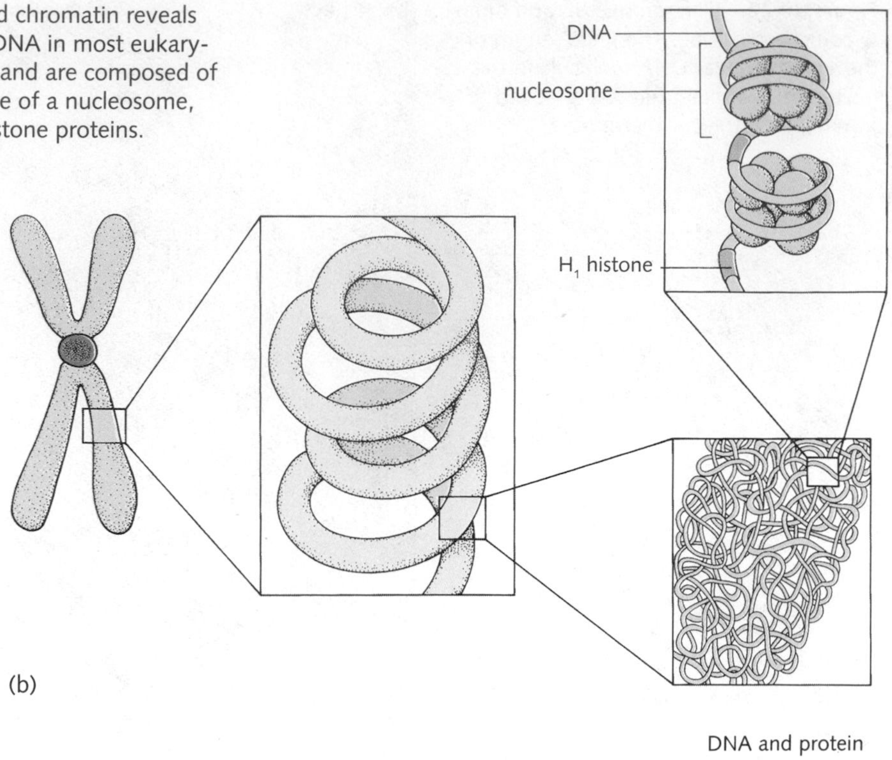

(b)

DNA

nucleosome

H₁ histone

DNA and protein

The DNA in a eukaryotic chromosome is tightly bound to protein, forming a material within the nucleus known as **chromatin.** There are two categories of proteins found in chromatin. **Histones** are proteins that contain large amounts of the extremely basic amino acids *lysine, arginine,* and *histidine. Nonhistone proteins* are distinguished only by the fact that they are not histones and bind very tightly to DNA.

If the nucleus of a single cell is gently disrupted and allowed to spill out on a thin carbon film, it is possible to stain or shadow the chromatin and examine it in the electron microscope. A typical chromatin preparation is shown in Fig. 20.23. The unraveling mass of material shows a distinct substructure that looks a bit like beads on a string. These beads, which are made up of DNA and histones, are known as **nucleosomes.**

Each nucleosome contains a strand of double-helical DNA wound twice around eight histone proteins. One of the reasons for the very basic nature of the histone proteins is now clear: the positive charges on the amino groups ($-NH_3^+$) found in lysine, arginine, and histidine help them bind to the DNA strands, which are negatively charged because of their phosphate groups.

What is the function of the nucleosome? Its primary task may be to manage the enormous mass of DNA by allowing it to fold into a chromosome (see Fig. 20.23). That's no small order. Each human cell has enough DNA to form a double helix about 10 *meters* in length, but it all has to fit within a nucleus that is generally smaller than 10 *micrometers* in diameter. This means that the DNA has to be folded by a factor of about 1 million just to fit into the nucleus, and the nucleosome is the first level at which that folding takes place.

There is some very recent evidence suggesting that the position of nucleosomes on some genes may be related to gene expression. At least in some cases, nucleosomes may be repositioned when a gene is turned on or turned off, although the full story is still being developed.

Nonhistone proteins do not form regular structures such as the nucleosome. The nonhistone chromosomal proteins are also much more diverse than the nucleosomes. They vary a great deal from one kind of cell to the next, and among organisms the variation is enormous. The nonhistone proteins include a great many DNA-binding proteins whose main function is the regulation of gene expression, a topic we will deal with next.

Seeing Genes at Work

Is a picture really worth a thousand words? In some cases, maybe a little more. In the case of molecular biology, sometimes a single picture makes everything clear. Oscar Miller, a scientist at the University of Virginia, has made an art of getting just the right picture. He has helped develop techniques to isolate DNA molecules gently enough so that they can be prepared for electron microscopy without breaking.

One of his most striking electron micrographs is shown here. It was made from nucleic acids isolated from a bacterium that was actively making protein. This one micrograph captures both transcription (RNA synthesis) and translation (protein synthesis). The picture also shows that prokaryotes carry out transcription and translation simultaneously: ribosomes actually begin to translate one end of an mRNA before the synthesis of the other end is complete. This would not be possible in eukaryotic cells, in which the nuclear envelope separates transcription (which occurs in the nucleus) from translation (which occurs in the cytoplasm).

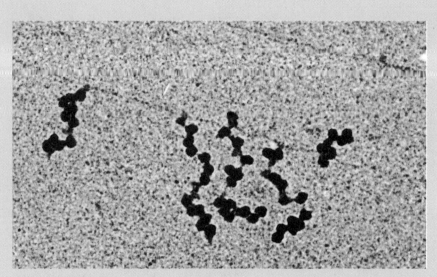

(Magnification factor: 172,000)

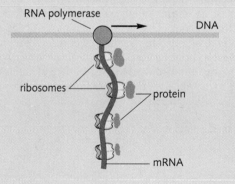

(a)

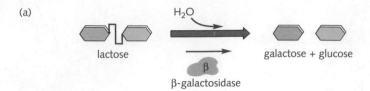

(b)

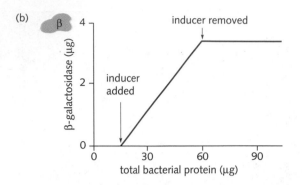

(c)

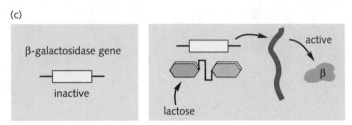

Figure 20.24 **(a)** β-galactosidase catalyzes the hydrolysis of lactose to produce galactose and glucose. This is the first step in the chemical pathways that utilize lactose as a source of food energy. **(b)** When lactose is added to a culture of *E. coli*, the amount of β-galactosidase in the cells quickly rises. This phenomenon is known as enzyme induction. **(c)** In the absence of lactose, the gene for β-galactosidase is inactive. By some mechanism, the presence of lactose activates the β-galactosidase gene.

THE CONTROL OF GENE EXPRESSION IN PROKARYOTES

Genetic studies tell us that even an organism as simple as *E. coli* has several thousand genes. It would be remarkably wasteful for each of these genes to be expressed at the same time. Instead, the products of most genes are adjusted relative to the need for them. One good example is found in the genes that enable *E. coli* to use lactose as food.

Lactose is a form of sugar found in milk, a food that frequently finds its way into *E. coli's* normal habitat, the human intestine. In order for *E. coli* to be able to utilize **lactose,** it must produce an enzyme that cuts this disaccharide into two simple sugars. The enzyme that does this is known as β-**galactosidase**.

When cells are grown on a medium that doesn't contain any lactose, they have no need of β-galactosidase. Not surprisingly, such cells have only very small amounts of the enzyme. However, when the same cells are transferred to a medium in which a lactose sugar is the only food source, the amount of the enzyme in the cells rises very quickly.

It seems almost as though the cell "knows" what kind of sugar is in the medium and then begins to make more of the appropriate enzyme. This phenomenon—a rapid increase in the amount of an enzyme in response to culture conditions—is known as **enzyme induction.** The substance that causes the increase (in this case, the sugar lactose) is known as the **inducer** of the enzyme. In effect, lactose turns on the β-galactosidase gene (Fig. 20.24). How does induction occur? How does the cell "know" that lactose is in the medium? The answers to these questions were provided in the late 1950s by two researchers at the Pasteur Institute in Paris, Francois Jacob and Jacques Monod. In a brilliant series of genetic studies, they proposed a scheme for the control of the gene. Their ideas were confirmed by a series of subsequent studies and now serve as a model for how many prokaryotic genes are regulated.

The *lac* Operon

The gene for β-galactosidase is found in a cluster of three genes, each of which is related to lactose metabolism: *z* is the gene that codes for β-galactosidase itself, and the *y* and *a* genes code for other enzymes that are also involved in lactose metabolism. Because these three genes code directly for the *structure* of proteins, they are known as *structural genes*. On one side of the three genes are two sections of DNA that are termed the **promoter (p)** and **operator (o)** sequences. The p and o regions are regulatory sequences that help control the expression of the three genes. The entire structure of regulatory and structural genes is known as an **operon** because the genes are "operated" together. Because these genes affect the metabolism of lactose, this region of the *E. coli* chromosome is known as the *lac* **operon** (Fig. 20.25).

The promoter (p) region is a segment of DNA to which RNA polymerase can bind. If the polymerase does bind to the p region, it can then move to the right until it encounters a special "start transcription" signal at the beginning of the z gene. The polymerase then produces a messenger RNA that contains a complementary copy of each of the three structural genes, and that mRNA can be

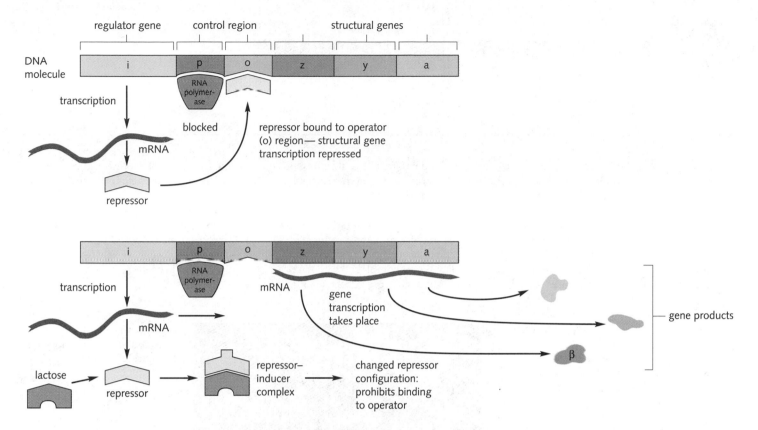

Figure 20.25 The *lac* operon. RNA polymerase binds to the promoter (p) region and transcribes an mRNA molecule that codes for the proteins specified by the *z, y,* and *a* genes (the *z* gene product is β-galactosidase). The repressor, a protein that binds to the operator (o) region, blocks transcription of the genes, preventing their expression. Lactose induces expression of the operon by binding to the repressor and unblocking the o site.

translated into proteins: one of them is β-galactosidase. If this is the case, what prevents the cell from filling up with the enzyme?

Another gene, known as the *i* gene, codes for a protein known as the ***lac* repressor** (*i*). The repressor is a DNA-binding protein, and it binds tightly to the operator (o) region of the gene. The effect of repressor binding is dramatic. Because RNA polymerase must pass through the o region to get to the structural genes, the presence of the repressor prevents the structural genes from being transcribed. In effect, the repressor turns off the operon—it represses it.

Now our problem is a bit different. With the repressor around, how does the cell make certain that β-galactosidase can be synthesized when it is needed? The repressor has *two* binding sites. One, of course, is for the o region of the operon. The other binding site is for lactose and similar sugars. When an inducer molecule such as lactose enters the cell, it binds to the repressor and causes a *conformational change* in the structure of the molecule.

The repressor–inducer complex is no longer able to bind to the operator, and it falls off the DNA back into solution. Now the operator region is open, and RNA polymerase is free to move from the promoter across it to allow gene transcription to take place.

The system regulates itself. When lactose is scarce, the repressor binds to the o region of the operon and very little of the enzyme is made. But when lactose is provided as the food source, it diffuses into the cell, binds to the repressor, and "turns on" the genes that enable the cell to use lactose for food. This automatic control system provides the cell with the right amount of lactose-metabolizing enzymes to suit the moment. The *lac* operon also has a positive control system that is sensitive to the overall energy needs of the cell and does not activate the operon if the cell has adequate supplies of other sugars for food, especially glucose.

Not all bacterial genes are like the *lac* operon. Some genes are expressed at low levels regardless of environmental stimuli. A few others are repressed, rather than

Grabbing DNA

In an important sense, DNA-binding proteins are the switches that turn genes on and off. Some of them, like the *lac* repressor, block RNA polymerase and prevent gene expression. Others, variously called "enhancers," "activators," or "transcription factors," bind to DNA, increase the efficiency of transcription, and turn gene expression on.

For many years, how these proteins bound to DNA was a mystery. The proteins themselves gave few clues as to how their presence on the DNA double helix could affect gene expression. That's no longer the case, largely because a number of laboratories have successfully crystallized DNA–protein complexes and determined their molecular structures using X-ray diffraction.

One such protein is known as CAP (catabolite gene activator protein). Under the proper conditions, CAP binds to the promoter region of the *lac*

operon and enhances transcription of the *lac* operon, resulting in gene activation. How does CAP turn the genes on?

Tom Steitz and his associates at Yale found that the CAP protein binds to the promoter region and produces a sharp bend in the DNA. The sharply bent DNA, now complexed with CAP, presents a much more attractive bind-

ing site for RNA polymerase and causes a dramatic increase in gene expression.

Do all DNA-binding proteins work like CAP? Not exactly. The study of these DNA–protein complexes is still in its infancy, but many proteins work differently. Given time, it's likely that a wide variety of activation mechanisms will turn up.

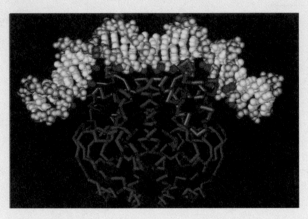

CAP–DNA complex. CAP binds to the specific DNA sequence of the *lac* promoter.

induced, when specific molecules from the environment are introduced. Other operons are regulated by positive control elements, which enhance transcription rather than restrict it. But the *lac* operon provides a general model for the manner in which a gene can be organized so that it responds to changes in the environment in specific ways.

GENE REGULATION IN EUKARYOTES

Gene regulation in eukaryotes is more complex than it is in prokaryotes. There are several reasons for this. First, even a simple eukaryote may possess more than 50,000 genes, making it especially important that genes be expressed only when they are needed. Second, the different cell types in complex organisms require vastly different patterns of gene expression. Cells as different as muscle cells and nerve cells must express very different groups of genes. Finally, the different cell types, which are produced during development in a complex organism are

themselves the product of differential gene expression. Development requires that gene expression be regulated in *time* and *space* (Fig. 20.26).

Regulation at Many Levels

Gene expression can be regulated at many different levels (Fig. 20.27). The *lac* operon was an example of transcriptional regulation. However, gene expression can also be controlled posttranscriptionally. Eukaryotic gene expression can be affected by the manner in which RNA molecules are processed before they leave the nucleus, by the stability of mRNAs in the cytoplasm, and by the rate at which mRNAs bind to ribosomes to begin protein synthesis. In fact, there are at least six different levels at which gene regulation can occur in eukaryotes (see Fig. 20.27). Which is the key? Every one. Researchers have found important examples of gene regulation that act at *each* of these levels.

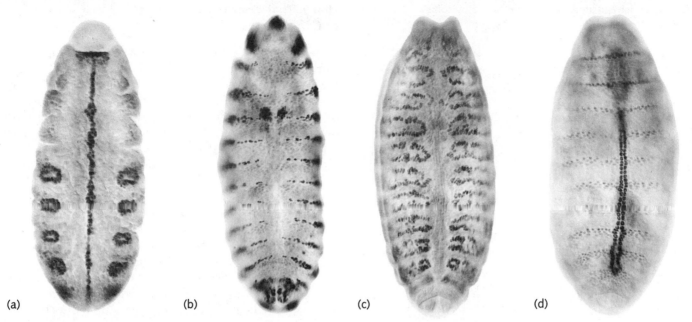

Figure 20.26 Position-sensitive expression of genes in the *Drosophila* embryo. A marker gene was placed in four different positions in the genome. Cells expressing the gene can then be stained. Depending on the control elements that surround the gene, it is expressed in **(a)** the trachea and tissues in posterior segments, **(b)** sensory organs, **(c)** muscle, or **(d)** the dorsal ectodermal stripe.

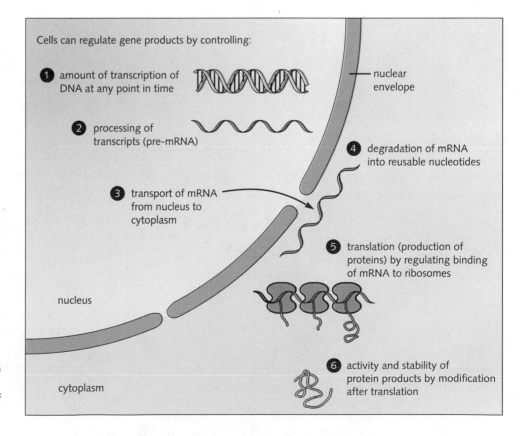

Figure 20.27 Gene expression can be regulated at many levels, including (1) transcription, (2) pre-mRNA processing, (3) transport of RNA from the nucleus, (4) stability of RNA in the cytoplasm, (5) translation, and (6) stability and activity of protein product.

Cells can regulate gene products by controlling:

1. amount of transcription of DNA at any point in time
2. processing of transcripts (pre-mRNA)
3. transport of mRNA from nucleus to cytoplasm
4. degradation of mRNA into reusable nucleotides
5. translation (production of proteins) by regulating binding of mRNA to ribosomes
6. activity and stability of protein products by modification after translation

nuclear envelope

nucleus

cytoplasm

Figure 20.28

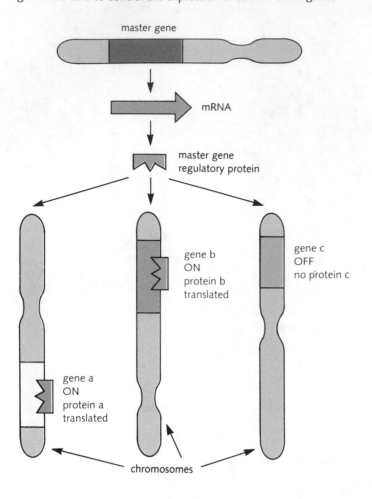

Figure 20.28 The regulatory proteins produced by some genes are able to control the expression of several other genes.

master gene

mRNA

master gene regulatory protein

gene b
ON
protein b
translated

gene c
OFF
no protein c

gene a
ON
protein a
translated

chromosomes

Genes That Control the Expression of Other Genes

The complexity of eukaryotic organisms requires that whole groups of genes be turned on and off in groups. We have already seen one example of this in the form of *X*-chromosome inactivation (Chapter 11). The genes in one of the two *X* chromosomes of a female mammal are inactivated as a group. Other genes may be inactivated by repressor molecules that bind to DNA and prevent transcription. In some cases, tightly condensed chromatin structures may prevent RNA polymerase from making contact with DNA, effectively turning off all of the genes within the condensed region.

Naturally, genes may be activated as well as repressed. Researchers have theorized for years that cells might contain "master genes" which activate clusters of genes that control complex developmental patterns (Fig. 20.28). It now seems that master genes do exist, and several such systems have been investigated. Immature muscle cells, for example, produce a regulatory protein known as *myoD1*. The expression of myoD1 activates a series of genes, which in turn produce a range of muscle-specific proteins required for further muscle cell development.

Researchers have shown that connective tissue cells (fibroblasts), which do not normally express muscle-specific genes, still have the capacity to respond to the myoD1 master gene. When pieces of DNA containing extra copies of the myoD1 gene were inserted into fibroblast cells, the cells began to produce muscle-specific proteins and to develop into muscle-like cells.

Intervening Sequences and the Control of RNA Processing

In the early 1970s, a number of techniques were developed that made it possible to isolate segments of DNA about the size of a single eukaryotic gene. One of the first experiments that several laboratories tried was to compare the DNA segment of a gene with the mRNA it produced. Naturally, scientists expected the DNA and the mRNA to be a perfect match in size. But a surprise was waiting. One of the first genes to be examined was the DNA segment that codes for *ovalbumin,* the major protein in the "white" of a chicken egg. The gene was discovered to be much larger than the mRNA. When a piece of single-stranded DNA containing the gene was allowed to base-pair with the ovalbumin mRNA, things got even stranger: the gene seemed to be filled with segments that had no matching sequence in the RNA.

The existence of **intervening sequences** in DNA, which had no corresponding segment in mRNA, was a complete surprise. And these **introns,** as they are now called, were found to be a common feature of eukaryotic

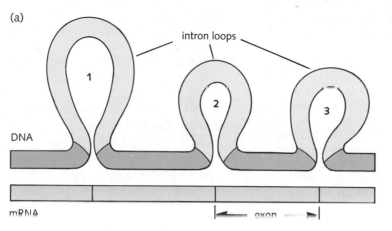

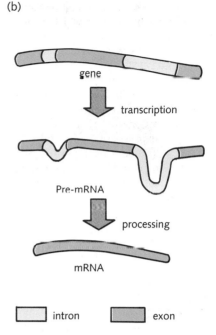

intron   exon

Figure 20.29 **(a)** Intervening sequences, or introns, were discovered when mRNA was hybridized to the DNA sequences from which it had been transcribed. Although complementary RNA/DNA sequences form a double-stranded hybrid, the regions of DNA that do not match the RNA sequence are visible as large loops. **(b)** Introns and exons are transcribed into a pre-mRNA. The introns are then removed as the pre-mRNA is processed into mature mRNA.

genes (although they are not universal). The regions that *are* complementary to mRNA are called **exons** (expressed sequences). What is the function of introns? We now know that most eukaryotic mRNAs are made as "pre-mRNAs" and then processed in the nucleus to remove introns and produce a mature mRNA molecule (Fig. 20.29).

It is also apparent that the splicing and processing of pre-mRNA is yet another level at which gene expression may be regulated. Several recent studies have shown that the same pre-mRNA molecules may be spliced in different ways, producing different mRNAs in different tissues. This makes it possible for a single gene to produce different gene products in the different tissues.

Genome Structure

In prokaryotes, closely related genes are often found together, like those of the *lac* operon, which allows them to be regulated as a group. While this is sometimes true for eukaryotes, it is quite common for related genes to be scattered in different locations around the genome, often on different chromosomes. Scattered genes can still be regulated as a group if they all contain similar regulatory elements that can respond to the same transcription factors, and this is often the case.

Eukaryotic genes are sometimes present in multiple copies. The genes for ribosomal RNA and the histone

proteins may be repeated hundreds or even thousands of times, making it possible for cells to produce large amounts of these products in a short period of time. Other genes, including those for contractile and cytoskeletal proteins, are present as members of *multigene families,* closely related groups of genes that serve similar purposes.

SUMMARY

The simultaneous development of biochemistry and genetics led to an expectation that genetics should have a molecular basis. Frederick Griffith discovered that an extract of killed bacterial cells was able to transform a strain of harmless bacteria into pathogenic organisms. Oswald Avery and his co-workers examined the transforming extract and showed that the transforming molecule was DNA, suggesting that DNA was the molecule of genetics. The structure of DNA was determined by James Watson and Francis Crick on the basis of an important X-ray diffraction pattern made by Rosalind Franklin. The Watson–Crick structure is a double helix, in which two polynucleotide strands are twisted around each other in antiparallel fashion. The strands are held together by hydrogen bonds, which produce specific pairs between the nitrogenous bases. The sequence of bases in DNA is copied to mRNA in a process known as transcription. This message is then translated as mRNA directs the synthesis of a protein.

The essential role of nucleic acids in information transfer can be seen in viruses, which infect host cells by inserting either DNA or RNA molecules. These viral genomes then direct the synthesis of new virus particles, disrupting the cells that they infect and releasing new viruses, which may infect other cells. In eukaryotic chromosomes, DNA is tightly bound to several types of proteins, including the histone and nonhistone chromosomal proteins. DNA is replicated in semiconservative fashion. Because most organisms contain far more genes than they will ever need to express at the same time, nearly all genes are regulated. One well-studied example of gene regulation is found in the bacterium *E. coli*, where three genes are operated as a group known as the *lac* operon. The expression of the operon is controlled by a repressor–inducer system.

STUDY FOCUS

After studying this chapter, you should be able to:

- Trace the history of how DNA was discovered as the genetic material and how its structure was deduced.

- Explain the "central dogma" of information transfer: DNA → RNA → protein.

- Give the details of the connection between genetics and the biochemical characteristics of an organism.

- Recount some of the latest discoveries in molecular biology, and describe its exciting, changing nature.

TERMS AND CONCEPTS

replication 397	reverse transcriptase 418
transformation 399	retroviruses 418
purines 403	plasmids 419
pyrimidines 403	chromatin 420
template 407	histones 420
transcription 408	nucleosomes 420
codon 409	*lac* operon 422
anticodon 411	introns 426
ribosome 411	exons 427
translation 413	

REVIEW

Objective Questions (Answers in Appendix)

1. Complementary base pairing in DNA happens between
   (a) adenine and guanine.
   (b) adenine and thymine.
   (c) cytosine and adenine.
   (d) thymine and guanine.

2. A strand of DNA serves as a _____ because it can guide the generation of a new _____ strand.
   (a) base pair; identical
   (b) biological strand; different helical
   (c) template; complementary
   (d) helix; exact

3. The coded information for synthesizing a protein is carried to the ribosome by
   (a) DNA.   (c) guanine.
   (b) mRNA.   (d) tRNA.

4. Pre-mRNA is found in the _____ of a eukaryotic cell.
   (a) cytoplasm only
   (b) ribosome
   (c) nucleus only
   (d) cytoplasm and nucleus

Discussion Questions

5. What is the molecule of inheritance? How was it discovered?

6. What is Chargaff's rule? What does it imply about the structure of DNA?

7. Describe the major features of the Watson–Crick model for the structure of DNA.

8. What are the basic differences among transcription, translation, and replication?

9. Describe the basic steps associated with the process of translation. Begin with the formation of an initiation complex.

10. What is a virus? What key characteristic distinguishes viruses from other forms of life?

READINGS

Portugal, F. H., and J. S. Cohen. *The Century of DNA: A History of the Discovery of the Structure and the Function of the Genetic Substance.* Cambridge, MA: M.I.T. Press, 1977. A comprehensive historical treatment of DNA.

Sayre, A. *Rosalind Franklin and DNA.* New York: W. W. Norton, 1975. A thoughtful analysis of Rosalind Franklin's role in the double-helix story.

Watson, J. D. *The Double Helix.* New York: Mentor Books, 1968. A lively and intensely personal account of the development of the double-helix model by one of those who developed it. No one interested in the nature of scientific discovery should overlook this book.

Watson, J. D., and F. H. C. Crick. "Molecular structure of nucleic acids: A structure of deoxyribose nucleic acid." *Nature* 171 (1953): 737. This is the article that announces its authors' discovery of the double-helical structure of DNA.

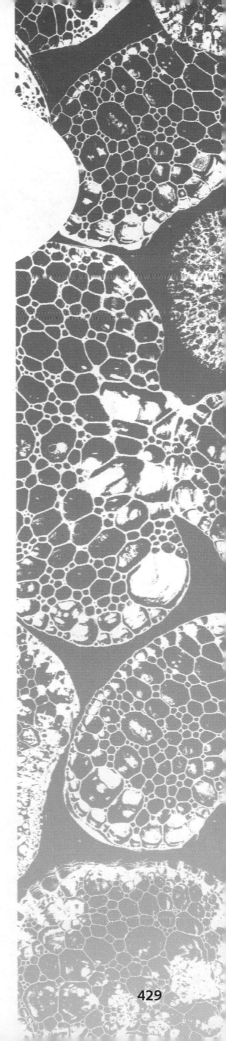

21

Biotechnology and Molecular Medicine

i nside a factory on the California coast, a culture of bacterial cells courses through miles of stainless steel pipes and circulates in giant vats. Hundreds of electronic devices and scores of technicians monitor and control the culture's temperature and chemical composition. What's so special about these bacteria? They have been genetically altered to churn out large quantities of human growth hormone, a vital protein that our bodies can produce only in minute amounts.

The device that nurtures those cells and collects their precious product is a *bioreactor*. Bioreactors are already producing proteins used in treating numerous debilitating (and often fatal) genetic diseases. Other proteins manufactured in this manner may open the door to effective therapies for conditions doctors once thought intractable: hardening of the arteries, cancer, and insulin-dependent diabetes, to name a few.

At the heart of the bioreactor are genetically altered bacteria that produce a most unlikely product—a human hormone. How did the hormone gene get into the bacteria? By a remarkable process known as genetic engineering, which enables researchers to cut and paste genes from one organism to another. The success of genetic engineering, even on a small scale, makes it clear that the science of biology will do more than study the living world of the next century—it will help to redefine it.

In this chapter, we will examine the tools and techniques that have made it possible to isolate and analyze genes and to transfer them from one organism to another. And we will introduce some of the important questions that this new technology will require society to deal with.

GENETIC ENGINEERING: READING AND EDITING THE BOOK OF LIFE

The information stored in our genes interacts with the environment to regulate just about every function of our physical bodies—and many aspects of our behavior, too. Genes, by governing the production of proteins, control a host of traits, ranging from hair and skin color to factors determining whether or not blood clots properly, how effectively the kidneys work, and how quickly cholesterol deposits form in our arteries.

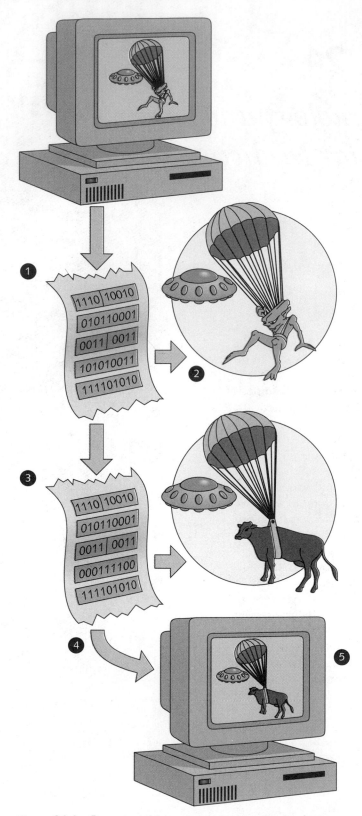

Figure 21.1 Reprogramming a computer program to change a game is a complex process. At least five steps, as illustrated here, would be necessary to make a small change in the appearance of the game. We can compare this simple job to the more complex task of modifying the genetic program of a living organism.

In the previous chapter, we saw how genetic information is encoded and expressed. We explored the way in which the base sequences of DNA transfer information to direct the synthesis of RNA and protein, and we examined some of the cellular machinery that regulates gene expression. As we have done in other fields of science, it's only natural that we would seek to use that basic knowledge to alter living things for our own purposes. In the past 15 years, that's exactly what has happened in molecular biology laboratories throughout the world.

Of course, we have been manipulating the genes of organisms *indirectly* for thousands of years—through experiments in plant and animal breeding. The common-sense approach of breeding only those individuals that best suit human needs has produced many of our most important crops and farm animals. The plant or animal breeder, however, has always been limited by the diversity of genetic material that existed within a species. In other words, if farmers wanted to breed a sheep with brown wool, they had to wait for a brown individual to appear, breed that individual with their existing flock, and work gradually over several generations towards the desired characteristic.

Genetic engineering is something different: it is the direct introduction of *new* genes, producing *new* characteristics, into an organism. This process is much more rapid than traditional breeding methods, and it allows researchers to combine genes from different sources, producing organisms that serve specific purposes. Now, for example, it is theoretically possible to find the genetic instructions that produce brown hair in goats—or even humans—and to transfer those instructions into sheep. This emerging power has earth-shaking implications for the future of our species and our relationships with other organisms. Because the technologies of genetic engineering are recent developments, our first task is to understand how these technologies work and what they are capable of doing. We will then highlight some of the exciting—and potentially troubling—issues that this technology raises.

Editing a Complex Code

Living organisms are complex systems, and we might well ask how it would be possible to "design" new organisms without having a complete understanding of their genetic systems. That's a fair question, and we might answer it by means of an analogy.

Video games are also complex systems (although much, much simpler than any living organism), and we might compare the lines of coded instructions in the programs for these games to the molecular specifications coded in an organism's DNA. As an example, let's imagine a computer game in which the player tries to capture aliens that parachute from a spaceship (Fig. 21.1). Let's suppose, for reasons that are not necessarily obvious, that we'd like to change the game so that *cows* parachute from the spaceship instead of aliens. What would we have to do?

First, armed with a general understanding of how a computer game works, we'd have to get a copy of the program for the game, written in the computer code that the machine interprets to produce the game. Somewhere in that code is the information that produces the image of an alien on the video screen. Our plan might work in five steps:

1. Extract the coded program.
2. Find the lines in the program that produce the images of aliens.
3. Change those lines so that they now produce images of cows.
4. Reinsert the altered program code into the computer.
5. Run the modified program to see if we have made a successful change.

As anyone who's ever written a computer program knows, every step of this project will be tricky. If we understand very little about how the program works, steps 2 and 3 will be particularly difficult. If the program is long, we might have to spend a great deal of time looking for the small portion that produces the images of aliens. Once we did find it, it might not be obvious how to change those lines to produce cows, and that might force us to take a shortcut. For example, by borrowing lines of program code from another video game that already has pictures of cows, we could put cows in our game without knowing exactly how the image of a cow is produced. Eventually, with luck and persistence, our modified game might be ready to play.

Changing a Genetic Code

In a certain sense, the information coded in DNA is similar to the program that runs our video game. As we have already seen, the elements "written" in DNA code, the genes, help to determine the characteristics of an organism. Remember what we had to do to modify the computer game: remove, modify, and reinsert the program code. To change one of the genes in a living organism, we could follow a scheme not very different from the way in which we modified our computer program (Fig. 21.2):

1. Extract the DNA code.
2. Find the genetic information for the characteristic we wish to change.
3. Insert new or modified genetic information into the RNA.
4. Reinsert the modified DNA into a living cell.
5. Test the modified cell to see if the new gene is expressed.

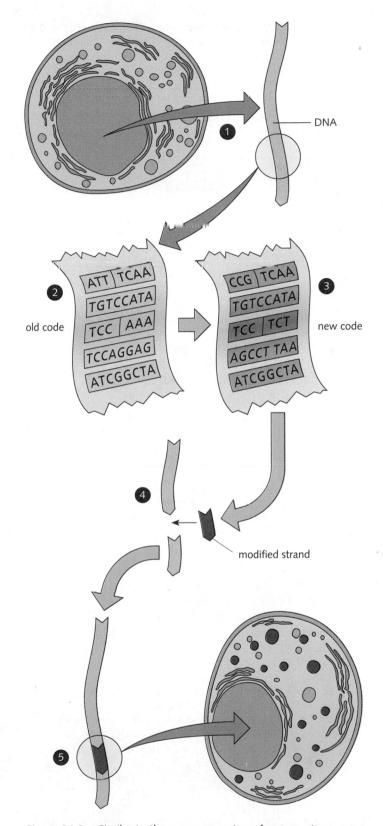

Figure 21.2 Similar to the reprogramming of a computer game, we can visualize genetic engineering as a multistep process involving the isolation, analysis, modification, and reinsertion of a DNA sequence, followed by the expression of that sequence in a transformed cell.

Figure 21.3 Restriction enzymes recognize specific sites in DNA and cut both strands of the double helix at those sites. The cutting sites of a few widely used restriction enzymes are indicated.

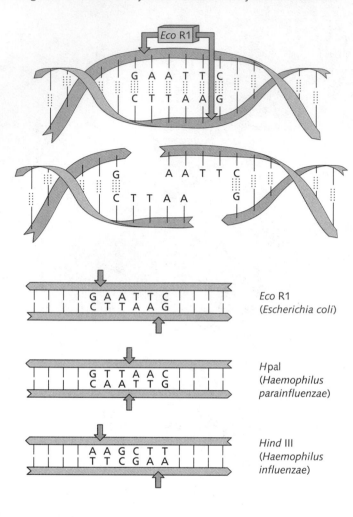

Eco R1
(*Escherichia coli*)

Hpal
(*Haemophilus parainfluenzae*)

Hind III
(*Haemophilus influenzae*)

RESTRICTION ENZYMES—TECHNICAL DETAILS
- More than 100 different restriction enzymes have been discovered.
- Each one cuts DNA at a specific base sequence.
- Eco R1 is a typical restriction enzyme. It is found in the bacterium *E. coli.*
- Eco R1 and many other restriction enzymes produce staggered cuts across the double helix (see figure).
- Most DNA sequences recognized by restriction enzymes are palindromes. (Palindromes read the same way in one direction as in the other. Names like "Otto," numbers like 193391, and phrases like "Able was I ere I saw Elba" are examples of palindromes, as are the three restriction cutting sites in the figure.)

In the case of the computer game, we'd make our changes on a video screen using software tools like an editor and a compiler (which changes the written code into a form that the computer can execute). Are there similar tools that can be used to edit DNA? The answer is yes. In the rest of this chapter, we will see what those tools are, and we will explore the ways in which they can be used to create genetically modified organisms by means of genetic engineering.

Analyzing DNA

Removing DNA from most cells is a simple task, usually accomplished by breaking the cells up and then chemically removing proteins, carbohydrates, lipids, and RNA until only DNA remains.

As we've already seen, in reprogramming a computer game, we find it much easier to look at just a few lines of code at a time. The same is true for DNA. DNA molecules are very long—much too long to handle easily. For the study of individual genes, therefore, DNA molecules must be cut up into smaller pieces. The pieces must be separated from each other and then analyzed by reading their base sequences.

Cutting DNA is easy. Special enzymes, known as **restriction enzymes,** cut DNA at specific base sequences (Fig. 21.3). This makes it possible to cut the DNA extracted from thousands or even millions of cells into precise pieces of manageable size.

These polynucleotide pieces are then separated from each other by means of a process known as **electrophoresis.** Small samples of a polynucleotide mixture are loaded on top of a porous gel. When an electric field is applied to the gel, polynucleotides are separated from each other on the basis of size—small fragments move more quickly than large ones (Fig. 21.4).

Gel electrophoresis is an extremely sensitive procedure. Not only can it be used to separate molecules as large as several hundred thousand bases, but also gels can be designed so that two fragments that differ in size by just a single base can be separated into different bands.

Reading the Sequence

The next step is to analyze the DNA sequences on each of the polynucleotide fragments. This can be done by using an RNA sequence to find a fragment with a complementary sequence (see Theory in Action, Finding a Gene: Southern Blotting, p. 436). It is also possible simply to read the base sequence from each of the fragments by using the technique known as **DNA sequencing.** This technique also involves gel electrophoresis, and it produces a

Figure 21.4   Polyacrylamide gel electrophoresis can be used to separate polynucleotide fragments of different lengths.
(a) The technical basis of electrophoresis. Although both large and small molecules move towards the positive end of the field, small molecules move more quickly through the gel and travel farther. Identically sized molecules move roughly the same distance and wind up in groups that appear as "bands" when the gel is stained.
(b) In the laboratory, electrophoresis is a quick and simple way to separate molecules. After electrophoresis is complete, the gel is stained to produce bands that reveal the migration distance of the polynucleotide fragments. The migration distance of each band is measured and compared with standards to give a rough estimate of the molecular weight of each fragment.

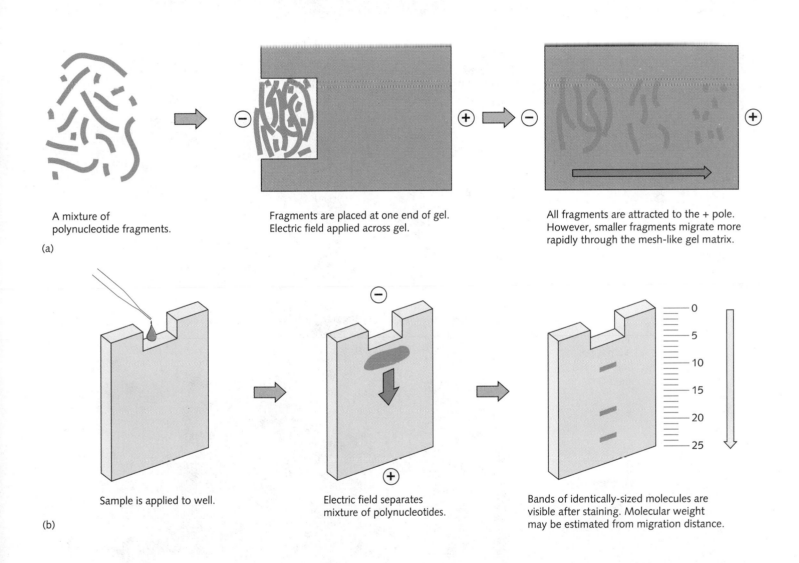

A mixture of polynucleotide fragments.

(a)

Fragments are placed at one end of gel. Electric field applied across gel.

All fragments are attracted to the + pole. However, smaller fragments migrate more rapidly through the mesh-like gel matrix.

Sample is applied to well.

(b)

Electric field separates mixture of polynucleotides.

Bands of identically-sized molecules are visible after staining. Molecular weight may be estimated from migration distance.

Figure 21.5 The DNA sequencing technique developed by Sanger.

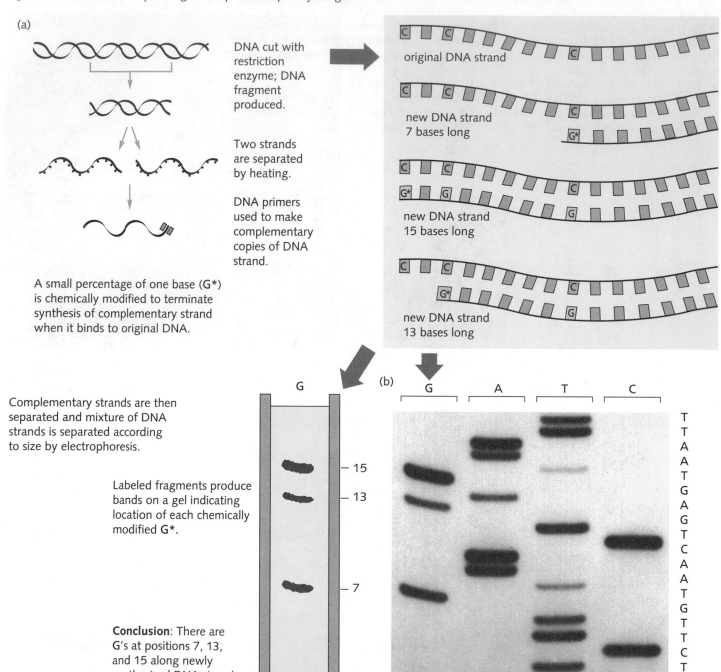

(a)

DNA cut with restriction enzyme; DNA fragment produced.

Two strands are separated by heating.

DNA primers used to make complementary copies of DNA strand.

A small percentage of one base (G*) is chemically modified to terminate synthesis of complementary strand when it binds to original DNA.

original DNA strand

new DNA strand
7 bases long

new DNA strand
15 bases long

new DNA strand
13 bases long

Complementary strands are then separated and mixture of DNA strands is separated according to size by electrophoresis.

Labeled fragments produce bands on a gel indicating location of each chemically modified **G***.

Conclusion: There are G's at positions 7, 13, and 15 along newly synthesized DNA strand.

(b)

G   A   T   C

T
T
A
A
T
G
A
G
T
C
A
A
T
G
T
T
T
C
T
T
C

DNA SEQUENCING—TECHNICAL DETAILS

The Sanger technique for DNA sequencing works like this:

- Restriction enzymes cut DNA into pieces of manageable size: a bit less than 200 bases.
- The paired strands are then separated by heating, and opposite strands are isolated.
- DNA replication enzymes are used to make a radioactively labeled complementary copy of one strand.
- In each of four test tubes used for reaction, a small percentage of one of the four bases added to the mixture has been chemically modified

so that it will terminate the growth of the new DNA strand, as shown in part a of the figure.

- After replication in each tube is complete, the newly synthesized strands are loaded onto a polyacrylamide gel with four wells, one for each base. Electrophoresis separates the strands according to size, and the banding pattern in each well is visualized by exposing photographic film against the gel.
- When all four wells are visualized side by side, it is possible literally to "read" the sequence by examining the banding pattern (see part b of the figure).

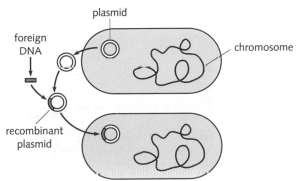

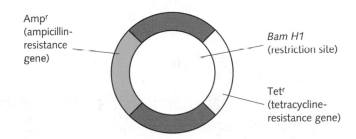

Figure 21.6 LEFT: Genetic engineering of a bacterium—in brief. Bacterial plasmids are used as "vectors" to bring foreign DNA into the cell. The plasmid is isolated, the foreign DNA is spliced into it, and the genetically altered plasmid is inserted into a new bacterium.

RIGHT: A typical plasmid, pBR322, which is widely used for genetic engineering. pBR322 contains a restriction site for *Bam H1* as well as a replication origin and two antibiotic resistance genes. The antibiotic resistance genes are useful markers to select cells that contain the plasmid.

PLASMIDS—TECHNICAL DETAILS

- An ideal plasmid for genetic engineering has two characteristics.
  1. A site where a restriction enzyme can make a single cut in the molecule, making it easy to manipulate in the lab.
  2. A *marker* gene to help identify cells that contain the plasmid. (The best markers are those that make the cell resistant to drugs normally toxic to a bacterium, such as *penicillin* or *tetracycline*.)
- When DNA fragments are mixed with plasmids that have also been cut with the same restriction enzyme, the ends of the DNA fragments and plasmids are joined by special enzymes to seal the plasmid and the foreign DNA.
- The combination of plasmid and DNA fragment is often called a **chimera.** (The word *chimera* in classical mythology refers to a female monster formed by a combination of a lion's head, a goat's body, and a serpent's tail.)

---

pattern of bands that can be "read" to determine the base sequence of the DNA fragment as shown in Fig. 21.5.

The development of techniques to read DNA sequences has been an enormous boon to molecular biology. In the space of a few afternoons, researchers can now take a large DNA fragment, cut it into smaller pieces, read the DNA sequence of each of them, and then reconstruct the complete sequence by "pasting" the results together. Complete DNA sequences for viruses have been determined in this way, and the sequencing of entire eukaryotic genomes is now actively under way (see Theory in Action, Mapping the Territory, p. 439).

GENE MANIPULATION

Having isolated, read, and analyzed the DNA code, now we must find a way to change it. This is usually done by inserting a new DNA sequence into a bacterium. The new sequence can come from anywhere. It may come from another bacterium or from a plant or mammalian cell, or it may even have been synthesized in the laboratory.

One might think that the easiest way to insert this new sequence into an organism would simply be to inject this "foreign" DNA into the bacterial cell. As it turns out, this

isn't always effective. Small fragments of DNA are quickly broken down and destroyed by most cells, so a more sophisticated strategy is needed. Researchers often sneak the foreign DNA into a cell by attaching it to a piece of DNA that the cell recognizes as its own.

The most useful vehicles for transferring DNA in this way are small DNA molecules known as **plasmids,** which are found in many bacteria (Fig. 21.6). Plasmids generally contain a *marker* gene that allows researchers to identify cells that carry plasmid DNA.

Researchers open the plasmid by cutting it with a restriction enzyme, then join the foreign DNA to it, and close the plasmid to form a **recombinant-DNA** molecule, so named because it was formed by recombining DNA from two sources. This recombinant plasmid is then inserted into a bacterium.

Gene Insertion

You might remember that the experiments of Griffith and Avery's group (Chapter 20) showed that DNA molecules could be taken up by bacteria, thereby *transforming* them permanently. Researchers take advantage of this same process by mixing recombinant plasmids with bacteria that do not contain plasmids.

Finding a Gene: Southern Blotting

One of the most difficult problems that molecular biologists must solve is how to find the genes that they are interested in studying. Let's suppose that we had taken DNA from frog cells, cut it into thousands of pieces with a restriction enzyme, and separated the fragments into thousands of bands on a gel. How could we now locate those specific pieces of DNA that contain the sequence for, say, α-globin? Since the

α-globin gene represents less than a millionth of the frog genome, this problem is aptly compared to looking for a needle in a haystack.

Looking for a needle in a haystack is impossible—right? Well, only as a figure of speech—and only if you try to find the needle using only your eyes and hands. If the needle were made of steel (most are), it would be attracted to a magnet. For that reason, if all that hay

could be tossed in front of a strong magnet, sooner or later the needle would stick to it. Thus with the right approach, it might actually be rather easy to find a needle in a haystack. This solution works, of course, only because the needle differs from hay in one important respect: it will bind to a magnet.

The problem of finding a single specific gene among so many lengths of DNA could be solved in a similar way—

(a) Nucleic acid hybridization. A DNA sample is separated, a probe, is added, and the probe base-pairs to complimentary DNA. **(b)** Southern blotting. Restriction fragments are separated by gel electrophoresis and then blotted to a sheet of filter paper. Once bound to the paper, the DNA is heated to separate the two strands of the helix. Then the paper is soaked in a solution containing a radioactively labeled probe, allowing the probe to hybridize to any fragment with a complementary sequence. The position of a labeled band can be compared with known standards to estimate the size of the band bound to the probe.

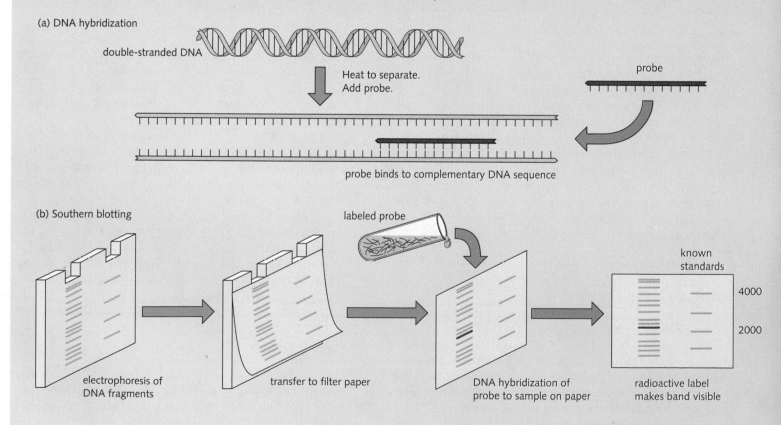

(a) DNA hybridization

double-stranded DNA

Heat to separate.
Add probe.

probe

probe binds to complementary DNA sequence

(b) Southern blotting

labeled probe

known standards

4000

2000

electrophoresis of DNA fragments

transfer to filter paper

DNA hybridization of probe to sample on paper

radioactive label makes band visible

if only we had a "magnet" that could bind to a particular base sequence. Well, it turns out that we do, and the figure shows how. If DNA fragments are heated, the double helix opens up, exposing both strands as shown in the figure. Now we can add our "magnet"—in the form of single-stranded RNA that researchers call a "probe." If that probe is complementary to any of the exposed regions on DNA, it will base-pair to one of the strands. The result is a DNA–RNA "hybrid" held together by hydrogen bonds.

Now, if we had labeled our single-stranded probe with a radioactive isotope, the hybrid could be located by testing for radioactivity. The process of nucleic acid hybridization is every bit as effective as the magnet would be in finding the needle and much more specific, since the base sequence of the RNA probe determines which DNA fragment it will bind.

Nucleic acid hybridization can be combined with gel electrophoresis to find our frog gene, as shown in Fig. 21.5. After the restriction fragments have been separated on a gel, they are transferred to a piece of treated paper that tightly binds DNA. Then a radioactive RNA probe (in this case, part of the mRNA sequence for α globin) is passed through the paper as the DNA is heated to open up the double helix. The probe will not bind to the paper—except when it finds a piece of DNA with a complementary base sequence. Finally, a piece of photographic film can be placed over the paper to detect the location of the radioactive probe and thereby identify the position of the DNA band containing the gene. This powerful technique is known as Southern blotting, after its inventor, Edward Southern.

About one bacterium in a million takes up a recombinant plasmid and is transformed. The rest are unaffected. The presence of the marker gene can then be used to "find" the transformed cells among the millions of bacteria that do not contain plasmids. If, for example, the marker gene produces resistance to an antibiotic, it is possible to "select" the cells that have taken up the plasmid in a single stroke, as shown in Fig. 21.7.

Cloning

Once the right cells have been selected, large numbers of them can be grown to produce an almost limitless supply of bacteria containing the recombinant-DNA molecule. This technique is sometimes referred to as "cloning" a DNA molecule. The term **clone** is used because of the way transformed bacteria are handled in the lab: the bacteria are spread on a culture plate in such a way that single cells produce round colonies of bacteria. Each of these colonies is a "clone," a group of cells formed by cell division from a single cell. For this reason, the cells of a colony are genetically identical and contain exactly the same recombinant-DNA molecules.

Genetic Engineering of Bacteria

There are several reasons why cloning DNA molecules into bacteria is useful. One of the original reasons for cloning was to get large amounts of a particular DNA fragment for study. However, a DNA synthesis technique known as PCR (see Theory in Action, PCR, p. 440) has now replaced bacterial cloning in most cases where large quantities of DNA are needed.

Genetically engineered bacteria can serve as living factories to make large amounts of human proteins that cannot be obtained in any other way. Despite the fact that humans and bacteria last shared a common ancestor at least 3 billion years ago, our genetic codes are still almost identical. As a consequence of that similarity, the cellular machinery of bacteria can transcribe and translate the information contained in many human genes into functional proteins.

Human growth hormone, which is used to treat patients suffering from pituitary dwarfism, was once scarce, precious, and even dangerous to use. (So little hormone is present in any pituitary gland that 70 cadavers were required to produce enough hormone to treat a single patient for a year. This placed recipients of the hormone at high risk for infection by certain viruses that affect the central nervous system.) This hormone is now widely available because it is mass-produced by recombinant bacteria that express the human growth hormone gene.

IN DETAIL:

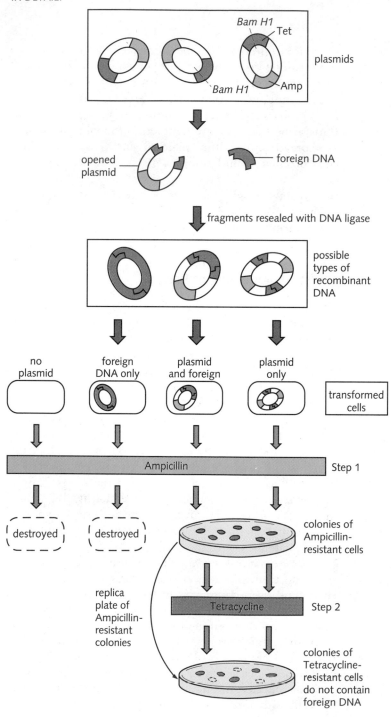

Figure 21.7 Genetic engineering of a bacterium in detail. Recombinant DNA is formed by mixing the foreign DNA with plasmid DNA molecules. After the "sticky ends" have attached, DNA ligase is used to seal the molecules together, producing a *recombinant-DNA molecule*. These recombinant-DNA molecules may then be taken up by bacteria. The researcher must then select the cells that have been transformed, using a two-step process:

1. The transformed bacteria are screened by plating them onto a petri dish that contains ampicillin. Bacteria that have not been transformed are not resistant to the antibiotic and are destroyed.

2. Two types of bacteria survive this treatment: those that contain only plasmid DNA, and those that contain foreign DNA as well. For identification of the colonies that contain foreign DNA, a "replica plate" is made by transferring a few cells from each colony to a plate containing tetracycline. Because of the position of the restriction site, a functional gene for tetracycline resistance can only exist in the absence of foreign DNA. Colonies that grow on the *amp* plate but not on the *tet* plate are therefore recombinant and are used for further study.

CELL TRANSFORMATION—TECHNICAL DETAILS

- Recombinant molecules are mixed in with a population of bacteria that contain no plasmids, under conditions that promote DNA uptake.
- About one bacterium in a million takes up a plasmid molecule and is transformed.
- Transformed cells are identified by means of a large dose of antibiotic. The cells that don't contain plasmids are killed by the drug. But those cells that contain the plasmid are resistant and therefore survive.
- By screening that group of surviving cells with the second antibiotic (see figure), we can determine which of the plasmid-containing cells also contains the foreign DNA.

Mapping the Territory

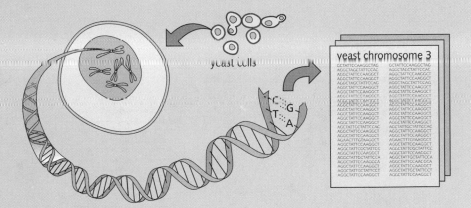

Chromosome III from yeast is the first eukaryotic chromosome to be completely sequenced.

In 1806 Merriweather Lewis and William Clark returned from a two-year expedition through the enormous territory of the Louisiana Purchase. The report they submitted to the president fired the imagination of the nation about its new western territories and began an era of exploration and settlement that changed the history of North America.

In many ways, the Lewis and Clark expedition was important not because it completely described the new territory but, rather, because it made it clear that this new land contained so much that remained to be explored.

Our understanding of the eukaryotic genome may have reached exactly this point. In 1992, a French team finished the entire DNA sequence of a eukaryotic chromosome, chromosome III of the yeast *Saccharomyces cervisiae*. Reading the 315,357 bases of DNA in this small chromosome was a big project—it took more than 3 years, and the scientific paper reporting the results had 147 authors!

When they analyzed the sequence, researchers found that the chromosome has 182 probable genes. Naturally, they compared the DNA sequences in these genes to the more than 30,000 gene sequences from other organisms that have been determined to date. Incredibly, only 66 of them showed any similarity at all to known genes.

What of the remaining 116? Right now they are unknown territory, raising new questions and exciting the imagination. The European group chose to work on yeast because its genome is so small—the human genome, which is the object of a worldwide DNA-sequencing effort known as the Human Genome Project, is 400 times as large.

With 15 chromosomes and 15 million bases to go, the job of sequencing the yeast genome should be completed by early in the next century. But even now the theme of this "expedition" is becoming clear. Determining the DNA sequence of a chromosome is nothing more than the first trip into new territory. It is the beginning, not the end, of a journey of exploration that will take us into the organization of life itself.

PCR

The first thing that a molecular biologist needs is plenty of DNA. DNA sequencing, cloning, and analysis all require significant amounts of sample. Unfortunately, nature doesn't always cooperate. Sometimes an interesting DNA sample may be found in a speck of blood or a fragment of bone that has too little DNA to be analyzed. Fortunately, there's a solution, and it goes by the name of polymerase chain reaction, or PCR.

PCR is such a simple technique that it can be automated, and run by a preprogrammed machine. In brief, researchers separate the two strands of a DNA sample by heating them to 70°C. Then they add small pieces of single-stranded DNA known as "primers" that mark the outer edges of the DNA sequence of interest. The temperature is lowered, allowing the primers to bind to DNA and serve as starting points for DNA polymerase to go to work and synthesize complementary copies of the sequence. Then the process starts over again, heating to separate the strands and adding new primers.

Because the entire DNA sequence is copied in the region between the primers, the amount of DNA is doubled with each round. In just a few hours, enormous quantities of DNA can be produced (hence, a "chain reaction"). PCR is now routinely used to amplify small quantities of DNA to be used for DNA fingerprinting and has even been used to produce workable quantities of DNA from Egyptian mummies more than 4000 years old.

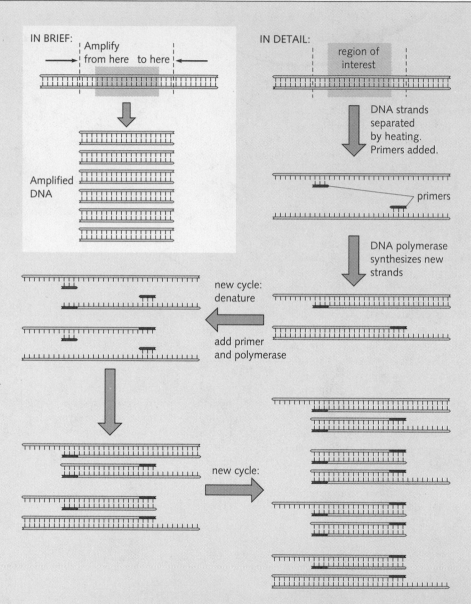

DNA amplification by polymerase chain reaction (PCR). Two short pieces of DNA, known as primers, mark the outer limits of the region to be amplified. After the two strands of DNA are separated, the solution is cooled and primers are bound to complementary regions; they serve as the starting points for DNA polymerase to extend the primers. The process is stopped by heating and recycled by the addition of more primer. By repeated cycles of heating and cooling, thousands of copies of this region between the primers are produced.

Other products that are already being synthesized in genetically engineered bacteria include insulin to treat diabetes, blood-clotting factors for hemophilia, and potential cancer-fighting molecules such as interleukin-2 and interferon.

Recombinant bacteria that express other foreign genes have potential uses in a wide range of applications (Fig. 21.8). Examples include strains of bacteria that carry enzymes to digest organic wastes, produce crop fertilizers, and break down cellulose fibers in wood to serve as a plentiful source of food.

Transforming Eukaryotes

Bacteria are "naturals" for recombinant-DNA technology, because plasmids' ability to transform bacterial cells makes the introduction of foreign DNA relatively easy. In fact, simple kits available over the counter today enable anyone with a high school understanding of biology and a minimal amount of equipment to transform bacteria. Can different techniques be used to achieve the same sorts of results in other organisms? The techniques vary, and none of them are simple, but the answer is a very definite yes.

Cloning systems similar to those used in bacteria have been developed in yeast by employing a plasmid-like minichromosome system. The transformation of yeast cells is now a routine laboratory procedure, and it is pos-sible to carry out genetic engineering in yeast in much the same way as in bacteria. Why should we be interested in transforming yeast? Although they can be cultured in large quantities, just as bacteria can, yeasts are eukaryotes. Therefore, yeasts are excellent model systems for experiments aimed at cloning and studying the workings of eukaryotic genes.

Yeasts are also valuable as factories for producing proteins unique to eukaryotic cells. The biochemical machinery of eukaryotes is more elaborate than that of bacteria. Mammalian cells often add "finishing touches" to proteins, adding sugars here and there, or cutting and folding polypeptide chains in specific ways. For that reason, putting certain human genes into bacterial cells is rather like asking a bicycle shop to build a racing car; some of the basic tools are available, but lots of necessary equipment is missing.

Yeasts have now been engineered to make human proteins that may be important in fighting cancer, and one lab has even engineered yeast cells to secrete *rennin*, an enzyme used by cheesemakers. Large amounts of rennin now being made cheaply by special yeast strains are a boon to the cheese industry. Other genetically engineered eukaryotic cells are being used to produce *tissue plasminogen activator*—a compound that helps clear arteries blocked during a heart attack—and *erythropoietin*—which controls the growth of red blood cells. A host of other proteins useful as drugs are now or will soon be produced in similar ways.

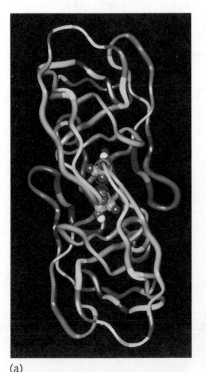

(a)

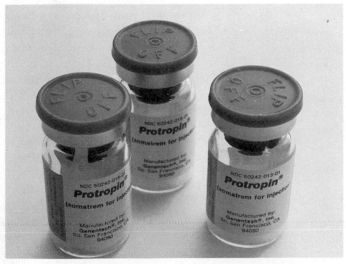

(b)

Figure 21.8 **(a)** Three dimensional structure of HIV protease, an enzyme produced by the virus that causes AIDS. Genetically engineered versions of this enzyme are being used to test drugs that may interfere with the action of the virus. **(b)** Vials of genetically engineered human growth hormone.

Figure 21.9 Genetic engineering has been successfully applied to higher organisms. TOP: A transgenic pig with higher levels of growth hormone produced the meatier pork chop. BOTTOM: Transgenic soybean plants containing genes for herbicide resistance on the left grow in marked contrast to those without the resistance gene on the right when exposed to the herbicide.

Transgenic Multicellular Organisms

Related techniques have also made it possible to insert genes into the eggs of fish, amphibians, reptiles, birds, and mammals. Several laboratories have now used such procedures to produce mammals that contain and express transplanted genes. Agricultural researchers have begun to experiment with genes that will increase the size and weight of farm animals, and medical researchers have transplanted the genes of pathogens such as the AIDS virus into mice for experimental study. Organisms whose genetic composition has been artificially modified by genetic engineering are known as **transgenic organisms.** Some transgenic animals and plants already show the promise of founding improved breeds of farm animals and more productive crop plants (Fig. 21.9). Other, more extraordinary feats of genetic engineering may completely revolutionize both the drug industry and the use of farm animals.

One of the most useful systems for transforming plants employs *Agrobacterium tumefaciens*, a bacterium that infects certain plants and causes tumors known as crown galls. The ability of *Agrobacterium* to cause tumors stems from a plasmid it carries. This plasmid, known as *Ti* (tumor-inducing), enters plant cells and becomes incorporated into chromosomes within the nucleus. Scientists have been able to use the Ti plasmid to transform plant cells by inserting new genes into single cells and then growing complete plants from the transformed cells (the technique is described completely in Chapter 33). One recent achievement of this technology is a plant containing microbial genes that direct the synthesis of raw materials for plastics previously obtained only from coal and oil.

Transgenic Milk: A Case Study

Transgenic animals can be produced in a number of ways. As a case study, we will examine the way in which one group of researchers has successfully engineered transgenic sheep that produce milk containing an enzyme useful in fighting certain forms of the human lung disease known as emphysema.

Why should researchers be interested in genetically engineering animals to produce desired proteins in their milk? There are two obvious reasons: (1) Mammary gland cells normally produce very large amounts of protein. (2) Those proteins are secreted into a liquid that is *meant* to be easily removed from the body.

To accomplish this feat, researchers first isolated the gene for the enzyme known as α-1-antitrypsin, hoping to produce enough of it to use in emphysema patients. Researchers at a Scottish firm (Pharmaceutical Protein Limited) then attached this gene to a promoter that is

Figure 21.10 The essential steps involved in making a transgenic sheep that will produce a human protein in its milk. Note that a number of intermediate and peripheral steps have been omitted for the sake of clarity.

1. The sequence of bases that codes for the desired human protein (for the case described in the text, α-1-antitrypsin) is isolated from human DNA.
2. A promoter region that controls the production of a protein normally found in sheep's milk (in this case, β-lactoglobulin) is identified and isolated from sheep DNA.
3. The human protein-coding region is joined to the sheep promoter forming chimeric DNA.
4. Sheep eggs and sperm, harvested from donor animals, are mixed together *in vitro* to produce a fertilized egg.
5. The chimeric DNA produced in step 3 is injected into the fertilized egg. One micropipette is used to hold the egg in place, while another, much finer microsyringe is inserted into the egg to inject the DNA. (The particular egg shown here is a mouse egg, but the procedure is the same for sheep.)
6. "Surrogate mother" sheep are primed for pregnancy with hormone injections.
7. The injected eggs, a few of which will be transgenic, are surgically implanted into the surrogate mothers.
8. Some proportion of the implanted embryos survive and are born. A small percentage of the females among those animals produce the human protein in their milk.
9. The transgenic animals now carry the chimeric DNA in all of their cells—including the germ line cells, which produce eggs and sperm. The most productive individuals are used in a breeding program to develop a line of transgenic, mammalian bioreactors.

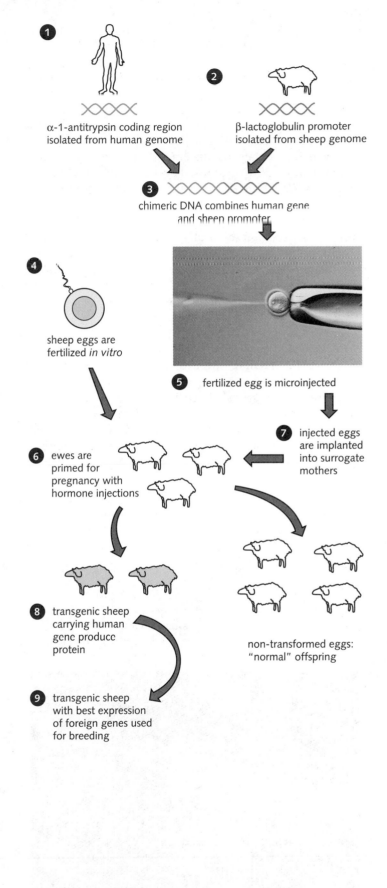

activated in mammary gland cells, hoping that the gene would be activated when milk is produced (Fig. 21.10).

The next step was to get the recombinant DNA into sheep. "Holding" laboratory-fertilized embryo cells with a micropipette, they injected dozens of copies of engineered DNA into each. The researchers then implanted these altered cells into the uterus of a surrogate mother, hoping that they would develop normally.

The odds of success in these ventures were not very high. In one series of experiments, 550 sheep eggs were injected, of which 439 survived the procedure. Only 112 of those successfully implanted in their surrogate mothers, and only 5 of those were, in fact, transgenic animals. Three of those matured into females who produced at least some of the desired protein in their milk, but only one produced enough of it to be useful commercially. That doesn't sound like much, but remember that the genetic alteration is now part of that ewe's DNA, meaning that the ability to produce genetically engineered milk can now be passed along to her offspring.

Needless to say, the effort and expense that went into producing this animal make her and her offspring especially valuable. The investments required to engage in this sort of work are very large, which makes it easy to understand why biotechnology companies are determined to protect their handiwork.

In the United States and many European countries, transgenic animals and plants can be protected by patent laws, which give their producers the right to control their use and to charge royalties. The first animal to be patented, in 1988, was the so-called Harvard mouse, into which cancer-causing genes had been transplanted (Fig. 21.11). Three more rodents, designed to help researchers study prostate disease, viral infections, and AIDS, were patented late in 1992, and many more applications for similar patents are now pending. The limits of patent control over living animals and plants are sure to be tested in the courts in the years ahead.

IS GENETIC ENGINEERING SAFE?

When several scientists first realized that it would soon be possible to carry out genetic engineering in living organisms, concern began to mount that this new technology should be carefully thought out before it was applied. In 1974 an informal conference of leading scientists from around the world was called to consider the safety question.

The scientists appreciated fully how little we knew about something as complex as the genome of a whole cell. They recommended that genetic engineering experiments be undertaken with great caution. They also advised that special care should be taken to ensure that engineered organisms did not escape the laboratory, in order to protect the environment against any unforeseen consequences of this new technology. Initially, the National Institutes of Health adopted guidelines requiring that genetic engineering be done in specially sealed labs and that the bacteria used for certain experiments be weakened genetically so that they could not live outside the laboratory. For several years, these guidelines closely regulated the flow of recombinant-DNA research.

In recent years, however, many of the original rules and guidelines have been relaxed. It has become clear that the burden of extra genes present in genetically engineered organisms actually make them grow more slowly than their natural counterparts and that there is little danger of creating a "super germ" any worse than the disease-causing microbes that are normally part of the environment. Evidence has also accumulated that exchanges of DNA molecules occur in nature all the time as microbes break down and release DNA fragments into the environment.[1] Therefore, the efforts of laboratory scientists have not broken any new barriers in terms of the transfer of DNA between different species of prokaryotes.

Nonetheless, concerns about the environmental hazards of genetically engineered organisms persist. Several organizations have opposed the release of such organisms into the environment. One early case involved a genus of bacteria called *Pseudomonas* that often grows on the leaves of fruit trees. Normally, these bacteria produce a protein that serves as a nucleating source for ice crystals, so normal *Pseudomonas* bacteria make plants upon which they grow susceptible to frost damage. Because frost damage causes the loss of millions of dollars worth of fruit every year (Fig. 21.12), this is a serious problem.

One biotechnology firm removed the gene for that protein, producing a strain of bacteria that does not produce those potentially damaging molecules. When crop plants are sprayed with the engineered bacterium, thereby displacing the frost-inducing strain, plants can be protected against frost damage. Because of the enormous savings it may permit, this engineered bacterium, which

Figure 21.11 The first patented transgenic animal. This mouse, patented by Harvard University, carries a number of human cancer genes spliced into its genome. It is used for laboratory studies on cancer.

1. The nature of the evidence is the large amount of DNA present in places such as digestive systems and sewage, where trillions of bacteria die and decompose. There is a rather high background rate of transformation in these environments, and this helps to account for the rapid spread of some genes from species to species.

is now in use, may become an important commercial product.

Opponents of this sort of project, however, argue that we cannot be sure what the ecological consequences of releasing the engineered bacterium will be. If it does penetrate the environment, will it affect the nucleation of ice crystals in the atmosphere and in turn reduce the amount of rainfall in certain areas? Although most scientists don't think that this is very likely, it is difficult to answer such objections with absolute certainty.

This controversy has already sharpened as the first transgenic organisms have been submitted for approval as food products. Certain consumer groups have voiced opposition to the use of genetically engineered hormones that increase milk production in cows, and the sale of meat from transgenic pigs and chickens will spark similar concerns. As we write this edition, no genetically engineered foods are yet on the market, although by the time you read this there may well be. The first step was taken in 1990, when a Dutch company received approval in England to proceed with the development of a transgenic strain of yeast designed to make bread rise faster. Then in 1992 in the United States, the Calgene Corporation requested approval for a genetically engineered tomato it named FLAVR SAVR®, produced using the antisense RNA technique (see Theory in Action, Antisense: Turning Off a Gene, p. 447).

The debate over the licensing and labeling of such products promises to be long, controversial, and enormously important. It is no exaggeration to say that every responsible citizen needs a clear understanding of genetic engineering to participate in the public decision-making process.

Figure 21.12 Frost-damaged citrus trees. A commercial product, consisting of genetically engineered bacteria, is now available to prevent frost damage. When sprayed on trees, these bacteria displace natural bacteria, which produce a protein that helps ice crystals to form.

THE NEW HUMAN GENETICS

The powerful techniques of molecular biology make it possible to carry out a whole range of new experiments involving human genetics. Those experiments, like their animal counterparts, offer enormous potential to alleviate or even eliminate the pain and suffering caused by many different diseases. They also raise a series of new ethical and social issues that we will have to deal with in the years ahead.

One of the most important tools in modern human genetics was developed as an outgrowth of the Southern blot technique (see Theory in Action, Finding a Gene). After researchers had worked with the bands on Southern blots for a few years, a fundamental fact began to sink in: *those bands are inherited.* That should have been obvious from the beginning, of course. After all, the position of a band on a blot is determined by its DNA sequence—a sequence that is inherited.

Therefore, we can think of the bands that hybridize to a particular probe as characters, just like red hair or blue eyes (Fig. 21.13a). Simple enough. In many cases, when lots of different individuals are tested with a probe, the pattern of bands on a Southern blot are identical. Very uninteresting for genetics. But in cases where some individuals show *different* bands for the same probe, things get very interesting indeed.

A character that differs among members of a population is said to be **polymorphic** ("many-shaped"). If the size of a restriction fragment differs among members of a population, that fragment is polymorphic, too.

Why should bands on a gel differ? Figure 21.13 shows why. A gene might lie between two restriction sites 4000 bases apart, as shown in Fig. 21.13(b). A probe for the gene would hybridize to a 4000-base-pair restriction fragment. However, in some individuals, random mutations—changes in DNA sequence that are described in more

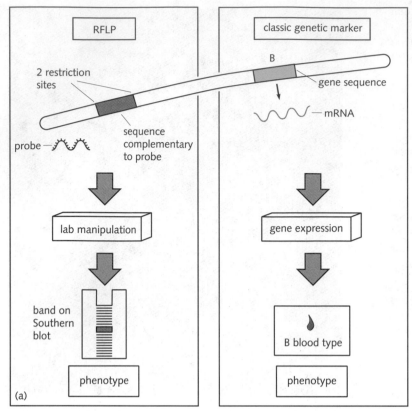

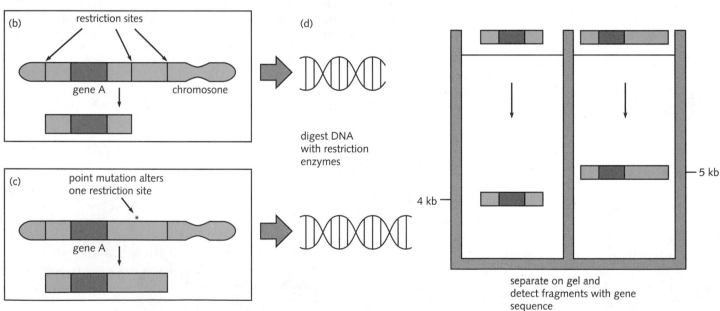

Figure 21.13 **(a)** Classic genetic markers are gene sequences that produce an *observable* phenotype, such as the B blood type shown here. Restriction fragment length polymorphisms (RFLPs) don't affect phenotype in the usual way. Instead, laboratory manipulation (restriction digestion of DNA followed by Southern blotting) is required to make RFLPs observable. Despite this, since the position of the band on the blot depends on the DNA sequence, a RFLP is every bit as useful in genetic analysis as a classic marker is.
(b) The size of a RFLP is determined by the distance between the two restriction sites that surround a DNA sequence complementary to the probe to be used (in this case, the probe is complementary to gene A).
(c) A point mutation that alters a restriction cutting site changes the size of a RFLP, even if it has no effect on the gene of interest.
(d) The change is easily detected by gel electrophoresis.

detail in the next chapter—could abolish one of those restriction sites. What happens then? The same gene becomes part of a larger restriction fragment (Fig. 21.13c), producing a different-sized band.

We could then say that this particular restriction fragment is polymorphic. There's no reason why we shouldn't consider those bands, which are inherited just like blood types, as ordinary phenotypic characters. We could say

that we had just discovered a restriction fragment length polymorphism, RFLP for short, usually pronounced "rif-lip."

For purposes of genetic analysis, a RFLP can be treated like any other trait. It is passed from one generation to the next and can be detected in DNA isolated from any tissue. Because each RFLP has a definite position, RFLPs on the same chromosome display *linkage*, and

Antisense: Turning Off a Gene

Genetic engineering makes it possible to insert a new gene into a cell and, from that cell, to produce a transgenic animal or plant with *added* properties. That's useful as far as it goes, but what about a situation where we'd like to turn an existing gene *off*?

Molecular biologists have been able to do exactly that by a technique known as antisense RNA (see figure). The idea is to insert into the plant cell a copy of the gene that is identical to the original gene in every way but one: the new gene is in the reverse orientation. The base sequence of RNA produced by the new gene (antisense RNA) is complementary to the normal mRNA; and when the two find each other in the cell, they will base-pair with each other to form double-stranded RNA, preventing either RNA molecule from directing protein synthesis. The molecular result is that the expression of the original gene is effectively blocked. As you might expect, this technique is widely used in the lab to study the effects of blocking expression of individual genes.

A perfect example of how this technology has been applied is a plant gene for flower color. In chrysanthemums, for example, genes code for a series of enzymes that produce pink-colored petals. By turning off one or more of those genes, plant breeders now claim to be able to produce whiter-than-white flowers. Keep your eyes on the local florist—this technique has already produced many flowers with novel colors.

You might want to watch the grocery store, too, because antisense RNA has recently been used to block the synthesis of an enzyme that causes tomatoes to overripen. Normally, when tomatoes ripen on the vine, they become sweeter and more flavorful, turn red—and be-

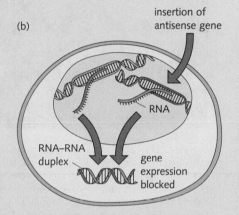

(a)

gene
mRNA

protein
product

(b)

insertion of
antisense gene

RNA

RNA–RNA
duplex

gene
expression
blocked

gin a process of softening that ultimately causes them to rot. For that reason, commercially raised tomatoes—which cannot be too soft while in transit and which should last as long as possible on grocers' shelves—are plucked while still green, hard, and fairly tasteless.

Enter Calgene Corporation, a biotechnology firm that identified and sequenced the gene that produces the enzyme that causes softening and rotting. Using the antisense approach, Calgene has managed to create a strain of tomatoes, called FLAVR SAVR®, in which the action of the softening gene has been blocked—without interfering with the action of other genes that

Antisense RNA. **(a)** The normal gene is expressed by means of mRNA, which is then translated into protein. **(b)** If a gene is inserted in the reverse orientation of the original gene, an "antisense" RNA is produced. The two RNAs, being complementary, will bind to each other to form double-stranded RNA, which cannot be translated. This blocks expression of the original gene. **(c)** The white chrysanthemum on the right is the product of an antisense gene that blocks pigment synthesis in the flower.

(c)

cause the fruits to become more flavorful. With luck, this trick should enable growers to ship full-flavored, vine-ripened fruit, rather than the tasteless supermarket tomatoes many of us have become used to.

Not every use of antisense RNA is this simple, of course. In some cases, the effects of the antisense gene are unpredictable. This is probably due to the fact that the site at which the new gene is inserted into host DNA, which cannot be controlled, is just as important as its DNA sequence. Despite this drawback, it is clear that antisense technology will play a major role in the new technology of genetic engineering.

DNA: The Ultimate Fingerprint?

Fingerprints are ideal for identifying individuals because they differ so widely from one person to the next. But fingerprints are useful only in those rare cases where a criminal leaves clean, complete prints that can be matched with police records. Today, however, molecular biology has developed a new tool that may become the ultimate weapon of criminology: the *DNA fingerprint.*

DNA fingerprinting makes use of the fact that certain DNA regions between genes are extremely variable from one individual to the next. When probes to these "hypervariable" regions are constructed, Southern blotting can be used to produce a pattern of RFLPs that differs markedly from one individual to the next.

If the probe recognizes a large number of fragments from around the genome, a single gel can produce a "fingerprint" that is unique for one individual. Scientists hoped that by using several probes, they could produce a gel pattern so specific that it could be distinguished from the pattern of any other individual in the world. DNA fingerprints can be prepared from cells found in a drop of blood or semen and even from fragments of skin caught under the fingernails of a crime victim.

In 1988 this technique came to the aid of a 27-year-old computer operator at Disney World in Florida who had been attacked, beaten, and raped in her home in Orlando in 1986. Although she caught a brief glimpse of her assailant's face during the rape, she faced the terrifying prospect of having to convince a jury that her brief eyewitness identification of the suspect, 24-year-old Tommie Lee Andrews, was absolutely certain.

Andrews claimed that he was at home during the evening the rape occurred, and he even produced a witness to substantiate that claim. It was a classic case of the victim's word against that of the accused. The prosecuting attorneys

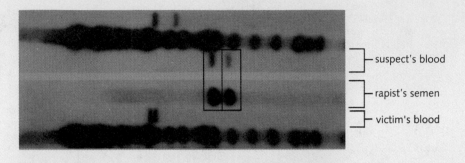

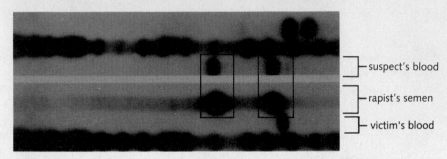

suspect's blood
rapist's semen
victim's blood

The first use of DNA fingerprinting in the United States is shown in these photographs. TOP: Two matching bands from DNA obtained from the blood of a suspect and semen left by the rapist. BOTTOM: To confirm the result, the test was run with a second restriction enzyme. This result also showed a match between the suspect's blood and semen left by the rapist. Neither sample matched the bands in the victim's DNA. The other lanes on the gels are control samples and marker bands.

sought the aid of a molecular biologist to perform the "DNA fingerprinting" test on a semen sample taken by police the night of the rape. When the DNA fingerprint of this sample was compared with a blood sample taken from Andrews, the results were conclusive: It was a perfect match. The jury returned a verdict of guilty, resulting in a jail sentence of 22 years for Andrews.

As the legal issues of this technique were explored, the Andrews trial turned out to be no more than the opening legal chapter on DNA fingerprinting. In several trials, defense attorneys correctly pointed out that there was not enough data to be absolutely certain that identical patterns could not appear in two different individuals. This could lead to a false DNA identification.

After 3 years of study, in 1992 the National Academy of Sciences attempted to settle the matter by issuing a report that recognized the scientific standing of DNA fingerprinting. The report proposed a new, conservative way to estimate the possibility that a DNA identification could be mistaken and proposed DNA testing standards for laboratories to meet. At this writing it remains to be seen whether or not courts throughout the country will agree that evidence meeting the new standards will be routinely admitted in court.

Only when uniform, court-approved standards are developed will DNA fingerprinting have the chance to fulfill its early promise as the ultimate weapon of criminal identification.

the rate of recombination between linked RFLPs can be used to construct genetic maps, as described in Chapter 10. More than 5000 different RFLPs have been discovered in the human genome, and more are added every week. The power of these new genetic markers is so great that they are forming the basis of a worldwide effort to map and sequence the entire human genome.

Testing for Genetic Disease

Sometimes RFLPs are located in or near disease-causing alleles. RFLPs closely linked to cystic fibrosis were crucial in pinpointing the location of that defective gene to chromosome 7 (see Theory in Action, Cystic Fibrosis: A Triumph and a Warning, p. 450). It is also possible to test for the presence of a defective gene simply by probing for a RFLP that is either in the gene or closely linked to it. Such tests are now available to detect the presence of more than 30 genetic disorders, and more are being developed every month. For better and for worse, tests for genetic diseases are much easier to develop than cures. Tests to determine the presence or absence of a defective gene are often available years ahead of treatments.

One of the first tests to be developed, for example, was based on a RFLP close to the allele for Huntington's disease, a dominant genetic disorder that causes degeneration of the nervous system and death. As noted in Chapter 11, Huntington's disease usually does not produce symptoms until middle age, so people are usually unaware that they carry the allele until after they have produced children. A test for the Huntington's allele can now be done with a small blood sample from which DNA is extracted to test for the telltale RFLP (Fig. 21.14). This test allows the children of Huntington's sufferers to learn whether or not they carry the deadly defective gene, but nothing can yet be done to halt the disease.

Genetic Testing: Ethical Issues

The availability of tests for Huntington's disease and cystic fibrosis are major medical breakthroughs. But the existence of such tests in the absence of effective treatment confronts the children of Huntington's sufferers and the potential carriers of certain other genetic diseases with a dilemma: to test or not to test.

On one hand, universal application of tests for inherited diseases might make it possible to reduce the incidence of many through a procedure known as **genetic screening.** If all potential carriers were to be tested, and if all agreed not to have children if they tested positive, the incidence of such diseases could be cut dramatically. A case in point is Tay–Sachs, a hereditary, degenerative disease of the nervous system that causes convulsions and

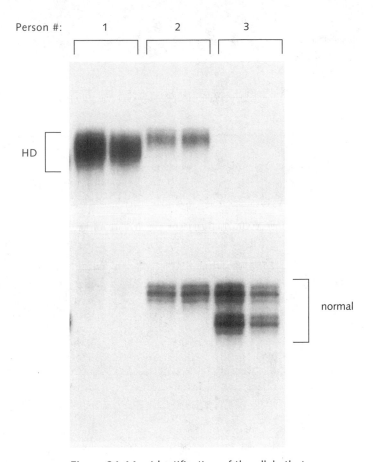

Figure 21.14 Identification of the allele that causes Huntington's disease (HD). DNA from three individuals has been analyzed on this gel. PCR (polymerase chain reaction) was used to amplify the DNA from a small region of the allele believed responsible for Huntington's disease. In normal versions of this gene, the region is relatively short ("normal" on the figure). However, in the HD allele, this region is greatly expanded, producing a band that runs more slowly on the gel (see "HD" on the figure). The three individuals shown are siblings. Person #1 has two copies of the HD allele, #2 has one HD allele and one normal allele, and #3 has two normal versions of the allele. Persons #1 and #2 will probably develop the symptoms of Huntington's later in life. (Gel kindly provided by Dr. Marcie MacDonald.)

Cystic Fibrosis: A Triumph and a Warning

A young couple nervously awaits the results of a pregnancy test at a hospital clinic. They seem at once more anxious and more hopeful than typical new parents, and for good reason. Both husband and wife are carriers of cystic fibrosis (CF), a deadly genetic disease. Normally, each child they conceive would have a one in four chance of suffering from CF. But if the woman is pregnant now, they can rest assured that their baby will be healthy. How do they know? A few weeks earlier, their eggs and sperm met in a laboratory dish, and the resulting fertilized eggs were cultured until they reached the eight-cell stage. At that point, a genetic test was performed on a single cell plucked harmlessly from each embryo. That test identified embryos that were free of the deadly defect, and two of those were placed in the woman's uterus. If either has implanted successfully, molecular biology will have freed their child from a gruesome and deadly inheritance.

This scenario, which occurs more and more frequently each year, is a testimonial to the power of human molecular genetics. CF affects roughly 1 child out of 2500 in the United States, and many of its victims die as children. It is the most common fatal genetic defect among people of European ancestry in the United States. For years researchers tried, unsuccessfully, to find the biochemical basis of CF, whose symptoms were diverse and puzzling. CF patients produce an abnormally salty sweat. They are unable to release normal amounts of digestive enzymes from the pancreas, which leads to digestive disorders. Their lungs clog with a thick mucus that makes it difficult for them to breathe and results in serious respiratory infections. Unable to find a single cause for all of these symptoms, many laboratories decided the only solution was to find the CF allele itself.

That allele was located by a combination of classic genetics and molecular biology. Researchers started by identifying a series of RFLPs in CF patients. Then they looked for the same RFLPs in their parents and grandparents, trying to find one that was always inherited together with the CF allele. The idea, of course, was to find one that was linked to the defective gene—in other words, located close to it on the same chromosome. Once that was accomplished, then they would use the RFLP probe to zero in on the allele itself.

In 1989 the strategy paid off. A group of researchers including Lap-Chee Tsui of Toronto and Francis Collins of Michigan announced that they had found the involved gene on the long arm of chromosome 7. The gene was large and complex, and it coded for a particular membrane protein: a transmembrane channel for chloride ions (see the figure). Nearly three-fourths of CF patients have a three-base deletion in this gene, causing the loss of a single amino acid and disrupting channel function.

What's happened since? In a general way, the isolation of a disease allele makes three things possible: testing for carriers of the allele, further study of the disease, and ultimately, curing the disease.

Progress against CF is now advancing rapidly on each of these fronts:

1. Tests are now available for carriers of the CF allele. As many as 1 adult in 25 carries one copy of the recessive allele, and most of these can now be identified by means of a Southern blot test. This should make it possible to screen prospective couples to find cases where both partners carry the allele and advise them of the risk of having children. Researchers are trying to perfect a test that will identify more than 99 percent of carriers.

2. Next is the development of an animal "model" for the disease. Research on CF has always been hampered by the fact that it was not possible to test therapies on animals before trying them on human patients or to study the disease prenatally, since CF often does considerable damage to the digestive system before birth. Beverly Koller's lab at the University of North Carolina has solved that problem by removing the mouse equivalent of the CF gene, engineering a DNA deletion, and then placing the defective gene back into embryonic cells. Mice with the engineered "CF" allele have defective chloride channels and develop intestinal blockages similar to those in human CF patients. The CF mice are now being made available to investigators around the world.

3. Curing a genetic disease is always the ultimate goal. Less than a year after the gene was isolated, researchers were able to "cure" the CF defect in cultured human cells by infecting them with a virus carrying a healthy copy of the involved gene. Another lab took the idea of "gene replacement" a step further by inserting the gene into the DNA of a virus that infects the respiratory passages of mice. The mice inhaled a mist containing the virus, and for more than 6 weeks they produced the normal human CF protein.

In 1993 a group of human volunteers with CF agreed to try the same approach, and trials were started using inhaled virus. The hope, of course, is that normal copies of the gene will be expressed in the air passageways of these patients, solving many of their respiratory problems and extending their lives. Will it ever be possible to *permanently correct* the CF allele? The ability to do this would eliminate the

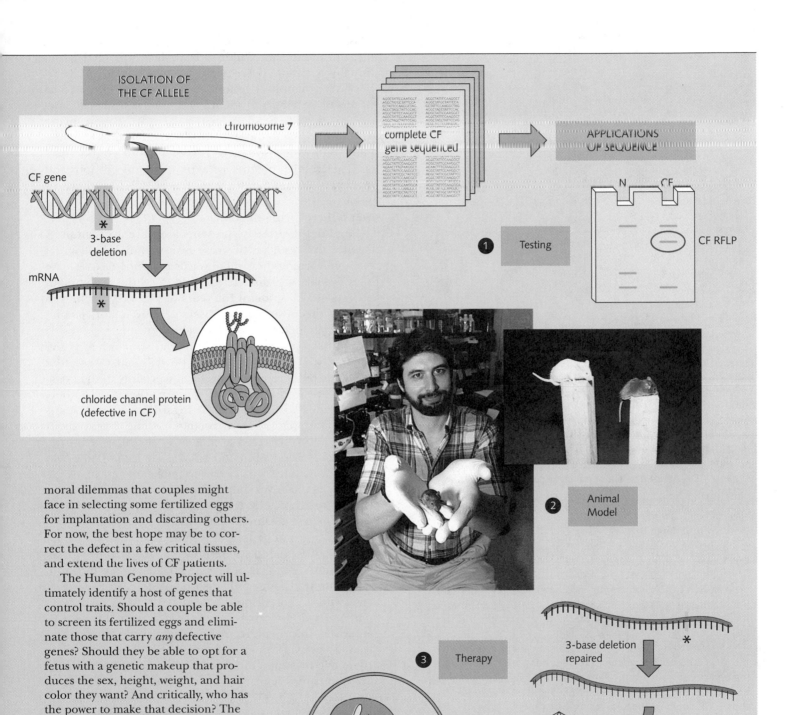

ISOLATION OF
THE CF ALLELE

chromosome 7

CF gene

* 3-base
deletion

mRNA

*

chloride channel protein
(defective in CF)

complete CF
gene sequenced

APPLICATIONS
OF SEQUENCE

1 Testing

N   CF

CF RFLP

2 Animal
Model

3 Therapy

3-base deletion
repaired   *

normal gene
placed in virus

virus used to insert
gene in CF cells

moral dilemmas that couples might face in selecting some fertilized eggs for implantation and discarding others. For now, the best hope may be to correct the defect in a few critical tissues, and extend the lives of CF patients.

The Human Genome Project will ultimately identify a host of genes that control traits. Should a couple be able to screen its fertilized eggs and eliminate those that carry *any* defective genes? Should they be able to opt for a fetus with a genetic makeup that produces the sex, height, weight, and hair color they want? And critically, who has the power to make that decision? The time is not far off when we will be able, theoretically, to redefine the human condition by manipulating our genes. We should consider the implications of this power sooner rather than later.

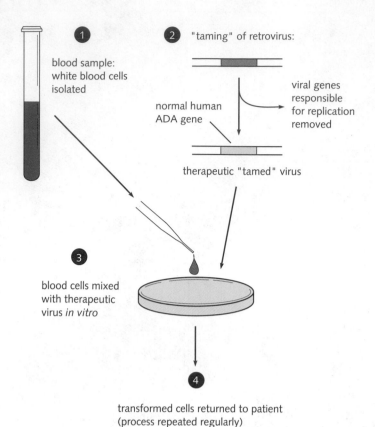

1
blood sample:
white blood cells
isolated

2 "taming" of retrovirus:

normal human
ADA gene

viral genes
responsible
for replication
removed

therapeutic "tamed" virus

3
blood cells mixed
with therapeutic
virus *in vitro*

4
transformed cells returned to patient
(process repeated regularly)

Figure 21.15 ADA therapy. ADA, a severe, often fatal, inherited immune deficiency, is caused by a defect in the gene that codes for the enzyme adenosine deaminase (ADA), resulting in the death of white blood cells, which are the body's defense against infection. Because these cells are continually produced during life, because they can be removed from a patient for manipulation, and because the disease is caused by a single defective gene, ADA deficiency is a prime candidate for gene therapy.

1. A blood sample is removed from the patient and a particular class of white blood cells is isolated.
2. A retrovirus known to infect white blood cells is "tamed" by removing the genes that enable the virus to replicate itself and replacing them with a working copy of the ADA gene.
3. The white blood cells isolated from the patient are mixed, *in vitro,* with retroviruses carrying intact copies of the ADA gene. Some, though not all, of the cells are successfully transformed, meaning that they incorporate the ADA gene into their own DNA.
4. The transformed cells are returned to the patient. Note that the ideal (and not yet possible) therapy for ADA requires finding and transforming the "stem cells" that live in bone marrow and produce white blood cells for the life of an individual. If those cells could be located, a single treatment of the sort described here might cure a patient for life. Because those stem cells have not yet been successfully isolated from humans, older and more mature white cells with much more limited life spans are used in the procedure. For that reason, the patient must go through this procedure every few weeks.

loss of sensory perception and usually results in death by the age of 4 years.

Because of the founder effect (see Chapter 13), the recessive allele that causes Tay–Sachs disease is most common in Jews of Eastern European descent. Careful screening, counseling, and education of individuals at risk to transmit the disease has had a dramatic and positive effect: the number of children born in the United States with Tay–Sachs has decreased by 90 percent since 1972.

Other attempts at genetic screening have not been so successful. Attempts to identify carriers of sickle-cell anemia in several U.S. states in the early 1970s were made without informed consent, with little education or counseling, and with insufficient procedures for confidentiality. The results of this screening produced job and insurance discrimination against perfectly healthy carriers of the allele and did not reduce the number of children suffering from the disease.

Huntington's disease presents a different situation. A negative test result in this case offers young adults great relief on learning they will never develop the disease. The unlucky souls who discover they do carry the allele might theoretically be grateful for years of advance warning to plan for the onset of the disease. They might, among other things, decide not to have children.

But many people who may one day succumb to Huntington's have decided that they do *not* want to take the test. Some of these people prefer uncertainty to the possibility of discovering that they *do* carry the allele. Others have decided to have children regardless of their status on the Huntington's test, seeing nothing tragic about bringing into the world a child who might develop the disease after living a full and normal life into middle age. These conflicting considerations make the issue of genetic testing far from simple. In one study, nearly 50 percent of the possible Huntington's carriers who were offered the test declined to take it.

The science of genetic screening is still in its infancy, and only a few such tests are now available. In the not-so-distant future, however, it may be possible to test individuals for hundreds of genetic disorders and even to identify those who are predisposed to ailments such as cancer, heart disease, and diabetes. As useful as it may be to gather that information, it also raises serious issues of individual liberty. Does an individual have the right to keep his genetic makeup confidential? Do employers or insurance companies have the right to screen out individuals who carry certain alleles? Is it anyone's business to say whether or not you should be able to pass your genotype, defective genes and all, on to the next generation?

These are just some of the questions that will be raised by the new biotechnology, which has given us powerful new tools and will force difficult choices on us in the years ahead.

Gene Therapy

You have probably already realized that the techniques of genetic engineering that are now being successfully applied to sheep, goats, cows, and mice could also work on another mammalian species: *Homo sapiens*. If taken to its ultimate potential, **gene therapy** could solve the problems posed by genetic screening, by replacing defective human genes with properly functioning copies.

The first known recipient of human gene therapy was a young girl who was 4 years old during her first treatment in 1990. She was born with ADA deficiency, a single-gene defect that causes serious (and often ultimately fatal) immune deficiency. By removing a sample of her white blood cells, adding normal copies of the vital gene, and returning the cells to her body, the gene therapy research team led by French Anderson at the National Institutes of Health has, according to her parents, miraculously enabled the child to enjoy the sort of good health that most children take for granted (Fig. 21.15).

Other efforts at gene therapy are in various stages of safety testing (see Theory in Action, Cystic Fibrosis) and will soon be ready for clinical trials. Along with the excitement about curing genetic disease, of course, this technology raises weighty questions about "designer genes" producing "designer people." Anderson, a pioneer (and champion) of gene therapy, asserts firmly that the technology should be used only to cure disease. But many other people—ranging from members of the general public to physicians—are already thinking about using human genetic engineering for "enhancement" purposes. We should remember, however, that much of what molecular biologists take for granted today was considered science fiction as recently as a decade ago. We should also remember that as we begin to alter our genes, we begin tinkering with the very essence of what it means to be human.

The molecular tools that make genetic engineering, genetic testing, and gene therapy possible are simply scientific techniques that enable us to manipulate DNA. But several prominent molecular biologists, including Erwin Chargaff (whose historic role in unraveling the secrets of DNA we described in the previous chapter), are concerned. Genetic engineering, Chargaff warned in a commentary to *Nature* magazine some years ago, "is in great danger of leading to various brutalities which society, once awakened, may not wish to tolerate." Chargaff and many of his colleagues agree that these technologies are not intrinsically "bad." Rather, their ultimate effects, good and bad, will be determined by how we—as individuals and as members of society—choose to use them. That's why all of us must prepare to make our decisions with fully informed judgment and fully open eyes.

SUMMARY

Genetic engineering is the intentional modification of genetic information in an organism. The technology of genetic engineering emerged from a number of scientific developments. They included the discovery of restriction enzymes, which cut DNA at specific sequences, and physical techniques like gel electrophoresis, which makes it possible to separate nucleic acid fragments on the basis of size. Southern blotting allows scientists to locate specific DNA sequences by means of nucleic acid hybridization, and sequencing techniques enable researchers to read the base sequence of isolated fragments. Molecular techniques also make it possible to recombine DNA fragments from different sources in the laboratory. These recombinant-DNA molecules can then be used to transform living cells. In bacteria, this can be done by joining foreign DNA to plasmids, small DNA molecules possessing marker genes that make it easy to select cells containing them.

Genetically engineered bacteria can now be used to produce dozens of useful products, including human hormones. Bacteria have also been modified by genetic engineering to perform special tasks, such as blocking the growth of natural bacteria that promote frost formation on citrus trees. Eukaryotic cells can also be transformed, and in some cases it is possible to produce transgenic organisms, animals, or plants that contain recombinant DNA. In other cases it has even been possible to join foreign genes to promoter sequences that result in the production of recombinant proteins in milk.

The new genetic technologies have allowed scientists to analyze small differences in DNA sequences between individuals. When these differences produce restriction fragments of different sizes, they are known as restriction fragment length polymorphisms, or RFLPs. RFLPs can be used to identify individuals by means of DNA fingerprinting and are also useful as genetic markers in gene mapping. In the very near future, it may be possible to employ "gene therapy" to treat or cure genetic disorders by inserting functional DNA sequences to replace or supplement nonfunctional ones. These technologies hold great promise, but they will also pose a new series of ethical and moral issues that societies and individuals will be forced to address.

STUDY FOCUS

After studying this chapter, you should be able to:

- Explain some of the tools of modern molecular biology and genetic engineering.

- Explain the techniques used to isolate and analyze DNA and to produce transgenic plants and animals.

- Discuss some of these ethical and scientific questions: Should engineered organisms be released into the environment? Is it ethical to attempt human gene therapy? Do humans have the right to alter their own reproductive cells or to control the course of evolution? As genetic testing becomes easier, what guidelines (if any) should govern its use? Why (or why not)?

REVIEW

Objective Questions (Answers in Appendix)

1. Restriction enzymes are used to
   (a) cut DNA at specific sites.
   (b) inhibit bacteria from infecting humans.
   (c) convert one enzyme to another.
   (d) cut RNA at specific sites.

2. A DNA molecule that has genes from different sources is a
   (a) chimera.   (c) replication intermediate.
   (b) operon.   (d) genome.

3. Insertion of foreign DNA into a bacterial plasmid creates
   (a) a minichromosome.   (c) a RFLP.
   (b) recombinant DNA.   (d) a clone.

4. *Agrobacterium tumifaciens* is useful in plant genetic engineering because of its ability to
   (a) fix nitrogen.
   (b) dissolve the plant cell wall.
   (c) join different restriction fragments.
   (d) produce a plasmid that integrates into host cell DNA.

Discussion Questions

5. Describe the use of restriction enzymes to prepare defined sequences of DNA. Would genetic engineering be possible without restriction enzymes? Why or why not?

6. If an individual research worker were able to determine the DNA sequences of 1000 bases of DNA in a single day, how long would that investigator require to sequence the entire human genome (approximately 6×10^9 base pairs)?

7. Imagine that you want to use genetic engineering to produce a complex human protein for medical purposes. The only catch is that you must have a product to market within 10 years. How would you go about choosing your production system? Would you choose bacteria, yeast, mammalian cells in culture, or intact animals? Explain your decision, and describe the steps in the process that you would have to perform.

8. Discuss some of the issues involved in the release of genetically engineered organisms into the environment.

9. Blood samples recovered at the scene of a crime are being used to identify a suspect. A license plate number noted at the scene has led to the arrest of the car's owner, Mr. J. Smith. He agrees to a DNA fingerprint analysis of his cells. The DNA laboratory finds that only 50 percent of the RFLP bands in Mr. Smith's blood match those of the unknown criminal. Mr. J. Smith claims he was at home when the crime was committed and that he had loaned his car to his brother, Mr. K. Smith. On the basics of the DNA test, is J. Smith's alibi believable? Should a DNA fingerprint analysis be performed on K. Smith? (*Hint:* Which relatives of any individual should show a 50 percent match of genetic material?)

READINGS

The legal status of DNA fingerprinting is still developing. Here are five recent articles that provide some of the essential background of the controversy.

Anderson, C. "FBI gives in on genetics." *Nature* 355 (1992): 663.

Chakraborty, R., and K. K. Kidd. "The utility of DNA typing in forensic work." *Science* 254 (1991): 1735–1739.

Leowontin, R. C., and D. L. Hartl. "Population genetics in forensic DNA typing." *Science* 254 (1991): 1745–1750.

Roberts, L. "Fight erupts over DNA fingerprinting." *Science* 254 (1991): 1721–1723.

Neufeld, P. J., and N. Colman. "When science takes the witness stand." *Scientific American* (May 1990): 46–54. A brief examination of the legal and ethical aspects of DNA fingerprinting in court.

Eijgenraam, F. "Yeast chromosome III reveals a wealth of unknown genes." *Science* 256 (1992): 730. A short summary of data from the first complete sequencing of a eukaryotic chromosome.

Oliver, et. al. "The complete DNA sequence of yeast chromosome III." *Nature* 357 (1992): 38–46.

Saltus, R. "Harvard wins gene-altered rodent patents." *Boston Globe* (Wednesday, December 30, 1992): 57.

Watson, J. D., M. Gilman, J. Witkowski, and M. Zoller. *Recombinant DNA.* 2d ed. New York: Scientific American Books, 1992. A thorough treatment of the range and detail of recombinant-DNA technology.

Mullis, K. B. "The unusual origin of the polymerase chain reaction." *Scientific American* (June 1990): 56–65. How did PCR, now so widely used in molecular biology, originate? The answer may surprise you.

Roberts, L. "To test or not to test?" *Science* 247 (1990): 17–19. A brief examination of the issues involved in cystic fibrosis screening.

Verma, I. M. "Gene therapy." *Scientific American* (November 1990): 68–72.

Weintraub, H. M. "Antisense RNA and DNA." *Scientific American* (January 1990): 40–47. An informative discussion of the theory and practice of the antisense technique.

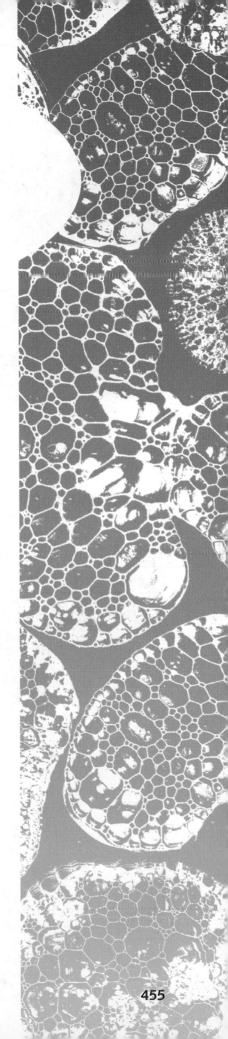

22

Molecular Evolution

*n*ear the end of the eighteenth century, Scotsman Mungo Park arrived in what is now Senegal, aiming to travel inland through West Africa and explore the Niger River. He began with 45 men, plenty of pack animals, and ample supplies. But no sooner had the expedition begun than his men began to die of malaria. Soon, Park's journal showed how routine death had become. "We made camp last night," a typical entry reads. "Three men died." Before the expedition was over, all of Park's men—like most Europeans who ventured into West Africa—had perished. Small wonder they called the area "White Man's Grave."

But that name itself reveals a mystery: as Europeans succumbed to malaria in droves, Africans went about their business. This is not to imply that local people never caught malaria. They certainly did. But many did not become as ill, and far fewer died. Even today, American and European visitors to Senegal gulp antimalarial drugs daily and slather insect repellent on every bit of exposed skin—for good reason. Yet tall, graceful, Senegalese, dressed in brilliant, flowing robes, travel from village to markets in the capital of Dakar and back again, apparently unconcerned about either mosquitoes or "the fever" those insects carry (Fig. 22.1).

A century after Mungo Park died and thousands of miles away, Dr. James Herrick of Chicago examined a young African-American patient with striking symptoms. The young man had a fever, chills, and numerous scars on his legs and thighs, some as big as silver dollars. His heart was enlarged and beat with an audible murmur, as though damaged. Then when Herrick made a routine blood smear, he saw something truly baffling. "The shape of the red cells was very irregular," he wrote in the *Archives of Internal Medicine*, "but what especially attracted attention was the large number of thin, elongated, sickle-shaped and crescent-shaped forms." (See Fig. 22.2.)

Within a few decades, scientists learned that *sickle-cell anemia*, as the syndrome was called, is inherited. They also learned that something was different about sickle-cell patients' hemoglobin, although they didn't know what caused the problem.

Figure 22.1 Many things have changed in Senegal since the beginning of the nineteenth century, but the hustle and bustle of the marketplace remains. Imagine the confusion of European visitors to the region at that time—visitors who almost invariably sickened and died of malaria, while local people went about their business. What mysterious protection did the indigenous people have that others lacked?

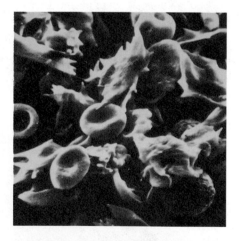

Figure 22.2 Red blood cells from a person with sickle-cell anemia.

Decades later still, the pieces of these seemingly separate puzzles fell into place—and merged into a single phenomenon. Sickle-cell anemia and Africans' mysterious resistance to malaria turned out to be phenotypic flip sides of the same molecular coin; they are both caused by the same change in a single nucleotide base pair within a single gene. The full story behind this evolutionary "devil's bargain" unites molecular biology, evolutionary biology, and medicine in a fascinating tale of molecular evolution.

MOLECULAR BIOLOGY MEETS EVOLUTIONARY THEORY

Modern biology has two main strategies for studying life. One approach, covered in Parts 2 and 3 of this book, stresses ecological and evolutionary interactions among organisms. Until recently, its practitioners treated genes essentially as black boxes. That attitude was sensible; even after the double helix was discovered, no one had a clue as to how the gene controlled processes at the organismal level. The other approach, described in the previous few chapters, concentrates on genes and molecules and treated *organisms* as black boxes. That approach, too, made sense; the evolution and behavior of multicellular organisms were just too complex to deal with at a molecular level.

But things are changing—and quickly. Ask almost any researcher today for the biological equivalent of "All roads lead to Rome," and he or she would probably suggest, "All studies lead to an understanding of molecular evolution." You've seen in the previous two chapters how gene action can be understood in chemical terms. Information coded in DNA is expressed by transcription to RNA and then translation into proteins, which help determine the phenotype of an organism. Thus any change in DNA has the potential to alter the phenotype of an organism. Conversely, a heritable change in an organism can be traced to changes in its genes.

Molecular biology now provides a host of tools to analyze the evolutionary process as it occurs in nature. The ability to read and manipulate DNA enables us to track the movements of alleles through populations and to move genes from one organism to another to study their effects. As a result, findings that would once have been considered "final results" in evolutionary experiments now serve as starting points for molecular studies. When biologists noted in the past, for example, that some species evolve and survive, while others stagnate and slip towards extinction, their findings simply puzzled theorists and either raised conservationists' hopes or dashed them. Today, those same observations inspire questions about how species' genetic characteristics influence their fates.

Mutations as Sources of Evolutionary Change

Darwin, Mendel, and early geneticists knew that heritable changes can occur spontaneously in both plants and animals, producing individuals called "sports" that were frequently used to breed new varieties (Fig. 22.3). The term **mutation,** from the Latin word meaning "to change," is applied to such heritable changes. Mutations can be passed on to future generations, they can occur in any gene, and they usually occur at random.

An early clue to the nature of mutations came from studies on the effects of radiation (see p. 275). In 1927, H. J. Muller found that flies irradiated with high levels of X-rays showed a dramatic increase in the incidence of mutations. Muller suggested that radiation caused physical changes in the molecules of genes. Later it was discovered that certain chemicals could also act as mutagens, agents that cause mutations. Today we know something Muller did not—that genes are made of DNA. Therefore, *mutations are heritable changes in DNA.*

Many nonscientists' feelings about mutations seem to be shaped by sci-fi thrillers such as *Godzilla* or *Toxic Avenger.* Most people think that mutations are always "bad," in no small part because "mutants" who carry them are invariably portrayed in films as either deformed individuals or giant insects bent on destroying Tokyo or New York. That general impression has been unintentionally fostered by the science of genetics itself. How? In classical genetics,

Figure 22.3   These mutants of black-eyed Susan have dramatically enlarged flowers.

the only mutations discovered were those that had observable (and often deleterious) effects—such as the useless wings of fruit fly mutants or the inability of bacteria to survive in their usual nutrient medium. In addition, of course, scores of awful human diseases, such as Tay–Sachs and Huntington's, are caused by mutations that interfere with normal gene function.

It may therefore surprise you to learn that mutations are essential to life as we know it. Why? Because mutations are a major source of genetic *variability* in a population, and inheritable variation (as Darwin knew) is essential to evolution. A population of organisms in which DNA flawlessly replicated itself, producing new generations without a single mutation, would lack the variation on which natural selection operates—and would therefore never evolve.

We now know that mutations occur constantly in nature. We also know that some mutations are beneficial, many are harmful, and many more don't make much difference at all to the organisms that carry them. Evolutionary change seems to require that mutations be *rare enough* so that problems caused by deleterious mutations don't become overwhelming, but *common enough* so that the potentially useful effects of beneficial mutations allow change and natural selection to continue.

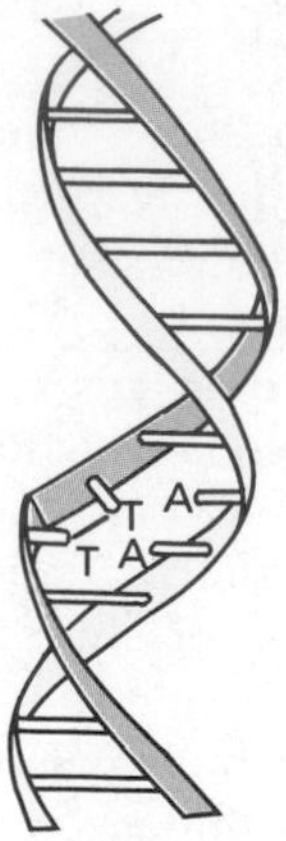

Figure 22.4 Thymine dimers are produced by ultraviolet light. DNA repair enzymes search for such dimers, remove them from DNA, and replace the missing bases. If such dimers are not removed, mutations result.

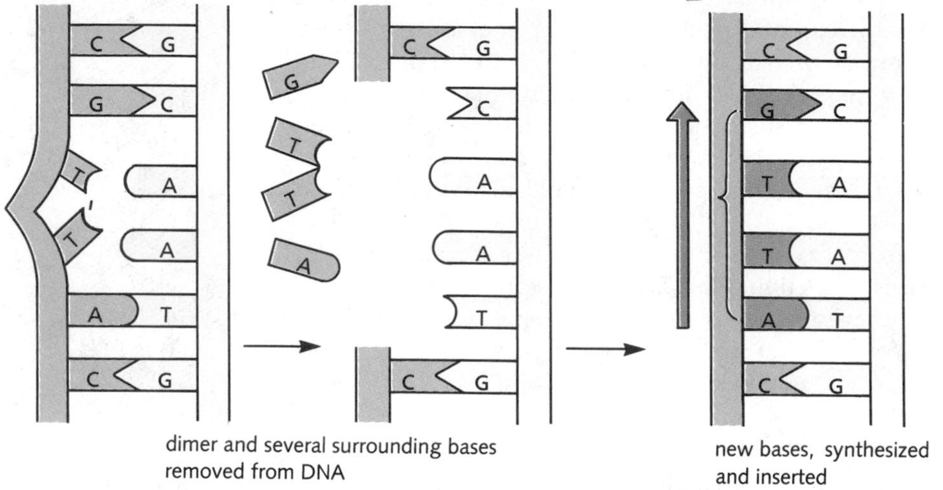

dimer and several surrounding bases removed from DNA

new bases, synthesized and inserted

Figure 22.5 Individuals with two copies of a recessive allele for defective DNA repair suffer from a disorder known as *xeroderma pigmentosum*. Unable to repair the damage done by ultraviolet light, these individuals suffer extensive skin tumors produced by exposure to sunlight.

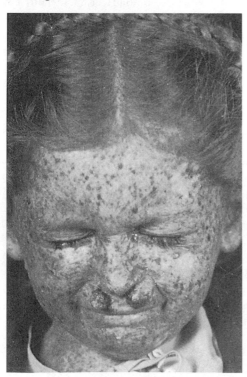

Sources of mutations Many naturally occurring mutations are caused by errors during the replication of DNA. When the double helix is duplicated, nucleotide bases that form the new strands are matched with preexisting strands according to the base-pairing mechanism discussed in Chapter 20. Although this matching process usually works well, mistakes occasionally occur when an incorrect nucleotide is inserted into one of the new DNA strands. Once such a mistake has been made, subsequent replication of the affected strand can pass that mistake along to succeeding generations of cells.

Some mutations, however, are caused by specific agents known as **mutagens.** Chemical mutagens are molecules that increase the rate at which mutations occur in DNA. Many chemical mutagens are molecules that can chemically displace correct nucleotides during DNA replication and therefore increase the rate of mutations dramatically. Others affect the function of DNA polymerase in such a way that the general error rate during replication is increased. Physical mutagens, such as ultraviolet light, X-rays, and gamma radiation, can also produce chemical changes in DNA, as shown by Muller's classic experiments with X-rays.

(a)

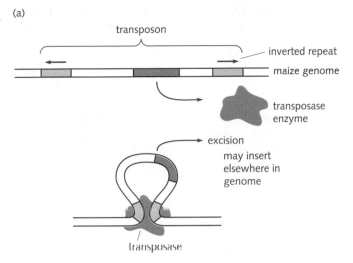

transposon

inverted repeat

maize genome

transposase enzyme

excision

may insert elsewhere in genome

transposase

(b)

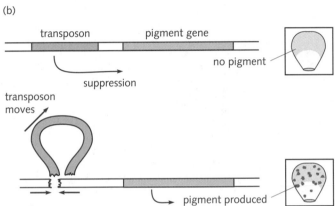

transposon

pigment gene

no pigment

suppression

transposon moves

pigment produced

Figure 22.6 **(a)** Transposable genetic elements, also known as transposons, have two essential elements. At either end of the transposons are short "inverted repeats" from 8 to 30 bases in length. These sequences enable a transposase enzyme, often produced by a gene within the transposon, to recognize the transposon sequence and catalyze its insertion or removal from the chromosome. **(b)** One transposable element in maize suppresses the expression of a pigment gene, producing pale-colored kernels. When the transposon moves from this position, the pigment gene is reactivated and a spot of red color in the kernel is the result. **(c)** The mottled colors of these maize kernels are the result of transposable genetic elements that have disrupted the gene coding for pigmentation enzymes. McClintock's studies provided the first proof that DNA segments could move from one position to another within the genome.

(c)

Ultraviolet light, for example, is strongly absorbed by DNA molecules. Energy absorbed from UV light can cause the formation of a covalent bond between two adjacent thymidine bases, producing a thymine dimer. Most cells have a repair mechanism that can remove the dimer and replace the damaged bases (Fig. 22.4). But if DNA replication takes place *before* the dimer can be excised, random bases are placed in the new DNA strand, and a pair of adjacent base substitutions results. Individuals lacking fully functional DNA repair mechanisms are extremely sensitive to sunlight and must take measures to protect themselves against exposure to help guard against a greatly increased risk of skin cancer (Fig. 22.5).

Transposons: "Jumping" genes The mutations described above involve changes in individual bases. But some mutations involve hundreds and even thousands of bases and result from large-scale movements of whole regions of a chromosome. Large sections of DNA may be *deleted, inverted,* or moved from one region of the genome to another (*transposed*). Because these mutations occur at

the level of the chromosome, they are sometimes referred to as chromosomal mutations.

Many chromosomal mutations are produced by transposable elements, short segments of DNA that are removed and then reinserted in the chromosome. These **transposable elements,** sometimes called **transposons,** were first discovered in maize (corn) by Nobel Prize winner Barbara McClintock during the 1940s and 1950s. McClintock's work, which took place well before the era of modern molecular biology, showed that certain genes could "jump" from one place to another in the maize genome (Fig. 22.6a). In one system she investigated, the instructions within a transposon suppress the expression of a neighboring pigment gene. If that transposon "jumps" out of position, the pigment gene is activated, producing a colorful speckling in the kernels of corn (Fig. 22.6b, c).

McClintock's work has produced a new appreciation among scientists of how flexible and dynamic chromosomes may be. We now know that transposons are found in many organisms, including other plants, bacteria, flies, yeast, and humans.

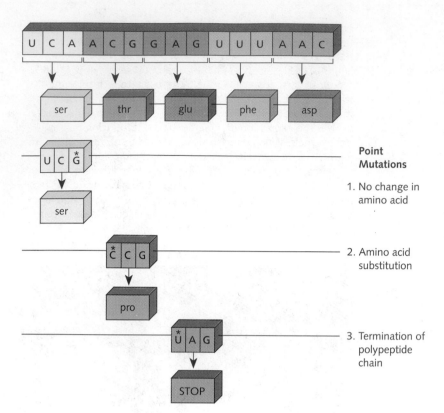

Figure 22.7 The effect of a substitution mutation depends on the location of the base change. In some cases, there is no change in the amino acid specified by the codon. In other cases, a different amino acid is specified, or a stop codon that causes premature termination of the polypeptide chain is produced, as shown.

Point Mutations

1. No change in amino acid

2. Amino acid substitution

3. Termination of polypeptide chain

EFFECTS OF MUTATIONS

Although mutations may be caused in different ways, they share a common feature: *they alter the base sequence of DNA molecules.* The effect of those alterations on the phenotype of an organism depends both on the *nature* of the change and on the *location* of that change with respect to the functional parts of genes.

Mutations that involve changes in only one or two bases are known as **point mutations,** because they exert their effect at a specific *point* in the gene. Base substitutions are point mutations involving a change in a single nucleotide base. What effect can a base substitution have? In some cases, it has no effect, but generally, the substitution *does* affect the amino acid specified by its codon (see genetic code chart, Fig. 20.12). Such changes can drastically affect the activity of a protein. A substitution may also change an ordinary codon into a stop codon, causing only the first part of the protein to be made (Fig. 22.7). Many genetic diseases (including colorblindness, sickle-cell anemia, and hemophilia) are the result of a few base changes in critical regions of important genes.

Point mutations can also involve the insertion or deletion of bases. If insertions or deletions occur in a protein-coding region, they can have major effects on gene products. Consider, for example, a gene in which a strand of 300 nucleotides codes for a polypeptide containing 100

amino acids. Remember that the genetic code is always read in groups of three bases, beginning with a start codon. Thus if a single base is added or deleted anywhere "downstream" of the starting point, that mutation causes a *shift* in the *reading frame* (Fig. 22.8). Such changes, known as **frameshift mutations,** disrupt the reading of every subsequent codon, radically change large portions of the genetic message, and often produce nonfunctional proteins.

Neutral Mutations

Many mutations, however, do *not* affect phenotype. They are called "neutral mutations," meaning that they don't change gene products in any important way. Recent studies using the techniques of molecular biology suggest that neutral mutations are common—in other words, base changes accumulate in those parts of the genome where they least affect protein function.

Take, as one example, introns—the intervening sequences within genes that are removed before mRNAs are translated. Evidence suggests that most mutations in introns don't affect the protein product. An evolutionary biologist might explain the result this way: Because intron mutations rarely alter phenotype, there is little or no pressure from natural selection on base changes in introns. Therefore, such changes can occur with little effect on an organism's chances for survival. This is why introns

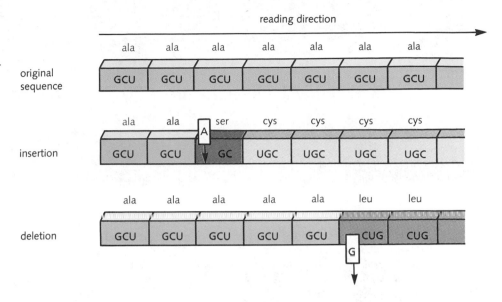

Figure 22.8 The insertion or deletion of a base can result in a frameshift mutation, in which the reading frame of the mRNA is shifted. This has a dramatic effect on the protein produced by the mRNA, changing most of the amino acids that are "downstream" from the point of the mutation.

are often regions of rapid genetic change and why introns taken from different individuals within the same species often differ from one another (Fig. 22.9).

This is not the case in exons—regions that actually code for the amino acid sequences of proteins. Exon mutations often do alter phenotype and therefore are subject to strong pressure from natural selection: substantial changes may be lethal.

TRACING THE COURSE OF EVOLUTION

Because DNA sequences (and the proteins they code for) change over time, we can compare proteins and genetic sequences to study the evolutionary relationships of plant and animal species. Using molecular information in this way is not easy to do, but the concept behind the procedure is quite simple. The more time that has passed since two species shared a common ancestor, the more differences should exist between their DNA (and the proteins coded for by that DNA). Conversely, the more alike the DNA and proteins of two species are, the more recently they diverged from one another. This approach has been used successfully by one group of researchers to identify the mainland ancestors of a unique group of Hawaiian plants (see Theory in Action, Molecular Links: Hawaiian Plants with Californian Roots).

Figure 22.9 Introns may serve as places where mutations can accumulate. When many mutations have altered the base composition of an intron, a final mutation at the border of an intron may alter the site that normally serves as the cutting point for RNA-processing enzymes. This may cause the entire intron to be added to the coding portion of the protein, introducing a substantial change in the amino acid sequence of the protein product.

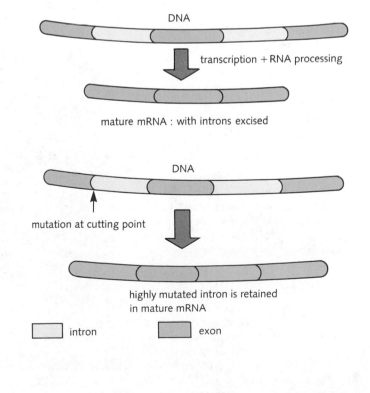

Molecular Links: Hawaiian Plants with Californian Roots

The Hawaiian Islands, like the Galapagos, were populated by colonists—animals and plants as well as humans. Biologists working on island fauna and flora often struggle to connect living island species with their mainland ancestors. Sometimes that job is straightforward; researchers determined many years ago that Darwin's finches sprang from a South American founding species. Other times, however, making those connections can be quite difficult.

One persistant challenge has been identifying the closest living mainland relatives of the unique Hawaiian plants known as silverswords (see photos).

Mount Haleakala on Kauai holds the dubious distinction of being the wettest place on Earth, and Maui's Haleakala crater receives less rainfall than a typical desert. Yet both these environments (and many others) are home to one or more of the 39 silversword species, which range in habitat from shrubby perennials to bushes, trees, and even vines.

But where did silverswords' ancestors come from? Most Hawaiian organisms "island-hopped" across the South Pacific from Australia or Asia. But silverswords most closely resemble an obscure group of plants called tarweeds from western North America. Tarweeds, though, live on mountaintops more than 2400 miles away over the open sea. No plant had ever been known to make that long a journey before, so botanists faced a real challenge in proving the connection between tarweeds and silverswords. Researchers tried one standard test—crossing living silverswords and living tarweeds with one another. But crosses set up according to physical appearance alone didn't work; none of the several attempts produced viable offspring. And randomly attempting to cross each of 99 extant tarweed species with 39 different silverswords was obviously out of the question.

The breakthrough came when botanists Donald Kyhos of the University of California at Davis, Gerald Carr of the University of Hawaii, and Bruce Baldwin, now of Duke University, compared DNA from silverswords and tarweeds. Specifically, they studied the DNA of the plants' chloroplasts, looking for similarities and differences. When they found what they thought was a close match between a silversword and a tarweed, they crossed the plants—and could hardly believe their eyes: the experiment produced healthy hybrid plants for the first time. An ancient link between silverswords and tarweeds had been proven, based on molecular clues! To evolutionary biologists, this was a first—the identification of close relatives of probably extinct continental species that founded a group of island species.

The two silversword species shown here are members of a diverse group of species that inhabit island ecosystems ranging from deserts to swamps. After silverswords' ancestors reached Hawaii—probably in the form of a single seed lodged in the feathers of a migrating bird—the archipelago's ecological patchwork allowed them to experience a dramatic adaptive radiation.

The Hawaiian silversword, *Argyroxiphium sandwichense macrocephalum* (LEFT) grows in dry volcanic craters. Its relative, *Dubautia laevigata* (RIGHT), belongs to a genus with members adapted to habitats ranging from sunbaked lava flows to swamps. These are but two of roughly three dozen members of the silversword alliance, unique to the Hawaiian chain.

Studies of chloroplast DNA suggested that *D. laevigata* was the Hawaiian species most closely related to the California tarweed, *Raillardiopsis muirii* (TOP). Crosses between these two species confirmed that relationship by producing viable hybrid plants (BOTTOM).

To compare data from many species at once, researchers use computer programs. The best of those programs take data from several different DNA regions (or several different proteins) from each species and use them to construct a host of possible evolutionary trees. The programs then test those trees—each of which relates the species differently—to determine which produce the best fit with the available data.

In general, molecular data agree quite well with evidence that paleontologists have found in the form of fossils. If we examine the differences in amino acid sequence between particular proteins in different species and compare the results to the length of time the species have been separate according to the fossil record, we get a strong correlation. Take, for example, the respiratory protein cytochrome c. There are 43 amino acid differences between human cytochrome c and that of yeast—two organisms separated by hundreds of millions of years of evolution. There are only 20 such differences between humans and bullfrogs—two vertebrates who shared a common ancestor more recently. Only 9 differences are found between humans and rabbits—both mammals, closer relatives still. Only a single difference exists between the cytochrome c of humans and that of rhesus monkeys. Finally, cytochrome c proteins from humans and chimpanzees—probably our closest living relatives—are *identical*.

Molecular Clocks: Evolutionary Timekeepers?

According to some researchers, molecular studies enable us to determine not only the *order* of evolutionary branching points but their *timing* as well. These researchers argue that neutral mutations—which are neither favored nor discriminated against by natural selection—act like "molecular clocks" that record moments of evolutionary time at a steady, predictable rate (Fig. 22.10).

Once again, though the notion is theoretically simple, turning theory into practice is more complicated than it first appeared. We now know that there is no single, universal molecular clock, because different parts of the genome change at different rates, and because the rate of neutral mutations may vary among different evolutionary lines. As a result, many original claims of molecular clock proponents have been challenged by researchers who study relationships among organisms in different ways.

Nonetheless, molecular data can provide valuable clues in situations where the fossil record is poor or absent. As you will see in Part 5, molecular data are used to study the very earliest events in the evolution of life—a period for which useful fossils are virtually nonexistent. If current interpretations of those data are correct, they suggest a dramatic restructuring of the traditional tree of life. At the other end of the evolutionary time scale, molecular approaches are also being applied to chart recent events in the history of *Homo sapiens*. There, too, the fossil record is spotty, perhaps because our primate forebears and relatives never flourished in very large numbers.

Molecular studies of human ancestors, though still controversial, are steadily gaining adherents. Molecular clock data gathered by the late Alan Wilson of Berkeley, for example, successfully challenged the once inviolate notion that human ancestors diverged from our closest relatives around 20 million years ago. Although debate continues, many researchers now agree that about 7 mil-

lion years seems closer to the mark. Other molecular sequence data, gathered by a team led by Maryellen Ruvolo of Harvard, suggests not only that humans are more closely related to chimps than to gorillas, but also that we are more closely related to the two living species of chimps than either chimps or humans are to any other great ape. This extensive molecular similarity led Jared Diamond of UCLA Medical School to suggest that both species of chimpanzee should be placed in the same genus as humans—a suggestion he made in the provocatively titled essay "The Third Chimpanzee" (Fig. 22.11).

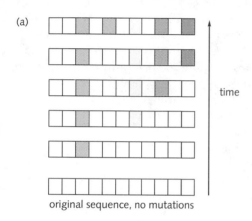

(a)

time

original sequence, no mutations

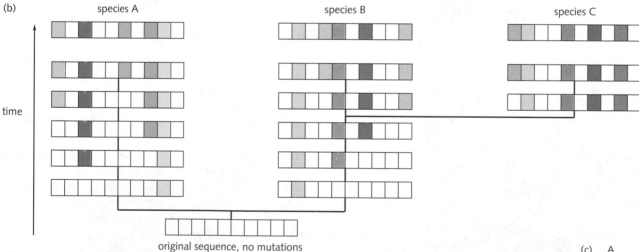

(b)

species A   species B   species C

time

original sequence, no mutations

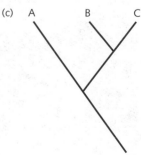

(c)   A   B   C

Figure 22.10   Molecular biology can be used to trace the evolution of several related species. **(a)** Because mutations accumulate over many generations, over time, a DNA sequence will become increasingly different from an ancestral sequence. In this example, a total of five base changes have cropped up over time. **(b)** If two species diverged in evolution some time ago (species A and B), the base changes they accumulate will be quite different. Species A and B differ in all ten of the positions shown. However, if two species share a recent ancestor (such as species B and C), they will share many of these base changes. Species B and C differ in only four positions. **(c)** By comparing DNA sequences between the two species, it is possible to construct a branching tree that shows the evolutionary relationships between the three species. This tree, based solely on molecular data, indicates that species B and C shared a more recent common ancestor with each other than they did with species A.

MOLECULAR MECHANISMS OF EVOLUTIONARY CHANGE

"All this relationship stuff is well and good," you might say, "but is there any evidence linking *specific* changes in DNA with evolutionarily significant changes in phenotype?" Could something as simple as a point mutation, for example, enable a population to overcome a survival challenge in its environment? Intriguingly, the most dramatic example of just such a situation occurs in our own species; it involves the case of malaria resistance described at the beginning of this chapter.

Point Mutations and the Case of the Sickled Cell

The full story of that phenomenon begins with molecular biology. The mutation involved in this case is about as small a change in DNA as you can imagine: a single base substitution in the gene for β-globin, one of the polypeptides in the hemoglobin molecules that fill our red blood cells. That point mutation causes a change in *a single* amino acid: at position 6 on the β-globin chain, the amino acid *glutamate* is changed to *valine* (Fig. 22.12). This simple change, from a charged amino acid to the nonpolar valine, makes the entire molecule—which is given the name *Hemoglobin-S*—a bit less soluble. Because of that solubility change, when the oxygen concentration surrounding Hemoglobin-S molecules drops below a certain level, many of them come out of solution and form bundles of tiny fibers within the red cells. These bundles of fibers push out against the inside of the cell membrane, producing the characteristic sickle shape.

But how does a mutation like this become common in a population? And what has this molecular phenomenon to do with either resistance to malaria or sickle-cell anemia? For the answers to those questions, we need to take

Figure 22.11 Application of molecular clocks to human evolution. **(a)** Older studies, based on paleontology and comparative anatomy, posited that humans diverged from other primates 20 million years ago and that chimps were most closely related to the other great apes. **(b)** New studies, including molecular clock data, moved the human–ape split up to about 7 million years ago. Parallel studies suggest that humans are more closely related to both living chimpanzee species than either chimps or humans are to other apes.

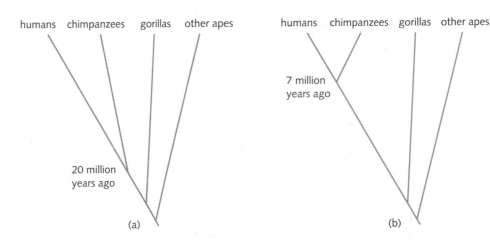

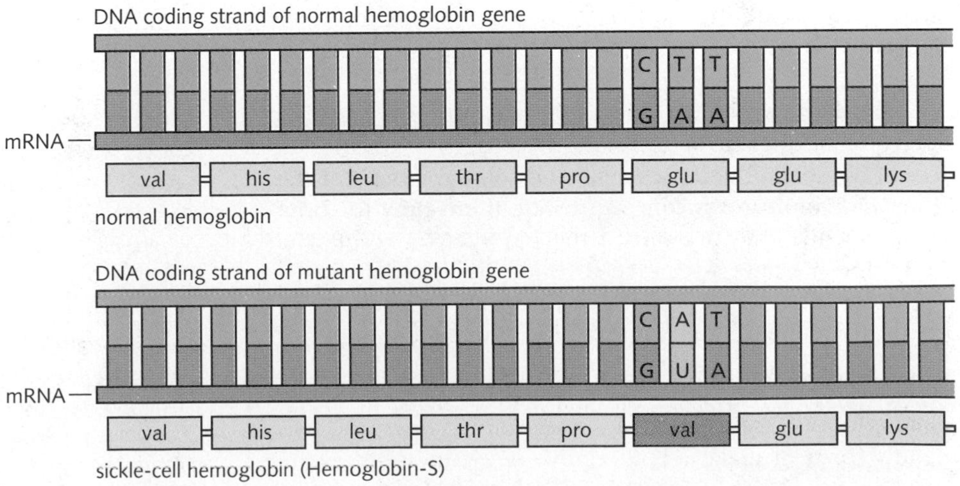

Figure 22.12 The base-pair substitution in the sixth triplet of the mutant hemoglobin gene causes the amino acid valine to replace glutamic acid, the amino acid normally found in hemoglobin. The result is sickle-cell hemoglobin (Hemoglobin-S).

Figure 22.13 The life cycle of the parasite *Plasmodium*, which causes malaria.

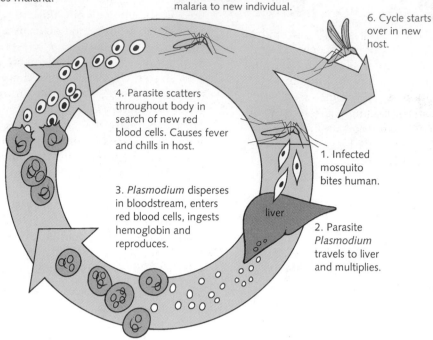

5. If female mosquito bites infected person, it carries the malaria to new individual.

6. Cycle starts over in new host.

4. Parasite scatters throughout body in search of new red blood cells. Causes fever and chills in host.

1. Infected mosquito bites human.

3. *Plasmodium* disperses in bloodstream, enters red blood cells, ingests hemoglobin and reproduces.

liver

2. Parasite *Plasmodium* travels to liver and multiplies.

an interdisciplinary approach uniting medicine, molecular biology, evolutionary biology, physiology, and history.

For a start, it is important to understand how malaria operates and how deadly a foe it can be (Fig. 22.13). Once transmitted to a human by the bite of an infected mosquito, the parasite—a member of the genus *Plasmodium*— heads straight for the liver, where it lives for a short while. It then emerges, disperses in the bloodstream, and searches out red blood cells. When it locates those targets, it enters them, feeds on the hemoglobin they contain, and reproduces furiously. Ultimately, the burgeoning parasite population shatters the host cell and scatters throughout the body in search of new red blood cells—causing fever and chills in the process. This cycle repeats over and over again.

Now here's the problem: because these parasites spend most of their lives *inside* our cells, they are hidden from our normal defenses against disease for most of the time. Our immune system can tackle them only for brief periods while they are out in the blood. As a result, the body doesn't have a chance to build up the defenses it normally uses to defeat disease. That's why a person who encounters malaria for the first time runs such a high risk of dying.

Hemoglobin-S and malaria Apparently, however, this deadly form of malaria is a relative newcomer—in evolu-

tionary terms, that is. For much of prehuman and early human history in Africa, it did not infect humans. Sometime during that period, the mutation that produces Hemoglobin-S occurred—probably at least three times in Africa and once or twice in Asia. As it turns out, people who carry only a single copy of this mutation have red blood cells filled with a roughly 50–50 mixture of Hemoglobin-S and normal hemoglobin. These heterozygotes for the Hemoglobin-S allele suffer no ill effects, so the mutation could persist and even spread through local populations with no problem. But because the mutation conferred no real benefit, it was probably not particularly common.

Then, however, the virulent modern form of malaria appeared, probably about 10,000 years ago. Malaria epidemics swept across the Old World tropics, and Hemoglobin-S was suddenly thrust to center stage. Why? Because *people heterozygous for the Hemoglobin-S allele had better chances of surviving their first malarial attack than those who didn't carry it.*

As Ronald Nagel of Albert Einstein College of Medicine explains, Hemoglobin-S heterozygotes are far more likely to survive their first malarial infections than people who don't carry that allele. Apparently, that 50–50 mixture of Hemoglobin-S and normal hemoglobin thwarts malaria parasites in several ways, two of which seem particularly important. When parasites infect red cells, they

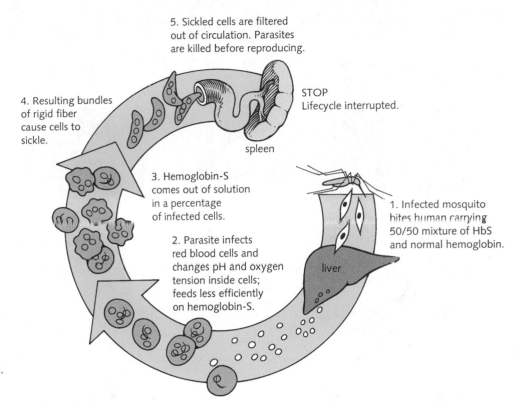

5. Sickled cells are filtered out of circulation. Parasites are killed before reproducing.

4. Resulting bundles of rigid fiber cause cells to sickle.

STOP
Lifecycle interrupted.

spleen

3. Hemoglobin-S comes out of solution in a percentage of infected cells.

1. Infected mosquito bites human carrying 50/50 mixture of HbS and normal hemoglobin.

2. Parasite infects red blood cells and changes pH and oxygen tension inside cells; feeds less efficiently on hemoglobin-S.

liver

Figure 22.14 When the mosquito, infected with the malaria-causing parasite *Plasmodium*, bites a human who is heterozygous for Hemoglobin-S, the parasite's life cycle is interrupted. Even if the parasite is not killed when the blood cells become sickled, *Plasmodium* does not feed well on Hemoglobin-S, and its growth and reproduction will be thwarted.

change both the pH and the oxygen tension inside, inducing the Hemoglobin-S to come out of solution. The resulting bundles of rigid fibers change the shape of the cells, causing them to sickle (Fig. 22.14). This sickling alerts the spleen that something is wrong; it then filters those cells out of circulation—killing the parasites before they can reproduce. What's more, malarial parasites don't seem to feed on Hemoglobin-S very well, so their growth and reproduction are impaired.

What all this means is that Hemoglobin-S heterozygotes have a vital advantage when infected with malaria: they are protected from the life-threatening first exposure to the parasite long enough for their immune system to "learn" how to handle later attacks. That, in turn, means that heterozygotes are more likely to survive and have children than people who lack their additional protection.

The result was a classic example of natural selection in action. Early on, the allele was uncommon enough that Hemoglobin-S heterozygotes usually married people who didn't carry the allele. Because heterozygotes had more children, and because half of those children were themselves heterozygotes, they were in turn protected from malaria with no ill effects. Over time, the frequency of the Hemoglobin-S allele rose. Soon the Hemoglobin-S allele was common across nearly all wet, tropical regions from Africa into Asia—a distribution that tracks the incidence of malaria (Fig. 22.15).

Sickle-cell anemia Sooner or later, however, the cost of this protection became evident. Remember that Hemoglobin-S was not an ideal solution to a problem that was "designed" with any sort of forethought; it was a random mutation, no more, no less. It just happened to make life difficult for the *Plasmodium* parasite without interfering too badly with hemoglobin's vital function—*in heterozygotes.* But as the frequency of the allele in populations rose, heterozygotes became common enough that they began having children with each other. Half the children from those marriages were also heterozygotes who shared their parents' molecular blessing. But by the laws of Mendelian genetics, one quarter of heterozygotes' offspring were *homozygous* for Hemoglobin-S. And those children had serious problems.

When a person's hemoglobin is composed entirely of Hemoglobin-S, red blood cells can sickle in droves under a variety of circumstances. As those sickled cells pass through the smallest passageways of the circulatory system, they get stuck in tight spots and clog the flow of blood. Pressure builds up behind the blockage, and small blood vessels can burst; internal bleeding occurs. Organs with rich blood supplies, such as the spleen, heart, and

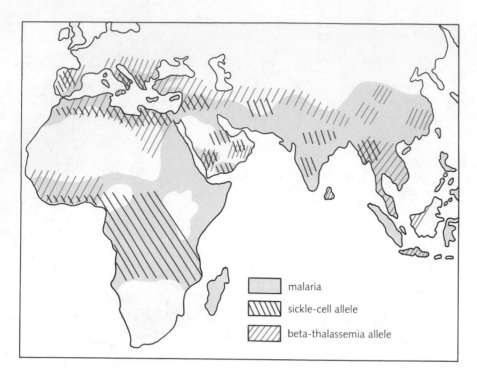

Figure 22.15 Distribution of the sickle-cell allele in human populations closely parallels the incidence of malaria.

malaria

sickle-cell allele

beta-thalassemia allele

liver, are repeatedly damaged by sickle-cell crises in which blood flow is blocked in many places at once.

These are the classic symptoms of sickle-cell anemia. People who suffer from the disorder have greatly reduced life expectancies, often succumbing to severe internal bleeding or blood clots in their lungs.

Thus, the negative effect of the Hemoglobin-S allele on homozygotes created a negative selective pressure that acts counter to positive effects on heterozygotes. The result is a life-and-death balancing act. The allele is common where malaria exists, because in such environments the positive effects of the allele outweigh its negative impact. Today, Hemoglobin-S is found in roughly 15–20 percent of Senegalese. In places where malaria is more common, the allele's frequency can reach 40 percent. Sickle cell is rare where malaria is absent, because the negative effects of the allele tend to remove its carriers from the population. The significant frequency of Hemoglobin-S allele among contemporary African-Americans (as high as 10 percent in some areas) has a straightforward, though repugnant, explanation: their ancestors were abducted into slavery only a few generations ago from malarial regions of Africa.

Gene Duplication: A Route to Diversity

The story of malaria and Hemoglobin-S offers dramatic proof that even a point mutation can have major effects on a population's survival in difficult environments. But in and of itself, that phenomenon doesn't explain how mutations might shape new species—organisms that function in fundamentally different ways than their ancestors. Molecular biologists, however, have uncovered other mechanisms that seem to have made just those sorts of striking phenotypic changes.

One of those mechanisms begins with a process called *gene duplication.* Sometimes, for reasons that we don't yet understand, a cell's DNA-copying machinery acts like an erratic photocopying machine; instead of producing only a single copy of DNA, it turns out two or more copies of entire sections of the genome. As a result, the progeny of the affected cell end up carrying several copies of numerous genes. Much of the time, apparently, those extra copies serve no function and begin to accumulate mutations at random. Over a long enough period of time, their original message may be garbled, and they can become the genetic equivalent of rusting old cars in an automobile graveyard—part of what is often called "junk DNA" (see Current Controversies: "Junk DNA," p. 473).

But numerous times in the history of life, those extra copies haven't been abandoned. Instead, they remain active, but because there are several copies "available," some of them can accumulate mutations that change their function *without depriving the organism of the original protein* (Fig. 22.16). It now seems clear to many researchers that the entire human genome was duplicated once, and then portions of it have been duplicated several times since then.

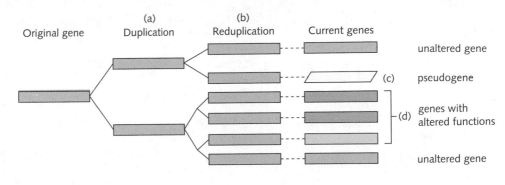

Figure 22.16 Gene duplication. As molecular biologists "read" DNA, they find evidence that genes and even entire genomes have been duplicated over time. Many researchers believe that duplications are the molecular basis for the evolution of new gene functions. **(a, b)** Once a gene is duplicated, some of the copies may undergo mutations in coding regions or promoters that change their function. Over time, such changes can cause the copies to degenerate into what are called "pseudogenes"—stretches of DNA that resemble genes but have lost all function **(c)**—or into new genes with different functions **(d)**.

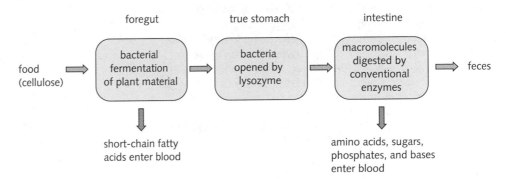

Figure 22.17 Lysozyme and the evolution of grazing mammals. Ruminants live on high-cellulose diets—despite the fact that no mammals produce cellulose-digesting enzymes—because bacteria in their stomachs digest the plant material for them. But then what? Bacterial digestion of cellulose alone frees up only certain nutrients for absorption by the host animal. Animals receive the full benefits of their diet only if they can digest the bacteria.

Lysozymes: Theme and variation One example of how gene duplication fueled the evolution of an animal group is no further away than your local farm. Most mammals that humans raise for milk or meat belong to the group known as ruminants. One reason these animals are so useful is that their digestive systems enable them to survive on foods like twigs, leaves, and grass that are high in cellulose. Why is that so exceptional an ability? Because no multicellular animals can produce the enzyme necessary to break down cellulose. Ruminants manage because their guts play host to cultures of cellulose-eating bacteria.

That's all well and good. But after the bacteria digest cellulose, the animals are left with rich cultures of bacteria—a stomach full of organisms protected by very tough cell walls. To make use of bacterial cellulose-digesting skill, the animals must, in turn, be able to digest the bacteria.

Ruminants accomplish that feat by producing a special stomach enzyme that breaks down bacterial cell walls (Fig. 22.17). Interestingly, that stomach enzyme is almost identical to another enzyme, called lysozyme, which is produced by an enormous variety of animals in just about every body secretion, including tears, saliva, and blood. Lysozyme protects eyes and other vulnerable body openings against bacteria by destroying their cell walls before they can cause any trouble. Because lysozyme is found in

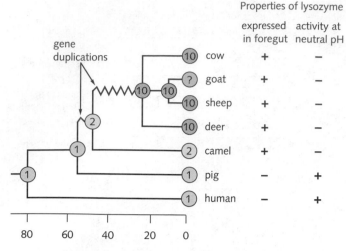

	Properties of lysozyme	
	expressed in foregut	activity at neutral pH
cow	+	–
goat	+	–
sheep	+	–
deer	+	–
camel	+	–
pig	–	+
human	–	+

Figure 22.18 The numerous genes that produce stomach lysozyme in ruminants have evolved through duplication and subsequent modification of a single gene found in nearly all animals. This chart, prepared by David Irwin of the University of Toronto and several colleagues, presents their hypothesis concerning the "evolutionary tree" of modern lysozymes. The small numbers in circles represent the number of copies of the lysozyme gene present at each stage, and the zigzag lines represent duplication events. (NOTE: Data and diagrams taken from an article by Irwin, Prager, and Wilson.)

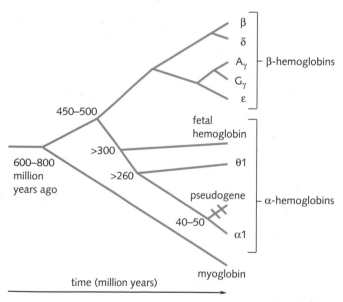

Figure 22.19 The globin gene family and its evolutionary history. It appears that all the known globin genes (like lysozymes) can be traced to a single common ancestral gene that has been duplicated and modified through time. Its current descendants include myoglobin (the oxygen-binding compound that makes red meat red) and the several forms of α- and β-hemoglobin that combine to form the four-part polymer found in human red blood cells. Note that the α-globins include fetal hemoglobin, a type of hemoglobin normally expressed only *in utero* (for reasons discussed in Chapter 36). [NOTE: Data and diagram adapted from Wen-Hsiung Li and Dan Graur (1991).]

such a wide range of animals, researchers believe that the original lysozyme gene arose very early in the history of life.

But "standard issue" lysozyme operates in liquids whose pH is close to neutral. If placed in the highly acidic stomach environment, it would "curdle," in much the same way that the proteins in milk curdle if poured into lemon juice. That is, lysozyme itself would curdle if it weren't digested by other stomach enzymes first!

To work in the stomach, therefore, the structure of lysozyme has to be altered in a way that protects it—while retaining its ability to digest bacterial cell walls. But how could that be done without compromising its vital original function? Apparently, through gene duplication.

David Irwin, now of the University of Toronto, used the tools of molecular biology described in the previous chapter to study lysozyme genes in numerous living ruminants. He found that all species have several copies of the gene. By determining the number of copies each species carries and comparing those data with the group's family tree, he decided that the gene was first duplicated in a common ancestor that lived roughly 55 million years ago. He also suggests that additional duplications occurred much later, somewhere between 25 and 40 million years ago (Fig. 22.18). Subsequent random mutations in introns, exons, and promoter regions of the duplicated genes ultimately produced the stomach enzymes of modern ruminants. In the meantime, the original gene retained its infection-fighting function.

Irwin hypothesizes that all these modifications took somewhere around 20 or 30 million years to occur, so fully functional stomach lysozyme probably first occurred between 25 and 40 million years ago. The increased ability to utilize difficult sources of plant food had very far-reaching evolutionary and ecological consequences; the ancestors of modern ruminants displaced earlier large plant-eating animals and soon became Earth's dominant large herbivores.

Gene families Lysozyme is far from the only example of gene duplication, which seems to be fairly common. Molecular biologists have identified several large groups of genes that each appear to have evolved from a single ancestral gene that was duplicated and modified over time. These groups of genes are called "gene families," and some of them contain as many as 200 members (Fig. 22.19). One such family contains the immunoglobulin genes, some of which you will learn about in Chapter 42. The proteins produced by these genes each perform slightly different tasks, but nearly all of them help cells recognize each other in some way. Other examples of this phenomenon include the genes for proteins found in muscle cells and multiple genes for common cellular proteins such as tubulin, which makes up the microtubules found in most cells.

THE IMPORTANCE OF GENETIC DIVERSITY

You've undoubtedly heard the saying "Variety's the spice of life." Molecular biologists who study evolution generally agree; genetic diversity seems vital, both to the evolutionary process overall and to the survival of many species. Circumstantial evidence to support that notion is abundant.

One still-unfolding story involves the state bird of Hawaii, a large handsome goose called the nene (pronounced nay-nay) that is found nowhere else in the world (Fig. 22.20). We know from fossil evidence that nenes once populated all the main Hawaiian Islands in great numbers—until humans arrived. The first settlers, Polynesians, hunted nene for food, and by the time the celebrated Captain Cook made landfall, only between 50,000 and 100,000 wild birds remained, and only on Maui and Hawaii. Thereafter, whalers killed the tasty birds by the hundreds, farmers usurped their habitat, and mongooses and rats introduced to the islands intentionally and by accident preyed upon eggs and young.

By 1938, only scattered nenes remained in the wild, along with a few dozen in a captive colony maintained by Herbert Shippman, a conservation-minded Hawaiian. By 1950, there were probably no more than 30 birds left anywhere in the world. After that, thanks to Shippman's efforts and a state-sponsored captive-breeding program, the nene population rose slowly to a few hundred. But the gene pool of the species had been pushed through what biologists call a "population bottleneck." It is estimated that all nenes alive today are the descendants of only 6 birds.

Today, conservation biologists are trying to reestablish the species in the wild—and are meeting with little success. According to Robert Fleischer of the National Zoo in Washington, most birds released into the wild haven't been reproducing successfully; many eggs don't hatch, and goslings' survival rate is very low. Researchers fear that the population bottleneck eliminated most of the species' genetic diversity, and DNA studies to date support that hypothesis. DNA fingerprints taken from the blood of living birds are much more similar to each other than those of normal, wild, nonendangered species.

In addition, using the PCR technique described in the previous chapter, Fleischer has been able to produce similar fingerprints from what is often called "dead DNA"— DNA fragments scavenged from museum specimens of nenes collected before the population bottleneck in the 1930s. Results from those investigations are still preliminary but also bear out the hypothesis; individuals from historic nene populations seem to have had more genetic variation than those living today.

But why should decreased genetic diversity create problems? You may know that several highly inbred breeds

Figure 22.20 The handsome nene, the state bird of Hawaii, was once numerous throughout the archipelago. Its modern population, though not especially small, was founded by as few as half a dozen individuals that survived a serious population bottleneck earlier this century. The species' limited genetic diversity seems to be hampering its survival when reintroduced to the wild.

of dogs have a high incidence of problems such as hip dysplasia. Animal breeders also know of a more general phenomenon called "inbreeding depression"; when animals are bred for too long to close relatives, their fertility falls, and they generally seem less robust and healthy.

Parallel experience with cheetahs—a species pushed through a bottleneck during the nineteenth century—is not encouraging. A recent study at the National Cancer Institute found that cheetahs have less than one-thirtieth the genetic variation of humans and wild mammal species (Fig. 22.21). And it has been known for years that efforts to breed cheetahs in captivity are rarely successful. Two reasons are actually visible; male cheetahs have sperm counts only about one-tenth those of domestic cats, and their sperm show more than twice the usual number of physical deformities such as malformed heads and bent tails.

Cheetahs' immune systems are also too similar to one another to protect populations from epidemics as they should. You will learn in Chapter 42 that the immune system's response to disease depends heavily on genetic diversity. But because the alleles of all living cheetahs are so similar, any disease that can outrun one animal's defenses can cause trouble for all of them. One recent epidemic of feline infectious peritonitis in the United States highlighted that problem. The disease killed barely 1 percent of domestic cats and scarcely affected lions exposed to it at the Wildlife Safari Park in Oregon. But the same disease killed fully 60 percent of the zoo's cheetahs.

That's why Fleischer and his colleagues working with nenes are so concerned. But they aren't just waiting for trouble to show itself; they're using molecular tools to try and make the best of a bad situation. Several of Fleischer's colleagues are capturing, marking, and sampling the DNA of the few birds that remain. They hope to identify descendants of any truly wild birds that may have survived over the years. If they can find any animals with DNA sequences not found among captive birds, they will use those individuals in a breeding program to try and increase the genetic variation in the captive colony. The results of their efforts will be closely watched by curators in zoos and conservationists in the wild, for they will have much to say about strategies for protecting rare and endangered species.

It is important to note that the relationship between population size and genetic diversity described here seems to hold for the kinds of organisms most of us usually think about: multicellular animals and plants that reproduce sexually and live in moderate- to large-sized populations. But a number of protozoans, fungi, insects, and plants can reproduce asexually. These organisms, which either never exchange genetic material with one another or do so only rarely, either survive handily with less genetic variation or have other ways of maintaining necessary diversity. There are also some insects—such as ants and bees—whose normal life history strategy involves a high degree of inbreeding.

Figure 22.21 Cheetahs, known as the fastest land animal alive, are seriously threatened by their lack of genetic diversity. Because the alleles of genes of all living individuals are so similar, their immune systems are impaired; they can accept skin grafts from one another without rejection and are highly susceptible to disease. Many cheetah sperm are malformed, and their fertility is very low.

"Junk DNA"

One phenomenon that was puzzling to molecular biologists for some time is the fact that multicellular plants and animals have a great deal more DNA per cell than bacteria. How much more? A typical prokaryotic cell contains about 4 million base pairs of DNA. Single-celled eukaryotes such as yeast have between 10 and 20 million base pairs. Typical animals and plants have several billion base pairs in their genome (humans have 6 billion). What is the function of all this "extra" DNA? Interestingly, a staggeringly large amount of it doesn't seem to code for proteins. This part of the genome—up to 95 percent of it by some estimates—is often called "junk DNA," because it seems to have no purpose.

In the human genome, for example, there are more than 300,000 copies of a 300-base-pair sequence called *alu*. That's almost 3 percent of all human DNA. It looks as though *alu* may have arisen as a result of random duplication of a particular DNA sequence, as though somehow, at some point, DNA polymerase acted like a broken photocopying machine, turning out a roomful of unneeded copies of a single page of a manuscript. This sizable chunk of DNA—like a very large number of other sequences found throughout the genome—doesn't appear to have any function at all.

Until quite recently, many researchers dismissed introns as "junk" too. After all, the sequences introns contain are snipped out of messenger RNA and aren't translated into protein. In some cases, introns may be unimportant, but indirect evidence suggests that they somehow affect gene expression.

One such example was turned up unintentionally by one group of genetic engineers described in the previous chapter. The first time they inserted chimeric DNA into their test animals, they used a sequence that included only the exons of the gene whose product interested them. (They created this intron-less artificial gene by making a DNA copy of the mRNA for that protein, thus eliminating the introns.) Although some of the desired protein was produced, yields were disappointingly low. All the other parts of their chimera—promoters and so forth—checked out fine. Then strictly on a hunch, they made a new construct using the entire gene, including introns. Gene expression increased dramatically, although no one yet understands why.

Other researchers, working so far primarily with single-celled organisms, have found that mRNA transcripts in some animals are "edited" after they are made. Sometimes, the editing changes are minor—the addition of a single base here or there, for example. But sometimes, mRNA is so extensively edited that its final form—the one that directs the synthesis of proteins—looks nothing at all like the DNA sequence from which it was originally transcribed. Sometimes, in fact, the "unedited message" carried in the original DNA sequence is virtually unrecognizable as a working set of instructions for protein synthesis. "If you were sequencing the human genome and found some of these original DNA sequences," asserts molecular parasitologist John Boothroyd, "you would assume that you'd made a mistake, or

that it was a cloning artifact, or junk. You wouldn't recognize it as a functional gene." Although this type of RNA editing has not yet been found in higher animals, it is difficult to spot and may have been overlooked.

These findings should remind us that we still know very little about how genes coordinate and control their activities. Simply because significant parts of the genome don't code for recognizable pieces of proteins doesn't mean they don't do anything important. We should also keep in mind the difference between "junk" and "garbage." Garbage is something we throw away altogether. Junk we might not use very often but keep squirreled away—because it might just come in handy some day.

Human Genetic Diversity

Homo sapiens is arguably the most successful large mammal on Earth. As you will learn in Chapter 29, our ancestors evolved in Africa and spread with remarkable speed to colonize every habitable continent. Until recently, far-flung human populations had little contact and therefore rarely exchanged alleles with one another. Over time, mutation, natural selection, and random genetic drift gave rise to an impressive amount of genetic diversity within our species.

Some of this diversity—the kind that comes to mind right away—is visible; people from different parts of the world have different eye shapes, nose sizes, hair types, distributions of body hair, and, of course, skin color. A far greater amount of "hidden" genetic diversity underlies those trivial cosmetic differences. In the context of human evolution, and in the world of academic medicine, that diversity is invaluable.

You have already seen how important it was for human populations in Africa that some individuals carried the Hemoglobin-S allele when malaria began to infect our species. Without that protection, the inhabitants of entire regions of Africa and Asia might have been wiped out. Intriguingly, malaria has been such a powerful selective force in so much of the world for so long that 20 different antimalarial mutations affecting hemoglobin have occurred in other populations.

The stories about this particular type of genetic diversity alone could easily fill an entire book; in one isolated population of Jews in Kurdistan, for example, there are 13 distinct mutations that protect against malaria while causing a blood condition called thalassemia. Other thalassemias are common around the Mediterranean, where they overlap the occurrence of the Hemoglobin-S allele. In addition, literally hundreds of other antimalarial mutations help various groups of people fight the parasite in other ways, including some that involve the immune system. And some of these, according to Ronald Nagel (Albert Einstein College of Medicine), offer valuable hints at new ways to help fight the disease in people who lack those genetic defenses. Closer to home, ongoing preliminary studies in San Francisco offer tantalizing suggestions that some people may have some kind of genetically based resistance to the human immunodeficiency virus (HIV) that causes AIDS.

Genetic Differences and Human Populations

Genetic differences among human populations can be invaluable research tools in other ways. Some of us (including one author of this text) carry an allele that predisposes us to atherosclerosis and heart disease. Stud-

ies of that unusual allele shed light on details of normal cholesterol metabolism—with enormous potential benefit to the general population (see Theory in Action, Cholesterol and Heart Disease, Chapter 37).

Similarly, recent data on Native Americans show surprising differences in the incidence of certain conditions among various tribes. The Sioux, for example, suffer higher rates of heart disease than Americans at large, while Arizona's Pima and Maricopa tribes have lower-than-average incidence of heart disease. But Pima and Maricopa people have the highest rates of diabetes *in the world*. Some of these differences may be caused by differences in diet and lifestyle, but others seem related to genetic differences among these groups. Ongoing studies not only may help affected individuals make health-appropriate lifestyle choices but also may lead to a better understanding of both conditions.

Unfortunately, misunderstandings about the nature and significance of human genetic diversity have caused a great deal of pain and suffering throughout history. There has been a resurgence in recent years of terms such as *racial purity* and *ethnic cleansing*. Molecular studies of human genetics, however, shows that these terms, in addition to being morally repugnant, are scientifically meaningless.

Molecular analyses of human DNA proves, for example, that alleles related to malaria resistance make a mockery of the concepts of "race" and "ethnicity." By studying segments of DNA adjacent to a particular gene, researchers can distinguish between otherwise identical mutations that occurred in different populations. In that way, they can tell that Hemoglobin-S alleles, which originally occurred in black African populations, are now widely distributed throughout fair-skinned populations of Italy, Greece, Spain, and Portugal. Thalassemia alleles have been followed in the same way; of those 13 mutations found in Kurdistani Jews, 1 crops up in India, 7 have been found in various Mediterranean peoples, and 1 was even found in a Moslem from Iran.

On a broader scale, however, the differences among humans that we can *see* (and thus usually think of) as human genetic diversity—the sorts of cosmetic variations that are often used to distinguish among so-called "races" (Fig. 22.22)—are caused by no more than 6 percent of the *total* genetic variation our species possesses. In other words, only 6 percent of the total human genetic diversity identified to date separates along those so-called racial lines. In similar fashion, only about 8 percent of our total genetic variation is divided among the more numerous and smaller populations that we often refer to as ethnic groups or nationalities. Where, then, do the vast majority of genetic differences that make us individuals reside? Roughly 85 percent of human genetic diversity occurs among people *within* any single population.

Figure 22.22 Human genetic diversity. A poorly understood combination of chance, genetic drift, and natural selection over time has resulted in the visible differences among groups of humans that we ordinarily think of under the heading of human genetic diversity. But molecular analyses reveal that there is much more variation—much of it invisible—between individuals *within* each of what we consider "nationalities," "ethnic groups," or "races" (see text).

Human population geneticists are emphatic on this point. One researcher, who is married to a woman of his own ethnic group, made the point in an interesting way. "There could well be more genetic differences between myself and my wife," he assures us, "than there are between me and a Kalahari bushman." British geneticist Steven Jones, of University College, London, concurred in a recent series of lectures. "The overall difference . . . between Africans and Europeans," Jones told his audience, "is no greater than that between different countries within Europe or within Africa. Individuals—not nations, and not races—are the main repository of human variation." Finally, in the words of Berkeley's Mary-Claire King, "The richness of human diversity has the splendid social effect of completely debunking racial mythology. One learns that race is a social construct, ultimately, and not a genetic one."

We as scientists can only hope, therefore, that increased awareness about the nature of genetic diversity will enable us to appreciate its value, both in our own species and in those with whom we share this planet. Over the next several chapters, as we survey the living diversity that mutations in DNA have produced, we hope to encourage an appreciation for the remarkable molecular diversity and ingenuity represented by the genetic endowments of living species—ingenuity whose survival value has been finely honed by natural selection over the entire 3.5 billion years since life began on Earth.

SUMMARY

Malaria, a debilitating parasitic disease that attacks human red blood cells, is prevalent in tropical regions throughout the world. Many individuals from these regions also carry an allele for sickle-cell anemia, a hereditary blood disorder. There is a connection between these two facts, a link that shows how the pressures of evolution can shape the molecular history of the human genome.

Mutations are heritable changes in DNA. Spontaneous mutations occur at random, and the frequency of mutation may be increased by physical or chemical mutagens or even by mobile genetic elements known as transposons. Mutations in DNA-coding sequences may have direct effects on gene products, altering the amino acid sequences of proteins, although some mutations are "neutral," having no direct effect on phenotype. The accumulated changes that mutations represent make it possible to trace the course of evolution by comparing the DNA sequences of different species. These comparisons not only reveal the relationships between species but also give clues that can be used to estimate the amount of time that has passed since two species were separated in the evolutionary past.

The case of sickle-cell anemia can be understood in evolutionary terms as a point mutation that changes a single amino acid in one of the polypeptides found in the oxygen-carrying protein, hemoglobin. Although this mutation reduces the solubility of hemoglobin, producing the symptoms associated with sickle-cell anemia, it also confers partial resistance to malaria, and this accounts for the prevalence of the sickle-cell allele in malaria-ridden environments.

Many genes, including those that produce lysozyme, are members of "gene families," groups of similar gene sequences produced by the process of gene duplication.

Genetic diversity within a population is essential to the long-term ability of a species to adapt to environmental change, a fact increasingly recognized by conservationists and zoologists. As a species, humans display a wide range of genetic diversity, reflecting our successes in colonizing diverse habitats throughout the planet.

STUDY FOCUS

After studying this chapter, you should be able to:

- Explain the molecular nature of sickle-cell anemia and account for the evolutionary pressures that have made the allele common among certain population groups.

- Outline the molecular mechanisms that underlie evolution.

- Describe the process of gene mutation, and explain how changes in DNA cause changes in amino acid sequence and ultimately in phenotype.

- Explain how transposable genetic elements can affect gene expression.

- Describe the strengths and also the uncertainties of using DNA sequences to trace the course of evolutionary change.

- Explain why a lack of genetic diversity in a species is a serious threat to its long-term survival.

TERMS AND CONCEPTS

mutation 457
mutagens 458
transposable elements 459

point mutations 460
frameshift mutations 460

REVIEW

Objective Questions (Answers in Appendix)

1. Sickle-cell anemia is
   (a) caused by an iron-poor diet.
   (b) characterized by increased susceptibility to malaria.
   (c) carried from one individual to another by mosquitoes.
   (d) a recessive genetic disorder.

2. Point mutations
   (a) occur more frequently in introns than in exons.
   (b) occur more frequently in exons than in introns.
   (c) always result in mutant phenotypes.
   (d) may be neutral, causing no change in phenotype.

3. Jumping genes (transposons)
   (a) do not affect protein synthesis.
   (b) are capable of moving from one place to another on a chromosome.
   (c) can cause bacterial infection.
   (d) are chemical mutagens.

4. Two organisms that diverged in evolution 20 million years ago are likely to have
   (a) no similar DNA sequences.
   (b) identical DNA sequences.
   (c) more differences in their DNA sequences than two organisms that diverged 5 million years ago.
   (d) fewer differences in their DNA sequences than two organisms that diverged 5 million years ago.

5. The lack of genetic diversity in cheetah populations is a threat to the future survival of the species because
   (a) the genetic similarity of mates increases the likelihood of recessive genetic defects in newborn animals.
   (b) the lack of genetic diversity makes the animals more susceptible to disease.
   (c) the lack of genetic diversity gives the species few resources with which to cope with environmental change.
   (d) of all of the above.

Discussion Questions

6. In the United States, sickle-cell anemia is most often associated with people of African ancestry. However, sickle-cell anemia is also common in the Middle East and across southern Asia. Why?

7. What is a mutation? Outline some of the factors that can cause mutations. Are all mutations irreversible?

8. Mutations may have both positive and negative effects. Explain. Are any mutations truly "neutral" in their effects? Why?

9. The position of a point mutation in a gene—intron or exon—helps determine what effect it has on phenotype. Why?

10. Discuss how molecular biology can be used to trace evolution.

11. What do genetic studies of human genetic diversity say about the concept of "race?"

READINGS

Prager, D. Irwin, and A. C. Wilson. "Evolutionary genetics of ruminant lysozymes." Submitted to *Animal Genetics*.

Diamond, J. *The Third Chimpanzee*. New York, NY: HarperCollins, 1992.

Cavalli-Sforza, L. L. "Genes, peoples, and languages." *Scientific American* 265 (May 1991): 104–110. The history of the human species can be traced by linguistic techniques as well as by biological ones, and the interplay between these studies makes for some interesting conclusions.

Culotta, E. "How many genes had to change to produce corn?" *Science* 252 (1991): 1792–1793. An interesting genetic account of the evolution of one of the world's most important plants.

Li, W.-H., and D. Grauer. *Fundamentals of Molecular Evolution*. Sunderland, MA: Sinauer Associates, 1991.

Noonan, D. "Genes of war." *Discover* (October 1990): 46–52. The story of Mary-Claire King's use of DNA fingerprinting to identify the remains of people who "disappeared" under the Argentine dictatorship.

Wilson, A. C. "The molecular basis of evolution." *Scientific American* 240 (April 1985): 164–173.

Kimura, M. "The neutral theory of molecular evolution." *Scientific American* 240 (May 1979): 98–126.

Herrick, J. B. "Peculiar elongated and sickle-shaped red corpuscles in a case of severe anemia." *Archives of Internal Medicine* 6 (1910): 517–521.

PART 5

Diversity of Life

These butterflies, pausing on leaves and flowers during their search for nectar, represent insects, the largest group of multicellular animals on Earth. They also demonstrate the intimate ecological and evolutionary relationships that bind animals and plants.

*b*iological diversity is one of our planet's greatest treasures. Each living species—from iceworms clinging to the edges of Arctic glaciers, to bacteria thriving in superheated, sulfur-laden water on the ocean floor, to mammals such as *Homo sapiens*—is a unique product of an evolutionary drama that began nearly 4.5 billion years ago.

Ever since the first life-forms emerged on the infant earth, the histories and fates of their descendants have been shaped by genetic change, chance, and natural selection. Lineages that have persisted have done so through both luck and adaptation, as their body structures, physiological processes, and reproductive strategies have been repeatedly put to the test. The species that "passed" that test survived. Those that failed became extinct. No one knows how many species have survived that test to date; scientists' estimates vary widely between 3 and 40 million living species, dependent on how the approximation is made.

The task of identifying current organisms, grouping and ordering them into categories that have scientific significance, and sorting out their evolutionary relationships is arduous. The task is also vital—especially because human activity is driving some species to extinction every day.

Part 5 begins with three introductory chapters. Chapter 23 discusses why and how organisms are named and classified. Chapter 24 gives an overview of the origins of life on Earth, and Chapter 25 describes how life evolved through the geologically defined periods in Earth's history. The latter chapters in the section divide life into five kingdoms and discuss representative organisms and their characteristics in each. Chapter 26 covers the kingdoms Monera and Protista, Chapter 27 discusses the Fungi and Plant kingdoms, and Chapters 28 and 29 cover the invertebrates and the vertebrates, respectively, within kingdom Animalia. Although the taxonomic discussion of plants has been separated from the discussion of animals for ease of learning,

It now remains to speak of animals and their nature. So far as we can, we will not exclude any of them, no matter how mean; for though there are animals which have no attraction for the senses, yet for the eye of science, for the student who is naturally of a philosophic spirit, and who can discern the causes of things, nature which fashioned them provides joys that cannot be measured.
— Aristotle, ca. 330 B.C.

we re-emphasize through all of these chapters that plants and animals have evolved together, a theme that was first discussed in Part 2, Organisms and Ecology.

23

Finding Order in Diversity: Biological Classification

ist hangs in the air as Rosemary Gillespie, the University of Hawaii's indefatigable "spider woman," leads her slicker-bedecked crew through the Pu'u Maka'ala forest reserve, 1200 m above sea level on the slopes of Kilauea volcano. Beads of moisture drip from tree fern fronds overhead and glisten on mosses underfoot. The setting is as close as reality gets to the Hollywood image of a forest primeval.

Gillespie is searching for spiders—but not just any spiders. She is gathering data on recently discovered varieties that may be evolving into new species. "What I'm most interested in finding," she instructs her crew, "are green spiny ones, humpbacked spinys, and gold bandy-legs." Spotting a few raised eyebrows at her pet names for the animals, she laughs and responds with a shrug, "There are lots of these things that aren't identified, so I've made up nicknames for them. What else do you do when you haven't proper names yet?"

"We're after these particular spiders," she continues, "because we're looking at the DNA of different populations, trying to find out which ones already look like different species and which ones might be headed that way. And they aren't too common here. Oh! Got one? Well done!" And off she runs down the trail.

Gillespie's work, aimed at identifying and categorizing unknown forms of life, epitomizes two related and vital research goals in modern biology. First, Gillespie is part of an informal global effort to catalog biological diversity—to name and describe the species with whom we share the earth. At the same time, she is observing and attempting to understand the processes of genetic change which give rise to that diversity.

TAXONOMY: THE SCIENCE OF NAMES

In their efforts to find order in diversity, biologists must begin with the basics: they must distinguish species from one another, assign each species a unique name, and arrange the vast numbers of living and fos-

sil organisms into logical groupings. **Taxonomy,** the science of naming organisms, has a long and colorful history that dates back to Aristotle (Fig. 23.1).

Over the years, as the number of known organisms grew, finding suitable names for them became a problem. Common names—of the sort Rosemary Gillespie relies upon as a temporary measure—were as vague and confusing then as they are today. "June bug," for example, refers to at least a dozen species of beetles in the United States alone, and the name "bluebell" is used for several different plants (Fig. 23.2). Conversely, a single species of American wildcat is known in different areas as the puma, the cougar, the catamount, the American panther, and the mountain lion.

Early scholars tried to standardize the system by giving each organism a name in Latin, the universal language of science. Unfortunately, these Latin names were difficult to use; they were changed regularly and could be up to 15

Figure 23.2 These three plants are only a sample of those commonly called "bluebells." Their scientific names and families are (LEFT) *Mertensia ciliata* (Boraginaceae), (CENTER) *Campanula rotundifolia* (Campanulaceae), and (RIGHT) *Clematis hirsutissima* (Ranunculaceae).

words long. One carnation-like flower, for example, bore the cumbersome name *Dianthus floribus solitariis, squamis calycinis subovatis brevissimus, corollis crenatis.* Biologists clearly needed a single, stable, concise way of naming things.

Binomial Nomenclature: *Genus* and *species*

Swedish botanist Carolus Linnaeus (1707–1778) devised a shorthand method of naming organisms called the system of **binomial nomenclature** (*bi* means "two"; *nome* means "name"; hence *binomial* means "two names"). The Linnaean system, which is now used around the world, gives each species a unique, two-part name. For the plant whose nine-word name is given above, for example, Linnaeus used the name *Dianthus caryophyllus.*

The first part of each name is a single word, usually a proper noun derived from either Latin or Greek, called the **genus** name (plural: *genera*). Because Latin was the language of science, many genera represent the common names used by the Romans. Similar cats were placed in the genus *Felis,* for example; dogs in the genus *Canis;* and horses in the genus *Equus.* Note that generic names begin with a capital letter and are italicized in print.

The second part of the name, usually a Greek or Latin adjective called the specific epithet, describes some significant characteristic of the species. The plant *Campanula rotundifolia,* for example, is a bellflower with rounded leaves (*campana* means "bell"; *rotund* means "round"; *folia* means "leaves"). Similarly, the name for our own species is *Homo sapiens* (*homo* means "man"; *sapiens* means "wise"). Note that the specific epithet is italicized but not capitalized. One of Rosemary Gillespie's favorite spiders, which she still calls "big step" after its style of walking, has a Latin name which means exactly that!

Table 23.1 *Classification of Common Organisms*

Taxon	Human	Fire Ant	Sunflower
Kingdom	Animalia	Animalia	Plantae
Phylum	Chordata	Arthropoda	Anthophyta
Class	Mammalia	Insecta	Dicotyledones
Order	Primates	Hymenoptera	Asterales
Family	Hominidae	Formicidae	Compositae
Genus	*Homo*	*Solenopsis*	*Helianthus*
Species	*sapiens*	*saevissima*	*annuus*

Higher Taxonomic Categories

Scientists also find it useful to group plants and animals into larger taxonomic categories or **taxa** (singular: *taxon*), each defined by a certain set of important characteristics. Animals belonging to our own *class,* Mammalia, for example, are identified in part by having body hair, being "warm-blooded," having a placenta, and nourishing newborns with milk from mammary glands.

Biological taxonomy uses a *hierarchical* system; that is, each category is nested within the category above it and contains lower categories within it. All of us use hierarchical systems regularly in daily life, because they provide a handy way of referring to groups of objects. If, for example, you ask someone for a "pen," he or she knows that you want one member of a diverse group of writing implements that includes ball-points and fountain pens—but not pencils.

Biologists refer to groups of organisms using a hierarchy that ranges from the largest divisions, called *Kingdoms,* down to species:

Kingdom
    Phylum (for animals) or Division (for plants)
        Class
            Order
                Family
                    Genus
                        Species

(A simple mnemonic device to help you remember these main divisions in the taxonomic hierarchy is the nonsense sentence "**K**ings **P**lay **C**hess **O**n **F**ine **G**reen **S**and.") Table 23.1 gives the classification of three familiar organisms according to this system. In some cases, however, researchers who assign organisms to these groups decide that these seven categories alone are not sufficient to demonstrate natural groupings among organisms; in those situations, additional categories are added. Phyla, for example, are sometimes divided into subphyla, orders are occasionally grouped into superorders or divided into suborders, and so on.

SYSTEMATICS: THE SCIENCE OF CLASSIFICATION

Working alongside taxonomy, the discipline of **systematics** tries to group organisms into higher taxa in ways that provide the most biologically relevant information. In Linnaeus's time, kingdoms and families were used simply to "pigeonhole" similar-looking organisms into groups. But the recognition that living species have evolved from earlier organisms posed a new challenge: to group organisms into taxa that represent lineages or lines of evolutionary descent. Thus, ideally, the hierarchy of taxa should represent "limbs," "branches," and "twigs" on the evolutionary tree of life. By this reasoning, species within a single genus resemble each other because they share a recent common ancestor. Similarly, members of a family represent a larger evolutionary lineage descended from common stock in the more remote past.

Creating taxa that accurately represent evolutionary history, however, is difficult. Until recently, most systematists aimed to group together those organisms that most closely resembled one another. To this end, systematists following a school of thought called *phenetics* examined as many anatomical and physiological characteristics as possible. Many characteristics—such as bone structure, tooth arrangement, body size, and claw size—were given equal weight. Other characteristics were considered more important. Those organisms that shared the greatest number of similar characteristics—in other words, those organisms that *looked* most similar to one another—were as-

sumed to be most closely related (Fig. 23.3a,b). But the phenomena of adaptive radiation and convergent evolution often make it difficult to distinguish organisms that look alike because they are closely related from those that look alike because they have adapted to similar niches. For this reason, classifications that rely exclusively on structural similarities do not always reflect evolutionary history.

Another school, *cladistics,* aimed specifically to create taxonomic groupings that accurately reflect evolutionary history. In other words, cladists seek to group organisms with their closest relatives, rather than with organisms that happen to resemble them because of convergent evolution. To do so, cladists divide characteristics used for classification into two groups: those characteristics shared by various subgroups of living organisms because they evolved from recent common ancestors, and those characteristics shared by larger groups of organisms because an ancient common ancestor had them. By treating recently evolved characteristics differently from very old

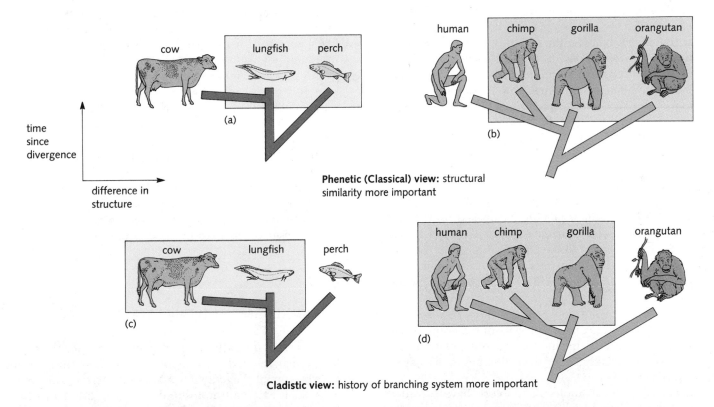

Phenetic (Classical) view: structural similarity more important

Cladistic view: history of branching system more important

Figure 23.3 **(a)** Though the lineage leading to lungfishes separated from other fishes hundreds of millions of years ago, both still live in water and look like "fish," so the phenetic (classical) approach groups them together. **(b)** Chimpanzees, gorillas, and orangutans retain many characteristics of early primates, whereas humans have undergone extensive changes in anatomy and behavior. Pheneticists therefore group apes in the family Pongidae and humans in their own family, the Hominidae. **(c)** Though cows and lungfishes look different because they have adapted to

different environments, they share more recent common ancestors with each other than either shares with a perch. Cladists thus group lungfish together with cows and separately from other fishes, which branched off from the main line of vertebrate evolution much earlier. **(d)** Humans, chimps, and gorillas share recent common ancestors, whereas orangutans branched off in the more distant past. Cladists therefore group humans together with chimpanzees and gorillas, our closest living relatives.

characteristics, and by checking the fossil record whenever possible, this scheme emphasizes evolutionary branching patterns while downplaying shared structural similarities (Fig. 23.3c,d).

Many modern evolutionary systematists attempt to combine the best features of both phenetic and cladistic approaches. This ambitious effort is fraught with conflict, because, as shown in Fig. 23.3, classification schemes that emphasize evolutionary relatedness often give rise to dramatically different patterns of relatedness than schemes that stress overall similarity. Perhaps the most equitable way to summarize the situation is to emphasize that *all* classification schemes are artificial attempts to impose a logical order on enormously complex assemblages of organisms. Every system of classification necessarily involves difficult subjective judgments about the importance of various characteristics.

In the end, therefore, we should look to each system not for "truth" in an absolute sense but for the insights each can give us into nature and relationships among living things. For that reason, the usefulness of a particular classification scheme depends on how its groupings are going to be used. To an ecologist, a lungfish is more like a perch than a cow, because the first two animals live in water and the third does not. In an ecological study, therefore, it makes sense to treat lungfishes like the fishes whose aquatic habitat they share. To physiologists or paleontologists, on the other hand, lungfishes more closely resemble mammals in certain details of body structure and some aspects of their physiology.

Molecular Systematics

As successful as systematics techniques based on physical features and fossils have been, they have certain limitations. The fossil record, for example, is patchy. Certain groups of organisms don't preserve well, so their fossil history is unclear. In addition, there are precious few fossils of the very first life forms, and the record of the most recent events in evolutionary history also leaves something to be desired. Finally, comparisons among visible features offer little help in sorting out relationships among organisms as different from one another as apes, algae, and bacteria.

Recent advances in molecular biology have provided systematists with new tools. Researchers can now consider molecular similarities and differences as further clues to relationships among organisms. As you learned in Part 4, all organisms (except some viruses) contain DNA, and a great many share other important molecules such as hemoglobin and chlorophyll. And as explained in Chapter 22, evolutionary changes in those molecules—both visible and invisible, adaptive and "neutral"—are caused by

changes in DNA and are reflected in proteins produced by changing genes.

Molecular systematics is still in its infancy, and there is widespread (and often severe) disagreement among researchers concerning the best classes of molecules to study (DNA, RNA, or proteins), which groups of molecules within those classes are most reliable (nuclear, mitochondrial, or chloroplast DNA; ribosomal RNA, or respiratory enzymes), and even which *parts* of those molecules (which nucleotide sequences or strings of amino acids) offer the best information. Similarly heated debates surround scientists' choices of particular mathematical formulae and computer algorithms to analyze the data obtained from molecular analyses.

Despite all those questions, the potential power of this approach makes it irresistible to many systematists. You would be hard-pressed to find a significant group of organisms whose history and relationships are not being reevaluated in light of molecular information. In some cases, new data have provided answers to previously unanswerable questions. Botanists working with the remarkable group of Hawaiian plants known as silverswords, for example, used studies of chloroplast DNA to identify a likely ancestor for the group among an obscure group of mountain plants in the western United States. Within that group of plants, molecular analyses are being used in efforts to sort out relationships among the unique species that inhabit various islands in the Hawaiian archipelago (see Theory in Action, Molecular Links Across the Pacific, Chapter 22).

In other cases—the immediate ancestry and history of *Homo sapiens*, for example—molecular data have shed light on some questions while doing little to clarify others (see Chapters 22 and 29). And in still other cases, molecular systematists have proposed sweeping (and controversial) revisions of existing taxonomic categories (see below).

THE KINGDOMS AND THE TREE OF LIFE

In Aristotle's time, the largest taxonomic categories (the kingdoms) seemed obvious: most common things were clearly Animal, Vegetable, or Mineral. A few odd organisms, such as bread molds and sponges, didn't fit neatly into those groups, but dividing lines were generally straightforward.

As biological knowledge grew, however, it became obvious that additional kingdoms were needed. Microscopes exposed the primary visible distinction in the living world: the division between *eukaryotes*—whose cells contain a variety of membrane-bound structures, including a nucleus—and *prokaryotes*, which lack a nucleus.

Molecular Dances with Wolves

In North Carolina's Alligator River National Wildlife Refuge, a pack of red wolves prowls. These animals were declared extinct in the wild in 1975; this group exists thanks to a captive breeding and release program of the U.S. Fish and Wildlife Service. A group of Florida panthers deep in Florida's Everglades National Park is in a similar situation. This population, estimated at 30–50 individuals, clings precariously to life while state and federal officials conduct a similar breeding program. But the futures of both these endangered animals are in doubt, due to studies in molecular genetics that question their identity as distinct biological entities.

Traditionally, red wolves have been known as *Canis rufus*—a designation implying that they are distinct from their close relatives: *Canus lupus*, the grey wolf, and *Canis latrans*, the coyote. Similarly, Florida panthers carry the scientific name *Felis concolor coryi*, in recognition that they represent one of eight

distinct subspecies of cougars known in North America.

But recent data collected from both nuclear and mitochondrial DNA seem to confirm some researchers' suspicions that red wolves are actually hybrids between grey wolves and coyotes. And RFLP analyses of DNA from Florida panthers have revealed that at least some of them carry genes traced to one or more Central or South American cougars that were set free in southern Florida 20 or 30 years ago.

If the results of these studies become widely accepted, red wolves and Floridian panthers could be in trouble. The programs that maintain them fall under the Endangered Species Act, which protects only recognized species and subspecies. A series of legal opinions has declared not only that hybrids aren't protected under the act but also that maintaining hybrids could jeopardize the integrity of the "proper species" for whom the act was written.

Thus if the red wolf is reclassified as a hybrid between two species, and if the Florida panther is declared a hybrid of two subspecies, the programs which protect and nurture them could be denied federal funding.

Some citizens' groups—including ranchers, hunters, and private landowners—press for strict enforcement of what they call the "letter of the law" of the Endangered Species Act. Several biologists call instead for case-by-case analysis that includes both expert scientific testimony and a clear understanding of population genetics. The "foreign" genes in the Florida panther population that could doom them legally, for example, may be their only hope of biological survival. If they are pushed through a population bottleneck that restricts their genetic diversity (see Chapter 22), hybrid vigor might prove to be a key to their survival.

Recent molecular evidence suggests that the Red Wolf, *Canis rufus* (LEFT), is not a "true" species, but rather a hybrid between the Grey Wolf, *Canis lupus* (CENTER), and the coyote, *Canis latrans* (RIGHT).

Table 23.2 *Major Characteristics of the Five Kingdoms*

Monera	Protista	Fungi	Plantae	Animalia
Prokaryotic; no membrane-bound intracellular structures	Eukaryotic; nucleus, mitochondria, some have chloroplasts	Eukaryotic; nucleus, mitochondria, but no chloroplasts; cell wall of chitin	Eukaryotic; nucleus, mitochondria, chloroplasts; cell wall of cellulose	Eukaryotic; nucleus, mitochondria, but no chloroplasts; no cell wall
Solitary, filamentous, or colonial	Mostly solitary, some colonial or multicellular	Some unicellular, most grow in thread-like branches	Multicellular, most sedentary on land	Multicellular, most motile
Aerobic or anaerobic	Mostly aerobic	Mostly aerobic	Strictly aerobic	Strictly aerobic
Autotrophic or heterotrophic	Autotrophic or heterotrophic	Mostly saphrophytic	Mostly photosynthetic autotrophs	Heterotrophic
Bacteria of diverse types, including cyanobacteria, formerly called "blue–green algae"	*Amoeba*, paramecia	Yeasts, molds, mushrooms	Mosses, ferns, flowering plants, seaweeds	Sponges, worms, snails, insects, mammals

Over the years, researchers have proposed classification systems involving anywhere from three kingdoms to twenty. One of the most commonly used, the five-kingdom system, is a modification of one originally proposed by R. H. Whittaker of Cornell University (Table 23.2). This system, which we have used as the organizing principle for our presentation because it is widely accepted, is often depicted as an evolutionary "tree" (Fig. 23.4).

This five-kingdom tree is "rooted" in the kingdom Monera, which contains all the prokaryotic organisms. The "trunk" of the tree is a motley collection of mostly single-celled eukaryotes lumped into the kingdom Protista (some multicellular algae are often placed here as well). The three main "branches" of the tree are defined mainly according to nutritional type: the kingdom Plantae contains predominantly photosynthetic primary producers with cell walls, the kingdom Fungi consists of heterotrophic decomposers and parasites with cell walls, and the kingdom Animalia includes a variety of heterotrophs without cell walls.

Although this tree is helpful in grouping organisms, the information it conveys about the evolution of living species is easily misinterpreted. First, because living kingdoms are stacked on top of one another, this tree can be misread to suggest that single-celled eukaryotes evolved from living monerans; and that plants, animals, and fungi were derived from protist species alive today. (They didn't, and they weren't.) Second, the location of living monerans and protists "beneath" multicellular groups can imply that those unicellular organisms are "living fos-

sils," and that they stopped evolving millions of years ago. (They aren't, and they certainly did not.)

More recently, extensive molecular studies of unicellular organisms have given rise to several alternative classification schemes that profoundly alter our family tree. As one example, Carl Woese and his colleagues at the University of Illinois suggest replacing the "trunk" of the tree with three very ancient branches that diverged 3.5 billion years ago. The three kingdoms represented by these branches are the Eubacteria ("true" bacteria), Archaebacteria ("ancient" bacteria), and Eukaryotes. (We will explain the differences among those groups in the next chapter.) Note also that this diagram looks more like a broadly branching "bush" than a "tree," and that the majority of living branches represent unicellular organisms (Fig. 23.5).

Issues in Taxonomy and Systematics

Although the workings of systematics and taxonomy may appear straightforward, some living things defy easy classification. Certain single-celled organisms can act as either plants or animals, depending on environmental conditions. Other organisms straddle the line between single-celled and multicellular life-forms. And viruses, which many biologists do not even consider organisms, fall outside this classification scheme altogether.

Even something as basic as the definition of a "species" can be problematic. Rosemary Gillespie—like

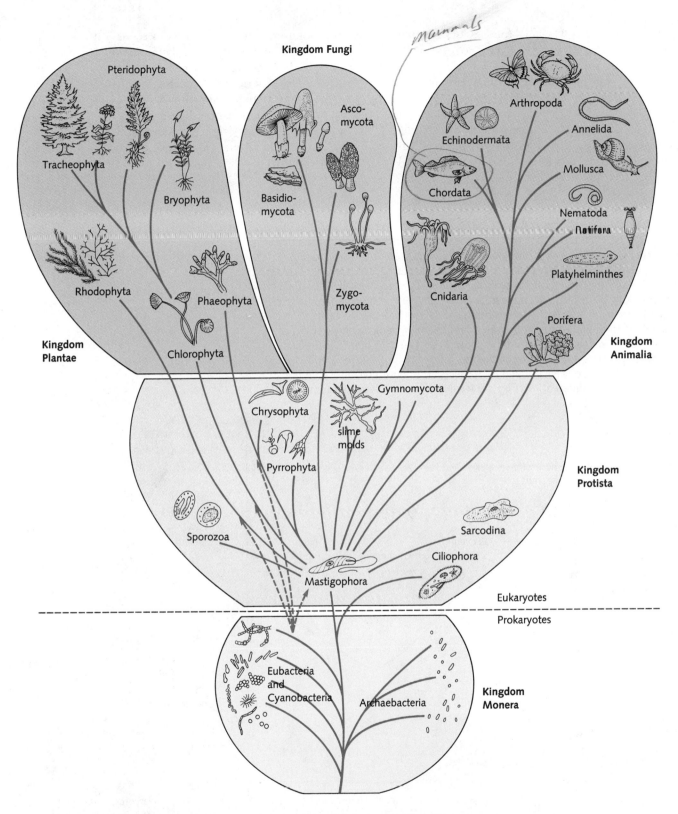

Figure 23.4 This schematic of the five-kingdom system depicts hypothetical evolutionary relationships. All prokaryotes are placed in the kingdom Monera. Unicellular eukaryotes are placed in the Protista, and multicellular eukaryotes are divided among the Fungi, Plantae, and Animalia. Note that although this system keeps all prokaryotes within the Monera, recent discoveries have led to the subdivision of the Monera into the Archaebacteria and Eubacteria. Some systematists feel that these divisions should be separate kingdoms.

many of her colleagues in Hawaii—focuses on isolated populations of organisms that may (or may not) be evolving into different species. But how does one define and distinguish species, subspecies, and populations in precise genetic terms? Many biologists have strong—and differing—opinions on the subject. In these days of vanishing habitats and organisms threatened with extinction, questions regarding the identity of species can change overnight from academic issues to political "hot potatoes." In these situations, as John Gittleman and Stuart Pimm wrote recently in *Nature*, "bad taxonomy can kill" if it denies a threatened group of organisms status as a full-fledged species (see Current Controversies, Molecular Dances with Wolves, p. 485). Given the renewed importance now being accorded to preserving biological diversity, modern taxonomists and systematists may find their work invested with unexpected responsibility.

Figure 23.5 Researchers following pioneer Carl Woese have split the "trunk" of the tree of life into three branches that diverged 3.5 billion years ago. One version of the resulting "bush of life" is shown here. One branch leads to most familiar living bacteria, the Eubacteria (or "true" bacteria). Another leads to organisms traditionally classified with Eubacteria on the basis of morphology but which now appear biochemically quite distinct. These Archaebacteria ("ancient" bacteria) are adapted to extreme environments: solfatara fields, where volcanic gases bubble through hot sulfurous water; and deep-sea vents, where geysers of superheated, mineral-laden water pour from the ocean floor. The third branch leads to the eukaryotes, all of which share important resemblances on the molecular level. Within this group, many organisms traditionally classified as "protists" turn out to be quite different from one another—far more different than we are from any other multicellular forms of life.

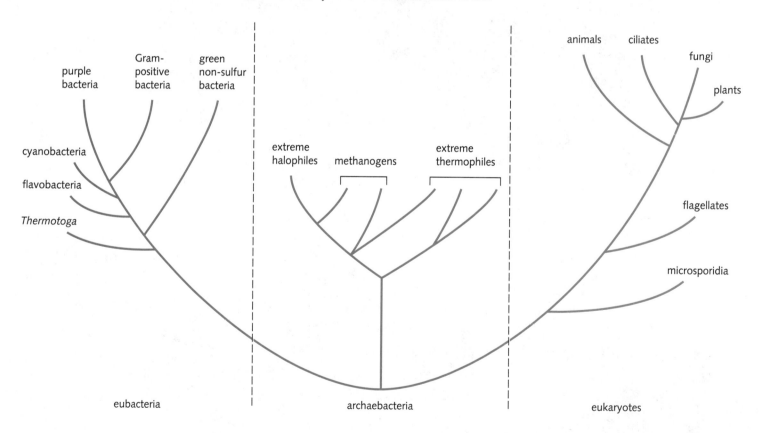

SUMMARY

Taxonomy and systematics work together in assigning names to organisms and grouping them into useful categories. Evolutionary systematics attempts to group organisms according to common ancestors they share. Our newfound ability to sequence DNA combined with techniques that analyze the molecular structure of proteins has created a new, powerful, and rapidly expanding molecular approach to systematics. Molecular analyses have solved some old taxonomic problems, offered new perspectives on existing debates, and generated new controversies concerning the structure of life's family tree.

STUDY FOCUS

After studying this chapter, you should be able to:

- Explain the scientific system for classifying and naming organisms.

- Show how different taxonomic approaches lead to different groupings of organisms.

- Outline the five-kingdom taxonomic system.

- Explain how traditional "evolutionary trees" can be misleading, and outline the points that alternatives make clear.

TERMS AND CONCEPTS

taxonomy *481*
binomial nomenclature *482*
genus *482*

taxa *482*
systematics *482*

REVIEW

Objective Questions (Answers in Appendix)

1. The classification of organisms into meaningful groups is the science of
    (a) taxonomy.   (c) evolutionary biology.
    (b) systematics.   (d) organismic biology.

2. The cladistic taxonomist would study _____ among organisms.
    (a) molecular differences
    (b) evolutionary history
    (c) similarity of function
    (d) similarity of appearances

3. The five-kingdom system classifies organisms according to _____ and _____ .
    (a) plants; animals
    (b) cellular organization; nutritional type
    (c) biochemical differences; nutrition
    (d) photosynthetic; heterotrophic characteristics

Discussion Questions

4. Devise two classification systems for the contents of a typical home workshop: tools, construction supplies, and various types of nuts, bolts, nails, screws, and fasteners. Devise one system based exclusively on the physical appearances of objects and a second system based solely on their function. What are the benefits and disadvantages of each system? Explain how you think the problems you encounter in devising this system parallel those involved in organizing classification schemes for living organisms.

5. The text emphasizes that all classification systems are necessarily "artificial." What does that mean? Which of the strategies for classifying organisms do you think comes closest to being a "natural" system? Why?

READINGS

See integrated list of readings for Part 5 after Chapter 29.

24

The Origins of Life

*Life is infinitely stranger than anything which the mind of man could invent.
We would not dare to conceive the things which are really mere common-
places of existence.*

—A. Conan Doyle (1891)

*i*n laboratory apparatus that looks like a giant pressure cooker, a researcher cultures bacteria whose natural habitat is superheated water
at the bottom of the sea. On a field expedition to hot springs in
Iceland, another scientist collects prokaryotic organisms that manage to
survive in water that would quite literally boil nearly any other organism—water at temperatures as high as 105°C.

These investigators and their colleagues are trying to answer a question that has perplexed and inspired scientists and philosophers for centuries: How did life first appear on the primordial earth? And how did
those earliest life-forms begin the long evolutionary process that has led
to the biosphere today? That inquiry is still incomplete, but geologists
and biochemists have come up with several plausible hypotheses about
the chain of events that culminated in the first living organisms.

CONDITIONS ON THE EARLY EARTH

The cloud of cosmic gas that formed the earth congealed and cooled sufficiently to produce the first solid rocks about 4.5 billion years ago (see
Chapter 1). During that process, volcanic activity produced an atmosphere that probably consisted mainly of water vapor (H_2O), carbon
monoxide and carbon dioxide (CO and CO_2), nitrogen (N_2), hydrogen
sulfide (H_2S), and hydrogen cyanide (HCN). Because the earth's surface
was still hot, water existed only as a gas; any rain hitting the ground
boiled right back into the atmosphere.

Sometime around 3.8 billion years ago, the planetary crust cooled
enough to enable water to remain on the surface in liquid form. Vast
thunderstorms raged across the planet, and ocean basins began to fill.
Then, somehow and somewhere in this inhospitable and inorganic landscape, the building blocks of life began to appear.

The Organic Soup: Life from Nonlife

Russian biochemist A. I. Oparin began theorizing about life's origins in 1924, but his ideas weren't tested experimentally until 1953. That's when American graduate student Stanley Miller built a flask that simulated conditions on the primordial earth by combining water, inorganic matter, and gases thought to have been present in the early atmosphere. Miller sterilized the equipment to kill microorganisms, excluded all traces of oxygen, and simulated lightning by generating sparks between tungsten electrodes.

Incredibly, within a week the raw materials of life assembled themselves like enchanted building blocks, creating the ordered structures of amino acids where before there had been only chaos. Several subsequent experiments using gas mixtures somewhat different from Miller's produced similar results, forming complex organic molecules in the absence of life.

From Molecules to Life

An amino acid stew, however, is far from a living cell. The jump from inert organic molecules to ordered, self-replicating biological systems is still the biggest conceptual gap in our understanding of life's origins. Recently, discoveries in molecular biology have led to hypotheses that offer a clearer picture of how that transition *might* have been made.

One line of thought points out that the organic molecules in these primordial "soups" are attracted to electrically charged clay and pyrite crystals. Because clay crystals have layered, repeating structures, organic molecules that stick to them are organized into nonrandom arrays. Furthermore, clay crystals are often interspersed with atoms of metals such as zinc and iron that promote certain chemical reactions. Held close to one another and to metal atoms in this way, small organic molecules often join together spontaneously to form larger molecules—including short stretches of RNA.

Guided by the structure of clay and agitated by waves or currents, large organic molecules may also organize themselves into different classes of spherical objects called *coacervate droplets* and *proteinoid microspheres* (Fig. 24.1). Both of these structures exhibit some properties of life: their structure is nonrandom, they are surrounded by a two-layered membrane, they accumulate and incorporate organic material, and they can "grow." Although these collections of molecules are not truly alive, many of them facilitate certain biochemical reactions within their ordered interiors. Several hypotheses link these "protocells" or "precells" with more organized structures that can direct more complicated biochemical reactions related to cellular metabolism (see Chapter 18).

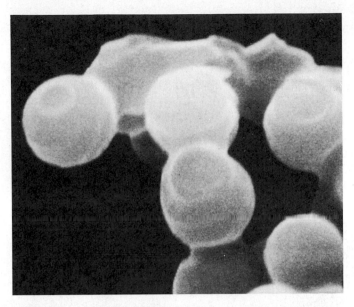

Figure 24.1 Proteinoid microspheres form spontaneously under conditions similar to those present on the early earth. Some microspheres act as natural chemical factories, encouraging reactions that incorporate materials from their surroundings. When they grow beyond a certain size, they become unstable and may split to form daughter spheres, as shown in this micrograph. (Magnification factor: approx. 14,000)

As you may realize, however, a major stumbling block remains. Living cells today all work the same way: information stored in DNA directs the assembly of various RNAs, which in turn direct the synthesis of the proteins that form cell structures and carry out cell functions. But while DNA codes for proteins, certain proteins (including several enzymes) are necessary for both making and reading DNA. Thus we are stuck with a classic "chicken-and-egg" problem. The current system requires all of its parts in order to operate: proteins can't be made without DNA, and DNA can't work without proteins. How could such a system have gotten started?

One hypothesis, offered by G. Cairns-Smith and J. Bernal as an alternative to the protocell concept, suggests that clay crystals might have served as templates or blueprints not only for the first proteins but also for polynucleotide strings as well. Other researchers have built on that foundation. Recent evidence indicates that once molecules resembling RNA were formed, some of them could have catalyzed their own reproduction. Several researchers hypothesize that this ability led to an "RNA world"—a self-perpetuating but precellular organic world based on the molecule that, until recently, biologists had treated as little more than a messenger (see Current Controversies, Was There an RNA World?). From that point on, evolutionary processes took over, although any scenarios about how the first true cells might have formed are still strictly conjecture.

Where might all this have taken place? Some origin-of-life theorists suggest that life began in one or another of these ways near the surface of the sea. (Darwin believed it happened in a "warm little pond" somewhere.) Others believe that intense asteroid bombardments of the early earth would have caused such enormous fluctuations in surface temperatures and sea levels that "life" at the surface would have been destroyed before it ever got going.

Still other researchers point out that the intense heat, pressure, and steady supply of high-energy inorganic compounds found in certain spots on the sea floor make those environments likely sites for critical reactions, even today (see Theory in Action, Hot Springs: Ancient Bacteria, Symbiosis, and the Origin of Life). Other investigators' proposals suggest similar locations on the earth's surface—in volcanic craters where superheated, sulfur-

laden water erupts from geysers and bubbles up through mineralized mud.

THE EARLIEST LIFE-FORMS

Regardless of how the first cells evolved, they must have been anaerobic—that is, they must have lived in the absence of oxygen—because there was no free oxygen on the early earth. They must also have been heterotrophs, depending on organic compounds formed around them, because they had not evolved ways to capture energy in any other manner. Finally, they would have been unicellular and prokaryotic, like many modern-day bacteria. Researchers have found fossils of microscopic organisms which fit this description in rocks that formed 3.5 billion years ago, barely a billion years after the earth solidified. This discovery surprised even the most avid aficionados of early life, for it pushed the origin of life much farther back in time than anyone expected.

The Road to Autotrophy

At first, life for these primitive bacteria would have been simple, for the oceans were well stocked with the complex organic molecules they needed to survive. But as early cells multiplied, they would have used up their chemical "foods," and intense natural selection would soon have favored individuals able to manufacture complicated molecules from simpler ones. The elaborate series of organic reactions you learned about in Part 4 would have evolved one step at a time.

Many metabolic pathways used by bacteria today probably evolved during this early period of the earth's history. Numerous living bacteria, in fact, still depend on *anaerobic fermentation* pathways to break sugars apart into alcohol and carbon dioxide or similar products in the absence of oxygen (see Chapter 18).

As early heterotrophs exhausted the finite supply of abiotically produced complex molecules, they would have experienced a shortage of energy to fuel their metabolic processes. The ensuing selection pressure to harness new energy sources ultimately led to the evolution of the earliest forms of bacterial photosynthesis—solar-powered reactions quite different from those carried on by modern plants. One type of bacterial photosynthesis, for example, takes in hydrogen sulfide (H_2S) instead of water (H_2O) and releases free sulfur instead of oxygen gas. These early photosynthesizers spread quickly. By around 3.5 billion years ago, several different types were growing together in layered, mat-like formations called **stromatolites** (*stroma* means "layer"; Fig. 24.2). Modern purple sulfur bacteria still use this kind of photosynthetic pathway today.

Figure 24.2 TOP: These fossils of ancient stromatolites are more than 3 billion years old. BOTTOM: These modern communities of photosynthetic bacteria exist only in extreme environments, such as these saline lagoons in Australia. Grazing pressure excludes these organisms from environments where plant-eating animals can survive.

Was There an RNA World?

DNA directs the assembly of RNA, which in turn directs the assembly of proteins. That is the "central dogma" of molecular biology, and that's how all living cells operate. Yet DNA replication, transcription, and translation all require the participation of several proteins. The system of information transfer in living cells today seems hopelessly dependent on *all* its parts; how could it ever have "bootstrapped" itself into operation? This mystery is far from solved, but studies by numerous researchers have suggested a possible explanation.

For some time, Norman Pace from Indiana University, Thomas Cech of the University of Colorado, Sid Altman of Yale, and Leslie Orgel of Scripps have obtained bits and pieces of RNA during experiments that simulate conditions on the primordial earth. Those ribonucleotide sequences aren't very long, but they appear in ways suggesting that, given enough time, significant stretches could have accumulated in a prebiotic environment.

Then independent experiments showed that certain forms of RNA do much more than serve as "messengers" for DNA. One study found that some stretches of ribosomal RNA can join amino acids into chains by itself—without the aid of protein enzymes. Other experiments demonstrated that some forms of RNA can copy themselves. Still other studies showed that some RNAs can "edit" other RNAs, adding and deleting ribonucleotides in nonrandom fashion. These experiments revived a hypothesis first suggested by Francis Crick and Leslie Orgel in 1968: the suggestion that RNA, not DNA, served as life's first central player. According to this hypothesis, a living world based on RNA could have started when RNA fragments that had been abiotically assembled acquired the ability to copy themselves.

You may recall that RNA is not as stable as DNA; because it lacks proofreading mechanisms, it would have "mutated" often. Early on, that would have been an advantage, as the first form of natural selection began. Those RNA variants most efficient at copying themselves and modifying each other would soon come to outnumber their more inert neighbors.

What next? Sooner or later, RNA could have "delegated" two vital functions to other sorts of molecules. One function, the "grunt work" of performing chemical reactions, was passed along to proteins that could have been assembled by RNA acting both as information source and as enzyme. The other function, information storage, was "delegated" to DNA—a polynucleotide that is more stable than RNA but still susceptible to mutation over time. Where could the first DNA have come from? Theoretically, primordial RNA could have created DNA with reverse transcriptase, an enzyme that was discovered only recently but that appears to be very old. Today, reverse transcriptase is utilized by retroviruses like HIV which work "backwards" to transcribe RNA into DNA. Ultimately, over time, RNA, DNA, and protein could have acquired the complicated and specialized functions described in Part 4.

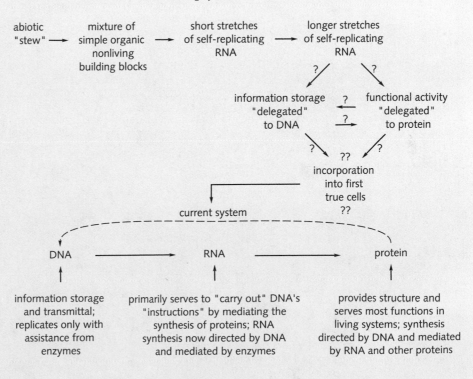

Schematic representation of one highly speculative scenario for the evolution of modern DNA-based living systems.

Hot Springs: Ancient Bacteria, Symbiosis, and the Origin of Life

Much of the deep ocean floor is nearly devoid of multicellular life. Water pressures reach several tons per square inch, and temperatures hover around freezing. The deep sea is pitch dark, so photosynthesis is impossible. Yet in this inhospitable environment, scientists have uncovered both a new class of ecosystem and clues to the origin of life.

In 1977, geologists aboard a deep-sea submersible near the Galapagos Islands stumbled upon a rich community that thrives around submarine hot springs. There, where geysers of mineral-laden water as hot as the core of a nuclear reactor erupt from the sea floor, the explorers saw animal life in abundance around the vents. No one had dreamed such a profusion of life

could exist on the sea floor, and at first, no one had any idea what energy source nourished these organisms in the absence of sunlight.

We know now that this community is powered by bacteria that utilize hot hydrogen sulfide (deadly poison to most eukaryotes) as a source of chemical energy to produce complex organic compounds. These chemosynthetic

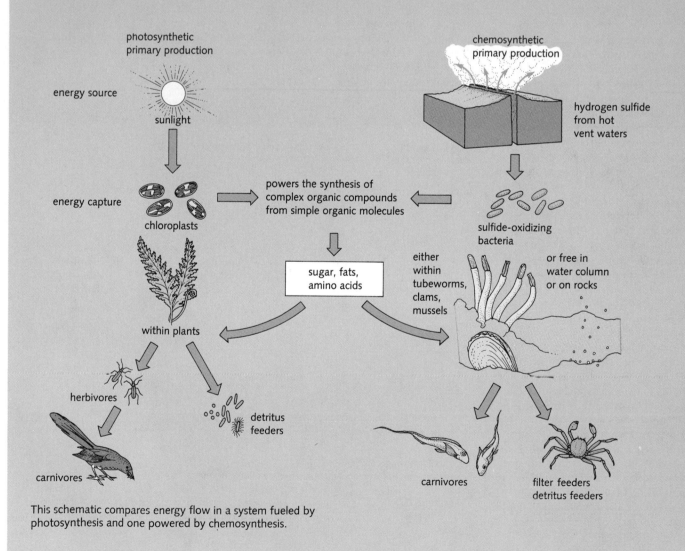

This schematic compares energy flow in a system fueled by photosynthesis and one powered by chemosynthesis.

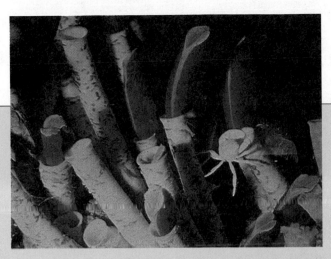

The deep-sea hot springs community includes scarlet tube worms a meter long, clams 30 cm in length, clusters of mussels, and scores of shrimps, crabs, and fishes.

bacteria, among the most ancient organisms alive today, replace photosynthetic plants as the base of the energy pyramid. Giant tube worms, clams, and mussels survive thanks to symbiotic relationships with bacteria that reside in their gills. Water passing over the gills supplies the bacteria with oxygen, carbon dioxide, and hydrogen sulfide; the bacteria, in turn, provide their hosts with organic carbon.

Several scientists note that all the conditions required for the chemical reactions leading to the origin of life exist at these vents and have existed there since the oceans were formed: high temperature, the necessary chemical menu, turbulent currents, and clay deposits. By duplicating these conditions, researchers have produced both amino acids and short stretches of RNA. Given millions of years, these building blocks could have produced the earliest living prokaryotes. The fact that the oldest living direct descendants of those first cells, the chemosynthetic archaebacteria, inhabit hot springs today lends additional support to this fascinating, though highly controversial, hypothesis.

Photosynthesizers Change the World

About 3 billion years ago, a new group of photosynthetic prokaryotes evolved. These organisms harnessed solar energy in much the same way that plants do today—and began to release free oxygen as a by-product (see Chapter 19).

It is difficult to convey just how revolutionary this metabolic innovation was and just how profoundly it transformed the nature of life on earth. Because we depend on oxygen, we think of that gas as essential to life. But oxygen is *not* essential to most life processes—only to the particular metabolic pathways used by modern eukaryotes. And, in fact, oxygen is such a highly reactive gas that it is *poisonous* to many forms of life because it oxidizes and destroys many organic compounds. Oxygen will kill any cell not protected by shielding compounds.

So imagine the scene 3 billion years ago: In an anaerobic world, one group of organisms started churning out this deadly gas in large quantities. For a time, geochemical activities shielded life from this toxic waste, because free oxygen rapidly combined with iron ions dissolved in the world's oceans. Over several million years, the resulting insoluble ferric oxide (rust) precipitated out as layers on the ocean floor. But once earth's reactive minerals were all oxidized, free oxygen began to accumulate in the atmosphere.

This was the first time that wastes produced by one group of organisms poisoned the biosphere for others. Anaerobes that did not evolve protection from toxic oxygen either perished or were excluded forever from living openly on the earth's surface or in open water in lakes and seas. These physiologically archaic organisms survive today only in habitats from which oxygen is excluded—places such as hot sulfur springs, deep mud, and cavities or wounds within the bodies of multicellular animals.

The production of free oxygen did, however, have a beneficial side effect. When atmospheric oxygen reached a concentration equal to 1 percent of its modern-day levels, the ionizing effects of ultraviolet (UV) radiation would have produced the beginnings of an ozone layer. As the ozone layer began shielding the earth from harmful UV rays, organisms could live safely near the sea's surface.

Evidence of an Aerobic World

About 2 billion years ago, a new sort of organism appeared—one whose fossils offer the first evidence of adaptation to higher oxygen levels. These organisms formed chains of small cells punctuated at intervals with larger, thicker-walled cells. The resulting colonies resemble modern prokaryotes called **cyanobacteria,** or *blue—green bacteria.* In modern blue—greens, similar thick-walled

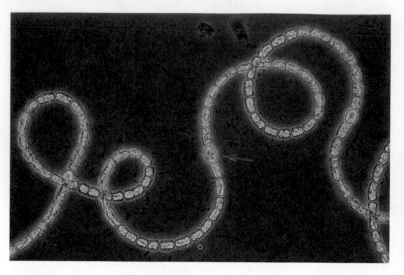
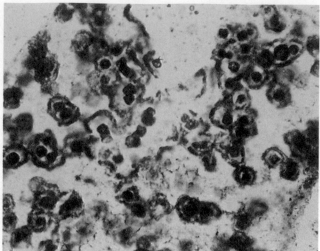

Figure 24.3 LEFT: This cyanobacterium (blue–green bacterium) shows the enlarged structures called heterocysts that protect its anaerobic, nitrogen-fixing enzymes from oxygen. RIGHT: This blue–green fossil of an ancient relative of modern cyanobacteria from the Gunflint Formation is evidence that prokaryotic organisms were evolving ways to tolerate increasing free-oxygen levels in the atmosphere.

cells, or **heterocysts** (*hetero* means "different"; *cystis* means "bladder or pouch"; hence *heterocyst* means "different-looking cell"), shield nitrogen-fixing enzymes from oxygen, enabling those anaerobic pathways to function in an aerobic world (Fig. 24.3).

Figure 24.4 This microfossil Acritarch is thought to exemplify the first eukaryotic cells.

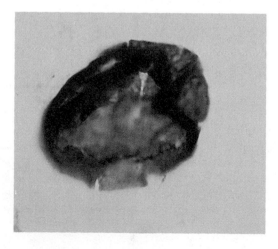

THE RISE OF EUKARYOTES

Between 2.5 and 1.4 billion years ago, a series of events set the pace for the rest of Earth's history. First, as oxygen concentrations continued to rise, organisms evolved physiological systems that not only *tolerated* oxygen but *depended* on it. Over time, new forms of aerobic metabolism evolved, ultimately developing in ways that extract 18 times as much energy from each sugar molecule as the older, anaerobic metabolism.

Second was the evolution of the first eukaryotes—organisms whose cells contain a nucleus surrounded by a double membrane and one or more classes of membrane-bound structures such as chloroplasts and mitochondria. The transition from prokaryotic to eukaryotic structure probably occurred among floating marine microorganisms whose remains comprise a class of microfossils called *Acritarchs* (from a Greek phrase meaning "of uncertain origin"; Fig. 24.4). Fossils of presumed eukaryotic cells have been dated at between 1.4 and 1.6 billion years old, which means that eukaryotic cells first evolved in, and were fully adapted to, an aerobic environment.

Third, within 200 or 300 million years, some eukaryotes "invented" sex, and the process of evolution accelerated to speeds never seen before. That evolutionary innovation has been supremely important in the history of eukaryotic life, because it enhances genetic variability from generation to generation. Here's why.

Prokaryotic cells (with some exceptions) reproduce by simple **binary fission;** they duplicate their genetic material and send half to each daughter cell. As a result, each

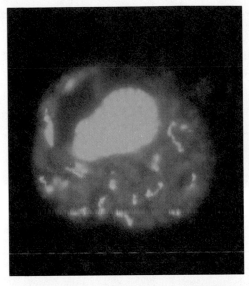
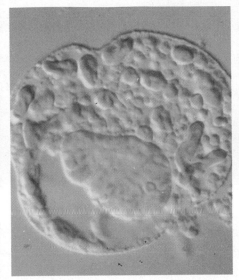
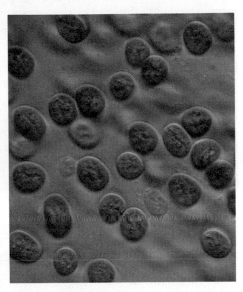

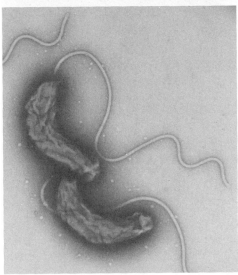

Figure 24.5 These micrographs show the evidence that leads many researchers to believe that chloroplasts and mitochondria originated as free-living organisms. Like the grin of the Cheshire cat, the DNA of these symbionts lingers as a clue to their possible independent origins. TOP LEFT: An alga, *Gyrodinium,* photographed in such a way that all its DNA fluoresces green. Why does this cell have DNA in both its nucleus and its chloroplasts? CENTER: The same alga, photographed in natural light to highlight its green chloroplasts. RIGHT: This blue–green bacterium, a descendant of the earliest photosynthesizers, looks much like the proposed free-living ancestors of modern chloroplasts. BOTTOM LEFT: This bacterium, *Bdelovibrio,* is thought to resemble the free-living ancestors of modern mitochondria.

daughter cell is a precise replica of its parent. Although this is a fast and effective means of cell division, it restricts genetic variation to mutations such as those resulting from errors in DNA replication.

The recombination and reshuffling of genetic material during sexual reproduction, on the other hand, provide many more chances for existing genes to combine in new ways (see Chapter 10). The evolution of sex thus vastly increased the genetic variation on which natural selection can operate. Once eukaryotes adopted sexual reproduction, new species began to appear far more rapidly than before. The first multicellular organisms evolved within a few hundred million years, and not long after that, a host of exotic multicellular animals appear in the fossil record, as you will see in the next chapter.

The Symbiotic Theory of Eukaryote Origins

The evolutionary origins of mitochondria and chloroplasts have puzzled biologists for decades. Both these organelles contain their own DNA and reproduce on their own when cells in which they reside divide. These organelles also bear an uncanny physical and biochemical resemblance to certain living prokaryotes such as the alga *Gyrodinium* and the bacterium *Bdelovibrio* (Fig. 24.5). These and other observations led to the conclusion that eukaryotes evolved from prokaryotes through a mechanism that neither Darwin nor most of his successors ever considered.

At some point, ancestral eukaryotes evolved a membranous envelope around their nuclei. Just how this happened is not clear. Then, according to a widely accepted theory championed by Lynn Margulis, a biologist at the University of Massachusetts, eukaryotic evolution continued when two or more simple organisms joined in cellular symbiotic associations that have lasted to this day. According to Margulis, mitochondria and chloroplasts don't just *look* like prokaryotes; they once *were* prokaryotes that took up residence inside other cells and lost their independence. Given certain chemical differences among the chloroplasts of living algae and higher plants, it is likely that these events happened not just once but several times, producing several distinct evolutionary lines (Fig. 24.6). This fascinating story of intracellular symbiosis reminds us that even the most basic components of our bodies—our cells—are collections of once-independent parts. Figure 24.7 offers a review of key milestones in the history of the earth up to the appearance of eukaryotic cells.

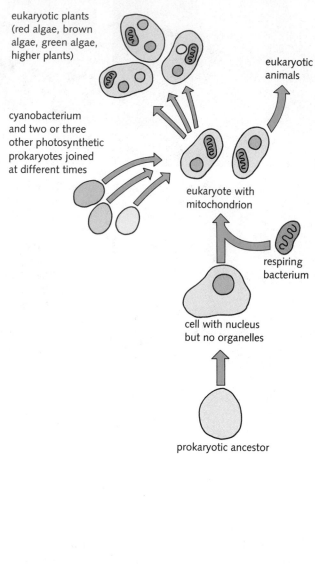

Figure 24.6 The symbiotic origin of eukaryotic cells. First, a host cell that had already evolved a nucleus combined with a bacterium to form an animal-like cell with mitochondria. This duplex cell (or another like it) was later joined by a photosynthetic prokaryote to form a unicellular alga with chloroplasts. These latter events probably happened several separate times.

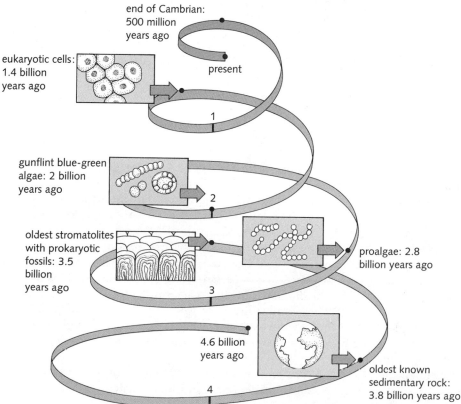

Figure 24.7 This schematic summarizes the timing of major events in early earth history. Note that life seems to have evolved roughly 1.1 billion years after the earth formed. The evolution of eukaryotic cells took more than 2 billion years longer.

SUMMARY

Life almost certainly began on the earth as, over billions of years, complex organic molecules formed abiotically and became organized in progressively more complex ways. Several theories have been postulated to explain how the first cells evolved, but the earliest life-forms must have been anaerobic heterotrophs that were unicellular and prokaryotic, similar to modern-day bacteria. The evolution of photosynthesis liberated free oxygen and radically changed the composition of the earth's atmosphere, allowing organisms to live safely near the sea's surface. As oxygen levels rose, new forms of aerobic metabolism evolved, and the evolution of the first eukaryotes was not far behind. Sexual reproduction supplied the genetic variation that allowed new species, including more complex multicellular organisms, to appear at a relatively rapid rate—within a few hundred million years.

STUDY FOCUS

After studying this chapter, you should be able to:

- Cite various hypotheses about the origin and early evolution of life.

TERMS AND CONCEPTS

stromatolites *492*

cyanobacteria *495*

heterocysts *496*

binary fission *496*

REVIEW

Objective Question (Answer in Appendix)

1. The earliest living cells
   (a) lived in aerobic conditions.
   (b) could get energy from other organic compounds.
   (c) were eukaryotes.
   (d) were autotrophs.

Discussion Questions

2. According to available evidence, what was the earth's earliest atmosphere like? What secondary atmosphere soon replaced it?

3. How could organic molecules have been formed abiotically early in the earth's history?

4. Why must the first organisms, or protoorganisms, have been heterotrophs? How could autotrophy have evolved?

5. Explain the symbiotic theory of the origin of eukaryotic cells.

6. Why do you think the "RNA world" hypothesis is so attractive to some scientists? Explain how recent discoveries in molecular biology have given this old idea new viability.

READINGS

See integrated list of readings for Part 5 after Chapter 29.

25

Charting Life's History on Earth

recall, for a moment, the scientific quests we've described on islands—from Darwin's classic observations in the Galapagos to cutting-edge work on molecular evolution in Hawaii. The power of those observations testifies that for studies aimed at revealing both the relationships among living species and the processes by which those species evolved, islands are perfect natural laboratories.

Why? Because islands (especially volcanic ones) tend to be geologically young, geographically isolated, relatively small, and ecologically diverse. Hawaii, for example, is barely the size of a pinhead on a world map, a mere 400,000 years old, and separated by 2000 miles of Pacific Ocean from the nearest continent; yet it has ecosystems with microclimates as diverse as some continents many times its size. The story of life in such a place, though rich and full of variation, is necessarily short and is invariably told by a fairly small cast of characters that are often clearly distinct from mainland species. That's all good news for evolutionary biologists.

Sorting out the diversity and history of life on a global scale, by contrast, is fiendishly difficult. The story began, as you saw in the previous chapter, more than 3.5 billion years ago. Geography—which is usually treated as a constant in studies spanning a few hundred thousand years or so—becomes a variable over such long periods, as you'll see in this chapter. And the global cast of characters is enormous, as you'll see in chapters to come. In short, a great deal has been going on, in a great many places, with myriad interactions, among countless players, for a very long time.

Not surprisingly, there's a good deal we don't know about the global history of life. But there are plenty of fascinating stories that we do know, and this chapter presents the most important of them. As it unfolds, you will see both successes and failures in the evolutionary adaptation of different organisms to the earth's physical and biological environments.

Connecting those stories are several common themes:

- **Life, which appeared soon after our planet formed, has been shaped by changes in physical environments ever since.** Sea levels have risen and fallen as glaciers advanced and retreated. Movements of the earth's crust have created continents, torn them apart, and smashed them into one another, creating and destroying oceans in the process. Periodic, major environmental fluctuations or cataclysms have caused mass extinctions that wiped out many species and made room for adaptive radiations among survivors.

- **Key evolutionary innovations in body structure and life cycle provide the raw materials that fuel adaptive radiations in major new groups of organisms.** Evolutionary systematists define major animal phyla and plant divisions by using the same physical and physiological traits that trace their histories. In this light, each major group of organisms can be seen as an evolutionary "experiment" based on a particular body plan.

- **The history of life has been profoundly influenced by long-term evolutionary interactions among organisms.** These interactions include competition between species exploiting similar resources; "arms races" between plants and herbivores, predators and prey, or parasites and hosts; and beneficial partnerships between two or more organisms. When those interspecific associations are tightly knit and involve stepwise reciprocal interactions, they demonstrate coevolution.

- **Organisms that have disappeared over the eons weren't "inferior" in any absolute sense to those that survived.** Although natural selection is a potent force, many species became extinct strictly by chance, not because they were poorly adapted. At any number of stages in Earth's history, chance could have produced a dramatically different planet.

CHANGING SCENERY: THE RESTLESS PLANET

Before we can understand the evolution of Earth's living groups, we must appreciate the fact that continents and oceans are not the static stage sets they seem to be during a human lifetime. On a geologic time scale, continents are constantly moving and changing. They dramatically alter both local and global climate as they do so.

Current awareness of continental history dates back to 1915 and a book entitled *The Origin of the Continents and Oceans* published by German scientist Alfred Wegener. Wegener proposed that land masses move slowly around the surface of the planet, sometimes colliding, sometimes pulling apart. He collected evidence from geography, botany, and geology to prove the existence of such movement, but he had no idea of the forces that could *cause* it.

Today, Wegener's notions have been expanded and elaborated into the theory of **plate tectonics,** or *continental drift*. According to plate tectonic theory, the earth's crust is divided into large, irregularly shaped pieces called *plates* (Fig. 25.1a). Forces in the earth's mantle create powerful convection currents of molten rock (Fig. 25.1b), which create new crust at certain boundaries between plates. At other boundaries, plates slip past one another or are forced over and under each other. (That's what's happening in earthquake-prone California.)

Biological Consequences of Continental Drift

Continental movements have been important in shaping life's history for several reasons:

1. Continents have not always been where they are today. Greenland and Antarctica, for example, have drifted from equator to pole, carrying their resident organisms from tropical to arctic climates.

2. Earth's land masses have repeatedly joined and separated from one another, alternately forming "supercontinents" and splitting up into many smaller pieces. The interiors of large continents experience greater fluctuations in weather than the interiors of smaller land masses or areas near major inland seas.

3. When previously separate continents join, they divide the seas that once flowed between them, dramatically altering ocean current patterns. Those currents, in turn, have powerful effects on global rainfall and temperature patterns.

The global geological forces that drive plate tectonics also influence climate directly by affecting the biogeochemical cycles discussed in Chapter 4. For reasons yet unknown, the forces that drive continental drift seem to fluctuate over millions of years. At certain times, major episodes of volcanic activity have expelled large quantities of carbon dioxide from the earth's interior into the atmosphere. Until this "extra" heat-retaining greenhouse gas was removed by either geochemical or biological means, it could have caused the earth's temperature to rise. Some researchers believe this mechanism was responsible for warm spells like that of the dinosaur era.

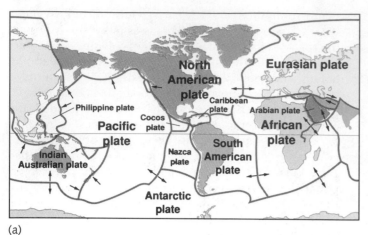

(a)

Figure 25.1 **(a)** The outlines of several major plates in the earth's crust. Note the intersection of plates along the coasts of Alaska and California; movement along these plates causes California's earthquakes and Alaska's volcanic eruptions.
(b) The incredible heat in the earth's core and currents in the molten mantle drive the movements of continental plates into and away from one another.

(b)

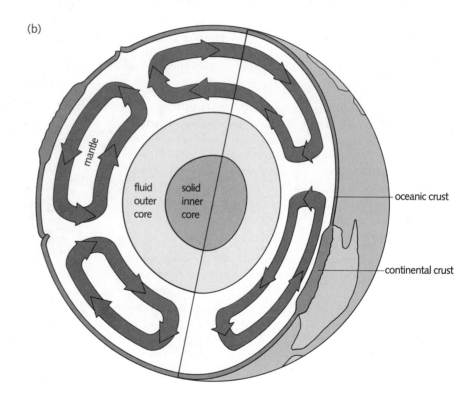

CHARTING EARTH'S HISTORY

The histories of both Earth and its life-forms cover an almost unimaginably long period of time. Both geologists and biologists, therefore, found it convenient to divide geological time into shorter segments—creating in the process a hierarchy of eras, periods, and epochs, as shown in Fig. 25.2. (In this book, we concern ourselves only with eras and periods; epochs are more fine-grained than our analyses.)

These divisions in time are of unequal length because they are not meant to be arbitrary measurements like seconds or hours. Instead, the beginnings and ends of these units are usually determined by major biological events. The boundaries between the Paleozoic and the Mesozoic eras (225 million years ago), and between the Mesozoic and the Cenozoic eras (65 million years ago), for example, mark two of the greatest mass extinctions in Earth's history. The boundary between the Proterozoic (Precambrian) and the Paleozoic, on the other hand, marks a

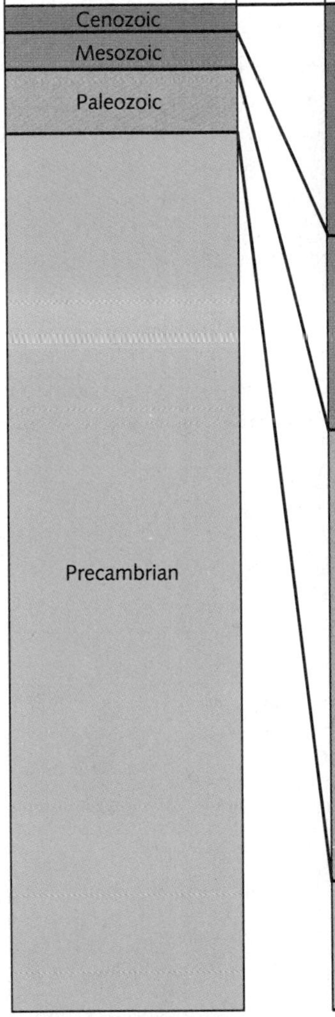

Relative Time Span		Era	Period	Epoch	Began (millions of years ago)	Length (millions of years)
Cenozoic		Cenozoic	Quaternary	Recent	0.01	
Mesozoic				Pleistocene	1.5	
Paleozoic			Tertiary	Pliocene	12	10
				Miocene	25	13
				Oligocene	34	9
				Eocene	56	22
				Paleocene	63	7
		Mesozoic	Cretaceous		135	70
			Jurassic		100	15
			Triassic		225	45
		Paleozoic	Permian		280	55
	Carboniferous		Pennsylvanian		310	30
			Mississippian		350	40
			Devonian		400	50
			Silurian		430	30
Precambrian			Ordovician		500	70
			Cambrian		570–600	70–100
		Precambrian	Proterozoic		2500	2000
			Archaeozic		4600	2000

← Late Cretaceous extinction: 50% of marine species, all dinosaurs

← Permian extinction: 50% of animal families, 96% of marine species

← Cambrian explosion

Figure 25.2 The geological time scale. Note that the beginning of each major era is marked by a major event in the history of life on Earth. Most of the boundaries between periods are marked by similar, though generally less dramatic, biological events.

great flowering of multicellular life. Note that we invoke the names of these divisions only for convenience; because there is no need for you to memorize them, they appear at the left-hand side of each figure concerned with geological history.

As this chapter covers major episodes of Earth's history, it will highlight major evolutionary developments, referring to specific plant and animal groups wherever they first appear. More detailed discussions of both plant and animal groups can be found in Chapters 26–29.

LATE PROTEROZOIC (OR PRECAMBRIAN) ERA

The Proterozoic era, stretching from about 2.5 billion years ago to about 570 million years ago was one of the longest periods in Earth's history (Fig. 25.3). For most of that time, life evolved slowly; as discussed in the previous chapter, it took nearly 2 billion years for the first eukaryotic cells to appear. But toward the end of that era, about 650 million years ago, two global trends fueled the evolution of multicellular organisms: a change in global climate and an increase in the amount of free oxygen in the atmosphere.

Dawn of Multicellular Life

In the oceans, the mid-Proterozoic saw a rapid diversification of single-celled plankton—the first great *adaptive radiation* of eukaryotes. But around 650 million years ago, a global cold spell triggered an ice age that wiped out many early plankton species. This may have been the world's first **mass extinction.** When the ice retreated and climate warmed, a second adaptive radiation of plankton began. This time, numerous multicellular animals called *metazoa* were among them.

This was the first time that multicellular, aerobic organisms *could* have evolved. The earliest metazoa had not yet evolved respiratory and circulatory systems and, therefore, depended on simple diffusion to supply oxygen to their cells. Because oxygen diffuses through thick, dense tissues very slowly at the low concentrations present during the early Proterozoic, clusters and many-layered sheets of aerobic cells could not have obtained the oxygen they required. But by the mid-Proterozoic, photosynthesis had produced enough free oxygen to raise atmospheric oxygen content to 10 percent of present levels—a concentration adequate to support thicker tissues of primitive, multicellular, aerobic organisms.

Much of what we know about early metazoans comes from excavations in the Ediacaran Hills of Australia. These hills have yielded a rich fauna of soft-bodied, multi-

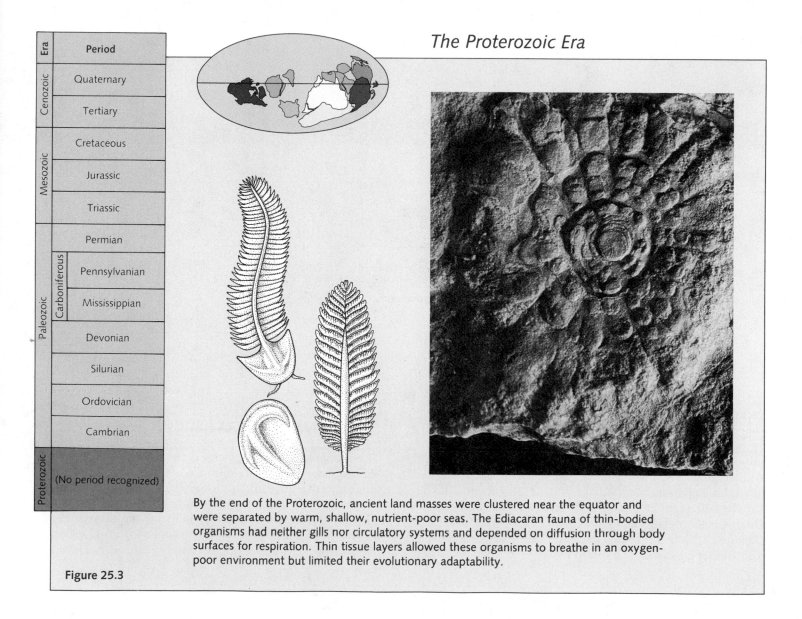

The Proterozoic Era

By the end of the Proterozoic, ancient land masses were clustered near the equator and were separated by warm, shallow, nutrient-poor seas. The Ediacaran fauna of thin-bodied organisms had neither gills nor circulatory systems and depended on diffusion through body surfaces for respiration. Thin tissue layers allowed these organisms to breathe in an oxygen-poor environment but limited their evolutionary adaptability.

Figure 25.3

cellular animals dated at 680–700 million years old. Some paleontologists believe that this *Ediacaran fauna* contained the forebears of modern jellyfish, certain worms, and their kin. But most researchers disagree, arguing that the Ediacaran fauna represents unsuccessful evolutionary "experiments" that left no descendants. In any case, these organisms probably lived in somewhat the same manner as many corals and jellyfish do today, depending on photosynthetic symbionts for energy in nutrient-poor seas.

By the end of the Proterozoic, about 570 million years ago, a few of the first fossil shells appeared, along with remnants of the first multicellular plants. From this time, too, date the first **trace fossils:** petrified tracks and burrows in sediment produced by animals that were not themselves preserved. These early shells and trace fossils indicate that other animals were also evolving—hard-shelled forms that would soon dominate the seas.

On land, however, not much was happening yet; life was still mostly an aquatic phenomenon. But by the end of the era, bacteria may have begun to colonize the soil, and primitive fungi and multicellular green algae may have ventured out onto moist beaches around shallow bays.

PALEOZOIC ERA: CAMBRIAN PERIOD

During the geologically active Cambrian period (Fig. 25.4), the giant supercontinent began to fragment and pull apart, creating many shallow seas and thousands of

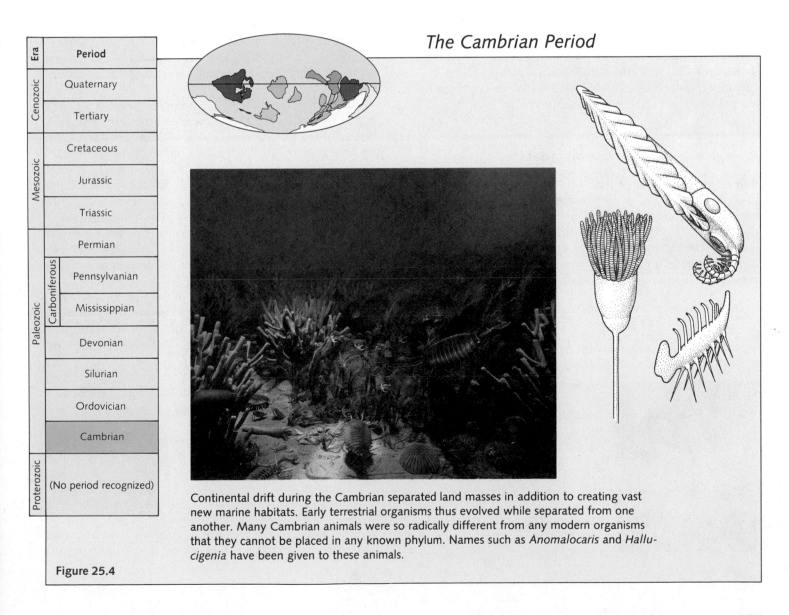

The Cambrian Period

Era	Period
Cenozoic	Quaternary
	Tertiary
Mesozoic	Cretaceous
	Jurassic
	Triassic
Paleozoic	Permian
	Pennsylvanian
	Mississippian
	Devonian
	Silurian
	Ordovician
	Cambrian
Proterozoic	(No period recognized)

Continental drift during the Cambrian separated land masses in addition to creating vast new marine habitats. Early terrestrial organisms thus evolved while separated from one another. Many Cambrian animals were so radically different from any modern organisms that they cannot be placed in any known phylum. Names such as *Anomalocaris* and *Hallucigenia* have been given to these animals.

Figure 25.4

miles of shoreline. Atmospheric oxygen concentrations were still rising and soon reached 20 percent of current levels.

The increased availability of oxygen enabled multicellular organisms to grow thicker tissue layers with greatly increased evolutionary flexibility. Under the pressures of natural selection, different cell groups within those thicker tissues began to specialize, adapting in structure and function in ways that enabled them to perform a variety of tasks impossible for generalized cells.

This biological division of labor led to the evolution of organ systems. Some excitable cells evolved into nerve cells, which in turn formed *nervous systems* that aided rapid communication between body parts. Other excitable cells evolved the ability to contract, forming *muscles* that enabled complex movement. *Gills* served as specialized respiratory structures, and *circulatory systems* distributed oxygen and food throughout the body.

This increased structural complexity and diversity fueled the Cambrian explosion, life's second series of experiments in body design. Some Cambrian body plans, as fantastic as any imaginary space creatures, were failures that rapidly vanished. Others, no less unusual, turned out to be evolutionarily successful and adaptable. These organisms founded several major phyla, many representatives of which are alive and abundant today.

Ecological Diversity in Cambrian Seas

Before the Cambrian, most animals lived by scavenging dead organic matter or by obtaining energy from photosynthetic symbionts. The rise in oceanic nutrient levels during the Cambrian, however, supported abundant growth of both unicellular and multicellular algae. That algal growth, in turn, allowed the evolution of herbivores with new and varied ways of feeding. We know, for example, that stromatolites suffered a dramatic decline in diversity during the Cambrian, probably because the first grazing herbivores had appeared (see p. 492).

As the earliest members of modern phyla emerged, they took their places in the first complex food webs. Many species assumed the roles of predators and began feeding on other animals. Emerging anatomical complexity fueled ecological interactions, as organisms occupied new niches and adjusted to emerging pressures of competition and predation. The development of gills and circulatory systems, for example, freed many animal groups from the need to exchange respiratory gases through body surfaces. That freedom, in turn, made it possible for emerging forms to cover their bodies with armor for protection against predators.

Modern Phyla First Known from the Cambrian

Many major living phyla probably trace their roots back to yet-unknown, soft-bodied ancestors that first evolved during the Proterozoic era. Unfortunately, the dearth of fossils from that time still shrouds these origins in mystery.

By the beginning of the Cambrian, however, several major lineages of **invertebrates,** animals without backbones, began leaving recognizable fossils (*in* means "without"; *vertebrae* means "skeletal elements of the backbone"; hence *invertebrate* means "without a backbone"). Though modern members of these groups did not evolve for millions of years, the basic body plans that appeared during the Cambrian represented evolutionary blueprints or archetypes for organizing body tissues. Within each phylum, variations on those blueprints could form new classes or orders, each of which could diverge in adapting to new ways of life—or disappear.

The Cambrian saw the debut of the first sponges, jellyfish, and sea anemones, several important groups of worms, and the ancestors of modern starfishes, sea urchins, snails, clams, and squid. But perhaps the best-known fossils from this period are the **trilobites,** the first representatives of the phylum Arthropoda, or "jointed-leg" animals, whose living members include crustaceans, spiders, insects, and a variety of other animals many people call "bugs."

Trilobites, the first animals to display the characteristics that made arthropods among the most successful of all invertebrate groups, rapidly took over the early Cambrian seas. At first, their speed, agility, keen vision, and armor made them deadly and invincible predators (Fig. 25.5). By the middle of the Cambrian, however, predators such as *Anomalocaris* and the giant sea scorpions called *Eurypterids* began to take their toll on the trilobite popula-

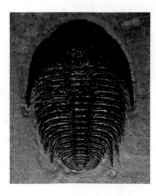

Figure 25.5 Trilobite fossils are the most common remains of primitive arthropods.

tion. Neither trilobites nor their predators survive today; many died out by the end of the Cambrian.

ORDOVICIAN PERIOD

The closing days of the Cambrian ushered in one of the largest mass extinctions on record; nearly 50 percent of existing animal families, including many trilobites, became extinct. The Ordovician period that followed saw dramatic adaptive radiations within each major group that survived that extinction (Fig. 25.6). Ancestors of modern snails, clams, and squid diversified extensively, as did aquatic arthropods. One group of early soft-bodied animals (not the corals of today) formed the first reefs.

The Ordovician also saw substantial evolution among multicellular marine algae, which by this time had probably already split into three major divisions: green algae, red algae, and brown algae. Because each division contains a different set of pigments that capture light for photosynthesis, the groups probably arose independently when early eukaryotic ancestors joined symbiotically with three different types of photosynthetic prokaryotes. By this time, too, the ancestors of modern green algae and the terrestrial plants to whom they are related had taken separate evolutionary paths, with the latter entering shallow, freshwater environments.

The Ordovician Period

Era	Period	
Cenozoic	Quaternary	
	Tertiary	
Mesozoic	Cretaceous	
	Jurassic	
	Triassic	
Paleozoic	Permian	
	Carboniferous	Pennsylvanian
		Mississippian
	Devonian	
	Silurian	
	Ordovician	
	Cambrian	
Proterozoic	(No period recognized)	

The Ordovician saw expansion of shallow marine habitats as oceans flooded vast land areas. The southern continent, Gondwanaland, drifted across the South Pole, where it underwent glaciation. Other major land masses straddled the equator where climates remained temperate. Ordovician reefs, though not formed by modern corals, filled the same ecological role as today's coral reefs and provided habitats for many other organisms.

Figure 25.6

SILURIAN PERIOD

Global climate and geological events during the Silurian period set the stage for two crucial evolutionary events: colonization of the land by plants and evolution of aquatic vertebrates with jaws. A global cold spell ended the Ordovician, causing the extinction of many marine organisms. A warm, moist climate followed, and that lasted through the Silurian.

Meanwhile, continental drift carried the southern supercontinent of Gondwanaland northward, toward other land masses near the equator (Fig. 25.7). Plentiful swamplike, freshwater habitats served as cradles for the evolution of both land plants and freshwater fishes. During the Silurian, several important groups of marine animals that had appeared much earlier evolved hard body parts and left clear fossil records for the first time.

Plants Colonize the Land

As life in the sea continued to evolve, competition and predation became fierce. Because terrestrial habitats were still devoid of life, any organism able to survive out of water could enjoy major selective advantages. But because life evolved underwater, both plants and animals encountered similar obstacles in adapting to terrestrial life. Plants were the first to make that leap successfully, so we will examine the problems they faced and the evolutionary adaptations that enabled their transition to land (Table 25.1).

Desiccation, the loss of body water to surrounding air, is the foremost obstacle to terrestrial life. Plants could not colonize the land, therefore, without a way to retain water. Most land plants have evolved waxy, waterproof coverings that retain water in tissues surrounded by dry air.

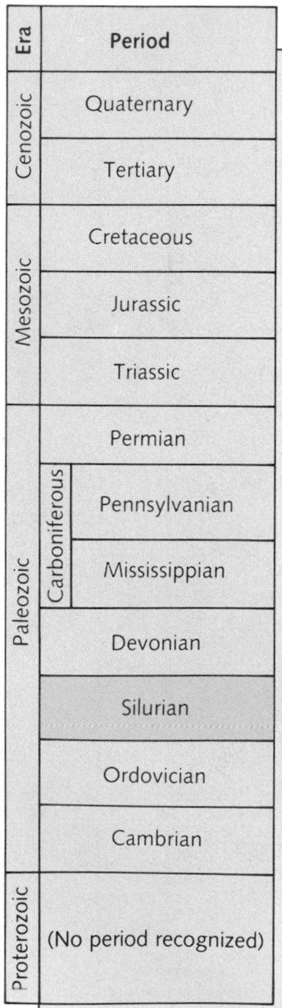

The Silurian Period

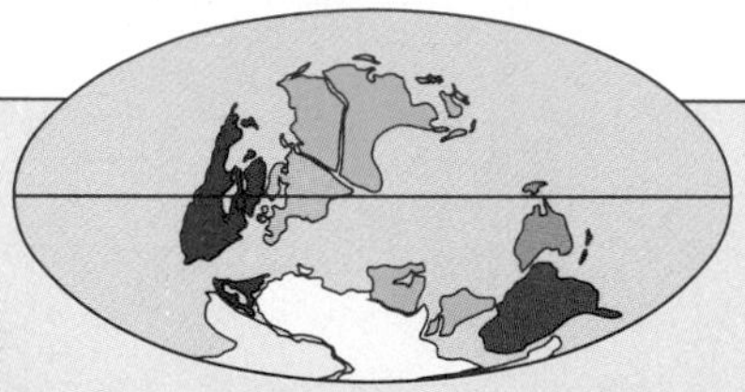

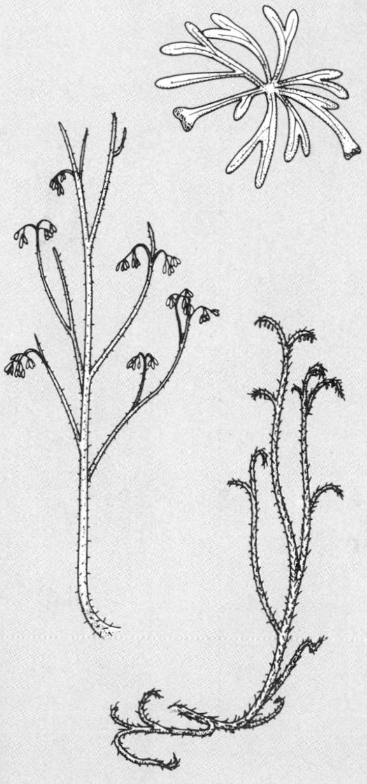

Era	Period
Cenozoic	Quaternary
	Tertiary
Mesozoic	Cretaceous
	Jurassic
	Triassic
Paleozoic	Permian
	Pennsylvanian
	Mississippian
	Devonian
	Silurian
	Ordovician
	Cambrian
Proterozoic	(No period recognized)

During the Silurian, many land areas were uplifted, pushing back shallow seas and creating moist, tropical, freshwater habitats. Predatory arthropods reached 2 meters in length, and numerous other strange animals inhabited the seas. During this period, the first multicellular plants emerged onto land.

Figure 25.7

Table 25.1 *Evolutionary Adaptations from Aquatic to Terrestrial Environment*

Multicellular Algae	Requirement	Terrestrial Plants
Medium supportive; no specialized support tissues	Support of plant tissue	Medium nonsupportive; specialized structural tissues
Occurs in most cells	Photosynthetic areas	Confined to portions above ground
Direct access to environmental water and minerals; no specialized transport system	Transport of water and dissolved nutrients	Elevated parts not in direct contact with water and minerals; transport of water and minerals required
By water	Transportation of gametes	By wind, water, animals
By water	Seed/spore dispersal	By wind, water, animals

This evaporation barrier is interrupted by pores that open or close as needed to allow exchange of carbon dioxide and oxygen for respiration and photosynthesis.

Transport of body fluids is another problem for terrestrial plants. Because aquatic plants can absorb water and nutrients through the leaf-like tissues used for photosynthesis, all their cells have ready access to the requirements for life. The leaves of land plants, on the other hand, stretch up to the sun while roots probe underground for water and nutrients. Because roots cannot carry on photosynthesis, and because most leaves cannot absorb water, land plants require transport systems to move these essential materials through their bodies. This function is served in modern land plants by two types of specialized **vascular tissues** that conduct water, inorganic nutrients, and the organic products of photosynthesis. These tissues evolved slowly as plants adapted to progressively drier environments. (We will have more to say about all these structures in Chapter 30.)

Physical support for body parts is another requirement for terrestrial life. Because living tissues are usually lighter than water, they are buoyed up by the medium around them. Aquatic plants can thus simply *extend* photosynthetic surfaces without having to hold them up. Land plants, however, have developed rigid supporting structures to hold their leaves up to the sun.

Reproductive and dispersal strategies must also adapt to conditions on land. Not surprisingly, early aquatic plants depended on reproductive systems that relied on water to support and carry gametes and young plants around. But on land, air currents can neither carry gametes over great distances nor disperse mature plants. For these reasons, plants could not survive and reproduce far from the water's edge without new, terrestrially oriented reproductive and dispersal strategies.

These changes occurred slowly in stages over millions of years. Each time a terrestrial adaptation was completed successfully, it initiated an adaptive radiation of plants able to colonize ever-drier habitats. Plants' first ventures onto land would have gone completely unopposed, of course, for there were as yet no herbivorous terrestrial animals.

The First Land Plants

The first land plants undoubtedly resembled modern green algae and probably began their invasion during the Ordovician. By the Silurian, they had already given rise to the two main branches of the plant kingdom defined by the presence or absence of vascular tissues (Fig. 25.8). The single surviving nonvascular plant group contains mosses and their kin; the several divisions of vascular plants include ferns and all the seed-producing plants—the most familiar multicellular, photosynthetic organisms.

The first vascular plants appeared around the beginning of the Silurian, about 430 million years ago. The stepwise development of vascular tissue, true roots, and water-retaining leaves rapidly led to the emergence of several early plant divisions, each with a distinct body plan. Like the Ediacaran animal fauna, some of these early plant groups were evolutionary "experiments" that failed to survive environmental changes and became extinct. The division Rhyniophyta—the first vascular plants known well from fragmentary fossil records—was one

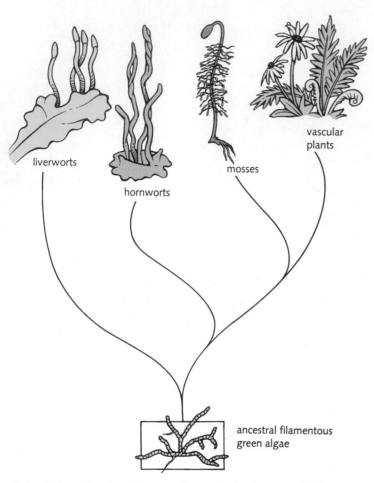

Figure 25.8 The first plant to colonize the land was probably a filamentous green alga. Soon after appearing, the lineage containing these ancestral land plants radiated to produce three lines of plants that lacked vascular tissue and the vascular plants that are both familiar and abundant on land today.

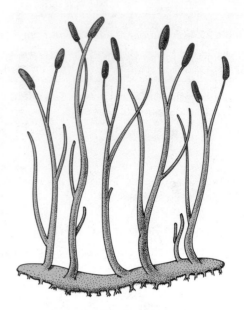

Figure 25.9 The earliest vascular plant species, such as *Rhynia,* lacked both true leaves and true roots.

such group (Fig. 25.9). Rhyniophyte's aboveground and below-ground parts were virtually identical; they had not yet differentiated into what in later plants are called *root* and *shoot.*

Three other early plant groups—club mosses, horsetails, and ferns—experienced major adaptive radiations early on but ultimately relinquished dominance to plants that evolved later. Both mosses and horsetails experienced wildly successful radiations that lasted for over 100 million years, from the beginning of the Carboniferous to the end of the Permian. Ferns were also enormously successful during the Carboniferous and Permian when tree-sized forms made huge forests.

The First Land Animals

After plants colonized the continents, animals could follow—assuming that they could evolve adaptations that dealt with the same problems faced earlier by plants: desiccation, the need to support their bodies, and the need for well-developed systems for the internal transport of water and essential nutrients.

Obviously, not all phyla were equally good candidates for a terrestrial existence. The body plans of sea anemones and jellyfish, for example, though eminently successful in the sea, were useless on land, for they offered natural selection no raw material from which to shape necessary terrestrial adaptations. Arthropods, on the other hand, had evolved tough external skeletons which—with minor evolutionary modifications—provided both waterproof body coverings and stiff, jointed appendages for walking, burrowing, or flying. It should come as no surprise, therefore, that among the first land animals were members of the same arthropod group that contained the early sea scorpions and horseshoe crabs (Fig. 25.10).

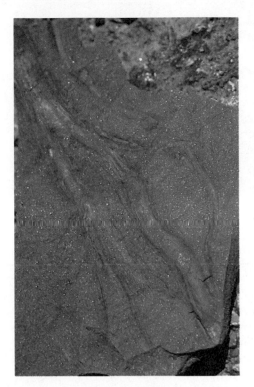

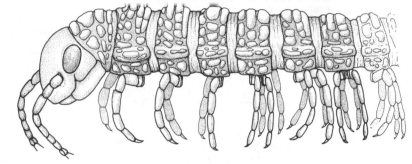

Figure 25.10 LEFT: These burrows, left in the soil of what is now the state of Pennsylvania during the Silurian period, are believed by some researchers to represent the first traces of multicellular animal life on land. No one knows just what sort of animal left these trace fossils. TOP RIGHT: This millipede fossil from the early Devonian period, similar to those that date back to the Silurian, is one of the oldest known full-body fossils of land animals. LOWER RIGHT: An artist's reconstruction shows what this arthropod might have looked like.

The Origins of Vertebrates

While major invertebrate groups diversified, another less ecologically conspicuous group of animals evolved—the ancestors of the first animals with backbones. The fossil record of those organisms is virtually nonexistent; they seem to have lacked hard body parts, and, as important as they are to vertebrate history, they don't seem to have played a major role in early ecosystems. Nevertheless, by the Silurian, vertebrates had come into their own and had founded a dynasty that would ultimately reshape much of the earth.

The ancestors of vertebrates, which are likely to have separated from the other lines of chordate evolution as early as the lower Cambrian, were probably shallow-water marine organisms. They had evolved a flexible series of cartilaginous units that wrapped around the nerve cord to protect it. Those units interlocked with one another to provide a rudimentary backbone.

The first fossil traces of vertebrates are tiny fragments of scale-like and tooth-like hard parts from the late Cambrian and Ordovician periods. From this evidence we infer that these animals were heavily armored, a trait that offered protection from formidable predators such as eurypterids. These fragments of armor, whose complexity is clearly seen under a microscope, indicate that vertebrates had been evolving for some time before they left any substantive fossils.

Jawless fishes: The first vertebrates The first vertebrates of which we have a reliable record are a curious group of fishes without jaws, commonly called **agnathans** (*a* means "without"; *gnathos* means "jaw"). Agnathans had mouths, of course, but those mouths did not have true *jaws*: the reinforced, hinged structures that serve the rest of us in feeding. The first agnathans, which became established in the Ordovician and diversified during the Silurian, exhibited a variety of odd shapes. Although jawless

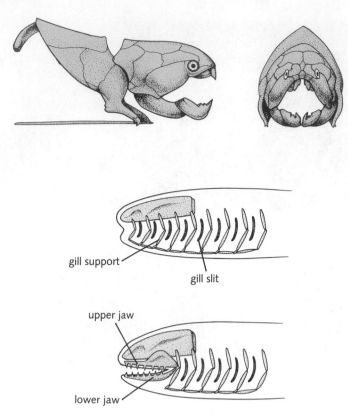

gill support

gill slit

upper jaw

lower jaw

Figure. 25.11 TOP: Many early jawed fishes were fearsome predators; this one reached a length of nearly 3 m. BOTTOM: Through many steps over millions of years, several of the anterior gill supports fused and changed shape to create the highly adaptable jaws that characterize most fishes and higher vertebrates today.

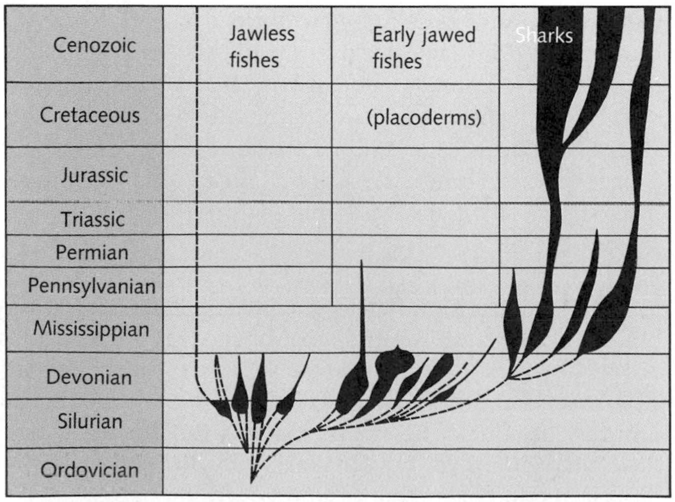

	Jawless fishes	Early jawed fishes	Sharks
Cenozoic			
Cretaceous		(placoderms)	
Jurassic			
Triassic			
Permian			
Pennsylvanian			
Mississippian			
Devonian			
Silurian			
Ordovician			

Figure 25.12 This chart illustrates the relative abundance (signified by the width of the bands) of three classes of fish-like vertebrates in the fossil record. For explanation, see text.

fishes thrived well into the Devonian, only about 45 species remain alive today.

Placoderms: The first jawed fishes Sometime during the middle of the Silurian, one group of primitive fishes evolved true jaws, apparently from the most anterior of the structures that previously supported gills (Fig. 25.11). Jawed fishes diversified so quickly that they appear in the fossil record fully developed. Beginning their diversification in the mid-Silurian, the first jawed fishes (called *placoderms* because of their plate-like armor) dominated the Devonian fossil record.

Evolutionary Innovations and Adaptive Radiations

Many popular accounts of evolution imply that advanced groups of modern organisms evolved either *from* more primitive living forms or at least *after* those primitive forms underwent their major adaptive radiation. One might get the impression, for example, that vascular plants evolved from modern green algae, or that jawed fishes evolved from living jawless forms. This is not the case.

Throughout history, adaptive innovations leading to advanced groups occurred in an evolving line *before* primitive members of that line diversified very much. In the case of the earliest fishes, for example, the ancestors of placoderms evolved jaws before the agnathans diversified, as Fig. 25.12 shows. And, as you will soon see, the first sharks and bony fishes appeared before placoderms reached their prime.

There are several possible explanations for this situation, but the powerful ecological principle of competition may be sufficient to account for it. It is likely that under certain circumstances, adaptation and diversification are stimulated by competition. If so, then competition with the first jawed fishes excluded agnathans from some niches and stimulated them to diversify and exploit other niches, where they temporarily became equal or superior competitors. In the long term, however, the advances in body plan evolved by jaw-bearing fishes provided them with an adaptive superiority that—together with the stresses of changing environments—ultimately spelled doom for their predecessors. Placoderms in turn, after flourishing in the Devonian, fell victim to a mass extinction, clearing the way for the higher fishes whose ancestors had emerged as early as the late Silurian.

DEVONIAN PERIOD

The Devonian period saw rapid evolutionary radiation of vertebrates in the sea and extensive colonization of the land by plants and animals (Fig. 25.13). During this geologically active time, Earth's continents once more drifted together, forming a single, vast land mass called *Pangaea*. This supercontinent, which persisted for 200 million years, became the stage for the evolution of all major lines of terrestrial plants and animals. Across Pangaea, Devonian climates were moist and comfortable, though cool. Ample rainfall created freshwater basins that cradled many new aquatic and soon-to-be-terrestrial organisms.

The Reign of Vertebrates Begins

The Devonian is often called the "age of fishes," thanks to the diversity of aquatic vertebrates during this period. As placoderms reached the peak of their adaptive radiation, three key innovations appeared among different lines that had already evolved jaws: paired appendages, lungs, and, ultimately, the two sets of limbs that characterize higher vertebrates. As each of these innovations appeared, it established a new body plan and sparked the adaptive radiation of one or more vertebrates classes (Fig. 25.14).

Paired fins The earliest fishes had only single *dorsal* or *ventral* fins, which acted as stabilizers, and tail fins, which propelled them forward. As a result, agnathans and most placoderms couldn't steer very well, either to chase prey or to escape predators. *Paired fins* added significant

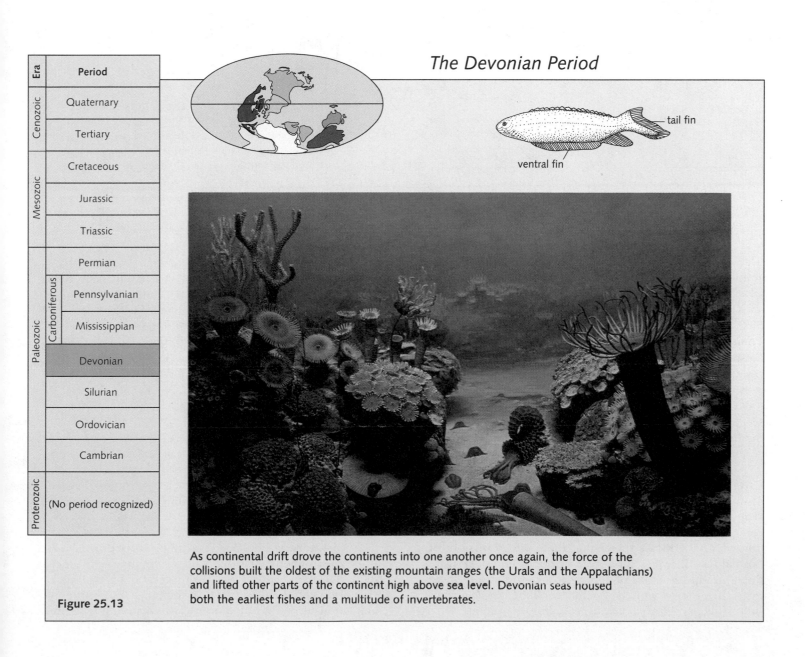

The Devonian Period

tail fin

ventral fin

Era	Period
Cenozoic	Quaternary
	Tertiary
Mesozoic	Cretaceous
	Jurassic
	Triassic
Paleozoic	Permian
	Pennsylvanian
	Mississippian
	Devonian
	Silurian
	Ordovician
	Cambrian
Proterozoic	(No period recognized)

Carboniferous (Pennsylvanian, Mississippian)

Figure 25.13

As continental drift drove the continents into one another once again, the force of the collisions built the oldest of the existing mountain ranges (the Urals and the Appalachians) and lifted other parts of the continent high above sea level. Devonian seas housed both the earliest fishes and a multitude of invertebrates.

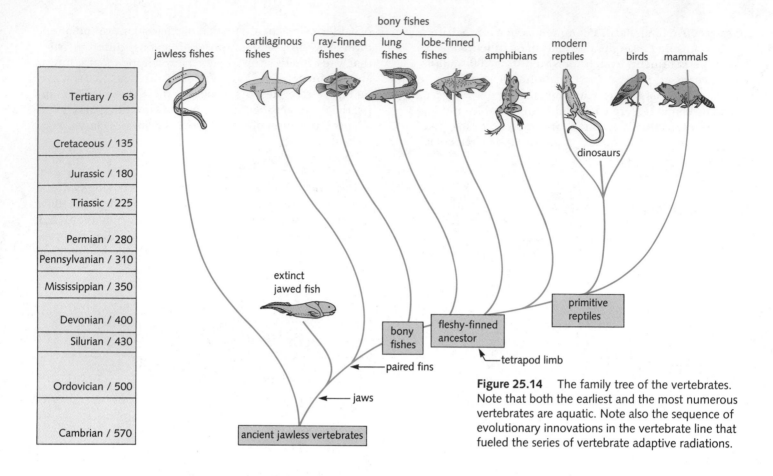

Figure 25.14 The family tree of the vertebrates. Note that both the earliest and the most numerous vertebrates are aquatic. Note also the sequence of evolutionary innovations in the vertebrate line that fueled the series of vertebrate adaptive radiations.

ability to steer and thus bestowed an enormous selective advantage.

The first vertebrates to possess both jaws and paired fins were the ancestors of modern sharks, skates, and rays. Their first adaptive radiation produced a host of species, most of which became extinct fairly rapidly. The class survived and radiated once again during the Cretaceous period, however, producing the modern species we know today. Another line of early placoderms combined two sets of paired fins with strong, lightweight bone to found the lineage that led both to modern fishes and to the ancestors of all terrestrial vertebrates.

Origins of the tetrapod limb One group of peculiar, primitive fishes of great evolutionary significance crawled around Devonian swamps on four stubby, fleshy fins (Fig. 25.15). They are called, with surprising clarity, "fleshy-finned fishes." X-rays reveal that these pudgy versions of normal fish fins were supported by bones homologous to the bones in the limbs of all terrestrial vertebrates. These were the first *tetrapods* (*tetra* means "four"; *poda* means "limbs"; hence *tetrapods* means "four-limbed animals").

Lungs Early *lungfishes* probably evolved in Devonian swamps where shallow, muddy water contained little oxygen. In these fishes, a pouch connected to the upper digestive tract evolved into a primitive lung that enabled its bearers to breathe air. Though common and diverse at first, lungfishes dwindled to a handful of species in parts of Australia, Africa, and South America.

Plants Invade the Land ...

At the beginning of the Devonian, terrestrial plants were probably rare, but mosses and vascular plants soon became common (Fig. 25.16). One or more lines that led ultimately to higher plants continued to evolve refinements in their vascular tissue, enabling them to grow taller and thicker stems.

At first, early land plants could reproduce only in very moist environments, and their life cycles involved dispersal by means of spores that were carried by the wind. Then, sometime during the Devonian, a new suite of adaptations ushered in a more fully terrestrial way of life.

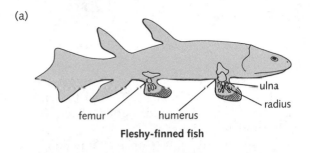

(a)

femur   humerus   ulna
                     radius

Fleshy-finned fish

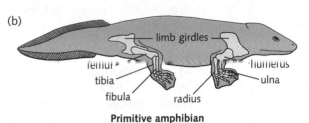

(b)

limb girdles

femur   humerus
tibia   ulna
fibula   radius

Primitive amphibian

(c)

Figure 25.15 **(a)** A generalized representation of a fleshy-finned fish, showing rough positions and skeletal supports of the paired front and rear limbs. **(b)** A generalized reconstruction of an early amphibian, showing the homology between the fin bones of fleshy-finned fishes and the limbs of more recent tetrapods. **(c)** *Ichthyostega,* one of the earliest amphibians.

The most important single innovation was the evolution of the **seed,** a self-contained, drought-resistant reproductive package that houses a dormant plant embryo and an ample supply of food within a protective seed coat.

The seed coat protects the infant plant from desiccation, enabling it to survive for weeks, months, or even years until conditions are right for germination. The stored food nourishes the embryo until it develops its own functioning roots and leaves. Of all plant reproductive strategies, those involving seeds are best adapted to terrestrial life. Seed plants remained rare for some time before giving rise to the major groups of modern plants that display several additional changes in their life cycles.

...And Vertebrates Follow

The first terrestrial vertebrates evolved from fleshy-finned ancestors through a range of intermediate forms. None of these can be unambiguously labeled *the* single ancestor of all land vertebrates, but it is clear that the bones of the tetrapod limb—which now carry the weight of all land vertebrates—were modified from those fins' original bony supports.

The first **amphibians**—ancestors of frogs, toads, and salamanders—climbed out of shallow swamps into moist terrestrial environments in tropical parts of Pangaea (see Fig. 25.15c). There, constant temperatures and high humidity made the transition from water to land less stressful than it would have been in colder or drier habitats. Possibly as a result of the value of being able to waddle

Figure 25.16 This fossil of a vascular plant known as *Baragwanathia* represents the oldest definitive evidence of primitive vascular plants. This specimen dates from the late Silurian or early Devonian period; its later Devonian descendants were up to a meter tall.

from pond to pond in search of food and open water, or as an adaptive response to competition or predation in shallow water, the descendants of early amphibians became progressively better adapted to terrestrial conditions.

These first terrestrial vertebrates faced the same problems as the first land plants: providing body support, avoiding desiccation, and "learning" to reproduce without standing in water. Early amphibians—like the first land plants—developed some, but not all, of the adaptations necessary for a fully terrestrial existence. The fleshy fins of their ancestors lengthened and strengthened into stocky limbs, each with five digits, and two skeletal "girdles" that provided those limbs with strong connections to the backbone. These major adaptations initiated the slow, steady amphibian radiation and spawned a complex and fascinating fauna that thrived for more than 130 million years. Most early amphibians seem to have been carnivores that fed on land-dwelling scorpions and spiders and on the very first insects that appeared at this time.

CARBONIFEROUS PERIOD

During the Carboniferous period, Pangaea remained intact while mountain building created a diversity of terrestrial environments ranging from moist, swampy lowlands to cooler and drier upland habitats. Toward the end of the period, several major glaciations affected the northern and southernmost land areas.

The Devonian ended with a mass extinction that claimed most jawless fishes, some placoderms, and many trilobites. Still, the major lines of higher plants and animals continued to evolve. The first forests, whose fossilized remains form most of our coal deposits and give the period its name, spread across Pangaea (Fig. 25.17). In and around those swampy forests, the amphibian adaptive radiation continued, encouraged both by the expansion of the great moist forests and by the emergence of a new type of prey in abundance: insects.

The Carboniferous Period

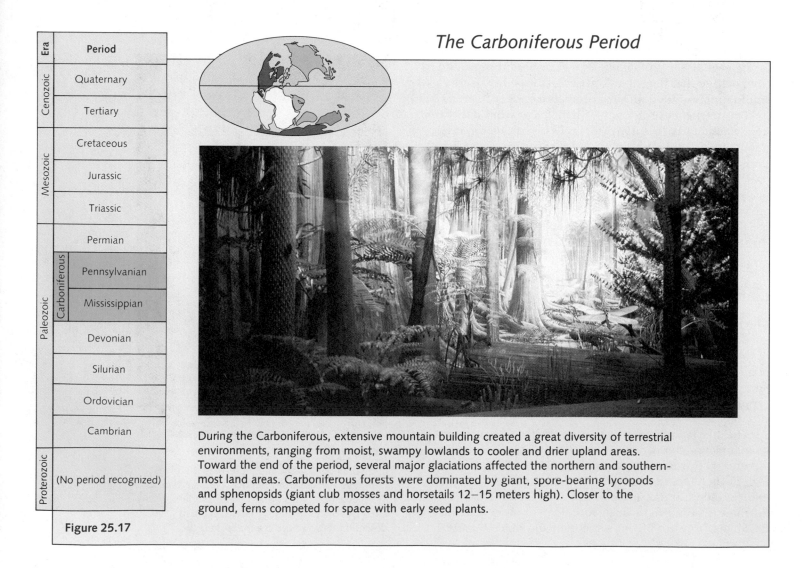

During the Carboniferous, extensive mountain building created a great diversity of terrestrial environments, ranging from moist, swampy lowlands to cooler and drier upland areas. Toward the end of the period, several major glaciations affected the northern and southernmost land areas. Carboniferous forests were dominated by giant, spore-bearing lycopods and sphenopsids (giant club mosses and horsetails 12–15 meters high). Closer to the ground, ferns competed for space with early seed plants.

Figure 25.17

The Success of Insects

The great quantity and diversity of Carboniferous land plants provided countless new ecological opportunities for insects, whose first representatives had appeared earlier. Although not as old as other arthropod groups, insects were far more successful, partly because they were the only flying organisms for over 100 million years. Insects' success had both positive and negative repercussions for other organisms. To plants, insects were predators that quickly became Earth's principal herbivores. To early carnivorous vertebrates, insects were prey that provided abundant, bite-sized animal food for the first time in history.

The First Reptiles

While amphibians diversified, reptiles, which first appeared during the Devonian, began their long-lived and enormously successful reign. The reptiles' adaptive radiation did not get under way until long after their amphibian predecessors had diversified, but the group radiated explosively from the second half of the Carboniferous into the Permian.

Like amphibians, reptiles seized upon insects as their primary food source and took advantage of stronger jaw muscles that enabled them to *bite* rather than just snap at prey. Early reptiles also evolved more agile limbs and stronger limb girdles. But the key innovation that spurred the reptilian radiation was the **amniotic egg,** a watertight egg produced by internal fertilization and wrapped in three protective membranes. Amniotic eggs serve land vertebrates much the same way seeds serve land plants— by making it easier to reproduce without being in standing water. Amniotic eggs protect developing embryos from desiccation, nourish them as they grow, provide a space for the storage and ultimate disposal of waste products, and enable the egg to exchange respiratory gases with the surrounding air.

CURRENT CONTROVERSIES

Hot-Blooded Dinosaurs?

Many body tissues, including muscles, operate well when body temperature is within a narrow range. The ability to move rapidly, either to capture prey or to avoid becoming prey, thus depends on a constant, optimal body temperature. Living reptiles are all ectotherms that make use of environmental conditions to regulate body temperature. To get warm, they bask in the sun. To avoid overheating, they shelter in the shade. This strategy works well in warm climates, but in very cold weather many ectotherms must become dormant.

Endothermic mammals and birds use insulation such as feathers, fat, and fur to retain metabolic heat. When it gets cold, endotherms metabolize food to keep warm. When it gets too hot, they sweat or pant to get rid of extra heat. Endotherms can perform muscular work for longer periods and under more widely varying environmental conditions than ectotherms. They can also grow faster. This vitality is expensive; an endotherm needs between *10* and *50 times* more food than a reptile of the same size.

Some paleontologists, led by Robert Bakker of the University of Colorado Museum, argue that dinosaurs were endotherms that regulated body temperatures as modern mammals do. This theory would account for dinosaurs' adaptive superiority and high growth rates, and it would explain how birds became endothermic.

Other paleontologists point out that dinosaurs were diverse and that not all were necessarily endothermic. It is not unlikely, says Yale's John Ostrom, that small, carnivorous dinosaurs were endothermic. Indeed, endothermy among the small dinosaurs that gave rise to birds would explain the role of the first feathers. But Ostrom and others point out that mammalian-style endothermy in large herbivorous dinosaurs is unlikely for practical reasons; they could never have eaten enough to survive. Elephants spend up to 15 out of every 24 hours stuffing 300-odd pounds of leaves and branches into their huge mouths. Could tiny-headed apatosaurs have gathered the daily *2000* pounds of conifers and lower plants needed to maintain endothermic metabolism in their enormous bodies?

Bakker's supporters point out that elephants use their large mouths for chewing and grinding. Dinosaurs, however, used their jaws only for *taking in* food; grinding was done by a gizzard-like arrangement in which stones that were swallowed were rubbed together by muscles inside the digestive tract.

Many researchers agree that dinosaurs had body temperatures higher than those of modern lizards. The debate over how they maintained those temperatures will continue for some time. Extinction, Ostrom says, ensures a certain amount of ambiguity. "When the dinosaurs died," he writes, "they took many of their secrets with them."

PERMIAN PERIOD

The Permian was a time of great environmental stress and innovation for all plants and animals, as geological changes in Pangaea produced cooler, drier climates. A mass extinction of extraordinary proportions claimed more than 50 percent of all terrestrial animal families and more than 95 percent of all marine species. Among its victims were the last trilobites, many amphibians, and nearly all non–seed-bearing plants.

The Rise of Seed Plants

Because the Permian's cooler, drier conditions spelled hard times for "lower" plants, they opened up new ecological opportunities for seed bearers by eliminating competition. By the end of the Permian, early seed-bearing plants had diversified into numerous lines, most of which are now extinct except for a few relic species (Fig. 25.18).

But one line of plants whose reign began during this period—the conifers—remained dominant for nearly 200 million years and is still important today.

The Rise of Reptiles

Driven by changes in climate and encouraged by diversifying terrestrial ecosystems, animal evolution proceeded rapidly. Reptiles made the most impressive progress, as the ancestors of modern reptiles, dinosaurs, and mammals appeared in quick succession (Fig. 25.19).

By the end of the Permian, terrestrial ecosystems were dominated by **therapsids,** a collection of curious animals that roamed across Pangaea and diversified into both carnivorous and herbivorous niches (Fig. 25.20). These animals were clearly intermediates in the evolution of mammals from reptiles. In fact, paleontologists differ as to whether they should be called transitional mammal-like reptiles or reptile-like early mammals.

The Permian Period

Era	Period		
Cenozoic	Quaternary		
	Tertiary		
Mesozoic	Cretaceous		
	Jurassic		
	Triassic		
Paleozoic	Permian		
	Pennsylvanian	Carboniferous	
	Mississippian		
	Devonian		
	Silurian		
	Ordovician		
	Cambrian		
Proterozoic	(No period recognized)		

During the Permian, the plates composing Pangaea continued to grind against one another, lifting larger and larger areas above sea level. The climate became cooler and drier, the great continental seas retreated, and many swamps dried up completely. Two plant relics from this period are ginkgoes (LEFT) and cycads (RIGHT). The ginkgo, thought by Western scientists to have been extinct for millions of years, was rediscovered in China where it had been maintained in horticulture.

Figure 25.18

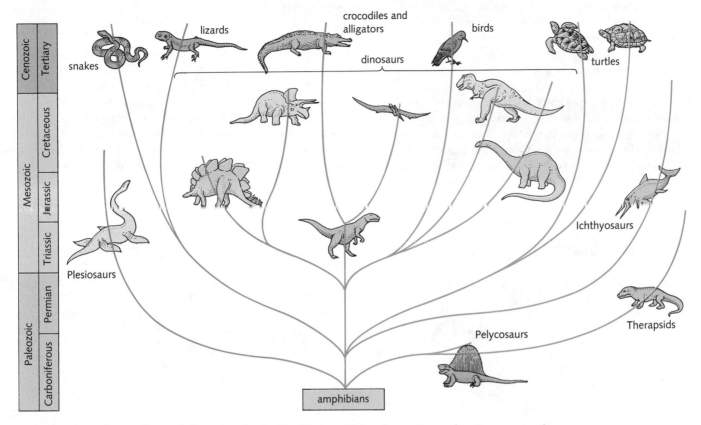

Figure 25.19 The reptilian radiation, showing both extinct and living forms. For explanation, see text.

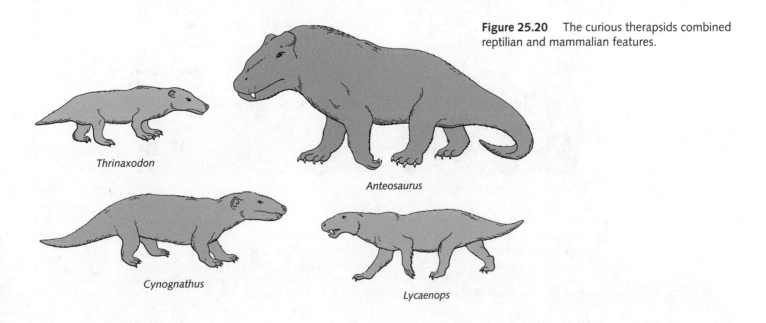

Figure 25.20 The curious therapsids combined reptilian and mammalian features.

In any case, because therapsids clearly maintained high internal body temperatures by metabolic means, they are described as **endotherms** (*endo* means "inside or interior"; *therm* means "heat"). In contrast, living reptiles and amphibians, all invertebrates, and most lower vertebrates are **ectotherms,** because their body temperatures rise and fall with the temperature of their external environment (*ecto* means "outside"). Endothermy is a useful, though energetically expensive, strategy that ultimately held the key to the success of mammals. [A heated controversy surrounds the notion that dinosaurs (or at least some of them) were endotherms. See Current Controversies, Hot-Blooded Dinosaurs?, p. 517.]

MESOZOIC ERA

The Mesozoic era, often called the "age of dinosaurs," spanned 183 million years and encompassed three periods: the Triassic, the Jurassic, and the Cretaceous (Fig. 25.21). Following the mass extinctions of the late Permian, the Mesozoic era saw major adaptive radiations among both marine and terrestrial animals and plants. At the same time, the first true mammals appeared. Because the evolutionary histories of all these groups are closely intertwined, we discuss the Mesozoic as a single unit.

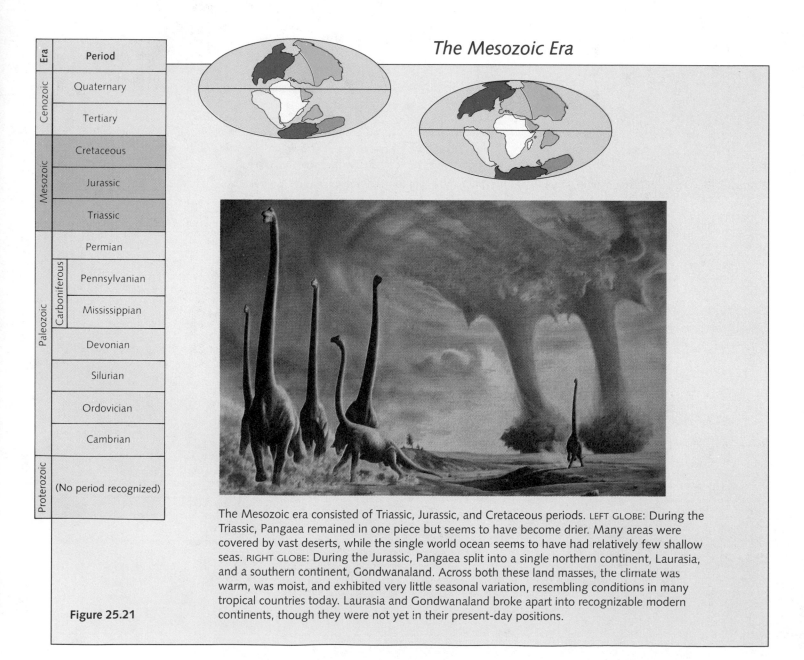

The Mesozoic Era

Era	Period
Cenozoic	Quaternary
	Tertiary
Mesozoic	Cretaceous
	Jurassic
	Triassic
Paleozoic	Permian
	Pennsylvanian
	Mississippian
	Devonian
	Silurian
	Ordovician
	Cambrian
Proterozoic	(No period recognized)

The Mesozoic era consisted of Triassic, Jurassic, and Cretaceous periods. LEFT GLOBE: During the Triassic, Pangaea remained in one piece but seems to have become drier. Many areas were covered by vast deserts, while the single world ocean seems to have had relatively few shallow seas. RIGHT GLOBE: During the Jurassic, Pangaea split into a single northern continent, Laurasia, and a southern continent, Gondwanaland. Across both these land masses, the climate was warm, was moist, and exhibited very little seasonal variation, resembling conditions in many tropical countries today. Laurasia and Gondwanaland broke apart into recognizable modern continents, though they were not yet in their present-day positions.

Figure 25.21

The Mesozoic era was a time of great geological changes that profoundly influenced the evolution of terrestrial organisms. The supercontinent of Pangaea remained intact through the Triassic but began to break up during the Jurassic. As the land mass fragmented, its pieces carried once-unified plant and animal populations into isolation and set the stage for new adaptive radiations.

Climatically, the Mesozoic was stable and warm; the dry warmth of the Triassic was followed by a warm, moist period that lasted into the Cretaceous. For reasons not completely understood, episodes of mass extinction terminated all three periods. The extinction at the end of the Triassic, for example, wiped out 35 percent of all animal families, including many primitive reptiles on land and several major marine groups. Toward the end of the Cretaceous, the rise of the Rocky Mountains and the Andes drained inland seas and cooled continental climates worldwide, with dramatic consequences for both plants and animals.

Ruling Reptiles

The reptilian adaptive radiation reached its peak during the Mesozoic era, and reptiles dominated both land and sea. The reptilian lineage, which had diverged back in the Permian, spawned two aquatic lines (ichthyosaurs and plesiosaurs) that are now extinct; the forebears of modern lizards, snakes, and turtles; and the ancestors of dinosaurs, crocodiles, and birds.

Dinosaurs were extraordinary animals. More than half were as large as or larger than any living land mammal. Evolving in an ecological arena already populated by therapsids and the first mammals, they quickly outstripped the competition to take over an incredible variety of niches. Dinosaurs ranged from chicken-sized

carnivores preying on even smaller animals to giant flesh eaters and scores of herbivores, large and small.

The most famous Jurassic reptiles were *Allosaurus*, *Ichthyosaurus*, *Apatosaurus* (formerly called *Brontosaurus*), and *Stegosaurus*. Their Cretaceous successors were the greatest of the dinosaurs: *Tyrannosaurus rex,* the king of the carnivores; *Triceratops* and the ankylosaurs, the heavily armored herbivores; *Pteranodon*, the great flying reptile; and the duck-billed dinosaurs.

Until the mid-1960s, most scientists envisioned dinosaurs as oversized versions of modern reptiles: plodding, pea-brained evolutionary misfits that survived simply because they were bigger than everybody else. Apatosaurs, estimated to have tipped the scales at 25 to 35 tons, were depicted slogging through swamps submerged up to their bellies, because it was assumed that their legs could barely support their weight.

But during the past 25 years, paleontologists have learned that old reconstructions of dinosaur skeletons were often wrong. Most dinosaur skeletons in museums, for example, show the animals' legs splayed out sideways like those of living reptiles and amphibians. In that configuration, leg bones would have splintered under the animals' massive bulk. More careful and extensive comparison of fossil bone and joint structure has shown that dinosaurs held their legs erect beneath them, as mammals do.

Furthermore, fossil footprints tell us that even the largest dinosaurs moved at respectable speeds, and that some lived in social groups like those of today's large, herbivorous mammals. One set of Cretaceous tracks shows a herd of 30 individuals traveling in such a way that larger animals on the edges of the group protected juveniles in the center. Also, dinosaur nests preserved with well-developed babies still inside suggest that certain dinosaurs incubated their eggs and cared for their young (Fig. 25.22).

Figure 25.22 Dinosaurs, such as these hypsilophodonts, may have used several techniques to maintain dependable body temperatures, and appear to have taken care of their young after hatching.

The Origins of Birds

In 1861 the discovery of a truly remarkable fossil caused remarkably little excitement. The fossil, named *Archaeopteryx,* had feathers and a wishbone, but it also had teeth in its beak and a long, bony tail. The animal was presumed to be an early bird that had evolved from the ancestors of dinosaurs and other reptiles at the beginning of the Mesozoic era.

In the early 1970s, however, John Ostrom of Yale University examined several *Archaeopteryx* fossils and noticed that their teeth, shoulders, forelimbs, pelvis, and tail looked more like parts of a dinosaur than those of a modern bird. Ostrom concluded that birds had evolved from a group of small, running, carnivorous dinosaurs. More recently, Sankar Chaterjee discovered another, older fossil he named *Protoavis,* whose characteristics were also intermediate between those of a small dinosaur and a bird. The question of how these first birds evolved the ability to fly is the subject of considerable debate and experimentation (see Current Controversies, How Does Half a Bird Fly?).

Such finds are valuable because they demonstrate a phenomenon known as **mosaic evolution,** in which different sets of traits evolve at different rates as organisms exploit new ecological opportunities. In-between forms such as these clearly document evolutionary links between extinct reptiles and living birds, and they refute the arguments of "creation scientists" who specifically deny the existence of evolutionary intermediates between major groups.

Figure 25.23   Painting of a late Cretaceous mammal.

The First Mammals

The promising radiation of the therapsids was cut short during the Triassic as these mammal-like reptiles were replaced by the ruling reptiles. The only survivors of the therapsid line were a group of small, insect-eating, furry, endothermic animals: the first true mammals (Fig. 25.23). They were probably nocturnal, feeding at night when small dinosaurs might have been sluggish. While the dinosaurs ruled the earth, these mammals seem to have cowered in the underbrush. But from these small beginnings, the entire mammalian radiation ultimately arose.

The Rise of Seed Plants

With all the attention lavished on dinosaurs, it is easy to forget that other important organisms were evolving along with them. Among the most important of those were the still-diversifying land plants. Dinosaurs and the first mammals lived in habitats dominated by conifers, cycads, ginkgoes, and a few extinct seed plants. Then during the Cretaceous, the true flowering plants (angiosperms) began to dominate the land.

Flowering plants radiated explosively during the mid-Cretaceous, their success fueled by seeds and other reproductive innovations. By the end of the Cretaceous, the coast and river valleys along North America's inland sea were carpeted with dense vegetation that probably resembled undisturbed areas near the Louisiana shore today. Redwoods, cypresses, and a variety of broad-leaved trees flourished in the wet, subtropical climate.

Coevolution of plants and animals   To understand the full significance of higher-plant evolution, we must consider the animals with whom the evolving plants shared the continents. The ancestors of flowering plants appeared toward the end of the Jurassic, just as reptilian herbivores were reaching their peak. Concurrently, the second adaptive radiation of terrestrial insects produced (among other groups) termites, ants, and bees. The advent of these animals made them part of the environment to which angiosperms adapted, and they became plants' partners in two major classes of coevolutionary relationships.

First, intense herbivory created powerful selection for the evolution of mechanisms to discourage plant eaters. Many recent plants evolved ways of "biting back" at predators; these plants synthesize and concentrate poisonous or distasteful chemicals in their leaves.

Equally important, reproductive habits of modern flowering plants coevolved with insects and mammals. Whereas the earliest true land plants were almost exclusively wind-pollinated, flowering plants as a group are pol-

How Does Half a Bird Fly?

The fossilized remains of both *Archae-opteryx* and *Protoavis* raise two thorny questions. Neither possessed the combination of muscle, bone, and feather development necessary for full flight. But if these animals couldn't fly, what purpose did their feathers serve? Furthermore, as their descendants lurched along the evolutionary road to full-fledged birdhood, how did they get off the ground? What good is half a bird that can almost—but not quite—fly?

The presence of feathers is relatively easy to explain. Long before they were sufficiently well-developed to use in flight, feathers could have provided efficient insulation for small, endothermic animals that had to keep warm at night. But two groups of investigators hold conflicting views about the way the first flying vertebrates left the ground.

In 1974 John Ostrom of Yale University suggested that the earliest birds, such as *Protoavis* and *Archaeopteryx,* were swift runners whose feathery forelimbs gradually evolved into net-like insect-catching devices. Then, once they began to resemble wings, the forelimbs would have enabled the animal to spring off the ground and maneuver to catch insects. From that point on, the

Reconstructions of *Archaeopteryx* illustrating two views of the evolution of flight in primitive birds.

animals could have begun taking longer and longer leaps and spending more and more time in the air. Ostrom recently declared that the insect-net part of the idea was probably wrong, but he still favors a "ground-up" origin of flight.

Others argue for a "tree-down" evolution of flight. According to this idea, the earliest birds were tree dwellers that first became airborne by gliding from

tree to tree. Sooner or later, they extended their range by flapping their wings.

Currently, these hypotheses are being tested both through sophisticated aerodynamic analyses worked out on computers and by physical models based on a combination of fossil evidence and modern calculations. Here, perhaps, is a paleontological question that can be solved by experiment. May

linated by animals that carry pollen from flower to flower. Bees are among the most important pollinators today, but moths, beetles, bats, and hummingbirds also play that role. The flowers that we admire look and smell as they do because they evolved in ways that enable them to attract pollinating animals.

While flowers were evolving with pollinators, seeds evolved with birds and mammals that both ate and dispersed them. Angiosperms wrap their seeds inside **fruits,** structures that are often both brightly colored and

tasty—and thus attractive to many birds and mammals. At least some of the hard-shelled seeds inside survive passage through the animals' digestive tracts and are scattered far from the parent plant.

These coevolutionary relationships provided flowering plants with two major advantages over their predecessors. Plants fertilized by animal pollinators, even if they were separated by large distances, could exchange genetic material without having to produce pounds of wind-carried pollen. And plants whose seeds were spread by animals could invest substantial energy in provisioning their embryos, despite the fact that extra food and protection made those seeds too heavy to be dispersed by wind like the spores of lower plants.

GREAT CRETACEOUS EXTINCTION

The end of the Cretaceous was literally the end of an era; the Mesozoic closed with one of the most dramatic mass extinctions in the history of life on land. Dinosaurs, many archaic birds, and most early land plants vanished. Interest in this Cretaceous extinction runs high, not only because dinosaurs are fascinating but also because if the dinosaurs had not been wiped out, mammals might never have undergone their adaptive radiation, and *Homo sapiens* might never have evolved!

Many explanations for the great dying are nearly as interesting as dinosaurs themselves. One proposal suggests that a giant meteor (or several meteors) struck the earth, raised enormous clouds of dust into the atmosphere, blocked solar radiation, and triggered a sudden, planetwide cold spell. But though dinosaurs and many gymnosperms disappeared, mammals, insects, and many nonreptilian marine animals survived. And scrutiny of the fossil record has shown that even groups which did become extinct died out very slowly, over millions of years. Rather than suggesting a sudden, global, catastrophic event, the fossil record paints a picture of prolonged changes in the global environment punctuated by a brief period of more rapid change.

Many researchers maintain that the Cretaceous extinction can be explained by long-term changes in climate caused by global geological events. As continents rose, inland seas retreated, taking with them their tempering effects on weather. Continental climates became both cooler and more variable; the hottest days became hotter and the colder days grew *much* colder. Those changes might have been accelerated by dust clouds cast up either by a long period of intense volcanic activity or by the impact of a large meteorite, but many researchers argue that such extreme mechanisms are not necessary.

Whatever problems the end of the Cretaceous posed for dinosaurs, however, it offered mammals and birds conditions under which their energetically expensive strategy of endothermic temperature regulation finally paid off. Endotherms spend more than 90 percent of the energy they take in just to maintain their body temperatures; while climates were warm and comfortable, this was an expensive strategy with few selective advantages. When temperatures dropped worldwide, however, endothermy became a great advantage.

CENOZOIC ERA

During the Cenozoic era, which spans the last 65 million years, the fragments of Pangaea acquired their modern continental shapes and drifted into their current positions (Fig. 25.24). As continents separated, ocean currents as we know them today began to flow, exerting profound effects on global climate. The first part of the Cenozoic was warm, but its middle period, from 50 to 20 million years ago, turned cooler and drier. The tropical and subtropical forests of earlier times were replaced in North and South America, Africa, and Asia, first by temperate forests and then by vast plains. The period climaxed with several major glaciations, the last of which ended only 10,000 years ago.

These climate changes had important effects on the era's rapidly evolving flora and fauna. The mass extinction that ended the Cretaceous wiped out the dinosaurs and decimated the lower plants, offering ecological and evolutionary opportunities to the mammals, birds, and flowering plants that survived.

Flowering plants were dominant on land by the end of the Mesozoic. One important family, the grasses, appeared early in the Cenozoic but did not diversify and spread widely until the cooling and drying trend began. Then, as tropical forests gave way to drier plains and savannahs, grasses and other quick-growing, rapidly reproducing plants flourished. Such plants, examples of ecological opportunists described in Chapter 5, quickly colonized the great plains and became major food sources for evolving mammals.

Mammals responded to the elimination of reptilian competitors by diversifying dramatically toward the end of the Cretaceous. The earliest mammals dispersed widely while Pangaea was still intact, though they remained small and ecologically inconsequential. Mammalian evolution thus began on the Mesozoic supercontinent and continued slowly for about 80 million years on the great land masses of Laurasia (the northern continent) and Gondwanaland (the southern continent). By the time the mammalian adaptive radiation finally began in force, at the beginning of the Cenozoic, all the continents had separated except Australia and Antarctica. We will pick up the story of mammals as a detailed evolutionary case study in Chapter 29.

The Cenozoic Era

Era	Period
Cenozoic	Quaternary
Cenozoic	Tertiary
Mesozoic	Cretaceous
Mesozoic	Jurassic
Mesozoic	Triassic
Paleozoic	Permian
Paleozoic (Carboniferous)	Pennsylvanian
Paleozoic (Carboniferous)	Mississippian
Paleozoic	Devonian
Paleozoic	Silurian
Paleozoic	Ordovician
Paleozoic	Cambrian
Proterozoic	(No period recognized)

As Cenozoic continents drifted to their modern positions, the supercontinent of Gondwanaland completed the breakup it began in the Jurassic. During the Cenozoic, however, new connections between continents—called land bridges—periodically appeared between Alaska and Siberia (across the Bering straits) and between eastern North America and Europe (through Greenland). Ultimately, Africa collided with Europe, and North and South America were joined through Central America.

Figure 25.24

SUMMARY

The great adaptive radiations in the Proterozoic, Cambrian, and Ordovician were evolutionary "experiments" during which many body plans and life cycles appeared. Several major invertebrate groups appeared, some of which diversified to found major animal phyla, and others of which became extinct due to competition, chance, and/or environmental change. The Cambrian also saw the first abundant growth of algae, whose success permitted the evolution of herbivores. Herbivores, in turn, became victims of early predators as the first complex food webs were established.

During the Silurian, arthropods diversified in the sea and some lineages emerged to become the first land animals; since then, arthropods have radiated into the most diverse phylum of multicellular organisms. The first vertebrates left virtually no fossil record. Their descendants—first, the jawless fishes; then, the jawed fishes and sharks—began to fill the seas by the end of the Silurian.

During the Devonian, evolving fishes split into two groups, the modern fishes and the fleshy-finned fishes. Among the ancestors of the latter were the forebears of the earliest terrestrial tetrapods—four-limbed amphibians that colonized the land. This period also saw the emergence of the first seed plants, whose reproductive methods were adapted to drier environments.

The Carboniferous saw a major diversification of insects, which fed on land plants and became food for carnivorous amphibians and the first reptiles. The evolution of the amniotic egg among reptiles served much the same purpose that seeds served for plants: it allowed those organisms to reproduce without standing water.

During the Mesozoic era, reptile diversity reached its peak, dinosaurs ruled, and the first birds and mammals appeared. During this time, too, the first flowering plants appeared; they have been coevolving with insects, birds, and mammals ever since. The era was ended by a mass extinction that wiped out the dinosaurs, setting the stage for the modern world.

The Cenozoic era saw major fluctuations in climate that drove the evolution of mammals, birds, and flowering plants as continents drifted into their current positions.

TERMS AND CONCEPTS

amniotic egg *517*   ectotherms *520*
therapsids *518*   mosaic evolution *522*
endotherms *520*

STUDY FOCUS

After studying this chapter, you should be able to:

- Using a geological time scale, trace the evolution of the first multicellular organisms.

- Describe the evolutionary events of the Cambrian and the Ordovician periods.

- Explain the adaptations that enabled aquatic plants to colonize the land.

- Trace the movement of arthropods onto land.

- Explain the early diversification of vertebrates, noting the key evolutionary innovations that fueled their adaptive radiations.

- Describe the climatic events of the Cenozoic, and explain how plants and animals responded to these global changes.

- Trace the evolution of higher plants, and explain their coevolution with modern insects and other pollinators.

REVIEW

Objective Questions (Answers in Appendix)

1. The Devonian period is called the age of
   (a) fishes.
   (b) amphibians.
   (c) reptiles.
   (d) birds.

2. The supercontinent from which major evolutionary lines originated is called
   (a) Mesozoic.
   (b) Devonian.
   (c) Pangaea.
   (d) Chondrichthyes.

3. Mammals and birds originated during the _____ period from reptile groups.
   (a) Devonian
   (b) Carboniferous
   (c) Permian
   (d) Triassic

4. Dinosaurs became extinct during the
   (a) Triassic period.
   (b) Jurassic period.
   (c) Cretaceous period.
   (d) Permian period.

5. Flowering plants, called angiosperms, began to dominate over gymnosperms during which of the following periods?
   (a) Devonian
   (b) Cretaceous
   (c) Permian
   (d) Triassic

6. The Cenozoic era began approximately _____ million years ago.
   (a) 5
   (b) 20
   (c) 30
   (d) 65

Discussion Questions

7. Why was the evolution of multicellular life tied to the production of oxygen by photosynthesis?

8. What is one explanation for the fact that jawed fishes appeared before the diversification of the agnathans (the jawless fishes) in the fossil record?

9. How did animals and plants of the Mesozoic era influence each other's evolution? Describe two classes of coevolutionary relationships that began during this time.

10. Why might cool global temperatures have seriously jeopardized the survival of large dinosaurs?

11. What problems did the first terrestrial plants and animals face when colonizing land? What plant and animal adaptations evolved as a result of these pressures?

12. Why was the evolution of the seed such an important innovation in terrestrial plants?

13. How is the evolution of modern mammals tied to the history of flowering plants?

READINGS

See integrated list of readings for Part 5 after Chapter 29.

26

Monera and Protista

O ver vast regions of Africa, neither a domestic cow nor a human village is to be seen. Why? Because a species in the parasite genus *Trypanosoma*—a unicellular organism that causes sleeping sickness—lives there, along with a biting fly that carries it from host to host. This deadly organism, although it belongs to one of the "oldest" groups of organisms on the tree of eukaryotic life, is relentlessly "modern" in keeping up with our bodily defenses. As you will learn later, trypanosomes' genes are far from fossilized; they change regularly, staying one jump ahead of both the human immune system and medical science.

Closer to home, in cities and prisons across America, a bacterium humans have known about for centuries has emerged from obscurity with a vengeance. *Mycobacterium tuberculosis*, the cause of a disease we in the developed world had all but forgotten, has, through a series of molecular tricks, evolved resistance to virtually every useful, commercially available antibiotic. Far more contagious than AIDS and much more rapidly lethal, this new tuberculosis threatens to become a serious health risk—and soon. Physicians are unable to foresee an easy solution to this newly recognized problem.

Sleeping sickness and multiple-drug-resistant tuberculosis are both caused by organisms that textbooks (including this one) often refer to as "primitive," "ancient," or "lower" forms of life. Both carry an important message: the fact that morphologically recognizable ancestors of living organisms appeared hundreds of millions or even billions of years ago does not imply that their descendants are either inferior or somehow "behind the times." As the "bush" of life made clear in Chapter 23, these organisms have been evolving in more or less their present form for a lot longer than we have. Not surprisingly, they are experts at adapting to challenges that environmental resistance throws at them. We overlook them both to our loss and at our peril.

Most unicellular life-forms don't cause disease, of course; many of them fulfill important roles in global ecosystems (Chapter 4). We selected disease-causing examples here to emphasize a general point: that they and their relatives play important and dynamic roles in the living world. Previous chapters have discussed aspects of bacterial physiology and genetics, and Chapter 42 will discuss the relationship between microorganisms and disease. Here we will introduce the basic groups of these organisms and explain some important details about their life cycles and ways of making a living.

BACTERIA: KINGDOM MONERA

Organisms similar to today's living bacteria were among Earth's first life-forms, and living members of the kingdom Monera are the most abundant organisms today (Fig. 26.1). A handful of soil contains billions of bacteria, and the bacteria in your mouth alone outnumber all the humans who have ever lived. Although there is ample microfossil evidence of early life, many important steps in early evolution were chemical events that left no fossil traces; therefore, biologists have tried to trace the links between ancient prokaryotes and their descendants by making biochemical comparisons among living forms. The advances in molecular biology over the past 20 years have led to extensive revisions of the classification of unicellular organisms.

Bacteria owe their success to an extraordinary metabolic diversity that makes the biochemical pathways of multicellular organisms look like carbon copies of one another. All plants, for example, depend on similar types of photosynthetic reactions. Bacteria, by contrast, exhibit several photosynthetic pathways that are radically different from one another. Bacteria reproduce rapidly by binary fission. The intestinal bacterium *Escherichia coli,* when grown in the laboratory, matures and is ready to divide in little more than 15 minutes. This short generation time generates genetic diversity in the absence of true sexual reproduction, helping bacteria adapt to changing environments.

The higher-level classification of monerans is both complex and under dispute, as discussed in Chapters 23 and 24. Many experts now believe that the ancient prokaryotes split early in their evolution into at least three separate lines or divisions: the *Archaebacteria,* the *Eubacteria,* and the *Cyanobacteria.* Some microbiologists, however, place the cyanobacteria within the eubacteria, and others favor giving the archaebacteria their own kingdom (Table 26.1). Until the dust settles, we will work within a revised version of traditional classification schemes.

"Ancient" Bacteria: Division Archaebacteria

The **archaebacteria** are called "ancient" bacteria (*archaic* means "old") because they appear quite similar to the first prokaryotes. Many bacteriologists place the archaebacteria in their own kingdom because they are chemically unique in several important ways—they have, for example, different RNA and chemically different cell walls from other bacteria.

Some archaebacteria, called **methanogens,** live in the anaerobic mud of swamps and marshes, where they produce methane gas as a metabolic by-product. Other methanogens live in the intestinal tracts of many organisms (including sheep, cows, and other mammals), where the methane they produce accumulates until it is belched from the mouth or exits from the other end of the gut. Some researchers believe that the methane these bacteria produce in the guts of farm animals contributes significantly to the greenhouse effect.

The members of another group, the **thermoacidophiles,** inhabit extremely hot, acidic habitats such as sulfur springs and deep-sea volcanic vents (*thermal* means "heat"; *philos* means "to love"; hence *thermoacidophile* means "heat and acid lover"; see Theory in Action, Hot Springs). The **halophiles** (*halo* means "salt"; hence *halophile* means "salt lover") are photosynthetic but thrive only in places such as the Dead Sea and Great Salt Lake, where the water is so salty that it is lethal to virtually all other organisms.

"True" Bacteria: Division Eubacteria

The **eubacteria,** or "true" bacteria, are numerous and ecologically diverse. Some are aerobic, others strictly anaerobic. Some are photosynthetic; others depend mostly on fermentation for their energy; and still others are dangerous parasites of multicellular eukaryotes. Major groups of eubacteria are shown in Table 26.1.

Many keys that aid in the identification of eubacteria rely on a staining color test known as the Gram stain, developed in the late 1800s by the Danish physician Christian Gram. Depending on the structure of their cell walls, bacteria appear either violet, in which case they are called *Gram-positive,* or red, in which case they are called *Gram-negative.*

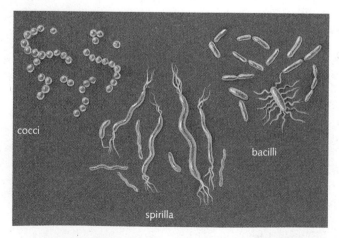

Figure 26.1 Moneran shapes: spherical cocci (singular: *coccus*), rod-shaped bacilli (singular: *bacillus*), spiral-shaped spirilla (singular: *spirillum*), and chain-forming streptobacilli and streptococci.

Table 26.1 *Bacterial Diversity –* MONERA

Division	Shape	Nutrition	Example
Archaebacteria			
Methanogens	Bacilli, cocci, or spirilli	Chemosynthetic or photosynthetic	Cellulose digesters in human and animal guts
Thermoacidophiles	Bacilli	Chemosynthetic	Rift-vent bacteria; bacteria of sulfer springs
Halophiles	Bacilli	Photosynthetic (using bacteriorhodopsin, not chlorophyll)	Pink bacteria of the Dead Sea and salt ponds
Cyanobacteria (some place these within the Eubacteria)	Cocci, bacilli, streptobacilli	Photosynthetic (using chlorophyll); nitrogen-fixing	Common in fresh water, brackish water, and marine habitats
Eubacteria			
Pseudomonads	Bacilli	Heterotrophic; chemosynthetic	Soil bacteria of decomposition; invaders of hot tubs
Spirochetes	Spirilli	Heterotrophic; free-living or parasitic	Parasite that causes syphilis
Enteric bacteria	Bacilli	Heterotrophic symbionts; in mammalian guts may be mutualistic, commensal, or parasitic	Normal gut symbionts; serious parasites causing sometimes-lethal diarrhea and typhoid
Chlamydias	Very tiny spheroids	Heterotrophic; parasitic	Urogenital infections in humans
Actinomycetes	Bacilli, streptobacilli	Heterotrophic; free-living or parasitic	Soil bacteria that produce streptomycin; nitrogen-fixing symbionts with plant roots; parasites that cause tuberculosis and leprosy

The "Blue–Greens": Division Cyanobacteria

The **cyanobacteria,** or "blue–greens," are common world-wide. Cyanobacteria carry on the same sort of photosynthesis as green plants, and some of them fix nitrogen in conspicuous, specialized cells called *heterocysts* (see Fig. 24.3, p. 496). Some cyanobacteria are solitary, floating forms; others form clusters; and still others live in mats, forming stromatolites similar to those their ancestors made more than 3 billion years ago.

Bacteria in Nature and Technology

Most bacteria are not only harmless to humans but also indispensable in the nutrient cycles of both natural and artificial ecosystems. Recall that bacteria oxidize the nitrogen-containing wastes of animals and keep the nitrogen cycle turning in both terrestrial and aquatic habitats. Furthermore, nitrogen-fixing bacteria are the biosphere's only link to the vast reservoir of nitrogen in the atmosphere around us.

Useful bacteria have been "domesticated" for centuries. Bacteria that ferment milk sugar (lactose) into lactic acid produce dairy products such as cheeses and yogurt. The bacteria of the nitrogen cycle are used to break down waste products in applications ranging from compost heaps to sewage treatment plants, aquaculture systems, and even your home fish tank. Bacteria are used in the medical industry as well, producing such modern antibiotics as erythromycin and streptomycin. Much more recently, our skills in manipulating the genetic material of bacteria have turned them into the workhorses of biotechnology. Recombinant bacteria are now being used to produce scores of important biological compounds, including insulin and other hormones that are difficult or impossible to synthesize in the laboratory.

Bacteria and the Human Body

Our intestines are filled with symbiotic eubacteria such as *E. coli*, most of which are either commensal forms (living without causing harm) or mutualistic forms (benefiting both themselves and their hosts). Some members of this "gut flora" help us digest such foods as fats; others synthesize and release vitamin K and a few of the B vitamins. (Recall the importance of gut bacteria to herbivores noted in Chapter 4 and revisited in Chapter 23.)

The human digestive system becomes accustomed to the normal activities of these microbial boarders. Anything that disrupts their activity or population, such as treatment with certain powerful, broad-spectrum antibiotics, invariably causes intestinal distress. Travelers' diarrhea occurs when unfamiliar strains of bacteria are introduced to the gut in drinking water, incompletely cooked food, or unwashed raw vegetables and fruits.

There are relatively few parasitic bacteria, but they cause some of the most deadly diseases known: bubonic plague, cholera, tuberculosis, bacterial pneumonia, diphtheria, and tetanus, to name just a few. Most bacterially caused diseases have been under control for years in industrialized nations, though they linger under unsanitary conditions in less developed countries. Diarrhea brought on by cholera bacteria is still one of the major causes of death among Third World infants. Bacteria are also responsible for several of the most common *venereal* or *sexually transmitted diseases* (STDs, in medical parlance), including gonorrhea (*Neisseria gonorrheae*), syphilis (*Treponema pallidum*), and chlamydial infections (*Chlamydia trachomatis*).

UNICELLULAR EUKARYOTES: KINGDOM PROTISTA

The Protista is by far the most diverse of the eukaryotic kingdoms; its members range from free-floating plankton to deadly parasites in mammals. The classification scheme we have adopted for this book defines the protists as unicellular, eukaryotic organisms. It further divides protists into three broad groups according to their nutritional habits: animal-like heterotrophs, fungus-like saprophytes, and plant-like autotrophs. Among the ancestors of these groups, we might expect also to find the ancestors of the three multicellular kingdoms.

The evolutionary systematics of the kingdom Protista is under a great deal of dispute. Because of differences between protists and higher plants and animals, it is not certain whether living protists and multicellular organisms share a common ancestor or evolved along completely separate lines. Because of differences among the protists themselves, it is not even clear that all living protists evolved from a single ancestor, as classic evolutionary trees imply (Fig. 26.2). According to some recent molecu-

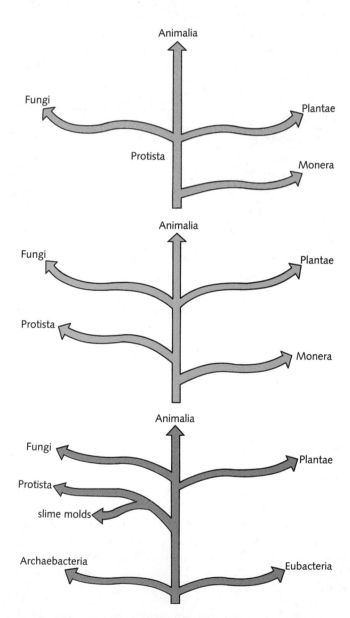

Figure 26.2 The evolutionary position of the unicellular kingdoms is still in dispute. The Protista are viewed by some as ancestors of all eukaryotes and by others as a separate line. Still others believe that slime molds, archaebacteria, and eubacteria should all be treated as belonging to separate kingdoms.

lar studies, protists—like bacteria—are far more biochemically diverse than their appearance suggests. For the sake of simplicity while new classification schemes for these organisms remain unsettled, we have chosen to retain the four large, traditionally recognized phyla. Many protozoologists, however, would consider these "phyla" to be more akin to subkingdoms and would divide the organisms within them into more than 35 distinct phyla.

Protists, though unicellular, are far from simple. Their single-celled condition hasn't kept them from evolving intricate relationships with their environment and with more recently evolved organisms. Whereas multicellular organisms have evolved specialized organs and organ systems, protists have evolved specialized structures ranging from complex organelles to locomotor structures protruding through their outer membranes.

Animal-Like Protists

Phylum Mastigophora The *mastigophora,* also known as **flagellates,** are among the simplest in appearance of the living protozoa (Fig. 26.3). Despite their structural simplicity, however, many modern flagellates are highly specialized. Some live symbiotically in the guts of termites, where they help these insects break down the complex molecules of cellulose (Chapter 4). Other mastigophorans are deadly parasites of humans and other animals. One species in the genus *Trypanosoma,* as mentioned earlier, causes African sleeping sickness, a disease serious enough even today to make much of Africa uninhabitable for both humans and livestock (Chapter 42).

Phylum Sarcodina Members of the phylum Sarcodina, commonly called **amoebae,** are also simple in appearance but ecologically diverse. Many, such as members of the genus *Chaos,* wander freely in wet habitats. Radiolarians and foraminiferans are abundant and ecologically important members of free-floating plankton communities, whose delicate shells are common in marine sediments.

Some amoebae, such as *Entamoeba histolytica,* are parasitic in mammals (including humans), where they can cause amoebic dysentery. After entering the body in contaminated food or water, these parasites live in the lining of the intestine, where the histolytic enzymes for which they are named cause major tissue damage (*histo* means "cell"; *lyse* means "break open"; hence *histolytica* means "breaker or destroyer of cells"). In severe cases, they can spread through the bloodstream to other tissues in the liver, the lungs, the skin, and even the nervous system.

Phylum Sporozoa Organisms belonging to the phylum Sporozoa have extremely complex life cycles, often involving two hosts and usually including some sort of spore-like phase. The best known of these **sporozoa** are members of the genus *Plasmodium,* which causes malaria in humans (Fig. 26.4).

Phylum Ciliata The phylum Ciliata contains another of the best-known protozoans, the genus *Paramecium.* Covered with countless, rapidly beating, hair-like *cilia,* members of this phylum may be free swimming or sedentary (Fig. 26.5). **Ciliates** such as *Paramecium* exhibit the most sophisticated organelles of all protozoans. Some are detritus feeders, others voracious predators.

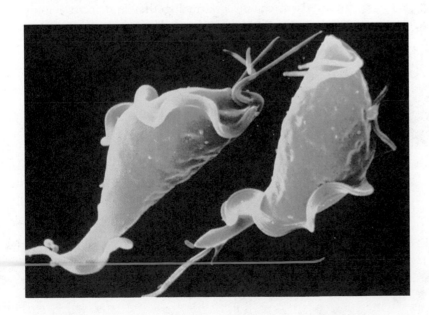

Figure 26.3 These typical mastigophorans bear a conspicuous flagellum, with which they propel themselves through their liquid environment.

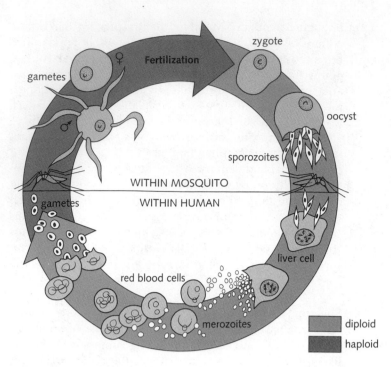

Figure 26.4 The complex life cycle of the malaria parasite demonstrates the coevolution of the parasite with both host and insect carriers. Starting with the left side of the cycle: The female mosquito bites an infected person and takes in protozoan gametes along with blood. Within the mosquito digestive tract, gametes unite and form a zygote. Structures called *oocysts* develop from zygotes. Thousands of small, spindle-shaped *sporozoites* develop within oocysts; some invade the mosquito's salivary glands after the oocysts burst. The mosquito then bites another person and infects that person with sporozoites, which enter liver cells and multiply to form *merozoites*. The liver cells burst; merozoites are released and enter red blood cells, where they multiply. Merozoites break out of red blood cells, release metabolic wastes, and cause fever in the victim. Some of these merozoites become gametes. The female mosquito then bites another infected person, and the cycle continues.

Fungus-Like Protists

The slime molds: Myxomycetes The phyla grouped among the fungus-like protists contain a variety of unicellular organisms with fungus-like characteristics (Fig. 26.6). They may be either saprophytic or parasitic on plants or animals. Among the most interesting of the fungus-like protists are the cellular slime molds. These curious organisms spend part of their lives as independent, free-living, amoeboid cells. When food becomes scarce, however, they secrete a chemical attractant that causes hundreds of cells to swarm together and form a multicellular, slug-like *plasmodium*. The plasmodium crawls around on its own for some time and then produces a tall fruiting body that releases spores when conditions become favorable.

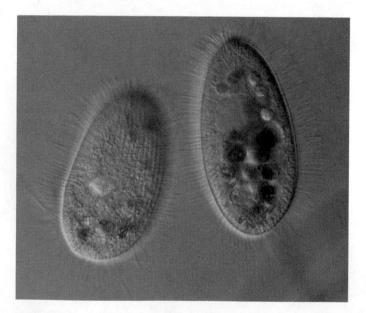

Figure 26.5 These marine ciliates clearly show the hair-like cilia that give the group its name.

Plant-Like Protists

Phylum Euglenophyta The phylum Euglenophyta contains *Euglena,* a genus that defies firm classification (Fig. 26.7). Some individual euglenoids have chloroplasts and photosynthesize as plants do; others lack chloroplasts and either absorb complex molecules to survive (as heterotrophs do) or absorb simpler nutrients from their environment (as fungi do). By manipulating green species in the lab, researchers can produce colorless forms lacking chloroplasts, thus crossing the traditional boundary between plant and animal kingdoms.

Phylum Chrysophyta The **chrysophytes,** which make up the phylum Chrysophyta, include the single-celled algae commonly called *diatoms, golden-brown algae,* and *yellow–green algae.* Many are solitary, single-celled members of both freshwater and saltwater phytoplankton communities. However, some form long floating chains, and others grow attached to larger objects underwater. Diatoms are among the most important aquatic primary producers. They often grow densely; a 5-gallon pail of seawater can hold several million diatoms. Many diatoms have spectacularly beautiful shells made of silica (Fig. 26.8).

Phylum Pyrrophyta The pyrrophytes are commonly called **dinoflagellates,** or "armored" flagellates (*dino* means "powerful"), because of the tough cellulose plates that encase their bodies. These protists swim through the water using a pair of long, whip-like flagella.

Dinoflagellates are second only to diatoms as aquatic primary producers and come in a wide variety of sizes and shapes. Some give off a bright, bluish light when disturbed in the water. The famous phosphorescent bay in Puerto Rico glows because of its dinoflagellate population. Other dinoflagellates live symbiotically within the tissues of many marine organisms, from corals to giant clams.

Several dinoflagellates produce powerful toxins that have the potential to cause illness or even death in higher organisms. *Red tides* are caused by the sudden, explosive growth, or *bloom,* of certain free-swimming dinoflagellates. A virulent form of seafood poisoning called ciguatera is caused by dinoflagellates that grow on the blades of bottom-dwelling algae in the tropics. Poisons manufactured by the dinoflagellates are picked up by herbivores and accumulate in the food chain through biological magnification.

THE VIRUSES

Somewhere between living cells and inanimate crystals exist the **viruses,** whose name means "poison" in Latin. Viruses, so small they are invisible under even the most powerful light microscope, are life stripped down to its barest essentials: a central core of nucleic acids surrounded by a coat of protein. Viruses are not "alive" the way other organisms are. Though other specialized parasites can *reproduce* only inside the cells of their hosts, a naked virus can carry on *no* metabolic processes of any kind on its own. Viruses "come alive" only by taking over the metabolic machinery of their hosts' living cells.

Viruses were discovered in the early 1900s by the Russian biologist Dimitri Iwanowski, who was studying a disease of tobacco plants. He discovered that he could transmit the disease by dripping juice from infected

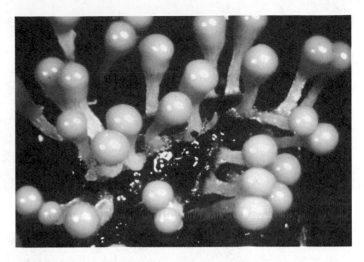

Figure 26.6 These fruiting bodies, or sporangia, are the spore-producing stage of a slime mold.

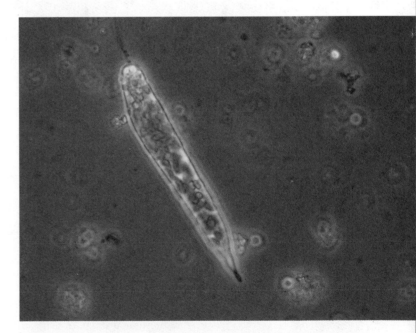

Figure 26.7 *Euglena,* a genus that blurs the boundary between autotroph and heterotroph. (Magnification factor: 160)

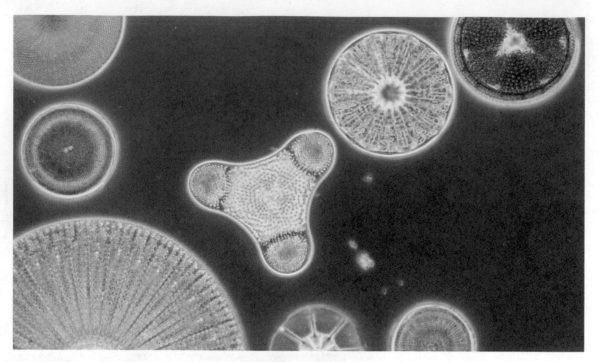

Figure 26.8 Chrysophytes, such as these diatoms, are vitally important primary producers in aquatic food chains.

Table 26.2 *Important Groups of Viruses*

Virus Type	Results of Infection
DNA Viruses	
Adenovirus	Respiratory illness; some animal tumors
Herpesvirus	Herpes (both oral cold sores and genital herpes); infectious mononucleosis; Burkett's lymphoma; chicken pox; shingles
Papavovirus	Warts; cervical cancer; tumors in monkeys
Parvovirus	In joint infection with adenoviruses: gastric and intestinal distress; possibly hepatitis A in humans
Poxvirus	Smallpox; cowpox
RNA Viruses	
Orthomyxovirus	Flu
Paramyxovirus	Mumps and measles
Picornavirus	Polio; common cold; intestinal ailments
Reovirus	Diarrhea
Retrovirus	Certain cancerous tumors and leukemias; AIDS
Togavirus	Yellow fever; equine encephalitis; German measles

leaves onto healthy plants. Even when he passed the juice through a filter so that not even a tiny cell fragment could pass through it, the filtered liquid could still transmit the disease. He concluded that an infectious organism (later called the tobacco mosaic virus), smaller than any yet known to science, must be causing the disease.

Viral Diversity

Viruses exist that can infect any kind of cell—animal or plant, prokaryote or eukaryote. The electron microscope has made viruses visible to us, and these small particles often have exotic and beautiful structures (Fig. 26.9). There are many "families" of viruses, all classified according to the nature of their nucleic acid core; some contain DNA, and others contain RNA (Table 26.2).

In either case, these macromolecules carry just enough information to enable viruses to take over their host cells and reproduce themselves. Viral diseases range from mild infections, such as the common cold, to debilitating and fatal diseases, such as AIDS. You will learn more about viruses and diseases in Chapter 42.

There are several competing theories about the evolutionary origin of viruses, but we know for certain that despite their simplicity, they cannot predate the oldest cellular forms of life. Because they can reproduce only with the aid of living cells, they must have originated well after the first prokaryotes.

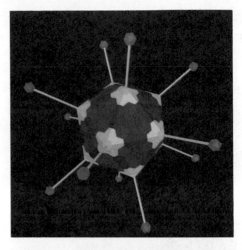

Figure 26.9 The crystalline structure of viruses is readily apparent whether reconstructed using computer graphics techniques (common cold virus, TOP LEFT), or visualized in electron micrographs (influenza virus, BELOW LEFT; polyoma virus, AT CENTER; and pox virus, AT RIGHT).

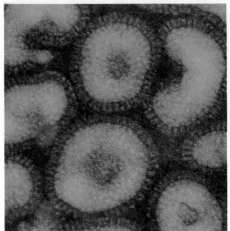

(Magnification factor: 460,000)

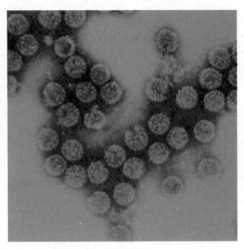

(Magnification factor: 120,000)

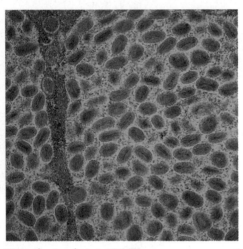

(Magnification factor: 59,000)

SUMMARY

The prokaryotic kingdom Monera is divided into three distinct evolutionary lines: the divisions Archaebacteria, Eubacteria, and Cyanobacteria. Many bacteria are essential participants in global nutrient cycles; others are commensal or mutualistic symbionts with higher organisms.

The unicellular eukaryotic kingdom Protista contains many diverse organisms; some resemble plants, others resemble animals, and still others resemble fungi. Many species are free living and occupy vital positions at or near the bases of aquatic food chains. Others are dangerous parasites.

The viruses, nature's genetic engineers, can survive and reproduce only by using their genetic material to take over the metabolic machinery of a host cell. Viruses parasitize organisms ranging from bacteria to humans and cause diseases ranging from the common cold to AIDS.

STUDY FOCUS

After studying this chapter, you should be able to:

• Describe the kingdoms Monera and Protista and the viruses.

TERMS AND CONCEPTS

archaebacteria *528*	amoebae *531*
methanogens *528*	sporozoa *531*
thermoacidophiles *528*	ciliates *531*
halophiles *528*	chrysophytes *533*
eubacteria *528*	pyrophytes *533*
cyanobacteria *529*	dinoflagellates *533*
flagellates *531*	viruses *533*

REVIEW

Objective Questions (Answers in Appendix)

1. Which group of bacteria are in the division Archaebacteria?
   (a) bacteria that produce methane gas
   (b) bacteria that have cells called heterocysts
   (c) bacteria that fix nitrogen
   (d) bacteria that live in aerobic conditions

2. Which division of bacteria contains the organisms that cause tuberculosis and the organisms that cause syphilis?
   (a) Cyanobacteria   (c) Eubacteria
   (b) Archaebacteria   (d) "blue–greens"

3. African sleeping sickness is a disease caused by which type of protist?
   (a) dinoflagellates   (c) plant-like protists
   (b) fungus-like protists   (d) animal-like protists

4. Viruses must have originated along the evolutionary "bush" of life
   (a) before the first prokaryotes.
   (b) after the first prokaryotes.
   (c) with the first eukaryotes.
   (d) with the first prokaryotes.

Discussion Questions

5. Why do we say that viruses fall outside all five kingdoms of life? What differentiates viruses from forms of life we call "organisms"?

6. Discuss the difference between the biological meanings of the words *primitive* and *ancient* and the negative associations those words usually carry in colloquial American usage. Do you think people in other cultures—where *old man* and *old woman* are terms of respect—have a different perspective on things that have been around for a long time?

READINGS

See integrated list of readings for Part 5 after Chapter 29.

27
Fungi and Plants

*t*here are a handful of multicellular species on which virtually all human life depends. Were these organisms somehow to disappear, global famine would result; if *Homo sapiens* didn't perish, our numbers would certainly decline precipitously.

Surprised? Most of us, removed from the daily business of food production, rarely think about where food comes from. But nearly the entire human race depends on a few species of grasses (mainly rice, corn, and wheat), on a few varieties of potatoes, and (to a lesser extent) on several types of beans. Without them, vast human populations—to say nothing of the domestic animals relished by meat eaters among us—would starve.

Our species is far from unique, of course. With the exception of photosynthetic and chemosynthetic bacteria, plants are the only autotrophs—organisms able to harness a nonliving energy source and assemble inorganic atoms and molecules into the organic building blocks of life. Virtually all multicellular animals (along with many heterotrophic bacteria and fungi) depend on plants directly or indirectly for both nutrients and energy (Chapter 4). Had plants not adapted to land, animals might never have done so. Had plants not coevolved with insects, birds, and mammals, human agriculture would have had a much harder time getting started—if it ever got going at all.

Because living plants and animals share common ancestors back at the dawn of life, our cells speak a common molecular language; genes from animals that are correctly cut and pasted into plant chromosomes can function much as they do in our own cells. (The same is true of fungal cells.) For that reason, the basic information you have learned about molecular genetics and cellular metabolism applies as much to plants and fungi as it does to us. But because the ancestors of plants split off from the ancestors of animals at least 2 billion years ago, many other plant features—from cellular structure to styles of reproduction—differ markedly from those of animals.

This chapter will trace the evolution and describe the unique characteristics of important groups within two kingdoms: the fungi and the plants. This chapter sets the stage for the more detailed discussion of plant structure and function that follows in Chapters 30–33.

KINGDOM FUNGI

Fungi are among the most important, diverse, and curious organisms on earth (Fig. 27.1). Although we barely mentioned them in Chapter 25, the ancestors of modern fungi probably began making their way onto land along with the plants (or perhaps before them) at least 570 million years ago. Some ecologically indispensable fungi are decomposers (saprophytes), and some are vital symbiotic partners of green plants. Other fungi are predators that capture nematodes, and still others are parasites that infest both animals and plants, causing serious diseases and millions of dollars worth of crop damage annually.

Some fungi, such as yeasts, are single-celled, but most are called "multicellular." Most fungi are not, however, multicellular in the same way as plants and animals, because, as a rule, individual fungal "cells" are not separated by complete cell walls, and many of the cell-like compartments have more than one nucleus. The predominant fungal body form is the **mycelium,** a tangled, cottony mass composed of branched filaments called **hyphae** (from the Greek *hyphe,* meaning "web"). The familiar structures we recognize as "mushrooms" are actually the fruiting bodies produced by much larger masses of mycelia, concealed in the soil, rotting wood, or decaying animal body in which the fungus is growing.

Fungi as Decomposers

Saprophytic fungi, which live on dead organic matter, are the most important decomposers of plant tissue in terrestrial environments, where they break down such complex organic molecules as cellulose and allow their constituent elements to be recycled. The hyphae of saprophytic fungi penetrate dead wood, leaves, and animal tissues and secrete digestive enzymes into the substrate around them. The hyphae then absorb the products of this digestion, which takes place outside their bodies.

Some saprophytic fungi live by digesting organic matter that is important to humans: many species spoil bread and fruit or attack and decompose paper, wooden structures, and natural fabrics. Humans use numerous fungi to produce beer and wine, make bread rise, flavor many cheeses, and produce penicillin and other antibiotics. A number of fungi—from common cooking mushrooms found in every supermarket to rare and costly truffles—are both edible and tasty.

Fungi as Symbionts

Symbiotic fungi enter into relationships with primary producers. Those associations enable the primary producers either to live in habitats they could never colonize alone or to function more efficiently in hospitable locations.

Lichens The 25,000 species commonly called **lichens** are actually symbiotic associations in which cells of either photosynthetic cyanobacteria or green algae are woven into a dense mat of fungal hyphae. Those hyphae are occasionally vividly colored, a trait that probably evolved because colored hyphae shield the photosynthetic partner from too much sun. Often, specialized hyphae penetrate the algal or bacterial cell to pick up nutrients more efficiently. The photosynthetic cells respond to their fungal hosts by producing certain substances they do not manufacture when growing alone.

This intimate association equips lichens to live in habitats where virtually nothing else can grow, such as on bare rock, on mountaintops above the timberline, and on the tundra. In such places, lichens are often the first colonists during succession, breaking down and enriching sterile rock and sand, thus paving the way for other, less hardy plants. When the photosynthetic partner is a cyanobacterium, the association can fix nitrogen, some of which is released to the immediate environment in organic form. For this reason, some researchers propose that lichens may have been among the very first forms of multicellular terrestrial life. Because terrestrial life did not exist in any significant quantity before the Silurian, early soils would have been totally devoid of the organic, nitrogen-containing compounds found in today's rich soils. In such inorganic habitats, nitrogen-fixing lichens would have been ideal pioneers.

Mycorrhizae Numerous fungi called **mycorrhizae** (*myco* means "fungus"; *rhizae* means "roots") enter into symbiotic associations with the roots of 80 percent of all higher plants (well over 200,000 species). In most mycorrhizal associations, fungal hyphae penetrate the outer cells of delicate plant roots. Part of the fungal mycelium then forms within the root cells, and the rest spreads out into the surrounding soil. Other mycorrhizae encase roots with mycelia but do not actually penetrate root tissue (Fig. 27.2).

Although these associations might seem parasitic, they are highly beneficial for both partners. Mycorrhizae help host plants take up copper, phosphorus, zinc, water, and other nutrients from the soil. In return, the plants provide the fungi with organic carbon. Mycorrhizal symbioses are widespread today; according to some estimates, these symbiotic fungi account for up to 15 percent of the weight of plant roots round the world. Plants with mycorrhizae seem to be more resistant than others to drought, nutrient shortages, cold, and possibly even the effects of acid rain. Mycorrhizae have been found on the earliest fossil plant roots, reinforcing speculation that they may have helped vascular plants to colonize inorganic Silurian soils.

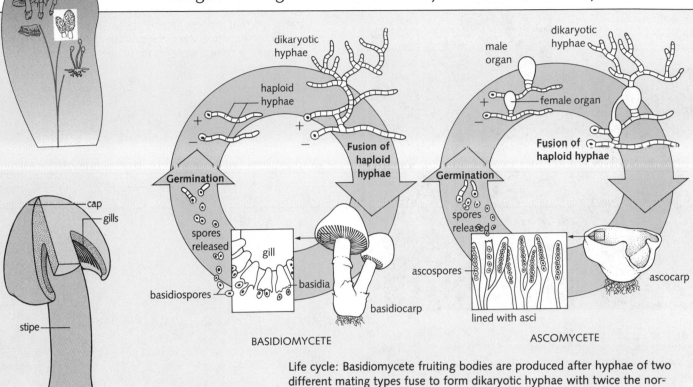

BASIDIOMYCETE

ASCOMYCETE

Anatomy: Familiar "mushrooms" are the fruiting bodies of basidiomycete fungi. A cap covers the reproductive structures located on the gills underneath. The stalk of the mushroom is called the *stipe.* A typical ascomycete fungus consists of a conspicuous cap or *ascocarp.* Both fruiting bodies are small compared with the *mycelium,* the much larger underground portion of the fungus.

Life cycle: Basidiomycete fruiting bodies are produced after hyphae of two different mating types fuse to form dikaryotic hyphae with twice the normal number of nuclei. The "mushroom" is a tightly woven mass of these dikaryotic hyphae. Reproductive structures called *basidia* form on the undersurface of the cap. Ascomycete fruiting bodies are formed after hyphae of two mating types meet. Some of these hyphae fuse to create *dikaryotic hyphae.* All three classes of hyphae (+, −, and dikaryotic) build the ascocarp. Across cells of the ascocarp surface, nuclei fuse to create the spore sacs, or asci, which produces haploid spores.

ABOVE LEFT: The basidiomycete *Amanita* is poisonous to mammals.
ABOVE RIGHT: Vividly colored lichens form a patchwork on rock surfaces, where they are often the first multicellular colonizers. LEFT: The fruiting bodies of an ascomycete.

Figure 27.1

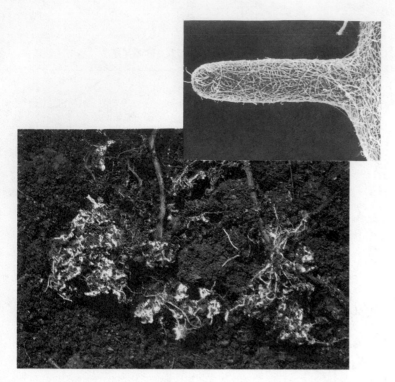

Figure 27.2 Many mycorrhizal fungi form a thick mantle of hyphae around plant root hairs, visible as cotton-like white masses in the large photo and seen on a single root hair in the inset photo.

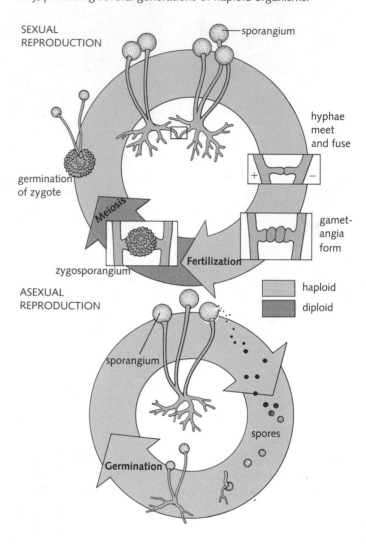

Figure 27.3 BELOW: Zygomycetes reproduce sexually when haploid hyphae of different mating types (simply called "+" and "−") meet. One cell from each hypha fuses to form the diploid zygote. Upon germination, the zygote immediately produces a sporangium that releases haploid spores that grow into the next generation of hyphae. Zygomycetes can also reproduce asexually, producing several generations of haploid organisms.

Main Groups of Fungi

The water molds: Division Oomycota   These organisms, **oomycetes,** are commonly called water molds, because they grow mainly in water or in wet soil. Some taxonomists classify them with fungi, while others consider them more closely related to algae. Many are parasites that cause mildew and blights on beets, grapes, tomatoes, and other crops. One oomycete, _Phytophora infestans,_ which causes late blight in potatoes, changed the course of European and American history. Between 1845 and 1860, a series of unusually cool, damp growing seasons in Ireland allowed _P. infestans_ populations to devastate the potato crop on which Irish peasant farmers depended for their livelihood. The resulting famine and outbreaks of disease forced unprecedented numbers of Irish families to emigrate, many to the United States.

Common molds: Division Zygomycota   This group of fungi, called **zygomycetes,** includes bread molds, saprophytes that live in soil and water, and parasites of animals, plants, and even other fungi. The common bread mold, _Rhizopus,_ is a zygomycete that can reproduce either sexu-

ally or asexually (Fig. 27.3). During sexual reproduction, zygomycetes produce a diploid zygote inside a thick-walled _zygosporangium_ that can survive inhospitable conditions by remaining dormant for months.

Sac fungi and yeasts: Division Ascomycota   This group, the **ascomycetes,** is also called the sac fungi because their spores are carried inside a sac, or _ascus_ (see previous page). The ascomycetes include edible saprophytes like truffles and pathogens such as those that cause Dutch elm disease. The fungal components of most lichens are ascomycete fungi. The single-celled yeasts, although dramatically different in appearance from other ascomycetes, are placed in this group because under certain circumstances they form spores.

In the course of competitive evolutionary interactions with other decomposers, many ascomycetes have evolved chemicals that inhibit the growth of bacteria around them. Cultured varieties of these fungi are widely used in the production of antibiotics.

Mushrooms: Division Basidiomycota What most of us think of as mushrooms are **basidiomycetes.** The familiar shape, however, is only the organism's fruiting body; the bulk of the mushroom is the mass of hyphae slowly digesting its way through organic matter beneath the surface. Many basidiomycetes are edible, although many others (such as *Amanita*) are hallucinogenic, deadly poisonous, or both. It is therefore unwise to hunt mushrooms to eat unless you are extremely knowledgeable. Furthermore, among many potentially hallucinogenic species, the difference between a hallucinogenic dose and a lethal one is very small.

The imperfect fungi: Division Deuteromycota These organisms, the deuteromycetes, are also called *fungi imperfecti,* or "imperfect fungi," because they seem to lack a sexual stage in their life cycle. For convenience, these 25,000-odd species are lumped together into this division, even though they are probably not closely related. From time to time, a sexual stage is discovered in this group and the species is moved to one of the other fungal divisions. Among the deuteromycetes important to humans are several *Penicillium* species that produce antibiotics. Other deuteromycetes are parasites that cause serious diseases in many plants and animals, including humans. One species, *Candida albicans,* causes a throat infection called candidiasis or "thrush." This infection strikes people with impaired immune systems, such as people with AIDS.

KINGDOM PLANTAE

Algae

As you learned in Chapter 25, the ancestors of multicellular algae probably split into three groups about 500 million years ago. Those groups usually recognized today are the chlorophytes, or green algae; the rhodophytes, or red algae; and the phaeophytes, or brown algae (Fig. 27.4). These divisions make intuitive sense, because each contains organisms that bear a different set of chlorophylls and *accessory pigments* that capture light for photosynthesis.

Figure 27.4 TOP: The green alga, *Ulva.* CENTER: Brown algae include giant kelps that form dense forests in cold-water seas. BOTTOM: *Chondrus crispus,* a common species of red algae, inhabits temperate rocky coasts.

But algal taxonomy is problematic for several reasons. First, physiological characteristics such as photosynthetic pigments don't leave fossil traces. Second, modern algae have changed extensively since the Paleozoic when they first appeared. Third, there are numerous apparent evolutionary connections among unicellular green algae, multicellular green algae, and higher plants.

As a result, the family tree of the plant kingdom is full of question marks. Many biologists, for example, classify single-celled green algae in the kingdom Protista. But because unicellular and multicellular algae are similar in so many ways, such a strategy would force us to add multicellular algae to that unicellular kingdom. In this book, therefore, we place all the green, red, and brown algae in the plant kingdom with higher plants, while acknowledging that some experts prefer to treat them differently.

Green algae: Division Chlorophyta

The more than 7000 species of living green algae vary enormously in size and complexity, from single-celled *Chlamydomonas* and its relatives through colonial forms such as *Volvox* to truly multicellular species such as *Ulva*, the common sea lettuce (Fig. 27.5). Although most chlorophytes are aquatic, a few can be called "semiterrestrial"; they can grow on land in very wet soil or on moist tree trunks. Like higher plants, green algae contain two types of chlorophylls—chlorophyll *a* and chlorophyll *b*—and a variety of accessory pigments called *carotenoids*.

The ancestors of modern chlorophytes and higher plants were probably much like *Chlamydomonas*: single-celled, motile, and exhibiting a primitive reproductive pattern. *Chlamydomonas* provides the simplest example of a life cycle that involves both a sexually reproducing stage that produces gametes and an asexual stage that produces spores. In *Chlamydomonas* the stage that produces spores, called the **sporophyte,** has two complete sets of chromosomes and is called **diploid.** The stage that produces gametes, called the **gametophyte,** has half that number of chromosomes and is called **haploid.** In *Chlamydomonas* both of these stages are single-celled. In *Ulva* and many other plants, both stages are multicellular, and this switching back and forth between haploid gametophyte and diploid sporophyte is called **alternation of generations.**

In algae and in many lower plants, male gametes must swim to join the female gametes. Because swimming requires water, plants employing this style of reproduction can grow only in places that are covered in at least a thin film of water during the plants' reproductive periods. Adaptive changes in water-dependent life cycles were crucial in enabling plants to colonize terrestrial habitats.

Brown algae: Division Phaeophyta

The brown algae, which obtain their color from accessory pigments called *xanthophylls* and *fucoxanthins,* include most plants commonly known as seaweeds: the "rockweeds" of the intertidal zone, the giant kelps of deeper water, and the floating "sargassum weeds." Each of these major groups creates a complex, three-dimensional habitat for many other organisms.

Although clearly not closely related to higher plants, the phaeophytes are among the most structurally complex of all marine algae. Many have specialized holdfasts that secure them to a rocky bottom and bladders, or floats, that hold their long stems upright in the water.

Red algae: Division Rhodophyta

The 4000-odd species of red algae are primarily marine and very diverse. Many have extremely complex life cycles. Red algae are most abundant in tropical waters, but some species (such as *Chondrus crispus,* commonly called "Irish moss") are found below the low-tide line along rocky temperate shores. All red algae possess chlorophyll *a* and varied accessory pigments called *phycocyanins* (*phyco* means "algae"; *cyan* means "blue–green") and *phycoerythrins* (*erythro* means "red"). Many red algae also contain chlorophyll *d*, which is not found in any other eukaryotes.

The accessory pigments that give these plants their red, violet, yellowish, or even brownish colors are very efficient at absorbing the blue and green wavelengths of light that penetrate to great depths in seawater. These wavelengths are not absorbed well by chlorophyll. By passing this energy on to the algae's photosynthetic machinery, the accessory pigments enable rhodophytes to live in deeper water than green algae.

Terrestrial Plants

As you learned in Chapter 25, the first land plants probably resembled modern green algae and apparently began making their way onto land about 500 million years ago. In adapting to truly terrestrial life, however, organisms that evolved in water faced four major adaptive challenges: *desiccation,* the loss of body water to surrounding air; the need to *transport body fluids* from place to place; the need to provide *physical support* for body parts; and the need to evolve *reproductive and dispersal strategies* that can function without standing water. It is useful to think of both the evolution and the taxonomy of living plants in relation to the step-by-step process of solving these adaptive challenges, because plants are classified largely according to structures that relate to these issues.

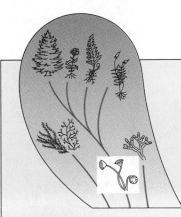

Green Algae: Division Chlorophyta

Ulva life cycle: Both haploid and diploid stages of the life cycle are multicellular. The diploid plant that produces haploid spores is called the *sporophyte*. These spores grow into multicellular, haploid plants called *gametophytes,* because they produce gametes. *Ulva*'s gametes are different from one another; larger ones are arbitrarily called "female," and smaller ones, "male." The condition of having different-looking gametes is called *heterogamy*. Fertilization forms a diploid zygote that grows into a multicellular sporophyte.

Chlamydomonas life cycle: This unicellular green alga is capable of sexual or asexual reproduction. Many individuals are haploid and reproduce asexually by cell division. Sometimes a mature haploid cell will produce identical haploid gametes, which fuse to form a diploid zygote. This zygote does not grow, but either develops a protective covering and rests, or divides to release haploid cells. In *Chlamydomonas,* the gametes look identical, a condition called *isogamy* (*iso* means "equal"; *gamy* refers to gametes; *isogamy* means "similar gametes").

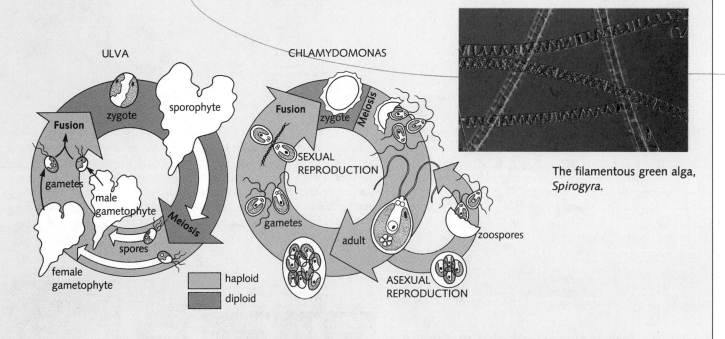

ULVA

CHLAMYDOMONAS

The filamentous green alga, *Spirogyra.*

sporophyte

zygote

Fusion

gametes

male gametophyte

Meiosis

spores

female gametophyte

Fusion zygote Meiosis

SEXUAL REPRODUCTION

gametes

adult

zoospores

ASEXUAL REPRODUCTION

haploid

diploid

LEFT: *Codium,* a genus of green algae common in temperate seas, often grows in rocky tide pools. RIGHT: *Acetabularia* is a green alga often found in shallow, warmer waters, such as those of the Mediterranean Sea.

Figure 27.5

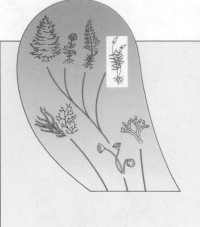

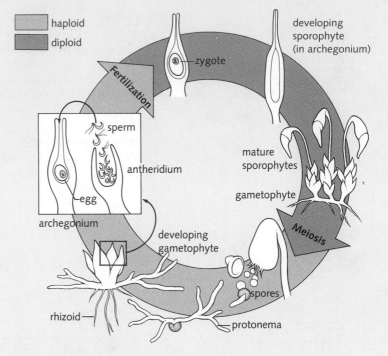

haploid

diploid

Fertilization

zygote

developing sporophyte (in archegonium)

sperm

antheridium

mature sporophytes

gametophyte

egg

archegonium

Meiosis

developing gametophyte

spores

rhizoid

protonema

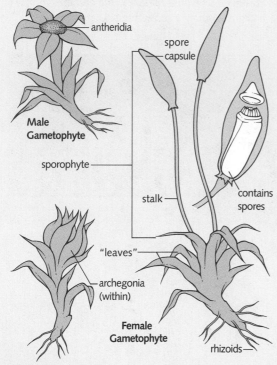

antheridia

Male Gametophyte

spore capsule

sporophyte

stalk

contains spores

"leaves"

archegonia (within)

Female Gametophyte

rhizoids

Life cycle: Sperm must swim to the archegonium to fertilize the egg. The diploid sporophyte grows from the archegonium, producing a spore capsule on a slender stalk. Within this capsule, dust-like, haploid spores are produced and released during dry, windy weather. Spores landing in favorable locations grow into algae-like *protonemata*, from which gametophytes sprout.

Anatomy: The haploid moss gametophyte has short stems, primitive vascular tissue, root-like rhizoids, and leaf-like photosynthetic surfaces one cell layer thick. Reproductive organs grow at the tips of these shoots. The male *antheridium* produces sperm, and each female *archegonium* produces a single egg.

CENTER: Many of the green moss gametophytes support the stalk and spore capsule of the sporophyte generation. Liverworts (LEFT) and hornworts (TOP) are other bryophytes common in moist environments.

Figure 27.6

544

In order to retain water in exposed leaf surfaces, most land plants evolved waxy, waterproof **cuticles** that cover leaf surfaces. This evaporation barrier is interrupted by openings called **stomata,** which open or close as needed to allow exchange of carbon dioxide and oxygen for respiration and photosynthesis. Another adaptation of major significance is the presence of **vascular tissues** specialized to conduct fluids. Water and inorganic nutrients are transported through the **xylem,** and the organic products of photosynthesis are carried through the **phloem.** These tissues evolved slowly as various groups of plants adapted to progressively drier environments. Roughly in parallel, land plants evolved rigid stems and other support structures to hold their leaves up to the sun (see Table 25.1, p. 509).

Reproduction and dispersal on land are also major challenges. Early multicellular plants adapted to aquatic or semi-aquatic habitats; like many modern algae, sperm swim to fertilize eggs, and gametes and young plants can be carried far from their parent plants by waves and currents. In many terrestrial habitats, those events are unlikely. Thus, as certain groups of plants entered drier and drier habitats over millions of years, they evolved both reproductive and dispersal strategies more fully adapted to the absence of water.

By around 400 million years ago, the two main branches of the plant kingdom had already diverged from one another: the single surviving nonvascular division **Bryophyta** and the divisions of vascular plants that we will discuss shortly. (NOTE: Plants known only as fossils are discussed in Chapter 25.)

Mosses, liverworts, hornworts: Division Bryophyta

This group, the **bryophytes,** is at once an ancient and successful division of more than 16,500 species and a collection of "also-rans" that never achieved the diversity of other plant groups. Viewed by some botanists as a primitive group close to the ancestors of "higher plants," they are seen by others as highly specialized forms that have become simplified during their long history.

Bryophytes, like all other plants, have a life cycle involving the alternation of generations, as shown in Fig. 27.6. Dust-like haploid spores, easily spread by the wind, germinate on moist surfaces and grow into thread-like *protonemata* that look remarkably like green algae. These protonemata ultimately give rise to the leafy green shoots we know as moss plants, the separate haploid male and female gametophytes that reproduce sexually to complete the life cycle.

Though well suited for life in wet terrestrial habitats, bryophytes have no true roots to take up water, no vascular tissue to transport fluids, and no cuticles to retain water. For those reasons, mosses cannot grow very tall; most are no more than 2 or 3 cm in height. Although bryophytes require standing water to reproduce, many species can enter a dormant state during droughts or subfreezing temperatures. Thus, though they probably evolved in the wet tropics, mosses are today the most abundant plants in both the Arctic and the Antarctic.

Club mosses and horsetails: Divisions Lycophyta and Sphenophyta

The club mosses (**lycophytes**) and horsetails (**sphenophytes**) were among the first vascular plants to combine conducting tissue, true roots, and leaf-like structures for efficient photosynthesis. Their reproductive cycles, however, still require standing water. After a slow start, both mosses and horsetails experienced wildly successful adaptive radiations that lasted for over 100 million years, from the beginning of the Carboniferous period to the end of the Permian. Today, however, they are represented by only 5 genera and about 1000 species worldwide (Fig. 27.7).

Figure 27.7 Living sphenophytes, though neither as large nor as widespread as their ancestors, retain their division's characteristic shape.

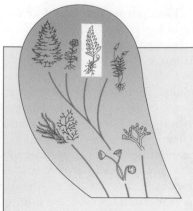

Ferns: Division Pterophyta

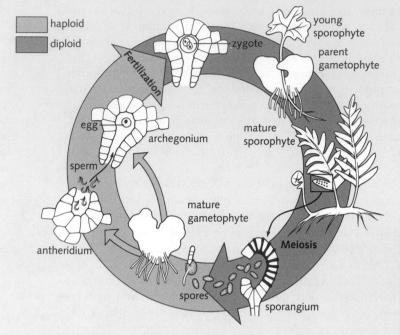

haploid
diploid

Fertilization

young sporophyte
zygote
parent gametophyte

egg
archegonium

sperm

mature sporophyte

antheridium

mature gametophyte

spores
sporangium

Meiosis

Life cycle: Haploid spores are released from sporangia to be carried on the wind. The spores germinate into small, haploid gametophytes. Sperm produced in antheridia swim to eggs in archegonia; the zygote produces a diploid sporophyte that grows out of and soon dwarfs its parent.

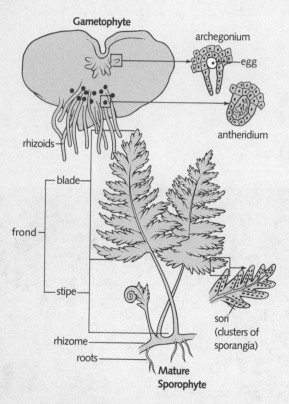

Gametophyte

archegonium
egg

antheridium

rhizoids

blade

frond

stipe

sori (clusters of sporangia)

rhizome

roots

Mature Sporophyte

Anatomy: Leafy fronds, composed of *blade* and *stipe,* belong to diploid fern sporophytes, which are larger and more independent than those of mosses. Fronds unfurl from creeping *rhizomes* that grow underground or along the surface. Clusters of sporangia called *sori* form on the undersides of mature fronds. Well-developed roots extend into the soil.

Living ferns range from small species common in moist-temperate environments to giant tropical tree ferns.

Figure 27.8

Ferns: Division Pterophyta Like bryophytes, ferns (**pter-ophytes**) can reproduce only with the aid of standing water (see Fig. 27.8). But fern sporophytes—with their better-developed vascular tissue, thicker and more re-silient leaves, and true roots—can survive under some-what drier conditions and may live for many years. Like club mosses and horsetails, ferns were enormously suc-cessful in the Carboniferous and Permian when their tree-sized representatives formed huge forests. Because they were better adapted to survive changing conditions, however, ferns are still both diverse (some 12,000 species) and widespread today.

Evolutionary Trends in Plant Reproduction

As you have seen thus far, all modern "higher" plants and many algae share reproductive cycles based on alterna-tion of generations. But within that similarity, a trend emerges as we survey plant evolution. Among single-celled aquatic forms, haploid gametophyte and diploid sporophyte are of roughly equal size and importance. As plants become increasingly adapted to terrestrial life, however, the sporophyte becomes dominant and the ga-metophyte becomes progressively smaller (Fig. 27.9). This early trend came to fruition with the evolution of seed plants.

The First Higher Plants: Gymnosperms

As we discussed in Chapter 25, the most important single innovation in the evolution of fully terrestrial plants was the evolution of the **seed,** a drought-resistant reproduc-tive package that houses a dormant plant embryo and an ample supply of food within a protective seed coat. Mod-ern seed plants display several other changes in their life cycles as well. Their haploid female gametophytes have been reduced to little more than a few cells, which live obligatorily on the diploid sporophyte. Male gameto-phytes, for their part, dwindled to dust-like **pollen grains** that are carried by wind or by animals to the female re-productive organs where fertilization occurs.

The earliest seed-bearing plants diversified into nu-merous lines around 225 million years ago. Most of those lines are now extinct, but they did leave some liv-ing descendants: the **gymnosperms,** or "naked seed" plants, so-called because their seeds aren't enclosed in well-developed fruits, as are seeds of flowering plants that evolved later on. In all these plants, haploid game-tophytes are reduced to a few cells that are totally de-pendent on the sporophyte for sustenance.

Cycads and ginkgoes: Divisions Cycadophyta and Ginkgophyta Two "naked seed" groups diversified suc-cessfully during the Permian period but left few living representatives. **Cycads** are slow-growing, tropical and subtropical palm-like plants with thick trunks that are rep-resented today by only about 100 species. **Ginkgoes** are represented by a single "living fossil," *Ginkgo biloba* (see Fig. 25.18).

Conifers: Division Coniferophyta Another gymnosperm group, the **conifers,** had its first great adaptive radiation around 250 million years ago, remained the dominant land plants for nearly 200 million years, and today is rep-resented by about 550 species. Vast coniferous forests cover much of the northern temperate high-altitude re-gions of the world.

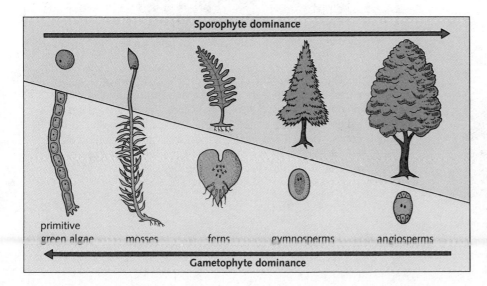

Figure 27.9 This schematic representation of trends in the life cycles of algae, lower plants, and higher plants shows the steadily increasing size of the diploid sporophyte (top part of figure) and the decreasing size of the haploid gameto-phyte (lower part of figure).

In conifers, male and female gametophytes are produced on separate, specially modified reproductive structures called **cones** (Fig. 27.10). The processes of fertilization and seed maturation in conifers proceed very slowly. Seeds are usually not released from the cone for nearly two years after pollination occurs.

Flowering Plants (Angiosperms): Division Anthophyta

Angiosperms, the true flowering plants, began to dominate the land during an explosive adaptive radiation that began somewhere around 100 million years ago. In this group, the trends we've been following in plant reproduction are carried even further. Not only are the male and female gametophytes reduced to a few cells living on the sporophytes, but also they complete their functions much more rapidly than gametophytes of conifers (Fig. 27.11).

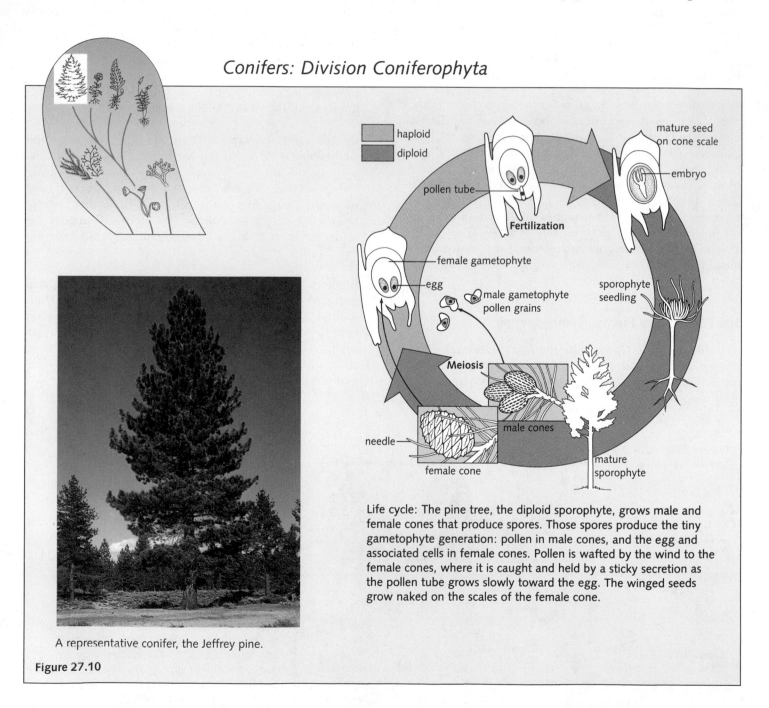

Conifers: Division Coniferophyta

Life cycle: The pine tree, the diploid sporophyte, grows male and female cones that produce spores. Those spores produce the tiny gametophyte generation: pollen in male cones, and the egg and associated cells in female cones. Pollen is wafted by the wind to the female cones, where it is caught and held by a sticky secretion as the pollen tube grows slowly toward the egg. The winged seeds grow naked on the scales of the female cone.

A representative conifer, the Jeffrey pine.

Figure 27.10

Flowering Plants: Division Anthophyta

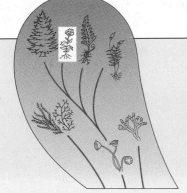

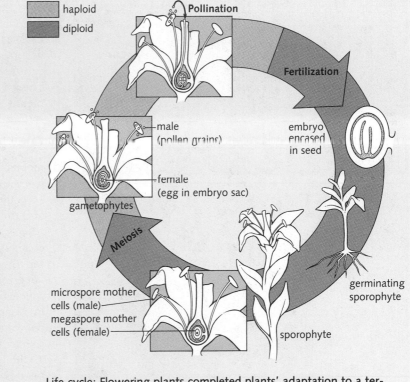

haploid
diploid

Pollination

Fertilization

male (pollen grains)

female (egg in embryo sac)

gametophytes

Meiosis

microspore mother cells (male)

megaspore mother cells (female)

embryo encased in seed

germinating sporophyte

sporophyte

Life cycle: Flowering plants completed plants' adaptation to a terrestrial existence in a world populated by insects, birds, and mammals.

ABOVE: This fossilized, radially symmetric flower belonged to an early angiosperm. RIGHT: Living angiosperms provide much of the color in our world today. Their flowers have evolved in such a way that they attract animal pollinators, and their fruits have developed in response to herbivores.

Figure 27.11

(a)

(b)

Many angiosperms can produce seeds in a few weeks. This ability to reproduce rapidly may have given early angiosperms a great advantage under intense grazing pressure, first from herbivorous dinosaurs and later from plant-eating insects and mammals.

Classes Monocotyledonae and Dicotyledonae Two major groups of angiosperms emerged during the Cretaceous. The *monocotyledons* include the lilies and the grasses and today comprise around 65,000 species (*mono* means "one"; *cotyledon* means "seed leaf"; hence *monocotyledon* means "bearer of a single seed leaf") (Fig. 27.12a).

The *dicotyledons* (*di* means "two"; hence *dicotyledon* means "bearer of two seed leaves") include most common trees and shrubs (Fig. 27.12b). Today they number about 160,000 species. We will examine angiosperm structures and their functions in great detail in Chapters 30–33.

Recall from Chapter 23 that both monocots and dicots experienced their adaptive radiations during and after the great radiation of insects that produced the ancestors of modern bees. In addition, many seed plant species evolved more or less concurrently with the emerging mammals. As a consequence, angiosperms developed coevolutionary relationships of two main types. Various species of flowering plants now depend for pollination on a wide range of insects—including mosquitoes, moths, beetles, and bees—on several types of birds, and even on a few bats. Those plants and others also depend on animals ranging from ants to fishes, birds, and mammals to disperse their seeds in various ways. Finally, of course, there are the complex interrelationships between plants and humans that we commonly call agriculture. According to some people who like to stretch rigid definitions, agriculture might be the most complicated example of recent plant–animal coevolution yet.

Figure 27.12   **(a)** These wild barley grasses are representative monocots. **(b)** The dogwood is a representative dicot.

SUMMARY

Fungi and plants are among the most important organisms on Earth. Some fungi are ecologically indispensable decomposers, and plants are autotrophs which are vital either directly or indirectly to virtually all multicellular animals.

Within the kingdom, fungi are very diverse. Saprophytic fungi decompose plant tissue and include species that spoil bread, others that produce beer and wine, and yet others that produce antibiotics. Symbiotic fungi maintain associations with primary producers and include lichens and mycorrhizae. The main groups of fungi are the water molds (oomycetes), the common molds (zygomycetes), sac fungi and yeasts (ascomycetes), mushrooms (basidiomycetes), and the imperfect fungi (deuteromycetes).

In this text, all the algae are grouped within the kingdom Plantae. The green algae (Chlorophyta) include single-celled forms like *Chlamydomonas,* colonial forms like *Volvox,* and multicellular species like *Ulva.* The life cycles of algae and other plants are characterized by alternation of generations, the switching back and forth between haploid and diploid stages. The brown algae and the red algae obtain their color from accessory pigments that absorb the blue and green wavelengths of light.

Terrestrial plants face many challenges, including desiccation, the need to transport body fluids, the need for physical support, and the need for reproductive and dispersal strategies. The bryophytes (mosses, liverworts, and hornworts) have no means for taking up, transporting, or retaining water. The lycophytes and the sphenophytes have conducting tissue and roots but still require standing water for reproduction. Pterophytes (ferns) also need standing water for reproduction but have a better-developed vascular tissue for survival in drier environments.

The development of seed was a significant innovation in the evolution of plants to a terrestrial environment. The gymnosperms (cycads, ginkgoes, and conifers) have "naked" seeds that lack a protective fruit. The flowering plants, or angiosperms, have seeds enclosed in a well-developed fruit and can reproduce rapidly. This division is represented by the monocotyledons (lilies and grasses) and the dicotyledons (trees and shrubs).

STUDY FOCUS

After studying this chapter, you should be able to:

- Name and describe the divisions of multicellular algae.
- Outline the life cycle of typical green algae.
- Describe the lower plants and their life cycles.
- Explain in detail the adaptations that enabled aquatic plants to colonize the land successfully.
- Trace the evolution of higher plants.

TERMS AND CONCEPTS

fungi *538*	haploid *542*
mycelium *538*	alternation of generations *542*
hyphae *538*	bryophytes *545*
lichens *538*	lycophytes *545*
mycorrhizae *538*	sphenophytes *545*
zygomycetes *540*	pterophytes *547*
ascomycetes *540*	seed *547*
basidiomycetes *541*	pollen grains *547*
sporophyte *542*	gymnosperms *547*
diploid *542*	conifers *547*
gametophyte *542*	angiosperms *548*

REVIEW

Objective Questions (Answers in Appendix)

1. In the species of *Chlamydomonas* that exhibits isogamy, the gametes
   (a) are produced by the sporophyte.
   (b) are diploid.
   (c) are the same size.
   (d) have the same number of chromosomes as the spores.

2. Most fungi produce cellular filaments called
   (a) hyphae.   (c) saprophytes.
   (b) mycelia.   (d) ascomycetes.

3. Mosses and liverworts are not considered to be completely adapted to land because they
   (a) require water to reproduce.
   (b) do not grow in soil.
   (c) have alternation of generations as part of their life cycle.
   (d) require warm, moist temperatures to grow and reproduce.

4. As plants became more adapted to land,
   (a) the haploid gametophyte predominated.
   (b) the diploid sporophyte predominated.
   (c) the haploid gametophyte and the diploid sporophyte predominated.
   (d) alternation of generations ended completely.

Discussion Questions

5. What characteristics distinguish the three divisions of modern algae from each other?

6. Why are fungi among the most important, diverse, and curious organisms on the earth? How is multicellularity in fungi different from that in plants and animals?

7. What trend emerged in the relative size of sporophyte and gametophyte as plants became increasingly well adapted to terrestrial life?

8. Why was the evolution of the seed such an important innovation in terrestrial plants?

READINGS

See integrated list of readings for Part 5 after Chapter 29.

28

The Invertebrates

On land we usually notice them when they bite us, buzz around our heads, or make noises late at night. In the sea, whether in a temperate zone, rocky tidal pool, or on a coral reef, they dazzle us with fairy-tale forms and shapes. Many are innocuous, many fill vital roles in Earth's varied ecosystems, and some cause serious problems for humans and other animals. Their life cycles vary almost beyond belief, from mundane, to bizarre, to downright gruesome.

These are the invertebrates, a motley collection of multicellular animals that share no common features to hold them into any single, formal taxonomic group. We lump them together because of what they lack: the internal support structure we call a vertebral column, or backbone. Arthropods alone outnumber all other multicellular animals. Insects are both essential allies and fearsome foes. Starfish and their kin look like leftovers from geologic ages past. Parasitic worms cause some ailments almost too horrible to mention.

We introduced the major invertebrate groups in order of their evolutionary appearance in Chapter 25. Here, we will study them in more detail, learn their life cycles, and see where they fit into the living world.

LIVING INVERTEBRATES

Ecologically and evolutionarily, invertebrates are enormously successful; they outclass members of our own phylum (Chordata) in length of presence on the earth and in both numbers of living species and numbers of living individuals.

Sponges: Phylum Porifera

Like their ancestors, modern-day sponges have simple body plans. Although multicellular, sponges are just barely so; they have neither organs nor well-differentiated tissues. They do have several different cell types, including wandering *amoebocytes* and flagellated *collar cells* that cooperate to assemble the sponge structure (Fig. 28.1). Sponges are so different from other metazoans in their organization that they probably evolved multicellularity independently. (Some taxonomists place the sponges in a separate subkingdom—the Parazoa—to emphasize their uniqueness.) Like many other primitive animals, sponges can reproduce either sexually, using eggs and sperm, or asexually, by budding.

Sponges: Phylum Porifera

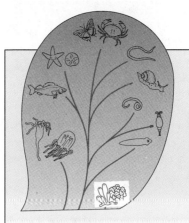

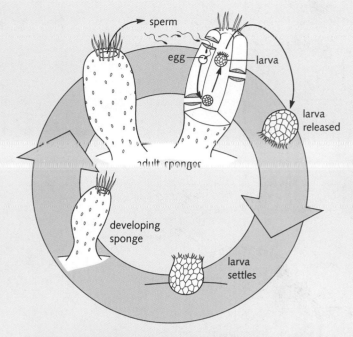

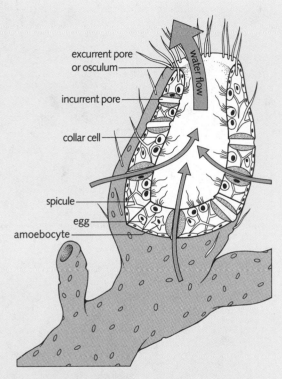

Life cycle: Sponges reproduce either asexually (by budding) or sexually. In sexual reproduction, sperm released by other sponges are drawn through the incurrent pores, and fertilization occurs internally. Zygotes grow into multicellular larvae that swim to a point of attachment and metamorphose into miniature adults.

Anatomy: Sponges propel water currents through their bodies using flagellated *collar cells.* Water enters through incurrent pores and exits through the *excurrent pore,* or *osculum.* This current allows sponges to breathe and to filter feed. Needle-like *spicules* ("little spikes") of calcium carbonate offer skeletal support to many sponges. Tough elastic fibers of *spongin* serve that purpose in other species.

A vase sponge.

Figure 28.1

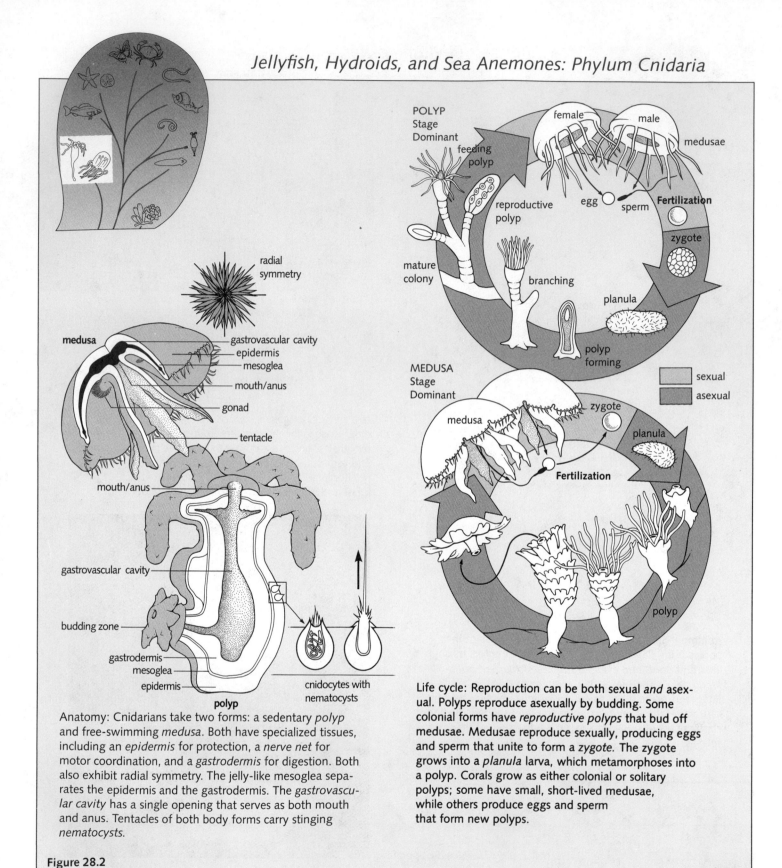

radial symmetry

medusa

gastrovascular cavity
epidermis
mesoglea
mouth/anus
gonad
tentacle
mouth/anus

gastrovascular cavity
budding zone

gastrodermis
mesoglea
epidermis

cnidocytes with nematocysts

polyp

POLYP Stage Dominant
female
male
medusae
feeding polyp
reproductive polyp
egg
sperm
Fertilization
zygote
mature colony
branching
planula
polyp forming

MEDUSA Stage Dominant
medusa
zygote
planula
Fertilization
polyp

sexual
asexual

Anatomy: Cnidarians take two forms: a sedentary *polyp* and free-swimming *medusa*. Both have specialized tissues, including an *epidermis* for protection, a *nerve net* for motor coordination, and a *gastrodermis* for digestion. Both also exhibit radial symmetry. The jelly-like mesoglea separates the epidermis and the gastrodermis. The *gastrovascular cavity* has a single opening that serves as both mouth and anus. Tentacles of both body forms carry stinging *nematocysts*.

Life cycle: Reproduction can be both sexual *and* asexual. Polyps reproduce asexually by budding. Some colonial forms have *reproductive polyps* that bud off medusae. Medusae reproduce sexually, producing eggs and sperm that unite to form a *zygote*. The zygote grows into a *planula* larva, which metamorphoses into a polyp. Corals grow as either colonial or solitary polyps; some have small, short-lived medusae, while others produce eggs and sperm that form new polyps.

Figure 28.2

Jellyfish, Hydroids, and Sea Anemones: Phylum Cnidaria

The phylum Cnidaria was the first to contain specialized tissues, including a simple nervous system called a *nerve net*. Cnidarians also have specialized sensory cells, muscles, and cells for catching and digesting food (Fig. 28.2). Their body plans exhibit **radial symmetry:** body parts are arranged like spokes of a bicycle wheel around a central mouth. Cnidarians don't have a digestive tract with separate mouth and anus. Instead, they have a *gastrovascular cavity* with a single opening that serves for both taking in food and expelling wastes.

Cnidarians may be either solitary, like many sea anemones and jellyfish, or colonial, like hard and soft corals (Fig. 28.3). Many cnidarians are predators that use stinging structures called **nematocysts** to paralyze or ensnare prey ranging from small shrimp to sizable fishes (*nema* means "thread"; *kystis* means "bag"; hence nematocyst means "thread in a bag"; see Fig. 28.2). Other cnidarians fall somewhere between predators and filter feeders, spreading web-like arrays of individual polyps, each of which contributes its tentacles to a network that snares passing plankton. Quite a number, including hard corals and some sea anemones, host symbiotic single-celled algae within their tissues. Many cnidarians exhibit a complex life cycle that alternates between asexual budding and sexual reproduction.

The cnidarian body plan is simple yet successful. Though most numerous and diverse in the tropics, cnidarians are common in temperate seas and are represented in freshwater habitats by *Hydra*. Hard corals, which secrete calcium carbonate skeletons, build coral reefs, the largest biologically formed structures in the world and one of the most important marine ecosystems.

Worm-Like Phyla

Modern worm-like animals actually belong to several very different phyla, including two simple ones that probably evolved by the middle of the Cambrian period: the Platyhelminthes, or flatworms, and the Nematoda, or roundworms. The earliest members of these groups were free living, making their way independently as herbivores and carnivores. Modern representatives range from various free-living forms to highly specialized parasites whose lives are irrevocably linked to those of other organisms.

Flatworms and roundworms exhibit **bilateral symmetry,** a body plan in which similar body parts occur on either side of a central dividing line. All bilaterally symmetrical organisms, such as ourselves, have left and right halves that are (at least externally) mirror images of each other. Some worms also evolved a linear "tube within a tube" digestive tract with a mouth at one end and an anus at the other.

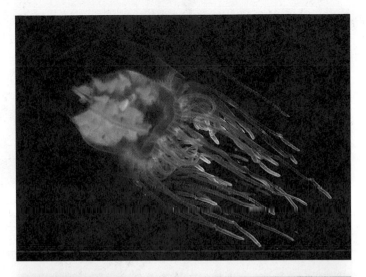

Figure 28.3   TOP: Jellyfish spend much of their lives as medusae; their polyp stage is short-lived and serves only to produce new medusae asexually. BOTTOM: This coral spends most of its life in the form of a colonial polyp.

Bilateral symmetry offers several adaptive advantages, and most higher invertebrates and all vertebrates exhibit this sort of body plan. Bilaterally symmetrical organisms have, in addition to a right side and a left side, a front (anterior) end and a rear (posterior) end. As organisms evolved more mobile lifestyles, they evolved specialized sense organs that detect both food and enemies. Over time, those sense organs accumulated at the front of the animal—a logical development considering that it is generally more useful to gather information about where you're going than to review where you've been. This progressive steady process of **cephalization** (*cephalos* means "head"; hence *cephalization* means "making of a head") ultimately led to development of the brain. Bilateral symmetry also makes directed movement more efficient.

Body cavities The evolving groups of bilaterally symmetrical animals diverged in the design of their body cavity or *coelom*. The most primitive, the *acoelomates* (*a* means "without"), have no body cavity. Others, the *coelomates*, have a fluid-filled cavity lined with tissue called *mesoderm*. This mesoderm forms structures called mesenteries, from which internal organs are suspended. Still other organisms, *pseudocoelomates*, have a simple body cavity that is lined with tissue other than mesoderm (Table 28.1).

Table 28.1 *Protostomes and Deuterostomes**

Protostomes	Deuterostomes
Cleavage: Division of Early Cells	
In protostomes, cells divide at an angle to one another, producing a *spiral* arrangement of cells. The ultimate fate of each cell is determined early in development.	In deuterostomes, the early cells divide vertically, producing a *radial* arrangement of cells. The fate of embryonic cells is not determined until much later in development. Each of these early cells can become part of almost any tissue.
Formation of the Coelom	
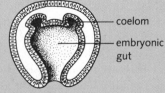	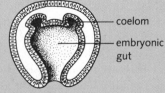
Protostomes are said to be *schizocoelous* (*schizo* means "split"). Here, solid masses of mesoderm tissue *split* to form the coelom.	Deuterostomes are said to be *enterocoelous* (*enteric* means "intestine"; hence *enterocoelous* means "intestine coelom"). The coelom is formed by buds that branch out and separate from the embryonic gut.
Development of the Gut	
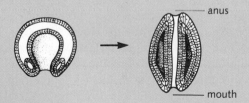	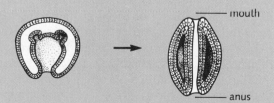
Protostome means "mouth first" (*proto* means "first"; *stoma* means "mouth"). The embryonic gut opening becomes the animal's *mouth*, and a new opening forms the anus.	*Deuterostome* means "mouth second" (*deutero* means "second"). The embryonic gut opening becomes the *anus*, and a new opening forms the mouth.

*Protostomes and deuterostomes are separated by differences in embryonic development. Because these events happen early in development and are constant across all members of a phylum, they are considered fundamental characteristics dividing groups of phyla.

Flatworms: Phylum Platyhelminthes

Flatworms (Fig. 28.4) are the simplest of the worm-like phyla and the first organisms on the evolutionary tree to exhibit not only specialized tissues but also tissues arranged into *organs*. Most have neither respiratory nor circulatory systems, but flatworms do have primitive nervous systems, excretory and osmoregulatory systems that get rid of wastes and control water balance, and reproductive systems with separate male and female organs.

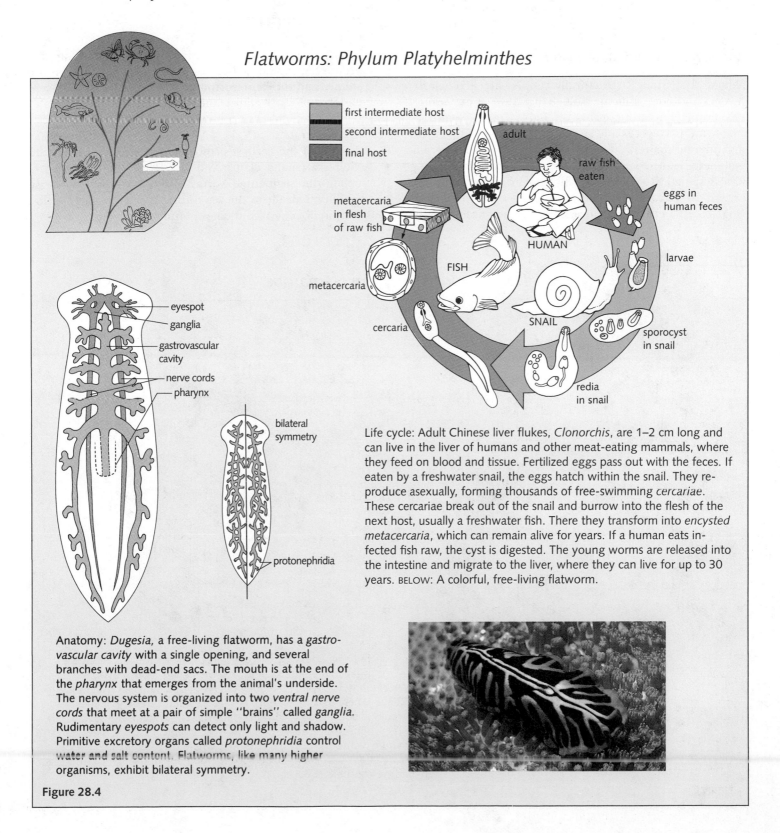

Flatworms: Phylum Platyhelminthes

first intermediate host

second intermediate host

final host

adult

raw fish eaten

HUMAN

eggs in human feces

larvae

FISH

SNAIL

sporocyst in snail

metacercaria in flesh of raw fish

metacercaria

cercaria

redia in snail

eyespot

ganglia

gastrovascular cavity

nerve cords

pharynx

bilateral symmetry

protonephridia

Life cycle: Adult Chinese liver flukes, *Clonorchis*, are 1–2 cm long and can live in the liver of humans and other meat-eating mammals, where they feed on blood and tissue. Fertilized eggs pass out with the feces. If eaten by a freshwater snail, the eggs hatch within the snail. They reproduce asexually, forming thousands of free-swimming *cercariae*. These cercariae break out of the snail and burrow into the flesh of the next host, usually a freshwater fish. There they transform into *encysted metacercaria*, which can remain alive for years. If a human eats infected fish raw, the cyst is digested. The young worms are released into the intestine and migrate to the liver, where they can live for up to 30 years. BELOW: A colorful, free-living flatworm.

Anatomy: *Dugesia*, a free-living flatworm, has a *gastrovascular cavity* with a single opening, and several branches with dead-end sacs. The mouth is at the end of the *pharynx* that emerges from the animal's underside. The nervous system is organized into two *ventral nerve cords* that meet at a pair of simple "brains" called *ganglia*. Rudimentary *eyespots* can detect only light and shadow. Primitive excretory organs called *protonephridia* control water and salt content. Flatworms, like many higher organisms, exhibit bilateral symmetry.

Figure 28.4

Today's free-living flatworms may be either scavengers or active predators in fresh water, saltwater, or very wet terrestrial habitats. Two classes within this phylum contain the **flukes** and the **tapeworms,** exclusively parasitic forms with complex life cycles. Infection with parasitic flatworms can cause devastating damage in humans.

Roundworms: Phylum Nematoda

Nematodes, though structurally simple, are ecologically diverse and adaptable animals that are the most numerous multicellular organisms on Earth today. Some are free-living herbivores and predators, others live on decaying organic matter, and still others are parasites of animals and plants. Parasitic nematodes cause debilitating illnesses such as trichinosis, whose painful symptoms are caused by larval nematodes that form cysts in muscles (Fig. 28.5). Because trichinosis can be acquired only by

eating the muscle tissue of animals that are themselves meat eaters, and because most mammals we consume are strictly herbivorous, humans acquire the disease almost exclusively by eating raw or incompletely cooked pork.

HIGHER INVERTEBRATES

By the middle of the Cambrian period, evolving animal groups had already split into two separate lines, the **protostomes** and the **deuterostomes,** distinguished by different patterns of early embryonic development. The details of these differences are summarized in Table 28.1 and later in Chapter 41.

Two of the earliest protostome phyla are the phylum Mollusca (snails, clams, and squid) and the phylum Annelida (the segmented worms). Both these groups diversified dramatically in the sea during the Ordovician period but did not colonize the land until much later.

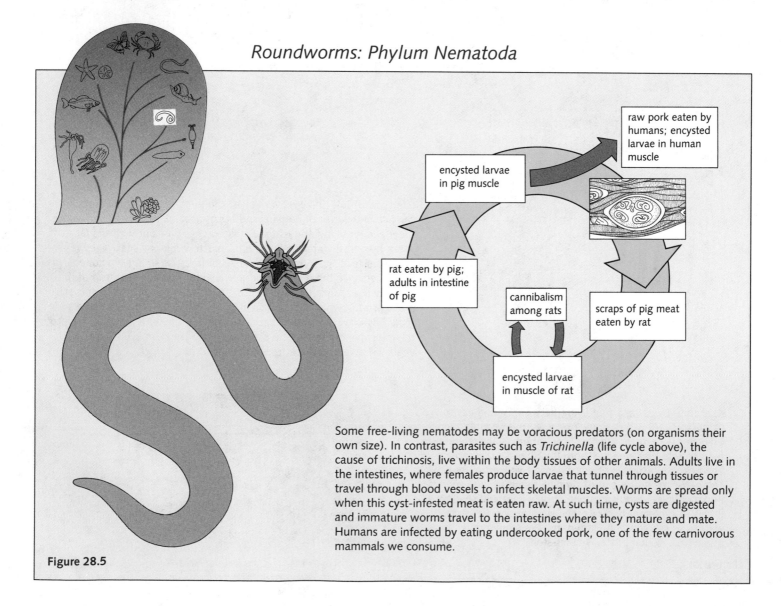

Roundworms: Phylum Nematoda

raw pork eaten by humans; encysted larvae in human muscle

encysted larvae in pig muscle

rat eaten by pig; adults in intestine of pig

cannibalism among rats

scraps of pig meat eaten by rat

encysted larvae in muscle of rat

Some free-living nematodes may be voracious predators (on organisms their own size). In contrast, parasites such as *Trichinella* (life cycle above), the cause of trichinosis, live within the body tissues of other animals. Adults live in the intestines, where females produce larvae that tunnel through tissues or travel through blood vessels to infect skeletal muscles. Worms are spread only when this cyst-infested meat is eaten raw. At such time, cysts are digested and immature worms travel to the intestines where they mature and mate. Humans are infected by eating undercooked pork, one of the few carnivorous mammals we consume.

Figure 28.5

Snails, Clams, and Squid: Phylum Mollusca

Molluscs were the first animals to evolve hard external coverings, an invaluable evolutionary advantage that helped early forms discourage predators and enabled more recent species to colonize the land. Members of this phylum ultimately diversified into several classes, each utilizing a variation on the phylum's basic body plan (Fig. 28.6). Because their shells are easily preserved, molluscan fossils are abundant.

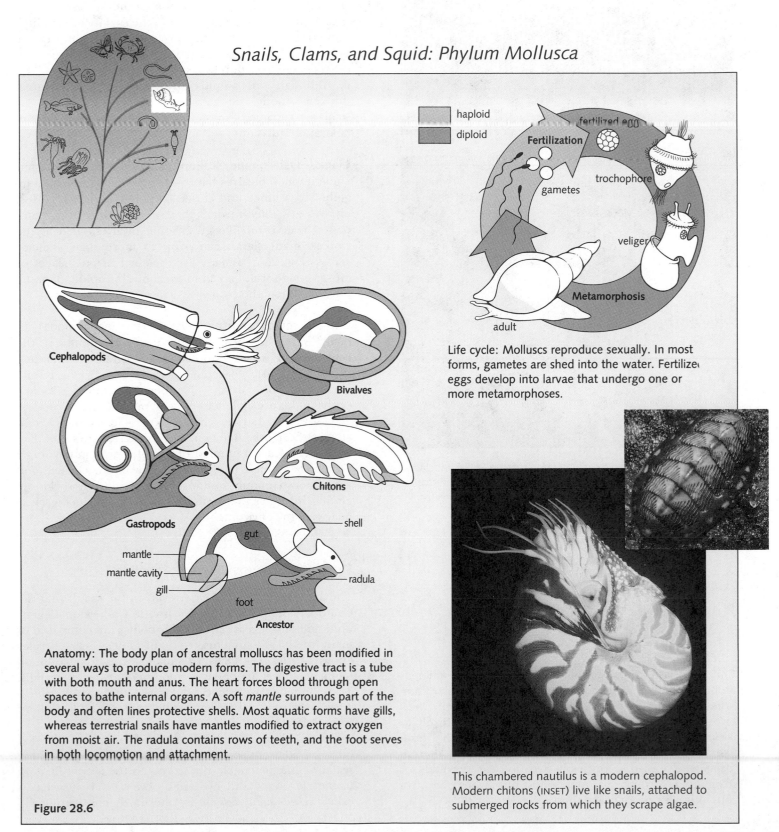

Snails, Clams, and Squid: Phylum Mollusca

haploid

diploid

Fertilization

fertilized egg

gametes

trochophore

veliger

Metamorphosis

adult

Life cycle: Molluscs reproduce sexually. In most forms, gametes are shed into the water. Fertilized eggs develop into larvae that undergo one or more metamorphoses.

Cephalopods

Bivalves

Chitons

Gastropods

shell

gut

mantle

mantle cavity

gill

foot

radula

Ancestor

Anatomy: The body plan of ancestral molluscs has been modified in several ways to produce modern forms. The digestive tract is a tube with both mouth and anus. The heart forces blood through open spaces to bathe internal organs. A soft *mantle* surrounds part of the body and often lines protective shells. Most aquatic forms have gills, whereas terrestrial snails have mantles modified to extract oxygen from moist air. The radula contains rows of teeth, and the foot serves in both locomotion and attachment.

This chambered nautilus is a modern cephalopod. Modern chitons (INSET) live like snails, attached to submerged rocks from which they scrape algae.

Figure 28.6

(a)

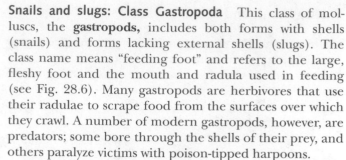

(b)

Figure 28.7 More living molluscs. **(a)** *Tridacna*, the giant clam, filter feeds and houses symbiotic algae in its mantle. **(b)** Octopi are the most intelligent predators among the cephalopods.

Chitons: Class Polyplacophora The **chitons** are ancient molluscs whose living members resemble their Cambrian ancestors (see Fig. 28.6). They are slow-moving herbivores that spend most of their time clamped firmly to rocks, where they scrape off encrusting algae with a rasp-like tongue, or *radula*. Organisms with feeding mechanisms such as this probably led to the downfall of stromatolites, which are now found only where herbivorous animals cannot reach them.

Snails and slugs: Class Gastropoda This class of molluscs, the **gastropods,** includes both forms with shells (snails) and forms lacking external shells (slugs). The class name means "feeding foot" and refers to the large, fleshy foot and the mouth and radula used in feeding (see Fig. 28.6). Many gastropods are herbivores that use their radulae to scrape food from the surfaces over which they crawl. A number of modern gastropods, however, are predators; some bore through the shells of their prey, and others paralyze victims with poison-tipped harpoons.

Although one would hardly expect to find great thinkers among the gastropods, even the lowliest sea snails are capable of mastering primitive learning tasks in laboratory situations.

Clams, oysters, and scallops: Class Bivalvia As their name implies, **bivalves** have two shells or "valves" held together by a hinge and closed by powerful muscles. Most bivalves are filter feeders that sift plankton from water passed over comb-like gills. Some bivalves, such as the tropical giant clams, can grow to more than a meter across (Fig. 28.7a). Many are attached in one place for life, although scallops can move about by rapidly clapping their shells together.

Octopi, squid, and nautilus: Class Cephalopoda In many ways the most advanced of the molluscs, the **cephalopods** (or "head-footed" molluscs), are also among the oldest. Fossilized cephalopods bear a striking resemblance to the living chambered nautilus. Today's cephalopods are agile, fast-moving predators with remarkably advanced nervous systems (see Fig. 28.6). Octopi, among the most intelligent of all invertebrates (see Fig. 28.7b), can learn a number of complex tasks in ways once thought restricted to the vertebrate brain. Cephalopod eyes are also among the most advanced in the world, operating in much the same way as ours, producing clear images of the environment.

Segmented Worms: Phylum Annelida

The **annelids** diversified extensively in the sea during the Cambrian period but did not invade freshwater habitats until millions of years later. Their bodies are composed of many identical segments, each of which bears bristles known as *setae* (Fig. 28.8). In aquatic forms, many of these segments have appendages modified to serve as gills. Some annelids are active predators, others are harmless scavengers, and still others use delicate gills to filter food from the water as they breathe.

The coelom of each annelid body segment is sealed off from adjacent segments by divisions, or septa. These sealed segments enable the worms to use a combination of muscle power and hydraulic power to create what is called a *hydrostatic skeleton*. By alternately contracting dif-

Segmented Worms: Phylum Annelida

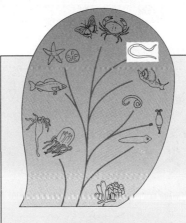

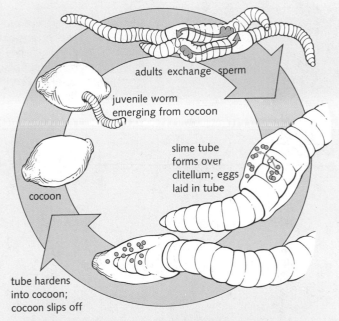

adults exchange sperm

juvenile worm emerging from cocoon

cocoon

slime tube forms over clitellum; eggs laid in tube

tube hardens into cocoon; cocoon slips off

Life cycle: Annelids reproduce sexually. Terrestrial species, such as earthworms, function as males and females at the same time. During mating, they exchange sperm. The clitellum secretes mucus that protects the sperm during the exchange and later secretes a slime tube that becomes the cocoon. After mating, worms separate and lay eggs, releasing stored sperm at that time. These eggs develop directly into adults. In marine forms, gametes are shed into the water; free-swimming larvae live in plankton before metamorphosing into adult form.

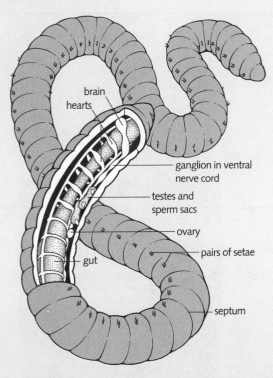

brain
hearts
ganglion in ventral nerve cord
testes and sperm sacs
ovary
pairs of setae
gut
septum

Anatomy: Each body segment is separated from the next by an internal partition. Most body parts, including hearts, blood vessels, and nerves, are repeated in many segments.

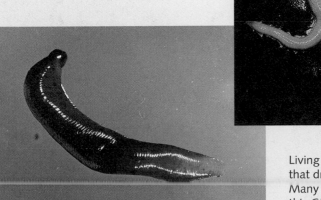

Living annelids: Some leeches (LEFT) are parasites that drink the blood of fishes and mammals. Many free-swimming marine annelids, such as this *Glycera* (ABOVE), are predators.

Figure 28.8

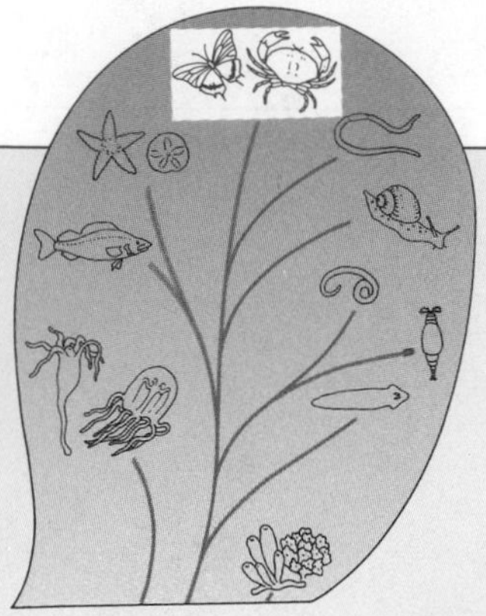

Spiders, Crustaceans, and Insects: Phylum Arthropoda

Life cycle: Lobsters, like most crustaceans, reproduce sexually. Eggs hatch into larvae that bear little resemblance to adults; these pass through four planktonic larval stages before metamorphosing into bottom-dwelling adult forms.

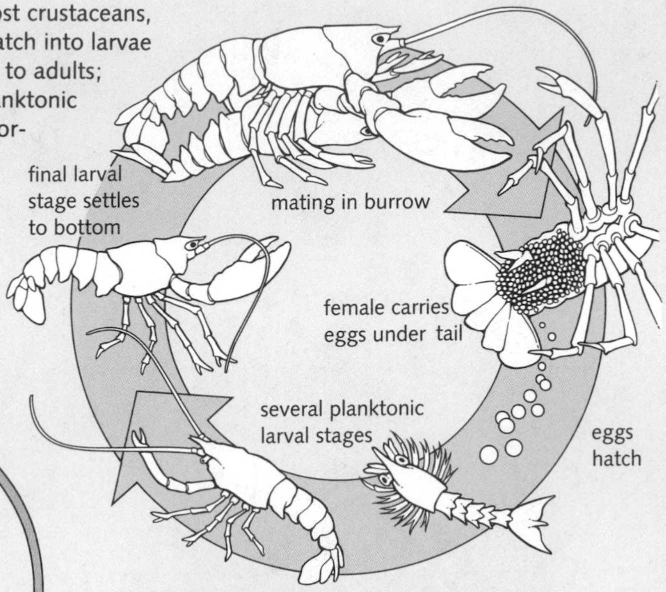

final larval stage settles to bottom

mating in burrow

female carries eggs under tail

several planktonic larval stages

eggs hatch

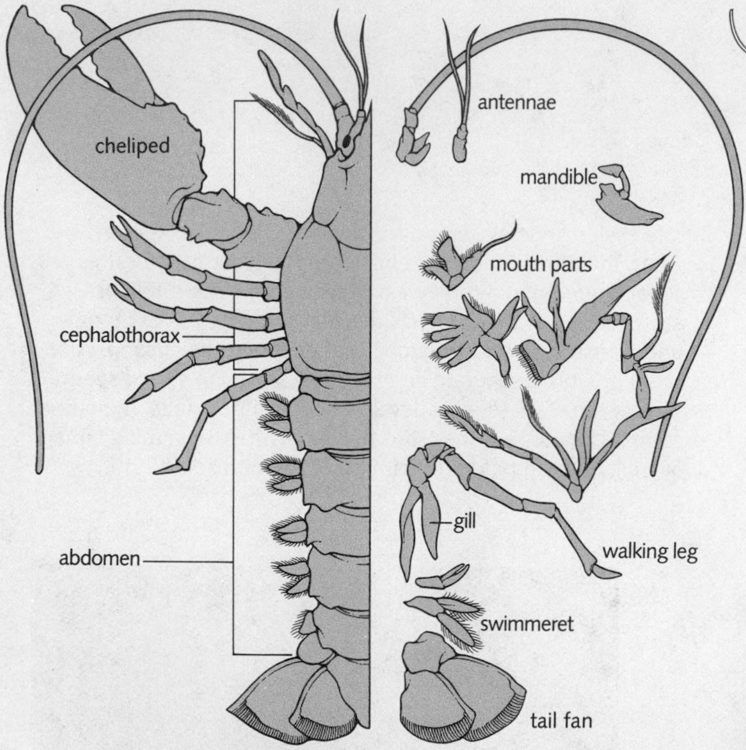

cheliped

antennae

mandible

cephalothorax

mouth parts

abdomen

gill

walking leg

swimmeret

tail fan

Anatomy: In crustaceans, such as lobsters, numerous body segments have fused, forming the cephalothorax and abdomen. The numerous appendages have specialized and perform various functions.

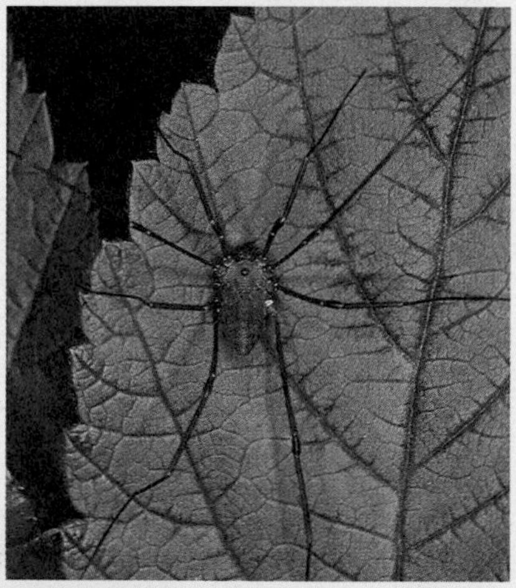

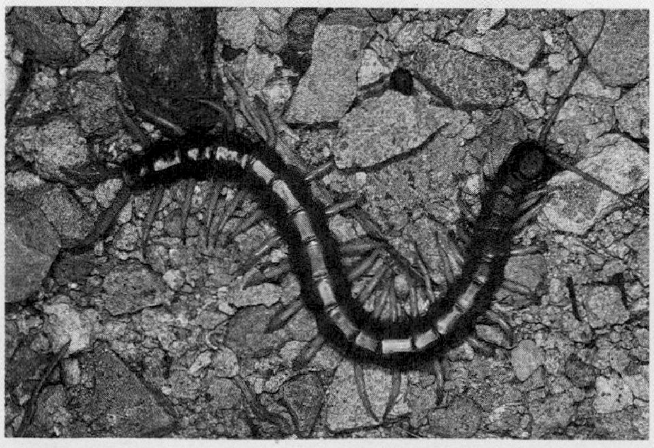

Arthropod body plans show the fusion of primitive segments during development. In centipedes (RIGHT), most visible body segments carry one pair of legs. The daddy-longlegs (TOP RIGHT) is an example of an organism whose embryonic segments have fused to form larger body regions.

Figure 28.9

ferent sets of muscles, annelids use the fluid yet incompressible contents of their segments to generate power for crawling, swimming, or burrowing, even though they lack any sort of hard skeletal support.

The First Arthropods: Phylum Arthropoda

The "jointed leg" animals, or **arthropods,** have enjoyed one of the longest-running and most spectacularly successful adaptive radiations in Earth's history. This phylum contains more species than any other group of organisms on Earth today. Between 800,000 and 900,000 arthropod species have been described, and estimates of the number of undiscovered species range from 2 million to as high as 40 million. Arthropods' basic body plan contains several key features retained from ancestral forms such as trilobites (Chapter 25) and several evolutionary innovations responsible for the group's success.

Like annelids, arthropods have bodies divided into a number of segments. In some, such as centipedes, those segments are readily visible externally. In others, such as crabs and beetles, a number of embryonic segments fuse to form larger body regions (Fig. 28.9).

Each arthropod segment carries a pair of the jointed appendages that give the phylum its name. As certain groups of arthropods evolved over time, several of these segments joined together; when that happened, the appendages associated with those segments were either lost or became specialized in ways that enable them to serve a variety of functions. Through the course of evolution, arthropod appendages have evolved into mouth parts that chew, flippers and legs that facilitate walking or swimming, and powerful claws that can seize and dismember prey.

All arthropods have hardened external skeletons, or *exoskeletons,* that are jointed in ways which enable legs and body segments to move. Exoskeletons are made from a mixture of proteins, a polysaccharide called *chitin,* and varying amounts of calcium salts. Exoskeletons bestowed many advantages on early arthropods; some of their benefits included protection, support, and hard structures to which their muscles could attach. Millions of years later, when the phylum invaded land, the exoskeleton not only served as a barrier to loss of water through evaporation but also provided stiff, jointed appendages for walking, burrowing, or flying.

Exoskeletons do have their limitations, however; arthropods can't grow inside such inflexible coverings. As a result, all arthropods must periodically shed their old skeletons and grow new ones, a traumatic process called *ecdysis,* or molting. During molting, the animal is completely immobilized, and for some time afterwards the new exoskeleton is too soft either to support the animal or to protect it. The animals behave accordingly; as the time for ecdysis approaches, arthropods seek secluded hiding places in burrows, under stones, or in piles of debris.

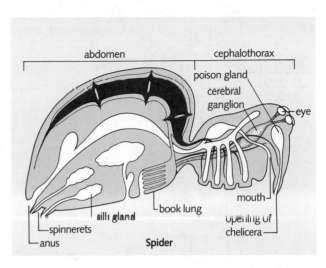

Figure 28.10 A typical spider's anatomy shows the fusion of body segments into cephalothorax (*cephalo* means "head"; *thorax* means "breastplate") and abdomen. Arachnid nervous systems have a large *cerebral ganglion* and several compound eyes. Silk for weaving webs or other traps is produced by glands called silk glands. Arachnids breathe primarily through respiratory organs called *book lungs,* whose blood-filled folds resemble the pages of a book. Spiders do not swallow prey whole; they inject digestive enzymes through openings made with their chelicerae and suck up the resulting liquid.

Arthropods as a group have well-developed nervous systems and an elaborate set of sensory cells and sense organs. The sophisticated compound eyes of many species enable them to see in color, and extraordinarily sensitive taste receptors on the feet and antennae of others can detect vanishingly small concentrations of chemicals given off by food. Arthropods also have well-developed and varied respiratory organs, circulatory systems, and excretory systems.

Zoologists believe that annelids, molluscs, and arthropods are closely related and evolved from a common protostome ancestor. Although modern members of these three great phyla differ widely from one another, they share enough larval and developmental characteristics to indicate close ancestral ties. Furthermore, chance and natural selection have kept alive several plausible descendants of a "missing link" between these phyla—animals called Onychophorans (see Theory in Action, *Peripatus: The Worm That Time Forgot,* p. 569).

Scorpions, spiders, and mites: Subphylum Chelicerata

Both living and extinct members of this subphylum are classified together because all possess mouthparts called *chelicerae,* often shaped into poisonous fangs or pincers, that evolved from the frontmost pair of appendages of the early arthropods (Fig. 28.10). **Chelicerates** have another pair of modified legs called *pedipalps* that are used in some species for feeding and in others only for grasping the partner during copulation. Chelicerates also have

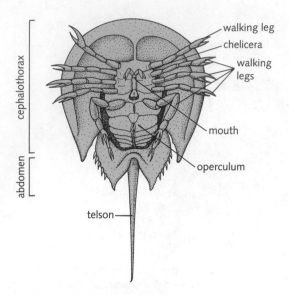

cephalothorax

walking leg
chelicera
walking legs

mouth

operculum

abdomen

telson

Figure 28.11 *Limulus*, the horseshoe crab, is a living chelicerate that closely resembles fossilized ancestors who were among the first hard-shelled animals.

four pairs of walking legs, in contrast to the three pairs evolved by insects 200 million years later.

One fairly old chelicerate still alive today is the common horseshoe crab or king crab, *Limulus* (Fig. 28.11), whose larvae look so much like trilobites that they are named after them.

Some of the earliest animals to crawl among the emerging plants were the detritus-feeding and herbivorous ancestors of today's scorpions, spiders, ticks, and mites. Modern terrestrial scorpions are venomous predators that carry their stings on their tails. There are about 1200 scorpion species alive today, most of which live in the tropics, in the subtropics, and in deserts everywhere. Spiders are ecologically important predators, particularly of insects. Mites can be either predators or parasites; the latter, which are serious agricultural pests of both plants and livestock, have also evolved intricate ecological relationships with both hosts and other parasites. Several kinds of ticks are important carriers, or vectors, of diseases such as Lyme disease and Rocky Mountain spotted fever.

Subphyla Crustacea and Uniramia While chelicerate arthropods were evolving on land, the ancestors of the two most important groups of modern arthropods, the subphyla Crustacea (modern shrimps, crabs, and lobsters) and Uniramia (modern centipedes, millipedes, and insects), were evolving in the sea. These subphyla combined well-developed nervous systems, adaptable exoskeletons, efficient respiratory systems (gills in crustaceans, air-filled tubes called *tracheae* in uniramians), and open circulatory

systems to produce a body plan that has proved adaptable to an infinite variety of ecological conditions.

Over the course of their evolution, crustaceans and uniramians developed innumerable specialized versions of the adaptable appendages that characterize all arthropods. It is on the basis of those appendages that members of these two subphyla are distinguished from one another. Both groups have jaws called **mandibles** that evolved from the appendages of the second or third body segment (rather than from the first, as did the chelicerae of the Chelicerata) (see Fig. 28.10). Beyond that similarity, however, lies a fundamental difference. Crustacean limbs have two branches and are called *bi*ramous, or two-branched. *Uni*ramian appendages, on the other hand, have only one branch.

Shrimps, crabs, and lobsters: Subphylum Crustacea The **crustaceans,** which include the shrimps, crabs, lobsters, copepods, and barnacles, evolved primarily in the sea, where they quickly became the ultimate aquatic arthropods. Living crustaceans include more than 26,000 species, the vast majority of which live in marine and freshwater habitats. Ecologically they can be detritus feeders, filter feeders, scavengers, herbivores, carnivores, or omnivores.

Many crustaceans, such as copepods and a variety of shrimps, are free swimming all their lives and don't grow much larger than a few millimeters in size. These are important herbivorous and carnivorous members of the zooplankton in both fresh water and saltwater, where they form critical links in many open-water food chains. Larger crustaceans, such as lobsters and crabs, have tiny, planktonic larvae but settle down to a bottom-dwelling existence as adults.

Centipedes, millipedes, and insects: Subphylum Uniramia The major **uniramian** adaptive radiations took place on land, after plants had paved the way. The very first members of the class Insecta were close behind chelicerates in their forays out of water, but insects as we know them did not definitively appear for another 100 million years. That evolutionary hesitation hasn't been a handicap in the long run, though. Biologists believe the subphylum Uniramia (including the mammoth class Insecta) contains more living species than all other groups of organisms combined.

Centipedes and millipedes: Classes Chilopoda and Diplopoda The centipedes (2500 modern species) and millipedes (10,000–50,000 modern species) were the earliest uniramia to evolve and leave fossil records of themselves. Of all the living arthropods, these classes show the most resemblance to their probable onychophoran-like ancestors. Modern centipedes are voracious carnivores whose first pair of walking legs has been modified into venomous fangs. Most millipedes, on the other hand, make their living scavenging through decaying vegetation.

Insects: Class Insecta   This group of organisms, the **insects,** has been enormously successful; they were the only flying animals for more than 100 million years after their great adaptive radiation. The class has many orders, the most significant of which are shown in Fig. 28.12. Insects are both numerous and ecologically important: ants, bees, and wasps alone comprise fully 30 percent of the total animal matter in tropical rain forests. More than 900,000 species of insects have been described, and estimates of yet-unknown species range from 2 million to 40 million or more.

Figure 28.12   RIGHT: Selected insect orders with their main characteristics. BOTTOM: Representative features of typical insects, shown for the common grasshopper.

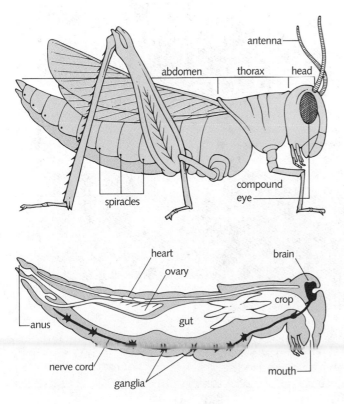

Order	Main Characteristics
Odonata	Mouthparts (biting); wings (2 pairs); metamorphosis incomplete *Examples:* dragonflies, damselflies
Orthoptera	Adult mouthparts (biting and chewing); membranous wings (2 pairs); metamorphosis incomplete *Examples:* grasshoppers, locusts
Isoptera	Mouthparts (chewing); wings (2 pairs) or wingless; social; employs division of labor; metamorphosis incomplete *Example:* termites
Heteroptera	Mouthparts (piercing, sucking); horny, membranous wings (2 pairs); metamorphosis incomplete *Example:* bedbugs
Coleoptera	Mouthparts (biting and chewing); membranous wings (2 pairs); armored exoskeleton; metamorphosis complete *Example:* beetles
Diptera	Mouthparts (sucking, piercing, lapping); wings (1 pair); metamorphosis complete *Examples:* flies, mosquitoes
Siphonaptera	Mouthparts (piercing, sucking); wingless; legs (jumping); metamorphosis complete *Example:* fleas
Lepidoptera	Tongue (long, coiled, for sucking); hairy body; wings (2 pairs) or wingless; metamorphosis complete *Examples:* butterflies, moths
Hymenoptera	Head mobile; eyes developed; mouthparts (chewing, sucking); membranous wings (2 pairs); metamorphosis complete *Examples:* bees, wasps, ants

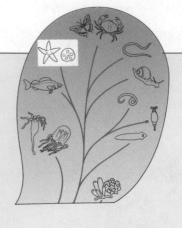

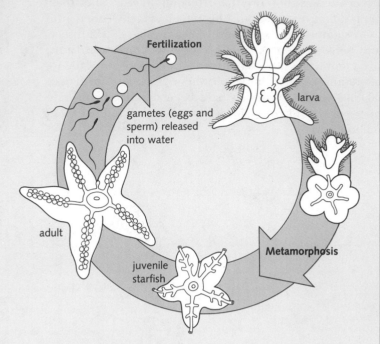

Fertilization

gametes (eggs and sperm) released into water

larva

adult

Metamorphosis

juvenile starfish

Life cycle: Echinoderms reproduce sexually, releasing eggs and sperm into the water. Free-swimming larvae go through complicated metamorphoses before settling down.

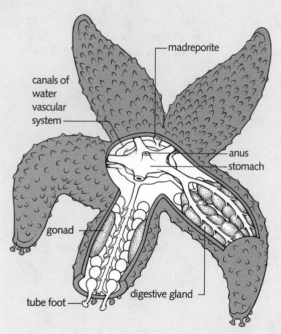

madreporite

canals of water vascular system

anus

stomach

gonad

digestive gland

tube foot

Anatomy: The water vascular system includes a *sieve plate* or *madreporite*, a series of canals, and tube feet. The arms contain digestive glands and reproductive organs.

Figure 28.13

Living echinoderms: A sea cucumber (LEFT) is a detritus feeder while most starfish (RIGHT) are carnivores that prey on bivalve molluscs. Sea urchins (CENTER) are grazers that scrape algae from submerged rocks.

In a very real sense, insects are humanity's most important and adaptable competitors for food and living space. Globally, billions of dollars each year are spent on controlling locusts and other agricultural pests; cockroaches, termites, and other household pests; and fleas and ticks that plague pets and livestock. Many of the reasons for insects' evolutionary and ecological success, as well as the symbiotic relationships they evolved with animals and flowering plants, are described in Chapter 25.

Starfish, Sea Lilies, and Sea Cucumbers: Phylum Echinodermata

The **echinoderms,** whose name means "spiny-skinned," are deuterostome animals that evolved during the Cambrian period (Fig. 28.13). Most echinoderms are radially symmetrical, although they seem to have evolved from bilaterally symmetrical ancestors. Echinoderms have never spread into freshwater environments but have always been abundant in the sea, where they live as filter feeders, detritus feeders, or carnivores.

The most peculiar characteristic of this phylum is the unique *water–vascular system,* a network of tubes connected to muscular, extensible suction cups called *tube feet.* Echinoderms use muscles in parts of the tube network to move water around inside, effectively employing hydraulic power to extend or retract their tube feet. Like hydraulic lifts used in garages, the water–vascular system can develop tremendous force over long periods of time. This sustainable power enables starfish to prey on clams and mussels by attaching scores of tube feet to the bivalves and pulling relentlessly until the shells gape open. The starfish then turns its stomach inside out into the mussel shell, spilling out digestive enzymes that digest the prey.

Although modern echinoderms have been evolving since the Cambrian period, many (particularly the filter-feeding "sea lilies" or "feather stars") look remarkably similar to their fossilized forebears (Fig. 28.14). Sea cucumbers are detritus feeders found in shallow marine habitats around the world and in "herds" of thousands on the ocean floor.

VERTEBRATE ORIGINS

As far back in time as the Silurian period, while major invertebrate groups diversified, another, far less physically and ecologically conspicuous group of animals evolved—those that ultimately led to the first animals with backbones. These peculiar animals left very little in the way of a fossil record and have left only a handful of odd descendants.

Chordates: Phylum Chordata

At first glance, you probably wouldn't place the most primitive living chordates in the same phylum with the rest of us vertebrates. Both groups are exclusively marine and have very simple body plans. The sparse fossil record is of little help in classifying very primitive chordates, which are discussed in this chapter because they are often called "invertebrate chordates"—as paradoxical as that may sound. The unusual animals called tunicates and

Figure 28.14 A crinoid, or feather star, from the phylum Echinodermata.

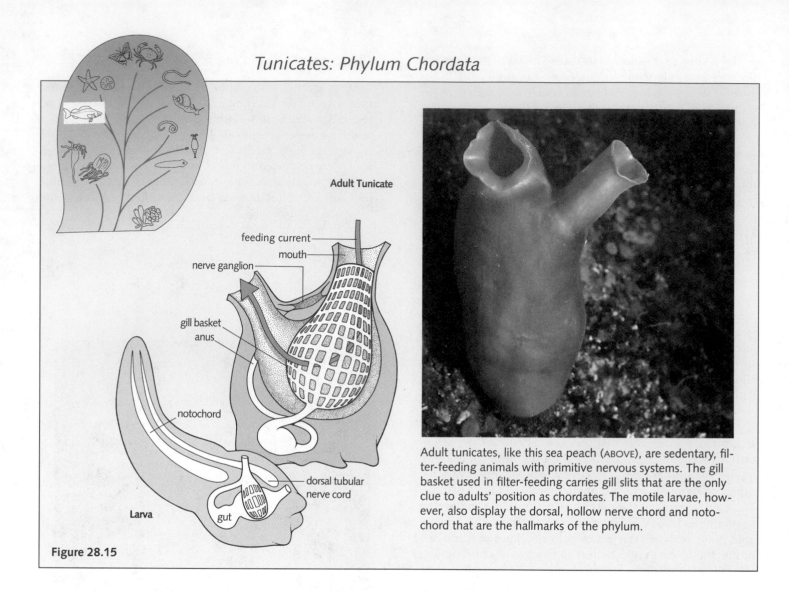

Adult Tunicate

feeding current
mouth
nerve ganglion
gill basket
anus
notochord
dorsal tubular nerve cord
gut
Larva

Figure 28.15

Adult tunicates, like this sea peach (ABOVE), are sedentary, filter-feeding animals with primitive nervous systems. The gill basket used in filter-feeding carries gill slits that are the only clue to adults' position as chordates. The motile larvae, however, also display the dorsal, hollow nerve chord and notochord that are the hallmarks of the phylum.

lancelets lack true vertebral columns (backbones), so they are certainly not vertebrates. They are placed in the phylum *Chordata* primarily because they share patterns of embryological development with higher chordates and because, for at least part of their lives, they possess the following features common to all **chordates:**

- A nervous system organized around a *dorsal hollow nerve cord,* a hollow tube of neurons that lies on top of (dorsal to) the gut
- A flexible supporting structure called a **notochord** that runs the length of the body between the gut and the nerve cord
- A series of **pharyngeal gill slits:** openings or pouches in the lining of the upper part of the digestive tract
- A tail that continues past the end of the digestive tract

The last three items on this list of chordate characteristics often apply only to the larvae or embryos of more advanced forms, such as our own species.

Tunicates: Subphylum Urochordata Most adult tunicates are stationary, filter-feeding organisms that look nothing like other chordates (Fig. 28.15). The free-swimming larvae of tunicates, however, are tadpole-like animals with all the expected chordate characteristics. Larval tunicates swim by using muscles in their tails to pull the flexible notochord from one side to the other and allowing it to spring back into shape. Sedentary adults are produced by metamorphosis. A few tunicate species that never metamorphose into sedentary forms as adults offer clues to the paths that early chordate evolution may have taken as far back as the Cambrian period.

Lancelets: Subphylum Cephalochordata Lancelets are small (usually under 5 cm), torpedo-shaped, superficially fish-like animals. Most of the few living species, which belong to the genus *Branchiostoma* (erroneously called *Amphioxus*), burrow into sandy or muddy areas of the continental shelves where they filter feed (Fig. 28.16).

Lancelets have many characteristics that link them to tunicate "tadpoles," but they have also evolved a few more advanced traits that link them to higher vertebrates. Like most higher chordates, lancelets have segmental muscles organized into chevron-shaped units on either side of the notochord, a distinct advance over the muscular organization of tunicates. By contracting units on opposite sides of the body alternately against the flexible but incompressible notochord, lancelets flex their bodies during swimming, much the way fishes do. But because lancelets lack the paired fins of fishes, they have poor directional control.

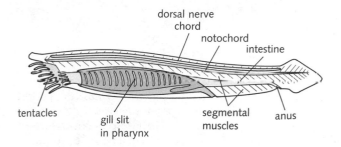

Figure 28.16 The lancelet, a simple, filter-feeding chordate, is a poor swimmer that spends most of its time partially buried in sediment.

THEORY IN ACTION

Peripatus: The Worm That Time Forgot

In 1826 the Reverend Lansdown Guilding published a description of an animal he couldn't quite place in the established scheme of things. It looked rather like a slug, yet it had many body segments and several dozen pairs of tiny feet with claws at their tips. It fed by shooting a sticky, gluey substance that entangled prey. At the base of each antenna it had a simple eye that much more closely resembled an annelid eye than the compound eyes of arthropods. Most of its body was soft and flexible, but the tips of its feet were encased in the rudiments of an exoskeleton.

Guilding called this organism *Peripatus* and created a new class for it among the molluscs, where he thought it belonged. Other zoologists thought it was an annelid; still others were certain it was a primitive arthropod. Today, most taxonomists place *Peripatus* and its relatives in their own phylum, Onychophora.

Onychophorans have survived since at least the Ordovician, and they were sufficiently widespread on the supercontinent to have left survivors in widely separate parts of the tropics.

Their host of transitional characteristics bridge the gaps between annelids and arthropods. Because onychophorans have survived nearly unchanged for millions of years, they—like *Limulus*—are often called living fossils. Yet in their other, far more advanced features,

the onychophorans remind us that evolutionary time never stands still. One pair of limbs near the head has evolved into jaws. And though some onychophorans lay eggs, others bear their young alive after a gestation period of 12 to 15 months.

The onychophoran *Peripatus*, an organism that displays a combination of characteristics found in molluscs, annelids, and arthropods.

SUMMARY

The major phyla of invertebrates—including the Porifera, Cnidaria, Nematoda, Platyhelminthes, Echinodermata, and Arthropoda—first appeared during the Cambrian period. Arthropods evolved first in the sea, but some lineages emerged and became the first land animals. Their descendants, which include shrimps, crabs, lobsters, insects, spiders, centipedes, and millipedes, have diversified to make the Arthropoda the most diverse phylum of multicellular organisms.

The first chordates, who share common roots with the ancestors of starfish and tunicates, left virtually no fossil record. Their descendants—first the jawless fishes, and then the jawed fishes and sharks—began to fill the seas by the end of the Silurian period.

STUDY FOCUS

After studying this chapter, you should be able to:

- Name and describe the phyla of the lower and higher invertebrates, and outline their life cycles.

- Explain arthropods' success, both in water and on land.

TERMS AND CONCEPTS

radial symmetry *555*	arthropods *563*
bilateral symmetry *555*	chelicerates *563*
cephalization *555*	mandibles *564*
protostomes *558*	crustaceans *564*
deuterostomes *558*	uniramian *564*
molluscs *559*	insects *565*
chitons *560*	echinoderms *567*
gastropods *560*	chordates *568*
bivalves *560*	notochord *568*
cephalopods *560*	pharyngeal gill slits *568*
annelids *560*	

REVIEW

Objective Questions (Answers in Appendix)

1. Sponges differ from other multicellular animals because they
   (a) can only reproduce asexually.
   (b) have complex body plans.
   (c) do not reproduce sexually.
   (d) lack organs and tissues.

2. The nervous system of cnidarians contains
   (a) a nerve net and sensory cells.
   (b) a nerve net only.
   (c) specialized gametes.
   (d) many-celled sense organs.

3. Bilateral symmetry is a characteristic of
   (a) flatworms.   (c) sponges.
   (b) cnidarians.   (d) Porifera.

4. Which of the features of embryonic development is characteristic of protostomes?
   (a) There is a radial arrangement of cells in embryo.
   (b) The fate of embryonic cells is not determined until late in development.
   (c) Coelom is formed by the splitting of the mesoderm.
   (d) The anus forms first, then the mouth.

5. Chordates share a common ancestor with the
   (a) centipede.   (c) mollusc.
   (b) lobster.   (d) tunicate.

Discussion Questions

6. What were the key innovations that made the arthropod body plan such an evolutionary success?

7. What makes the organization of sponges so different from that of other metazoans? Why is the sponge considered just barely multicellular?

8. Why are the nematodes probably the most numerous multicellular organisms on the earth today?

9. What characteristics contributed to the success of arthropods as the first terrestrial animals?

10. What shared features place tunicates and lancelets in the phylum Chordata?

READINGS

See integrated list of readings for Part 5 after Chapter 29.

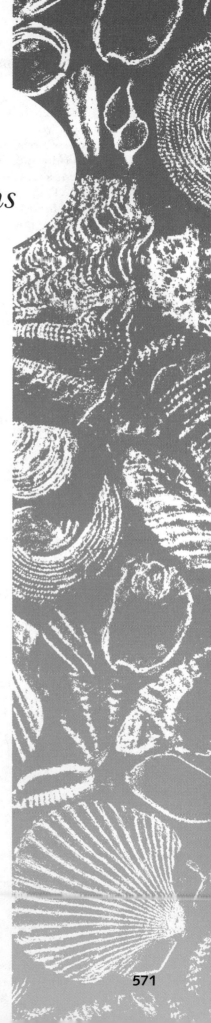

29

The Vertebrates:
From Fishes to Humans

"*f*ate makes our relatives," a wise man wrote, "choice makes our friends." Conversely, we might add, choice unmakes friendships, but relatives stay relatives. That's why family histories are so fascinating; we are who we are (at least in part) because of who our ancestors were—whether we like it (or them) or not. Those of us with no prior interest in Native American, Polish, or Middle Eastern history suddenly become fascinated on discovering that our ancestors were Cherokee or Arapaho, or that they hailed from Minsk or Jerusalem. Connecting ourselves to the past helps bring the past alive. Have we a knight in the family, perhaps? Or a despot? A famous rabbi, priest, or wilderness explorer?

Biologists feel a similar kinship with all living organisms. As you now know, every organism that has ever lived on Earth—from bacteria, to dinosaurs, to spotted owls, to Uncle Harry and Aunt Sophie—is linked to every other organism by the invisible record written in DNA. That common history makes for common problems, common solutions to those problems, and common destinies. That's why we've emphasized that to know other life-forms is to know ourselves and our environment better.

This chapter covers what even the most parochial among us see as *real* family history—the story of our species' closest kin: the vertebrates. In addition to sharing invisible molecular ties, these animals share with us our entire basic body plan: our backbones, our two pairs of limbs, the basic design of our heads, nervous systems, and digestive tracts. Our group is an ancient one whose origins we traced in Chapter 25. We will now look more closely at the surviving leaves and fruits on the family tree, examining close relatives in tropical forests and more distant (but still recognizable) relatives in the depths of the seas. Some we will be proud of, others we might not care for. But all of them are kin.

INNOVATION AND ADAPTIVE RADIATION

The evolution of terrestrial vertebrates from aquatic forebears, and ultimately from ancestors shared with invertebrate chordates, was both marked and fueled by four significant "innovations" in body plan: true jaws, paired fins, lungs, and the two sets of limbs which characterize the four-limbed terrestrial vertebrates known as tetrapods. Jaws came first, appearing in the now-extinct placoderms. The next innovation added paired fins. While that might not seem like much at first glance, paired fins add enormously to an aquatic animal's ability to steer and turn more efficiently, to accelerate and stop suddenly, and even to hover in place in the water. That addition of paired fins to the evolving vertebrate body plan thus bestowed an enormous selective advantage and set in motion the adaptive radiation of the first vertebrate group to survive in large numbers today.

Jawless Fishes: Agnatha

The jawless fishes, whose ancestors were the first vertebrate members of the phylum Chordata, were quite successful early on, but virtually all vanished by the end of the Devonian period. Only two jawless fishes, lampreys and hagfishes, remain alive today.

Lampreys (Fig. 29.1) retain many primitive features of their forebears but have adapted to specialized niches as predators of far more recent organisms. They have most peculiar larval forms, whose appearance and behavior serve as strong reminders of the connections between invertebrate chordates and vertebrate chordates. Lamprey larvae are pink, worm-like organisms that look like an imaginary cross between fishes, lancelets, and tunicates. They bury themselves in the mud of rivers and streams and filter feed for several years before metamorphosing into adults.

Lampreys: Phylum Chordata

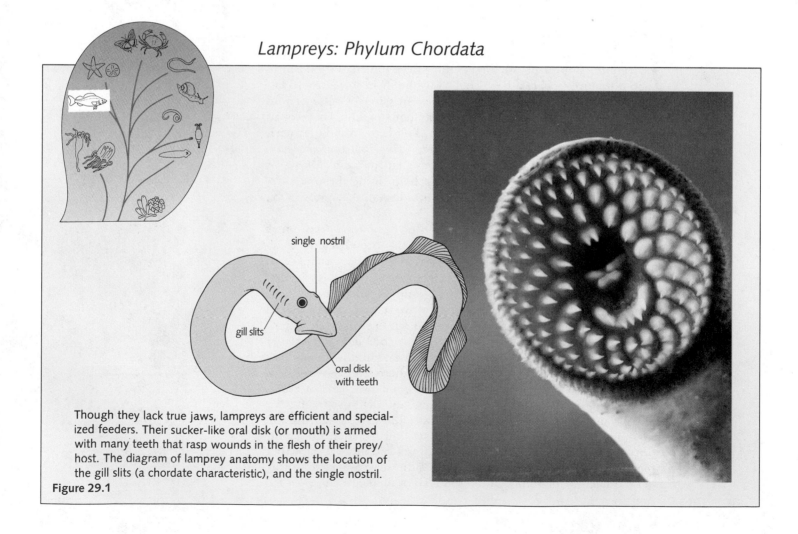

single nostril

gill slits

oral disk with teeth

Though they lack true jaws, lampreys are efficient and specialized feeders. Their sucker-like oral disk (or mouth) is armed with many teeth that rasp wounds in the flesh of their prey/host. The diagram of lamprey anatomy shows the location of the gill slits (a chordate characteristic), and the single nostril.

Figure 29.1

Hagfishes, too, are evolutionary relics. They burrow through the mud of the continental shelf, where they hunt annelid worms and scavenge pieces of dead and dying vertebrates.

Cartilaginous Fishes: Class Chondrichthyes

The first vertebrates to possess both jaws and paired fins were the **cartilaginous fishes,** so called because their skeletons were made of firm but resilient cartilage (Fig. 29.2). The scientific name for this group is class Chondrichthyes (*chondra* means "cartilage"; *ichthyes* means "fishes").

Although the first adaptive radiation of cartilaginous fishes didn't produce many species and didn't last very long, the group didn't fall into extinction. Instead, it survived to diversify again hundreds of millions of years later, during the Cretaceous period. That radiation produced modern sharks, skates, and rays.

Modern sharks and their kin are often referred to as "primitive" fishes, because they lack true bone and because their paired fins aren't as versatile as those of more recently evolved groups. But remember that when biologists use the word *primitive* in this context, they only mean that the animals involved appeared in more or less their present form a long time ago. They *don't* mean that those

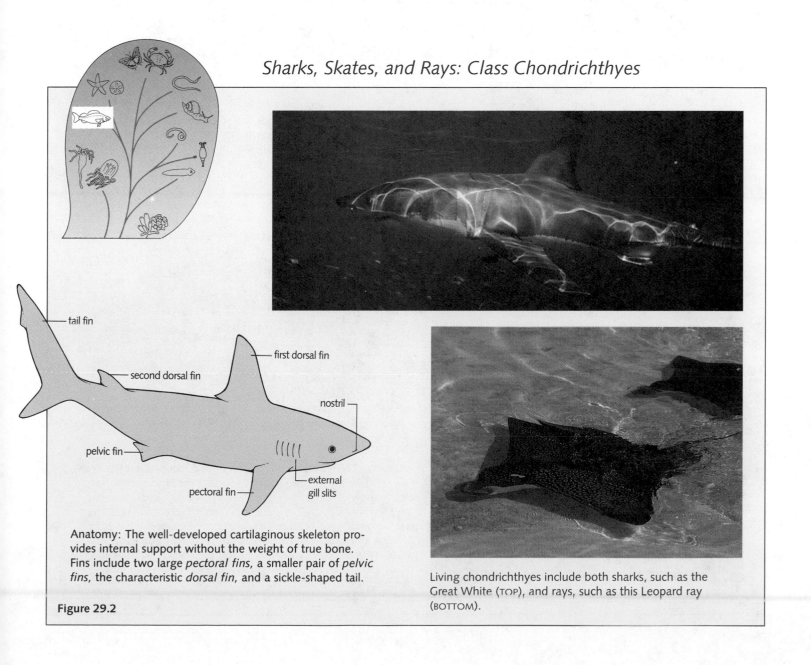

Sharks, Skates, and Rays: Class Chondrichthyes

tail fin

second dorsal fin

first dorsal fin

pelvic fin

nostril

pectoral fin

external gill slits

Anatomy: The well-developed cartilaginous skeleton provides internal support without the weight of true bone. Fins include two large *pectoral fins*, a smaller pair of *pelvic fins*, the characteristic *dorsal fin*, and a sickle-shaped tail.

Figure 29.2

Living chondrichthyes include both sharks, such as the Great White (TOP), and rays, such as this Leopard ray (BOTTOM).

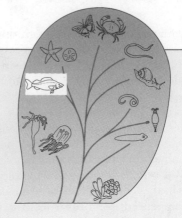

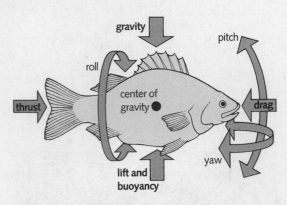

Maneuvering through a three-dimensional medium such as water is no easy matter; to capture prey efficiently, fishes must control roll, pitch, and yaw, just as airplane pilots do. The single fins of the earliest fishes provided only rudimentary steering ability; paired fins in more advanced species permit precision maneuvering.

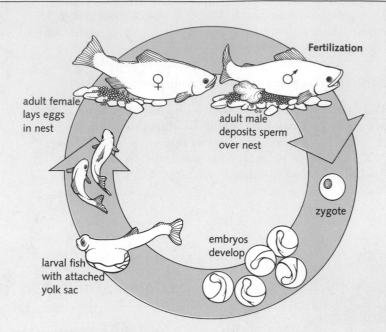

Life cycle: In most fishes, sexes are separate, fertilization is external, and eggs hatch into free-swimming larvae. There are also many species with internal fertilization that bear their young alive.

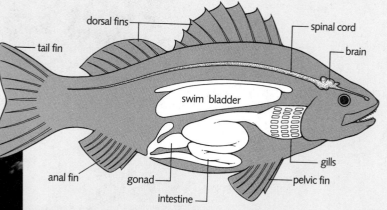

Anatomy: Ray-finned fishes have three or four unpaired fins and two sets of highly mobile paired fins. LEFT: The Nassau grouper clearly shows the highly mobile paired pelvic fins and lateral pectoral fins (not depicted in anatomy diagram). Most fishes of this class also have gas-filled swim bladders that counter the weight of bone and muscle and allow the fish to hover effortlessly. Fish anatomy varies greatly with ecological specialization. This carnivorous species has jaws armed with sharp teeth and a short digestive tract. Herbivorous fishes have jaws specialized for rasping and long intestines to digest plant tissue.

Figure 29.3

animals are inferior or "backwards" in any way. (If they were, they would have been extinct long ago!) So although modern sharks and their relatives have retained such primitive features as bulky fins and a cartilaginous skeleton, they have been evolving for nearly 400 million years. Armed with reasonably good eyesight and an exceptionally keen sense of smell, they are among the planet's most efficient and sophisticated predators. In addition, many can detect the minute electric currents generated by the muscles of distant potential prey.

Bony Fishes: Class Osteichthyes

Another group of early fishes acquired not only paired fins but also a skeleton made of strong, lightweight bone; these animals founded the lineage of **bony fishes,** formally called the class Osteichthyes (*osteo* means "bone"; Fig. 29.3). By the middle of the Devonian period, bony fishes split into two distinctive lines: the *ray-finned fishes,* the most diverse group of vertebrates alive today, and the *fleshy-finned fishes,* from which the ancestors of land vertebrates arose.

Ray-finned fishes: Subclass Actinopterygii A modest early radiation of ray-finned fishes during the Devonian left behind only a few primitive species, such as gars and sturgeons (Fig. 29.4). Two hundred million years later, however, the main line of ray-finned fishes finally began the spectacular adaptive radiation that produced the more than 30,000 species alive today. Many modern fishes are extraordinarily agile swimmers that use their paired fins to maximum advantage in ways that enable them to dart, hover, pivot, swim in reverse, and stop on a dime.

They use this agility in feeding, avoiding predators, and navigating around such complex structures as coral reefs.

Fleshy-finned fishes: Subclass Sarcopterygii Only a few species of fleshy-finned fishes survive today as relics of the final early vertebrate innovations. Early *lung fishes* probably evolved in shallow Devonian swamps where the water contained little oxygen. In these fishes, a pouch connected to the upper digestive tract evolved into a primitive lung that enabled its bearers to breathe air. Though lungfishes were common when they first appeared, there are now only a few species left in Australia, Africa, and South America.

The **coelacanth** is the sole survivor of a primitive group of fleshy-finned fishes that crawled around Devonian swamps on four stubby, fleshy fins (see Fig. 25.15). This group of odd animals was thought to have been extinct for hundreds of millions of years, until a living specimen was captured by a fishing trawler in the Pacific Ocean in 1938. Since then, several dozen specimens, practically identical to their fossil forebears, have been captured or sighted from submersibles.

The living coelacanth shown in Fig. 29.5 is thus of enormous interest to students of vertebrate evolution. X-rays reveal that these pudgy versions of normal fish fins were supported by bones homologous to the bones in the limbs of all terrestrial tetrapods. Although the term *living fossil* is problematical—because living coelacanths have certainly had the opportunity to evolve somewhat over the millions of years since their ancestors were fossilized—this animal offers us a remarkable glimpse of the stages in limb development ultimately leading to terrestrial vertebrates.

Figure 29.4 One of the most archaic of living bony fishes is the ancient and sedentary sturgeon.

Figure 29.5 The living coelacanth, a fleshy-finned fish whose group was long thought extinct, was rediscovered in the Pacific Ocean by local fishers.

Figure 29.6 The midland mud salamander is representative of the class Amphibia.

Frogs, Toads, and Salamanders: Class Amphibia

As described in Chapter 25, vertebrates adapting to life on land faced the same sorts of problems as those faced by the first terrestrial plants: providing body support, avoiding desiccation, and adjusting to forms of reproduction that don't require standing water. And like the living descendants of the first mosses and lycopods, living amphibians display a range of adaptations that include some, but not all, adjustments necessary for a fully terrestrial existence (Fig. 29.6).

Like lower plants, most amphibians lack a waterproof barrier to prevent the loss of body fluids and therefore lose water rapidly through their skin in dry air. (That's not altogether bad if they are living in wet places; many amphibians respire through their skin, an ability that is lost when skin is made waterproof.) The adults of several living amphibian species have adaptations that minimize desiccation, so they can survive on land for limited periods, even in fairly dry areas. But the majority of modern amphibians must still return to water to lay their eggs and must spend at least part of their lives as fully aquatic larvae. For that reason, although adults might seem admirably terrestrial, most species can survive and reproduce over the long term only in wet (or at least seasonally wet) habitats (Fig. 29.7).

Lizards, Snakes, and Turtles: Class Reptilia

Reptiles, as discussed in Chapter 25, were the first truly terrestrial vertebrates. They combined better resistance to desiccation in the form of a dry, scaly skin with still stronger and more efficient skeletal support systems and the fully terrestrial **amniotic egg** (Fig. 29.8). Together with internal fertilization, these adaptations enabled them to become the dominant (and largest) vertebrates both on land and in the sea for hundreds of millions of years. The extinct dinosaurs are discussed in Chapter 25.

Living reptiles belong to four orders that include, respectively, crocodiles, caimans, and alligators; turtles and tortoises; snakes, lizards, and most other common reptiles; and a single relic species, the tuatara, whose closest relatives disappeared over 100 million years ago. Modern reptiles are ectothermic animals that depend on environmental heat sources and sinks to keep their body temperatures within a suitable operating range. Although this habit limits reptiles in some ways, it does permit them to subsist on far less food than endotherms of similar sizes.

Birds: Class Aves

Birds, which evolved either from truly ancient stem reptiles or from more recent dinosaur ancestors (Fig. 29.9),

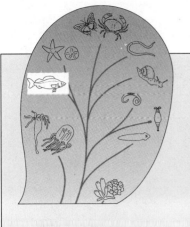

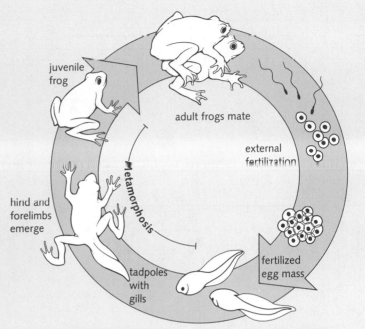

Life cycle: The typical cycle includes a fully aquatic larval stage, or *tadpole*. Frog and toad tadpoles are herbivorous filter feeders whose internal anatomy changes as profoundly as their external anatomy as they metamorphose into carnivorous adults.

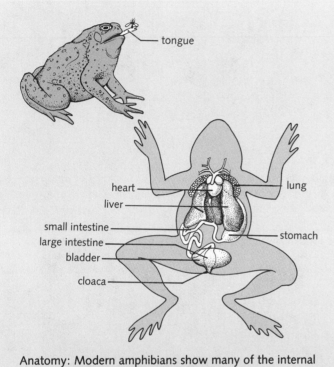

Anatomy: Modern amphibians show many of the internal organs common to higher terrestrial vertebrates.

Many modern amphibians have become highly specialized in adapting to specific habitats. ABOVE: Eggs of the Surinam toad sink into the protective skin of their parent's back, where they develop. LEFT: Spadefoot toads have turned water-permeable skin into an advantage in desert habitats. They burrow deeply at the end of rainy season and stay buried for up to 10 months, using their skin to extract water from moist soil as the roots of plants do.

Figure 29.7

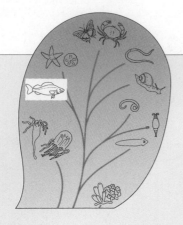

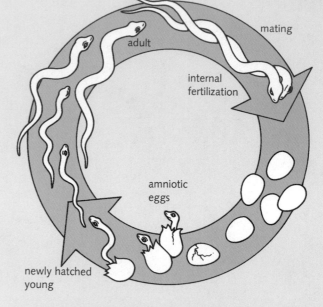

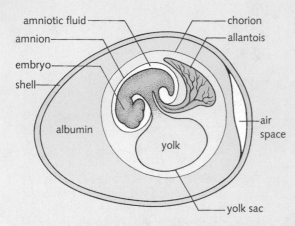

The amniotic egg, which is found in both reptiles and birds, protects embryos from desiccation, and allows for nutrition, respiration, and excretion within a shell that may be either tough and leathery (as in turtles) or brittle (as in birds) but is always porous. Inside the shell are three layers of protective membranes: the amnion, chorion, and allantois.

Reptiles, like all truly terrestrial vertebrates, have internal fertilization. Most species lay eggs that develop externally, although the females of some species retain eggs within their bodies until they hatch and then "give birth" to already wiggling young.

Reptilian diversity includes (TOP LEFT) an iguana, *Dipsosaurus dorsalis*, (TOP RIGHT) a crocodile, *Crocodylus johnstoni*, and (BOTTOM RIGHT) a milksnake, *Lampropeltis triangulum sinaloa*.

Figure 29.8

Birds: Class Aves

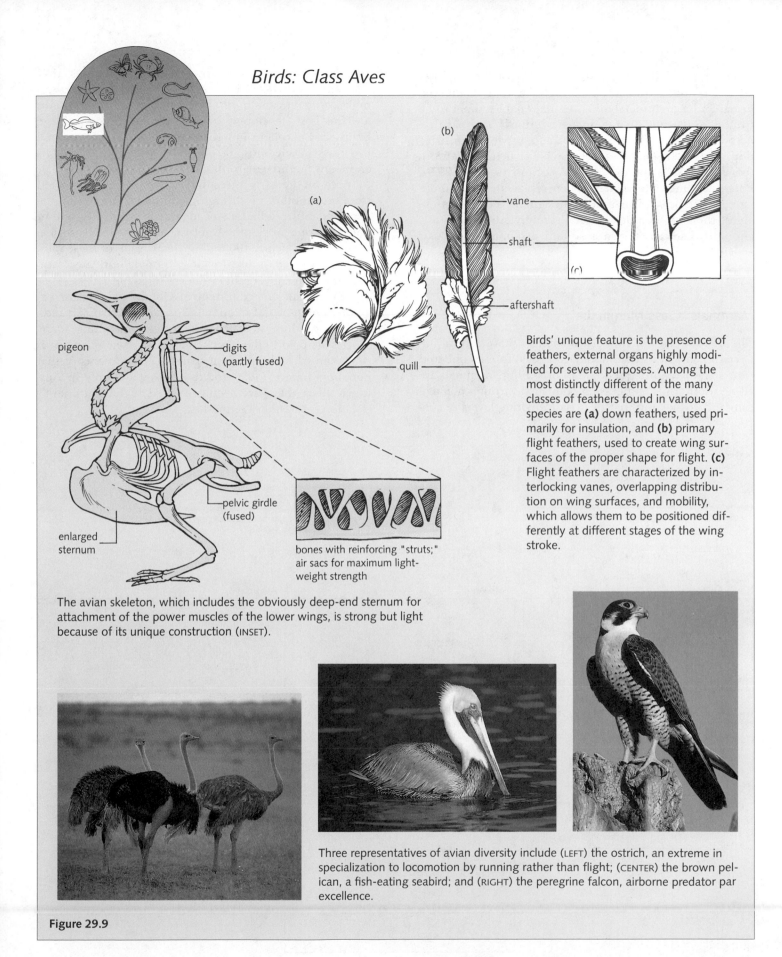

vane

shaft

aftershaft

quill

(a)

(b)

(c)

pigeon

digits
(partly fused)

pelvic girdle
(fused)

enlarged
sternum

bones with reinforcing "struts;"
air sacs for maximum lightweight strength

Birds' unique feature is the presence of feathers, external organs highly modified for several purposes. Among the most distinctly different of the many classes of feathers found in various species are **(a)** down feathers, used primarily for insulation, and **(b)** primary flight feathers, used to create wing surfaces of the proper shape for flight. **(c)** Flight feathers are characterized by interlocking vanes, overlapping distribution on wing surfaces, and mobility, which allows them to be positioned differently at different stages of the wing stroke.

The avian skeleton, which includes the obviously deep-end sternum for attachment of the power muscles of the lower wings, is strong but light because of its unique construction (INSET).

Three representatives of avian diversity include (LEFT) the ostrich, an extreme in specialization to locomotion by running rather than flight; (CENTER) the brown pelican, a fish-eating seabird; and (RIGHT) the peregrine falcon, airborne predator par excellence.

Figure 29.9

are characterized by scaly skin, feathers of several types used for both insulation and flight, a four-chambered heart, and a high metabolic rate that enables them to be truly endothermic, as mammals are. Birds also have remarkably lightweight bones, thanks to a style of construction that incorporates a reinforced structural network that looks like a sponge in cross section (see Fig. 29.9). Potentially heavy teeth have been replaced with a more lightweight beak. The unique design of bird lungs enables them to extract oxygen rapidly and efficiently enough to sustain the intense and prolonged muscular activity necessary for flight.

Mammals: Class Mammalia

The 4100 species of modern **mammals** share several important characteristics. Mammals possess four-chambered hearts and a diaphragm to aid in breathing. Mammals are also endothermic, generating body heat through metabolism and conserving that heat via insulating layers of sub-cutaneous fat and body hair. With two exceptions, they (or rather we) bear live young and feed newborns on milk secreted by the mother. Mammalian nervous systems are the best developed in the animal kingdom; large brains and excellent muscular coordination have helped this class invade most of the earth's terrestrial habitats and many marine environments.

Because the early ancestors of mammals were isolated from one another on diverging continents, three simultaneous adaptive radiations of mammals occurred—one in Laurasia, one in South America, and one in Australia. Separated for nearly 100 million years, each of these radiations produced similar-looking, though unrelated, animals through convergent evolution (Fig. 29.10).

On the ancient continent of Gondwanaland, for example, the group of mammals known as marsupials radiated to produce a spectacular fauna, including many species adapted to the same niches as their familiar North American or African counterparts. The remnants of that radiation survive today, for the most part, only in Australia, where they have been preserved by that continent's isolation from other continents.

North America

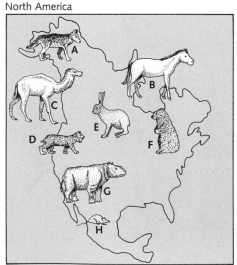

South America

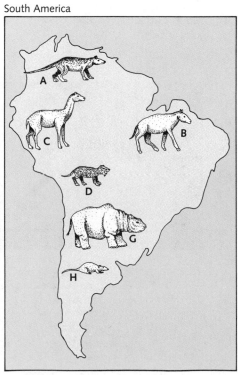

Australia

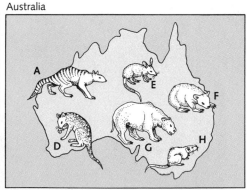

Figure 29.10 Cenozoic mammalian radiations. Mammalian ancestors dispersed across Pangaea were isolated on wandering continents. Convergent evolution among the independently evolving faunas of Laurasia (including North America), South America, and Australia produced both remarkable ecological look-alikes and bizarre, one-of-a-kind animals.

Egg-laying mammals: Subclass Prototheria In Australia and parts of New Guinea live two most peculiar animals: the duck-billed platypus (Fig. 29.11) and the echidna, or spiny anteater. These strange creatures represent a fascinating mixture of mammalian and reptilian traits. **Prototherians** are clearly endothermic, but they do not regulate their body temperature as effectively as other mammals. Most peculiarly, they lay eggs but suckle their young after the eggs hatch.

Marsupials: Subclass Theria, infraclass Metatheria Isolated on the continents formed from Gondwanaland, the first **metatherians, marsupials** such as kangaroos, evolved along a path distinct from, yet parallel to, the main mammalian line (Fig. 29.12). These pouched mammals were once common across all the southern continents that had formed Gondwanaland, but significant numbers of species survive today only in Australia. Nearly all the marsupials that roamed South America during the early Cenozoic era became extinct when that continent ultimately collided with North America. Marsupials do not fare well in competition with placental animals; and when North American mammals entered South America via the Central American land bridge, most local mammals—including many fossil species discovered by Darwin—became extinct. The only South American marsupials to make the trip north successfully were the opossums.

Marsupials are better at endothermic temperature regulation than prototherians, but they have somewhat lower metabolic rates than other mammals. Their young are born alive but emerge from the womb at an extraordinarily premature stage of development. Kangaroos, for example, which ultimately grow to the size of an adult human, emerge from the birth canal while still little more than embryos, only 2 to 3 cm long. Each marsupial infant crawls blindly and instinctively up its mother's belly and down into her pouch. There it attaches to a teat that then swells in its mouth so the baby cannot drop off. The infant hangs on for weeks or months (depending on the species) until it is ready to begin independent life.

Placental mammals: Infraclass Eutheria The **eutherian,** or "true," mammals diversified rapidly at the beginning of the Cenozoic era, producing dozens of orders, 17 of which survive today. Many extinct mammals were giants by today's standards; there were once beavers as large as modern bears, relatives of elephants that would dwarf living species, and ground sloths 6 m (nearly 20 ft) tall!

Eutherians are often called **placental** mammals, after the highly developed reproductive organ that connects the developing fetus to the mother during pregnancy. Eutherian young are born in a much more advanced stage of development than marsupials. And though they depend on the mother's milk for varying lengths of time, many are able to stand and walk just hours after their birth.

Figure 29.11 Prototherian mammals such as this duck-billed platypus are evolutionary intermediates between reptiles and mammals. The platypus does not represent the ancestors of modern forms, but it undoubtedly preserves several characteristics of those ancestors.

Figure 29.12 Marsupials, such as this red kangaroo, look much like familiar placental mammals but differ from our closer relatives in reproductive physiology and metabolic rate. Marsupials are "born" as little more than embryos that complete their development in their mother's pouch.

The evolution of modern mammals has been intimately linked to the history of flowering plants. The great herds of bison, buffalo, wildebeest, gazelle, and other mammalian herbivores evolved in direct response to the widespread success of grasses during the Cenozoic. Among the changes seen in the evolution of modern horses, for example, was the development of molar teeth that grind the tough leaves of grasses into digestible pulp. Our own order, the Primates, owes most of its basic character to millions of years spent evolving in the branches of angiosperm forests. And as we will see shortly, our immediate ancestors probably emerged from the trees in response to ecological opportunities in the spreading grassy savannahs of Africa.

The biology of eutherian mammals could easily fill several books; mammalian physiology will feature prominently in Chapters 34–46. Here we will concentrate on the evolution of the singular species known as *Homo sapiens*.

PRIMATE AND HUMAN EVOLUTION

Human beings, *Homo sapiens* (*homo* means "man"; *sapiens* means "wise"), are members of the order **Primates.** Our closest living relatives are the chimpanzees and gorillas of Africa. We can trace our line of descent back through a host of intermediate forms to tree-dwelling ancestors that lived in the forests of Africa some 65 million years ago. Because no other statements of evolutionary fact have caused such an uproar from Darwin's time to the present, we will examine the events in human evolution closely.

Figure 29.13 The earliest primates, like other primitive mammals, seem to have resembled the diminutive living tree shrew.

The earliest primates were small, insect-eating, tree-dwelling mammals that probably looked a great deal like today's tiny tree shrews (Fig. 29.13). Even though several primates ultimately abandoned the trees for a ground-dwelling existence, many of the characteristics we consider quintessentially human first evolved as adaptations to life in trees.

1. Highly mobile digits (fingers and toes); strong, grasping, flexible hands; and freely movable arms.

2. A complex visual system involving large, mobile, forward-pointing eyes connected to enlarged visual-processing centers in the brain. This combination (eyes with overlapping fields of view and sophisticated information processing in the brain) endows primates with excellent *binocular vision* (*bi* means "two"; *ocular* means "eyes") that permits precise perception of depth and distance.

3. Complex central nervous systems based on enlarged brains that can take in and evaluate large amounts of environmental information quickly and efficiently. Information flow to and from the brain is channeled largely through a well-developed spinal cord protected by the backbone.

4. A trend toward bearing only one offspring per pregnancy. This trend is reflected in the reduction in the number of nipples throughout the order. Primitive primates have several pairs, as do most other mammals; humans and other higher primates have only a single pair.

5. A complex yet flexible social system in which parents and relatives care for infants and juveniles for long periods of time. As you will see in Chapter 46, higher primates often exhibit a number of strikingly human behavior patterns.

Division of the Primate Line

Soon after the primate line appeared, it split into three groups (Fig. 29.14). There were two groups of **prosimians** (*pro* means "before"; *simians* means "monkey"), or "lower primates," including the relatively primitive tarsiers and lemurs, pottos, and galagos. These are the primates whose origins lie closest to the base of the family tree.

The third group, **anthropoids** (*anthro* means "human"; *anthropoid* means "human-like"), or "higher primates," first appeared in the fossil record between 40 and 36 million years ago. Their origins among the lower primates are still uncertain, but anthropoids clearly split into two groups as the continents of Africa and South America drifted away from one another. The *New World monkeys* of Central and South America have *prehensile* (long and grasping) tails that they can use much like a fifth hand for grasping

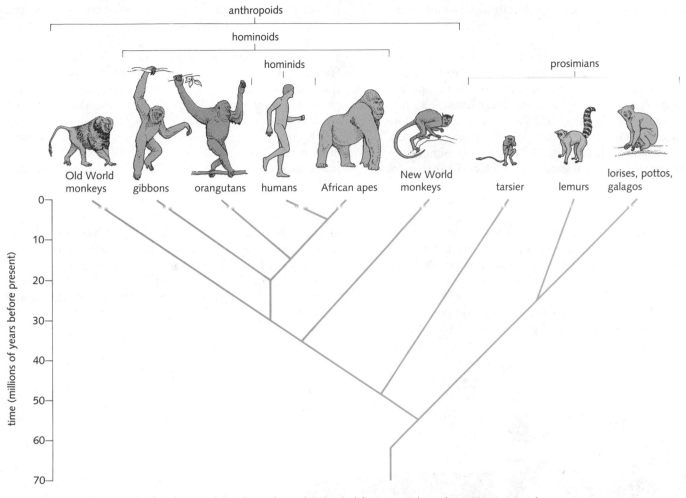

Figure 29.14 The primate family tree shows how the order divided first to produce the prosimians and anthropoids and then again to produce the New and Old World monkeys and great apes. Paleontologists still argue over which group of lower primates contained the ancestors of the anthropoids.

branches. The *Old World monkeys* and *great apes* often lack tails; and when tails do occur, they are not prehensile.

Hominoid lines   Humans, extinct human ancestors (the family Hominidae), and the living great apes (the family Pongidae) together make up a group of anthropoids called the **hominoids.** Early hominoids quickly evolved into several species that spread out over most of Africa, Europe, and Asia by about 26 million years ago. Between 10 and 20 million years ago, two early branches off this line led to the gibbons and orangutans of Asia (Fig. 29.15). These apes have remained in the trees that sheltered their ancestors. Other ancient hominoids—the ancestors of both the African apes and humans—began to spend more and more time on the ground as the Cenozoic climate changed and Africa's tropical forests gave way to broad savannahs.

Figure 29.15   Though the orangutan may look more "human" than several other apes, its shared ancestor with *Homo sapiens* was further back than the ancestor we share with African apes.

Unfortunately, this stage of hominoid evolution occurred in parts of Africa where researchers have not yet found good fossil sites from this time period. For that reason, there is little fossil evidence to date the final splitting of the human line from that of the apes. Based on that limited evidence, paleontologists once suggested that the human line split off from the apes as far back as 25 million years ago. As discussed in Chapter 22, however, biochemists can estimate the date of that split using the "molecular clock" approach. By examining similarities and differences among the genes and blood proteins of chimpanzees, gorillas, and humans, they have calculated that the hominid–pongid (living ape) split occurred somewhere between 4 and 9 million years ago. Although the data remain controversial, the general consensus (after 15 years of intense debate) is that humans and chimps diverged from human ancestors no more than 5–7 million years ago.

Hominid radiation Precisely what happened next to the hominid line is not clear, but the fossil record reveals that human-like organisms appeared with increasing frequency between 2.5 and 3.5 million years ago. When the very first of these specimens were found more than 20 years ago, they were assembled into a sort of evolutionary "ladder" along which ape-like species led to progressively more human-like forms with larger brains and more erect posture. You may recall that this is much the same view that people once had of horse evolution.

But beginning in the 1970s, a series of exciting discoveries threw this simple scheme into disarray. These finds made it clear that between 1.5 and 2.5 million years ago there was a whole cluster of hominid species, several of which must have lived virtually side by side. Four of the more primitive species from this time period are usually classified in the genus *Australopithecus* (Fig. 29.16), whereas the two more advanced forms are placed in our own genus, *Homo*. The new wealth of hominid fossils had turned the simple evolutionary ladder into a bush with several branches, just as similar finds had done for the story of the horse.

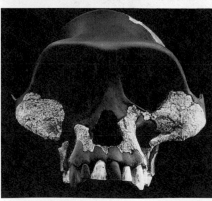

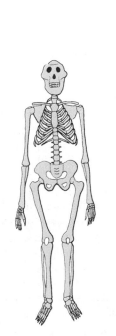
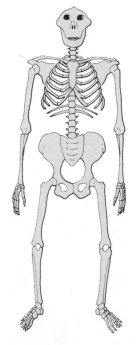

Figure 29.16 All these species of australopithecines lived on the ground, stood between 1 and 1.5 m tall, and weighed around 20 kilograms. CLOCKWISE FROM TOP: The skulls of *Australopithecus robustus*, *A. afarensis*, and *A. boisei*. RIGHT: Representative skeletal structures of *A. afarensis* and *A. boisei*.

Australopithecus afarensis   *Australopithecus boisei*

Australopithecines

In 1974 Donald Johansen discovered the skeleton of a female hominid dated at between 3 and 3.5 million years old (Fig. 29.17). The find was spectacular. In a field where most specimens are reconstructed from isolated teeth and skull fragments, this skeleton was nearly 40 percent complete and included enough bones to yield a great deal of information about this woman's appearance. Johansen and his co-workers named the skeleton "Lucy" because the night of the discovery they held a party at which they played the old Beatles tune "Lucy in the Sky with Diamonds."

Lucy is an excellent example of a transitional organism between apes and humans. From the shape of her pelvis and leg bones, we know that she walked *bipedally*—that is, erect on two legs (Fig. 29.18). But her stance was intermediate between the knock-kneed, uneven shuffle of chimpanzees and the more straight-legged and graceful modern human walk.

Lucy was assigned to the species *Australopithecus afarensis*, the species many researchers believe to be the common ancestor of all other hominids. But the relationships between the other three species of *Australopithecus* (*A. boisei*, *A. africanus*, and *A. robustus*) and the early members of the genus *Homo* (*H. habilis* and *H. erectus*) are unclear. The situation was further confused when Alan Walker of the Johns Hopkins University medical school found a fairly intact skull—now referred to as KNM-WT 17000—in a fossil bed dated 2.5 to 2.6 million years of age. The discovery of this skull provoked suggestions that there may have been a sudden adaptive radiation of the hominid line between 3 and 3.5 million years ago, producing either three or four parallel branches. No one knows what evolutionary or ecological event might have sparked this sudden proliferation, and experts are still arguing over who gave rise to whom (Fig. 29.19). (Even the names of

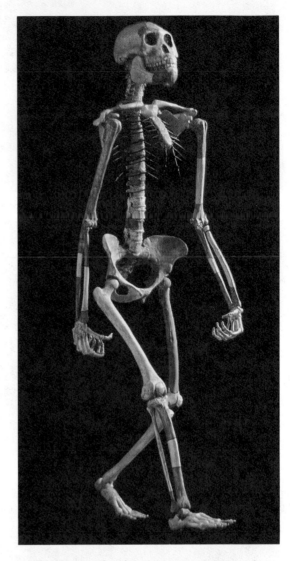

Figure 29.17 This dramatic reconstruction of Lucy, one of the most nearly complete fossil hominids ever discovered, has been assigned to the species *A. afarensis*. This skeleton is between 3 and 3.5 million years old.

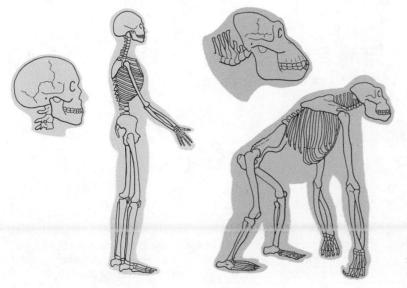

Figure 29.18 Differences and similarities in skull shape and general skeletal structure between humans and chimpanzees.

these hominids are still in dispute; the fossils we refer to as *Australopithecus robustus*, for example, are assigned by some authorities to the genus *Paranthropus*.) Because several of these species existed at the same time, however, one obviously could not have evolved into another.

Though much is often made of the "humanness" of australopithecines, there is ample evidence that they actually occupied the same sort of ecological niche in African savannahs that baboons do today. Australopithecines had not yet learned to hunt; rather, they were still the prey of

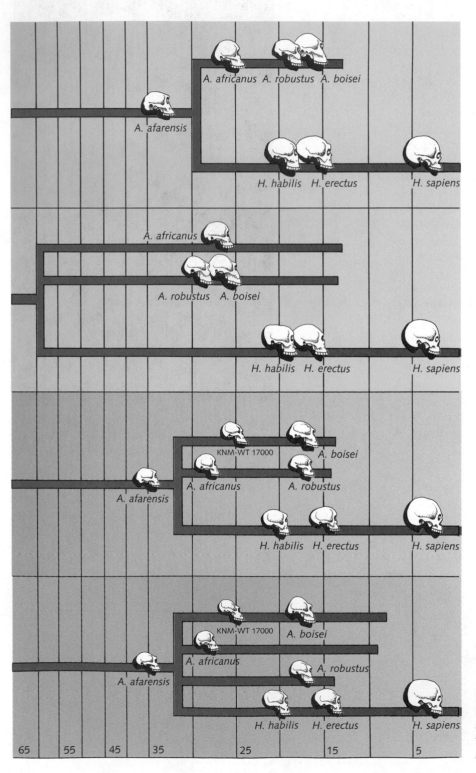

Figure 29.19 Four alternative models of early hominid evolution, all of which can be supported, and none of which can be conclusively disproved, through existing fossil evidence. Adapted from *Discover* magazine, September 1986.

carnivorous mammals with whom they shared the grasslands. South African paleontologist C. K. Brian has found holes in 2–3-million-year-old hominid skulls that match precisely with the bite of a leopard's canine teeth. By 1 million years ago, these sorts of injuries vanish, but it is not certain whether hominids had learned to hunt on their own or had simply grown more adept at avoiding other hunting species.

Evolutionary Trends in Hominids

Although their evolutionary interrelationships are uncertain, the early hominids exhibit clearly the evolutionary trends that separated them from the great apes:

- A continued *lack* of specialization in jaw and tooth structure, indicating an omnivorous diet. Studies of wear patterns in fossil teeth show that both meat and vegetables were consumed.
- The continual modification of lower spine, pelvis, and leg bones to permit more efficient bipedal walking.
- A phenomenal increase in brain size, from 280–400 cm³ in chimpanzees to 440–530 cm³ in australopithecines, 775–1100 cm³ in *Homo erectus,* and 1300–1400 cm³ in modern *Homo sapiens.*

Of all these changes, the extraordinarily rapid increase in brain size is the most difficult to explain. Certain scholars believe that at some time in the history of the hominid line, mental development in the slowly enlarging hominid brain reached a crucial milestone and made possible the development of a primitive culture. At that point, mental ability and social organization themselves became influential factors in hominid life, fueling ever-more-rapid evolutionary increases in mental abilities. This is the process that Harvard biologist Edward O. Wilson calls *gene–culture coevolution.*

Genus *Homo*

Around 1.9 million years ago, the first member of our own genus appeared in the form of *Homo habilis* (*habilis* means "able"). Many (though not all) authorities believe that *H. habilis* was the first hominid to use stone tools (Fig. 29.20). Despite this ability and despite a broad range covering much of eastern and southern Africa, this species lasted only about 500,000 years, a short span compared with the life of the hominid line.

Judging by the fossil record, *Homo habilis* was replaced around 1–1.5 million years ago by *H. erectus.* This more modern-looking, more intelligent species not only used tools but also regularly employed fire. *H. erectus* was also a great traveler: the species migrated far from Africa into Europe and across India into Asia. Two of the first fossil hominids ever discovered in Asia, "Java man" and "Peking man" (now "Beijing man"), belong to this species.

Figure 29.20 *Homo habilis* may have been the first hominid to use stone tools. These flaked stone tools are artifacts of Neanderthal culture found from southern Africa through western Europe.

We know a great deal about this species because of a nearly complete specimen of a young *H. erectus* male found in Kenya by a team of British anthropologists led by Alan Walker and Richard Leakey (Fig. 29.21). Contrary to previous assumptions that all early hominids were small, this skeleton shows that *H. erectus* could have stood nearly 2 m tall. At 880 cm³, his brain was significantly larger than that of australopithecines, but his spinal cord was so much smaller that some researchers feel it would have seriously limited neuromuscular control and coordination.

This information is consistent with what we know about *H. erectus* culture. Members of the species apparently lived mostly by gathering food and scavenging meat from carcasses left by larger carnivores, although they may have done some hunting on their own. They quickly developed the hand axe but never went any further in tool development, even though they survived for more than a million years.

The first *Homo sapiens* Just when, where, and how fully modern humans—usually referred to as the subspecies *Homo sapiens sapiens*—originated is still uncertain. Among the questions that remain are the number and extent of migrations made by modern humans, and the relationship those wandering populations had with scattered groups of *Homo erectus* and earlier populations of *Homo sapiens.* Two opposing camps are still bitterly locked in

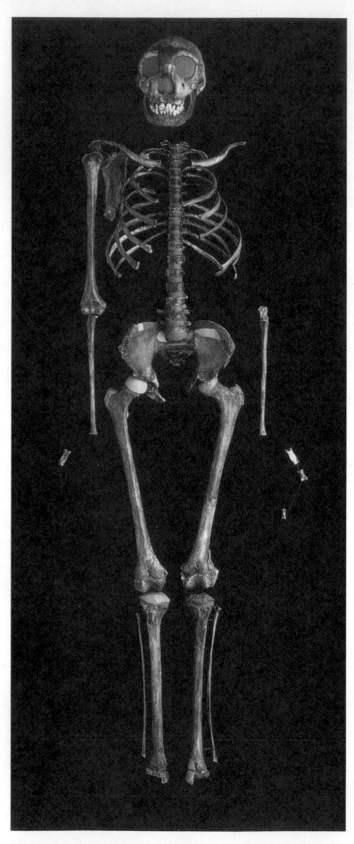

Figure 29.21 This nearly complete specimen of a young *Homo erectus*, often called the Nariokotome boy, shows that his species could have stood nearly 2 m tall.

intense (and often nasty) debate. Because it would be difficult and confusing to lead you through two or three conflicting hypothetical scenarios at once, we will stick to one "party line" here and confine the debate to Current Controversies, Molecules, Mandibles, Modern "Man," and "the Mother of Us All," p. 590.

The earliest fossils of our own species suggest that *Homo sapiens* originated in or near Africa about 500,000 years ago and spread to Europe around 70,000 years ago. Because the first of these European fossils was found in Germany's Neander Valley, they were called Neanderthals. They are now generally recognized as a subspecies of *Homo sapiens: Homo sapiens neanderthalensis*. *Neanderthals* were mentally similar to modern humans, though by modern standards they looked rather brutish, for their skulls carried massive, bony ridges that jutted out above their eyes. Neanderthals were heavyset and muscular, though no more so than certain modern-day weight lifters.

Despite their physically primitive appearance, Neanderthals had clearly begun to develop a sophisticated and multidimensional culture. They buried their dead provisioned with weapons, food, and flowers, a behavior usually taken to signal belief in life after death. Neanderthal skeletons showing extensive healing of serious wounds give evidence that these early humans cared for their sick and injured. Whether they did so because they relied on one another for survival or because they were emotionally attached to each other is not clear.

Even so, many authorities believe that Neanderthals, like several australopithecines before them, were a 300,000-year evolutionary dead end. They were replaced by a wave of modern humans that swept out of Africa into Europe and Asia in the Neanderthals' footsteps.

Cro-Magnons The first skeletally modern humans— *Homo sapiens sapiens*, or **Cro-Magnons**—seem to have appeared in Africa about 100,000 years ago. What were apparently small bands of Cro-Magnon wanderers headed into Europe and Asia, where they quickly replaced the resident Neanderthals about 35,000 years ago. Some students of the human molecular clock, in fact, believe that the entire human race descended from no more than a handful of Cro-Magnons. These scholars use their data to remind us how closely related all humans are today.

The Cro-Magnons were physically weaker than Neanderthals but had a much more advanced culture. Whether they interbred with the Neanderthals or wiped them out cannot be determined from the fossil record. The Cro-Magnons made a wide variety of tools out of stone, bone, and ivory. The paintings they left behind in caves are remarkable for the use of color, form, and texture (Fig. 29.22). These paintings, together with what look like ritual implements, suggest sophisticated religious practices related to hunting. The Cro-Magnons were skilled hunters who survived the great ice ages of

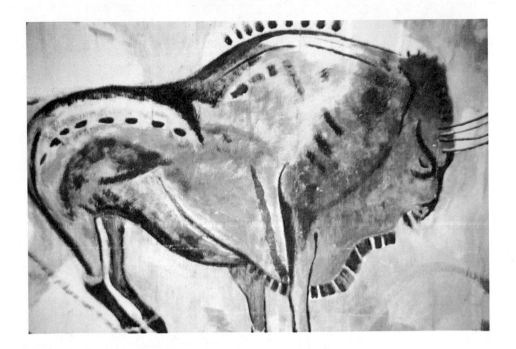

Figure 29.22 The Cro-Magnon culture left behind many cave paintings, such as this image of a bison. These paintings are interpreted as artifacts used in religious rituals aimed at appeasing animal spirits and ensuring success in dangerous hunts for large game.

the late Cenozoic era by feeding on game as large as 3-ton ground sloths, mammoths, mastodons, and woolly rhinoceroses.

Most of this early human history occurred in the Old World. Although Cro-Magnon people spread rapidly throughout Europe and Asia, there were none in the Americas until about 12,000 years ago. At that time, an ice-free period gave travelers the opportunity to cross over a land bridge from Siberia through western Canada southward into the American plains. These ancestors of the American Indians were called the Clovis people, because their stone tools were first discovered near the town of Clovis, New Mexico.

Hunting with distinctive stone blades and building homes of mammoth skin and bones, the Clovis people quickly spread across the Americas. Within 1000 years they had colonized North America and had expanded from Canada all the way to Tierra del Fuego.

Just at the time the Clovis people were expanding their range, 75–80 percent of the large mammals in the New World abruptly disappeared. Many biologists believe that the skillful slaughter of large mammals by the Clovis people caused this wave of extinction about 11,000 years ago, though the "overkill" hypothesis cannot be conclusively proved. Opponents point out that modern "primitive" peoples such as the Australian aborigines are not particularly adept at hunting big game and argue that a change in climate was the main culprit in the extinction.

The theory's supporters point to fossils of bison preserved with spear tips between their ribs (Fig. 29.23) and to numerous bones at fossil campsites as proof that the Clovis people hunted large animals extensively at precisely the time of the great extinctions. If early humans did contribute to the demise of other mammals, it was the beginning of humanity's impact on the environment and the start of the largest mass extinction since the end of the Cretaceous period.

Figure 29.23 Fossils such as this one, showing a Clovis spear point between the ribs of a bison, are taken by some as evidence that early humans were responsible for the disappearance of many large mammals, even in prehistoric times.

Molecules, Mandibles, Modern "Man," and "the Mother of Us All"

There are few issues in the history of life that truly intrigue the general public, but the history of the peculiar species named *Homo sapiens* certainly does. Ever since Darwin's time, paleontological, archaeological, and anthropological finds concerning human evolution have been trumpeted in the press and have stirred dispute. Not surprisingly, recent efforts to apply molecular techniques to the human family tree have been both fascinating and controversial.

On the one hand, studies of specific allele frequencies in various groups confirm with uncanny precision certain movements of recent human populations that are documented in written history. In a letter to *Science*, for example, population geneticists Richard Lewontin and Daniel Hartl cite several examples: current frequencies of specific alleles among Spaniards bear witness to the Moorish conquest of 500 years ago; certain southern Italian populations—particularly those in Sicily—hold certain alleles in common with Greeks and North Africans; northern Italians, by contrast, exhibit German alleles that can be traced to the Lombard invasions.

All well and good. But can molecular studies tell us more about what we really need to know: events in prehistory for which we have (at best) myths or (at worst) no clue at all? The answer to that question (at least, for the present) depends largely on which members of the scientific community you speak to.

The last point of agreement is that *Homo erectus*, our immediate predecessor, appeared in Africa somewhere around a million years ago. Not too long after, early humans spread across much of Europe and Asia (see map at right). From there on, however, consensus rapidly breaks down.

One group of anthropologists, for example, builds an argument based on intensive analysis of the fossil record, including careful scrutiny of fossilized jawbones, or mandibles. These scientists contend that widespread populations of *H. erectus* across the Old World slowly evolved—in parallel with one another and with substantial interbreeding, but nonetheless independently—into modern *Homo sapiens* (see figure a). That is, populations of *H. erectus* that reached Asia—represented by remains found in Java and Beijing, among other places—were the founding fathers and mothers of indigenous Asian peoples. Other *H. erectus* populations that reached Europe, on the other hand, gave rise to Neanderthals, which in turn evolved into modern Europeans.

But an outspoken band of "molecular historians" disagrees. One group of those researchers builds its case on studies of mitochondrial DNA, which is passed strictly from mothers to their children without any genetic contribution from fathers. According to these researchers, their data, taken from 189 people of various races scattered around the globe, suggest that *H. sapiens sapiens* originated in Africa, remained there for some time, and then spread out across the Old World, somewhere between 150,000 and 200,000 years ago. Thus, they feel, this new stock of *H. sapiens* replaced preexisting populations of archaic humans, rather than interbreeding with them (see figure b).

They further claim to have traced the mitochondrial lineages of all living humans to a single female—unfortunately dubbed "mitochondrial Eve" or "the mother of us all"—who lived in Africa between 150,000 and 200,000 years ago (see figure c).

At this writing, several warring camps are still exchanging heavy intellectual artillery fire over these hypotheses. Molecular historians have won out over paleontologists in previous skirmishes (see text), but the outcome of this battle is still far from clear. Keep an eye on newspapers and science magazines for ongoing developments.

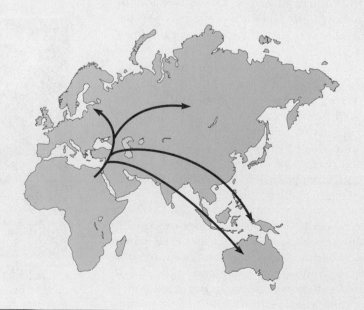

(a) One group of anthropologists argues that far-flung local populations of *Homo erectus* evolved slowly and independently into modern humans—while engaging in enough interbreeding with other groups to maintain our identity as a single species.

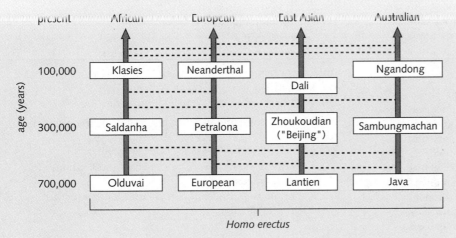

(b) The main point of molecular researchers is that *H. sapiens sapiens* originated in Africa rather recently and supplanted preexisting groups completely.

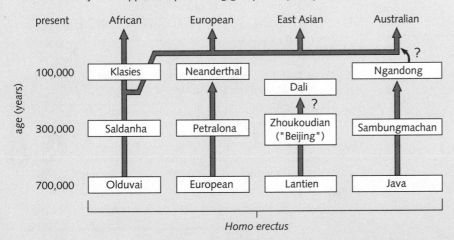

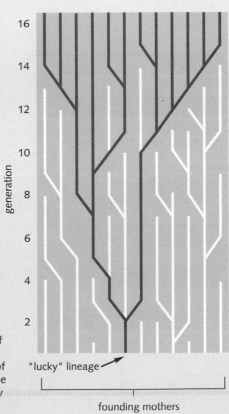

(c) This hypothetical diagram traces humankind to a single mito-chondrial lineage but does *not* suggest that a single female be-gat our entire species. As the diagram shows, mitochondrial lineages can disappear by chance, much like the family name of a father who has no sons disappears in his lineage, whether or not he has any daughters. Similarly, the mitochondrial lineage of a mother disappears if she has no daughters, whether or not she has any sons. Thus a single mitochondrial lineage can ultimately spread throughout a population.

SUMMARY

The bony fishes split into two groups, ray-finned fishes (most modern fishes) and fleshy-finned fishes (lungfishes and coelacanths). Ancestors of the fleshy-finned stock also gave rise to terrestrial tetrapods, four-limbed amphibians that emerged onto land. Reptiles and birds both display amniotic eggs that enable true terrestrial reproduction.

Mammals are endothermic organisms that have four-chambered hearts, bear live young, and nourish newborns with mother's milk from mammary glands. Three separate mammalian lines evolved: (1) Prototherians suckle their young but lay eggs like reptiles. (2) Marsupials give birth to tiny live young that continue development attached to a nipple in the mother's pouch. (3) Eutherians develop in the womb with the aid of a placenta; many can walk within hours of birth.

Primates, the order that includes *Homo sapiens,* is characterized by mobile digits, flexible appendages, binocular vision, trends toward fewer offspring per pregnancy, progressively larger brains, and complex social systems. The primate line is divided into three groups: two prosimian groups (tarsiers and lemurs) and the anthropoids (higher primates). Hominoids include humans, extinct human ancestors, and living apes, such as gibbons, orangutans, and chimpanzees. The first human ancestors to approach the cultural complexity of modern humans were the Neanderthals, members of a *Homo sapiens* subspecies. The Cro-Magnons, *Homo sapiens sapiens,* were the first members of our own subspecies and had a culture sufficiently advanced to include religious rituals.

STUDY FOCUS

After studying this chapter, you should be able to:

- Describe the major classes of fishes, amphibians, and reptiles, and explain the emergence of mammals.

- Discuss the mammalian adaptive radiation that took place during the Cenozoic era.

- Trace early primate history and the evolution of *Homo sapiens.*

TERMS AND CONCEPTS

cartilaginous fishes *573*	placental *581*
bony fishes *575*	primates *582*
amniotic egg *576*	prosimians *582*
mammals *580*	anthropoids *582*
prototherians *581*	hominoids *583*
metatherians *581*	Cro-Magnons *588*
marsupials *581*	
eutherian *581*	

REVIEW

Objective Questions (Answers in Appendix)

1. Mammals share all of the following characteristics except that they
   - (a) are exothermic.
   - (b) have body hair.
   - (c) bear live young.
   - (d) nurse their young.

2. The Cenozoic era began approximately _____ million years ago.
   - (a) 5
   - (b) 20
   - (c) 30
   - (d) 65

3. Primate evolution began
   - (a) on the savannah.
   - (b) near the seacoasts.
   - (c) in the forest.
   - (d) on the desert.

4. An early primate might have resembled today's
   - (a) kangaroo.
   - (b) tree shrew.
   - (c) platypus.
   - (d) chimpanzee.

5. Early hominids exhibited all of the following evolutionary trends except
   - (a) an increase in brain size.
   - (b) a modification of lower spine, pelvis, and leg bones for efficient bipedal walking.
   - (c) a continued specialization of jaw and tooth structure.
   - (d) highly mobile digits.

Discussion Questions

6. Why is the skeleton "Lucy" considered a transitional organism between apes and humans?

7. About 11,000 years ago, 75–80 percent of the large mammals in the New World disappeared. What are two possible explanations for this extinction?

8. What evidence leads scientists to believe that Cro-Magnons engaged in religious practices?

9. How is the evolution of modern mammals tied to the history of flowering plants?

10. The duck-billed platypus exhibits the reptilian trait of laying eggs. Is the duck-billed platypus a reptile or a mammal? What characteristics determine how it is classified?

READINGS

(Integrated list for Part 5)

Gould, Stephen Jay. *Eight Little Piggies* (1993), *Bully for Brontosaurus* (1991), *The Panda's Thumb* (1987), *The Flamingo's Smile* (1985). New York: Norton. If you never read any other essays on natural history, don't miss these collections of Gould's entertaining and scholarly essays.

Wilson, Edward O. *The Diversity of Life*. Cambridge, MA: Belknap Press of Harvard University Press, 1992. A practical follow-up to the material covered in this section. Argues the importance of diversity and the need to understand and preserve all species.

Gittleman, J. L., and S. L. Pimm. "Crying wolf in North America." *Nature* 351 (1991): 524–525.

Wayne, R. K., and S. M. Jenks. "Mitochondrial DNA analysis implying extensive hybridization of the endangered red wolf *Canis rufus*." *Nature* 351 (1991): 565–568.

Gould, Stephen Jay. *Wonderful Life: The Burgess Shale and the Nature of History*. New York: Norton, 1989. The Burgess shale is a Canadian rock formation that contains some of the oldest and best-preserved animal fossils on the planet. The discoverer of these fossils originally tried to place them into existing animal phyla, often unsuccessfully. More recent research suggests that the Burgess shale contains a record of evolutionary novelty outside of existing phyla, and Gould makes the most out of this, drawing some fundamental conclusions about contingency and the nature of history. The title of the book is drawn from the Jimmy Stewart movie *It's a Wonderful Life*.

Eldredge, Niles. *Life Pulse*. New York: Facts on File, 1987. A sprightly, well-documented presentation of modern evolutionary theory and the process of science in evolutionary biology.

Meyers, Norman. *A Wealth of Wild Species: Storehouse for Human Welfare*. Boulder, CO: Westview Press, 1983. Wild species are worth preserving for their own sake, but Meyers makes a compelling argument that our planet's biological resources should be saved for our own self-interest as well.

Wegener, A. *The Origin of the Continents and Oceans*, 4th German ed., 1929. Translated by John Biran. Dover Publications: New York, 1966.

PART 6

Plant Systems

Plants, like these towering poplars in North Carolina, dominate much of the earth's landscape.

*f*rom the towering trees of temperate forests to the gentle blades of grass that color the rolling prairie, plants have successfully colonized almost every corner of the planet. In our effort to study and understand living things, it is all too easy to assign plants a passive role, relegating them to the background with respect to the ongoing drama of animal life. As animals, we tend to equate life with movement. The fact that plants are often stationary has led many people to the unfortunate conclusion that plants are less than alive. The great irony of this view is that the environment of Earth has been modified by plants and their relatives in a way that has made it hospitable to other organisms, including ourselves.

The atmosphere, one of Earth's most precious resources, is the direct result of the process of photosynthesis carried out by plants and their relatives. Plants were essential in forming the biosphere, and they continue to be needed for its maintenance and good health. The central position of plants in the biosphere is reason enough to be curious about plant biology. The utility of plants as providers of food in our daily lives is another. But there is a third, more subtle reason why we should be interested in plants: they help to expand our understanding of the possibilities of living organisms.

With few exceptions, plants do not gather food, move about, or struggle directly with predators. Despite these differences from animals, plants are successful. Arguably, they are *the* most successful form of life on Earth. Part of this success arises from the ability to convert sunlight into chemical energy. But the abilities of plants to respond to challenges, to reproduce

under adverse conditions, and to respond to biological and meteorological adversity should prove a powerful reminder that there is more than one road to evolutionary success.

In Part 6, we explore some of the adaptations that have made plants so different from other organisms and yet so successful. Chapter 30 begins with a general analysis of

> *In the dooryard fronting an old farm-house near the white-wash'd palings, / Stands the lilac-bush tall-growing with heart-shaped leaves of rich green, / With many a pointed blossom rising delicate, and with perfume strong I love, / With every leaf a miracle—and from this bush in the dooryard, / With delicate-color'd blossoms and heart-shaped leaves of rich green, / A sprig with its flower I break.*
>
> —Walt Whitman, 1865

plant structure and function, while Chapter 31 specifically covers reproductive strategies among plants. Chapter 32 explains how plants obtain nutrients, and Chapter 33 discusses the mechanisms through which plants respond to their environment.

30

Plant Structure and Function

*i*magine that you are asked to write a science fiction novel about the future earth. If you take the course of evolution into account, you might imagine a world in which many new kinds of organisms have evolved and many have become extinct. You might imagine an earth in which reptiles, or birds, or mammals have become extinct. You might even imagine the extinction of human beings. But a world without plants is almost beyond comprehension. Plants are the basis of nearly all life on the land. They provide the material that enters the terrestrial food chain at the bottom, making it possible for so many other forms of life to live on earth.

Plants depend on the sun for energy. Their need for sunlight is so great that it comes to dominate the structure, the life cycle, and even the shape of plants. In this chapter, we will explore the basic structure and reproductive patterns of plants, and we will examine some of the ways in which plants reconcile the demands of their environment with the need for sunlight (Fig. 30.1).

WHAT IS A PLANT?

We will use the common term *plants* to refer to the members of the kingdom **Plantae** and to all algae, whether they are classified as Plantae or Protista. These organisms share a number of common features (Fig. 30.2). The kingdom Plantae is composed of many divisions of plants that differ from each other in the ways in which they have met environmental challenges. (Plant taxonomists, for historical reasons, use the term *division* rather than *phylum* when they classify plants. For our purposes, however, *division* and *phylum* are interchangeable.) Although we will not examine each division in the kingdom, we can explore some of the ways in which plants are organized by examining representative members of the most important groups.

In earlier chapters, we have considered the evolution of plants, and we have seen the vital roles that plants of all types play in the earth's

Figure 30.1 Plants dominate the landscape of the Monteverde cloud forest in Costa Rica, employing a wide variety of survival strategies.

many ecosystems. In this chapter, we will begin to examine the organisms themselves and to explore the ways in which plants have adapted to the many environments on the earth.

We discussed the process of photosynthesis in some detail in Chapter 19, but we should also consider the fact that photosynthesis exerts a powerful influence on plant adaptation and evolution. Their dependence on photosynthesis for food and energy means that plants must have access to water and carbon dioxide, and it also requires that as many cells of the plant as possible be directly exposed to sunlight.

The Need for Sunlight

Whether they are found on land or in the water, plants require sunlight to provide the energy needed for photosynthesis. Every species of plant displays some adaptation shaped by the need to ensure a steady supply of solar energy in its particular environment. Plants living underwater are found close to the surface where the light is most intense. Many algae contain organelles that sense the presence of light and direct the cell to swim toward it. Larger underwater plants, including many forms of algae known as seaweeds, produce gas-filled sacs that enable them to float near the surface. Those plants living deeper in the water have special accessory pigments that help them harvest those wavelengths of sunlight that penetrate more deeply.

Figure 30.2 At the cellular level, the common features of plants include cells with nuclei, cell walls, and the green photosynthetic pigments chlorophylls a and b.

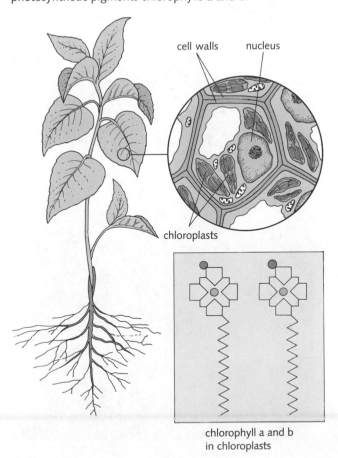

cell walls   nucleus

chloroplasts

chlorophyll a and b
in chloroplasts

Figure 30.3 The three principal needs of plants are illustrated in the adaptations of these three species. TOP: Some water plants, like the feather boa kelp, stay near the surface of the water with the aid of gas-filled sacs, ensuring that they receive plenty of sunlight. CENTER: The airfoil-like wings of these maple tree fruit increase the chances for reproductive success by spreading their seeds over great distances. BOTTOM: The thick, stubby stems and fruits of *Opuntia englemanii* help to conserve water in the Utah desert.

Plants on the land have body shapes that maximize the amount of light they are able to absorb. The thin, rigid stalks and broad leaves of land plants enable them to gather light efficiently. Trees, the largest plants, grow towards the sky where they have unimpaired access to enormous amounts of sunlight. Plants growing in the shade of these giants increase the amount of light-capturing chlorophyll in their chloroplasts, and they gradually change the orientation of their leaves during the day. This maximizes the amount of light they are able to absorb.

The Need for Water

Plants, like all living things, need water. For the plants that live in water, this problem solves itself, and all plants living on the land have evolved ways to ensure a steady supply of water. Water is the means by which plants absorb mineral nutrients from the soil; plants also require water for internal transport. Plants' need for water is similar to that of animals, with two important differences: For plants, water is one of the raw materials of photosynthesis, so it is used up at a high rate during carbohydrate production. Second, keeping up a steady supply of water is complicated by the need to absorb light energy; sunny places are also places where living tissues can dry out. This presents the plant with an important challenge: how to absorb as much light as possible *and* avoid severe water loss.

The Need to Reproduce

All organisms must ensure that their offspring have a reasonable chance to survive and begin a new generation. Plants, however, face special challenges (Fig. 30.3). Plants do not have nervous systems, and they are not able to run away from predators or pests. Because nearly all plants live in fixed positions, they must also manage to find mates without being able to move around. Therefore, they have evolved strategies for dealing with these prob-

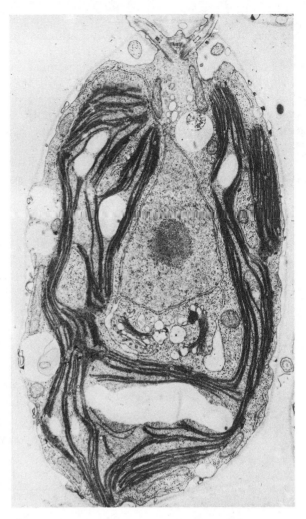

Figure 30.4 *Chlamydomonas,* one of the most common of the freshwater algae.

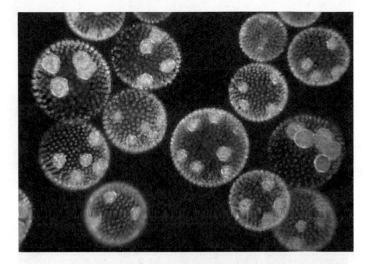

Figure 30.5 TOP: *Volvox,* a colonial organism closely related to *Chlamydomonas,* which displays a limited degree of cellular specialization. Small daughter colonies are seen within each mature colony. (Magnification factor: 12) BOTTOM: *Ulva,* a multicellular alga that displays even greater specialization than *Volvox.*

lems that are essentially passive. An important part of such strategies is a reproductive pattern enabling each individual to produce large numbers of offspring.

PLANT STRUCTURE: FROM WATER TO LAND

Plants Restricted to Wet Environments

The simplest plants are the **algae.** Most algae are unicellular like the typical green alga *Chlamydomonas,* shown in Fig. 30.4. *Chlamydomonas* is a photosynthetic autotroph, and each cell contains a single cup-shaped chloroplast. There are two flagella at one end of the cell, enabling it to swim rapidly through the freshwater ponds and lakes where most species of *Chlamydomonas* live. *Chlamydomonas*

is surrounded by a tough cell wall. Under favorable conditions, *Chlamydomonas* cells can grow rapidly by mitosis; like many algae, they are also capable of sexual reproduction, as we will see in the next chapter.

Other green algae, such as *Volvox,* have individual cells that are similar to *Chlamydomonas.* However, in *Volvox* these cells are associated with each other to form a more complex organism in which a degree of cell specialization develops. More complex algae include *Ulva,* the "sea lettuce." (See Fig. 30.5 to compare *Volvox* and *Ulva.*) These plants, which are found along rocky seacoasts, contain a number of specialized cell types, including *holdfast cells* that help attach the plant firmly to the rocks. Many biologists believe that the presence of specialized cells in these organisms suggests that they are similar to the first organisms to evolve towards an exclusively multicellular lifestyle.

Figure 30.6 Mosses (TOP) and liverworts (BOTTOM) are bryophytes. The lack of specialized vascular tissue confines these plants to damp environments.

In a way, the success of *Ulva* in the absence of such specialized tissues points out one of the strengths of the algae. Because algae are small and simple, nearly every cell of an alga is in direct contact with the environment. Therefore, algae have no need to develop specialized tissues to control the flow of water and nutrients. Most algae live in direct contact with water, that means that their exchange of water and other nutrients with the environment occurs on a cell-by-cell basis.

The simplicity of the algae also places severe limits on the places where they are able to live. Although some

species (including a few representatives of *Chlamydomonas*) are able to live in the soil, and others (including the algae that associate with fungi to form lichens) are able to live on dry land, for the most part algae must go where the water is. Terrestrial habitats are beyond their reach. Those algae that are able to survive on land do so only in very special environments where the supply of water is ensured. The other types of plants differ from the algae in displaying adaptations that have enabled them to develop a degree of independence from the water. The greater that independence, the more firmly they are committed to the terrestrial lifestyle.

Plant Life on Land

Plants living on land face a completely different set of problems from those that live in saltwater or fresh water. The problems that confront plants living on land include the following:

- *The lack of buoyant support.* Whereas water supports aquatic plants, the lower density of air requires supporting tissues for all but the smallest land plants.

- *The tendency of land plants to desiccate, or dry out, owing to water loss from evaporation.* Plants exposed to the air face the possibility of disastrous water loss from exposed tissue.

- *Rapid changes in temperature and humidity.* Plants on land face a daily environment that is much less stable than that which water plants face.

- *The extraction of nutrients.* The nutrients used by water plants are taken directly from the external environment, but land plants must acquire nutrients that are present in the soil in solid form.

- *The scarcity of water for reproduction.* Gametes must be exchanged in an environment that often lacks the water that exposed reproductive cells might need to survive.

Land plants have adapted to these problems by evolving a greater degree of complexity in their cells and tissues. The evolutionary patterns of land plants show a clear trend towards establishing independence from water; the details of that evolution are obvious in the very first division of the Plantae, the Bryophytes.

Bryophytes, the mosses, liverworts, and hornworts, are true land plants and have structures very much like stems and leaves (Fig. 30.6). They are often covered with a waxy coating, or cuticle, that helps them retain water when the weather is dry, an important adaptation for life on land. Bryophytes are anchored to the soil by thin hair-like structures known as *rhizoids*. Rhizoids are not considered "true roots" because **true roots** contain **vascular tissue,** which is specialized for carrying water. The lack of vascular tissue is one of the most important characteristics separating the bryophytes from "higher" plants.

Water is absorbed through thin cell walls in the rhizoids and then moves from cell to cell by means of osmosis. This cell-to-cell passage of water works well over short distances, but it cannot move water very far against the force of gravity, as vascular tissue can. This is one of the main reasons why these organisms are unable to grow very far from their sources of water—most bryophytes are relatively small. In addition, the sperm cells of bryophytes are propelled by flagella and must swim through water to reach the egg cells. Therefore, they are capable of sexual reproduction only when the areas in which they grow are very damp. Their inability to move water efficiently and their need for a moist environment in which to reproduce confines these organisms to places that are very moist. Bryophytes are found on damp forest floors and at the edges of brooks and streams.

The Vascular Plants

Vascular plants, which possess specialized water-conducting tissues, are known as **tracheophytes.** This term comes from specialized cells called **tracheids** that are key components of water-conducting tissue. The existence of specialized water-conducting tissue is just one of the specializations found in tracheophytes. Others include the development of *true roots,* containing vascular tissue, that gather water from the soil and pass it through conducting tissues bundled within the main stalk of the plant. The early tracheophytes formed the great forests of the ancient earth, and their ability to move water against the force of gravity enabled them to grow several meters tall. The tracheophytes, including club mosses, ferns, and seed-bearing plants, are true land plants that have decreased their dependence on the presence of water (Fig. 30.7). This independence is not complete, for the reproductive stages of the life cycle of some members of the group still require the presence of water.

Gymnosperms and Angiosperms

The seed-bearing plants, a subdivision that includes the **Gymnosperms** and the **Anthophyta,** show the greatest degree of cellular specialization. All cone-bearing seed plants, including pine and fir trees, are gymnosperms. The Anthophyta, or flowering plants, are sometimes referred to as *Angiosperms,* a name derived from an older classification scheme. Because the angiosperms are the predominant form of plant life on land, we will concen-

Figure 30.7 Two large fronds of *Polypodium scolopendria,* a representative fern. The vascular tissue of ferns and other tracheophytes gives them a degree of independence from water. The structures on the undersides of the fronds are *sori,* clusters of spore-producing tissue.

trate our attention on the structure and function of the tissues of flowering plants. These plants have true roots, stems, and leaves, and they have developed an extensive network of conducting tissues that moves water, minerals, and nutrients throughout the organism.

The angiosperms themselves are divided into two classes on the basis of the number of *cotyledons,* or seed leaves, they possess. **Monocotyledons** (also known as **monocots**) contain a single seed leaf, and **dicotyledons (dicots)** contain two. Although this sounds like a trivial distinction, other structural differences between monocots and dicots carry over into the organization of stems, leaves, and flowers, as shown in Fig. 30.8.

seedlings	stems	leaves	flowers

Dicots

two seed leaves

vascular tissue arranged in a ring

veins usually netlike

floral parts in 4s or 5s

Monocots

single seed leaf

scattered vascular tissues

veins usually parallel

floral parts in 3s

Figure 30.8 Some of the principal structural differences between monocots and dicots.

Figure 30.9 Plants contain dermal tissue, vascular tissue, and ground tissue. Their general arrangement is illustrated in this diagram.

PLANT ORGANIZATION

The most abundant land plants are the angiosperms; more than 200,000 species are known. As different from each other as many of these species are, we can make some useful generalizations about their structure. The body of a plant contains distinct regions known as **roots, stems,** and **leaves.** The roots are typically anchored in soil, providing stability and acting as a conduit for water and mineral nutrients. The stems rise above the ground. They contain supporting tissue to keep the plant rigid as well as conducting tissue to carry water and nutrients from one end of the plant to the other. Leaves are the principal organs of photosynthesis and are commonly arranged in such a way that they are able to capture the maximum amount of sunlight.

Three tissue systems are found throughout the body of the plant (Fig. 30.9). **Dermal tissue** is the outer covering that protects the plant from the environment. The dermal layer is continuous from the roots to the leaves. Plants also contain a system of **vascular tissue** specialized for conducting fluid. The vascular system runs throughout the plant and enables its various parts to specialize in different functions. Finally, the vascular system itself is embedded in **ground tissue,** which provides strength and support and also contains most of the cells that are active in photosynthesis.

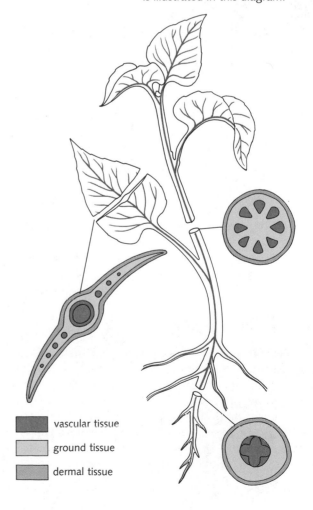

vascular tissue

ground tissue

dermal tissue

Plants and Medicine

A young leukemia patient resting peacefully in a cancer ward is thousands of miles away from the exotic island of Madagascar, which lies off the eastern coast of Africa. She and her hopeful parents, who have been told that the odds are good she will recover from the deadly cancer, may not know the location of that island, and they may not have heard of the Madagascar periwinkle (*Catharanthus rosea*), a plant with white and pink flowers native to the island. Nonetheless, the little girl may owe her life to this flower and to a compound extracted from it known as *vincristine*. Vincristine is a cellular poison that breaks down the mitotic spindle and makes it impossible for cells to complete cell division. Cancer cells, which divide rapidly, are particularly vulnerable to vincristine, and in the past decade or so the drug has become a powerful weapon in the medical arsenal with which patients are winning battle after battle in the war against childhood cancer.

Vincristine is just one example of a human dependence on plants that is much less obvious than our dependence on plants as sources of food. Plants are one of our most valuable sources of medicines, old and new. In many cases, well-known drugs are responsible for the effectiveness of folk remedies. Hindu folk medicine has made wide use of *Rauwolfia serpentina*, the snakeroot plant, as a treatment for circulatory and nervous disorders. From a careful analysis of the effects of this plant, researchers extracted *reserpine*, a drug often essential to the treatment of high blood pressure. The use of willow bark as a pain remedy is effective because the plant contains a close chemical relative of aspirin. The flowering foxglove contains *digitalis*, one of

the most effective heart medicines known. It is estimated that more than 25 percent of the drugs prescribed for use in the United States contain plant materials as their active ingredients, and that percentage would be much larger if synthetic chemicals first discovered in plants were included in the total.

The possibilities of plants in medicine are almost limitless. For years, biologists have collected rare plant species from around the world with the hope of finding new medicines. The search has been long and costly, but it has also been rewarding. Sometimes success in that search comes from the commonplace: the polyacetylene poisons in marigolds hold great promise as antitumor drugs. But just as frequently, it comes from the exotic, as with *curare*, a potent muscle relaxant derived from the plant poisons used by Amazon natives on their blowgun darts.

Ironically, just as we develop the biochemical techniques to make the most of these botanical gifts, we may be on the verge of destroying the world's greatest storehouses of plant chemicals. The tropical rain forests, including the Amazon, are vanishing at an astonishing rate. As one acre after another is cleared for the quick gain of timber or for cultivation, the wild forests become ever smaller and the chances for discovery of rare plant species diminish. Sadly, the land beneath the forest is of little use for farming. So much of its nutrient content is tied up in vegetation that once it has been cleared, the land will support farming for no more than a few years. Recent estimates are that more than half the land cleared for the Amazon forest has now been abandoned. The farms and ranches come and go, but once a plant species has been lost, its potential for medical application is gone forever.

Flowering foxglove (*Digitalis purpurea*) is the source of digitalis, a powerful medicine used to treat heart disease.

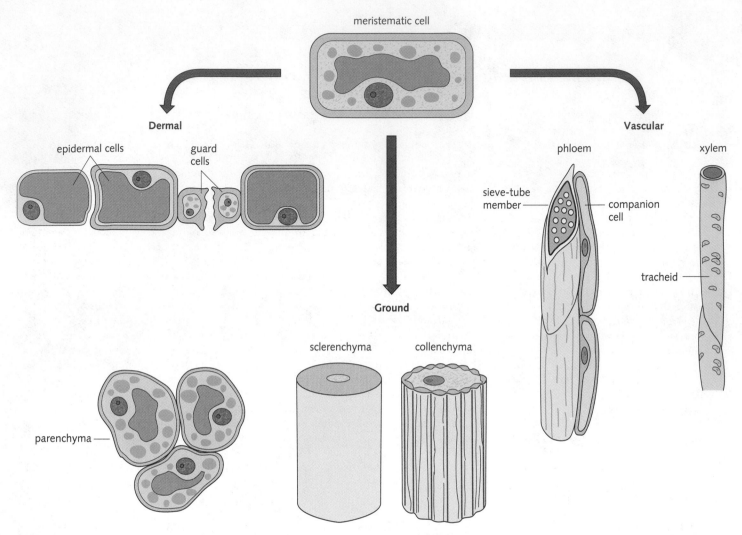

Figure 30.10 Meristematic cells develop into the cells that make up dermal, ground, and vascular tissue. Several important examples of cell types in each of these stylized drawings of tissues are shown.

Plant Cells: Growth and Development

Unlike animals, most plants continue to grow and increase in size throughout their lives. Even the ancient trees of the Pacific Northwest, some several thousands of years old, continue to put out new growth each year. How do they do it? The answers are found in **meristems,** regions of plant tissue in which cells remain "forever young," constantly growing and producing new tissue.

As we will see, apical meristems are found at the tips of roots and stems, where they are responsible for the **primary growth,** or elongation of the plant. Lateral meristems produce **secondary growth,** the thickening of stems, branches, and roots.

Growth in the meristems produces cells that make up the three tissue systems. In general, when a meristematic cell divides, one of the daughter cells remains part of the meristem, while the other moves into the body of the plant. The cells that leave the meristem may then continue to grow and divide as they develop into more specialized cells, a process known as **differentiation.**

Plant Cell Types

Each plant tissue contains a number of cell types, as shown in Fig. 30.10. Dermal tissue, for example, is remarkably diverse. **Epidermal cells** form much of the outer covering of the plant body, which in most plants is only one layer thick. This covering may also contain **guard cells,** which regulate the flow of gases in and out of leaves, and cells with tiny hair-like projections that provide protection to the plant.

The ground tissue of most plants consists primarily of **parenchyma cells,** which are thin-walled and often form

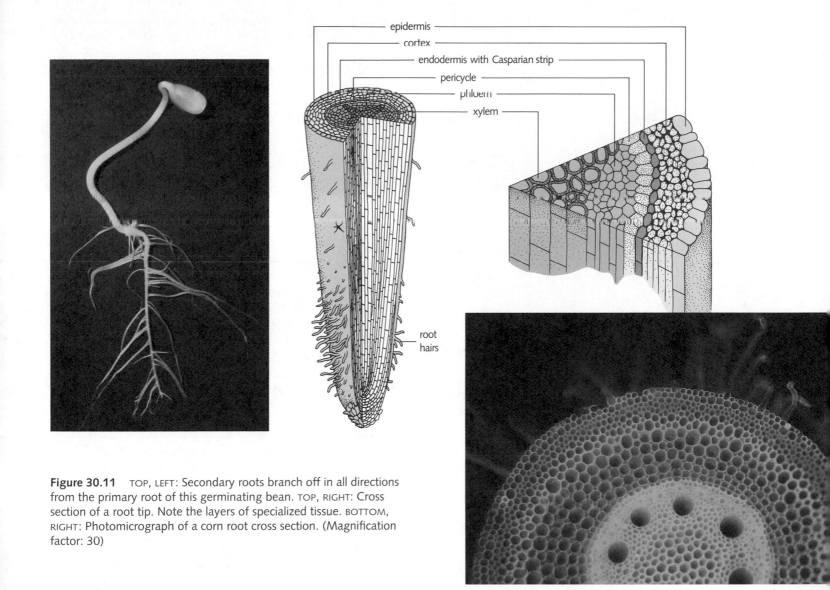

epidermis
cortex
endodermis with Casparian strip
pericycle
phloem
xylem

root hairs

Figure 30.11   TOP, LEFT: Secondary roots branch off in all directions from the primary root of this germinating bean. TOP, RIGHT: Cross section of a root tip. Note the layers of specialized tissue. BOTTOM, RIGHT: Photomicrograph of a corn root cross section. (Magnification factor: 30)

the large masses of tissue in stems, roots, leaves, and, sometimes, fruits. Parenchyma cells in leaves are very active in photosynthesis. The ground tissue may also contain **collenchyma cells** and **sclerenchyma cells,** which have thickened cell walls. These cells are important in strengthening the ground tissues and provide much of the support for the thin-walled parenchyma cells. Unlike collenchyma cells, sclerenchyma cells have hardened, wood-like cell walls and often are dead at maturity. Fine linen cloth is made mostly from sclerenchyma fibers of the flax plant.

The vascular tissue of plants performs the vital role of conducting fluids from one part of the organism to another. The principal vascular tissues are **xylem,** the water-conducting tissue, and **phloem,** the food-conducting tissue. A number of different cell types are associated with each of these tissues, as shown in Fig. 30.10.

THE ROOTS: STARTING FROM THE GROUND UP

Plants extend their roots into the soil, where these organs gather minerals and water and provide mechanical support for the portions of the plant above the ground. The best soils for plants are composed of loosely packed particles, and as much as 50 percent of total soil volume may consist of nothing more than water and air. Plant roots grow into these empty spaces in the soil and collect the water and dissolved minerals that accumulate there. When a seedling begins to grow, it extends a **primary root** into the soil, and numerous **secondary roots** branch off from the primary root as it grows (Fig. 30.11). The size of the root system can be astonishing: A classic study of a single 4-month-old rye plant showed that the total surface

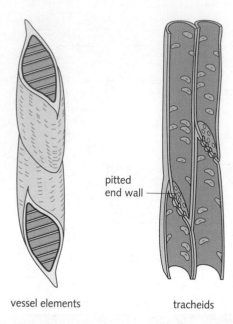

pitted
end wall

vessel elements   tracheids

Figure 30.12 The structural organization of xylem. Tracheids are the most basic conducting tissue. Water moves from the empty chamber of one tracheid to the next through *pits,* areas bounded by a thin primary cell wall. Like tracheids, vessel elements are hollow. However, the boundaries between adjacent vessel elements are generally perforated, forming a continuous tubule through which fluid moves easily.

area of its root system was more than 600 square meters, 130 times greater than the combined surface areas of stems and leaves.

Root Structure

Part of the outer surface of a root is covered with tiny projections called **root hairs,** which extend from the outer surface of the root and make direct contact with the soil. The water and dissolved minerals absorbed into the root are first taken in through the root hairs. The root hairs are part of the outermost layer of cells in the root, the **epidermis.** The epidermis surrounds a spongy layer of cells known as the *cortex,* and near the center of the root is another tissue layer called the *endodermis,* the innermost cells of the cortex (Fig. 30.11). The endodermis surrounds the vascular tissue of the root, and the walls of the endodermal cells contain a thin, waxy layer called the **Casparian strip,** which prevents water from moving between the cells of the layer. Therefore, to enter the vascular tissue, water must pass directly through these endodermal cells. This allows the endodermal cells to control the rate at which water enters the root.

Inside the endodermis is a cylinder of cells known as the **pericycle.** During rapid root growth, dividing cells from the pericycle are the sources of new tissue as roots send out lateral branches. Inside the pericycle are the two vascular tissues called xylem and phloem. These tissues are the major fluid-conducting systems in the plant—in effect, they are its circulatory system. The specialization of functions in a plant requires that the movement of material should occur in two directions: water and dissolved minerals must be brought up from the roots to the rest of the plant, and the products of photosynthesis (sugars and other nutrient molecules) must be transported from their production sites in the leaves to the rest of the plant. *Xylem* is the tissue that conducts water and minerals up from the roots; *phloem* circulates a nutrient-rich sap throughout the plant.

Because of osmosis, water is able to cross biological membranes to enter a cell whenever the concentration of dissolved material is greater within the cell than outside of it (in the terminology of osmosis, whenever the cell is *hypertonic* to its surroundings). In plants, this same situation occurs whenever the soil is damp. Osmosis causes a net movement from moist soil into root hair cells. This movement of water dilutes the cytoplasm of these cells, and water then continues to move by osmosis to adjacent cells in the root. This osmotic transfer of water from one cell to another in the cortex region of the root moves water into the roots and into the vascular system of the plant.

Xylem

Xylem tissue extends from the roots of a plant up through its stem into the leaves and branches. The tissue is composed of *tracheids* and **vessel elements.** Tracheids are a series of elongated cells that grow as tight bundles and then gradually die, leaving only the interconnected tracheid cell walls. Water can pass from one tracheid cell to another through *pits* that are separated by extremely thin cell walls. The hollow *vessels* are formed from individual *vessel elements* (Fig. 30.12). Vessel elements join end to end to form a continuous hollow tube through which water can pass easily. In smaller plants, osmotic pressure from the roots is sufficient to force water up through this network of hollow tracheids and vessels. You can see this for yourself if you snip off the end of a healthy blade of grass and watch fluid accumulate on the cut end of the blade.

In larger plants, however, osmotic pressure is not enough to overcome gravity and force water all the way from the ground to the upper branches. There are two other forces in operation that combine to do the job. The first of these is known as **capillary action.** Water molecules have a strong *cohesive* character, meaning that they are attracted to each other (by hydrogen bonding), and water molecules can also be attracted to other molecules, a

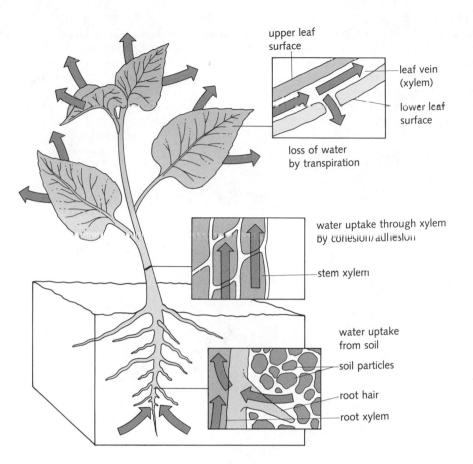

upper leaf surface

leaf vein (xylem)

lower leaf surface

loss of water by transpiration

water uptake through xylem by cohesion/adhesion

stem xylem

water uptake from soil

soil particles

root hair

root xylem

Figure 30.13 The consumption of water in the leaves, by transpiration and photosynthesis, produces an upward tension or "pull" that ultimately moves water from the roots into the upper regions of the plant. As noted, the forces of *cohesion, adhesion,* and *tension* draw water upwards through the vascular system.

process called *adhesion.* The combination of cohesion and adhesion helps water to rise in a small tube-like opening. The smaller the diameter of the tube, the greater the capillary action and the higher the level to which water can rise.

A second, and more important, force that lifts the water molecules in larger plants is generated by the plant's own use of water. A great deal of water is used up by photosynthesis in leaf cells, and still more water is lost from these cells by evaporation from cell surfaces into air spaces within the leaf. The evaporation of water from leaves, a process known as **transpiration,** can release as much as 100 liters of water a day from a large tree.

When water is lost in this way, the water content of the cells surrounding the vascular tissue drops, and osmotic pressure moves water out of the vascular tissue into the leaf. Just like a locomotive pulling a train hundreds of cars long, the movement of water in the leaves then "pulls" water upwards through the vascular system all the way from the roots (Fig. 30.13). How can this possibly work? The key can be found in the combined forces of cohesion, adhesion, and tension.

The "pull" or *tension* generated in the leaves applies a force to water in the vascular system. The strong *cohesion*

of water molecules transmits this force throughout the vascular system. When these forces are combined with *adhesion,* the attraction of water for the walls of the xylem vessels, enough combined force is produced to lift water even to the tops of the tallest trees.

Phloem

Phloem tissue is similar to xylem in that it forms a continuous network reaching from the roots to the leaves of the plant. However, unlike xylem, phloem is composed of living cells, and it is the principal food-conducting tissue of the plant. Phloem tissue is made up of **sieve tubes,** and individual phloem cells are called *sieve tube members* (Fig. 30.14). Unlike xylem cells, the individual cells within the phloem tissue remain alive, and each sieve tube contains a thin layer of cytoplasm. As they mature, sieve tube members lose ribosomes and a functional nucleus. The roles these organelles play in other cells may be taken over by *companion cells,* which lie next to sieve tube members and are connected to them by *plasmodesmata,* small passageways that permit molecules to flow from the cytoplasm of one cell to that of the next.

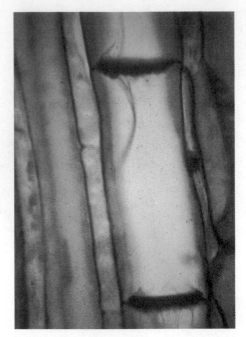

Figure 30.14 LEFT: Photomicrograph of a sieve tube member. Nutrients move from cell to cell through the sieve plates which separate phloem members. BELOW: Companion cells are adjacent to sieve tube members and perform some of the metabolic tasks required to keep the sieve tube members alive (DIAGRAM AT LEFT). Phloem vessels can be collected for scientific study by exploiting an insect pest: aphids are allowed to attach to a stem and begin collecting phloem sap. Then their bodies are sliced off with a razor blade, leaving the mouth parts inserted into the plant. Drops of sap are then collected from the feeding tube.

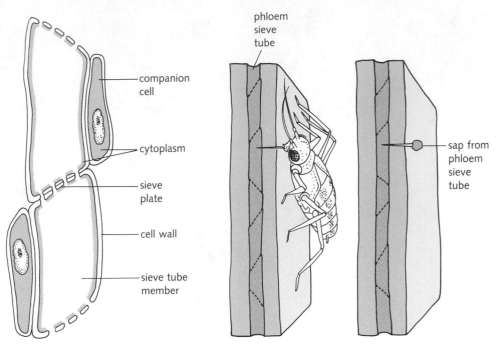

companion cell

cytoplasm

sieve plate

cell wall

sieve tube member

phloem sieve tube

sap from phloem sieve tube

Phloem sap, the material that flows through these interconnected sieve tubes, has a completely different composition from xylem sap. Phloem sap is filled with sugars and other nutrients. Radioactive tracer experiments show that the sugars produced in photosynthesis quickly show up in the phloem sap but not in the xylem sap. These and other experiments suggest that phloem sap is used to move the products of photosynthesis throughout the plant.

The phloem system can move material very quickly in both directions, but this process, known as *translocation,* requires the input of energy. There are a number of theories about what force might be behind the movement of material through the phloem system. One suggestion, known as the **pressure–flow hypothesis,** relies on a combination of active transport and osmotic pressure. If sugars are actively transported into the sap at one location, osmotic pressure produces a flow of water that follows these sugars. If sugar is then actively absorbed at another location (the roots, for example), the active uptake of the sugar into the roots again causes water to follow it. The

movement of sugar and water produces a steady flow between tissues that produce sugar and tissues that consume sugar, as shown in Fig. 30.15.

Xylem and Phloem?

Why do plants need two different systems, one for water (xylem) and another for nutrients (phloem)? The reasons are simpler than you might think. A typical plant picks up water in the soil through its roots and produces nutrients from photosynthesis in its leaves. To supply water, the plant must have a vascular system capable of moving massive amounts of water up from the roots. Could the same system be used to transport nutrients?

Probably not. The water transport system in xylem is moving in the wrong direction, away from the roots and towards the leaves. Nutrients generally need to be moved in the *opposite* direction and might never reach cells in the roots unless there were a second system capable of bringing them downward from the leaves. Phloem is that sys-

Figure 30.15 The pressure–flow hypothesis for the transport of nutrients in phloem. **(a)** In the upper part of the plant (near the leaves), sugars and other nutrients are actively pumped into the phloem. As a result of increased sugar concentration in this region, water moves by osmosis into the upper regions of the phloem system. In the lower part of the plant (the roots), sugar is pumped out of the phloem for use by neighboring cells. Once again, osmosis causes water to move, but this time it moves out of the phloem and into the root cells. **(b)** Because water moves into the phloem in the leaves and out in the roots, a net flow is produced, running from the tissue that produces nutrients to the tissues that consume them.

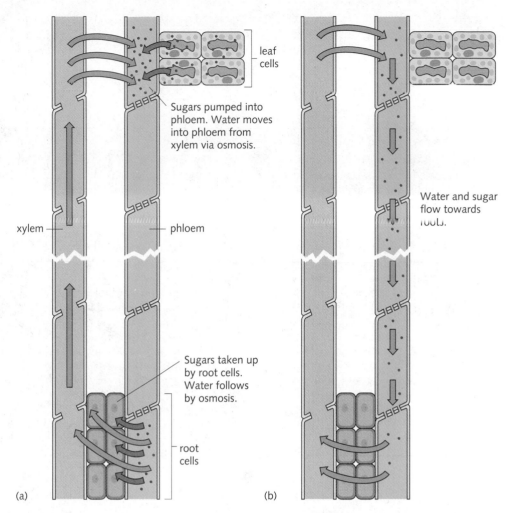

leaf cells

Sugars pumped into phloem. Water moves into phloem from xylem via osmosis.

Water and sugar flow towards roots.

xylem — — phloem

Sugars taken up by root cells. Water follows by osmosis.

root cells

(a)   (b)

tem, and Fig. 30.15 illustrates how it works in concert with xylem to transport nutrients where they are needed, even in a direction opposite the major flow of water.

Water Movement into the Roots

If plants had to depend on the presence of enough water around their roots to acquire water by osmosis, they would be able to thrive only in wet, muddy soil. However, the cells of the root cortex help the osmotic process along. These cells pump minerals into the root across their own cell membranes by *active transport*. Mineral ions are pumped into the root cortex cells by protein pumps in the cell membrane that use energy in the from of ATP. As the cells of the root cortex accumulate dissolved nutrients, the concentration of dissolved material within the cells increases. This increase eventually reaches the point at which the concentration of dissolved material inside the cell is greater than its concentration in the soil, producing an osmotic gradient that causes water to flow into the roots. The flow of water follows the active transport of minerals by the cortex.

After water passes through the root cortex, it moves into the cells of the endodermis, which form a cylinder that encloses the water-transporting vascular tissues of the root. Active transport also takes place in the cells of the endodermal layer. These cells pump minerals into the central region of the root surrounding the vascular layer. As we saw earlier, because the Casparian strip prevents water and minerals from leaking back into the cortical region, the water is "trapped" inside the central region of the root, a zone of water-transporting tissue known as the **vascular cylinder.** No water can leave or enter the vascular layer without passing through the endodermal cells. Hence, these cells can regulate the movement of water and dissolved minerals into the vascular tissues of the root. The continuing osmotic pressure

Frozen in Time

The thick, nutrient-rich phloem sap found in the vascular tissues of plants is their lifeblood. Phloem carries nutrients throughout the organism, and interrupting the flow of sap is one of the surest ways to injure an otherwise healthy plant. As biologists, we can be grateful for another property of the thick, resin-laden sap from certain trees: the fact that it occasionally leaks from the plant and over time can be transformed into a mineralized, gem-like material known as *amber*. The oils and resins found in the sap of pine trees make it a particularly rich source of amber. As droplets of sap dry and gradually oxidize, they acquire the hard texture and yellow–brown color of amber.

Material trapped within the sticky sap is sealed off from the environment as amber forms, and it can be preserved for hundreds or even millions of years. This has made amber particularly useful to biologists and geologists who are trying to learn about the earth's past. Insects trapped on the surfaces of ancient pine trees by large drops of sticky sap more than a hundred million years ago have been locked in those amber gemstones ever since (see the accompanying figure). Such specimens have given evolutionary biologists a rare look at ancient living things in a detail unmatched by ordinary fossils.

Author Michael Crichton, in his best-selling book *Jurassic Park,* even speculated that amber containing fossilized mosquitoes could serve as a source of DNA molecules from the blood of dinosaurs. As improbable as this fictional saga might seem, several laboratories have indeed recovered DNA from amber and are now using it to trace the course of evolution.

Besides insects, amber frequently contains small bubbles, and these bubbles have given scientists a chance to determine the composition of the ancient atmosphere. Robert Berner of Yale University and Gary Landis of the U.S. Geological Survey crushed samples of amber that were from 25 to 40 million years old and analyzed the contents of gases released from small bubbles within each stone. The preliminary results of their studies seem to indicate that the ancient atmosphere might have consisted of as much as 30 percent oxygen, considerably more than the 21 percent it exhibits today. More studies are needed to confirm these results, but they illustrate the scientific importance of phloem sap, transformed to amber, as a priceless record of the past.

Amber containing a beautifully preserved midge, so detailed that even a tiny mite on its back is visible.

Figure 30.16 The movement of water from root hairs into the vascular cylinder. Endodermal cells are embedded in the waxy Casparian strip, preventing water from moving between the cells. This enables these cells to regulate water movement into the vascular cylinder.

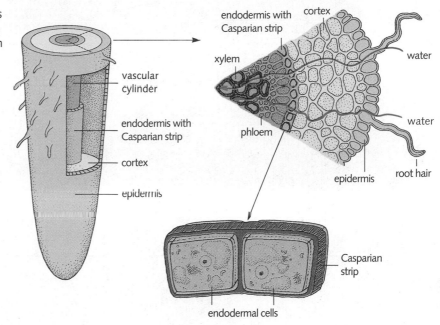

forces water into the root system and creates a root pressure that is the starting point for the flow of water through the system (Fig. 30.16).

The sensitivity of root hairs to osmotic effects is one of the reasons why saltwater flooding is so severe a threat to crop plants. High concentrations of salt in seawater can cause osmosis to drive water *out of the root hairs* rather than into them. If the water is salty enough and the situation persists, the cells of the root die of water loss, killing the plant. A similar situation occurs in houseplants and in fields that have been artificially irrigated for many years. Unlike rainwater, river water and well water contain small quantities of dissolved salts. As irrigation water evaporates, much of this salt is left behind in the soil, making it more and more difficult for plants to grow. This gradual accumulation of salt can ruin soils that have been carelessly irrigated for long periods of time.

Plants that live in salty environments such as salt marshes or coastal dunes are able to endure salt concentrations in the soil that would dehydrate and wilt most other plants. Such plants are known as *halophytes* ("salt plants"). Halophytes are particularly interesting to plant breeders, who would like to transfer their salt tolerance to crop plants, making it possible to grow food in coastal ar-

eas where salty soil has limited agriculture. Like other plants, halophytes absorb water via the active transport of salt and other dissolved minerals into their root tissue. When the concentration of dissolved salts is higher there than in the surrounding soil, water follows into the root cortex by osmosis. The key adaptation of halophytes is that their root cortex cells are able to tolerate much higher salt concentrations than the roots of most other plants, enabling them to continue to draw water when other plants have wilted.

Growth Patterns in Roots

Plant growth occurs primarily in regions known as *meristems,* in which rapidly dividing, unspecialized tissue provides a reservoir in cells that can develop into the more specialized tissues of the plant. Meristematic regions near the tips of roots or stems are known as **apical meristems.**

As a root grows, cells produced in the meristem enlarge and elongate and then begin to develop into the various tissues of the root. This process produces a *zone of elongation,* followed by a region known as the *zone of matu-*

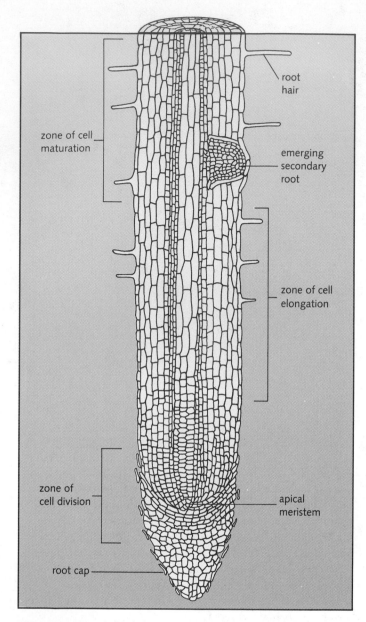

Figure 30.17 Growth in roots is localized. Distinct regions, including a *zone of cell division* (where cells are rapidly dividing), a *zone of cell elongation* (where cells increase in size), and a *zone of maturation* (where cell development takes place) can be recognized.

ration (Fig. 30.17). At the tip of the root, some of the cells produced by the meristem join the **root cap,** a sheath of cells that protects the meristem against injury from the soil particles that it pushes aside as the root grows.

Root growth is not necessarily limited to the apical meristem. In most plants, *secondary roots*, which contain apical meristems of their own, branch off from the main root, increasing root system complexity. Cells in the *pericycle* of the root begin to divide, and these cells form a new lateral meristematic region that produces cells that push through the endodermis and penetrate the cortex of the existing root. The new root grows through the epidermis of the existing root and forms an independent branch root that is completely connected—xylem, phloem, and epidermis—to the existing root structure.

THE STEMS: MAINTAINING THE FLOW

In the vascular plants, stems connect the leaves that carry on photosynthesis and the roots that absorb water and nutrients. Stems may run only a few centimeters, or they may extend tens of meters into the air. Either way, they furnish support, water, and nutrients for the other tissues of the plant. Xylem and phloem tissues pass through the stems and directly link all parts of the plant, allowing water and nutrients to be carried between the roots, stems, leaves, and fruits. In most plants, stems contain distinct **nodes** where leaves are attached and **internode** regions that connect the nodes. Small **axillary buds** are found where leaves attach to the nodes. These buds contain undeveloped meristem tissue that can produce new stems and leaves (Fig. 30.18).

Stems are surrounded by a layer of epidermal cells that have thick cell walls and a waxy coating for protection. In monocots, **vascular bundles** of xylem and phloem are embedded in ground tissue (Fig. 30.18). In dicots, these vascular bundles are also embedded in ground tissue, but the bundles are arranged in a ring surrounding the center of the stem. Ground tissue inside the ring of vascular bundles is known as **pith;** ground tissue outside the vascular bundle forms the **cortex** of the stem.

Growth Patterns in Stems

Like growing roots, growing stems contain active meristematic regions. The apical meristem is found just beneath the apex of the growing stem, and it is followed by zones of elongation and maturation. In some plants, including many common trees, the apical meristem is the

Figure 30.18 TOP LEFT: In monocots, bundles of vascular tissue are scattered in the ground tissue of the stem. TOP RIGHT: This cross section of a mature monocot stem (corn) shows the vascular bundles, ground tissue, and epidermis. (Magnification factor: 50) BOTTOM LEFT: In dicots, the bundles are arranged in a ring around the central pith. BOTTOM RIGHT: In a mature dicot, as shown in this cross section of a basswood trunk, the pith, cortex, and epidermis are clearly visible. The vascular bundles are now seen as distinct bands of xylem and phloem surrounding the pith. (Magnification factor: 18) BELOW: Many stems produce axillary buds at nodes, which then give rise to lateral branches.

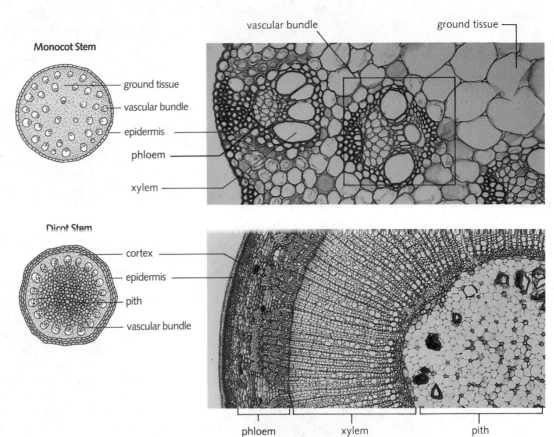

only region of the stem capable of increasing in length. This is why tree branches remain at the same height as a plant matures. Other plants, including corn, contain meristematic regions between the leaf nodes, allowing rapid growth as several meristems contribute to stem elongation. The **terminal bud** of a growing woody stem is covered by thick scales shielding the meristem and the leaves that develop around it. **Lateral buds** develop at the points of connection between leaves and the stem, and these serve as the starting points for lateral branches that arise from the stem.

The growth that occurs at apical meristems is known as primary growth (Fig. 30.19). Primary growth is responsible for elongation of stems and roots. In larger plants, particularly those that live for many years (known as *perennials*), secondary growth is also important to plant development. Secondary growth results in a *thickening* of stems and roots in the *lateral* direction. The tissue region that contains the cells responsible for such growth is known as the **lateral meristem,** or **cambium.** Secondary growth is particularly pronounced in the stems of woody plants and trees.

The patterns of primary and secondary growth are important in dicots, but secondary growth does not occur in many monocots. Although few monocots grow large enough to be called trees, those that do (most notably the palm tree) display only primary growth. This is one reason why the trunks of palm trees are of nearly uniform diameter (Fig. 30.19).

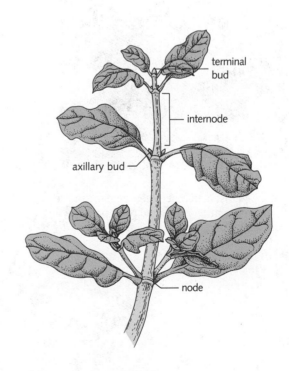

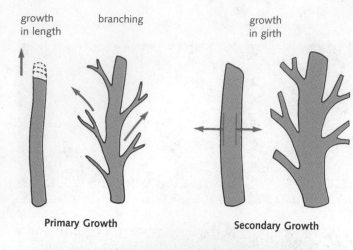

growth in length   branching   growth in girth

Primary Growth   **Secondary Growth**

Figure 30.19 TOP: The distinction between primary and secondary growth. Stem elongation is referred to as primary growth. Secondary growth refers to the increase in stem diameter. BOTTOM: Like most monocots, palm trees display only a limited amount of secondary growth because their growing tissue is localized at the bases of growing scale-like leaves.

Growth Patterns in Tree Stems

The vascular bundles in dicots are arranged in a ring, phloem tissue near the epidermis and xylem near the center of the stem, as shown in Fig. 30.20. As the stem of a woody plant grows, lateral meristems develop at the zone between xylem and phloem in each vascular bundle until they form a layer of **vascular cambium** tissue that completely encircles the stem. This growing layer of cells gives rise to tissues known as **secondary xylem** and **secondary phloem.** As seasons come and go, the vascular cambium continues to produce secondary xylem and secondary phloem cells, causing the stem to become thicker and thicker. This growth presents no special difficulties for the xylem cells; new layers of secondary xylem are simply added to older ones. Continued activity in the cambium, however, pushes the phloem outward, causing the tissue to split and fragment as the diameter of the stem increases. If this process went unchecked, serious breaks in the outer covering of the stem would occur. However, the *cortex* of the stem contains another meristematic tissue known as **cork cambium,** which forms an outer covering of cork that grows to cover places where the phloem tissue has been split by the growth process.

The **bark** of a tree is composed of all the tissues external to the vascular cambium. The **wood** of a tree is formed by the production of new xylem tissue by the cambium, whereas the vascular cambium itself is at the boundary between wood and bark. In other words, *wood is secondary xylem.* In temperate climates the pattern of wood growth follows a yearly cycle. In early spring, when water is generally in good supply and growing conditions are ideal, the cambium forms xylem cells that are quite a bit larger than those formed as the days begin to cool and shorten at the end of the summer. This seasonal variation in the formation of wood produces the **annual tree rings** exposed when you cut through a piece of wood. The annual rings that this process produces reveal more than the age of a tree. By analyzing the thickness and mineral composition of tree rings from many trees, scientists are able to get some idea of what growing conditions were like many hundreds of years ago. Severe weather conditions, including droughts, are clearly reflected in the thinner rings that result from reduced wood growth.

Reaction to Injury

Unlike animals, trees cannot run from predators or take active measures to defend themselves from being eaten. Instead of flight, trees rely on their own growth patterns to react to serious injuries and control predators. When a branch is broken and the wood of a tree is exposed to the air, chemicals in the wood begin to oxidize, producing a series of molecules related to *phenol,* a ringed six-carbon compound toxic to nearly all organisms. The phenol compounds cause a noticeable discoloration in the wood, tinting it in shades of dark blue, red, or green. These compounds help to sanitize the wound, killing fungi and bacteria that might otherwise begin to attack the exposed wood.

Another danger posed by a wound is the loss of fluid: the vascular cambium produces sap-carrying phloem, and were it to continue growing with no change, sap might flow continuously from the wound, weakening the plant. Instead, the cambium at the site of the wound produces far fewer fluid-transporting cells than normal. In their place, the cambium forms a layer of dense-walled cells that cover the wound and are able to resist attack from pathogenic organisms.

Finally, the growth pattern of the tree is adjusted in a way that walls off the injured and infected area. Vascular tissue is plugged on either side of the wound, thick new cell walls surround the injured area, and the tree begins to structure its new growth around the wound. The growth patterns of trees allow them to deal with injury by *compartmentalizing* it: walling it off, limiting damage, and continuing to grow and reproduce.

First year

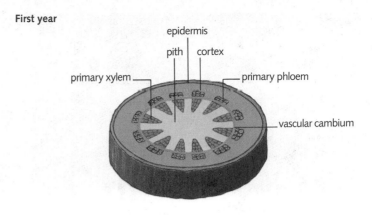

Second year

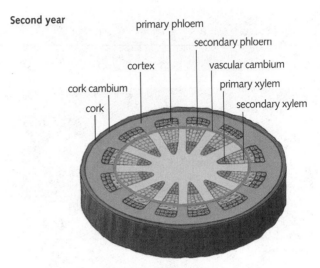

Figure 30.20 RIGHT: In woody plants, new layers of vascular tissue are produced every year. TOP: In the first year of growth, xylem and phloem tissues are formed from the vascular cambium. CENTER: In the second year, these primary xylem and phloem tissues are pushed apart by new layers of secondary xylem and secondary phloem produced by the vascular cambium. BOTTOM: This process repeats itself every year, producing a series of alternating rings as the result of slower growth during the colder months. BELOW: The spectacular growth of woody trees is recorded in the annual rings of this trunk cross section.

Fourth year

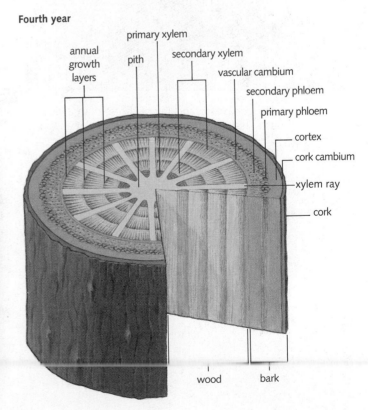

Figure 30.21 The complex branching of the veins of a dicot leaf (TOP) contrasts sharply with the simpler pattern of a monocot (BOTTOM).

THE LEAVES: HARVESTING SUNLIGHT

The leaves of a plant are the primary organs of photosynthesis. Most leaves consist of long thin *blades* attached to the nodes of the stems by a stalk called a *petiole* (Fig. 30.21). Some leaves may be divided and may consist of several interconnected parts, or *leaflets.* Leaves are covered by an outer layer of *epidermal tissue,* which protects them from the environment. The epidermal cells are coated with the **cuticle,** a thick waxy coating that protects the tissue against insect invasion and water loss. Because of their need to exchange gases, leaves cannot be completely sealed off from the environment. Therefore, leaves contain small openings known as **stomata** (singular: *stoma*), most commonly on the leaf's lower surface, which allow gases to flow into the tissue of the leaf itself. The stomata are surrounded by *guard cells* that regulate the passage of gases by opening and closing the stomata.

Regulation of Gas Exchange

At first it might seem that leaves would keep their stomata open at all times, to allow carbon dioxide to enter the leaves, where it could be used for photosynthesis. But this would cause a problem. Carbon dioxide must be dissolved in water in order to enter the photosynthetic cells of the leaf. Therefore, the surfaces of these cells must be kept wet all the time. This means that on a bright sunny day, which is ideal for photosynthesis, transpiration occurs; water evaporates from the cell surfaces within the leaf and is lost through the stomata. We have already seen that transpiration is one of the forces that drives the transport of water from roots to leaves. However, if the stomata were kept open all the time, the water losses due to transpiration would be so great that very few plants would be able to take in enough water to survive. Therefore, plants are faced with a serious problem. They must keep their stomata open just enough to allow photosynthesis to take place, but not so much that they lose an excessive amount of water.

The regulation of the guard cells of the stomata is not entirely understood. It is clear that the stomata respond to fluid levels within leaf cells by closing when fluid levels begin to drop and opening when the amount of water is adequate. One way in which this may occur is represented in Fig. 30.22. The cell walls that surround the guard cells contain cellulose microfibrils oriented in a pattern that allows the cells to lengthen, but not to thicken, when they swell under osmotic pressure. When the cells are swollen by water uptake, the paired cells increase in length, producing an opening between the two cells. Therefore, when water moves into the guard cells by osmosis, the

(a)

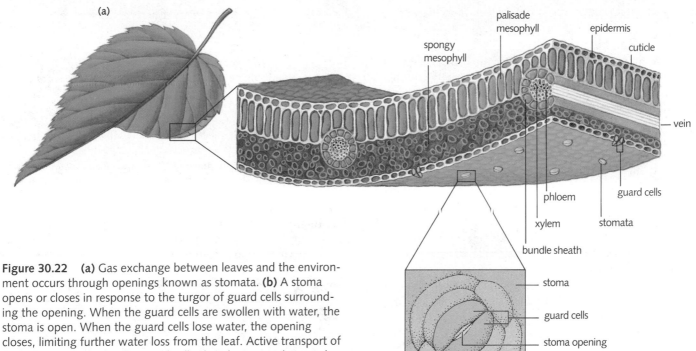

palisade
mesophyll

spongy
mesophyll

epidermis

cuticle

vein

phloem

xylem

bundle sheath

guard cells

stomata

stoma

guard cells

stoma opening

Figure 30.22 **(a)** Gas exchange between leaves and the environment occurs through openings known as stomata. **(b)** A stoma opens or closes in response to the turgor of guard cells surrounding the opening. When the guard cells are swollen with water, the stoma is open. When the guard cells lose water, the opening closes, limiting further water loss from the leaf. Active transport of potassium ions (K^+) by the guard cells also plays a regulatory role. Osmosis causes water to enter or leave the guard cells, following the direction of potassium transport. A stoma from a tobacco leaf opens, allowing gas exchange to take place. (Magnification factor: 343) **(c)** A scanning electron micrograph showing the internal structure of a turnip leaf. Stomata are present on both surfaces of this leaf, although they are more numerous on the underside. (Magnification factor: 57)

(c)

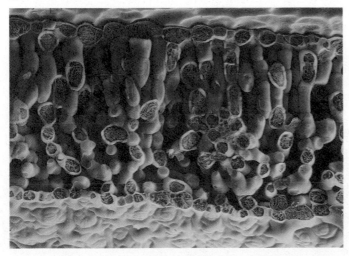

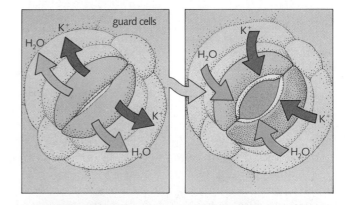

guard cells

K^+

H_2O

K^+

H_2O

H_2O

K^+

H_2O

K^+

(b)

stomata are forced open. When water is depleted and lost from the cells, they shrink a bit and the guard cells gradually close the stomata.

The actual control of osmotic pressure in the guard cells is somewhat more complex. The guard cells are able to take up potassium ions (K^+) by active transport across their cell membranes. This accumulation of potassium makes the cytoplasm slightly hypertonic with respect to the surrounding medium, and osmosis drives water into the cells. The resulting increase in *turgor,* or water pressure, within the cells causes them to swell and open the stomata. Several factors seem to affect the active transport of K^+. *Light* stimulates the potassium pump, which causes the stomata to open at dawn as photosynthesis speeds up, allowing CO_2 to enter the leaf. High CO_2 levels within the leaf and water loss cause the stomata to close, whereas low CO_2 levels within the leaf cause the stomata to open. Superimposed on all of these factors in many plants is a 24-hour cycle, or **circadian rhythm,** of stomatal opening and closing. Plants that have been grown in normal day–night cycles and then placed in a dark chamber continue to open their stomata during the daytime and close them at night for several days in the absence of any environmental cues. This internal timing mechanism affects the degree to which the stomata respond to outside influences.

Internal Leaf Tissue

The xylem and phloem tissues found in leaves are direct continuations of the tissues that run through the roots and stems. In leaves these tissues are bound together to form a *vascular bundle* that appears as the *leaf vein* easily seen in most plants. These bundles run more or less parallel to each other in monocot leaves (grasses such as corn are good examples). They follow a more pronounced branching pattern in dicots such as tomato and bean plants.

The ground substance of the leaf is a tissue known as **leaf mesophyll.** The mesophyll cells, packed with chloroplasts, perform most of the plant's photosynthesis. A typical dicot leaf contains an upper layer of *palisade cells,* which form a tightly packed row of cells just below the epidermis. Beneath the palisade cells is a much looser layer of cells known as the *spongy mesophyll.* The air spaces between cells in this region make them especially suitable for gas exchange, and the carbon dioxide that enters through the stomata generally enters these cells first. Moisture is supplied to the mesophyll cells by osmosis from the vascular bundles that run through the mesophyll tissue. C4 plants have a specialized tissue around the vascular bundle known as the *bundle sheath,* and the pathways of carbon fixation in C4 plants are divided between mesophyll and bundle sheath cells.

SUMMARY

Plants include single-celled algae as well as the terrestrial plants that dominate the temperate regions of the earth. Plants require sunlight, water, and carbon dioxide, and the need for these three resources influences their structure and lifestyle. The simplest plants, algae, absorb water directly from the environment.

The vascular plants have evolved specialized water-conducting tissues, and the ability to move water through such tissues has enabled them to colonize areas in which water is relatively scarce. Other adaptations of land plants include specialized supporting tissue, protection against water loss, the ability to extract nutrients from the soil, and reproductive systems that can function effectively in dry environments.

Seed-bearing plants include the Gymnosperms and the Anthophyta, or flowering plants. The structural organization of flowering plants involves three tissue systems: dermal tissue, ground tissue, and vascular tissue. These tissues are found throughout the body of the plant, in roots, stems, and leaves. Plants continue to grow throughout their lifetimes, and this growth is concentrated in meristems, which contain rapidly dividing cells capable of developing into the many specialized cells of the adult plant.

The primary function of roots is to absorb water from the soil. Water is drawn into the roots by osmosis and raised into leaves and stems against the force of gravity by a combination of forces generated by the cohesion of water, adhesion to the walls of conducting tissue, and tension produced by water use in the leaves. The roots contain another conducting tissue, known as phloem, which serves as a transport system that passes nutrients throughout the organism. The movement of nutrients from one region of the plant to another through phloem may be explained by means of the pressure–flow hypothesis.

The stems of plants contain the same two conducting tissues, xylem and phloem, arranged in a pattern that identifies the plant as a monocot (one seed leaf) or dicot (two seed leaves). In woody plants the most rapid rates of cell growth are found in the vascular cambium, a tissue that forms between the xylem and phloem, laying down tissue known as secondary xylem and secondary phloem.

The leaves, attached to the stems, are a plant's primary organ of photosynthesis. A rapid rate of photosynthesis can be maintained because special guard cells on the underside of the leaf control openings known as stomata, which regulate the passage of air into the space within the leaf.

STUDY FOCUS

After studying this chapter, you should be able to:

- Describe the problems plants face in a terrestrial habitat, and indicate some survival strategies unique to plants.

- Outline the classification of plants and plant-like groups.

- Explain how plants draw water from the soil and raise that water to the tallest parts of the plant.
- Explain the pressure–flow hypothesis for the movement of nutrients through phloem.
- Explain the growth processes of plants and the processes that produce specialized plant tissues.

TERMS AND CONCEPTS

vascular tissue *600*	Casparian strip *606*
tracheids *601*	pericycle *606*
monocots *601*	capillary action *606*
dicots *601*	transpiration *607*
dermal tissue *602*	sieve tubes *607*
ground tissue *602*	pressure–flow
meristems *604*	hypothesis *608*
primary growth *604*	vascular bundles *612*
secondary growth *604*	cambium *613*
xylem *605*	stomata *616*
phloem *605*	

REVIEW

Objective Questions (Answers in Appendix)

1. Mosses have not completely adapted to land because
   (a) they need water in order to survive.
   (b) they need water for reproduction.
   (c) they have true roots.
   (d) every cell is in direct contact with soil nutrients.

2. Which of the following is a seedless, vascular plant?
   (a) fern   (c) flowering plant
   (b) pine tree   (d) liverwort

3. Flowering plants called angiosperms are divided into
   (a) gymnosperms and monocots.
   (b) bryophytes and gymnosperms.
   (c) monocots and dicots.
   (d) bryophytes and anthophyta.

4. Plant cells increase in size by
   (a) conduction.
   (b) meiosis.
   (c) differentiating into specialized tissues.
   (d) elongation.

5. Meristematic tissues are capable of
   (a) rapid conduction of nutrients throughout the plant.
   (b) rapid conduction of water throughout the plant.
   (c) regulating the flow of gases into leaves and stems.
   (d) developing into any of the cell types found in mature tissue.

6. The main photosynthetic area of the leaf is composed of
   (a) xylem.   (c) stomata.
   (b) phloem.   (d) mesophyll.

Discussion Questions

7. "Plants are organisms that eat sunlight." How valid is this generalization? How useful is it in explaining the forces that have governed plant structure and function?

8. What were some of the special challenges faced by plants as they evolved from an aqueous environment to a terrestrial one? What rewards were available to the first plants to inhabit the land successfully?

9. The vascular plants are sometimes referred to as the "higher" plants. Is this nomenclature justifiable? In what sense are they "higher" than algae and bryophytes? What is the difference between a true root and a rhizoid?

10. Although the high salt concentration found near the seashore is toxic to most plants, a few species thrive in such environments. How does the root epidermis differ between salt-tolerant and typical plants in terms of cellular activities and transport mechanisms?

11. How does transpiration pull function? Does it work in concert with capillary action or antagonistically to it?

12. Explain the pressure–flow hypothesis of nutrient movement through phloem. The development of a large fruit, such as an apple, requires that substantial quantities of nutrient be transported into the fruit. Explain how the pressure–flow hypothesis would redirect nutrient flow into a developing fruit.

13. The loss of water from the leaves of a plant on a hot day causes the cells to lose water pressure and shrink slightly. What direct effect does this have on the guard cells of the stomata? Does this change increase or decrease the rate of water loss?

READINGS

Raven, P. H., R. F. Evert, and S. E. Eichorn. *Biology of Plants.* 5th ed. New York: Worth Publishers, 1992. The latest edition of a comprehensive and highly readable text on plant biology.

Dussourd, D. E. "The vein drain; or how insects outsmart plants." *Natural History* 99 (February 1990): 44–49. A brief photographic study of how insects tap into xylem and phloem to help themselves to a quick meal.

Stewart, D. "Green giants." *Discover* 11 (April 1990): 61–64. A beautiful study of giant sequoias and how their internal systems meet the challenges of their great size in terms of fluid transport, mechanical support, and wind resistance.

Feldman, L. J. "The habits of roots." *BioScience* 38 (October 1988): 612–618. A study of root structure and function, featuring some striking images of delicate root structures.

Bateman, G., ed. *Flowering Plants of the World.* New York: Mayflower Books, 1978. An authoritative, thorough, and beautiful review of flowering plants. To be admired as much for its striking color drawings as for its value as a reference text.

31

Reproduction in Flowering Plants

*t*he evolution of flowering plants is a relatively recent event. Although we might tend to think of the last 100 million years as the "Age of Mammals," or even the "Age of Humans" (if we confined ourselves to the last 4 million years), we would do better to call it the "Age of Flowering Plants." The rise of the angiosperms, the flowering plants (division Anthophyta), has been one of the most important events to occur in the last 100 million years of the earth's history. From their first appearance in the Cretaceous, flowering plants have risen to ecological dominance. With the exception of isolated conifer forests, flowering plants are found nearly everywhere on the planet where plant growth on land is possible.

The rapid rise of these plants has led evolutionary biologists to wonder what features made them such fearsome competitors to the gymnosperms that had dominated the earth's landscape for the previous 100 million years. Flowering plants are not significantly different from gymnosperms in terms of their ability to draw water from the soil or to produce large stems and efficient leaves, and each type of seed-bearing plant is able to reproduce well in both dry and wet environments. The anthophytes, it seems, have a secret weapon that has given them domination over the planet. That secret weapon involves one of the most important functions of any living organism—reproduction.

ASEXUAL AND SEXUAL REPRODUCTION

The essential advantage enjoyed by the flowering plants over the gymnosperms may be the very characteristic that separates these two groups: unlike the gymnosperms (the "naked seed" plants), the flowering plants surround their seeds with a layer of tissue known as **fruit,** formed from the reproductive tissues of the flower itself.

Why should the formation of fruit be such a key characteristic in the ability of flowering plants to profit from natural selection? A hint

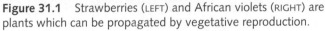

Figure 31.1 Strawberries (LEFT) and African violets (RIGHT) are plants which can be propagated by vegetative reproduction.

may be found in the **flowers,** the plant reproductive organs that add so much beauty to our world. Fruit and flowers serve as lures for insects, birds, and mammals. The adaptations by which fruit and flowers interact with animals solve two problems that plants face: finding mates and dispersing their offspring. Being rooted in place, plants cannot move about to exchange reproductive cells with others of the same species, and they face enormous problems in dispersing the thousands of seeds they are capable of producing. Associations with animals, which can carry pollen *and* disperse seeds over great distances, provide a way to solve both of these problems. These associations with animals may have given flowering plants the critical advantage that enabled them to cover the planet.

Asexual Reproduction

Vegetative reproduction in nature Angiosperms, the "highest" of the "higher" plants, are distinguished from the most advanced animals, the vertebrates, in many ways. One of the most interesting of these is the ability they share with other plants to reproduce *asexually.* This process is known as **vegetative reproduction,** and its re-

sults are familiar to any gardener. Many plants can grow from a detached leaf, stem, or bundle of roots. Why is it so easy to regenerate a plant from a small cutting? Remember that the key to plant growth is the presence of undifferentiated, actively dividing cells in regions known as **meristems.** Meristematic cells are capable of developing into any plant tissue, including reproductive cells, and their presence gives any portion of the plant the chance to regenerate into a complete organism. Little by little, the growing meristematic cells of the detached tissue form new tissues that replace the parts from which the cutting was taken, and eventually a completely new plant, genetically identical to the original, is formed.

The formal name for this type of asexual reproduction is **fragmentation,** because the "parent" plant is fragmented to produce the vegetative offspring. In nature, vegetative reproduction is an important means of propagation in a number of species (Fig. 31.1). One famous example is the spider plant (*Chlorophytum* sp.), which produces *runners,* slender lateral shoots capable of developing buds of their own. If the *adventitious* roots (roots that appear in an unusual or unexpected place, such as the end of a lateral shoot) of the buds find soil, they grow into an independent organism that can be detached from the original plant. Strawberries can propagate in much the same way, and thick beds of strawberries are produced by carefully allowing a few plants to spread their runners in every direction to establish new plants.

Many grasses also propagate by sending runners along the surface of the soil, and thick clumps of dune and desert grasses that are formed as lateral runners take root and develop into separate plants. Because plants produced by vegetative processes are genetically identical to their parents, clusters of plants formed in this way are *clones,* groups of organisms produced by cell division from a single ancestral cell—in these cases, by division of the single zygote cell from which the original plant was formed.

Vegetative reproduction in horticulture The capacity of plants for vegetative reproduction has been widely used by humans to propagate the plants we grow for our own purposes. Many common houseplants, such as African violets (Fig. 31.1), can be propagated from single leaves, and fruit trees are often produced from stem cuttings. As you may know, potatoes produce small buds known as "eyes," which can grow into complete potato plants if they are planted.

Many fruit trees, including apples and pears, are propagated exclusively from cuttings of preexisting stock that have been rooted in nutrient solutions and grown into small saplings before they are transplanted to the orchard. The ability of plants to grow under such conditions even makes it possible for two different plants to be grafted together. The careful joining of buds from one plant to the rootstock (stems and roots) of another makes it possible for a single apple tree to yield five or six varieties of apples (the stems of several varieties are grafted to a single tree) and also enables farmers to combine the best aspects of two completely different varieties of plants.

The ability to combine characteristics has been particularly important to the wine industry. In the late nineteenth century a combination of diseases destroyed thousands of acres of vineyards in France, threatening to wipe out wine production. Fortunately, it was discovered that the Concord grape, which is native to North America, was immune to these diseases and flourished in the soil of France. The fruit of the Concord grape, however, does not make a very good wine; therefore, winegrowers attempted to graft the shoots of European grape varieties onto North American rootstock. The attempt was successful, and today most of the grapes grown for wine throughout the world are combinations of European and American grape species grafted together to produce the best possible plant for all purposes.

In recent years, tissue culture techniques have been widely used to produce plants from single cells. We will discuss plant cell culture more fully in Chapter 33, and 33 we have already seen how it may be used in plant genetic engineering (Chapter 21). 21

Sexual Reproduction

Sexual reproduction in the angiosperms takes place in flowers, which are specialized shoots containing reproductive organs. Like sexual processes in other organisms, reproduction in the flowering plants involves the formation of *gametes* (specialized reproductive cells) and the exchange of genetic information when gametes fuse to form a new organism. Beyond that point, however, diversity is the hallmark of angiosperm reproduction. One of the major themes in the evolution of higher plants has been the progressive modification of reproductive organs and of the reproductive process itself. We often tend to think of evolutionary success in terms of survival. Catch phrases like "survival of the fittest" contribute to a predisposition to view natural selection in terms of the ability of a group of organisms to compete successfully for survival against predators. Yet evolutionary success is measured most directly by the number of offspring that an organism is able to pass along to the next generation. One powerful strategy for assessing evolutionary success, then, is to focus on the reproductive process itself: the efficient distribution of gametes and the placement of seeds into environments where they are likely to grow. The sexual processes of flowering plants are case studies in the success of this evolutionary strategy.

ALTERNATION OF GENERATIONS

The Principle of Alternation

When we examined the life cycles of other plant groups earlier in the text (Chapter 27), we saw a pattern of alternating haploid (n) and diploid ($2n$) generations. This *alternation of generations,* in its simplest form, involves a **sporophyte,** a diploid plant that produces haploid *spores* by means of *meiosis,* and a **gametophyte,** a haploid plant that produces haploid gametes by *mitosis. Fertilization* occurs when two gametes fuse to form a diploid *zygote,* which then grows to form a new sporophyte. The most basic type of alternation of generations—one wherein the gametophyte and sporophyte are distinct, independent plants—is shown in Fig. 31.2. In some plant groups, notably the mosses, the haploid gametophyte is the larger and more prominent phase of the life cycle. In flowering plants, however, the sporophyte dominates, and the gametophyte has become a tiny organism that lives briefly and almost invisibly within the tissue of the larger plant. 27 10

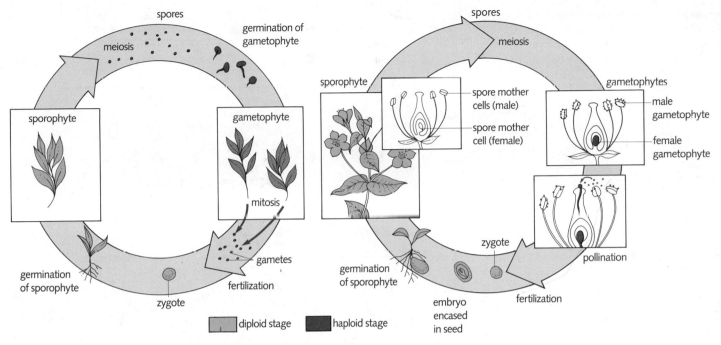

Figure 31.2 The alternation of generations in plant reproductive cycles. In principle, the sporophyte and gametophyte of a life cycle may be represented by free-living plants, as shown in the idealized cycle at the left side of the diagram. However, in flowering plants (RIGHT), the sporophyte stage of the cycle is the dominant form of the plant, and the male and female gametophytes exist only as small groups of cells enclosed within the reproductive organs.

Alternation of Generations in the Angiosperms

Anthophytes produce two types of spores, a large **megaspore** and a smaller **microspore.** In many plants, both spore types are produced in a single flower. Meiosis in the sporophyte produces each type of haploid spore from a diploid **spore mother** cell, as illustrated in Fig. 31.2. These spore cells then go through several rounds of cell division to produce the female and male *gametophytes.* Fertilization occurs when a haploid nucleus from the male gametophyte and one from the female gametophyte fuse to produce a diploid zygote. This zygote then develops into an embryo that is encased in a seed, awaiting further development into a complete diploid sporophyte plant.

The details of this life cycle are complex; this description is only a simple blueprint. It is necessary to fill in many details to complete a picture of reproduction in the flowering plants. Nonetheless, it is important to appreciate the extent to which alternation of generations has been modified in the evolution of higher plants. Most dramatic has been the reduction in the role of the gametophyte. It is an almost invisible stage of the life cycle that is completely hidden within the plant's reproductive organs.

THE DETAILS OF PLANT REPRODUCTION

The Organs of Reproduction

The structures within a flower are actually modified leaves, and a flower can be viewed as a small, truncated shoot with four distinct leaf layers. We generally do not think of flowers as leaves, an indication of how greatly they have been modified. The four distinct parts of a typical flower are **sepals, petals, stamens, and carpels** (Fig. 31.3).

In many flowers, the *sepals* are truly leaf-like, enclosing the flower bud during its early stages of development and retaining the green color of chlorophyll throughout the life of the flower. Taken together, the sepals form the *calyx,* the outermost enclosure of the flower. The *petals* of the flower may be brightly pigmented, giving the flower its characteristic color and helping to attract insects, birds, and other animals that may aid in the reproductive process. The actual reproductive organs are found in the two remaining portions of the flower. The male portions of the flower are the *stamens,* which produce **pollen.** The stamens consist of thin filaments that emerge from the stem of the plant and terminate in the **anthers,** enlarged sacs where pollen is developed and released. The female

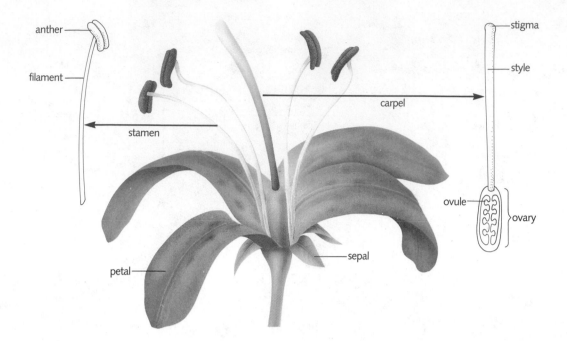

Figure 31.3 An idealized flower, showing the four principal structures of a complete flower: sepals, petals, stamens, and carpels.

Labels: anther, filament, stamen, petal, sepal, stigma, style, carpel, ovule, ovary

portions of the flower are the *carpels*. A single flower may have one or several carpels, each of which has a broad base containing an **ovary,** within which are **ovules** containing the female gametophyte. The diameter of the carpel narrows as it leads upward through a slender portion known as the **style** and terminates in a sticky **stigma.** Taken together, the stamens are known as the *androecium*—literally, the "house of man." In the same vein, the carpels collectively are known as the *gynoecium*—the "house of woman."

In a sense, the flower shown in Fig. 31.3 is an ideal one. It contains the four principal flower parts, and its male and female tissues are perfectly developed, making it a complete flower. The flowers of many plants are not complete, and we will consider them shortly.

Formation of Spores and Gametes

The female tissues: megaspores and eggs The formation of female megaspores begins in the ovary of the flower, within the swollen base of the carpel. On the inner wall of the ovary, one or several clusters of cells have developed into *ovules*. As the ovule matures, a small cluster of cells develops within it, known as the *nucellus*. The nucellus is surrounded by a protective *integument* that leaves a single opening, the *micropyle*, through which fertilization will occur.

One large cell known as the *megaspore mother cell* appears within the nucleus. This diploid cell undergoes meiosis to produce four haploid megaspores. Three of

these megaspores disintegrate in most species, leaving a single haploid megaspore. In lower plants, the mature spore would now begin to divide and to lead an independent existence as the gametophyte stage of the life cycle. In the seed-bearing angiosperms and gymnosperms, however, the gametophyte stage consists of a small group of cells derived from this single surviving megaspore.

Typically, the megaspore, which is now more properly called the *megagametophyte*, passes through three rounds of mitosis, producing first two, then four, and finally eight nuclei that are not separated by cytokinesis (Fig. 31.4). The megaspore now contains eight nuclei—four at the end of the cell closest to the micropyle and four at the opposite end. One nucleus from each group moves to the center of the cell, where they are known as the *polar nuclei*. Cell wall formation now occurs around the three nuclei closest to the micropyle forming an **egg cell** and two *synergids*. Cell walls also form around the three cells at the opposite end, which are then called the *antipodals*, completing the formation of the megagametophyte. Anticipating the role it will soon play in the reproductive

Figure 31.4 Details of reproduction in a representative angiosperm. Pollen develops within pollen sacs located at the tips of the anthers. The embryo sac develops within the ovule at the base of the carpel. A double-fertilization event occurs in angiosperms, producing a diploid zygote and a triploid endosperm mother cell. Both cells form critical tissues in the seed. ◊

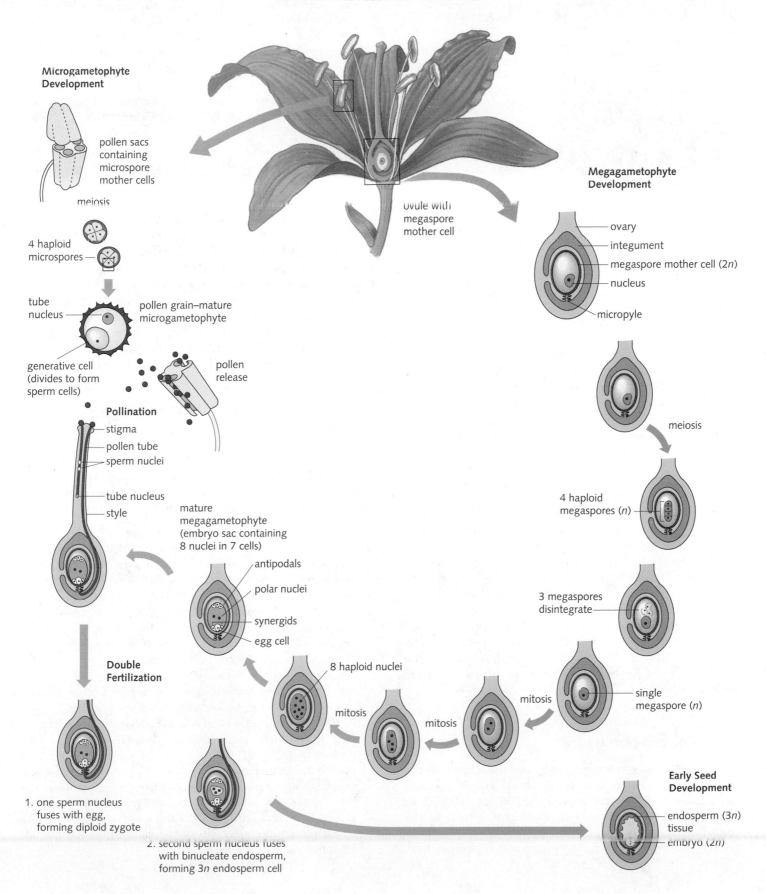

Microgametophyte Development

pollen sacs containing microspore mother cells

meiosis

4 haploid microspores

tube nucleus

pollen grain–mature microgametophyte

generative cell (divides to form sperm cells)

pollen release

Pollination

stigma
pollen tube
sperm nuclei

tube nucleus
style

mature megagametophyte (embryo sac containing 8 nuclei in 7 cells)

antipodals
polar nuclei
synergids
egg cell

Double Fertilization

1. one sperm nucleus fuses with egg, forming diploid zygote

2. second sperm nucleus fuses with binucleate endosperm, forming 3n endosperm cell

8 haploid nuclei

mitosis

mitosis

mitosis

ovule with megaspore mother cell

Megagametophyte Development

ovary
integument
megaspore mother cell (2n)
nucleus

micropyle

meiosis

4 haploid megaspores (n)

3 megaspores disintegrate

single megaspore (n)

Early Seed Development

endosperm (3n) tissue
embryo (2n)

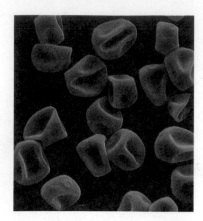

Figure 31.5 Pollen grains of timothy, a common grass. The protective coating around pollen grains enables them to survive until they make contact with a receptive flower.

process, we call the seven-cell, eight-nucleus gametophyte the **embryo sac.**

The male tissues: microspores and pollen *Microspores* are the smaller spores, and they develop within the anthers of the flower from diploid *microspore mother cells.* It's a bit confusing to speak of both the "male" and "female" spores as deriving from "mother" cells, but bear in mind that the production of spores is *not* a sexual process. The microspore mother cells are contained in chambers within the anther known as **pollen sacs,** which may contain as many as 1000 other mother cells. Each microspore mother cell passes through meiosis and produces a group of four haploid microspores (Fig. 31.4).

Each of these microspores then develops into a **pollen grain.** A mature pollen grain is surrounded by a series of thick carbohydrate and protein coats, and the details of pollen wall construction vary so widely that it is possible to identify most plants from their pollen grains alone! Within the developing pollen grain, a single round of mitosis produces two nuclei. One of these, the *generative cell,* will produce two *sperm cells* to fertilize the female gamete; the other, the *tube nucleus,* will form the *pollen tube,* a structure that enables the generative nucleus to reach the female gametophyte. The binucleate pollen grain is the male version of the gametophyte stage of the plant life cycle.

Pollination and Fertilization

Mature pollen grains are released from the anthers as the pollen sacs split open, allowing the loose pollen grains to be scattered on the surface of the anther (Fig. 31.5). From the anthers the pollen grains are spread to the stigma by a variety of methods, including insects, birds, and air currents. If the pollen grains are delayed in find-

ing their way to the stigma, their thick coating allows the cells within the pollen grains to survive many days of dry weather and high temperature, and their ability to survive outside the sporophyte for prolonged periods is one of the key elements of the independence of seed plants from water. However, pollen cells cannot survive indefinitely, and in many species, successful pollination is a matter of delicate timing between the release of pollen and the development of the female gamete.

Initial events: the growth of the pollen tube The surface of the stigma is covered with a moist, sugar-containing secretion that activates and nourishes the pollen cells that fall upon it. The cells within the pollen grain break through the wall of the grain and form a pollen tube, which begins to grow into the tissue of the stigma and down through the style (Fig. 31.4). The generative nucleus divides to form two sperm nuclei, and these two nuclei follow the tube nucleus down the stigma toward the ovule. The three nuclei of the growing pollen tube represent the complete microgametophyte stage of the angiosperm cycle. The pollen tube passes through the micropyle to enter the ovule through one of the synergids, where the two sperm nuclei are released from the pollen tube.

True fertilization occurs when one of these nuclei fuses with the *egg cell* to produce a diploid zygote, the first cell of a new sporophyte plant. Strictly speaking, the female gametophyte produces a single gamete cell, the egg cell of the embryo sac (megagametophyte), and the male gametophyte produces two gametes, the sperm nuclei of the pollen tube (microgametophyte).

Double fertilization—embryo and endosperm In plants and animals that reproduce sexually, the fusion of male and female gametes produces a diploid zygote that then develops into a new organism. This is the case in flowering plants. However, flowering plants also undergo a unique *double-fertilization* event in which the second sperm nucleus of the pollen tube fuses with the two polar nuclei to form a **triploid (3n) endosperm** cell (Fig. 31.4). Because of its role in forming endosperm, the cell containing the two polar nuclei is sometimes called the *endosperm mother cell.* The formation of endosperm tissue is unique to the flowering plants, and it is one of the most important evolutionary novelties of the angiosperms. The diploid zygote will develop into an embryo that will produce a mature plant. The triploid endosperm cell will develop into a rich, nutrient-laden, living endosperm tissue that will nourish the embryo during its early growth. The remaining cells of the embryo sac, a synergid and three antipodals, disintegrate early in embryonic development.

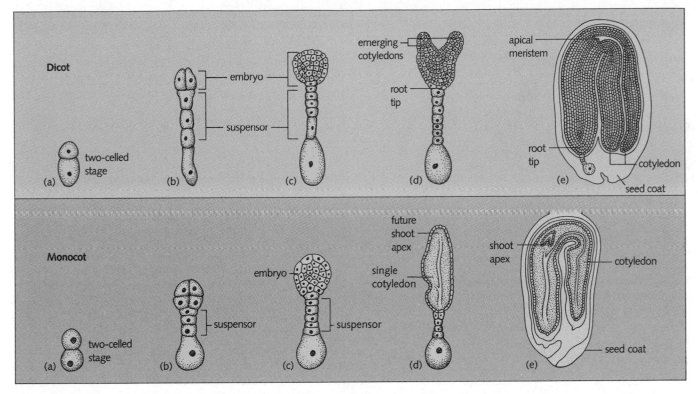

Figure 31.6 The formation of embryonic tissues in dicots (TOP) and monocots (BOTTOM). The basic pattern of seed leaves is established early in development and is obvious well before seed development is complete.

Embryonic Development

Immediately after the double-fertilization event produces diploid and triploid nuclei within the embryo sac, the process of seed formation begins. Although seeds are complex structures, each of the major tissues of the seed is formed from cells that are already present in the ovary at the time of fertilization. The *zygote* develops into an *embryo*, the *endosperm mother cell* produces nutrient-storing *endosperm*, the wall of the *ovary* develops into *fruit*, and the *integument* that surrounded the nucellus and embryo sac develops into a *seed coat*.

The first cell division of the zygote produces two cells. One develops into the embryo itself, and the other forms a structure known as the *suspensor*, which attaches the embryo to the nucellus. As mitosis continues and the embryo increases in size, the primary tissues of the plant are produced, and vascular tissue appears as the embryo begins to take shape. Gradually the first leaves of the plant, its "seed leaves" or cotyledons, appear. You will recall that the two major groups of the flowering plants, the *monocots* and *dicots*, are distinguished on the basis of the number of cotyledons that appear in the seed. Dicot embryos, which produce two cotyledons, quickly assume a heart-

shaped appearance as the cotyledons emerge from the embryo (Fig. 31.6). The apical meristem takes shape between the cotyledons, and the enlarging embryo may gradually bend to fit within the confines of the seed coat. Monocot embryos assume an elongated, cigar-like shape that may also become bent to fit within the seed.

As the tissues of the embryo mature, nutrients from the plant flow into the developing seed, filling the cotyledons and the endosperms with proteins, carbohydrates, and lipids that will be available for the growth of the young plant. When embryonic development is complete, water is removed from the seed until the remaining moisture constitutes less than 10 percent of the seed mass. The coating of the desiccated seed hardens, and it is ready for release into a potentially hostile environment (Fig. 31.7).

Fruits

Angiosperm seeds are not borne exposed on the surface of the reproductive tissues, as gymnosperm seeds are, but are enclosed within the protective tissues of the ovary. As

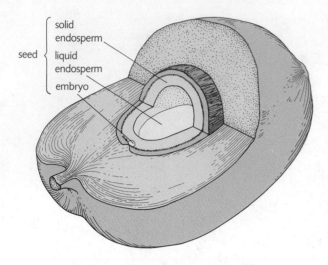

Figure 31.7 Coconuts and a cross section of a typical coconut. The "milk" of the coconut is liquid endosperm tissue. The thick protective coating of the coconut enables it to survive harsh treatment.

the seed matures, the ovary walls surrounding the seed go through a series of transformations until they ripen to form a fruit. Fruits differ widely from one species to another, and the size, type, and structure of the fruit surrounding a seed play a key role in the process of *seed dispersal.*

Fruit types   **Simple fruits** are those produced from a single carpel or from several carpels that fuse as the fruit is formed. The ovary wall surrounding a simple fruit may be fleshy, like that of the grape or the tomato, or tough and dry, like a bean pod. In some fruits, such as the peach or cherry, the inner wall of the ovary is rigidly attached to the surface of the seed (Fig. 31.8). In others, such as the maple, the dry fruit forms an aerodynamic surface that helps the seed float gracefully when it is released from the parent plant. **Aggregate fruits** are formed from multiple ovaries within a single flower. These ovaries are held tightly together but do not fuse. Blackberries and raspberries are examples of aggregate fruits. **Multiple fruits** are formed from several separate flowers clustered tightly together. The fruits that are formed from these tightly clustered flowers, which include the pineapple, consist of distinct, individual sections derived from the separate flowers. Table 31.1 summarizes the defining characteristics of the three types of fruits.

In everyday speech we tend to divide edible plants into the categories of *fruits* and *vegetables.* Apples, pears, oranges, and bananas, which are commonly called "fruits," are all cases where the term is properly applied. However, many "vegetables," including peas, corn, and beans, are actually fruits in the biological sense: They are seeds enclosed within a ripened ovary. We may think of fruits as tasting sweet, but the biological definition of the term has nothing to do with taste, sweetness, or even edibility.

PLANTS AND ANIMALS—FLOWERS, POLLEN, AND SEEDS

Early in this chapter we stressed the fact that the evolutionary history of flowering plants, as well as their basic structure and organization, reflects a pattern of association with animals. Many plants depend on animal assistance in two of the most important tasks of the reproductive process: pollination and seed dispersal. Evolutionary theory tells us that we cannot expect animals to aid in plant reproduction for altruistic reasons. Evolutionary pressures imply that animal assistance will be available only if plants have developed adaptations that make playing a role in pollination and seed dispersal beneficial to the animals. That is, it must improve their chances for survival, as well as the plants'.

Flower Specialization

The basic flower structure illustrated in Fig. 31.4 is the exception rather than the rule for many species of flowering plants. Many plants produce *incomplete flowers* in which either the carpel or the stamens are missing (Fig. 31.9). Such flowers are usually described as *staminate* (if only the stamens are present) or *carpellate* (if only the carpel is present). In *monoecious* organisms, staminate and carpellate flowers are found in different portions of the same plant. This is the case in corn, wherein pollen is released from the staminate flowers (the tassels of the plant) to fertilize the carpels (the silk and kernels), which then bear aggregate fruits (ears of corn). *Dioecious* organisms are those that produce staminate and carpellate flowers on separate plants, such as the American holly or buckthorn.

Flowers, complete or incomplete, differ from each other in a wide variety of ways. Flowers are classified according to the positions of the stamen and anthers, the number of stamens, the symmetry of petals and sepals surrounding the reproductive structures, and a host of physical characteristics including color, scent, and shape. As we will see, many of these differences can be understood in terms of the process of pollination.

Figure 31.8 The peach is a simple fruit. Aggregate fruits, like raspberry, are produced by multiple ovaries in a single flower. Multiple fruits, including the pineapple, are formed from clusters of separate flowers.

Table 31.1 *Types of Angiosperm Fruits*

Fruit Type	Definition	Examples
Simple	Single ovary or fused ovaries from a single flower	Pea, bean, cherry, grape, banana, peach
Aggregate	Many unfused ovaries from a single flower	Blackberry, raspberry
Multiple	Separate ovaries from adjacent flowers	Fig, pineapple

Figure 31.9 Many plants produce specialized flowers. LEFT: Corn produces separate male and female flowers. CENTER: The stamens of the Scarlet passion flower are fused into a single structure. RIGHT: The sunflower is a composite structure, actually consisting of hundreds of smaller flowers.

Techniques of Pollination

Wind pollination The simplest technique of pollination is dispersal by the wind. Gymnosperms rely on this technique, and they produce enormous quantities of pollen to compensate for the slim chance that any given pollen grain will fall upon receptive female reproductive tissue. As you might expect, wind pollination is most effective when a large group of plants from a single species grow side by side, increasing the density of the pollen released and, hence, the chances for success. Corn is a wind-pollinated angiosperm, and gardeners appreciate the fact that a successful corn crop cannot be grown from just a few plants. Corn does well only when a large number of plants are grown side by side in a pattern that maximizes the likelihood of successful wind-aided pollination. Many grasses (corn has evolved from a tropical grass) and a few large trees, including the oak, are wind-pollinated.

Insect pollination Evolutionary biologists have long speculated that the dependence of many plants on insects for pollination may have come about as a natural consequence of adaptations that increased the success of wind pollination. One of the key problems for a wind-pollinated plant is ensuring that the stigma (the tip of the carpel) is sticky enough to catch and hold the pollen grains that chance to fall upon it. These sticky secretions may have been just sweet enough to attract the interest of foraging insects. As the insects traveled from flower to flower to find more of the sticky fluid, they may have inadvertently bumped into the anthers of several flowers, covering their bodies with powdery pollen. In this way, insects may have begun to scatter pollen from one flower to another. As chancy as this process may sound, the fact that insects tend to visit one flower right after another made it much more likely that pollen would arrive by means of their wanderings than via the random stirrings of the wind. This process would have produced a selective pressure for plants to develop more effective means of luring insects to their flowers. Plants possessing new adaptations ensuring that insects could not visit a flower *without* scattering pollen would have a tremendous advantage

in terms of reproductive efficiency—exactly the sort of situation in which natural selection acts most effectively.

The direct spread of pollen from one plant to another by an animal is called **vector pollination** (the insect is the *vector*, the carrying agent). Because hardworking honeybees move quickly from one flower to another, pollen picked up by a bee has a good chance of being deposited directly in another flower. Much less pollen is wasted than in wind-aided dispersal.

Bees are the most important insect pollinators. (See Theory in Action, The Grand Masquerade: Plants as Mimics, p. 633.) Worker bees visit flowers repeatedly, gathering sweet secretions known as *nectar* as well as protein-rich pollen. Bees have sharp, well-developed visual systems. They are capable of distinguishing color in the short-wavelength range of the spectrum, meaning that they are particularly sensitive to yellow, green, blue, and violet colors. Their range of sensitivity extends well into the ultraviolet, a region where our own eyes cannot detect light (Fig. 31.10). Flowers that attract bees are often strongly scented and brightly colored, enabling bees to locate and visit even widely scattered flowers regularly. The long funnel-like openings of many bee-pollinated flowers ensure that the insect must brush against the anthers to gather nectar and thereby force the animal to scatter pollen on the stigma as it feeds.

There are more than 15,000 known species of bees, and many bee species are specific pollinators of only certain types of plants. For such insects, there is strong evolutionary pressure toward efficient pollination. An adaptation that enables a bee species to pollinate its flowers more efficiently increases the number of offspring of that plant species, thereby increasing the potential food source for the bee species.

Bird pollination   Many flowers are specifically adapted to pollination by birds. In North America, the most common bird pollinators are the many members of the hummingbird family (Fig. 31.11). The metabolic demands of hummingbirds are so great that they must obtain a steady supply of water and sugary nutrients in order to survive, and bird-pollinated flowers satisfy these needs effectively. The straw-like bills of hummingbirds enable them to drink large amounts of watery nectar from cavities deep within the flower. Meanwhile, anthers brush against their heads and feathers, loading them down with pollen that will be transferred to the next flower on their feeding tour. The storage of nectar in deep, narrow cavities prevents bees from drinking it, ensuring that pollination will be species-specific. Hummingbirds do not have well-developed senses of smell, and therefore bird-pollinated flowers generally do not have the lovely scents that we associate with bee-pollinated flowers. However, birds can see the red end of the spectrum much better than insects, so many bird-pollinated flowers are extremely colorful.

Pollination by other animals   Moths and butterflies are also involved in pollination relationships with a series of plants. These flowers attract their insects via a combination of scents and colors not unlike those of bee-pollinated flowers. However, their nectar is generally found at the base of a long, thin tube, where bees cannot reach it. Moths and butterflies insert their long, thin tongues

Figure 31.11 The ruby-throated hummingbird, an active bird pollinator.

Figure 31.12 A cave-dwelling nectar-eating bat, pollinating a banana flower.

through these tubes and gather the nectar as they hover over the flower.

A few flowers are pollinated by small, pollen-eating bats (Fig. 31.12). The poor sense of sight of bats requires that these flowers, which include several species of cactus, be powerfully scented so as to attract the animals when they emerge at night. Flower-feeding bats may lack teeth and have long, brush-like tongues that quickly gather nectar and pollen.

Pollination and self-incompatibility   Many plants are capable of self-pollination, and a few have flowers that are structured in a way that makes self-pollination almost unavoidable. In some species, however, there are barriers to self-pollination, even in cases where stamen and carpel are found in the same flower. The most interesting of these are genetic systems that induce **self-incompatibility,** meaning that one plant is unable to pollinate either itself or genetically similar plants. The barriers to self-fertilization are not completely understood, but they seem to act on markers present on the surface of the pollen grain. If the pollen grains carry marker proteins that interact with similar recognition proteins on the stigma, the result is that the entry of the pollen into the flower is blocked. Only pollen grains with different markers are allowed to enter the flower and complete fertilization. Incompatibility systems ensure that the full potential of sexual reproduction, including the ability to produce new combinations of genetic information, is utilized in plant breeding. Incidentally, these protein markers on pollen may be the very same ones that cause hay fever in humans.

Techniques of Seed Dispersal

Seed dispersal raises problems that are similar to those of pollination. In each case, the plant must develop a way to release a small object and increase the odds that that small object will find its way to a distant goal—a stigma of the same species, or a fertile spot to grow—all without the ability to intervene directly itself.

Wind dispersal   The most direct agent of seed dispersal is the wind. Small seeds can be easily scattered in the wind. Large seeds are sometimes encased in a wing-like structure that causes them to spin and twirl, gaining more distance, as they float to the ground. Others, such as dandelions and milkweed seeds, contain feather-like attachments that allow them to float on the slightest breeze, scattering seed and fruit for miles on a moderately windy day (Fig. 31.13). Westerners are familiar with the tumbleweed (*Salsola kali*) plant, whose bushes break their attachments to the roots as the plant dies, causing them to

The Grand Masquerade: Plants as Mimics

Some plants are able to employ subtle adaptations in flower structure, even to the point where they mimic the flowers of other species. The normal relationship between plant and pollinator is *mutualism:* Each organism benefits, and each organism makes a contribution. But could a plant exploit the insect by luring it without providing nectar as its contribution to the mutual relationship? Ordinarily, one might say no, because that would produce an evolutionary pressure to avoid the nectarless flower (and avoid starving to death!). However, if a species developed that looked enough like a genuine nectar-containing flower, insects might be unable to tell the two species apart.

Such a situation seems to have developed with two common plants in Western Europe—the bellflower (genus *Campanula),* which produces nectar, and the red orchid *(Cephalanthera rubra),* which does not. Despite differences in shape and color, the reddish orchid and the violet bellflower attract the same species of bees. Each flower is pollinated by the bees, and this first generated confusion among scientists who could not understand why the bees could not distinguish one flower from the other. L. Anders Nilsson of Sweden solved this problem by showing that the flowers do indeed resemble each other in the ultraviolet region of the spectrum. The orchid mimics the bellflower's color in the ultraviolet, where the bee's visual system is the most sensitive. Thus the mimic gets a "free ride" on the mutualistic relationship between the bees and the bellflower.

A few plants have even developed adaptations that mimic means to satisfy drives other than hunger. One genus of orchids *(Ophrys)* has a flower structure that looks like the abdomen of a female bee and releases a fragrance nearly identical to the sexual attractant she

produces! When the males emerge in the springtime, they attempt to mate with the flower. Such attempted breedings, known as *pseudocopulation,* do not produce offspring for the bees, but they spread pollen between flowers as male bees engage in one frustrating

Mimicry in plants. The markings of this orchid flower resemble a female bee, attracting male bees in a way that ensures pollination.

Figure 31.13 Milkweed seeds are dispersed by the wind and can be spread for miles under the right conditions.

tumble along the dry western plains in stiff winds. The plant retains its seeds until it breaks loose; they are then scattered as the dried bushes are blown aimlessly along the prairie.

Water dispersal Plants that grow in lakes and streams, including the water lily and water hyacinth, produce seeds that trap small air pockets, enabling them to float after they are shed. These floating seeds are carried by currents and surface winds, which disperse the seeds throughout the wet environment. One of the most remarkable seeds is the coconut. This monocot seed contains a *liquid* endosperm layer (the "milk" of the coconut), and it is light enough to float in seawater within its protective coating for many weeks. Volcanic islands in the Pacific, which may form in a matter of days, are quickly colonized by coconut palms. Water dispersal is clearly the reason for the success of this species in these remote locations.

Dispersal by animals The success of flowering plants in recruiting animals for pollination is mirrored by similar successes in seed dispersal (Fig. 31.14). In one sense, these two adaptations are related to the same structure, the flower. Many evolutionary biologists speculate that the development of an ovary, which tightly encloses the female gametophyte, was necessary to prevent the ovule tissue from being eaten by insects in search of nectar. After a successful fertilization, that ovary was available for

further modification. As we have seen, ripened ovaries are the fruits that enclose the mature seeds of flowering plants.

When a seed is enclosed within a rich, tasty fruit, it immediately becomes an object of interest to plant-eating animals. Apples, pears, grapes, and cherries are eaten by a wide variety of animals, and although the fruits provide these animals with rich sources of nourishment, the seeds within them pass through the animals' digestive systems unharmed. In many cases, in fact, the coatings of seeds are so tough that they will not allow germination until the seeds have been weakened by digestive enzymes. Chewing helps prepare for germination the seeds of the common grass timothy, which is widely used as hay for farm animals. Timothy seeds are bundled in a tight cob at the tip of the plant, and the seed surfaces are *scarified* (roughed up) as the hay is eaten. The manure of horses provides a rich source of potential fertilizer for the scarified seeds, and farmers appreciate the fact that spreading such manure on barren fields is an excellent way to plant them with prefertilized timothy.

When fruits are eaten by an animal, the ability of the animal to move guarantees that the seed will be deposited some distance from the parent plant. The bright colors produced by such fruits when they are fully ripe can be seen as "signals" to animals that they are ready to be eaten, helping to make certain that seeds are not dispersed until their development is completed. (See Current Controversies, Is Sex Doomed? Apomixis.)

Is Sex Doomed? Apomixis

One of the most widely known flowering plants is *Taraxacum officinale*. This organism thrives in areas of high light intensity, including frequently mowed lawns. It is exceptionally hardy, blooms early in the spring, and produces hundreds of seeds just a few days after blooming. The dandelion, as *Taraxacum officinale* is more commonly known, is one of a handful of plants that have adapted well to twentieth-century suburban life. It has survived even the most vigorous attempts to exterminate it. Could the dandelion be an example of how sexual recombination shuffles genes to achieve the greatest degree of fitness in a population? Not really. Beneath the fields of glistening yellow flowers, the humble, successful dandelion conceals a dark secret: it is one of a handful of plants that have abandoned sex.

For years, evolutionary biologists have been puzzled about why sexual reproduction is so widespread. Although sexual recombination does produce new combinations of genes, the constant and random mixing also makes it more difficult for a beneficial gene to be passed on to the next generation. Asexually reproducing organisms face no such problems; offspring are identical to parents, so beneficial genes take hold in a population quickly. The dandelion seems to have taken a first step along the road to producing identical offspring by abandoning the sexual process.

Although dandelions seem to be typical composite flowers, the pollen they produce is sterile. Furthermore, the female megagametophyte mother cell does not undergo meiosis to produce a haploid gametophyte. Instead, the diploid egg cell begins to develop into an embryo and quickly produces seeds that are genetically identical to the mother plant. The technical name for this process in plants is *apomixis*. Interestingly, although fertilization of the egg does not occur, pollination is still important to development in dandelions: the application of pollen to the carpel seems to serve as a signal that development should proceed. Thus the dandelion has discarded the gametophyte stage of the plant life cycle along with sex. The significance of a "higher" plant going asexual has not been lost on biologists. One might also say that dandelions have discarded our concept of *species* as well. Our definition of this term depends in part on the ability of organisms to interbreed. Because dandelions do not interbreed under any circumstances, they have led biologists to wonder whether the species concept can still be applied to them.

It is possible that dandelions are onto a good thing. They may have discovered that sex is too much trouble, too chancy, too random, and too conservative in the evolutionary sense. They retain the appearance of normal flowers and the need for a pollination signal as evolutionary remnants of processes that they have not yet completely discarded, suggesting that their not-too-distant ancestors at one time reproduced by means of a typical sexual cycle.

Only time will tell whether the dandelions are just an interesting experiment or the harbinger of things to come. Certainly we need not yet conclude that the sexual processes of flowering plants, which have filled the world with such beauty and provided so much inspiration, are about to be pushed off the scene. Just further motivation, perhaps, for those homeowners who resolutely set the lawn mower blades a little lower and cut off every little yellow flower they can reach.

Dandelions in a suburban lawn.

Figure 31.14 Fruits play an essential role in seed dispersal. LEFT: The eastern fox squirrel assists in the dispersal of apple seeds after eating the fruit of the apple. RIGHT: The domestic sheep carries away the burrs of Bido-bio.

Many plants produce seeds that are covered with tiny hairs or barbs, allowing them to stick to feathers or fur. Eventually, animals stop and take the time to remove these burrs from their bodies, but by the time they do, the seeds may have been carried for miles. The saliva that moistens such seeds as the animals pull them out may even help to activate the germination process.

SUMMARY

Flowering plants are the dominant form of plant life on land, and this dominance may well be the result of their reproductive abilities. The presence of meristematic tissue makes it possible for most plants to reproduce asexually, a fact that has been used by humans for plant propagation.

The basic reproductive cycle of flowering plants is an alternation of generations in which a diploid sporophyte plant produces a haploid gametophyte. The male gametophyte, the pollen grain, is formed within organs known as anthers. The female portion of the flower, the carpel, gives rise to the female gametophyte, the embryo sac.

Pollen released from the anthers falls on the stigma at the tip of the carpel and forms a pollen tube that grows through the female tissue of the carpel to reach the female gametophyte. A double-fertilization event in the ovary produces a diploid zygote cell, which produces the new plant embryo, and a triploid cell that forms the food-storing endosperm tissue of the seed. The embryo then develops into a form in which it is prepared for germination, endosperm grows and is packed with stored food material, and the seed is encased in the mature wall of the ovary to form a fruit.

Many flowering plants are dependent on animals for pollination and seed dispersal. Although pollination may occur simply by wind dispersal, pollination by animal vectors, such as bees, birds, and bats, is widespread and of great economic importance. The interactions of plants with their pollinators place selective pressures on both and have resulted in elaborate relationships between plant and animal. Seed dispersal systems may make use of physical forces, such as wind or water, or may take advantage of the movements of animals to disperse seeds.

STUDY FOCUS

After studying this chapter, you should be able to:

- Explain the relationship between plant reproductive processes and the life cycles of individual species.

- Trace how the alternation of generations is accomplished in various species, and identify the gametophyte and sporophyte stages of each cycle.

- Explain the "double-fertilization" event that takes place in flowering plants, and trace the tissues that are formed from each of the fertilization events.

- Describe the details of flower structure, seed formation, and the reproductive cycle in a number of flowering plants.

- Explain the phenomenon of coevolution of plants and animals, especially as it occurred among the flowering plants.

TERMS AND CONCEPTS

REVIEW

Objective Questions (Answers in Appendix)

1. A gametophyte is
   (a) haploid.
   (b) the plant produced when gametes join.
   (c) produced from zygotes.
   (d) diploid.

2. The female gametophyte develops within the floral organ called the
   (a) sepals.   (c) stamens.
   (b) petals.   (d) carpels.

3. Which of the following will develop into the male gametophyte?
   (a) anther   (c) carpel
   (b) microspore   (d) ovary

4. In addition to the fusion of male and female gametes, which produces a zygote, flowering plants have a second sperm nucleus that
   (a) fuses with a second haploid female, which develops into a second zygote.
   (b) is ejected from the plants once the first fertilization occurs.

(c) becomes the protective covering of the seed.
(d) fuses with the binucleate endosperm mother cell to form a triploid endosperm cell.

5. The bright colors of flowers can best be understood as
   (a) heat-regulating devices.
   (b) camouflage to distinguish floral organs.
   (c) attractive markings to assist vector pollination.
   (d) safeguards against self-fertilization.

Discussion Questions

6. How do sexual and asexual reproduction differ? What are the principal forms of asexual reproduction in flowering plants?

7. What are the major parts of the flower? What are the differences between complete and incomplete flowers?

8. What are the general features of the angiosperm life cycle? Can pollen grains properly be considered gametes? Why or why not?

9. Sketch the formation of the female gametophyte from a single megaspore cell. Which cell(s) of the gametophyte can properly be considered a gamete and why?

10. Many flowers have special adaptations that allow only one species of pollinating insect to enter. To a plant, adaptations that restrict pollination to a single vector species would present a selective advantage. Why?

READINGS

Coen, E. S., and E. M. Meyerowitz. "The war of the whorls: genetic interactions controlling flower development." *Science* 353 (1991): 31–37. The four floral organs are produced by the interactions of three sets of genes. This key paper explains what is known about how these genes interact.

Haring, V., et al. "Self-incompatibility: a self-recognition system in plants." *Science* 250 (1990): 937–941. A detailed description of the genetic and molecular basis of self-incompatibility.

Barrett, S. C. H. "Mimicry in plants." *Scientific American* 257 (September 1987): 76–83. Plants can mimic animals as well as other plants. This article describes some remarkable examples.

Corbet, S. "More bees make better crops." *New Scientist* 23 (July 1987): 40–43. A case study of the value of a robust local bee population in promoting plant productivity.

Stiles, E. W. "Fruit for all seasons." *Natural History* 93 (August 1984): 43–53. A study of seed dispersal by means of animal vectors.

32

Plant Growth and Nutrition

i f an alien civilization were to study our planet by means of highly detailed photographs taken from space, it would quickly note that one form of life covers more of the planet's surface than any other. That form of life does not have a nervous system and, in most cases, does not even have the ability to move about. The dominant form of life on land is the flowering plant. All too often we humans suffer from a curious prejudice whereby we fail to consider plants as living organisms. It is all too easy to ascribe the characteristics of animal life to life in general and to forget that in many ways, plants represent the most successful evolutionary experiment of all.

In this chapter we will explore some of the ways in which plants cope with a fundamental need of all living things: to obtain materials for growth and nutrition. Because plants lack a nervous system, we have a tendency to assume they are unable to respond to their environment—after all, that's what a nervous system is for. But we must remember that plants are involved in playing out an entirely different kind of survival strategy than animals are. Because most plants are rooted in place, they cannot search about for nutrients. Plants must obtain nutrients from the soils in which they find themselves. Their success is an important lesson in the diversity of living things.

GERMINATION AND DEVELOPMENT

The seed of a flowering plant consists of an outer coating that encloses and protects a plant embryo arrested at a critical point of development, and endosperm tissue that serves as a storage reservoir of food for the developing plant. In most monocots, the endosperm tissue of the seed remains separate from the embryo and serves as stored food waiting for the embryo to begin rapid growth. In many dicots, however, nearly all of the endosperm is taken up by the seed leaves of the developing embryo, and very little endosperm is left by the time the seed is mature. In such plants the stored food is concentrated in the cotyledons themselves (Fig. 32.1).

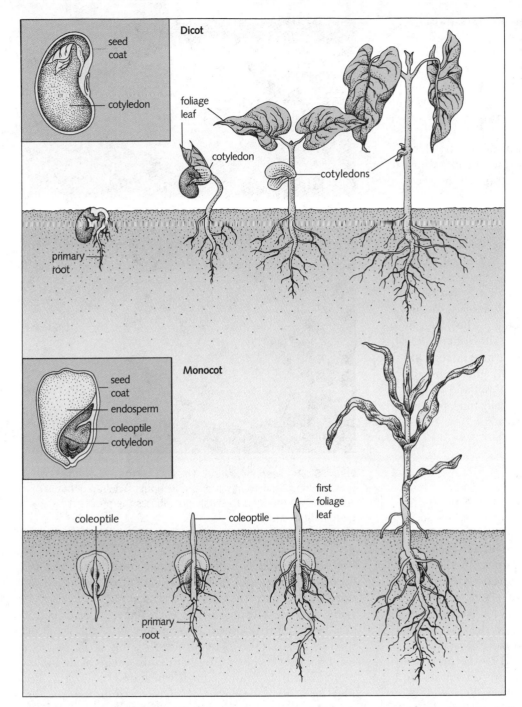

Dicot

seed coat

cotyledon

foliage leaf

cotyledon

cotyledons

primary root

Monocot

seed coat
endosperm
coleoptile
cotyledon

coleoptile

coleoptile

first foliage leaf

primary root

Figure 32.1 Seed germination in dicots and monocots. The coleoptile, shown in the diagram of monocot germination, is a protective sheath covering the growing tip of the plant.

Dormancy

Most seeds lie dormant for anywhere from a week to a year after they are produced. Cells within the seed have been dehydrated (seed water content is very low: 5–20 percent of the total weight of the seed), and most cellular processes have stopped. Although the cells within a seed are alive, the dehydration that removed nearly all water from the seed has left the cells of the embryo almost in a state of suspended animation. For chemicals to react, they first must come into contact with each other. The drastic reduction that occurs in the amount of free water when a seed develops makes it nearly impossible for chemicals to move around, and it all but stops the normal metabolic processes of the cells in the seed.

Many seeds can survive extended storage. Lotus seeds more than 2000 years old have been sprouted successfully, and seeds of the arctic tundra lupine (*Lupinus arcti-*

cus) recovered in the Yukon, and estimated by radioactive dating to be at least 10,000 years old, have also been germinated successfully. Despite such remarkable examples, very few seeds can survive indefinitely. Most seeds can last for only a few years without special types of storage, and viable seeds that are more than 100 years old are very rare. The seed is alive, but it cannot last forever in "suspended animation."

In many species, seeds cannot sprout unless they have undergone an extensive dormant period during which normal metabolic processes virtually stop. Other seeds cannot sprout unless they have been nicked or scraped or have had their tough outer coatings damaged in a fire. There are several reasons why seed dormancy is so common. Dormancy aids in the dispersal of seeds, and makes it more likely that the offspring will find a place to develop where they need not compete with the established parent plant. In the temperate regions of the world, extended dormancy can help ensure that a seed sprouts only in the spring. In seeds that are likely to be eaten by animals, a tough outer coating makes it more likely that the seeds will sprout only after the digestive system has weakened their tough outer coating. By that time the seed is likely to have passed out of the animal and to have been deposited in rich manure, which will help provide nutrients for the early stages of its growth.

Germination

Germination is the reactivation of growth in a plant embryo that leads to the plant's sprouting from its seed. Many factors control germination, and they vary with the species, but three factors are important in the germination of nearly all seeds: water, oxygen, and temperature.

In order to begin using the food stored within it, the seed must absorb a large amount of water, a process known as **imbibition** (literally, "drinking"). The rapid influx of water causes the seed to swell, and the coat of the seed may actually burst under the pressure this swelling creates (Fig. 32.2). As cells are activated by the presence of water, they begin to metabolize food stored within the seed, and this creates a demand for oxygen. The seeds require a supply of oxygen, which is plentiful in loose, damp soil. Finally, many species begin germination only at certain temperatures. Seeds from temperate areas of the world often germinate at temperatures as low as 2°C or 3°C, wheras plants from warmer climates may require temperatures no lower than 20°C.

The Beginning of Growth

The first structure to emerge from the seed is the primary root. The root quickly penetrates the soil and begins to absorb the water the developing embryo needs. As this

Figure 32.2 Seed imbibition. The lima bean seed on the bottom has been soaked in water for 12 hours. The rapid intake of water causes the seed to swell and initiates the process of germination.

occurs, the seed is rapidly appropriating and using the food that was stored in the endosperm or in the cotyledons. Seeds carry a store of extracellular digestive enzymes that are activated in the early stages of germination and help to break down the complex molecules making up the stored food. The process of digestion makes a brief flood of carbohydrates and amino acids available for plant growth. In grains such as wheat and barley, the *aleurone layer,* a single layer of cells surrounding the endosperm, releases these digestive enzymes and makes the food energy stored in the endosperm available for germination. The covering of the seed and the aleurone layer together are known as *bran;* they are removed when white flour is prepared from wheat seeds. The embryo is usually removed as well, although whole wheat flour contains the embryo, which is rich in proteins and oils.

The developing plant quickly establishes directional growth under the soil: Its roots are directed downward, and it begins to send a shoot towards the surface of the ground. We have seen that root apical meristems are protected with a root cap. In monocot grasses, the apical meristem of the first shoot is protected by the coleoptile as it pushes through the soil, and the first vegetative, or *plumular,* leaves, distinct from the cotyledons, do not emerge until after the stem has left the soil. In dicots, a number of different mechanisms are employed to protect the meristem. In some the cotyledons remain closed around the first true leaf until ground is broken. In others the stem bends into a loop that penetrates the soil and drags the top of the stem up into the air after it (see Fig. 32.1). Once the growing region of the stem, the meristem, has been moved above the soil, the stem straightens out and the plant begins to grow upward towards the sunlight.

NUTRIENT REQUIREMENTS

Like all plants, a growing seedling requires carbon dioxide and water to support the process of photosynthesis. As we have noted, these two components are available in the atmosphere and the soil, and plants have developed efficient ways of gathering them as they grow. Photosynthesis makes it possible to harness the energy of the sun to produce high-energy carbon compounds, satisfying the major requirements of most plants for a steady source of energy.

Both plants and animals, however, have nutritional needs that carbohydrates alone cannot satisfy. As they grow, all organisms must find the chemicals necessary to produce new proteins and nucleic acids. The synthesis of these compounds requires a source of reduced nitrogen compounds and other chemical constituents. You will recall that many proteins also contain sulfur atoms, and the prosthetic groups of many enzymes include metals such as zinc, iron, and manganese. The chlorophyll molecule contains a magnesium atom. Nucleic acids include phosphorus. Calcium is used within many cells to control the rates of biochemical pathways. These elements are just a few of the essential nutrients that plants must obtain in order to grow and reproduce.

Plant Chemical Composition

The first step in determining which nutrients are required for plant growth is to analyze the chemical composition of healthy plants. In nonwoody plants, anywhere from 70 to 95 percent of the total weight is water. Most of the remaining dry weight is attributable to organic compounds. Carbon, hydrogen, and oxygen account for 95 percent of dry weight, but 13 other elements are present in amounts large enough to measure.

The average amounts of 16 elements found in significant quantity in typical crop plants are shown in Table 32.1. A list such as this is useful, but it represents only an average. There is tremendous variation within individual plants: Calcium averages about 1.7 percent of dry weight in alfalfa but only 0.4 percent in corn. A white oak leaf contains as much as 10 times the dry weight percentage of manganese as does a blade of alfalfa. Furthermore, individual species vary with the type of soil in which they have been raised. Plants grown in the iron-rich soils near mining operations contain much more iron than plants of the same species grown in more typical soils. Finally, the chemical composition of a plant tells us only which elements are present, not which ones are essential for the growth of the plant.

Table 32.1 *Chemical Composition of Typical Crop Plants*

Element	Form in Which Element Is Available	Percent of Dry Weight in Tissue
Carbon	CO_2	45
Oxygen	O_2, H_2O, CO_2	45
Hydrogen	H_2O	46
Nitrogen	NO_3^-, NH_4^+	6
Potassium	K^+	1.5
Calcium	Ca^{2+}	1.0
Magnesium	Mg^{2+}	0.5
Phosphorus	$H_2PO_4^-$, HPO_4^{2-}	0.2, 0.2
Sulfur	SO_4^{2-}	0.1
Chlorine	Cl^-	0.010
Iron	Fe^{3+}, Fe^{2+}	0.010
Boron	H_3BO_3	0.002
Manganese	Mn^{2+}	0.005
Zinc	Zn^{2+}	0.0020
Copper	Cu^+, Cu^{2+}	0.0006
Molybdenum	MO_4^{2-}	0.00001

Plant Nutrients

Early in the twentieth century, plant scientists attempted to determine which elements were essential for plant nutrition by growing large numbers of plants in nutrient-poor soil and then watering the plants with solutions containing different combinations of mineral nutrients. This technique identified a number of nutrient elements, including copper, zinc, and manganese, whose importance had not been suspected before that time. More recently, plant biologists have used a liquid growth technique known as **hydroponic culture** to determine a plant's needs for small amounts of nutrients (Fig. 32.3). The roots of plants grown this way are immersed in defined nutrient solutions. Because plant root cells need oxygen for respiration, the culture medium must be bubbled with oxygen.

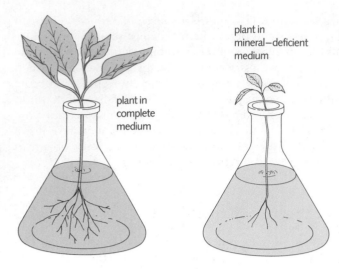

Figure 32.3 The nutritional requirements of plants can be determined by hydroponic culture, growth in a defined liquid medium.

Studies with nutrient cultures have helped to show that a few essential elements must be present in relatively large amounts for plants to thrive. They include oxygen, carbon, hydrogen, nitrogen, potassium, calcium, magnesium, phosphorus, and sulfur. These nine elements, which are known as **macronutrients,** are common in plant tissue. Potassium and sodium, for example, are the major salts found in intracellular and extracellular fluids, respectively, in plants. Calcium is required for the production of cellulose-containing cell walls. The remaining essential nutrient elements are known as **micronutrients.** The roles played by micronutrients, which may be present in plants in amounts less than 10 parts per million of dry weight, are less obvious. The micronutrients include boron, chlorine, copper, iron, manganese, molybdenum, and zinc. Many of these elements are required in key cellular enzymes (manganese, for example, is needed for the water-splitting apparatus of photosynthesis), although the actual cellular roles of some of the nutrients remain obscure.

Deficiencies in Mineral Nutrients

It is very rare for any soil to lack a certain nutrient completely. Plants may thrive even in nutrient-poor soils, in part because plant roots contain specific transport systems for many nutrients. One indication of this is the fact that root tissue grown in nutrient solutions may contain 50 to 1000 times the concentrations of sodium, potassium, and calcium found in the nutrient solution itself.

Active transport processes move these ions across the cell membranes of root hairs and into the general circulation of water flow into the plant. Remember that in order to enter the xylem, each mineral must first pass through the ring of endodermal cells embedded in the impermeable Casparian strip. This process allows the endodermal cells to control which ions are added to the general circulation of the plant. Cells in other plant tissues are able to concentrate ions by a similar mechanism, ensuring that each cell has an adequate supply of nutrient ions.

However, even the most efficient transport system has its limitations, and in many cases the soil content of a few nutrients is so low that the plant is unable to concentrate them in its root system. A persistent deficiency of any of the major nutrients eventually causes serious problems for the plant (Fig. 32.4). Magnesium deficiencies, for example, limit the ability of the plant to synthesize chlorophyll, because one magnesium atom is needed for each chlorophyll molecule. Magnesium-deficient plants develop **chlorosis,** a yellowing of the leaves, in the older portions of the plants because damaged chlorophyll molecules cannot be replaced quickly enough to maintain the normal green color. By contrast, plants lacking copper synthesize chlorophyll in normal amounts, but they can synthesize neither the electron transport components in mitochondria nor the chloroplasts, which require copper atoms. This places several limitations on their ability to produce and use energy, and copper-deficient plants are severely stunted in growth.

Table 32.2 summarizes the major functions of the essential nutrients in plants and the symptoms that result from a prolonged deficiency of those nutrients.

Figure 32.4 Symptoms of magnesium deficiency in tomato leaves.

Table 32.2 *Essential Nutrients in Plants*

Nutrient	Chemical Form	Need in Plant	Symptoms of Nutrient Deficiency
Macronutrients			
Carbon	CO_2	Organic molecules throughout plant	
Oxygen	O_2, H_2O	Organic and inorganic molecules throughout plant	These three nutrients, CO_2, O_2, and H_2O, are so basic to plant life that major deficiencies result in the death of the organism. Therefore it is not appropriate to speak of the effects of a carbon or hydrogen deficiency.
Hydrogen	H_2O	Organic and inorganic molecules throughout plant	
Nitrogen	NO_3^-, NH_4^+	Found in proteins and nucleic acids	Paling and loss of green color from leaves (chlorosis), which may develop red or purple color
Potassium	K^+	Required for enzyme activities and regulation of cell ionic balance	Chlorosis; brown spots along edges of leaves; general weakening of plant
Calcium	Ca^{2+}	Required for cell wall synthesis, as an enzyme cofactor, and for cell division	Stunting of growth in meristemic (growing) regions; most soils have so much Ca^{2+} that deficiency is rare
Magnesium	Mg^{2+}	Part of the chlorophyll molecule; required for protein synthesis and as an enzyme cofactor	Chlorosis develops between the veins; patches of red or purple color appear on older leaves
Phosphorus	$H_2PO_4^-$, HPO_4^{2-}	Forms part of nucleic acids and phospholipids, ATP, and other energy-storing intermediates	Stunted growth; deep green color develops in older leaves
Sulfur	SO_4^{2-}	Required in many proteins and key compounds	General chlorosis and yellowing of leaves
Micronutrients			
Chlorine	Cl^-	Important in maintaining cellular osmotic balance and in some reactions of photosynthesis	Wilting; stunting of root growth; reduced fruit
Iron	Fe^{3+}, Fe^{2+}	Important part of many enzymes, including the cytochromes; required for chlorophyll synthesis	Chlorosis of younger leaves
Boron	H_3BO_3	Important in nutrient transport within plant	Thickened darker leaves; stunted meristemic growth
Manganese	Mn^{2+}	Required in enzymes of the Krebs cycle and for oxygen release during photosynthesis	Small, speckled dead (necrotic) spots on leaves
Zinc	Zn^{2+}	Required for auxin synthesis and protein synthesis	Dramatically reduced leaf size and chlorosis between veins; reduced flowering
Copper	Cu^+, Cu^{2+}	Enzyme cofactor; required for photosynthetic electron transport	Withered leaf tips and darkened leaf color
Molybdenum	MO_4^{2-}	Required for nitrogen fixation	Chlorosis, including mottling and wilting of leaves

FULFILLING THE NUTRIENT NEEDS OF PLANTS

Soil and Its Nutrients

Plants obtain nearly all of their nutrients from the soil in which they grow. **Soils** vary widely in their chemical composition as well as in their ability to hold water, and this accounts in part for the great differences in native plant growth in various parts of the world (Fig. 32.5). The surface of the earth contains all of the naturally occurring chemical elements. Most of these are locked in the rocks that make up the majority of the earth's crust. The action of wind and water breaks many of these rocks up into smaller particles, and the soluble minerals within the rocks are dissolved by rainwater and washed away. This weathering process gradually produces the small inorganic particles from which soil is formed.

As we saw in Chapter 3, soil contains organic materials as well, and these are formed first from microorganisms, including bacteria, protozoa, fungi, and simple plants. Many of these organisms contribute to the breaking up of solid rocks, as pressures from their growing roots and rhizoids split the largest particles of the crust into smaller and smaller fragments. Gradually, the combination of *physical* and *biological* processes produces a complex soil containing organic and inorganic molecules, as well as a mixture of minerals that may include essential plant nutrients.

Soil particles vary in size and composition. Sandy soils contain quartz (SiO_2) particles that are classified as fine (20–200 μm). (Refer to the Appendix for a discussion of metric units.) Smaller (2–20 μm) quartz particles are known as *silt;* soils containing a mixture of sand, silt, and clay particles are known as **loam.** Soils that are too sandy are unable to hold water, whereas soils that are mostly clay do not allow water to penetrate. The best loamy soils contain a mixture of sand and clay that allows rainfall to penetrate and then holds much of the water in the moist soil for many days.

Clay particles are covered with negative (–) charges, and this is one of the reasons why clay is able to hold the highly polar water molecule and to bind large amounts of positively charged ions (+). The importance of this property can also be seen in Fig. 32.6. Many of the essential elements needed for nutrition are cations (positively charged), and clay particles hold these elements until they are absorbed by root cells.

Soil Fertilization and Management

In some regions of the world, soils are so rich that they contain ideal nutrient mixtures for a wide variety of crop plants. This is true in much of the American Midwest, one of the most productive farming regions of the world. In many places, however, the soil may lack one or two essential nutrients. For thousands of years, farmers have

Figure 32.5   Soils vary widely in their ability to support plant life. Rich layers of humus, soil packed with decaying organic matter, are major nutritional sources for plants in many environments.

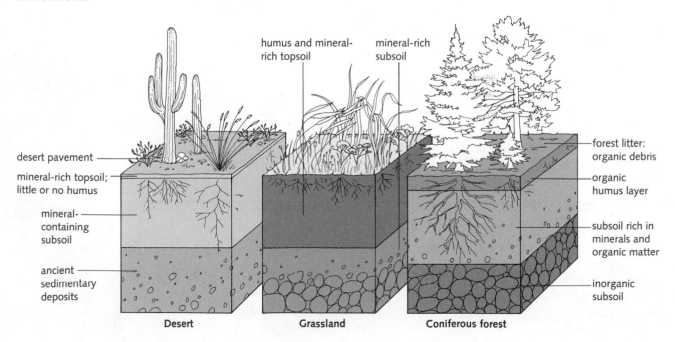

desert pavement

mineral-rich topsoil; little or no humus

mineral-containing subsoil

ancient sedimentary deposits

humus and mineral-rich topsoil

mineral-rich subsoil

forest litter: organic debris

organic humus layer

subsoil rich in minerals and organic matter

inorganic subsoil

Desert   **Grassland**   **Coniferous forest**

sought to improve the condition of their soils by adding essential nutrients in the form of **fertilizers.** Manure—human and animal—is one of the oldest and most widely used forms of fertilizer. Manure is rich in organic and inorganic nutrients. By adding animal manure to heavily farmed fields, it is possible to return to the soil many of the micronutrients and macronutrients that plants remove during their growth process.

Many plants require large quantities of macronutrients. Harvesting these plants may permanently remove macronutrients from the soil and deplete its nutritional value, reducing crop yields in following years. One of the most important food crops, corn, removes large amounts of nitrogen, phosphorus, and potassium from farmland. To use the same land for producing corn year after year requires that the soil be carefully managed and that most nutrients be replaced when the next crop is planted.

One way to replace such nutrients is by using artificial fertilizers (Fig. 32.7). Some of the elements in fertilizers are obtained from the mining of mineral-rich rock formations, and others are synthesized by the chemical industry. Many fertilizers used in the United States are particularly rich in the macronutrients nitrogen, phosphorus, and potassium. Commercial fertilizers, including those used for lawn and garden applications, are commonly labeled with the percentage content of these three nutrients. A bag of garden fertilizer labeled "20-10-5" contains 20 percent nitrogen, 10 percent phosphorus, and 5 percent potassium.

These fertilizers are useful in replacing the nutrients that crop plants remove, but they must be used with great

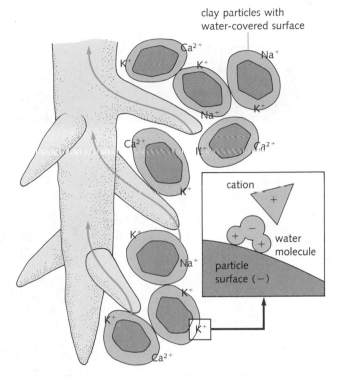

Figure 32.6 The surfaces of clay particles hold negative charges, and this enables them to bind positively charged mineral ions as well as water. These properties enable clay-containing soil to bind moisture and nutrients.

Figure 32.7 The proper use (and overuse) of chemical fertilizers. LEFT: The fertilization of tomatoes contributes to the growth of this valuable cash crop. RIGHT: The overuse of fertilizers may inhibit plant growth and lead to barren, wind-eroded fields.

Solar-Powered Marine Animals

The nutritional relationships between plants and animals are so obvious that they scarcely need to be mentioned. Plants are the ultimate food sources for nearly all forms of animal life. In most cases this means that, directly or indirectly, animals eat plants and extract chemical energy from them. As humans, we may believe that the invention of farming represents a fundamental advance, made possible by human intelligence and the development of culture, in harnessing plants for our own uses. As important as farming is, however, we were not the first species to develop it. Corals, for example, have developed associations in which algae are trapped in small chambers within the coral animals. Nutrients released from the algae give the corals much of their food supply and provide much of the energy required to build great coral reefs throughout the world. Does this mean that "farming animals" must lead stationary lives, in the manner of coral, in order to be successful? Not at all.

Consider, for example, the sea slugs, a group of invertebrates known by the more scientific name of nudibranchs. These flattened, shell-less animals are found in warm waters throughout the world feeding on corals, hydroids, and sea anemones. But as William Rudman,

a researcher in Sydney, Australia, has shown, a few nudibranchs have something else going for them: They have developed large colonies of algae that produce their food. One species, the "blue dragon" (*Pteraeolidia ianthina*), is found off the coast of Australia. Early in

The blue dragon (*Pteraeolidia ianthina*) gets its color and much of its nourishment from colonies of algae that live within it.

its life the blue dragon grazes on hydroids that themselves contain colonies of algae. As the blue dragon grows, it gradually develops colonies of its own, possibly by trapping some of the algae from its food in pouches within the nudibranch gut. Algae find points within the gut in which they can remain and thrive, releasing nutrients into the animal's digestive system. The blue dragon is remarkably well adapted to taking advantage of the crop that it grows within itself. Large outgrowths known as *cerata* protrude from its body. These cerata increase the animal's surface area, and thin, algae-filled networks of tubes linked to the animal's gut run just beneath the surface of its skin. Given the photosynthetic resources of a plant, the blue dragon assumes at least some of the structural characteristics of a plant. The delicate clusters of cerata, which furnish nutrients to the animal almost as leaves do to a plant, are arranged so that they seldom shade each other. For its part, the blue dragon makes the most of this association. Once it has acquired a colony of algae, an individual does not seem to return to normal feeding but appears content to bask its nearly transparent body in the sunshine and enjoy the nourishment that comes from within.

care. Overfertilizing, a mistake many backyard gardeners make, can kill crop plants by putting too high a concentration of salts into the soil. The intensive use of fertilizers can also affect the groundwater. When large amounts of nitrogen- and phosphate-containing fertilizer are used near wetlands and streams, runoff from the fields may contaminate the water with large amounts of these two nutrients. Because the growth of algae in fresh water is often limited by the availability of these nutrients, the

sudden increase in nutrients may cause a "bloom"—the sudden increase in the population of algae—that upsets the natural balance of the freshwater ecosystem.

Symbiosis and Plant Nutrition

Mycorrhizae Plants live by their leaves and by their roots. The leaves, of course, are the main reasons why we

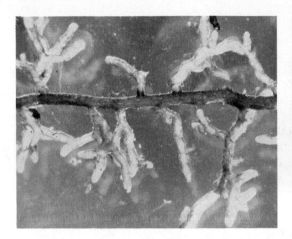
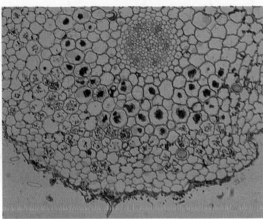

Figure 32.8 LEFT: The roots of many plants are covered with mycorrizhae fungi, which perform many of the functions of root hairs. Mycorrhizae are essential for the growth of several species of plants. RIGHT: This cross section of an orchid root shows the filamentous processes of mycorrhizae fungi inside the cells of the cortex.

consider plants autotrophs. Leaves are the major sites of the photosynthetic activity that provides plants with an independent source of energy. Many plants, perhaps as many as 80 percent of terrestrial plants, have root systems that are dependent on other organisms. The most widespread example of this is a symbiotic relationship with the fungi known as **mycorrhizae** (*myco* means "fungus," *rhizae* means "roots"; see Fig. 32.8).

In many common forest trees, including oaks and pines, these fungi grow in close association with root tips and penetrate the internal tissues of the roots. They take over the functions of root hairs and provide the plant with moisture and mineral nutrients. In a few cases, there is good evidence that the mycorrhizae also manufacture hormones that affect the plant itself and stimulate the growth of new root tissue. Many of the mushrooms that spring up in shaded areas on the forest floor are merely the reproductive structures of these complex fungi, which cover the roots of the very trees that shade them.

In many types of orchids, the dependence on mycorrhizae has become extreme. Unless their seeds are infected by fungi of the genus *Rhizoctonia*, the orchids cannot germinate. The forest trees that commonly associate with mycorrhizae often cannot be grown in the soil of the plains and prairie, not because the plants themselves have any difficulty with the environment but because their mycorrhizal associates cannot adapt to the soil conditions there. The relationship between mycorrhizae and the plants they infect is a perfect example of *mutualism,* the form of symbiosis in which both species benefit. The fungi seem to be more efficient at gathering mineral nutrients from the soil than most plant roots are, and the plant in turn provides the fungi with organic nutrients.

Nitrogen fixation Earlier in the text we considered another important example of symbiosis involving root tissue. The nitrogen-fixing bacteria *Rhizobium* (Fig. 32.9)

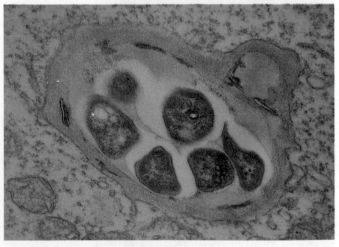
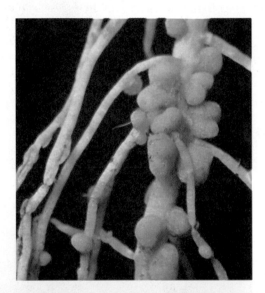

Figure 32.9 TOP: Nodules containing nitrogen-fixing bacteria on the roots of peas. BOTTOM: This thin section through the root system of a pea plant shows portions of six nitrogen-fixing *Rhizobium* bacteria.

live in nodules of the roots of legumes such as peas and soybeans. The roots provide these bacteria with a stable protective environment in which they are supplied with food, and the bacteria in turn use their ability to "fix" nitrogen gas from the atmosphere (N_2) into reduced nitrogen compounds such as ammonia (NH_3). Plants must have reduced nitrogen in order to synthesize proteins, and as we have seen, it often must be provided by adding nitrogen-rich fertilizers to the soil. Plants that can associate with nitrogen-fixing organisms, however, can make their own useful nitrogen fertilizer. Its relationship with these bacteria has made the soybean one of the world's leading protein-producing crops. As you might expect, some of the major research efforts in plant science are directed towards finding ways to produce nitrogen-fixing relationships in such other important crop plants as wheat, corn, and rice.

At first, it seemed as if these research efforts were genuine long shots, but the surprising news from plant research laboratories is that some of these attempts are already halfway to success—they have been able to coax nonlegume plants to produce the nodules that can accommodate nitrogen-fixing bacteria. Researchers used enzymes to dissolve much of the cell wall surrounding root cells and then infected the plants with nitrogen-fixing bacteria with the aid of other compounds that produce holes in the plant cell membrane.

The results of these efforts are bacteria-laden root nodules (Fig. 32.10) in which bacteria remain for the lifetime of the plant. Laboratories throughout the world are working on the second stage of these efforts, which may prove more elusive—"turning on" the nitrogen-fixing activities of bacteria so that the nodules will "feed" the plants with fertilizer. Nonetheless, the first step—establishing a stable relationship between plant and bacteria—has been taken, and the potential benefits of making that relationship work are very great.

Meat Eaters

Though all higher plants are photosynthetic autotrophs, a few species find ways to supplement their diets. Many of these plants are found in nitrogen-poor soil, so the occasional capture of small organisms enables them to obtain badly needed nitrogen compounds. The prey of choice for such plants is small insects that can be lured and trapped by specially modified leaves (Fig. 32.11).

The best-known **insectivorous** ("insect-eating") plant is the Venus-flytrap, which is native to North America and is found primarily in the Carolinas. The bilobed leaves of the Venus-flytrap are an open invitation for flying insects to land and sample the juices that moisten their inner

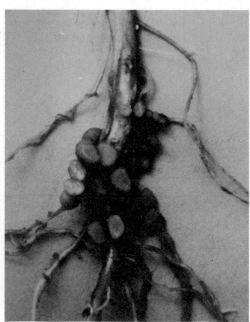

Figure 32.10 LEFT: The oilseed rape plant is the source of canola oil. RIGHT: Preliminary experiments have induced the plants to form root nodules, like these in a soybean plant, that accommodate nitrogen-fixing bacteria.

Figure 32.11 Insectivorous plants. LEFT: The Venus-flytrap is about to digest an insect that has become trapped in its clamp-like leaf. CENTER: The sundew has captured a number of insects in its sticky droplets. RIGHT: The pitcher plant has caught a mosquito in its cup-shaped leaf.

surfaces. When an insect is unfortunate enough to step on the leaves, however, they snap shut in a reaction so fast that the insect is caught inside. The response of the plant is so swift that it almost seems to have a nervous system. Actually, the steps of the insect on the leaves stimulate hair cells that activate a very rapid proton-pumping system in the cells within the leaf. The sudden pH change that the pumping produces causes a rapid movement of water into the tissues where the two lobes of the leaf are joined. As these cells expand with water, they force the jaws of the plant's trap shut. The animal trapped inside is then digested by enzyme-rich solutions secreted from the epidermis of the leaf, and the breakdown products are absorbed directly into the plant.

Few plants can match the mechanical perfection of the Venus-flytrap, but there is more than one way to trap a bug. The pitcher plant, also a North American native, has a large curved leaf covered with tiny downward-pointing bristles. When a fly enters the plant to sip the juices that moisten its inner surface, the direction in which the bristles lie forces the fly to move deeper and deeper. The close quarters of the leaf make it impossible for most insects to fly away, and when they finally reach the bottom of the pitcher-shaped leaf, they are digested in a pool of enzymes and absorbed into the leaf. These plants illustrate not only the remarkable adaptations that plants have evolved but also one of nature's great ironies. Many plants suffer the fate of being eaten by insects, but many others have been able to put insects to use for their own benefit. The most benign example is insect pollination, but the insectivorous plants have added a new chapter to

the story: They have invited the bugs to dinner and then served them as the main course.

SUMMARY

The seeds of higher plants are structures that are specialized to suit the reproductive patterns of their particular species. In many plants, a pattern of seed dormancy is essential to survival. Dormancy is ensured by the dehydrating of seed tissue, enabling seeds to survive for months or years without apparent changes. Food is stored within the seed, either as endosperm or in the seed leaves of the plant embryo. Germination, the activation of a seed, is often prompted by imbibition. As it begins to grow, the developing plant sends out a shoot towards the surface of the ground as well as a series of roots that push through the soil in search of water and nutrients.

Although they obtain much of their chemical energy from the products of photosynthesis, plants have important nutritional needs in terms of the chemicals, organic and inorganic, that they must obtain from the environment. The exact chemical needs of plants can be determined by several techniques, including hydroponic culture. Deficiencies in any essential nutrients may affect plant growth and cause such conditions as chlorosis, a loss of leaf color. The ability of plants to absorb nutrients is highly dependent on the nature of the soil in which plants grow. The best soils are mixtures of clay and sand that retain some water while allowing excess water to drain. Soils that are lacking in certain nutrients may be supplemented by the addition of fertilizers, although these bring with them their own set of problems and concerns. Many plant nutritional needs are met by close associations with other types of organisms, including nitrogen-fixing bacteria and the fungi known as mycorrhizae. A select group of plants is able to meet its nitrogen needs by trapping and digesting insects.

STUDY FOCUS

After studying this chapter, you should be able to:

- Cite some basic physiological principles that govern plant behavior, growth, and development.

- Explain why dormancy is important in seeds, and describe the initial events of seed germination.

- Explain the relationship between soil composition and plant growth and development.

- Describe the close relationship between plants and other organisms that provide plants with nutrients and moisture.

- Explain the demands that growing commercial crops places on soil resources, and explore some of the ways in which these resources are replenished.

TERMS AND CONCEPTS

germination *640*
imbibition *640*
hydroponic culture *641*
macronutrients *642*
micronutrients *642*
chlorosis *642*

soils *644*
loam *644*
fertilizers *645*
mycorrhizae *647*
insectivorous *648*

REVIEW

Objective Questions (Answers in Appendix)

1. A dormant embryo is called a(n)
   (a) endosperm.
   (c) seed.
   (b) cotyledon.
   (d) gametophyte.

2. The type of soil most likely to help plants grow contains
   (a) many negatively charged particles.
   (b) a mixture of sand and silt.
   (c) few air spaces and is tightly packed around the roots.
   (d) mostly silt.

3. Mycorrhizae are
   (a) vascular bundles of phloem.
   (b) plants that have a mutually beneficial association among legumes and nitrogen-fixing bacteria.
   (c) structures on roots that have nitrogen-fixing bacteria.
   (d) fungi that have a mutually beneficial association with a root.

4. A symbiotic nitrogen-fixing relationship is exemplified by
   (a) *Rhizobium.*
   (c) orchids.
   (b) mycorrhizae.
   (d) fungi.

5. Which of the following statements about commercial nitrogen fertilizers is *not* true?
   (a) Nitrogen runoff from overfertilized fields can cause water pollution.
   (b) Excessive nitrogen fertilization can harm plants by causing them to absorb too much water.
   (c) Nitrogen fertilizers are often formulated with potassium and phosphorus.
   (d) Some crops remove much more nitrogen from the soil than others.

6. The folding of the leaf lobes of the Venus-flytrap is a response to changing
   (a) light patterns.
   (c) wind speed.
   (b) water pressure.
   (d) wind direction.

Discussion Questions

7. What critical events precede the germination of a seed? Why is dehydration effective in increasing the dormant period of most seeds?

8. Most plants are considered to be autotrophs, even though they have (as we have seen in this chapter) extensive mineral needs that must be met by soil nutrients. How can organisms with so many nutritional needs be considered autotrophs? (You may want to look up the definition of *autotroph* for this question.)

9. What are some of the macronutrients necessary to plant growth? Which of these are provided by the raw materials used for photosynthesis? Which are not?

10. Why might an excess of nitrogen-rich fertilizer cause "bulging" of plant roots (excessive water loss and death of root hair cells)?

11. Carnivorous plants are often able to grow on very poor soil that has serious nutrient deficiencies. How might the ability to trap and digest insects be useful under such conditions?

READINGS

Rudman, W. B. "Solar-powered marine animals." *Natural History* 96 (1987): 50–53.

Mandoli, D. F., and W. R. Briggs. "Fiber optics in plants." *Scientific American* 251 (August 1984): 90–98. The germinating plant uses fiber optics to sense the direction of sunlight.

Heslop-Harrison, Y. "Carnivorous plants." *Scientific American* 238 (February 1978): 104–115. A fascinating look at plants that trap animals for food.

33

Control of Plant Growth and Development

*i*n many cultures throughout the world, plants are not considered living things. Consciously or subconsciously, we sometimes regard plants as part of the background, the scenery, a landscape that supports life but is distinctly passive. One of the reasons for this belief may be that plants do not seem to display the rapid, coordinated responses to their environment that animals do. If we spend a few quiet hours in a deep forest, our attention at first may be attracted to the great diversity and activity of animal life. It is impossible not to notice the rapid movements of insects, the characteristic songs of birds, or the quiet stealth of mammals as they seek their prey. Everywhere animals seem to possess the ability to respond to their surroundings, quickly and decisively. Behind this activity, the trees and lawn beneath may seem inanimate and unresponsive. Yet just as surely as animals, plants can and do respond to their environment.

If we neglect the ways in which plants react to their surroundings, it may be only because the pace and scale of that response are more subtle than in many animals. As we shall see, this does not make that response any less effective in ensuring the organism's survival, and it does not make it any less important to other organisms, ourselves included. In this chapter we will examine some of the ways in which higher plants produce an organized and coordinated response to their surroundings and examine characteristics that have made these organisms so successful in colonizing the land surface of the planet.

CONTROL OF PLANT GROWTH PATTERNS

When a plant germinates and begins to develop, an obvious need arises to coordinate activities in different parts of the growing organism. The tissue in the root must grow quickly enough to provide moisture, support, and nutrition—but not so quickly that all the resources of the plant are used up in root tissue growth. Similarly, the tissues that produce leaves must grow quickly enough to ensure a rate of photosynthesis that will enable carbohydrates to flow back into the roots. In

Figure 33.1 Although the growth of a plant is indeterminate, it is not random. The growth patterns of trees are often so specific that it is possible to distinguish two species merely by their branching patterns.

Figure 33.2 Geotropism, the ability to respond to the force of gravity, enables plants to orient themselves correctly even in the most difficult terrain.

order to do so, the leaves and stems that carry out photosynthesis must bend toward the sunlight; in this way the plant will find as much solar energy as possible. The plant must respond to the time of day, so that growth and metabolic cycles can be regulated efficiently; it must respond to the time of year, so that reproduction and growth can be completed during the most favorable weather; and it must survive the ravages of predators such as insects that eat its most productive tissues.

In animals, many of these problems are solved by the *nervous system,* a group of tissues and organs specialized to carry messages from one end of an organism to the other. Plants lack nervous systems, yet they perform many of these functions every bit as efficiently as animals.

The Nature of Plant Growth

The growth of most plants is *indeterminate*—that is, there are no absolute restrictions on the precise size a plant will reach or the shape it will assume. This does not mean, however, that plant growth is random. In fact, plant growth is generally subject to a number of controls and restrictions, some of which are so precise that they produce floral and leaf structures identical in all members of a species. These controls are the effects of influences that regulate where and to what extent growth can occur. They are the reason why it is possible to tell, even from a great distance, whether a large tree is an oak or a maple (Fig. 33.1). Controls on growth rates and patterns are important to survival because they help mold the shape of the organism to suit important factors in the environment such as light, gravity, wind direction, and moisture.

Responses to the Environment

For thousands of years humans have marveled at the ability of plants to respond to stimuli from the environment. These responses of plants to factors in the environment are known as **tropisms,** a term derived from a Greek word that means "to turn."

Geotropism Every seedling has the ability to sense and respond to the force of gravity, a property known as **geotropism** (Fig. 33.2). Geotropism affects stems and roots differently, directing root growth toward the source of gravitational attraction and stem growth away from it. The differences between root growth and stem growth provide an example of how tropisms can be either positive or negative. Emerging roots exhibit *positive geotropism,* stems *negative geotropism* (they turn against the force of gravity). Besides helping a developing seedling find soil and air, geotropism also helps larger plants respond to major disruptions such as storms. Thanks to geotropism, small trees blown over in wind storms are able to return slowly to a vertical growth pattern.

Phototropism   Plants respond to light, a property known as **phototropism.** When the amount of light hitting one side of a plant is greater than the amount hitting the other side, the plant turns toward the light. If the light source is fixed, the plant gradually orients itself such that its leaves are perpendicular to the angle of illumination. The phototropic response is so quick that young seedlings turn toward a source of light in a matter of just a few hours (Fig. 33.3).

Thigmotropism   Plants are also affected by touching solid objects, a response known as **thigmotropism.** Climbing plants, including ivy and pole beans, are able to sense when their growing shoots make contact with a solid object (Fig. 33.4). When they do, the rate of growth on the side making contact slows down and the shoot curls around the solid object, attaching to it tightly. This tropism enables these plants to find solid objects and cling to them for support. (See Theory in Action, How Does a Plant Know You Touched It?, p. 655.)

Besides these well-defined tropisms, plants can respond to a host of other environmental factors, including the length of the day and the season of the year. Many plants, including such well-known species as the morning glory, open their flowers only during the daytime. Others, such as the cereus cactus of the American Southwest, open only at night. Many plants respond so precisely to seasonal changes that they are said to contain *biological clocks* that adjust biochemical processes going on within them to the length of day and the time of year.

PLANT HORMONES

One of the ways in which plants solve the problems of survival is by using chemical messengers that carry signals from one part of the organism to another (see Theory in Action, Chemical Defense Mechanisms: How Plants Fight Back, p. 656). These messengers ensure that all tissues of the organism can act in concert, enabling even the largest plants to respond directly to changing conditions in the environment. The messengers that perform this task are chemicals known as **hormones** (Fig. 33.5). A hormone is a substance that is produced, usually in small quantities, in one part of an organism and affects the physiology of another part. Hormones are accurately described as "chemical messengers." Plant hormones can be produced by several different tissues, including the cells found in leaves, fruits, and seeds.

Figure 33.3   These corn seedlings, bending towards a light source, display positive phototropism.

Figure 33.4   Thigmotropism, the ability to respond to contact, enables the tendrils of this squash plant to obtain support from a solid object.

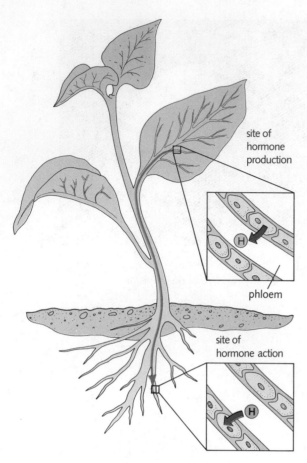

Figure 33.5 Hormones are chemical messengers which may be produced in one region of a plant and affect cells in another region.

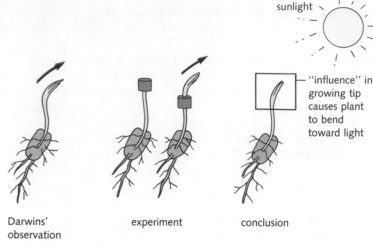

Figure 33.6 Charles and Francis Darwin discovered that the tip of a growing plant plays a special role in phototropism. If the tip was covered with a dark band, a seedling would not bend towards the light even if the rest of its stem was illuminated. Covering a region below the tip had no effect—the seedling still turned towards the light.

Plants are able to use hormones to regulate their rate and direction of growth, to control the time at which they produce flowers and drop leaves, and even to coordinate the functions associated with germination. A hormone does not necessarily affect every cell of an organism in the same way. In fact, many cells cannot respond to a hormone message at all. In order to respond to the message carried by a particular hormone, a cell must contain a **receptor** for that hormone. Receptors are molecules to which hormones bind, forming a *receptor–hormone complex* that then affects cellular metabolism. Cells cannot respond to a hormone unless they contain the proper receptor. Those cells that do contain the receptor are known as *target cells,* and it is to such cells that the hormonal message is directed. The nature of the response depends on the amount of hormone that reaches the target cell, and it may also be influenced by the presence of other hormones that affect the same cell.

Auxin

In the late 1880s Charles Darwin and his son Francis carried out some simple experiments on phototropism. They were interested in understanding why growing plants seemed to bend toward the light and in whether there was a specific cause for phototropic growth. The Darwins' experiment showed that the tip of the growing plant was the key to phototropic behavior (Fig. 33.6). If the tip of the plant was covered, the plant did not bend toward the light even if the rest of the stem was illuminated. Covering a region below the tip had no effect. Could the tip be releasing a substance (the Darwins called it an "influence") that retarded growth on the illuminated side and accelerated growth on the side away from the light?

In the late 1920s Frits Went showed that the "influence" was a diffusable chemical. His first hunch was that

How Does a Plant Know You Touched It?

If you touch an animal (most animals, anyway), it reacts. It may flinch, or run, or purr, or growl, but there's little doubt that it is aware of contact. The animal's sensory nerves, of course, carry the sensation of contact to its brain, and the nervous system reacts quickly to the stimulus.

Plants can also react to touch, a phenomenon known as *thigmotropism*. Although these reactions are slow, they do occur. As many as 80 percent of plant species show changes in growth patterns as a result of touching and mechanical stimulation. Just a few soft touches, for example, are enough to cause a plant like *Arabidopsis* to produce shorter, stouter stems (see figure). So sensitive are some plants that as one researcher found, even the slight contact caused by measuring the plants with a ruler was enough to cause a marked inhibition of growth.

Until a few years ago, the mechanism of thigmotropism was unknown. Now, however, Janet Braam and Ronald Davis of Stanford University may have found the key to this intriguing mystery. They found that slight stimulation of *Arabidopsis* plants increased the production of mRNA from five different genes as much as 100-fold. What do these *touch-induced* (TCH) genes do? They don't have an answer yet, but they have discovered that three of the genes code for proteins that closely resemble calmodulin, a calcium-binding protein associated with regulatory pathways in the cell.

If expression of the TCH genes changes the level of cellular calcium, then calcium itself may be the messenger that carries the message "We've been touched" throughout the cell. The next time you touch a plant, remember that you may be setting off a chain of events that affect mRNA expression, protein synthesis, cellular calcium levels, and possibly even the growth of the plant.

Two 6-week-old *Arabidopsis* plants. The plant on the left was touched twice daily, and the plant on the right was not touched. Touching the plant stimulates the production of mRNA from several genes, and these may be responsible for the inhibited growth of the touched plant.

Chemical Defense Mechanisms: How Plants Fight Back

Most insects eat plants. The relationship between insects and plants, therefore, would seem to be mostly one-sided. The bugs eat green plants as fast as they can, and the plants, unable to swat or squirm or run away, seem to be unable to do anything to stop them. But plants have evolved a number of weapons that fight back.

Plants produce a whole range of chemicals that are deposited in their tissues seemingly to poison the insects that might otherwise eat them. Plants of the mustard family (Cruciferae) produce a chemical known as *sinigrin*, which contains the toxic compound lylisothiocyanate. By removing this compound from mustard plants (which few insects will even attempt to eat) and infusing it into celery leaves, we can study the effect of the chemical on some insects that normally eat celery plants. One such insect is *Papilo poly-*

xenes, the black swallowtail butterfly. Larvae of the butterfly that are placed on treated celery leaves grow very slowly when the dose of the chemical is low and are killed outright when the dose is raised to the level found in mustard.

Some scientists have suggested that chemicals that are produced by one species and are able to affect the growth, development, or population levels of another species should be called *allochemicals*. Sinigrin is just one example of a toxic allochemical. There are many more. But plants can be subtle in their defenses as well, and in many cases they can achieve their ends without directly poisoning their insect predators. Insects generally go through several distinct cycles of growth. Most butterflies, for example, hatch first to a series of larval stages that do most of the feeding; the final larval stage forms a pupa;

and the pupa develops into the mature butterfly. Each stage of the cycle is triggered by *hormones*, chemical messengers released from special organs within the insect's body.

Recently, Isao Kubo and other scientists were investigating the effects of a devastating plague of locusts in Kenya when they discovered that one plant species, a bugleweed (*Ajuga remota*), had survived the devastation. To figure out why, they fed extracts of the plant to a number of insects. When the time came for the insects to develop into pupae, they all grew several extra head capsules, blocking the development of their mouthparts; they starved and died. The plant extract caused this by producing an allochemical that was a slightly modified version of the insect's own development hormones. The plants had sabotaged the inner workings of their predators!

---

the *coleoptiles* (thin sheaths that cover the tips of rapidly growing grasses) might contain a chemical influence that affected plant growth. To test this hypothesis, he stripped the tips off growing oat seedlings and set them on small blocks of agar to allow fluid from the tips to diffuse into the agar (Fig. 33.7). Then he placed one of the agar blocks next to the tip of a seedling whose own tip had been cut off and placed the seedling in the dark. Within a few hours, the growing seedling began to bend away from the agar block as if responding to a light source. The agar was carrying a growth-promoting substance that it had picked up from the oat seedlings. Went named this substance **auxin,** a term derived from a Greek word meaning "to increase," because auxin increased the rate of plant growth. Before long, Went's auxin was identified as *indoleacetic acid* (abbreviated IAA).

Auxin acts on different receptors at different concentrations to induce several different effects. We will examine some of the more important ones.

Auxin promotes cell growth. It is synthesized in the apical bud, near the tip of the growing seedling, and transported down the stem. As Went's experiments showed, a difference in light intensity across the stem results in an unequal distribution of auxin. The accumulation of auxin on the side of the stem away from the light causes the cells on that side to elongate more quickly than those on the side the light shines on. This rapid elongation causes the stem to bend toward the light: the response we recognize as phototropism. We are not sure exactly how the difference in auxin concentration comes about, but it is clear that the apical meristem itself is the source of the difference, as shown by the Darwins' original experiment. The actual effects of auxin are subtle, and they depend on the concentration of the hormone as well as on the tissue responding to it. As shown in Fig. 33.8, the same concentration of auxin has different effects on roots, buds, and stems, and very high concentrations of auxin can actually inhibit growth.

Figure 33.7 An experiment performed by Frits Went on the nature of the phototropic response. Coleoptile tips were placed on agar blocks so that material from the tips could diffuse into the blocks. When such a block was then placed next to a growing shoot, the shoot bent away from the side touching the block, suggesting that the block had absorbed a growth-promoting substance from the coleoptile tip.

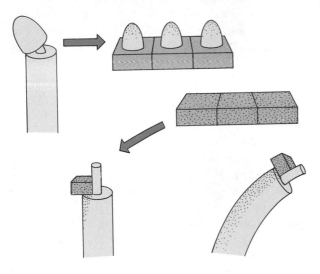

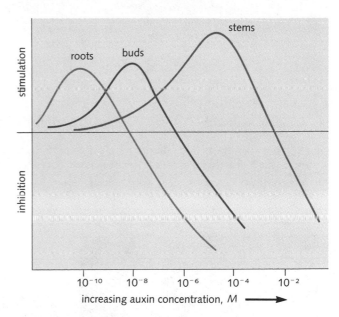

Figure 33.8 The effects of different auxin concentrations on growth in roots, buds, and stems. Note that the same concentration of auxins may stimulate stem growth and inhibit root growth.

Auxin is the hormone that produces geotropism. When a plant is tilted on its side, auxin accumulates at high concentrations in the bottom side of the main stem. As a result, cells on this side begin to grow more quickly, righting the plant and allowing it to grow in a normal upward pattern. Incidentally, auxin does not accumulate on the lower side of the stem because gravity "pulls" it down. (Auxin is a small molecule, and the force of gravity is not great enough to concentrate it in the lower part of a horizontal stem.) As in the case of phototropism, we still have not determined by what mechanism the plant produces the unequal distribution of auxin that accounts for geotropism.

The effect of auxin on roots provides an interesting extension of this story. Note in Fig. 33.8 that the same concentrations of auxin that *stimulate* growth in stems *inhibit* growth in roots. Roots exhibit a behavior exactly opposite to that of stems. If placed horizontally, roots turn away from light and downward toward the force of gravity (Fig. 33.9). Measurements of auxin concentration, however, show that it gathers on the *same* sides of both roots and stems: toward the force of gravity and away from the light. Auxin in the roots is released from the root apical meristem. The pattern of hormone release is the same in root and stem, but the patterns of *target cell response* differ (Fig. 33.10).

Figure 33.9 Germinating corn seeds show the effects of positive geotropism as they orient root growth in the direction of gravity.

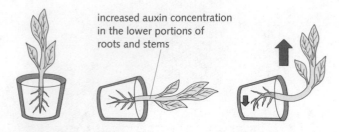

increased auxin concentration in the lower portions of roots and stems

Figure 33.10 The effects of geotropism are different on roots and stems. This is due to the differing sensitivities of roots and stems towards auxin.

Figure 33.11 Apical dominance controls the shape of these spruce trees, limiting the rate of growth in branches closest to the apical meristem.

High concentrations of auxin inhibit the growth of lateral buds. For quite a long time, gardeners have known that the growing tip of a plant inhibits the development of lateral branches but that when the tip is cut off, rapid lateral branching occurs. Horticulturists call this effect **apical dominance:** The tip, or apex, of the branch tends to dominate the growth pattern (Fig. 33.11). If the tip of a growing branch is cut off, lateral buds begin to grow rapidly.

A carefully controlled concentration of auxin can promote root growth. This discovery is exploited when cuttings are used to root new plants. When a little bit of auxin is carefully applied to a plant cutting, it promotes root formation and increases the chances that the cutting will be successful. Commercially available rooting and propagation products enable home gardeners to take advantage of this effect.

More than 200 different compounds have been discovered that mimic auxin's effects. Molecules that are chemically similar to IAA have many of the same effects as auxin. Some of these are much stronger than IAA itself, and the most potent compounds affect plants in an unexpected way. Because high concentrations of auxin can inhibit growth, the most potent synthetic auxin-like molecules are *toxic* to fast-growing plants. These compounds have been used as weed killers, or **herbicides,** since the late 1940s. Herbicides are very useful in agriculture, because killing off plants that compete with a desired crop can increase crop yields. One of the most widely used herbicides is a compound known as 2,4-D (2,4-dichlorophenoxyacetic acid). Because 2,4-D is more effective against dicots than against monocots, farmers can spray high levels of the herbicide on monocot crops such as corn and, without badly damaging the corn, eliminate a fair share of the dicot weeds that would otherwise compete with the corn seedlings.

But herbicides are not without their dangers. A mixture of 2,4-D and another auxin-like chemical was used as *Agent Orange,* a chemical defoliant widely sprayed in Vietnam to deny enemy troops the use of jungle foliage as cover (Fig. 33.12). One of the chemical by-products of synthesizing these compounds is dioxin, and dioxin was present in trace amounts in Agent Orange. Dioxin is a potent carcinogen (cancer-causing chemical), and its presence in Agent Orange has been a continuing concern to the Vietnamese population as well as to American servicemen who fought in Vietnam.

Table 33.1 summarizes the functions, origins, and targets of auxin and several other plant hormones.

Gibberellin

For many years, rice farmers in Japan had been aware of a disease that weakened their crops by causing the rice plants to grow unusually tall. The giant plants were fragile and often failed to produce rice. They called the disease the "foolish seedling" disease. Eiichi Kurosawa, a Japanese biologist, showed that the rapid growth was due to a fungus called *Gibberella* that grew on the plants. The fungus produced a soluble compound that Kurosawa named **gibberellin.** Gibberellins are actually a family of about 60 closely related compounds. They are similar in chemical structure to the steroid lipids we described in Chapter 15. Gibberellins are also made by higher plants (including rice), and the effect of the fungus was due to its ability to synthesize a chemical that mimics the actions of a class of naturally occurring plant hormones.

Gibberellin seems to regulate the rate at which internode regions of plant stems elongate. Applying gibberellin to a growing plant can result in a rapid increase in the length of stems between branch nodes (Fig. 33.13). Gibberellin can cause dramatic increases in size among dwarf plants, often allowing them to reach near-normal sizes. Gibberellins are also important in seed germination in some plants. In barley, for example, as the seed takes up water, special tissues release gibberellin, which diffuses throughout the seed. A target tissue in the seed responds to the gibberellin message by releasing proteolytic enzymes, which begin to break down the stored endosperm tissue in the seed—something the embryo requires for a fast start on growth. Gibberellins, therefore, can cause sprouting in barley. The brewers of beer have noticed this: Barley seeds are sprayed with gibberellin so that they will all sprout at once to produce the uniform barley malt that is the basis of the best beers.

Figure 33.12 Agent Orange, the chemical defoliant used in Vietnam, contained a mixture of chemicals mimicking the effects of IAA (indolacetic acid).

Table 33.1 *Major Classes of Plant Hormones and Their Effects on Plant Cells and Tissues*

Hormone	Functions	Origin	Target
Auxins (including IAA)	Stimulate cell growth; and stem and root elongation, control geotropism and phototropism	Meristems in roots and shoots	Cells in roots, stems, and leaves
Gibberellins	Stimulate stem elongation and bud and fruit development	Chloroplasts in leaf tissue	Stem cells
Cytokinins	Stimulate cell division and growth	Several plant tissues	Growing tissue in roots, stems, and leaves
Ethylene	Stimulates ripening of fruit; affects (+ or −) development of leaves and fruit	Root tissues; aging leaves	Fruits and flowers
Abscisic acid	Inhibits growth, promotes dormancy	Several plant tissues	Stem tissue and buds

Figure 33.13 Gibberellin dramatically increased the rate of leaf elongation in the bird's nest fern at the right.

Cytokinins

Cytokinins are plant hormones that were discovered in attempts to improve the growth of plant cells in culture. Carlos Miller, a researcher in the laboratory of Folke Skoog at the University of Wisconsin, was testing the ability of nucleotide-containing mixtures to stimulate growth when he took a bit of herring sperm DNA off the shelf and added it to his cultures. It caused a rapid increase in the rate of plant cell division, but pure DNA had no such effect. Instead, Miller and Skoog showed that one of the breakdown products of DNA was a potent plant hormone. The compound these investigators discovered belonged to a class now known as cytokinins (because of their effect on cell division). Their chemistry is similar to that of *adenine,* one of the bases in DNA.

Cytokinins seem to complement many of the effects of auxin. Cytokinin applied to a lateral bud causes the bud to grow, even in the presence of a high concentration of auxin (which would normally inhibit growth of the bud because of apical dominance). Recent experiments show that the ratio between auxin and cytokinin concentrations determines the rate of tissue growth. This is one of the main reasons why the apex itself is not subject to the inhibitory influences of auxin. When applied to clumps of cells in culture, cytokinins tend to cause the development of stems, whereas auxin promotes the growth of roots. When the balance between the two is just right, a mixture

of roots and stems is formed. Xylem sap contains cytokinins that are manufactured in the roots and transported upward to the rest of the plant. The balance between cytokinins and auxin seems to be one of the major elements that controls the shape and growth of a plant. This precise control of cellular behavior by the interplay of different hormones is something we will see again when we discuss animal hormones in Chapter 39.

Abscisic Acid

Although many plant hormones seem to be geared toward the stimulation of plant growth, there are cases in which it is important for a plant to limit or restrict its rate of growth. Plants that live in temperate areas of the world must prepare for a winter season in which temperatures are low and light levels reduced. Their survival requires that they conserve material resources gained during the warm seasons of the year and protect them against the ravages of winter weather. **Abscisic acid,** a hormone produced in a variety of plant tissues, has a striking ability to inhibit growth. Abscisic acid is released from cells within a limb bud in early fall. It slows growth in the meristematic tissues of the bud and promotes the development of thick, scaly coverings that will serve as protection for the bud throughout the winter.

The name *abscisic acid* was originally coined because of a belief that this substance played a role in abscission, the seasonal loss of leaves, flowers, and other plant parts. As luck would have it, although this hormone is present in leaves about to be lost from the plant, it plays little role in the process of abscission itself.

In woody tissues, abscisic acid slows down the rates of both primary and secondary growth, preparing the plant for a reduction in the amount of photosynthetic product that occurs as the leaves are shed. Abscisic acid also plays a role in the arresting of cell growth that occurs during seed formation. Although the seeds of many plants are formed by rapid growth of the embryo and ovary tissue, once the seed is mature, abscisic acid slows down the rate of cell growth and prepares the seed for the extended dormant period that may be required for germination.

What happens to seeds if abscisic acid is not present? Mutants of corn that lack the ability to produce this hormone are unable to stop the growth of embryos in the seeds they produce. As a result, their seeds germinate immediately, right in the cob, before they have a chance to be dispersed.

Ethylene

In the nineteenth century the major source of indoor lighting was illuminating gas. Use of the gas seemed to have some important effects on indoor plants: Growth was stunted, stems swelled, and leaves fell off. Fruit on the indoor plants began to ripen much more quickly than normal. Gradually the effect was traced not to the gas itself but to one of its minor components, **ethylene.** Ethylene is produced by plants themselves, and ethylene production is under the control of auxin. In fact, many of the effects of auxin are actually brought about by the ethylene that auxin causes to be produced. Ethylene is produced as a chemical trigger during fruit formation, and this explains its ability to control the ripening process.

Commercial producers of fruit exploit these properties. Many commercial crops, including lemons and tomatoes, are generally picked unripe and then packed for shipment under conditions wherein circulating air removes any ethylene the cells of the fruit release. This prevents the fruit from ripening during shipment. Then, just before delivery to stores, the fruit is treated with synthetic ethylene and ripens rapidly (Fig. 33.14). Although this treatment has commercial value, it doesn't always duplicate the quality of vine-ripened fruit, in part because light absorption also plays a role in natural ripening for which ethylene does not compensate.

Figure 33.14 Tomatoes are generally harvested while still green and then ripened by the application of ethylene gas.

PHYTOCHROME: THE REGULATION OF PLANT GROWTH

The major events in the life of a plant include germination, flowering and reproduction, and senescence. In nature these events are coordinated with the seasons of the year. Given the central role that *light* plays in the life of a plant, it would be surprising if light did not play a major role in regulating the timing of these events in plant development. The fact that all plants, tropical or temperate, experience daily cycles of light and darkness provides a powerful physiological reason why plants regulate their own cycles of activity to make the best possible use of sunlight hours. Not surprisingly, most plants have developed mechanisms by which they can respond to the length of the day as well.

The Discovery of Phytochrome

Light affects the *form* of a plant as well as the direction of its growth. If a bean seedling is allowed to sprout in darkness, its stem grows very quickly and its leaves remain folded and undeveloped. When the seedling is later exposed to light, it rapidly unfolds and enlarges its leaves and slows down its rate of stem elongation to assume a more normal appearance (Fig. 33.15). Light actually serves two purposes in this situation. First, light is required for the conversion of chemical intermediates to chlorophyll. Second, light causes a change in the shape and growth pattern of the plant, a process known as **photomorphogenesis.** (In animals the growth and shaping of

Figure 33.15 Photomorphogenesis, the influence of light on plant form, is dramatically illustrated in this photograph of bean seedlings raised in darkness (LEFT) and in light (RIGHT).

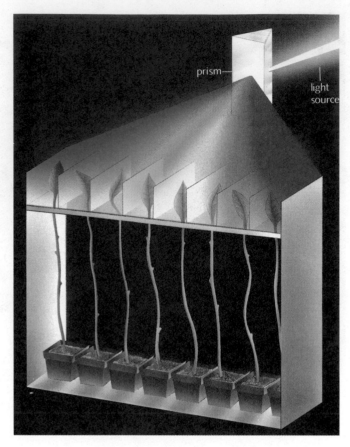

Figure 33.16 The experimental apparatus designed by Hendricks and Borthwick to study photomorphogenesis enabled them to determine that a pigment absorbing in the far-red region of the spectrum was responsible for the effect. This pigment is now known as phytochrome.

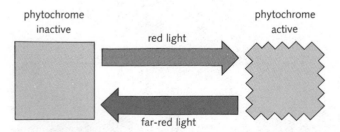

Figure 33.17 Phytochrome is converted into an active form by red light and can be converted back to its inactive form by far-red light.

body parts may have nothing to do with exposure to light and are called *morphogenesis.*)

Photomorphogenesis is not caused by photosynthesis. In 1954 two scientists, Hendricks and Borthwick, who worked at the U.S. Department of Agriculture Experimental Station in Beltsville, Maryland, performed experiments to determine which wavelengths of light were the most effective in inducing photomorphogenesis. The very best wavelengths were in the red region of the spectrum (about 660 nm). Blue light, which drives photosynthesis very effectively, had absolutely no ability to induce photomorphogenesis. The two scientists did other experiments that produced a puzzling result: Far-red light could completely reverse the effects of red or white light. The wavelength that was most effective at this reversal was 730 nm—so far into the red region that most people cannot even sense its presence.

The scientists realized that these experiments could be explained if a single pigment molecule were responsible for photomorphogenesis. The absorption of light in the red region (at 660 nm) should convert the pigment to an active form that triggers photomorphogenesis. However, this new, activated form should absorb light in the far-red region (730 nm), converting the pigment back from the active form to the inactive form and explaining the reversal effect. They proposed that the pigment, which was still theoretical, be called **phytochrome** (*phyto* means "plant"; *chrome* means "color"). It would work like a switch, red light (660 nm) switching it on and far-red light (730 nm) turning it off (Fig. 33.16).

These scientists then learned that nearly 10 years earlier two other scientists in the same lab had shown that red light (660 nm) was most effective in causing lettuce seeds to germinate. They had also shown that the effects of this light could be reversed by far-red light (730 nm). Hendricks and Borthwick began to believe that they had discovered one of the fundamental mechanisms by which plants respond to light. And they were right.

Phytochrome has now been isolated and closely studied, although the links between phytochrome and development are still obscure. Phytochrome is a pigment–protein complex with a molecular weight of 120,000, and it does indeed undergo a shift between two forms when it absorbs light, as shown in Fig. 33.17.

Photoperiodism

Besides the germination of seeds and the control of photomorphogenesis, phytochrome is involved in a host of plant responses to light. Phytochrome controls the growth and development of chloroplasts, the synthesis of chlorophyll, the release of hormones such as gibberellin, and even the repositioning of leaves as the angle of sunlight changes. But perhaps the most important response that phytochrome controls is **photoperiodism,** a physiological response of a plant to changes in the *photoperiod,* or length of the day. Accurate measurement of the photoperiod is particularly important to plants, for it is the best clue to the passing of the seasons.

Plants are able to time their flowering with some precision, and a stroll through the woods will remind you that some plants flower exclusively in the spring, others only in the fall. Different species clearly have some sort of mechanism to sense the length of the day and adjust their growth and flowering pattern accordingly. In fact, experiments reveal that plants fall into three groups in terms of photoperiodism (Fig. 33.18). **Long-day plants** flower only when the daylight lasts longer than a certain minimum; they include plants that flower in late spring and early summer, such as hollyhocks and wheat. **Short-day plants** flower only when the daylight period is less than a certain amount; they include soybeans, poinsettias, and ragweed, which flower in the late summer or early fall. A great many plants are unaffected by day length, and they are known as **day-neutral plants.**

At first, you might be tempted to think that plants would respond to changes in the amount of daylight to which they are exposed. However, experiments in controlled growth chambers yielded a surprising result: It was actually the length of the *dark* period that was critical. What the "long-day" plants actually responded to was a "short-night" period of darkness (Fig. 33.19). It might be more appropriate to use the terms *long-night plants* and *short-night plants,* but established terminology doesn't yield easily to new discoveries. In any case, the crucial question is how a plant senses the length of the day and responds by flowering.

Phytochrome seems to be at the heart of the photoperiodic response. Cockleburs flower when they are exposed to a dark period of more than 9 hours. However, that dark period must be uninterrupted. If even a flash of light occurs in the middle of the dark period, flowering does not occur. The wavelength of the flash of light that is most effective in preventing flowering is 660 nm. And when the flash of white or red light is followed by a flash of far-red light (730 nm), the plant flowers normally. This is powerful evidence that phytochrome controls the photoperiodic response.

Figure 33.18 TOP: Wheat (*Triticum aestivum*), a long-day plant. BOTTOM: Sierra primrose (*Primula suffrutescens*), a short-day plant.

Florigen

If different parts of the plant are covered when photoperiodic experiments are carried out, it soon becomes evident that the leaves are the source of the response. When

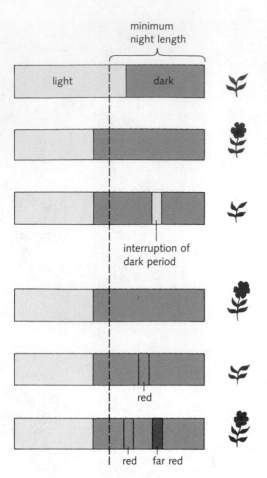

minimum
night length

light — dark

interruption of
dark period

red

red   far red

Figure 33.19 A summary of experiments that showed the involvement of phytochrome in flower production. Cockleburs flower when they are exposed to a dark period longer than 9 hours. If that dark period is interrupted by a flash of red light, flowering does not occur. However, if that flash is followed by a period of far-red illumination, flowering occurs. The ability of far-red light to reverse the effects of the flash is indicative of the involvement of phytochrome.

even a single leaf of a cocklebur is covered so that it receives a sufficient dark period, the whole plant flowers. This seems to implicate a hormone—one that is produced in the leaf under phytochrome control and then travels throughout the plant system to induce flowering. Botanists have suggested the name *florigen* ("flower maker") for the hormone, and there is additional evidence that florigen exists. To date, however, florigen has not been isolated, and therefore it is not yet possible to say with certainty that the flowering response is caused by a single phytochrome-controlled hormone.

AUTUMN: A CASE STUDY

In the temperate areas of the world, as summer ends the days begin to grow shorter and shorter. The change in photoperiod is sensed by the *phytochrome* system, and several processes are set in motion by the response to activated phytochrome. The *auxins* that were produced in the leaves begin to diminish, and the plant increases production of *ethylene* and *abscisic acid.* Phytochrome turns off the pathways for chlorophyll synthesis in the leaves. Because no new chlorophyll is synthesized, existing chlorophyll molecules are gradually bleached and destroyed by the bright fall sun, leaving only the accessory pigments in the leaves. These pigments, including yellow and orange *carotenes* from the chloroplasts and the reddish *anthocyanins* found in cellular vacuoles, become visible as chlorophyll fades, and the lovely colors of autumn are revealed (Fig. 33.20).

In time the appearance of *hydrolytic enzymes,* which break down thin cell walls, causes a weakening of the point where the leaf is attached to the stem. This zone of weakness is called the **abscission layer.** The growth of stems begins to slow down, and the apical meristem develops a thick, waxy *terminal bud* that will help it to survive the winter. Little by little, hormones shut down the metabolic activities of the leaf until it breaks off at the weakened abscission layer and drifts to the ground. This

Figure 33.20 Changes in pigment biosynthesis are responsible for the brilliant colors of autumn, illustrated in this photograph of a sugar maple forest near the Montreal River in Ontario.

pattern is called leaf **senescence,** and it is vital to the plant. Senescence enables the plant to shed tissue that would be difficult to protect through a long cold season and allows the plant to conserve its resources for the winter. Senescence has been a difficult process to investigate, but the programmed senescence of various tissues, including leaves, fruit, and flowers, is an important feature of the normal life cycle of flowering plants.

TOMORROW'S PLANTS

Plant Cell Culture

Today's botanists are experimenting with a number of techniques that promise an unprecedented ability to develop new varieties of plants. In recent years scientists have discovered how to produce complete plants from a few cells taken from another plant. Isolated cells are grown in a broth rich in nutrients and hormones, and groups of cells gradually form a small clump known as a *callus.* Each callus is then transferred to the surface of sterile agar (also rich in nutrients and hormones), and a small *plantlet* begins to grow with fully developed tissues. As the new plant increases in size, it can be transferred to soil and eventually grown outdoors.

In addition, proper growth conditions can produce plant **protoplasts,** cells that lack cell walls (Fig. 33.21). Scientists have developed techniques whereby protoplasts from different plants or even different species can be fused, forming hybrid cells with new combinations of genes and bypassing the natural breeding process entirely. The ability to grow and manipulate plant cells in this way gives scientists an opportunity to alter the genes of a single cell and then grow a complete organism from that altered cell.

Plant Genetic Engineering

In order to manipulate the plant genetic system, one first needs a way to get DNA molecules inside plant cells. There are several methods that can be used to insert DNA molecules into plant cells, and transgenic plants are now routinely produced in laboratories throughout the world.

Some of the tools used to construct transgenic plants are biological in nature. Two main classes of DNA viruses infect plant cells, and each holds promise as a tool for genetic engineering. At the present time, however, the organism of greatest interest to plant molecular biologists is the soil bacterium *Agrobacterium tumefaciens.* **Agrobacterium** infects several types of plants and produces large tumor-

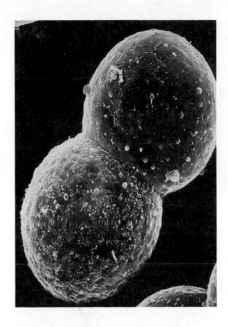

Figure 33.21 The fusion of plant protoplasts, shown in this electron micrograph, is an essential part of techniques for plant gene manipulation. (Magnification factor: 320)

like growths of plant tissue known as *galls.* During an *Agrobacterium* infection a small piece of DNA, a *plasmid,* from the bacterium enters the plant cell and transforms it into a gall cell.

This plasmid, known as *Ti,* contains about 200,000 base pairs of DNA. Recombinant-DNA techniques can be used to splice Ti DNA to another segment containing a gene that is to be inserted into plant cells. This altered Ti plasmid can then be used to infect plant cells with the DNA segment. Different strains of Ti can be constructed so that they do not cause galls, and this allows Ti to be used as a vehicle for inserting new genes into plants. These techniques make it possible to isolate a particular gene, insert it into a single plant cell, and then use that cell to grow a complete organism containing the genetic modification (Fig. 33.22).

The power of these techniques is apparent in a series of experiments in which the Ti plasmid was used to carry the luciferase gene into tobacco cells. Luciferase, an enzyme found in fireflies, produces light from the energy available in ATP. After infecting tobacco cells with the plasmid, Steven Howell and his co-workers in San Diego were able to grow mature tobacco plants from the transformed cells. These plants expressed the luciferase gene, giving them a most un-plant-like quality: they glowed in the dark like fireflies (Fig. 33.23)!

These experiments illustrate that there is no insurmountable barrier to the transfer of DNA from the animal kingdom to the plant kingdom. Several labs have shown that it is possible to use plants to produce biologically active proteins like insulin and glucagon, and it has even been possible to produce antibody polypeptides in

plants. Thus plant products may one day be used to aid the immune system in the fight against disease.

We should note that there are many plant species that cannot be transformed with the Ti plasmid, and for a time it was not possible to carry out genetic engineering on these species. However, new techniques have been developed that now make it possible to transform virtually any species of plant. In many cases, researchers use physical techniques to insert foreign DNA into protoplasts. One procedure begins by coating tiny metal beads with DNA. These microscopic beads are then loaded into a "particle gun" and fired into plant cells (Fig. 33.24). In a large number of cases, the beads penetrate the cells and release their DNA; the cell is transformed as the foreign DNA is integrated into the plant genome.

Figure 33.22   LEFT: These corn seedlings, growing in sterile agar, have each been cloned from single cells in culture. BELOW: Foreign genes can be introduced into plant cells by genetic engineering. One useful technique involves the Ti plasmid from *Agrobacterium tumefaciens*. DNA can be joined to the Ti plasmid used to transform *Agrobacterium tumefaciens,* which in turn may be used to transform plant protoplast cells in culture. Cells containing the DNA may be used to regenerate a complete plant, as shown.

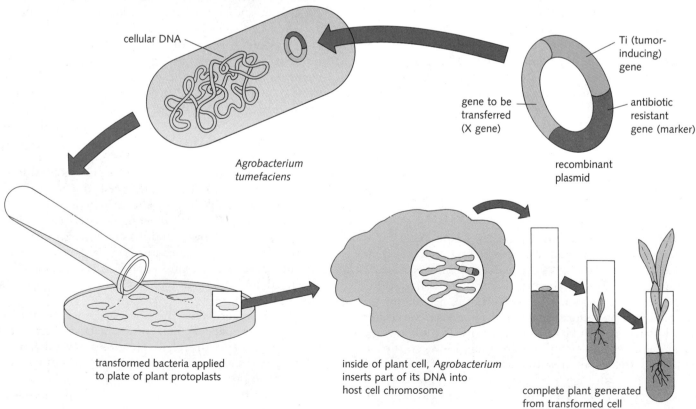

cellular DNA

Agrobacterium tumefaciens

Ti (tumor-inducing) gene

gene to be transferred (X gene)

antibiotic resistant gene (marker)

recombinant plasmid

transformed bacteria applied to plate of plant protoplasts

inside of plant cell, *Agrobacterium* inserts part of its DNA into host cell chromosome

complete plant generated from transformed cell

Figure 33.23 A genetically engineered tobacco plant containing the gene for luciferase, the protein that causes fireflies to glow.

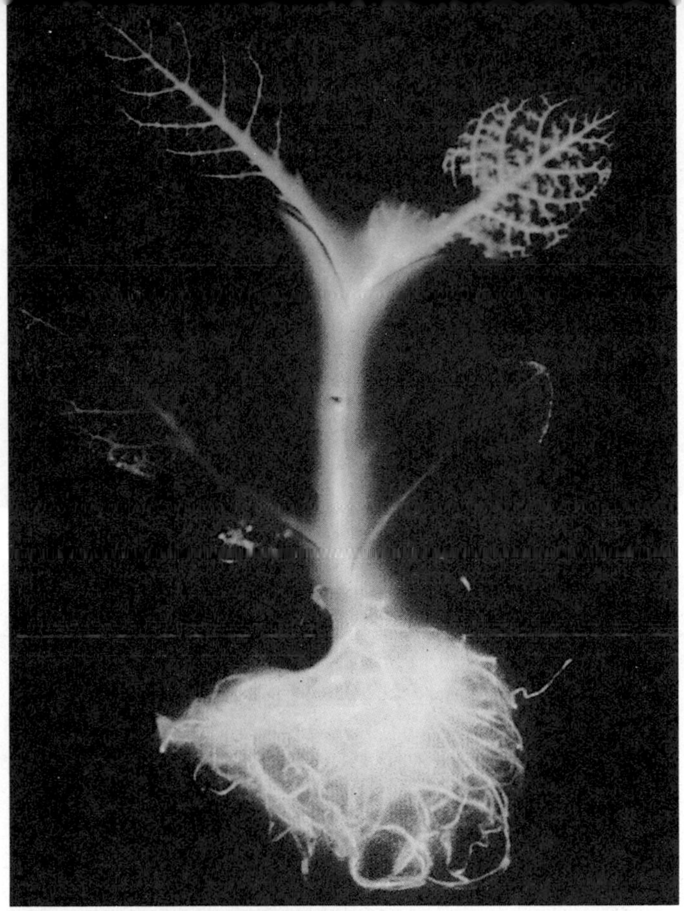

Figure 33.24 The production of transgenic plants is now a routine procedure. BOTTOM LEFT: The "particle gun" method transforms plant cells by shooting DNA-coated metal beads into plant cells. When hundreds of cells are bombarded with DNA-coated beads, a few of them incorporate the foreign DNA and become transformed. BOTTOM RIGHT: Examples of transgenic plants include **(a)** herbicide-resistant cotton plants (nonresistant plant on right); **(b)** spoilage-resistant tomatoes (nonresistant tomatoes on right); **(c)** high-starch potatoes (dark stain shows starch content; engineered potato on left).

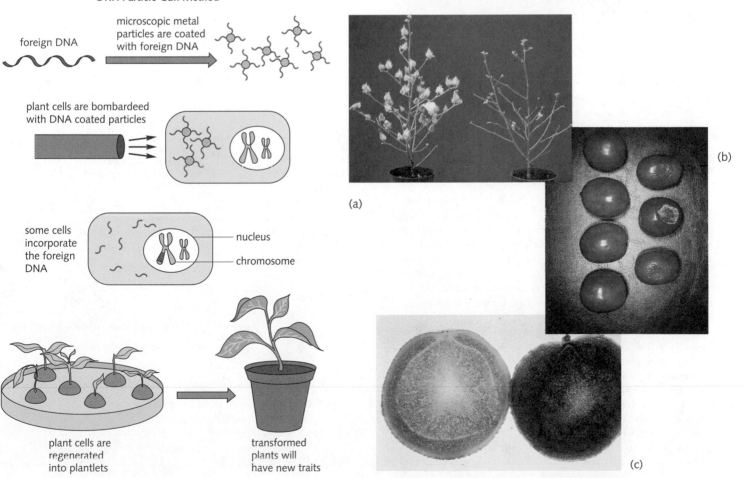

DNA Particle Gun Method

foreign DNA

microscopic metal particles are coated with foreign DNA

plant cells are bombardeed with DNA coated particles

some cells incorporate the foreign DNA

nucleus

chromosome

plant cells are regenerated into plantlets

transformed plants will have new traits

(a)

(b)

(c)

The New Harvest

As shown in Fig. 33.24, molecular biology has produced an extraordinary yield of transgenic plants designed for a wide variety of purposes. The ease with which plants can be cultivated makes them ideal for the production of genetically engineered compounds such as antibodies, antibiotics, plastics, and hormones. Materials produced in this way may be on the druggist's shelves as you read this chapter, and it is clear that more are on the way. Food plants, modified to improve their nutritional value or shelf life, will soon be on the tables of people throughout the world.

Despite the great promise these advances hold, we must exercise care in the choice of such experiments and in our attempts to release genetically engineered material into the environment. The construction of new plant varieties may have unforeseen biological and economic consequences, and one of the key questions facing scientists and nonscientists alike in the coming years will be how to regulate such technologies.

SUMMARY

The ability of a plant to survive depends on the existence of systems that control and coordinate events in different parts of the organism. Many of these responses are known as tropisms. Common tropisms include geotropism, phototropism, and thigmotropism—the ability to respond to gravity, light, and touch, respectively. Many tropisms are caused by hormones, substances that are made in one region of the plant and affect cellular behavior in another part. The first plant hormones to be discovered were the auxins. Auxins affect the rate of cell growth, and the accumulation of auxins in different amounts on either side of a stem is the major cause of responses to light and gravity. Chemicals similar to auxins have been used as herbicides (weed killers). Other plant hormones include gibberellins, which regulate the rate of stem elongation and also play an important part in seed germination; cytokinins, which help to regulate cell division; abscisic acid, which inhibits growth under certain conditions; and ethylene, which is important in the ripening of fruit.

Phytochrome is a pigment–protein complex that helps to regulate the timing of important light-sensitive events such as flowering and photomorphogenesis. Phytochrome conversions between active and inactive forms are used by plants to sense the length of the day and to control seasonal responses such as flowering and reproduction.

Emerging techniques, including plant cell culture, depend on the informed use of plant hormones to stimulate cell growth and to promote the development of new tissue in plants grown by cloning cells isolated from other individuals. The application of new techniques in molecular biology has now made it possible to insert genes into plant cells, altering the genetic constitution of individual cells, and then use those cells to produce new plants. These techniques have already produced a host of transgenic crop plants, and more are on the way.

STUDY FOCUS

After studying this chapter, you should be able to:

- Define the term *hormone*, and explain how hormones act in controlling plant growth and development.

- Cite some methods and results of experimental plant physiology.

- Describe some techniques of plant cell culture and genetic engineering, and discuss their possible implications.

- Appreciate the complexity of plants as living organisms that respond to their environment in sophisticated and complex ways.

- Explain the central role that light plays in the regulation of plant growth.

- Describe the basic techniques used to produce transgenic plants, and give some examples of how plants have been modified by these techniques.

TERMS AND CONCEPTS

tropisms *652*	ethylene *661*
geotropism *652*	photomorphogenesis *661*
phototropism *653*	phytochrome *662*
thigmotropism *653*	photoperiodism *663*
hormones *653*	long-day plants *663*
receptor *654*	short-day plants *663*
auxin *656*	day-neutral plants *663*
apical dominance *658*	senescence *665*
herbicides *658*	protoplasts *665*
gibberellin *659*	*Agrobacterium* *665*
cytokinins *660*	

REVIEW

Objective Questions (Answers in Appendix)

1. When a seedling rights itself, this is characterized as a _____ response.
   - (a) righting
   - (b) phototropic
   - (c) geotropic
   - (d) thigmotropic

2. The processes that allow plants to respond to changing conditions in the environment are most directly influenced by the
   (a) concentration and type of minerals in the soil.
   (b) plant species.
   (c) concentration of hormones.
   (d) amount of shade.

3. Darwin and his son showed that
   (a) plant hormones regulate growth.
   (b) plants demonstrate a phototropic response.
   (c) plants are affected by the force of gravity.
   (d) plants exhibit geotropism.

4. Auxins are active in producing
   (a) cell elongation.   (c) growth of lateral buds.
   (b) seeds.   (d) fruit ripening.

5. Auxins, gibberellins, and cytokinins are all
   (a) hormones.
   (b) types of flowering plants.
   (c) types of soil nutrients.
   (d) types of organic fertilizers.

6. A plant that is day-neutral
   (a) will only flower when the amount of daylight exceeds the amount of darkness.
   (b) will only flower when the amount of darkness exceeds the amount of daylight.
   (c) is unaffected by photoperiodism.
   (d) will not flower.

7. *Agrobacterium tumefaciens* is a
   (a) plant virus.
   (b) nitrogen-fixing bacterium.
   (c) bacterium that produces plant tumors known as galls.
   (d) transgenic plant.

Discussion Questions

8. What is a hormone? Why is it justified to classify auxin as a hormone?

9. To many plant biologists, Went's experiments with agar blocks proved the existence of plant hormones (in general) and of auxin (in particular). What else did the experiments show about the nature of the chemical message?

10. Is phytochrome a hormone? What are some of the major differences in actions and effects between phytochrome and auxin?

11. Why would it perhaps be more appropriate to refer to "long-day plants" as "short-night plants"?

12. What basic steps are required to produce transgenic plants that express genes from other organisms?

READINGS

Gasser, C. S., and R. T. Fraley. "Transgenic crops." *Scientific American* 266 (June 1992): 62–69. A clear and readable review of the explosion of genetically engineered crop plants.

Moffat, A. S. "High-tech plants promise a bumper crop of new products." *Science* 256 (1992): 770–771. A brief and up-to-date summary of genetically engineered plant products that are close to being brought to market.

Oeller, P. W., et al. "Reversible inhibition of tomato fruit senescence by antisense RNA." *Science* 254 (1991): 437–439. Technical details of the production of a genetically engineered tomato with a long shelf life.

Salisbury, F. B., and C. W. Ross, eds. *Plant Physiology*, 4th ed. Belmont, CA: Wadsworth Publishing, 1991. A thorough and authoritative text covering the full range of plant physiology.

Braam, J., and R. W. Davis. "Rain-, wind-, and touch-induced expression of calmodulin and calmodulin-related genes in *Arabidopsis*." *Cell* 60 (1990): 357–364. A scientific report on the influences of mechanical stimuli on plant gene expression. A technical paper, but still quite readable.

Rosenthal, G. A. "The chemical defenses of higher plants." *Scientific American* 254 (January 1986): 94–99. A marvelous description of some of the extraordinary chemical defense mechanisms of higher plants.

Sisler, E. C., and S. F. Yang. "Ethylene, the gaseous plant hormone." *BioScience* 33 (1984): 233–238. A concise and well-illustrated review of the role of this hormone in ripening and other processes.

PART 7

Animal Systems

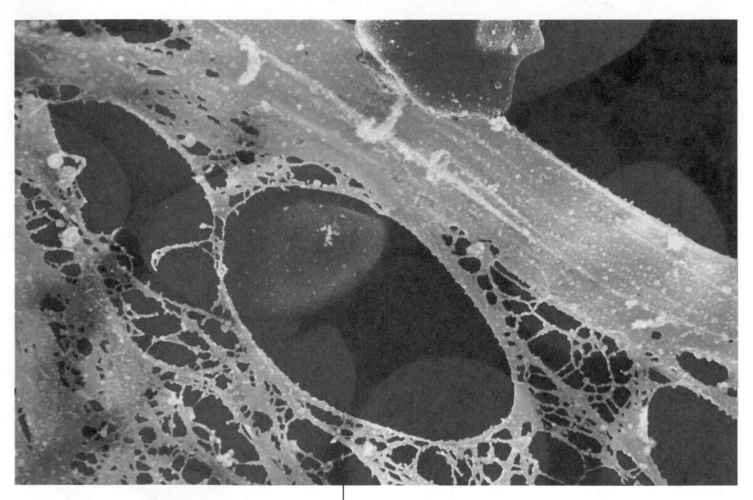

After an injury, a net of fibrin and other proteins forms a clot that seals the wound and halts the loss of blood.

*a*complaint that has been heard from generations of students is that "the more we learn the less we seem to know!" Although it is often true that answering one question results in our posing new ones, it is also true that each question science is able to answer enriches our understanding of nature. A scientist works with the inner belief that all knowledge is worth having and, therefore, that the newer, deeper questions that our studies raise will be that much more profound and satisfying to answer.

Anatomy and physiology—fields devoted to the structure and function of the systems of living organisms—are excellent examples of sources of never-ending questions. Thousands of years ago, biologists realized that it was appropriate to consider an organism as a number of systems that could be studied separately. As time passed, classic Greek and Roman texts were updated by scientists in the Middle Ages who compared the older writings with what they were able to discover on their own. With few exceptions, however, it was impossible to find a link between any of these systems and the chemicals of which living things are made.

Even so, there were hints of the processes at work within an organism. The electrical activity of nerve and muscle tissue, as well as the acids found in the digestive system, suggested that basic principles of physics and chemistry were at work within living organisms. Engineers pointed out that hydraulics could be used to explain the flow of fluid in the circulatory system, and little by little, the field of physiology evolved in an effort to

link each action of a particular system with a series of cellular and molecular events.

Today, physiology has fully developed the connections with chemistry and physics that were only suspected a century ago. We discuss a number of systems in this section using the cellular and molecular perspective gained in Part 4. Where appropriate, we will use a broader perspective to examine how

We must not conceal from ourselves the fact that the causal investigation of the organism is one of the most difficult, if not the most difficult problem which the human intellect has attemped to solve, and that this investigation, like every causal science, can never reach completeness, since every new cause ascertained only gives rise to fresh questions concerning the cause of this cause.

—Wilhelm Roux, 1894

similar systems function in a variety of different animals and how such systems have been shaped by evolution. Chapter 34 starts by examining how multicellular organisms are organized and how their systems work together to maintain each animal's internal environment. Chapters 35–42 each cover a major body system, while Chapters 43–46 form a subunit on the various aspects of neurological physiology.

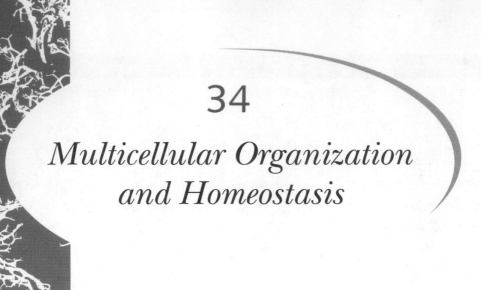

34

Multicellular Organization and Homeostasis

*i*magine that you are visiting Africa's Serengeti Plains in late May. The grasslands are parched and brown, and more than a million wildebeest are on the move, plodding toward greener pastures in Kenya.

Imagine now that you are watching a single animal in this huge herd. He moves mechanically, as if in a trance. Internally, unconsciously, his actions and metabolic processes are geared to scarcity. His measured steps expend the least possible energy. With no food in his gut, his body mobilizes and burns stored energy from fat deposits, distributing it to organs and muscles. Lacking water on today's long march, he retains body fluids by producing as little urine as possible.

Suddenly, a lioness springs from the underbrush. Alarmed, the wildebeest rouses from his stupor, all senses on alert. Energy in the form of glucose pours into his muscles. He gallops away with a grace and speed startling for an animal his size. The lioness strikes and tears his skin with her claws but fails to knock him down. She cannot keep up with his frantic pace.

Several minutes later, once out of danger, the wildebeest relaxes. His metabolism returns to normal, and he lapses into the steady pace of his migratory style. Already, blood in damaged tissues around his wound has thickened enough to stop dripping. Already, bacteria that entered the wound are being surrounded and destroyed by his body's defenses. And already, he has resumed his 200-mile journey toward the northwest.

MAINTAINING THE BALANCE

If you could somehow have watched these events from the perspective of individual cells, you would have seen dozens of remarkable acts of communication, coordination, and control among cells belonging to different tissues. Those acts are neither more nor less remarkable than similar phenomena that occur within your own body or, for that matter, within the body of an ant.

The cells of every organism are at once independent entities and interdependent parts of a larger whole. Each and every cell, whether a nerve cell, a muscle cell, or a skin cell, must respond to changes in its immediate environment within the body. Each must successfully balance the inputs and outputs on which its metabolic activity depends. Yet despite cells' individual requirements, they must also function together as members of the greater cellular community that is the entire organism.

One of the remarkable accomplishments of multicellular organisms is the creation of an internal environment that is carefully controlled. Cells in the wildebeest's brain, for example, must be kept at constant temperature, supplied with energy in the form of glucose, bathed in fluid with a constant concentration of water, and cleansed of their waste products. Those conditions must not change, regardless of droughts, floods, famines, heat waves, or cold snaps. Failure at any of these tasks, even for a few minutes, would lead to permanent injury or death of the entire organism.

Maintaining constant internal conditions in the face of rapidly changing external environments is no small accomplishment. The major systems of the body must rely on a foolproof system of checks and balances that monitors and guides the activities of every cell.

Starting in this chapter, we will examine how the major body systems of animals succeed at this daunting task. We will begin here by providing a brief overview of the tissues and major body systems of multicellular animals. Earlier in this text (Chapters 26–29), we examined some of the basic differences in structure and physiology in the many diverse groups of the animal kingdom. Although we will include a number of different species in our treatment of animal physiology here, we will most often focus on that species of particular interest to us, *Homo sapiens* (Fig. 34.1).

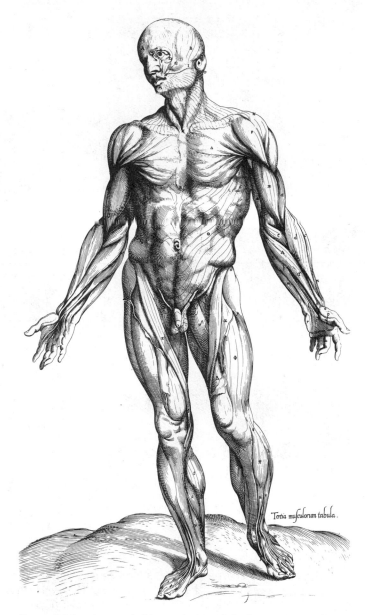

Figure 34.1 A remarkable rendering of the human muscular system, by Vesalius.

THE RANGE OF CELLS AND TISSUES

The Diversity of Cells

Humans begin life as a single cell, a zygote formed by the fusion of sperm and egg. Over the course of a few weeks, that single cell divides into thousands and ultimately millions of cells in the developing embryo. The embryo develops three cell layers (called **germ layers**), the *ectoderm, mesoderm,* and *endoderm,* and from these the rest of the body is produced. Gradually, the cells in these three layers undergo a process called *differentiation* ("becoming different") in which they are transformed into the astonishing variety of cells found in the adult.

- **Ectoderm** develops into the cells of skin and the nervous system.
- **Mesoderm** develops into the cells of the organs found in between ectoderm and endoderm, including muscles, blood vessels, the reproductive tract, and the kidneys. The mesoderm also provides the outer layers of the organs of the digestive and respiratory systems.
- **Endoderm** produces the cells that form the lining of the digestive and respiratory systems and the glands (liver and pancreas) that are outgrowths of these systems.

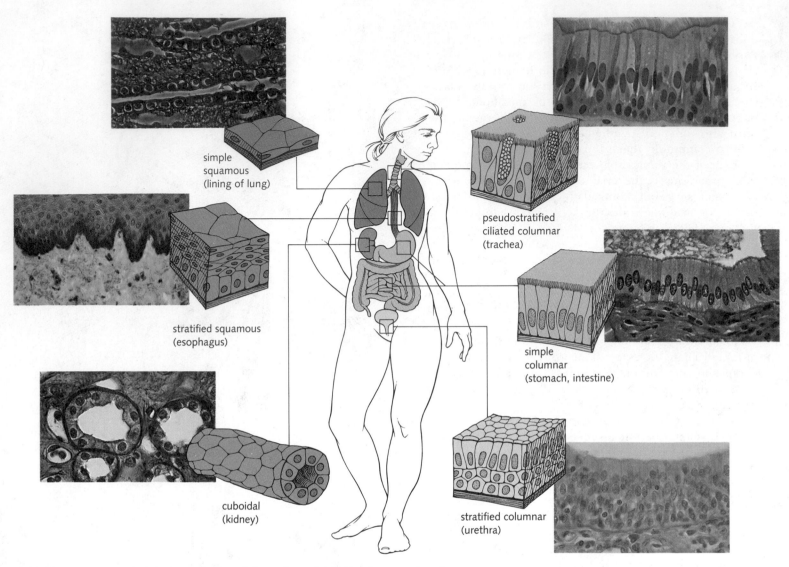

Figure 34.2 Epithelial tissues are found throughout the body. They are placed into categories based on the structural organization of the cells of which they are made. Photomicrographs of six major types of epithelial tissue. Clockwise (from the upper left): simple squamous epithelium (lining of lung), pseudostratified columnar epithelium (trachea), simple columnar epithelium (stomach), stratified columnar epithelium (urethra), simple cuboïdal epithelium (lining of kidney), and stratified squamous epithelium (esophagus).

As different cell types are produced, they do not take up random locations within the embryo. Rather, they become organized into groups that perform specialized functions. These groups of similar cells, which are called **tissues**, are found throughout the body, and they represent a basic level of organization just above the cell itself.

Traditionally, anatomists (biologists who study the organization of the body) have divided tissues into four types: **epithelial, connective, nervous,** and **muscle.** These categories are based on the characteristics of the tissue itself, not on the germ layer from which it was formed.

Within each of these four categories there is a great range of diversity. Cells as diverse as blood cells and bone cells, for example, are considered different types of connective tissue.

Epithelial Tissue

Epithelial tissues (Fig. 34.2) form surfaces around and within the body and often act as barriers that divide the body into distinct compartments. The most obvious epithelial tissue is *skin,* which covers the body and serves as

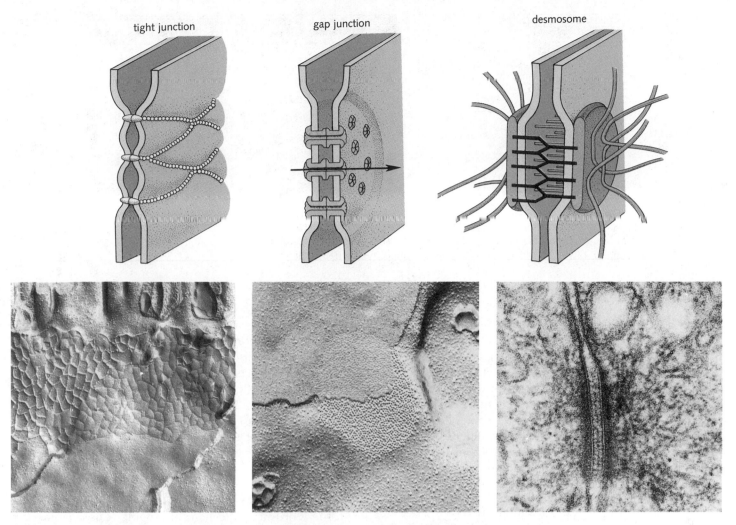

Figure 34.3 Diagrammatic representations of three major types of cell junctions in a simple epithelium. Tight junctions block the passage of materials in the extracellular space between cells, gap junctions form tiny, regulated channels through which small molecules may pass from cell to cell, and desmosomes provide for strong mechanical attachment between adjacent cells. The three major types of cell junctions are shown in the electron micrographs. LEFT: Tight junctions consist of a web-like series of sealing elements between the membranes of two adjacent cells. CENTER: Several hundred gap junctions are seen as a cluster of small particles in the center of this micrograph. Each particle is an individual connecting unit that allows material to pass from one cell to the next. RIGHT: A single desmosome consists of extracellular attachments and thickened cell membrane regions anchored to filaments in the cytoskeleton.

its first line of defense against the outside world. Epithelial tissues that are derived from endoderm form the linings of the digestive and respiratory systems, while mesoderm produces the epithelial tissues that cover body organs and line body cavities and blood vessels. *Secretory glands*, found in tissues derived from ectoderm, endoderm, and mesoderm, are usually formed by invaginations (infoldings) of epithelial tissue.

Epithelial tissues are classified according to the shapes of the cells that form them. Figure 34.2 shows examples of *squamous, cuboidal,* and *columnar* epithelia. A **simple epithelium** consists of a single layer of cells; a **stratified epithelium** consists of two or more layers of cells. **Pseu-**dostratified epithelium** appears to have layers of cells, yet all the cells actually touch a basement membrane.

Cell junctions   In order for tissues to form, the cells that compose them must be held together by some means. The cells of many tissues must also have some means of communication with the cells adjacent to them. **Cell junctions,** specific attachments between adjacent cells, perform these tasks. Junctions are found in many nonepithelial tissues as well, and they can be grouped into three general types (Fig. 34.3).

Tight junctions seal off the external space between two cells, preventing material from leaking between

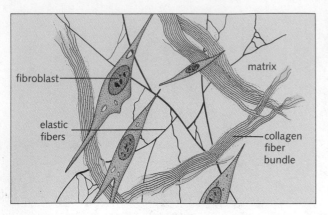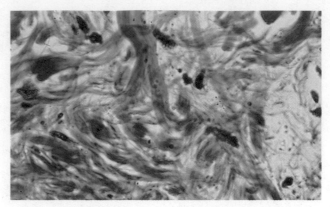

Figure 34.4 LEFT: Loose connective tissue consists of fibroblasts in a matrix of elastic fibers and tough bundles of collagen. Fibroblasts help to produce this matrix and are embedded in it. RIGHT: Photomicrograph of fibroblasts in the collagen-rich connective tissue just beneath the human scalp.

them. Tight junctions are particularly important in epithelia that line fluid-filled cavities, because they prevent the movement of fluid between cells. Tight junctions between cells of the intestine keep its contents from leaking across the intestinal wall into body cavities.

Gap junctions provide a means of intercellular communication. They are small, pore-like protein channels that connect the cytoplasm of one cell to that of its neighbor. Compounds such as salts, sugars, and ions have been shown to pass from one cell to the next through gap junctions. They also allow electrical signals and metabolites to pass between cells, enabling groups of cells to function as a coordinated unit such as heart muscle.

Recent studies have shown that gap junctions are quite sophisticated: they may open or close in response to certain cellular signals, including calcium ion and hydrogen ion (pH) concentration. Gap junctions between the contractile cells of the heart are the principal reason why the heart muscle is capable of contracting as a single unit.

Adhering junctions, which are also known as **desmosomes,** are strong mechanical attachments between adjacent cells. These junctions do not block the movement of material between cells. Desmosomes are attached to the *cytoskeleton* of each cell involved in the junction, and this enables the junction to act almost as a "spot weld"—a point at which two cells are cemented together. Desmosomes are particularly common in tissues that are subject to mechanical stress, such as skin, where they prevent the tissues from breaking apart.

Connective Tissue

Connective tissue gets its name from the fact that it makes up the basic support structures of the body; it connects the body's other tissues in a manageable framework. If you place your hand around your upper arm as you flex

your elbow, you will feel many of the basic types of connective tissues: bones, ligaments, and tendons.

One of the distinguishing characteristics of this tissue is the fact that cells of connective tissues generally produce an **extracellular matrix,** a layer of material that surrounds the cells. The extracellular matrix may be liquid or solid, loose or dense, flexible or rigid. The unique characteristics of the different types of extracellular matrices are often responsible for the different properties that we associate with various types of connective tissues.

Supporting tissue One fundamental role of connective tissue is providing mechanical support. **Loose connective tissue** surrounds major organs, blood vessels, and epithelial tissues. The support provided by its loose matrix of fibrous proteins is indispensable in forming the structure of an organism. In loose connective tissue the most common cell type is the **fibroblast** (Fig. 34.4). The fibroblast produces fiber-like proteins that are woven into a com-

Figure 34.5 Adipose tissue is a form of connective tissue.

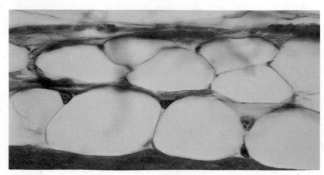

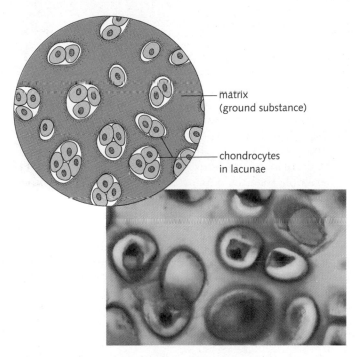

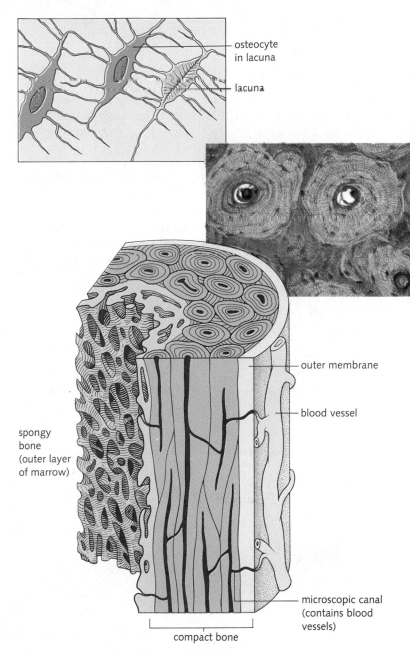

Figure 34.6 Cartilage is produced by chondrocytes, which secrete a tough, flexible extracellular matrix around spaces called lacunae ("holes") in which the chondrocytes remain. The photomicrograph of cartilage shows clusters of chondrocytes confined to isolated lacunae within the tissue matrix.

plex extracellular bundle. The most important of these proteins, **collagen,** forms large, tough fibers that are found throughout the body. It is the single most abundant protein in the vertebrate body. In some tissues, the layers of collagen are denser and form the basis for connections between muscles and bones and other points subjected to mechanical stress.

Adipose tissue, or **fat,** is another type of connective tissue widely distributed throughout the body. Each fat cell contains a single large droplet of fat stored for possible future use (Fig. 34.5).

Cartilage is a form of connective tissue in which an extremely dense matrix containing collagen fibers is laid down in precise orientation (Fig. 34.6). Later in development, elastic fibers are laid down with the collagen, and the result is a tough, resilient, springy tissue that can support great weight and still remain flexible. In some vertebrates, notably sharks and rays, the entire skeleton is made of cartilage. In adult humans, cartilage is found at the endpoints of many bones, in the knee, and in flexible structures such as the ear and the tip of the nose.

Bones are strong and rigid, and they provide support and protection for the vertebrate body and its organs. Bone is produced by cells that first lay down a matrix of cartilage and then gradually *mineralize* that matrix by filling it with crystals of calcium phosphate and calcium carbonate (Fig. 34.7). Bone is not a solid matrix, however;

Figure 34.7 Bone is produced by bone cells called osteocytes, which surround their lacunae with a dense, calcium-rich matrix. In spongy bone, the organization of the matrix is loose and open. In compact bone, the tight matrix is penetrated by a series of microscopic canals through which blood vessels and nerves are able to reach the osteocytes. The photomicrograph of human compact bone clearly shows the system of microscopic canals that bring nutrients to osteocytes. It is important to remember that bone is a living tissue, filled with cells that carry on an active metabolism.

it is penetrated by thousands of tiny microscopic canals through which blood vessels flow. The cells that produce the matrix remain alive and active within it, constantly dissolving old matrix and laying down new matrix. This activity enables broken bones to heal rapidly (see Fig. 34.7).

The inner portion of most large bones, the **bone marrow,** is the principal blood-forming tissue of the body. Most of the cellular components of the circulatory system are produced in bone marrow.

Circulating tissue The white and red blood cells that are found in blood are also derived from connective tissue. Unlike the other connective tissues we have described, they have a liquid matrix, and the tissue itself flows throughout the body. Blood cells in their liquid matrix form key tissues of the *circulatory system* and the *immune system.* We will discuss these two systems in greater detail later in this section.

Nervous Tissue

Neurons, the cellular units of the nervous system, carry out a variety of tasks, all of which involve some form of communication. The actions of neurons enable information to pass rapidly from one end of a large organism to another, making it possible for animals to respond rapidly to changes in their environments. The messages carried by neurons may relay information about the environment, informing the body that an object is hot or cold, for example, or they may pass commands along to other organs, instructing a muscle to contract or a gland to secrete.

Neurons are derived from ectodermal tissue that develops into the nervous system of the embryo. As an organism develops, neurons grow throughout the body, producing a network of cells and nerve fibers (Fig. 34.8).

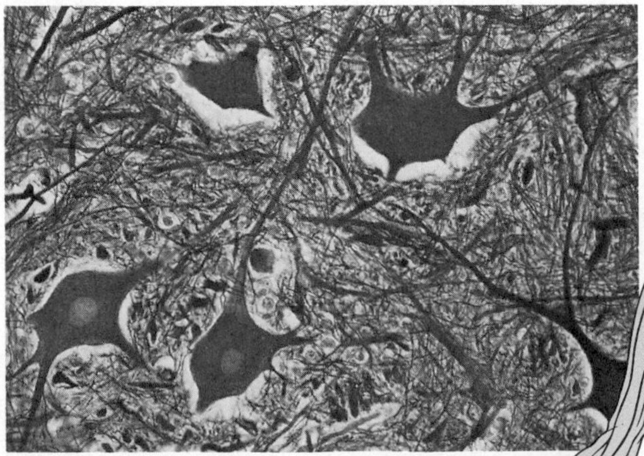

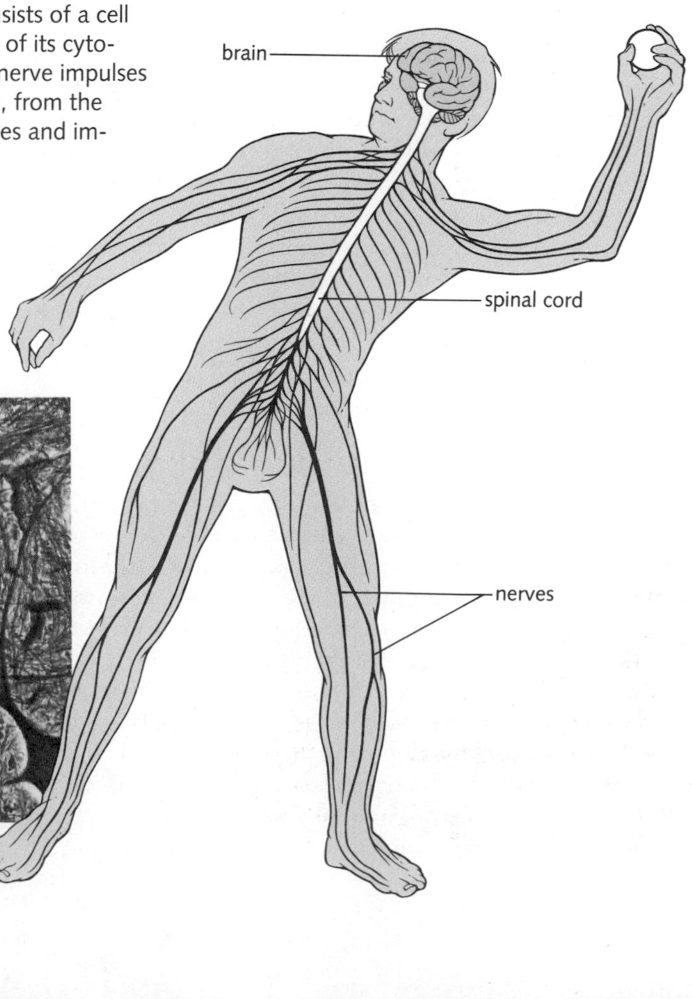

Figure 34.8 Nervous tissue is specialized for the conduction of electrical impulses. Nerve cells vary in size, and they include some of the longest cells in the body. A typical neuron consists of a cell body, containing the cell nucleus and a large portion of its cytoplasm, and a number of cell processes, along which nerve impulses may travel to or from the cell body. This nerve tissue, from the spinal cord, consists of a complex jumble of cell bodies and impulse-carrying cell processes.

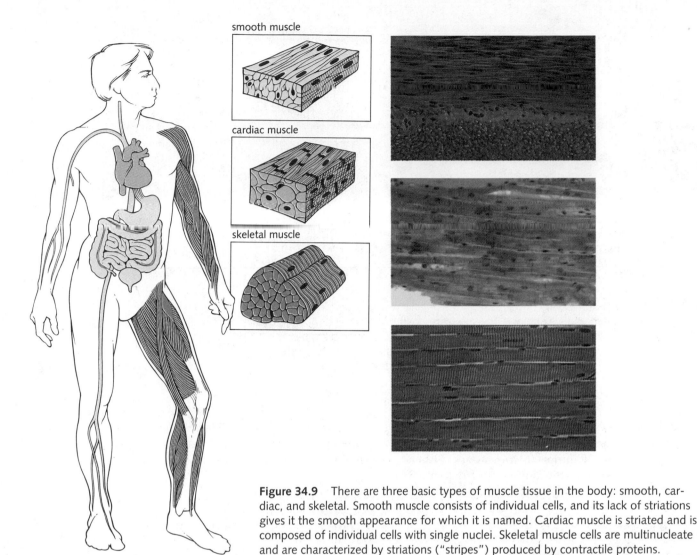

smooth muscle

cardiac muscle

skeletal muscle

Figure 34.9 There are three basic types of muscle tissue in the body: smooth, cardiac, and skeletal. Smooth muscle consists of individual cells, and its lack of striations gives it the smooth appearance for which it is named. Cardiac muscle is striated and is composed of individual cells with single nuclei. Skeletal muscle cells are multinucleate and are characterized by striations ("stripes") produced by contractile proteins.

Muscle Tissue

Muscle cells are specialized for contraction. Contraction requires energy, and muscular tissue is one of the major users of food energy in the body (Fig. 34.9). Their ability to use food energy rapidly means that muscle tissues are major sources of heat. Because only a fraction of the food energy can be converted into movement, the difference is given off as heat.

Smooth muscle tissue is found throughout the body, and it is most common in tissues where muscular contractions are not under conscious control. Smooth muscle cells surround arteries and veins, as well as the walls of the digestive system and the reproductive tract. Smooth muscle gets its name from the fact that the spindle-shaped cells of this tissue have a cytoplasm that appears smooth and unbroken.

Skeletal muscle tissue forms the large muscles of the limbs and trunk that we often associate with physical fit-

ness and muscular development. **Cardiac muscle** is found in the heart, and it combines many of the properties of smooth muscle and skeletal muscle. Both skeletal and cardiac muscle show a characteristic striped pattern (known as striations) produced by an orderly, repeating array of contractile proteins in the muscle cytoplasm. Although cardiac muscle functions largely without voluntary control, most skeletal muscles are under conscious control.

ORGANS AND ORGAN SYSTEMS

Throughout the body we find examples of different types of tissues organized in ways that enable them to perform a single function or a series of related functions. These groupings of tissues are known as **organs.** Many organs

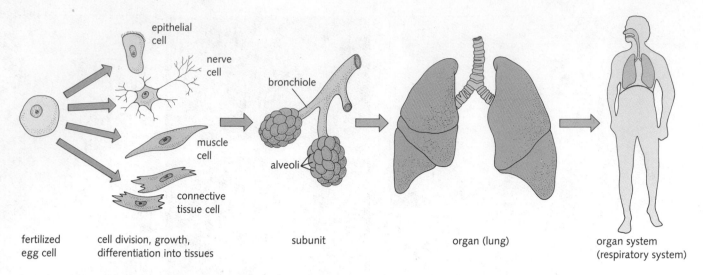

epithelial cell

nerve cell

bronchiole

muscle cell

alveoli

connective tissue cell

fertilized egg cell

cell division, growth, differentiation into tissues

subunit

organ (lung)

organ system (respiratory system)

Figure 34.10 Each system of the body develops from the original fertilized zygote. Specialized cell types are formed during development. The specialized cells of the respiratory system form a host of tissues and structures, such as the alveoli in which gas exchange takes place. Many such tissues make up a complete organ, such as the lungs. And finally, the complete respiratory system includes a series of organs that work in concert during breathing.

are small, such as the *pituitary*, a gland about twice the size of a pea that is located within the skull and is responsible for the production of important hormones. Other organs are very large, such as the body's most expansive organ, the *skin*, which has a surface area of about 2.0 m^2 in the average adult human.

Organs themselves can be grouped into **organ systems,** a higher category comprising groups of organs that are physically or functionally related (Fig. 34.10). For example, the *mouth, stomach,* and *large intestine* are separate and distinct organs that form part of the *digestive system,* a group of organs that digests food. The many different muscles throughout the body can all be thought of as separate organs, but taken together, they form part of the *muscular system,* the most massive of all organ systems in most vertebrates.

Major Organ Systems

In this book we will group the organs and tissues of the body into 10 major organ systems. These groupings are not absolute, and some organs are functional parts of two or more systems. The diaphragm, or breathing muscle, may be placed in the muscular or the respiratory system, for example, and the mouth serves both the respiratory and the digestive systems. Furthermore, it would be wrong to think of the organ systems as units that function in isolation from each other. Detailed physiological work has shown very clearly that links among the organ systems

are important in the daily workings of nearly every organ in the body. Table 34.1 describes and illustrates the 10 major organ systems of the human body.

STUDYING THE HUMAN BODY

The Basic Body Plan

Anatomists use a series of standardized terms to help them describe the positions of tissues and organs. These terms are more precise than our everyday references to "front," "top," and "side," and we will use them in the chapters that follow. Figure 34.11 illustrates how three imaginary planes may be passed through the body to produce the three views (sagittal, frontal, and transverse) that are used to show body structures.

The basic features of the human body are similar to those of other vertebrates. We have an internal skeletal system (endoskeleton) and a dorsal nerve cord (spinal cord). Our bodies exhibit bilateral symmetry, which means that each half is a mirror image of the other half. Even so, that symmetry is not absolute—many internal organs, such as the heart, liver, pancreas, stomach and intestines, are arranged asymmetrically. The human body has three major cavities: dorsal, thoracic, and abdominopelvic (Fig. 34.12). We will refer to these cavities frequently to pinpoint the location of organs and systems.

Table 34.1 *The Major Organ Systems and Their Functions*

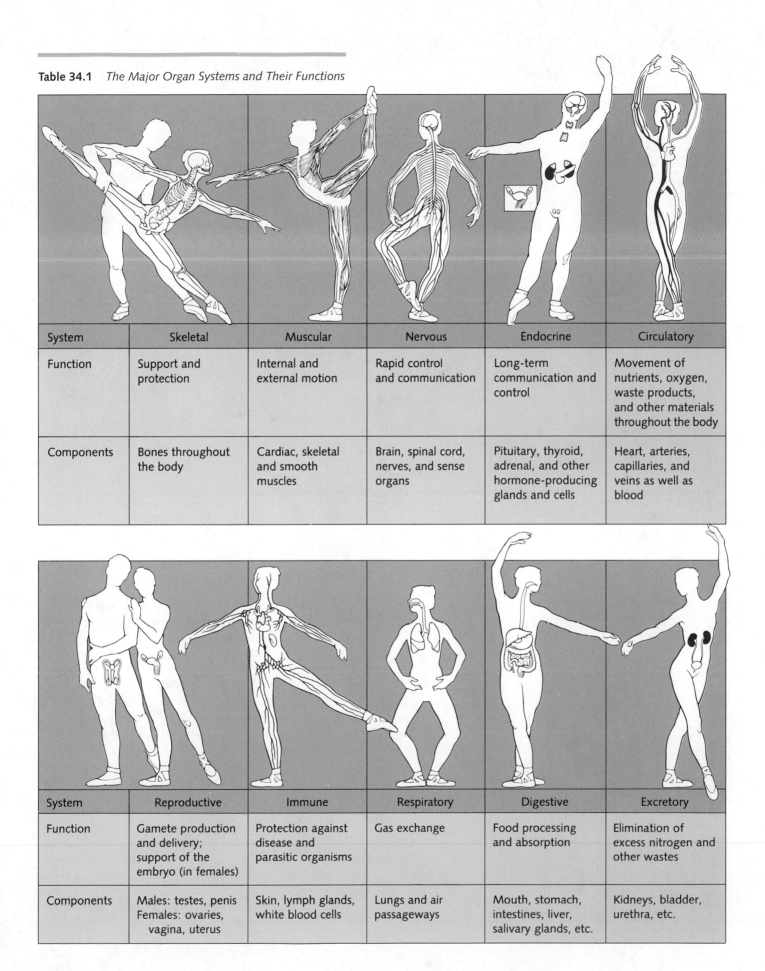

System	Skeletal	Muscular	Nervous	Endocrine	Circulatory
Function	Support and protection	Internal and external motion	Rapid control and communication	Long-term communication and control	Movement of nutrients, oxygen, waste products, and other materials throughout the body
Components	Bones throughout the body	Cardiac, skeletal and smooth muscles	Brain, spinal cord, nerves, and sense organs	Pituitary, thyroid, adrenal, and other hormone-producing glands and cells	Heart, arteries, capillaries, and veins as well as blood

System	Reproductive	Immune	Respiratory	Digestive	Excretory
Function	Gamete production and delivery; support of the embryo (in females)	Protection against disease and parasitic organisms	Gas exchange	Food processing and absorption	Elimination of excess nitrogen and other wastes
Components	Males: testes, penis Females: ovaries, vagina, uterus	Skin, lymph glands, white blood cells	Lungs and air passageways	Mouth, stomach, intestines, liver, salivary glands, etc.	Kidneys, bladder, urethra, etc.

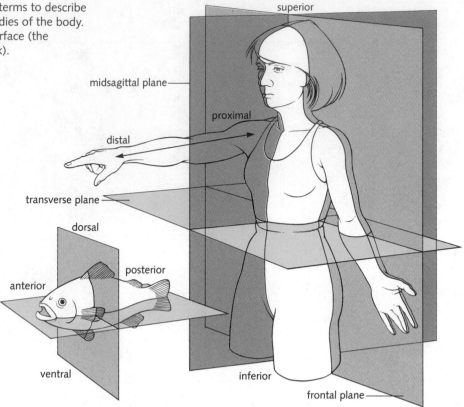

Figure 34.11 Anatomists use a consistent set of terms to describe a series of imaginary planes that help to orient studies of the body. Not labeled in the human figure are the ventral surface (the front of the body) and the dorsal surface (the back).

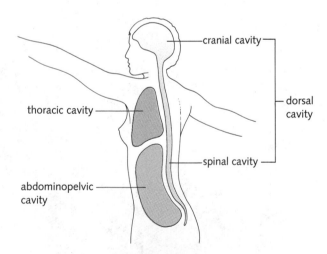

Figure 34.12 The human body contains three major cavities: dorsal (cranial and spinal), thoracic (chest), and abdominopelvic (abdomen).

HOMEOSTASIS

Maintaining the Internal Environment

The body is an organized collection of cells, tissues, and organs whose individual efforts make it possible for the whole organism to survive and function in a particular environment. Therefore, the whole organism depends on the proper functioning of individual cells. This dependence is mutual, of course, because the individual cells also depend on the organism for their survival. The cells and tissues of the body cannot survive on their own, and their individual needs for food, oxygen, waste removal, and protection can be met in nature only if they remain part of the complete organism. Another way to describe the situation is to say that cells require a specific *environment* that only the organism can provide.

The immediate environment of most cells is determined by the *extracellular fluid* in which they are bathed. Its temperature, nutrient and oxygen content, and salt concentration must be maintained within narrow ranges for cellular activities to continue. The internal environment is regulated by a series of automatic mechanisms that are part and parcel of the body's organs and systems (Fig. 34.13). These mechanisms work to counterbalance changes in the environment and to keep the cellular environment consistent.

This regulatory process is called **homeostasis,** which means "keeping things the same." Homeostatic mechanisms work to maintain the constancy of the internal environment. Homeostasis is the central theme of *physiology*, the study of the function of the body's organs and systems. As we will see, it is possible to understand the functions of every system in the body in terms of homeostasis—the maintenance of an acceptable cellular environment.

The Concept of Feedback

Homeostatic systems must respond to changes in the local environment, and then they must counteract those changes by adjusting their own activities. Homeostatic mechanisms are found not only in nature but also in simple machines and everyday appliances. Many of these mechanisms operate by means of a **feedback** principle, in which conditions existing in the environment are "fed back" to the system and used to regulate its activity. At its simplest, a feedback system consists of four basic elements:

1. the *controlled system*, the system regulated to maintain a set level of a particular variable,

2. a *receptor* that monitors the level of the regulated variable,

3. a *processing center* that integrates information about the set level of the variable and the activity of the controlled system, and

4. the *output*, the effect or product of the controlled system.

Feedback systems may be either positive or negative.

Negative feedback   In **negative feedback** systems, output from the system *inhibits* the system itself. A thermostatically controlled furnace is an excellent example of how negative feedback can control a system.

The heat receptor (thermometer) in the thermostat responds to the household temperature and passes that information along to a processing center, in which the actual temperature is compared with the desired temperature set by the homeowner. If the temperature is at the desired level, no action is taken. If the temperature is too low, an electrical signal is sent to the furnace (the controlled system), which then begins to produce heat (the output).

The furnace burns continually until the household temperature rises above the set point in the thermostat. At that point, the receptor "reports" the higher tempera-

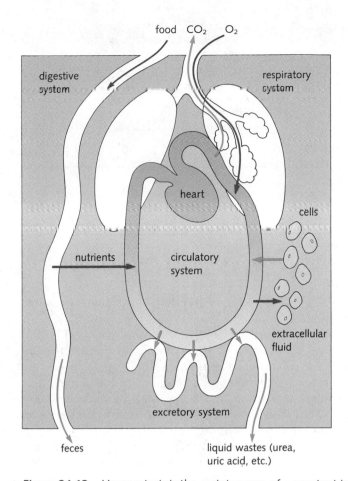

Figure 34.13   Homeostasis is the maintenance of a constant internal environment. Homeostasis demands a variety of systems to regulate temperature, salt and water balance, nutrient flow, and other conditions within an organism.

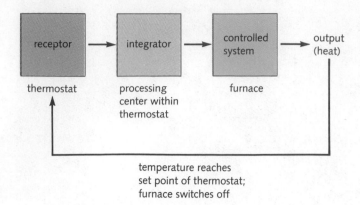

Figure 34.14 The basic elements of a feedback system are found in an ordinary household furnace and thermostat. A feedback system is one that regulates its own output. This block diagram illustrates how *heat,* the output of a furnace, feeds back to stop the production of excess heat, regulating the system to maintain a constant temperature.

ture to the processing center, and the furnace is turned off (Fig. 34.14).

This is a negative feedback control system: The furnace produces an effect (heat) that tends to turn the furnace off over time. Negative feedback mechanisms are widely used to stabilize, or maintain, a particular variable (such as temperature) at a set point. If the variable moves away from the set point, systems are stimulated that tend to return it to the set point.

The body itself regulates temperature by a mechanism that is remarkably similar to a household heating system. The body contains a series of receptors in several organs, especially in the part of the brain called the *hypothalamus,* that monitor the temperature of the blood and compare it to the normal set point of 37°C.

If body temperature drops below this level, cellular activities that use food energy are stimulated and the body produces more heat. One of the most noticeable of these activities is *shivering.* The muscular contractions associated with shivering release heat, causing the body temperature to rise. When temperature receptors in the hypothalamus no longer sense a difference from the set point, they begin to inhibit the production of more heat.

If body temperature rises too far above the set point, sweating is stimulated and the evaporation of sweat cools the body surface. This cooling effect lowers body temperature until it returns to the set point and the *sweating* reaction stops.

Positive feedback Whereas negative feedback acts to stabilize a situation, **positive feedback** tends to intensify a particular effect. During childbirth, for example, the pressure of the baby's head against the wall of the uterus produces a stimulus that increases the contractions of smooth muscles surrounding the uterus. These contractions cause the head to be pushed harder against the wall, and this in turn causes still stronger contractions. The

positive feedback in this case (each contraction produces a stimulus that *increases* the strength of the next contraction) serves an important physiological purpose. It helps force the baby through the uterus into the birth canal. Once birth is completed, the stimuli are gone and the contractions become less intense until they finally stop.

Homeostasis in Action: Eat, Drink, and Be Merry

Water balance Homeostatic systems normally operate so smoothly that we are scarcely aware of their existence. Every action we take has consequences that might affect the internal environment, and it is only because our internal organs regulate that environment so closely that we can carry out everyday functions such as eating and drinking.

When you exercise strenuously on a hot, dry day, for example, you lose body moisture in the form of sweat. If this loss of water continued unabated, your body would soon be in trouble for several reasons. But that doesn't happen, because your body's homeostatic mechanisms swing into action (Fig. 34.15).

In addition to the temperature sensors mentioned earlier, the hypothalamus also contains cells that are sensitive to the concentration of water in the blood. As you lose body moisture, causing your blood to become more concentrated, the hypothalamus does two things at once. On one front, it creates in your consciousness a feeling of thirst, so you "know" you ought to take a drink. At the same time, the hypothalamus causes the neighboring pituitary gland to release a substance known as antidiuretic hormone (ADH). ADH molecules pass through the bloodstream to the kidneys, the organs whose activities control the amount of water removed from the blood to be eliminated from the body as urine. ADH *inhibits* the removal of water from the bloodstream, enabling the body to conserve water.

Then, when you finally get around to taking your long-awaited drink, you might take in as much as 1 or 2 liters of fluid over the course of an hour. Most of that water is quickly absorbed across the wall of the digestive system into the bloodstream. But 2 liters of water added to the blood would dilute it so much that the equilibrium between the blood and the cells of the body would be disturbed. Large amounts of water would diffuse across blood vessel walls into the tissues. The cells of the body would swell with the excess water. Capillaries might be blocked by swollen red blood cells, and the skin would become swollen and puffy as water diffused into it from the bloodstream.

Needless to say, this doesn't happen, because the same homeostatic system intervenes. When the water content of the blood rises, as it might after a large drink, the pituitary releases less ADH. With lower ADH levels, the kidneys quickly remove water from the bloodstream, forming large quantities of urine and restoring the blood to its original salt concentration.

Thus the system sets both upper and lower limits for blood water content: a water deficit stimulates the release of ADH, causing the kidneys to conserve water; whereas an overabundance of water causes the kidneys to eliminate the excess (Fig. 34.16).

Note that the system works even more quickly when the excess fluid contains alcohol, because alcohol inhibits the production of ADH. For that reason (as you may have noticed), drinking alcohol results very rapidly in a need to urinate.

Sugar balance Without regulation of blood sugar, even the simple act of eating a snack could cause problems for certain cells in the body. Many tissues absorb glucose (sugar) directly from the bloodstream for energy production. For that reason, a sudden surge in blood sugar could completely upset the metabolism of certain cells. Conversely, a decrease in blood sugar between meals would starve other cells—such as brain cells—that depend on blood glucose for nearly all of their energy. This is not an abstract problem. If the body didn't have the means to prevent rapid changes in sugar concentration in the blood, eating a candy bar or drinking a glass of lemonade could double the concentration of blood sugar.

But the body does have a sugar-control mechanism. Blood absorbs glucose and other sugars as it passes through vessels in the small intestine. From the intestine it passes to the liver, where excess sugar is removed from the bloodstream and stored in liver cells, under the control of the hormone *insulin*. The insulin-stimulated absorption of sugar by the liver ensures that sweet foods do not affect blood sugar levels in the general circulation.

Figure 34.15 A homeostatic system in action.

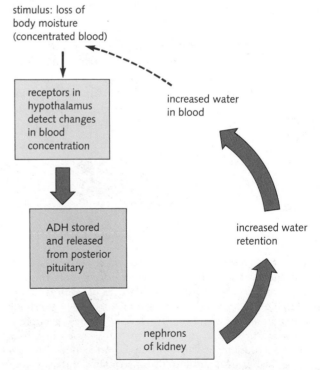

Figure 34.16 Water balance is controlled by the hypothalamus, the pituitary gland, and the kidneys. When the blood becomes too concentrated, the hypothalamus causes the pituitary to release ADH (antidiuretic hormone), which *inhibits* water removal by the kidneys. When the water content of the blood is too high, ADH release is inhibited, and the kidneys act to remove excess water. This is a classic negative feedback system.

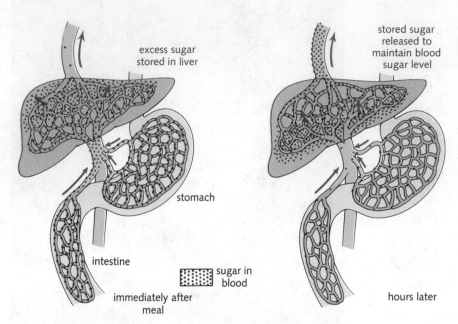

excess sugar
stored in liver

stored sugar
released to
maintain blood
sugar level

stomach

intestine

sugar in
blood

immediately after
meal

hours later

Figure 34.17 Blood sugar level is regulated by the action of the liver. Immediately after a meal, sugar-laden blood from the small intestine passes through the liver, where enough sugar is removed and stored to prevent a sudden rise of sugar in the general circulation. Between meals, stored sugar is released from the liver to maintain a constant sugar level in the blood. The uptake and release of blood sugar is controlled by a series of hormones and will be investigated more fully in Chapter 39.

 As blood sugar levels fall, on the other hand, the liver releases its carbohydrate stores, once again keeping blood sugar levels constant (Fig. 34.17). The liver's strategic position in the circulatory system and its ability to regulate blood sugar make it another critical organ of homeostasis. We will explore the regulation of blood glucose in Chapter 39.

SUMMARY

The human body develops in a manner similar to the bodies of other vertebrates. Three cell layers—ectoderm, mesoderm, and endoderm—are produced in the early stages of embryonic development. The wide variety of organs and tissues of the body are formed from these three germ layers. Tissues, groups of cells that perform similar functions, can be grouped into four categories: nervous, muscular, epithelial, and connective. Tissues can be further grouped into distinct organs, and groups of organs in turn form a series of organ systems. Interactions between adjacent cells occur at the level of the cell membrane, and specialized cellular junctions mediate these interactions in a way that assists in the formation of complex, interconnected tissues. The interdependence of the cells in a large organism requires that the internal environment of the organism be closely regulated. Homeostasis, the maintenance of a regulated cellular environment in the face of changing external conditions, is one of the important tasks of nearly every system in the body. Each organ system plays a role in homeostasis. A key part of any homeostatic process is feedback, in which the condition of one aspect of the internal environment directly affects tissues or organs that help to regulate it. The regulation of water balance, in which a series of tissues interact to regulate the concentration of dissolved material in the blood, is a classic example of a feedback system. Similar systems exist to regulate the levels of blood sugar and other compounds within the circulation, as well as such diverse variables as body temperature and hormone levels.

STUDY FOCUS

After studying this chapter, you should be able to:

- Explain the cellular organization of tissues and organs in multicellular organisms.
- Describe the major human organ systems.
- Describe the concept of homeostasis, and provide some examples of feedback regulation.

ectoderm *673*
mesoderm *673*
endoderm *673*
epithelial tissue *674*
connective tissue *674*
nervous tissue *674*
muscle tissue *674*
tight junctions *676*
gap junctions *676*
adhering junctions *676*

extracellular matrix *676*
fibroblast *676*
adipose tissue (fat) *677*
cartilage *677*
bones *677*
organs *679*
homeostasis *683*
negative feedback *683*
positive feedback *684*

REVIEW

Objective Questions (Answers in Appendix)

1. Select the sequence that reflects the correct order of levels of organization in living organisms.
   (a) organs–tissues–cells–systems
   (b) systems–tissues–organs–cells
   (c) organs–cells–tissues–systems
   (d) cells–tissues–organs–systems
   (e) systems–organs–tissues–cells

2. Epithelial tissues are categorized by *function* as
   (a) supportive and fibrous.
   (b) protective and glandular.
   (c) stratified.
   (d) squamous, cuboidal, and columnar.

3. _____ is a flexible tissue that surrounds major organs and contains fibroblasts.
   (a) Epithelial tissue
   (b) Smooth muscle tissue
   (c) Loose connective tissue
   (d) Bone tissue

4. In mammals, the _____ governs control of body temperature and the _____.
   (a) pituitary gland; sugar-control mechanism
   (b) hypothalamus; thirst mechanism
   (c) adrenal cortex; hunger mechanism
   (d) pituitary; reproductive mechanism
   (e) adrenal cortex; thirst mechanism

5. The sugar-control mechanism in humans is located in the
   (a) liver.
   (b) kidneys.

(c) large intestine.
(d) gallbladder.
(e) small intestine.

Discussion Questions

6. Describe the major organ systems of the human body. Which systems seem to have overlapping functions?

7. Do you think that single-celled organisms have homeostatic systems? Why or why not?

8. Describe the major types of tissues. What relationship do they have to the three germ layers that were formed in the embryo?

9. What are the differences between a tissue and an organ? Between an organ and a system?

10. Describe your understanding of the term *homeostasis*. Give several examples of homeostatic control mechanisms that exist throughout the body.

11. What critical homeostatic roles are played by the liver and the kidneys when we eat a single large meal and also drink a large amount of liquid?

12. Some recent experiments have shown that when embryos are treated with substances[1] that block movement through gap junctions, they fail to develop properly. What does this imply about the role that gap junctions might play in normal development?

READINGS

Fox, S. I. *Human Physiology*, 4th ed. Dubuque, IA: Wm. C. Brown, 1993. An excellent and up-to-date textbook on human physiology.

Schmidt-Nielsen, K. *Animal Physiology: Adaptation and Environment*, 4th ed. New York: Cambridge University Press, 1990. An excellent and comprehensive textbook on comparative animal physiology.

Fawcett, D. W. *A Textbook of Histology*, 11th ed. New York: Saunders, 1986. A classic textbook that is the standard reference work on the microscopic anatomy of cells and tissues.

1. Antibodies against the major protein of the gap junction.

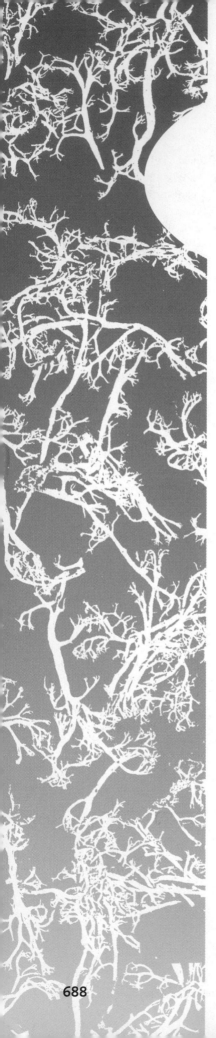

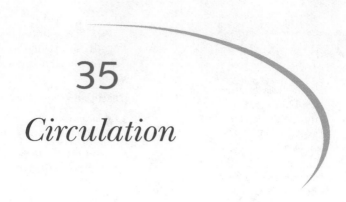

35

Circulation

"●

i was about 47 years old at the time. I was at work, and I collapsed one day. There are 22 minutes out of my life that I don't remember. I had gone into cardiac arrest."

That was the experience of one man who survived a heart attack and who—after open-heart surgery—is now leading a healthy, more or less normal life. (He happens to run the Boston Marathon regularly.)

But half a million other Americans who suffer heart attacks each year are not so lucky. They die. In fact, the average American has one chance in four of dying from a heart attack.

The occurrence of heart disease, as we will see shortly, is related to several aspects of modern life, both in the United States and in many other developed nations. But the severe consequences of heart attacks bear grim witness to the importance of this small, muscular organ, about the size of a clenched fist. They also testify to the importance of the circulatory system, whose function depends utterly on the heart.

THE IMPORTANCE OF INTERNAL TRANSPORT

To understand why the circulatory system is so important, we must backtrack a bit to consider the requirements of the individual cells of which all multicellular organisms are composed. Each of those cells must have a constant supply of oxygen and nutrients and a method for removing wastes.

When an organism consists of a single cell, this exchange can be carried out by the processes of transport and diffusion across the cell membrane. Some simple multicellular organisms, such as *Hydra*, can also get by without a specialized system to transport nutrients and dissolved gases through their bodies (Fig. 35.1). *Hydra* is a simple animal composed of two layers of cells. Food is captured and brought into a large digestive cavity in the center of the organism. Water flows into

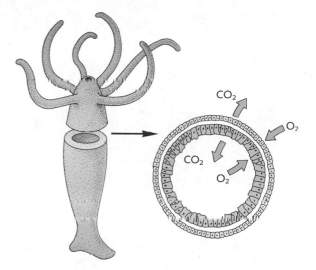

Figure 35.1 In a small organism, gas exchange can take place directly with the environment. *Hydra,* shown here, consists of two cell layers. Oxygen and carbon dioxide diffuse between the environment and each cell layer.

and out of the digestive cavity, permitting every cell in the organism to come into direct contact with the environment.

But although diffusion is efficient on the microscopic scale, in larger organisms it does not meet the needs of thousands of cells, many of which have no direct contact with the environment. Furthermore, in most multicellular organisms of any size, specialized groups of cells are organized into functional units known as organs. One organ may specialize in the processing of food, another in the exchange of gases with the environment, and another in the elimination of waste material. These organs pose a serious problem: oxygen, carbon dioxide, nutrients, and other substances must be carried from one place to another so that these organs can serve the needs of the whole organism.

A **circulatory system** is a group of organs and tissues that do just that by *circulating* a fluid substance through the body of an organism. The fluid may carry dissolved respiratory gases, food, chemical messages, waste materials, and even living cells. The fluid that moves through a circulatory system is generally called **blood,** and the tubules through which blood passes are known as **blood vessels.**

Circulatory systems in the animal kingdom can be divided into two broad categories: **open circulatory systems,** in which the circulating fluid percolates throughout the body of an organism and bathes nearly all the cells of the body, and **closed circulatory systems,** in which the circulating fluid is confined to blood vessels.

Open Circulatory Systems

Open circulatory systems are found in most molluscs and arthropods. The blood, or *hemolymph,* that circulates in these organisms moves through open ended vessels that empty into the body cavity or **hemocoel.** As blood flows through the hemocoel, it carries cellular waste products to the excretory system and nutrient materials from the digestive system.

One of the best-studied open circulatory systems is found in the common grasshopper. The major organ of the grasshopper circulatory system is a large **heart** that lies along the back of the body (Fig. 35.2). This dorsal heart is a muscular tube that collects blood near the rear of the animal and squeezes it toward the front of the animal by a series of regular contractions.

Blood leaving the heart courses through the tissues of the body and gradually flows toward the rear of the animal, completing the circuit. Blood reenters the heart through small openings called *ostia,* where one-way valves prevent blood from leaking back during a contraction of the heart muscle. The flow of blood carries minerals and nutrients throughout the body and transports waste products to the excretory organs. In insects the blood does not carry oxygen (insects have a special respiratory system in which tubules called *tracheae* carry the air directly to the tissues).

Closed Circulatory Systems

The simplest forms of closed circulatory systems are found in echinoderms and annelids, which include sea cucumbers and earthworms, respectively. More complex systems are found in vertebrates. The circulatory system of the earthworm is a good example of the general plan of a closed system (see Fig. 35.2). Blood is pumped into large, muscular vessels from a series of five *aortic arches.* These arches are the equivalent of the single pump, or heart, that forces blood through the more advanced vertebrate circulatory systems.

As blood flows from the arches, it passes through smaller and smaller vessels, finally moving into a fine meshwork of **capillaries**—small vessels whose walls are seldom more than one cell thick. Diffusion between the cells of the body and the blood takes place across the capillary walls. It is estimated that no cell in the human body is farther than 0.1 mm from a capillary.

Gradually, blood flows out of the capillary system into larger and larger vessels, which ultimately return the blood to the aortic arches to be pumped through the system again. A closed system produces higher pressures than an open system does, allowing blood to circulate more rapidly. Why is rapid circulation useful? The slow

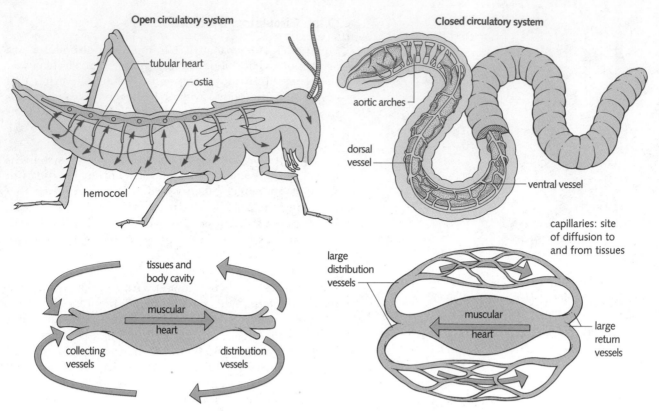

Figure 35.2 LEFT: The grasshopper has an *open circulatory system*. Blood is forced through the animal's muscular heart and percolates directly through the tissues. Blood returns to the heart through the *ostia*. RIGHT: The earthworm has a *closed circulatory system*. Blood is forced from the heart and passes into the body by means of vessels and capillaries. The restriction of blood flow to these vessels in a closed system allows the circulatory system to produce a higher pressure than an open system, enabling blood to flow more rapidly.

circulation of blood in the open systems of insects cannot supply enough oxygen to meet the needs of the active tissues of the body, which is one of the reasons why insects have a separate system of tracheae to distribute oxygen. By contrast, the rapid flow in many closed systems, including that of the earthworm, makes it possible for blood to carry respiratory gases as well as food and waste material.

THE VERTEBRATE CIRCULATORY SYSTEM

The closed circulatory systems of vertebrates have several important features in common: they all contain a single muscular heart with multiple chambers, and they all contain a number of special features that integrate circulation with gas exchange in the organs of *respiration*.

Vertebrate blood is itself a living tissue, and it contains a range of specialized cells. Its functions include (1) the transportation of nutrients, gases, and wastes; (2) the maintenance of a proper internal environment; and (3) the protection of the organism against disease.

Vertebrate Blood Vessels

Vertebrate circulatory systems contain three distinctly different types of blood vessels: arteries, veins, and capillaries (Fig. 35.3).

Arteries are vessels that carry blood away from the heart. Arteries are thick-walled vessels composed of four layers of tissue: an *endothelium* (a layer of flattened cells that line the artery wall), an elastic layer, a thick layer of smooth muscle, and an outer layer of connective tissue. These four layers are tough and elastic. Because blood is pumped from the heart in powerful spurts, blood in the arteries closest to the heart is under increased pressure. The elasticity of artery walls allows these vessels to expand

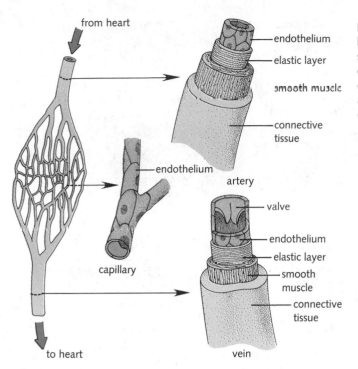

from heart

endothelium
elastic layer
smooth muscle
connective tissue

artery

endothelium

capillary

to heart

valve
endothelium
elastic layer
smooth muscle
connective tissue

vein

Figure 35.3 The basic structure of arteries, capillaries, and veins. Arteries and veins are multilayered structures that include layers of elastic tissue and connective tissue, as well as a layer of smooth muscle. The walls of the smallest capillaries, by contrast, are only one cell thick, bringing blood into close contact with the surrounding tissues.

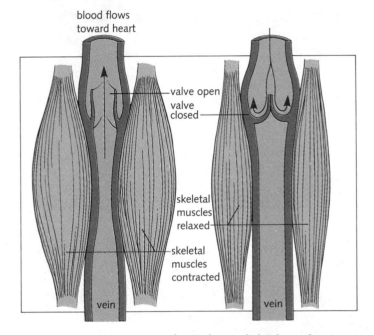

blood flows toward heart

valve open
valve closed

skeletal muscles relaxed

skeletal muscles contracted

vein

vein

Figure 35.4 Many veins are located near skeletal muscles. When these muscles contract, they help to force blood through the veins. Veins contain one-way valves which prevent the backflow of blood.

to accommodate the surges of blood, and by the time blood flow reaches the smallest vessels, it has become smooth and continuous.

As blood moves away from the heart, the arterial system branches into smaller arteries, **arterioles,** which lead directly to the smallest vessels of the circulatory system, the **capillaries.** Capillaries are blood vessels whose walls, as we have said, are only one cell thick. These tiny vessels form an intricate network that brings blood throughout the tissues of the body, enabling materials to diffuse across the capillary wall in both directions: into the blood and out of it. The interior of a capillary may be as small as 10 μm (just big enough for red blood cells to slip through in single file).

Near the end of the capillary circulation, blood begins to collect into larger vessels called **venules,** which channel blood into larger **veins** that lead eventually to the heart. The walls of veins are similar to arteries; they are composed of three layers—an endothelium, a layer of smooth muscle, and an outer layer that contains both and connective tissue. Because the venous blood has traveled a great distance from the heart, the source of fluid pressure in the system, blood passes through the veins under much less pressure than it does through arteries. Many veins have **valves** that prevent blood from flowing in a reverse direction. In some vertebrates, veins are located near skeletal muscles so that when a muscle contracts, blood is forced through the venous system (Fig. 35.4).

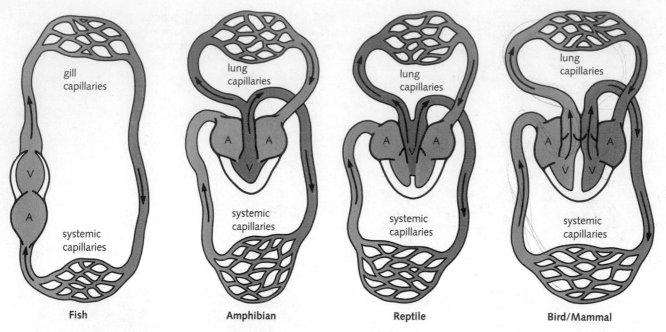

Figure 35.5 Circulatory patterns in fish, amphibians, reptiles, birds, and mammals. Oxygen enters each circulatory system in lung or gill capillaries and is carried to the tissues in the systemic capillaries.

The Vertebrate Heart

The center of the vertebrate circulatory system is the heart. Heart structure differs among major vertebrate groups, and comparisons among those variations offer clues to the evolutionary process that gave rise to the mammalian heart.

The simplest vertebrate heart is the *two-chambered heart* found in fish. Blood collects in a single large vein, the *sinus venosus*, and enters the first chamber, the **atrium.** It is pumped into the second chamber, the muscular **ventricle,** which forces it into a single large artery, the *conus arteriosus*. (The words *atrium* and *ventricle* are derived from Latin. The atrium was the room by which one entered a Roman house, and the word *ventricle* is derived from a word that means "to depart.")

The pattern of blood flow in fish is known as **single circulation,** because the blood passes through the entire circulatory system after each pass through the heart. One difficulty with this pattern of blood flow is that blood must pass through two systems of capillaries, one right after another. The first capillary network is in the gills, where the blood is oxygenated; the second is in the cells and tissues of the rest of the body. The passage of oxygen-poor blood through the gills reduces fluid pressure, slowing the rate at which oxygen-rich blood then moves through the rest of the body, limiting the efficiency of oxygen delivery.

The circulatory systems of higher vertebrates show a trend toward separation of oxygen-rich and oxygen-poor blood, which improves the efficiency of oxygen delivery and helps to support higher metabolic rates in body tissues. Amphibians have a *three-chambered heart* that directs part of the circulatory flow to the *lungs,* making it possible for these animals to use their lungs to exchange gases directly with the air. As shown in Fig. 35.5, all the blood that passes through the lungs is returned to the heart, where it is mixed with blood returning from the systemic circulation and pumped back out through a vessel called the **aorta.**

In amphibians, the aorta splits into two main vessels, one that goes to the lungs and one that enters the systemic circulation. Blood flowing in this system moves in a pattern known as **double circulation,** because there are two basic pathways that blood leaving the heart may enter: the **pulmonary circulation** leading to the lungs and the **systemic circulation.**

The circulation of blood in reptiles is similar to that in amphibians, but an important difference illustrates the beginning of an evolutionary trend. In most reptiles, the single ventricle is partially divided by a muscular wall, as shown in Fig. 35.5. This wall helps to separate the blood that enters the right and left atria, and experiments have shown that as a result, there is very little mixing of oxygen-rich blood from the lungs with oxygen-poor blood from the systemic veins.

Birds and mammals seem to have completed the evolutionary experiment begun in reptiles. Their ventricles are completely divided into two separate chambers, a right and a left ventricle, so that there is no chance for oxygen-rich blood to mix with oxygen-poor blood. These **four-chambered hearts** can be thought of as two separate pumps. The right atrium collects blood from the veins, and the right ventricle pumps it into the lungs. The left atrium fills with blood from the lungs, and the left ventricle pumps this oxygen-rich blood into the systemic circulation.

This four-chambered heart ensures that (1) *all* the blood that returns to the heart is pumped through the lungs where gas exchange takes place and (2) *all* the oxygen-rich blood returning from the lungs is immediately pumped into the systemic circulation. This double circulation is particularly well suited for animals that are very active and have a high demand for energy, because it pumps fully oxygenated blood into the circulatory system at high pressure and high velocity. Therefore, it is not surprising to find that birds and mammals (warm-blooded animals with high metabolic rates) possess circulatory systems that exhibit this pattern.

The Structure of the Human Heart

The human heart is a cone-shaped muscle a bit larger than a fist (Fig. 35.6). It is located in the thoracic cavity, in a space between the lungs, just to the left of the midline of the body. The heart is contained within a loose sac known as the *pericardium* ("around the heart"). A thin film of fluid between the surface of the heart and the pericardium serves to prevent friction during the heartbeat. The wall of the heart muscle is composed of an outer layer, the *epicardium;* a thick middle layer of muscle tissue, the *myocardium;* and an inner lining, the *endocardium.*

The work of the heart is done largely by powerful contractions of the myocardium. In the ventricles, myocardial tissue forms column-like projections known as *papillary muscles.* The myocardium contracts in a regular fashion, providing most of the force that causes blood to flow through the circulatory system.

The heart muscle pumps about 70 milliliters of blood with each contraction, and it beats about once a second, sleeping or waking, without taking even a few minutes' rest, during an entire lifetime. From the time your heart started beating before birth, to the end of an average lifetime, it will have pumped roughly 45 million gallons of blood through your body.

The heart is divided into right and left halves. Each side of the heart consists of two chambers: an atrium through which blood enters the heart and a ventricle

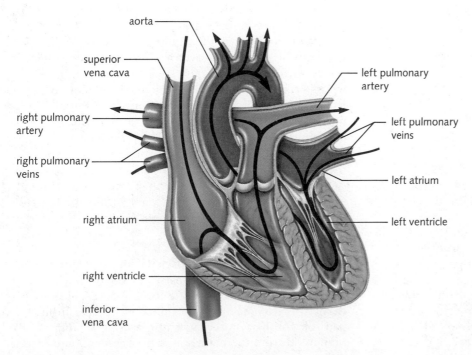

aorta

superior vena cava

right pulmonary artery

right pulmonary veins

right atrium

right ventricle

inferior vena cava

left pulmonary artery

left pulmonary veins

left atrium

left ventricle

Figure 35.6   The basic pattern of blood flow through the valves and chambers of the human heart.

Figure 35.7 The opening and closing of cardiac valves is crucial to the pumping action of the heart. **(a)** As the heart muscle relaxes, blood flows through the atria and into the ventricles. **(b)** As the heart muscle contracts, the valves between atria and ventricles close and **(c)** blood is forced from the ventricles into the major vessels leading to the lungs and body.

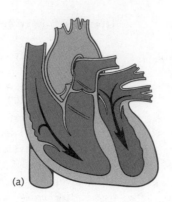 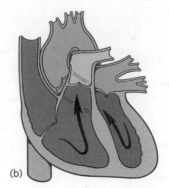 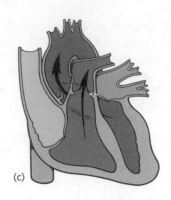

(a) (b) (c)

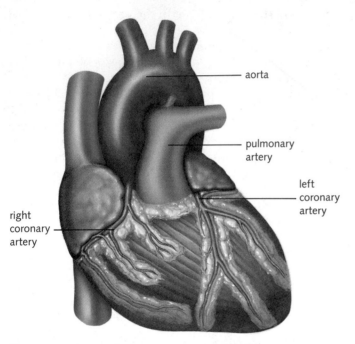

aorta

pulmonary artery

left coronary artery

right coronary artery

Figure 35.8 Nourishment and oxygen are supplied to the cells of the heart through the coronary arteries.

Blood Flow Through the Heart

Blood enters the heart through the left and right atria, filling each of these chambers. When the heart muscle contracts, blood moves from the atria into the ventricles (Fig. 35.7). Contraction of the ventricles does not force blood back into the atria, because special flaps of tissue form *valves* across the passageways between the atria and the ventricles. The **atrioventricular valves,** or AV valves, are easily pushed open when blood flows through from the atria to the ventricles. But when the ventricles contract and blood is forced in the other direction, these valves are forced shut. Another set of valves, the *pulmonary* and *aortic valves,* are located where blood leaves the ventricles to enter the lungs and the systemic circulation, respectively. (These valves owe their name, *semilunar valves,* to their half-moon shape.)

When blood leaves the right side of the heart, it passes through the **pulmonary arteries** into the lungs, where it releases carbon dioxide and picks up oxygen, a process we will consider in detail in the next chapter. When it leaves the left side of the heart, it passes through a thick, muscular artery known as the aorta. This major vessel branches off to send blood to every branch of the systemic circulation. The two sides of the heart represent the two major circuits of the mammalian circulatory system: the right side of the heart pumps blood through the lungs (the *pulmonary circulation*), while the left side pumps oxygen-rich blood through the body (the *systemic circulation*).

Blood supply to the heart muscle is critical Although an enormous volume of blood passes through the chambers of the heart, the heart muscle itself gets very little nourishment or oxygen from the blood it pumps. Instead, a pair of **coronary arteries** (right and left) (Fig. 35.8) branch off from the main systemic circulation and feed a

through which it leaves. Although blood flows between the two sides of the heart during embryological development, a thick *septum* gradually develops that closes completely at birth, preventing blood flow between the sides. In some infants, the septum does not close properly, allowing blood to flow between the left and right sides of the heart. This produces a mixture of oxygen-rich and oxygen-poor blood that limits the efficiency of the circulatory system. If this problem is not corrected, these babies are permanently weakened by the delivery of relatively oxygen-poor supplies of blood to their cells.

network of tiny capillaries that permeate the heart muscle. These vessels are relatively small, and any blockage of the circulation through the coronary arteries can deprive the heart muscle of oxygen. When blockage of the coronary arteries is extensive enough, cardiac muscle cells begin to die, causing a heart attack.

Control of Heartbeat

Although the heart beats as a single unit, all the cells of the heart do not contract at exactly the same time. Rather, contraction begins in the atria and spreads to the rest of the muscle in a wave-like pattern. Unlike skeletal muscles, the nervous system does not cause each cardiac muscle cell to contract. Instead, individual cardiac muscle cells have the innate ability to contract in a regular, rhythmic cycle.

Each cardiac muscle cell is connected to neighboring cells by a series of *gap junctions* (described in Chapters 16 and 34) that allow small molecules and ions to pass directly from one cell to another (Fig. 35.9). When several cardiac muscle cells are grown together in a culture dish, they form gap junctions with each other and gradually begin to beat as a unit. Each contraction is accompanied by the flow of current (in the form of ions) across the muscle cell membrane. Because these ions can also flow through gap junctions, the contraction of one cell begins an ion flow that starts a contraction in the neighboring cell that gradually spreads from cell to cell.

Specialized cardiac muscle cells speed the movement of impulses throughout the heart. These impulses originate from a region of the heart known as the **sinoatrial node** (otherwise known as the **SA node**) located in the upper corner of the right atrium (Fig. 35.10). The sinoatrial node is a source of a wave of contraction in the heart that spreads through both atria via the gap junctions linking cardiac muscle cells. This region is known as the **pacemaker** of the heart. The wave of excitation initiated at the pacemaker causes the two atria to contract together, helping to force blood into the ventricles.

The impulse does not spread directly to the ventricles, however, because there is a region of connective tissue that does not conduct impulses. Instead, the impulse is picked up by a special bundle of conducting fibers (the atrioventricular bundle of His) that originate at an area known as the **atrioventricular node (AV node).** These fibers conduct the impulse to the lower regions of the ventricles, which then contract smoothly to force blood up and out of the ventricles through the semilunar valves. It takes about one-tenth of a second for the impulse to travel from the AV node to the muscular walls of the ventricles, so there is a delay between atrial contractions and ventricular contractions.

Figure 35.9 Electron micrograph of the boundary between two cardiac muscle cells. The arrows point to two regions that contain gap junctions. These junctions allow electrical current to flow directly from one cell into the next, enabling the heart muscle to contract as a unit.

Figure 35.10 The heartbeat begins in a region of the heart muscle known as the *sinoatrial node* (also called the *pacemaker*). The contraction spreads through the heart muscle like a wave. Conducting fibers that originate at the *atrioventricular node* spread the contraction to the ventricular region of the muscle.

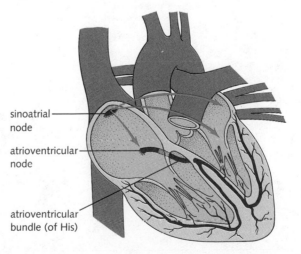

sinoatrial node

atrioventricular node

atrioventricular bundle (of His)

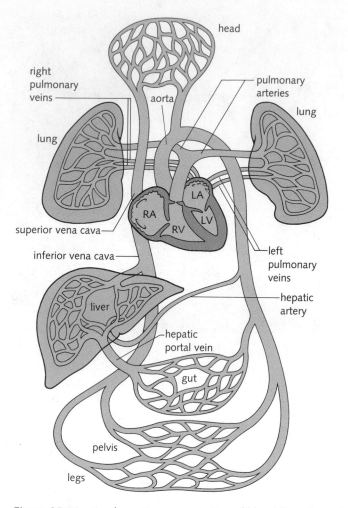

Figure 35.11 A schematic representation of blood flow through the human body. Throughout most of the body, arterial blood passes through a single bed of capillaries before it returns to the venous circulation. However, blood passing out of the capillaries associated with the gut is collected in the *hepatic portal vein* and then passes through a second set of capillaries in the liver. The *hepatic portal circulation* ensures that nutrients absorbed from the gut will pass through the liver before they reach the general circulation.

In some forms of heart disease, the AV node does not initiate an impulse properly. In its place, other groups of cells may begin to initiate independent contractions, and the usually smooth wave of contraction may become broken up into a series of partial contractions. Blood pumping becomes much less efficient, and the inadequate oxygen supply may cause a feeling of weakness. This condition can be remedied by an *artificial pacemaker,* a tiny electronic device that is surgically installed near the AV node and initiates normal contractions by sending out regular electrical pulses to simulate the pulses of a healthy pacemaker.

At rest, the normal heart rate varies from 60 to 80 beats per minute. During anxiety or vigorous exercise, the heart rate increases; sometimes it approaches 200 beats per minute. Although the nervous system does not initiate each individual heartbeat, it can influence heart rate. Nerves that originate in the central nervous system terminate at the SA node. These nerves release chemical messengers that can increase or decrease the heart rate.

The Cardiac Cycle

The two-stage cycle pattern of the heartbeat moves blood through the heart's chambers efficiently (the pumping of the atria "primes" the ventricles, so they are full of blood when they contract), and it also produces the characteristic sounds of a heartbeat. If you listen to a heartbeat through a stethoscope, you will notice that it is a double sound: described as "Lub-dub . . . lub-dub."

The sounds are produced by the closing of heart valves. "Lub," the first sound, occurs when the AV valves slam shut at the beginning of a ventricular contraction; "dub" follows when the semilunar valves leading from the ventricles close after blood has been ejected. The heart is completely at rest for about 0.4 second before a new cycle of contraction begins. Physicians can detect many heart problems, especially those involving the valves, by listening to the heartbeat for abnormal or extra sounds.

CIRCULATORY PATTERNS IN THE HUMAN BODY

The Pulmonary Circulation

The mammalian circulatory system is organized in a way that enables it to pump the entire blood volume through the lungs before returning it to the rest of the body. This part of the system is known as the pulmonary circulation. Blood passes from the right ventricle to the lungs via the pulmonary arteries and returns to the left atrium via the pulmonary veins (Fig. 35.11).

One interesting aspect of this configuration is that it creates an exception to our usual impressions of the circulatory system. Unlike most other arteries and veins, the pulmonary arteries carry *oxygen-poor* blood, and the pulmonary veins carry *oxygen-rich* blood. In the lungs, the pulmonary circulation is routed through a network of tiny capillaries that allows gas exchange to take place between the blood and the air that has been inhaled.

The Systemic Circulation

Blood leaving the heart flows through the largest artery in the body, the aorta, which has an opening of about an inch in diameter. The major arteries that lead to the upper and lower parts of the body branch off from the aorta, and they in turn give rise to many smaller vessels that carry blood to individual organs and tissues.

In the organs of the body, arteries branch and rebranch until arterioles lead to capillaries. The capillaries form profuse networks throughout the tissues, where the thin capillary walls facilitate the exchange of oxygen, nutrients, hormones, carbon dioxide, and waste products.

Blood that has passed through the capillary networks is collected by the venules that lead into the larger veins. These veins lead into two major vessels, the **superior vena cava,** which collects blood from the head, neck, and arms, and the **inferior vena cava,** which collects blood from the rest of the body. These large vessels lead directly to the right atrium of the heart.

Control of Blood Circulation Patterns

It may be tempting to assume that the circulatory system is just so much passive plumbing through which the blood is pumped like cooling fluid circulating through an engine. It is not. The circulatory system is a flexible, dynamic collection of elements that can quickly accommodate the changing demands of the body.

Capillaries are often grouped into clusters called *capillary beds* that carry blood from an arteriole to a venule. The rate at which blood may enter the capillary bed is partially controlled by a **precapillary sphincter,** a control region encircled with smooth muscle (Fig. 35.12). As long as the smooth muscle is contracted, very little blood can enter the capillary bed. However, other control factors in the body (including the nervous system) may cause the sphincter to open, allowing blood flow to increase through the capillary bed and changing the pattern of circulation.

Several tissues in the body regulate blood flow by means of precapillary sphincters. In many other tissues, blood flow is controlled by contractions of the smooth muscles that surround small arteries and arterioles. Al-

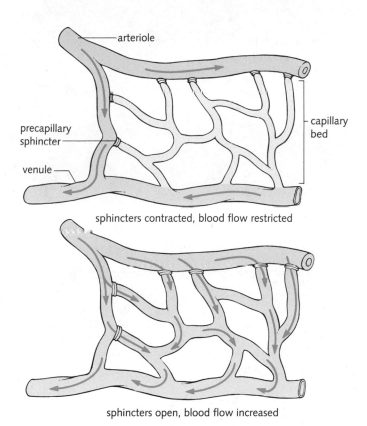

Figure 35.12 Blood flow to muscles and other tissues may be controlled by the opening and closing of precapillary sphincters.

though skeletal muscles have a rich blood supply, when a muscle is called on to work very hard, its capillary beds open and the muscle fills with blood, increasing the rate at which oxygen is carried to the tissue. This increased flow of blood during heavy exercise causes a noticeable increase in the size of the muscle, an effect that weight lifters call "pumping up."

The contractions of smooth muscles surrounding arterioles in the skin produce changes in circulatory patterns that are easy to see. When a light-skinned person is frightened or nervous, these arterioles contract, limiting blood flow to the skin—such a person may literally "pale" in fear. At times when the body temperature is too high, these smooth muscles relax and the face, "blushing" as a result of increased blood flow, becomes reddish. The small intestines, where food absorption takes place, also have a rich blood supply controlled in this way. After a full meal, the capillary beds in the small intestine open, allowing more blood to flow through them so that water and digested food can be absorbed as rapidly as possible. When absorption is complete, the sphincters close again, restricting the blood supply and changing the overall flow pattern of the circulatory system.

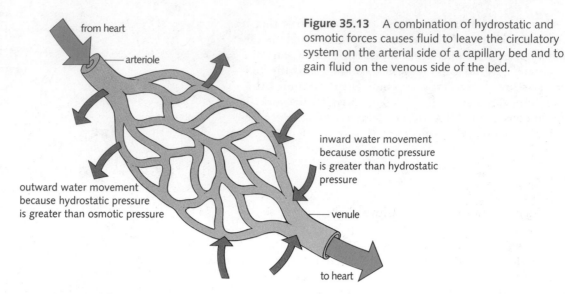

Figure 35.13 A combination of hydrostatic and osmotic forces causes fluid to leave the circulatory system on the arterial side of a capillary bed and to gain fluid on the venous side of the bed.

Material Exchange in the Capillaries

The major vessels of the circulatory system—the large arteries and veins—are the ones that most easily attract our attention. However, for most of the cells of the body, the important work of the circulatory system is done across the capillary wall. The extensive network of capillaries makes contact with an enormous surface area of body tissues. This brings the blood into close contact with nearly every cell in the body. The thin wall of the capillary allows materials to leave the blood and also permits molecules in the **interstitial fluid** ("in-between" fluid) that surrounds the cells to enter the blood.

Two important forces act in opposition across the capillary wall (Fig. 35.13). As oxygenated blood enters the circulatory system from the heart, it is under significant hydrostatic pressure. This pressure is great enough to force water and small molecules across the membranes that surround arterioles and capillaries, and there is a substantial loss of fluid from the blood into the interstitial fluid. Blood pressure drops dramatically as circulation continues through the capillaries, so the rate of water loss begins to diminish.

Another force complicates the picture, however, and that is *osmosis*. **Plasma,** the liquid portion of the blood, is a thick liquid containing a rich assortment of dissolved salts, sugars, and proteins. Interstitial fluid contains many of these materials, too, but it lacks the proteins.

Thus plasma is *hypertonic* with respect to interstitial fluid, and the forces of osmosis tend to drive water into the blood. As long as the hydrostatic pressure is higher than the osmotic pressure, water and solutes are lost from the capillaries to the surrounding tissue. But when hydro-static pressure drops off in the capillaries and venous circulation, osmosis drives fluid back into the plasma.

This movement of fluid out of the circulatory system and back in again has a positive effect: it produces a current of fluid movements that helps move materials between the circulation and the tissues of the body, helping the circulatory system do its work. It also tends to keep the composition of the blood stable. Despite all the movement of water, both ends of the system roughly balance out, so that more than 99 percent of the fluid volume that leaves the heart returns again in the venous circulation.

Blood Pressure

Each beat of the heart exerts a force known as **blood pressure** against the walls of the blood vessels. Blood pressure is a hydraulic force, and it can be studied and analyzed in the same way that forces on other fluids are. Because the heart is not continuously contracted, blood pressure rises and falls with the contraction cycle. The expansion and contraction of arterial vessels with the rise and fall of pressure constitute a **pulse.**

Arterial blood pressure   When the ventricles of the heart contract (a state known as **systole**), blood pressure reaches a maximum in the aorta and the major arteries. When the ventricles relax (a state known as **diastole**), pressure drops to a minimum in these vessels. Blood pressure is therefore expressed as two numbers: systolic pressure and diastolic pressure.

Blood pressure measurements are made with a *sphygmomanometer,* which is connected to a pressure cuff. This device measures blood pressure in units of millimeters of mercury, abbreviated "mm Hg." The pressure of the atmosphere at sea level is 760 mm Hg, equal to a column of mercury 760 mm high. A blood pressure reading of "100" *exceeds* atmospheric pressure by 100 mm Hg.

The person taking a blood pressure reading listens with a stethoscope for the sound of blood flowing through an artery in the arm. The cuff is gradually inflated until the pulse disappears completely. This happens when the pressure of the cuff cuts off the flow of blood through the surface artery. Then the pressure in the cuff is gradually released. When the blood flow is audible again, the pressure of the cuff is noted. This is the **systolic pressure,** the maximum pressure exerted by a heart contraction. However, the blood still flows in spurts—flowing only when it exceeds the pressure of the cuff. The pressure is then lowered still further until a change in sound indicates that the blood is flowing constantly; the blood pressure at this point is the **diastolic pressure.**

Detecting the second sound is difficult and requires considerable experience. Knowing both systolic and diastolic pressure is important, however, because the difference between the two can be critical in diagnosing hypertension. In a young adult, for example, normal blood pressure is roughly 120 mm Hg systolic and 80 mm Hg diastolic; this is expressed as 120/80.

Blood pressure is always measured in the same place (at the inside of the elbow joint with the cuff on the upper arm) because actual blood pressure varies from one part of the circulatory system to another. Blood pressure is highest in the major arteries closest to the heart muscle. That pressure steadily drops as blood moves through arteries, arterioles, and capillaries. Pressure is lowest in the venous part of the system (Fig. 35.14).

Factors affecting blood pressure Blood pressure in a healthy individual can be influenced by a number of factors. As the heart rate increases, pressure increases with it, because more blood is being forced into the system. Contraction of the smooth muscles that line the arteries and veins also increases blood pressure by reducing the volume of the circulatory vessels.

Blood pressure is influenced by the volume of blood in the circulatory system. Normally, blood volume remains relatively constant. But when more fluid is added to the blood, its volume increases and blood pressure rises. When fluid is lost, on the other hand, blood volume decreases and blood pressure falls. As you will see in Chapter 39, several organs controlled by hormonal feedback loops regulate the fluid content of blood.

All of these mechanisms affecting blood pressure and blood volume operate in healthy individuals to ensure that blood flow to various parts of the body can be increased or decreased in response to the changing de-

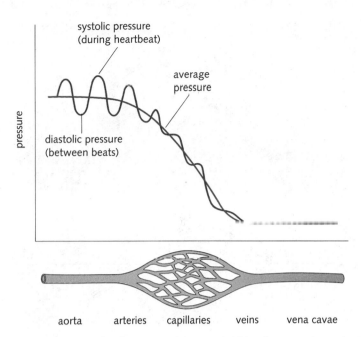

Figure 35.14 Blood pressure drops as the blood moves through the circulatory system.

mands of exercise and physiology. When any of these regulatory systems fail, however, the results can be quite serious.

DISORDERS OF THE CIRCULATORY SYSTEM

Any medical disorder that affects heart muscle can be fatal, for the body cannot survive an interruption in blood flow that lasts longer than a few minutes. *Diseases of the circulatory system are the number one cause of death* in the United States and several other developed nations.

Hypertension

Hypertension, the technical term for excessively high blood pressure, is a serious medical problem. The heart must pump blood against the diastolic pressure in the arteries, so as that pressure rises, the work load on the heart increases. Heart muscle must work nearly twice as hard to pump the same volume of blood in an individual with a blood pressure of 180/140 as in someone with a blood pressure of 100/70.

Both numbers are important, however. A large increase in *systolic pressure* may mean that the circulatory

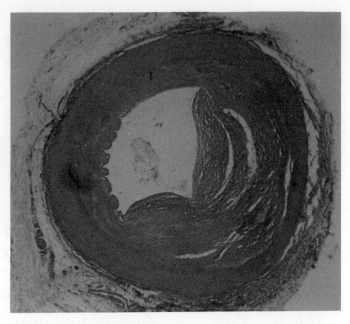

Figure 35.15 Severe atherosclerosis has produced fatty deposits large enough to block more than 60 percent of the area of this coronary artery.

system has become less elastic. Normally, the major arteries are elastic enough to expand slightly as the ventricles contract, moderating the pressure increase as blood surges out of the heart. Sometimes, however, arteries "harden" as a result of the buildup of fatty tissue on the arterial inner surface. This condition, known as *atherosclerosis,* is discussed in more detail below.

The extra load that high blood pressure imposes on the heart has its consequences: the heart muscle fatigues easily, can be more easily damaged, and has less reserve capacity to draw on if sudden demands are made on the body. For these reasons, physicians often try to control high blood pressure with a number of treatments. The usual recommendations for mild to moderate hypertension include some regular exercise and a reduction of salt in the diet.

What has salt to do with blood pressure? Eating too much salt increases the amount of sodium ion in the blood. In some individuals, this increased sodium load causes the body to retain more fluid, increasing both blood volume and blood pressure.

Other recommendations for controlling blood pressure include prescription drugs that help to lower blood pressure by reducing blood volume and a reduction in fatty foods to reduce atherosclerosis.

Atherosclerosis

Many problems in the circulatory system are produced by a disorder known as **atherosclerosis,** or "hardening of the arteries." In atherosclerosis, the normally smooth and unobstructed inner surface of the body's arteries is lined with fatty deposits and clusters of tissue produced by abnormal cell growth (Fig. 35.15). We will discuss the nutritional and metabolic factors that *cause* atherosclerosis in Chapter 37; here we will concern ourselves only with its *effects* on the circulatory system.

When atherosclerosis proceeds too far, the obstruction it creates within arteries can have two serious consequences. First, as we have already seen, it can seriously impede the flow of blood and increase blood pressure. Second, it increases the risk that a normally harmless blood clot can block the artery altogether. This condition is serious everywhere, but it is most dangerous in the brain and in the coronary arteries, the vessels that feed blood to the heart muscle itself.

Heart Attack

Atherosclerosis in coronary arteries is extremely dangerous because the constantly beating heart muscle requires an equally constant supply of oxygen and nutrients in order to survive. As coronary arteries begin to narrow, blood flow may become insufficient to support high rates of heart muscle activity, and the heart may be seriously weakened and possibly damaged.

When a coronary vessel closes completely, part of the heart muscle may begin to die for lack of oxygen, causing intensive pain and failure of the heart to pump sufficient blood to the rest of the body. This is known as a **myocardial infarction,** or **heart attack.** When a heart attack strikes, *immediate* medical attention is necessary to provide anticlotting drugs that can quickly increase the flow of blood to the heart. CPR (cardiopulmonary resuscitation) is necessary if the heart stops beating due to oxygen deprivation.

Many heart attacks occur not when atherosclerosis completely seals an artery but when a small blood clot, called a **coronary thrombus,** becomes lodged in a partially obstructed region of the coronary artery. As the clot blocks the flow of blood through that vessel, the heart attack begins. Medical scientists believe that many such clots form elsewhere in the body and drift through the circulatory system until they become trapped in a small vessel or passageway.

Risk factors in heart disease There is no single, simple cause of heart disease. Rather, there are several known risk factors that—especially in combination with one an-

EKGs

Every time the heart beats, each of its cardiac muscle cells contracts. The electric currents produced by the impulses that stimulate those contractions are strong enough to be detected at the surface of the body. An *electrocardiogram* (EKG) is a procedure that produces a recording of the electrical activity of the heart muscle from a series of electrodes placed at the surface of the skin. These electrodes are typically placed at the wrists and the left ankle.

A normal EKG (*below, left*) shows five distinct components, or waves, labeled P, Q, R, S, and T. The P wave is produced by the electrical impulse that initiates contraction in the atria—the first event in the contraction cycle. The Q, R, and S waves result from similar impulses in the ventricles. The T wave is generated by postcontraction recovery in the ventricles. The pattern as a whole gives an experienced health worker an overall view of the health of the heart muscle.

A comparison of a normal EKG (*right, top*) with several abnormal recordings appears below. *Atrial fibrillation* (*right, middle*) is a more serious condition in which uncoordinated impulses spread throughout the atria, and the P wave all but disappears from the EKG. *Ventricular fibrillation* (*right, bottom*) results from a similar condition in the ventricles. A heart in ventricular fibrillation is ineffective in pumping blood. Because the ventricles do the lion's share of pumping, ventricular fibrillation is fatal unless it is quickly corrected by administration of drugs and electric shock.

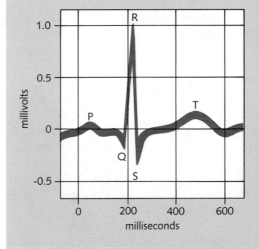

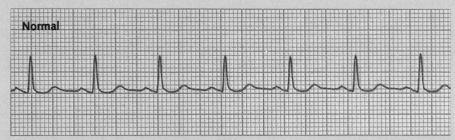

Normal

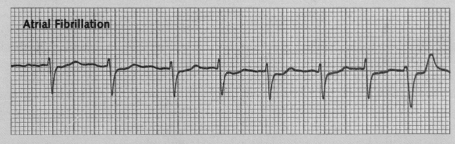

Atrial Fibrillation

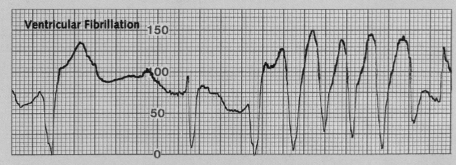

Ventricular Fibrillation

other—increase the likelihood of heart attack. Given the severity and prevalence of heart disease in the United States, it is important that everyone understand these risk factors and take appropriate action.

Certain risk factors are genetic; individuals may inherit from either of their parents a predisposition to heart disease in general or to atherosclerosis in particular. Some of these problems can be treated with medication, but to the extent that they are genetically controlled, they persist throughout life.

Several other risk factors, however, are produced by certain patterns of behavior and can therefore be eliminated by appropriate changes in those behaviors.

First, it is now clear that eating too much saturated fat and **cholesterol** leads to atherosclerosis. Second, cigarette smoking enormously increases a person's risk of coronary disease, although the mechanism by which it does so is not completely understood. There is also a high degree of correlation between obesity and lack of exercise and the incidence of heart disease. High blood pressure further compounds the risk, because it increases the amount of work the heart must do and makes even the slightest interruption in coronary blood flow that much more serious.

For these reasons, the best advice physicians can give for avoiding heart attacks are reminders to exercise regularly, to avoid fatty foods and cigarettes, and to control body weight.

Stroke

A drifting blood clot need not block a coronary artery to cause serious problems. Such a clot may move into the blood vessels leading to the brain, where it may become stuck and block the flow of blood to a group of nerve cells. If the clot manages to block blood flow completely, the brain cells served by the vessel begin to die for lack of oxygen, a condition known as a **stroke.** Strokes can happen without warning, although they are made more likely by the same conditions that have been identified as favoring heart attacks: smoking, atherosclerosis, and high blood pressure.

Strokes vary in their effect on individuals, depending on the region of the brain affected. If the stroke affects brain cells that control movement, a patient may become paralyzed. If blood flow to the visual center of the brain is blocked, blindness may result.

Because strokes are localized, their effects may be confined to one side of the body; temporary or permanent loss of the use of one leg or one arm is not uncommon. Each half of the brain controls motor functions on one side of the body, and the localized nature of a single blood clot accounts for the localized effects of a stroke.

THE LYMPHATIC SYSTEM

We noted earlier that roughly 99 percent of the blood volume returns from capillary beds to the venous system. This implies that about 1 percent of the blood volume *does not* return. What happens to it? The system is set up in such a way that a bit more fluid leaves the capillaries on the arterial side than reenters on the venous side. This extra fluid enters the interstitial fluid, and over time, the volume of interstitial fluid increases. If nothing is done to counteract the flow, large amounts of excess fluid build up in the tissues of the body.

Fortunately, a special system exists to collect the fluid. Thin-walled tubes known as **lymph vessels** absorb the excess fluid and move it slowly through the body. The lymph vessels are part of the **lymphatic system,** and the fluid that moves through the vessels of the lymphatic system is known as **lymph.**

The lymphatic system does not contain a muscular pump like the heart to move fluid through it. Instead, lymph vessels contain valves like those found in veins, which ensure that fluid can move through the system in only one direction. Lymph flows slowly through this system, driven by the osmotic pressure of the blood and by occasional mechanical pressure from nearby skeletal muscles, until it empties back into the general circulation by flowing into the subclavian veins in the upper portion of the chest (Fig. 35.16).

Lymph vessels do not merely return excess fluid to the circulation. They also flow near the cells that line the intestines, where they pick up fat from the digestive tract (Chapter 37). Lymph vessels also play an important role in the movement of white blood cells to aid in the immune response. It is essential that the flow of lymph not be blocked. *Edema,* a swelling of the tissues that is due to excess fluid accumulation, can occur when the lymphatic vessels do not function properly.

A parasitic disease caused by a roundworm can destroy and block lymph vessels, resulting in a grotesque swelling known as *elephantiasis.* Another cause of swelling is nutritional. If the diet does not contain enough protein, the protein concentration of the blood falls. Less fluid may return to the circulation by the process of osmosis, increasing the load on the lymphatic system. Beyond a certain point, the system cannot cope with the added work, and the tissues of the body swell—an ironic consequence of prolonged malnutrition.

BLOOD COMPOSITION

For certain purposes, it is convenient to consider blood a fluid and to imagine it flowing through the circulatory system like water through a garden hose. But we must never forget that blood is a tissue, that it contains billions of living cells, and that these cells are absolutely vital to the functioning of the circulatory system.

Plasma

The **plasma,** the fluid in which these cells are found, is 90 percent water. Nearly 7 percent of blood plasma is composed of plasma proteins. There are three major kinds of plasma proteins: **globulins,** which are produced by the immune system and help protect the body against infection; **fibrinogen** and **prothrombin,** which control the clotting process; and **albumins,** which stabilize the plasma by binding fats and other insoluble molecules and, with the other plasma proteins, regulate the osmotic character of the blood. We have already seen that blood is hypertonic with respect to interstitial fluid, and the plasma proteins are the principal reason why.

Blood plasma is slightly basic (its pH is usually about 7.4). If this pH is allowed to vary by as much as a single pH unit, serious problems develop. The solubility characteristics of dissolved proteins, carbohydrates, and salts are all pH-dependent, and the entire blood function of transporting materials can come to a halt if the pH changes rapidly. Vigorous exercise (which increases blood concentrations of carbon dioxide and lactic acid) and hyperventilation (which decreases blood levels of carbon dioxide) have the potential to upset blood pH. Plasma proteins (especially albumin) help to buffer the blood against these and other influences and to maintain its pH within allowable limits.

Salts dissolved in blood plasma include potassium, sodium, calcium, and magnesium (positively charged ions) and chloride, phosphate, carbonate, and sulfate (negatively charged ions). The concentrations of these ions vary within remarkably narrow limits, and large changes in ion concentration are indications of serious medical conditions.

Glucose is present in blood plasma as a result of digestive activities. A great deal of glucose obtained in a meal is absorbed by the liver and stored in the form of glycogen. As the cells of the body absorb glucose from the circulation, the liver provides more of this sugar to keep the level of blood glucose within a narrow range. Precise control of blood sugar is especially important for cells in the brain, which obtain all of their energy from glucose carried in the bloodstream. The control of blood sugar is a complex process that involves the close coordination of activities in the liver, digestive system, nervous system, and endocrine system. It will be discussed in detail in Chapter 39.

Lipids (fats) are carried in blood plasma bound to proteins, forming **lipoproteins.** One fatty compound of

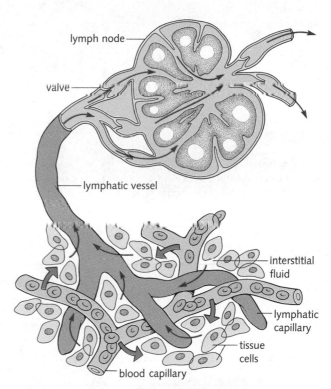

Figure 35.16 Fluid lost from the circulatory system is collected into the vessels of the lymphatic system. This fluid passes through lymph nodes and is eventually returned to the circulatory system. The lymphatic system plays a vital role in a number of physiological activities including the immune system.

Artificial Hearts

If the heart is nothing more than a pump, then it should be possible to replace it with a mechanical device that pumps blood. This concept has been the inspiration for a number of research teams who have sought to develop an artificial heart to replace worn or damaged human hearts—and thereby to prolong the lives of patients suffering from severe heart diseases.

A ventricular assist device, used to assist the heart in pumping from the left ventricle.

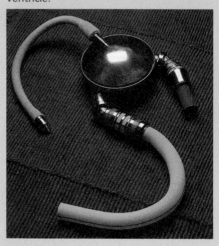

The first successful replacement of a human heart with a mechanical device was done in 1982 by surgeon William DeVries. DeVries removed the heart of Barney B. Clark, a patient who was suffering from acute heart failure, and replaced it with the Jarvik-7, a mechanical heart designed by William Jarvik. The heart was implanted into the chest and was driven by compressed air. Initially, the implant into Clark was successful; he regained consciousness and was able to talk and move about his hospital room. However, Clark suffered a number of seizures and strokes, and he died 112 days after insertion of the heart.

The Jarvik-7 helped one patient to survive for 620 days, and it was used in almost 200 patients as a temporary substitute for their own hearts while they waited for a heart transplant. Problems with clotting, infection, and seizures led the U.S. Food and Drug Administration to withdraw approval for the Jarvik-7 several years ago.

This does not mean that the effort to develop an artificial heart is over. The success of the Jarvik-7 has led a number of companies to develop ventricular assist devices that help a weak-

ened heart to pump blood. These devices are implanted up against the left ventricle, the chamber that pumps blood directly to the body (see Fig. 35.6). Driven by compressed air or electric motors, these pumps squeeze the left ventricle in a way that assists the heart muscle's own efforts to force blood into circulation.

Ventricular assist devices are now widely used to help restore circulation in individuals with weakened heart muscles. In one case, a woman survived more than 5 years with a ventricular assist device until a suitable heart transplant was available. In other cases, the support provided by the device is sufficient to enable the heart to heal itself and regain enough strength to make a transplant unnecessary.

More than 65,000 Americans a year could benefit from devices that might replace or assist the heart. The success of ventricular assist devices is an indication that bioengineering is on the right track, and it may be only a matter of time before those patients can count on help from a machine that may literally make the difference between life and death.

---

great medical interest is the steroid cholesterol. Cholesterol is carried in two different forms through the blood: in *high-density lipoproteins (HDLs)* and in *low-density lipoproteins (LDLs)*. A certain amount of cholesterol is essential to the body; it is used in the construction of many cell membranes. Excess cholesterol, however, particularly in the form of LDLs, leads to atherosclerosis.

Blood Cells

Nearly 45 percent of the volume of human blood is cellular. Blood contains a range of cell types that performs a variety of functions (Fig. 35.17). These living components

of blood are of three types: erythrocytes (red blood cells; *erythro* means "red"), leukocytes (white blood cells; *leuko* means "white"), and thrombocytes (platelets).

Red blood cells The majority of blood cells, called **red blood cells** or **erythrocytes,** are shaped like biconcave disks, and they lack nuclei, which are extruded during development (Fig. 35.18). Erythrocytes are produced by cells in the bone marrow that undergo rapid cell division. These cells develop into erythrocytes gradually, producing large amounts of **hemoglobin** (the reddish, oxygen-carrying protein) and gradually losing their internal cellular organelles.

As they develop, these cells go through a final, irreversible step: their nuclei are extruded and destroyed. When released from the bone marrow into the circulation, erythrocytes are really concentrated solutions of hemoglobin and other proteins surrounded by a specialized cell membrane. A new red cell has an average life span of about 4 months, meaning that roughly 1 percent of the body's erythrocytes must be replaced every day.

The number of red cells in the blood does not vary much from day to day, but it can be affected by environmental demands on the circulatory system. Individuals who participate in demanding aerobic exercises (those that stimulate the heart rate and place extra demands on the oxygen-carrying capacity of the blood) produce erythrocytes more rapidly than average, resulting in a higher concentration of red cells in the blood. People who live at high elevations, where the air is much thinner, show similar changes in their red cell content. When we consider the respiratory system, we will look at the details of the interaction between oxygen and hemoglobin.

Although mammalian erythrocytes lose their nuclei during development, the erythrocytes of most other vertebrates—including birds, reptiles, amphibians, and fish—do not. These cells retain nuclei throughout their useful life span, although they suffer much the same loss of cellular organelles as mammalian erythrocytes do.

White blood cells Vertebrate blood includes a number of other cell types that do not contain hemoglobin. These are known as **white blood cells,** or **leukocytes,** to contrast them with the reddish color of the erythrocytes. There are about 500 erythrocytes for every leukocyte. Like erythrocytes, leukocytes are formed in the bone marrow, and they are released into the bloodstream as fully developed cells with active cytoplasmic organelles. There are five ba-

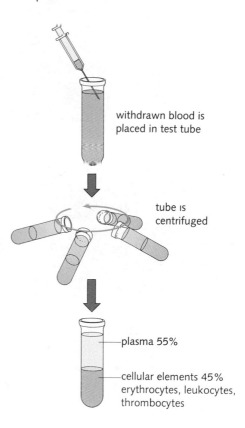

Figure 35.17 Centrifugation separates cellular elements from blood plasma.

withdrawn blood is placed in test tube

tube is centrifuged

plasma 55%

cellular elements 45% erythrocytes, leukocytes, thrombocytes

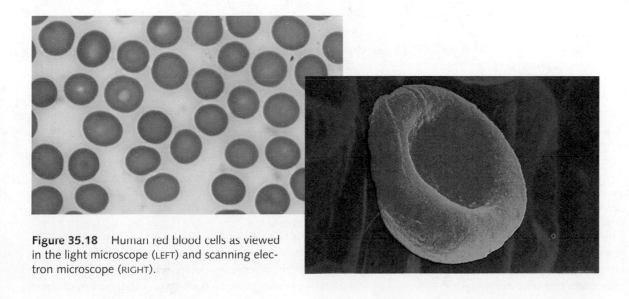

Figure 35.18 Human red blood cells as viewed in the light microscope (LEFT) and scanning electron microscope (RIGHT).

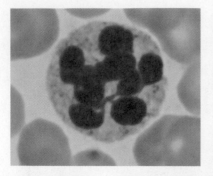
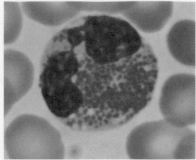
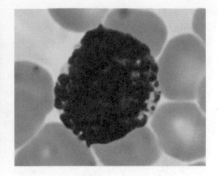
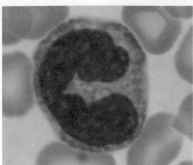
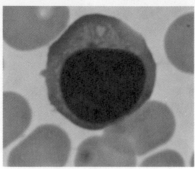

Figure 35.19 Five basic types of *leukocytes,* or white blood cells (CLOCKWISE, FROM UPPER LEFT): neutrophil, eosinophil, basophil, lymphocyte, and monocyte.

sic types of white blood cells, and they can be distinguished from each other by their size, the shape of their nuclei, and the types of granules visible in their cytoplasm (Fig. 35.19).

Leukocytes are important components of the **immune system,** and they function in a number of ways to guard the body against attack from invading bacteria, viruses, and eukaryotic parasites. Some of the cells, including granulocytes and macrophages, deal with foreign cells by engulfing and destroying them, much as an amoeba might dispose of a small cell it happened across. Such cells are aptly called **phagocytes** ("eating cells"). Other cells, known as **lymphocytes** (because they are common in lymph fluid), deal with infection by producing special proteins known as antibodies that attach to and help destroy foreign cells. These leukocytes do not occur only in the blood; they are found in much greater numbers in the lymphatic system and circulating through various places in the body. They are also important in variations of the immune reaction, such as allergies, in which special chemicals are released from immune system cells in response to exposure to certain foreign molecules.

Although leukocytes are found in other tissues, circulating with the blood is an ideal way for them to do their job. It allows them to visit all portions of the body rapidly, enabling the blood to serve as a second line of defense against disease and infection. The operation of the immune system is much more complicated (and more interesting) than this suggests, and we will examine the func-tions of the immune system in Chapter 42.

Platelets and blood clotting As we have seen, blood is more than just a liquid. It can respond to changes in the circulatory system in a sophisticated way. Blood is even able to stop leaks in the circulatory system. We have all had minor cuts and scrapes that bleed for several seconds or minutes and then stop bleeding and begin to heal. The first stage in that process is the formation of a **blood clot,** a fibrous mass of material that forms automatically from molecules in the blood itself and stops the loss of fluid (Fig. 35.20).

For many years, physicians believed that blood clotting was caused by exposure to the air. This was a natural enough assumption, given the fact that most visible wounds are open to the air. However, a great deal more is involved. One of the most important molecules in the clotting reaction is **fibrinogen,** a large protein produced by the liver and always present in normal plasma. Fibrinogen polymerizes to produce **fibrin** during clot formation. Fibrin forms a net-like meshwork that traps erythrocytes and platelets and actually creates the clot.

But if fibrinogen is present all the time, what controls its conversion to fibrin? When tissues are torn at the site of a wound, an enzyme called **thromboplastin** is released. This enzyme catalyzes the conversion of a soluble protein called **prothrombin** into another enzyme called **thrombin.** And thrombin, in turn, catalyzes the reaction in which fibrinogen is converted into fibrin.

The clotting process is aided by fragments of cells known as *thrombocytes* or **platelets.** Platelets are small pieces of cells called **megakaryocytes** that are found in the bone marrow. As these cells grow, small pieces of their

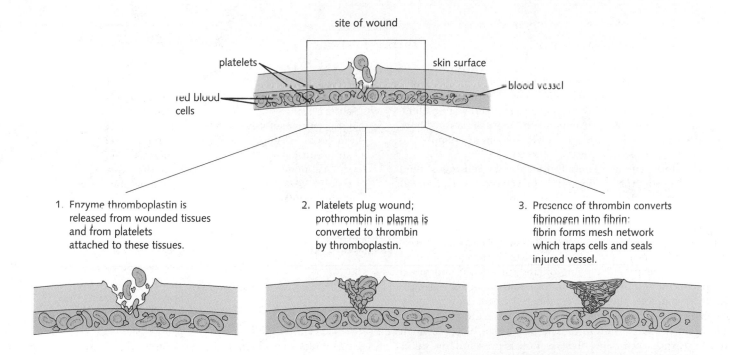

site of wound

platelets

red blood cells

skin surface

blood vessel

1. Enzyme thromboplastin is released from wounded tissues and from platelets attached to these tissues.

2. Platelets plug wound; prothrombin in plasma is converted to thrombin by thromboplastin.

3. Presence of thrombin converts fibrinogen into fibrin: fibrin forms mesh network which traps cells and seals injured vessel.

cytoplasm are broken off and released into the circulation to become **platelets.** Platelets are essential to the clotting reaction. Their surfaces are quite sticky, and they easily attach to the broken tissue at the edges of a wound. There they seem to release thromboplastin, and the platelets serve as anchoring points for the network of fibrin that envelops the blood cells around a wound.

All of this may give you the impression that blood clotting is complicated. It is indeed. Blood clotting is caused by a sequence of reactions, each of which produces a product that catalyzes the next reaction in a *cascade* of chemical events, resulting in the production of large amounts of clot-forming fibrin. The cascade method has a number of advantages. The most obvious is that a few molecules at the beginning of the cascade can ultimately produce millions of fibrin molecules at the end, because each step *catalyzes* the next step. This increases the power of the blood clotting system, as well as its sensitivity to small injuries.

One problem with such a system, however, is the fact that every component of the cascade must be present and must function perfectly if the clotting reaction is to be successful. This makes the system very vulnerable to genetic disorders that affect even one element of the pathway. **Hemophilia** is an example of such a genetic disorder. As we saw in Chapter 11, hemophilia is a human disease that results from a defective protein in the clotting pathway. Individuals who suffer from hemophilia are unable to produce blood clots quickly enough to stop even minor bleeding, and they must take great care to avoid injury. Hemophiliacs are usually treated with extracts of clotting factors prepared from normal whole blood.

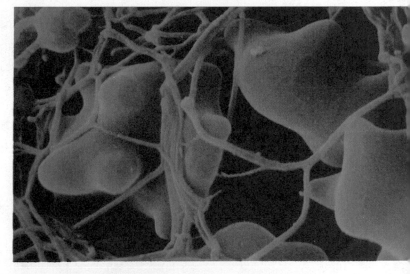

Figure 35.20   Blood clotting is triggered by the release of thromboplastin from broken tissues and platelets. This sets in motion a series of reactions that form the clot, stopping further bleeding and beginning the healing process. The scanning electron micrograph is of red blood cells tangled in the fibrin meshwork of a blood clot.

In small animals, the exchange of gases with the environment can occur directly through cell membranes. Nutrients and waste products can diffuse directly from cell to cell and can be eliminated directly into the environment. In larger organisms, however, this is not possible, and specialized systems are required to allow for the internal transport of material.

Circulatory systems generally function by moving a fluid through the body of an organism. In vertebrates a muscular heart moves blood through a closed circulatory system. Fish have a two-chambered heart that pumps oxygen-poor blood into the gills (where gas exchange occurs), from which vessels lead to the rest of the body. Amphibians and reptiles have three-chambered hearts in which there is a partial separation of blood pumped to their gas-exchanging organs (the lungs) and blood pumped to the rest of the body. In birds and mammals this separation is complete: their four-chambered hearts support completely different pulmonary and systemic circulations. The human heart is a large, hollow muscle that forces blood through atria into paired ventricles. Reverse flow is prevented by valves that separate the chambers, and contractions are under the control of a specialized bundle of cells (the pacemaker) in a region of the muscle known as the SA node.

With each contraction of the heart, a surge of blood under high pressure is pumped into the aorta, the large artery from which every other vessel in the systemic circulation receives blood. Blood pressure is maintained in the arteries, whose thick elastic walls are surrounded by smooth muscle. As blood flows into arterioles and capillaries, its pressure is diminished, and the lowest pressures in the circulatory system are found in the veins that return the blood to the heart. The liquid portion of the blood is known as plasma and consists of a complex mixture of salts, organic compounds, and specific blood proteins. The cellular portion includes erythrocytes (red blood cells), leukocytes (white blood cells), and platelets (subcellular particles that play an important role in the clotting reaction).

STUDY FOCUS

After studying this chapter, you should be able to:

- Analyze the process of diffusion at both the cellular level and the tissue level in order to understand the problems an organism encounters as it exceeds a certain size.

- Explain how organisms in various phyla have evolved ways to deal with the problem of circulation.

- Describe the details of the human circulatory system, its organs, and its cells.

- Explain how the circulatory system integrates its function with the larger systems of the body.

- Describe some of the most serious disorders of the circulatory system.

TERMS AND CONCEPTS

atrium *692*	myocardial infarction *700*
ventricle *692*	coronary thrombus *700*
aorta *692*	cholesterol *702*
atrioventricular valves *694*	stroke *702*
sinoatrial (SA) node *695*	lymph *702*
atrioventricular (AV) node *695*	plasma *703*
systole *698*	erythrocytes *704*
diastole *698*	hemoglobin *704*
atherosclerosis *700*	leukocytes *705*
	platelets *706*

REVIEW

Objective Questions (Answers in Appendix)

1. Which type of circulatory system is found in insects?
   - (a) open system
   - (b) closed system
   - (c) open or closed system, depending on the size of the insect
   - (d) open system when the insect is immature and closed system when the insect reaches adulthood

2. Which type of blood vessel carries blood away from the heart?
   - (a) vein
   - (b) venule
   - (c) artery
   - (d) superior vena cava

3. Which of the following conditions is otherwise known as "hardening of the arteries"?
   - (a) atherosclerosis
   - (b) myocardial infarction
   - (c) stroke
   - (d) hypertension

4. Which of the following elements in mammalian blood is a living cell containing a nucleus?
   - (a) erythrocyte
   - (b) leukocyte
   - (c) platelet
   - (d) a and b only

5. Which of the following statements correctly describes the human lymphatic system?
   - (a) It serves as a conduit between the interstitial fluid and the venous fluid.
   - (b) It contains vessels similar to arteries.
   - (c) It contains tiny muscular pumps that move fluid through the system.
   - (d) All of the above statements are correct.

Discussion Questions

6. What are the basic differences between a closed and an open circulatory system? Give an example of an organism with each type of system.

7. Most of the devices that have been used to assist weakened hearts in recent years have been designed to help pump blood out of the left ventricle. Of the four chambers of the heart, why is this the most practical location for an assist device?

8. What evolutionary trends are apparent in the development of the mammalian heart? What key characteristic of the four-chambered heart enables it to deliver oxygenated blood to the tissues more efficiently than the two-chambered heart?

9. What key role do the valves play in the functioning of the human heart? What problems might result if one of the valves failed to close? How would this affect the heart's ability to pump blood?

10. What is the difference between the pulmonary circulation and the systemic circulation?

11. Why is the level of blood plasma proteins important in preventing fluid loss when blood passes through the capillary beds?

READINGS

Lawn, R. M. "Lipoprotein (a) in heart disease." *Scientific American* 266 (June 1992): 54–60. A detailed treatment of the relationship between lipoproteins and atherosclerosis. You may wish to read this article after you have read the sections in Chapter 37 on lipids and lipoproteins.

Golde, D. W., and J. S. Gasson. "Hormones that stimulate the growth of blood cells." *Scientific American* 259 (July 1988): 62–70. The formation of cellular components of the blood is a fascinating process. This article explores some of the controls behind it.

Robinson, T. F., S. M. Factor, and E. H. Sonneblick. "The heart as a suction pump." *Scientific American* 254 (June 1986): 84–91. A lucid treatment of the mechanical principles behind the action of the heart muscle in pumping blood.

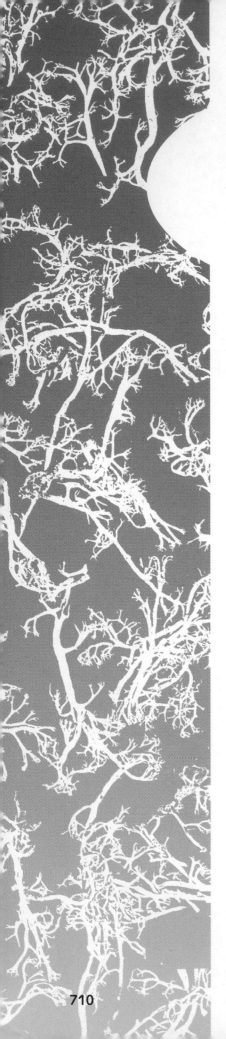

36

Respiration

a limp body is pulled from the wreckage of an automobile, and the rescuers who have hurried to the scene wonder whether there is still a life to save. One pulls a tiny mirror from her medical kit and holds it under the victim's nostrils. The mirror fogs—there is hope. With not a moment to lose, they begin to administer first aid and skillfully load the victim through the open ambulance doors.

This situation reflects something we often take for granted: the connection between life and the earth's atmosphere—that is, the process of respiration. **Respiration** is the exchange of gases with the environment: the release of carbon dioxide and the uptake of oxygen.

The process of respiration is so vital that it serves as a universal sign of life, so persistent that it punctuates every other physical activity. The need for an organism to exchange these gases stems from the chemical processes that are at work in living cells. Respiration at the level of the organism, the subject of this chapter, is nothing more than a reflection of cellular respiration, a process we examined in Chapter 18.

THE CELLULAR ROOTS OF RESPIRATION

For hundreds of millions of years, Earth's atmosphere has been rich in oxygen. Multicellular organisms evolved in the presence of this gas, and their metabolic pathways use oxygen to generate a constant supply of energy in the form of ATP (Chapter 18). As carbon-containing food molecules are broken down, cells must find acceptors for the electrons released from them. Oxygen is a powerful electron acceptor, and its abundance makes it a logical last step in the electron transport chain of respiration.

So urgent and constant is the need for energy that many cells, especially those of the brain, cannot survive more than a few minutes without oxygen. This cellular demand for oxygen creates a need for oxygen on the part of the whole organism. At the same time, the breakdown of energy-rich carbon compounds in food results in the production of carbon dioxide, a waste product that must be eliminated from the body. Thus all multicellular organisms must constantly exchange oxygen and carbon dioxide with their environment (Fig. 36.1).

The term *respiration,* in addition to its meaning at the cellular level, is also applied to this organismal process of *gas exchange* with the environment. In this sense, a **respiratory system** is a group of cells, tissues, and organs that exchanges gases with the environment and helps distribute those gases within the organism.

Respiration in Water and Air

To satisfy the respiratory needs of cells, oxygen must first pass from the gas phase of the atmosphere into the liquid phase that exists within living organisms. Oxygen and carbon dioxide are both soluble in water. For that reason, animals in typical aquatic environments are surrounded by a supply of oxygen that can diffuse directly into cells to support respiration. Similarly, carbon dioxide can diffuse out of living cells into the surrounding water. Terrestrial animals, which exchange gases directly with the atmosphere, face one additional complication: they must supply a moist surface for gases to dissolve into the liquid phase before these gases can enter living cells.

Many small organisms, whose ratio of surface area to total volume is large, rely on simple diffusion across their body surfaces to meet their cells' needs. This is possible not only for many aquatic animals but also for those terrestrial species whose habits enable them to keep their outer surfaces moist.

But in many larger organisms—both aquatic and terrestrial—the body cells' demand for gas exchange exceeds the diffusion capacity of the organisms' outer body surfaces. Over time, different groups of animals have evolved various kinds of respiratory systems through which gas exchange with the environment takes place. Not surprisingly, the design of these systems varies with the organism's size, its body structure, its evolutionary history, and the nature of the environment in which it lives. In this chapter we will look at some of the special problems that confront the respiratory systems of various organisms, and we will examine in detail the human respiratory system.

THE VARIETY OF RESPIRATORY SYSTEMS

Every respiratory system has two basic elements: a system to exchange gases with the environment and a system to distribute those gases throughout the organism. Although respiration itself, which occurs at the cellular level, is very much the same from one cell to the next, respiratory systems are not at all the same in different organisms.

Well-developed respiratory systems usually work by bringing either air or water into close contact with moving body fluid. It is not surprising, therefore, that the most advanced respiratory structures are found in animals that have well-developed circulatory systems.

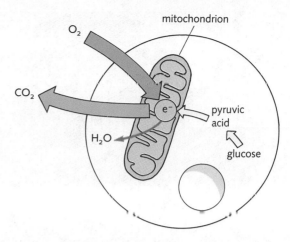

Figure 36.1   Respiration begins at the cellular level. The mitochondria of eukaryotic cells require a steady supply of oxygen to convert the chemical energy in glucose and other foods into biologically useful forms.

Exchange Through Body Surfaces

The point where gases enter and leave an organism is known as the **respiratory surface.** The simplest respiratory surface is the outer covering of an organism—its skin, or *integument.* Many small aquatic organisms exchange gases through their skins, and so do some terrestrial animals such as the earthworm.

The earthworm illustrates several specializations required in a relatively sizable organism that carries out gas exchange across its skin. First, there is a close coupling between the skin surface and the circulatory system (Fig. 36.2). The body of the earthworm is too large for oxygen to diffuse into tissues deep within, and a network of fine capillaries just below the skin allows the oxygen and carbon dioxide that pass through the skin to move directly into or out of the circulation.

Second, the skin must be kept moist to enable oxygen gas to dissolve in it and then diffuse into the body. For this reason, the earthworm's skin contains a series of glands that cover its body with a thick, wet mucus. (If you've picked up a worm, you know this is true.)

Gas exchange through the integument is not limited to invertebrates. A number of fishes acquire some oxygen directly through their skin, and the thin, moist skin of amphibians is particularly well suited to gas exchange. Frogs fulfill a large fraction of their respiratory needs directly through the skin.

During one stage of their lives, reptiles and birds carry out gas exchange through their integument as well. The active, growing embryo in an egg needs a constant supply of oxygen, which it obtains through thousands of tiny pores that pass through the eggshell. As the embryo

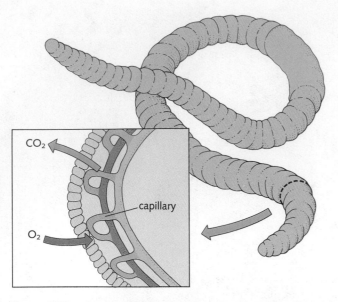

Figure 36.2 Gas exchange in earthworms takes place directly through the skin. The process is made more efficient by a circulatory system that carries blood close to the surface of the body, and by a thin, moist skin which assists in the diffusion of oxygen and carbon dioxide.

develops, a tissue known as the *chorioallantois* extends from the body of the growing embryo and covers the entire inner surface of the shell. Richly endowed with blood vessels, the chorioallantois is separated from the eggshell pores by a thin extracellular membrane that helps to block the entry of bacteria into the egg. Gas exchange occurs across this membrane and completely answers the respiratory needs of the embryo until its lungs become functional, usually just a few days before hatching.

Plants can be considered skin-breathing organisms, too. Although we usually pay more attention to their photosynthesis (a process not found in animals), plants *do* respire. The nonphotosynthetic tissues of a plant derive most of their energy from aerobic respiration, as does the entire plant at night. Gas exchange takes place by means of simple diffusion through the "skin" of the plant.

Exchange Through Gills

Gills are organs specialized for gas exchange in water. Gills may be internal or external, they may function in saltwater or fresh water, and they may be exposed at the surface of an organism or protected by shells or parts of the skeleton. All gills, however, solve the problem of gas exchange by bringing oxygen-poor blood into close contact with oxygenated water so that gas exchange can occur between them. The rate of gas exchange depends in part on the size of the surface over which blood and water are in close contact, so gills of all types have one thing in common: thousands of tiny projections that increase their surface area. By doing so, gills enable their bearers to live an active life supported by a high rate of respiration and gas exchange.

The gills of the common starfish, an echinoderm, are delicate stalks of tissue that project from the surface of the animal directly into seawater (Fig. 36.3). To protect them from injury, the organism is covered with spine-like projections that extend past the gills. These projections make the surface of the starfish rough to the touch.

Some vertebrates also have external gills. Many have gill clusters on either side of the head that look almost like feathers (Fig. 36.3). The tiny structures within the gills enable the cluster to expose an enormous surface of the circulatory system for gas exchange.

Fishes have two large banks of internal gills that develop from the lining of the throat. These gills are protected behind two bony plates, or gill covers, that also help control the movement of water across the gills. Some fast-swimming bony fishes, such as tunas, simply swim through the water with their mouths and gill covers open wide. As these fishes move, water enters the mouth, flows across the gills, and leaves through the openings behind the gill covers. More sedentary fish species must pump water across their gills. They first suck water in by opening their mouths with the rear ends of their gill covers pressed against the sides of their body. Then they close their mouths and force the water past the gills and out past the relaxed gill covers. Both techniques produce a steady flow of oxygen-rich water across the gill.

Gill tissue has a rich blood supply that comes directly from the heart. This oxygen-poor blood passes into large **gill arches** that support rows of tiny **gill filaments.** Within each filament, there are two vessels—one through which blood enters the gill and another through which it leaves. Between these vessels are dense networks of thin-walled capillaries packed into hundreds of wafer-thin projections called **lamellae.** As blood travels through the lamellae, the forces of diffusion cause them to release carbon dioxide and to absorb oxygen from the water.

Note that blood is pumped through gill filaments and lamellae in a direction opposite to the flow of water past the gills (Fig. 36.4). This pattern is known as **countercurrent flow,** because the blood and the water move in opposite directions.

This countercurrent flow pattern dramatically increases the efficiency of the gill system over that of a configuration in which water and blood move in the same direction. In fact, countercurrent flow is so efficient in this regard that it crops up again and again among many

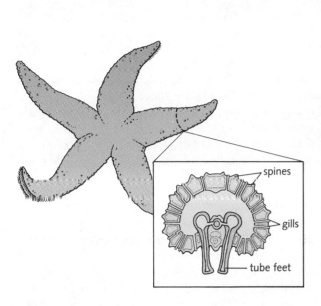

Figure 36.3 LEFT: Gas exchange in the starfish takes place in feathery gills which exchange oxygen and carbon dioxide with the surrounding seawater. Spiny projections from the surface of the animal provide protection for the gills. TOP: The external gills of this tiger salamander larva provide a large surface area for gas exchange.

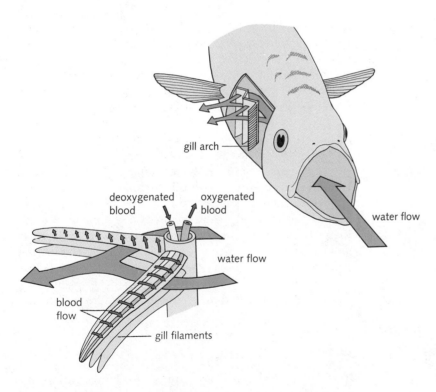

Figure 36.4 The gill systems of fishes allow blood and water to come into close contact. Notice that blood flows through the gill filaments in a direction opposite to the flow of water.

animals and in different physiological systems throughout the body.

Why is this arrangement so efficient? Recall that diffusion is a passive, physical process. Molecules diffuse from areas of high concentration to areas of low concentration. In doing so, they move at rates proportional to the difference in concentration between the two areas.

Now imagine the diffusion of oxygen in an imaginary gill through which blood and water flow in the same direction. Here oxygen-poor blood meets oxygen-rich water so, at first, oxygen diffuses rapidly into the blood. But the rate of net diffusion soon slows down as the oxygen concentration rises in the blood and falls in the water.

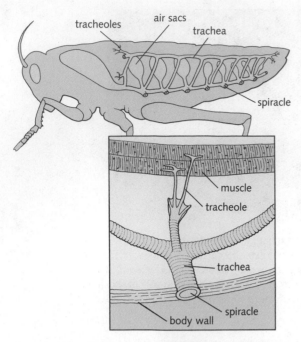

Figure 36.5 The tracheal system of insects allows air to come into close contact with oxygen-utilizing tissues such as muscle.

Figure 36.6 Air-filled swim bladders in fishes help to regulate buoyancy. Air sacs connected to the oral cavity might have played a role in the evolution of lungs.

In a countercurrent flow system, oxygen-poor blood first "encounters" water that has already lost some of its oxygen. But because there is still more oxygen in the water than in the blood, net diffusion occurs anyway. And from that point on, as the oxygen content of the blood increases, it encounters water that has a progressively higher oxygen content (see Fig. 36.4). The architecture of the gill system thus ensures that concentration differences between blood and water enable gas exchange to occur across the entire length of the capillary.

Exchange Through Tracheae

Terrestrial animals face a different set of problems from those of animals living in water. On the one hand, the at-

mosphere is a much richer source of oxygen than water. But on the other hand, these animals face the challenge of exchanging gases across a gas–liquid interface rather than a liquid–liquid interface. Because oxygen must enter solution to diffuse into the body, terrestrial animals must keep their respiratory surfaces moist. Obviously, such surfaces are also potential sites for water loss.

The largest group of terrestrial animals, the arthropods, respire in a unique way (Fig. 36.5). Their tough outer shell of chitin is nearly impermeable to gases. Instead, air enters insects and other terrestrial arthropod bodies through a series of openings known as **spiracles.** This air then travels through a series of dead-end passages known as **tracheae** (singular, *trachea*) that carry oxygen to tissues throughout the body.

Tracheae, in turn, lead into tiny **tracheoles** that make direct contact with individual cells, especially those (such as muscle cells) that have high oxygen demands. Oxygen diffuses directly into the cells from the interior of the tracheoles, and carbon dioxide diffuses in the opposite direction. In smaller insects, the only force that moves air through the system is diffusion, but large insects can assist respiration by moving their abdomen. This helps to pump air back and forth mechanically through the system.

Remember that insects have open circulatory systems, which are less efficient than closed ones. The main reason why insects can make do with an open system is that their tracheal systems carry oxygen directly to their tissues. Although this system works well for insects, the nature of its tracheal system places severe restrictions on the size of an organism. As an animal increases in size, it becomes more and more difficult for fresh air to penetrate the deepest passageways of the system. In larger insects, the tracheal system becomes less capable of meeting peak respiratory demands—and this is one of the principal reasons why insects are seldom longer than a few centimeters. Giant insects are common in science fiction movies, but the limitations of the tracheal system have (thankfully) prevented their emergence in the real world.

Lungs

Terrestrial vertebrates carry out gas exchange in internal organs known as **lungs.** Although there is no fossil record of the development of the lung (soft tissues are seldom preserved in fossils), we can get some idea how lungs may have developed by looking closely at present-day fishes. Many fishes have internal air sacs known as *swim bladders* that help them adjust their buoyancy for efficient swimming (Fig. 36.6). Some fishes also have sacs in the mouth region that can be filled with air as they swim to the surface. A few fishes, including the common goldfish, can trap air in these sacs and extract some of the oxygen from them directly into the bloodstream.

The Ice Fish: Blood Without Hemoglobin

The low solubility of oxygen in water is the principal reason why oxygen-carrying proteins such as hemoglobin are necessary to support the gas-exchange activities of the respiratory system. What might happen if an organism lived in an environment where its own oxygen demands were lower and the solubility of oxygen in water was greater? We need not speak hypothetically in answering this question. An organism known as the ice fish (scientific name, *Mallotus villosus*) lives in the cold waters of the polar regions, where the average water temperature is less than −4°C. It is similar to most fish in every respect but one: Its blood is clear and colorless—it has no hemoglobin. How does this remarkable animal survive with its pale, colorless blood?

Low temperature produces two effects that enable the ice fish to transport oxygen without an oxygen-carrying protein. First, *low* temperatures *lower* the animal's metabolic rate, so its oxygen demands are much less than they would be at higher temperatures. This means that the respiratory system may supply less oxygen. Second, *lower* temperatures *increase* the solubility of gases in water. This is why warm bottles of soda tend to pop or even explode. The CO_2 used to create the fizz in soda is quite soluble when the drink is cold. But when a warm bottle is opened, the gas comes out of solution so quickly that the drink may foam out of its container. At the temperatures in which the ice fish lives, oxygen solubility is so much greater than normal that dissolved oxygen alone is sufficient to

meet the respiratory needs of the organism. Why the ice fish apparently lost the ability to synthesize hemoglobin is a question you might wish to ponder in terms of the selective pressures shaping evolutionary change.

The head of an ice fish shows the ghostly hue produced by its colorless blood.

The combination of an oral cavity and an internal sac in ancient fishes might easily have developed into a gas-exchanging system that would have enabled these fishes to survive in shallow, muddy waters where the dissolved oxygen content is low. Developments such as these might have been the first steps in the evolution of lungs. The development of the lung made it possible for vertebrates to thrive on land.

In most modern terrestrial vertebrates, air enters each lung through a single large tube (also called a trachea) that branches into thousands of tiny passageways. Those passageways lead to terminal sacs known as **alveoli** (singular, *alveolus*), where gas exchange takes place. Human lungs have more than 300 million alveoli and their combined surface area is more than 75m²—more than three times the area of a bowling alley. The flow of blood through a closed circulatory system is routed through the lungs, allowing blood to be used as the oxygen-transporting medium (Fig. 36.7).

In most vertebrates, including mammals, lungs are dead-end passageways. Air enters the lungs as the chest

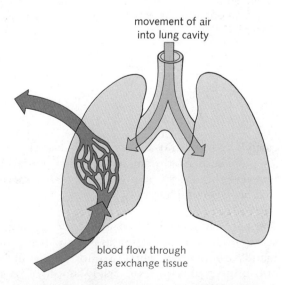

movement of air into lung cavity

blood flow through gas exchange tissue

Figure 36.7 The lungs of modern terrestrial vertebrates are sacs into which air is drawn by breathing. The circulatory system absorbs oxygen in the lungs and transports it to the rest of the body.

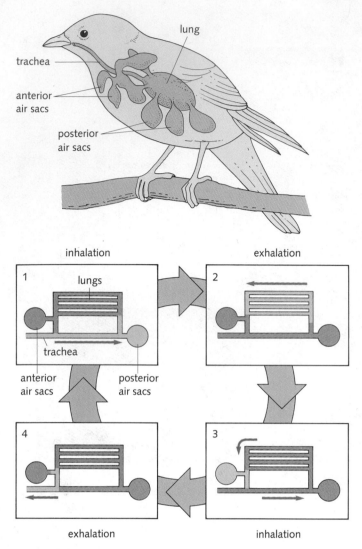

Figure 36.8 labels: lung, trachea, anterior air sacs, posterior air sacs

inhalation — exhalation

1 — lungs — trachea

2

4

3

anterior air sacs — posterior air sacs

exhalation — inhalation

Figure 36.8 Air sacs connected to the lungs of many birds provide for a one-way system of air flow, which results in highly efficient breathing.

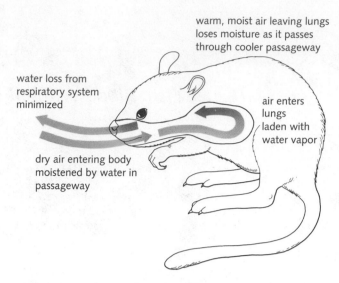

warm, moist air leaving lungs loses moisture as it passes through cooler passageway

water loss from respiratory system minimized

air enters lungs laden with water vapor

dry air entering body moistened by water in passageway

Figure 36.9 The nasal passageways of the kangaroo rat help to minimize moisture losses associated with breathing.

cavity expands, and air leaves as the cavity decreases in size. Because of this in-and-out flow pattern, fresh air entering the lungs is always mixed with some residual air that has been depleted of oxygen.

The avian lung—increased efficiency Although soaring sea gulls may make flight look easy, flying actually demands tremendous physical effort from all birds. That effort, in turn, requires a steady supply of oxygen. Interestingly, the gas-exchange system in birds is more efficient than that in other terrestrial vertebrates because air flows through the lungs in a one-way cycle.

The efficient gas-exchange system in birds works as follows: Fresh air is first routed to a series of sacs behind the lungs. This fresh air then passes directly into the lungs and through a series of small passageways, from where it is exhaled. Because air flows through the lungs in a one-way cycle, the gas-exchange surfaces in avian lungs are always presented with air containing the maximum oxygen content of 21 percent. The air-sac–lung system of a bird satisfies that demand while minimizing weight, helping to produce an efficient "flying machine" (Fig. 36.8).

Conserving Moisture

When dry air moves across the moist surfaces of an animal's respiratory system, some of that moisture evaporates and may be lost. This problem reaches an extreme in desert animals that live in an arid climate characterized by high temperatures and low humidity.

The kangaroo rat illustrates the way in which the respiratory systems of many desert animals are adapted to limit water losses in breathing (Fig. 36.9). Its upper air passages are covered with tightly packed hairs moistened with mucus and other secretions from the epithelial tissue beneath them. As the dry desert air moves into the body, it is moistened by water from these passageways, and by the time the air arrives in the lungs, it is humid enough so that the moist membranes across which gas exchange takes place do not dry out.

Every time the animal inhales, evaporation cools the upper passageways of the system. When it exhales, the warm, moist air from the lungs passes through the cooler

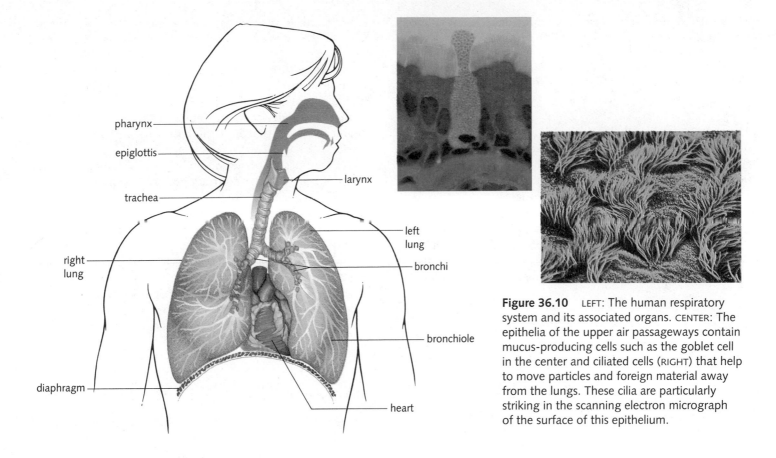

Figure 36.10 LEFT: The human respiratory system and its associated organs. CENTER: The epithelia of the upper air passageways contain mucus-producing cells such as the goblet cell in the center and ciliated cells (RIGHT) that help to move particles and foreign material away from the lungs. These cilia are particularly striking in the scanning electron micrograph of the surface of this epithelium.

passageways of the upper part of the system, allowing most of the moisture in the exhaled air to condense on the lining of the upper system. This arrangement limits the loss of moisture, conditions the air that is inhaled, and enables the animal to exist in an environment where water is at a premium.

Mammals that are not desert dwellers are less efficient at water conservation than the kangaroo rat, because their nasal passages are not so specialized for holding moisture. The same basic process of water conservation, however, does occur in most organisms, including humans.

THE HUMAN RESPIRATORY SYSTEM

Inhaled air enters the human respiratory system through the mouth and nose (Fig. 36.10). The nasal passageways are specially adapted to filter and moisten the air as it passes into the body. Air then passes through the **pharynx** and into the **trachea,** which leads to the lungs. The accidental entry of food or drink into the lungs is prevented by the **epiglottis,** a small flap that closes over the entrance of the trachea during swallowing. The upper part of the

air passageway, called the **larynx,** contains the **vocal cords,** which produce sounds by vibrating as air passes between them.

In the chest the trachea divides into two tubes, the **bronchi,** one of which enters each lung. The bronchi are supported by rings of cartilage, which prevent the passageways from collapsing. Within the lungs, the air passageway divides further into smaller bronchi, then into **bronchioles,** and finally into the alveoli, where gas exchange takes place. The rich blood supply of the alveoli makes them ideal sites for gas exchange.

As air moves through the pharynx and larynx, it picks up moisture from *mucus* lining the epithelial tissue of the passageway (see Fig. 36.10). This helps to keep the gas exchange surface within the alveoli moist. This mucus also traps inhaled particles of dust or smoke. Such foreign material is forced out of the respiratory system by the movement of *cilia* lining the passageways.

These ciliated cells are among the first to be damaged by environmental assaults such as smoking. Nicotine, one of the components of cigarette smoke, works like an anesthetic on the cilia in the respiratory tract, reducing their efficiency in clearing the lungs. A smoker's morning cough is an attempt to clear the respiratory passages of accumulated mucus—an early warning that the cilia are not working properly and that the respiratory system is headed for trouble.

Breathing

The process of **breathing,** or *ventilation,* is the movement of air into and out of the lungs. The lungs are not directly connected to any muscle system, and the only force that moves air into the lungs is air pressure. The body exploits atmospheric pressure every time we inhale.

The lungs are sealed in the **thoracic cavity.** They lie within two **pleural sacs** that border the **diaphragm,** a large dome-shaped muscle separating the thoracic cavity from the abdominal cavity. The thoracic cavity can be expanded in two ways: by contractions of the diaphragm muscle that expand the lower portion of the cavity, and by contractions of the *external intercostal muscles* between the ribs that raise the rib cage and enlarge the cavity from side to side. Because the thoracic cavity is sealed, this creates a partial vacuum within the pleural cavity, a vacuum that is rapidly filled with inrushing air. Figure 36.11 shows the expansion of the chest that occurs during inhalation, as well as a mechanical model that illustrates the physical principles at work.

In contrast to inhalation, exhalation is largely a passive process: the muscular contractions that expanded the thoracic cavity cease, and the highly elastic components of the lung and the tissue surrounding it retract and return to their original shape. As the thoracic cavity returns to its original position, air is expelled.

This is an efficient system, but it depends on maintaining the thoracic cavity around the lungs intact. A puncture wound in the chest, even if it does not touch the lungs, may admit air into the thoracic cavity and make breathing impossible—one of the reasons why chest wounds are always serious.

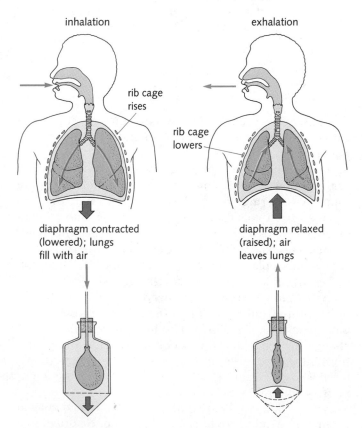

Figure 36.11 Inhalation. The lungs inflate due to atmospheric pressure after movements of the diaphragm and rib cage expand the chest cavity.

The Movement of Air

A normal breath at rest brings about 500 milliliters of air into the lungs. A person at rest breathes about 10 times a minute, so air is moved into the lungs at a rate of 5000 milliliters per minute. But the anatomy of the respiratory system means that not all of this inhaled breath actually reaches the alveoli for gas exchange. About 150 milliliters of each breath remains within the trachea, bronchi, and bronchioles, where gas exchange does not take place. This means that, at rest, only the first 350 milliliters of a breath actually have a chance to reach the alveoli, a condition that reduces the effective respiratory volume. During times of stress or physical exertion, breathing becomes much deeper; a single deep breath can bring in as much as 3500 milliliters of air.

GAS EXCHANGE

The Alveoli

Table 36.1 illustrates what happens to fresh air as it passes through the respiratory system. By the time it is exhaled, air has lost about 30 percent of its oxygen. In contrast, it contains 100 times as much carbon dioxide. The exchange of these two gases takes place in the alveoli.

The alveoli are extremely thin; their walls are only one cell thick. Each alveolus is surrounded by a delicate network of thin-walled capillaries that bring blood into close

Table 36.1 *Changes in Composition of Respiratory Gases*

Gas	Inhaled Air % (by volume)	Inhaled Air Pressure (mm Hg)	Exhaled Air % (by volume)	Exhaled Air Pressure (mm Hg)	Alveolar Air % (by volume)	Alveolar Air Pressure (mm Hg)
O_2	20.71	157	14.6	111	13.2	100
CO_2	0.04	0.3	4.0	30	5.3	40
H_2O	1.25	9.5	5.9	45	5.9	45
N_2	78.00	593	75.5	574	75.6	574

contact with the surface of the alveolar wall. Inhaled air enters each alveolus, filling the tiny sac with oxygen. As oxygen-poor blood rushes over the surface of an alveolus, oxygen from within the sac dissolves in the film of liquid coating the alveolus and diffuses through the capillary wall into the blood. Carbon dioxide from the blood, in turn, diffuses across the alveolar wall into the air, which is then exhaled (Fig. 36.12).

The movement of oxygen and carbon dioxide across the alveolus is a matter of simple diffusion. *There are no active transport processes associated with gas exchange in the lungs.* By the time it leaves an alveolus, the bloodstream has passed carbon dioxide into the atmosphere in exchange for oxygen. The enriched blood flows back to the heart through the pulmonary veins. This oxygen-rich blood is then pumped through the aorta to the rest of the body.

The Role of Hemoglobin

Although oxygen diffuses rapidly into solution, the oxygen-carrying capacity of most liquids is low. A liter of water in equilibrium with atmospheric pressure contains only about 3 milliliters of dissolved oxygen. If blood carried only 3 milliliters of O_2 per liter, an enormous volume of blood would be required to meet the respiratory needs of the average human: about 600 milliliters O_2 per minute. Fortunately, this is not the case. Human arterial blood has a much higher oxygen content: roughly 200 milliliters O_2 per liter. The increased capacity is possible because only a small amount of the oxygen is actually

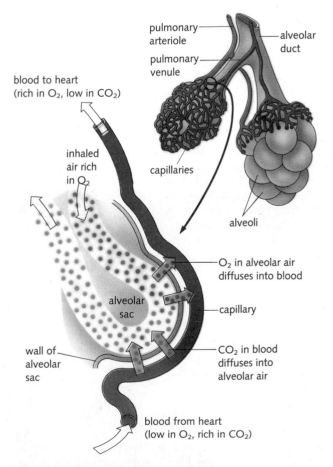

Figure 36.12 Gas exchange takes place in alveoli across the thin walls of the alveoli and capillaries.

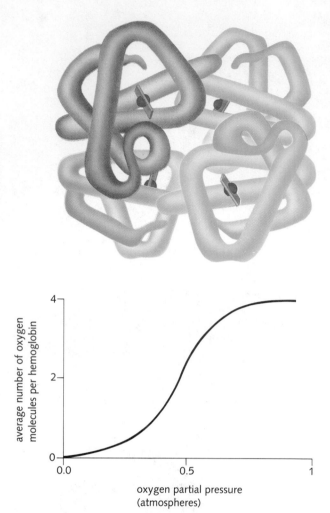

Figure 36.13   TOP: Oxygen is carried within red blood cells by hemoglobin, a protein with four oxygen-binding sites. Each subunit in this diagram of the hemoglobin molecule contains a single oxygen binding site. BOTTOM: The binding of oxygen to hemoglobin follows a sigmoid-shaped curve indicating that the affinity of hemoglobin for oxygen increases after one or two oxygen molecules have bound to the protein.

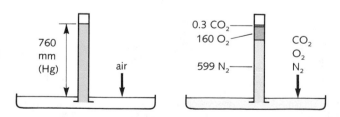

Figure 36.14   Atmospheric pressure at sea level is equivalent to a column of mercury 760 mm high. Oxygen accounts for 160 mm of this pressure, and nitrogen for 599 mm. Therefore, the partial pressure of oxygen in sea-level air is said to be 160 mm.

dissolved in water. More than 98 percent of the oxygen carried by blood is bound to a carrier molecule called **hemoglobin** (Fig. 36.13).

Hemoglobin is an oxygen-carrying protein contained within the red blood cell. The hemoglobin molecule consists of four polypeptide chains, and each of these chains contains a **heme** group. Heme is a ring-like, iron-containing prosthetic group (as discussed in Chapter 15) that binds an oxygen molecule. Each hemoglobin can bind a total of four oxygen molecules, one at each of its four heme groups (see Fig. 36.13).

Hemoglobin accounts for as much as 31 percent of the total weight of the red blood cell. Oxygen binds to hemoglobin molecules as red blood cells pass through the alveoli, and it is released as they pass through capillary networks in body tissues.

The binding of oxygen to hemoglobin causes a slight change in the color of the molecule. Fully oxygenated hemoglobin is bright red, and even partially deoxygenated hemoglobin is a much deeper bluish red. These differences account for the fact that venous blood is a much deeper, more bluish color than the bright red blood that flows through the arteries.

Gas Exchange Between Blood and Tissues

Why does hemoglobin bind oxygen in the alveoli and release it in the tissues? In order to understand this fascinating process, we must first understand a few terms and concepts related to the behavior of gases in mixtures and in solution.

Partial pressure   Generally, normal atmospheric pressure is 760 mm of mercury, which means that the pressure of the air produces a force equivalent to the weight of a column of mercury 760 mm high (Fig. 36.14). Only about 21 percent of the atmosphere is oxygen, so only 21 percent of atmospheric pressure—about 160 mm Hg—is due to oxygen. This value is known as the **partial pressure** of oxygen. Partial pressure is the pressure that can be directly attributed to one gas in a mixture of gases such as air.

When a container of water is opened to the air, oxygen from the air diffuses into the water until an *equilibrium* is reached at which the number of oxygen molecules entering the water is matched by the number leaving it. Because the dissolved gas is in equilibrium with the gas found in the air, we speak of the dissolved oxygen as also having a partial pressure of 160 mm Hg.

Hemoglobin and oxygen   Hemoglobin serves as a good carrier of oxygen in living systems because of the manner in which it binds oxygen. As it happens, oxygen molecules attach loosely, rather than permanently, to each of

The Bends

Although nitrogen constitutes 79 percent of the atmosphere, our discussion of the respiratory system has all but ignored the existence of this gas. In most respects there is no reason to be concerned with nitrogen; at normal partial pressures it is not known to participate in any physiological process. However, nitrogen does dissolve in the blood, and as its partial pressure increases, the amount of dissolved nitrogen in the blood increases too. This can cause serious problems for deep sea divers, who must breathe pressurized air to keep their lungs from collapsing under the force of scores of meters of water.

Under these unusual pressures, large amounts of nitrogen gas do dissolve in the blood. When a diver rises to the surface rapidly, the pressure falls quickly enough to cause nitrogen to come out of solution. The diver's circulatory system is blocked by millions of tiny nitrogen bubbles in the capillaries, causing pain, paralysis, and sometimes death. This serious condition, which is known as *the bends*, can be avoided by rising to the surface slowly, allowing excess nitrogen to leave the blood gradually through the lungs.

Hours of breathing nitrogen-rich air under high partial pressure can affect the nervous system and cause another problem known as *nitrogen narcosis*, in which the diver may seem drunken and silly, losing the ability to make sound judgments. Some of the first divers to experience this condition reported seeing mermaids and other exotic sights, leading some to describe the effect as the "rapture of the deep." The danger of nitrogen narcosis is one of the reasons why professional divers work in pairs and limit the time they spend beneath the water.

---

the four binding sites in hemoglobin. These loosely bound oxygen molecules are thus free to dissociate under certain physical conditions.

The property of hemoglobin that makes it so useful to the body is that it binds oxygen when the O_2 partial pressure is high and releases oxygen when the O_2 partial pressure is low (Fig. 36.15). In the alveolus, the O_2 partial pressure is high, so hemoglobin passing through picks up oxygen. In the capillary networks in other body tissues, however, the O_2 partial pressure is low. There oxygen dissociates from hemoglobin, so it is free to diffuse across the capillary wall into the tissues.

The need for oxygen to diffuse into and out of the blood can cause problems under certain circumstances. At high altitudes, for example, atmospheric pressure is lower than at sea level. Therefore, the partial pressure of oxygen is low, and less oxygen is available to diffuse across the alveolus. **Hypoxia,** a deficiency of oxygen in the tissues, may result. To compensate, an individual breathes harder at high altitudes (meaning that a larger volume of fresh air is inhaled). However, at levels above 20,000 ft, diffusion cannot move enough oxygen into the bloodstream to satisfy metabolic demands, and special breath-

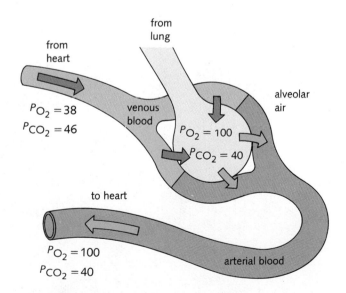

Figure 36.15 Because the lungs are not completely emptied with each breath, oxygen in the alveolus has a partial pressure (P_{O_2}) of 100 mm. The movement of gases between capillaries and alveoli gives blood leaving the alveoli different partial pressures of oxygen and carbon dioxide than the blood that enters the alveoli.

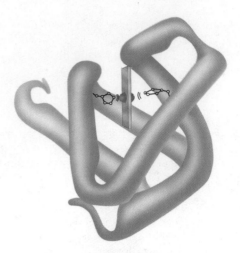

Figure 36.16 Oxygen is stored in muscle and other tissues by binding to myoglobin, a protein containing a single oxygen-binding site.

stream to satisfy metabolic demands, and special breathing equipment is usually necessary.

During pregnancy, mammals are confronted with another interesting problem: Oxygen must be transferred from the blood of the mother to the blood of the developing fetus. The circulations of mother and offspring are separate, but they come into intimate contact in the capillary beds of the *placenta,* where oxygen and nutrients diffuse from the maternal circulation to that of the fetus. If the hemoglobin molecules of mother and fetus were identical, the rate of oxygen transfer would be very slow; the best that could be achieved would be a 50/50 equilibrium between the two circulations.

As a result, in most species the red cells of the fetus contain a different form of hemoglobin, which has a higher affinity for oxygen than does adult hemoglobin. This fetal hemoglobin thus picks up oxygen at the same partial pressure at which the maternal hemoglobin releases it. For this reason, transfer of oxygen from mother to fetus is rapid and effective. Shortly before the animal is born and starts to breathe for itself, the body begins to synthesize the adult form of hemoglobin.

Skeletal muscle cells of the body face a similar problem. They need to draw large amounts of oxygen from the blood, and they do so in much the same way. These cells draw oxygen from the circulation and maintain their own stores of oxygen in the tissue. **Myoglobin,** a protein very similar to hemoglobin, is stored in skeletal muscle cytoplasm (Fig. 36.16). Each myoglobin molecule binds a single molecule of oxygen, allowing a reserve of oxygen to be built up within the muscle cell. This reserve is released when the partial pressure of oxygen within the cell drops, as it does during vigorous exercise.

The binding curve for myoglobin shows that it, like fetal hemoglobin, has a greater affinity for oxygen than does adult hemoglobin. This ensures that myoglobin will bind to oxygen at partial pressures low enough for hemoglobin to release it. Thus oxygen flows into muscles efficiently. The reddish color of myoglobin, incidentally, is one of the major reasons why the meaty muscle tissue of beef is a deep, bright red. Conversely, the "white meat" of a turkey has very little myoglobin.

The Transport of Carbon Dioxide

Carbon dioxide is produced during the oxidation of glucose and other food molecules. In active tissues, large amounts of carbon dioxide are present and the partial pressure of CO_2 rises, prompting CO_2 to diffuse into the bloodstream. As was the case with oxygen, only small amounts of CO_2 can dissolve directly in the blood. But unlike oxygen, only a small amount of CO_2 (about 11 percent) binds directly to hemoglobin, occupying the sites previously taken by oxygen. Some CO_2 molecules (about 8 percent) dissolve directly in the blood plasma. But the majority of CO$_2$ molecules (81 percent) undergo a two-step chemical reaction as they enter the bloodstream:

$$CO_2 + H_2O \rightarrow H_2CO_3$$
$$H_2CO_3 \rightarrow HCO_3^- + H^+$$

In the first equation, carbon dioxide dissolved in water reacts with a water molecule to form carbonic acid. This chemical reaction is catalyzed by an enzyme known as *carbonic anhydrase* and is found within the red blood cell. In the second equation, carbonic acid dissociates to produce the negatively charged bicarbonate ion and to release a positively charged proton. The proton may then bind to hemoglobin, displacing oxygen which can then be released into the tissues. The binding of protons to hemoglobin helps to release oxygen where it is most needed, and to prevent the blood from becoming highly acidic.

Bicarbonate ion is readily soluble in water, so large amounts of carbon dioxide can be carried by blood returning to the heart. The release of bicarbonate and hydrogen ions slightly lowers the pH of the blood (you will remember that an increase in the hydrogen ion concentration lowers the pH of a solution). However, it turns out that deoxygenated hemoglobin has a high affinity for hydrogen ions, so most of these ions are bound to hemoglobin. Therefore, blood returning to the heart is only slightly more acidic than blood in the arterial circulation. When deoxygenated blood flows through the alveolar capillaries, bicarbonate is converted to CO$_2$ that rapidly diffuses out of the blood into the alveoli and is exhaled.

CONTROL OF RESPIRATION

Breathing, or ventilation, is caused by muscular contractions under the control of the nervous system. Although breathing occurs automatically, we override that unconscious control when we want to blow up a balloon, play the flute, or hold our breath underwater. This sometimes gives us the mistaken impression that breathing is purely voluntary. It's not. After only a few seconds of hard running, athletes begin to puff to "catch their breath." That adjustment to the increased respiratory demands of the body is automatic. You may be able to hold your breath for a minute or so, but before long your body "forces" you to breathe. How does this happen?

A **breathing center** located in the *medulla* (a portion of the brain just above the spinal cord) controls breathing. Nerves carry impulses from the breathing center to the diaphragm and to the muscles of the rib cage. These impulses produce a cycle of contraction and relaxation that ventilates the lungs.

How does the breathing center react to an increased need for oxygen? Cells in the center are particularly sensitive to the partial pressure of CO$_2$ in the bloodstream. As the content of CO$_2$ in the blood rises, so does the CO$_2$ content of the breathing center. The center reacts by stimulating the breathing muscles. If the CO$_2$ content of the blood reaches critical levels, the impulses from the breathing center become so powerful that we cannot override them, no matter how strong our determination is.

In most cases, the sensitivity of the breathing center serves the body well, because an increased CO$_2$ partial pressure is a good indication that the body needs oxygen. However, there are some circumstances in which the breathing center can be tricked, and those cases can be fatal. Passengers in hot air balloons at high altitudes often have to be *told* to begin breathing pressurized oxygen, because they do not feel the effects of oxygen deprivation. The thin air at high altitudes starves the body for oxygen, but because CO$_2$ does not build up in the bloodstream, the breathing center does not react.

A similar problem occurs in divers who "hyperventilate" by rapid breathing before diving underwater. Rapid breathing decreases the carbon dioxide levels of the blood so much that these divers may run out of oxygen, pass out, and then drown underwater before the breathing center senses a buildup of CO$_2$.

DAMAGE TO THE RESPIRATORY SYSTEM

Subverting the System: Carbon Monoxide

Few molecules are as dangerous to the respiratory system as **carbon monoxide** (CO), a colorless, odorless gas produced when fuel is burned under conditions wherein oxygen availability is limited. Carbon monoxide is given off by gasoline and diesel engines and also by wood and charcoal stoves that limit air flow.

The carbon monoxide molecule is similar to the oxygen molecule in size and shape, and it attaches very tightly to the oxygen-binding sites of hemoglobin. But unlike oxygen and carbon dioxide, CO does not dissociate easily from the oxygen-carrying protein. This allows a relatively small number of CO molecules to "crowd out" oxygen and dangerously limits the oxygen-carrying ability of the respiratory system. In CO poisoning, the tissues of the body are deprived of oxygen, and death may occur within minutes.

Carbon monoxide poisoning is possible whenever the CO concentration of the air reaches significant amounts. Therefore, engines and stoves that produce CO should always be used in open, well-ventilated places where their exhaust fumes cannot accumulate. The first symptoms of carbon monoxide poisoning are drowsiness, headache, and disorientation. The best first aid is plenty of fresh air and concentrated oxygen, if available, to increase the O$_2$ partial pressure and force more O$_2$ onto hemoglobin.

Saving a Life: The Abdominal Thrust

Dozens of people die each year as a result of an unfortunate aspect of the structure of the respiratory system: the fact that air and food enter the body by means of the same passageway. When an individual manages to swallow a piece of food that is just a bit too large, that food may become lodged in the upper part of the throat, blocking the air passageways. In most cases a quick burst of air from the lungs—a cough—dislodges the food, but sometimes the coughing action is not powerful enough to move the food. Air is sealed off from the lungs, and death may follow from suffocation in as little as four minutes. The speed of a "café coronary," as these sudden attacks of suffocation have been called, leaves no time to summon trained medical personnel. Fortunately, a technique for dealing rapidly with the problem has been developed. Originally named the *Heimlich maneuver* after its inventor, this *abdominal thrust* technique takes advantage of the residual air that is always contained within the lungs and the elastic properties of the chest cavity.

The abdominal thrust should be attempted only when there is clear evidence that the air passageways have been completely blocked and the patient is in serious distress. Even if an individual is choking violently, if she or he is able to breathe or speak, the air passageways are not blocked and there is no need for the maneuver. The figure shows how the maneuver is applied. The rescuer reaches around the patient from behind and joins hands together to make a small fist just below the rib cage. The rescuer then drives that fist inward and upward while trying to lift the patient, compressing the thoracic cavity and forcing air against the food fragment. If the force of the maneuver is great enough, the food pops out of the throat, freeing the breathing passage.

The abdominal thrust technique depends on a sudden compression of the chest cavity to produce the pressure needed to dislodge an object from the breathing passageways.

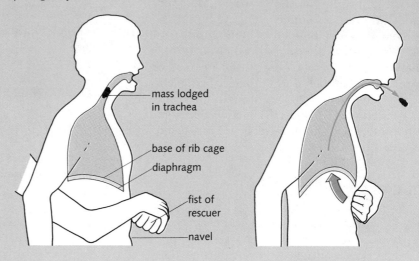

mass lodged in trachea

base of rib cage

diaphragm

fist of rescuer

navel

Because the CO molecule binds directly to the oxygen-binding site of hemoglobin, CO-poisoned blood has the same bright red color as fully oxygenated blood. Emergency squad workers are often struck by the fact that the victims of carbon monoxide poisoning have a flushed, "healthy" color; it is caused by this mechanism.

Smoking and Lung Damage

The upper passages of the respiratory system keep foreign particles from entering the delicate alveoli in which gas exchange takes place. Persistent exposure to foreign particles may eventually overwhelm the ability of the system to cleanse itself and may begin to damage the air passageways (Fig. 36.17). Remarkably, the major source of such damage is self-inflicted; it is smoking.

When tobacco is burned and the smoke inhaled, a mixture of nicotine, carbon monoxide, ash particles, and dozens of dangerous organic chemicals finds its way into the lungs. Persistent smoking, even in moderate amounts, can lead to *bronchitis*, a chronic inflammation of the bronchi; to *emphysema*, serious damage to the elastic tissue of the lungs; and potentially to the most serious consequence of all, *lung cancer.*

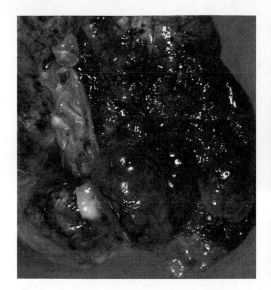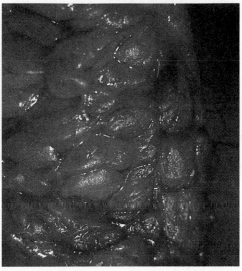

Figure 36.17 Part of the damage done by smoking is shown in this comparison of the lungs of a smoker (LEFT) with the healthy lungs of a nonsmoker (RIGHT).

Lung cancer is now the leading cause of death from cancer for both men and women in the United States. Over 160,000 people will develop lung cancer each year in the United States during the 1990s, and very few will survive it; the death rate is greater than 87 percent after 5 years. The cancer cells develop as solid masses that interfere with lung activity. Very quickly, cells can break off from the original tumor and spread throughout the body, causing dozens of secondary tumors and often bringing a painful death.

In addition to its own carcinogenic properties, cigarette smoking also increases the smoker's risk of developing lung cancer as the result of exposure to other environmental hazards. All of us who live in major metropolitan areas continually inhale scores of potentially noxious airborne particles. Normally, the self-cleaning mechanism in the respiratory tract traps these particles and moves them up, out of the lungs and into the pharynx, where they are swallowed. But because cigarette smoke paralyzes this transport system, smokers' lungs cannot clean themselves efficiently and particles accumulate in the lungs and remain for long periods.

Smoking is also a danger to nonsmokers. In 1993, the Food and Drug Administration listed the "secondhand smoke" inhaled by those who live or work with smokers as a cause of lung cancer. By recognizing the dangers of secondhand smoke, the FDA has lent a sense of urgency to efforts to provide smoke-free environments for students, workers, and young people.

The great tragedy of lung cancer is that it is almost completely preventable. More than 90 percent of all cases of lung cancer are caused by smoking. If you don't smoke and avoid secondhand smoke as much as possible, you greatly diminish your chances of contracting this terrible disease.

Emphysema

The breathing process can be affected by anything that makes it more difficult for air to enter the lungs. **Emphysema** is a condition that often develops in older people who have been exposed to conditions that weaken the respiratory system, including smoke, dust particles, and pollutants.

Recall that the elasticity of healthy lung tissue produces much of the pressure differential that drives air out of the lungs. The lung tissue of an emphysema patient, however, has lost its normal elasticity; when expanded, it does not "snap back" well, if at all. The lowered efficiency that results from emphysema reduces the volume of air that is exhaled and the amount of air that can be inhaled. In advanced cases of emphysema, the lungs are able to bring in so little air that patients must be administered air enriched in oxygen.

Asthma

Asthma is an allergic condition that we will discuss in Chapter 42. The allergies of an asthma patient may trigger an involuntary reaction in which smooth muscle cells around the airways contract, reducing the diameter of the passageways. Because this increases the resistance to airflow (for the same reason that water flows more slowly through a narrow garden hose), less air is able to flow to and from the lungs during a normal breath. To counteract this and allow them to breathe normally, asthma sufferers often take medications that help to relax the muscles surrounding the air passageways.

The process of respiration, in which cells oxidize carbon-rich food molecules to obtain chemical energy, requires the presence of oxygen to act as an electron acceptor. In smaller organisms, this oxygen can be obtained directly from the environment by the inward diffusion of oxygen from surrounding water or air, and carbon dioxide is diffused directly outward. In larger organisms, however, diffusion is not adequate to the task, and a specialized respiratory system makes possible gas exchange with the environment. Respiratory exchange may occur through a moistened integument, through gills, through tracheae that conduct air directly to the tissues, or through lungs that allow gas exchange to occur into a circulating blood, which carries oxygen to the tissues.

The human respiratory system is typical of those of terrestrial mammals. Air is inhaled through a series of passageways and enters the lung, driven by a partial vacuum created within the chest cavity when the diaphragm muscle contracts. Air enters the lungs and finds its way into hundreds of thousands of tiny alveoli, where gas exchange takes place between the air and the circulating blood. Oxygen passing across the walls of the alveoli enters the fluid phase of the blood, and most of it is taken up by hemoglobin, the oxygen-carrying protein contained within red blood cells. Oxygen is bound to special sites within the hemoglobin molecule. This binding is loose enough so that when the blood passes into tissue regions where little oxygen is found, oxygen molecules released from hemoglobin are able to diffuse into the surrounding area. The carbon dioxide that is a waste product of respiration in the tissues also enters the blood, and most of it returns to the lungs as bicarbonate ions, where it produces carbon dioxide to be exhaled. The oxygen-carrying system may be subverted by a molecule such as carbon monoxide, which competes with oxygen for binding sites on hemoglobin.

STUDY FOCUS

After studying this chapter, you should be able to:

- Discuss the relationship between respiration at the cellular level and the respiratory system.

- Explain some of the mechanisms and organs that have evolved to meet the challenges of gas exchange.

- Trace the operation of the human respiratory system, and describe its connection with the circulatory system.

- Cite the properties of hemoglobin, and describe how hemoglobin facilitates oxygen transport through the system.

TERMS AND CONCEPTS

REVIEW

Objective Questions (Answers in Appendix)

1. Which of the following statements about diffusion is true?
   (a) Oxygen cannot diffuse through the moist external membranes of most animals.
   (b) Oxygen exchange occurs in individual cells by diffusion in both small and large organisms.
   (c) Diffusion does not occur at the cellular level in large terrestrial organisms.
   (d) a and b only

2. All respiratory systems have
   (a) a way to exchange gases by diffusion.
   (b) a tracheal system.
   (c) countercurrent flow.
   (d) a system using lungs.

3. The most efficient gas-exchange systems are found in
   (a) reptiles.   (c) humans.
   (b) mammals.   (d) birds.

4. Hemoglobin contains _____ heme units, each of which can combine with _____ of oxygen.
   (a) four; one molecule   (c) two; one molecule
   (b) eight; one molecule   (d) two; one atom

5. Most of the carbon dioxide transported in the blood is carried in what form?
   (a) bound to hemoglobin
   (b) bound to myoglobin
   (c) dissolved as CO_2 in blood plasma
   (d) dissolved in blood as bicarbonate

Discussion Questions

6. If respiration occurs at the cellular level, why is it necessary for an organism to have a respiratory *system*?

7. Describe the various types of respiratory structures found in animals. How are they similar? How are they different?

8. Why is an effective respiratory system often associated with a circulatory system?

9. An animal can suffocate in any atmosphere that lacks oxygen. Why is brief exposure to concentrated carbon monoxide much more dangerous than brief exposure to concentrated carbon dioxide?

10. If a small cage of mice is pumped full of 100 percent nitrogen gas (N_2), the animals seem to take no notice of it. Gradually, they lose consciousness, and then will die if oxygen is not supplied. If 100 percent carbon dioxide is used instead, the mice react immediately and try desperately to get out of the chamber, obviously aware that they are suffocating, until they are rescued by the addition of oxygen. Why should the animals be aware of suffocation in one case but not in the other? (This question has an answer in terms of physiology, which you should be able to come up with right away. It also has an answer in terms of evolution, which you may need to think about for awhile.)

READINGS

Houston, C. S. "Mountain sickness." *Scientific American* 267 (October 1992): 58-66. High altitude means less oxygen available for the respiratory system to absorb from the atmosphere. The effects of high altitude, subtle and far-reaching, are explored in this article.

Feder, M. E., and W. W. Burggren. "Skin breathing in vertebrates." *Scientific American* 253 (November 1985): 126-132. Vertebrates have lungs or gills. But this does not mean that skin breathing does not take place in vertebrates. This well-written article describes the adaptations in a variety of vertebrates that allow for skin breathing.

Bramble, D. M., and D. R. Carrier. "Running and breathing in mammals." *Science* 219 (1983): 251-255. A summary of physiological differences in the respiratory systems of mammals associated with their active lifestyles.

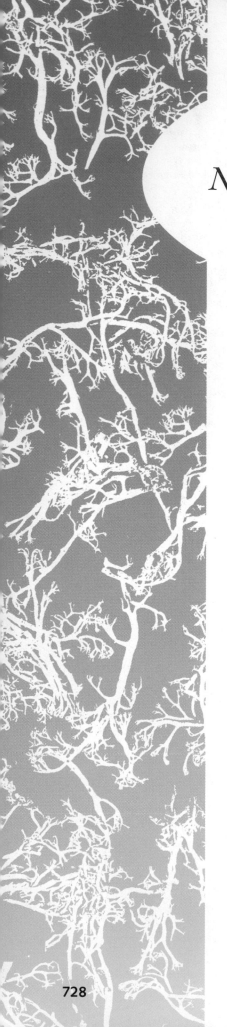

37

Nutrition and Digestion

One must eat to live, and not live to eat.
—Molière, *Amphitryon* (1668)

You can find them in America's remote rural hamlets and tumultuous urban ghettos, in Somalia in Africa, and in Bangladesh in Asia: children whose lives will be tragically shortened or permanently impaired by lack of proper energy and nutrients. Yet in an equally universal range of places are children and adults whose lives are in peril from too much or the wrong sort of food: chronically overweight individuals whose hearts, blood vessels, liver, and kidneys struggle daily with problems caused by misdirected, overzealous eating.

Molière, with his classic combination of wit and insight, captures all too well these extremes of the perpetual human preoccupation with food. Today, more than three centuries after he wrote those lines, the human race is still struggling, not only with the demons of malnutrition and obesity but also with a host of other medical problems that result from an improper diet.

Even those of us who are healthy are constantly bombarded with a confusing mix of scientific and pseudoscientific advice on nutrition. We have been advised by physicians on the basis of scientific evidence to add fiber to our diets and to reduce our intake of fats. We are pressed by various "health food" manufacturers to replace "refined sugar" with "complex carbohydrates" and to swallow megadoses of "essential vitamins and amino acids." And everyone seems to have a new diet plan that promises weight loss with no change in either eating patterns or exercise habits.

A thorough exploration of all these important, daily concerns about nutrition is beyond the scope of this chapter. We can, however, touch on several of them and lay sufficient conceptual groundwork for you to pursue further answers on your own.

NUTRITIONAL NEEDS OF HETEROTROPHS

Organisms need to eat for two main reasons. The *generation of chemical energy* is the principal use to which most food is put. As we saw earlier, carrying on the activities of life requires energy in the form of ATP.

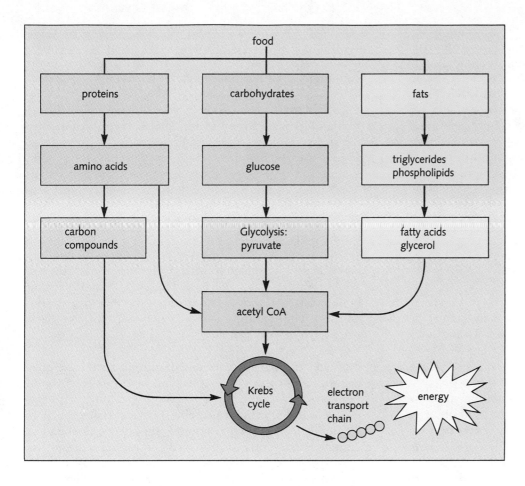

Figure 37.1 Food is a source of raw materials for the body, but it is also a source of energy. Food in the form of proteins, carbohydrates, or fats can be broken down and used as a source of cellular energy.

That ATP can be produced from the breakdown of glucose and other food molecules in pathways known as *glycolysis,* the *Krebs cycle,* and *oxidative phosphorylation* (see Chapter 18). Every cell has energy needs that must be met by a constant supply of food molecules entering these chemical pathways.

Animals' diets must also provide them with numerous *molecules needed for growth and maintenance.* Given only a simple source of chemical energy, such as sugar, cells can synthesize some of these compounds. But there are many other molecules that animal cells cannot synthesize and that must be provided in the diet. These essential nutrients vary from one species to the next, and finding diets that provide all essential nutrients, along with the proper amount of food energy, is a major focus of research in animal nutrition.

Food Energy and Calories

We measure the energy available in a food by measuring the amount of heat that can be released from it when it is completely oxidized. This heat is expressed in units known as calories. A **calorie** is the amount of heat energy needed to raise the temperature of a gram of water 1°C. One thousand calories is a kilocalorie. (Practicing nutritionists usually use the term *Calorie* to represent *1 kilocalorie,* but in this text we will adhere to scientific usage to avoid confusion.)

Every organism needs a constant supply of energy just to maintain its basic cellular activities. This background level of activity is known as the **basal metabolic rate,** and it represents the minimum activity required to sustain vital functions such as breathing and nervous system activity. The basal metabolic rate for average-sized humans is about 1500–1600 kcal per day.

Nearly all foods can provide usable energy. In Chapter 18 we traced the chemical pathways followed when glucose was broken down to release energy used in the formation of ATP. We chose glucose because it is the molecule on which the pathways of cellular metabolism are based. However, the same pathways can be used to obtain chemical energy from nearly any food molecule. Fats, proteins, and carbohydrates can all be broken down into simpler molecules, which then enter the same basic chemical pathways (Fig. 37.1). In this limited sense, basic energy requirements can be satisfied by any source of calories.

A lack of sufficient calories in the diet is known as **undernourishment.** The effects of chronic undernourishment can be tragic. Starved for sources of energy, the body begins to break down its own macromolecules to provide ATP, causing a loss of muscle and other tissue and a drop in the efficiency of several major organ systems. Prolonged undernourishment is common in areas such as Ethiopia and the Sudan that have suffered over time from severe famines, but it is also found in parts of many supposedly wealthy countries, including the United States. Because of its ability to cause permanent damage to the skeleton and the nervous system, long-term undernourishment of children can lead to stunted growth and decreased mental ability.

Food and Essential Nutrients

Most animals, humans included, cannot manufacture all of the compounds needed to maintain life. Those molecules that an organism cannot synthesize, along with a variety of minerals, are known as *essential nutrients,* and a healthful diet must include them.

Determining which nutrients are essential and the amounts in which they are required is not a simple task, especially in a difficult organism to experiment with, such as the human. However, nutritional scientists now believe that roughly 50 compounds are essential to human life. Given these 50 compounds in proper amounts and an adequate source of chemical energy, the human body can synthesize each of the thousands of compounds found in living cells and tissues.

However, if one or more of these essential nutrients are missing, the body is said to suffer from **malnourishment,** or malnutrition. Malnourishment cannot be compensated for by excesses of other nutrients; it can be corrected only by a diet that contains proper amounts of the missing nutrients.

Water   Water can be thought of as the most important of all essential nutrients. In fact, animals deprived of both food and water die of thirst long before they would have begun to suffer the ill effects of hunger. More than half of the total body weight of most animals is water—in some animals, that figure is as high as 90 percent. Water is needed in and around every cell of the body, and it makes up the major portion of blood, lymph, and other body fluids. Water is required for processing food in the digestive system, for eliminating waste through the kidneys, and for temperature regulation in animals that sweat or pant when they are hot.

If the total water content of an animal's body drops below the minimum level needed for normal functions, the animal is said to be *dehydrated.* If dehydration is not promptly treated, it can lead to problems with the circula-

tory, respiratory, and nervous systems. Under typical circumstances, the average human must take in at least a liter of water a day to avoid dehydration. Under desert conditions, 2 liters or more may be required.

Carbohydrates   The major sources of food energy in the human diet are carbohydrates. **Simple carbohydrates** are the sugars found in fruits and honey and in refined sugar. **Complex carbohydrates** include the starches in potatoes, grains, and vegetables. Enzymes in the digestive system break down complex carbohydrates to release simple sugars. These are absorbed directly into the circulation, making carbohydrates an immediate source of food energy for the body.

Many carbohydrate-rich foods also contain cellulose-based *fibers* that we cannot digest because we lack the enzymes to break down cellulose. However, these fibers add bulk to the food that moves through the digestive system, and a certain amount of fiber is necessary for the digestive system to work properly.

Proteins   Proteins are essential nutrients because they are sources of the *amino acids* used by the body to build new proteins. There are 22 different amino acids commonly found in cellular proteins. Plants can synthesize all 22 amino acids as long as they are supplied with chemical energy and a source of nitrogen. Thus plants do not require these compounds as nutrients. Animals, however, can synthesize only about half of these 22 amino acids and must acquire the rest in their food. Those amino acids that animals cannot synthesize are called **essential amino acids.**

Nutritionists recommend that an average-sized adult eat a diet that contains at least 60 grams of protein per day. However, the need for protein includes *quality* as well as *quantity,* for humans must derive eight essential amino acids from the protein they eat. If one or more of these essential amino acids are lacking in the diet, **protein deficiency** results. Figure 37.2 shows that when even one of the essential amino acids is lacking in the diet, the overall level of new protein synthesis in the body declines, despite an abundance of other nutrients. For this reason, a diet deficient in *any* of the essential amino acids dramatically reduces the amount of new protein that can be synthesized. Amino acids that cannot be used for protein synthesis are then used as food energy—a waste of nutrient value in one of the most important types of food.

Protein-rich foods from animal sources, such as eggs, fish, meat, and milk, are said to contain *complete* proteins, which means that they include a complete set of essential amino acids.

Many protein-rich plant foods, however, are nutritionally *incomplete* because they lack some of the essential amino acids. Corn, for example, is deficient in *lysine,* an essential amino acid. Beans have plenty of lysine, but they

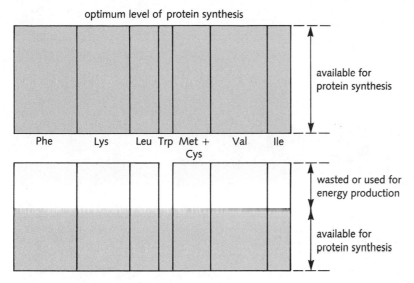

optimum level of protein synthesis

Phe | Lys | Leu | Trp | Met + Cys | Val | Ile

available for protein synthesis

wasted or used for energy production

available for protein synthesis

Figure 37.2 Humans require eight essential amino acids in the diet. Because each of these amino acids is needed for protein synthesis, a deficiency in any one of them lowers the overall level of protein synthesis and may cause malnutrition, even if the other amino acids are present in excess.

lack adequate amounts of *methionine* and *cysteine,* also essential amino acids. Vegetarians must therefore take care to balance their food sources in order to obtain a complete supply of the essential amino acids. Fortunately, this can be done quite simply. A diet made up exclusively of either beans or corn produces protein deficiency. However, a diet that contains ample portions of both supplies all of the essential amino acids and hence supports normal levels of protein synthesis.

Protein deficiencies produced by diets that lack complete sources of protein cause serious problems in many areas of the world where meat and dairy products are scarce. One such problem is *kwashiorkor,* a severe protein-deficiency disease found in several Third World countries (Fig. 37.3). It strikes children (who need large amounts of protein for growth) particularly hard. Kwashiorkor, which is characterized by weakness, stunted growth, skin inflammation, and anemia, is caused by diets based on only a single vegetable source of protein: corn

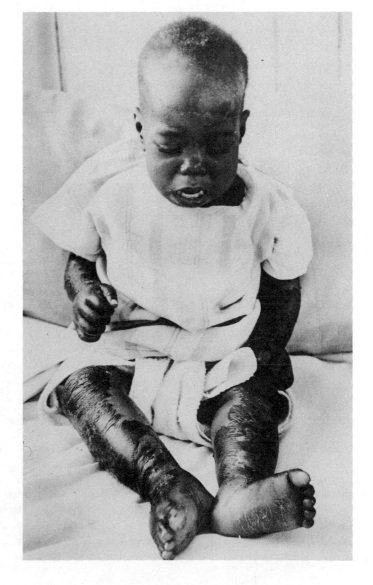

Figure 37.3 This child has suffered from kwashiorkor, a nutritional disorder caused by a diet lacking in one or more of the essential amino acids. The disease is common in parts of the world where diets contain only a single source of protein, such as rice or corn.

Cholesterol and Heart Disease: Research and Changing Perspectives

A 60-year-old man, a veteran of two coronary bypass operations, sits in a chair, connected by tubes from both his arms to an experimental laboratory device. As his blood flows slowly through the apparatus, a chemical filter removes the excess cholesterol that threatens to clog his few functioning coronary arteries. Elsewhere, a 6-year-old girl receives in a single operation both a new heart to replace the one crippled by fatty deposits and a new liver that doctors hope will keep the vessels of her replacement heart intact.

Both of these individuals, one 60, the other only 6, are victims of *atherosclerosis* (Chapter 35), a disease that contributes to nearly half of all deaths in the United States each year. Theirs are extreme cases, for they carry defective genes that interfere with the homeostatic control mechanism that regulates blood cholesterol levels.

"Normal" blood cholesterol concentrations among Americans range from 100 to 200 milligrams per deciliter. The 60-year-old man, who carried a single copy of the defective gene, had cholesterol levels ranging from 300 to 450 milligrams per deciliter. The 6-year-old girl, who carried *two* defective genes, had levels in excess of 600 milligrams per deciliter. Levels this high are rarely, if ever, seen outside the 1 in 500 individuals with such a defective gene.

Atherosclerosis occurs when fatty deposits, laden with cholesterol, accumulate in the walls of arteries throughout the body. The medical effects of high cholesterol are chillingly clear. In villages in Japan and the former Yugoslavia, where the mean cholesterol level is 160, the heart attack rate is extremely low: fewer than 5 per 1000 men in 10 years. But in Finland, where the mean cholesterol level is 265, the incidence of heart attack is 14 times higher.

The incidence of heart attack and the mean cholesterol levels in the United States are midway between these two extremes. It is quite clear that high cholesterol levels—along with other contributing factors, such as obesity, lack of exercise, smoking, and stress—lead to elevated risk of heart attacks.

But cholesterol is not simply a "bad" compound; it is an essential part of cell membranes and a vital step in the biosynthetic pathways of several important hormones. How, then, does excess cholesterol cause such serious problems?

Cholesterol is present in foods derived from animals, including meat, eggs, and dairy products (see table). The body can also manufacture cholesterol, primarily in the liver. Cholesterol itself is insoluble in water, and it therefore circulates in the blood combined with other fats and protein in either of two forms: as *low-density lipoprotein*, or LDL, and as *high-density lipoprotein*, or HDL. It is the LDL form of cholesterol that gets deposited in arterial walls; HDL particles seem to pick up cholesterol from such places and carry it to the liver for disposal.

The workings of the cholesterol regulatory machinery in the body were described by Michael Brown and Joseph Goldstein in a series of elegant experiments that earned them a Nobel Prize. They discovered that the membranes of many cells contain specialized receptor molecules that bind to cholesterol-containing HDLs and LDLs and carry these lipoproteins into the cell. There, cholesterol can be incorporated into the cell membrane, converted into hormones, stored, or metabolized. In healthy individuals, a negative feedback loop within liver cells "informs" the cells' metabolic pathways that enough cholesterol is present, removes excess

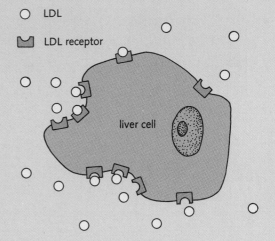

Receptors for low-density lipoproteins (LDLs) on the surface of liver cells help to regulate LDL levels in the bloodstream. When the receptor system does not operate properly, high LDL levels may produce atherosclerotic deposits that threaten the circulatory system.

cholesterol from circulation, and shuts down the synthesis of more cholesterol.

The 60-year-old man and the 6-year-old girl have defective genes for those receptors. Unable to detect and monitor cholesterol in the bloodstream, the liver cells of such individuals are not only unable to *remove* excess cholesterol but they are also "tricked" into continually producing and releasing more, even if blood cholesterol levels are already high. Thus these individuals cannot control their cholesterol levels with diet alone; their bodies continue to synthesize the compound even when none is taken in.

How is a gene defect in the cholesterol receptor related to cholesterol in the diets of normal individuals? Brown and Goldstein have suggested that a diet rich in cholesterol provides the liver with more cholesterol than it can possibly use. No longer requiring cholesterol for its own metabolism, the liver cell shuts down the synthesis of LDL receptors. Because the liver is subsequently unable to measure and react to the true cholesterol concentration in the blood, blood cholesterol rises. In a subtle but dangerous way, therefore, a diet high in cholesterol causes symptoms that mimic the genetic disease.

Nationwide surveys have shown that possibly 50 percent of Americans have dangerously high cholesterol levels caused mainly by diets containing too much fat. Many—though not all—of these individuals are at increased risk of heart attack and would benefit from lowering their blood cholesterol. There is little question that the American diet contains far too much fat: a typical American consumes more than the equivalent of an entire stick of butter in fat and cholesterol each day. These facts have led researchers to assemble recommendations about cholesterol control that range from diet to drugs:

1. Individuals over 20 years of age should have their cholesterol levels checked once every 5 years.

2. Those with LDL cholesterol levels above 130 milligrams per deciliter should be instructed in ways to cut back their dietary intake not only of cholesterol itself but also of saturated fats that the body converts into cholesterol. Some dietary fats, including those found in olive oil and the "omega-3" polyunsaturated fats found in fish oils, seem to lower LDL and raise HDL in a beneficial way.

3. Those individuals with LDL levels above 160 milligrams per deciliter should be considered at high risk and

should be monitored closely. If these individuals do not respond to changes in diet, their physicians should consider treatment with one of the recently developed drugs that control cholesterol uptake and synthesis.

It is important to note that, scientifically, the last word on cholesterol is not yet in. We do not fully understand the relationship between the different classes of lipoproteins and atherosclerosis, and despite the clear statistical connection between cholesterol levels and atherosclerosis, physicians recognize that the connection is not absolute. Some individuals are able to sustain high levels of cholesterol without apparent damage, and some others with relatively low cholesterol levels nonetheless suffer heart attacks. Despite this uncertainty, health experts continue to stress the importance of lowering cholesterol levels for most individuals.

Learning how to eat a satisfying meal without all that fat is the first step in controlling high cholesterol. Another helpful step involves eating foods high in soluble fiber, which binds cholesterol-containing bile salts. As a last resort for those whose cholesterol levels are resistant to dietary measures alone, supplemental drug therapy may help. Combined therapy with several classes of drugs has been shown to lower cholesterol levels by as much as 40 percent, and there is clear evidence that lowering cholesterol reduces the risk of heart attack.

Cholesterol Content of Common Foods

Food	Cholesterol (mg)
Fruits/vegetables	0
Grains	0
Milks (1-cup serving)	
Whole milk	33
Yogurt (whole milk)	29
Low-fat milk	18
Low-fat yogurt	14
Cheeses (1-oz serving)	
Cheddar	30
Processed American	27
Swiss	26
Ice cream (1/2 cup)	30
Butter (1 tsp)	12
Margarine, all vegetable (1 tsp)	0
Creams (1 tbsp)	
Whipped	20
Sour	8
Half and half	6
Meats (3-oz serving)	
Veal	86
Lamb	83
Beef	80
Pork	76
Chicken	39 to 63
Eggs (1 large egg)	
Yolk	274
White	0
Fish (3-oz serving)	
Shrimp	128
Lobster	72
Clams	50
Oysters	38
Fish fillet	34 to 75

Source: Adapted from *Cholesterol Information Sheet* (Rosemont, Ill.: National Dairy Council,

in Africa, rice in Asia. The cure for kwashiorkor is a change in diet to include complementary sources of proteins. The addition of soybean products—such as tofu—to a rice diet can supply the missing nutrients.

Fats Although animals can synthesize many fats and lipids from a few starting compounds, they do require a source of unsaturated fatty acids. These **essential fatty acids** are used both to synthesize certain lipid components of cell membranes and to produce the important fatty acid–like hormones known as *prostaglandins.*

Fatty acids that can supply these requirements are found in most plant and animal foods, so long-term deficiencies in these nutrients are not common. In fact, the total fatty acid dietary needs of a human can be supplied by as little as 60 milliliters of unsaturated vegetable oil per day—about 4 tablespoons.

Despite the fact that very little fat is required in the diet, recent trends in developed countries have led to the consumption of more and more fat. Roughly 40 percent of the calories in the diet of an average American are now derived from fat. The overabundance of fat in the diet may cause a number of problems. For instance, fat intake may crowd out ingestion of other foods, leading to long-term protein and vitamin deficiencies. Diets containing excessive amounts of saturated fats also tend to increase the levels of cholesterol and other dangerous forms of fat in the blood, often leading to heart disease (see Theory in Action, Cholesterol and Heart Disease: Research and Changing Perspectives). High-fat diets are also associated with increased incidence of breast and colon cancers. Although the ideal level of fat in the diet is hard to determine, most dietary experts believe that the fat content of the typical diet should be reduced to the point where only 20–30 percent of one's daily calories are derived from fat.

Vitamins Animals also require more than a dozen small organic molecules known as **vitamins.** The term was coined as a contraction of "vital amines," because *thiamine,* the first vitamin to be discovered, belonged to the chemical family known as the amines. Vitamins are usually defined as *complex organic compounds needed in small amounts in the diet.* This distinguishes them from proteins and other nutrients that are needed in much larger amounts. Most vitamins are involved in chemical reactions as *coenzymes* or *cofactors,* which help to catalyze chemical reactions. Just a handful of molecules can catalyze hundreds or thousands of such reactions.

Although vitamins are needed in small amounts, vitamin deficiencies can have serious—even fatal—consequences. Vitamin B_1, or thiamine, for example, is present in many meats and grains, including the hulls that surround the grains of natural brown rice. The need for this vitamin was first appreciated in the nineteenth century when prisoners in the Far East fell sick with a disease known as *beriberi.* Prisoners afflicted with beriberi were weak and nervous and suffered from *edema* (swelling in the tissues) as well as heart failure. The fact that none of the jail guards caught the disease suggested that it was not a contagious illness. The jail wardens were also proud to point out that they fed the prisoners the very best rice on the market: polished white rice, from which the brown husks had been removed.

A clue to the source of this mysterious ailment came from an unlikely source: the behavior of chickens kept in the prison compound. These birds suffered from symptoms similar to those of the human prisoners—loss of muscle coordination that caused them to fall repeatedly when they tried to walk. The fact that the chickens had been fed table scraps from the jail led to the idea that the prisoners' food might be responsible for beriberi. In fact, when the ailing chickens were fed table scraps containing ordinary brown rice, they quickly improved.

It became clear that the disease was caused by the absence of thiamine, vitamin B_1, in the polished white rice. Because rice hulls contain plenty of thiamine, all that was necessary to cure the disease was to switch from white to brown rice.

One vitamin after another has been discovered over the years, and at least 14 vitamins are now recognized. Wherever possible, nutritional scientists have developed a *recommended daily allowance* (RDA) for each vitamin. The RDA serves as a guide to developing a diet that exceeds the minimum amount of each vitamin required for normal body function.

As shown in Table 37.1, 10 of the vitamins are water-soluble and 4 are fat-soluble. The body is not able to store any of the water-soluble vitamins; if the diet contains more of these vitamins than can be used, the excess is removed from the blood by the kidneys and excreted in the urine.

Excess amounts of fat-soluble vitamins, on the other hand, can be stored in the fatty tissues of the body. Although this allows the body to develop storage reserves of these vitamins, it also poses a potential danger. Vitamins A, D, and K are toxic (poisonous) in high concentrations, and the use of excessive numbers of vitamin pills can cause these compounds to build up to the point where they become a serious medical problem. For this reason, the excessive use of vitamin supplements (except under the supervision of a physician) should be avoided.

Minerals The *inorganic* nutrients known as minerals are usually required in very small amounts. There are at least 14 **essential mineral nutrients** (Table 37.2), many of which are required for the body to build specific tissues. Deficiencies of these and other minerals can result in stunted growth, muscle weakness, or anemia, depending on what is lacking and the severity of the deficiency.

Table 37.1 *Vitamins*

Vitamin	Food Sources	Function	Needed Daily	Results of Vitamin Deficiency
Water-Soluble Vitamins				
Vitamin B$_1$ (thiamine)	Yeast, liver, grains, legumes	Coenzyme for carboxylase	1.5 mg	Beriberi, general sluggishness, heart damage
Vitamin B$_2$ (riboflavin)	Milk products, eggs, vegetables	Coenzyme in electron transport (FAD)	1.8 mg	Sores in mouth, sluggishness
Niacin	Red meat, poultry, liver	Coenzyme in electron transport (NAD)	20 mg	Pellagra, skin and intestinal disorders, mental disorders
Vitamin B$_6$ (pyridoxine)	Dairy products, liver, whole grains	Amino acid metabolism	2 mg	Anemia, stunted growth, muscle twitches and spasms
Pantothenic acid	Liver, meats, eggs, whole grains, and other foods	Forms part of coenzyme A, needed in Krebs cycle	5-10 mg	Reproductive problems, hormone insufficiencies
Folic acid	Whole grains and legumes, eggs, liver	Coenzyme in biosynthetic pathways	0.4 mg	Anemia, stunted growth, inhibition of white cell formation
Vitamin B$_{12}$	Meats, milk products, eggs	Required for enzymes in red cell formation	0.003 mg	Pernicious anemia, nervous disorders
Biotin	Liver and yeast, vegetables, provided in small amounts by intestinal bacteria	Coenzymes in a variety of pathways	Unknown	Skin and hair disorders, nervous problems, muscle pains
Vitamin C (ascorbic acid)	Citrus, tomatoes, potatoes, leafy vegetables	Required for collagen synthesis	45 mg	Scurvy: lesions in skin and mouth, hemorrhaging near skin
Choline	Beans, grains, liver, egg yolks	Required for phospholipids and neurotransmitters	>700 mg	Not reported in humans
Fat-Soluble Vitamins				
Vitamin A (retinol)	Fruits and vegetables, milk products, liver	Needed to produce visual pigment	1 mg	Poor eyesight and night blindness
Vitamin D (calciferol)	Dairy products, fish oils, eggs (also sunlight on skin)	Required for cellular absorption of calcium	0.01 mg	Rickets: bone malformations
Vitamin E (tocopherol)	Meats, leafy vegetables, seeds	Prevents oxidation of lipids in cell membranes	15 mg	Slight anemia
Vitamin K (phylloquinone)	Intestinal bacteria, leafy vegetables	Required for synthesis of blood clotting factors	0.03 mg	Problems with blood clotting, internal hemorrhaging

SOURCE: N. S. Scrimshaw and V. R. Young, "The requirements of human nutrition," *Scientific American*, 1976.

Calcium phosphate, for example, is the major component of bone, and large amounts of both calcium and phosphorus are required in the diet when new bone is being formed.

Iron is needed as a component of a number of important proteins, including hemoglobin and myoglobin, which bind oxygen, and of the cytochromes, which are involved in electron transport. Iron deficiencies are particularly common in premenopausal women, owing to loss through menstruation, and often result in chronic anemia. Even college-age women should pay particular attention to the amount of iron in their diets.

Most bodily fluids contain sodium, potassium, and chlorine, and a proper supply of these minerals is important to many electrically excitable tissues, including muscle and nerve.

Table 37.2 *Essential Mineral Nutrients*

Mineral	Food Sources	Function	Needed Daily	Results of Mineral Deficiency
Calcium	Milk, cheese, legumes, dark green vegetables	Bone formation, blood-clotting reactions, nerve and muscle function	800 mg	Stunted growth, weakened bones, muscle spasms
Phosphorus	Milk products, eggs, meats	Bones and teeth, ATP and related nucleotides	800 mg	Loss of bone minerals
Sulfur	Most foods containing proteins (derived from sulfur-containing amino acids)	Used in formation of cartilage and tendons	Provided by sulfur-containing amino acids in diet	Deficiency of sulfur-containing amino acids
Potassium	Most foods	Acid–base balance, nerve and muscle function	2500 mg	Muscular weakness, heart problems, death
Chlorine	Salt	Acid–base balance, nerve and muscle function, water balance	2000 mg	Intestinal problems, vomiting
Sodium	Salt	Acid–base balance, nerve and muscle function, water balance	2500 mg	Weakness, diarrhea, muscle spasms
Magnesium	Green vegetables	Enzyme cofactors, protein synthesis	350 mg	Muscle spasms, stunted growth, irregular heartbeat
Iron	Eggs, leafy vegetables, meat, whole grains	Hemoglobin, electron transport enzymes	10 mg	Anemia, skin lesions
Fluorine	Drinking water, seafood	Structural maintenance of bones and teeth	2 mg	Tooth decay, bone weakness
Zinc	Most foods	Enzyme cofactor	15 mg	Fever, vomiting, diarrhea, nausea
Copper	Meats, drinking water	Required for hemoglobin-synthesizing enzymes	2 mg	Anemia
Manganese	Whole grains, egg yolks, green vegetables	Required for several enzymes	3 mg	No reported effects in humans
Iodine	Seafood, milk products, iodized salt	Thyroid hormone	0.14 mg	Goiter (enlarged thyroid)
Cobalt	Meat, liver, milk products	Part of vitamin B_{12}	Required in the form of vitamin B_{12}	No reported effects in humans

SOURCE: N. S. Scrimshaw and V. R. Young, "The requirements of human nutrition," *Scientific American*, 1976.

Sulfur is a component of many proteins, and iodine is used in very small amounts for the synthesis of *thyroxine*, a vital hormone produced by the thyroid.

One of the most critical minerals for animals living on land is sodium. The large amount of sodium in the body is an evolutionary reflection of the abundance of sodium chloride (common salt) in seawater, where life first began. On land, however, sodium is hard to come by. Land plants, which need to conserve water, contain very little sodium. Therefore herbivores often have chronic sodium deficiencies, and this can influence their behavior in the wild. Farmers often provide blocks of salt for their horses and cattle, and hunters know that a block of salt placed in the woods will draw wild animals from miles around. For example, salt blocks are so effective in drawing deer that their use in hunting has been outlawed in most states.

Humans have the same need for salt as other land animals, and ancient civilizations valued salt highly. The discovery that salt could be obtained by boiling seawater predates civilization, and many cultures have incorporated the refining and use of salt into their cultural and religious ceremonies.

The successes of human culture in developing means of purifying salt as a mineral supplement, however, have had a serious side effect. Probably because our mammalian ancestors needed salt, our taste buds find it eminently desirable. Salt enhances the flavor of dozens of foods, and this has led to an excess of sodium in the typical diet in some countries. The average American takes in roughly 25 times the amount of sodium needed for a healthy diet, and this extra sodium can cause several problems.

The immediate effect of sodium is to raise blood pressure and increase stress on the kidneys, which remove excess sodium from the blood. Over the long term, these effects can increase the likelihood of heart disease and stroke. A high level of dietary sodium is one of the major causes of circulatory disease in the United States and other industrialized countries.

Excesses of other minerals can also cause serious problems. An excess of iodine in the diet can depress thyroid activity, large amounts of iron can produce liver damage, and high levels of magnesium in water supplies can cause chronic diarrhea. As with the other major classes of nutrients, a healthful diet contains adequate, but not excessive, amounts of each of the essential minerals.

A BALANCED DIET

How can you be sure that the food you eat contains enough nutrients, vitamins, and minerals without providing too many calories in the form of fat? The simplest way is to eat a balanced diet, containing a variety of foods. The U.S. Department of Agriculture (USDA) recommends that individuals should build their daily menu around a "pyramid" of food from six different groups, as shown in Fig. 37.4. At the base of the pyramid are carbohydrate-rich foods such as bread, cereal, rice, and pasta, of which the USDA recommends from 6 to 11 servings per day. Fruits and vegetables are next, followed by lesser amounts of protein-rich foods like milk, cheese, beans, and meat. At the tip of the pyramid are fats, oils, and sweets, which should be used only sparingly. The basic idea behind these recommendations is that individuals should limit their intake of fatty foods and eat from a wide variety of food groups every day.

Balancing Nutrient Input and Requirements

As we have seen, the basal metabolic rate for an average-sized human is about 1500–1600 kcal per day. This figure represents no more than the basic energy needed to remain alive with minimal activity. Another way of thinking about the basal metabolic rate is to note that it is roughly the amount of energy one would burn while spending the whole day in bed.

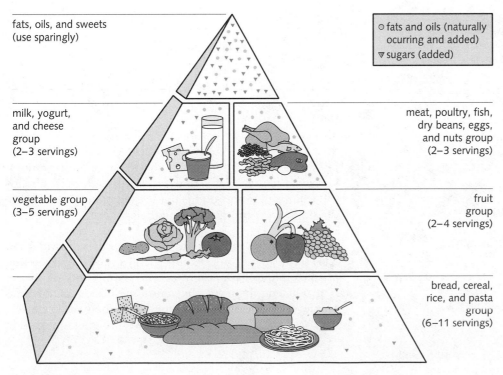

fats, oils, and sweets
(use sparingly)

milk, yogurt,
and cheese
group
(2–3 servings)

vegetable group
(3–5 servings)

○ fats and oils (naturally
ocurring and added)
▽ sugars (added)

meat, poultry, fish,
dry beans, eggs,
and nuts group
(2–3 servings)

fruit
group
(2–4 servings)

bread, cereal,
rice, and pasta
group
(6–11 servings)

Figure 37.4 The elements of a well-balanced diet are illustrated in this food pyramid. The pyramid is built around the principle of minimizing consumption of fatty foods and emphasizing the use of carbohydrate-rich foods in the daily diet.

The Hundred-Watt Bulbs

The basal metabolic rate for humans ranges from 1300 to 1800 kcal per day. Ultimately, whether it is used for internal movement, breathing, circulation, or philosophical thought, nearly all of this energy is lost as heat. This gives the average human an energy output equal to that of a 100-watt light bulb. This may not sound like much, but architects realize how important those heat sources can be. When building a 1000-seat theater, for example, the designer must deal with the fact that 1000 humans produce a steady heat output of 100 megawatts! That's roughly equivalent to the heat output of a large barbecue fire roaring in the middle of the theater. When the audience laughs or cries, the energy output is even higher, because emotional involvement increases metabolic rate. Air conditioning may be a product of our modern society, but it is made necessary by one of the oldest forces—the metabolic fire of life itself.

Table 37.3 *What It Takes to Burn Calories*

Food	Activity	Calories/minute	Time to Burn (minutes)
Cheeseburger 470 Calories	Resting	1.1	427
	Walking	5.5	85
	Swimming	10.9	43
	Running	14.7	32
Milkshake 318 Calories	Resting	1.1	289
	Walking	5.5	58
	Swimming	10.9	29
	Running	14.7	22
Corn, 2 pats butter 170 Calories	Resting	1.1	155
	Walking	5.5	31
	Swimming	10.9	16
	Running	14.7	12
Corn, unbuttered 70 Calories	Resting	1.1	63
	Walking	5.5	13
	Swimming	10.9	6
	Running	14.7	5

Physical activity, however, requires energy in *excess* of the basal rate. The normal daily activities of walking, lifting, and doing household chores may consume several hundred kilocalories in addition to the basal rate; vigorous activity may add more than a thousand kilocalories. Swimmers and long-distance runners may need more than 4000 kcal a day. Table 37.3 presents the amount of energy required for a number of physical activities in terms of common foods.

Obesity

Naturally, it is a rare day when the body takes in exactly enough food to match precisely the number of calories burned. When more food is eaten than can be converted into energy, the body stores the excess as glycogen in the liver and muscles. This ready reserve of carbohydrate can be used quickly when the body demands more energy. However, if the body's reserves of glycogen are already full when still more food is made available, the body converts that food energy into an efficient form for long-term storage: fat.

The energy required to produce fat is provided by ATP, so we should not delude ourselves into thinking that it is possible to gain weight only by eating fatty foods. Fat is produced whenever the amount of food taken in is greater than the amount of energy the body is capable of using. Therefore we can gain weight by eating any food to

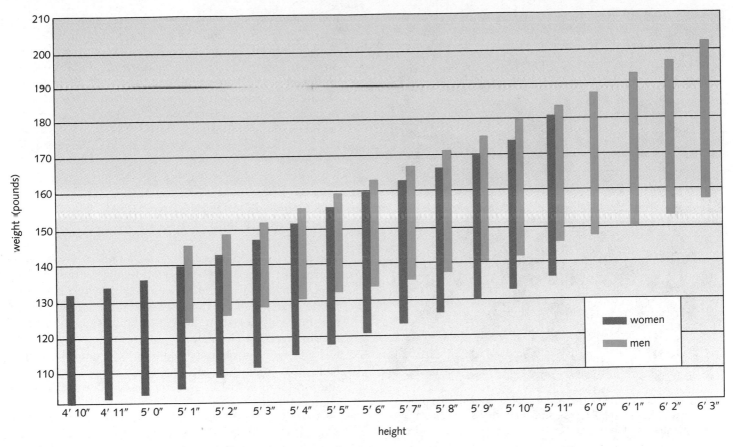

Figure 37.5 Ideal weights for individuals of various heights and ages. The "ideal" weight is actually a range that varies with body type and muscular development, as well as bone structure.

excess, including proteins, simple sugars, complex carbohydrates, and fat.

Individuals are said to be **obese** when their weight exceeds the ideal weight for someone of their size by more than 20 percent (Fig. 37.5). Obesity is a serious health problem. In addition to limiting physical activity, extra weight places an unnecessary strain on the heart and circulatory system.

Losing Weight

The causes of obesity seem simple. Too much food is taken in, and too little is burned up in daily exercise. The cure for obesity seems equally simple: more exercise and better control of diet. When a person's daily activity *exceeds* his or her food intake, the first sources of extra energy to be called into action are glycogen reserves stored in liver and muscle cells. When these reserves are depleted, the body takes fat molecules out of storage and

uses them for energy. If activity exceeds food intake on a regular basis, a slow but steady reduction in fat reserves takes place and weight is lost.

There is, of course, a lot more to it. Many people, despite their best efforts, try and try again and fail to lose weight. Nutritional scientists have noticed that an individual's appetite for food and personal energy level seem to help maintain a *set point* of body weight. When food intake is restricted, basal metabolic rate slows down and the dieter feels tired and sluggish. This lowering of the metabolic rate causes fewer calories to be burned, and weight loss becomes more difficult. Some scientists have suggested that the set point may be genetic or that it may be determined by feeding habits in early infancy.

There is no complete solution to the difficulties many of us have in losing weight, but the most successful weight control programs couple a moderate decrease in food intake with increased exercise. Over time, the physical activity (whether cycling, running, walking, or dancing) seems to alter the set point and give the dieter a feeling of per-

Figure 37.6 Proper exercise is an essential element to a program of weight control. Exercise also strengthens the circulatory and respiratory systems.

sonal fitness and pride that makes weight control much easier (Fig. 37.6).

Eating Disorders

The personal and social pressures to lose weight can be so great that serious psychological disorders are produced in efforts to control obesity.

Individuals suffering from **anorexia nervosa** have developed such a strong aversion to food that they may be unable to eat without vomiting. Anorexics have a distorted perception of their own bodies, which sometimes makes them interpret the presence of normal amounts of body tissue as a symptom of obesity. Such images may lead these individuals to lose weight to the point where their healthy tissue begins to break down and their very lives are threatened.

A related eating disorder is **bulimia,** in which individuals force themselves to vomit after eating high-calorie foods. In the early stages, sufferers may rationalize their behavior as helping to control weight, but this rationalization masks serious health problems. Vomiting the acidic contents of the stomach gradually destroys tooth enamel, causes a loss of salts from the body (which may produce

heart failure), and may rupture the esophagus, causing death. There is no doubt that these eating disorders are partly the result of social pressures that place a premium on a slim appearance. They require immediate medical and psychological attention. Unfortunately, the frequency of eating disorders among young women and men seems to be increasing, and the disorders also have been diagnosed in young children.

OBTAINING FOOD: THE UNIVERSAL PREOCCUPATION

The need for food affects every aspect of the life of an organism. The need to ensure an adequate supply of food affects physiology; it influences behavior and reproduction, helps determine the structure of the nervous system and the skeletal system, and defines the tasks performed by the digestive, circulatory, and endocrine systems. The way an organism obtains food also determines how it affects other organisms, what pressures its species places on the local and global ecosystems, and which organisms it competes with.

Our analysis of digestion and nutrition would be much simpler if we could make generalizations that held true across each of the major taxonomic groups—if just one type of digestive system were found in each phylum, for example. But this is not the case. Although some aspects of the organization of any system are limited by whether an animal is an arthropod, a mammal, or an echinoderm, within most major groups there are organisms that differ enormously with respect to habitat, diet, behavior, and other factors.

Feeding Techniques

Success or failure in obtaining food is one of the great tests of survival. As we saw in Chapter 4, the rewards for success in the quest for food have fueled the evolution of a tremendous variety of ways of obtaining and processing different types of foods (Fig. 37.7).

Most familiar animals are carnivores that eat other animals, herbivores that eat plants, or omnivores (such as ourselves) that eat both plants and animals. But many aquatic organisms, such as clams, are filter feeders that trap food particles as they strain enormous volumes of water. Others, such as mosquitoes, are fluid feeders that consume either the blood of an animal or the sap of a plant. And many ecologically important organisms are detritus feeders that eat decaying organic material.

Because of this great diversity in feeding strategies, we cannot carry out a phylum-by-phylum analysis. We can, however, illustrate a few common "themes" that emerge in a study of digestive systems.

Cellular Digestion

Digestion is such a basic process that it is found in organisms that consist of nothing more than a single cell. A number of protozoans, including *Paramecium* and *Amoeba*, take in food particles directly from the environment. *Amoeba* engulfs particles directly, enclosing them in part of the cell membrane to form a food vacuole. *Paramecium* uses a ciliated groove to sweep particles into a specialized region of the cell membrane known as the *cytopharynx*, where food vacuoles are formed (Fig. 37.8).

Once inside the cell, food particles must be broken down into usable form. Food vacuoles fuse with *lysosomes* to form digestive vacuoles. Lysosomes are organelles laden with enzymes capable of breaking down many large molecules (Chapter 16). These enzymes split polymers in food into monomers that are released, along with other small molecules, into the cytoplasm where they can be utilized by the rest of the cell. (Note that lysosomes are not restricted to protozoans; certain cells in multicellular organisms, such as the white blood cells of humans, handle large food particles in the same way.)

Material that cannot be broken down in lysosomes is released from the cell. In *Paramecium*, the release of indigestible material occurs at a specialized region known as the *anal pore*.

Animals with Gastrovascular Cavities

The simplest digestive systems have a single opening to the exterior. Food is taken into the cavity, it is broken down by enzymes released from cells lining the cavity, and the breakdown products are absorbed into the tissues of the organism. Indigestible material is then released from the cavity via the same opening through which it entered.

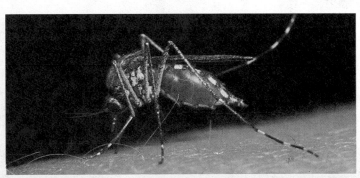

Figure 37.7 There's more than one way for an animal to acquire food, as demonstrated here by (CLOCKWISE FROM TOP LEFT) a brown pelican with fish; a mosquito sucking blood; a spiny sun star attacking a green sea urchin; and a yellow rat snake devouring its prey.

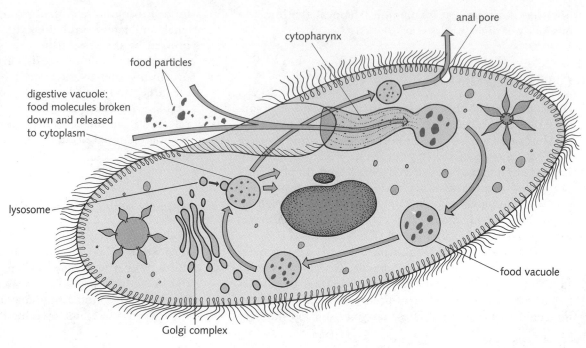

anal pore

cytopharynx

food particles

digestive vacuole: food molecules broken down and released to cytoplasm

lysosome

food vacuole

Golgi complex

Figure 37.8 The formation of food vacuoles in *Paramecium*. The movement of food particles through this organism is an example of the pathways followed by food material in many other cells. The primary work of chemical digestion is done by lysosomal enzymes.

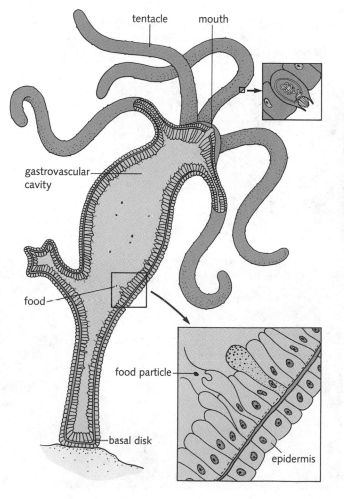

tentacle   mouth

gastrovascular cavity

food

food particle

basal disk

epidermis

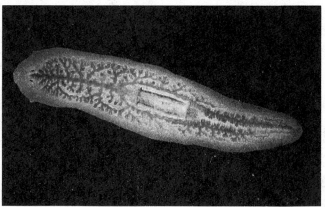

Figure 37.9 Digestion in *Hydra*. LEFT: Food is broken down in the gastrovascular cavity, and digestion is completed within the cells lining the cavity. RIGHT: The extensive gastrovascular cavity of *Planaria* is stained red in this photomicrograph. The intricate branchings of this cavity allow food to diffuse to all parts of the animal.

One organism that digests food in this way is *Hydra,* shown in Fig. 37.9. *Hydra* disables its food by stinging it with *nematocyst* cells on its tentacles. Then it pushes its paralyzed prey through its mouth opening into its **gastrovascular cavity,** where digestive enzymes break the food down. This *extracellular digestion* begins the breakdown of food, and *intracellular digestion* completes it, as the partly digested material is taken up by the cells lining the cavity. Material that cannot be used for food, including the mineral skeletons of some small organisms, is ejected from the cavity through the mouth.

A gastrovascular cavity is also found in flatworms such as *Planaria,* and it functions in much the same way (see Fig. 37.9). The cavity is highly branched, extends throughout the length of the organism, and enables food to diffuse to all parts of the animal. This is why the word *vascular* is used to describe the cavity.

Animals with Alimentary Canals

Many organisms have evolved an **alimentary canal,** or **gut,** a tubular passageway that is open at both ends and thus allows a one-way flow of material to take place. Food enters at the **mouth** and is digested, and the products of digestion are absorbed through the canal. Undigested material is expelled from the terminus of the gut, the **anus** (Fig. 37.10).

This one-way movement of food through the alimentary canal allows for a degree of specialization that is not possible in digestive systems with only one opening. Specialized organs have evolved along the length of the canal, forming a kind of "dis-assembly line" in which food is broken down in an orderly and systematic fashion. As food enters the canal, for example, specialized sharp mouth parts, bills or teeth, can break the food into small pieces. In some animals, especially birds, this process continues in a muscular compartment called the *gizzard.*

The early stages of digestion occur in another specialized portion of the canal called the *stomach,* in which food may remain for several minutes or hours. The final chemical breakdown and absorption of nutrients occurs in the *intestines.* In most organisms this is the longest portion of the canal, reflecting the high priority placed on absorbing every possible nutrient molecule. Finally, undigested material is collected in the *rectum* and is periodically expelled through the *anus.*

Associated with the alimentary canal are a series of glands that produce the enzymes needed to break down the different macromolecules found in food. As food passes through the canal, these enzymes are released in sequence, breaking down food in stepwise fashion and preparing it for the absorption that occurs in the intestines.

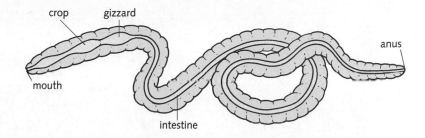

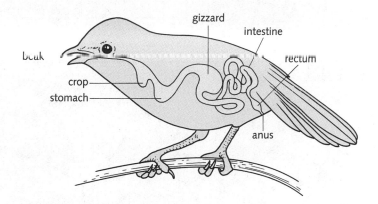

Figure 37.10 The digestive systems of many animals are organized around an alimentary canal that allows one-way movement of food from a mouth to an anus. The alimentary canals of the earthworm and a bird contain a number of specialized organs that process the food as it moves through the digestive system.

Alimentary canals in many organisms have valve-like rings of smooth muscle, called *sphincter muscles,* between different compartments. These valves manage the flow of material through the system, ensuring that food is kept in each compartment long enough for chemical processes to be completed before the food enters the next portion of the canal.

THE HUMAN DIGESTIVE SYSTEM

The overall plan of the human digestive system reflects the fact that we are omnivores. This system, called the **gastrointestinal tract,** is illustrated in diagrammatic fashion in Fig. 37.11. It is essentially a tube between 6 and 10 m in length and composed of specialized digestive organs.

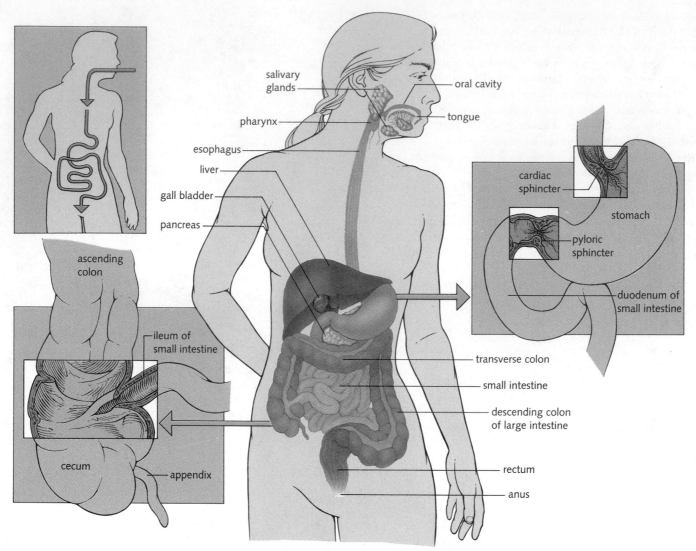

labels on figure:
salivary glands
oral cavity
pharynx
tongue
esophagus
liver
gall bladder
pancreas
cardiac sphincter
stomach
pyloric sphincter
duodenum of small intestine
ascending colon
ileum of small intestine
transverse colon
small intestine
descending colon of large intestine
cecum
appendix
rectum
anus

Figure 37.11 The human digestive system. After being ground and chewed in the mouth, food passes down the esophagus to the stomach, where partial digestion occurs over a period lasting from 1 to 5 hours. The final stages of digestion and absorption take place in the small intestine, and excess water is resorbed into the body in the large intestine. A number of organs, including the liver and pancreas, produce secretions that are essential for digestion.

A cross section of the tract shows that its wall is composed of four distinct layers (Fig. 37.12). The **mucosa** lines the interior of the tube. It consists of a layer of epithelial cells, some of which release mucus, digestive enzymes, ions, and water into the tract; a thin layer of connective tissue; and an equally thin layer of muscle.

Beneath the mucosa is the layer of **submucosa,** containing major blood and lymph vessels and a network of nerve fibers.

Next is the **muscularis externa,** a layer of muscle fibers oriented in two directions. Most of the muscle fibers are arranged in a circular pattern around the tract such that their contractions squeeze the tube-like lumen, making it thinner. A few of the fibers are oriented at right angles to the circular fibers; their contractions shorten the tube.

Contractions of these two types of muscle squeeze food through the tract.

Finally, the gastrointestinal tract is surrounded by an outermost layer of connective tissue known as the **serosa.** This pattern varies a bit from one region of the tract to another, but the same four tissue layers are found throughout the system.

Over the next few pages we will trace the pathway followed by food as it moves through the digestive system, and we will examine the ways in which each major organ of the system contributes to the overall absorption of nutrients.

The Oral Cavity

Food processing begins in the **mouth (oral cavity).** The lips are covered with thin, almost transparent skin that allows the color of capillaries to show through. This gives them their reddish color but also makes them susceptible to moisture loss during dry weather—one of the reasons why the lips may become dry and cracked. Food is not absorbed in the mouth, so the inner lining of cells in the oral cavity is not specialized for absorption. The upper portion of the mouth is the **palate.** The palate is hard and bony near the front of the cavity, but the **soft palate** near the rear is a flexible covering that is important in separating breathing and swallowing activities.

Teeth are anchored in special sockets in the bones of the jaw and are connected to the underlying bone by a network of blood vessels and nerve fibers that enter through the root. Teeth are living tissue despite the tough coating of mineralized enamel on their surfaces.

Several basic types of teeth can be distinguished: *incisors,* whose sharp edges are capable of cutting directly through meat; *cuspids* and *bicuspids,* which have one or two points (cusps) to grasp and tear food; and *molars,* whose large flat surfaces make them ideal for grinding food into fine particles. Human tooth structure is midway between that of an herbivore (in which molars predominate) and a carnivore (in which incisors and cuspids are most common), reflecting the mixed human diet of both meat and vegetable material (Fig. 37.13).

The oral cavity contains three pairs of large **salivary glands** that produce *saliva,* a fluid that initiates the process of chemical digestion of carbohydrates. The release of saliva is under the control of the nervous system, and often just the scent of food increases the production of saliva (this is the biological basis of the expression "mouth watering"). An adult produces from 1 to 2 liters of saliva in a day, and this fluid serves a critical function in beginning the digestive process.

Buffering chemicals, including *bicarbonate ions,* in saliva maintain the pH of the oral cavity near neutral,

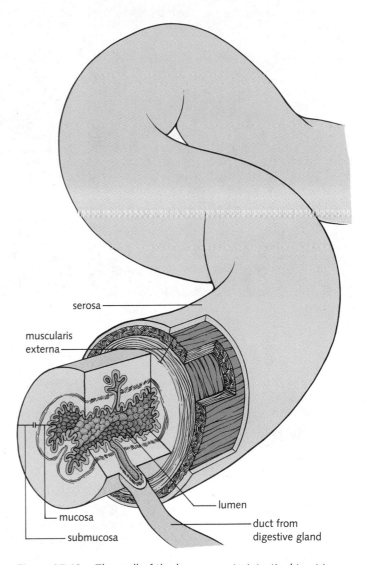

Figure 37.12   The wall of the human gastrointestinal tract is composed of four layers: an outermost *serosa;* a *muscularis externa,* a layer of smooth muscle fibers; a *submucosa,* which contains blood and lymph vessels; and finally an inner *mucosa.*

Figure 37.13 The teeth of an animal reflect its diet. LEFT: The flattened surfaces of an herbivore's teeth are used for grinding plant material. RIGHT: The sharp surfaces of a carnivore's teeth are useful in cutting flesh.

even when acidic food is eaten. The enzyme *salivary amylase* breaks the bonds in starch, releasing disaccharides. *Mucin,* a glycoprotein, gives saliva its somewhat slippery character. Mucin secretions help food slide easily through the smaller openings of the digestive system as the food is swallowed.

The **tongue** is a large muscle that covers the floor of the mouth. *Taste buds,* clusters of sensory cells that react to the chemical composition of food, are found on the surface of the tongue. There are two distinct sets of muscle fibers within the tongue, giving it tremendous flexibility in movement. This allows the tongue to be used for manipulating food and swallowing, as well as for such nondigestive functions as talking, singing, and whistling.

During chewing, the tongue mixes food with saliva and gradually forms it into a ball called a **bolus,** which is swallowed via the combined action of the back of the tongue and the upper portion of the throat, the **pharynx.**

Swallowing

The pharynx (generally called the throat) serves as the passageway for both food and air. Separating the functions of breathing and swallowing is a critical matter, because the entry of just a small amount of food or water into the breathing passages can cause a blockage with serious consequences. (See Theory in Action, "Saving a Life," in Chapter 36.)

The channeling of food and air to the proper passageway is accomplished by the combined actions of several sets of muscles in the lower portion of the pharynx. When a bolus of food is swallowed (Fig. 37.14), the soft palate is forced upward, temporarily sealing off the nasal passageway from the pharynx (that's why it is nearly impossible to swallow and breathe at the same time). Nerves in the pharynx sense contact with the bolus and stimulate the **glottis,** the opening between the vocal cords, to close, helping to seal off the air passageway. The movement of the bolus itself down the throat pushes a small flap of tissue, the **epiglottis,** over the glottis to protect it further from the entry of food or water. As the bolus is swallowed, a circular band of muscle, the **upper esophageal sphincter,** relaxes to allow food to enter the esophagus, the tube that leads to the stomach.

The Esophagus

Food moves through the **esophagus** into the stomach. Because we humans eat in an upright position, one might assume that gravity moves food down the esophagus into the stomach. However, that is not the case. Humans can swallow effectively while lying down, while standing on their heads, and even in the weightless environment of space. Grazing animals such as horses generally eat in positions that require their food to move through the esophagus *upward,* against the force of gravity.

Food is forced through the esophagus by a series of muscular contractions, known as **peristalsis,** in the wall of the tube (Fig. 37.15). The circular muscles surrounding the esophagus contract just behind the bolus while the muscles just ahead of it relax. This produces a wave of muscular contractions that forces food into the stomach

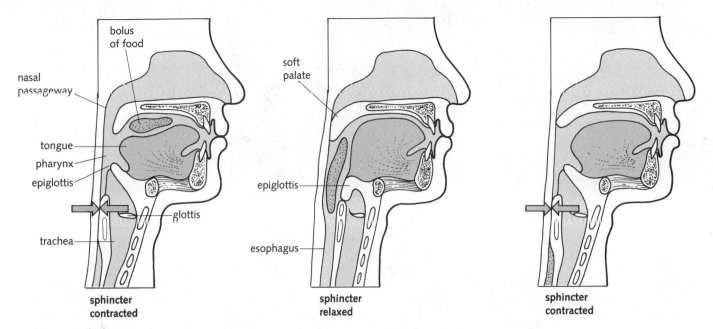

Figure 37.14 Swallowing is a coordinated muscular reaction that traps food into a mass called a *bolus* and then forces it towards the back of the throat past the soft palate. As it passes the palate, a muscle relaxation opens the upper esophageal sphincter, allowing food to pass through and enter the esophagus. Once the bolus has passed, the sphincter closes again.

regardless of the direction of gravity. In humans a typical peristaltic wave takes about 8 seconds to reach the stomach. Peristalsis then continues down the wall of the stomach, mixing the food with digestive fluids and moving it forward.

The wall of muscle surrounding the esophagus is particularly thick at the point where it enters the stomach, a region known as the **cardiac sphincter.** This sphincter maintains a seal between the stomach and the esophagus, although the seal is not perfect. Excess pressure on the abdomen or in the stomach can force fluid from the stomach up past the sphincter into the esophagus, and the acids in stomach fluid cause a painful sensation known as *heartburn* when they enter the esophagus. Eating too much food creates such pressure and is a major cause of heartburn.

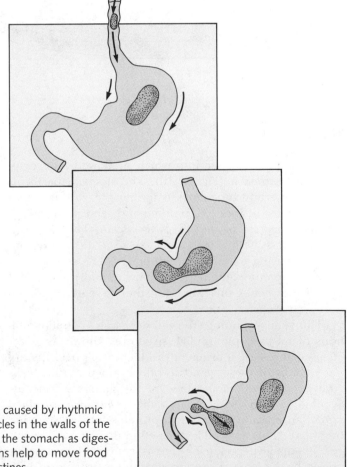

Figure 37.15 Peristaltic waves, caused by rhythmic contractions of the smooth muscles in the walls of the stomach, force material through the stomach as digestion proceeds. Similar contractions help to move food through the esophagus and intestines.

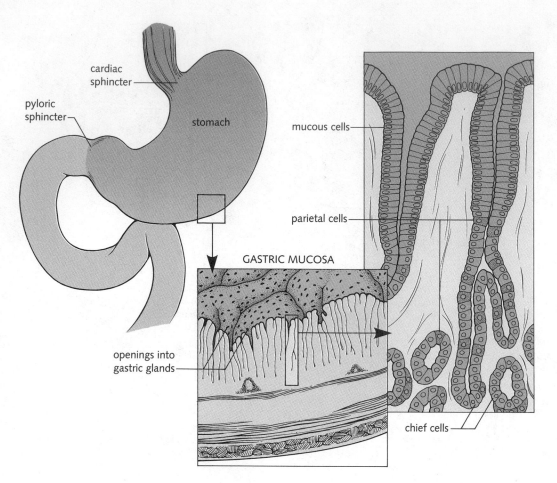

cardiac
sphincter

pyloric
sphincter

stomach

mucous cells

parietal cells

GASTRIC MUCOSA

openings into
gastric glands

chief cells

Figure 37.16 This detailed drawing of the lining of the stomach illustrates the mucous cells, which secrete mucus to protect the stomach lining, and the gastric glands, groups of cells that secrete a variety of digestive enzymes. The parietal cells secrete hydrochloric acid, producing an acidic environment necessary for the chemical digestion of many foods. The chief cells release pepsinogen, which helps to break down proteins.

The Stomach

The large muscular sac into which swallowed food empties is the **stomach** (Fig. 37.16). The stomach performs several important functions: Its size enables it to take in a large amount of food at one time and send it, bit by bit, through the rest of the digestive system. This, in turn, makes it possible for us to eat a few large meals a day instead of having to nibble continuously. The stomach releases a potent combination of hydrochloric acid and digestive enzymes that work together to start the digestion of proteins.

The stomach lining, the *gastric mucosa,* contains millions of microscopic, pit-like structures known as *gastric glands* that release four major products. Cells near the upper regions of these glands produce mucus. Other cells (the *parietal cells*) at the base of the glands secrete hydrochloric acid, while cells called *chief cells* produce *pepsin,* an enzyme that cleaves proteins into polypeptide fragments. This enzyme is released in the form of *pepsinogen,* an inactive precursor that is converted into active pepsin by the low pH produced by the hydrochloric acid. (Pepsin also works best in an acidic environment.) The

lower portion of the stomach glands contains cells that secrete *gastrin,* a hormone that enters the circulation and then stimulates the cells lining the stomach to secrete still more digestive enzymes and HCl.

Gradually, the combined action of these secretions and the muscular contractions of the stomach wall, which are sometimes violent enough to be felt (and even heard), break the food down into smaller and smaller particles. As digestion in the stomach continues, the food becomes a nutrient-rich, milky slurry known as **chyme.** Chyme is forced toward the bottom of the stomach, where the pyloric sphincter controls the rate at which it passes from the stomach into the duodenum.

Ulcers Why don't the powerful enzymes and acids released into the stomach digest the stomach wall? The answer is that sometimes they do! In most cases, a thick layer of mucus protects the stomach lining against digestion by its own enzymes and acids. However, the stomach wall may become damaged, and tissue may be attacked by the digestive enzymes. As the tissue breaks down, cellular damage produces a wound in the wall of the stomach or intestine known as a **peptic ulcer.** The damage may become so severe that bleeding into the digestive system results, and extensive bleeding can become life threatening. Foods that stimulate the production of digestive fluids, including the caffeine found in coffee and tea, can aggravate the damage caused by a peptic ulcer and should be avoided to help the tissue heal.

The Duodenum

As the stomach fills with chyme, it gradually releases small amounts of material through the **pyloric sphincter** into the **duodenum,** the upper portion of the small intestine. The pressure of chyme against the sphincter causes it to open and close in a rhythmic fashion, timed to coincide with waves of peristaltic contractions across the stomach. Gradually, over 3 to 6 hours, the entire contents of the stomach empty into the small intestine, in portions of about 50 milliliters (just under 4 tablespoons) at a time.

Most chemical work of the digestive process is done in the duodenum, and the breakdown of food molecules in this portion of the small intestine is the final preparation for absorption, which takes place in the remainder of the intestine.

Two major glands, the liver and the pancreas, produce secretions essential to this process (Fig. 37.17). These secretions are added to the chyme through an opening at the *duodenal papilla.* Hormones triggered by the presence of food in the stomach help ensure that secretions from the liver and pancreas are released only when food passes into the duodenum.

Pancreatic secretions The **pancreas** is a double gland. As we saw in Chapter 34, the *endocrine* portion of the pancreas produces several hormones, two of which are glucagon and insulin. Each helps regulate the level of sugar in the blood. The *exocrine* pancreas, which makes up 99 percent of the pancreas's mass, produces a range of powerful digestive enzymes that break down proteins, fats, and carbohydrates.

Most of these enzymes are released from the cells of the pancreas in inactive forms. This prevents them from digesting each other or the tissues of the pancreas itself. They are activated by the environment of the intestine, and they rapidly begin to break down food molecules in the duodenum.

Other pancreatic cells produce *sodium bicarbonate,* which makes the secretions of the pancreas strongly alkaline (high pH). The release of bicarbonate neutralizes the strongly acidic chyme from the stomach. This pH change inactivates pepsin and activates the enzymes from the pancreas and the duodenum.

The release of these materials from the pancreas is timed to coincide with the movement of food into the duodenum, enabling the pancreas to store its digestive

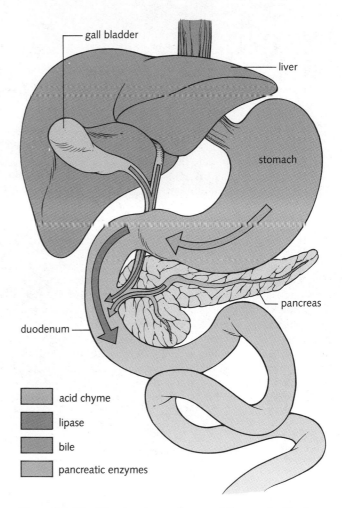

acid chyme

lipase

bile

pancreatic enzymes

Figure 37.17 The duodenum is one of the most critical regions of the digestive system. Secretions from the liver and pancreas enter the digestive system in the duodenum, where they mix with acidic chyme from the stomach.

Alcohol: Metabolism and Illness

The liver is capable of detoxifying a wide range of drugs and poisons, and one of these is ethyl alcohol. Alcohol has been a part of human culture for thousands of years. Mesopotamian pottery dating to 4200 B.C. shows details of the fermentation process. The technique of distilling a beverage to increase its alcoholic strength has been known for nearly 2000 years. Alcoholic drinks are a part of popular culture throughout the world. Alcohol should be recognized as a powerful and addictive drug. At least 12 million Americans are active abusers of alcohol. Alcohol contributes directly to high rates of suicide, murder, assault, abuse, violence, traffic accidents, and mental illness. As if these difficulties were not enough, al-cohol has long-term destructive effects on the body itself.

Alcohol is absorbed rapidly across the lining of the stomach, which is one reason why drinking on an empty stomach has an immediate effect. When a drink is mixed with food, the alcohol concentration in the stomach is lower and absorption into the bloodstream is less rapid. Alcohol's most immediate effects are on the nervous system. It lowers brain activity, dulls reflexes, and interferes with sensory input. High doses of alcohol can be fatal.

Alcohol is broken down to acetaldehyde by enzymes in liver cells. Ultimately, acetaldehyde can be converted to a compound that enters the Krebs cycle (Chapter 18). In this limited way, alcohol can supply energy like other foods. Acetaldehyde, however, is a toxic chemical that can damage mitochondria and upset the metabolism of liver cells. Long-term heavy drinking causes repeated damage to liver cells. The liver enlarges with swollen cells and fat globules. Patches of dead and dying cells appear throughout the organ, scar tissue builds up, and blood flow through the liver is impaired—a condition known as *cirrhosis*.

The effects of cirrhosis extend throughout the body. They include internal bleeding, abdominal swelling, and sexual disorders. There is no cure for cirrhosis, and the only way to arrest the ongoing damage is to stop drinking alcohol.

---

enzymes until they are needed. How is this timing accomplished? The release of acidic chyme from the stomach to the duodenum stimulates the duodenum to release hormones. These hormones, in turn, cause the pancreas to dispense its mixture of bicarbonate and digestive enzymes. The neutralization of stomach acid that occurs when pancreatic fluid enters the duodenum inhibits further release of duodenal hormones, making the whole process self-regulating.

Liver secretions   Unlike the pancreas, the liver does not produce digestive enzymes. It does, however, produce a fluid that is absolutely vital to the successful digestion of fatty foods. This fluid is known as **bile.** Bile is a mixture of salts and lipids (including *cholesterol*) that aids in the digestion of fats. The liver also secretes *bilirubin,* a breakdown product of hemoglobin, into the bile to be eliminated from the body.

The best way to explain the function of bile is to compare it with a soap or detergent. Fats from foods are in the form of thick globules by the time they reach the small intestine. Like droplets of grease, these globules do not dissolve in water, so the fat-digesting *lipase* enzymes produced by the pancreas are unable to reach many of the fat molecules.

The bile salts and phospholipids in bile, however, contain both *hydrophilic* (water-soluble) and *hydrophobic* (water-insoluble) parts (see Chapter 15), just as detergents do. The hydrophobic portions of the bile compounds interact with fat molecules, while the hydrophilic portions interact with water. This enables bile compounds to break fat droplets into smaller and smaller pieces, a process known as *emulsification*.

Gradually, fats are emulsified into tiny lipid droplets known as *micelles* that are subject to direct attack by pancreatic lipase. Without bile to break fats into small droplets, as much as 90 percent of the fat in food would pass through the intestine undigested.

The secretions of bile from the liver accumulate in a sac known as the **gallbladder.** As a meal is digested, bile from the gallbladder passes through the same duct as pancreatic secretions to enter the duodenum. The gallbladder is not a vital organ, for the movement of bile into the digestive system is unimpaired by its absence. Many animals, in fact, have no gallbladder, and in humans it must be removed if it becomes blocked or infected.

Remarkably, the total content of bile salts in the body at any one time is only about 50 percent of the amount needed to emulsify the fats in a large meal. The digestive system copes with this undersupply of bile salts by rapidly *recycling* the bile salts it does have. As bile is used up to help emulsify fats, the bile salts are rapidly absorbed in the small intestine, where fat itself is also absorbed and returned to the blood. The liver retrieves the bile salts and cholesterol from the blood and resecretes them as new bile, which is added to the digestive tract to emulsify the next batch of fatty food moving into the duodenum. This rapid recycling makes a limited resource go a long way.

The Small Intestine

By the time food passes through the duodenum into the rest of the small intestine, nearly all the important *chemical* work of digestion has been completed. What remains is the task of absorbing the nutrient molecules now available within the small intestine.

The rate at which material can be absorbed across a barrier (such as the wall of the small intestine) depends on the surface area of that barrier. All other things being equal, the larger the surface area, the greater the rate of absorption. For this reason, the inner surface of the small intestine is wrinkled and folded, and these folds are covered with finger-like projections called villi (Fig. 37.18). Although these foldings produce a tremendous increase in surface area, the actual surface is even larger. Epithelial cells lining the intestine are covered with **microvilli—** tiny projections of the cell membrane that dramatically increase its surface area.

Nutrient molecules pass into these cells from the intestinal tract via a number of mechanisms. Free fatty acids, simple lipids, and a few simple sugars (such as fructose) diffuse across the epithelial cell membranes in response to a concentration gradient: their high concentration within the intestine drives them by passive diffusion into the epithelial cells. Other molecules, including glucose, nucleic acids, amino acids, and even vitamins, are taken into the epithelial cells by active transport. As a result of osmotic pressure, a great deal of water follows the movement of nutrients into cells lining the small intestine.

Nutrient molecules don't remain long within the epithelial cell but are rapidly transported from the intestine to the bloodstream. Each villus of the small intestine has a fine network of capillaries that lies just below a single layer of epithelial cells. The close contact makes it possible for nutrient molecules to move into the bloodstream within a matter of minutes after the first food arrives.

Fats are transported by a different mechanism. Although most fat is taken into epithelial cells in the form of fatty acids, very few fatty acids leave those cells. Instead, epithelial cells convert fatty acids into triglycerides that form lipid droplets within the endoplasmic reticulum. These droplets, known as **chylomicrons,** are released from the inner surfaces of the epithelial cells.

Chylomicrons are prevented from entering the capillaries by a thick *basement membrane* of polysaccharide molecules. However, there is no basement membrane surrounding the *lacteals* (extensions of the lymphatic system), and chylomicrons pass into them rapidly. This allows the lymphatic system to absorb nearly all of the fat molecules taken up in digestion. After transport by the lymphatic system, fat molecules eventually enter the venous circulation.

Material is mixed in the small intestine by rhythmic contractions of the rings of smooth muscle that surround the organ. It is interesting to contrast these contractions, which are known as **segmentation movements,** to the peristaltic waves that move along the digestive tract, propelling its contents. Segmentation contractions squeeze the food and move it *back and forth,* which helps to ensure that all the contents of the intestine eventually come into contact with the absorptive wall.

The capillaries that absorb nutrients in the villi of the small intestine do not lead to the general circulation. Instead, they converge into larger and larger vessels that drain into the hepatic portal vein, leading to the liver (Fig. 37.19).

The Liver and the Hepatic Portal Circulation

Functionally, the liver is one of the most complex organs in the body. Its functions include the detoxification of drugs and alcohol, the formation of urea from nitrogen-containing wastes, the synthesis of several important blood plasma proteins, and, as we have seen in this chapter, the emulsification of fats.

The importance of the hepatic portal circulation that delivers blood to the liver cannot be overemphasized. If the circulatory system in the small intestine emptied directly into the general circulation, the levels of sugars, amino acids, and nucleic acids in the blood would rise dramatically every time food was consumed and would fall precipitously a short time thereafter. A quick snack or candy bar would cause, among other things, a rush of sugar into the circulation. This might not seem like much of a problem, but many of the cells of the body, including those of the nervous system, are extremely sensitive to the level of sugar in the blood.

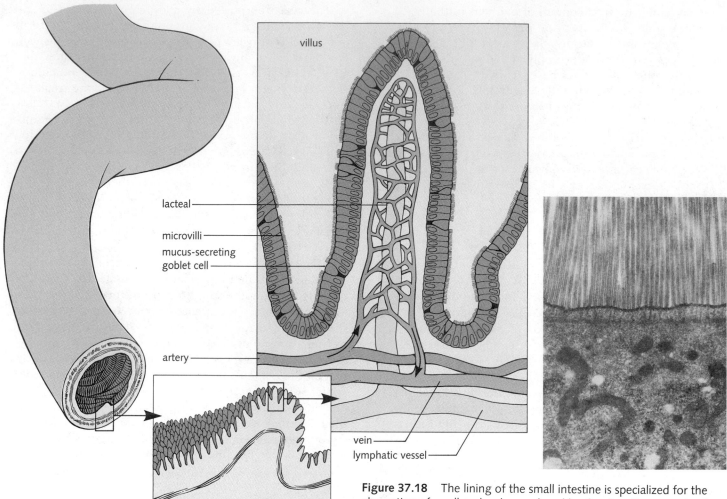

Figure 37.18 The lining of the small intestine is specialized for the absorption of small molecules produced by the digestive process. The blood and lymph vessels enable these absorbed molecules to be rapidly transported to the rest of the body. RIGHT: The electron micrograph shows one of the main reasons why the cells lining the small intestine absorb material so efficiently. Each cell contains thousands of microvilli, which greatly increase the surface area of the cell membrane.

The primary organ that regulates blood sugar is the liver. During the digestion of a large meal, the liver removes much of the sugar that comes to it via the hepatic portal vein. This ensures that the concentration of sugar in the blood does not rise rapidly. Then, when the meal is digested and the absorption of sugar from the small intestine has ceased, the liver gradually releases its store of carbohydrate to keep blood sugar levels from falling.

In this way, the liver acts as a sort of gatekeeper between the digestive system and the circulatory system, allowing only an appropriate amount of sugar and other nutrients to enter the general circulation and doling out its reserves of those compounds when the rest of the body needs them.

The Large Intestine

The large intestine, or **colon,** is about 1.3 m long and consists of three main parts: the *ascending colon* lies along the right side of the body, the *transverse colon* crosses the top of the abdomen from right to left, and the *descending colon* goes down the left side of the body.

The small intestine dispenses its chyme into the colon in an area known as the *cecum* (Fig. 37.20), where a small finger-like pouch, the *vermiform appendix*, is found. In rodents and some primates, the cecum is a large organ in which bacterial action breaks down cellulose, releasing usable carbohydrates and vitamins. In humans, however, the appendix is not an important organ and generally

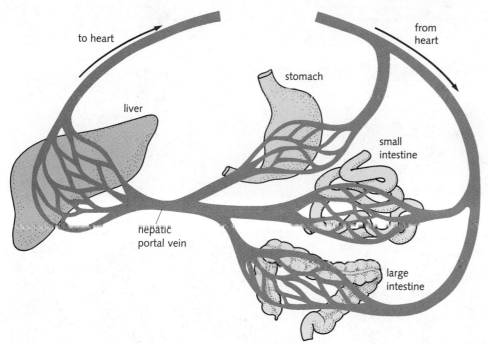

to heart

from heart

liver

stomach

small
intestine

hepatic
portal vein

large
intestine

Figure 37.19 Veins leaving the digestive system do not pass directly back into the general circulation. Instead, they collect into the hepatic portal vein, ensuring that newly absorbed food passes through the liver before it enters the general circulation. The *hepatic portal system,* as this circulation pattern is known, allows the liver to control the entry of food molecules into the bloodstream.

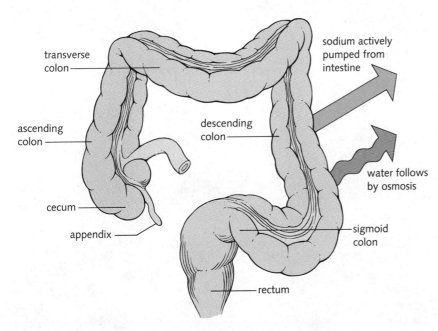

transverse
colon

sodium actively
pumped from
intestine

ascending
colon

descending
colon

cecum

appendix

water follows
by osmosis

sigmoid
colon

rectum

Figure 37.20 The large intestine is the final absorptive organ of the digestive system. As material passes through it, sodium is actively pumped out of the digestive tract into the circulatory system. Osmosis causes water to flow with the sodium, conserving water and producing a concentrated feces.

merits our attention only when it has become infected, a condition known as *appendicitis,* and needs to be surgically removed.

The 500 milliliters of chyme that enter the colon on average every day contain indigestible material that has neither been digested nor absorbed, large amounts of intestinal bacteria, and a substantial amount of water.

Throughout the large intestine, cells lining the tract actively transport sodium from the chyme into the blood. As sodium ions move in this direction, osmotic forces cause most of the remaining water to follow. Gradually, nearly all of the remaining water is removed, and what remains are dense *feces* concentrated in the last few centimeters of the tract. The descending portion of the colon enters the

Table 37.4 *Functions and Secretions of Digestive and Accessory Organs*

Organ	Function	Secretion
Digestive Tube		
Mouth	Chops and grinds food	
Pharynx	Channeling of food and air to proper passageway	
Esophagus	Propels food to stomach	
Stomach	Stores and churns food Releases HCL and enzymes to initiate proteolytic breakdown	HCl, activates enzymes, breaks up food, kills germs Gastrin, stimulates HCl secretion Pepsin, cleaves proteins Peptidase cleaves proteins
Small intestine	Completes digestion Absorbs nutrients	Sucrases cleave sugars Amylase cleaves starch and glycogen Lipase cleaves lipids Nuclease cleaves nucleic acids
Large intestine	Reabsorbs water and sodium ions Stores wastes	
Appendix	No known digestive function Contains cells of the immune system	
Rectum	Holds waste before expulsion	
Anus	Opening for waste elimination	
Accessory Organs		
Tongue	Mixes food with saliva	
Salivary glands	Moisten food	Salivary amylase cleaves starch Mucin makes saliva slippery
Liver	Synthesis of blood plasma proteins Regulates blood sugar Formation of urea Secretes bile Detoxifies drugs and alcohol	Bile aids in lipid digestion
Gallbladder	Stores bile	
Pancreas	Adds digestive enzymes Regulates blood glucose levels Neutralizes stomach acid	Bicarbonate neutralizes stomach acid Trypsin and chymotrypsin cleave proteins Carboxypeptidase cleaves peptides Amylase cleaves starch and glycogen Lipase cleaves lipids Nucleases cleave nucleic acids Glucagon and insulin regulate level of blood sugar

rectum, a muscular passageway wherein the feces may be stored for a time before they are eliminated by defecation.

When water is not properly removed from the large intestine, the result is *diarrhea,* the elimination of a large volume of watery feces, which is unpleasant at best and life threatening at worst. Many forms of diarrhea result from infections that impair the ability of the large intestine to pump salts out of the intestinal tract. The result is that salts and water are both lost from the body in large amounts. If left unchecked, this can lead to dehydration and to a loss of salts so severe that they literally cannot be replaced. In many areas of the world, diarrhea is the leading cause of death among infants and young children.

Timing

In digestion, like politics, timing is everything. How does the digestive system coordinate the release of its secretions so that they are produced only when they are needed? Some responses of the digestive system to food are obvious, such as the way in which your mouth waters with saliva when the scent of your favorite food wafts by. The sight or scent of food can also trigger the release of hydrochloric acid and pepsinogen in the stomach, causing the stomach to "churn" even when it is empty.

As you recall, when food enters the stomach, receptors in the stomach wall react to the pressure by secreting the hormone *gastrin,* which causes the cells lining the stomach to secrete more acid (Table 37.4). The release of gastrin helps to ensure that just the right amount of acid is released. A large meal causes the secretion of more gastrin, which stimulates the release of more acid. Just as the stomach reacts to a volume of food, the duodenum reacts to the presence of chyme by releasing the hormones *secretin, cholecystokinin,* and *enterogastrone.* Secretin causes the pancreas to secrete bicarbonate (which neutralizes the acidic chyme), and cholecystokinin stimulates the gallbladder to contract and release bile. Enterogastrone travels through the bloodstream to the stomach where it inhibits muscular activity in the stomach, ensuring that chyme is not forced too quickly into the small intestine. Working together, these hormones ensure that the organs of the digestive system release the proper sequences of secretions to break down even the most complex lunch into a series of sugars, amino acids, and lipids that can be absorbed in the small intestine. In addition, one of these hormones, cholecystokinin, is also a *neural transmitter substance* that acts on the hypothalamus to reduce appetite. This is an important signal that eating should stop (Fig. 37.21).

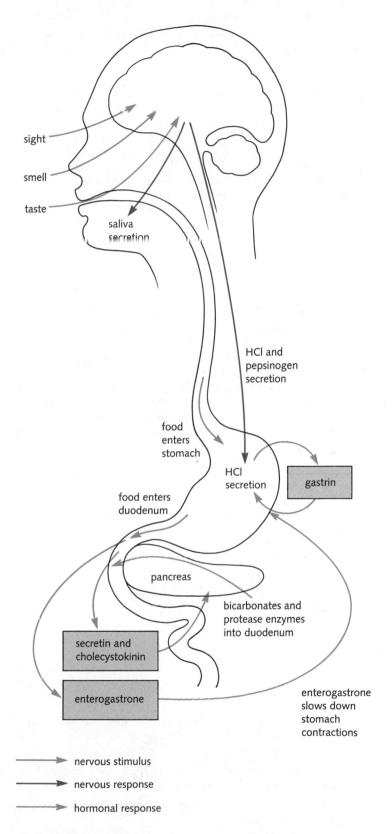

Figure 37.21 The digestive process is controlled by the senses, signals from the brain, and hormones secreted by the stomach and intestines.

Animals are heterotrophs that eat a varied diet to obtain the nutrients they need to survive. On a cellular level, food molecules are needed to provide a source of cellular energy and the building blocks from which new molecules can be synthesized as needed for growth and repair. Food for energy can be derived from carbohydrate, fat, or protein. However, food for growth and repair requires a balanced diet that includes appropriate amounts of carbohydrates, fats, and proteins; small amounts of essential organic molecules known as vitamins; and a number of essential minerals.

Digestive systems are specialized for the intake, processing, and absorption of food. Simpler animals have simple digestive systems, beginning with those organisms in which digestion is a cellular process. More complex organisms have gastrovascular digestive systems, in which food enters and waste leaves by the same opening. Higher animals, including vertebrates, have digestive systems built around alimentary canals, which have two openings.

Food enters the human digestive system through the mouth, where it is chopped and ground by the teeth, and is then swallowed into the esophagus, where it is forced into the stomach by means of peristaltic waves of muscle contraction. The stomach releases digestive enzymes in an extremely acidic (low-pH) environment. As food enters the small intestine, it is joined by secretions from the liver and the pancreas, and these digestive juices complete the final chemical breakdown of material. Within the small intestine, epithelial cells transport nutrients into the circulatory system and the lymphatic system. The large intestine removes sodium ions and water from the contents of the gastrointestinal tract, producing dense feces containing very little water.

STUDY FOCUS

After studying this chapter, you should be able to:

- Describe the basic organs and tissues of the digestive system.

- Appreciate how food is broken down into small molecules that can be used for cellular activities.

- Explain the interactions between the digestive system and other systems of the body, including the nervous system and the circulatory system.

- Describe the basic requirements of the human diet, and explain how to establish a sound program of nutrition.

- Understand the medical dangers of high cholesterol levels in the bloodstream.

TERMS AND CONCEPTS

calorie 729	mucosa 744
basal metabolic rate 729	palate 745
undernourishment 730	pharynx 746
malnourishment 730	esophagus 746
vitamins 734	peristalsis 746
essential mineral	chyme 748
nutrients 734	duodenum 749
gastrovascular cavity 743	pancreas 749
alimentary canal 743	gallbladder 750
mouth 743	chylomicrons 751
anus 743	

REVIEW

Objective Questions (Answers in Appendix)

1. If deprived of nutrients, an organism will suffer first from a lack of
   (a) water.
   (b) carbohydrates.
   (c) proteins.
   (d) fats.

2. An immediate source of food energy for the body is
   (a) protein.
   (b) fat.
   (c) lipid.
   (d) carbohydrate.

3. Which figure is closest to an appropriate daily caloric intake for a sedentary adult?
   (a) 500 kcal
   (b) 1000 kcal
   (c) 1500 kcal
   (d) 2000 kcal

4. Peristalsis is the movement of food through the digestive tract by
   (a) the pressure of water which enters the digestive tract by diffusion.
   (b) the contraction and relaxation of the esophageal sphincter.
   (c) secretions of mucus.
   (d) wave-like contractions of the smooth muscle.

5. Gastric glands are responsible for the production of
   (a) trypsin.
   (b) hydrochloric acid.
   (c) pepsinogen.
   (d) b and c only.

Discussion Questions

6. Outline the basic nutritional requirements for humans. How many general categories of nutrients are in the minimal list?

7. Drinking a large amount of liquid after a heavy meal can often cause "heartburn," a burning sensation in the lower part of the esophagus. From what you know about the functioning of the stomach and the causes of heartburn, explain why.

8. Explain the concept of "complete" proteins and how it is relevant to individuals who choose a vegetarian diet.

9. What mechanisms operate to protect the lining of the stomach from digesting itself? What happens when these mechanisms fail?

10. When an individual goes for an abnormally long period of time without eliminating feces, the individual is said to suffer from constipation. One of the common remedies for constipation is a laxative solution containing magnesium salts. Magnesium salts move into the large intestine but are absorbed by the intestinal wall very slowly, leaving large amounts of these salts in the digestive tract. Explain how the presence of these hard-to-absorb salts might affect water movements in the large intestine and how this might contribute to the effect of the laxative.

11. Emergency treatment of individuals suffering from chronic diarrhea includes the intravenous administration of saline (water + salt) solution and oral administration of drugs that stimulate salt transport in the intestinal tract. Explain why *both* of these treatments are necessary to save the lives of such individuals.

12. Individuals hiking or mountain climbing in very cold weather generally bring high-fat foods that contain the largest possible number of calories per gram, thereby reducing the weight of their gear. What special difficulties would an individual whose gallbladder had been removed face on such an expedition?

READINGS

Lawn, R. M. "Lipoprotein in heart disease." *Scientific American* 266 (1992): 54–60. Lipoproteins transport cholesterol and are associated with heart disease. This interesting report explores the genetic and molecular story of lipoproteins in the circulatory system and helps to explain why the cholesterol story is not quite as simple as researchers once thought.

Livingston, E. H., and P. H. Guth. "Peptic ulcer disease." *American Scientist* 80 (1992): 592–598. The powerful digestive enzymes released in the stomach are capable of destroying the walls of the digestive system itself, leading to ulcers. This well-illustrated article explores some of the causes and cures of peptic ulcer disease.

Scrimshaw, N. S. "Iron deficiency." *Scientific American* 265 (October 1991): 46–52. For a detailed examination of the biological and social effects of a specific nutrient deficiency, this is one of the very best articles available.

Uvnas-Moberg, K. "The gastrointestinal tract in growth and reproduction." *Scientific American* 261 (July 1989): 78–83. A very clear account of how the digestive system adjusts to changes brought about by pregnancy.

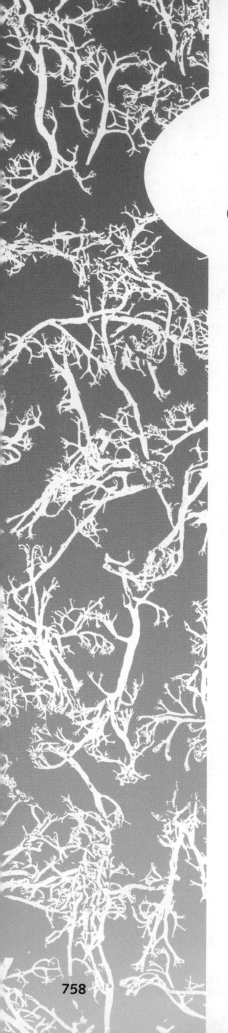

38

Regulation of the Cellular Environment

"Water, water every where,
Nor any drop to drink"

—Samuel Taylor Coleridge,
The Rime of the Ancient Mariner

*t*he mariner in Coleridge's great poem found himself adrift at sea and lacking water to drink. The fact that seawater cannot support human life has been known since prehistoric times, and the mariner knew that he could easily die of thirst in the middle of the ocean. He might even have known that if he *did* drink seawater, he would die of dehydration far sooner than he would if he drank no water at all! Not surprisingly, he found the situation ironic.

With plenty of time on his hands, the mariner could well have wondered about the same questions that occur to us today when we think of other animals that live in and around the ocean. How can seabirds, mammals, and marine fishes thrive on the same saltwater that would cause us to die of thirst? For that matter, how do desert animals—from rattlesnakes to camels—survive in intensely hot places with little water or none at all?

In this chapter we will seek the answers to these and several other questions pertaining to water balance, the regulation of salt, and the control of body temperature. All of these are variations on the great theme of organismal physiology: the need to maintain constant internal conditions in the face of a changing external environment.

WATER, WASTES, AND TEMPERATURE

As you'll recall from several earlier chapters, the physiological systems of multicellular animals have been shaped by the requirements of life at the cellular level. Among those requirements are three whose functions are intertwined: the *excretion* of metabolic wastes, the *control* of salt and water balance within the body, and the *regulation* of body temperature.

It is a chemical fact of life that cellular metabolism produces molecules that must be removed from the body. As cells utilize nutrients, they produce many small molecules that are toxic to the organism and are released as *wastes*. One major waste product, carbon dioxide, is removed by the respiratory system.

But the handling of many other waste molecules is more complex. Furthermore, when animals ingest food they may also consume other substances, ranging from complex toxic compounds synthesized by plants to simple sodium salts. Many of these molecules and ions, if allowed to accumulate in the body, disrupt the delicate balance required to maintain the internal environment within its narrow functional limits.

At the same time, animals need a sufficient supply of water in their bodies. The task of maintaining that supply is doubly complicated. For one thing, the elimination of many waste products requires the simultaneous elimination of at least some water. And water is also required by two other homeostatic functions: *respiration* and the *regulation of body temperature*.

We will begin our exploration of these physiological balancing acts performed by multicellular animals by examining a universal problem of all animals, the elimination of toxic nitrogenous wastes.

Toxic Ammonia

Several important metabolic pathways, such as the breakdown of amino acids, release nitrogen-containing compounds. When the amino group (—NH$_2$) is removed from an amino acid, a process known as *deamination*, ammonia (NH$_3$) is formed. A water molecule consumed in the process provides the extra hydrogen that converts NH$_2$ to NH$_3$ (Fig. 38.1).

For this reason, the complete metabolism of proteins releases ammonia—a potent poison—into the body. Few cells can survive high concentrations of ammonia, so it must be eliminated from living tissues as quickly as possible or converted to less toxic compounds. The elimination of metabolic wastes, such as ammonia, from the body is known as **excretion.**

Dealing with Ammonia: Three Strategies

Animals excrete nitrogenous wastes in one of three ways: as ammonia, as uric acid, or as urea.

Ammonia excretion The simplest strategy, and the one used by most single-celled organisms, is to let ammonia diffuse out into the environment. In fact, because ammonia is quite soluble in water, many aquatic invertebrates simply allow ammonia to diffuse passively out of their tissues into the surrounding water.

Freshwater fishes, whose skin and scales present formidable barriers to ammonia diffusion, eliminate ammonia in two ways. First, they lose ammonia continually across their permeable gill membranes. Second, their kidneys collect ammonia from the bloodstream and release it regularly in thin, very dilute *urine*. Anyone who keeps pet fishes should know that the water in a fish tank quickly becomes loaded with ammonia. In successful aquaria and

Figure 38.1 When amino acids are metabolized, they are *deaminated:* the amino groups are removed, producing ammonia. Because ammonia is toxic at high concentrations, it is essential that it be removed or converted to a less dangerous form.

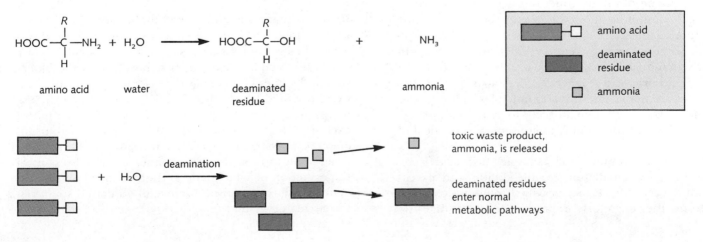

Terrestrial Animal	Nitrogen Waste Excreted
birds insects reptiles	uric acid
mammals some amphibians	urea

Figure 38.2 LEFT: Land animals usually convert ammonia to less toxic compounds, such as uric acid or urea. Urea is highly soluble and can be eliminated as urine. Uric acid is much less soluble, and is eliminated as a thick, sticky paste. ABOVE: The large amounts of uric acid left by these seabirds on coastal rocks have produced *guano* deposits.

in natural ecosystems, beneficial bacteria convert ammonia into far less toxic nitrate. Filling the tank with underwater plants, which absorb ammonia as a source of nitrogen, also helps to minimize ammonia poisoning.

Uric acid   Land animals, on the other hand, do not have the luxury of allowing ammonia to diffuse away. Furthermore, because most terrestrial animals must also conserve body water, they can't afford to flush ammonia away in a constant stream of dilute urine. But because ammonia is so toxic, it cannot be allowed to accumulate and become concentrated anywhere in their bodies.

One solution, used by reptiles, birds, and insects, is to remove ammonia from the circulation in the liver and convert it to **uric acid** (Fig. 38.2). Uric acid is much less toxic than ammonia, and it is also much less soluble. This low solubility, though it might seem to be a disadvantage, actually offers an opportunity to conserve a great deal of water.

In birds and reptiles, for example, uric acid is removed from the circulation by the kidneys and passed through ducts to the **cloaca,** a cavity through which material from the excretory, reproductive, and digestive

systems leaves the body (*cloaca* is the Latin word for "sewer"). The lining of the cloaca absorbs water from this fluid, causing the uric acid to precipitate in the form of white crystals. As more and more uric acid precipitates, it leaves behind more water than the cloaca can resorb. The nearly dry, milky-white paste called **guano** that birds finally excrete is composed primarily of uric acid. The insolubility of uric acid makes these familiar calling cards of birds almost impossible to wash away.

Urea   In mammals and some amphibians, the liver removes ammonia from the bloodstream and produces **urea.** The liver returns urea, which is highly soluble in water, to the blood. Urea is then removed from the bloodstream by the kidneys and concentrated to form an excretory fluid known as **urine.** Although urine can contain quite a high concentration of urea, this method of excreting nitrogenous waste still requires significant amounts of water to keep urea in solution. The kidneys, in addition to removing nitrogenous wastes, also control the excretion of salts, water, and other metabolites, thus maintaining the requisite balance of water and solutes in the circulatory system.

EXCRETORY SYSTEMS

The simplest solutions to the problems of excretion and water balance are found in unicellular organisms that eliminate wastes by simple diffusion. Some multicellular aquatic organisms, particularly those (such as jellyfish) with thin, gelatinous bodies, can also eliminate their nitrogenous wastes by simple diffusion.

As an organism increases in size, however, the volume of its tissues (that produce ammonia) increases more quickly than its surface area (across which diffusion might take place). Thus even among aquatic animals, greater body size usually precludes the elimination of ammonia by simple diffusion alone.

For that reason, most larger organisms have specialized **excretory systems** that control water balance and handle the excretion of nitrogenous wastes. These systems may take a variety of forms. A freshwater organism without a rigid cell wall must deal with a persistent inflow of water due to osmosis. In pure distilled water, the influx of water can be so rapid that a cell may double its volume in only a few minutes. One cellular solution to the problem of excess water is the **contractile vacuole.** The contractile vacuole gets its name from its behavior. Under a microscope we can watch the transparent organelle grow larger and larger, then suddenly contract as it releases its contents from the cell. How does the contractile vacuole help to solve the problem of water balance?

Osmosis causes water to flow into the cell, diluting the concentration of dissolved material in the cytoplasm (Fig. 38.3). The vacuole is surrounded by smaller vesicles that swell with fluid that is *isotonic* to the cytoplasm. Solutes are pumped out of these vesicles (a process that requires energy), producing a fluid that is nearly pure water. These vesicles then dump their contents into the contractile vacuole, causing the vacuole to swell as it fills with water. When the vacuole is swollen with fluid accumulated in this way, a sudden contraction of the cytoplasm around it forces its contents outside the cell, releasing the accumulated fluid. Furthermore, although nitrogenous wastes from such organisms can diffuse directly through the cell membrane, some of the wastes may be eliminated with the water in the contractile vacuole.

Flatworms (platyhelminths) do not have circulatory systems, but they do have an excretory system. A network of *excretory canals* runs throughout the body, opening to the exterior through small pores known as *nephridiopores*. Water and nitrogenous wastes are pumped into the tubules throughout the body, and fluid is moved through the system by ciliated cells found at the ends of bulb-like passages in the system. The constant movement of cilia in these cells gives the impression of a flickering fire, and

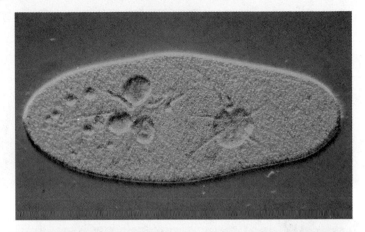

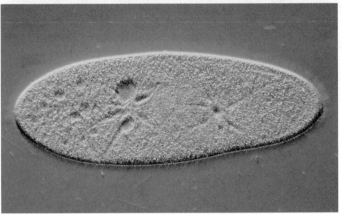

Figure 38.3 The contractile vacuole of *Paramecium* is an organelle that collects water from the cytoplasm (TOP) and expels it to the exterior (BOTTOM).

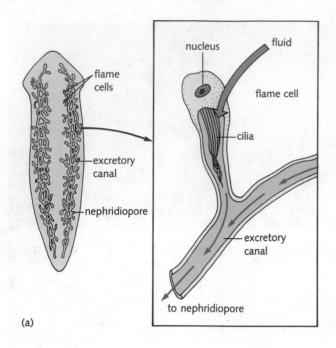

(a)

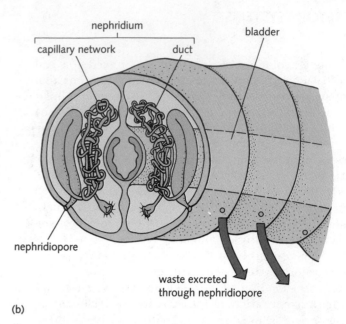

(b)

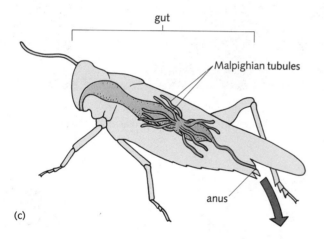

(c)

Figure 38.4 Excretory systems vary widely from one organism to another. **(a)** The flame cells of flatworms propel fluid through a network of excretory canals. **(b)** Earthworms contain pairs of excretory organs in each segment known as nephridia that remove wastes from the bloodstream and the coelomic fluid and pass them to the exterior. **(c)** Insects contain a network of Malpighian tubules that collects wastes and directs them into the digestive system.

they are known as **flame cells** (Fig. 38.4a). The flame cell system is one of the earliest structures that arose in the evolution of an excretory system.

Earthworms have excretory systems that take advantage of their closed circulatory systems by processing metabolic wastes collected by the circulating blood. In each segment of the worm's body, there is a pair of excretory organs called **nephridia** (Fig. 38.4b). Each is composed of a tangle of capillaries that surrounds a duct of tissue leading to the outside of the body. Each *nephridium* collects metabolic wastes and excess water from the coelomic fluid and sends these to a bladder-like reservoir, which is periodically emptied to the exterior through a nephridiopore. By working in concert with the circulatory system, these isolated organs are able to excrete wastes from every segment of the earthworm's body.

Insects, which have open circulatory systems, remove nitrogenous wastes through a system of organs known as **Malpighian tubules.** These tubules, whose mode of operation resembles that of the flatworms' tubular system, thread throughout the posterior half of the body. The movement of fluid through the animal's open circulatory system helps to sweep waste materials towards the Malpighian system. Unlike the excretory canals of the flatworm, however, these tubules do not exit directly to the exterior. The Malpighian tubules empty into the digestive system just before the hindgut (Fig. 38.4c). Water and salts are resorbed by the lining of the gut, leaving very dry feces that pass out through the anus.

Vertebrates eliminate nitrogenous wastes and maintain water balance by means of **kidneys,** paired organs located in the abdominal cavity. Kidneys have a rich blood

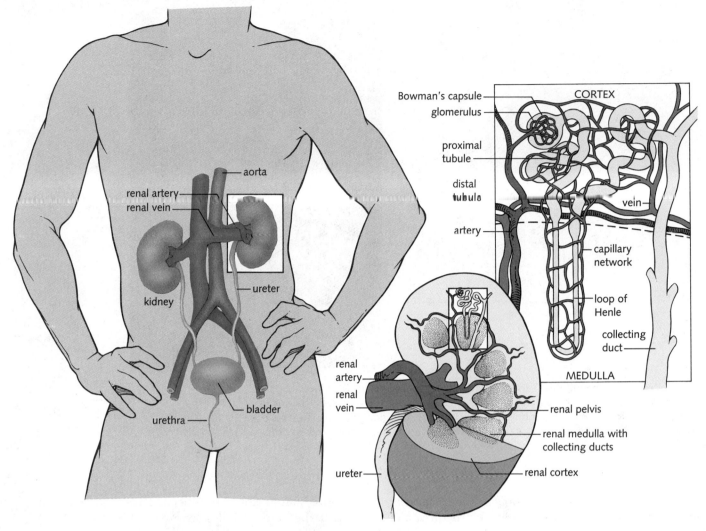

Figure 38.5 The human excretory system. Paired kidneys remove wastes from the circulation and concentrate them in urine that is stored in the bladder prior to elimination. A cross-sectional view shows the internal organization of a kidney, as well as the structure of a nephron, one of the kidney's functional units. Roughly 1 million nephrons are present in a human kidney.

supply, and their functions include the removal of urea or uric acid from the bloodstream. Vertebrate kidneys function in several different ways related to their environments. The need to conserve water has led to a variety of adaptations we will explore later in this chapter.

THE HUMAN EXCRETORY SYSTEM

In the strictest sense, a complete inventory of the human excretory system includes the *lungs,* through which water vapor and carbon dioxide are eliminated during breath-

ing, and the *skin,* which eliminates water and salt during sweating. The lungs and skin do not *regulate* water loss in response to water balance, however, so we will confine our attention here to the excretory and regulatory functions of the kidney.

The Kidney

Human kidneys are paired organs located near the posterior wall of the abdominal cavity (Fig. 38.5). Each adult kidney is about the size of a fist and weighs roughly 170 grams. Each kidney is connected to the circulatory system by a large **renal artery** that brings blood into the organ

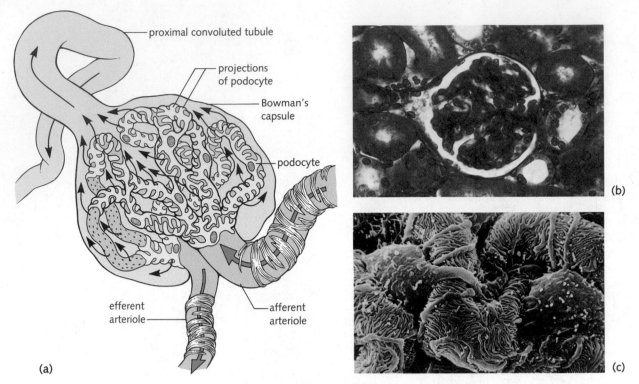

Figure 38.6 **(a)** The glomerulus is the site at which the primary filtrate is produced. Fluid passes from the bloodstream through a filtration barrier between the processes of podocytes, and enters the proximal convoluted tubule. **(b)** The meshwork of glomerular capillaries is enclosed within Bowman's capsule. **(c)** This scanning electron micrograph shows a web of podocytes covering the capillaries of a glomerulus.

and by a **renal vein** that returns blood to the circulatory system (*renal* is an adjective we apply to structures and functions associated with the kidney).

In cross section, the kidney shows two distinct regions: the *renal cortex*, the tissue near the outer edge of the organ, and the *renal medulla*, which lies inside the cortex and borders a cavity known as the *renal pelvis*.

The kidney removes water and waste products from the blood and forms urine. Urine passes through tiny **collecting ducts** that drain into the renal pelvis, which empties into vessels called **ureters.** The ureters lead to the **bladder,** where urine is stored. Urine is eliminated from the body by passing through the **urethra** to the outside.

The Nephron

Each human kidney contains roughly 1 million functional units known as **nephrons.** Figure 38.5 illustrates the basic structure of a typical nephron and its relationship to the flow of blood through the kidney.

The close association of capillaries to the collecting tubes within each nephron might lead you to conclude that the function of the nephron is simple: blood passes through the capillaries, wastes and some water are re-

moved, and urine is produced. That is indeed what happens, but four distinct steps are involved:

1. Formation of a primary filtrate
2. Resorption of the filtrate
3. Concentration of the urine
4. Tubular secretion

Formation of the filtrate Blood entering the nephron first passes through a roughly spherical network of thin capillaries known as a **glomerulus.** The glomerulus is surrounded by a cup-shaped structure known as **Bowman's capsule** (Fig. 38.6a, b).

Here a fluid called the **primary filtrate** is produced through an essentially mechanical filtration process. The walls of glomerular capillaries are exceptionally "leaky." Because the arterial blood entering the glomerulus is under fairly high pressure, a substantial proportion of the blood plasma seeps through the walls of the glomerulus like water through the walls of a leaky garden hose. This filtration process is far from haphazard, of course; cells called *podocytes* ("foot cells") cover the capillary walls with millions of tiny projections (Fig. 38.6c). A filter-like extracellular layer extends between these projections, and the

The Artificial Kidney

The role the kidneys play in purifying the blood is essential to life. Fortunately, we have two kidneys, so if one is affected by injury or disease, a functional kidney remains. The capacity of this organ is so great that one kidney can easily perform all of the filtering and concentrating functions the body requires. If both kidneys are lost or damaged, there are only two ways to keep an individual alive. The best solution is a kidney transplant. A kidney is taken from a healthy donor and carefully implanted in the body of the patient suffering from kidney failure. If the transplanted tissue is a close match to the tissues of the patient, or if medication can control the immune system's natural response to the foreign tissue (called rejection), the kidney may survive as a permanent part of the recipient's body.

In many cases, however, an appropriate donor is not available or surgery is not advisable. Medical technologists have developed artificial kidney machines that purify the blood by simulating the pathway that normal blood takes through the kidney. The accompanying figure shows the basic principles of the artificial kidney. Blood is removed from the body through a tube inserted in the arm and pumped through tubing in a chamber filled with a solution isotonic to blood. The blood moves through special dialysis tubing, which contains tiny pores. These openings allow salts and small molecules to pass through the walls of the tubing. When blood laden with nitrogenous wastes moves through the dialysis machine, these wastes diffuse away into the chamber, and purified blood is returned to the body.

Ingenious as it is, the artificial kidney is not a perfect solution to the problem of kidney failure. It is expensive and the process is time-consuming, occupying several hours a day as often as three times a week. The need to insert needles for the removal of blood is irritating to the skin and may ultimately lead to infection or emotional depression on the part of the patient. The ideal solution when a kidney transplant is not possible would be an internal artificial organ to take the place of the kidney. Work towards developing such an artificial kidney continues.

A kidney dialysis machine allows wastes to diffuse out of the bloodstream across an artificial membrane.

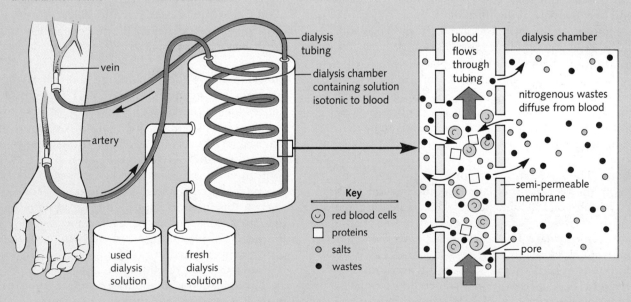

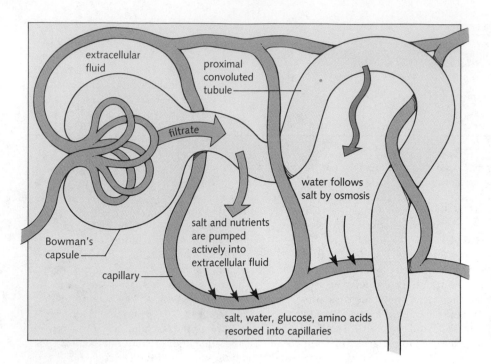

Figure 38.7 Most of the primary filtrate is resorbed in the proximal convoluted tubules. Salt is actively pumped into the extracellular fluid and resorbed into the capillaries. Osmosis causes water to follow the flow of salt, greatly reducing the volume of the filtrate.

Labels in figure: extracellular fluid; proximal convoluted tubule; filtrate; Bowman's capsule; capillary; salt and nutrients are pumped actively into extracellular fluid; water follows salt by osmosis; salt, water, glucose, amino acids resorbed into capillaries

primary filtrate is formed by passage through this layer. This primary filtrate passes into Bowman's capsule, flows through the tubules of the nephron, and eventually is collected, along with fluid from thousands of other nephrons, to form urine.

Blood cells and platelets cannot pass through the tiny openings in the membrane. Neither can large protein or lipoprotein molecules. Smaller molecules (including salts, amino acids, and sugars) do pass through the membrane and become part of the filtrate.

The removal of the primary filtrate from the blood, however, is only part of the story. Blood leaving the glomerulus passes through two other places where dense capillary beds surround the tubule before it returns to the venous circulation. Why should the primary filtrate run through a series of tubes that are so closely associated with capillaries? The answer to this question is the key to understanding the function of the kidney.

Resorption   The rate at which primary filtrate is removed from the blood is very large: as much as 125 milliliters per minute, or 180 liters per day. That's nearly three times the body weight of an average-sized adult. Fortunately, we don't urinate 180 liters per day. Average daily urination is much closer to 1.5 liters—less than 1 percent of the filtrate's original volume. This reduction is accomplished by a process of **resorption** in which approximately 70 percent of the primary filtrate is returned to the blood.

Resorption begins in the first portion of the tubular system leading from the glomerulus, the **proximal convoluted tubule** (*proximal* means "close to" the glomerulus;

convoluted means "twisted"). The walls of the tubule actively transport salt molecules and nutrients from the filtrate within to the extracellular fluid surrounding it. This energy-requiring transport removes most salts from the primary filtrate. When these salt molecules are pumped from the tubule, water follows by osmosis. The salt, nutrients, and water moving into the fluid are then *resorbed* by capillaries that follow the tubule pathway. Urea is not resorbed, however, so its relative concentration within the tubule increases as water is lost. This process alone returns to the blood nearly 65 percent of the water from the primary filtrate (Fig. 38.7).

Concentration   The remaining fluid still has about the same total concentration of liquid and dissolved substances as the blood. If the tubule were now to lead directly to the bladder, the concentration of dissolved material in the urine would be the same as in blood plasma. If urine were always to be released at the same concentration as plasma, the kidney would not be able to get rid of excess water and could not conserve water when it was in short supply. Terrestrial animals, because they must conserve water, have a mechanism built into the structure of the nephron that enables them to produce a concentrated urine.

Concentrating the filtrate might seem like an easy job; all that is needed is a system to remove water from the filtrate. But water is not pumped out of the nephron directly. Instead, in a process resembling the operation of plant root hairs, cells lining the nephron move water from the filtrate by generating an osmotic gradient across the tubule wall. The kidney concentrates urine by actively

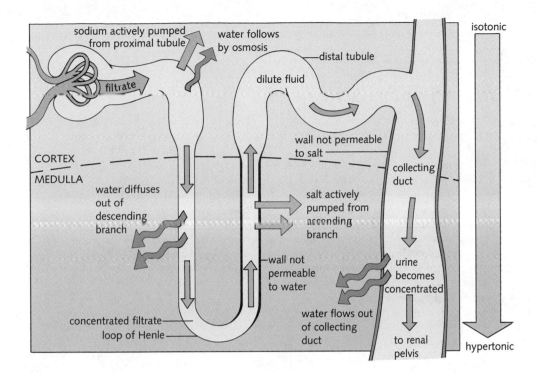

Figure 38.8 Fluid and ion movements occur between the filtrate and its surroundings as the filtrate moves through the nephron. The fact that the wall of the collecting tubule is not permeable to salt allows the force of osmosis to draw water out of the filtrate, forming a more concentrated urine. This countercurrent flow pattern, in which fluids move through different segments of the tubule in opposite directions, is responsible for the kidney's great efficiency.

transporting salt ions and allowing water to follow the salt by osmosis.

But that process alone is not powerful enough to concentrate urine sufficiently. In order for animals to survive in habitats where fresh water is in short supply, their excretory systems must be able to produce urine that is much more concentrated than their blood. To accomplish that feat, the mammalian kidney amplifies the abilities of its membranes to create an osmotic gradient by using a **countercurrent multiplier system.** In order to understand how this system works, you must understand the peculiar configuration of the tubule system.

Down the line from the proximal convoluted tubule, urine passes through a hairpin loop called the **loop of Henle.** Fluid enters the descending portion of this loop and descends from the cortex into the medulla. After making a sharp turn, it then rises through the ascending portion of the loop, back into the cortex, and into the distal convoluted tubule.

The ascending and descending portions of the loop of Henle have different properties with respect to the movement of water and salt (Fig. 38.8). The ascending portion actively pumps salt from the urine into the extracellular fluid around it. Because the wall of this portion of the loop is impermeable to water, water cannot follow the salt here. The net result of this process, therefore, is to increase the salt concentration outside the tubules.

The activity of the ascending portion of the loop then creates a zone of high salt concentration that includes the entire medullary region of the kidney. The flow of water through the loop, combined with continued pumping of salt across the ascending tubule wall, sets up a sharp concentration difference in the extracellular fluid. The saltiest portion is near the bottom of the loop of Henle, and the least salty portion is in the cortex. The strength of this gradient is increased by the geometry of the loop itself. This is a countercurrent multiplier system, in which a reverse flow process increases the effect of ion pumping by one part of the system.

Why should this be of any significance? Look at Fig. 38.8 and you will see that both the descending portion of the loop and the collecting ducts pass through the medulla. As urine flows through the *descending* loop of Henle, the walls of which are permeable to water, water leaves the tubule by osmosis into the surrounding region of high salt concentration. The urine thus decreases in volume and becomes more concentrated.

Because salts are pumped out of the *ascending* portion of the loop, the urine becomes less concentrated but stays at roughly the same volume. You might think that nothing has been accomplished: the fluid is no more concentrated than it was when it entered the system. But the "purpose" of the loop is not to concentrate the urine in the loop itself; it is to establish a concentration gradient in the extracellular fluid surrounding the tubule.

What function does that gradient serve? As dilute fluid leaves the loop, it enters the **distal convoluted tubule** and then flows into a **collecting duct.** The collecting duct flows back down through the extracellular gradient that is set up by the loop. The walls of the duct are

Dealing with Nitrogen: Another Style of Life

The liver is capable of detoxifying a wide range of harmful substances. Animals excrete nitrogenous waste by releasing concentrated wastes in forms such as urine that are easily eliminated from the body. Do plants face the same problem? What happens to the metabolic wastes of a plant?

The fact that plants don't eliminate nitrogen in the same ways as animals is really a reflection of a profound difference in internal metabolism between plants and animals. Animals are able to

synthesize only a few of the amino acids they require; the rest must be obtained through the diet. Therefore, most animals are unable to do anything useful with the ammonia (NH_3) molecules that are released when amino acids are used as food. Most plants, however, do not take up proteins as food. Plants are able to make most of their own amino acids, and the chemical pathways by which they do this *require* nitrogen to complete the synthesis of the amino acid. Therefore, plants have no need

to eliminate nitrogen produced by metabolic pathways. Instead, they recycle much of the nitrogen that they produce, use it to synthesize their own amino acids, and never experience that great need common to so many animals at regular intervals: to empty the bladder.

---

not permeable to salt. They are, however, permeable to water. During this final trip through the medulla, therefore, the urine loses a great deal of water through the tubule walls and becomes more and more concentrated as it approaches the end of the tubule. By the time the fluid leaves the duct and passes toward the kidney pelvis, its concentration of dissolved material may be as much as four times that of blood plasma.

The remarkable ability of the loop of Henle to produce a water-concentrating salt gradient means that a terrestrial animal need devote only a small amount of water to eliminating nitrogenous waste. Nevertheless, a certain amount of water must be used to dissolve urea and other waste products. In humans, this minimal volume is about 400 milliliters of water per day. This means that even under conditions where water intake is limited, this amount of water is lost from the body each day. This is one of the serious problems facing humans and most other mammals in dry environments.

Secretion Besides filtration and resorption, there is a third process at work in kidney tubules. **Secretion** is the active transport of specific molecules from capillaries into the tubules. Unlike filtration, secretion is very specific. Secretion pumps potassium (K^+) and hydrogen (H^+) ions into the filtrate. Kidney tubules also secrete a number of foreign substances, which include chemicals, ammonia, and drugs, into the filtrate. Secretion is the process that "clears" many drugs and poisons from the circulation, enabling them to be excreted in the urine.

Urination Let us now turn our attention briefly to the concentrated urine that leaves the collecting duct. This urine flows into the renal pelvis and is carried by the ureters into the bladder. The bladder is a flexible organ surrounded by smooth muscle. The bladder can comfortably accommodate as much as 800 milliliters of urine, a bit more than two cans of soda or beer. The bladder wall begins to stretch when the bladder accumulates about half this amount (400 milliliters). The extension of smooth muscle causes nerve cells in the wall of the organ to communicate a sensation to the central nervous system, and the individual becomes aware that the bladder is full. The bladder is emptied through the *urethra* when a sphincter of muscle at the boundary between the bladder and the urethra is relaxed.

Control of Water Balance

Conserving water when conditions demand it is only one role of the kidney. As we all know, when water intake is greater than normal, the kidney is able to produce large amounts of dilute urine. This prevents body fluids from becoming diluted when liquid intake is high. How can the same system that concentrates urine when water is in short supply produce a dilute urine when too much water is present?

ADH and the collecting ducts One key to the versatility of the kidney in maintaining water balance is the respon-

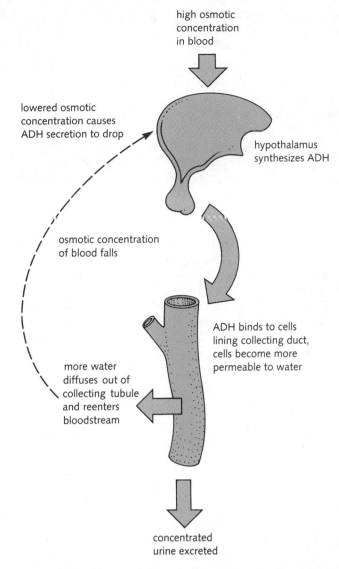

Figure 38.9 The regulation of water balance by the antidiuretic hormone, ADH.

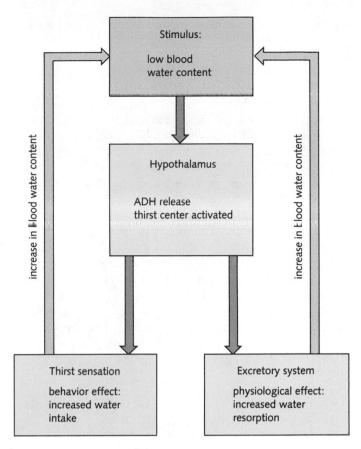

Figure 38.10 Feedback pathways affecting thirst and water loss as they regulate blood water content.

siveness of the collecting ducts to a hormone known as **antidiuretic hormone,** or **ADH.** A diuretic is an agent or drug that causes the urine to be more dilute and increases the rate at which water is removed from the body. An *anti*diuretic hormone, then, acts to concentrate the urine and conserve water.

The system works as follows. ADH, when present, acts on the wall of the collecting ducts to increase their permeability to water. The more permeable the walls of the duct, the more water leaves the tubules to reenter the bloodstream, and the more concentrated the resulting urine becomes. When ADH is not present, the wall of the tubule is much less permeable; more water remains in the tubule, and a dilute urine is produced. Thus the concentration of ADH determines how much water is extracted from the tubule and how much is retained in the body (Fig. 38.9).

The concentration of ADH in the bloodstream is controlled by the *hypothalamus* and the *posterior pituitary,* two extremely important regions of the brain that we will discuss in more detail in Chapter 39. Cells in the hypothalamus monitor the osmotic concentration of the blood. When the concentration of material in the blood is too high (and thus the concentration of water too low), the hypothalamus synthesizes ADH and causes it to be released into the blood from the posterior pituitary. When ADH binds to the cells lining the collecting ducts within the kidney, it stimulates the resorption of water from the urine. When the concentration of dissolved material in the blood is too low (when there is excess water), the release of ADH is inhibited, preventing the retention of water. The system thus responds to changes in the blood's water balance, acting to conserve water when it is in short supply and to remove water when there is too much of it.

Thirst Changes in the water content of blood affect the central nervous system as well as the kidneys. The sensation of *thirst* seems to be produced in the hypothalamus by the same factors that increase the release of ADH (Fig. 38.10). In this way the hypothalamus regulates water balance, controlling the excretory system to reduce water loss and modifying behavior to increase water intake. Experiments with laboratory animals have shown that artificial stimulation of certain brain regions causes an insatiable thirst that leads the animals to drink continually.

Anyone who has ever stood near the restrooms at a baseball park on a hot day knows that there is a relationship between beer and urination. The casual drinker may think that this is due to the amount of liquid he has drunk, but this is only partly true. A heavy drinker feels "hung over" the next morning, and one of the symptoms of the hangover is a powerful thirst. The need to visit the restroom and the hangover thirst are related. Alcohol inhibits the release of ADH, and this increases the rate at which water is lost from the kidneys. Excessive drinking therefore causes a great deal of water to be lost, producing an unusually large amount of urine and dehydrating the body. Consider the case of two fans sitting side-by-side at the ball park, one drinking beer and the other drinking an equal volume of soft drink. The nonalcohol drinker will not only see more of the game (fewer visits to the restroom) but will also feel better in the morning!

Aldosterone Although ADH and drinking behavior both help regulate the concentration of water and salt in the blood, they do not directly control the total volume of blood. A second hormone system acts in concert with ADH to help regulate blood volume. When blood volume is low, blood pressure falls, causing certain cells in the kidney to release an enzyme known as **renin.** Renin converts *angiotensinogen*, an inactive hormone, to its active form, *angiotensin*. Angiotensin has two effects. First, it causes smooth muscle surrounding blood vessels to constrict, increasing blood pressure. Second, it stimulates the **adrenal cortex** to release **aldosterone.** This hormone affects the distal convoluted regions of the tubule and increases the resorption of sodium and the secretion of potassium. Because this action increases the sodium resorption from the distal tubule, the salt concentration of the blood increases. This in turn increases the amount of water retained in the blood in the kidney.

High blood pressure, on the other hand, turns off this system, leading to the excretion of salt and a lowering of blood pressure. Figure 38.11 summarizes the effects of this feedback system.

Figure 38.11 The regulation of blood volume by aldosterone, a hormone produced in the adrenal cortex.

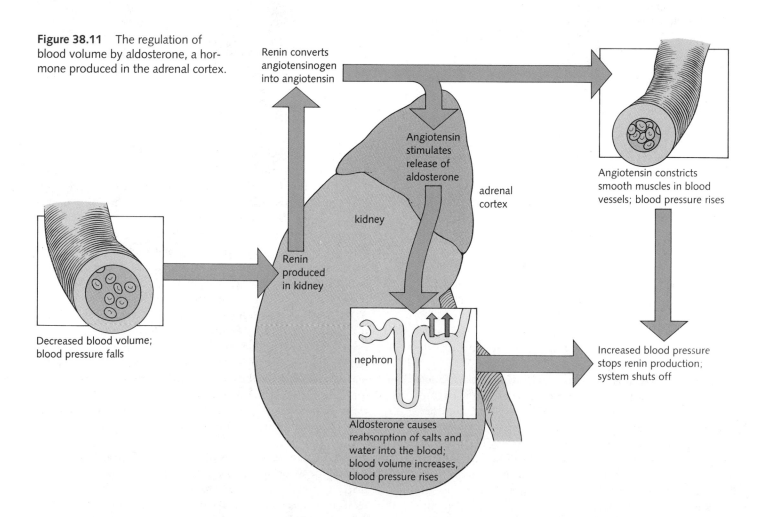

Renin converts angiotensinogen into angiotensin

Angiotensin stimulates release of aldosterone

adrenal cortex

kidney

Renin produced in kidney

Decreased blood volume; blood pressure falls

nephron

Aldosterone causes reabsorption of salts and water into the blood; blood volume increases, blood pressure rises

Angiotensin constricts smooth muscles in blood vessels; blood pressure rises

Increased blood pressure stops renin production; system shuts off

Acid–base regulation The normal pH of arterial blood is 7.4. Venous blood has a slightly lower pH (about 7.35) because of its higher content of carbon dioxide, which reacts with water to produce carbonic acid and bicarbonate:

$$CO_2 + H_2O \rightleftharpoons \underset{\text{carbonic}}{H_2CO_3} \rightleftharpoons \underset{\text{bicarbonate}}{HCO_3} + H^+$$
$$\underset{\text{acid}}{}$$

The presence of these bicarbonate ions buffers the blood against a rapid decrease in pH when hydrogen ions (H^+) are produced by normal metabolic processes. If there were no mechanisms to eliminate H^+, however, the limited buffering capacity of the blood would soon be overwhelmed. Although the majority of pH regulation occurs in the lungs, the kidneys prevent the blood from becoming too acidic by secreting excess H^+ into kidney tubules. Some H^+ recombines with bicarbonate (HCO_3^-) in the tubule to produce water and CO_2, which is eliminated through the lungs. If blood pH is very low, then HCO_3^- may not be available, and the excess H^+ may be combined with ammonia or phosphate salts and excreted in the urine (Fig. 38.12). The H^+-secreting ability of the tubule wall is so great that the pH of tubule fluid may drop as low as 4.5, signaling an H^+ concentration almost 1000 times higher than that of the blood.

WATER BALANCE IN OTHER ANIMALS

As we saw at the beginning of the chapter, humans have wondered for thousands of years how some animals can survive in salty environments. Our own kidneys enable us to survive in a variety of environments, conserving water when we have little to drink and excreting excess water when we drink more than we need. Animals in different environments display a variety of adaptations that enable them to maintain a proper balance of water and salt.

Aquatic Animals

Freshwater fishes live in an environment of nearly pure water. Therefore, they tend to gain water by osmosis. Although their bodies are coated with a thick mucus that tends to reduce the influx of water, water still enters through the gills. Without some means of counterbalancing the influx of water, the cellular salt concentration would be diluted to a dangerously low level. The kidneys of freshwater fishes remove large amounts of water from the blood and release a dilute, ammonia-containing urine. The kidneys resorb nearly all salts into the bloodstream, enabling the animal to conserve essential minerals (Fig. 38.13).

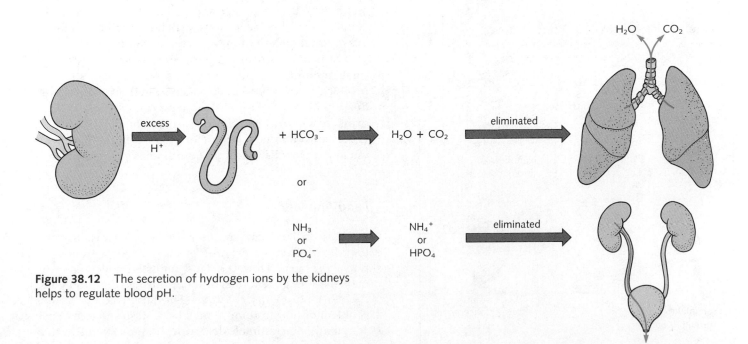

Figure 38.12 The secretion of hydrogen ions by the kidneys helps to regulate blood pH.

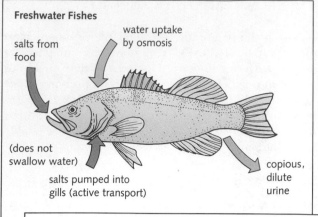

Freshwater Fishes

salts from food

water uptake by osmosis

(does not swallow water)

salts pumped into gills (active transport)

copious, dilute urine

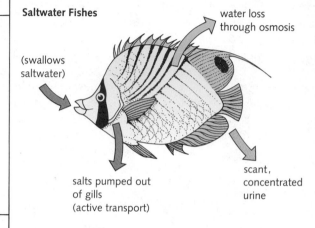

Saltwater Fishes

water loss through osmosis

(swallows saltwater)

salts pumped out of gills (active transport)

scant, concentrated urine

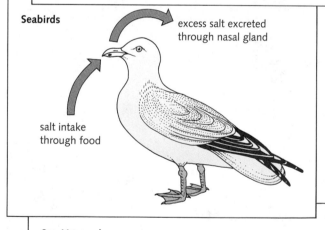

Seabirds

excess salt excreted through nasal gland

salt intake through food

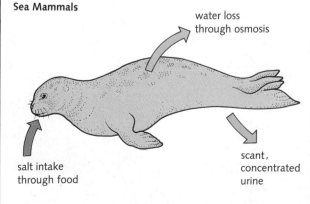

Sea Mammals

water loss through osmosis

salt intake through food

scant, concentrated urine

Figure 38.13 Freshwater fishes must retain salts and release a dilute urine, while saltwater fishes must deal with an over abundance of salts. They produce a dense, concentrated urine and eliminate some salt by active transport from the gills. Seabirds use very little water in the elimination of nitrogen wastes and remove excess salt from the bloodstream by means of salt glands. Sea mammals produce an extremely concentrated, salty urine, freeing them from the need to drink fresh water.

Saltwater fishes must deal with the opposite problem: **dehydration.** In spite of the fact that they are surrounded by water, the salt concentration of seawater is a bit higher than that of their body fluids. Therefore, these fishes tend to lose water to their environment by osmosis. They compensate for this in two ways. First, their kidneys produce a dense, concentrated urine that contains very little water. Second, specialized cells in their gills remove salt from the bloodstream by means of active transport and release it to the outside. By using their gills to excrete some salt, these fishes are able to maintain a proper water balance (Fig. 38.13).

Seabirds, like other birds, produce a thick, uric acid–laden guano to remove nitrogenous waste, which contains very little water. This ability to conserve water helps to minimize their water loss. However, the high concentration of salt in seawater would still produce dehydration were it not for special glands that pump salt out of the bloodstream (Fig. 38.13). These salt glands enable seabirds to drink seawater and extract the excess salt. The glands dump their secretions just above the bill, and crystals of dried salt are often observed in this region.

Sea mammals drink little or no seawater. They obtain small amounts of water from the food they eat, and they produce an extremely salty, concentrated urine that minimizes water loss (Fig. 38.13). Why can't land mammals drink seawater? Our kidneys are unable to produce a urine concentrated enough to remove both the excess salt and metabolic wastes that our bodies generate.

Land Animals

Animals on land face another set of problems. Because they live in air, water gain or loss through osmosis is not a problem. However, land animals in dry environments lose large amounts of water during respiration and therefore must conserve water. They also face the serious problem of how to remove metabolic wastes in a dry environment. Some animals deal with these problems by presenting tough barriers against water loss, such as the insect's chitinous exoskeleton and the reptile's leathery skin. Others have evolved lifestyles that minimize water loss. Thin-skinned amphibians live in or near the water. The delicate mosquito, which loses water easily, searches for prey only in humid weather (Fig. 38.14). A mosquito contains less than 3 microliters of water—about one-twentieth of a drop. If a mosquito were to take to the air in dry weather for more than a few minutes without finding a meal, its water loss from increased respiration would be fatal.

The most extreme adaptations of land animals for water conservation are seen in desert animals such as the kangaroo rat (Fig. 38.15), which is found in the American Southwest. As we saw in Chapter 36, the long nasal passageways of this animal help to minimize water loss during respiration. To escape the sweltering desert heat, this animal spends most of the day deep in its cool burrow, emerging only at night. Its kidneys hold the world record for efficiency, producing urine 25 times more concentrated than the kangaroo rat's blood. Not surprisingly, the nephrons of the rat have extremely long loops of Henle, providing a very long surface for water resorption. The large intestines of this desert dweller are equally efficient, extracting so much water that the feces it produces are nearly dry. With so many adaptations for water conservation, how much water does the kangaroo rat need to drink? None! The kangaroo rat obtains most of its water from a diet of dry seeds. You will remember from Chapter 18 that the oxidation of glucose produces carbon dioxide and *water*. This metabolic water, produced by the oxidation of food carbohydrates, is enough to meet the kangaroo rat's needs. Efficiency personified.

TEMPERATURE REGULATION

Regulation of fluid balance is an important component of internal homeostasis, but other factors are also controlled. One of these is the temperature of the body. Our review of basic chemistry emphasized the fact that chemical reactions, including enzyme-catalyzed ones, are affected by the temperatures at which they occur. The wide variety of local environments (and local temperatures) on the earth has made it important for many animals to control their internal body temperatures, a process called **thermoregulation,** whether they live in the Arctic or in the jungle.

Chemical reactions generally occur more quickly at higher temperatures, so the basal metabolic rate increases as the temperature increases. This phenomenon leaves an animal with two "choices." If it allows the external environment to influence its internal temperature, its metabolic rate will change as the external temperature changes. The alternative is to adjust heat-producing processes within the body so that a relatively constant

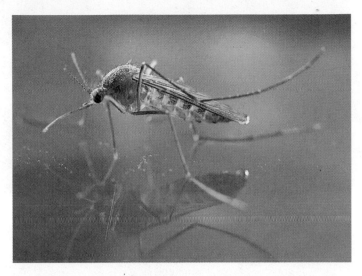

Figure 38.14 Water loss is a critical problem for small insects like the mosquito.

Figure 38.15 The kangaroo rat, an animal so well adapted to life in the arid desert that it does not need to drink water.

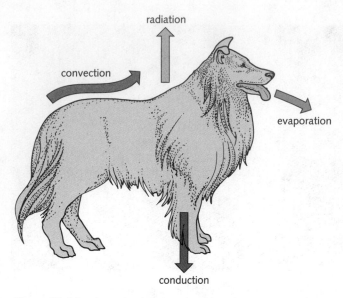

radiation

convection

evaporation

conduction

Figure 38.16 An animal may gain or lose heat in a number of ways. Rapid panting on a hot day helps this dog to lose heat by evaporation.

internal temperature is maintained while the external temperature varies.

Every organism has a source of heat available to it: its own metabolism. As we noted in Chapter 18, the metabolic pathways that are used to produce ATP from glucose capture only about 40 percent of the total available energy in the glucose molecule. The remaining 60 percent is lost as heat, and when ATP is used to do cellular work, much of the energy available in ATP is also released as heat. These facts of chemistry enable organisms to control the sources of heat production within their own bodies.

An organism may gain heat from or lose heat to its environment in a number of ways, as illustrated in Fig. 38.16. (1) Heat gain from *radiation* occurs whenever an animal absorbs direct or indirect sunlight. Because all warm objects emit infrared radiation, radiation may also cool an organism. (2) *Conduction,* the direct transfer of heat between two bodies, occurs whenever an animal's temperature is different from its surroundings. Conduction may either heat or cool an animal, depending on the temperature of the environment. (3) Warm air is lighter than cool air, and this physical fact results in *convection,* in which rising warm air produces air currents. Convection currents increase the efficiency of cooling by bringing cooler air in contact with the surface of an organism. Convection may also occur in water. (4) Finally, *evaporation,* which involves water changing from a liquid to a gas phase, consumes a great deal of energy and is very effective in surface cooling.

Ectotherms

Most organisms are **ectotherms**—that is, their body temperatures are largely determined by their interactions with the environment in which they live. Virtually all invertebrates are ectothermic, as are all amphibians, most fishes, and most reptiles. Although such animals are commonly called "cold-blooded," this term is really a misnomer; the body temperatures of such animals may actually be quite high.

For small animals living in the water, ectothermy is almost unavoidable. The density of water is much greater than that of air, and as a result, much more energy is required to change the temperature of water a few degrees than is required to change air temperature by the same amount. In addition, the temperature of most bodies of water is relatively constant, so animals living in them can depend on a stable environmental temperature. This relative stability enables fishes and other organisms to adapt to the temperatures of their surroundings and be free of the need to regulate their own temperatures.

One consequence of this adaptation is that many ectothermic organisms can live only in narrow temperature

Figure 38.17 Many ectotherms help to control their body temperature by their behavior, swimming on hot days and basking in the sun on cool ones.

ranges. Anyone who has neglected the temperature controls on a tank of tropical fish has found that even slight variations in water temperature can mean the difference between life and death.

Land ectotherms, including amphibians and reptiles, face a far more challenging environment. In many terrestrial habitats, daily variations in air temperature can be as large as 25°C. These variations affect ectotherms dramatically, for they must adjust their behavior in ways that keep their body temperatures within a suitable operating range.

Many active ectotherms can, in fact, control their body temperatures remarkably well by moving around; they seek shade or dip into ponds on hot days and bask in the sun on cool ones (Fig. 38.17). Still, these animals are slow and sluggish during cool nights and must take great care to lower their body temperature on warm days. On cool, sunny days in the spring and fall, it is not unusual to find snakes curled up on rocks in the sunshine until their body temperature increases to the point where they are able to be more active.

Endotherms

Endotherms are animals that regulate their body temperature to a large extent by physiological means and maintain a more or less constant internal temperature regardless of environmental fluctuations. Endotherms generally maintain a body temperature greater than the temperature of their environment. Actual body temperatures vary from one endothermic species to the next, but the normal human body temperature of 37°C is a typical value.

Maintaining a constant temperature gives endotherms a metabolic advantage over other organisms. Biochemical pathways and the enzymes that catalyze them often have their maximum efficiencies at temperatures near this point, and this enables endothermic organisms to remain active even in cold environments. Birds and mammals are endotherms.

Adaptations for temperature regulation Endotherms have evolved many adaptations that help them maintain a constant body temperature. Mammals that live in cold environments have thick coatings of insulating fur and body fat to minimize the rate at which heat is lost to the outside. The blubber of aquatic mammals like whales and dolphins is such an adaptation, enabling them to cope with the problems of living in a cool ocean environment where the heat capacity of water makes thermoregulation especially difficult.

Birds, for their part, have several layers of feathers that serve to insulate their relatively small bodies and prevent heat loss. Birds from particularly cold environments have an especially thick layer of down feathers; humans have exploited the insulating ability of down for many years.

Other organisms have adaptations to hot environments. Horses and many other animals cool themselves

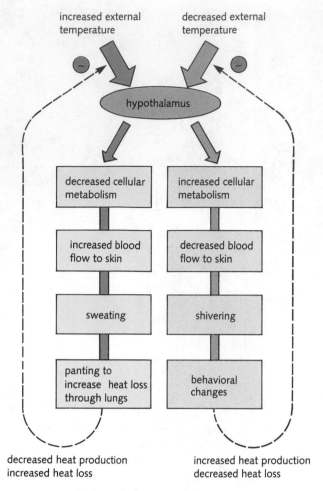

increased external temperature

decreased external temperature

hypothalamus

decreased cellular metabolism

increased cellular metabolism

increased blood flow to skin

decreased blood flow to skin

sweating

shivering

panting to increase heat loss through lungs

behavioral changes

decreased heat production increased heat loss

increased heat production decreased heat loss

Figure 38.18 The hypothalamus coordinates the regulation of body temperature.

by sweating, others by panting. Some of these adaptations are behavioral. A bird lifts its wings to increase the surface area available for heat loss. Burrowing animals stay in the cool ground on hot days.

Thermoregulation in Humans

Humans, like other mammals, maintain a constant internal temperature regardless of local environmental temperature. The system of temperature regulation is so precise that a change in body temperature of even a few degrees is usually a signal of serious trouble.

Body temperature can be regulated by adjusting two key variables: the rate of heat production and the rate of heat loss or gain with respect to the environment. For the systems and organs of the body to work effectively, there must also be a temperature-sensitive center that coordinates and controls these heat-producing and heat-loss functions.

In humans, this center is the region of the brain known as the *hypothalamus* (Fig. 38.18). The hypothalamus has two sources of temperature information: input from temperature receptor neurons in the skin and input from receptors in the hypothalamus itself that monitor changes in blood temperature. These two sources of information alert the hypothalamus when a change in temperature occurs. This enables the body to respond quickly to the two types of temperature stress: cold and heat.

Response to cold The hypothalamus reacts to low temperatures by coordinating activities of both the nervous system and the endocrine system. Blood vessels near the skin constrict, limiting heat loss from the body. There is a gradual increase in skeletal muscle tone which may eventually lead to shivering, helping to produce heat. The body releases two hormones, epinephrine and thyroxine, that increase the body's metabolic rate: cells throughout the body break down carbohydrates more quickly, fat reserves are mobilized, and sugar is released into the bloodstream from the liver. These changes produce heat in the tissues of the body, tending to elevate body temperature.

To minimize heat loss through the skin, certain smooth muscles surrounding blood vessels contract, restricting blood flow to the limbs. This is one of the reasons why your fingers and toes may seem extremely cold on a winter day. This response is not without risk. As reduced blood flow lowers the temperature of the extremities, they may approach the freezing point. In very cold weather, ice crystals may form in the tissues, causing a condition known as **frostbite** (Fig. 38.19). At first, one experiences little pain in tissue numbed by frostbite, even though the growing ice crystals rupture cells and break blood vessels. (Freezing deadens nerve cells that would

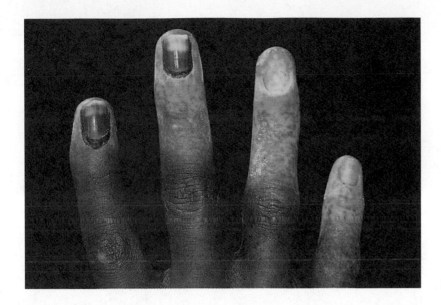

Figure 38.19 Frostbite.

otherwise warn of tissue damage.) But later, as the tissue thaws, the pain may be intense and there is danger of permanent injury. One should avoid frostbite by leaving the cold immediately when tissue goes numb. Frostbitten tissue must be warmed slowly to minimize further injury as ice crystals melt.

The hypothalamus may also initiate **shivering,** an involuntary response that helps produce heat throughout the body. Some of the major sources of internal heat are the muscles of the body. Muscles make up as much as 40 percent of the total body mass, and they are major energy users. Shivering is a series of uncontrolled skeletal muscle contractions. These contractions do little mechanical work, because the body itself often remains stationary. Nearly all of the energy released by shivering contractions is converted into heat. A few minutes of vigorous shivering can increase the heat output of surface muscles by a factor of five or six! The cold response may also include "goose bumps," which result from a contraction of the adductor muscle in the root system of a hair. In fur-coated animals, the contraction of adductor muscles causes some of their fur to stand out straight and increases the amount of air trapped within an already thick fur coat, making it an even better insulator. In humans, the heat produced by the formation of goose bumps is not important. Our goose bumps may be only a reminder of the thicker hair that some of our evolutionary antecedents bore.

Response to heat High body temperature causes the hypothalamus to initiate a series of responses that slow down heat production and increase the rate at which heat is lost from the body. The same chemical responses that stimulate metabolic activities under cold stress are now reversed to check the breakdown of fats and carbohydrates. These changes may cause one to feel sluggish and tired on a very hot day.

In addition, circulation to the skin increases to maximize heat loss, and the increased blood flow may give the hands and faces of fair-skinned people a flushed, reddish appearance. The skin begins to produce *sweat,* a secretion of water and salts that cools the body by its evaporation. Breathing, which results in heat loss through the lungs during gas exchange, increases to some extent in humans. In those mammals unable to sweat, such as dogs and cats, temperature loss through the lungs and via the tongue is increased by rapid *panting.*

In extreme cases, the body may lose the ability to maintain its own temperature. Body temperature may rise so high that cells in the skin are unable to function, producing a life-threatening condition known as **heat stroke.** A heat stroke patient ceases to sweat and has hot, dry skin. He or she may collapse or hallucinate. Immediate first aid to lower the body temperature is necessary to save a victim of heat stroke, and the most effective measure is often to place the victim in a tub of cool water.

Homeostasis and Thermoregulation

Thermoregulation is an example of a homeostatic mechanism that regulates a single variable of the internal environment—the temperature. Each aspect of temperature response, whether to cold or to heat, can be understood as part of a negative feedback loop.

In the case of cold, low temperatures initiate a series of responses that raise the body temperature. Once it has been elevated, further increases in temperature have a negative feedback effect on the hypothalamus.

A similar situation exists when the body responds to high temperatures. In each case, the body responds to the external temperature as though it maintained a specific temperature, or "set point," in its "biological thermostat" within the hypothalamus. Although biologists do not fully understand the nature of this thermostat, it is clear that one does exist, at least in a functional sense.

An interesting exception to the homeostatic tendency of the body to maintain a fixed temperature is **fever,** a long-term elevation in body temperature caused by illness. Fever can be understood as a response in which the body changes its thermostatic "set point" because of special circumstances. Normally, as body temperature begins to rise, the hypothalamus initiates a series of responses that slow cellular metabolism to reduce heat production. When it is faced with disease, however, the last thing the body needs is a reduction in cellular activity while the cells of the immune system are working to defeat an infection. The 2° or 3°C elevation in body temperature associated with fever, therefore, is part of an integrated defense mechanism against disease. Allowing body temperature to rise for a few days while the immune system works to deal with the infection may make us miserable, but it is an effective way of combating disease.

The kidney is made up of hundreds of thousands of functional units known as nephrons. The filtration process involves the formation of a primary filtrate that contains much of the dissolved material from the bloodstream. This filtrate passes through the convoluted tubule system, in which much of the water and dissolved material is resorbed into the bloodstream. This process is assisted by the osmotic gradient created by the loop of Henle. The amount of water resorbed into the bloodstream is regulated by the amount of antidiuretic hormone (ADH), which is released from the posterior pituitary.

Temperature regulation in warm-blooded (endothermic) and cold-blooded (ectothermic) organisms is achieved by a number of mechanisms, including control of the overall metabolic rate. The internal temperature of an organism is determined by its ability to respond to rapid changes in its own internal activity and in the temperature of the external environment.

STUDY FOCUS

After studying this chapter, you should be able to:

• Explain the role of the excretory system in maintaining water balance, and cite the differing demands placed on the system in various environments.

• Describe the structure and physiology of the human excretory system and its relationship to other systems in the body.

• Discuss the process of thermoregulation.

TERMS AND CONCEPTS

SUMMARY

Organisms living in water must deal with the forces of osmosis. In fresh water they must retain salts and deal with the inward pressure of osmosis; in salt water they excrete salt and maintain an osmotic balance with the water around them. Animals living on land have evolved ways to conserve water and to eliminate waste products. The ammonia wastes produced by protein breakdown are especially difficult to deal with because the ammonia that is formed in these pathways is toxic. Ammonia may be converted into less toxic compounds and removed from the body by the excretory system. Excretory systems vary tremendously from one group of organisms to another. Flatworms contain flame cells that move wastes to the exterior; earthworms have nephridia in each segment; and insects have a system known as Malpighian tubules that adds metabolic wastes directly to the digestive system. In vertebrates, kidneys remove metabolic wastes and excess water from the circulation and produce urine, which is excreted.

Objective Questions (Answers in Appendix)

1. Deamination refers to the
   (a) breakdown of ammonia.
   (b) failure of the kidneys to keep a water balance in the body.
   (c) filtering of amino acids through the liver.
   (d) removal of an —NH$_2$ group from an amino acid.

2. Terrestrial organisms must excrete nitrogenous wastes in the form of urea or uric acid because
   (a) ammonia cannot be excreted in crystalline form
   (b) ammonia is toxic.
   (c) too much water would be needed for removal of ammonia.
   (d) of all of the above.

3. During the process of resorption,
   (a) dissolved solutes are returned to the bloodstream.
   (b) specific molecules are transported from the bloodstream to the tubules.
   (c) urea is removed from the tubules.
   (d) water is produced as an end product.

4. During the process of filtration,
   (a) the cellular components of the blood pass into the primary filtrate.
   (b) salt and other small molecules pass through the filtration barrier into the primary filtrate.
   (c) water is removed from the filtrate.
   (d) proteins are actively pumped from the blood into the filtrate.

5. When blood volume is high, blood pressure rises. This results in
   (a) renin, angiotensin, and aldosterone levels all falling.
   (b) renin, angiotensin, and aldosterone levels all rising.
   (c) renin levels rising and angiotensin levels falling.
   (d) renin and angiotensin levels falling but aldosterone levels rising.

6. The kangaroo rat relies mostly on _____ in order to keep a healthy water balance.
   (a) food containing water
   (b) living close to a stream
   (c) oxidative processes
   (d) searching for prey only in humid weather

7. It is possible to conduct a laboratory experiment in which the temperature of blood is slightly elevated in the vessels leading to the hypothalamus. If this were done in a mammal, which of the following would be the likely result?
   (a) an increase in overall body temperature
   (b) a decrease in body temperature
   (c) no change in temperature, but increased kidney output
   (d) no change in temperature, but decreased kidney output

Discussion Questions

8. Describe the difference between the functions of the two capillary beds (glomerulus and net around tubules) in each nephron.

9. Some "crash" diets call for protein-rich food and the complete elimination of carbohydrates from the diet. People on these diets often complain of frequent urination and increased thirst. What is the likely source of these symptoms?

10. At what stage of the excretory process in the kidney are foreign substances removed from blood?

11. If you did not drink any water for 24 hours, would you expect your level of ADH to increase or decrease? Why?

12. Animals with an extreme need to conserve water often have larger kidneys than related organisms that inhabit damp environments. The increased size of their kidneys corresponds to an increase in the distance between the glomerulus and the bottom of the loop of Henle. How does this structural change assist in water conservation?

13. Explain the key mechanisms of thermoregulation in humans, and indicate which mechanisms are subject to negative feedback loops.

READINGS

Underwood, B. A. "Bee cool." *Natural History* 99 (December 1990): 50–58. See note for Heinrich entry below.

Heinrich, B. "Thermoregulation in winter moths." *Scientific American* 256 (March 1987): 104–111. In this text we did not discuss temperature control adaptations in insects. If you're interested, you'll find these two fascinating articles well worth reading.

Strickler, E. M., and J. G. Verbalis. "Hormones and behavior: The biology and thirst of sodium appetite." *American Scientist* 76 (1988): 261–268. An excellent discussion on how the body balances salt and water intake.

Smith, H. W. *From Fish to Philosopher.* Boston, MA: Little, Brown, 1953. Practically a "cult classic" among physiology students, this book traces the lessons of vertebrate evolution by using kidney structure and function as a key. Take the time. It's a fascinating study.

39

Chemical Communication

Your brain requires a steady, uninterrupted supply of glucose to function properly; if the flow of this vital sugar stops—even for a few moments—brain activity decreases or even stops. Yet you can get along perfectly well by eating just three times a day, and most people can fast for several days with no lasting ill effects (Fig. 39.1). How does your body maintain a steady supply of glucose to the brain over a 24-hour period when food intake is restricted to three brief meals a day?

An inquisitive toddler touches the outside of a barbecue grill containing red-hot coals. Almost instantaneously, before he is even aware of what has happened, he jerks his hand away. A second or two later, he feels pain and begins to cry, but long before that he has removed his fingers from danger, preventing more extensive damage to his body. How did that immediate withdrawal reaction occur before the child knew he was in pain?

These are just two examples of the need for coordination and interaction among the trillions of cells that make up your body. In order for you to survive, those cells must coordinate their activities in ways that enable them to serve not only their own needs as individual cells but also your needs as a multicellular organism.

Long-term, sustained responses to environmental conditions—such as the regulation of blood glucose level—require steady, dependable coordination of cellular- and organ-level activities over periods ranging from a few minutes to days, weeks, months, and even years. *Short-term, rapid responses to events in the environment*—such as the hand's quick withdrawal from extreme heat—require swift reactions and precise coordination over periods ranging from milliseconds to seconds, minutes, or hours.

THE BODY'S VITAL MESSENGERS

Both types of responses we have described require that cells or groups of cells somehow affect the actions of other cells, and that, in turn, requires **communication.** The bodies of most multicellular animals, from insects to humans, contain two major organ systems specializing in

Figure 39.1 A protest hunger strike. The brain requires a constant level of glucose in the bloodstream. Chemical messengers within the body make it possible to maintain this level even during fasting.

communication: the **nervous system** and the **endocrine system.**

The nervous system generally handles messages that must be delivered quickly, that generally (though not always) produce rapid responses, and that need not last for very long periods of time. These sorts of messages enable animals to catch prey (or to avoid being preyed on) and to detect and respond in a timely fashion to events in the world around them. As we will see when we discuss the nervous system in Chapters 43 and 44, a typical nerve cell transmits messages at great speed over long distances by means of *electrochemical impulses*. These messages are carried quickly across the tiny gaps that separate adjacent nerve cells by chemical messengers known as *neurotransmitters*.

The endocrine system generally handles messages that can be delivered more slowly but that must have long-lasting effects. Such messages enable young animals to grow, mature animals to repair damage caused by disease and injury, and all of us to regulate body water content and blood nutrient levels. These messages are carried by chemical signals known as **hormones,** a term derived from a Greek verb that means "I arouse." Because hormones often cause dramatic changes in the cells they affect, and because they can influence the activity of millions of cells at once, their name is quite appropriate. Hormones travel through the bloodstream and can diffuse through body tissues to reach most cells of even large organisms.

The functions of these two communication systems overlap a great deal. In many situations, animals must make long-term changes in physiology and behavior in response to environmental conditions that can be detected only by the nervous system. Birds, for example, may need to prepare for migration as days lengthen or shorten. And when an animal is threatened, it may need to respond immediately, with reactions so swift that they must be mediated through the nervous system. At the same time, however, that animal may need to mobilize its resources for longer periods of intense physical exertion to battle intruders or to flee over long distances.

In these circumstances where the functions of the two communication systems overlap, information is received by the nervous system, acted on by the nervous system, and simultaneously translated into the chemical language of the endocrine system. It should not surprise you, therefore, to learn that there are critical links between the two systems in several places, notably in a region of the brain known as the *hypothalamus*. Once you learn more about the nature and function of the body's chemical messengers, you will also discover that several compounds act both locally within the nervous system as neurotransmitters and in the body at large as hormones.

THE ENDOCRINE SYSTEM

In 1849, a researcher named A. A. Berthold removed the testes from several immature roosters and noted that the castrated birds failed to develop the comb and wattles normally found in mature males. Furthermore, they showed no interest in female birds. This by itself was no real surprise; it had long been a common practice among farmers to castrate some of their roosters to minimize fights among males scrambling to establish mating territories in the barnyard.

But Berthold took his experiment one step further. He transplanted testes back into the abdominal cavities of castrated birds and found that male plumage and male mating behavior developed in a matter of weeks. Berthold knew that the transplanted testes had no direct connection to any organ system in the recipient birds. So how had those free-floating gonads had any effect on the birds?

Berthold concluded that the testes were releasing a substance or group of substances into the bloodstream that produced male characteristics and male behavior in the birds. Those substances, of course, were male sex hormones. We now recognize Berthold as the first scientist to demonstrate definitively the activity of an *endocrine gland*.

The Principle of Chemical Signaling

The phenomenon Berthold observed is widespread, not only in the animal kingdom, but in both simple and complex members of other kingdoms as well. The cellular slime mold *Dictyostelium discoidium*, for example, offers a

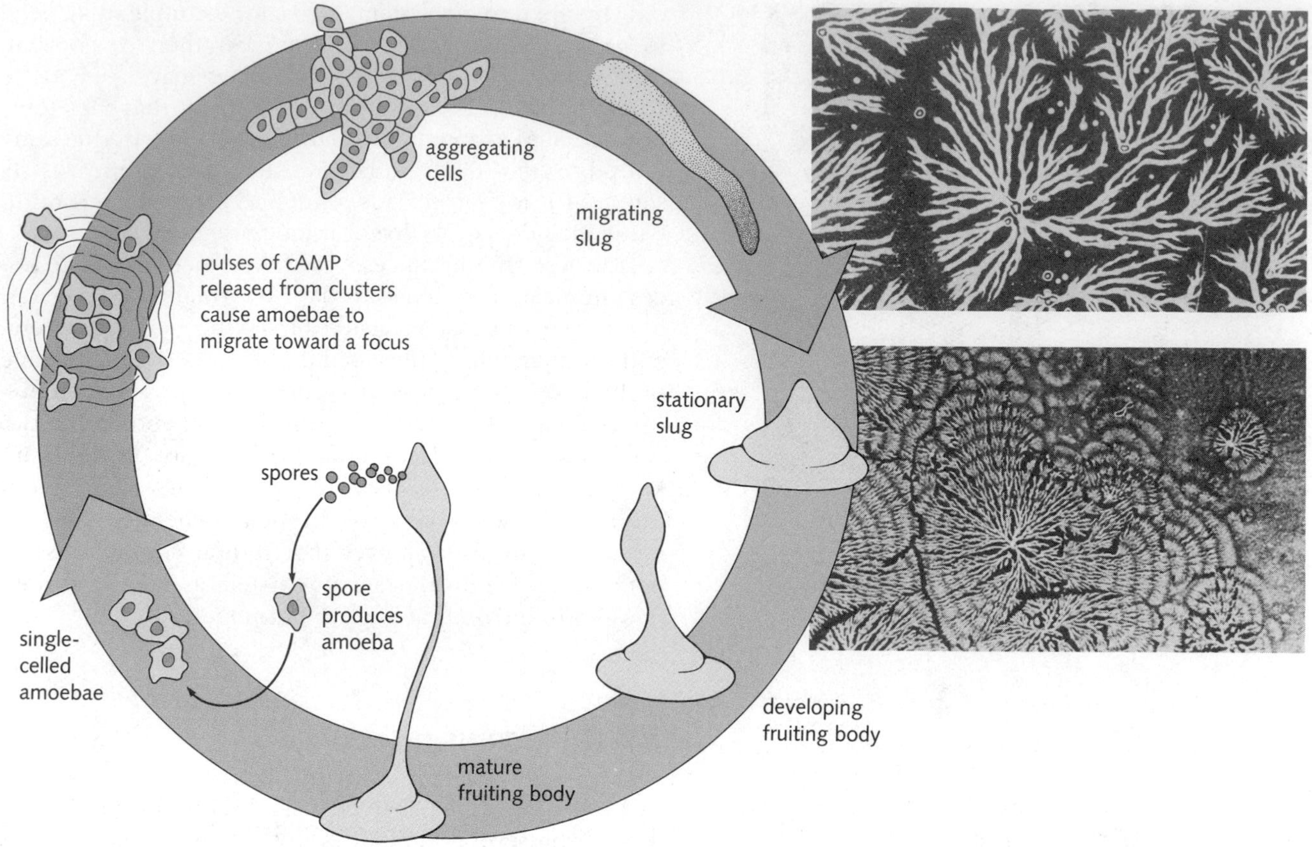
aggregating cells
migrating slug
pulses of cAMP released from clusters cause amoebae to migrate toward a focus
stationary slug
spores
spore produces amoeba
single-celled amoebae
developing fruiting body
mature fruiting body

Figure 39.2 LEFT: The life cycle of *Dictyostelium discoidium,* the cellular slime mold. The aggregation of individual amoeba-like cells to form a migrating slug occurs in response to chemical signals. RIGHT: The mass migration of thousands of *Dictyostelium discoidium* cells forms branch-like patterns leading to clusters where migrating slugs will form.

perfect example of primitive chemical communication in action (Fig. 39.2). *Dictyostelium* spores, which are scattered on the surface of the soil, give rise to individual ameboid cells that forage independently of one another. When the food supply available to the ameboid cells begins to fail, however, the individual cells aggregate into clusters that take on the appearance of a thick slug that moves along the surface of the soil. After moving several centimeters, it and the cells form a fruiting body that matures to release new spores and begin the cycle again.

This striking convergence of independent cells into a single, organized, multicellular structure requires the coordination of thousands of cells. How is it accomplished? By the release of chemical messages. As their nutritional situation worsens, individual cells send out pulses of a chemical known as **cAMP** (cyclic adenosine monophosphate), which diffuses through the moist soil surface and attracts other cells. As a few cells unite to form a cluster, they begin to release cAMP in coordinated pulses, and still more cells are drawn toward the center of the pulse. These small cAMP molecules carry the essential signal that enables the many cells of *Dictyostelium* to act as a single organism. Each cell can produce the chemical message, and each can respond to it. If *Dictyostelium* did not have this ability, it is difficult to see how it could regulate a life cycle involving a multicellular phase.

The relatively simple chemical message used by *Dictyostelium* offers clues to the origin of more complicated hormonal control systems. In this slime mold, the message "calls" independent cells together and primes them to act as a single unit. Endocrine systems in more complex organisms provide the communication links necessary to produce coordinated responses in widely separated cells.

Hormonal Control in Insects

Insects, with their complex cycles of growth and larval development, provide both interesting and important examples of hormones in action. As you recall from Chapter 28, insects are covered by a tough exoskeleton that serves 28

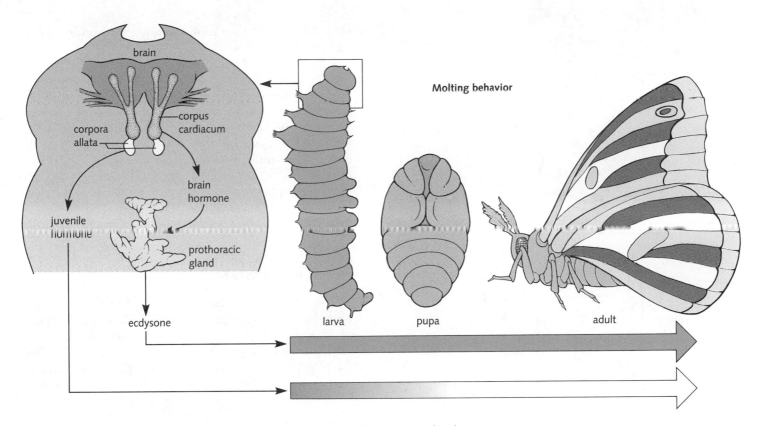

Figure 39.3 Changing concentrations of brain hormone, juvenile hormone, and ecdysone regulate molting and maturation in insects.

several important functions. But this external skeleton also creates a serious problem: Insects cannot grow unless that exoskeleton is periodically discarded and replaced in a process called **molting** or **ecdysis.**

Molting is a complex process that involves coordination of many cells throughout the organism. During a molt, the animal's epidermis partially digests the existing exoskeleton. The insect then splits the weakened shell, pulls its body out, and grows a new, larger exoskeleton to replace the old one. Because the exoskeleton covers virtually every body surface of the organism (including part of its gut), and because molting is accomplished all at once, the process must be precisely orchestrated.

This coordination of metabolic functions among cells scattered across the body is precisely the sort of need that is well served by chemical signals. The specific signal that organizes the molting process is a hormone called **ecdysone,** produced and released by an organ called the *prothoracic gland,* which is located just behind the head of the larva. Early experiments showed that molting could be prevented if this gland were surgically removed. Conversely, extracts prepared from the gland induce molting when injected into other larvae.

The production of ecdysone is controlled, in turn, by a substance known as **brain hormone,** produced and released by a small portion of the brain known as the *corpus cardiacum* (Fig. 39.3). Thus, although ecdysone actually controls the molting process itself, the production of brain hormone determines the timing of the molt.

There is, however, an important aspect of insect development that cannot be explained solely by the system we have described so far. Many insects, such as butterflies, pass through several larval stages, each of which ends in a molt. During the first several molts, the larva simply grows larger. But at some point, the nature of the molt changes, and a *pupa* is formed instead of a larger larva. The development of adult tissues takes place within the pupa, and ultimately an adult insect emerges from it.

If all molts are triggered by ecdysone, what makes the final molt differ from those that precede it? The answer to this question is found in a different region of the brain: the paired structures known as the *corpora allata.* These small organs produce a substance known as **juvenile hormone (JH),** which controls the effect that ecdysone has on the larva. When JH is present during an ecdysone burst, the next stage of development is another

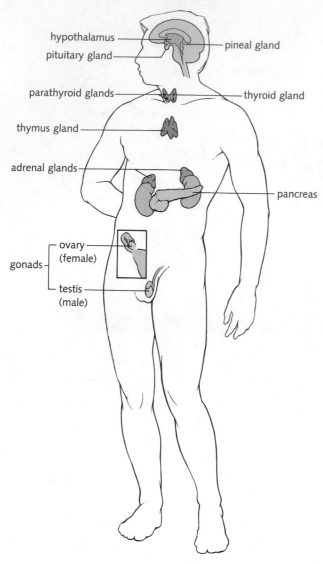

Figure 39.4 The organs of the human endocrine system.

larva. During the final molt, however, JH is not released, and an adult develops (see Fig. 39.3).

Over time, biologists have learned to exploit this system for help in the control of insect pests. Because reproduction occurs only in the adult stage, any compound that mimics the action of juvenile hormone can prevent the final molt and keep an insect in its larval stage indefinitely. One strategy for controlling manure flies in stables today, for example, is to add JH-like chemicals to the manure. Fly larvae grow in the manure but never molt into mature flies; hence the reproductive cycle is broken and the fly population controlled.

COMPONENTS OF THE ENDOCRINE SYSTEM

The word *endocrine* comes from two Greek words: *endo*, which means "inside," and *krinein*, which means "to divide or separate." As the word implies, **endocrine glands** release their secretions *inside* the body—in most cases into the bloodstream. At one time endocrine glands were referred to as *ductless glands*, because they do not have ducts or passageways that lead outside the body or into the digestive tract. The term *endocrine* distinguishes these glands from **exocrine glands,** such as sweat glands or digestive glands, that release their products outside the body or into the digestive tract.

The basic structure of the endocrine system is different from that of other organ systems. The human endocrine system, for example, is made up of a series of apparently independent organs found in various places throughout the body. These scattered organs do not have a common developmental origin and do not respond to the same types of stimulation. They are small, specialized glands, each of which produces a particular product or group of products for release. The major human endocrine organs are shown in Fig. 39.4. They include the **thyroid,** the **parathyroids,** the **endocrine pancreas** (islets of Langerhans), the **adrenals,** the **thymus,** the **pituitary,** and the **gonads: ovaries** in females, **testes** in males.

At first the isolation of endocrine organs from one another is puzzling: How can widely scattered organs form a single system? But if you keep in mind the fact that the endocrine system functions by means of chemical messages, the separation seems much less problematic. Even though the glands are scattered throughout the body, their ability to synthesize and release hormones into the bloodstream permits them to act in concert, to respond to subtle changes in the body, and to maintain the stability of the body's internal environment.

The Pineal Gland

The tiny human *pineal gland* is located in the midbrain region of the skull (the word *pineal* is derived from the gland's resemblance to a small pine cone). The Greeks had noted the existence of the pineal gland as early as the fourth century B.C., and Descartes (the seventeenth-century French philosopher and mathematician) suggested that the gland was "the seat of the rational soul." The precise function of the pineal gland in humans is unknown, but experiments on other vertebrates have provided a number of valuable clues.

The reproductive activities of many vertebrates are geared to the seasons of the year, ensuring that mating and the birth of offspring occur under optimal conditions. When the pineal gland is surgically removed from such animals, their ability to remain in synchrony with the seasons is often lost. The pineal seems to adjust an organism's physiology to environmental changes.

The pineal produces a hormone known as *melatonin*. In some vertebrates (fishes and salamanders), daily surges of melatonin adjust skin color by controlling the dispersal of pigment granules in skin cells, making the animals less readily visible as their environment changes from bright light to total darkness. Melatonin is also produced in mammals, including humans, but it does not regulate skin color in these organisms. Nonetheless, the onset of darkness brings about an increase in melatonin production that lasts until the first light of morning.

Some scientists have found evidence that they believe suggests that the nervous system can "tell" the pineal gland when darkness has begun even while the organism is asleep! Interestingly, although the human pineal is buried deep within the skull, in many reptiles it is located at the very top of the brain

case. There it develops as a miniature "third eye," complete with transparent lens and a layer of rudimentary photoreceptors. Thus, at least in these ancient vertebrates, the pineal was capable of sensing the presence and absence of light directly.

A number of effects have been suggested for melatonin, and there is some

Migrating animals, including these geese, must precisely sense the seasons of the year for their movements to be successful.

evidence linking its levels to an internal "clock" that regulates daily cycles of sleeping and waking and is responsible for the disorientation known as "jet lag" experienced by travelers who pass through several time zones in just a few hours. Other experiments have suggested that the deep depression many people feel during the darkest days of winter, especially near the polar regions where only 5 or 6 hours of daylight occur, results from more than a romantic longing for the rebirth of spring. The diminished hours of daylight may cause such high melatonin levels that normal cycles are disrupted.

The best therapy for such bouts of depression, not surprisingly, may be larger doses of light. A number of recent studies have shown that high levels of artificial light can reverse winter's seasonal depression. The actual experiments vary, but the one common factor is that the individuals under study worked and lived either in natural light or bright artificial light for at least 12 hours per day. Researchers have not yet determined the mechanism behind the benefits of increased lighting, but they may very well act against seasonal depression by going directly to the source of the problem—the pineal gland.

In mammals with seasonal mating patterns, the changes of season may influence melatonin in a way that makes the gonads shrink during the times of the year when mating is not appropriate. Does melatonin have similar sexual effects in humans? There is no evidence for seasonal variation in human mating patterns. But it may be significant that some studies show that human melatonin levels decline as much as 75 percent during the transition from childhood to early puberty. A signal for sexual maturation? Only time and more research will tell.

Type of hormone	steroid hormones	polypeptide hormones	amino acid derivatives
Structure	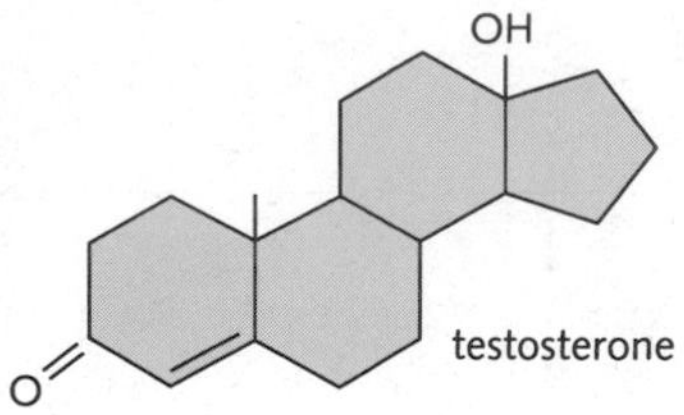 OH / O= testosterone	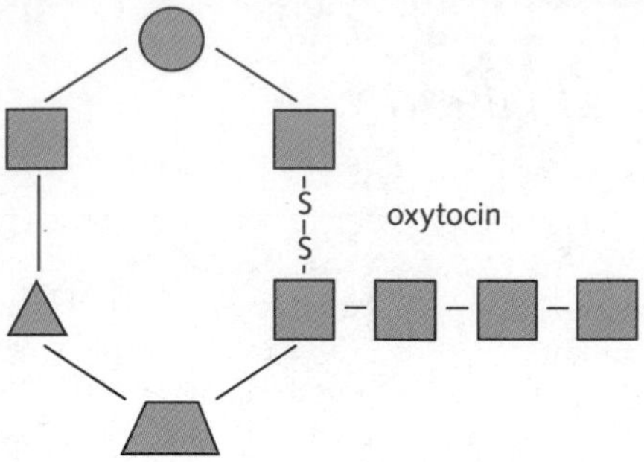 oxytocin S–S	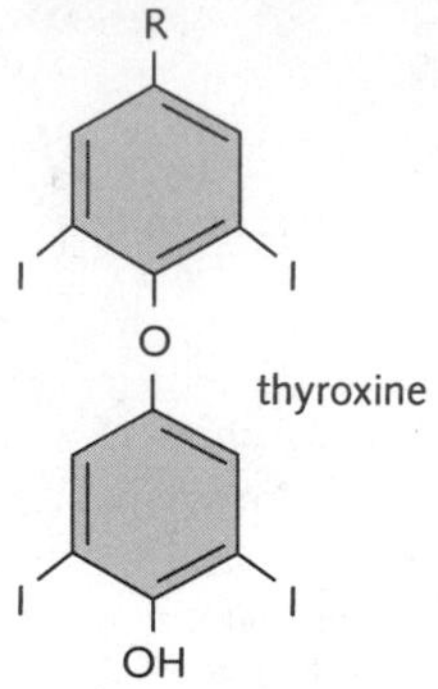 R / O / OH thyroxine
Other examples	estrogen, ecdysone	epinephrine / norepinephrine	glucagon / brain hormone

Figure 39.5 Chemical structures of several major classes of hormones.

TYPES OF HORMONES

Hormones secreted by endocrine glands are biochemical messages destined for distant targets throughout the body. Hormones' ability to travel through the blood enables them to reach virtually any tissue. It is important to keep in mind, however, that the term *hormone*, like the term *vitamin*, has a functional rather than a chemical definition. Hormones range from compounds as simple as ethylene (in plants) to simple peptides, lipids, and complex glycoproteins (in humans).

Ecdysone, for example, is a **steroid hormone.** As shown in Fig. 39.5, its chemical structure is similar to that of other steroids, including the important human sex hormones *testosterone* and *estrogen.* Steroids are lipids, and steroid hormones have all the characteristics of lipid molecules, among them the ability to pass easily across the lipid bilayers that form cell membranes.

Many hormones are **proteins** or **polypeptides;** one of these is the brain hormone that regulates ecdysone production in insects. Polypeptide hormones may range in size from a few amino acids to several hundred, and many are complex glycoproteins with molecular weights exceeding 20,000. *Oxytocin,* which is released from the pituitary gland to induce labor and promote milk production, is a representative polypeptide hormone (see Fig. 39.5).

Another important class of hormones called **amino acid derivatives** consists of chemically modified versions of common amino acids. Among the best-known amino acid derivatives is *thyroxine,* which is synthesized by the thyroid gland from the amino acid tyrosine, as illustrated in Fig. 39.5.

THE NATURE OF HORMONE ACTION

In studying hormones and their activities, researchers long ago realized that any theory of hormone action had to explain four intriguing phenomena.

First, hormones exert powerful effects in very small amounts. Even when we say that a hormone is present at "high" levels, the concentration of that hormone within the organism is still usually less than one part per million. Somehow, the effects of relatively few hormone molecules are amplified to cause many far-reaching changes.

Second, a single hormone can have different effects on several different processes in the same cell. For example, the hormone *epinephrine (adrenalin)* stimulates the conversion of glycogen to glucose in a cell, while simultaneously slowing down the production of glycogen from glucose.

Third, a single hormone may affect only a single cell type or it may affect several cell types. And a hormone that affects several cell types may affect them in similar ways or it may affect them differently. *Insulin,* for example, both promotes the conversion of glucose to glycogen in liver cells and stimulates the conversion of glucose to fats in adipose cells.

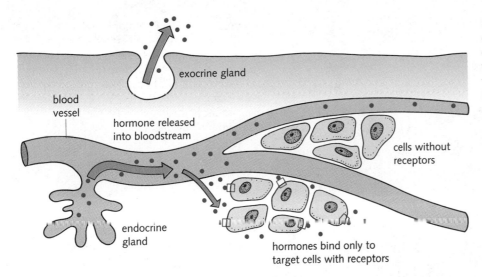

Figure 39.6 Exocrine glands release their secretions outside of the bloodstream. An endocrine gland, however, releases hormones into the circulation. Cells with receptors enabling them to respond to a particular hormone are known as target cells.

Fourth, different hormones can affect processes in the same cell differently. Often, one or more hormones act to speed up a particular reaction, whereas other hormones act to slow it down.

The Importance of Receptors

The ability of a particular cell to respond to a hormone depends on whether that cell has a **receptor,** a molecule that binds to the hormone and directs a cellular response to it. Cells containing hormone receptors are referred to as **target cells** for the hormone that affects them. If cells in different tissues carry receptors for a certain hormone, that hormone can affect all those cell types. Cells that lack such receptors, on the other hand, are not influenced by the hormone. This phenomenon explains how hormones can act throughout the body and yet influence only certain groups of cells. It is this ability that enables the endocrine system to regulate a range of body activities (Fig. 39.6).

It should not surprise you to learn that hormones as different as simple amino acids and steroids exert their effects through different types of receptors and act in very different ways. Before we examine the effects of several specific hormones and study how they are produced and regulated, it will be helpful to learn how different classes of hormones influence cells.

Steroids

As lipids, steroid hormones pass through cellular membranes easily. Once they are inside, the ability of steroid hormones to influence a cell depends on the presence of receptor proteins that are usually found within the nucleus. If a cell contains the appropriate receptor, hormone and receptor bind to each other to form a **hormone–receptor complex,** as shown in Fig. 39.7.

The shape of the complex enables it to bind very tightly to specific regions of chromatin, and that is where the hormone exerts its effect. Hormone–receptor complexes can either activate or deactivate specific genes or groups of genes within the nucleus. Note that neither the receptor protein nor the hormone alone possesses this regulatory ability; only the *combination* of hormone and receptor activates or deactivates the genes.

Through this type of direct effect on gene expression, steroid hormones can cause fundamental changes in the molecular biology of target cells. The binding of a single hormone-receptor complex to chromatin may increase by a factor as great as 1000 the rate at which certain mRNA molecules are produced. For example, an increase in mRNA levels occurs in the chicken oviduct following introduction of the steroid hormone estrogen. The effect is also highly specific—only some genes are activated.

Peptides and Their Derivatives

Many other classes of hormones, including polypeptides and amino acid derivatives, do not enter their target cells. Instead they bind to receptor molecules located on the cell surface. These receptors, which form part of the cell membrane, are very specific; in most cases, they allow one, and only one, hormone to bind.

The binding of a hormone molecule at the cell surface triggers a sequence of biochemical events that affects

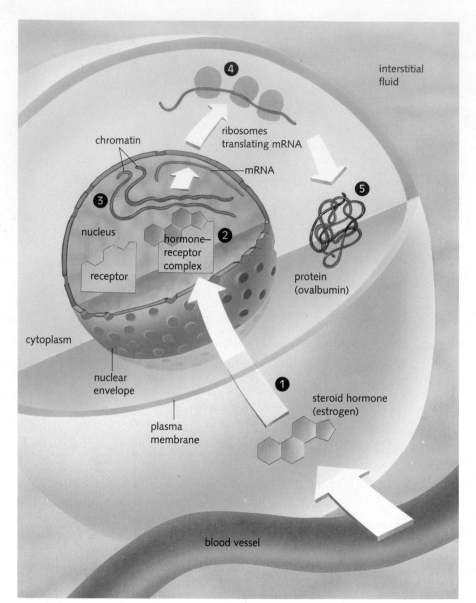

Figure 39.7 Steroid hormones enter their target cells (1) and bind to a protein receptor molecule. This forms a hormone–receptor complex (2) that can bind to chromatin (3) and alter the pattern of gene expression (4). In chickens, the steroid hormone estrogen results in a 1000-fold increase in the amount of ovalbumin mRNA (5) produced by cells lining the oviduct.

the target cell. Figure 39.8 shows an example of such a sequence of events. The binding of a hormone to its receptor causes it to form a complex with a so-called *G protein*, a membrane protein that binds *GTP* (guanosine triphosphate). Together, the G protein and the receptor cause a change in **adenylate cyclase,** an enzyme bound to the inner surface of the cell membrane.

When the hormone-binding site in the receptor is empty, adenylate cyclase is *inactive*. But the binding of the hormone to its receptor switches the enzyme to its active form, and it begins to catalyze a chemical reaction in which cAMP is formed from ATP:

$$ATP \longrightarrow cAMP + P \sim P$$

By activating an enzyme such as adenylate cyclase, a single hormone molecule can stimulate the synthesis of several hundred cAMP molecules in a very short period of time. Why is this important? cAMP functions as a **second messenger,** a molecule that carries the signal from the primary messenger (the hormone) into the cell.

Second Messengers as Amplifiers

The power of this system is further amplified because the activation process does not end with the production of cAMP. Many proteins, including certain enzymes called **protein kinases,** are activated by cAMP. Thus each acti-

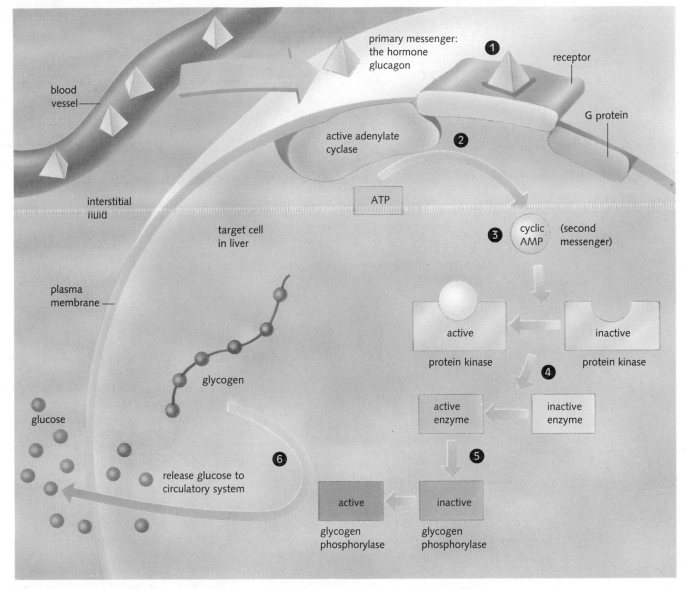

Figure 39.8 Polypeptide and protein hormones, such as glucagon, bind to receptors at the cell surface (1). Hormone binding causes the production of a second messenger (2), such as cyclic AMP (cAMP) (3). cAMP activates protein kinases (4) which may phosphorylate enzymes (5) that in turn control the activity of other enzymes. In the case of glucagon, this "cascade" of events results in the activation of a large number of enzyme molecules that break down the storage carbohydrate glycogen (6) and release glucose into the circulation.

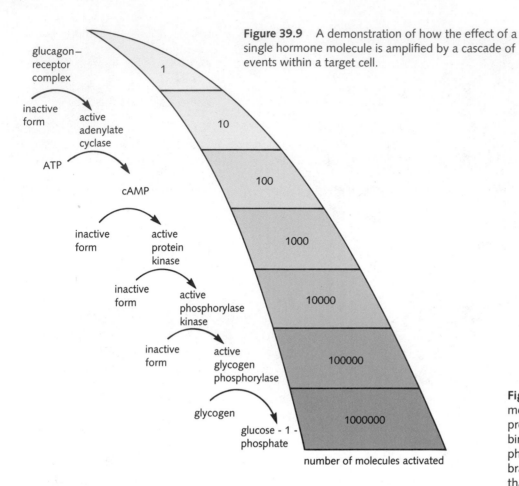

Figure 39.9 A demonstration of how the effect of a single hormone molecule is amplified by a cascade of events within a target cell.

glucagon–
receptor
complex

inactive form → active adenylate cyclase

ATP

cAMP

inactive form → active protein kinase

inactive form → active phosphorylase kinase

inactive form → active glycogen phosphorylase

glycogen

glucose - 1 - phosphate

1
10
100
1000
10000
100000
1000000

number of molecules activated

Figure 39.10 The binding of some hormones to their receptors (1) activates a "G protein" (so named because it contains a binding site for GTP, guanosine triphosphate). The G protein activates a membrane-bound phospholipase enzyme (2) that cleaves a membrane lipid, phosphatidyl inositol, into two parts: inositol triphosphate (3) and diacylglycerol (4). Each of these compounds may then act as a second messenger (5).

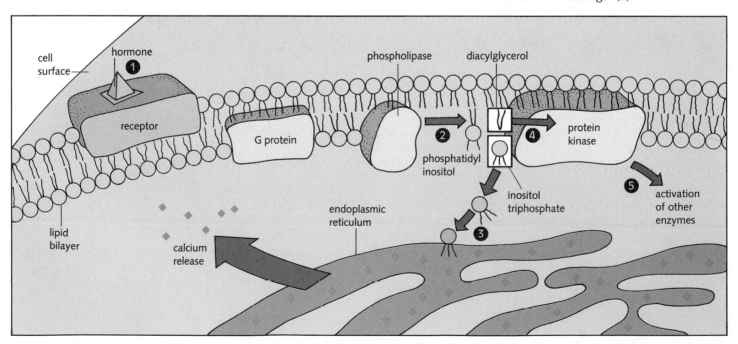

cell surface

hormone ❶

phospholipase

diacylglycerol

receptor

G protein

phosphatidyl inositol ❷

protein kinase ❹

❺ activation of other enzymes

lipid bilayer

calcium release

endoplasmic reticulum

inositol triphosphate ❸

vated protein kinase molecule can activate hundreds of other enzymes.

Ultimately, a single hormone can set in motion a "cascade" of events that amplifies by several orders of magnitude the message from a single hormone molecule (Fig. 39.9). This amplification process is one of the reasons why small amounts of a hormone can cause dramatic changes in body chemistry.

At the same time, hormones that participate as intermediates in the cascade can affect other hormones in different ways. Protein kinase, for example, simultaneously activates enzymes that catalyze the conversion of glycogen to glucose and *de*activates other enzymes that catalyze the conversion of glucose to glycogen. Thus the activation of adenylate cyclase can have different effects on different reactions in the same cell.

cAMP is one of the best-understood second messengers, but it is not the only one. Many hormones cause the release of calcium ions (Ca^{2+}) into the cell after they bind. The rapid change in calcium concentration then affects a wide variety of cellular functions through a calcium-binding protein called **calmodulin.** Because calmodulin binds to—and therefore influences the properties of—several enzymes and cellular proteins, the activities of those cellular components are indirectly controlled by calcium level.

The mechanism by which certain hormones affect calcium levels has only recently come to be understood. Hormone binding stimulates a membrane-bound enzyme (called phospholipase *c*) to cleave **phosphatidyl inositol,** a lipid found in the cell membrane, into two fragments (Fig. 39.10). One fragment, *inositol triphosphate,* causes calcium to be released from storage in the endoplasmic reticulum. The other fragment, *diacylglycerol,* remains in the membrane and activates a protein kinase that then activates a series of cellular enzymes by phosphorylation. A number of hormones, including *antidiuretic hormone* (ADH) and *serotonin,* release calcium as a second messenger by means of phosphatidyl inositol.

CONTROL OF THE ENDOCRINE SYSTEM

For many years scientists were puzzled by the fact that endocrine glands were so widely separated. In addition, many hormones seem to have overlapping functions, whereas other hormones seem to work in opposition to each other (Table 39.1). For example, the secretions of at least three glands directly affect the uptake and release of sugar: insulin and glucagon from the pancreas, glucocorticosteroids from the adrenal cortex, and adrenalin from the adrenal medulla. If each endocrine gland did, in fact, work independently, coordinated control of body chemistry would be next to impossible.

Hormonal Control Systems: Two Techniques

Despite their physical separation, the organs of the endocrine system do work together. The activities of several endocrine glands (such as the thyroid) are coordinated by a "master gland," the pituitary, through a process called **feedback control.** Several other glands (such as the pancreas) function more independently, but their hormones work together in pairs or groups in a complementary fashion. Before giving specific examples of these control systems in the body, let us briefly examine two analogies that help clarify the way each of these control mechanisms works.

Negative feedback loops Negative feedback control operates in much the same way as a thermostat in a home heating system. In such a system, when the temperature falls below a certain level, the thermostat turns on the heat. Then, when the system has raised the temperature above a certain level, the thermostat turns the heat off. Depending on the design of both the heating system and the thermostat, temperature may fluctuate a bit during the regulatory cycle, but it neither rises too high nor falls too low.

Later in this chapter you will see how a negative feedback system involving the pituitary gland controls blood concentrations of thyroid hormone. Then, in Chapter 40, you will see how a more complex version of the same system regulates the hormones of the mammalian reproductive system.

Complementary hormone action As Table 39.1 shows, two or more hormones often influence the same biochemical pathways in opposite ways. Such "opposing" actions regulate body chemistry in much the same way in which the brake and accelerator pedals in a car control its speed. Theoretically, a good driver can control the speed of a car on an open highway by using only the accelerator. (Especially if you drive a standard shift, you can, in fact, get quite proficient at this.) But in the more complex maneuvers of driving around town, even a good driver relying solely on the accelerator would quickly get into trouble; there are too many situations that require the opposing action of the brake to slow things down.

In much the same way, several endocrine regulatory functions rely not only on the negative feedback control of a single hormone but also on the complementary effects of two opposing hormones. This sort of control system regulates levels of both calcium and glucose in the bloodstream.

Table 39.1 *The Endocrine Organs and the Hormones They Secrete*

Organ / Hormone	Type of Hormone	Target Tissue(s)	Effects
Posterior Pituitary			
Oxytocin	Polypeptide	Breasts, uterus	Stimulates milk release, uterine contractions
Antidiuretic hormone (ADH) (vasopressin)	Polypeptide	Kidney, blood vessels	Regulates water balance, blood vessel constriction
Anterior Pituitary			
Melanin-stimulating hormone (MSH)	Polypeptide	Pigment cells in skin	Function in human unclear
Follicle-stimulating hormone (FSH)	Protein	Gonads	Stimulates gonads of both males and females; necessary for gamete cell development
Luteinizing hormone (LH)	Protein	Gonads	Stimulates gonads to produce sex hormones (estrogen and testosterone)
Prolactin	Protein	Breasts	Stimulates milk production
Thyrotropin	Protein	Thyroid	Stimulates thyroid activity
Corticotropin (adrenocorticotropic hormone: ACTH)	Polypeptide	Adrenal cortex	Stimulates adrenal cortex to release its endocrine secretions
Somatotropin (growth hormone: GH)	Protein	All growing cells	Stimulates growth, increases metabolic rate
Thyroid			
Thyroxine	Iodinated amino acid derivative	All cells	Stimulates metabolic activity
Calcitonin	Polypeptide	Bone cells	Inhibits Ca^{2+} release from bone
Parathyroid			
Parathyroid hormone	Polypeptide	Bone and digestive tract	Stimulates Ca^{2+} release from bone and Ca^{2+} uptake from digestive system
Adrenal Cortex			
Corticosteroids (cortisol)	Steroid	Many tissues	Stimulates carbohydrate metabolism
Mineralocorticoids (aldosterone)	Steroid	Kidney and blood	Stimulates sodium retention
Adrenal Medulla			
Adrenalin, noradrenalin (epinephrine, norepinephrine)	Amino acid derivative (modified tyrosine)	Muscles, liver, circulatory system	Flight-or-fight response

The Pituitary Gland

The **pituitary gland** is a small structure about the size of a large bean located beneath a region of the brain known as the **hypothalamus.** It is found near the center of the skull, in one of the most protected locations in the body. Interestingly, the pituitary does not develop as a single structure during embryological development. Instead, its *anterior* and *posterior* portions are assembled from two tissues of completely different origin (Fig. 39.11). As befits

their separate origins, the two parts of the pituitary function as completely independent glands and work in distinctly different ways.

Each part of the pituitary, however, is closely tied to specific regions within the hypothalamus. Because of these connections, we now know that the pituitary does not function by itself as the supreme regulatory gland it was once thought to be. Instead, the hypothalamus and pituitary together provide vital links between the nervous system and other endocrine glands.

Organ / Hormone	Type of Hormone	Target Tissue(s)	Effects
Pancreas			
Glucagon	Polypeptide	Liver and other cells	Stimulates glycogen breakdown, increases blood sugar
Insulin	Polypeptide	Liver and other cells	Stimulates glycogen synthesis, decreases blood sugar
Pineal			
Melatonin	Amino acid derivative	Nervous system, gonads	Regulates light/dark cycles; function in humans not entirely clear
Ovaries			
Estrogen	Steroid	Cells throughout body	Female sexual characteristics, development of uterine lining
Progesterone	Steroid	Uterus	Growth of uterine lining
Testes			
Testosterone	Steroid	Cells throughout body	Male sexual characteristics, sperm development
Thymus			
Thymosin	Polypeptide	T-lymphocytes	Stimulates growth and development of T-lymphocytes
Digestive (GI) Tract			
Gastrin	Polypeptide	Stomach lining	Stimulates HCl production
Cholecystokinin	Polypeptide	Pancreas	Stimulates release of digestive enzymes and bicarbonate
Secretin	Polypeptide	Pancreas	Stimulates release of digestive enzymes and bicarbonate
Enterogastrone	Polypeptide	Stomach lining	Inhibits smooth-muscle contraction and HCl release
Kidney			
Erythropoietin	Glycoprotein	Bone marrow	Stimulates production of red blood cells
Heart			
Atrial Natriuretic Hormone	Polypeptide	Kidneys	Stimulates removal of sodium from blood

The Posterior Pituitary and Its Hormones

The **posterior pituitary** grows down from the base of the brain as an extension of the hypothalamus and consists, in large part, of specialized **neurosecretory cells.** As their name implies, these cells act both as nerve cells and as hormone-producing cells. Their cell bodies—where hormones are produced—are located within the hypothalamus. From these cell bodies, long processes run several centimeters through the stalk that attaches the pituitary to the brain. Those cell processes end in a net-

work of branches that are wrapped in a dense network of fine capillaries.

Because neurosecretory cells are "rooted" in the brain, they are under direct control of the nervous system. Hormones produced in the cell bodies of these cells are transported down through the cell processes into their terminal branches. When the appropriate region of the hypothalamus is stimulated, the neurosecretory cells release these hormones into the capillary network, which carries them into the general body circulation where they have a variety of effects.

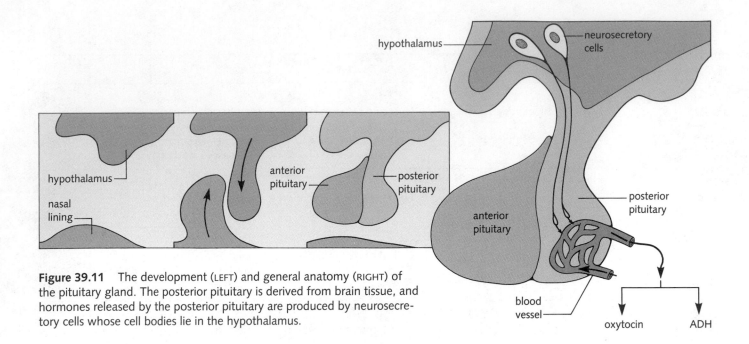

Figure 39.11 The development (LEFT) and general anatomy (RIGHT) of the pituitary gland. The posterior pituitary is derived from brain tissue, and hormones released by the posterior pituitary are produced by neurosecretory cells whose cell bodies lie in the hypothalamus.

The posterior pituitary produces two polypeptide hormones: **antidiuretic hormone (ADH)** and **oxytocin.** ADH affects the kidneys, where it stimulates water resorption. (A *diuretic* is anything that stimulates water loss through the kidneys, resulting in increased urine production. Because ADH does the opposite, preventing water loss, it is known as an *antidiuretic*.) As we saw in Chapter 38 (excretion), when blood becomes too dilute, the hypothalamus responds by not releasing ADH, allowing the kidneys to remove water from the blood. When the blood becomes too concentrated, the hypothalamus causes the release of ADH from the posterior pituitary, preventing further water loss. This self-regulating system forms a classic feedback loop that regulates the concentration of water in the blood with great precision.

Oxytocin, the other posterior pituitary hormone, has two important effects in women. It stimulates *labor*, the smooth-muscle contractions in the uterus that lead to childbirth. Physicians who feel that they must "induce" labor for one reason or another administer injections of oxytocin. Oxytocin is also important after childbirth, when its release causes milk to flow toward the nipples in the mammary glands. This "letdown reflex" is familiar to any woman who has nursed a baby.

The Anterior Pituitary and Its Hormones

The **anterior pituitary,** which forms from the lining of the mouth, grows up to meet the posterior pituitary as it descends from the base of the brain. Because of this dif-ference in developmental origin, the anterior pituitary does not have the same sort of direct neural connection with the hypothalamus that the posterior portion has. Researchers tried to find such a direct nerve link for many years but were unable to find one. Yet there was abundant evidence of some functional link with the brain, because electrical stimulation of the hypothalamus caused the release of several hormones from the anterior pituitary.

Endocrinologists found a clue to the operation of this system in a pattern of blood circulation around the pituitary. A tiny arteriole circulates blood through a dense bed of capillaries that make intimate connections with a small section of the hypothalamus. These capillaries flow together to form a larger vessel but then, instead of leaving the brain immediately, divide again to form a second network of capillaries that surround and penetrate the anterior pituitary. This is a most unusual pattern, and it suggests an unusual sort of connection between the hypothalamus and the pituitary.

Releasing hormones We now know that a group of neurosecretory cells in the hypothalamus produces small molecules (some are merely tripeptides) known as **releasing hormones.** Under appropriate stimulation from the brain, neurosecretory cells discharge hormones into the surrounding capillary network from which they are carried directly to the second capillary bed in the anterior pituitary. There these chemical messengers bind to receptors on anterior pituitary cells, where they regulate the release of other hormones (Fig. 39.12).

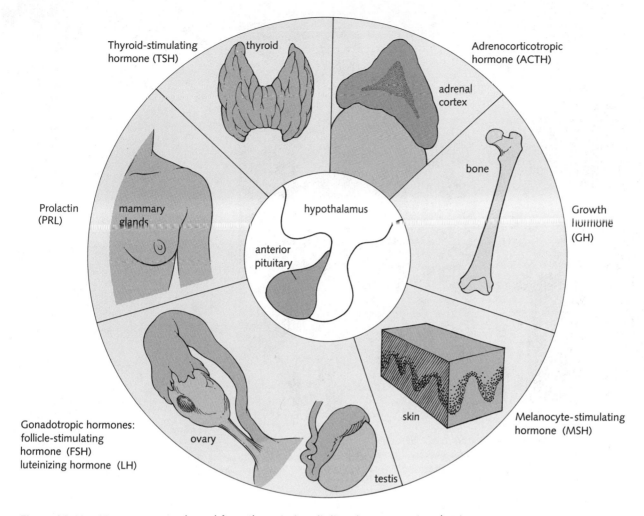

Figure 39.12 Hormones are released from the anterior pituitary in response to releasing hormones produced in the hypothalamus.

Image labels (clockwise from top):
- Thyroid-stimulating hormone (TSH) — thyroid
- Adrenocorticotropic hormone (ACTH) — adrenal cortex
- bone
- Growth hormone (GH)
- Melanocyte-stimulating hormone (MSH) — skin
- testis
- Gonadotropic hormones: follicle-stimulating hormone (FSH) luteinizing hormone (LH) — ovary
- Prolactin (PRL) — mammary glands
- hypothalamus
- anterior pituitary

Anterior pituitary hormones The anterior pituitary produces seven of the most important hormones in the body. Several of these are known as **tropic hormones** because their target organs are several other endocrine glands. A thorough understanding of these hormones requires knowledge of their target glands as well, so we will return later in this chapter, and again in Chapter 40, to discuss in detail their part in the regulatory process.

Thyrotropic hormone (known as **TSH** because it is also called thyroid-stimulating hormone), is a glycoprotein hormone with target cells only in the thyroid gland itself. When TSH binds to receptors in the thyroid, it increases the release of thyroxine, the major thyroid hormone.

Follicle-stimulating hormone (FSH) and **luteinizing hormone (LH)** are called **gonadotropins** because their main targets are the gonads. These glycoprotein hormones are named for their effects in the female, although the same two hormones are present in males, where they perform slightly different functions.

Adrenocorticotropic hormone (ACTH) controls the activity of the adrenal cortex. This hormone, a peptide consisting of 39 amino acids, seems to be released in response to the level of circulating *corticosteroids* in the blood. The other part of the adrenal gland, the adrenal medulla, is not directly affected by any of the pituitary hormones. Instead, signals for the release of adrenalin from the medulla come directly from the nervous system.

Prolactin (PRL), as its name implies, affects the development of the breasts in females and the production of milk. When levels of prolactin drop too low, milk formation in the breasts stops. Males also produce prolactin, but it is not clear whether the hormone plays any functional role in the male.

Figure 39.13 LEFT: The famous showman P. T. Barnum and "General Tom Thumb," a pituitary midget who starred in Barnum's circus shows for many years. RIGHT: A pituitary giant. Abnormal levels of growth hormone (GH) are responsible for both conditions.

Growth hormone (GH), or somatotropin, has target cells throughout the body. This small protein (191 amino acids) exerts powerful effects on cell growth. The hormone is *anabolic*—that is, it switches cellular metabolism in favor of reactions that *build up* larger molecules such as proteins and complex carbohydrates. The levels of growth hormone control the rate at which the body increases in size, and its effects on the growth of the skeletal system help determine the ultimate size to which an individual grows. Abnormally low levels of GH produce midgets; abnormally high levels of the hormone produce "pituitary giants" (Fig. 39.13).

Melanocyte-stimulating hormone (MSH) is also produced by the anterior pituitary. In lower vertebrates this hormone causes an increase in skin pigmentation. Its function in humans is not clear.

MAJOR ENDOCRINE GLANDS AND THEIR HORMONES

The Thyroid

The thyroid gland produces several hormones, but the most important of these is **thyroxine** (Fig. 39.14). Thyroxine is synthesized from the amino acid *tyrosine* in a reaction that requires *iodine*. This is the main reason why iodine is needed in the diet. Thyroid cells contain a powerful active transport mechanism that can pick up trace amounts of iodine from the bloodstream and concentrate it in specialized storage proteins.

Nearly every cell in the body is a target cell for thyroxine. This hormone helps to regulate metabolic rate—the rate at which cells use food and oxygen—and also the rate at which they grow. Increased levels of thyroxine stimulate greater metabolic activity; lower levels depress the metabolic rate.

Maintaining the right level of thyroxine is a critical job. When levels rise abnormally high, a syndrome known as **hyperthyroidism** results, with symptoms that include irritability, weight loss, high blood pressure and increased pulse rate, and bulging eyes (caused by an increase in pressure of the fluid within the eye). When too little thyroxine is produced, the result is **hypothyroidism:** lowered heart rate and blood pressure, sleepiness, loss of energy, and weight gain.

Hypothyroidism is often caused by a lack of iodine in the diet. As the thyroid tries to remedy this by increasing in size and attempting to extract even more iodine from the bloodstream, a noticeable swelling of the gland results, a condition known as *goiter.* The cure for this kind of goiter is to increase the amount of iodine in the diet, which causes the gland to return to normal size. Enlarged thyroids are not uncommon today in the inland areas of developing countries where seafood, which tends to be rich in iodine, is not available. Goiter was once common in parts of the American Midwest until it became standard practice to add small amounts of iodine to table salt.

Underactivity of the thyroid gland early in life can produce one form of dwarfism. A deficiency in thyroxine lowers metabolic rates in the cartilage tissue that forms the ends of growing bones, and a permanent decrease in stature results. Thyroid dwarves also suffer from mental retardation (cretinism) caused by the low metabolic rate of the brain as it develops.

The thyroid produces another major homeostatic hormone known as *calcitonin,* which affects calcium levels in the blood. Calcitonin does not control calcium levels alone but rather works together with another calcium-regulating hormone produced by the parathyroid glands and discussed in more detail later in this chapter.

Feedback control of thyroxine levels

The medical conditions caused by too much or too little thyroxine emphasize an important point that applies to other hormones as well: regulating hormone levels is critical to normal growth, development, and body maintenance. We can now examine the control of thyroxine as an example of the way negative feedback makes possible this sort of precise regulation.

The feedback control system that regulates thyroxine concentration consists of three parts: the thyroid itself, the hypothalamus, and the anterior pituitary (Fig. 39.15). As we have said, most cells in the body are target cells for thyroxine. In the vast majority of these cells, thyroxine stimulates an increase in metabolic activity. But thyroxine also *inhibits* cells in both the hypothalamus and the anterior pituitary. Thus when the blood thyroxine level drops, metabolic activity throughout the body slows a bit. The

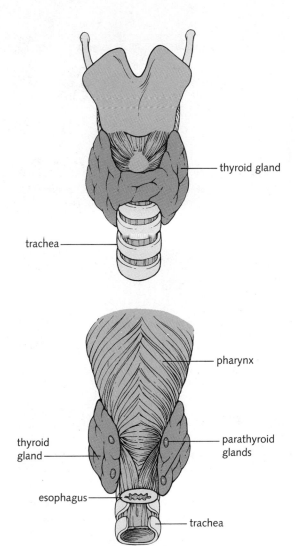

Figure 39.14 The thyroid and parathyroid glands.

hypothalamus responds to this drop in activity and also to the lowered level of thyroxine by producing **thyroid releasing hormone (TRH).** This TRH is carried through the special capillary network we described earlier to the anterior pituitary, where it causes the release of thyroid-stimulating hormone (TSH). TSH, in turn, stimulates the thyroid to release thyroxine, raising the level of thyroxine in the bloodstream and increasing metabolic activity throughout the body.

But because thyroxine inhibits both the hypothalamus and the anterior pituitary, an increase in thyroxine concentration causes the hypothalamus to decrease its production of TRH, thus decreasing TSH release by the anterior pituitary and ultimately reducing the further release of thyroxine. This elegant self-regulating system ensures that the level of thyroxine is appropriate to maintain normal metabolic activity.

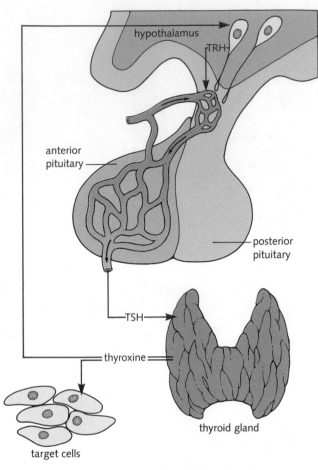

Figure 39.15 Thyroid activity is regulated by a feedback loop involving the hypothalamus and the anterior pituitary.

The Parathyroids

The **parathyroids** are four tiny glands embedded in the posterior surface of the thyroid. One might expect that these glands are related to the major function of the thyroid, and medical workers believed this for years, but this notion has proved to be mistaken. These glands produce a hormone known as **parathyroid hormone (PTH),** a polypeptide containing 83 amino acids.

PTH acts to increase blood calcium concentrations by stimulating the release of calcium from several sources. The major calcium reservoirs are the bones, where deposits of calcium phosphate make up the greatest portion of the tissue. Increased PTH levels cause calcium to be released from this reservoir into the bloodstream. PTH also converts vitamin D into its active form, which increases the rate at which calcium ions are absorbed from the intestines.

The effects of PTH are counterbalanced by **calcitonin,** a hormone produced by the thyroid, in an example of the

second type of hormonal control mechanism discussed earlier. Calcitonin's effects are precisely the opposite of those of PTH. Calcitonin *reduces* blood calcium levels by preventing calcium removal from the bones. As the other cells of the body take up calcium, blood levels of the ion drop.

The release of both calcium-regulating hormones is controlled directly by the glands that produce them. Each gland responds directly and independently to calcium levels in the blood. The regulated release of these two complementary hormones maintains precise control over the level of calcium in the blood. Precise control is important, because several important cell types require steady calcium levels for their metabolic activities. Nerve cells are particularly sensitive to calcium levels, and muscle contraction and blood clotting require calcium ions. Diseases that result in the overproduction of PTH (or the underproduction of calcitonin) can severely weaken the skeleton by causing a chronic loss of calcium.

The Pancreas and Control of Blood Glucose

The **pancreas** is a large gland located in the abdominal cavity between the stomach and the duodenum (Fig. 39.16). Most of the pancreas is an *exocrine gland* that produces digestive enzymes and releases them through ducts that empty directly into the digestive tract. Scattered throughout the pancreas, however, are small clusters of *endocrine* cells known as the **islets** of the pancreas. Islet tissue contains several different types of cells and produces a number of important hormones. The best known of these are two important hormones that control the level of glucose in the bloodstream, **insulin** and **glucagon.**

To appreciate fully the complex and complementary functions of these hormones, recall for a moment the design and activity of the digestive tract. Remember that after a large meal, blood leaving the intestines carries large amounts of glucose that fuel the metabolism of all the body's cells from neurons to skin cells. Between meals, on the other hand, the intestine adds little or no glucose to the blood passing through it, although the body's other cells keep utilizing that vital energy source. Thus if there were no "sugar buffering system" in the body, blood glucose levels would rise and fall cyclically (Fig. 39.17), with the sort of catastrophic results seen in severe cases of diabetes mellitus.

But there *are* tissues that can serve as "sugar buffers." After meals, the liver can remove glucose from the blood and convert it into *glycogen,* a polymer that may contain thousands of glucose molecules that can be stored for future use. The liver can also use excess glucose to produce lipids that are released into the bloodstream and stored by fat cells. Muscle cells use glucose to fuel their contractions, and a wide variety of other cells use glucose as an

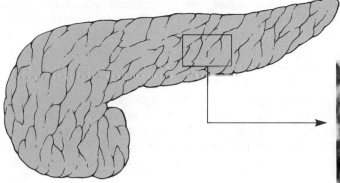

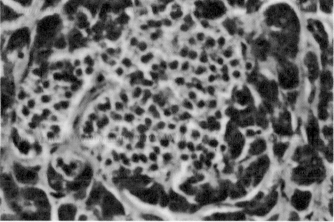

Figure 39.16 LEFT: The pancreas. The endocrine tissues of the pancreas, located in the Islets of Langerhans, produce insulin and glucagon. RIGHT: An Islet of Langerhans is visible in the center of this light micrograph, surrounded by exocrine tissue.

energy source. Between meals, both liver and adipose cells reverse their activity, converting glycogen and fat, respectively, back into glucose. As a result, glucose concentrations in the blood remain quite stable.

These complementary activities (and several related functions) are controlled by the complementary actions of insulin and glucagon. Not surprisingly, these hormones have several important targets—the liver, muscle tissue, and adipose cells.

Insulin, produced by islet cells known as *beta cells,* is released whenever a large amount of sugar is absorbed into the bloodstream. In short order, insulin stimulates the uptake of sugar by all three types of target cells. In both liver and muscle cells, insulin-stimulated cells allow sugar molecules to pass across their cell membranes, convert them into glycogen, and store them. In adipose cells, insulin increases the storage of fat. Insulin also stimulates the production of proteins and fats, and it simultaneously inhibits their metabolic breakdown, thus increasing growth in many cell types. Releasing insulin at just the right time and in just the right amount also prevents extra sugar from being lost through the kidneys. All these activities, in concert, ensure that the valuable chemical energy in sugar is either utilized or stored for future use.

Glucagon, a hormone produced by the *alpha cells* of the islets, is released when blood sugar levels begin to drop. This hormone causes liver cells to break down glycogen and release the resulting glucose into the bloodstream. Thus a proper balance of insulin and glucagon, released at the appropriate times, keeps blood sugar levels relatively constant and makes certain that sugar is effectively used and stored by the body.

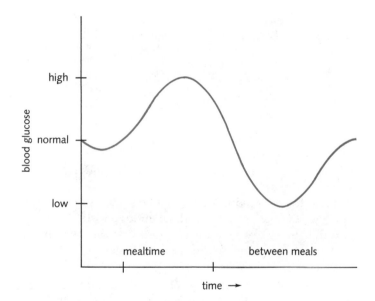

Figure 39.17 Blood leaving the intestines after a meal is loaded with sugar from food absorption. Between meals, blood sugar tends to fall as cells throughout the body use the sugar for energy.

Diabetes Unfortunately, we have direct evidence of the importance of this balance: Important medical complications occur when it fails. By age 65, nearly 3 Americans out of every 100 suffer from a metabolic disease known as **diabetes mellitus.** In adult onset diabetes, the pancreas usually still produces some insulin but not enough for the body to respond properly. In juvenile onset diabetes, on the other hand, the pancreas does not make insulin. This

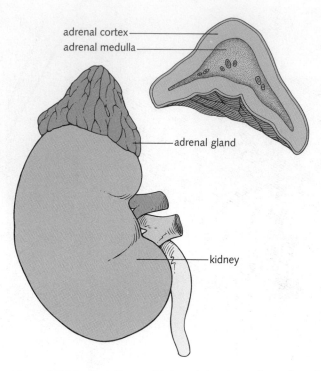

Figure 39.18 Location and internal structure of an adrenal gland.

latter, more severe form of the disease sometimes runs in families, and there may be an inheritable tendency to develop the disease.

Most forms of adult onset diabetes are caused by the failure of the pancreas to produce enough insulin on cue, although there are rare forms caused by the inability of the body to respond to the insulin molecule. Either of these conditions spells trouble. When a diabetic eats a meal, sugar is not taken from the blood into the liver and other tissues as quickly as it should be. The kidneys remove much of the excess sugar, and a great deal of water along with it. This produces one of the principal symptoms of diabetes: production of a large amount of sugar-containing urine. The Greeks noticed that bees would gather around the urine of diabetics; hence the name *diabetes mellitus* (from Greek words meaning "passing-through" and "honey"). Even today, a high sugar level in the urine is often the first symptom of diabetes noticed by physicians.

In addition to surges of blood sugar after eating, all diabetics can experience dangerously low blood sugar levels between meals, because their tissues have not built up stores of sugar. Some cells and tissues suffer local episodes of malnutrition as a result, and damage to parts

of the body can accumulate over many years. Muscle cells, for example, are nearly impermeable to glucose in the absence of insulin, so muscle wasting and weakness are typical symptoms of the disease.

Fortunately, most adult onset diabetics are able to make some insulin, so some cases are relatively mild. In many people, it can be controlled by a careful diet that avoids sudden surges of sugar into the blood. In more severe cases, the only effective therapy is the administration of small, carefully controlled doses of insulin. This enables most diabetics to live quite normally, but it is not a perfect solution. It is impossible by this means to fine-tune the level of blood sugar the way a healthy pancreas does, and fluctuations of therapeutic insulin can cause damage to several tissues.

Juvenile onset diabetes is almost always insulin-dependent, meaning the patients must rely on injections of insulin to survive.

The Adrenal Glands

The **adrenal glands** are paired organs that sit atop the kidneys (the name *adrenal* means "next to the kidney"). Each adrenal can be thought of as two separate endocrine glands: The outer part of the gland, the **adrenal cortex,** is made up of tissue with a completely different function from that of the inner portion, the **adrenal medulla** (Fig. 39.18).

The adrenal cortex   The adrenal cortex makes up about three-quarters of the total mass of each adrenal gland. The cells of the cortex produce a wide variety of steroid hormones essential to normal body function. For this reason, removal of the adrenal cortex is fatal. These hormones, known as **corticosteroids,** are synthesized from cholesterol, and they fall into three broad classes: **glucocorticoids, mineralocorticoids,** and **sex hormones,** which include testosterone and estrogen.

Mineralocorticoids affect salt transport in the kidneys. In the absence of these hormones, the kidneys excrete dangerously high amounts of sodium.

Glucocorticoids regulate certain reactions to stress and infection, such as inflammation and wound healing, and also elevate blood sugar levels in ways similar (although not identical) to glucagon. **Cortisol,** the most important glucocorticoid present in humans, activates chemical pathways in adipose and muscle cells that produce glucose from fats and proteins. By complementing the effects of insulin in these tissues, cortisol makes more glucose available as a source of ready energy. Stress increases the release of cortisol and other glucocorticoids from the adrenals. During periods of difficulty, when we

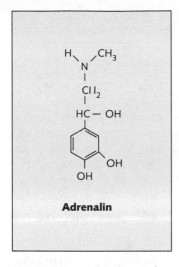

Figure 39.19 The chemical structure of adrenalin, which is synthesized from the amino acid tyrosine.

Figure 39.20 Adrenalin release is part of the "flight-or-fight" response, which plays a role in the actions of predator and prey.

experience physical danger or psychological stress, the release of cortisol makes extra energy available. Cortisol also suppresses some functions of the immune system, including the inflammatory response. This may be one of the reasons why prolonged stress increases our susceptibility to disease.

The adrenal medulla The adrenal medulla does not produce steroid hormones. It produces a series of hormones by chemically modifying the amino acid tyrosine. The best known of these are the compounds **adrenalin** and **noradrenalin** (Fig. 39.19). The word *adrenalin* comes from the same Latin roots as *adrenal* and means "near the kidney." (Greek also has its fans in the scientific community, hence the synonyms **epinephrine** and **norepinephrine.**) Historically, scientists working on the endocrine system have used the term *adrenalin,* whereas scientists working on the nervous system call the same compound epinephrine. Adrenalin is synthesized by the cells of the medulla and stored until it is needed. Small amounts of adrenalin are released on a regular basis, but a flood of the hormone can be produced in response to signals to the adrenals from the nervous system.

Why should the nervous system have a direct connection to the adrenal medulla? The target cells for adrenalin are found throughout the body. When adrenalin enters the bloodstream, it travels quickly to these cells and causes a number of immediate reactions. The heart rate increases, blood pressure rises, sugar reserves are released from the liver, blood is diverted from the intestine and directed toward skeletal muscles, the pupils of the eyes widen, and metabolic rates throughout the body increase. These changes prepare the body for vigorous physical activity. They are part of what is often called the *"flight-or-fight"* response (Fig. 39.20).

The nervous system triggers the release of adrenalin in response to fear or anger, and the set of responses that adrenalin brings about prepares the body to act. The pallid skin and rapid pulse you may feel in stressful situations are a result of this response. Adrenalin release may be thought of as a survival mechanism. It is one of the reasons why the strength and determination of a trapped wild animal are to be feared and one of the reasons why humans are often capable of feats of great strength and endurance in life-or-death situations.

The Gonads

The **gonads,** ovaries in females and testes in males, produce large amounts of steroid hormones. The most common of these are the **estrogens** in females and

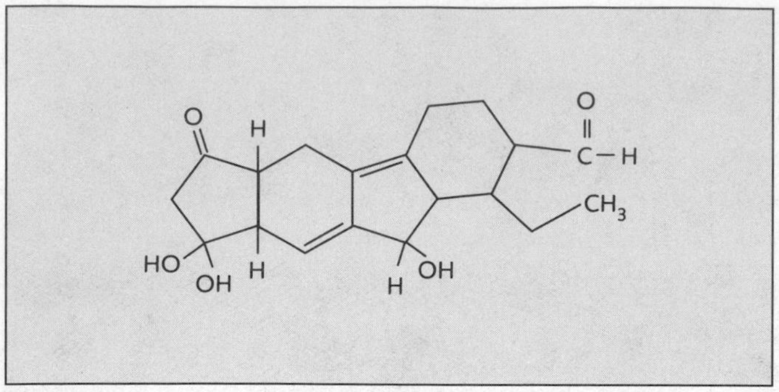

Figure 39.21 The structure of a representative prostaglandin: PGE2.

testosterone in males. These sex hormones are responsible for regulating the reproductive process and producing the sexual characteristics that distinguish males from females. Because they are closely related to the reproductive process, it is best to discuss their functions in conjunction with the reproductive activities they control. We will therefore explore these hormones fully in Chapter 40.

The Thymus

The **thymus** is a gland located near the center of the chest. It reaches its maximum size about the time of puberty and becomes significantly smaller with age. The gland is generally considered part of the immune system and is important in the maturation of *lymphocytes,* specialized white blood cells that help to fight infections. It is included in a listing of the endocrine glands because it produces **thymosin,** a polypeptide hormone, which stimulates the development of certain lymphocytes.

Other Endocrine Glands

A number of organs that are usually *not* associated with the endocrine system produce hormones. The kidneys, for example, produce **erythropoietin,** a hormone that stimulates the development of erythrocytes (red blood cells) in the bone marrow. In Chapter 37 we saw how a series of hormones were produced by cells in the digestive system, helping to regulate the release of digestive enzymes. Recent studies have shown that the heart is also an

endocrine organ (see Theory in Action, The Heart Is an Endocrine Gland Too).

Prostaglandins

The definition of a hormone is a fairly general one, and it leaves open the possibility that other types of chemical messengers may be discovered as we learn more about individual organs and tissues. More than 50 years ago, a group of researchers discovered that a component of semen was able to produce contractions in the uterine muscles of the female. Gradually this substance was purified and shown to be a chemical derivative of fatty acids (Fig. 39.21). Assuming that the material was contributed by the prostate gland (one of the glands that helps to produce the fluid portion of semen), they named the substance **prostaglandin.** This name has persisted, but it is a misnomer: The prostaglandins found in semen are actually synthesized in the *seminal vesicle.* Prostaglandins produced in the lining of the uterus during childbirth stimulate the smooth muscle contractions that occur during labor.

After the first group of prostaglandins in semen had been characterized, researchers began to look elsewhere for these hormones, and prostaglandins were discovered in great numbers. The list of tissues and organs that produce them has grown steadily, and it is now clear that prostaglandins can raise or lower blood pressure, cause the air passageways leading to the lungs to expand or contract, control several aspects of the immune response to foreign organisms and substances, and even regulate the level of pain sensed by the nervous system. Many of these effects are recent discoveries, and there is great hope that prostaglandins and medications controlling their metabolism will be useful in treating a number of difficult medical conditions.

Though all of the data are not yet in, it appears that prostaglandins exert their effects on target cells by binding to a specific receptor molecule and producing a second messenger within the cell. Unlike other hormones, prostaglandins are not stored within cells but, rather, are manufactured and released in response to cell injury or other hormones. They generally act locally, near the site of release. The redness and irritation associated with many injuries are partly a result of prostaglandin release.

Until prostaglandins were discovered, the action of aspirin, the most commonly used drug in many Western countries, had been a mystery. There is good evidence that aspirin is a strong inhibitor of prostaglandin synthesis, and this action may help to dampen the threshold for pain sensation. It also explains the anti-inflammatory action of aspirin and the ability of aspirin to reduce the tendency of blood to clot. Both of these reactions involve prostaglandin hormones.

The Heart Is an Endocrine Gland Too

Not very long ago, it was possible to summarize the workings of the endocrine system with a list of seven or eight glands and a dozen or so important hormones that they produced. In recent years, however, new evidence has forced us to alter this simple description of the endocrine system. Not only has the discovery of prostaglandins shown that tissues throughout the body are capable of producing hormones with powerful effects, but recent research suggests that many organs that nominally are part of other systems also have endocrine functions.

The heart, an organ that is often assigned the purely mechanical role of pumping blood for the circulatory system, is one of the latest organs shown to have an endocrine function. With its two atria filling with blood on every stroke, the heart is in an ideal position to sense the volume of blood in the circulatory system. When too much blood swells the atria, the blood volume should be reduced; when too little blood rushes in, the volume of blood should be increased. The endocrine function of the heart brings these changes about.

Tiny granules are found in certain cells within the atria. These granules contain a polypeptide hormone that is released when the walls of the atria are stretched, either artificially or by an unusually large volume of blood. This polypeptide hormone, called atrial natriuretic factor, or *ANF* (*natriuretic* means "sodium eliminating"), acts on the kidneys to stimulate the removal of sodium from the blood. Water follows the flow of sodium, and the blood volume is decreased. The hormone also acts on the smooth muscle cells that line many blood vessels, helping to lower blood pressure, and it affects receptors in the hypothalamus, inhibiting the release of ADH (antidiuretic hormone). This remarkable system not only brings to light a new hormone produced by a "nonendocrine" organ but also suggests that many other endocrine mechanisms have yet to be discovered lurking in systems and organs.

SUMMARY

Cellular communication is mediated by two major systems, the nervous system and the endocrine system. Messages sent by the endocrine system are carried in the bloodstream by chemical messengers called hormones. Target cells that possess specific receptor molecules respond to hormone messages. The ability of cells to send and respond to chemical messages is found in many organisms, even unicellular forms such as the slime mold. A hormonal system that we analyzed as being representative coordinates the molting process in insects by means of a hormone known as ecdysone.

The major chemical classes of hormones include steroids, proteins, or polypeptides, and amino acid derivatives. Steroid hormones affect their target cells by binding to protein receptor molecules, forming a hormone–receptor complex within the cell nucleus. Steroid hormone–receptor complexes bind to specific sites in chromosomal DNA, where they activate specific genes and change the cellular pattern of RNA transcription. Other classes of hormones bind to receptors at the cell membrane and cause a system of proteins in the membrane to produce second messengers.

The major organs of the human endocrine system are the thyroid, parathyroids, pancreas, adrenals, gonads, thymus, and pituitary. The hormones produced by these glands affect a variety of cellular functions in tissues throughout the body, regulating blood sugar levels, metabolic activity, and growth processes. The pituitary, which consists of anterior and posterior portions, regulates several tissues by means of its hormone products, and it also controls the secretions of other glands via a series of tropic hormones. The release of tropic hormones is further controlled by releasing hormones produced in the hypothalamus. A final class of hormones are the prostaglandins, which are produced by cells throughout the body and have powerful local and long-range effects.

STUDY FOCUS

After studying this chapter, you should be able to:

• Explain the concept of chemical messengers, and describe the roles they play in integrating the activities of an organism.

scribe the major organs and hormones of the human endocrine system.

ive and explain specific examples of hormone action.

Examine the modes of hormone action at the cellular level, and discuss the diversity of hormone effects.

• Explain the interaction of the hypothalamus with the anterior and posterior pituitary.

• Explain the importance of second messengers in amplifying hormone effects.

TERMS AND CONCEPTS

hormones   *781*
cAMP   *782*
ecdysone   *783*
juvenile hormone   *783*
endocrine glands   *784*
exocrine glands   *784*
steroid hormone   *786*
target cells   *787*
hormone–receptor
   complex   *787*
adenylate cyclase   *788*

second messenger   *788*
protein kinases   *788*
feedback control   *791*
releasing hormones   *794*
tropic hormones   *795*
insulin   *798*
glucagon   *798*
diabetes mellitus   *799*
adrenalin   *801*
prostaglandin   *802*

REVIEW

Objective Questions   (Answers in Appendix)

1. Which organ or tissue is both an endocrine and an exocrine gland?
   (a) pineal gland
   (b) heart
   (c) pancreas
   (d) parathyroid gland

2. Pain resulting from the action of prostaglandins may be relieved by
   (a) aspirin.
   (b) oxytocin.
   (c) somatotropin.
   (d) thyroxine.

3. In animals, hormones are chemical messengers that are transported through _____ and have _____ effects.
   (a) the bloodstream; stimulating or suppressive
   (b) the ducts; stimulating
   (c) the lymphatic system; only stimulating
   (d) the nervous system; stimulating or suppressive

4. Neurosecretory cells in the hypothalamus produce
   (a) tropic hormones.
   (b) mineralocorticoids.
   (c) corticosteroids.
   (d) releasing hormones.

5. The action of thyroxine on the hypothalamus–pituitary system is an example of
   (a) positive feedback.
   (b) negative feedback.
   (c) end product inhibitor.
   (d) the second messenger effect.

Discussion Questions

6. What makes a hormone a hormone? What requirements does a molecule have to meet to be classified as a hormone?

7. Why are hormones associated with homeostasis? In what respect do hormones contribute to homeostasis?

8. How is the method of action of steroid hormones different from that of other hormones?

9. Describe the basic types of second messenger systems. Why does a second messenger system enhance the effect of a single hormone molecule? Do steroid hormones have the equivalent of a second messenger?

10. In what respect does the link between the nervous system and the release of adrenalin by the adrenal medulla increase the effectiveness of a nervous response?

11. Some individuals suffer from a genetic disorder in which the hypothalamus is unusually sensitive to the action of thyroxine. How is this oversensitivity likely to affect their overall metabolism? Explain your answer.

READINGS

Lienhard, G. E., J. W. Slot, D. E. James, and M. M. Muckler. "How cells absorb glucose." *Scientific American* 266 (January 1992): 86–93. Insulin and glucagon regulate the absorption and release of glucose by cells thoughout the body. But this hormone-controlled system also involves specialized proteins that allow glucose to cross cell membranes, and these proteins are the focus of this excellent article.

Weissmann, G. "Aspirin." *Scientific American* 264 (January 1991): 84–90. How does aspirin work? Many of its effects are explained by the fact that it inhibits prostaglandin synthesis. But that's only part of the story. Aspirin has other modes of action as well, and these are explored in this article.

Atkinson, M. A., and N. K. MacLaren. "What causes diabetes?" *Scientific American* 263 (July 1990): 62–70. Diabetes is one of the major health problems in the United States and throughout the world. This article incorporates the latest information on the causes and effects of this disorder.

Cantin, M., and J. Genest. "The heart is an endocrine gland." *Scientific American* 254 (February 1986): 76–81. An excellent review of experiments that led to the identification of atrial natriuretic factor, ANF, a hormone produced by the heart.

Fellman, B. "A clockwork gland." *Science* 85 (May 1985): 76–81. A lively discussion of the roles played by the pineal gland in a variety of vertebrates, including humans.

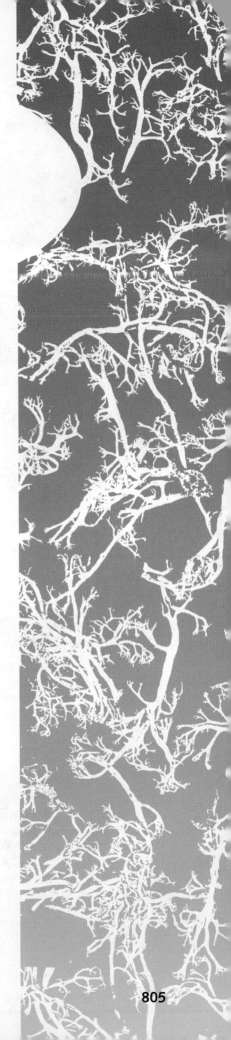

40

Reproduction

It is in this way that everything mortal is preserved, not by remaining the same, which is the prerogative of divinity, but by undergoing a process in which the losses caused by age are repaired by new acquisitions of a similar kind . . . ; it is in order to secure immortality that each individual is haunted by this eager desire and love.

—Diotima to Socrates in Plato's *Symposium*

*a*n elaborately shaped orchid flower entices a male bee with colors and odors that mimic the female of his species. A peacock spends his entire life dragging an impossibly large tail behind him—all for the few brief moments when he succeeds, through its size and color, in attracting the attention of a peahen. And two humans—after more than a decade of adolescent role playing and earnest searching—collapse, exhausted, in each others' arms.

The drive to reproduce is a fundamental part of life on earth. Among bacteria and plants it occurs as matter-of-factly as growth; in humans it is often accompanied by an addictive mixture of angst and ecstasy. But in every case, the act of reproduction is an organism's link with both past and future and is the key to its species' survival.

FORMING A NEW GENERATION

Among many single-celled organisms, reproduction is as simple as cell division; a lone bacterium can divide with staggering speed in a culture flask, filling it with billions of cells in a few days. Organisms that undergo **asexual reproduction** give rise to offspring that are genetically identical to themselves, sometimes blurring the distinction between parent and offspring. Asexual reproduction is highly efficient, for every member of an asexual population, even isolated individuals, can reproduce.

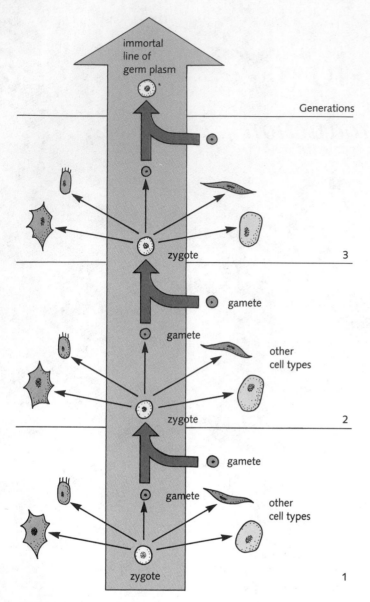

immortal
line of
germ plasm

Generations

zygote

3

gamete

gamete

other
cell types

zygote

2

gamete

gamete

other
cell types

zygote

1

Figure 40.1 Of all the cells in an adult, only the reproductive cells have a chance to become part of a new generation. These germ line cells are part of an unbroken succession of reproductive events that links each organism with its ancestors.

Among many multicellular creatures, however, **sexual reproduction** becomes so complex that its requirements govern many aspects of life from birth to death. At first glance, sexual reproduction seems not only more complicated than asexual reproduction but less efficient as well. Because two individuals are required for reproduction, but only one can actually produce eggs, sexual reproduction seems to cut the number of potential offspring per individual in half. Why, then, is this pattern of reproduction so widespread?

The classic view holds that sex has become common because the shuffling and recombination of alleles of genes during meiosis, combined with the addition of "new" alleles from another individual, produces new—and potentially useful—combinations of genetic information in offspring. This genetic diversity, as we have discussed in earlier chapters, produces variability in a population which may be useful over evolutionary time as that population faces competition and environmental challenges. Although this is probably true, some biologists view it as a secondary benefit of—rather than the primary impetus for—the evolution of sex.

Although Diotima spoke of love, procreation, and immortality from a philosophical point of view, her thoughts closely parallel a new hypothesis on the importance of sexual reproduction that is based on our increased understanding of molecular biology. We have known for some time that every body cell of a multicellular organism must eventually die, for no organism is immortal. Even when cells (other than cancer cells) are removed from animals and reared under perfect culture conditions, the cell line eventually becomes senile and dies. One reason for this unavoidable mortality is that errors in DNA replication and damage from environmental mutagens accumulate in somatic cell DNA throughout an organism's lifetime. Though some of these errors can be repaired before mitosis, others cannot, and both cell lines themselves and the organisms in which they exist age as damage mounts.

The cells of the reproductive system, however, are endowed with a potential that no other cells possess: to become part of the next generation. In that restricted sense, these cells have a shot at immortality. Recognizing this possibility, biologists sometimes refer to the cells of the reproductive system as being part of the **germ line**—a line of cells that are passed from one generation to another (Fig. 40.1).

Furthermore, researchers have recently learned that the pairing and recombination of chromosomes during meiosis offers an opportunity for a fundamentally different sort of DNA repair than normally occurs during mitosis. For this reason, as germ cells are produced during meiosis, even serious, double-stranded damage to DNA can often be repaired. In effect, this powerful repair mechanism—together with the addition of new genetic material from another parent—does indeed repair the "losses caused by age" to DNA. In the process, it offers cells of the germ line the closest thing on earth to immortality.

The unique position of the reproductive system in the formation of the next generation leads to some interesting paradoxes. Though they are essential to the survival of the species, the reproductive organs form the only system in the body that is not essential to the survival of the individual. An injury to the reproductive system is usually not life-threatening, and it is even possible to remove the entire system without reducing an individual's chances for survival. Yet functioning reproductive systems do much more than produce gametes; their hormones (in addition to preparing the body physically for reproduction) influence the nervous system, producing patterns of behavior tending to ensure that every individual will do its best to pass its genetic material on to the next generation.

THE MALE REPRODUCTIVE SYSTEM

In humans and other mammals, the male reproductive system has the task of producing and delivering **sperm cells,** the male gametes. Sperm are produced in the **testes** and suspended in a fluid known as **semen,** which is produced by glands that line the reproductive tract. Semen is stored in the reproductive tract and released into the fe-

male reproductive system through the male copulatory organ, the **penis,** during sexual intercourse.

The Human Testes

Reproductive systems include *primary sexual organs*, which produce sperm and eggs, and *secondary sexual organs*, which are important to the reproductive process but do not actually produce gamete cells. In the male, the primary sexual organs are the testes (Fig. 40.2). Although they develop within the abdomen, human testes descend at birth into a sac known as the **scrotum.** Sperm cannot be produced at normal body temperatures, and the external location of the testes lowers their temperature by 2°–3°C, just enough to allow sperm to be produced.

The testes are composed of tightly coiled tubes known as **seminiferous** ("sperm-bearing") **tubules.** The combined length of human seminiferous tubules (which is nearly 250 m long) is lined with sperm-producing cells.

Figure 40.2   The reproductive system of the human male.

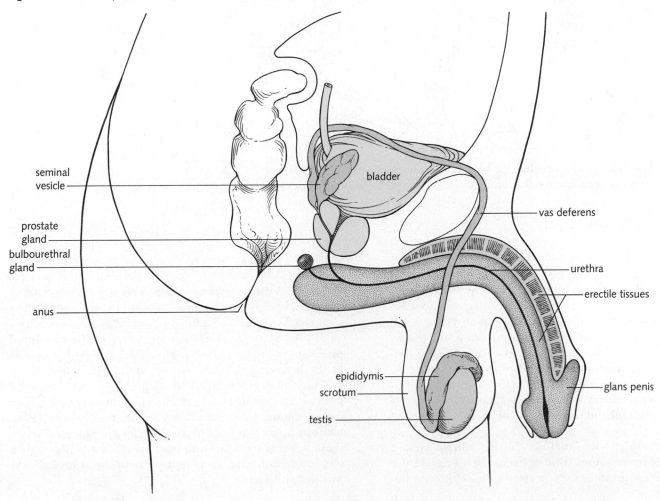

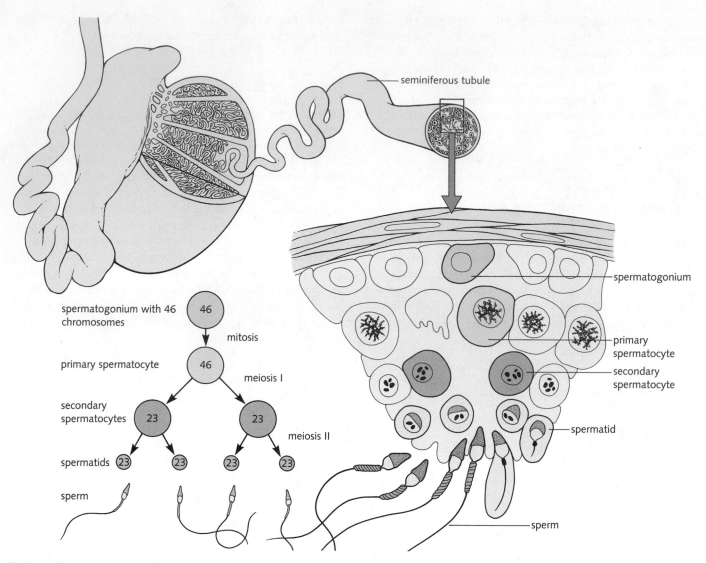

Figure 40.3 Sperm are produced within the testes in the seminiferous tubules. Meiosis takes place as mature sperm cells are produced from spermatocytes.

Sperm development along the walls of these tubules is a continuous process. Cells called **spermatogonia,** found near the periphery of the tubules, divide by mitosis, providing a steady source of new cells that are available for sperm development. A few of these spermatogonia move away from the wall of the tubule and increase in size, becoming **primary spermatocytes.** Meiosis now begins, and each primary spermatocyte passes through two meiotic divisions. The two cells formed by the first meiotic division are known as **secondary spermatocytes,** and the four cells formed by the second meiotic division are known as **spermatids.**

At the completion of meiosis the haploid spermatid has only 23 chromosomes (one of each pair), whereas the diploid primary spermatocyte possessed 46. However, the spermatid is not yet ready for the reproductive process. It must first pass through a complex developmental process in which the appearance of the cell undergoes a complete change. Its nuclear chromatin is condensed into a compact *headpiece*, and a long, powerful *flagellum* develops from the centriole. The Golgi apparatus of the spermatid produces at the tip of the sperm a flattened vesicle known as the **acrosome.** The acrosome contains a collection of enzymes that will help to break through the protective layers surrounding the egg cell. The complete process of development, from spermatogonium to mature sperm, takes approximately 72 days (Fig. 40.3). Released when their development is complete, sperm travel through a system of collecting ducts that leads to the rest of the reproductive system.

The production of sperm in humans is a continuous process that occurs in "waves" of development throughout the testes. Within any single portion of the tubule, nearly all of the primary spermatocytes divide at the same time. In a nearby segment, the secondary spermatocytes may be dividing. These waves of development sweep through the tubules, ensuring that a continuous supply of mature sperm is available at all times. In a healthy male, a single drop of semen contains between 5 and 10 million active sperm (Fig. 40.4).

Many mammals do not reproduce year-round, as humans are capable of doing, but instead have seasonal reproductive cycles. In males of such species, sperm development throughout the tubules is coordinated in such a way that sperm develop together in preparation for the breeding season.

The testes contain two other cell types: **Sertoli cells** are found within the seminiferous tubules, where they aid in sperm development by providing nutrients. **Interstitial cells of Leydig** (*interstitial* means "in between") are found between the tubules and are not directly involved in sperm production, but they produce *testosterone*, the male sexual hormone that is necessary for sperm development.

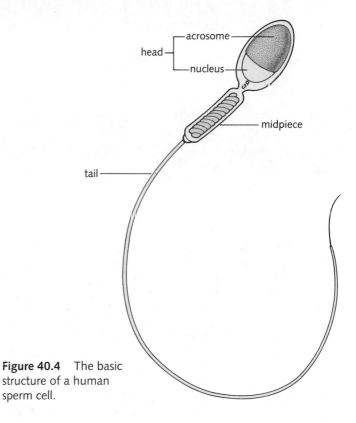

Figure 40.4 The basic structure of a human sperm cell.

The Male Reproductive Tract

The other organs and tissues of the mammalian male reproductive system are primarily concerned with the storage and delivery of sperm cells. They include the **epididymis,** through which sperm are collected from the testes and channeled to the **vas deferens,** which leads into the lower portion of the abdomen. The **seminal vesicle** secretes fluids into the vas deferens, and the **prostate** and **bulbourethral glands** secrete fluids directly into the **urethra,** the tube through which urine is passed from the body. These fluids produce semen, a suspension of salts, nutrients, and sperm.

As shown in Fig. 40.2, just beneath the prostate gland the vas deferens merges with the urethra. During sexual intercourse, semen is released into the urethra and leaves through the tip of the penis.

Male Sex Hormones

The testes are directly affected by two hormones produced by the anterior pituitary: **LH** (luteinizing hormone) and **FSH** (follicle-stimulating hormone), both of which were first discovered (and named) in females. These two hormones are the major **gonadotropins** (gonad-controlling hormones) in males as well as females (Fig. 40.5). **Gonadotropin-releasing hormone** (GnRH), produced by the hypothalamus, controls the release of FSH and LH from the pituitary. The target cells for LH

are the interstitial cells of the testes. LH stimulates interstitial cells to produce the steroid hormone **testosterone.** Testosterone and FSH together act on spermatogonia and induce the development of sperm.

In humans, testosterone receptors are found in a variety of cell types, so this hormone, in addition to directing the development of sperm, has several body-wide effects. The rapid increase in testosterone at puberty induces the development of **secondary sexual characteristics.** These include an increase in facial and body hair, a deepening of the voice, and a pattern of physical development that includes greater stature and a tendency to accumulate muscle mass. Testosterone also affects the nervous system, leading to the development of the male sex drive. In many other mammals the intensity of the male sex drive is directly—and exclusively—dependent on the level of circulating testosterone. In humans, however, the connection between hormones and sexual behavior is complicated by many psychological factors.

Because of testosterone's stimulatory (anabolic) effects on muscle development, young athletes are often tempted to take supplemental doses of this drug to increase muscle mass. But because of this hormone's complex effects on many body tissues—including the central nervous system—such steroid treatments pose serious hazards to both physical health and mental well-being.

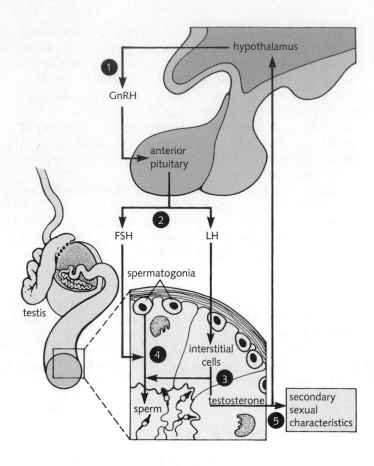

Figure 40.5 1. The hypothalamus produces GnRH (gonadotropin-releasing hormone). GnRH acts on the anterior pituitary. 2. GnRH stimulates the release of FSH (follicle-stimulating hormone) and LH (luteinizing hormone) from the anterior pituitary. 3. In males, the target cells for LH are the interstitial cells of the testis. LH stimulates them to produce the male sex hormone, testosterone. 4. FSH *and* testosterone are both necessary for sperm-producing cells (spermatogonia) in the testes to develop into sperm. 5. In addition to promoting sperm development, testosterone also produces secondary sexual characteristics in males. Testosterone also feeds back to the hypothalamus, where high levels of testosterone inhibit the release of GnRH, forming a negative feedback loop that stabilizes testosterone levels.

Puberty in Males

In young boys, sexual organs are immature. Beginning during embryonic development, the testes produce very low levels of testosterone—not enough to affect secondary sexual characteristics, but enough to direct the proper differentiation of male reproductive structures. Then, sometime between the ages of 11 and 14, a period known as **puberty,** the reproductive tissues develop and the sex organs become fully developed.

We do not understand for certain how puberty begins, but it is likely that during childhood the hypothalamus is extremely sensitive to feedback inhibition by testosterone. During early life, therefore, the hypothalamus does not secrete gonadotropin-releasing hormone (GnRH). Then, at the onset of puberty, the sensitivity of the hypothalamus to testosterone decreases, and GnRH is released. This releasing hormone stimulates the secretion of FSH and LH from the pituitary. LH in turn stimulates the interstitial cells of the testes to produce testosterone. At that point, spermatogonia begin to mature, secondary sexual characteristics develop, and the individual becomes sexually mature.

THE FEMALE REPRODUCTIVE SYSTEM

The female reproductive system also produces gametes, but the mammalian pattern of reproduction places special demands on the female system: offspring develop within the body of the female and are born alive. If fertilization occurs, the zygote becomes implanted in a portion of the reproductive tract and develops into an embryo. The mammalian female reproductive system, therefore, must not only produce mature eggs but also prepare the lining of the *uterus* to receive and nurture the developing embryo. This task requires a variety of complex reproductive organs and an equally complex system of hormonal controls to synchronize the system's activities.

The Ovaries

The primary female sex organs, the **ovaries,** develop from the same embryonic tissues as the testes do in males. The ovaries retain their abdominal location, although, like the testes, they are associated with a system of ducts that leads to the exterior (Fig. 40.6). The basic internal organization of the ovaries is quite different from that of the testes, as shown in Fig. 40.7, which reflects the fact that a woman produces fewer than 500 egg cells in her lifetime, contrasting sharply with the millions of sperm found in every drop of semen.

The female reproductive cells, or **oocytes,** develop from cells known as **oogonia,** the equivalents of spermatogonia in the male. Unlike spermatogonia, however, oogonia complete their process of mitosis before a woman is born. At birth the ovary contains about 400,000 **primary oocytes,** each of which is contained within a single layer of cells known as a *primary follicle,* ready to begin its first meiotic division. Prophase I of meiosis is not completed, however, until just before a mature egg is released from an ovary.

Roughly once a month in a mature woman, a single follicle and its primary oocyte begin to enlarge. The cells of the follicle, known as *granulosa cells,* divide rapidly as the follicle enlarges, and they secrete the *zona pellucida,* a dense layer of material that surrounds the primary oocyte. The first meiotic division takes place within the follicle, producing a **secondary oocyte** and a much smaller cell with very little cytoplasm, the *first polar body.* The secondary oocyte enters prophase of the second meiotic division, and it is in this state that the oocyte is released from the ovary. **Ovulation,** the release of the secondary oocyte, occurs when a mature follicle ruptures at the surface of the ovary, allowing the oocyte to escape (Fig. 40.8). Fluid movements sweep it into the **fallopian tube,** where cilia and smooth-muscle contractions slowly direct the oocyte towards the *uterus.* The second meiotic division, which produces the ovum and a secondary polar body, does not occur unless the secondary oocyte is fertilized by a sperm cell.

The Female Reproductive Tract

If fertilization occurs, the fallopian tubes gradually deliver the ovum into the **uterus,** a thick, muscular organ where a fertilized ovum implants and begins embryonic development. The lining of the uterus, a tissue known as the

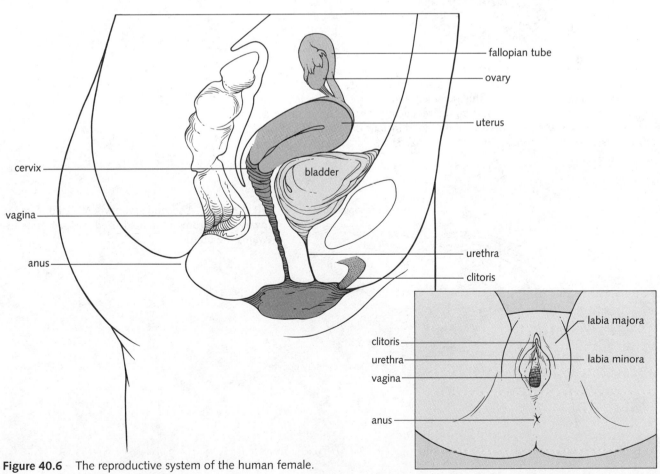

Figure 40.6 The reproductive system of the human female.

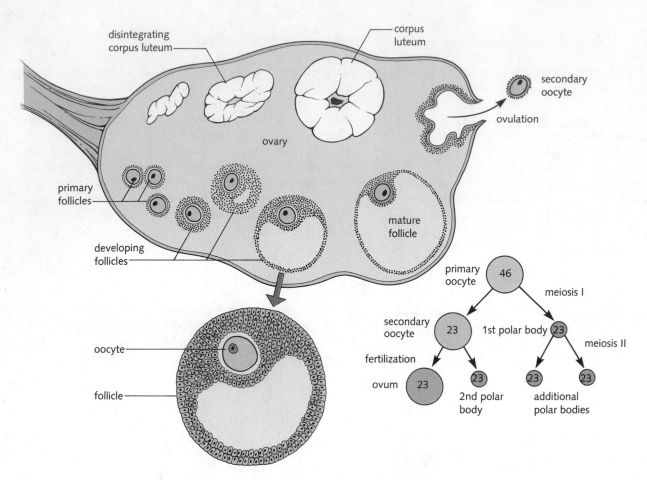

Figure 40.7 A diagrammatic view of the internal organization of the ovary. Oocytes develop within structures called follicles. The development of a secondary oocyte within its follicle is shown as a sequence of stages, beginning at the lower left and moving counterclockwise. After ovulation, the release of a secondary oocyte, the remaining follicular tissue forms a structure known as the corpus luteum. Note that the second meiotic division, which produces the ovum, only occurs when the secondary oocyte is fertilized.

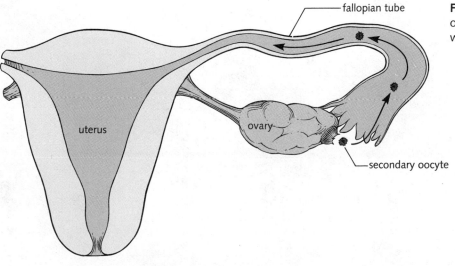

Figure 40.8 A secondary oocyte released from the ovary travels down the fluid-filled fallopian tube towards the uterus.

For Want of a Receptor . . .

We often equate testosterone and estrogen with "maleness" and "femaleness," respectively, when thinking about the secondary sexual characteristics that these primary sex hormones help to produce. However, there is more to the process of sexual development than just the hormones themselves. Consider the case of a genetic abnormality known as *TF syndrome*. Individuals with this disorder are seemingly normal females. They have no apparent medical problems until puberty, when it becomes apparent that something is wrong. Although they seem to mature sexually as young women, they fail to menstruate. It is at this point that patients with TF syndrome often seek help from a physician.

If a karyotype is performed, the physician discovers that the young woman is actually a genetic male: 46XY. The gonads within the patient's abdomen are testes, not ovaries. A check of blood hormone levels shows that estrogen levels are low but that the blood contains a normal level of testosterone for a young man.

The initials TF stand for *testicular feminization*. Individuals with TF syndrome have spent their childhoods believing themselves to be females, and there is no reason for that conviction to change once the syndrome is diagnosed. Like about 10 percent of married women, they are unable to have children. They remain infertile, but they can lead perfectly normal lives as women.

TF syndrome results from a genetic defect in the receptor protein for testosterone. Testosterone, like all steroid hormones, binds to receptor proteins that are present in target cells. Although the testes of a TF patient produce large amounts of testosterone, potential target cells throughout the body lack the receptor. Therefore, they cannot respond to the hormone, and male secondary sexual characteristics do not develop. TF syndrome emphasizes the central role that hormones play in sexual development—and reminds us that hormones cannot act without their receptors.

---

endometrium, has a rich blood supply that provides the ideal environment for early development. The uterus opens into the **vagina,** which leads to the exterior of the body. The connection between the uterus and the vagina, a muscular opening known as the **cervix,** secretes mucus that may block the passageway during pregnancy, helping to isolate the developing embryo. The external opening of the vagina is covered by folds of skin known as the **labia majora** and **labia minora.** As we will see later, the physiology of the male and female increases the likelihood that sexual intercourse will lead to successful fertilization.

Female Sex Hormones and the Reproductive Cycle

Human eggs are released approximately once every 28 days, so the female reproductive tract must prepare for a possible pregnancy at roughly 28-day intervals. Because this preparation requires the coordination of many organs and tissues, a chemical control system is necessary to ensure that all the events involved occur in their proper sequence. Naturally enough, the endocrine system provides this control.

It is easiest to understand the female reproductive cycle when we begin as we did with males—at puberty. Puberty begins in girls usually between the ages of 10 and 14 (on the average, about a year earlier than in boys). At that time, an increase in GnRH secretion by the hypothalamus initiates the release of FSH from the pituitary (Fig. 40.9a). Under the influence of FSH, a few of the many thousands of immature primary follicles begin to grow. Usually, one follicle soon takes the lead and the others stop growing, allowing only a single oocyte to develop to maturity at one time. [Sometimes two follicles develop and, if fertilized, the ova yield nonidentical (fraternal) twins.]

At the same time, joint action of FSH and LH causes the granulosa cells of the follicles to secrete **estrogen,** the first of two principal female steroid sex hormones. Estrogen produces female secondary sexual characteristics,

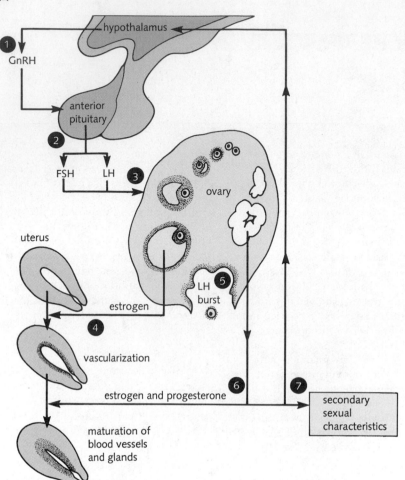

(a)

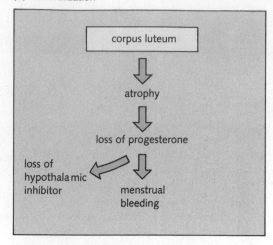

(b) no fertilization

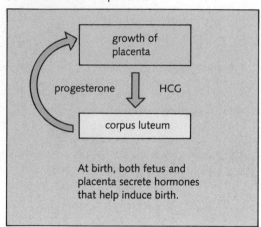

(c) fertilization and implantation

At birth, both fetus and placenta secrete hormones that help induce birth.

Figure 40.9 **(a)** 1. The hypothalamus produces GnRH. GnRH acts on the anterior pituitary. 2. GnRH stimulates the release of FSH and LH from the anterior pituitary. 3. LH and FSH *both* stimulate the maturation of a follicle containing an oocyte. 4. The cells of the follicle wall produce increasing amounts of estrogen, which stimulates development of the endometrium (the lining of the uterus). 5. A burst of LH causes *ovulation:* the wall of the follicle bursts and releases the oocyte. 6. The walls of the ruptured follicle, now known as the corpus luteum, produce estrogen and progesterone, which complete the maturation of the uterine wall. 7. Estrogen and progesterone produce secondary sexual characteristics in females. These hormones also feed back to the hypothalamus, where high levels of estrogen and progesterone inhibit the release of GnRH. Therefore, FSH and LH are not released until estrogen and progesterone levels drop as the corpus luteum disintegrates. **(b)** The drop in estrogen and progesterone levels produces menstruation (a sloughing off of the endometrium) *unless* fertilization and implantation take place. **(c)** If fertilization was successful, the developing embryo and the wall of the uterus produce hormones that keep the corpus luteum functioning and keep the endometrium intact.

including the development of breast tissue and mammary glands, the appearance of hair under the arms and in the genital region, and a typically female pattern of bone and muscle growth in which the pelvis widens and deposits of body fat appear on the hips and thighs.

At the same time, estrogen stimulates extensive cell division in the lining of the uterus. The uterine endometrium develops an extensive blood supply and large numbers of glands. Meanwhile, the follicle continues to grow until, for some reason that is not completely understood in humans, a surge of luteinizing hormone is released from the anterior pituitary. This LH surge stimulates ovulation.

Then, as the oocyte begins its descent through the fallopian tube, the follicle that released it changes into a new structure called the **corpus luteum.** The corpus luteum secretes some estrogen, but it is primarily responsible for the secretion of **progesterone,** a second female sex hormone that causes the maturation of the blood vessels and glands in the lining of the uterus. For this reason, progesterone is essential in the final preparation of the uterus for implantation of the embryo.

The Conclusion of the Cycle

During roughly two days following its release into the reproductive tract, the secondary oocyte is available for fertilization by sperm cells deposited into the reproductive tract via sexual intercourse. What happens next depends on the presence or absence of viable sperm during this period.

If a viable sperm cell fertilizes the oocyte, a fertilized zygote forms. If that zygote is successfully implanted in the uterine lining, pregnancy results.

If no sperm are present, the unfertilized oocyte is soon flushed out of the uterus into the vagina, the uterine lining disintegrates, and a new menstrual period results.

Menstruation: When fertilization does not occur Following ovulation, the lifetime of the hormone-producing cells of the corpus luteum is limited. If no fertilization and implantation occur within seven to ten days after ovulation, the corpus luteum begins to disintegrate, and its production of both estrogen and progesterone falls (Fig. 40.9b). This fall in progesterone levels has two effects.

First, because the enlarged uterine lining is a hormone-sensitive tissue, the decrease in estrogen and progesterone levels causes much of the newly developed tissue to be resorbed. The rest of this tissue detaches from the uterine lining, and as its blood vessels rupture, menstrual bleeding begins. This is why the onset of a "period" usually signals that a pregnancy has not occurred.

Second, the continued drop in circulating levels of estrogen and progesterone frees the hypothalamus from the inhibition caused by these hormones. The hypothalamus then releases GnRH, which stimulates the pituitary to secrete FSH and LH, and the entire cycle begins again.

In the absence of fertilization, this cycling of the reproductive system continues at regular intervals until **menopause,** when follicle development and the release of oocytes cease. Like the time of puberty, the time of menopause is highly variable; it usually occurs between the ages of 43 and 55.

Pregnancy: When fertilization does occur If fertilization occurs, the zygote begins cell division and forms a **morula,** a ball of 30–60 cells by the time it reaches the uterus. Gradually, a cavity develops in the middle of this ball of cells, and the embryo is then known as a **blastocyst.**

The blastocyst implants in the lining of the uterus and grows rapidly just within the lining. The presence of the embryo causes changes to occur in the uterine cells that surround it. At this time, tissues from the embryo join with those of the mother to produce the organ called the **placenta** (Fig. 40.9c). In addition to facilitating the exchange of nutrients, gases, and waste products between developing fetus and mother, the placenta itself serves as a vital, though temporary, endocrine gland.

The first major hormone secreted by the placenta, **human chorionic gonadotropin (HCG),** is essential to continuation of pregnancy. The action of HCG maintains the corpus luteum and its vital secretion of progesterone. High levels of progesterone, in turn, simultaneously preserve the uterine lining and inhibit the hypothalamus from releasing more GnRH during the early part of the pregnancy. Later on in pregnancy, the placenta itself begins to synthesize progesterone, ensuring that the uterine lining is maintained during pregnancy (Fig. 40.10).

This maintenance of the uterine lining gives a woman her first hint that she may be pregnant: Her monthly menstrual period does not occur. Home pregnancy tests generally contain chemicals that test for the appearance of HCG in the urine, often making it possible to detect a pregnancy in its early stages.

Timing of the menstrual cycle The average length of the human menstrual cycle is 28 days. The actual length of the cycle is determined by a number of factors, including the rate at which the corpus luteum disintegrates following ovulation, the speed with which a new follicle develops, and the responsiveness of the hypothalamus to circulating estrogen levels in the bloodstream. Because these individual factors are variable, the length of the menstrual cycle is variable too. Only about 30 percent of women have menstrual cycles that match the 28-day

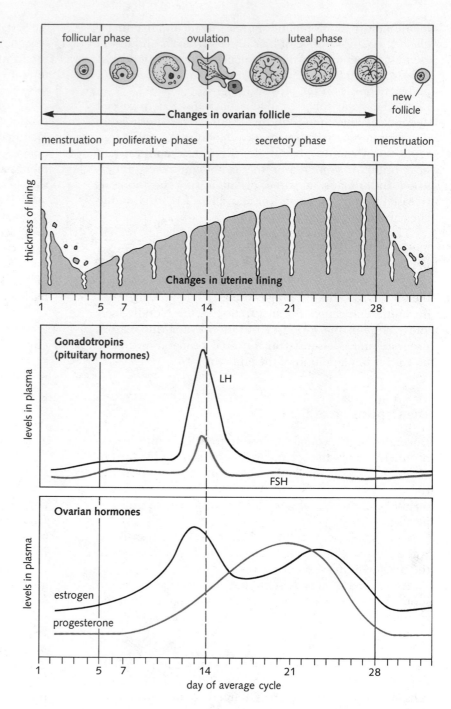

Figure 40.10 *Changes in follicle structure, the uterine lining, and hormone levels associated with the menstrual cycle.*

average; others vary from 21 to 48 days, and in many, the length of the menstrual cycle fluctuates from month to month (Fig. 40.11).

Hormones, Behavior, and Reproduction in Other Mammals

In most mammals, there are closer and much more definite connections between sex hormone concentrations,

sexual behavior, and the release of oocytes. As oocytes are prepared for release, larger amounts of the female sex hormones are released. These hormones prepare the reproductive tract for pregnancy and also cause the animal to become behaviorally receptive to mating, a condition known as "heat" or **estrus.** The **estrous cycle** is this sequence of endocrine, physical, and behavioral changes.

Higher primates are almost unique among mammals in that they do not exhibit an estrous cycle. Humans and other higher primates are physiologically and psychologi-

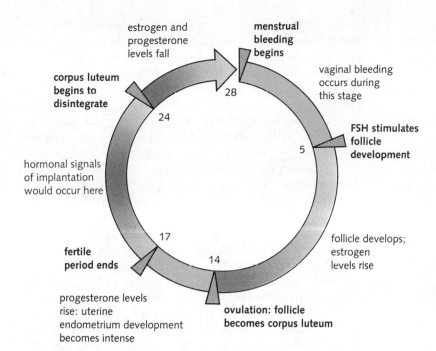

estrogen and
progesterone
levels fall

**menstrual
bleeding
begins**

**corpus luteum
begins to
disintegrate**

vaginal bleeding
occurs during
this stage

28

24

**FSH stimulates
follicle
development**

5

hormonal signals
of implantation
would occur here

follicle develops;
estrogen
levels rise

17

**fertile
period ends**

14

progesterone levels
rise: uterine
endometrium development
becomes intense

**ovulation: follicle
becomes corpus luteum**

Figure 40.11   A simplified view of events in the
menstrual cycle.

THEORY IN ACTION

Other Reproductive Cycles

The human menstrual cycle and the pattern of behavior that goes with it are all but unique in the animal world. Very few mammals, for example, are receptive to mating throughout the year, as humans are. More typical is a pattern closely linked to the seasons in which mating is timed to bring offspring into the world at the most favorable time of the year. In many domesticated animals there is no longer any practical connection between their mating cycles and the seasons, yet they still are receptive to mating only at certain times.

Domestic dogs and cats, for example, mate only during the few days immediately before and after ovulation—a period known as *estrus*, or "heat." Dog breeders recognize this time by noting a swelling of the female's genitalia and the release of a bloody discharge from her vagina. Some amateurs mistakenly take this discharge as the equivalent of a menstrual period and assume it means that the time of greatest fertility has passed. This is not the case. Instead, bleeding results from the rapid growth and enlargement of the uterine lining just prior to ovulation. In such animals, therefore, vaginal bleeding marks the beginning rather than the end of a period of high fertility.

North American deer have one of the most intricate of all sexual cycles. Male deer pass through a yearly cycle in which the level of testosterone in circulation rises and falls rhythmically. Testosterone stimulates the growth of *antlers*, a male secondary sexual characteristic important in the social ritual combat that precedes mating in many groups of deer. Throughout the summer the antlers, encased in a soft layer of skin known as *velvet*, grow rapidly un-

til the mating season begins in the autumn. As the season begins, the deer shed the velvet from their antlers, marking the point at which they are ready to mate with females. By the end of the mating season, another set of hormones causes the antlers to fall off and be lost.

Some of the most efficient reproductive cycles are found in cats and rabbits. In these animals, the activity in the nervous system that is caused by the act of sexual intercourse itself triggers the release of gonadotropic hormones from the pituitary. These hormones cause ovulation to occur within a few hours of intercourse, virtually ensuring that the newly deposited sperm will encounter oocytes. The expression "breeding like a rabbit" really does have a scientific basis!

cally capable of mating at all times, although the release of mature oocytes and the preparation of the uterine lining occur at regular intervals.

SEXUAL INTERCOURSE

The reproductive system is more than a collection of organs and tissues: It is associated with behavior integral to the psychology of both males and females. Part of this is a strong attraction for members of the opposite sex. When the presence of the opposite sex induces sexual excitement, the nervous system causes a series of subconscious changes that facilitate sexual intercourse.

In females, smooth muscle around the opening of the vagina relaxes and the walls of the vagina secrete a fluid that lubricates the passageway and has the right combination of pH, salts, and nutrients to help sperm cells survive. As sexual excitement increases, hollow cavities in the **clitoris,** a small structure (about 2 cm in length) near the front of the vaginal opening, fill with blood, causing it to become erect.

In the male, sexual excitement causes blood vessels leading from the penis to partially close, restricting the flow of blood. Although its exit from the penis is restricted, blood continues to flow into the organ at its normal rate. Thus spongy tissue in the penis fills with blood and it becomes stiff and erect, which enables it to pass between the labia of the female into the vagina (Fig. 40.12). During sexual intercourse, or **coitus,** the penis is inserted into the vagina, the feelings of pleasure in both male and female may rise and fall gradually, eventually leading towards a climax known as **orgasm** (Fig. 40.13).

The tip of the penis is richly endowed with sensory receptors, making it one of the most sensitive areas of the body. As the penis is thrust back and forth within the vagina, mechanical stimulation brings the male to climax. At a point where sexual pleasure intensifies, smooth muscle surrounding the male reproductive tract contracts, expelling as much as a teaspoonful or more of semen into the vagina. The release of semen is known as **ejaculation.** The contraction of smooth muscles during the period of ejaculation is involuntary, so it cannot be controlled by the male.

Orgasm in the female also occurs at the height of sexual excitation, and it is accompanied by smooth-muscle contractions in the vagina and uterus that are believed to assist the entry of sperm into the uterus. Female orgasm is not associated with a distinct physical event equivalent to male ejaculation, but like the male orgasm, it is associated with feelings of satisfaction and relaxation. The sensation of orgasm is often centered in the *clitoris*, and orgasm may begin with a series of powerful contractions of the uterine and vaginal walls, which spread a warm, pleasurable sensation throughout the pelvis. Note that orgasm in human females is not connected with the release of oocytes in any way and is not necessary for fertilization to occur. It is possible for the male to penetrate the reproductive tract, ejaculate sperm, and fertilize an egg without inducing climax in the female.

CONTRACEPTION

The biological organization of the entire reproductive system facilitates mating and reproduction. Although there are only one or two days a month when an oocyte is ready for fertilization, the fact that sperm cells may live within the reproductive tract for several days makes it possible for fertilization to occur a day or two after intercourse.

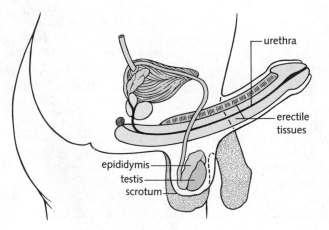

Figure 40.12 During sexual excitement, hollow tissues in the penis fill with blood, producing an erection.

urethra

erectile tissues

epididymis
testis
scrotum

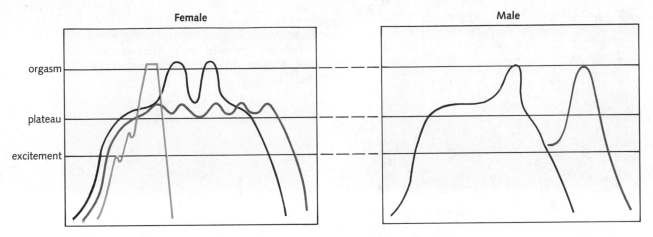

Figure 40.13 Levels of sexual excitement in both male and female usually rise quickly and reach a plateau. Sexual excitement may then rise again, reaching a climax that may be repeated several times in the female. In the male, orgasm is generally followed by a refractory period before a second peak of sexual excitement is possible. The different tracings show several possible outcomes for sexual arousal.

This fact, coupled with uncertainty as to the actual day of ovulation, means that the range of *possible* fertility for an individual woman may span as many as 10 days a month. Within a given year, 85 percent of fertile women who engage in regular sexual intercourse without using **contraceptive** (conception-preventing) techniques will become pregnant.

The simplest and most reliable technique, of course, is **abstinence.** Pregnancy does not occur without the entry of sperm into the reproductive tract. **Withdrawal,** in which the penis is withdrawn just prior to male orgasm, is one of the least reliable contraceptive techniques, because several drops of semen (containing millions of sperm cells) are often released before orgasm without the male being aware of it.

A technique of timing intercourse, sometimes known as the **rhythm method,** is reliable if a woman is able to determine the exact point at which she ovulates each month and avoids intercourse for several days before and after ovulation. This is often possible, because ovulation is accompanied by a slight change in body temperature. However, the irregularity of the menstrual cycle in many individuals makes this method uncertain, at best.

Some of the most useful contraceptive methods involve a physical barrier, such as a **condom** or **diaphragm,** that prevents sperm and oocyte from meeting. The reliability of these methods is increased when a **spermicide** is placed in the vagina to destroy any sperm that may escape through the barriers. The added ability of condoms to block the passage of many disease-causing organisms is an equally important benefit. Because traditional condoms

cover only the penis, a number of "female condoms" have now been developed that can be used by a woman directly and provide more complete protection for her genital region.

In the mid-1960s, an alternative method of contraception became available for the first time—the so-called **birth control pill.** Artificial birth control pills contain synthetic forms and estrogen and progesterone. Taking the pill on a daily basis keeps a steady high level of estrogen in the circulation. After 21 days, the woman either stops taking pills or takes "blanks" containing no hormone for 7 days. This causes estrogen levels to drop low enough to trigger a normal menstrual period but not low enough to stimulate the release of FSH. Because FSH release does not occur, a follicle does not develop, and pregnancy is prevented because no oocytes are released. The hormone-based birth control pills have a very high reliability; fewer than 1 percent of women who use the pill on a regular basis become pregnant in a given year.

In 1991, researchers developed a new method of administering hormonal birth control. Tiny silicon-based capsules containing synthetic progesterone are embedded under the skin, where they gradually release hormone into the bloodstream. These implants last for 5 years or more and are better than 99 percent effective in preventing pregnancy.

RU-486, a drug developed in France, has been used in Europe to terminate pregnancy, although it is not a contraceptive. RU-486 blocks the action of progesterone, the hormone that normally maintains the lining of the uterus. By breaking down the uterine lining, this drug

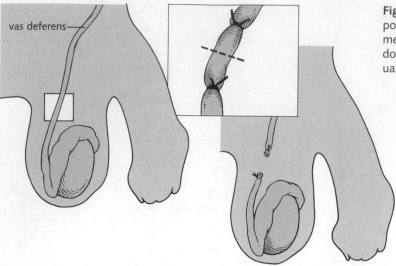

vas deferens

Figure 40.14 A surgical vasectomy. Removing a portion of the vas deferens is a safe and effective method of birth control for males. The operation does not affect sex drive or the ability to have sexual intercourse.

effectively prevents a developing embryo from implanting in the uterus, resulting in the expulsion and destruction of the embryo.

Research is also under way on male birth control pills that block sperm development. **Gossypol,** a compound first extracted in China from cotton seeds, has shown promise, and it may eventually be marketed as the first birth control pill for men.

The most widely used method of birth control in the United States (used by approximately 24 percent of couples) is **surgical sterilization.** Whether in the male or the female, the goal of sterilization is to prevent the delivery of sperm or egg cells to the reproductive tract. In women, this can be accomplished by closing the fallopian tubes that lead from the ovaries to the uterus, making it impossible for eggs to reach any area accessible to sperm. However, **tubal ligation,** as this procedure is called, is a significant surgical operation that is usually done under general anesthetic and therefore carries with it some risk.

The anatomy of the male makes a similar medical procedure much easier. A small incision is made in the scrotum near each testis, and a small portion of the vas deferens is removed. This procedure, known as a **vasectomy,** prevents sperm cells from leaving the testes and causes the male reproductive tract to produce semen that is normal in every respect except for the absence of sperm cells (Fig. 40.14). One disadvantage of any surgical sterilization is the fact that it cannot be easily reversed. Therefore, either tubal ligation or vasectomy is most appropriate for couples who have decided not to have children or any more children.

Another contraceptive method that has shown some promise is the **intrauterine device,** or **IUD.** An IUD is a small metal or plastic structure that is implanted in the uterus by a physician. Because of its size and shape, it remains within the uterus indefinitely. The method by which an IUD prevents pregnancy is not fully understood, but it probably interferes with implantation by slightly disturbing the uterine lining. Although the IUD is reliable and does not require daily attention (as the pill and barrier methods do), some IUD designs have proved harmful to the women using them, producing uterine irritations, infections, and even sterility. These problems have led to a decrease in the use of IUDs, although a well-designed and properly inserted IUD is a safe and effective method of birth control.

Table 40.1 summarizes the rationale, relative effectiveness, and problems associated with each major method of birth control.

Infertility

Pregnancy cannot occur unless both the male and the female reproductive systems are functioning properly. Roughly 10 percent of the couples in the United States are **sterile**—unable to have children. Another 10 percent are **infertile**—able to have only one or two children when they would like to have more. One common cause of sterility or infertility is a low sperm count in semen. Some men produce defective sperm that are unable to fertilize ova, but many others produce too few sperm to be effective in sexual intercourse. Low sperm counts do not necessarily consign a couple to sterility. Many physicians employ a technique in which the sperm from several ejaculates are concentrated in the laboratory to produce semen with a near-normal sperm count. This concentrated semen can then be used to inseminate

Table 40.1 *The Effectiveness of Widely Used Contraceptive Techniques*

Method	Rationale	Effectiveness*	% Used by	Problems
The pill	Synthetic estrogen and/or progesterone prevents follicle development	94–97%	18.5	Must be taken regularly; no barrier against diseases
Female sterilization (tubal ligation)	Fallopian tubes are surgically closed	99.6%	16.6	Sometimes irreversible; no barrier against diseases
Condom	Sheath covering penis traps semen; prevents passage of fluid between sexual partners	86%	8.8	Must be applied immediately before sex; some condoms may leak
Male sterilization (vasectomy)	Vas deferens are surgically closed	99.8%	7	Sometimes irreversible; no barrier against diseases
Diaphragm	Flexible membrane covers cervix and prevents sperm from entering uterus	84%	3.5	Must be applied before sex
Rhythm	Timing intercourse to avoid fertile periods	80%	1.8	Irregular cycles may limit effectiveness
IUD (intrauterine device)	Foreign object interferes with implantation	94%	1.2	Must be inserted by doctor; some risk of infection
Sponge	Diffusable spermicide kills sperm	72–82%	0.7	Must be inserted before sex; more effective if used with barrier device
Spermicidal foams and jellies	Sperm-killing chemicals placed in vagina	79%	0.6	Must be inserted before sex; more effective if used with barrier device

SOURCE: Data from National Center for Health Statistics. Derived from surveys of U.S. women aged 15–44.
*The percentage of women who avoid pregnancy during one year of use.

the female, and it is often successful in producing a pregnancy.

In other cases, physical obstructions in the vas deferens or the fallopian tubes prevent the delivery of reproductive cells, and these problems can be corrected by surgery. Occasionally, low progesterone levels make the uterine endometrium less receptive to implantation than it should be. This condition can be corrected by the administration of oral doses of progesterone. In many women, a lack of sensitivity to LH and FSH prevents the development of a mature follicle each month, making conception impossible. A number of synthetic "fertility drugs" that mimic the effects of these gonadotropins have been developed, and they often induce the oocytes to mature. Sometimes the drugs are too effective: one of their side effects is the possibility of multiple births resulting from the maturation of several follicles at once.

Endometriosis, a disorder produced by fragmentation of the uterine lining, has become an increasingly serious problem affecting fertility in recent years. Fragments of the endometrium (the lining of the uterus) break off and become lodged in other tissues such as the fallopian tubes or ovary. These scattered fragments are affected by hormones regulating the menstrual cycle in the same way as the rest of the uterine lining, producing periodic bleeding and disruption throughout the abdomen. The cause of endometriosis is not clear, but it is most common among women who have put off pregnancy until their thirties. Although extreme cases may require surgical removal of the ovaries, many cases of endometriosis can be successfully treated by administration of the hormones contained in oral contraceptives.

SEXUALLY TRANSMITTED DISEASES

The general pattern of mammalian reproduction includes internal fertilization. This pattern is effective and efficient, but it also creates a perfect opportunity for the spread of specialized parasites from one individual to another. The diseases caused by such organisms are known as **venereal diseases** (from Venus, the Greek goddess of love) or **sexually transmitted diseases (STDs).**

The Spread of STDs

Over the past 25 years, substantial changes in sexual behavior have taken place throughout the United States and other Western countries. The introduction of reliable contraceptive techniques has enabled people to engage in sexual intercourse without fear of unwanted pregnancy. However, contact with multiple sexual partners provides an ideal environment for the spread of STDs (Table 40.2).

Adding to these problems is the fact that none of the STDs produce a lasting immunity to reinfection. Although several STDs can be cured if recognized early, others cannot. Furthermore, several STDs—even those that can be cured if detected early—may cause sterility and other sorts of irreparable internal damage before their symptoms become apparent. For this reason, the only sensible method of dealing with STDs is to take every precaution necessary to *prevent* them.

Syphilis One of the most serious sexually transmitted diseases is **syphilis,** which is caused by a spirochete, a corkscrew-shaped bacterium known as *Treponema pallidum.* The spirochete is a delicate organism that doesn't survive well outside the body; it dies quickly when exposed to air or sunlight. Yet because the organism lives in the moist membranes of male and female reproductive tracts, sexual intercourse provides a perfect opportunity for the spirochete to be passed from one individual to another.

When a person is first infected with the spirochete, a hard sore known as a *chancre* forms at the point where the organism passed through the skin. The sore disappears after a few days or weeks, but this does not mean that the disease has passed. Instead, it has entered a *latent* phase in which it silently spreads throughout the body. It will appear in the reproductive tract, allowing it to be passed to other sexual partners. In addition, it will gradually infect the circulatory system and the nervous system and may

Table 40.2 *The Most Common and Most Serious Sexually Transmitted Diseases*

Disease	Organism	Mode of Transmission	Remarks
Syphilis	Syphilis spirochete (bacterium)	Sexual intercourse; infection occurs through penis or vagina	First sign is a small sore, or chancre; may cause brain damage, blindness, or death; 130,000 new cases per year
Gonorrhea	*Neisseria gonorrhoeae* (bacterium)	Sexual intercourse; infection in urinary tract	Produces burning sensation during urination; may infect fallopian tubes and cause sterility; 1,400,000 cases per year
Herpes	Herpes type II (virus)	Sexual intercourse	Painful inflammation of genital area; cannot be cured; 500,000 cases per year
Chlamydia	*Chlamydia trachomatis* (bacterium)	Sexual intercourse	Urinary inflammation; may cause sterility; difficult to detect; 4,000,000 cases per year
Genital warts	Human papilloma virus (virus)	Sexual intercourse	Painful and unsightly; linked with cervical cancer; 1,000,000 cases of warts; 14,000 cases of cervical cancer
Hepatitis B	Hepatitis B virus (HBV)	Sexual intercourse	Attacks the liver, producing liver malfunction; difficult to treat, causing 5,000 deaths annually; 300,000 cases per year
AIDS	Human immunodeficiency virus (HIV)	Sexual intercourse; intravenous drug use; transfusion with contaminated blood	Causes death by destroying immune response; 35,000 cases reported; 200,000 total U.S. cases and perhaps 1.5 million infected

Source: Data from Centers for Disease Control and Prevention. All numbers are from the United States. Incidences expressed are *new* cases per year, so the actual number of continuing cases in the population may be higher.

In vitro *Fertilization*

Although human fertilization normally occurs within the female reproductive tract, fertilization is a cellular event, and there is no technical reason why it cannot occur in a laboratory dish. In 1978 this possibility became a reality when a baby girl named Louise Brown was born in England. Her parents had been unable to conceive because of a blockage in her mother's fallopian tubes. This blockage prevented oocytes from reaching the uterus and made her infertile. Physicians solved this problem by surgically removing a secondary oocyte from the surface of Mrs. Brown's ovaries. This oocyte was then mixed with Mr. Brown's sperm, and fertilization took place, closely monitored under a microscope.

The fertilized ovum began to divide and, a few hours later, was carefully removed from the dish and placed in Mrs. Brown's uterus. It successfully implanted in the lining of the uterus, and normal development took over. Nine months later Louise was born. The process that the Browns helped to pioneer is called *in vitro* fertilization (*in vitro* means "in glass"). *In vitro* fertilization is now a routine procedure that has helped produce thousands of children for couples. The technical ease with which *in vitro* fertilization can be accomplished is stunning, and it is the product of our increasing knowledge of human reproduction.

As successful as it has been in helping childless couples, *in vitro* fertilization has already raised new ethical problems. It is possible for a fertilized ovum to be implanted in the uterus of someone other than the donor of the ovum, and courts have already been asked to rule on the validity of contracts in which individuals have agreed to bear children for others. In one case, identical twins were born 10 months apart because one of the embryos was kept frozen in liquid nitrogen until the other had been born. Divorce courts have already been asked to determine whether one spouse can prevent the implantation of frozen embryos into another in order to prevent "new" children from adding to a child support award.

Like many advancing technologies, *in vitro* fertilization is already challenging us to develop an ethical framework for considering whether, and under what circumstances, we want to do some of the things science has made possible.

eventually cause death. To make matters worse, the spirochete can be passed from a mother to her unborn child, so children may be infected with the disease before birth (congenital syphilis). In an attempt to guard against this, blood tests for syphilis are required in many states before a marriage license is granted.

Syphilis infections can generally be cured in their early stages by massive doses of such antibiotics as penicillin, but in its later stages, the disease can cause permanent physical damage that cannot be repaired by drugs.

Gonorrhea Another sexually transmitted disease caused by a bacterium is **gonorrhea.** This disease is produced by *Neissereia gonorrhoeae* bacteria, which live in the reproductive and urinary tracts, giving rise to one of the principal symptoms of the disease in males: a painful irritation associated with urinating. Blood and pus are often discharged in the urine as the infection persists. In some individuals, "clap" (as gonorrhea is sometimes known) does not cause any other serious problems. In most women, for example, the disease is virtually without symptoms until the infection produces *pelvic inflammatory disease*, which may result in sterility or even death in some cases. The bacterium can infect a newborn child during its passage through the vagina at birth, and it often causes blindness. This is one of the reasons why it is now standard practice to medicate the eyes of all newborns.

Until recently, gonorrhea—once detected—responded readily to such common antibiotics as penicillin. Unfortunately, however, in several parts of the world where prostitution is widespread, active prostitutes have adopted the practice of regularly dosing themselves with antibiotics as "preventive medicine." You already know enough biology to guess what has happened: these circumstances have bred several strains of antibiotic-resistant gonorrhea organisms that are slowly spreading across the world. This dangerous development may make this disease more difficult to control in the future.

Chlamydia A serious infection known as **chlamydia** is caused by the bacterium *Chlamydia trachomatis* and can be passed on through sexual contact. The symptoms of chlamydia infection are subtle, and individuals—particularly females—often carry the disease for long periods without being aware of it. The bacterium infects the urinary tract and can produce a painful sensation during urination. The results of long-term infection can be serious, and chronic infection is one of the common causes of sterility in women: the bacterium causes severe inflammation of the pelvic area and produces irreversible damage to the reproductive tract. Because of its long asymptomatic period and the ease with which it is transmitted, chlamydia is becoming distressingly common among sexually active individuals. Some studies suggest that as many as 20 percent of college-age women in some areas of the United States are infected with chlamydia.

Herpes   Syphilis and gonorrhea are the "classic" sexually transmitted diseases, but several other disorders have caused increasing concern in recent years. **Herpes** is a family of viruses that cause a number of ailments in various locations in the body. *Herpes type I* is often responsible for cold sores that frequently appear on the lips. But Herpes virus, especially *Herpes type II,* can also infect the genital area. It causes painful, persistent infections that appear as small red blisters on the skin within a few days after infection and generally vanish within two to three weeks.

The virus itself, however, does not disappear. It remains latent, infecting the skin and nerve cells in the genital area. Visible herpes infections may then reappear without warning, brought on by stress, illness, exposure to sunlight, or even the excitement of sexual intercourse. Active virus particles are released from these painful sores, and by means of these particles the infection can be passed on to a sexual partner. A number of treatments, including the antiviral agent *acyclovir,* can reduce the severity of the blisters, but as yet nothing can cure the infection. Childbirth poses an additional hazard to women infected with herpes. The virus can be passed on to the newborn child as it passes through the vagina during birth, causing, in the most severe cases, brain damage or death.

Genital warts   In recent years **genital warts,** which are caused by viral infections, have become an increasing medical problem. Although many warts can be treated with special drugs or surgical removal, they are a painful and unsightly problem. The viruses that are responsible for genital warts are spread through sexual contact, and researchers have now established a clear link between genital warts and several forms of cervical cancer.

Hepatitis B   *Hepatitis* is a term used to describe infections of the liver, and there are several forms caused by different infectious agents. Hepatitis type B is caused by a sexually transmitted virus known as HBV. Once a person is infected, there is no cure, although the infection can be treated and many infected individuals recover from the symptoms of the disease. Still, HBV kills more than 5000 people per year by destroying liver function, and it may be responsible for scores of death from liver cancer. Nearly 300,000 individuals become infected with HBV each year, nearly all from sexual contact. There is an effective vaccine against HBV that is recommended by health authorities to sexually active gay men and heterosexuals with multiple partners.

AIDS   In 1980 and 1981, a number of patients who were suffering from a series of unusual infections appeared at hospitals in New York and San Francisco. The physicians who treated them realized that the immune systems of these patients had all but stopped functioning, making them susceptible to pathogenic organisms that normally are not able to cause disease. Eventually this disorder came to be known as **AIDS (acquired immune deficiency syndrome).**

As physicians and public health officials followed the spread of AIDS, it soon became apparent that this disease, for which the long-term mortality rate is very close to 100 percent, can be transmitted by sexual activity from members of either sex to members of either sex. AIDS will be discussed more fully in Chapter 42, but one aspect of the disease should be stressed here: AIDS, which is currently incurable, is *caused by a virus that can be transmitted by sexual contact.*

42

AIDS can also be spread by transfusions with infected blood and by the sharing of hypodermic needles with an infected person. The rapid spread of AIDS, which may exceed 500,000 cases in the United States by 1995, along with the fatal nature of the disease, has added a new imperative to the prevention of sexually transmitted diseases.

The Prevention of STDs

Researchers are actively working on treatments for each of these diseases, but prevention is always better than cure. Two strategies are effective in the control of sexually transmitted diseases. The first is a reduction in the number of one's sexual partners. The danger of contracting a disease transmitted by sexual contact increases in direct proportion to the number of sexual partners, and a population in which the number of sexual partners is limited provides the fewest opportunities for the spread of such diseases.

The second preventive approach is a change in the most commonly used types of contraception. The pill, surgical sterilization, and IUDs offer no barriers to the

Is Sex the Cornerstone?

From a purely biological point of view, one of the most remarkable aspects of human sexuality is the fact that men and women are receptive to mating throughout the year. What evolutionary pressures could have led to the loss of seasonal estrus in our species? Is it possible that the loss of estrus was a consequence of selective pressure brought about by the beneficial aspects of being able to mate at any time?

Some biologists have speculated that the loss of estrus is at least partially responsible for the basic social fabric of even the most primitive human communities. Human offspring are born in a condition in which they require a greater amount of care for a longer period of time than the offspring of any other organism. If humans mated only during a distinct breeding season, these

biologists reason, there would be little incentive for males to remain with their mates once the season was over. The loss of estrus, however, enables sexuality to form a cohesive bond between males and females, increasing the likelihood that males will remain with their mates and assist in the process of forming a community in which the young can be raised to maturity.

This may not be a flattering way to think about the relations between the sexes: that men must be lured into social groups by the promise of sexual relations. It does, however, suggest a way to explain the evolution of the menstrual cycle to replace estrus. Females who were able to attract the presence of a male through their continuous sexual receptivity were more likely than other females to be successful in raising

their own offspring. Similarly, males who were willing to participate in the formation of family units were more likely to pass their genes into the next generation, inasmuch as their presence increased the chances that their children would survive.

Although this theory is pure speculation, a few scientists have suggested that sexuality may have been the source of the rapid evolutionary success of the human species. By providing the bond that held individuals together in social groups, sexuality may have produced a situation wherein human culture was able to flourish. How remarkable to think that it may not have been our big brains or our dexterous hands that were the keys to our evolution, but rather our constant attraction to the opposite sex!

entry of disease-causing organisms, including the virus that causes AIDS. The only method that does is the condom, and societies in which the use of condoms is highest (Japan, for example) have the lowest rates of sexually transmitted diseases. Although the condom is not 100 percent safe, it does provide a barrier against the entry of viruses and bacteria and helps prevent the spread of diseases associated with sexual contact. Public health officials agree that proper use of condoms will help to stop the spread of sexually transmitted diseases.

SUMMARY

The reproductive system passes the genetic heritage of each species into the future by producing gamete cells from which offspring are formed. The primary sex organs of the male reproductive system are the testes. Sperm cells are produced within

the seminiferous tubules of the testes and suspended in a fluid known as semen. Semen is delivered into the female reproductive tract by the remaining organs of the system. Sexual development in the male is under the control of pituitary gonadotropins, which are themselves regulated by a releasing hormone derived from the hypothalamus. LH stimulates the production of testosterone, the male sex hormone, by the interstitial cells of the testes. Testosterone and FSH promote the maturation of sperm cells, and testosterone has the added effect of producing the male secondary sexual characteristics.

The female reproductive system produces secondary oocytes in its primary sex organs, the ovaries. Under the influence of the endocrine system, a single oocyte matures each month within a cluster of cells known as a follicle. Development of a follicle is stimulated by the pituitary hormones FSH and LH. The follicle cells release estrogen, a female sex hormone, which is responsible for the production of secondary sexual characteristics. Ovulation, the release of the mature oocyte, occurs when the follicle has completely developed, leaving behind the corpus luteum. The oocyte travels down one of the fallopian tubes towards the uterus. If it is not fertilized, disintegration of the corpus luteum causes the uterine lining to be sloughed off in a process known as menstruation. If fertilization occurs, the resultant zygote becomes implanted in the lining of the uterine wall. The implantation of a zygote prevents the corpus luteum from disintegrating and triggers a series of hormonal changes.

Sexual intercourse is a complex process involving a series of emotional and physiological changes that lead to feelings of intense pleasure in both male and female. Fertilization can be prevented by a series of contraceptive methods, including surgical, chemical, and barrier techniques. The act of sexual intercourse also creates an opportunity for disease-causing organisms to be passed from one partner to another.

STUDY FOCUS

After studying this chapter, you should be able to:

- Explain in detail the physiology of the human male and female reproductive systems.

- Describe the interactions between the endocrine system and the reproductive system, and outline the extent to which the reproductive system depends on endocrine messages for its normal functioning.

- Discuss the most widely used methods of artificial contraception, citing the pros and cons of each.

- Identify the prevalence and mode of transmission of sexually transmitted diseases.

TERMS AND CONCEPTS

REVIEW

Objective Questions   (Answers in Appendix)

1. When blood testosterone levels fall below normal,
   (a) LH production in the pituitary ceases.
   (b) sperm production increases.
   (c) GnRH release ceases.
   (d) the pituitary releases LH.

2. What is the sequence of the developmental stages for sperm cells (starting with the earliest form)?
   (a) spermatogonia, sperm, spermatocytes, spermatids
   (b) sperm, spermatogonia, spermatocytes, spermatids
   (c) spermatogonia, spermatocytes, spermatids, sperm
   (d) spermatids, spermatocytes, spermatogonia, sperm

3. Ovulation occurs immediately following a sharp rise in the blood in the levels of
   (a) progesterone.　　　(c) LH.
   (b) estrogen.　　　　　(d) testosterone.

4. Which of the following is *not* a steroid hormone?
   (a) testosterone
   (b) progesterone
   (c) follicle-stimulating hormone
   (d) estrogen

5. The immediate cause of menstruation, the breakdown of the uterine lining, is
   (a) a drop in estrogen levels.
   (b) implantation.
   (c) an increase in estrogen levels.
   (d) an increase in FSH levels.

6. Which birth control method is the *least* effective?
   (a) rhythm　　　　　　(c) condoms
   (b) birth control pill　　(d) IUD

Discussion Questions

7. Describe the basic characteristics of the human male and female reproductive systems.

8. List the major male and female secondary sexual characteristics. Which hormones are responsible for these differences?

9. Some scientific writers have suggested that germ plasm is *immortal*. Although this cannot be strictly true, explain why the statement makes some sense.

10. Efforts to develop a male birth control pill have focused on drugs that would block the actions of FSH. Explain why it would be undesirable to block the actions of LH.

11. Long-term progesterone implant devices are placed just under the skin and release low levels of progesterone for serveral years. Would you expect individuals with these implants to have a regular menstrual cycle? Why or why not?

READINGS

"Special issue: The Science of Sex." *Discover* 13 (6) (June 1992): If you are curious about the biology of sex, this issue is for you. Ten articles in this issue deal with topics ranging from the evolution of the female orgasm to chromosome testing in Olympic athletes.

Adler, J., et al. "Safer sex." *Newsweek* (December 9, 1991): 52–61. This article in a popular newsweekly is a well-written summary of the issues surrounding sexual behavior and the spread of STDs.

Aral, S. O., and K. K. Holmes. "Sexually transmitted diseases in the AIDS era." *Scientific American* 264 (February 1991): 62–69. STDs have been rising rapidly in minority populations in the United States. This very readable article presents some disturbing statistics and suggests a range of solutions to counter these threats to health.

DeBuono, B. A., et al. "Sexual behavior of college women in 1975, 1986, and 1989." *New England Journal of Medicine* 322 (1990): 821–825. Although this is a primary scientific article in a leading journal, it is quite readable. It also carries an important message. The sexual behavior of college students in the United States has changed in the past two decades, but it has not changed enough to prevent the rapid spread of STDs. This article is well worth reading by anyone who is concerned about public health, and it is required reading in K. R. M.'s biology class at Brown University.

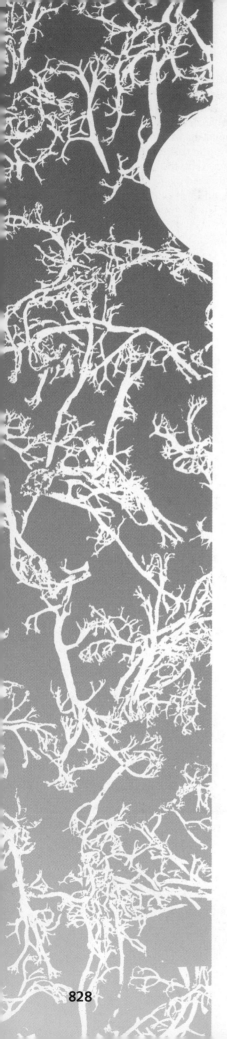

41

Embryology and Development

The most important event in your life is not birth, nor death, nor marriage. It is gastrulation.

—Lewis Wolpert

*e*very day of the year, in every corner of the earth, countless acts in an unseen drama play themselves out on the stage of life. Certain scenes in that drama unfold just beneath the quiet surfaces of ponds and streams. Other acts are set within the confines of eggshells. And still others transpire shrouded in the darkness of a mother's womb.

Yet despite their individual differences, all these dramas are but variations on a theme that unites all animal life: the formation of a new organism from a single, fertilized egg cell (Fig. 41.1). This orderly process of **development**—which begins with the union of sperm and egg and continues throughout the life of the organism—is still the heart of one of biology's greatest mysteries: how does the complexity of a multicellular organism emerge from a process that begins with a single cell?

At the core of that mystery is a phenomenon that is at once both elementary and baffling. Every cell descended from a fertilized egg contains the same set of genes. Yet somehow, during development, each individual's genome directs cells to change in form and function in scores of different ways. The net result, if no errors occur, is a new individual composed of highly specialized cells, each of which performs a different task.

THE PATTERNS OF DEVELOPMENT

Life begins with the fusion of sperm and egg to form a **zygote,** which then develops into an **embryo.** Embryos of different species follow a variety of developmental patterns. We will learn the most about our own development by studying the embryology of our closest relatives, other chordates.

Development in the lancelet: a small egg gives rise to a small organism that resembles the adult.

Development in the frog: medium-sized egg develops into a tadpole, tadpole metamorphoses into a frog.

Figure 41.1 These vertebrates employ a range of reproductive strategies. Vertebrate egg sizes range from the small to the very large. Egg types include those that may or may not be protected by an outer cover. The eggs of most mammals develop inside the mother's body.

Birds and reptiles produce eggs with very large amounts of yolk. During their long developmental period, they are protected with a tough outer shell.

Mammals develop within the bodies of their mothers. This pattern enables mammals to produce small eggs which develop over a long period.

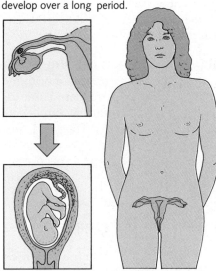

The Egg: Protecting and Nurturing the Embryo

The different types of *ova,* or egg cells, that animals produce reflect a variety of evolutionary responses to two overriding requirements of development: the protection of the helpless embryo and the provision of adequate nutrients and energy to support intense cellular activity. In one way or another, we can understand all of the different types of chordate ova as adaptations in response to these demands.

The ancestors of terrestrial vertebrates were small aquatic organisms, most nearly represented today by such animals as the lancelet *Branchiostoma.* These simple creatures produce small eggs that remain in water as they develop. The eggs contain a modest amount of stored food material known as yolk. The tiny lancelets that hatch from these eggs develop quickly enough to escape the notice of most predators, and only a small amount of yolk is necessary to support development.

Amphibians such as frogs also reproduce in the water, but there is much more yolk in their eggs to support a longer period of development. The egg first develops

into an immature or *larval* stage known as a tadpole. Later, the tadpole goes through a rapid developmental period known as *metamorphosis* and then emerges as a mature frog.

Reptiles, birds, and mammals have become independent of water for reproduction, but they have done so in ways that emphasize their evolutionary ties to our aquatic ancestors. Birds and reptiles produce large fluid-filled eggs provided with enormous amounts of stored yolk and enclosed in tough covering for protection. As the embryo develops within the egg, it produces special tissues that invade the mass of yolk and draw nutrients from it.

Most mammals exhibit a different adaptation: they produce small eggs with very little stored yolk, but embryos produced from these eggs develop and obtain nourishment while protected within fluid-filled membranes in the body of the mother. These mammals have evolved an organ known as the **placenta,** a connection that allows food and oxygen to be passed from mother to embryo.

Animals that lay eggs outside the body of the mother are said to be *oviparous,* a term derived from the Latin words for "egg" and "to give birth." Those that nourish their young inside their bodies are *viviparous* ("to give birth" "alive"). The eggs of a few vertebrates, including many common snakes, develop in a pouch within the mother's body but live off their yolk supplies and receive no further nourishment directly from the mother. These species are said to be *ovoviviparous.*

EARLY DEVELOPMENT

The single-celled zygote develops into a complex, multicellular organism through a process called **differentiation,** during which genetically identical cells develop along different pathways. Ultimately, the single fertilized egg produces the highly specialized cells that make up tissues as diverse as muscle, nerve, skin, and bone. But development is more than just cellular differentiation; embryonic tissues must grow within the limits of a set *pattern.* The organs must develop in certain positions, and the proportions of the limbs must be carefully scaled. The events in different parts of the embryo must be precisely timed.

Embryonic development can be divided into six distinct stages. We will review these stages one at a time and will follow the major structural changes that occur as the embryo develops.

Gametogenesis	Formation of gametes (egg and sperm)
Fertilization	Fusion of egg and sperm
Cleavage	The first rounds of cell division
Gastrulation	Formation of the primary germ layers
Neurulation	Formation of the nervous system
Organogenesis	Formation of the major organ systems

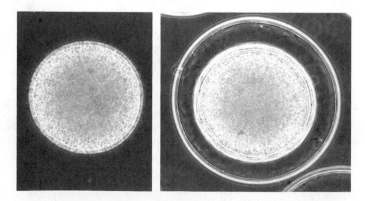

Figure 41.2 LEFT: The cortex of a sea urchin egg contains thousands of cortical granules. RIGHT: These granules are released when fertilization takes place, quickly forming a tough fertilization membrane that prevents other sperm from entering the egg.

Gametogenesis

Development begins with **gametogenesis,** the formation of sperm and egg cells. As we saw in Chapters 10 and 40, the genetic events of meiosis are nearly identical for male and female gametes. Other cellular events that occur during gametogenesis, however, are quite different, for sperm and egg cells are specialized for reproduction in very different ways.

While the sperm cell develops a motile flagellum, the cytoplasm of the larger egg cell is loaded with extensive supplies of nutrients and raw materials to support the early growth of the embryo. In most vertebrates, yolk proteins are synthesized in the liver and carried to the ovaries through the bloodstream. *Follicle cells,* which surround the oocytes, absorb these proteins from the bloodstream and pass them to the developing egg. Furthermore, the active transcription of oocyte DNA fills the cytoplasm with messenger RNA molecules, some of which are stored and used to direct protein synthesis only after the egg has been fertilized.

The egg is surrounded by the *vitelline envelope,* known in mammals as the *zona pellucida.* This layer of glycoprotein serves not only as a protective covering but also as a species-specific recognition site to which sperm can bind.

Fertilization

When a sperm cell makes contact with the glycoprotein layers surrounding the egg, receptors on the membranes of both sperm and egg cell are activated, with virtually immediate and dramatic results.

Response of the sperm Contact causes the *acrosome,* a structure at the tip of the sperm, to fuse with its cell membrane and to release enzymes that begin to dissolve a pathway through the layers surrounding the egg. The sperm cell penetrates through this opening and fuses with the egg cell membrane, releasing its nucleus into the egg cytoplasm. The haploid gamete nuclei then fuse to form the new diploid nucleus of the zygote.

Response of the egg In some organisms, the reaction of the egg cell to the first sperm contact is spectacular. In sea urchins, fertilization activates thousands of *cortical granules* scattered around the periphery of the egg (Fig. 41.2). These granules fuse with the cell membrane and release their contents to the exterior. A rapid reaction between the granule contents and the glycoprotein coating of the egg cell produces a *fertilization membrane* that surrounds the egg in a matter of a few seconds. This fertilization membrane completely blocks the entry of any other sperm, preventing *polyspermy,* or multiple fertilization.

Mammalian ova do not form fertilization membranes, but fertilization causes rapid changes in the zona pellucida. These changes mask sperm-binding sites and ensure that only a single sperm cell is able to fertilize the ovum.

In some organisms, changes in the structure of egg cytoplasm occur at the point where the sperm cell enters the ovum. These changes have profound effects on further development. In frog eggs, for example, the penetration of the sperm causes a movement of the cytoplasm that changes the distribution of pigment granules in the egg cell (Fig. 41.3). This produces a lightly pigmented region known as the gray crescent *opposite* the point where the sperm entered. The position of the gray crescent indicates the anterior–posterior axis of the embryo: the direction in which the head and tail of the tadpole will emerge from the egg.

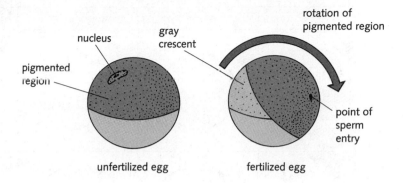

Figure 41.3 Sperm entry into the frog egg causes a movement of pigmented cytoplasm, producing a distinct region called the gray crescent *opposite* the point of sperm entry.

Cleavage in the Embryo: Cell Division Begins

After fertilization, the zygote goes through rapid cell divisions that partition the single cell into individual cells known as *blastomeres*. In species where the egg cell is large, these first divisions can be dramatic, and early embryologists chose the word **cleavage** ("splitting") to emphasize that fact. Differences in the distribution of yolk affect the way in which a zygote begins development (Fig. 41.4).

In eggs with small amounts of yolk, including those of most mammals, sea urchins, and the lancelet, *complete cleavage* occurs, dividing the zygote into cells of roughly equal size. Complete cleavage also occurs in amphibians. However, a complication is introduced by the fact that amphibian eggs contain a higher concentration of yolk at one end of the cell, which is designated as the **vegetal pole.** The opposite, yolk-poor end is called the **animal pole.** At first, cleavage in amphibians produces cells of equal size. Before long, however, a more rapid rate of cell division at the animal pole of the egg produces cells much smaller than those formed at the vegetal pole.

By contrast, in eggs with large amounts of yolk—including those of birds, reptiles, and insects—cell division can occur only at the surface of the egg, a process known as *superficial cleavage.*

The blastula   As cleavage proceeds, the number of cells in the zygote doubles with each round of mitosis. Gradually, the dividing cells form a **morula,** a tight mass of cells named for its resemblance to the mulberry fruit (the Latin word *morula* means "mulberry") (Fig. 41.5). In *Branchiostoma*, the cluster of cells gradually develops a hollow core. This stage of development is known as the **blastula** (the Greek word *blastos* means "bud"), and the cavity enclosed by the single layer of cells is called the *blastocoel.* A

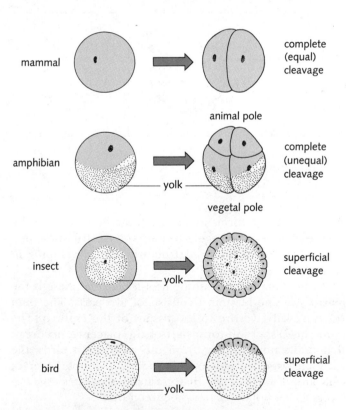

Figure 41.4   In small eggs (amphibians and placental mammals), complete cleavage follows fertilization. Insect eggs undergo superficial cleavage around the periphery, and superficial cleavage of the large eggs of birds and reptiles occurs in a small disk above the massive yolk.

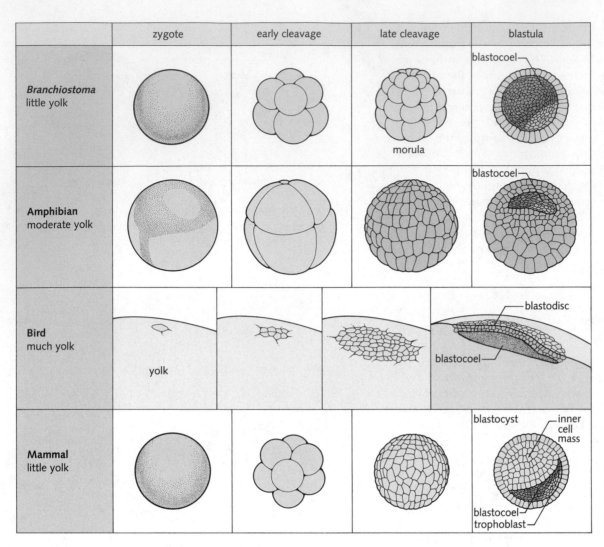

	zygote	early cleavage	late cleavage	blastula
Branchiostoma little yolk			morula	blastocoel
Amphibian moderate yolk				blastocoel
Bird much yolk	yolk			blastodisc / blastocoel
Mammal little yolk				blastocyst / inner cell mass / blastocoel trophoblast

Figure 41.5 Structures equivalent to the blastula, a hollow ball of cells, form in all vertebrates.

similar structure forms in all vertebrate embryos, but the different patterns of egg size and yolk distribution prevent a blastula of uniform structure from developing in all species.

In mammals, the equivalent of a blastula phase is the **blastocyst.** The blastocyst consists of two parts. The *inner cell mass* will give rise to the tissues of the embryo. The outer *trophoblast* will form part of the placenta, an organ that draws nourishment from the mother. In birds, the early cleavages produce a **blastodisc,** a double layer of cells atop the yolk mass that gradually separates into two layers with a hollow *blastocoel* between them.

Gastrulation and Germ Layer Formation

Development in chordates produces three germ layers (Fig. 41.6) that ultimately form organs and tissues. The outermost layer, the **ectoderm** ("outer skin"), eventually forms the skin and the nervous system. The inner layer, the **endoderm** ("inner skin"), forms the lining of the digestive tube and the organs of the digestive system. The layer between the other two is the **mesoderm** ("middle skin"), source of muscle and bone tissue, the circulatory system, and other internal organs and tissues.

During the developmental stage known as **gastrulation,** the first steps are taken to produce these layers and to establish the basic body plan of the embryo. By the time gastrulation is complete, the cells that will produce the three germ layers can be clearly identified, and the embryo is known as a **gastrula.** Details of gastrulation differ among embryo types, but in each case they involve massive migrations of cells inward and on the surface (see Fig. 41.6). The simplest form of chordate gastrulation, exemplified by *Branchiostoma*, involves an inward movement of cells from the surface of the blastula. Like a soft rubber ball squeezed in at one side, the cells at one end of the blastula begin to fold in to form a new cavity, the **archen-**

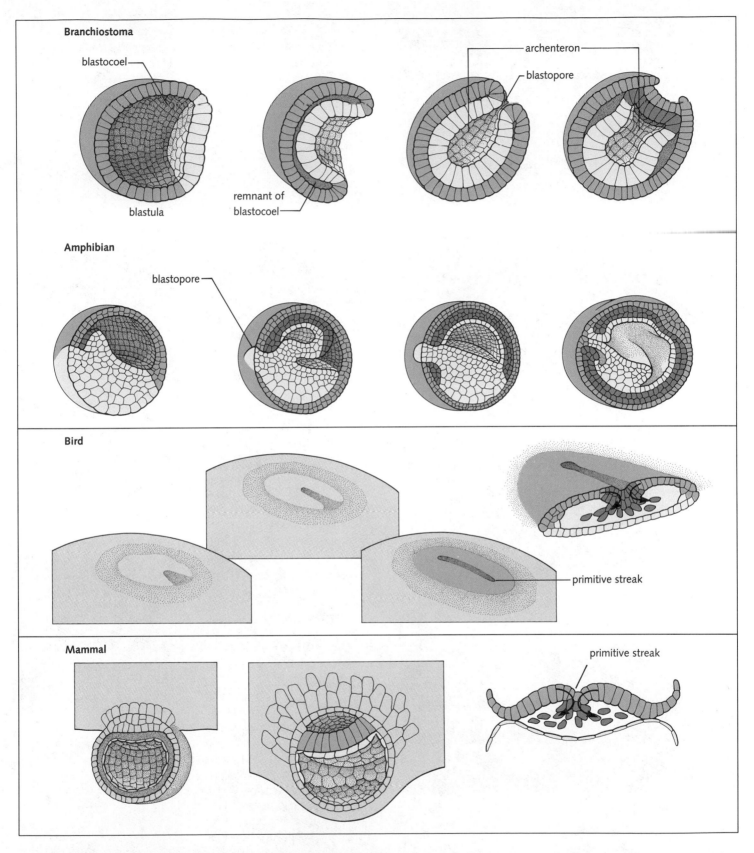

Branchiostoma

blastocoel

blastula

remnant of blastocoel

archenteron

blastopore

Amphibian

blastopore

Bird

primitive streak

Mammal

primitive streak

Figure 41.6 Gastrulation involves the inward migration of cells from the blastula. When gastrulation is complete, the three germ layers, ectoderm, mesoderm, and endoderm, have been formed. In lower chordates, mesoderm develops from early endoderm. In birds and mammals, mesoderm is formed by the inward migration of cells between ectoderm and endoderm.

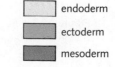

endoderm

ectoderm

mesoderm

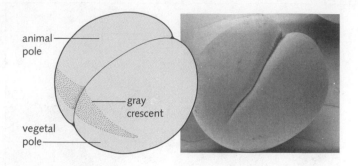

animal
pole

gray
crescent

vegetal
pole

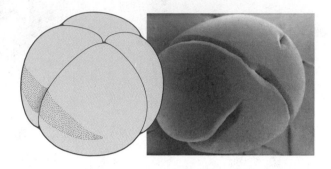

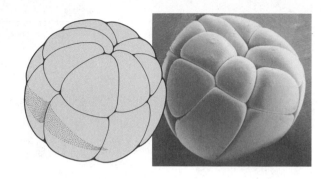

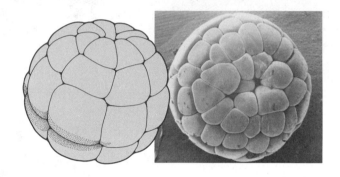

blastocoel

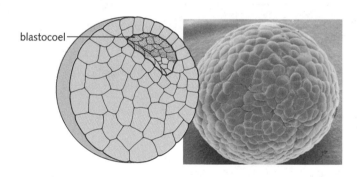

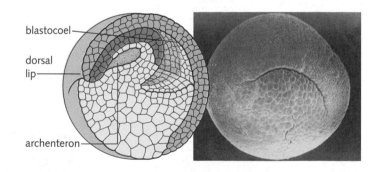

blastocoel

dorsal
lip

archenteron

archenteron

yolk
plug

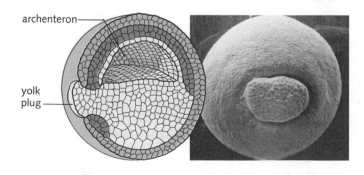

neural fold

notochord

gut
cavity

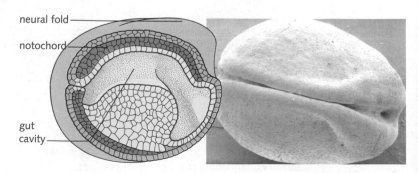

Figure 41.7 Blastulation and gastrulation in amphibians.

teron, that is open to the exterior through the **blastopore.** The *archenteron* (from the Greek words for "primitive" and "intestine") will form the cavity of the digestive system of the embryo. The early gastrula consists of two distinct cell layers, the ectoderm and endoderm, surrounding the archenteron. Within a matter of hours the mesoderm forms between these two.

Gastrulation in amphibians also involves the inward movement of cells. A slit-like blastopore forms at the site of the gray crescent, and cells migrate inward to produce an endoderm-lined archenteron. Further cell movements produce mesodermal layers that give rise to the notochord and internal organs.

In birds and mammals, gastrulation is different. Rather than moving inward toward a blastopore, certain embryonic cells migrate through a zone known as the **primitive streak,** which leads to the interior of the embryo. During gastrulation, movement of these cells toward and through the primitive streak is rapid, resulting in the immediate formation of mesoderm between the two existing cell layers. By the time gastrulation is complete, a small disk of three cell layers has been formed. Figure 41.7 shows, in greater detail, the blastula and gastrula stages of development in amphibians.

Neurulation

One of the most important steps in chordate development is **neurulation,** the development of the nervous system. During the formation of the three germ layers, a block of mesoderm begins to develop into the **notochord.** (All chordates, at some stage in their development, possess a notochord; in vertebrates, the notochord is eventually replaced by the spinal column.) As the notochord begins to take shape, there is a noticeable change in the ectoderm just above it, in the region known as the *neural plate.* Along the length of the notochord, ectodermal cells thicken and change shape in a way that produces paired cell ridges along the length of the organism. Gradually, these ridges fuse to produce a **neural tube** from which the spinal cord is formed.

As neurulation proceeds, clusters of ectodermal cells break off from the folding neural plate and occupy the region between the neural tube and the epidermal ectoderm (Fig. 41.8). These cells, *neural crest cells,* migrate throughout the body and are some of the most remarkable cells in the embryo. Some neural crest cells develop into sensory neurons, others form the *glial* cells that surround some nerve fibers, others become secretory cells in the adrenal medulla, and still others form the melanin-containing pigment cells that are responsible for skin coloration.

In humans, unsuccessful neurulation occurs in about 1 birth in 200, causing a birth defect known as *spina bifida.* This incomplete closure of the neural tube may leave the nervous system open to the exterior of the body. Spina bifida can be corrected by a surgical procedure that encloses the neural tube with skin to guard against infection.

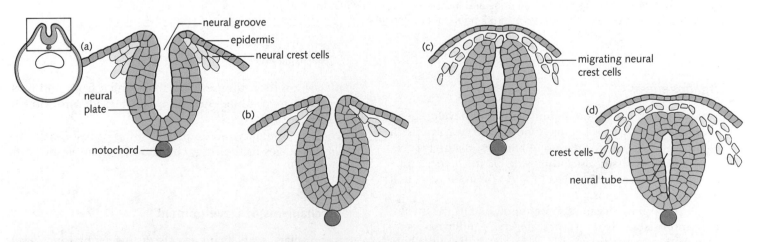

Figure 41.8 Neurulation. Ectodermal tissue forms a hollow tube, which gives rise to the spinal cord and the brain. A few cells break off during neurulation to form neural crest cells, which migrate to various locations throughout the body.

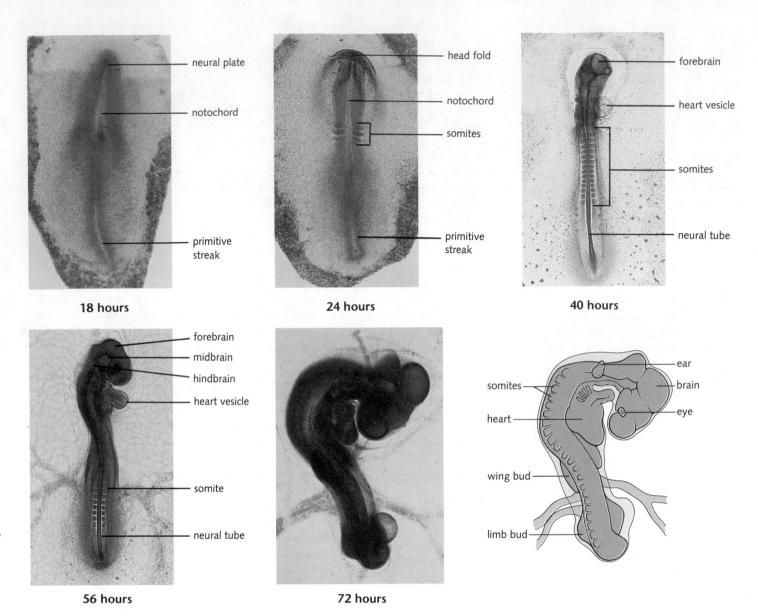

18 hours

- neural plate
- notochord
- primitive streak

24 hours

- head fold
- notochord
- somites
- primitive streak

40 hours

- forebrain
- heart vesicle
- somites
- neural tube

56 hours

- forebrain
- midbrain
- hindbrain
- heart vesicle
- somite
- neural tube

72 hours

- somites
- heart
- wing bud
- limb bud
- ear
- brain
- eye

Figure 41.9 Photographs of embryonic development in the chick. Note the formation of the first somites, blocks of tissue from which bone and muscle will form. The diagram clearly shows the positions of major structures after 72 hours of development.

Organogenesis

Even before neurulation is complete, organ development, or **organogenesis,** has begun. One of the key events in organogenesis is the formation of **somites,** distinct blocks of mesodermal tissue on either side of the neural tube (Fig. 41.9). These somites develop into bone, muscle, and certain internal organs.

At this point, organ development begins to involve cells and tissues from several layers. Epidermal ectoderm and nervous tissue, for example, both contribute to the formation of the eyes. Similarly, ectoderm and mesoderm form the *limb buds* from which front and back limbs develop, while endoderm and mesoderm form the pancreas.

As a summary, a comparison of the major developmental stages in the frog, chick, and human is shown in Fig. 41.10.

Mechanisms of Development

As you have seen, many events during embryonic development are rapid and dramatic. In order for the systems

Figure 41.10 A summary of major developmental stages in the frog, chick, and human.

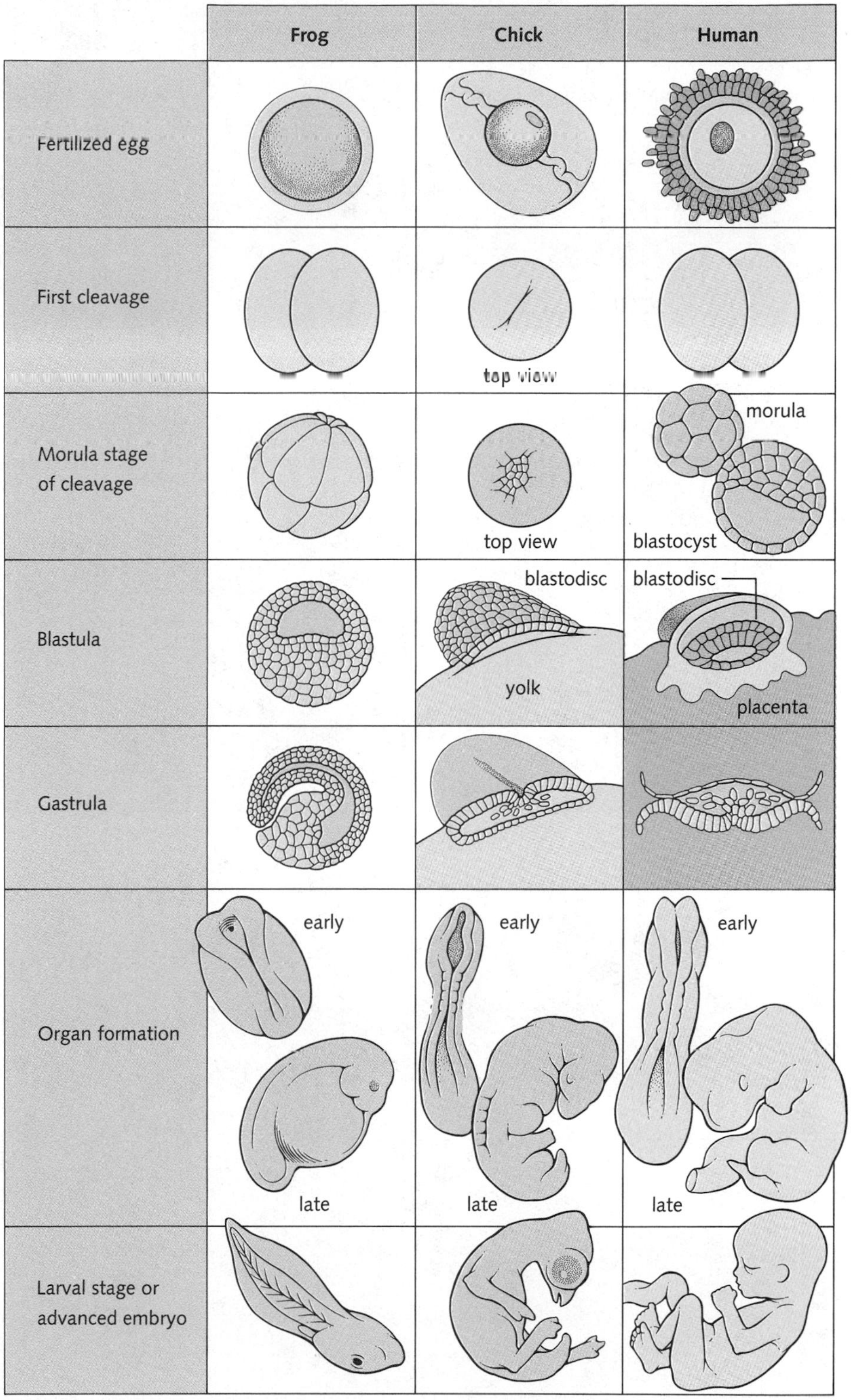

	Frog	**Chick**	**Human**
Fertilized egg			
First cleavage		top view	
Morula stage of cleavage		top view	morula / blastocyst
Blastula		blastodisc / yolk	blastodisc / placenta
Gastrula			
Organ formation	early / late	early / late	early / late
Larval stage or advanced embryo			

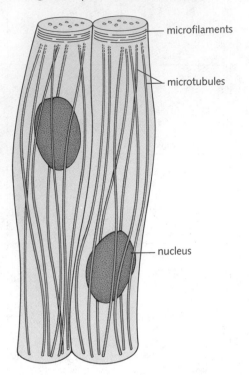

Figure 41.11 Individual changes in cell shape are responsible for neurulation. Cell elongation is the first in a series of changes that produce the neural tube.

microfilaments

microtubules

nucleus

of developing embryos to achieve their final form and function, hundreds or thousands of cells must perform specific operations in a highly coordinated fashion. Although there is much about this coordination that we do not understand, there are several developmental mechanisms and control systems that we can observe and describe, at least superficially.

The relatively undifferentiated cells of early embryos undergo several processes during which they become organized into recognizable body systems. These processes, which occur at the cellular level, are often called cellular mechanisms of development.

Neurulation, for example, involves a series of coordinated and carefully timed *changes in cell shape*. Ectodermal cells at the ends of the neural plate first become elongated and then constricted at one end to produce the neural tube (Fig. 41.11). These changes in shape involve microtubules that run the length of the cells; constriction of a ring of actin microfilaments acts like a purse string to draw one end of the cell together.

Other developmental processes involve the *migration* of cells through the embryo. Cells of the neural crest, for example, which all originate at the edges of the

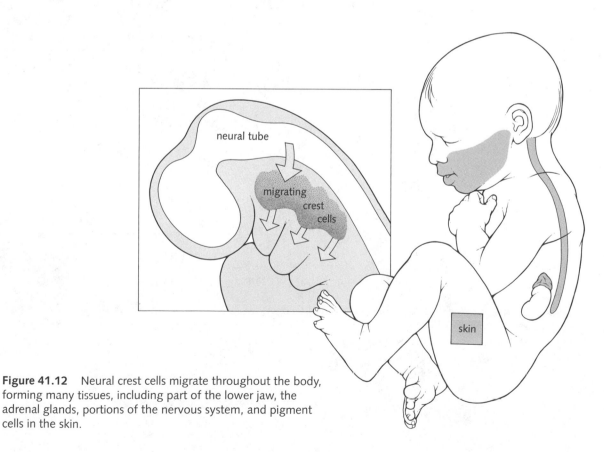

neural tube

migrating crest cells

skin

Figure 41.12 Neural crest cells migrate throughout the body, forming many tissues, including part of the lower jaw, the adrenal glands, portions of the nervous system, and pigment cells in the skin.

neural plate, migrate throughout the body to end up in an amazing variety of organs and tissues. As illustrated in Fig. 41.12, neural crest cells become part of the nervous system and the adrenal glands. Pigment cells in the skin are derived from neural crest, as are teeth and the bones and connective tissue of the lower portion of the head.

Neural crest cells do not move to these destinations randomly but follow several well-defined migration routes through the embryo. Many details of this phenomenon are not completely understood, but it is clear that these routes are marked by specific cell surface molecules. These molecules enable migrating cells to recognize the proper migration pathways.

Development also involves rapid *cell division, cell differentiation,* and even programmed *cell death*. The limbs of vertebrates, for example, originally form as paddle-like structures flattened on their ends. Individual fingers and toes form as some parts of the limb bud continue to grow while the death of cells between them creates openings in the original structure. As shown in Fig. 41.13, the principal developmental difference between the hind limbs of ducks and those of chickens is the degree of cell death between the digits.

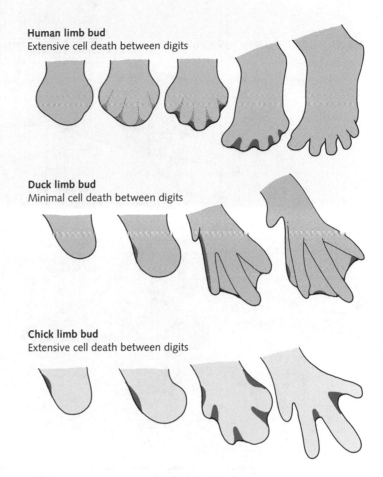

Human limb bud
Extensive cell death between digits

Duck limb bud
Minimal cell death between digits

Chick limb bud
Extensive cell death between digits

Figure 41.13 As limb buds develop, a genetically determined pattern of cell death helps to sculpt the individual digits.

HUMAN EMBRYOLOGY

We can now follow this entire process in detail by tracing human development from start to finish.

Gametogenesis and Fertilization

As we saw in Chapter 40, human sperm and ova are highly specialized cells produced by meiosis and differentiation. Although both gametes are haploid, the differences between the formation of sperm and ova are very great, reflecting the differences in their roles in the reproductive process.

Sperm are specialized for service as actively moving carriers of genetic information. A sperm cell's sole mission is to locate the ovum, bind to its outer surface, and deliver the sperm's haploid nucleus. Sperm are therefore little more than "stripped down" chromosome carriers. Their genetic material is condensed and packaged into a streamlined head, which is attached to an energy-generating midpiece and a long flagellum. Each sperm carries an acrosome whose enzymes help it penetrate the zona pellucida when it contacts the ovum.

The tasks of the ovum, on the other hand, are more complex. In addition to providing its own haploid nucleus, the ovum must provide almost all of the cytoplasm for the zygote, including ribosomes, transfer RNAs, and preformed messenger RNAs. The ovum also contains a store of food material in the form of yolk, which will provide energy for the zygote and support its development until it is able to draw nourishment from the mother. Thus, although the human ovum is quite small compared to the eggs of amphibians or birds, it is many thousands of times larger than sperm. The process of fertilization, as described earlier in this chapter, generally occurs in the fallopian tubes. The fusion of a sperm with an ovum causes the nucleus of the ovum to complete meiosis, and the haploid male and female *pronuclei* join to form the diploid zygote nucleus.

The First Month of Development

Human development follows a pattern that is generally typical of mammals. The first cell division typically occurs

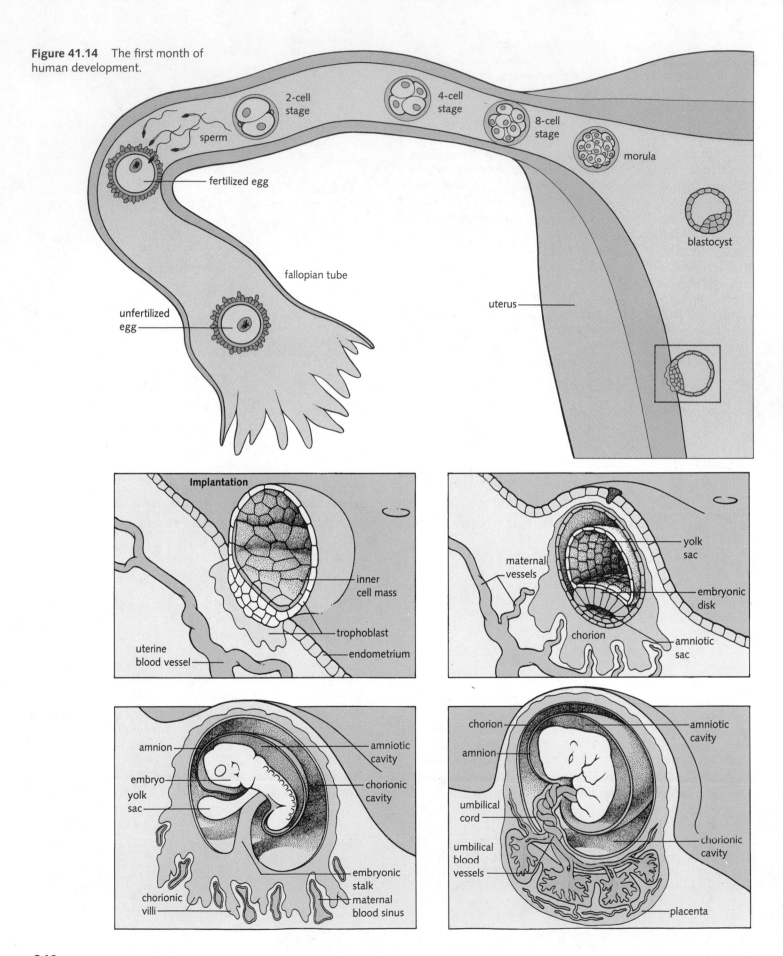

Figure 41.14 The first month of human development.

2-cell stage

4-cell stage

8-cell stage

sperm

morula

fertilized egg

blastocyst

fallopian tube

uterus

unfertilized egg

Implantation

inner cell mass

uterine blood vessel

trophoblast

endometrium

maternal vessels

yolk sac

embryonic disk

chorion

amniotic sac

amnion

amniotic cavity

embryo

chorionic cavity

yolk sac

chorionic villi

embryonic stalk

maternal blood sinus

chorion

amniotic cavity

amnion

umbilical cord

umbilical blood vessels

chorionic cavity

placenta

within 36 hours of fertilization (Fig. 41.14). Other cell divisions follow as the zygote moves down the fallopian tube, and within 4 days the *morula,* a ball of about 75 cells, has formed. Gradually the morula develops an internal cavity; when this occurs, the embryo has reached the *blastocyst* stage, and cell division slows down. In a sense, the embryo waits for the next critical step, **implantation,** which involves the binding of the blastocyst to the *endometrium,* the lining of the uterus.

Cell surfaces in the embryo carry glycoprotein molecules that bind to molecules on cells lining the endometrium. This enables the blastocyst to attach tightly to the endometrium. Once the embryo has made this attachment, cell division resumes as rapid growth begins again.

As with other mammals, the blastocyst contains the inner cell mass that eventually grows into the new individual—in this case, a human being. The inner cell mass is surrounded by a series of tissues that make direct contact with the uterus and already help to channel nourishment to the rest of the embryo. The outer tissues, initially known as the trophoblast, penetrate the tissues of the endometrium. As they develop, they form a distinct layer known as the **chorion.**

After implantation, the embryo literally invades the tissues of the endometrium, and within 2 weeks after fertilization it is totally embedded in the uterine wall. In addition to making direct contact with the mother's cells, the trophoblast has another important function. By releasing **human chorionic gonadotropin (HCG),** it prolongs the lifetime of the corpus luteum. The corpus luteum, in turn, prevents menstruation by continuing the secretion of progesterone.

As Fig. 41.14 shows, the implanted embryo goes through a number of dramatic changes in tissue organization. First, the inner cell mass produces an embryonic disk with two cell layers. Next, gastrulation occurs, resulting in the formation of three germ layers. As you will note, the embryo is already associated with two cell-lined *extraembryonic* ("outside the embryo") sacs: the **yolk sac** and the **amniotic sac.** In birds, as you recall, the yolk sac contains yolk that nourishes the embryo during development. Embryos of placental mammals, though they have very little yolk, form an empty yolk sac from which embryonic blood cells are formed, retracing their evolution from organisms that produced yolk-bearing eggs.

Shortly after gastrulation, the amniotic sac expands so that it surrounds the entire embryo. The yolk sac and another extraembryonic sac, the allantois (which, in birds and reptiles, stores metabolic wastes), are squeezed together in a thin stalk of tissue that passes through the growing amniotic sac to connect the embryo to the mother. This is the beginning of the umbilical cord.

Gradually, blood vessels from the developing embryo grow through the umbilical cord, and a new organ, the placenta, is formed from both maternal and embryonic tissue. Figure 41.15 shows how the circulation of blood from mother and embryo is organized within the

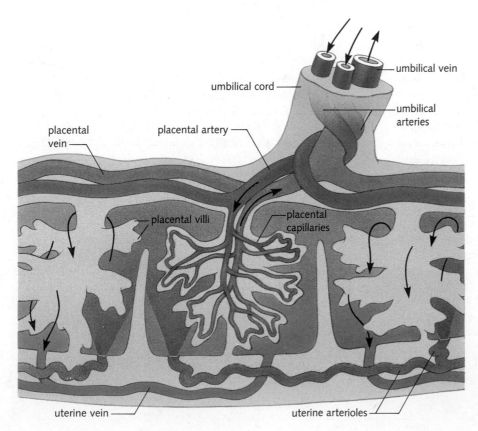

Figure 41.15 A schematic view of the placenta. Fetal and maternal blood do not mix within the placenta. Nonetheless, they do come into very close contact, allowing oxygen, carbon dioxide, food, and waste products to pass between mother and child.

Figure 41.16 Neurulation in the human embryo. After four weeks of development, the neural tube is completely enclosed.

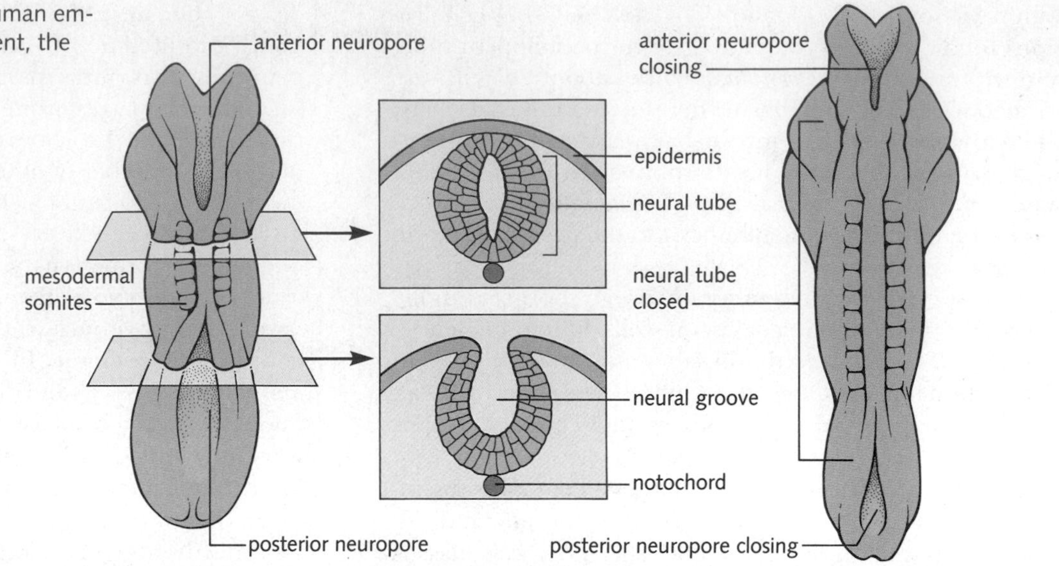

Labels (left diagram): anterior neuropore; mesodermal somites; posterior neuropore

Labels (center diagrams): epidermis; neural tube; neural groove; notochord

Labels (right diagram): anterior neuropore closing; neural tube closed; posterior neuropore closing

placenta. Tiny *villi* containing embryonic blood vessels extend into open *sinuses* of maternal blood. Although the two circulations come into very close contact, they normally do not mix. Food, oxygen, carbon dioxide, metabolic wastes, drugs, antibodies, and viruses cross the placental barrier, but whole cells generally do not.

The production of two parallel but separate circulatory systems is an important feature in the development of placental mammals. It prevents many infections from spreading through the circulatory system and enables both mother and developing embryo to be prepared for the day when the connection between them will be broken. In essence, the placenta serves the developing embryo as its principal organ of excretion, nourishment, and respiration.

In the fourth week of development, two prominent ridges of ectoderm begin to form running the length of the embryo's dorsal side. These ridges roll up to form a deep, hollow cleft, and neurulation begins (Fig. 41.16). This process is similar to that in other vertebrates, and by the end of the fourth week the neural tube has formed and become enclosed within the embryo.

By this time all the major cell layers of the embryo have been formed, and most of its major organs are either present or well on their way to formation. Small buds have formed where the limbs will emerge, and the embryo is about half a centimeter in length. All this has happened in about 25 days; the mother's menstrual period

has been delayed by only about 10 days, and she may have begun to wonder whether she is pregnant.

The First Trimester

For reasons of convenience, physicians often break the process of human development into three **trimesters,** 3-month periods of development. During the first trimester the embryo takes on the distinct appearance of a human child, both internally and externally (Fig. 41.17).

Because all body systems are established during the first trimester, this is a particularly sensitive time for development, and a number of external factors can affect the embryo. *Rubella* (German measles) infections can cause the heart to develop improperly, injure the eyes, or cause deafness. Drugs, such as the tranquilizer *thalidomide,* can prevent normal limb development, resulting in crippling malformations.

Thalidomide was used briefly in the early 1960s. It proved safe in all tests on animals and was released for sale in Europe. Its sale was blocked in the United States, however, because its safety in pregnant animals had not been established. This proved to be a fortunate precaution. The drug prevents normal limb formation, and women who took the drug and who were carrying fourth- and fifth-week embryos (the time when the limb buds form) gave birth to babies with severe limb malformations. Small flipper-like structures were produced in place

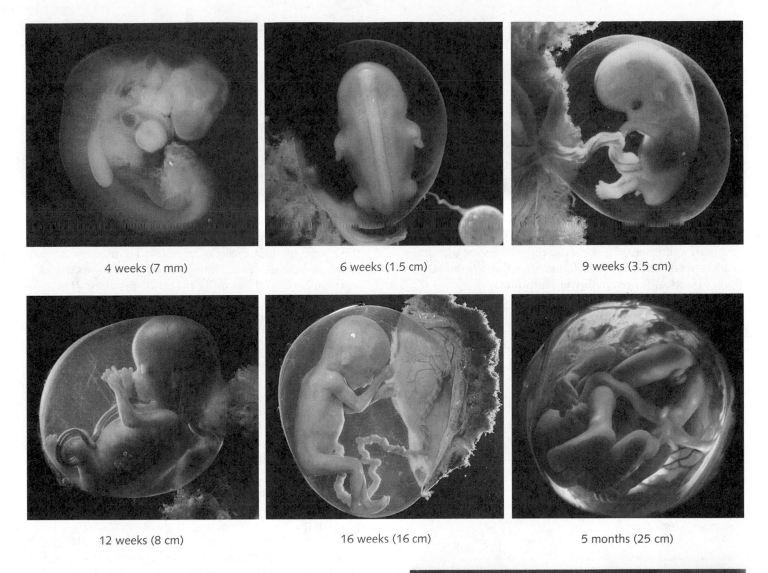

4 weeks (7 mm)

6 weeks (1.5 cm)

9 weeks (3.5 cm)

12 weeks (8 cm)

16 weeks (16 cm)

5 months (25 cm)

of hands or feet, and thousands of children were born handicapped for life.

By the end of 8 weeks, the embryo is officially known as a **fetus.** Its skeleton and muscle systems have developed to the point where it begins to make a few halting movements, and the sexual organs have begun to develop. (It is now possible to tell a boy from a girl.) By the end of 12 weeks, the fetus is nearly 10 cm long. For centuries, the end of the first trimester was known as the time the fetus "quickened" in the womb (became large and active enough for its movements to be noticed by the pregnant woman).

The Second Trimester

During this phase, existing organs and systems grow and mature. The fetus is covered with a layer of soft hair (the

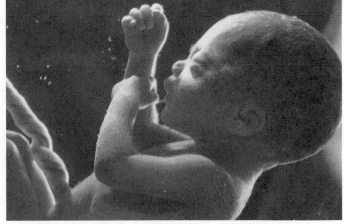

9 months

Figure 41.17 Photographs made of human development within the uterus at different periods.

lanugo, a term derived from a Latin word meaning "down"). By the end of the fifth month, the beating of the fetal heart is loud enough to be heard outside the mother with a stethoscope. The fetus's lungs are enlarging, its digestive system is almost ready to function, and it shows reflexes in response to sudden changes in light or loud sounds.

After 6 months of development, all body systems are formed and the fetus is nearly ready to be born. By this point it is about 35 cm long and weighs between 500 and 800 grams. A baby born at the end of the second trimester (or about 26–27 weeks) has a chance of surviving. The greatest problems faced by prematurely born infants at this stage are temperature regulation (because they are so small) and breathing (because their lungs are not quite ready to take over gas exchange for the whole body). Such infants suffer from chronic respiratory distress.

Despite these difficulties, the care of premature infants is improving almost yearly, and intensive-care wards at major hospitals are now able to save a large fraction of babies born near the end of the second trimester.

The Third Trimester and Birth

The third trimester sees the most dramatic weight gains of the developmental period. A mother is well aware that

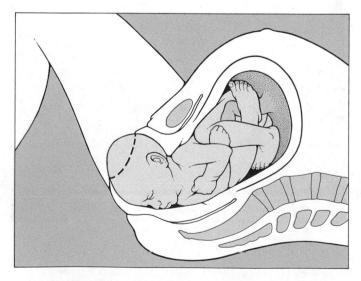

Figure 41.18 The usual position of a baby as it passes from the uterus through the birth canal during delivery.

her baby has gone from 600 grams to an average of 3000 grams during those last 3 months. During this period, all the major organ systems develop to the point where they are prepared to function independently of the mother. By the end of the eighth month, development has progressed so far that an early birth presents little danger. The major change physicians watch for at this stage is any sign that the placenta might be degenerating and ceasing to function properly.

Finally, approximately 266 days after it has begun, the process of embryonic development is complete, though it is not clear exactly how the body "knows" that it is time to send the fetus along to a life of its own.

Slowly at first, the large smooth muscles surrounding the uterus begin a series of rhythmic contractions known as **labor.** These contractions are triggered by **oxytocin,** a hormone released from the posterior pituitary. (It is possible to "induce" labor by injecting oxytocin intravenously, and this is sometimes done for medical reasons.) Other muscles gradually dilate (enlarge) the opening of the cervix, and little by little it becomes large enough (10 cm) for the fetus to pass through it.

The contractions of labor continue, becoming stronger and closer together. When the cervix has dilated to a full 10 cm and the contractions occur at intervals of 2–3 minutes, birth is ready to begin. The stronger contractions burst the amniotic sac, and the fluid within it rushes through the cervix and out of the body through the vagina. Through the action of repeated contractions, the fetus is forced, usually head-first, out of the uterus and through the cervix into the vagina, or birth canal (Fig. 41.18).

The baby is said to have *crowned* when its head is visible in the cervix. At this point, more contractions generally take from 2 to 90 minutes to force the baby through the cervix and out through the birth canal, still attached to its mother via the umbilical cord. The cord is just long enough to allow the infant to pass completely through the birth canal while still receiving nourishment and oxygen from the mother.

As the newborn makes contact with the outside world, it may cough or cry to loosen some of the fluid that has filled its lungs, and then it takes its first independent breath. Blood vessels leading to the placenta begin to close, and the placenta itself detaches from the uterine wall and is released, traveling through the birth canal as the **afterbirth.**

In larger hospitals, the placenta and part of the umbilical cord often are recovered—the skin that covers them is useful in treating burn patients. (The cells of the placenta are specialized, in a sense, to invade the tissues of another human without triggering the immune reaction. Hence they do not trigger such a reaction so readily as other tissues that might be used to cover burns.)

By nursing her child soon after it is born, a mother helps bring the birthing process to conclusion (Fig. 41.19). Hormonal feedback caused by nursing releases more oxytocin, which now restricts blood flow to the uterus. This helps slow down any bleeding that might have been induced when the placenta pulled out of the uterine wall.

In addition, the first secretions from the breast after birth contain not milk but **colostrum,** a fluid rich in nutrients that contains a special class of antibodies to help protect the infant from infection for many weeks.

Nursing a newborn has many advantages for both mother and child. Human milk is the perfect food for a baby, and it is always sterile, is always the right temperature, and doesn't require a special trip to the store. In addition, there is little chance that it will provoke the allergic reaction to feeding that sometimes occurs when cow's milk is substituted.

Maturation: Development Continues

Human development does not end with birth, for the cells in many body tissues continue to proliferate and differentiate for weeks, months, or even years. The immune system is only partially mature at birth; it is not really functional for several weeks into infant life. Actually, it is not *fully* functional until about age 2½ years. The nervous system continues to develop as it interacts with the environment, a fact that emphasizes the importance of providing infants with stimulating and interesting surroundings. Clearly, the reproductive organs are not fully functional until puberty. And the growth of the muscular and skeletal systems continues for 16 to 18 years after birth until the full adult height is reached.

It might be tempting to say that human development is complete at that point (or shortly thereafter). And it is true that beyond a certain point, most cells in most tissues stop dividing and differentiate into their final specialized form. From that time on, many cells of the brain and major nerves, for example, normally divide only when damaged. But in several body tissues, development never stops. In many tissues—including those of the liver, the skin, and the lining of the digestive system—continuous production of new cells is required to keep up with normal wear and tear. Bone marrow, too, continues to produce several different classes of blood cells throughout life.

You will see shortly that the processes of cellular reproduction, growth, and differentiation are regulated by several sorts of intercellular messengers. Normally, the actions of these messengers maintain body-wide homeostasis in mature individuals. If any cells escape this regulatory control system for some reason, one or another form of cancer may develop.

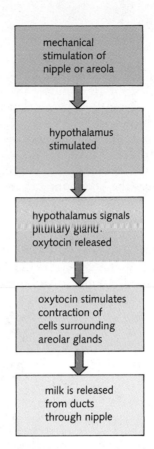

Figure 41.19 Successful nursing requires the involvement of the nervous system and the endocrine system to release the milk produced by exocrine glands in the breast.

Aging

For a number of years after physical maturity is reached, body systems continue to operate at peak efficiency. But after several decades, the flexibility of the skeletal system decreases, muscle strength ebbs, the skin becomes wrinkled and more transparent, and the reproductive system becomes less active. In females **menopause,** which usually occurs before or around age 50, marks a complete halt in the ability to ovulate and conceive children. Researchers often mark age 60 as the beginning of **senescence,** or old age.

Despite intense investigation into the cellular basis of aging, this process is still poorly understood. Explanatory theories include the accumulation of cellular waste products over time, the accretion of errors in DNA replication, and a preprogrammed genetic cycle of physical decline.

Whatever the ultimate causes may be, old age does not have to be a period of inactivity. Average life ex-

pectancy in developed countries is approaching 75 years, and it is clear that people are capable of enjoying life and making important contributions to society for each and every one of those years. With attention to good health, including a sensible diet and regular exercise, older people can make the last phase of the human life cycle very rewarding (Fig. 41.20).

COORDINATION OF DEVELOPMENT

Developmental processes as complex as those we have been discussing could not occur if individual cells—or even groups of cells—operated independently. Clearly, the behavior of embryonic cells and the fates of their descendants must be controlled and coordinated with great precision.

Yet each cell in an embryo contains exactly the same genes as all other cells around it. Each cell thus contains all the information it needs to develop into any of the tissue types found in the adult. This statement raises questions that are at once simple and profound. How do various cells "know" what sort of tissue to become? And if all cells are genetically identical, how does the process of controlled differentiation begin?

If you think about this problem for a moment, you will realize that the fate of an embryonic cell depends on the ways in which some of its genes are turned on while others are turned off. For that reason, research on the cutting edge of developmental biology today focuses on the patterns and mechanisms of gene expression during development. Before we delve into that fascinating research, however, we need to consider briefly the organisms chosen for this work and the strategy used by many researchers to investigate the intricate pathways of development.

The complexity of vertebrate embryos has led researchers to look for simpler organisms in which to apply the tools of genetics and molecular biology. One such organism is the fruit fly, *Drosophila melanogaster,* for which a rich library of genetic information has been accumulated over many years.

Figure 41.20 Four generations in the same family. Good diet, exercise, and attention to medical care help to ensure that every stage of human life can be full and rewarding.

Development in *Drosophila*

Development in *Drosophila* differs from the vertebrate pattern. After several rounds of nuclear division, the nuclei of the zygote move toward the periphery of the embryo. Cell membranes then separate the individual nuclei, forming a monolayer of cells known as the **blastoderm** (Fig. 41.21).

Fertilized Egg	Nuclear Division	Nuclear Migration	Blastoderm Formation		Gastrula	Early Organ Formation	Larva

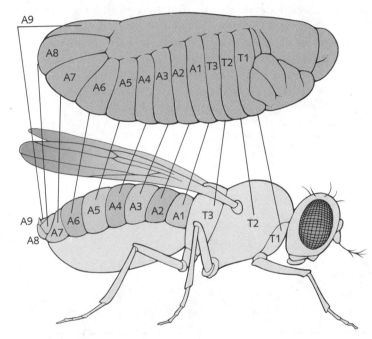

Next, during gastrulation, the three germ layers are produced and the embryo becomes fragmented longitudinally into a series of distinct *segments* from which the adult body is derived. A normal embryo contains 15 segments: 3 head segments, 3 thorax segments, and 9 abdomen segments.

This embryo hatches from the egg into a *larva* that retains these 15 segments and, after a period of intense feeding, growth, and molting, passes through a *pupal* stage and becomes an adult fly by the process of *metamorphosis.*

The limbs of the adult fly are not derived directly from the segments of the larva but arise from a series of small pouches in the larval epithelium that are known as **imaginal disks.** Disks on either side of the body produce the eyes and antennae, the three pairs of legs, and the wings. Experiments in which cells were transplanted from one part of the embryo to another have shown that in the blastoderm, all cells are already committed to their particular segments—their developmental fates have already been determined. Once this discovery was appreciated, investigators began to search for genes that controlled the pattern of segmentation and cell determination in *Drosophila.*

Figure 41.21 TOP: Stages in *Drosophila* development. BOTTOM: There is a direct correspondence between segments in the embryo and those in the adult fly.

Caenorhabditis: The World's Best-Known Embryo?

There is another organism, little known to the public, that plays an important role in developmental research. In the 1960s, British scientist Sidney Brenner suggested that the tiny roundworm *Caenorhabditis elegans* might be an ideal species in which to study development.

This organism, less than a millimeter in length, affords several experimental advantages. It can be grown easily on culture plates in the lab. Its generation time is less than three days, so several generations can be studied within a month. And its size and transparency make it possible to observe under a laboratory microscope all the embryonic events that characterize this species.

J. E. Sulston and his associates have painstakingly followed every single cell in the *C. elegans* embryo and established a complete *lineage map* for the organism, a road map for the ancestry of each and every one of the (exactly!) 959 cells in the adult (Fig. 41.22). This unique achievement—*C. elegans* is the only organism with such a complete map—has made it possible to isolate single genes that control the fates of individual cells in the embryo. For example, exactly 131 cells undergo "programmed" cell death during the development of a *C. elegans* embryo. Two genes, *ced-3* and *ced-4,* are involved in this process. Mutations that inactivate either of these genes result in the survival of nearly all of these "programmed-to-die" cells. The ability to trace the effects of genes on individual cells is one of the reasons that *C. elegans* is now an important experimental system for studies of development.

Developmental Anomalies as Research Tools

As you may recall from our discussion of cholesterol in Chapter 37, it is often easier for researchers to "get a handle" on a physiologically complex system if they can compare its proper functioning with a case in which some part of the system is missing or malfunctioning. For that reason, one important research strategy is to examine cases in which normal developmental processes go awry. In such situations, either naturally occurring mutations or experimental manipulations produce developmental anomalies: Siamese twins, animals with two heads, or individuals with malformed or duplicated limbs (Fig. 41.23). You will see as our discussion proceeds how important these unfortunate individuals have been in expanding our understanding of development.

Master Control Genes

Thus far we have considered genes largely as isolated "bits" of information; one codes for eye color, another codes for hair color, others control blood type, and a few working in tandem color human skin anywhere from alabaster to ebony. Yet the body is formed not from independent parts but from tissues whose growth and dif-

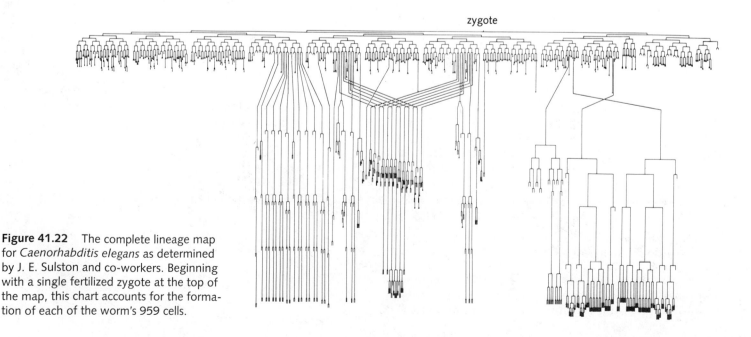

Figure 41.22 The complete lineage map for *Caenorhabditis elegans* as determined by J. E. Sulston and co-workers. Beginning with a single fertilized zygote at the top of the map, this chart accounts for the formation of each of the worm's 959 cells.

ferentiation are clearly coordinated in some way. How is that coordination accomplished?

As far back as 1915, researchers hypothesized the existence of some sort of "master control" genes. Such genes would somehow regulate the activity of hundreds or even thousands of other genes, turning them on and off in sequence to produce the patterns we know as a fully formed embryo.

The first evidence for such genes appeared in *Drosophila,* where certain bizarre mutations produce differences far more dramatic and far-reaching than changes in eye color or bristle number. One such gene is a mutation that results in *fushi tarazu,* which means "too few segments" in Japanese. Flies with the *fushi tarazu* mutation have half the usual number of segments, and they die before they are able to hatch from the egg.

Homeotic mutants Another series of mutant genes, known as *homeotic mutations,* transforms the body part produced on one segment into another body part that is normally found on a different segment. One such mutation produces legs on the head where antennae should be; another substitutes antennae in place of wings, and still another turns mouthparts into feet (Fig. 41.24).

Dozens of homeotic genes have been found, and most of these have been mapped to two clusters in the genome. One cluster affects the head and anterior thorax segments, whereas the other cluster controls the abdomen and posterior thorax. Researchers have even been able to isolate a homeotic mutant gene, insert it into the germ line of normal flies, and turn it on at will, essentially redesigning the fly in the process.

We are slowly learning how these genes exert their effects; at least some of them code for proteins that are found within cell nuclei, suggesting that their function is to bind to DNA and regulate the expression of other genes. These homeotic genes may bind to DNA sites throughout the genome, activating (or deactivating) whole groups of genes at once.

The homeobox sequence Intriguingly, a surprising biochemical similarity seems to link many—though not all—homeotic genes. When researchers bound labeled mRNA molecules to *Drosophila* chromosomes, they discovered a short piece of DNA, about 180 bases in length, whose nucleotide sequence is almost identical from one gene to the next. Because of its presence in these homeotic mutants, Swiss scientist Walter Gehring coined the name *homeobox* to refer to this piece of DNA.

Are homeoboxes in fruit fly genes special cases that mark master control genes only in flies? Definitely not. Sequences closely related to the *Drosophila* homeobox have now been found in yeasts, sea urchins, and frogs, and in 1984 they were found in humans as well. It is now

Figure 41.23 Siamese twin girls who were born with a single heart muscle.

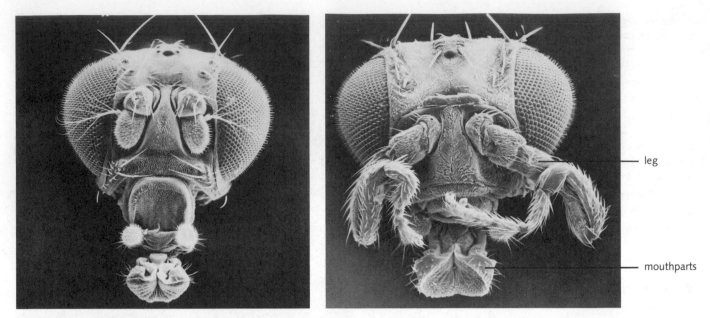

leg

mouthparts

Figure 41.24   Compare the normal fly (LEFT) with the mutated fly (RIGHT). The fly shown in the scanning electron micrograph on the right possesses a homeotic mutation that produces legs where its antennae should be. (Magnification factor left: 105; Magnification factor right: 135)

suspected that genes containing the homeobox DNA sequence control the timing of gene expression in a wide variety of animals and even in plants.

The universality of this sequence is suggestive; the homeobox genes function as master control elements for a variety of steps in development of nearly all organisms. Recent experiments in mammals have confirmed this view. When mouse homeobox genes are disrupted, the result is a complete disruption of organ development, affecting the heart, thymus, major blood vessels, and throat. The remarkable molecular similarity of homeobox sequences offers the tantalizing suggestion that at least some aspects of genetic control are shared among organisms that took separate evolutionary paths more than 600 million years ago.

Sex determination in humans   One of the most interesting of all developmental processes is the way in which male and female embryos develop their differences. Recently, researchers in Britain have discovered a single, short DNA sequence that acts as a master developmental switch: embryos that have the gene develop into males, and those that lack it mature as females.

The story is a fascinating one. For the first six weeks of human development, male (XY) and female (XX) em-

bryos are identical both structurally and biochemically. The tissues that ultimately form the gonads of both sexes (testes or ovaries) develop in the lower abdomen near two different duct systems. Then, as if on cue, the testes of male embryos begin to produce testosterone, stimulating the development of the male reproductive tract. In females, the other duct system develops, producing the female reproductive tract. Much later in life, at puberty, the processes we discussed in Chapter 40 complete the differentiation of mature males and females.

British molecular biologists Robin Lovell-Badge and Peter Goodfellow call the 35,000-base-piece DNA they have isolated *Sry* (sex-determining region Y). What is so special about *Sry*? To begin with, male mice contain a very similar region, as do many other animals. A strain of XY mice has been discovered that is biologically female and lacks the *Sry* sequence, strongly hinting that *Sry* is necessary for male development. Finally, Lovell-Badge and Goodfellow transplanted a piece of DNA containing the *Sry* region into female (XX) mouse cells. The result? Some of these transgenic female mice developed into "males" with normal testes, leaving little doubt that the *Sry* sequence can initiate the male developmental pattern.

The discovery of *Sry* does not mean that sex determination is fully understood. It does seem that female development occurs when the *Sry* gene is absent, and male development takes place when it is present.

Is there a "femaleness" factor, too? And what controls the activation of the *Sry* gene? Only more research on these intriguing candidates for master control genes in humans will tell.

The Influence of Egg Cytoplasm

As fascinating and powerful as master genes may be, however, their presence alone cannot explain the formation of patterns in early embryos. All cellular descendants of the fertilized ovum, after all, contain the same genes. Thus even master control genes themselves must be differentially activated in different embryonic cells in order to direct those cells along various developmental paths. The question thus becomes, "What controls the controllers?" It turns out that many key influences on the earliest stages of development are contained within the highly structured cytoplasm of the egg cell.

At the turn of the century, Hans Spemann showed that one region of the frog egg, the gray crescent, was absolutely essential to development (Fig. 41.25). By carefully dividing a two-cell embryo, he showed that when each half of the embryo contained part of the gray crescent, two normal tadpoles developed. However, any cell that contained none of the gray crescent failed to develop and produced only an undifferentiated ball of cells. Even simpler manipulations of certain eggs have equally dramatic results; merely centrifuging frog or trout eggs for 4 minutes within half an hour of fertilization produces "Siamese twins."

These classic experiments suggest that some sort of key developmental influence resides in the egg cytoplasm. More recent studies using molecular techniques have shown that the egg cytoplasm of many organisms contains a presynthesized store of mRNA molecules that are not translated until development begins. These molecules direct the synthesis of a host of proteins that are not coded by the embryonic nuclei but nonetheless direct many aspects of early development.

This situation has interesting consequences. Embryonic cells usually do not start to synthesize large amounts of their own mRNA until gastrulation. For that reason, any preformed messages that are present in the egg cytoplasm at fertilization can exert powerful influences on early embryonic cellular activities. Furthermore, the differential distribution of mRNA during early cell division can induce those cells to develop along different paths. This influence is not restricted to mRNA, of course; differential distribution of any other substance that affects either protein synthesis or gene expression could have the same effect.

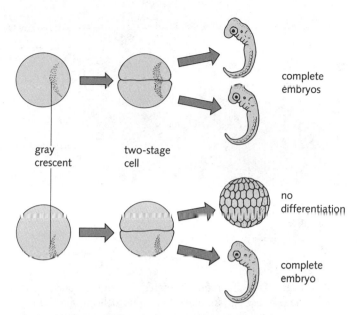

Figure 41.25 Hans Spemann's classic experiments with divided embryos showed that material in the gray crescent was necessary for complete development.

Recent research has shown that there is, in fact, a gradient of various compounds in early *Drosophila* embryos and that these compounds affect—either directly or indirectly—the expression of homeotic genes in different embryonic segments.

Embryonic Induction

No cell is an island; in any multicellular environment, the way a cell behaves very much depends on its surroundings. Thus, although preformed material in the egg cytoplasm is important in the earliest stages of development, it is not long before interactions among the cells of the embryo itself become critical in determining developmental changes. Several classes of experiments have demonstrated that various embryonic tissues exert powerful effects on one another.

During neurulation, how do the ectodermal cells "know" that they will become part of the nervous system? An experiment suggests one possibility. If cells derived from the gray crescent are removed from the dorsal lip of a frog blastula and transplanted inside a second blastula, they give rise to a block of mesodermal tissue that produces a second notochord. When neurulation occurs, two nerve tubes form—one above the original notochord and

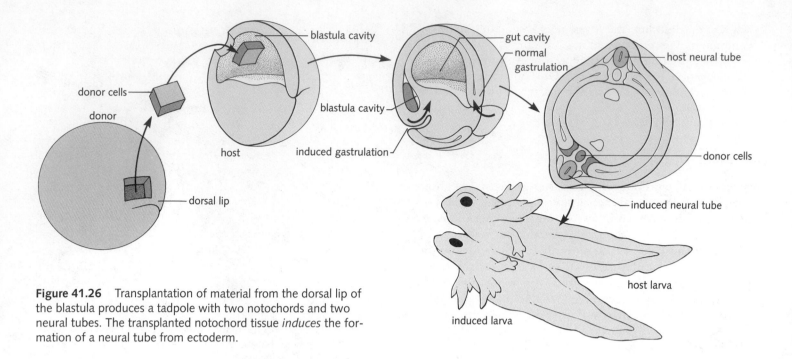

Figure 41.26 Transplantation of material from the dorsal lip of the blastula produces a tadpole with two notochords and two neural tubes. The transplanted notochord tissue *induces* the formation of a neural tube from ectoderm.

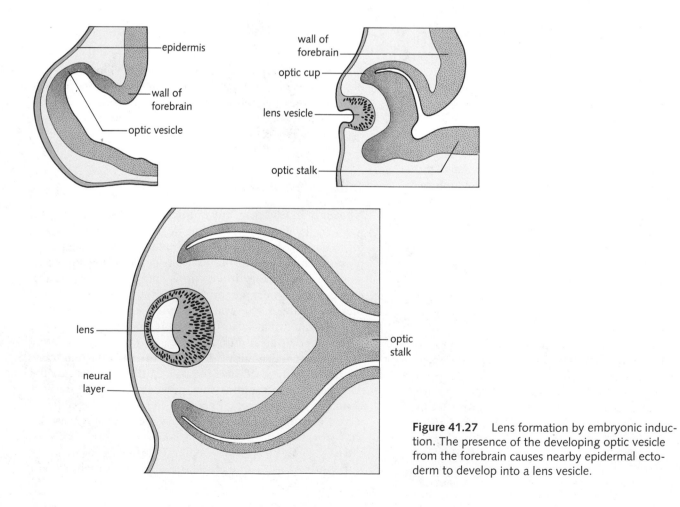

Figure 41.27 Lens formation by embryonic induction. The presence of the developing optic vesicle from the forebrain causes nearby epidermal ectoderm to develop into a lens vesicle.

one above the transplanted tissue (Fig. 41.26). What has happened? The ectoderm above the transplanted mesodermal tissue would normally have produced skin. But somehow the presence of notochord mesoderm at that point in development *induces* nearby ectoderm to develop into nerve tissue.

This process of **embryonic induction** is not an isolated phenomenon. It is a general pattern of development, and it is repeated in the formation of many different structures. Another well-known example is the *induction* of the lens of the eye by the **optic cup,** shown in Fig. 41.27. A lobe of the developing brain called the *optic vesicle* grows close to the ectoderm at the anterior end of the embryo. Gradually, this vesicle folds in to form the optic cup. This optic cup then seems able to induce *any* piece of ectoderm, *anywhere* on the body, to form a lens. In one experiment, an optic cup transplanted just beneath the skin in the belly region of a frog induced belly ectoderm to form a lens and a complete eye structure on the ventral surface of the tadpole!

In some tissues, induction occurs as the result of more complex signals, including hormones, which are quite specific and affect only certain target cells and tissues. But induction can also occur in response to very subtle cues, such as a change in local pH, a change in salt concentration, or the release of simple carbohydrates or other simple compounds. Although the most complex of these interactions have not been explored, there are a few cases in which we have begun to understand the complexities of induction.

One compound that seems to play an important role in inducing changes in embryonic tissue is *retinoic acid,* a derivative of vitamin A. One series of experiments demonstrated the effects of this compound on the developing limb bud of chick embryos. When a piece of filter paper soaked in retinoic acid is placed next to the front part of a limb bud, it causes the bud to produce a duplicate (though "mirror image") set of digits (Fig. 41.28). Further experiments showed that a gradient of retinoic acid occurs naturally in the limb bud and that this gradient normally controls the differentiation of different parts of the wing.

Intriguingly, retinoic acid also affects the development of limb buds in mammals. As we will see shortly, this simple compound even has inducing effects on several other classes of cells in mature individuals.

THEORY IN ACTION

Hen's Teeth!

Embryonic induction is not limited to tissues of the same species. Hans Spemann and Oscar Schotté had shown this in a classic experiment in 1932 when they switched the oral ectoderm of a frog and salamander embryo. The mesoderm of each embryo induced the ectoderm to produce a mouth, but the type of mouth reflected the origin of the ectoderm. The salamander larva had a frog mouth, and the tadpole had salamander teeth. We might explain this by saying that the mesoderm "told" the ectoderm to make a mouth, and the ectoderm "obeyed" by making the only kind of mouth it could.

Much more recently, Kollar and Fisher used the same technique to play a revealing trick. They transplanted epithelium from the jaw-forming region of a chick embryo onto the oral mesoderm of mouse embryos. What happened? The chicken tissue, responding to the instructions of the mesoderm, produced teeth. Hen's teeth!

Remarkably, although chicken cells have not formed teeth for millions of years, they still retain the genetic information to do so. All that was necessary was an induction signal from the mouse mesoderm to "make teeth." The fallout from this experiment is as much literary as scientific. The phrase "scarce as hen's teeth" will probably survive this assault, but it will never carry quite the same impact.

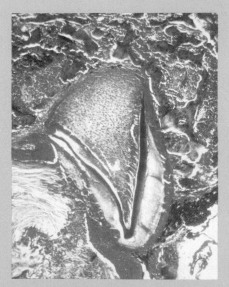

A tooth produced by chicken ectoderm transplanted next to the jaw mesoderm of a mouse.

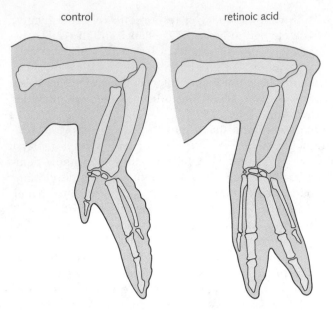

control   retinoic acid

Figure 41.28 The application of retinoic acid to the front edge of this developing limb bud has produced two "mirror image" sets of digits.

Further Control During Maturation

As we mentioned earlier, when an animal reaches maturity, rapid growth and differentiation give way to homeostasis in most tissues. Throughout the body, cells communicate with each other constantly, both through cell surface molecules and through more mobile chemical messengers such as peptide hormones and retinoic acid. Somehow, this collective interaction not only determines how many of each cell type there should be but also controls how much of which products are made by which cells. A number of studies in various systems have shown that this sort of communication is mediated by both positive and negative controls: some messengers stimulate cells to grow or to produce certain products, and others induce them to stop dividing or halt synthesis.

As we mentioned earlier, however, there are several tissues in which growth and differentiation continue throughout adult life. Not unexpectedly, these tissues are controlled by the same general sorts of intracellular messengers that regulate embryonic development.

The various types of circulating blood cells, for example, all have limited life spans. For that reason, they must be produced throughout life. But the various kinds of blood cells are not always needed in the same relative numbers. Injury may create a need for more red blood cells, for example, or infection may increase the demand for certain types of lymphocytes.

Where do new and replacement blood cells come from and how is their production controlled? It turns out that all classes of blood cells are produced by a single marvelously versatile class of **stem cells** (Fig. 41.29). These stem cells—which may either circulate in the bloodstream or settle in the bone marrow, spleen, or liver—are potentially immortal; they continue to divide and reproduce for the life of the organism.

Through a complex process involving several stages, stem cells are induced by a variety of circulating growth factors to produce one or another type of blood cell. When more red blood cells are required, for example, the kidney releases *erythropoietin*, and under its influence, stem cells produce more erythrocytes. The production of the various classes of white blood cells is controlled by a series of similar growth factors, several of which have been identified.

CANCER: WHEN CELLS REBEL

Cancers are some of humanity's oldest enemies and they are still some of the most formidable. Signs of cancer have been found in 5000-year-old Egyptian mummies, and it claims the lives of nearly half a million people in the United States each year. Nearly 1 person in 3 will develop some form of cancer during his or her lifetime, and present statistics indicate that 50 percent of those will die from the disease within 10 years after diagnosis.

The origin of the use of the Latin word *cancer* (which means "crab") to describe the disease is obscure. Some scholars have suggested that the name is derived from the crab-like shape of some cancerous tissue, and others believe that the intense pain caused by certain forms of the disease led some to compare it to the grip of a crab's claw.

Cancers have myriad causes; they can be triggered by exposure to any of hundreds of noxious chemicals called **carcinogens,** by radiation, and by certain viruses. Clusters of cancer cells called **tumors,** or **neoplasms** ("new cells"), can appear in skin, blood, bone marrow, internal organs, or brain. Some tumors linger for years, growing slowly or not at all, whereas others spread with frightening speed. Retinal cancer and acute leukemia strike mostly young children; breast cancer, skin cancer, lung cancer, and many others attack primarily adults.

Because of this variety of causes and effects, it seemed for years that "cancer" was actually a collection of more than 100 different diseases, each caused by a different carcinogen and each associated with a different set of symptoms. But over the past few decades, the techniques of molecular biology have proved that beneath its diversity *cancer is a disease of the genes.* There is now good evidence that at the root of all cancers lie genetic changes that transform normal body cells into tumor cells.

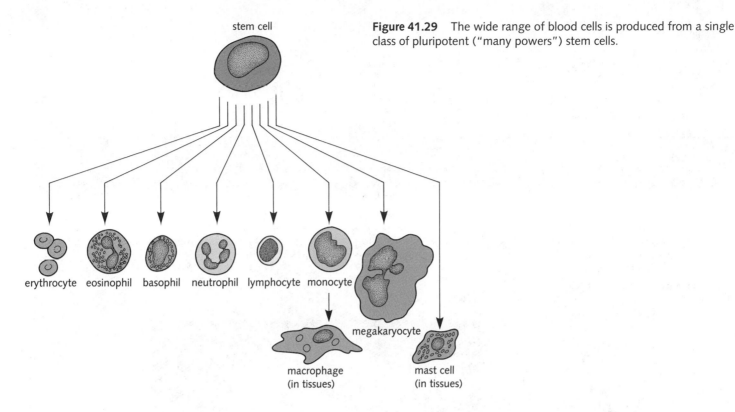

stem cell

erythrocyte eosinophil basophil neutrophil lymphocyte monocyte

megakaryocyte

macrophage
(in tissues)

mast cell
(in tissues)

Figure 41.29 The wide range of blood cells is produced from a single class of pluripotent ("many powers") stem cells.

Failure to Control Cell Growth

A cancer arises when the genetic material in a single cell is damaged in a manner that frees the cell from normal constraints on growth and reproduction. Certain cancers appear when cells fail to respond to the signals that normally prevent mitosis. Other cancers arise when cells act as though they were constantly stimulated to divide. And still other cancers are caused by cells that fail to heed body messengers that normally induce them to differentiate and stop reproducing.

Once transformed in this manner, a cancer cell is like a parasite reproducing out of control, spawning millions of similar cells that differ from the normal parent tissue by as little as a single gene. This is one thing that makes cancer so difficult to treat. There is no foreign organism that can be made the target of special drugs or medicine; the disease-causing cells are the cells of the victim, growing in the absence of normal controls.

Oncologists—physicians and scientists who study and treat cancers—classify tumors into two groups differing in behavior and in medical consequences. **Benign tumors** (from the Latin word for "kind") are those in which the growth of cells is restricted to the area of the tumor itself. Despite their name, benign tumors can cause serious problems, particularly when they develop in delicate areas such as the brain, where tumor growth can deprive normal cells of nutrients and put pressure on blood vessels and nerves. But the cells of benign tumors tend to stay together in one mass, often making it possible to remove them neatly and completely through surgery.

The cells of cancerous or **malignant** ("evil") **tumors,** on the other hand, tend to break off from the original mass, enabling them to **metastasize,** or spread through the body. This property makes malignant tumors especially dangerous. If such a tumor is not removed before it undergoes metastasis, dozens of tumors may spring up throughout the body, blocking the movement of materials, disrupting the functions of vital organs, and ultimately killing their host.

Cancer Traced to Altered Genes

The nature of the genetic changes that cause cancer and the mechanisms by which their altered gene products free cells from growth restraints are being brought to light by four separate lines of inquiry that have converged only in the past 20 years.

Metastasis: How Does a Cancer Cell Invade?

If cancer cells were not capable of metastasis, spreading throughout the body, cancer would not be such a serious medical problem. Localized tumors can be removed by surgery, but invasive ones spread to scores of places before they can be detected. Their ability to spread through the body is remarkable. A metastasizing cell must break its connections to other tumor cells, penetrate into the bloodstream, evade the immune system, and find a suitable place to grow. What gives malignant cancer cells this deadly ability? Researchers have tried to find out by analyzing individual cells derived from malignant tumors. One of the first findings was that some cells from a tumor are much more invasive than others.

These most invasive cells are armed with a formidable bag of tricks. In the bloodstream they attract macrophage cells and induce them to release an enzyme that helps the tumor cells break through blood vessel walls. Researchers have discovered that even if the malignant cells are detected by the immune system, the most invasive among them have cell surface molecules that enable them to avoid destruction by turning off the immune response. Finally, the most dangerous cancer cells produce a substance known as *tumor angiogenesis factor* (TAF). TAF affects nearby blood vessels and causes them to grow in the direction of a new tumor, providing a blood rich in nutrients to support further growth. The discovery of these tricks could not be called good news in the battle against the disease, but they have led to experimental strategies that may help to control metastasis by directly attacking the most invasive, and therefore the most dangerous, cells.

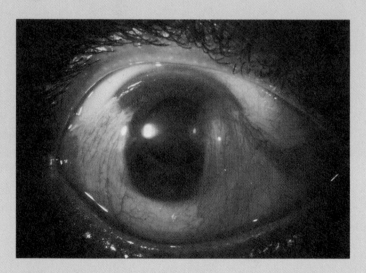

The rich growth of blood vessels associated with this tumor makes its active growth possible.

Studies of inherited cancer Over a century ago, in 1886, a British physician reported that survivors of *retinoblastoma,* a particularly virulent form of eye cancer, bore an unusually high percentage of children who were also struck by the disease. Then in 1910, it was noted that retinoblastoma survivors also exhibited a much higher-than-normal incidence of an otherwise rare form of bone marrow cancer later in life. These findings alerted researchers that some forms of cancer have a genetic basis, although many decades passed before techniques to explore this lead were developed. (See Theory in Action, A Gene That Suppresses Tumors, p. 858.)

Studies of environmental carcinogens Researchers have believed for decades that radiation and certain chemicals can cause cancer, and it is now clear that they do so by causing changes in DNA. Radiation, from the ultraviolet light in ordinary sunlight to the gamma radiation re-

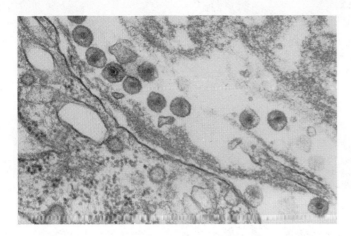

Figure 41.30 LEFT: These retrovirus particles are capable of producing cancer in the tissues they infect. BELOW: RSV and related retroviruses contain four key genes. One of these, the *src* gene, is essential for causing cancer.

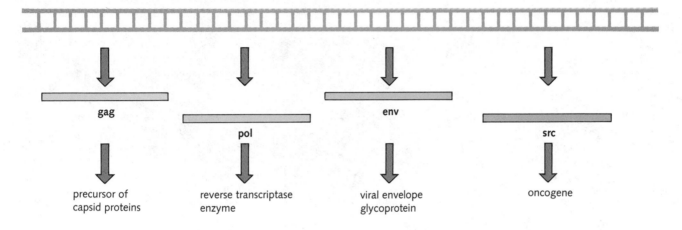

leased in a nuclear explosion, can cause molecular changes in DNA. Similar effects are noted with carcinogenic chemicals. Hundreds of chemicals have been shown to cause mutations, and nearly all of these are also able to produce cancers under the right conditions.

Studies of gene transfer using tumor cells Using a procedure similar to that used by Oswald Avery and his colleagues (Chapter 20), several investigators transferred pieces of DNA from animal and human tumor cells into healthy cells growing in cell culture. Some of the healthy cells were transformed into cancer cells that caused tumors when injected into healthy mice. The "instructions" that transformed healthy cells into cancer cells must have been carried from one cell to another by pieces of DNA.

Studies of cancer-causing viruses In 1910, Peyton Rous discovered that a virus could produce tumors in chickens and that the virus could be isolated and used to spread the disease to other chickens (Fig. 41.30). This virus, called the *Rous sarcoma virus* (RSV), belongs to a large family of oncogenic viruses, known as **retroviruses,** that

contain RNA as their genetic material instead of DNA. After infecting a cell, retroviruses produce a DNA copy of their genome, which is inserted into the host cell's DNA. Most retroviruses produce diseases of other sorts (Chapter 42), but RSV and more than 40 other members of its family have been shown to cause cancer in animals.

Although the significance of this work by Rous was not immediately recognized (he received the Nobel Prize decades later), genetic analysis of RSV and other oncogenic viruses during the 1970s showed that most contain a critical gene that causes cancer when inserted into the host genome. Appropriately, these cancer-causing DNA sequences were named **oncogenes.** The oncogene from RSV, discovered by Raymond Erickson and Marc Collett at the University of Colorado, is known as *src* [pronounced "sarc" (Fig. 41.30)].

At first, it was thought that oncogenes were merely viral genes that produced tumors. However, *src* and most of the more than 60 viral oncogenes that have since been discovered turn out to be nearly identical to genes found in normal human cells. These "cellular oncogenes" are normal parts of the human genome and are similar to

A Gene That Suppresses Tumors

Although the oncogene hypothesis gives us a basis to say that cancers are caused by mutations in a few cellular genes, there are relatively few places where workers have been able to pinpoint the exact nature of the change that leads to the transformation into a cancer cell. But those few important exceptions may teach us a great deal about the genetic changes responsible for cancer.

Retinoblastoma is an inherited form of cancer. It is caused by a single recessive allele and therefore is expressed only in individuals with two copies of the allele. In retinoblastoma, a tumor originates within the retina of the eye and grows rapidly, often damaging the eye and causing blindness. Earlier genetic studies had concluded that the retinoblastoma allele was *dominant* for the *tendency* to develop the tumor, and this finding provided some insights into how the disease begins. Most often, children inherit the gene defect from only one parent. Then, when a mutation damages the other, normal gene in

one of the hundreds of thousands of retina cells, a tumor develops.

Nearly 25 years ago, Jorge Yunis at the University of Minnesota found a patient with the tumor who also had a deletion on chromosome 13. Soon

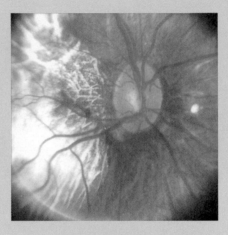

A retinoblastoma tumor is visible as the pale oval in this photograph of a human retina taken through the front of the eye.

other patients were found to have similar deletions (though not all of them did), and a link between the tumor and the loss of material on chromosome 13 seemed to be clear. Little by little, the affected region on that chromosome was narrowed down until, in 1986, two research groups in Boston were able to isolate the entire retinoblastoma allele. Now that this allele has been isolated, the really interesting work will begin. Already experiments have shown that abnormalities in the retinoblastoma allele are associated with other kinds of tumors. Gradually, researchers should be able to develop a test for the gene deletion that causes retinoblastoma, enabling them to look for something more revealing than a deletion on chromosome 13. The retinoblastoma allele belongs to a class of genes known as tumor suppressors (see text), which normally prevent the uncontrolled growth of cells. Researchers are eager to analyze the molecular details of this fascinating gene, whose *absence* allows a cancer to develop.

---

genes found in many other species ranging from yeasts to chimpanzees. Clearly, to have retained a common form through hundreds of millions of years of independent evolution, oncogenes must have very basic and important functions.

Common focus: Genetic change Today, the many lines of cancer research have come to a single conclusion, namely, that genetic changes are responsible for cancer. These changes may be the result of new genes introduced into normal cells (viruses and gene transfer), mutations that alter existing genes (carcinogens), or inherited gene defects (inherited cancers), but there is little doubt that

cancers are caused by alterations in a crucial set of genes that regulate cell growth.

Oncogenes and Tumor Suppressors

Which genes are responsible for tumor formation? It is too soon to be sure that we are aware of all the possible ways in which the growth of normal cells can go awry, but a great many cancers are caused by changes in two broad classes of genes. The first of these are the oncogenes.

As we have noted, viral oncogenes are often similar or identical to oncogenes found in normal cells. What are these genes doing in those normal cells? No universal pattern characterizes oncogenes, and we are not yet sure of the function of each and every one. Nonetheless, some general patterns have become clear.

Many oncogenes have been shown to code for proteins that play important roles in regulating cell functions. Some, including *src,* code for protein kinase enzymes, the molecular targets of many powerful peptide hormones (Chapter 39). Thus oncogene products may regulate the activities of scores of other enzymes. Some oncogenes code for products that regulate the timing of the cell cycle, and still others seem to code either for growth factors that stimulate cell division or for the membrane proteins that serve as receptors for those growth factors.

How are oncogenes related to cancer? It is possible for a *point mutation*—a substitution involving only a single base among hundreds in each gene—to turn a normally functioning oncogene into one that produces a cancer. This means that certain oncogenes could create cancer by causing cells (1) to produce too much of a growth-stimulating substance, (2) to act as though a growth stimulator were present even when it isn't, or (3) to *ignore* the message from a *growth-inhibiting* or *differentiation-inducing* substance. Preliminary work shows that every one of these mechanisms is associated with a form of cancer.

The second class of genes are known as **tumor suppressors,** indicative of the fact that they normally prevent the formation of tumors. The retinoblastoma allele (see Theory in Action, opposite page) is an altered allele tumor suppressor, as is the allele responsible for neurofibramatosis, an inherited tendency to develop benign tumors that was dramatized in the movie *The Elephant Man*. More than a dozen tumor suppressors have been identified, and there is no doubt that many more are waiting to be discovered.

Some tumor suppressors code for receptors to growth-inhibiting signals, so that if both copies of the suppressor gene in a cell become defective, the cell cannot respond to these signals, and uncontrolled growth may result. The functions of many other tumor suppressors are still unknown.

One of the most interesting tumor suppressor genes is known as *p53* (because it codes for a protein with an approximate molecular weight of 53,000). Scientists originally thought that *p53* might be an oncogene, because mutated copies of it were discovered in a wide variety of cancers. It is now clear that *p53* may be the single most important checkpoint in the cell cycle, and that is why so many cancers result from *p53* mutations.

The *p53* gene product seems to bind to nuclear DNA, and in some unknown way, it prevents cells containing damaged or defective DNA from entering mitosis. Thus, their DNA can be repaired before the cells are allowed to divide. When both copies of the *p53* gene are missing or damaged, a cell with damaged DNA has no such check on growth and may become cancerous. Some studies indicate that individuals who inherit one defective copy of the *p53* gene may be at very high risk of developing cancer, because their cells retain only one functional copy to protect them against the proliferation of cells with genetic damage.

Treatment for Cancer

Despite intensive medical effort, drugs and other cancer treatments have had only limited success. For solid tumors, including breast cancer and prostate cancer (two common causes of cancer death among women and men, respectively), the best hope lies in early diagnosis. If these tumors can be removed by surgery before they have metastasized, the chances for a complete cure are excellent.

Other approaches represent attempts to take advantage of the fact that many cancer cells pass through rounds of cell division much more often than most normal cells. Therefore, if a poison can be targeted to destroy rapidly dividing cells, it may be able to kill most or all of the cells in a malignant tumor while killing relatively few normal cells. This approach is known as **chemotherapy.** Medical researchers have developed a host of drugs that act on such cells, and in a few cases the results have been spectacular. Several forms of childhood leukemia, a cancer of white blood cells, now have cure rates approaching or surpassing 80 percent.

In other cases, the results have been less encouraging, and many of the drugs have serious side effects. Because they tend to poison *all* rapidly dividing cells, many healthy cells in important organs are affected by the chemotherapy, and the patient may become quite ill as a result. However, there have been dramatic improvements in the design and testing of such drugs.

Many cancers can also be treated via **radiation therapy:** tumor regions are irradiated with high-intensity radiation to help kill the rapidly dividing cancer cells. This treatment can be effective because rapidly growing cells, which must replicate their DNA more frequently than other cells, have less time to repair radiation-induced damage. Therefore, they tend to accumulate serious mutations and to eventually die when the number of genetic defects induced by the radiation becomes too great. Radiation therapy is used for localized tumors.

Currently, several experimental cancer treatments focus on trying to get the body's own immune system to fight malignant cells. Because the immune system recognizes and destroys any proteins not normally found in the body (Chapter 42), and because many tumor cells display unusual proteins on their surfaces, it should theoretically be possible to "beef up" the body's own defenses to attack cancer cells. In 20–30 percent of the cases in which the

Conservation and Cancer Biology

The forests of the American Northwest contain a variety of trees, many of which produce the strong, sturdy lumber valued by craftsmen and builders. Many other trees, such as the small, slow-growing Pacific yew *(Taxus brevifolia),* are considered "trash" trees by lumbermen and have been routinely cleared out of the way to make room for their larger, fast-growing cousins. Barely 30 years ago, the Pacific yew remained an unimportant species surviving in scattered pockets throughout the Northwest and occasionally cultivated around houses and gardens as ornamental shrubbery.

The obscurity of *Taxus brevifolia* ended in 1964, when a government-sponsored survey of plants for anticancer activity collected samples of the plant and tested them on cancer cells. Extracts from the bark of the plant slowed the growth of cancer cells. The active compound in the extracts was given the name **taxol** and has been intensively studied by researchers ever since.

Taxol works by stabilizing microtubules, cytoskeletal structures that make up the mitotic spindle. Taxol-treated cells are unable to rearrange their microtubules in the ways needed to divide successfully, and therefore they cannot complete mitosis. The same principle seems to slow or stop the growth of cancer cells, including those in breast tumors. In one study, it shrunk tumors in 48 percent of patients with advanced stages of breast cancer, an impressive result.

The Pacific yew tree *(Taxus brevifolia),* source of the drug taxol.

Taxol has not yet been synthesized in the laboratory, and until it is, the Pacific yew will remain the best source of this powerful drug. And there's the problem. Nearly 30 pounds of bark are needed to extract a gram of taxol, and roughly three trees supply only enough taxol to treat a single patient for a year. Not surprisingly, the value of this "trash" tree has skyrocketed, and federal agents have begun to arrest "tree poachers" who hoped to sell yew bark stolen from federal land.

The issues faced by conservationists, medical researchers, and cancer patients now point to a serious dilemma: should existing yew forests be harvested to provide as much taxol as possible to treat today's patients, or should yews be protected until their numbers can be increased to supply enough taxol to meet the potential demand? However this issue is dealt with, the case of taxol has already taught a valuable lesson in biology. The extinction of any species, even a "trash" plant, is an incalculable loss for humankind. We can be thankful that enough of the Pacific yew population remained to give researchers a chance to discover this extraordinary drug.

most promising of these treatments was tried, it was spectacularly successful, but in most cases its effects were either negligible or insufficient. **Immune therapy** seems most successful with skin cancers called **melanomas,** but other types of cancer are highly resistant to immune therapy. Researchers are working hard to find out why this is true.

Ultimately, researchers studying the genetic basis of cancer hope to devise a new generation of cancer treatments based not on the brute-force methods of radiation and chemotherapy but on drugs specifically targeted at the mechanisms by which oncogenes release cells from normal growth controls. Unfortunately, because much of this research is still in its infancy, no one anticipates the discovery of such a cure in the immediate future.

There are some promising leads, however. One tantalizing series of experiments at Boston's Children's Hospital has recently shown that the application of retinoic acid can cause a certain type of cancer cell to differentiate and stop growing, thus arresting, if not permanently curing, the cancer. Intriguingly, at an early stage in the differentiation process of those cancer cells, researchers detected the expression of genes containing a homeobox.

Prevention and Control of Cancer

The fact that viruses have been linked to certain types of cancer should not be taken to mean that cancer can be acquired through contact with infected individuals. Only in the case of a single type of virally caused cancer of the cervix has human cancer been shown to be communicable, and in that case sexual intercourse is the only known means of transmission. There is no evidence that prolonged personal contact with any other class of cancer patient increases the risk of contracting the disease. Close contact with a cancer patient, in fact, is one of the most beneficial things that friends and relatives of a victim can do to aid in that patient's recovery.

There is strong evidence that nearly all forms of human cancer are triggered by exposure to specific agents in the environment (Table 41.1). As an example, let's look at two serious cancers of the digestive system. Stomach cancer is common in Japan but rare in the United States (where it occurs at about one-fifth its rate of incidence in Japan). Intestinal cancer, on the other hand, is much more common in the United States than it is in Japan (where it occurs at about one-quarter of the U.S. rate). A scientist might examine this information and suggest that the differences in cancer rates could be due either to environmental factors or to differences in genetic makeup between Americans and Japanese.

It is easy to choose between these two possibilities. Hundreds of thousands of Americans are of Japanese ancestry, and public health statistics for these Japanese-Americans show an interesting trend. The first generation of immigrants has cancer rates very close to those displayed by the homeland Japanese population. The second generation, those born in the United States, have rates midway between the homeland Japanese and general U.S. rates, and third- and fourth-generation Japanese-Americans have rates for these two types of cancer that match those of the general U.S. population.

It is clear that genetics does not play an important role in these cancers. General environmental agents, such as air and water, are probably not to blame either; this is suggested by the fact that first-generation immigrants exhibit a cancer rate matching that of their homeland. Studies of the lifestyles of this population group suggest that the most important factor in altering the rates for these two types of cancers is diet: the change from traditional Japanese food to typical American fare. Clearly,

Table 41.1 *The Most Potent Chemical Carcinogens*

Carcinogen	Exposure	Principal Cancers
Aflatoxin B1	Dietary	Liver
4-Aminobiphenyl	Occupational	Bladder
Arsenic	Occupational	Skin, lung
Asbestos	Occupational	Lung
Benzene	Occupational	Leukemia
Benzidine	Occupational	Bladder
Bis(chloromethyl) ether	Occupational	Lung
Chlornaphazine	Medical	Bladder
Cigarette smoke	—	Lung, bladder, pancreas
Diethylstilbesterol (DES)	Medical	Vagina
Estrogens	Medical	Corpus uteri
High-energy radiation	Medical, occupational, wartime	Various
Mustard gas	Occupational	Bladder
2-Naphthylamine	Occupational	Bladder
Ultraviolet light	Sunlight	Skin
Vinyl chloride	Occupational	Liver, brain (?)

SOURCE: Mathew S. Meselson, *Chemicals and Cancer*. Boulder, Colo.: Colorado Associated University Press, 1979.

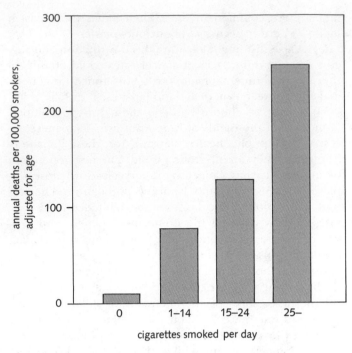

Figure 41.31 Cigarette smoking is one of the best-studied cases of a direct link between a behavior and cancer. This graph shows the rising incidence of lung cancer deaths with increased smoking.

each form of diet seems to have its drawback—and the best solution would be to identify the best (and worst) aspects of each diet.

Armed with these kinds of data, health researchers have tried to identify dangerous chemicals in the diet or in the immediate environment that might be responsible for cancer. The goal of this research is to identify chemicals we encounter in our everyday lives that may be responsible for the many cases of cancer that seem to have no direct cause. The best way to combat this dread disease is to eliminate its causes, and for many years health researchers have been trying to do just that. Interestingly, many of the most potent carcinogens do not cause cancer until changed by one of the body's enzymes into a carcinogenic form.

The Clearest Cause of Cancer

Some carcinogens in the environment are present in such low concentrations that their effects can scarcely be noted. But one carcinogenic agent has an effect so strong that it cannot be ignored: cigarette smoking. Lung cancers, nearly 90 percent of which are caused by smoking, are now the leading cause of death from cancer among men and women in the United States. A decision to avoid smoking and secondhand smoke, wherever possible, lowers the risk of lung cancer to 10 percent of that of a smoker and nearly ensures that an individual will not suffer from this disease (Fig. 41.31).

Given such evidence, it is difficult to understand why the habit of smoking, with its deadly consequences, persists at all. The best advice that the authors of this book can give you is to avoid smoking and exposure to secondhand smoke and not add your name to the list of lung cancer statistics.

SUMMARY

In vertebrates, all patterns of development follow a basic plan by which gastrulation results in the formation of the three primary germ layers: ectoderm, mesoderm, and endoderm. The details of this process differ dramatically, influenced by the size of the ovum and the mechanism by which it is protected and nurtured during development. The event that initiates development is fertilization, the fusion of sperm and egg. Cleavage occurs, and the zygote rapidly becomes a mass of cells known first as a morula and then as a blastula. Gastrulation is a process of cell migration in which the inward movement of the blastula surface establishes the internal germ layers of the embryo. Neurulation is the formation of a primitive nerve tube from ectoderm, an event that is completed after the major cell layers of the embryo have been formed.

Human development proceeds via a pattern typical for placental mammals: after the blastocyst implants in the uterine wall, an embryonic disk forms that is surrounded by outer cell layers that gradually invade the surrounding uterine tissue. As gastrulation and neurulation proceed, the extraembryonic membranes and placenta are formed, protecting the embryo and enabling it to draw nourishment from the mother.

Increasingly, biologists have turned to molecular models for development, including those derived from work with *Drosophila,* in which the techniques of classical genetics and molecular biology have been used to produce large numbers of developmental mutants. Analyses of these mutants have shown that certain DNA sequences, including one known as the homeobox, are found in many developmental genes.

Cancer is a cellular disease that results from the loss of control over cell growth in a multicellular organism. Tumors that develop in normal tissue may be either benign or malignant, depending on whether they tend to invade neighboring tissues and organs, a process known as metastasis. Cancers seem to be the result of heritable changes (mutations) in cellular DNA that lead to the loss of control over growth. Studies on oncogenic viruses have suggested that a certain set of genes, the oncogenes, are particularly important in the regulation of cell growth, and much cancer research now focuses on understanding what these genes do and how they are controlled.

STUDY FOCUS

After studying this chapter, you should be able to:

- Explain the process of embryonic development in representative vertebrates.

- Compare different types of embryos.

- Examine in detail the process of human embryonic development.

- Pose some of the fundamental questions of developmental biology, including the issue of gene activation in early development.

TERMS AND CONCEPTS

REVIEW

Objective Questions (Answers in Appendix)

1. The three primary germ layers (ectoderm, mesoderm, and endoderm) are formed during
   (a) gastrulation.   (c) cleavage.
   (b) neurulation.   (d) embryonic induction.

2. The yolk sac in the human embryo
   (a) contains a yolk.
   (b) is not present.
   (c) produces the mesoderm.
   (d) gradually disappears as the fetus matures.

3. The lining of most of the digestive system originates from
   (a) mesoderm.   (c) endoderm.
   (b) ectoderm.   (d) chorion.

4. In vertebrates, the notochord is replaced by
   (a) the digestive tube.   (c) somites.
   (b) the spinal column.   (d) limb buds.

5. Which of the following is *not* a cause of cancer?
   (a) radiation   (c) bacterial infection
   (b) oncogenic viruses   (d) mutation

Discussion Questions

6. Describe the basic vertebrate body plan, and distinguish among the three primary germ layers. Describe how the layers are formed in amphibian, bird, and mammalian embryos.

7. What are the primary means by which the endocrine system is signaled that a pregnancy has begun? Why do these signals prevent ovulation until the pregnancy is completed?

8. Which aspects of the trophoblast and the placenta are of particular interest to researchers who want to learn how to prolong the survival of skin grafts and other tissue transplants?

9. The primary signal for male sexual differentiation is testosterone. In the absence of testosterone, the female sexual pattern develops. Why is the production of estrogen by the embryonic ovary initially of little significance in sexual development?

10. Over the years, researchers have identified a number of families in which the individuals have a much higher-than-normal incidence of breast cancer, almost as if a tendency to develop this cancer were inherited. More recent studies have shown that many members of these families share a defective *p53* gene. Explain how this fact ties in with their inherited tendency to develop cancer.

11. A number of oncogenic viruses *do not* contain oncogenes. Several of them, however, do contain DNA sequences that act as very strong promoters of transcription. Explain how the insertion of a strong promoter next to an existing cellular gene might produce cancer.

READINGS

Daly, D. "Tree of life." *Audubon* (March/April 1992): 76–85. A superb article on the scientific history of taxol and the issues surrounding conservation of the Pacific yew.

Liotta, L. "Cancer cell invasion and metastasis." *Scientific American* 266 (February 1992): 54–63. A well-illustrated review of the mechanisms by which cancer cells cross barriers within the body to invade other tissues.

Cherfas, J. "Sex and the single gene." *Science* 252 (1991): 782. A one-page summary of experiments in which the *Sry* sequence was used to change the sex of *XX* mouse embryos from female to male.

Koshland, D. E. "Cancer research: Prevention and therapy." *Science* 254 (1991): 1089. This is a one-page editorial statement that serves as the introduction to an issue of *Science* devoted to cancer research. Seven excellent research reviews are found on pages 1131–1177 of the same issue.

Marx, J. "How embryos tell heads from tails." *Science* 254 (1991): 1586–1587. One of the fundamental unresolved questions in embryology is how embryos develop polarity. This article summarizes some interesting recent research.

Hoffman, M. "The embryo takes its vitamins." *Science* 250 (1990): 372–373. Retinoic acid, vitamin A, plays a key role in shaping the tissues of the developing embryo. This short article summarizes some recent developments in the vitamin A field.

Meselson, Mathew S. *Chemicals and Cancer.* Boulder, CO: Colorado Associated University Press, 1979. A brief, concise pamphlet summarizing the evidence associating certain chemicals with cancer.

42

Immunity and Disease

*i*n different parts of the world, four children suffer: one with measles, one with the flu, one with sleeping sickness, and one with AIDS. Their prognoses are dramatically different. The first will recover quickly and never contract measles again. The second will recover but will probably battle influenza (the "flu") several times in later years. The third will fight a long and difficult battle, perhaps winning, perhaps not. And the fourth, at least so far, is doomed to an early death (Fig. 42.1).

Why can the body fight off some diseases? Why do some diseases, such as measles, never recur? How does AIDS so completely overwhelm the body that a young life is brought to an early end?

We know today that each child described above is struggling with a **pathogen:** an organism or agent that infects our bodies and causes disease. We also know that our body battles these invaders with the **immune system,** an assortment of cells and tissues that identifies alien invaders and attacks them. The immune system also "remembers" infections; after a particular invader is defeated once, it is dealt with rapidly and more efficiently if it tries to strike again (*immunis* means "exempt," *immunity* means "freedom from infection"). Those may seem like simple statements of fact, yet our knowledge about disease is relatively recent. And our understanding of how the immune system responds to pathogens is newer still—and is growing rapidly.

DISEASE THROUGH THE AGES

For most of history, diseases were blamed on supernatural causes. By the sixteenth century, physicians began to understand conditions that *spread* disease but had no clue about what *caused* illness. The first useful theory attributed illness to *miasma,* or "bad air," generated by earthquakes, unburied corpses, sexual activity, or the wrath of God. Once miasma existed, the theory went, it transmitted disease from the sick to the healthy; it hovered around infected persons, houses, and towns and could be carried in clothes, bedding, and baggage. Though incorrect, this concept inspired sensible responses to epidemics. Contact

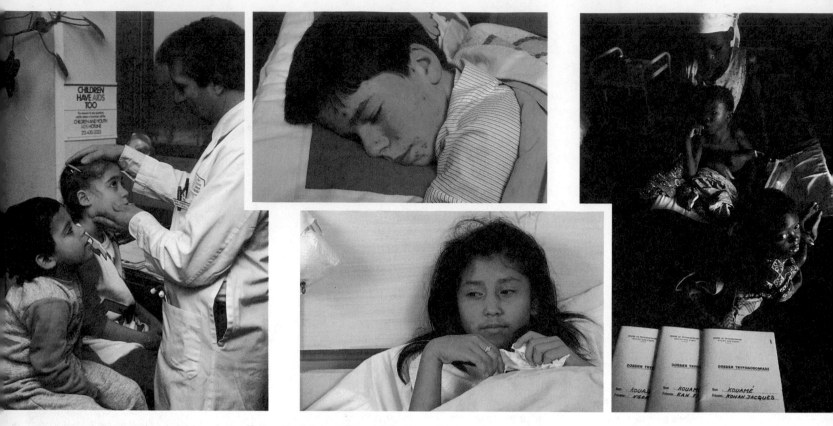

Figure 42.1 These children are suffering from four different infectious diseases: (FROM LEFT TO RIGHT) AIDS, measles, flu, and sleeping sickness.

with sick people was avoided, infected houses were fumigated, and contaminated bedding was burned. The microbial basis of infectious disease remained unknown for centuries, however, and medical breakthroughs usually followed leaps of intuition rather than methodical investigation.

The First Vaccination: A Shot in the Dark

One such leap was made in 1796, by Edward Jenner, an English physician. In Jenner's day, **smallpox** terrorized Europe, leaving millions dead in its wake. Those who survived the disease were left with disfiguring scars, but they also gained a mysterious protection; they would never get smallpox again. Somehow, they had become immune—resistant to further bouts with the deadly ailment. At the same time, a similar but far milder disease called **cowpox** occurred among milkmaids and dairy hands. Oddly enough, contracting cowpox seemed to confer immunity to smallpox.

Without understanding what he was doing or why, Jenner introduced to Europe a procedure that seems to have been first invented in the Middle East or Asia: he infected a patient with cowpox to see whether he could produce immunity to smallpox. To do so, he took fluid and pus from a cowpox lesion on a milkmaid and mixed them into a shallow cut he made in the arm of an 8-year-old boy, Jamie Phipps. Several weeks later, he performed the same procedure, but this time using material from a smallpox patient. Fortunately, this unorthodox procedure worked. Jamie did not come down with smallpox because he had become immune to the disease.

Society's response to Jenner's work was mixed. Many hailed the procedure, but numerous physicians—who rightly mistrusted a procedure they didn't understand—lobbied against it. After all, not only was the immune system unknown, but also bacteria and viruses hadn't even been discovered yet! In fact, the term **vaccination** was originally a derogatory term that literally meant "encowment" (*vacca* means "cow") (Fig. 42.2).

Jenner prevailed, and by the beginning of the nineteenth century, thousands had been vaccinated against smallpox. Today we know that vaccination works because cowpox and smallpox are caused by similar viruses; the immune system "remembers" exposure to either in a way that helps it head off subsequent attacks by both. We now have vaccinations of many kinds that can protect us against a host of diseases—such as polio and smallpox—but not yet against others—such as herpes and AIDS. Before we can discuss how and why vaccination works in some cases and fails in others, we need to know the sorts of organisms that make us sick and the ways in which the immune system fights them.

PATHOGENIC ORGANISMS: AGENTS OF DISEASE

The world around us teems with microscopic organisms that enter the body with every breath and through even the tiniest scrape or cut in the skin. Most of the time, most of these organisms pose no threat; they interact with our tissues in harmless or even beneficial ways. But some organisms are pathogens that can produce disease, either by causing direct physical damage to tissues, by releasing toxic chemicals, or by taking over cells' genetic machinery to use for their own ends.

The connection between pathogens and disease became clear during the late nineteenth century, as scientists associated particular sources of contagion (such as polluted water) with specific diseases (such as cholera). Pioneering microbiologists, including Louis Pasteur and Robert Koch, paved the way for a scientific understanding of several diseases. Bacteria, the first group of pathogens to be identified, were shown to cause cholera, anthrax, tuberculosis, and bubonic plague. Later, during the early decades of the twentieth century, researchers identified viruses, the causes of such diseases as smallpox, polio, the flu, and AIDS.

What Makes a Pathogen?

To cause disease, microorganisms must enter their hosts' bodies, manage to survive and reproduce, and ensure that their progeny are transmitted to other hosts. As important as this is, just because an organism enters a host *doesn't* necessarily mean that it causes disease; many kinds of bacteria live in the digestive tracts of most animals, and some of them are essential to proper health and nutrition.

Many potential pathogens don't normally cause disease because they are either kept out of the body or prevented from surviving inside by the immune system. Such pathogens cause problems only when natural defenses are compromised by old age or malnutrition or by immune-suppressing drugs or diseases such as AIDS. In addition, host species differ in their ability to repel different

Figure 42.2 A cartoon ridiculing Jenner's cowpox vaccine for smallpox.

Figure 42.3 A stop-action photo of this person sneezing shows the dispersal of droplets carrying possible pathogens.

invaders, and microorganisms differ in their ability to invade various hosts. For that reason, microorganisms that cause serious disease in some organisms are harmless to others.

Interestingly, the actual symptoms of disease reflect pathogens' evolutionary strategy for spreading their progeny around. Many viruses and bacteria of the nose, throat, and respiratory passages force hosts to sneeze and cough, thereby broadcasting pathogens in aerosol sprays (Fig. 42.3). Intestinal pathogens responsible for cholera and dysentery cause rampant diarrhea that releases them onto the ground or into surface water where they may be ingested by more potential hosts. And rabies can actually change its victims' behavior—inducing them to attack and bite other animals, thereby spreading the saliva-borne pathogen.

Some organisms, such as the protozoan *Toxoplasma gondii*, straddle the line between pathogens and harmless symbiotic partners. *Toxoplasma* is common in domestic cats and sheep and is easily picked up by humans who handle cat feces or eat raw lamb. Somewhere between a third and a half of Americans are "infected" by *Toxoplasma*. Luckily, it causes no disease symptoms in most healthy individuals. But if a pregnant woman not *already* infected becomes infected *while* she is pregnant, serious complications for the fetus may result. (Hence the folk wisdom that pregnant women should avoid cats if they don't already own one!) And people whose immune systems are impaired by AIDS can be ravaged by a virulent disease called toxoplasmosis.

Pathogen Diversity

Virtually every group of organisms, including bacteria, is plagued by pathogens of one kind or another. The most common pathogens are viruses, bacteria, protozoa, fungi, and worms (Table 42.1), but nearly every major taxonomic group of organisms includes at least a few pathogenic species.

Viruses are tiny bags of genetic material (either DNA or RNA) surrounded by a protein coat. Devoid of any cellular machinery for respiration or protein synthesis, viruses aren't fully alive on their own; to reproduce, they must enter living cells and usurp control of those cells'

Table 42.1 *Pathogens*

Pathogen	Examples	Diseases	Notes
Viruses			
DNA viruses	Herpes simplex II	Genital herpes	Spread by sexual contact
	Variola virus	Smallpox	Spread by personal contact (airborne)
RNA viruses	Rhinovirus	Common cold	More than 50 types known
	Flu virus	Influenza	New strains arise every few years
	Measles virus	Measles	Preventible by vaccine
	HIV	AIDS	Spread by blood or sexual contact; fatal
Bacteria	*Mycobacterium tuberculosis*	Tuberculosis	Treatable with drugs; a limited vaccine available
	Mycobacterium leprae	Leprosy	Causes long-term infections
	Streptococcus pneumoniae	Bacterial pneumonia	Before antibiotics, nearly 1/3 of patients died; now 95% recover
	Corynebacterium diphtheriae	Diphtheria	Serious childhood disease in nineteenth century; now preventible by vaccine
	Yersinia pestis	Bubonic plague	Still present in western United States; carried by prairie dogs
	Treponema pallidum	Syphilis	Spread by sexual contact
	Vibrio cholerae	Cholera	A leading cause of infant death in Third World
Protozoa	*Trypanosoma*	African sleeping sickness	Spread by tsetse fly
	Plasmodium	Malaria	Prevalent in tropics; a serious health problem
Flatworms (Platyhelminthes)	*Schistosoma mansoni*	Schistosomiasis	Disease spread by coinfection in snails; a very serious health problem in rice-growing areas
	Taenia saginata	Beef tapeworm	Spread by contaminated meat

biochemical machinery. Ultimately, this "genetic terrorism" causes infected host cells to die.

Viruses infect cells by binding to the cell membrane and either injecting genetic material or "tricking" the host cell into taking them in. Once inside, DNA viruses may immediately begin using host cellular machinery to produce viral RNA. By contrast, RNA viruses called **retroviruses** first use an enzyme called **reverse transcriptase** to create a DNA copy of their RNA-based genetic material. After that point, both DNA viruses and RNA viruses may do either of two things:

1. *The virus may produce an active infection.* In this case, viral DNA replicates and directs the host cell to synthesize new viruses. Some viruses do this slowly, gradually budding off new virus particles from the cell surface while the host cell survives. Other viruses overwhelm cells by forcing them to create hundreds of new viruses rapidly. The host cell bursts, releasing viruses that attack neighboring cells. Rhinoviruses, flu viruses, and measles viruses follow this course of action.

2. *The virus may produce a latent infection.* In such cases, instead of replicating immediately, viruses "splice" their DNA (or a DNA copy of their RNA) into host cell DNA. There, the viruses effectively "disappear" for weeks or years. We still do not know much about the molecular events that control latency and activation. Some latent viruses, such as herpes, seem to be activated by psychological stress. Others, such as the human immunodeficiency virus (HIV), may be triggered by other pathogens. Once acquired, latent viral infections stay with a host for life; there is no known way to remove viral DNA without killing the host cells.

Whenever free virus particles are released, they may be passed from one host to another, spreading the infection. Some viruses, such as those that cause colds and flu, survive outside the body and are spread by casual contact, coughs, and sneezes. Other viruses, including those that cause herpes, AIDS, and certain types of cervical cancer, cannot survive outside the body; these must be spread through direct contact or through the exchange of body fluids such as blood, mucous secretions, or semen. We will examine the transmission and action of HIV at the end of this chapter.

Bacteria are not all pathogenic; in fact, only a few members of this diverse group cause harm. Those few, however, cause very serious diseases, including diphtheria, tuberculosis, and syphilis. Pathogenic bacteria may attack and destroy cells directly, they may grow so quickly that they deprive host cells of necessary nutrients, or they may release poisonous waste products or **toxins** that injure host tissues.

Protozoa cause several serious diseases. The genus *Trypanosoma* includes flagellates that live in hosts' blood and release poisons. One species, carried by the tsetse fly, causes **African sleeping sickness** in humans and cattle. This disease is virulent enough to make it almost impossible to raise cattle in parts of Africa. Not only have these parasites resisted efforts to develop vaccines for either livestock or humans, but also they are nearly invulnerable to the body's defenses. We will explain why this is so shortly. **Malaria** is caused by *Plasmodium,* which spends part of its life cycle in the *Anopheles* mosquito. Infections are spread when an infected mosquito injects saliva containing the malaria organism as it begins to bite. *Plasmodium* reproduces within red blood cells and causes them to burst, producing fever and chills.

Most multicellular pathogens are simply known as **parasites.** Some parasites, such as tapeworms, live within the digestive systems of their hosts, where they absorb food and release eggs into fecal material. Others, including schistosomes or flukes, live within host tissues and have complicated life cycles involving two or more hosts, which may include humans, snails, and freshwater fishes. The use of human feces for fertilizer in Asia distributes the eggs of this species widely in such agricultural areas as flooded rice paddies.

NONSPECIFIC DEFENSES AGAINST INFECTION

The body's first defense against pathogens is simple: it tries to keep them out. It does so with a shield of physical, chemical, or biological barriers called **nonspecific defense mechanisms,** so named because they repel a wide variety of organisms.

The Skin

The largest barrier against infection is the **skin** (Fig. 42.4). Usually, pathogens can enter the body only where skin is broken or in places where skin is exceptionally thin and delicate, such as the mucous membranes of the nose, mouth, or genitals. Nowhere is skin's protective value more apparent than in serious burn victims, who face major risks of infection. The best sanitary precautions of modern hospitals cannot match the protective value of healthy skin, so medical workers try to grow or graft skin onto burned areas as quickly as possible.

The skin is not just a passive physical barrier; it is also an active part of the immune system. Skin cells respond to infection by creating scales, rashes, scabs, scars, and blisters. **Sebaceous glands** produce oily **sebum,** which inhibits bacterial and fungal growth. The pH of skin, influenced by sweat and oil glands, is acidic (pH 3 to 5) and prevents the growth of many potential pathogens. Skin

Figure 42.4 Skin is the primary barrier against infection. An outer layer of epidermis lies just above the dermis, which contains sensory nerves, glands, and hair follicles. Subcutaneous tissue (literally meaning "under the skin") lies below the dermis.

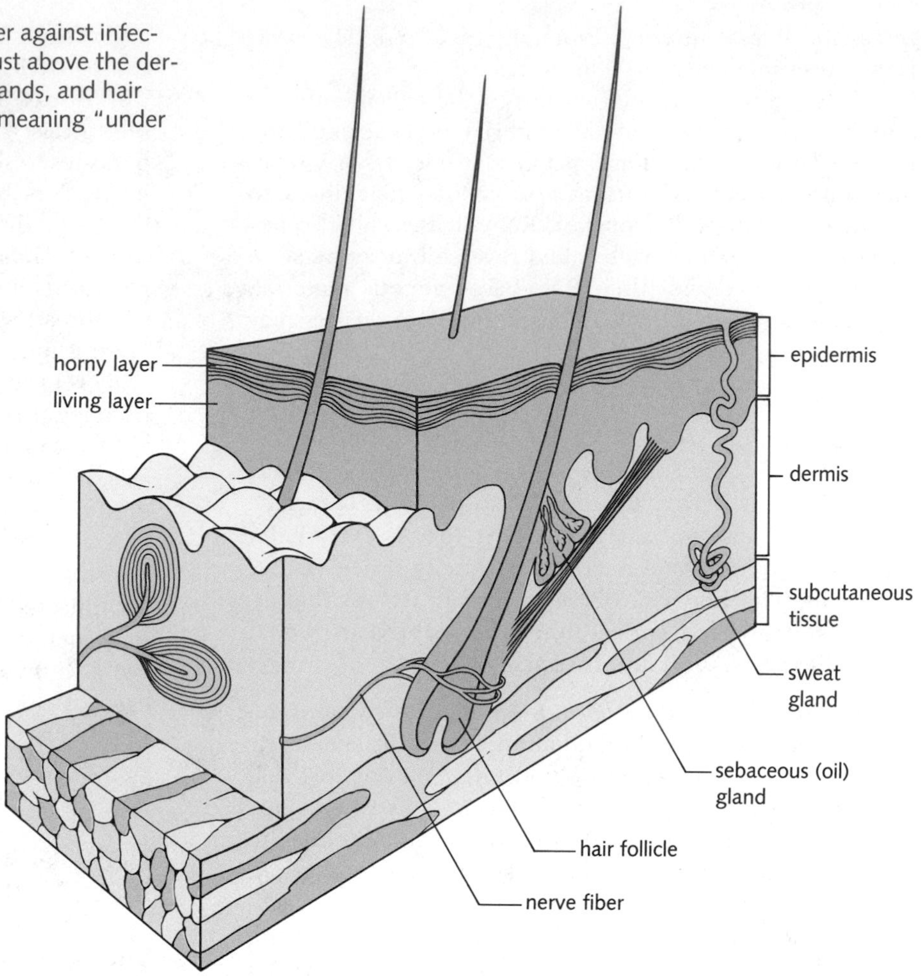

horny layer

living layer

epidermis

dermis

subcutaneous tissue

sweat gland

sebaceous (oil) gland

hair follicle

nerve fiber

cells also produce several messenger molecules that influence other cells in the immune system.

Enzymatic Defenses

Where skin is thin and delicate, other nonspecific defenses are used. Respiratory passageways are coated with mucus that traps inhaled microorganisms. Ciliated cells lining the passageways push that mucus towards the alimentary canal and the digestive system, where the trapped microorganisms are destroyed by digestive enzymes. Most bacteria that enter the mouth or urogenital openings are killed by an enzyme known as **lysozyme,** which breaks down their cell walls. Lysozyme is also secreted into tears, which protect the surface of the eye.

Inflammation

When skin is cut or torn, bacteria enter the wound and grow, spread, and release toxins. But this invasion is challenged by another line of nonspecific defense, called the **inflammatory response,** because it can cause the skin to turn a flaming color (Fig. 42.5). **Mast cells** near the skin surface release a substance called **histamine** that causes blood vessels near the wound to expand. Fluid leaks into the wounded area from these vessels, and the wound begins to swell. Histamine and other compounds released from the wound site attract and direct the actions of several types of white blood cells.

Phagocytes—cells that can engulf and destroy foreign cells—swarm over the infected area, engulfing and digesting bacteria. **Monocytes,** small roundish cells found throughout the circulation, are attracted to the wound, where they change into larger cells called **macrophages** that also engulf bacteria. Toxins produced by bacteria kill many of these defending cells, and as hours go by, fluid filled with dead bacteria and white blood cells builds up. Increased cellular activity in the wound area may raise its temperature several degrees. It is reddened, hot, and tender to the touch. Bit by bit, however, the infection is brought under control and the wound begins to heal.

SPECIFIC DEFENSES: THE TARGETED RESPONSE

The most powerful and complex activities of the immune system are targeted directly against specific pathogens. The life-and-death nature of these battles at the cellular level justifies the militaristic analogies often used to describe the sequence of events that characterizes this battle:

1. *Encounter and recognition:* The first task is to recognize a pathogen as "nonself"—as an invader that doesn't belong in the body. This is perhaps the immune system's most astonishing ability. It can recognize nearly any foreign substance in the body and can therefore attack virtually any pathogen.

2. *Mobilization:* After identifying an invader, the immune system prepares a specific immune response against it. This involves stimulating the reproduction and activation of several classes of cells that work together in confronting invaders.

3. *Attack:* The activated immune system attacks foreign cells or tissues and disposes of them.

4. *Cease-fire:* Once the invader is disposed of, the cellular "army" is "demobilized." Failure to do this can result in serious problems at the hands of misguided former defenders that run amok, damaging body tissues.

5. *Memory:* After an attack, the immune system "remembers" the defeated pathogen. If that same pathogen invades again, the system is primed for rapid action.

Figure 42.5 Cellular events associated with a simple splinter wound. Even a small wound brings about a dramatic response that we recognize as inflammation.

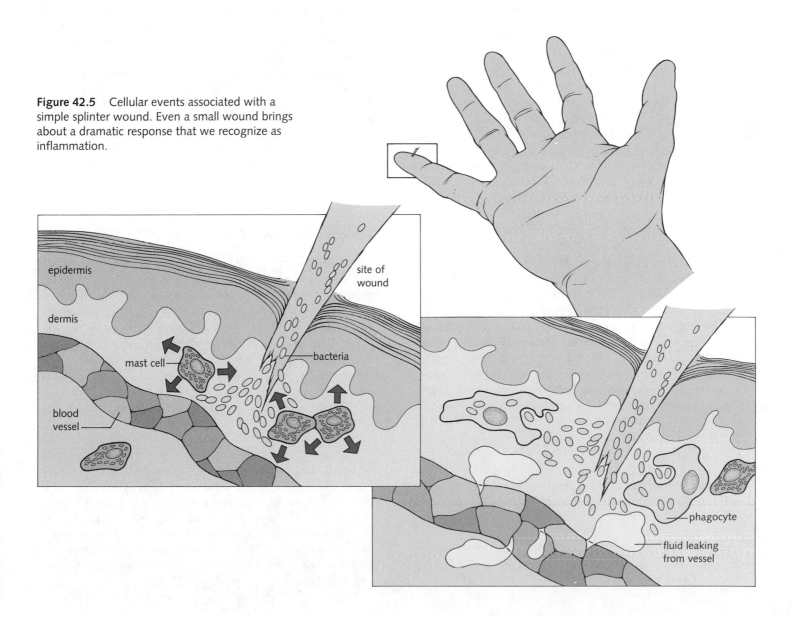

Steps 1–3 occur quickly and powerfully, so the host does not usually get sick again.

Each stage of the specific immune response involves several different classes of interacting cells. As yet, none of those stages are completely understood. We will explain what we do know about this system by first introducing the major participants. We will then show how those components work together, and we conclude by observing how the system's activity explains its response to the diseases mentioned in the chapter opener.

Distinguishing Self from Nonself: Antigens, Antibodies, and MHC

In order for the immune system to battle foreign invaders, it must somehow tell the difference between "self"—cells and molecules that belong in the body—and "nonself"—cells and molecules that come from somewhere else. This distinction is critical because, as you will soon see, the army of immune defenders wields a powerful arsenal of chemical and cellular weapons. If those weapons are turned on the body's own cells by mistake, a debilitating or fatal autoimmune disease—such as rheumatoid arthritis, multiple sclerosis, or diabetes—may result.

How does the system distinguish between self and nonself? The answer revolves around three types of molecules found on cell surfaces: antigens produced by pathogens, MHC markers carried by nearly all cells, and antibodies produced by certain cells of the immune system.

Antigen is a general term referring to any foreign substance that can be recognized by the immune system. Antigens are usually proteins, glycoproteins, or complex carbohydrates that form part of the surface coat of viruses, bacteria, or other pathogens. From the body's perspective, antigens serve as warning signs that alert the immune system to a foreign presence.

MHC markers are proteins produced by a region of the genome known as the **major histocompatibility complex (MHC complex),** because they determine which cells are "compatible" with one another and which are not. Because of extensive recombination in this part of the genome, no two individuals (except identical twins) have exactly the same MHC markers.

MHC markers identify body cells as self. While patrolling the body for invaders, your immune system ignores any cells that carry your personal MHC markers and targets cells that carry foreign markers. The greater the differences between the MHC markers of the foreign cells and those of the body, the stronger and more effective the response.

Antibodies are proteins manufactured by the immune system that attach, or "bind," to specific antigens. There are several major classes of antibodies, each of which interacts with different molecules and cells of the immune system. In general, however, antibodies serve two crucial functions:

1. Certain types of antibodies are released into blood, mucus, and saliva, where they bind directly to antigens on the surface of pathogens. In some cases, this binding by itself can "neutralize" a virus; in other cases, it marks the invader for destruction by other parts of the immune system.

2. Certain classes of antibody molecules remain attached to certain types of cells in the immune system, enabling those cells to recognize and attack pathogens.

At any given time, your body contains an astronomical number of different antibody molecules, each of which is manufactured specifically to bind to one particular antigen molecule. That, in fact, is where the word *antigen* comes from: "*anti*body *gen*erator." By recent estimates, the immune system of every human being can respond to as many as 10 billion different foreign proteins. Even immunologists find that hard to believe; they have tried repeatedly to find compounds that *won't* function as antigens—in other words, compounds that antibodies cannot recognize. Remarkably, they failed. Even artificial proteins—molecules that have never before existed on earth—are always recognized by the immune system. As you will see shortly, antibodies and MHC markers work together to target the immune response.

Antibody Structure and Function

All antibodies are proteins that share certain characteristics. Each has a **variable region** that recognizes a specific antigen, and each has a **constant region** that controls how the molecule interacts with other parts of the immune system (Fig. 42.6). Functionally, variable regions act like "glue" that attaches the antibody to the antigen. The constant regions at the other end of the antibody molecule act like a "signal flag" which announces, "There's an enemy here—do something!" As you will see shortly, this signal directs the responses of other components of the immune system.

One class of antibody called IgG, for example, is a Y-shaped molecule composed of two identical polypeptides called **light chains** and two other identical polypeptide **heavy chains.** Both the light- and heavy-chain polypeptides form part of each variable region and the constant region. IgG contains two **combining sites** formed from the variable regions. These combining sites bind to a specific antigen. The binding is remarkably specific: an antibody can distinguish between two molecules that differ by only a few atoms.

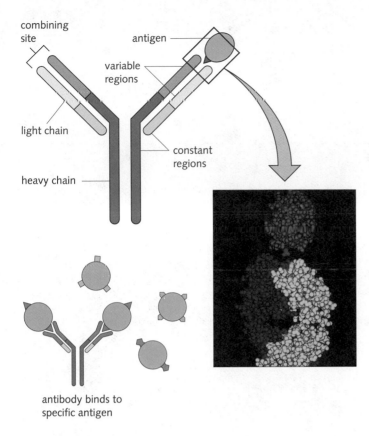

Figure 42.6 Antibodies are composed of heavy and light chains. The variable region of the molecule contains an antigen-binding, or combining, site. BELOW RIGHT: A computer-generated model based on experimentally determined atomic positions shows the close fit between an antigen and the combining site of the antibody. The critical portion of the antigen recognized by the antibody is shown in red. BELOW LEFT: An antibody can distinguish between antigen molecules that differ only slightly.

CELLS AND ORGANS OF THE IMMUNE SYSTEM

The immune system must guard the entire body against infection. Like undercover police, therefore, its cells cannot be confined to a single location; they must travel throughout the body so that no infection goes unnoticed. So it makes sense that the principal components of the immune system are circulating blood cells.

Lymphocytes: Munitions Plants, Foot Soldiers, and Commanders

Several of the most important components of the immune system are blood cells called **lymphocytes.** As their name implies, lymphocytes are found in lymphatic vessels and lymph nodes, as well as in the spleen, thymus, and general circulation (Fig. 42.7).

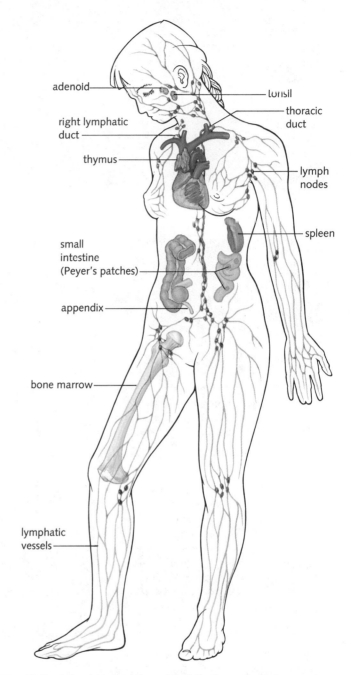

Figure 42.7 Lymphocytes, the cells of the immune system, are found throughout the body. Lymphocytes are present in large numbers in the lymphatic system, bone marrow, thymus, spleen, tonsils, adenoids, blood, and digestive system.

Millions of Antibodies from a Few Genes

The immune system can make antibodies against almost anything. Research workers who have purified molecules from biological systems know that if they inject those purified molecules into a rabbit or a mouse, the animal will make antibodies in response to the injection.

The ability of the immune system to manufacture an almost limitless variety of antibodies raises an interesting question. Because antibodies are large proteins assembled according to genetic instructions, shouldn't the genomes of mammalian cells be dominated by millions of genes coding for different antibodies? Putting it another way, how does an organism with an immune system manage to find room for anything else in its DNA?

This question baffled researchers. But in 1976, scientist Susumu Tonegawa made an interesting discovery. Using a DNA copy of the mRNA that coded for the variable and constant regions of an antibody, he discovered that the corresponding DNA in an embryonic cell of the same species was in several pieces. Why would DNA that was in one piece in the adult cell be in pieces in an embryonic cell? The conclusion was that some sort of DNA rearrangement must have taken place as the cells developed.

Tonegawa and other scientists have shown how that rearrangement occurs. Each developing cell contains several gene regions that code for the constant regions of antibody chains. There are also several hundred gene regions that code for the variable regions of the chains, and several more that code for the D and J regions that connect the two in the finished antibody molecule. As shown in the accompanying figure, cutting and splicing events during development produce a final immunoglobulin gene containing an apparently random collection of C, D, J, and V segments. The genes for antibodies are assembled almost as though by shuffling a deck of genetic segments present in each cell.

Because these events occur almost at random, the final immunoglobulin gene is different in every cell. Because millions of cells are involved, this process takes a major step towards guaranteeing that hundreds of millions of different possible antibody genes are created, so the system is ready to form antibodies against any antigen.

But that's not the end of the story. Immunologists once thought that antibodies produced by a stimulated clone of B-cells were identical to those produced by the cell that founded the clone. It has recently become clear, however, that as a stimulated B-cell develops, it gives rise to a variety of clones which produce antibodies that fit the antigen better and better as time goes on!

How does this occur? The best explanation seems to be that stimulation

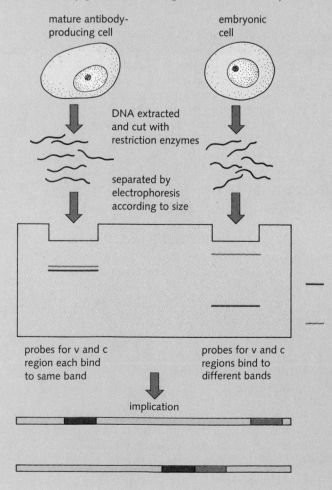

Susumu Tonegawa used DNA hybridization and gel electrophoresis to discover that antibody genes move during the course of development.

mature antibody-producing cell

embryonic cell

DNA extracted and cut with restriction enzymes

separated by electrophoresis according to size

probes for v and c region each bind to same band

probes for v and c regions bind to different bands

implication

somehow induces an extremely high rate of mutation in the DNA regions that code for most variable portions of the antibody combining site. Because of these rapid mutations, the daughter cells of a single B-cell produce a range of slightly different antibodies. Some of these antibodies fit the antigen better than the original antibody, and some of them fit it more poorly. Those B-cells producing antibodies that bind the antigen most tightly are stimulated to divide more rapidly than those that bind less tightly.

This process of mutation and antigen stimulation allows the immune response to "evolve" during the lifetime of a single clone, making the response more versatile and more effective. The fine-tuning continues throughout the proliferative phase of clonal selection and ends when mature plasma cells are formed.

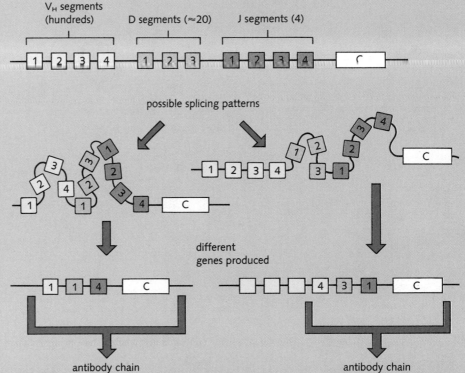

V_H segments (hundreds)   D segments (≈20)   J segments (4)

possible splicing patterns

different genes produced

antibody chain   antibody chain

Complete antibody chain genes are assembled from four different segments. Cutting and splicing brings together one copy of each of the V, D, J, and C segments to form the complete gene. Because the exact pattern of splicing is a chance event, it happens in a slightly different way in each cell, ensuring that the immune system as a whole will be able to produce thousands of completely different antibody chains.

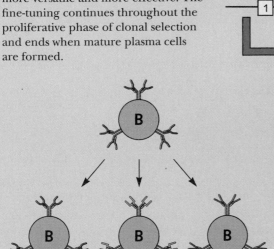

mutation produces variation in binding sites

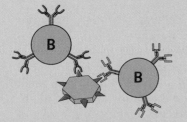

these B-cell clones are not stimulated

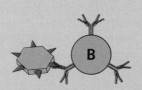

stimulated to grow and divide

Small mutations in the antibody-producing genes produce a B-cell population that is constantly selected (and improved) for its ability to bind antigen. Those B-cells whose antibodies fit the antigen best are stimulated to grow and divide.

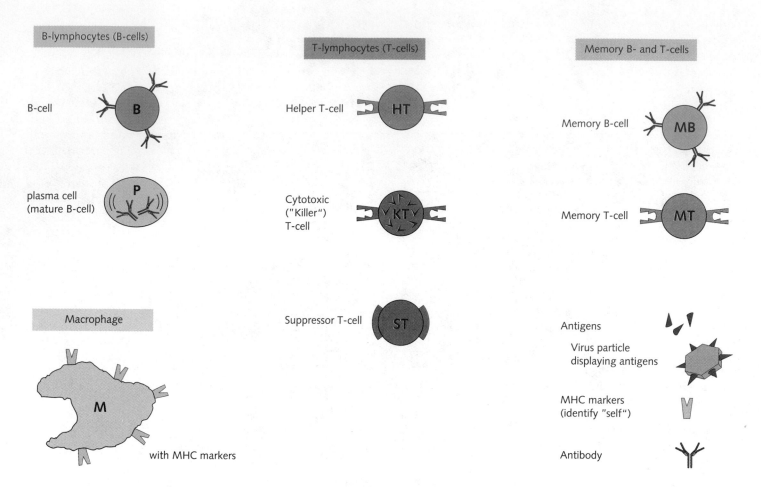

B-lymphocytes (B-cells)

B-cell — B

plasma cell (mature B-cell) — P

Macrophage

M

with MHC markers

T-lymphocytes (T-cells)

Helper T-cell — HT

Cytotoxic ("Killer") T-cell — KT

Suppressor T-cell — ST

Memory B- and T-cells

Memory B-cell — MB

Memory T-cell — MT

Antigens

Virus particle displaying antigens

MHC markers (identify "self")

Antibody

Figure 42.8   Visual key to cells, markers, and other elements involved in the immune system.

Functionally, the link between immune system and lymphatic system makes sense. Why? Remember from Chapter 35 that plasma leaks from capillaries into body tissues, drains into lymph vessels, and passes through lymph nodes before returning to the general circulation. This pattern ensures that a few bacteria or virus particles from any infection are swept along with the flow and make contact with lymphocytes. Once that contact is made, the bodywide immune response begins.

Lymphocytes, like other blood cells, originate from stem cells in bone marrow. These stem cells grow, divide, and differentiate into different classes of blood cells under the control of several hormones. As we introduce the various classes of these wandering immune cells, notice the icon used to signify each cell type in Fig. 42.8 and throughout this chapter.

There are two main classes of lymphocytes: **B-lymphocytes (B-cells)** and **T-lymphocytes (T-cells).** Although both B-cells and T-cells originate in bone marrow, T-cells mature only after they have passed through the *thymus gland,* which is where the T (for "thymus-derived") comes from. These two cell types are nearly identical in appearance but differ in function.

B-cells   B-cells function as cellular munitions factories that produce antibodies, the immune system's chemical weapons. They are responsible for what is known as the **antibody-mediated immune response.** Each mature B-cell produces only a single type of antibody, which is targeted against a specific antigen. One B-cell, for example, might produce an antibody that binds to a protein in the coat of the measles virus. Another B-cell might produce an antibody that binds to carbohydrates on the cell membrane of a trypanosome. This might seem like a hit-or-miss strategy, but the population of B-cells is large and diverse. This diversity ensures that the immune system contains at least a few cells able to recognize virtually any antigen the body encounters.

Young B-cells make numerous copies of the particular antibody they are genetically programmed to produce. Many of those copies are moved to the cell surface. There, one end of each molecule is anchored in the membrane, enabling the antibody to act like a receptor for its particular antigen. Ultimately, mature B-cells make many more copies of that antibody and release them into the blood and other body fluids. Because antibodies circulate through blood and other body fluids (which were once called "humors"), the production of antibodies is sometimes known as the **humoral immune response.** We will describe how this process works when we describe the immune system in action.

T-cells There are several types of T-cells, each of which performs an important function. T-lymphocytes mount a direct cell-to-cell campaign, or **cell-mediated immune response.** Like B-cells, T-cells have antibody-like molecules which recognize antigens and act as receptors on their cell surfaces. Unlike B-cells, however, T-cells do not secrete these molecules.

Helper T-cells, often referred to as the "command centers" of the immune system, work together to guide and direct the immune response. **Killer T-cells,** also called cytotoxic T-cells, are often referred to as the "foot soldiers" of the immune system, because they follow the directions of other cell types to attack pathogens and infected body cells directly. **Suppressor T-cells** moderate the immune response after infections are brought under control.

Memory B- and T-cells Once the immune system has been exposed to a specific pathogen, certain groups of B- and T-cells "remember" that encounter. These cells remain in the body for long periods of time, ready to spring into action should the same invader make a second appearance.

Macrophages

Still other cells, the macrophages or "big eaters," act essentially like roving military police. These cells cruise around the body, maneuvering in and out of tissues while searching for troublemakers. If they do encounter pathogens, macrophages engulf them and then help mobilize and direct the immune response by "alerting" T-cells to their presence.

THE IMMUNE SYSTEM IN ACTION

Now that you have met the major players, we can examine how they work together responding to a pathogen. Suppose that a few virus particles invade your body. At first, those viruses go about their business unimpeded; some of them float around in body fluids, while others enter your cells and begin replicating. Sooner or later, however, circulation and lymph movement cause chance encounters between those viruses and various branches of the immune system. Then the real battle begins. (Refer to the stages in Fig. 42.9 as you proceed through the immune process.)

Stage 1: Encounter and Recognition

If macrophages encounter free-floating virus particles, they engulf and partially digest them (stage 1a). Those macrophages then take pieces of virus coat proteins (which act as antigens) and combine them with newly synthesized MHC markers. The antigen-MHC complexes are then moved to the cell surface and "presented" on the outer surface of the cell membrane. In this way, macrophages use these antigens to alert the other cells of the immune system to the presence of pathogens.

Meanwhile, if a B-cell encounters an antigen that binds to its particular antibody, that B-cell becomes activated (stage 1b). The activated B-cell then digests the antigen into smaller pieces and places those pieces on the outer surface of its cell membrane in association with an MHC marker.

There are now two different types of cells spreading an alarm: "This particular pathogen has entered the body—help!"

Stage 2: Mobilization

If a macrophage presenting an antigen encounters a helper T-cell which recognizes that antigen, the cells lock together (stage 2a). The macrophage then secretes an **interleukin**—a messenger molecule that activates the helper T-cell. The T-cell itself then begins to secrete other interleukins that stimulate other T-cells to multiply.

Meanwhile, if a stimulated helper T-cell recognizes the antigen presented by an activated B-cell, those cells bind together (stage 2b). The helper T-cell then secretes different interleukins, which stimulate the B-cell to proliferate into a clone of cells. Because all the members of that clone are descended from the original stimulated B-cell, all of them will produce antibodies matched to the same antigen.

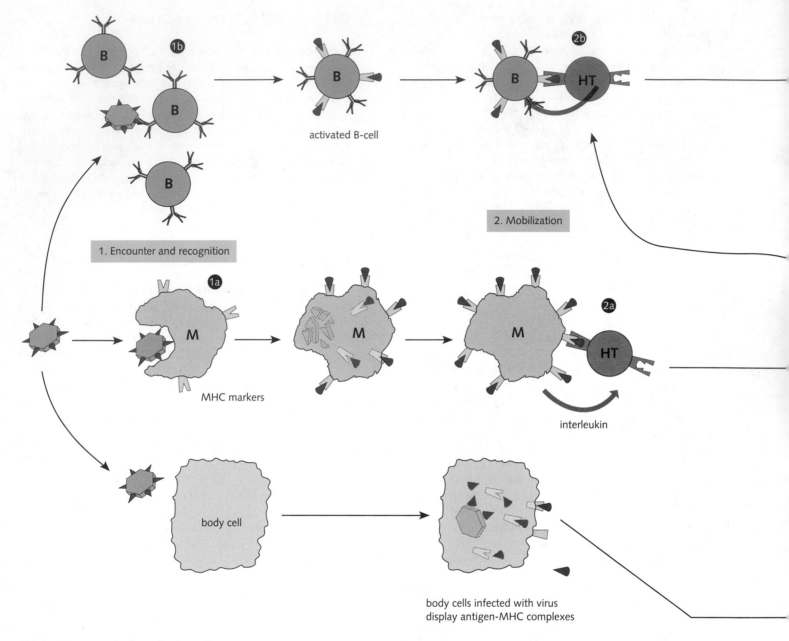

Figure 42.9 Antibody and cell-mediated immune responses combine to combat an invading virus. For this figure, we have arbitrarily divided the complex series of events into five stages and labeled them as encounter, mobilization, attack, cease-fire, and memory. Note also that several chemical messengers and processes have been omitted here for simplicity. **(1a)** Any macrophage that encounters an antigen first engulfs and then digests it. That macrophage then displays pieces of the foreign protein combined with MHC markers on its cell surface. **(1b)** Among the many B-cells in circulation, a few are stimulated by the binding of a foreign protein (antigen) to antibody-like molecules on their cell surfaces. These activated cells digest the protein and display pieces of it on their cell surface. **(2a)** Among the many helper T-cells in the body, a few carry antibody-like molecules that bind to the pieces of antigen on the macrophage surface. The macrophage secretes a messenger that activates the helper T-cell to multiply. **(2b)** Stimulated helper T-cells bind to the pieces of antigen on B-cell membranes and secrete a growth factor that stimulates those specific B-cells to proliferate and form activated clones. As B-cells proliferate, further secretions from helper T-cells induce them to mature and secrete antibodies to the activating antigen. **(3a)** Many newly produced cells in this activated clone mature into plasma cells and secrete antibodies to the antigen originally encountered. **(3b)** Viruses cross-linked by antibodies are singled out to be engulfed by phagocyte. **(3c)** Meanwhile, helper T-cells secrete chemical messengers that stimulate the multiplication of those killer T-cells that have also been activated by contact with antigens from the invading virus. **(3d)** Killer T-cells destroy those body cells infected with active virus. This sacrificial maneuver prevents further viral replication. **(4)** Suppressor T-cells halt the actions of both plasma cells and other T-cells, preventing their self-activating chain reactions from getting out of control. **(5)** Some B-cells and T-cells from activated clones differentiate into memory cells. These memory cells can respond rapidly to future infection for nearly the entire life span of the individual.

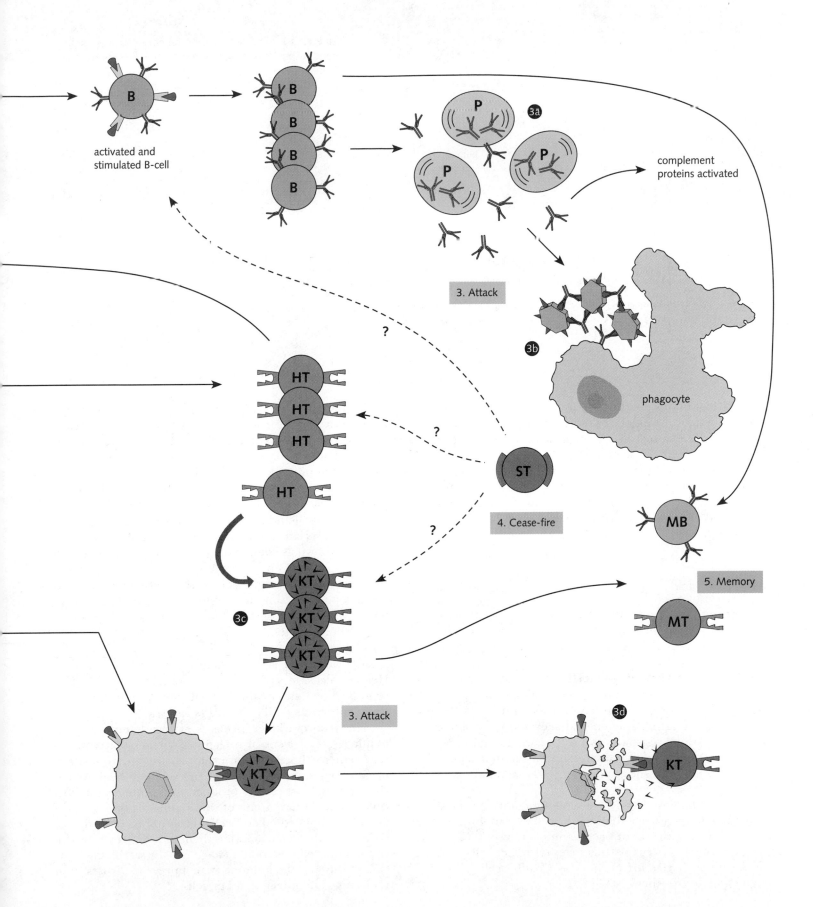

activated and
stimulated B-cell

complement
proteins activated

3. Attack

phagocyte

4. Cease-fire

5. Memory

3. Attack

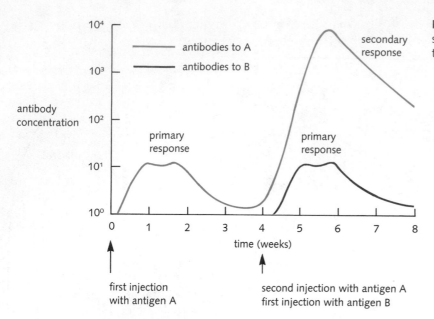

Figure 42.10 The presence of memory cells makes the second response to an antigen much more powerful than the first.

Note that in this process, the antigen activates the specific groups of lymphocytes able to make antibodies against it. For this reason, we might say that the antigen "selects" those cells. This interaction between antigen and immune system is known as the **clonal selection model.** Clonal selection enables the immune system to "economize" by producing large numbers of only those lymphocytes tuned to fight the particular pathogen at hand.

Stage 3: Attack

At this point, different parts of the immune system operate in two parallel lines of attack to exterminate the invading viruses.

B-cells launch the antibody-mediated response

Many stimulated B-cells mature into **plasma cells** specialized to synthesize and release antibodies targeted at the antigen that activated the original cell (stage 3a). This growth and differentiation takes between 5 and 10 days and is the reason for the time lag between infection and the production of large amounts of antibody.

Free antibodies perform several functions. Because antibodies have two combining sites, they can *cross-link* antigen-carrying viruses or bacteria into a network. Cross-linked bacteria or virus particles cannot infect cells, and they are easy targets for phagocytes ("eating cells") that engulf and destroy them (stage 3b).

In addition, the constant region of antibody molecules—the area we compared to a "signal flag"—carries "instructions" to other molecules and cells in the body on how to handle the cross-linked package. If the antigen is a free-floating protein, the antibody bound to it may single it out for excretion by the kidney. If the antigen is part of a virus or bacterium, the constant region attracts phagocytes that engulf and destroy the invader.

Finally, certain antibodies activate a group of blood proteins known as **complement proteins.** When these circulating proteins encounter an antibody bound to the surface of a cell, they attach to the cell membrane and create a pore, causing the cell to burst. Complement proteins thus help to destroy foreign cells too large to be eaten by phagocytes.

T-cells launch the cell-mediated response

Helper T-cells, in addition to encouraging the reproduction and maturation of activated B-cells, also stimulate activated killer T-cells (stage 3c).

The action of killer T-cells is extremely important in fighting several types of infections. Viruses that have already entered host cells, for example, are safe from attack by antibodies, but they have an Achilles' heel. While viruses cause host cells to produce viral proteins, the host cells often "display" some of those antigens on their surfaces. If stimulated killer T-cells recognize those antigens, they attack the cell displaying them and destroy it by disrupting the cell membrane (stage 3d). By sacrificing body cells in this way, the immune system prevents viruses from replicating and infecting other cells.

Killer T-cells are vital in fighting other pathogens as well, particularly certain bacteria, protozoans, and multicellular parasites. In these cases, certain groups of T-cells recognize the foreign MHC markers carried by those organisms.

Stage 4: Cease-fire

Once the infection is under control, the immune response must be halted to keep it from escalating out of control. This cease-fire response is accomplished by slowing down the rate of lymphocyte cell division and limiting the production of antibodies (stage 4). The function is performed by *suppressor T-cells*, which were once thought to be a separate group of cells. Recent research has shown that a separate class of suppressor T-cells may not exist, although there is no doubt that the suppressor function is real and that it is vital in keeping the immune response in check.

Stage 5: Memory

Even after the immune response has moderated, groups of antigen-stimulated B- and T-cells remain. These **memory cells** are stored in the spleen, where they are ready to respond rapidly should the body ever encounter the same pathogen again (stage 5). The existence of these memory cells makes a second response of the immune system to the same antigen swifter and much more powerful than the first (Fig. 42.10). This is the basis of *permanent immunity* to certain diseases.

WHEN THE IMMUNE SYSTEM WORKS AND WHEN IT FAILS: CASE STUDIES

Although the immune system is activated when a foreign organism enters the body, the results of that activation vary greatly. In some cases, the system deals with a pathogen effectively; in other cases, it cannot do so. Let us return to the first three children in the opening scene of this chapter—one suffering from measles, one from the flu, and one from sleeping sickness—and consider why the system works in some cases and fails in others.

Measles: Recovering from Disease with Permanent Immunity

When the body first encounters the viruses that cause smallpox and measles or the bacteria that cause tuberculosis and diphtheria, it produces a powerful immune re-

sponse that leaves behind large clones of memory cells. Those memory cells are sensitized to very specific antigens associated with those pathogens. If pathogens carrying those identical antigens enter the body again, the memory cells produce antibody so quickly that the invaders are destroyed before they can produce disease (Fig. 42.11a).

Luckily for us, many pathogens (including the virus that causes measles) change very little, even over long periods of time. For that reason, the measles virus your body encounters today will be virtually identical to one you may first have encountered during a childhood illness. Thus, memory cells "on file" in your body can prevent reinfection. Of course, this immune memory also explains why vaccines can protect us against certain diseases: as long as the right antigens are present in a vaccine, the body's response to that vaccine will leave behind memory cells that prevent infection.

Louis Pasteur (without understanding the specifics of memory cells, of course) more or less understood what the immune system did, and he applied his insight to several diseases for which no mild form (such as Jenner's cowpox) existed. Pasteur discovered that many pathogens could be altered to limit their ability to produce disease without affecting their ability to stimulate the immune response. This strategy is the basis for vaccines against diseases such as measles, whooping cough, and polio. It is also possible to produce some vaccines by killing the pathogen before injection.

Today, researchers are learning how to use techniques of molecular biology to identify individual proteins from pathogens that trigger a good immune response. Those proteins can then be cloned for use as vaccines by themselves. Alternatively, they can be inserted into the genome of a harmless microorganism that is then cultured and injected into the body. This fascinating technique holds great promise for controlling some of the world's most persistent diseases.

Influenza: Recovering from Disease with Temporary Immunity

The second child will mount a strong immune reaction against the flu and will probably recover completely within a few days. She will produce a clone of memory cells against antigens carried by that particular strain of flu virus and will therefore be protected from reinfection by that strain.

But the viruses that cause the flu and the common cold, unlike those that cause measles and smallpox, mutate rapidly in ways that alter the proteins in their outer coat. As a result, every few months (in the case of colds) or every few years (in the case of flu) new strains arise

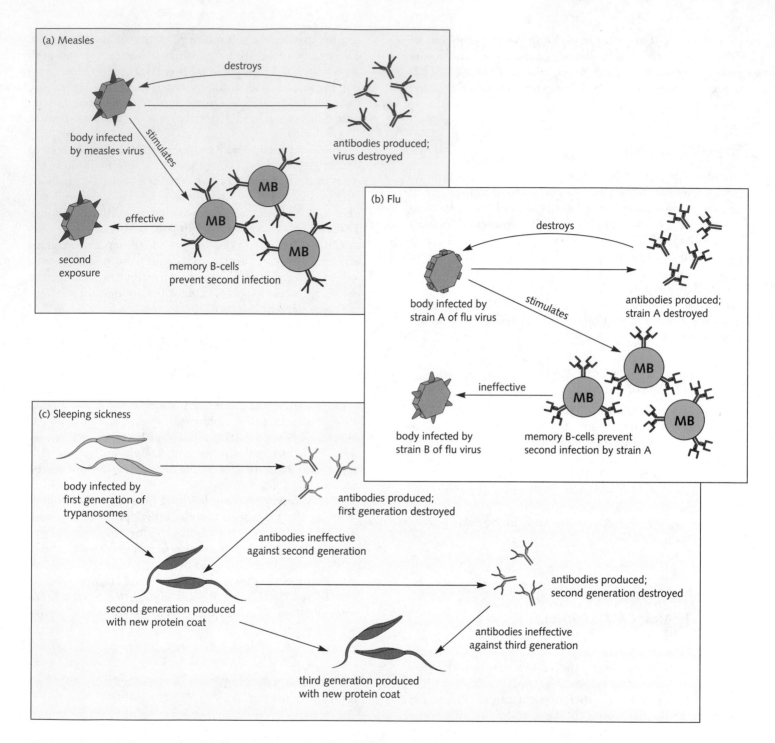

Figure 42.11 **(a)** The memory cells of the immune response are effective in preventing a second infection by measles virus. **(b)** However, the coat proteins of flu virus undergo frequent mutations, so that memory cells may not be effective against a second infection by a different strain of the flu virus. **(c)** Trypanosomes undergo rapid rounds of mutation as they grow in the body, ensuring that every time the immune response rises to destroy the parasite, a few cells with different coat proteins evade the response and continue the infection.

that carry completely different coat proteins. Memory cells specialized to respond to the "old" viral strains are useless against these new strains (Fig. 42.11b). That's why we cannot acquire permanent immunity to the flu or the common cold. That is also why it has been impossible to develop a flu or cold vaccine that works for a lifetime—or even for a decade. Instead, new vaccines are produced every year from new flu strains that seem to present a threat.

Trypanosomes: Pathogens that Defeat the Immune System

Our third patient has a much harder time with his illness, sleeping sickness. Why? Because trypanosomes employ more than 100 sets of genes to produce the proteins that coat their outer surfaces, thereby escaping the immune response.

Here's how their strategy works. The patient's immune system produces a powerful response against the pathogen only a few days after infection begins. But just as the immune system gears up to attack the trypanosomes that invaded the body, a new generation of pathogens turns on a different set of genes to make a different set of surface proteins. Because the original immune response was targeted against a specific set of antigens, which have now been replaced by new ones, it is as though the invaders had disappeared; the body must start from scratch to mount a response against the new antigens. By the time that new response is geared up, the trypanosomes switch coat proteins again, always staying one jump ahead of the body's defenses (Fig. 42.11c). That's why this patient will struggle through fever after fever, suffering brain inflammation, damage to the nervous system, and, all too often, death.

DISORDERS OF THE IMMUNE SYSTEM

Allergies

When the immune system overreacts to an antigen in the environment, an **allergy** results. The offending molecule can be part of a pollen grain, an animal hair, the toxin from a bee sting, or even house dust. We still do not know why some people suffer from allergies and others do not, but we understand in principle how the allergic response proceeds.

The functioning immune system is always capable of producing an inflammatory response when antigens are discovered. These reactions are mediated by a particular class of antibodies that activate **mast cells** (Fig. 42.12). Ac-

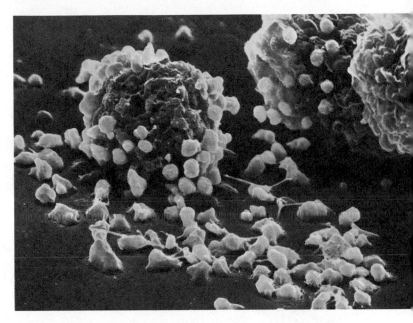

Figure 42.12 Mast cells releasing histamine granules. (Magnification factor: 3000)

tivated mast cells, in turn, release *histamine granules* that initiate the inflammatory response described earlier. Although there is no effective cure for the runny nose and itchy eyes of hay fever, many individuals can get some relief by taking *antihistamines,* drugs that counteract the effects of histamines. (Note in this context that cold medications containing antihistamines do *not* fight infection. If you take antihistamines when you have a cold or the flu, all they do is minimize the inflammatory response in nose and sinuses to help you breathe while your immune system goes after the viruses.)

The smooth-muscle reactions mediated by histamine can be dangerous. If a strong allergic reaction develops in breathing passageways, they can contract so that the movement of air is significantly decreased. This condition, known as **asthma,** may make breathing almost impossible. Asthma attacks are usually triggered by exposure to a particular substance to which the sufferer is allergic, so the best strategy is to avoid exposure to that substance. Immediate relief from a severe asthmatic attack can often be provided by a mist of *epinephrine* or similar compound (usually from a pocket-sized inhaler), which relaxes smooth muscles in the passageways, easing airflow and permitting more normal breathing.

Recognizing Self and Autoimmune Disease

As mentioned earlier, a properly functioning immune system distinguishes self from nonself. Normally, the **thymus gland** "teaches" T-cells to discriminate between those

LS ____
N ____
LL ____

Monoclonal Antibodies

When an antigen stimulates the immune response, antibodies are made against specific sites on the antigen molecule. These sites can be very small—often just three or four amino acids in the case of a protein antigen—and they are distributed along the exposed surfaces of the molecule. Because an individual protein may have hundreds of such sites along its surface, hundreds of different B-cells may be stimulated to produce antibodies when the immune system reacts to that protein. This leads to a *polyclonal* immune response, because many clones of B-cells develop in response to the antigen, each clone producing a different antibody in response to a different site.

Because the production of an antibody is a cellular event, it is possible to select a *single clone* of B-cells that will produce *monoclonal* antibodies against a single site on the antigen—a perfect supply of exquisitely specific antibodies.

That was the thinking of Cesar Milstein of Great Britain as he followed suggestions first made by George Kohler in Switzerland. Milstein and Kohler knew that two problems would arise in growing single clones of antibody-producing cells: selecting the individual single B-cells or plasma cells that might react to an antigen from the millions around them that did not, and coaxing those cells to grow and divide in culture. Milstein developed a technique to do both, and it is now widely used in research work.

First, the antigen is injected into a mouse, and after a few days, millions of lymphocytes are removed. Somewhere among these millions are a few cells stimulated by the antigen. In order to ensure that the antibody-producing cells grow and divide efficiently, the investigator mixes the lymphocytes with cells of a mouse *myeloma*, a cell line produced from a tumor of the antibody-producing plasma cell. The mixture of cells is treated under conditions wherein cell *fusion* occurs, producing hybrid cells between the mouse B-cells and the myeloma cells.

These "*hybridoma*" cells are now "immortal," meaning that they may grow almost without limit in culture. The investigator grows individual clones of the hybridomas and tests them, one at a time, for reactivity with the original antigen. The clones that test positive are set aside and grown under conditions in which they produce large amounts of antibody.

These antibodies are called *monoclonals*, because they are produced by a single clone of cells. Because monoclonal antibodies react to one specific site on the antigen, they can be used to detect individual antigens on proteins, virus particles, and tumor cells. Because of this extreme specificity, monoclonal antibodies are already used in a test for the AIDS virus and in targeting drugs to cancer cells.

inject mouse with antigen "A"

The preparation of monoclonal antibodies by the fusion of mouse lymphocytes with myeloma cells.

lymphocytes

myeloma cells

mix cells—some fuse transfer to specific medium

Hybridoma cells

unfused cells die; fused (hybridoma) cells grow

hybridoma cells cultured in separate wells for analysis of antibodies

positive clones used to produce large quantities of monoclonal antibodies

proteins that belong in the body and those that don't. According to current theory, the enormous diversity of T-cells present at birth includes many that could attack the body's own proteins if they survived. But before any damage can occur, the thymus gland singles out those T-cells and eliminates them.

How does the thymus accomplish this feat? It contains specialized cells that absorb a representative sample of the body's proteins, digest them, and display their component peptides to the T-cell population passing through. Any developing T-cell carrying receptors that bind to one of these "self" peptides is killed in the thymus. By some time after birth, therefore, all the T-cells that might attack our own tissues are eliminated.

When this ability to recognize the body's own molecules is lost, the immune system's power to destroy can be turned against the body's own tissues to cause an **autoimmune disease.** One such disorder is **myasthenia gravis,** in which victims grow weaker, lose control of voluntary muscles, and may become crippled. This disease is caused by an immune system response against molecules that form part of the neuromuscular junction. In **multiple sclerosis,** the immune system attacks the protective myelin sheaths around nerve cells; whereas in **rheumatoid arthritis,** a misguided immune response destroys cells in the body's joints. One theory holds that these autoimmune diseases are caused by renegade macrophages that pick up and display self molecules instead of nonself molecules, eventually triggering a destructive T-cell response.

Sometimes, infection can trigger autoimmune diseases, as in the case of **scarlet fever,** a disease that results when toxin-producing strains of *Streptococcus* infect the throat. The infection itself isn't serious, but "mistakes" by the immune system can cause permanent damage. Apparently, antigens carried by certain strains of the bacterium resemble proteins found on the surfaces of heart muscle cells. Therefore, in some cases of scarlet fever, the immune system response to the bacteria inadvertently attacks heart muscle, producing a complication known as **rheumatic fever.**

Transplant Rejection

Although not really a "disorder" of the immune system, the rejection of transplanted tissue is a result of immune responses that physicians wish they could avoid. The cell-mediated immune response recognizes foreign MHC markers on the cells of tissues and organs transplanted from one individual to another and mounts an attack on those foreign tissues. Physicians have tried two strategies to make transplants "take": finding a "compatible" donor (usually a family member whose MHC markers are very similar to those of the recipient) and suppressing the function of the immune system with drugs such as *cyclosporin.*

AIDS

The fourth child described earlier suffers from a disease that first appeared in tropical Africa, where diseases new to science are often discovered. This disease is unique; it is close to perfection in terms of pathogen evolution. Not only can it avoid the defenses of the immune system, but it also attacks that system directly, lives inside it, and destroys it. This is the retrovirus known as **HIV,** the *human immunodeficiency virus.*

HIV went unrecognized in central Africa for years, both because its victims died of other, already recognized diseases and because sophisticated medical care was usually wanting. It spread slowly to Europe and to Haiti, and then to the United States, Latin America, and Asia. There the first isolated cases—a few in Europe and one or two in the United States in the mid-1970s—baffled physicians. Patients were dying of infections from microorganisms they should have fought off easily, and no one could imagine why.

Finally, the virus found its way into two Western populations whose behaviors ensured its explosive spread: the gay community, where sexual freedom was a way of life, and the legions of intravenous drug users, whose ritual sharing of needles spread infected blood with deadly speed and efficiency. At about the same time, the disease began to spread among heterosexually active individuals throughout central Africa.

As discussed in Chapter 2, these recognized modes of HIV transmission—sexual activity and infected blood—seriously interfered with society's ability to control the disease. Because AIDS remains incurable, and because it presents a threat to the health of many Americans, we will try to correct some of the misconceptions about AIDS by reviewing what is now known about its history and mode of action.

The History of AIDS in the United States

In the early 1980s, physicians noticed patients with unusual infections and a form of skin cancer so rare that most had never seen a case before. These infections, including pneumonia caused by *Pneumocystis carinii* (a protozoan), were produced by organisms that cause disease only in extremely rare cases. Purple skin cancers known as *Kaposi's sarcoma* normally appear only in elderly patients with impaired immune systems or in patients whose immune systems are suppressed with drugs such as cyclosporin. Yet here they were occurring in otherwise apparently healthy men.

Thoughtful physicians recognized that these were **opportunistic infections,** successful only because the immune system of the patient was impaired. The first six recognized cases of the disease were described in

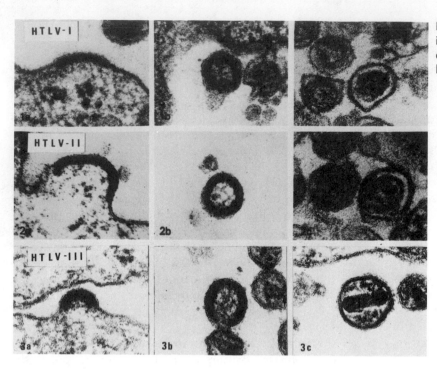

Figure 42.13 Original electron micrographs identifying the three known types of AIDS virus particles budding from cells grown in laboratory culture. HTLV III has been renamed HIV.

Los Angeles in 1981, and the name **AIDS** (for **acquired immune deficiency syndrome**) was suggested for the disorder.

Because the disease was first reported among gay men, some physicians assumed that something about these men's lifestyles impaired their immune response. But epidemiologists soon recognized that it was spread through blood (by the sharing of contaminated needles) and sexual intercourse. In 1983, Luc Montagnier and associates at the Institut Louis Pasteur in France identified HIV as the cause of AIDS. This discovery was quickly confirmed by a team led by Robert Gallo at the National Institutes of Health in the United States (Fig. 42.13). Research on the disease is now progressing around the world; we have learned a great deal about the biology of AIDS very quickly.

The Biology of HIV Infection

HIV is a retrovirus that infects lymphocytes and epithelial cells throughout the body. Many researchers call HIV the most "insidious" parasite ever to evolve. Not only does it avoid the defenses of the immune system, but it also subverts and breaks down the lymphocyte communication network in a manner that destroys the system.

HIV is composed of a glycoprotein envelope surrounding a dense, cylindrical core. In the core are two molecules of viral RNA that contain at least eight genes (Fig. 42.14). Some of these genes code for the envelope proteins, some code for the core proteins, one directs the synthesis of the enzyme reverse transcriptase, and three control the activity of the virus.

HIV especially targets T-cells, because one of its envelope proteins binds to a receptor molecule on the T-cell membrane. This attachment causes the viral core to be taken inside the cell. There the virus synthesizes reverse transcriptase, the enzyme that makes DNA copies of its RNA genome. Some of those copies insert themselves into the host cell DNA and stay there *permanently,* while other copies remain free in the cytoplasm (Fig. 42.15).

Viral DNA may remain inactive for widely varying periods of time. When activated, it directs the synthesis of viral RNA and proteins that are assembled into new virus particles and budded off into the bloodstream. Active HIV infection can kill not only the cell it has infected but also many other T-cells. Because they carry the critical viral protein on their cell membranes, infected cells bind to healthy T-cells, killing them *en masse.*

HIV also infects macrophages, within which they can "hide" for long periods without triggering immune responses. The virus also seems able to grow and reproduce indefinitely inside a macrophage without killing the host cell. By hiding inside monocytes that cross the blood–brain barrier, HIV can also enter the central nervous system, where it causes serious brain damage.

Difficulty in tracing and predicting the spread of HIV arises from its long, variable latent period, which ranges from a few weeks to 10–15 *years.* During that time, an individual may have no warning that the disease is developing. When the virus is first activated, signs of immune suppression may be subtle or invisible; they include night sweats, swollen lymph nodes, a fungal infection of the throat called "thrush," and reduced numbers of helper T-cells in the blood.

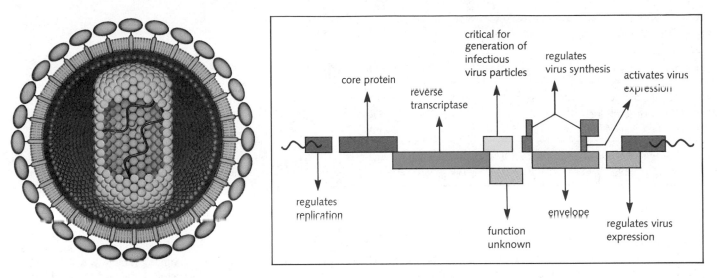

Figure 42.14 LEFT: A schematic model of the HIV retrovirus, showing its membrane-like outer coat and internal core structure. RIGHT: The HIV genome.

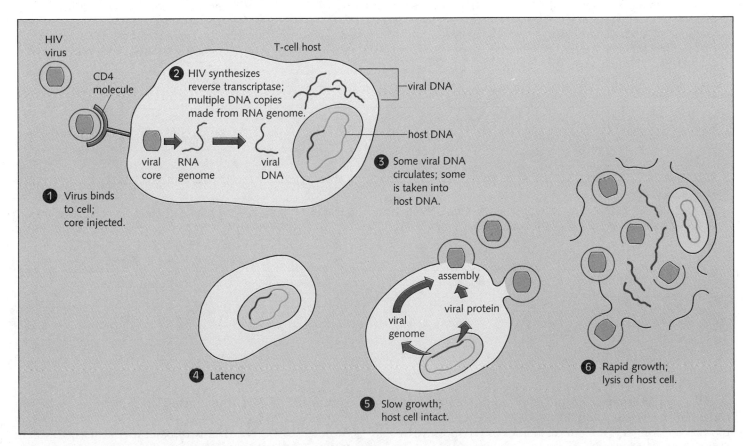

Figure 42.15 Infectious cycle of the HIV retrovirus. The RNA genome of the virus is copied into DNA by reverse transcriptase, and becomes part of the host cell DNA. The viral genome may remain latent within the cell for long periods of time. It may also direct the synthesis of new virus particles, or it may grow rapidly and destroy its host cell.

Table 42.2 *Patterns of HIV Infection in the World*

Pattern 1	Pattern 2	Pattern 3
Homosexual/bisexual men and intravenous drug abusers (IVDA) are the major affected groups	Heterosexuals are the main population group affected	More recent introduction with spread among persons with multiple sex partners
Period When Introduced or Began to Spread Extensively Mid-1970s or early 1980s	Early to late 1970s	Early to mid-1980s
Sexual Transmission Once predominantly homosexual; over 50% of homosexual men in some urban areas infected. Increasing rates of infection among women. New cases among college-age individuals indicate that heterosexual transmission is increasing among sexually active adolescents	Predominantly heterosexual; up to 25% of the 20- to 40-year age group in some urban areas infected, and up to 90% of female prostitutes. Up to 50% of men currently serving in military units in some central African states reported to be HIV-positive. Homosexual transmission not a major factor	Both homosexual and heterosexual transmission documented. Generally low prevalence of HIV infection in rural areas. Situation changing rapidly in certain Asian countries such as India and Thailand; in large cities, such as Bangkok, transmission increasing among the estimated 1 million prostitutes and their clients. International "sex tourism" worrisome
Parenteral Transmission Intravenous drug abuse accounts for the next largest proportion of HIV infections, even the majority of HIV infections in southern Europe. Transmission from contaminated blood or blood products not a continuing problem, but tens of thousands of persons infected before 1985	Transfusion of HIV-infected blood is major public health problem. Nonsterile needles and syringes account for undetermined proportion of HIV infections	Not a significant problem at present in rural areas; some infections in recipients of imported blood or blood products. Situation changing rapidly in large Asian cities; in Thailand, 50% of IVDA (many of whom are prostitutes) are HIV-positive
Perinatal Transmission Documented primarily among female IVDA, sex partners of IVDA, and women from HIV endemic areas	Significant problem in areas where 5 to 15% of women are HIV positive	Currently not a problem
Distribution Western Europe, North America, some areas in South America, Australia, New Zealand	Africa, Caribbean, increasing areas in South America	Asia, the Pacific Region (minus Australia and New Zealand), the Middle East, Eastern Europe, some rural areas of South America

ORIGINAL SOURCE: P. Piot, F. A. Plummer, F. S. Mhalu, J. Lamboray, J. Chin, and J. M. Mann, AIDS: An international perspective, *Science* 239 (February 1988):576. Updated in 1993 from various sources.

The average time from infection to full-blown AIDS seems to be 8 years; some do not develop the disease for 16 years (or longer). Long-term studies of HIV-positive individuals indicate that between 95 and 99 percent of infected individuals ultimately develop AIDS. There is as yet no test that can tell for certain when or how soon AIDS will develop in an infected individual.

How AIDS Is *Not* Transmitted

Although AIDS is a deadly disease, it is *not* easy to catch. HIV is *not* transmitted via food or water or through coughing, sneezing, dry kissing, hugging, or sharing clothing, bedding, or eating utensils. Everyday contact with AIDS victims, including shaking hands and engaging in close conversation, poses no risk of infection. There is neither purpose in nor need of quarantining or isolating HIV-positive individuals or AIDS patients.

There is no evidence that AIDS can be transmitted by mosquitoes. Mosquito-borne diseases are extremely host-specific; the mosquito that transmits malaria cannot carry yellow fever, and vice versa. Furthermore, mosquitoes transmit malaria only because the disease-causing organism enters the insect's salivary glands. There is no evidence that HIV either lives in the mosquito's system or enters its salivary secretions. Because the insect does not actually inject blood from a previous meal into new victims, there is no way for a mosquito bite to transmit HIV.

How AIDS *Can* Be Transmitted

At first, medical authorities and the media referred to "high-risk *groups*," by which they meant gay and bisexual men and intravenous drug users. It is now clear that AIDS is not restricted to any specific group of people but can be contracted by anyone who engages in specific high-risk *behaviors*. Those behaviors are clear: unprotected sexual intercourse of any kind and the sharing of hypodermic syringes (Table 42.2).

Sharing of syringes The AIDS virus is transmitted through needles shared among intravenous drug users. Between 50 and 75 percent of American intravenous drug users are infected with HIV. Because AIDS is also spread through sexual activity, sexual partners of intra-venous drug users are at risk and—if sexually active themselves—may serve as conduits for transmitting the virus into non-drug-using populations.

Transmission from mother to fetus The AIDS virus can be passed from an infected mother to child, either during the birthing process or across the placenta during fetal development; risk of such transmission appears to range from 25 to 40 percent. Because globally the predominant mode of transmission is through heterosexual intercourse, a staggering number of children are infected with the virus while in the womb.

Sexual transmission The initial prevalence of AIDS in the United States among non-drug-using gay men, many of whom practice anal intercourse, misled many people into thinking that AIDS is not transmitted through vaginal intercourse. Globally, however, heterosexual intercourse is the dominant mode of transmission, accounting for 60 percent of all new HIV infections. Internationally, the ratio of infected females to infected males is close to 1:1 (Fig. 42.16).

Anal sex may be more effective than vaginal sex in transmitting the virus. It also appears that the presence of other sexually transmitted diseases in either partner increases the risk of transmission during vaginal intercourse. The statistical risk of transmission in a single sexual act is small—somewhere between 1 in 100 and 1 in 1000. But many non-drug-using sexual partners of HIV-infected individuals have become infected, some of them through a single sexual encounter. The virus has clearly

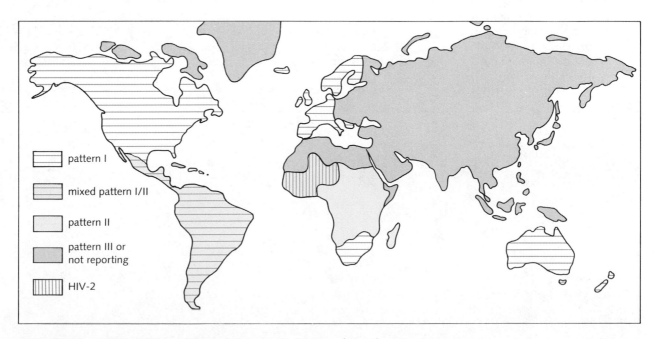

Figure 42.16 Global patterns of HIV infection. Latin America, formerly pattern 1, is now classified as a new category, mixed pattern 1/2.

Protecting Yourself Against AIDS

The best protection against contracting AIDS from sexual activity is not to have sex at all. The next best strategy is to form a monogamous relationship with another uninfected individual and to have no sexual activity outside of that relationship. For many individuals, this would mean simply abstaining from sex before marriage.

Sexually active individuals should know that the risks of unprotected sexual intercourse are well established and that the magnitude of those risks increases with the number of sexual partners. Most contraceptive methods, including the pill and IUDs, do *not* prevent the spread of HIV or other sexually transmitted diseases.

Only *condoms,* which prevent contact with semen, seem to be effective in preventing the transmission of HIV during intercourse, and they are effective only if used with great care during all sexual contact. The use of condoms by sexually active heterosexuals and homosexuals, together with limiting the number of sexual partners, may be the only way to slow the spread of the disease until a cure or treatment is found.

It is true that the statistical chance of getting infected with AIDS through

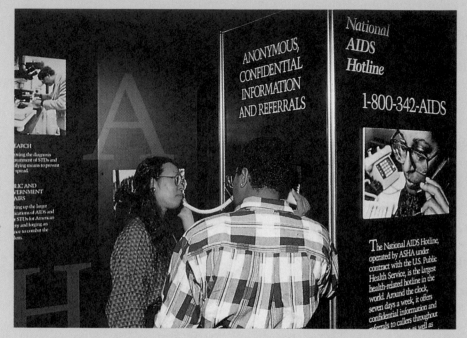

Although AIDS cannot be cured, it can be prevented. AIDS education is a critical weapon in the fight against the disease.

a single act of unprotected intercourse with a member of the general population is very low. But because of the virus's long latency period, as some

AIDS experts say, if you have sex with an individual, you are sharing in that individual's entire sexual history for the past 15 years.

been transmitted both from male to female and from female to male.

It does not appear that the virus is spreading explosively through the American population, though its continuing spread in Africa, Southeast Asia, and Japan is alarming. At this time, although most cases of AIDS are still found in members of the original high-risk groups, heterosexual transmission among sexually active adolescents is rising to levels that should cause serious concern among college-age people. Because of the virus's long latency period, because of uncertainty about the rate of transmission through heterosexual intercourse, and because of an almost total lack of knowledge about the sexual activities of the American public, prudent epidemiologists advise caution.

Current estimates of Americans infected with HIV range from 500,000 to nearly 2 million. By the end of 1992, the number of officially diagnosed cases of AIDS in the United States surpassed 270,000, and the number of AIDS-related deaths topped 160,000. By 1993, AIDS surpassed both cancer and heart disease in terms of the number of years of life lost before age 65. Globally, one person somewhere in the world is infected every 15–20 seconds.

Transfusion-borne AIDS   Before 1985, HIV infection was transmitted to hemophiliacs and surgical patients through transfusions of infected blood and blood products. Such cases have been nearly eliminated by two concurrent strategies.

1. The nation's blood supply is screened for antibodies to HIV, and public health officials assure us that the blood supply is safe. Because not all infected persons produce antibodies to the virus at all times, however, and because of the latency between infection and first antibody production, it is possible for blood that tests antibody-negative to carry HIV. Research and development are currently in progress to develop a commercial test for minute quantities of viral RNA. This technique will further minimize the risk of transmission through transfusion.

2. The blood industry is doing everything it can to discourage potentially infected individuals from donating blood.

Note that because licensed health care providers use sterile, disposable equipment when handling blood, it is impossible to contract AIDS when giving blood at a licensed blood-donating facility.

Tests, Vaccines, and Cures

The procedure commonly called the "AIDS test" is actually a test for antibodies *against* the AIDS virus. The presence of those antibodies does not mean that an individual is immune to the virus. Rather, it implies that she or he has been exposed to the virus and that the immune system has attempted to deal with the infection. A negative result for this test is a reasonably safe indicator that a person has not been infected with the virus.

As of this writing, no *cure* for AIDS has been found; there is no treatment that eliminates the virus from infected individuals. According to the American Foundation for AIDS Research, current treatments for HIV/AIDS involves the four complementary approaches described next.

Antiviral therapy   Drugs that hinder viral activity are being used to suppress HIV multiplication and spread through the body. The first of these was AZT, a modified thymidine molecule that blocks the synthesis of viral DNA by reverse transcriptase. Because AZT is not incorporated by most body cells in their DNA synthesis, it is tolerated by many patients. But AZT does interfere with DNA synthesis in the bone marrow of some patients, causing serious side effects. Several other drugs (including ddI, ddC, and d4T) are being used in conjunction with AZT and interferons in efforts to combat viral resistance to single drugs and in hopes of achieving synergistic action.

Immunotherapy   Several chemical messengers used in the immune system—interferons and lymphokines—are under study as ways to protect or stimulate the immune system against HIV. Tests are also under way on so-called therapeutic vaccines that enhance the body's natural response to the virus.

Treatment of opportunistic infections   The greatest success has been achieved with drugs aimed at opportunistic infections. One drug, pentamidine, is effective in preventing and treating *Pneumocystis* pneumonia; drugs aimed at other ailments are currently under study.

Protective vaccines   Unfortunately, the genes that code for HIV's protein envelope exhibit very high rates of mutation. This means that new strains of the virus are continually emerging, a fact that makes protective vaccine development extremely difficult.

DISEASE, IMMUNOLOGY, AND THE FUTURE

Spurred in part by advances in molecular biology and in part by the battle against HIV, our understanding of disease and the immune system has made unprecedented leaps over the last few years. According to the World Health Organization, smallpox has been eradicated from the face of the earth, and polio and many other childhood diseases are under control.

Other diseases continue to be problems, though, both nationally and worldwide (Table 42.3). Importantly, immunological research offers hope not only to people threatened by infectious disease but also to those with certain types of cancer, diabetes, and autoimmune diseases. Why? Because all those conditions—and experimental treatments for them—involve functions or malfunctions of the immune system.

In addition, pathogens have a nasty habit of evolving resistance to drugs and immune responses. Malaria and sleeping sickness continue to resist permanent control,

and some bacteria have developed resistance to penicillin and other common antibiotics. New strains of multiple-drug-resistant tuberculosis—resistant to nearly all antibiotics that don't have serious side effects on humans—present a growing threat to public health. And we haven't even discussed the scores of parasitic diseases that plague tropical countries. For all these reasons, the frontiers of molecular immunology and parasitology are likely to remain among the most challenging and exciting fields in the biological sciences.

Table 42.3 *Losses of Human Life to Epidemics*

Past Epidemics	Effects
Bubonic plague	One outbreak of the plague (1347–1350) killed between 17 million and 28 million people, one-third to one-half of Europe's population
Influenza ("the flu")	Killed 22 million people worldwide in 1917–1918
Smallpox	Killed 400,000 Europeans in the nineteenth century alone
Polio	Infected 400,000 Americans, of whom 22,000 died between 1943 and 1956

Present Epidemics	Effects
Measles	Fairly well controlled in industrialized countries, but kills at least 1.5 million people annually in the Third World
Tuberculosis	Kills at least 500,000 people annually worldwide
AIDS	Over 270,000 officially reported cases and 160,000 deaths in the United States alone
Diarrhea (largely cholera)	Kills 10 million people annually in Third World countries
Malaria	Kills 1.2 million people annually in Third World countries

SUMMARY

All organisms must cope with the diversity of pathogenic organisms that can cause disease. The primary barrier against infection is the skin, the single largest organ in the human body. The skin forms a tough and effective barrier, and defenses that exist just beneath the skin come into play if the skin is broken.

The immune system acts when the skin is breached. Cells from the immune system play a key role in the inflammatory response, which deals with such invasions. The specific immune response, mediated by T- and B-lymphocytes, recognizes and attacks pathogens based on interactions between antigens, antibodies, and MHC markers. The antibody-mediated immune response, triggered by foreign proteins called antigens, results in the formation of antibodies. The stimulation of B-cells occurs by clonal selection of the cells that react with a new antigen. The cell-mediated immune response includes a direct cell-to-cell process in which activated T-cells destroy large cells or tissue transplants from another individual.

The immune response can be manipulated by vaccination, a process in which a weakened or altered pathogen (or an antigen from a pathogen) is injected into a healthy individual to stimulate the immune response against that pathogen. Vaccinations have helped conquer many of the world's most dreaded diseases, including polio and smallpox.

Malfunctioning immune systems can produce allergies and autoimmune diseases, in which some aspects of the immune response seem to be directed against an individual's own cells. The importance of the immune response is illustrated tragically by AIDS, a viral disease that impairs the immune system to the point where death from other infections follows. AIDS is transmitted by blood (especially via the sharing of hypodermic needles) and during unprotected sexual intercourse.

After studying this chapter, you should be able to:

- Explain why certain organisms produce disease while others do not.

- Explain the essential features of the immune response and the components and functions of the immune system.

- Describe various disorders of the immune system, paying particular attention to AIDS.

TERMS AND CONCEPTS

pathogen *865*

parasites *869*

histamine *870*

phagocytes *870*

monocytes *870*

macrophages *870*

antigen *872*

antibodies *872*

B-lymphocytes *876*

T-lymphocytes *876*

humoral immune
   response *877*

helper T-cells *877*

killer T-cells *877*

suppressor T-cells *877*

plasma cells *880*

complement proteins *880*

allergy *883*

thymus gland *883*

autoimmune disease *885*

HIV *885*

AIDS *886*

REVIEW

Objective Questions (Answers in Appendix)

1. Which of the following immunological defenses in vertebrates is highly specific?
   (a) skin
   (b) inflammatory response
   (c) humoral immune response
   (d) mucous membranes

2. Which of these symptoms is *not* a manifestation of the inflammatory response?
   (a) phagocytosis by white blood cells
   (b) increased blood flow at location of skin cut
   (c) production of specific antibodies
   (d) release of histamines

3. What are phagocytes?
   (a) particles released by mast cells in response to pathogens
   (b) circulating white blood cells that engulf invading bacteria
   (c) antibodies in body secretions
   (d) blood proteins that attach to a cell membrane and puncture it, causing it to burst

4. The incubation period of the AIDS virus ranges
   (a) from 1 to 6 months.
   (b) from 24 hours to 3 weeks.
   (c) from several months to decades.
   (d) over an unknown period.

5. Certain viruses are called retroviruses because
   (a) their genetic code is carried on their DNA.
   (b) they can make DNA from an RNA template.
   (c) they cause T-cells to move in reverse.
   (d) they cause B-cells to attack T-cells.

Discussion Questions

6. Using material from this chapter and Chapter 2 together, explain why some scholars argue that society's understanding of and reactions to disease are at least as important as medical practices in controlling epidemics. How has public reaction to the AIDS epidemic helped or hindered our scientifically guided efforts to contain and cure this disease?

7. Explain how, without having a genome of infinite size, the immune system manages to manufacture antibodies against a seemingly endless parade of antigens it has never seen.

8. Explain how the immune system might act as agents of natural selection in affecting the evolution of bacteria, viruses, and multicellular parasites over time. What are some strategies that parasites have evolved to evade the immune system?

9. Many people have trouble understanding that, even though AIDS is a communicable disease that is fatal, it is not easily transmitted from one person to another. Can you explain?

READINGS

Diamond, Jared. "The mysterious origin of AIDS." *Natural History* (September 1992): 24–29. A brief description of how diseases are transmitted, with a more in-depth look at the history of AIDS.

Gibbons, Ann. "Exploring new strategies to fight drug-resistant microbes." *Science* 257 (August 21, 1992): 1036–1038. An illuminating article on microbial resistance to antibiotics and what researchers are doing about it.

Krause, Richard M. "The origin of plagues: Old and new." *Science* 257 (August 21, 1992): 1073–1077. An interesting article about the nature of epidemics.

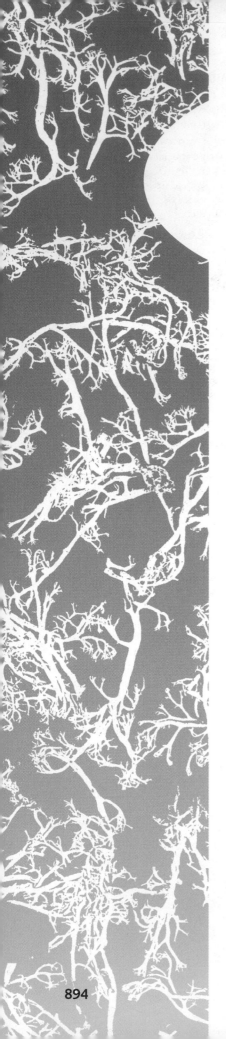

43

Nervous Control

The mind of man is capable of anything—because everything is in it, all the past as well as the future.

—Joseph Conrad, *Heart of Darkness*

*i*t's a desperate scene that plays itself out in endless variations every day across the nation: An addict buys a fleeting glimpse of paradise in a vial of crack cocaine. The first high, users say, is like nothing they have experienced before—euphoria, an overwhelming sense of self-confidence, and a combination of physical, emotional, and sexual pleasures that overshadows all delights they have ever known.

But this ticket to an ephemeral Eden, they soon discover, is a true devil's bargain. The effect of crack on the mind is so great that even a single dose can prove irrevocably addicting. You may have seen crack's effects on the psyche explained on television or in magazine articles. One user compared the cocaine high to "being Adam and having God blow life into your nostrils." But when the brief high wears off, reality returns. The "superman" or "wonderwoman" dissolves once again into the same, tired body. Depression sets in. The real pleasures of everyday life—friends, parents, children, music, learning, food, even sex—pale in comparison to the drug-induced high. With frightening speed, crack undermines rational thought; for the addict, money to get high becomes more precious than life itself. Lying, cheating, stealing (from relatives, if necessary), and even committing murder seem preferable to being deprived of the drug.

THE MIRACLE OF MIND

Let us take a moment to frame this question in biological terms. What is the human mind that a simple compound extracted from a plant can wreak such havoc upon it? What is pleasure? What is pain? Why do we respond to events around us as we do? And what, for that matter, are the thoughts that are running through your mind as you read this?

Any search for the human mind leads to the brain, an organ awe-inspiring in its complexity. We know fairly well what the brain is composed of; it contains more than 100 billion nerve cells, each of which makes as many as 1000 contacts with other nerve cells and receives input from 10,000 more. We are familiar with the list of things the brain does; within its physical substance, thoughts materialize, emotions fly, moods dance, creativity springs, and dozens of automatic systems monitor and control the body's life-sustaining processes. We have traced the physical "wiring patterns" of nerve cells into and through the brain in exquisite detail; accumulated information on the anatomy and interconnections of brain areas fills many volumes far larger than this one.

Yet when asked how the miracle of the conscious mind emerges from this mass of neurons, most neurobiologists simply shrug; the human nervous system is a whole that is far greater than the sum of its individual parts. We have yet to understand fully the workings of our neural connections or to decipher the messages those networks create and carry.

Of course, neurobiologists don't give up after admitting they don't have all the answers. Though the workings of the intact brain are still a mystery, those aspects of the nervous system that have been illuminated serve as fascinating and useful objects of study. We may not know how nerve impulses form thoughts, but we do know enough about brain chemistry to understand how crack and other drugs affect normal neural function. We do not yet know how you recognize a photograph of your mother, but we are learning enough about the brain's visual centers to guide us in designing computers that can see.

The human nervous system, of course, is among the most complex in the animal kingdom. In addition to the experimental difficulties this complexity presents, ethical considerations restrict the sorts of experiments we can perform directly on one another's brains. Luckily, however, our shared evolutionary history with less complex animals unites our nervous system with theirs through bonds of common chemistry and physiology. For that reason, researchers have learned a great deal by studying the tiny brains of sea snails and the giant nerve cells of squids. It is largely through work on these less complicated systems that neurobiologists have accumulated the knowledge and techniques that enable us to study our own nervous systems today.

WHY NERVOUS SYSTEMS?

You have seen how the body's chemical communication system uses hormones to mediate long-term processes such as digestion, metabolism, growth, and reproduc-

tion—processes that occur over seconds, months, or even years. But even the simplest multicellular organisms have more immediate needs that hormonal control systems alone cannot meet. Animals must constantly gather and analyze information about their environment and respond immediately and appropriately if they are to maintain homeostasis and survive in changing, and often hostile, environments.

Even single-celled organisms respond to a variety of stimuli. The undifferentiated cytoplasm of *Amoeba* somehow mediates that protist's negative responses to strong light, and specialized organelles in many other single-celled organisms detect such stimuli as light, noxious chemicals, and dissolved compounds emitted by food. Locomotory organelles such as flagella or cilia respond to these stimuli by propelling the organism either towards a stimulus or away from it.

But this type of simple response is inadequate for large, multicellular animals that must be able to pass information from cell to cell within their bodies. The complex behaviors of higher animals also require more detailed sensory information and more coordinated movement than simple organelles can provide. So just as higher organisms evolved specialized tissues and organs for respiration and digestion, they evolved multicellular nervous systems.

Components of Nervous Systems

The principal components of the nervous system proper are the many classes of nerve cells or **neurons,** several of which we will examine in detail in the next three chapters. A fully functioning nervous system, however, also depends on many other, nonneural cell types. Nerve cells throughout the body, for example, are intimately associated with specialized **glial cells,** or **neuroglia** (from the Greek word for "glue"). Neuroglia were once thought to form only passive, supporting structures in the nervous system, but we know now that they are intimately involved with the metabolic processes of neurons they surround.

The responses and behaviors that neural activity triggers also depend on cells that themselves are not neurons. A tiger's decision to lunge at prey, for example, is carried out by muscle cells that produce movement. Other neurons exert their effects through tissues of the endocrine, respiratory, circulatory, and excretory systems. Collectively, nonneural cells whose actions are under neural control are known as *effectors,* and the organs they form, such as muscles and glands, are called *effector organs.*

Types of neurons Neurons can be grouped into several classes according to their location and function in the chain of information processing and response.

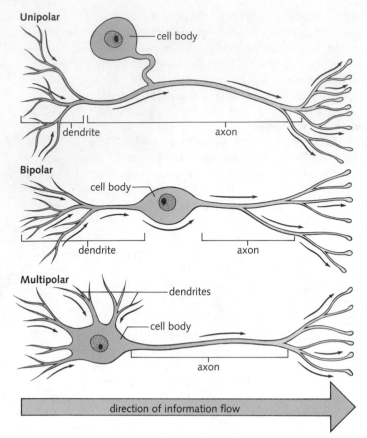

Unipolar
cell body
dendrite   axon

Bipolar
cell body
dendrite   axon

Multipolar
dendrites
cell body
axon

direction of information flow

Figure 43.1 A unipolar neuron, bipolar neuron, and multipolar neuron classified according to structure.

- **Sensory neurons** respond to light, heat, pressure, or chemicals in the environment and translate that information into the electrochemical language of the nervous system.

- **Interneurons** collect and relay information gathered from sensory neurons. Certain interneurons found in the spinal cord simply connect sensory neurons with the next link in a neural chain. Other interneurons in the eye and brain process and compare information from a variety of receptors, evaluate the incoming information, and determine the appropriate response(s).

- **Motor neurons** carry the "directions" issued by interneurons to the muscles that generate appropriate movements of the limbs, heart, lungs, gut, or other parts of the body.

Two other functional terms are often used to describe nerve cells: **afferent neurons** (*affere* means "to carry to")

carry information towards the brain. **Efferent neurons** (*effere* means "to carry out") conduct information from the brain or spinal cord to effector organs. Motor neurons are efferent neurons, as are neurons that control nonmuscular effectors such as glands.

Neurons can also be classified into three broad groups—unipolar, bipolar, and multipolar—according to the arrangement of their main cellular parts: the **cell body,** the finely branched **dendrites** (**dendron** means "tree"), and the long, slender **axon** (from the Greek *axis*) (Fig. 43.1). Usually, one or more dendrites pick up information and carry it towards the cell body, while the axon conducts information away from the cell body. Often, the far end of the axon divides into one or more branches that distribute the information either to other neurons or to effector cells.

Axons rarely run singly through the bodies of higher vertebrates; they are usually gathered together with accompanying glial cells into bundles called **nerves.** Some nerves contain only a few neurons, others contain thousands. Although most individual axons carry information in only one direction, a single nerve can contain both afferent and efferent neurons.

Neural circuits No neuron functions in isolation. Neurons connect and communicate with each other and with effector cells through specialized structures known as **synapses** (*synapsis* means "to clasp"). You can think of synapses as "connectors" that transport information from one neuron to the next. Synapses tie neurons together into *neuronal circuits,* collections of cells that work together to process information and to orchestrate such complex functions as coordinated movement, thought, and creativity (Fig. 43.2).

One of the simplest neural circuits, the **reflex arc,** is responsible for actions such as the knee-jerk reflex in humans. This reflex, initiated by a quick, sharp tap below the kneecap, requires no conscious activity and occurs faster than you could intentionally move your leg (Fig. 43.3). The sudden stretching of the upper thigh muscle is detected by sensory neurons called *stretch receptors.* The responses of those stretch receptors are relayed through the spinal cord to motor neurons that cause the stretched muscle to contract. Simple, unconscious reflexes such as this one are essential in nearly all coordinated movement.

The far more intricate neuronal circuits in the brain have thus far defied our efforts to understand their function as clearly as we understand the knee-jerk reflex. But scientists studying the smaller nervous systems of insects and snails, and the more accessible neuronal networks of mammalian eyes are uncovering tantalizing clues to the actions of ever more complex circuits, as you will see in this chapter and the next. Before examining what we know about complex neural circuits, however, we must first describe how individual neurons work.

Figure 43.2 Simple neural circuits. LEFT: In this circuit, a sensory neuron connects directly to a motor neuron, which directly stimulates muscle cells. Circuits this simple are common in invertebrates, less common in higher vertebrates. RIGHT: In most higher animals, specialized sensory cells gather information, one or more interneurons process that information, and several motor neurons allow for more precise control of effector organs.

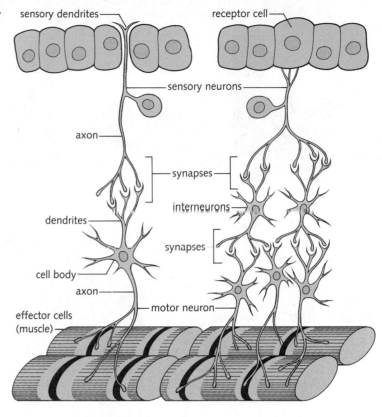

Figure 43.3 The knee-jerk reflex is controlled by one of the simplest neural circuits in humans. A sudden stretch of the upper thigh muscle stimulates a stretch receptor that, in turn, stimulates a motor neuron in the spinal cord. That motor neuron stimulates the thigh muscle to contract, extending the leg. At the same time, the stretch receptor causes an inhibitory neuron to prevent firing in the motor neuron leading to muscles that flex the leg.

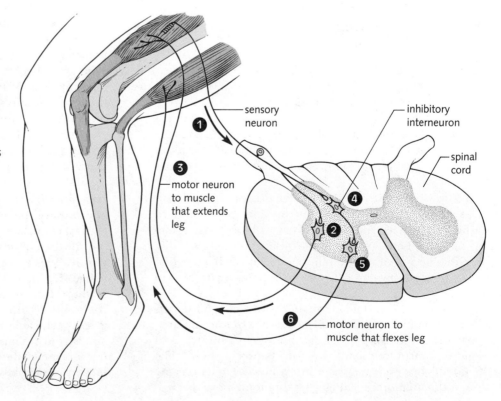

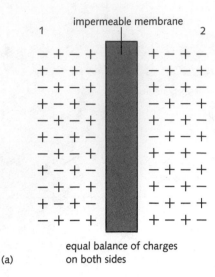

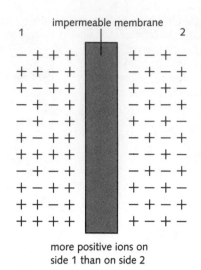

Figure 43.4 **(a)** A membrane separating two solutions containing equal numbers of positively and negatively charged ions. No electrical potential exists here. **(b)** This membrane separates solutions with different relative concentrations of positive and negative ions. An electrical potential exists across this membrane.

(a) equal balance of charges on both sides

(b) more positive ions on side 1 than on side 2

HOW NEURONS WORK

Neurons transmit information by using specialized proteins in their plasma membranes to create and manipulate electric currents. But nerve cells do *not* transmit messages by sending electric currents down their axons the way telephones send electrical pulses through wires. To understand the way neurons work, we must understand a few basic facts about electrical phenomena in living tissue.

The Difference Between Potential and Current

Electric current is defined as any flow of charged particles. Normally we think of current as a flow of negatively charged electrons through a conducting material such as a wire. But in living systems, electric current can be generated by the movement of either positive ions (such as Na^+, K^+, and Ca^{2+}) or negative ions (such as Cl^-), all of which are common in both cytoplasm and extracellular fluid.

Electrical potential is a force that causes charged particles to move and is measured in volts (V). Standard flashlight batteries, for example, have an electrical potential of 1.5 V between their poles. Batteries develop that potential because they are put together in such a way that electrons are drawn away from one pole and attracted to the other.

We can create a similar situation by placing an impermeable membrane between two solutions of ions in water. If each solution contains equal numbers of positive and negative ions, no potential exists between them (Fig. 43.4a). If, however, we place more positive ions on one side of the membrane, an electrical potential exists across

the membrane (Fig. 43.4b). Note that electrical potentials exist in these situations in the absence of any current flow. When charged ions *do* flow in response to an electrical potential, they move in a way that reduces that potential. That's why batteries run down when used.

MEMBRANE PUMPS, LEAKS, AND ELECTRICAL POTENTIALS

Nerve cells develop electrical potentials because specialized proteins in their plasma membranes use energy from ATP to pump sodium ions out of the cell as they pump potassium ions into the cell. The action of this protein, which is called the **sodium–potassium pump,** raises the concentration of sodium ions outside the cell to ten times the concentration of sodium ions inside. At the same time, the pump moves potassium ions into the cell.

But many of the potassium ions pumped into the cell do not remain inside because the nerve cell membrane is not absolutely impermeable. Instead, the membrane is slightly "leaky" to sodium ions and far more leaky to potassium ions. Because the membrane leaks potassium, and because potassium is more concentrated inside the cell than outside, those potassium ions tend to diffuse out. In the equilibrium that results, there are both more positive ions and more sodium ions outside the cell than inside (Fig. 43.5). For this reason, we say that the interior of the cell is negative with respect to its immediate surroundings. (Under normal circumstances, the membrane is less permeable to negatively charged chloride ions and almost totally impermeable to large, negatively charged organic molecules found inside the cell. These ions therefore cannot cross the membrane to balance out the charges.)

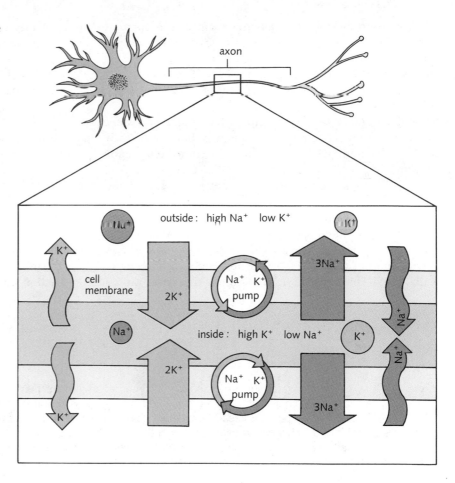

Figure 43.5 Ion pumps, leaks, and the potential across a nerve cell membrane.

This electrical potential across the nerve cell membrane is called the **resting potential,** because it is characteristic of nerve cells at rest. By placing one very fine electrode inside a nerve cell and another electrode alongside the cell, we can measure that resting potential, which turns out to be about −70 millivolts (mV). (That's 7×10^{-2}, or 0.070, V, compared with the battery's 1.5 V.) The minus sign indicates that the inside of the cell is negatively charged with respect to the outside of the cell.

If the interior of the nerve cell becomes *more negative* than it normally is at equilibrium—in other words, if the resting potential changes from −70 mV to, for example, −100 mV—we say that the cell has been **hyperpolarized** (*hyper* means "more"). The entry of negatively charged chloride ions into the nerve cell, for example, would hyperpolarize it.

If the interior of the cell becomes *less negative* than it normally is at equilibrium—in other words, if the resting potential changes from −70 mV to, for example, −40 mV—we say that the cell has been **depolarized** (*de* means "less"). Allowing positively charged sodium ions to enter a cell would depolarize it.

Ion Channels and Excitability

The existence of a resting potential is not unique to nerve cells. What distinguishes nerve (and muscle) cells from other cells is their ability to create and transmit disturbances in that resting potential to carry information. This ability depends on two classes of pores in the cell membrane that are called **gated ion channels** because they open and close like gates in a fence. (Note that these pores exist in addition to, and operate independently of, the sodium–potassium pump.)

Electrically controlled channels are pores that open and close when the electrical potential across the cell membrane changes. At a normal resting potential of −70 mV, for example, the channels that control the entry of sodium into the cell are closed, and virtually no sodium can flow passively across the membrane. If the cell becomes sufficiently depolarized, however, these electrically gated channels open and allow sodium to pass through readily (Fig. 43.6a).

Small depolarizations do not affect these channels, but if the neuron becomes less negative than it normally is at equilibrium to a level called the **threshold,** many sodium channels spring open at once. Electrically controlled channels are important in the generation of the nerve impulse, as we will see shortly.

Chemically controlled channels are pores that open only when activated by compounds called **neurotransmitters,** chemical messengers that transmit information across synapses (Fig. 43.6b). Neurotransmitters and chemically gated channels interact in much the same way as hormones and their receptors. The only difference is that hormones generally travel through the bloodstream before reaching their receptors, whereas neurotransmitters and their receptors are separated only by the small distance across a synapse. As we noted in Chapter 39 and we will see again shortly, several compounds, such as adrenalin (also called epinephrine), can act both as circulating hormones and as neurotransmitters.

THE ACTION POTENTIAL: ELECTRICALLY GATED CHANNELS IN ACTION

Neurons with long axons carry information by means of **action potentials**—brief changes in the membrane's elec-

Figure 43.6 **(a)** Electrically gated sodium channels are composed of proteins that span the cell membrane and can switch back and forth between different shapes. At normal resting potential, the shape of these proteins keeps the channel closed. When the interior of the cell depolarizes sufficiently, the channel proteins shift into a different shape, opening the channel. **(b)** Chemically gated channels are normally closed. When bound to their specific neurotransmitter molecules, however, they change conformation to open.

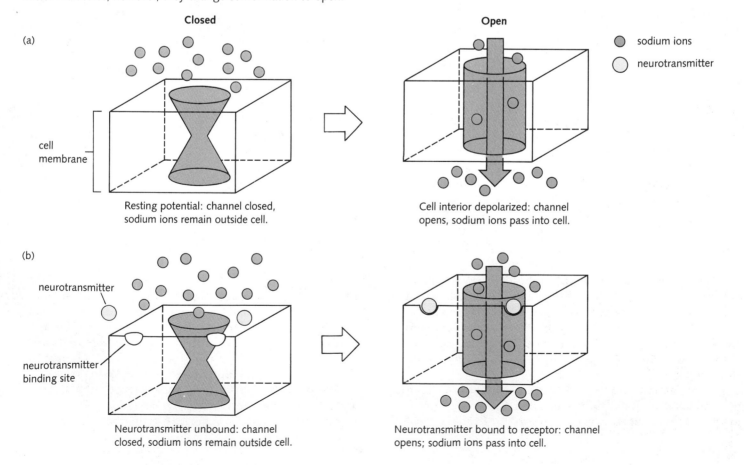

Closed   Open

sodium ions
neurotransmitter

(a)
cell membrane

Resting potential: channel closed, sodium ions remain outside cell.

Cell interior depolarized: channel opens, sodium ions pass into cell.

(b)
neurotransmitter
neurotransmitter binding site

Neurotransmitter unbound: channel closed, sodium ions remain outside cell.

Neurotransmitter bound to receptor: channel opens; sodium ions pass into cell.

trical potential that sweep along the axon. Although action potentials are also called *nerve impulses,* they are not pulses of electricity that travel through the axon; they are disturbances in the resting potential that move along the membrane like ripples passing along the surface of a quiet stream. Once an impulse is initiated, it doesn't need to be "pushed" in any way. It is **self-propagating,** which means it travels down the axon with no further input of energy from the cell.

Depolarization and Threshold

The action potential starts when part of a nerve cell is depolarized to threshold. As soon as threshold is reached, large numbers of electrically controlled sodium channels spring open, and positively charged sodium ions rush into the cell. Sodium ions move for three reasons: they are attracted by the negative charges inside the cell, they are repelled by the positive ions outside the cell, and they diffuse from an area of high sodium concentration to an area of low sodium concentration.

Remember that when current flows because of an electrical potential, the potential decreases. As sodium ions rush into one area of a cell, therefore, the resting potential in that region drops to zero and then actually reverses briefly, changing from −70 mV to +40 mV in about 2 milliseconds (Fig. 43.7). Then, almost as quickly, the membrane potential reverses again and returns to −70 mV, as the sodium channels close and other channels open, allowing positive potassium ions to rush out.

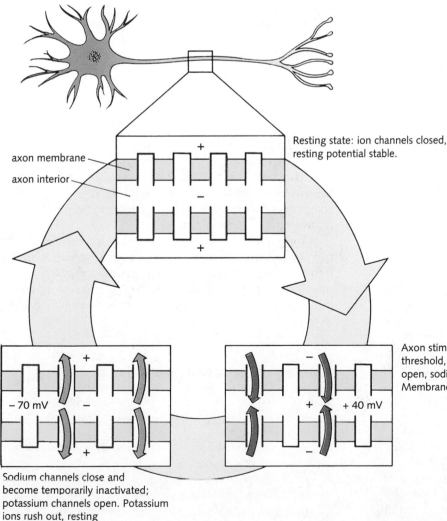

Figure 43.7 Events at the site of an action potential. Before stimulation, all ion channels are closed and the membrane is at resting potential. When the axon is stimulated to threshold, sodium channels open, and sodium ions rush in. The membrane potential reverses. When sodium channels close and become temporarily inactivated, potassium channels open and the resting potential is restored.

Resting state: ion channels closed, resting potential stable.

axon membrane

axon interior

Axon stimulated to threshold, sodium channels open, sodium rushes in. Membrane potential reverses.

+ 40 mV

− 70 mV

Sodium channels close and become temporarily inactivated; potassium channels open. Potassium ions rush out, resting potential restored.

	sodium (Na⁺)	potassium (K⁺)
ion flow		
closed channel		
open channel		

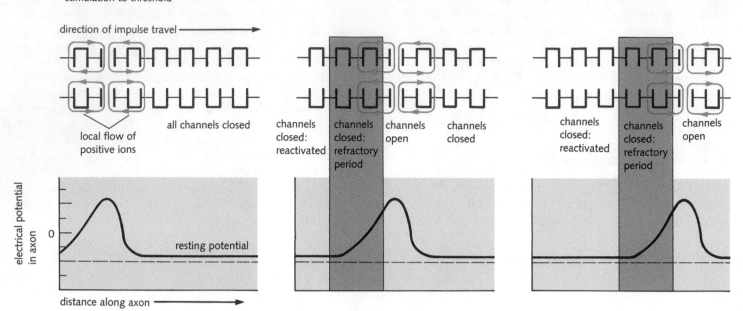

(a) Schematic of ion flow in axon at instant of stimulation to threshold

(b) Ion flow 1 millisecond after stimulation

(c) Ion flow 2 milliseconds after stimulation

direction of impulse travel ⟶

all channels closed

local flow of positive ions

channels closed: reactivated

channels closed: refractory period

channels open

channels closed

channels closed: reactivated

channels closed: refractory period

channels open

electrical potential in axon

0

resting potential

distance along axon ⟶

Figure 43.8 Propagation of the action potential. **(a)** On stimulation to threshold, inward movement of positive ions creates an electric current that depolarizes nearby parts of the membrane. **(b)** The newly activated channels open, allowing sodium to enter farther down the axon. At the same time, the first channels to open close and become inactivated. **(c)** As channels open and close in sequence along the axon, they create a traveling wave of electrical disturbance.

After sodium channels have opened and closed, they cannot be activated for a brief but finite period of time called the **refractory period.** This sequence of depolarization and repolarization constitutes the action potential.

The Traveling Wave: An All-or-Nothing Event

The action potential travels in one direction along the axon because of the way sodium channels work. As sodium ions rush in, these moving charged particles create a flow of current that depolarizes nearby parts of the axon (Fig. 43.8a). That depolarization is strong enough to open the electrically controlled sodium channels in the adjacent section of the axon (Fig. 43.8b). Because sodium channels *behind* the disturbance cannot reopen for a few milliseconds (their refractory period), the wave does not spread in two directions but travels "one way" down the axon (Fig. 43.8b,c).

Because all the electrically gated sodium channels of a neuron have the same threshold for opening, either all the channels in a region open in response to a depolarization or none of them do. As a result, action potentials are **all-or-nothing events;** either they happen or they

don't. Additionally, all action potentials generated in a particular neuron are exactly the same size. If the stimulus that produces an action potential is larger than threshold or lasts for a long time, it doesn't generate larger action potentials but instead gives rise to a series of many impulses, one after another.

Compared with the total number of ions near the membrane, the actual number of ions involved in an action potential is small. This fact and the constant activity of the sodium–potassium pump "reset" the resting potential so quickly that the axon is ready to fire again as soon as the brief refractory period is past.

Summary of the Action Potential

1. When part of an axon is depolarized past a critical *threshold,* electrically gated sodium channels in the membrane open.
2. Sodium ions rush into the cell, first depolarizing it and then reversing its potential.
3. Local current flow depolarizes adjacent areas of the membrane. This causes additional sodium channels to open and induces the wave of depolarization to travel down the length of the axon.

Figure 43.9 LEFT: Impulse transmission in a myelinated axon. RIGHT: An unmyelinated axon.

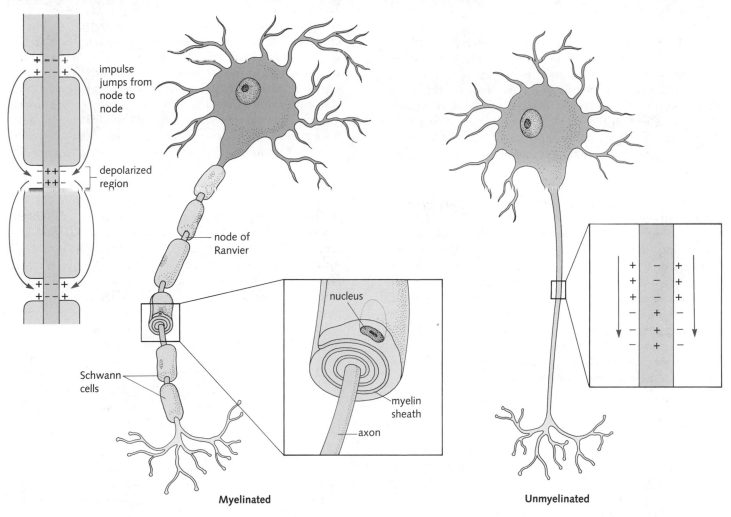

impulse jumps from node to node

depolarized region

node of Ranvier

Schwann cells

nucleus

myelin sheath

axon

Myelinated

Unmyelinated

4. As the peak of the wave passes a region, potassium ions rush out of the cell. Sodium channels close and cannot be reopened for a brief period.
5. The area returns to resting potential.

The Role of Myelin

The axons of many vertebrate neurons are wrapped in an insulating **myelin sheath** formed by a special class of glial cells called **Schwann cells.** Each Schwann cell wraps the nerve with a sheath that covers about 1 mm of its length. The glial cells space themselves along the axon in a manner that leaves a small section of membrane, the **node of Ranvier,** exposed between one section of the sheath and the next (Fig. 43.9).

This arrangement greatly speeds the conduction of action potentials along an axon for the following reason.

Electric current always flows in closed loops (between the poles of a battery, for example), so current entering an axon must leave it somewhere. In unmyelinated axons, the current entering the neuron at the site of the action potential flows through very tight local loops. Thus the action potential moves down unmyelinated axons in rapid but very small steps that result in conduction speeds of between 0.5 and 2 m per second.

In myelinated axons, on the other hand, the regions wrapped by the myelin sheath have very high resistance to current flow. Current entering the axon at one node of Ranvier can exit only at the next node, where it activates the electrically gated channels. The action potential thus "jumps" quickly from one node to the next—a process called **saltatory conduction,** after the Latin word for "jump"—rushing down the axon. The increase in speed is substantial; myelinated axons can conduct impulses at speeds up to 120 m per second!

SYNAPSES: SITES OF NEURAL CONTROL AND INTEGRATION

As we have seen, action potentials have definite advantages for carrying information along axons. But a series of action potentials in an axon is neither more nor less than a collection of identical signals. Because those impulses do not change in size, shape, or speed, regardless of recent activity in the nerve, they cannot be responsible for the sorts of changes in neural activity that make learning and memory possible. And action potentials cannot interact with other action potentials to process sensory information. Where, then, do these critical events occur?

As we mentioned earlier, individual neurons are connected into neural circuits by synapses. It is at these synapses that virtually all the essential work of comparing, integrating, and processing neural information occurs. Changes in the response of synapses to incoming action potentials under different conditions allow us to compare sensory stimuli, evaluate the meaning of those stimuli, and determine the appropriate course of action. And it is becoming clear to scientists that changes over time in the structure and function of certain types of synapses underlie the processes of learning and memory. For that reason, we will examine synapses closely.

Electrical Synapses

The simplest way for neurons to connect to one another is through direct electrical connections called **electrical synapses.** These connections, which are also called **gap junctions,** allow the free flow of ions between cells. Cells connected by electrical synapses are tied together into a single, continuous electrical unit.

Transmission of impulses across electrical synapses is essentially instantaneous, a feature that makes these connections valuable in certain situations. Electrical synapses linking the giant axons of lobsters, for example, enable those animals to respond with all possible speed to escape predators. (The *chemical* synapses you will learn about, by contrast, introduce a delay of several milliseconds between neurons.) Electrical synapses do have limitations, however. They have relatively little ability to process information so they are thought to play only minor roles in learning and other flexible types of behavior. It should not surprise you that in the nervous systems of animals that engage in complex, modifiable types of behavior, a different sort of neural connection is involved.

Chemical Synapses

Chemical synapses, of which there are several types, are both structurally and functionally more complex than electrical synapses. Action potentials do not simply pass across chemical synapses as they do across electrical synapses. Instead, when an action potential reaches a chemical synapse, it is converted into a chemical signal: a pulse of neurotransmitter molecules that diffuse across the synapse. On the far side of the synapse, that chemical signal is converted back into an electrical signal. Once you understand how this process works, you will have a glimpse into the mechanism that allows virtually all of the information processing that occurs in the nervous system.

A schematic representation of a simple chemical synapse is shown in Fig. 43.10. The neuron through which the action potential arrives, the **presynaptic neuron,** is separated from the **postsynaptic neuron,** the cell on the other side of the synapse, by a gap called the **synaptic cleft.** Inside the presynaptic neuron, *synaptic vesicles* store neurotransmitter molecules that will be released into the synapse when the presynaptic neuron is stimulated. In the membrane of the postsynaptic neuron, a host of chemically gated ion channels act as receptors for neurotransmitter molecules.

The sequence of events that occurs when an action potential reaches this type of synapse is as follows:

1. When an action potential arrives at a synapse, it triggers the opening of electrically gated calcium channels in the presynaptic cell near the synapse.

2. Calcium ions outside the cells diffuse inward, drawn to the cell's negative interior.

3. The presence of calcium ions inside the presynaptic cell causes synaptic vesicles to fuse with the cell membrane and release neurotransmitter molecules into the synaptic cleft.

4. Neurotransmitter molecules diffuse across the synaptic cleft and bind with receptor sites on the postsynaptic membrane.

5. The opening of various chemically gated channels allows either positive or negative ions into the postsynaptic cell, which causes either a local depolarization or hyperpolarization.

6. Transmitter remaining in the synapse is eliminated, either by chemical reactions that inactivate it or by transport mechanisms in the presynaptic membrane and glial cells that remove it. This clears the synapse for the arrival of the next impulse.

What happens to the postsynaptic membrane depends both on the specific transmitter(s) released at a synapse and on the character of the chemically controlled channels it binds to. This is because different transmitters can have different effects on the postsynaptic cell. Although neurobiologists once believed that a single neuron could

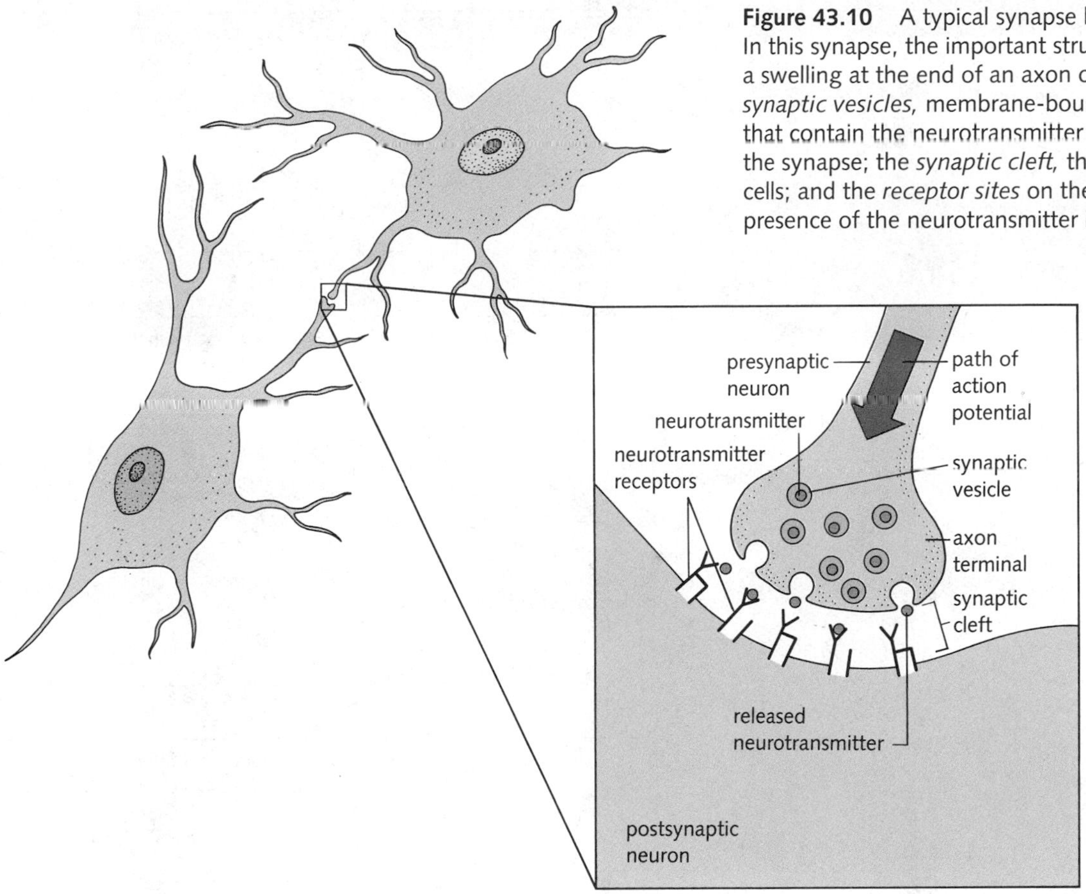

Figure 43.10 A typical synapse between two vertebrate neurons. In this synapse, the important structures are: the *synaptic bouton,* a swelling at the end of an axon of the presynaptic neuron; the *synaptic vesicles,* membrane-bound sacs in the presynaptic neuron that contain the neurotransmitter which carries information across the synapse; the *synaptic cleft,* the gap between the two nerve cells; and the *receptor sites* on the postsynaptic cell that detect the presence of the neurotransmitter in the synaptic cleft.

manufacture and release only a single transmitter, we now know of several neurons that use at least two. For the sake of clarity, we will look first at a relatively simple chemical synapse that uses a single transmitter.

A Simple Synapse:
The Neuromuscular Junction

The best-understood of all synapses is the **neuromuscular junction,** the connection between motor neurons and muscle cells (Fig. 43.11). (As you will see in Chapter 45, muscle cell function depends on membrane action potentials that are virtually identical to nerve impulses.) Neuromuscular synapses are large and simple, and they rely on a single neurotransmitter, a compound called **acetylcholine (ACh).** Stimulation of a neuromuscular junction by the neurotransmitter acetylcholine nearly always excites the muscle cell.

When an action potential reaches a neuromuscular junction, the presynaptic cell releases acetylcholine into the synaptic cleft. The transmitter diffuses across the cleft, binds to its receptors, and then opens channels for ions in the muscle cell membrane. Because the neuromuscular junctions release a substantial amount of acetylcholine, enough channels are usually opened by a single action potential to depolarize the muscle cell to threshold. At that point, the muscle cell membrane generates an action potential that causes that cell to contract (Fig. 43.11).

Neurotransmitter released into the synapse cannot continue to affect the postsynaptic cell for long, because the muscle cell should be stimulated only when its motor neuron is actively firing. For that reason, transmitter must be removed from the synapse or inactivated after it has performed its function. In synapses using acetylcholine as a transmitter, this is accomplished by a compound called *acetylcholinesterase,* an enzyme that destroys virtually all

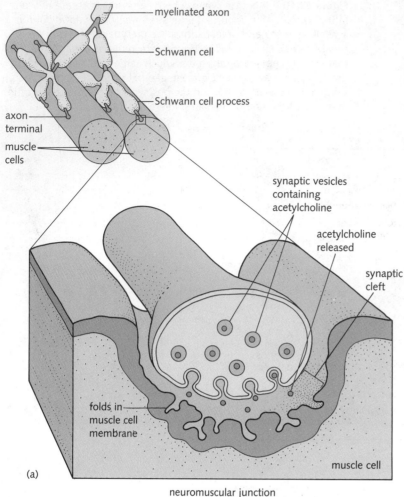

(a)

neuromuscular junction

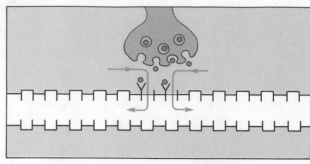

Stimulated nerve ending releases acetylcholine; acetylcholine-controlled channels are opened.

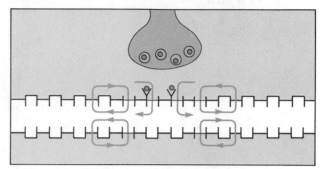

Channels opened chemically allow ion flow; this depolarizes membrane; electronically-controlled channels nearby are opened.

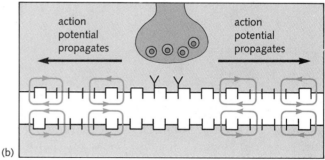

(b)

Figure 43.11 **(a)** The structure of a neuromuscular junction between a motor neuron and a major body muscle under conscious control. **(b)** Schematic summary of events at a neuromuscular junction. When an action potential arrives, the release of acetylcholine nearly always initiates an action potential in the muscle.

free molecules of ACh within a few hundred microseconds. The vital role played by acetylcholinesterase is dramatically demonstrated by the action of several nerve gases such as *sarin* and insecticides such as *parathion* that interfere with its activity. This interference results in the accumulation of acetylcholine in neuromuscular junctions, wildly unregulated postsynaptic activity, and loss of muscular coordination. Exposure to sufficient doses of acetylcholinesterase inhibitors can cause death from respiratory arrest by preventing relaxation of the diaphragm and other muscles essential to breathing.

Synapses Between Neurons

Synapses between nerve cells function similarly to neuromuscular junctions, but they are both more complicated and more flexible.

Summation: The key to information processing   A single action potential arriving at a typical synapse between two neurons usually does not initiate action potentials in the postsynaptic cell. Instead, each impulse creates a small, temporary, local change in potential called a **postsynaptic**

potential. If that change is a depolarization, it brings the postsynaptic cell closer to threshold (and hence closer to firing), so it is called an *excitatory postsynaptic potential* or **EPSP.** If the change is a hyperpolarization, which makes it more difficult for the postsynaptic cell to fire, it is called an *inhibitory postsynaptic potential* or **IPSP.**

EPSPs and IPSPs differ from action potentials in two important respects. First, they are not all-or-nothing events. They are **graded potentials** whose size depends on the amount of transmitter that binds to the postsynaptic membrane; the more transmitter, the larger the postsynaptic potential it produces (up to some maximum value). Second, IPSPs and EPSPs do not travel; they decrease in intensity with increasing distance from the synapse, and they disappear with time (Fig. 43.12).

Graded potentials are important because their effects add up in a process called **summation.** Remember that most nerve cells receive synaptic input from thousands of other neurons (Fig. 43.13). A single EPSP at any one of these synapses usually doesn't trigger an action potential, but many EPSPs arriving simultaneously can cause the postsynaptic cell to fire. Similarly, rapid, repeated stimulation of a few synapses can also add up to cause firing.

When a neuron receives both simultaneous EPSPs and IPSPs, it responds according to the algebraic sum of those positive and negative inputs. This ability to combine excitatory and inhibitory inputs for a net result underlies all processing of information in the nervous system.

You might wonder at this point just how IPSPs and EPSPs allow integration and control. As an example of a phenomenon with which you are already familiar, recall from Chapter 35 that the heart's natural "pacemaker," the SA node, can be induced either to speed up or to slow down the heartbeat. This adjustment occurs as a result of interactions between excitatory and inhibitory transmitters released onto the excitable cells of the SA node. Excitation of the SA node causes the heart rate to increase, and its inhibition causes the heart to slow down.

Similarly, the ability of chemical synapses to compare inputs from different presynaptic neurons enables sensory processing centers in the brain to weigh and interpret input from many sensory receptors. As we will see during the discussion of the visual system in the next chapter, this sort of comparison allows the eye and brain to judge distance, to perceive movement, and to see in color. Similar interactions appear to underlie the processing of auditory information and information gleaned from other senses. Far more complicated versions of similar processes underlie thought and other higher mental functions.

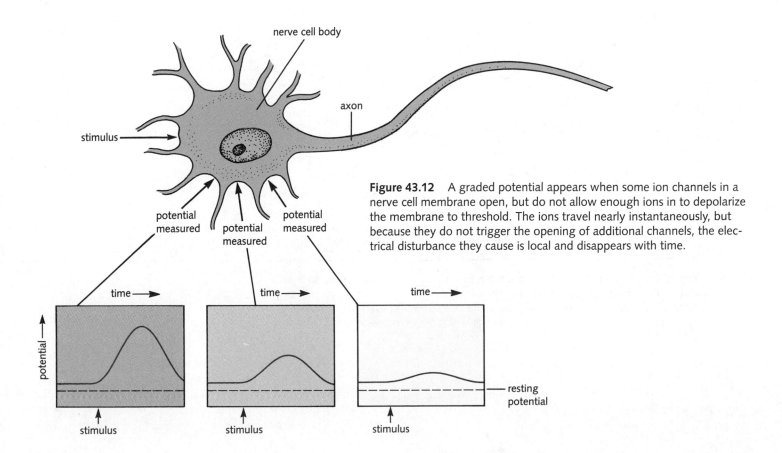

Figure 43.12 A graded potential appears when some ion channels in a nerve cell membrane open, but do not allow enough ions in to depolarize the membrane to threshold. The ions travel nearly instantaneously, but because they do not trigger the opening of additional channels, the electrical disturbance they cause is local and disappears with time.

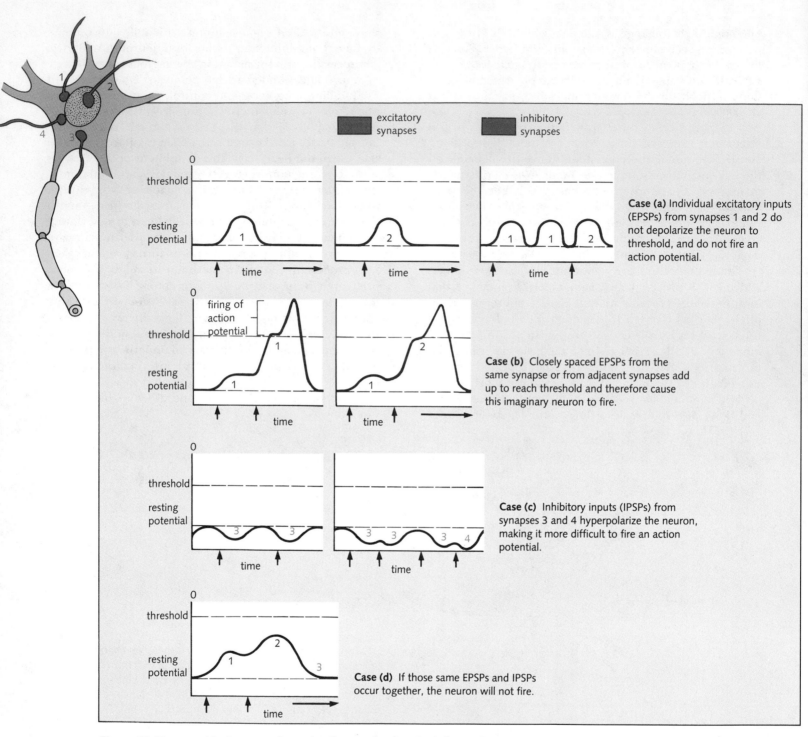

excitatory synapses

inhibitory synapses

Case (a) Individual excitatory inputs (EPSPs) from synapses 1 and 2 do not depolarize the neuron to threshold, and do not fire an action potential.

Case (b) Closely spaced EPSPs from the same synapse or from adjacent synapses add up to reach threshold and therefore cause this imaginary neuron to fire.

Case (c) Inhibitory inputs (IPSPs) from synapses 3 and 4 hyperpolarize the neuron, making it more difficult to fire an action potential.

Case (d) If those same EPSPs and IPSPs occur together, the neuron will not fire.

Figure 43.13 LEFT: Most nerve cells receive thousands of contacts from other neurons, which make synapses all over their cell bodies, dendrites, and axons. RIGHT: Graded potentials produced by synapses add up in time and space. The imaginary neuron receives four synaptic inputs, two of which are excitatory and two of which are inhibitory.

A Diversity of Transmitters

For many years, researchers thought that the entire nervous system employed only three neurotransmitters: acetylcholine, epinephrine (adrenaline), and norepinephrine (noradrenaline). This assumption was natural enough, because these three transmitters act at many synapses throughout the nervous system. Acetylcholine, for example, is found in the brain, in many other synapses throughout the nervous system, and in the neuromuscular junctions of skeletal muscles. Epinephrine and norepinephrine—which we have already encountered in their roles as hormones—also act as neurotransmitters that can either stimulate or inhibit a variety of body functions. (We will discuss these compounds again shortly with the major divisions of the nervous system.)

Recent studies of synapses in the brain and other parts of the central nervous system, however, have shown that there are a host of other neurotransmitters in different parts of the brain. Two such compounds, gamma aminobutyric acid (*GABA*) and the amino acid *glutamate,* are important transmitters at more than 50 percent of brain synapses. *Dopamine* is another transmitter that is vital to the function of several brain centers related to movement, arousal, and the emotions.

Still other compounds affect synapses but do not carry impulses from one neuron to another. These *neuromodulators* regulate synaptic activity by affecting transmitter synthesis, storage, or release, by affecting the ability of transmitters to bind to their receptors, or by influencing the fate of transmitters in the synapse. Neuromodulators, together with the diverse transmitters found in the brain, affect mood, attention, learning, sleep, and the sensation of pain. Interestingly, certain compounds that are well known as hormones—such as vasopressin (ADH)—seem to act both as neurotransmitters and neuromodulators. Vasopressin, for example, the principal function of which is the regulation of body fluid volume, also has a powerful influence on the processes of learning and memory through its actions on various brain centers.

Although a detailed discussion of these compounds and their actions is beyond the scope of this book, a few important points are worth noting. First, several psychiatric disorders such as manic depression and schizophrenia, which have traditionally been treated as strictly "emotional" phenomena, have relatively recently been linked to imbalances in specific neurotransmitters and neuromodulators. This information has spawned research into chemical therapies to treat such disorders. Second, evidence is accumulating that certain of these imbalances are caused by mutations in specific genes that control either the production and release of these compounds or the receptors that respond to them. These findings, in turn, provide clues to genetic influences on behavior of the sort that we will discuss in Chapter 46. Finally, it is by aiding, replacing, or interfering with neurotransmitters and neuromodulators that mind-altering drugs—from hallucinogens to stimulants and antidepressants—exert their influence (see Theory in Action, Drugs, Synapses, and the Brain, p. 910).

Endorphins: Pain and the brain Although many drugs can produce psychological and physical dependence, certain compounds known as **opiates,** such as opium, morphine, and heroin, are particularly addictive. The reason for this phenomenon is startlingly simple, but it puzzled researchers for many years. The epidemic of drug abuse that began in the 1970s fueled many studies on drug action that grew into one of the most fascinating scientific detective stories in the recent history of brain research.

Neuroscientists had suspected since the 1950s that opiates somehow acted directly at receptor sites on specific neurons in the brain. This hypothesis made sense for several reasons. Many opiates are extraordinarily powerful; a single dart coated with an opiate called etorphine, for example, can bring down a charging bull elephant. Yet even a slight change in the three-dimensional structure of an opiate can render it completely ineffective. And finally, there are several drugs that effectively block the action of opiates. All these phenomena suggest that opiates, like naturally occurring neurotransmitters and hormones, work by binding to three-dimensionally specific receptors on cell membranes.

Then, in 1972, Candice Pert and Solomon Snyder of Johns Hopkins University proved this hypothesis by showing that radioactively labeled opiates did, in fact, bind to specific neural receptors. Furthermore, those binding sites are located in specific brain areas that control perceptions of pleasure and chronic pain.

But why should the human brain contain receptors for compounds manufactured naturally by poppies? Several researchers around the world hypothesized that if brain receptors for opiates existed, there had to be a class of yet-undiscovered, opiate-like compounds that are normally found in the brain. After a great deal of intense (and competitive) work, three laboratories converged on the trail of opiate-like compounds they called **endorphins** (a contraction of "endogenous morphine" or "morphine within") and **enkephalins** (literally, "in the head"). These compounds, produced in the brain under conditions of stress or injury, serve a variety of functions associated with relief from pain. During pregnancy, for example, women produce eight times the normal quantities of endorphins, perhaps to cope with the added stress of pregnancy, labor, and delivery. Endorphin secretion during intense and prolonged physical activity also produces the blissful state that long-distance runners call "joggers' high."

Finally, all the pieces of the puzzle had fallen into place. Morphine, opium, and heroin bind to the brain's receptors for natural pain-relieving compounds, flooding the brain with "pleasure" signals and thus inhibiting re-

Drugs, Synapses, and the Brain

The use of mind-altering or *psychoactive* drugs is part of nearly every human culture, for better and for worse. South American jungle tribes induce healing trances by drinking or smoking plant extracts that cause hallucinations. Artists and writers ranging from Samuel Taylor Coleridge to Arthur Conan Doyle and Timothy Leary have credited drugs with enhancing sensitivity and spurring creativity. And legions of unfortunate addicts testify to the extraordinary power of such drugs as opium and heroin to cause not only short-term changes in mood but also long-term alterations in personality and physiology.

Psychoactive drugs alter brain function either by overstimulating synapses (causing rapid, uncontrolled firing of postsynaptic neurons) or by turning synapses off (blocking the normal flow of information through the brain). Each of these actions can be accomplished in several ways.

Synapses can be overstimulated by drugs that mimic the actions of transmitter molecules, cause transmitters themselves to be released in the absence of normal stimulation, or interfere with inactivation of transmitters in the synapse. Amphetamines (speed), for example, act as stimulants by causing rapid release of the transmitter noradrenaline at certain brain synapses. Nicotine, a major active component in cigarette smoke, acts as a stimulant because it mimics the effects of acetylcholine at other synapses.

Synapses can also be "turned off" by substances that block receptor sites on the postsynaptic membrane, block the synthesis or storage of transmitters in the presynaptic cell, or prevent the normal release of transmitters. Valium, of-

Drugs that act at synapses are often quite specific in their effects. This schematic, for example, shows the action of an antidepressant that blocks the action of the transmitter norepinephrine but allows another transmitter, serotonin, to act unhindered.

ten prescribed as an antianxiety drug, enhances the effects of GABA, an inhibitory transmitter at many brain synapses. It is also commonly used in hospitals to interrupt severe epileptic seizures because of the same inhibitory effect. Several commonly used, major tranquilizers, such as chlorpromazine, attach to postsynaptic receptors for both acetylcholine and noradrenaline and block the activity of those transmitters.

Although many psychoactive drugs (including nicotine and alcohol) can produce both psychological and physical dependence, certain compounds (such as opium and crack cocaine) are particularly addictive. This is because these drugs' short-term chemical effects that flood the brain with "feel-good" signals also cause significant long-term changes in synaptic function. Crack, for example, causes a sudden increase in the release of dopamine onto neurons in brain areas that control emotions, the sensation of pain, and the general sensation of "pleasure." This stimulation, however, causes a drop in the amount of dopamine stored in presynaptic neurons for later release. After a few doses of crack, stimuli that would normally cause pleasure no longer do so, and the absence of dopamine in these brain areas between "hits" leads rapidly to feelings of depression. Soon, the only way to feel "normal"—not to mention happy—is to rely continually on the drug. This vicious dependency has made crack a serious social and medical problem.

sponse to pain. But that flood of stimulation takes its toll. In some cases, the repeated artificial stimulation with opiates causes either a decrease in the production of naturally occurring transmitters or a decline in the number of receptor sites. Consequently, when addicts stop taking drugs, their brains are deprived of both natural *and* injected "feel-good" signals. The result is the agonizing pain and mental agony of withdrawal, which lasts until the brain recovers its normal function.

Changes at the Synapse: Learning and Memory

> You have to begin to lose your memory, if only in bits and pieces, to realize that memory is what makes our lives. Life without memory is no life at all. . . . Our memory is our coherence, our reason, our feeling, even our action. Without it, we are nothing. . . .
>
> —Luis Buñuel

How do we learn? How do we remember past events? These are among the most fundamental and exciting questions in neurobiology and psychology. Neurobiologists have long assumed that memory resulted from changes in the number, structure, and function of brain synapses, but they had no way to study the changes they imagined. Today, tissue culture, molecular genetics, and recombinant-DNA technology are ushering in a new era of molecular neurobiology. The proteins that form certain gated channels and the genes, mRNAs and tRNAs, that control them have been isolated, identified, and cloned. This allows neurobiologists to examine changes in synaptic function at a level it was impossible to achieve even five years ago.

Functionally, there are at least two different kinds of memory. **Short-term memory** enables us to remember information for a few minutes or hours; we use this facility when we memorize phone numbers for immediate use or "cram" for an examination. Some (but not all) information stored in short-term memory ultimately enters **long-term memory,** where it remains accessible for days, months, or years. How might changes in synaptic structure and function make these processes possible?

Experience and synaptic function   When similar thoughts or actions are repeated, certain groups of synapses are stimulated repeatedly. We now know that this kind of continued stimulation can affect the release of transmitter from the presynaptic cell, the expression of genes that control transmitter production, and even the genes that control the production of gated channels. Thus synapses that are repeatedly stimulated may end up making or releasing more (or less) transmitter than similar synapses that are not stimulated. In the same way, repeated stimulation may either increase or decrease the number of

functional chemically gated channels in the postsynaptic neuron, thereby increasing or decreasing that neuron's responsiveness to neurotransmitters. Additionally, certain neurotransmitters and neuromodulators, through their actions on various cell receptors, act as second messengers and affect long-term metabolic processes in cells just as hormones do (see Chapter 39).

Experience and change in neural structures   Experience and the brain activity it generates can also affect the expression of genes that control nerve growth and survival. In many cases, both in developing brains and in mature brains, neurons that are stimulated grow and expand their synaptic connections, while those that are not stimulated stay the same, regress slightly, or wither away and die. Those changes in nerve growth within our brain structure clearly play some role in creating the physical records that store our memories.

FROM NEURON TO BRAIN: THE EVOLUTION OF NERVOUS SYSTEMS

Nervous systems have evolved steadily during the history of multicellular animals (Fig. 43.14). The simplest multicellular animals, such as *Hydra*, have diffuse **nerve nets** with no defined control center. In many of these animals, the system can conduct action potentials in either direction.

As bilateral symmetry became more pronounced, animals developed a tendency to move in one direction. The front end of the animals became increasingly important, and both sensory cells and neural relay centers became concentrated there. At the same time, various neurons (such as sensory receptor cells) became more specialized, different parts of the nervous system became specialized for different functions, and the number of interneurons in neural circuits increased rapidly.

The accumulation of sensory neurons and interneurons in the head region, a process called **cephalization,** led to collections of nerve cell bodies called *ganglia,* from which one or two *nerve cords* ran down the length of the body. In most invertebrates, cephalization progressed to a certain point and then stopped, leaving several secondary segmental ganglia distributed along the animals' nerve cords. In some animals, such as flatworms, these ganglia have retained so much control over essential body functions that decapitated animals can survive long enough to grow a new head. In others, such as arthropods, ganglia along the nerve cord coordinate sensory input and muscular activity.

In the line leading to vertebrates, however, cephalization continued, resulting ultimately in the evolution of the vertebrate brain, a three-part swelling at the anterior end of a dorsal nerve cord. In the simplest vertebrates,

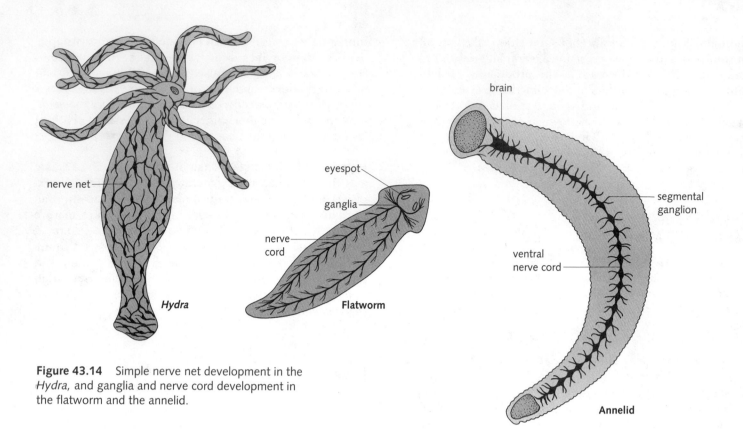

Figure 43.14 Simple nerve net development in the *Hydra*, and ganglia and nerve cord development in the flatworm and the annelid.

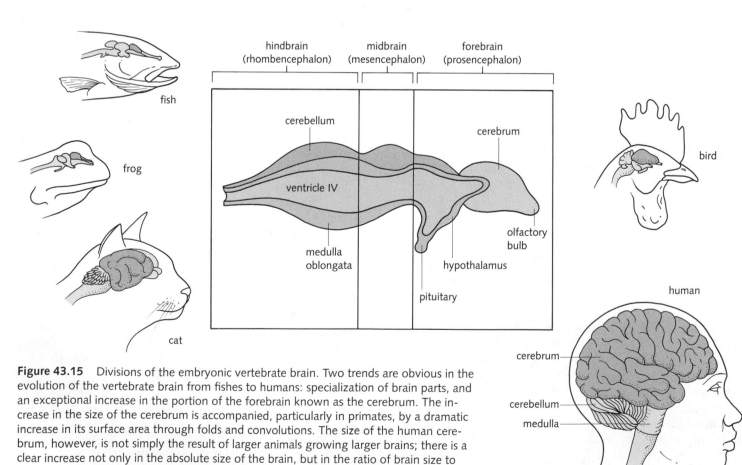

Figure 43.15 Divisions of the embryonic vertebrate brain. Two trends are obvious in the evolution of the vertebrate brain from fishes to humans: specialization of brain parts, and an exceptional increase in the portion of the forebrain known as the cerebrum. The increase in the size of the cerebrum is accompanied, particularly in primates, by a dramatic increase in its surface area through folds and convolutions. The size of the human cerebrum, however, is not simply the result of larger animals growing larger brains; there is a clear increase not only in the absolute size of the brain, but in the ratio of brain size to body weight. Living mammals and birds have brains about 15 times larger than those of most similarly sized lower vertebrates.

such as fishes, those three original regions are easy to see: the **hindbrain** (*rhombencephalon*), the **midbrain** (*mesencephalon*), and the **forebrain** (*prosencephalon*). Even in higher primates, whose brains are much larger, those three regions are visible during early development (Fig. 43.15).

In fishes and other lower vertebrates, the forebrain is concerned primarily with detecting and integrating information from the chemical senses of smell and taste. In higher vertebrates, including humans, the forebrain not only integrates incoming sensory information about the environment but also initiates voluntary movement and performs the mysterious processes we call "thinking."

THE HUMAN NERVOUS SYSTEM

The human nervous system is customarily divided into two main parts according to location and function (Fig. 43.16). The **central nervous system (CNS),** housed almost entirely within the bony protective structures of the skull and vertebral column, consists of the brain and spinal cord. The **peripheral nervous system (PNS),** distributed throughout the body, consists of afferent and efferent fibers and sense organs.

The peripheral and central nervous systems interact constantly. The CNS integrates information about internal and external environments that is provided to it by the sense organs and afferent neurons of the PNS. Once the CNS has "decided" on a course of action, it operates by distributing "instructions" to effector organs ranging from glands to blood vessels to fingers and toes through the efferent neurons of the PNS.

Both divisions of the nervous system are therefore essential for either *voluntary* actions (such as putting on a sweater before you go out into the cold) or *involuntary* activities (such as shivering when you don't dress warmly enough).

The Peripheral Nervous System

The peripheral nervous system contains both afferent and efferent neurons. The afferent portion includes *somatic sensory neurons* in all major sense organs (except the eyes) as well as *visceral sensory neurons,* which are isolated sensory cells that detect blood pressure, the contents of the digestive system, and the relative positions of various body parts. The efferent portion of the peripheral nervous system is divided functionally into the voluntary or **somatic nervous system** and the involuntary or **autonomic nervous system,** depending on whether the responses they mediate are under voluntary or involuntary control.

It is important to remember that the PNS and CNS interact constantly and that many body functions are under both voluntary and involuntary control. Breathing, for example, is normally controlled involuntarily. But during activities ranging from speaking to swimming, we can exert considerable voluntary control over the same muscles. Even heartbeat, which Western physicians consider an entirely involuntary function, can be controlled consciously by those trained in Eastern techniques of yoga and meditation.

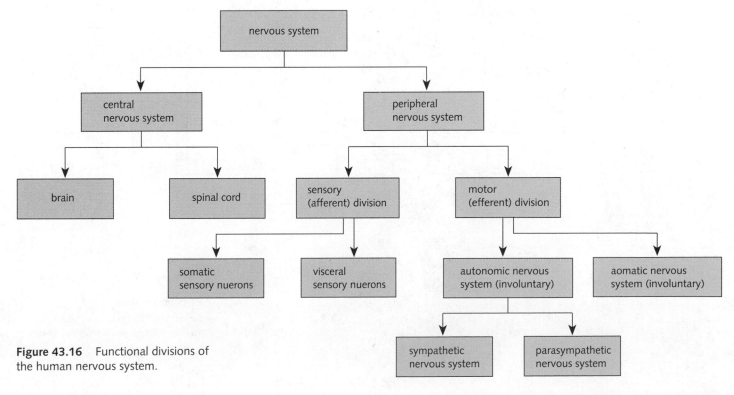

Figure 43.16 Functional divisions of the human nervous system.

The somatic nervous system The somatic nervous system includes most sensory neurons from the skin and skeletal muscles and most motor neurons leading to skeletal muscle. This division of the nervous system is responsible for both voluntary movement and reflex arcs of the sort described earlier.

The autonomic nervous system The autonomic nervous system is subdivided into the **sympathetic division** and the **parasympathetic division**. Sympathetic and parasympathetic neurons make synapses with glands and with smooth and cardiac muscles, where they release different neurotransmitters that often (but not always) have complementary (opposite) effects (Fig. 43.17). Specific effector organs controlled by autonomic neurons usually mediate long-term physiological processes such as secretion of saliva, narrowing or widening of blood vessels, and secretion and muscular activity in the digestive tract. The control functions of the autonomic nervous system thus overlap with those of many hormones (see Chapter 39).

The sympathetic division prepares the body for intense physical exertion, the same "flight-or-fight" response mediated by adrenaline. In fact, when activated by the CNS through fear, anger, or the perception of danger, the sympathetic division simultaneously stimulates the release of adrenaline (epinephrine) and reinforces that hormone's actions by increasing heart rate, widening capillaries in major muscle groups, mobilizing glucose from the liver for quick energy, and slowing down digestive activity. At the same time, the sympathetic system releases noradrenaline onto its effector organs.

Figure 43.17 The autonomic nervous system, showing the different paths taken by sympathetic and parasympathetic neurons and their complementary effects.

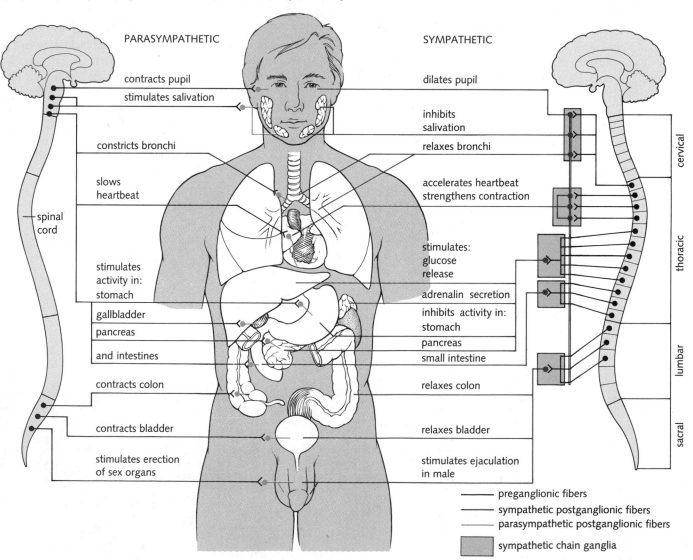

The parasympathetic division, on the other hand, helps the body adjust to the relaxed state one enters after a large meal; it slows down the heart rate, decreases blood flow to the muscles and skin, and increases the secretory and circulatory activity in the digestive tract. Parasympathetic neurons release acetylcholine onto their effector organs.

The Central Nervous System

The brain and spinal cord share several physical characteristics (Fig. 43.18). Both are protected by bony coverings and cushioned by three layers of connective tissue called **meninges.** Both brain and spinal neurons have cells organized in such a way as to produce regions called

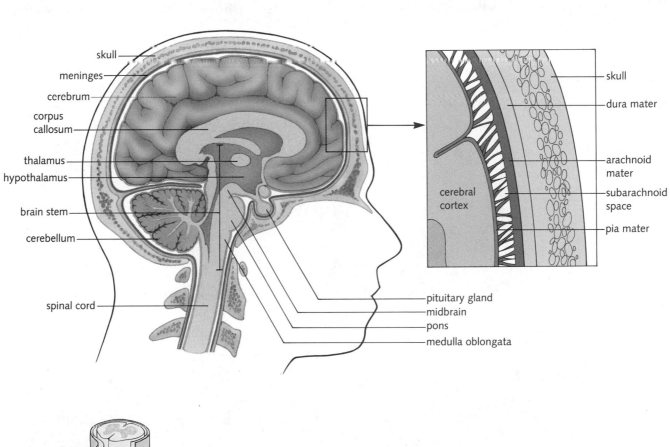

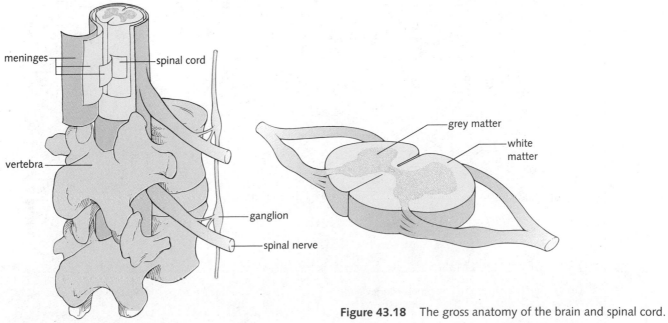

Figure 43.18   The gross anatomy of the brain and spinal cord.

gray matter and **white matter.** The glistening white matter—located in the interior of the cerebral cortex and on the exterior of the spinal cord—is composed predominantly of long axons whose color is due to their myelin sheaths. The gray matter—found on the outer layer of the forebrain and in the center of the spinal cord—is composed predominantly of cell bodies.

Both brain and spinal cord have internal cavities: the four *ventricles* of the brain and the *central canal* of the spinal cord. These cavities are lined with ciliated cells and filled with **cerebrospinal fluid,** which circulates through the CNS carrying oxygen, glucose, white blood cells, and hormones. Cerebrospinal fluid picks up these substances from blood capillaries, but not in the same way as other tissues do. Brain capillaries are not as "leaky" as other capillaries are. They form a highly selective membrane called the **blood–brain barrier.** The barrier does not permit many large proteins, including certain hormones and some drugs to cross, but it does allow the passage of other drugs, such as nicotine, alcohol, and cocaine.

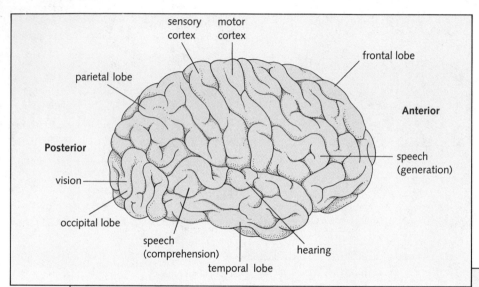

Figure 43.19 Specializations of the cerebral cortex.

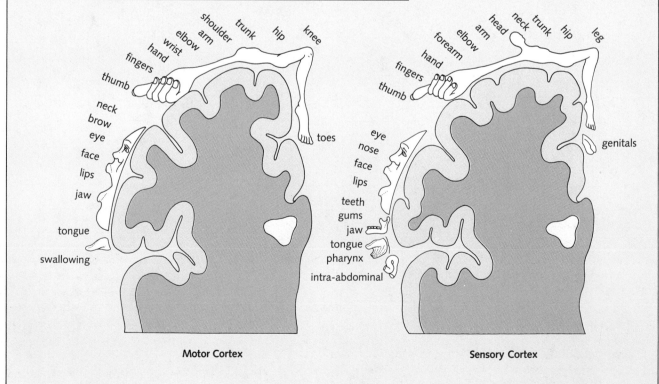

The Brain

The forebrain: Cerebrum and thalamus As the embryonic human brain develops, the leading edge of the forebrain enlarges to form the **cerebrum,** the largest part of the brain and the part that has changed most through evolution. The surface of the cerebrum, the **cerebral cortex,** is gray matter because it contains the cell bodies of the cerebral cells. Interestingly, gray matter in primitive brains is internal, as it is in the spinal cord. It seems likely that relocating the cell bodies on the outside was important because of the phenomenal increase in the number of cells in the cerebrum: many cells could be added to the outside of the brain without disrupting the axons that led to and from them.

The human cortex is folded so extensively to increase its surface area (and hence the number of cells it can hold in the confined space of the skull) that if it were laid out flat, it would cover an entire square meter. Different parts of the cerebral cortex are specialized for different functions (Fig. 43.19). Sensations from specific body areas are "felt" by well-defined regions, and commands to muscles that move those body parts originate from other cortical areas.

The cerebrum is divided into two halves, or **hemispheres,** connected by a narrow band of axons called the **corpus callosum.** Oddly enough, the sensations and motor areas of the two sides of the body are reversed in the cerebrum: the left side of the body is represented on the right side of the brain, and vice versa. Certain broad classes of higher cerebral functions seem to be localized in different hemispheres as well. The right hemisphere seems to govern creativity and artistic ability, while the left hemisphere plays the more dominant role in analytical, verbal, and mathematical ability.

A part of the forebrain just behind the cerebrum grows into a smaller structure called the **thalamus,** an important "relay center." Here, incoming messages from both olfactory and visual receptors and from many other parts of the nervous system are relayed to the cerebral cortex.

During the development of the thalamus, two buds from the thalamus grow outward to form the neural portions of the eyes. Another, smaller thalamic area grows upward to form the *pineal body,* a structure often called the "third eye" that functions in regulating day/night cycles of sleep and wakefulness. Still another part of the thalamus grows down to produce the **hypothalamus** and **posterior pituitary**—the vital links between the nervous system and the hormonal system that we considered in Chapter 39.

The midbrain The midbrain in humans does not enlarge proportionately as much as the cerebrum; it functions predominantly to relay incoming information from eyes and ears to the cortex and to connect the cerebrum with other parts of the brain. This region contains parts

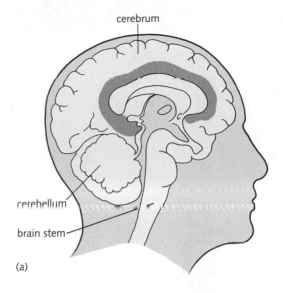

cerebrum

(a)

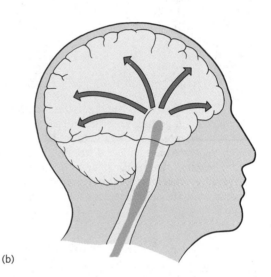

cerebellum

brain stem

(b)

Figure 43.20 **(a)** The limbic system. **(b)** The reticular formation.

of two small but indispensable functional units, the **limbic system** and the **reticular formation** (Fig. 43.20a,b).

The limbic system controls the emotions. Stimulation of various parts of the limbic system can evoke anger, hostility, or pleasure. The reticular formation monitors sensory input to the brain and determines which incoming stimuli the cortex attends to or ignores on a moment-by-moment basis. Thus it is the reticular formation that "wakes up" the rest of the brain when it detects certain noises or odors. Electrical stimulation of sensory areas in the cortex alone does not wake a sleeping individual, but

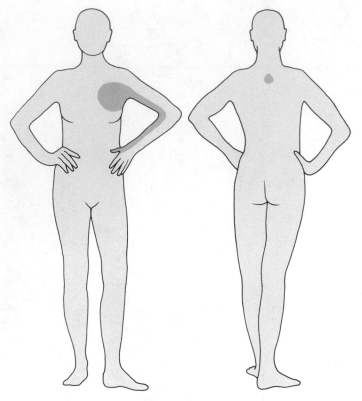

Figure 43.21 Patterns of nerve supply to the skin. Neurons leave the spinal cord in groups; branches of these spinal nerves supply both skin and internal organs. Because healthy internal organs do not generate pain signals, discomfort in those organs is often mistakenly reported by the brain as originating in the skin area served by the same nerve as the organ in question. The colored regions show the surface areas confused with pain from the heart.

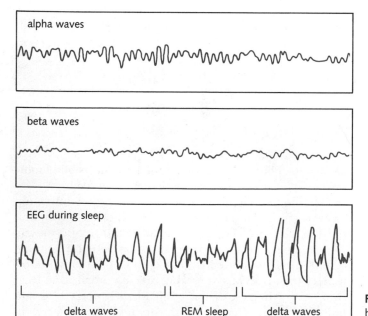

Figure 43.22 Electroencephalograms (EEGs) can be used to monitor brain activity during states of wakefulness and sleep.

stimulation of the reticular formation generates waves of impulses throughout the cortex, stimulating it to wakefulness. Other parts of the reticular formation suppress those "wake-up" centers and make sleep possible.

The hindbrain The hindbrain develops into three distinct regions: the **cerebellum,** the **pons,** and the **medulla oblongata.** Each of these regions acts as a neural "switchboard," controlling the flow of nerve impulses between the brain and the rest of the body. The medulla, for example, controls a number of functions, including breathing, blood pressure, heart rate, and coughing.

The cerebellum coordinates body movements by performing several sophisticated integrating functions. The cerebellum regulates general equilibrium (balance) by interpreting information from the organs of balance within the ear. It also controls muscle tone, both at rest and during action; this is essential to control muscles' resistance to stretch and their ability to stabilize joints. The cerebellum also coordinates the actions of opposing muscle groups by determining when to start and stop contraction and how much force to apply at each moment. To accomplish all these functions, the cerebellum processes input from skin, eyes, ears, stretch receptors in skeletal muscle, and the cerebrum.

The Spinal Cord

Perhaps nowhere in the body is the heritage of our segmented invertebrate and lower vertebrate ancestors more evident than in the structure of the spinal cord. Here, at regular intervals, efferent neurons leave the CNS and afferent neurons enter along what are known as **spinal nerves.** Some afferents and efferents synapse directly with one another, such as those of the knee-jerk reflex that we discussed earlier. Others connect through interneurons that travel along the spinal cord to and from the brain. The pattern in which spinal nerves connect with receptors in the skin reflects the fate of tissue derived from embryonic segments during development (Fig. 43.21).

It is interesting that the brain tends to confuse certain kinds of sensory information from nerves originating from the same body segment. This is particularly true if certain neurons from a segment are regularly stimulated and others are not. The heart and the region of skin indicated in blue in Fig. 43.21, for example, are served by nerves that enter the spinal cord together. During a heart attack, the brain can interpret heart pain as pain in the left armpit and the inner surface of the left arm, areas that are stimulated more often. This is known as *referred pain.*

SO ELEGANT AN ENIGMA: THE BIOLOGY OF THE MIND

Neurologists and psychologists have long been challenged by the difficulty of determining how the brain functions. For many years, researchers have been able to pick up traces of the brain's electrical activity through electrodes applied to the scalp. Typically, these patterns have been displayed in the form of an **electroencephalogram,** or **EEG.** Because the brain exhibits different patterns of activity during different states of wakefulness and sleep, EEGs have been useful in categorizing the levels of sleep and in investigating certain kinds of sleep disorders (Fig. 43.22).

Today, however, medical science has far more specific ways to see what is going on inside the brain. **CAT** scans (**c**omputer-**a**ssisted **t**omography) take a series of finely focused X-rays of the brain from many different angles, analyze the amount of radiation absorbed by different parts of the brain, and convert that information into a series of "maps" that represent visual cross sections of the anatomy of the brain (Fig. 43.23).

PET scans (**p**ositron **e**mission **t**omography) go a step further than EEGs and CAT scans by monitoring and offering physicians a record of the physiological activity of the brain. By administering doses of short-lived radioactive isotopes that are metabolized by the brain, physicians can create maps of relative metabolic activity in the brain during different activities (Fig. 43.24).

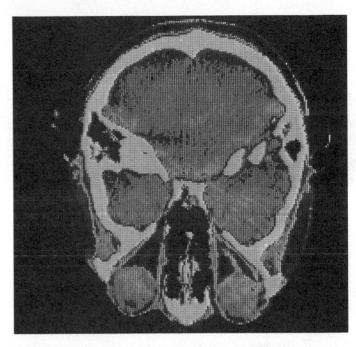

Figure 43.23   CAT scan of a normal human brain.

Figure 43.24   LEFT: PET scan patient. RIGHT: PET scan.

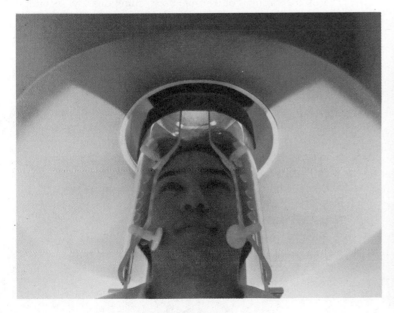

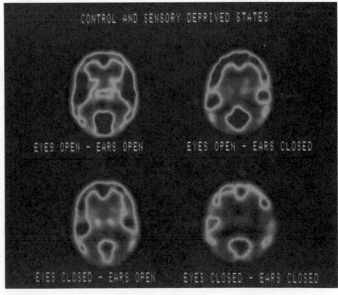

Finally, the more recently developed **BEAM** scans (**b**rain **e**lectrical **a**ctivity **m**apping) use the sort of sophisticated data processing techniques used in CAT and PET scans to translate the jagged lines of EEG recordings into a color map of brain function. With these and other tools at their disposal, both basic and clinical researchers hope to continue unraveling the mysteries of the intact brain.

SUMMARY

The nervous system receives information through sensory neurons, carries and processes information through interneurons, and directs responses by stimulating muscles and glands called effectors. Neurons carry information as action potentials, waves of electrical disturbance that travel along axons. Action potentials are made possible by ion pumps and ion channels in the nerve cell membrane. Neurons communicate at synapses, where electrical activity is converted into pulses of chemical neurotransmitters. Neurotransmitters may either excite or inhibit the next neuron in the circuit.

The simplest nervous systems are diffuse nerve nets with few specialized cells. Among higher invertebrates and vertebrates, however, axons are gathered into afferent and efferent nerves and central nerve cords. Collections of cell bodies and synapses called ganglia serve as information processing centers. As higher vertebrates evolved, the largest of these—the cerebral ganglion—grew into a highly developed brain. The human nervous system is divided functionally and structurally into *central* and *peripheral nervous systems*. Functionally, the peripheral nervous system is divided into the voluntary or somatic nervous system and the involuntary or autonomic nervous system. The autonomic nervous system is further divided into sympathetic and parasympathetic divisions.

STUDY FOCUS

After studying this chapter, you should be able to:

- Describe the principal components of the nervous system and how these components are interconnected.

- Understand the basic facts about electrical phenomena in living tissue.

- Explain how the nervous system transmits information by describing the mechanisms by which action potentials are generated and chemical synapses receive and transmit stimuli.

TERMS AND CONCEPTS

REVIEW

Objective Questions (Answers in Appendix)

1. When a neuron is not transmitting an impulse, a(n) _____ is present across its cell membrane.
   (a) impermeable barrier to K⁺ ions
   (b) action potential
   (c) resting potential
   (d) depolarized current

2. When a stimulus comes to a neuron cell membrane, the membrane becomes _____ permeable to _____ ions.
   (a) more; sodium   (c) less; potassium
   (b) less; sodium   (d) more; potassium

3. An evolutionary trend in the nervous system of invertebrates that is seen in a more advanced form in vertebrates is
   (a) accumulation of ganglia in the head region.
   (b) development of a ventral nerve cord.
   (c) formation of a diffused nerve net.
   (d) increasing control of body functions by secondary ganglia.

4. The phrase "all or none" as it relates to the action potential means that the
   (a) membrane generates a complete action potential or does not generate any action potential.
   (b) movement of nerve impulses is self-propagated.
   (c) nerve impulse doesn't lessen or end as it moves from the origin of the stimulus.
   (d) movement of the nerve is by saltatory conduction.

5. The blood–brain barrier does not allow the passage of
   (a) alcohol.   (c) cocaine.
   (b) nicotine.   (d) all hormones.

Discussion Questions

6. Explain the series of events that result in the generation and propagation of an action potential.

---

7. It is tempting to compare action potentials traveling down axons to pulses of electric current traveling through insulated wires. Why is this not an accurate comparison?

8. Neurons are rarely connected to one another by direct electrical connections. Why? How are chemical synapses important to the functioning of nervous systems?

9. Explain the important differences between graded potentials and action potentials. What are the advantages and disadvantages of each?

10. How do the central and peripheral nervous systems differ from one another? How are they similar?

11. Compare and contrast the activities of the sympathetic and parasympathetic divisions of the autonomic nervous system. Although the actions of these two systems are often described as "antagonistic," it is more accurate in some ways to describe them as complementary. Why?

READINGS

Alkon, D. M. *Memory's Voice: Deciphering the Mind–Brain Code.* New York: HarperCollins, 1992.

Holloway, M. "Unlikely messengers: how do nerve cells communicate?" *Scientific American* (December 1992): 52. This article identifies two recently discovered neurotransmitters, ATP and nitric oxide, and concludes that there may be many more neurotransmitters as yet unknown.

Premack, D. *Gavagai! or the Future History of the Animal Language Controversy.* Cambridge, MA: M.I.T. Press, 1986.

Restack, R. M. *The Mind.* New York: Bantam Books, 1988.

Sacks, O. *The Man Who Mistook His Wife for a Hat and Other Clinical Tales.* New York: Harper and Row, 1985.

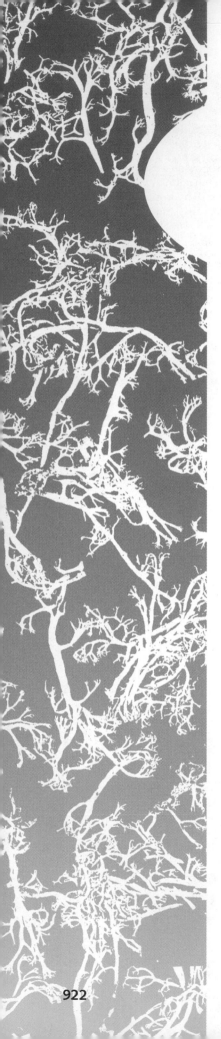

44

The Sensory System

If the doors of perception were cleansed, every thing would appear to man as it is, infinite.
For man has closed himself up 'til he sees all things through narrow chinks of his cavern.

—William Blake

*M*ake believe for a moment that you are in your favorite corner of the natural world—a tropical beach, a Colorado mountain ridge, or a lush New England forest. Now immerse yourself in that setting as completely as possible. Imagine the warmth of the sun on your skin, the feather touch of the breeze on your face, the tang of salt air, the bracing scent of pine needles, or the subtle odors of moss and wet soil. Conjure up the images of wildflowers blazing in an alpine meadow, butterfly wings flashing in the sun, or crickets chirping. Savor these sensations. Yes, this is your favorite spot, just as you remember it.

If you think the way most of us do, the collection of sensory images you've just conjured up *is* that place for you. Your perceptions *are* your world, for perceptions are all of the world any of us can ever know. Each of our senses gathers information about part of our immediate environment, and together they inform us of changes that occur around us.

Sensory systems, human and nonhuman, are of incalculable importance in daily life. In this chapter, we will explore the form and function of human senses and the senses of other animals. In the process, we will learn to view the senses as evolutionary adaptations to the advantages and disadvantages of various sensory modalities in different environments.

THE WORLD, THE SENSES, AND REALITY

The world we create from our perceptions—the environment that is so real to us—is an insular and idiosyncratic "reality" based on only a fraction of the phenomena present in the environment. The total reality of that spot you've chosen is far more complex and varied than human senses can fathom; there are sounds we cannot hear, objects and colors we cannot see, smells and tastes of which we are completely unaware. And altogether beyond our sensory capabilities are events we can describe and detect with instruments but cannot experience directly. The earth's magnetic field envelops us. The ground beneath us carries countless faint vibrations. And underwater, both inanimate objects and living or-ganisms emit minute electric currents that surround them with faint electromagnetic auras as personal as our fingerprints.

All these phenomena are potential sources of environmental information to animals with appropriately specialized sensory receptors. In some cases, animal senses are more acute than, but basically similar to, our own. Dogs track odors of which we are unaware. Bees can see ultraviolet light. Bats and dolphins "see" by emitting and responding to sounds far above our range of hearing. But other animals have senses that are completely different from anything we possess. Certain fishes communicate by generating electric currents, for example, and many birds navigate in part by detecting the earth's magnetic field.

Thus, if we accept the metaphor of our senses as windows on the world, then we must also say that we perceive the world only "through a glass, darkly." We can devise microscopes and telescopes, extending that glass's visual range with lenses. But the glass remains stubbornly in place, coloring our view of the world so intimately that we can conceive of different world views only abstractly.

THE NATURE OF SENSORY STIMULI

Information about the environment is carried by **sensory stimuli**—environmental factors to which human or animal sensory receptors can respond. Many sensory stimuli are in the form of energy and contain information encoded in variations in the form, frequency, or intensity of that energy. We see vibrantly colored flowers in a field, for example, because their petals reflect different wavelengths of light energy from the grass around them (Fig. 44.1). We hear the annoying buzz of mosquitoes because the vibrations of their wings create sound waves of a particular intensity and frequency.

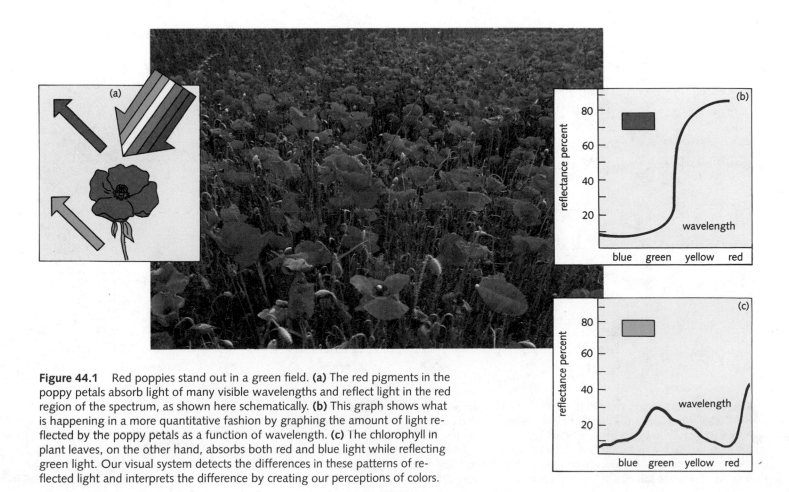

Figure 44.1   Red poppies stand out in a green field. **(a)** The red pigments in the poppy petals absorb light of many visible wavelengths and reflect light in the red region of the spectrum, as shown here schematically. **(b)** This graph shows what is happening in a more quantitative fashion by graphing the amount of light reflected by the poppy petals as a function of wavelength. **(c)** The chlorophyll in plant leaves, on the other hand, absorbs both red and blue light while reflecting green light. Our visual system detects the differences in these patterns of reflected light and interprets the difference by creating our perceptions of colors.

Other sensory stimuli are composed of matter: atoms or molecules. Because many flowers release fragrant compounds that travel through the air, you can distinguish between a rose and a dandelion with your eyes closed. Similarly, information about the differences between saltwater and fresh water, and between sugar solution and milk, is carried by atoms and molecules as well as by the liquids' visual appearance.

ESSENTIALS OF SENSORY FUNCTION

For animals to respond appropriately to environmental stimuli, their sensory systems must perform two interrelated tasks. First, they must convert, or *transduce*, sensory stimuli into the electrochemical "language" of the nervous system. Then they must process the resulting neural information in a manner that enables the central nervous system to identify and evaluate important stimuli. As you will see, part of this essential sensory processing often takes place in the sensory organs themselves, but a great deal occurs in higher brain centers.

Sensory Transduction

Transducing stimuli into neural activity is the task of *receptor cells,* specialized neurons that detect stimuli of one type or another. Each type of receptor cell is typically specialized to detect only one particular kind of stimulus; photoreceptors respond to light, whereas taste receptors respond to chemicals. Upon receipt of an appropriate stimulus, the receptor cell generates a **receptor potential,** a graded potential like those discussed in Chapter 43. In most types of receptors, the magnitude of this receptor potential varies with the intensity of the stimulus.

For sensory information to be acted on, it must be transmitted to the central nervous system for processing. But as you recall from the last chapter, graded potentials (including receptor potentials) cannot carry information very far from their point of origin. The sensory system handles this problem in two different ways.

Some receptor cells, such as the stretch receptors in muscle, convert receptor potentials directly into action potentials. When such receptors are stimulated, they fire impulses that travel down long axons wired into the central nervous system. Note that because all action poten-

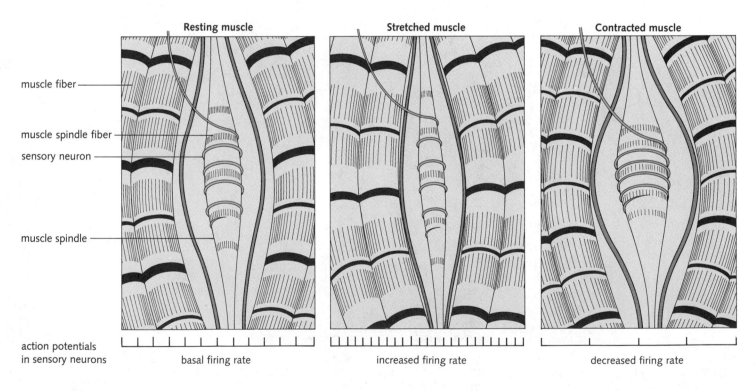

Figure 44.2 Stimulation and firing rate in stretch receptor cells of vertebrate skeletal muscle. Each vertical line in the graphs below the diagram indicates a single action potential. The horizontal axis indicates time. In a muscle at rest, muscle spindles are slightly stretched, and fire impulses slowly. In a muscle stretched beyond its resting length, the receptors increase their basal firing rate. In a muscle that has just contracted, the receptors decrease their firing rate. NOTE: Muscle spindles in higher vertebrates are more complex.

tials produced by a particular cell are identical, receptors cannot communicate the intensity of the stimulus by changing the *size* of those impulses. Instead, they encode the intensity of stimulation in the *rate* at which they fire.

Stretch receptors called **muscle spindles**, for example, are seldom totally inactive, for even in a muscle at rest they are slightly stretched. Resting spindles produce action potentials at a slow, steady rate called their **basal firing rate** or **rate of spontaneous activity**. When the muscle around them stretches, they fire more often, and when the muscle around them contracts, their basal firing rate decreases (Fig. 44.2).

This pattern of increases and decreases in basal firing rate is common throughout the nervous system. The pattern enables individual neurons to signal either an increase or a decrease in stimulation from a relatively stable resting state.

Other receptor cells, such as hair cells (which respond to motion and vibration), produce only graded potentials when stimulated. The cell bodies of these receptors (which usually lack axons) synapse directly onto one or more interneurons that produce action potentials in response to those graded potentials. In the hair cells of the ear, only a single interneuron is involved (Fig. 44.3). In the eye, there may be as many as five interneurons between individual photoreceptors and higher processing centers. These peripheral interneurons play a very important role in sensory processing, as we will soon see.

Information Processing in the Brain

It is said that Bertrand Russell, the eminent English philosopher and mathematician, once visited his dentist in great distress. "Where does it hurt?" the dentist asked Russell, ever the insightful philosopher, replied crisply "In my mind, of course."

Biologically, as well as philosophically, Russell was quite correct. Whether sensory receptors are located in our teeth or in our eyes, they do nothing more than transduce stimuli into neural activity. The actual process of sensory perception—the response to a toothache or the creation of visual images—always takes place in the brain. This is true because the "raw data" gathered by receptor cells must be integrated and processed by interneurons in two ways at once.

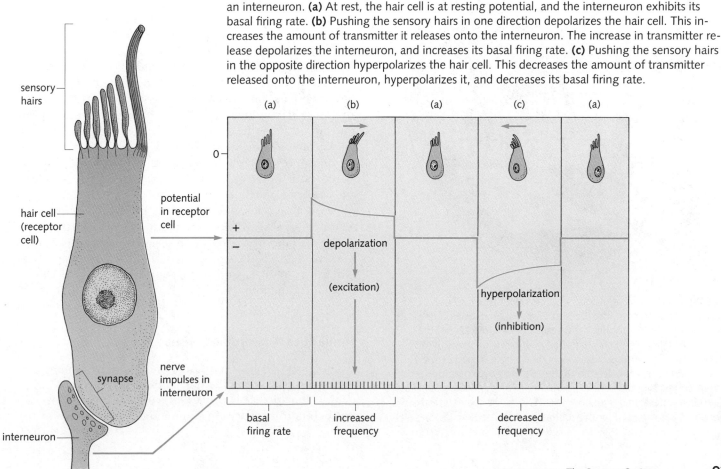

Figure 44.3 Conversion of graded receptor potentials from a hair cell into action potentials by an interneuron. **(a)** At rest, the hair cell is at resting potential, and the interneuron exhibits its basal firing rate. **(b)** Pushing the sensory hairs in one direction depolarizes the hair cell. This increases the amount of transmitter it releases onto the interneuron. The increase in transmitter release depolarizes the interneuron, and increases its basal firing rate. **(c)** Pushing the sensory hairs in the opposite direction hyperpolarizes the hair cell. This decreases the amount of transmitter released onto the interneuron, hyperpolarizes it, and decreases its basal firing rate.

First, the sensory system (including the brain) must extract important information—called the **signal**—from irrelevant information—called **noise.** A parent searching for a lost child in a crowded park, for example, strains to pick out that child's voice (the signal) from among all others in the loud throng (the noise). Second, animals are constantly bombarded with far more information than their brains can possibly pay attention to at once. The CNS, therefore, must selectively attend to certain stimuli and ignore others. Harried parents searching for a child may focus so much attention on the auditory search that they do not notice they have badly stubbed their toe until the little tyke is safely in hand. We can also generate that kind of concentration consciously; when intent on reading, for example, we often screen out such stimuli as a radio playing in the background.

It is also at the level of the CNS that moods, emotions, experience, and expectations color our ultimate perceptions, which differ not only between individuals but also between specific moments within each individual's life (Fig. 44.4). The odor that causes a child to respond "Ugh! Spinach!" might make the same individual salivate with anticipation several years later. A sunny room at a temperature of 70° Fahrenheit (21° Celsius) might be perceived as uncomfortably warm by an eskimo in winter, as pleasant by a Bostonian in May, and as chilly by a Bahamian in July. What is the "truth" about these stimuli? Is the odor of spinach pleasant or repugnant? How warm is a 70° room? The only truthful answer, thanks to the vagaries of perception, is, "It depends." Information processing is bewilderingly complex, and though we are beginning to understand some of the neural mechanisms that make perception possible, others remain a mystery.

THE DIVERSITY OF SENSORY SYSTEMS

Senses and Environments

Animal senses have evolved over time as adaptations to the nature of sensory stimuli in different environments, for each environment has physical characteristics that affect both energy and molecules that carry sensory information. Our environment, for example, consists of the media that surround us: the air around our bodies and the earth beneath our feet. Other animals may be immersed in the clear water of a mountain stream, the turbid water of a swamp, or the salty water of the sea.

Each of these media allows only certain environmental stimuli to pass. Air transmits visible light very well, for example, and conducts sound waves relatively efficiently. But air passes little or no electrical energy and carries only a limited assortment of small molecules detectable via the sense of smell. Water, on the other hand, carries sound both faster and farther than air, and it dissolves and carries a much wider range of chemicals, including sugars, amino acids, and large proteins. Water—especially seawater—is also an excellent conductor of electricity, but it absorbs (and hence fails to transmit) many wavelengths of visible light. Not surprisingly, animal sense organs have evolved in ways that relate to the environment in which they must function.

Figure 44.4 Look for a moment at this scene on an island shore, devoid of human life. Or is it? Look again. Now look still again, focusing on the space between the gnarled trees on the left. Isn't that Napoleon standing in a characteristic pose? Notice how what you see in this case is very much determined by what you are looking for. Notice also that you probably would not see the Napoleonic silhouette unless you had been exposed to similar images in the past; this is just one example of the way current perceptions are affected by expectations and experience.

Similarities Among the Senses

Cataloguing all the evolutionary adaptations among the senses and explaining the mechanisms by which the information they detect is processed would fill several books. But behind differences in stimuli and in the nature of receptor cells, many underlying similarities unite all the senses. For each sense, there is a fascinating story of environmental information, the evolutionary adaptation of receptor cells to detect that information, and neural processing in the CNS to make the use of that information possible. In this chapter, we have selected the visual system to serve as a detailed example of these major principles, though we will briefly examine other human and animal senses.

VISION: A MODEL OF SENSORY FUNCTION

Flowers, leaves, fruits, and the iridescent wings of tropical butterflies all differ in the way they absorb and reflect the electromagnetic energy of the sun. From honeybees searching for nectar-bearing flowers to fishes engaged in courtship dances, most animals depend heavily on patterns of reflected light to provide information about their environments and about each other. The elegance of design that enables vertebrate eyes to serve these purposes so impressed Darwin that he discussed them in a chapter of *On the Origin of Species* that he called "Difficulties of the theory: organs of extreme perfection and complication."

To understand why eyes are so complicated, and to appreciate the functions of visual systems, we will examine the characteristics of visual stimuli in nature and will explore the evolutionary responses of animal visual systems to the demands of vision in different environments. Because the visual system is among the best understood of the sensory systems, we will also study briefly the first stages of visual information processing. This examination will serve as a specific example of the way chemical synapses and neural wiring patterns work together to process information in the nervous system. It is important to remember that although we do not have the time or space to examine each of the other senses in this much detail, all senses perform tasks analogous to these.

Light and the Visual Environment

Visible light is our name for the portion of the electromagnetic spectrum that functions as an adequate stimulus for vision in humans. Describing the physical nature of light is difficult, for it has properties of both waves and particles. Physicists recognize both sets of properties and describe light simultaneously in terms of waves and in terms of packets of energy called **photons.** Our visible spectrum ranges from wavelengths of about 400 nanometers (nm), which we perceive as violet, to about 700 nm, which we perceive as deep red.

With the exception of the few organisms that produce their own light, natural objects are visible only by virtue of the sunlight they reflect. "White" objects reflect roughly equal amounts of all visible wavelengths, whereas "colored" objects reflect more light of certain wavelengths than of others (as we saw in Fig. 44.1).

But the light that illuminates objects in nature varies enormously at different times and in different places. Between midday in June and a moonless night, for example, the intensity of natural light varies by a factor of 10^{15}. And between dawn and midday, from summer to winter, and in different habitats, the color of available light changes enormously (Fig. 44.5).

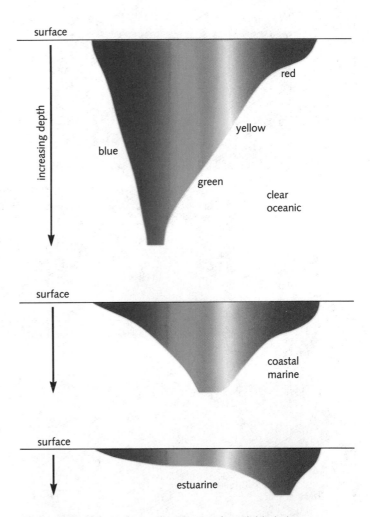

Figure 44.5 The color and intensity of available light varies greatly among different aquatic habitats. Seawater absorbs long-wave and extreme shortwave light and thus acts as a turquoise color filter placed over sunlight. Water along temperate seacoasts and in freshwater lakes carries dissolved and suspended materials that absorb light of short wavelengths; the light that persists is thus colored greenish-yellow. The minimal light available in many rivers, swamps, and marshes is deep reddish-brown.

Because of these changes in natural illumination, the physical stimulus the eye receives from an object at different times and in different places also changes greatly. Visual systems are therefore confronted with the formidable task of providing constant, reliable information under highly variable conditions; a ripe strawberry should look red regardless of when or where we see it. The eye and brain accomplish this task with such accuracy that we rarely even notice the necessary adjustments in progress.

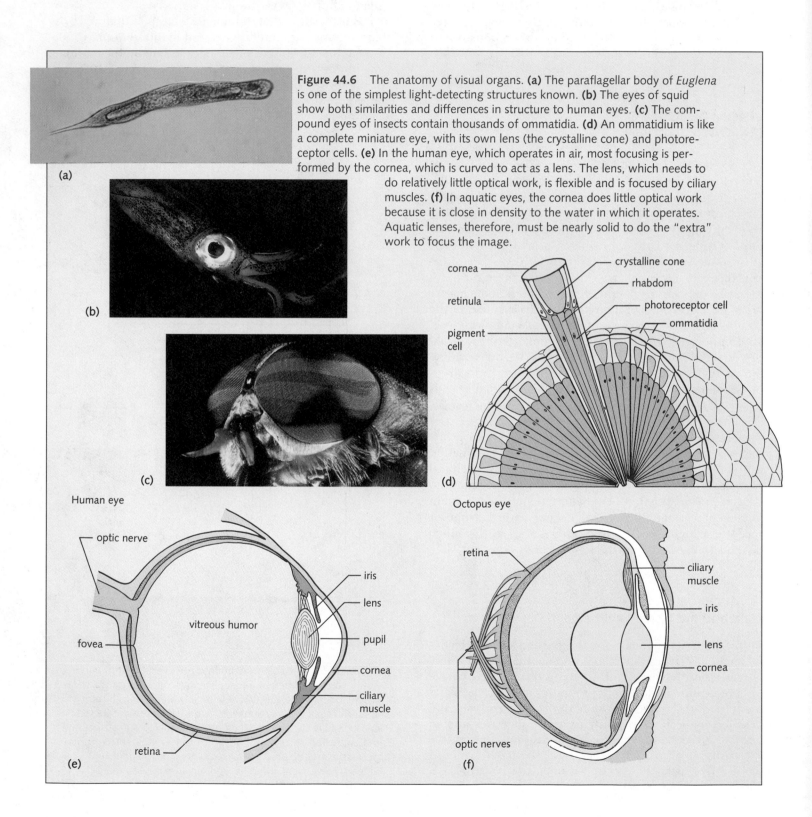

Figure 44.6 The anatomy of visual organs. **(a)** The paraflagellar body of *Euglena* is one of the simplest light-detecting structures known. **(b)** The eyes of squid show both similarities and differences in structure to human eyes. **(c)** The compound eyes of insects contain thousands of ommatidia. **(d)** An ommatidium is like a complete miniature eye, with its own lens (the crystalline cone) and photoreceptor cells. **(e)** In the human eye, which operates in air, most focusing is performed by the cornea, which is curved to act as a lens. The lens, which needs to do relatively little optical work, is flexible and is focused by ciliary muscles. **(f)** In aquatic eyes, the cornea does little optical work because it is close in density to the water in which it operates. Aquatic lenses, therefore, must be nearly solid to do the "extra" work to focus the image.

(a)

(b)

(c)

(d)

cornea
retinula
pigment cell
crystalline cone
rhabdom
photoreceptor cell
ommatidia

Human eye

optic nerve
fovea
vitreous humor
retina
iris
lens
pupil
cornea
ciliary muscle

(e)

Octopus eye

retina
optic nerves
ciliary muscle
iris
lens
cornea

(f)

Structures and Functions of the Eye

Light-sensitive organs have evolved independently in a wide variety of animals, and they vary from the simple "eyespots" of protozoa to the intricate visual sense organs and processing centers of higher vertebrates. One of the simplest such structures, the paraflagellar body of *Euglena,* is shown in Fig. 44.6a. The most highly developed eyes among invertebrates are found in the cephalopod molluscs and the arthropods (Fig. 44.6b,c). The **compound eyes** of arthropods are very different from ours, for they consist of dozens to thousands of identical units called **ommatidia** (Fig. 44.6d). Each ommatidium is a complete miniature eye with its own lens-like structure and *retinula,* or "little retina." Each retinula, in turn, contains eight or more light-sensitive **photoreceptor cells**.

It is difficult to imagine how the thousands of separate images produced by ommatidia are combined by the insect brain into a coherent picture of the world. Behavioral studies have shown that although insect eyes cannot provide so fine-grained an image of the world as our eyes can, they can detect movements far more rapid than those our visual systems can perceive. Many insects also have well-developed color vision.

Cephalopod eyes, on the other hand, though they evolved completely independently from those of vertebrates, show remarkable convergence in structure with vertebrate eyes (Fig. 44.6e,f). Both cephalopod and human eyes have a clear outer covering, or **cornea,** and a **lens** that aids in focusing the image on a **retina** lined with photoreceptor cells. Both eyes also contain a heavily pigmented **iris** that opens and closes under the control of circular muscles to regulate the amount of light entering the eye. The major optical structures of aquatic and terrestrial eyes work differently because water is more dense than air and because the focusing ability of a lens depends on the difference in density between the lens and the medium in which it operates.

For eyes (such as our own) that operate in air, the lens-shaped cornea actually does a good deal of the "work" in focusing light on the retina. Because the cornea is so important in this respect, irregularities in its shape can cause the visual distortions called **astigmatism.** Sitting behind the cornea, our lens is responsible mostly for "fine-tuning" the focus of the visual image. This process, known as **accommodation,** adjusts the eye to enable us to focus on either nearby or distant objects. In this situation, a moderately dense, yet flexible lens can perform all the necessary optical work through changes in shape. If either the lens or the eye as a whole is improperly shaped, however, the lens may not be able to accommodate sufficiently to bring all images into proper focus. In "nearsightedness" (**myopia**), nearby objects are seen clearly but distant objects are blurred. In "farsightedness" (**hyperopia**), the reverse occurs (Fig. 44.7). All of these problems can be corrected through the use of proper eyeglasses or contact lenses.

For eyes that operate in water, the difference in density between that much denser medium and the cornea is minimal, so the cornea does relatively little focusing itself. For this reason, a highly curved, very dense lens is necessary. Aquatic lenses are so dense, in fact, that they cannot change shape as ours do. Instead, these lenses focus by moving back and forth in front of the retina, in much the same way as we might focus the inflexible glass lens of a camera or microscope.

Figure 44.7 Myopia (nearsightedness) and hyperopia (farsightedness) result when an improperly shaped lens fails to focus images properly on the retina.

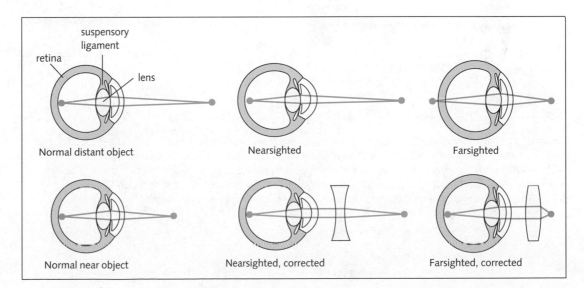

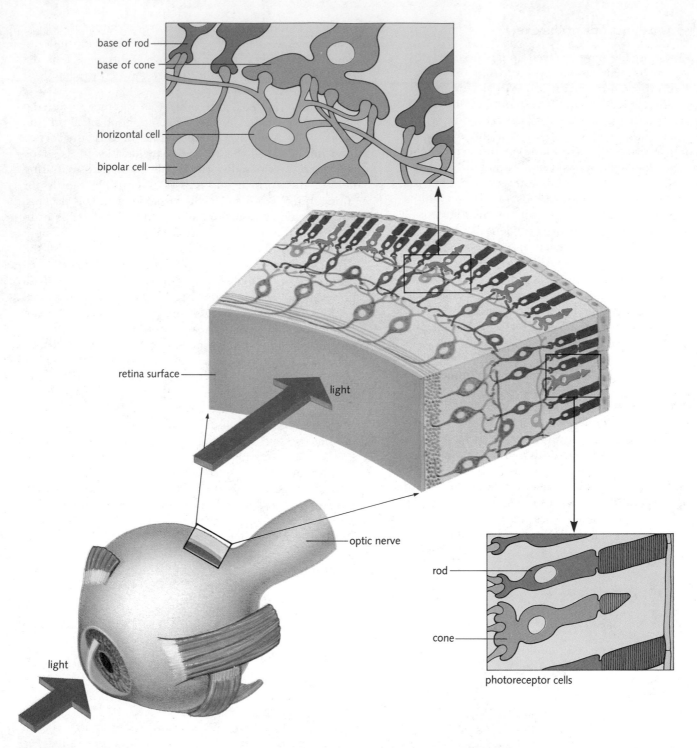

base of rod

base of cone

horizontal cell

bipolar cell

retina surface

light

optic nerve

rod

cone

photoreceptor cells

light

Figure 44.8 Vertebrate photoreceptor cells synapse with several bipolar and horizontal cells. Both bipolar and horizontal cells receive input from many photoreceptors. Several bipolar cells, in turn, synapse with each ganglion cell. These three types communicate using only graded potentials. The axons of ganglion cells (which do produce action potentials) lead out of the eye through the optic nerve. The dendrites of a retinal bipolar cell gather information from an area called the cell's receptive field.

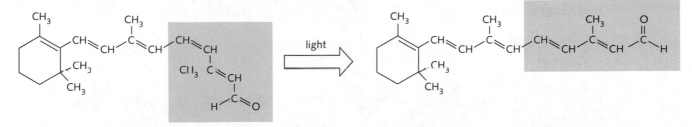

Figure 44.9 The effects of light absorption on visual pigments.

The Retina

The vertebrate retina is composed of two layers, the *neural retina* and the *pigment epithelium* (Fig. 44.8). The neural retina consists of two classes of photoreceptors, the **rods** and **cones;** several classes of glial cells; and four classes of interneurons, the *horizontal cells, bipolar cells, amacrine cells,* and *ganglion cells.* Cones, which function well only in fairly bright light, are responsible for the sharp, full-color visual images we depend on through most of the day. Cones are concentrated in the **fovea,** the part of the retina we instinctively direct at whatever we wish to see most accurately. Rods, sensitive enough to detect the absorption of a single photon of light, are responsible primarily for "black and white" vision in very dim light, although there is evidence that they participate in color vision at low light intensities.

The four classes of interneurons relay information from photoreceptors to higher visual centers. They also process that information, extracting clues to color and form in a manner we will soon examine. This processing is possible because each retinal interneuron gathers information from a significant area of the retina called its *receptive field.* The axons of the ganglion cells traverse the retina, gather together, and exit the eye through the **optic disk** to form the **optic nerve.**

The pigment epithelium behind the neural retina contains pigment that prevents light that is not absorbed by photoreceptors from scattering and degrading the image. In nocturnal animals such as cats, however, the epithelium also contains a reflective layer called the *tapetum.* The tapetum reflects light onto the photoreceptors, giving them a "second chance" to absorb the limited nighttime light available. This tapetum creates the "eye shine" you see when a cat's eyes are illuminated by a spotlight in the dark.

The pigment epithelium is also important in providing both physiological and physical support to the neural retina. If the photoreceptor layer pulls away from this supportive epithelium, creating a condition known as **detached retina,** retinal damage and impairment of vision can result.

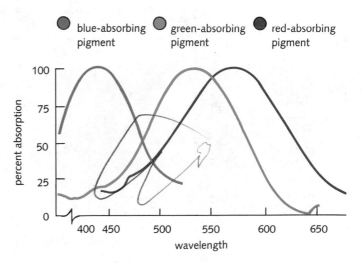

Figure 44.10 The absorption spectra of visual pigments from human rods and cones. Each pigment has a characteristic absorption spectrum with a well-defined peak, referred to as its wavelength of maximum sensitivity.

The Photoreceptors

Photoreceptors owe their light sensitivity to a class of compounds called **visual pigments,** which are composed of a protein group or **opsin** combined with a derivative of vitamin A_1 or A_2 called **retinal.** When visual pigment molecules absorb light, they change shape (Fig. 44.9) in a way that alters the resting potentials of the cells that contain them. After the shape changes, the opsin and retinal separate—a process called *bleaching* that temporarily inactivates the visual pigment. Reactivation requires that opsin and retinal be rejoined and returned to their original shape.

Visual pigments are concentrated in the region of the receptor cells known as the outer segment. In vertebrate photoreceptors, light absorption by visual pigments in these outer segments causes a graded *hyperpolarization* of the receptor, but it does not initiate an action potential. (Fig. 44.10)

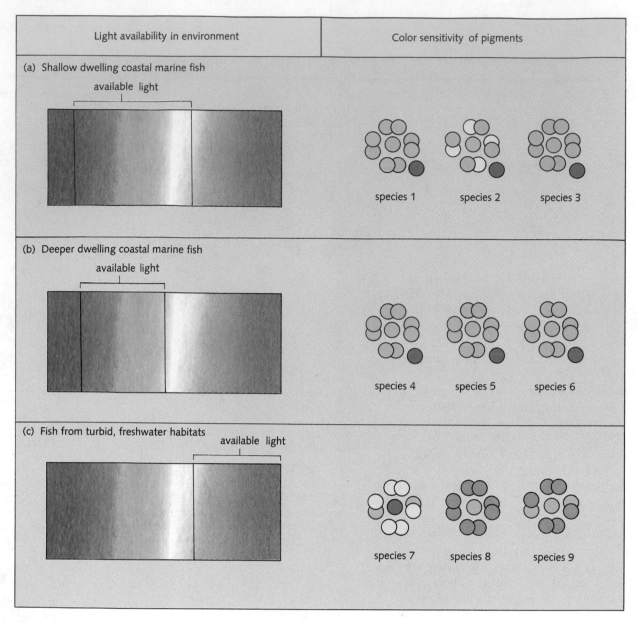

Figure 44.11 Visual pigments and environments. Aquatic animals living where light is dim have evolved visual pigments whose sensitivities match the color of available light. The colored dots for each species represent the colors to which their visual pigments are maximally sensitive.

Most animals have a single class of rod photoreceptors, all of which contain the same visual pigment. Diurnal (day-active) animals may possess as many as four classes of cone cells (humans have three), each of which contains a different visual pigment. All visual pigments absorb light of most visible wavelengths to some degree, but each particular visual pigment is maximally sensitive to light of certain wavelengths (as shown in Fig. 44.10). Extensive neural processing of signals from different classes of cone cells enables the visual system to tell objects apart on the basis of both brightness and color.

Although several mammals, such as cats and other nocturnal species, don't see color well, most other animals (including insects, fishes, turtles, birds, and monkeys) have well-developed color vision. This is undoubtedly because color vision provides a great deal of important visual information to them about their environment. So useful is color vision, in fact, that it seems to have evolved independently in different groups of animals at least four times. Some animals, such as bees, can see wavelengths invisible to humans because their visual pigments absorb ultraviolet light. Other animals, such as turtles and birds, have better color vision than we do; they can see more subtle differences in colors because they have more classes of cone receptors. Many species, particularly aquatic ones, have evolved visual pigments attuned to the colors of light present in their specific environments (Fig. 44.11).

Processing of Visual Information

> But to determine . . . by what modes or actions [light] produceth in our minds that phantasm of colours is not so easie.
>
> —Isaac Newton

The absorption of light by photoreceptors is the first step in a long series of events that ultimately results in vision. The complexity of information processing involved in these events is mirrored in the intricacy of neural networks in the retina and higher visual centers. Here we will investigate two phenomena: light adaptation and color vision. Like the other material discussed in this section, each of these visual phenomena has analogues in other sensory mechanisms.

Adaptation One of the most remarkable properties of the visual system is its capacity to operate over an enormous range of light intensities. Faced with widely varying conditions of light and dark, the eye adjusts its sensitivity through a process known as **sensory adaptation** (not to be confused with evolutionary adaptation). By comparison, photographic emulsions used in camera films—which cannot adapt—are restricted to a much smaller range of light intensities. To avoid over- or underexposures, photographers must select the proper high-light or low-light film for each situation and set their cameras carefully to control the amount of light that enters.

By contrast, adaptation—the "exposure control" in our eyes—proceeds rapidly and automatically through several processes that overlap in time. When you enter a darkened theater from a sunlit street, for example, the pupils of your eyes immediately expand to allow more light to enter. Visual pigment bleached in strong light is slowly but steadily regenerated. Finally, photoreceptors and interneurons throughout the system increase their sensitivities. All of these processes are reversed when you leave the darkened area and emerge once again into brighter light.

Color vision The three classes of cones in the human retina respond differently to light of different wave-

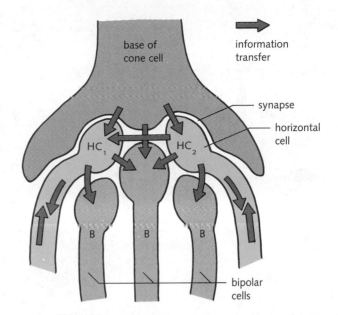

Figure 44.12 Information transfer within a cone cell synapse. The horizontal cell dendrite marked HC_1 can be influenced not only by the receptor cell but also by the horizontal cell labeled HC_2. The bipolar cell dendrites labeled B are influenced by the photoreceptor and by both horizontal cells at the same time.

lengths, making it possible for us to tell objects apart on the basis of the wavelengths of light they reflect. We call this ability color vision. Much of the neural wiring that makes color vision possible is right in the retina itself in the form of hookups among cone cells, horizontal cells, bipolar cells, and ganglion cells. At the bases of cone cells, for example, are large synapses that contain dendrites from both horizontal cells and bipolar cells (Fig. 44.12). Several of these dendrites can both receive and transmit information within this synapse. And that information may either excite or inhibit the cell to which it is transferred.

Bipolar cells and horizontal cells are usually stimulated by photoreceptors to which they attach directly. But horizontal cells, once stimulated, often cause inhibition at other synapses into which they send their dendrites. For this reason, bipolar cells can receive two different

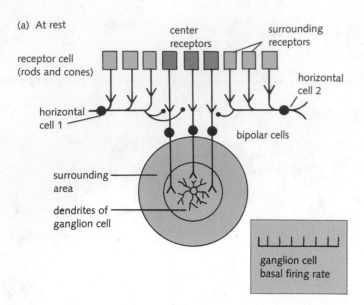

(a) At rest

center receptors
surrounding receptors

receptor cell (rods and cones)

horizontal cell 2

horizontal cell 1

bipolar cells

surrounding area

dendrites of ganglion cell

ganglion cell basal firing rate

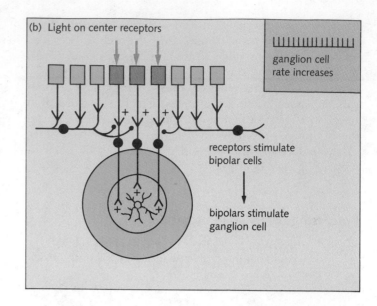

(b) Light on center receptors

ganglion cell rate increases

receptors stimulate bipolar cells

bipolars stimulate ganglion cell

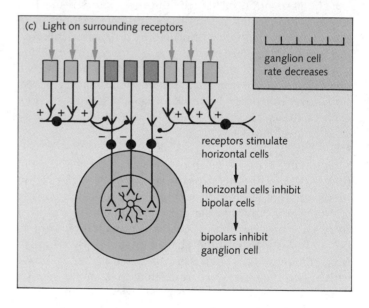

(c) Light on surrounding receptors

ganglion cell rate decreases

receptors stimulate horizontal cells

horizontal cells inhibit bipolar cells

bipolars inhibit ganglion cell

Figure 44.13 **(a)** Neural wiring and information processing in the retina. Let us assume that we are monitoring the response of the single ganglion in the center of the figure as shown. Note that in these schematic illustrations, we show functional (rather than anatomical) connections between cells. Excitatory connections are shown by "+," and inhibitory connections are shown by "−." **(b)** Light on the "center" receptors stimulates bipolar cells, which cause our central ganglion cell to increase its basal firing rate. **(c)** Light on surrounding receptors stimulates horizontal cells 1 and 2, which inhibit those bipolar cells, lowering the ganglion cell's firing rate.

kinds of input: they can be stimulated directly by certain photoreceptors, and they can be inhibited by horizontal cells stimulated by other photoreceptors (Fig. 44.13a).

This kind of wiring enables certain bipolar and ganglion cells to function as "spot detectors." Light falling on photoreceptors in the center of such a ganglion cell's receptive field stimulates bipolar cells. Those bipolar cells in turn stimulate ganglion cells, which respond by increasing the rate at which they fire action potentials (Fig. 44.13b). If a light falls on photoreceptors outside that central area, on the other hand, those photoreceptors stimulate horizontal cells that *inhibit* central bipolar cells and ultimately *decrease* the rate at which the ganglion cell fires (Fig. 44.13c).

These responses make it easier for the visual system to detect a spot of light against a background. And if several

of these "spot detectors" are hooked together in groups, they function as "edge detectors" that help the visual system detect the edges of large objects.

This sort of information processing is called *opponent processing*, because inputs from different cells are hooked up in opposition to one another. A type of opponent processing is also involved in color vision, as the demonstration shown in Fig. 44.14 illustrates. This coloropponent interaction is the first step in a long line of processes that ultimately results in the perception of a full-color image. Step by step, in ways that we still do not completely understand, all the neurally coded data about color, shape, and movement are "assembled" as the information passes through several brain centers to the visual cortex. Precisely where and how visual perception occurs is still unknown.

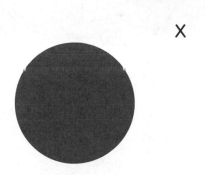

X

Figure 44.14 Demonstration of opponent color processing. Place this illustration next to a piece of plain white paper under a strong, white light (not a cool-white fluorescent). Stare intently at the "X" in the center of the figure for 2–3 minutes, and then quickly turn to look at the white paper. You will see an afterimage of colors complementary to those in the figure. Looking at a strong red light produces a green afterimage because red- and green-sensitive mechanisms in the visual system are wired in opponent fashion to one another. Similarly, looking at a strong yellow stimulus—which stimulates red- and green-sensitive cells together—produces a blue afterimage because those cells are hooked up in opposition to blue-sensitive cells.

Note that any sort of defect in this chain—from the visual pigments that first absorb light to the layers of neural processing that occur in higher neural centers—can result in visual deficiencies. The most common deficiencies in color vision (often labeled colorblindness) result when an individual is missing one or more of the cone visual pigments.

Lessons from vision applied to the other senses We have just outlined the basic functions of the visual system and have seen a few examples of how structures that serve these functions have evolved to meet different needs in different environments. All of these points should be viewed not simply as information about the visual system in particular but as principles demonstrated in all sensory systems.

In all species, sense organs and sensory receptor cells have adapted in ways that enable them to function in particular environments and for particular purposes. Similarly, opponent processing of neural signals is important because it allows higher brain centers to compare input from different sources. It is found not only in the visual system but also in many other senses and throughout the central nervous system. As just one example, opponent processing enables the auditory system to compare the sounds heard by each ear, thereby enabling the hearer to judge the location of a sound source.

AUDITION

After vision, hearing is probably the most important sense for humans and other primates. For numerous insects and several mammals, sound is the primary sensory channel for finding food, avoiding being made into food by others, and for communicating as well (see Chapter 46). Sound is a superb carrier of long-distance messages and, unlike light, can travel through soil and cloudy water and can move around tree trunks and foliage with little loss of energy.

What Is Sound?

Sound is a mechanical disturbance created in a gas, liquid, or solid by a vibrating or moving object. Physicists describe two components of sound waves, far-field sound and near-field sound.

Far-field sound consists of pressure waves and is so named because it can travel over long distances. When a loudspeaker produces sounds, for example, its cone vibrates, alternately pushing and pulling the air molecules next to it (Fig. 44.15). This repeated alternation of high and low pressure is transmitted to nearby air molecules, spreads outward from the sound source, and forms the stimuli we detect with our ears. What we perceive as the

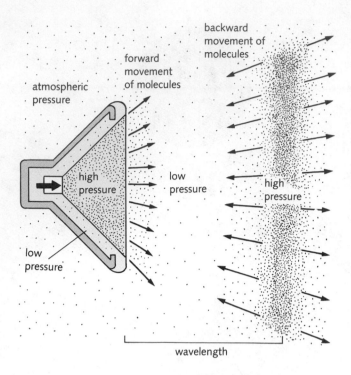

Figure 44.15 The generation of sound by a loudspeaker. Movement of air molecules is shown by arrrows, and areas of high and low pressure are shown by the amount of space between dots.

pitch of the sound depends on the frequency of its vibrations; the higher the frequency, the higher the pitch we perceive.

Sound frequency is expressed in units of cycles per second, or **hertz (Hz).** The human ear can detect sounds ranging in frequency from about 16 Hz to about 20,000 Hz. Dogs, by comparison, can hear frequencies up to 40,000 Hz. (That's why dog whistles work; they produce sound within dogs' hearing range but above ours.) Other species' hearing abilities extend even further; moths and bats, for example, hear frequencies of up to 100,000 Hz. We call sound above our hearing range **ultrasound.**

Near-field sound consists of the actual back-and-forth movement of molecules in the sound path. Near-field effects are rarely important to terrestrial animals because, as its name implies, near-field sound falls off rapidly with distance in air. Although we have shown the movement of air molecules in Fig. 44.15, such movements are usually undetectable more than a few inches from the speaker. Just about the only place a human can experience near-field effects is in a nightclub; if you walk up to a large, floor-mounted loudspeaker playing at very high volume, roll up your sleeve, and put your arm in front of the speaker cone, you will feel the hairs on your arm move back and forth during loud bass passages. Aquatic animals often experience near-field effects, however, because even near-field sound carries over long distances in water.

Figure 44.16 The anatomy of the human ear.

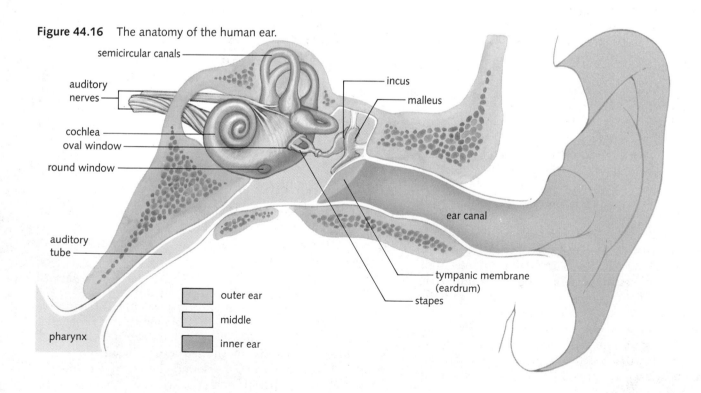

Hearing in Air: The Human Ear

The human ear uses both physical and neural elements to transduce pressure waves in air into sequences of neural impulses (Fig. 44.16). The **outer ear,** which is shaped like a funnel, channels pressure waves into the **ear canal,** amplifying them as it does so. At the end of the ear canal, the pressure waves set the **tympanic membrane** (or "eardrum") in motion. The vibrations of the eardrum are transmitted through the **middle ear,** a linked system of three bones (the **malleus** or "hammer," the **incus** or "anvil," and the **stapes** or "stirrup") to the **oval window.** Vibrations of the oval window in turn create pressure waves in the fluid-filled **cochlea** of the **inner ear.**

Within the long, coiled cochlea, vibrations of different frequencies are spread out along the **basilar membrane,** which separates the inner ear's two fluid-filled compartments. On the basilar membrane sits the **organ of Corti,** a structure that contains the **hair cells.** These hair cells are sensory receptors connected to the neurons that leave the ear to form the auditory (cochlear) nerve. As the basilar membrane vibrates, the hair cells brush against the **tectorial membrane,** bending their sensory hairs as shown in Fig. 44.17. The method by which hair cells transduce vibrations into neural impulses is illustrated in Fig. 44.3.

Sound in Water: The Lateral Line Sense

In water, near-field sounds are common, although pressure ripples from water currents make it hard to draw a distinction between touch and hearing. Confounding the distinction between ears and other organs are the **lateral lines** of fishes and amphibians—organized collections of hair cells clustered in open-ended canals just beneath the scales or skin surface. These receptors enable aquatic vertebrates to detect minute movements of water relative

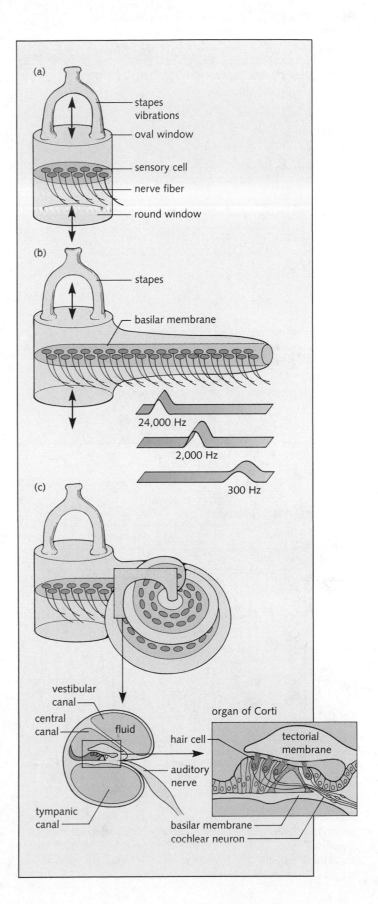

Figure 44.17 These schematics show how the cochlea is constructed and how vibrations of the oval window set the basilar membrane in motion. **(a)** This simplified version of the inner ear, which resembles the simple ears of certain reptiles, shows the relationships between the oval window, basilar membrane, and round window. The flexible round window allows the container to vibrate. **(b)** In this more complex system, the basilar membrane has stretched out. Because vibrations of different frequency are localized along different parts of this membrane, more accurate discrimination among sounds is possible. This development is characteristic of advanced reptiles and birds. **(c)** Inner ears with coiled cochleas, such as this one, are unique to mammals.

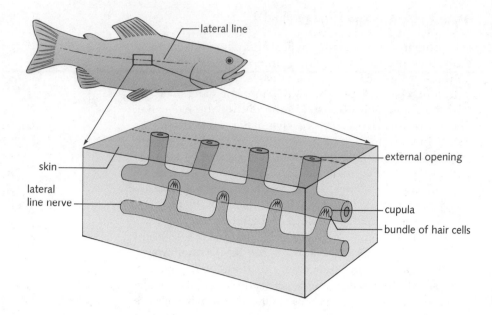

Figure 44.18 The lateral line of fishes. Near-field sounds and water movements conducted through the canals bend the hair cells and send sensory information to the brain. This system allows fishes great sensitivity to water-borne vibrations and currents.

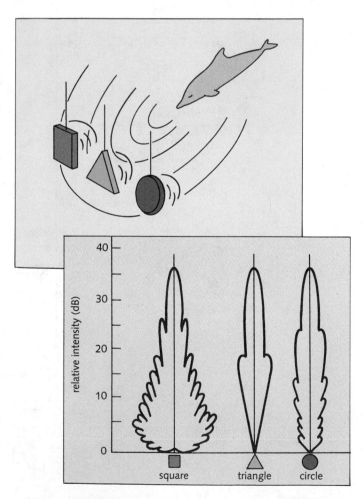

Figure 44.19 Echo patterns from three test objects during discrimination trials with an experimentally blindfolded bottlenose dolphin. Detecting the differences in sound reflection patterns from these objects, the dolphin correctly selected the circular target in 90 percent of the trials.

to their bodies (Fig. 44.18). This ability helps schooling fishes maintain their relative positions in their schools and enables many predatory fishes to detect the movements of nearby prey.

Echolocation

Two groups of mammals, the bats and the cetaceans, whales and dolphins, have evolved remarkably sophisticated systems for **echolocation**—a method of using echoes of their own cries to locate and identify prey and to navigate around obstacles. Bats emit pulses of ultrasound, whereas dolphins emit clicks that contain both audible and ultrasonic frequencies. So sophisticated is this echolocation system that bats and dolphins can navigate through mazes and identify inanimate objects even when blindfolded (Fig. 44.19). In nature, both bats and dolphins also use their sensitive auditory systems to identify sounds generated by moving prey.

BALANCE AND ACCELERATION

Nearly all organisms, aquatic and terrestrial, need to know which end is up or, more precisely, which end is down; they need to keep track of their body's position relative to the pull of gravity. In all organisms, this function is served by yet another receptor organ using hair cells as the sensory receptors. In many invertebrates, spherical organs called **statocysts** are lined with hair cells and equipped with either sand grains or other small particles. The particles settle to the lowest part of the chamber, where their pressure stimulates the hair cells (Fig. 44.20).

The human sense of balance or **equilibrium** is mediated by hair cells located in two fluid-filled chambers within the inner ear, the **utricle** and the **saccule** (Fig. 44.21). The cilia of these hair cells are covered with a layer of jelly-like material studded with particles of calcium carbonate. Because the jelly-like mass is denser than the surrounding fluid, it presses down on the hair cells beneath it, exerting pressure that changes whenever the head changes position. Those changes in pressure deform the hair cells' cilia, causing changes in neural activity that are reported to the brain.

The **semicircular canals,** a set of three mutually perpendicular, fluid-filled tubes, enable the brain to monitor precisely any sudden movements of the head. At the base of each canal is a tuft of hair cells; their cilia are embedded in a stiff yet pliable gelatinous mass called the cupula. When the head is moved suddenly, the fluid in the canals tends to lag behind, bending the cupula and stimulating the hair cells.

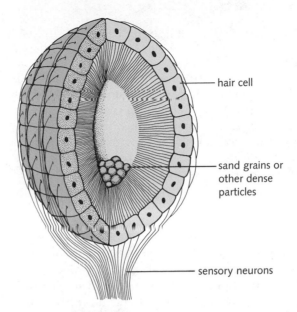

Figure 44.20 A statocyst is the balance organ of many invertebrates.

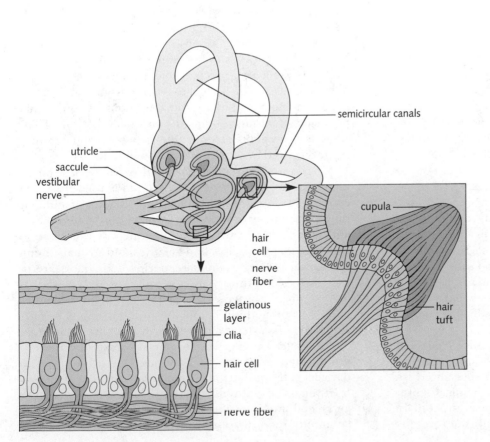

Figure 44.21 The organs of balance and acceleration detection in humans. Movements or changes in position of the head cause shifts in the pressures exerted by cupulae on hair cells beneath them. The utricle and the saccule detect position and the semicircular canals detect balance and sudden head movements.

In addition to serving the purpose of preserving balance and equilibrium in the brain, information from the semicircular canals provides sensory input to the *vestibulo-ocular reflex,* a response involving the muscles that position the eyes. The vestibulo-ocular reflex automatically adjusts the position of the eyes when the head is moved suddenly.

These sensory systems constantly gather information about movement and acceleration, whether we want them to or not. All of us have experienced times when we would have preferred to ignore that sensory input—times when the provocative movement of a car, boat, or plane has led to the intense discomfort of motion sickness. Researchers do not all agree on the specific processes that generate motion sickness, but most implicate the receipt of conflicting or "mismatched" cues about motion from different senses. When you are in the cabin of a tossing ship, for example, your senses of balance and acceleration insist that you are being tossed up and down, but your eyes report that things around you are stable. At other times, the mismatch may be produced by motion that causes the utricle and saccule to generate information that conflicts with reports from the semicircular canals. Large amounts of alcohol can alter the responses of the semicircular canals and can make a moderately bumpy ride in an automobile a great deal more unpleasant than it otherwise would be.

THE CHEMICAL SENSES

More than a century ago, French naturalist Jean Henri Fabré found his den invaded by male peacock moths. That morning, he had placed a newly emerged female moth into a wire-gauze cage, and as night fell he discovered "coming from every direction and apprised I know not how . . . forty lovers eager to pay their respects to the marriageable bride born this morning. . . ." And more than a century before that, Napoleon sent a brief note to Josephine from the battlefront. "Ne te lave pas. Je reviens," he wrote; "Don't wash. Coming home."

Both of these behaviors reflect animals' ability to detect and identify a wide variety of chemical substances ranging from individual inorganic ions—such as sodium and chloride—to complex organic compounds—including amino acids, hormones, and other proteins. Chemical senses are vital to animals in many ways; they make it possible to detect and identify potential mates by their odors, to locate prey, and/or to detect predators.

The senses that make these discriminations possible are usually divided into three categories: **olfaction** (smell), **gustation** (taste), and a variety of specialized receptors grouped together as the **common chemical sense.** In order for any of these receptors to detect an atom or compound, it must interact in solution with the receptor cell membrane. For this reason, all chemical sense receptors require a moist environment, which is usually provided by a mixture of liquid and mucus secreted by supporting cells and glands.

All these senses, though based on the actions of different receptors, are linked by neural processing in the CNS in such a way that the sensations they produce interact and overlap. Though served by distinctly different receptors from taste, for example, olfactory stimuli play an important role in determining the flavors of foods we eat. You must have experienced, for example, the major apparent loss of taste that accompanies a head cold. In this situation, your taste buds are functioning perfectly well, but your olfactory receptors are blocked by nasal congestion. Furthermore, the body's internal common chemical sense receptors can influence what odors and tastes animals find attractive. Olfactory centers in the brains of rats are excited by the odor of food when the rats are hungry. Yet if those animals are injected with sugars that satisfy the body's caloric needs, the olfactory centers of their brains show far less activity even when normally attractive food odors are present.

Olfaction

The sense of smell is the animal kingdom's long-distance chemical sense. Olfactory receptors are stimulated by **odors**—minute concentrations of chemicals carried to the receptors through air or water. The sensitivity of some olfactory receptors is extraordinary. Fabré's male moths responded to a sex attractant released by the female from as far away as 11 km. To locate females from that distance, males had to respond to concentrations as low as one molecule of attractant in 10^{15} molecules of air. That's roughly equivalent to being able to taste a single grain of sugar dissolved in an 8-ounce glass of water!

The moths' sex attractant is an example of a **pheromone,** an important class of compounds used in animal communication. As you will learn in Chapter 46, animals ranging from arthropods to primates use pheromones and other odors in urine and glandular secretions to locate and identify mates and relatives and to mark home territories. Among mammals, specific body odors are often vital to bonding between mothers and newborns. Many human mothers, for example, can identify blankets and items of clothing belonging to their infants by odor alone. Within a few weeks of birth, nursing infants can differentiate between their mothers' breasts and those of other lactating women. And although most Americans today feel differently from Napoleon about the sexual attractiveness of natural body odor, our use of perfumes and deodorants testifies to the importance of odor in communication. Curiously, many popular scents include musk, a powerful sex attractant distilled from the body secretions of other mammals.

Olfactory receptors in invertebrates may be located in a variety of places; moths generally carry them on their antennae (Fig. 44.22). In vertebrates they are found exclusively in the moist epithelium within the nasal passages (Fig. 44.23). Olfactory receptors are stimulated when odor molecules interact with receptor proteins in ways that alter their membrane potentials. Olfactory cells convert those receptor potentials directly into action potentials, which travel through the olfactory nerve to the olfactory bulbs of the central nervous system.

Gustation

Our sense of taste is based on four types of taste receptors that are sensitive to sweet, bitter, salty, and sour compounds. But much as the visual system combines the responses of three types of cones to produce innumerable colors, the chemical sense somehow combines these primary sensations with olfactory clues to register a nearly endless variety of flavors. Several hypotheses attempt to explain the evolution of the four basic tastes, and there are clear advantages to detecting certain components in edible objects. Sweet foods, for example, contain sugar, an easy source of available energy. Bitter tastes are often associated with secondary compounds in plants that may be toxic. And salt is essential to balancing the levels of body ions.

Human taste receptors are grouped into **taste buds** that line grooves in the surface of the tongue (Fig. 44.24). Although young humans are equipped with as many as 10,000 taste buds, that number declines to as few as 3000 with age. (This phenomenon may account for the increased appetite of older individuals for salt and certain spices.) Other animals may carry their taste receptors in surprising locations. In arthropods they may be on mouthparts, antennae, or walking legs; in fishes they may be scattered over the entire body surface.

Ever since the Romans discovered the first artificial sweetener nearly 2000 years ago, the food industry has invested a great deal of effort in determining how various molecules stimulate taste receptors. We now know that many compounds interact with sensory cells by fitting into specific receptor sites on the receptor cell membrane. Food chemists, or "flavorists," as they are sometimes called, have become adept at synthesizing compounds that "fool" taste buds by mimicking the shapes of natural food components. Such compounds as saccharin and aspartame (NutraSweet®), for example, are used to stimulate sugar receptors without providing the calories of sugar.

Figure 44.22   Olfactory receptors. Each of the hairs on this Australian moth's antennae carries many still finer hairs. Each of those contains the animal's exquisitely sensitive olfactory receptors.

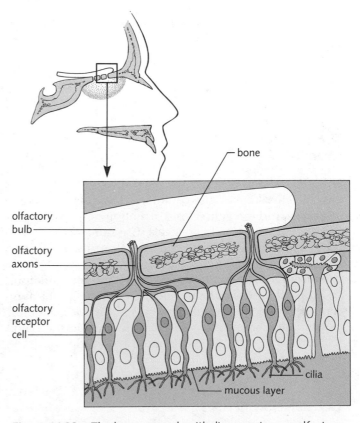

Figure 44.23   The human nasal epithelium carries our olfactory receptors. The cilia of these cells, which must be kept moist, are protected by a layer of mucus.

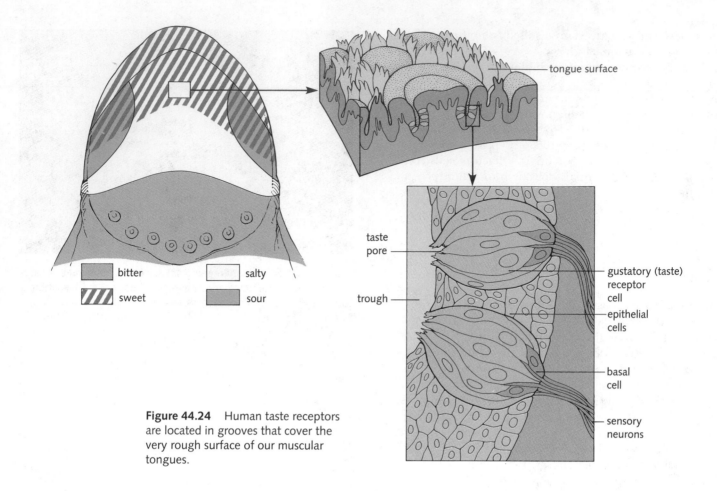

Figure 44.24 Human taste receptors are located in grooves that cover the very rough surface of our muscular tongues.

Labels in figure:
tongue surface
taste pore
trough
gustatory (taste) receptor cell
epithelial cells
basal cell
sensory neurons

bitter / salty
sweet / sour

The Common Chemical Sense

Other chemical receptors scattered throughout the body are collectively called the common chemical sense. Some of these produce generalized irritation responses when stimulated by noxious chemicals. Chlorine gas and cigarette smoke, for example, stimulate mucus production and choking responses in the respiratory tract. Other chemical receptors monitor the body's internal condition and report directly to higher neural centers. Chemical receptors within the carotid artery, for instance, monitor oxygen and carbon dioxide concentrations in the blood (Chapter 35), and others monitor blood glucose levels (Chapter 37).

THE GENERAL SENSES

The general senses, another group of miscellaneous receptors, are often broken into two groups: exteroceptors and interoceptors. **Exteroceptors** monitor external conditions impinging on the body and are responsible for the sensations of pain, warmth, cold, and light touch (Fig. 44.25). These receptors initiate appropriate reflexes when stimulated. The stimulation of cold receptors, for example, produces body responses to conserve heat. Pain is reported by a variety of free nerve endings in the skin that respond to mechanical, chemical, or thermal stimuli. Certain injuries remain painful for some time because tissue damage releases compounds such as histamine (Chapter 42), prostaglandins, and bradykinin, which bind to pain receptors and amplify the neural response to pain. Aspirin, one of humanity's oldest pain relievers, has been found to work by partially blocking prostaglandin production.

Interoceptors distributed throughout the internal organs report such sensations as cramps, hunger, and thirst, along with the need to defecate or urinate.

Proprioceptors, interoceptors imbedded in muscles, tendons, joints, skin, and connective tissues, report on the relative positions of body parts and on their move-

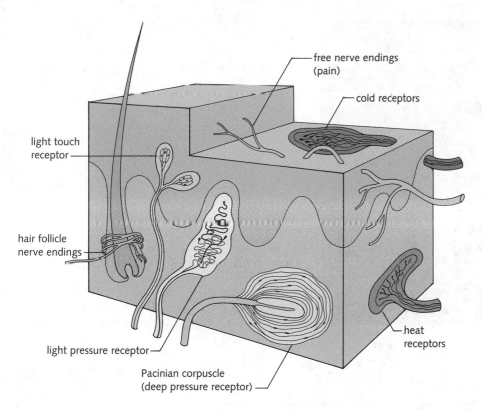

light touch receptor

free nerve endings (pain)

cold receptors

hair follicle nerve endings

light pressure receptor

Pacinian corpuscle (deep pressure receptor)

heat receptors

Figure 44.25 Our skin contains a host of receptors that provide information about our immediate environment. Pacinian corpuscles respond to pressure and touch. The end bulbs of Krause are believed to respond to cold, while the Ruffini corpuscles probably respond to heat. Free nerve endings mediate responses to pain. Hair-follicle nerve endings respond to displacements of body hairs by touch or air movement.

ments (Fig. 44.26). Well-known proprioceptors are *stretch receptors* that detect the length and tension of muscles and *Pacinian corpuscles* that detect pressure deep in body tissues.

SPECIAL SENSES OF ANIMALS

In addition to developing familiar senses differently from the way our species has, many animals have evolved senses totally different from any in the human repertoire.

Infrared Detection

A number of snakes—pit vipers, for example—have heat-sensitive pits located on either side of their snouts (Fig. 44.27). The *infrared receptors* in these pits assemble, and report to the snake's brain, a thermal map of the environment much like the visual map produced by the visual system. Snakes use this sense to track warm-blooded prey, such as small rodents, in the dark.

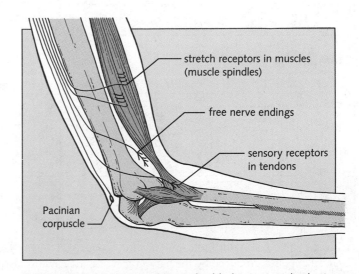

stretch receptors in muscles (muscle spindles)

free nerve endings

sensory receptors in tendons

Pacinian corpuscle

Figure 44.26 Proprioceptors, imbedded in various body tissues, continuously monitor the amount of stretch in our muscles and the relative positions of our joints.

Figure 44.27   Pit organs, infrared receptors of snakes, can detect the body heat of their small-mammal prey.

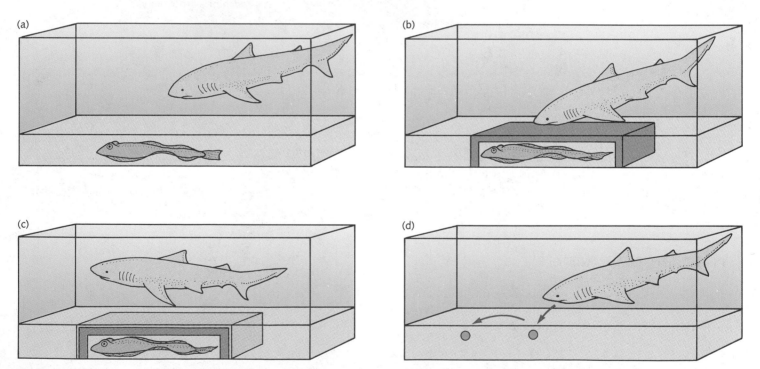

Figure 44.28   **(a)** A shark easily finds live fish buried out of sight by homing in on the tiny electrical impulses produced by the prey's breathing movements. **(b)** The shark detects just as easily a fish covered with agar—which blocks scent cues but not electric currents. **(c)** An agar chamber, which blocks scent cues, covered with an electrically insulating film successfully hides the fish. **(d)** A shark dives for electrodes that simulate the electric field of a living fish in the absence of any scent cues.

Electroreception

Sharks and several other fishes can detect minute electric currents in the water around them. The electroreceptive cells that make this sense possible, the *ampullae of Lorenzini,* are evolutionarily related to hair cells and are concentrated in a series of canals and pits around the animal's snout.

Electrodetection is a very useful ability for saltwater predators, because the essential life processes of most animals generate small electric currents. The exchange of ions across the gills of fishes as they breathe, for instance, generates electric currents, as do the movements of their respiratory muscles. By homing in on these currents, sharks are able to detect even prey that are completely buried in the sand (Fig. 44.28). In addition, sharks apparently use these receptors to help them navigate across large stretches of open ocean.

The exquisitely sensitive electrodetecting sense of deep-water sharks has made them a serious hazard to long-distance undersea cables. Those cables need to have amplifiers at regular intervals along their length. Unfortunately, these amplifiers generate small electric currents in the seawater around them that attract sharks and encourage them to attack the cables!

Two other groups of fishes from turbid habitats in Africa and South America generate their own electric fields by using specially modified muscle tissue. Such fishes can hunt, navigate around obstacles, and communicate with one another by detecting changes in these electric fields.

SUMMARY

By responding to natural phenomena, the senses of humans and other animals gather information about the environment. Though the specifics of sensory processing vary among senses, the principles of sensory function are always similar. Specialized neurons called receptor cells transduce such natural phenomena as light energy and sound waves into graded receptor potentials. Receptor potentials are converted—either by receptor cells or by interneurons—into action potentials that are processed by the CNS, where perception occurs. Animals often have abilities very different from ours because of the design of their sensory systems; bees can see ultraviolet light, for example, and dogs can detect ultrasound. Other animals have senses for which there are no human equivalents, such as sharks' ability to detect electric current.

Receptor cells of the human eye, which are stimulated when they absorb photons, fall into two classes: cones, of which there are three types, and rods. Each type of cone is maximally sensitive to light in one part of the visible spectrum. By comparing the signals from these types of cones, our visual system is able to distinguish among objects on the basis of color.

Hearing, the senses of balance and acceleration, and the lateral line sense of fishes and amphibians are all mediated by sense organs based on receptors known as hair cells. Hair cells produce generator potentials when their projecting cilia are deflected.

STUDY FOCUS

After studying this chapter, you should be able to:

- Explain the roles and functions of sensory systems, human and nonhuman.

- Describe how information is processed by using the visual system as an example, and identify the similarities that unite all senses.

- Explain how the physical characteristics of different environments influence the transmission of sensory information.

TERMS AND CONCEPTS

receptor potential	*924*	lateral lines	*937*
basal firing rate	*925*	echolocation	*938*
compound eyes	*929*	statocysts	*939*
retina	*929*	equilibrium	*939*
accommodation	*929*	olfaction	*940*
rods	*931*	gustation	*940*
cones	*931*	pheromone	*940*
visual pigments	*931*	taste buds	*941*
sensory adaptation	*933*	exteroceptors	*942*
hertz (Hz)	*936*	interoceptors	*942*

REVIEW

Objective Questions (Answers in Appendix)

1. In the compound eye of arthropods, there are dozens to thousands of complete miniature eyes known as
   - (a) foveas.
   - (b) rods.
   - (c) ommatidia.
   - (d) cones.

2. In humans, chemical sense receptors of airborne substances are primarily found in the
   - (a) semicircular canals.
   - (b) olfactory epithelium.
   - (c) nerve endings in the skin.
   - (d) oval window.

3. Mechanical vibrations of the air are converted into neural impulses by sensory receptors in the ear called

(a) hair cells. (c) olfactory receptors.
(b) horizontal cells. (d) ear canals.

4. When our ears respond to different _____ of sound waves, we are perceiving the _____ of the sound.
   (a) lengths; frequency (c) pitches; frequency
   (b) frequencies; pitch (d) lengths; pitch

Discussion Questions

5. How do the differences in structure between the human eye and a fish eye reflect the different operational demands of vision in air and water?

6. Explain why the stimuli available to the chemical senses of taste and smell are different for terrestrial and aquatic animals. What sorts of compounds could a lobster "smell" that we can only taste? Why?

7. Choose any single sense and explain how differences in receptor cell and/or sense organ structures among different animals result in different sensory capabilities.

8. Describe one animal sense that humans lack altogether.

READINGS

Downer, John. *Super Sense.* New York: Henry Holt, 1988. An intriguing, well-illustrated, and broad-ranging survey of the sensory capabilities of animals ranging from humans to bats and electric fishes.

Schnapf, J. L., and D. A. Baylor, "How photoreceptor cells respond to light." *Scientific American* (April 1987): 40–47.

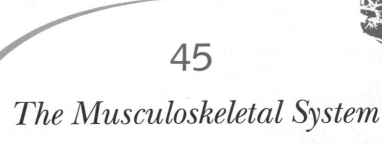

45

The Musculoskeletal System

To move things is all mankind can do, and for such the sole executant is muscle, whether in whispering a syllable or in felling a forest.

—Sir Charles Sherrington

*a*n eagle soars in graceful arcs on an updraft, controlling its glide with subtle adjustments of its wings. A cheetah chases a gazelle, expertly avoiding obstacles and adjusting to irregularities in the terrain beneath his feet as he races along at speeds up to 95 km (60 miles) an hour. A ballerina pirouettes on point, her body exquisitely balanced and positioned. And an athlete, all his muscles toned and working in harmony, sprints with cat-like grace towards a high jump. All these feats—flying, running on four legs, and balancing on a few toes—are accomplishments of the vertebrate *musculoskeletal system*, the bones and muscles that together enable animals to move efficiently and gracefully.

THE WONDER OF CONTROLLED MOVEMENT

Muscle tissue shortens when stimulated and by itself can generate the force necessary to pump blood or to move food through the intestines. But to permit movement through the environment, muscular force alone is of limited use. Jellyfish, for example, contract vigorously, but they are hardly high-speed swimmers. An evolutionary step in the right direction was taken by annelid worms, which—though they lack rigid body parts—use hydraulic principles to operate **hydrostatic skeletons** that enable them to crawl and burrow efficiently (Fig. 45.1).

But for animals to run, they must have rigid body parts to push against the ground. To fly or swim, they must apply force against air or water. Although muscle can supply the *force* to those body parts, only a skeleton can supply the necessary *support*. One highly successful support system is the external skeleton, or **exoskeleton,** of arthropods

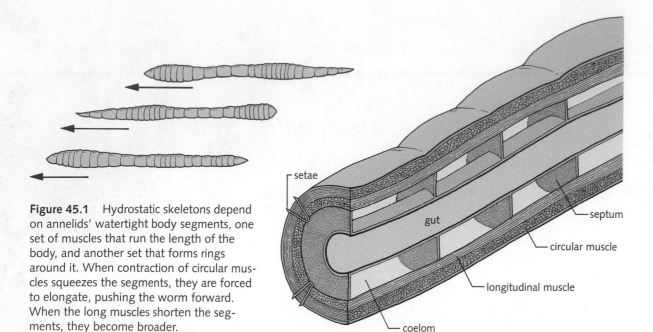

Figure 45.1 Hydrostatic skeletons depend on annelids' watertight body segments, one set of muscles that run the length of the body, and another set that forms rings around it. When contraction of circular muscles squeezes the segments, they are forced to elongate, pushing the worm forward. When the long muscles shorten the segments, they become broader.

Labels in figure: setae, gut, septum, circular muscle, longitudinal muscle, coelom

Table 45.1 *Relative Advantages of Exoskeletons and Endoskeletons*

Advantages	Disadvantages
Exoskeleton	
Provides good support for small animals	Too heavy to support large animals
Readily adaptable into wings, flippers, or claws	Must be shed during growth
Allows wide variety of movements	
Protects soft tissues from damage and desiccation	
If broken, can be replaced at next molt	
Endoskeleton	
Protects brain and some internal organs	Exposes soft tissues to mechanical damage and desiccation
Allows for steady growth	If broken, cannot be replaced and must be repaired
Provides good support per unit weight; can carry large animals	

(Fig. 45.2; see also Chapter 28). The alternative solution in vertebrates is the **endoskeleton,** a system of internal bones and joints (Fig. 45.2). Both of these systems of bones, joints, and muscles are called musculoskeletal systems, and both strategies have advantages and disadvantages (Table 45.1).

If rigid skeletons were all fashioned from one piece, of course, they would be useless; both endoskeletons and exoskeletons are composed of individual hard parts connected by **joints** that allow them to move relative to one another. Muscles are attached to bones by strong **tendons.** Many joints are actually held together both by ligaments attached to bones on either side of the joint and by muscles and tendons that stretch across the joints and act to stabilize them. As you can see in Fig. 45.2, bones and joints act as a mechanical system of levers and hinges that translate muscle contraction into body movement.

Because muscle tissue can generate force only by *shortening,* it can *relax,* but it cannot forcibly extend itself. In order to move body parts back and forth, therefore, muscles in both insects and vertebrates are arranged in **antagonistic pairs** that pull in opposite directions across skeletal joints. Many complex joints, such as the knee and shoulder joints, are actually operated by **antagonistic groups** of muscles, rather than by single pairs. Around the human elbow joint, for example, the primary (but not the only) muscles involved are the biceps and triceps groups. The biceps, which bends the elbow, acts as a *flexor* muscle. The triceps, which straightens the arm, acts as an *extensor.*

STRUCTURE AND FUNCTION IN MUSCLE TISSUE

As we saw in Chapter 34, there are three basic types of muscle tissues in vertebrates. All three contract in essentially the same way, but under a microscope they can be distinguished from one another by the organization of their cellular parts. (These three muscle types are illustrated in Fig. 34.9.)

Skeletal Muscle

Skeletal muscle, primarily responsible for voluntary movement, is also called **striated muscle** because of the alternating light and dark bands, or striations, that are produced by the organization of molecules within its cells. Vertebrate skeletal muscles are controlled by motor neurons through the specialized synapses called neuromuscular junctions that we discussed in Chapter 43. Because these synapses are strictly excitatory, stimulation of motor neurons always leads to muscle contraction. (There is no way, in other words, that a neural command can induce skeletal muscle to relax.)

Skeletal muscle is divided into two subtypes. **Red muscle,** also called *slow-twitch* muscle, can work for long periods without fatigue. These muscles, which obtain their energy primarily through aerobic respiration, contain many mitochondria. The dark coloration of red muscle results from biochemical and structural adaptations to its need for oxygen; these tissues contain a substantial quantity of dark, oxygen-storing myoglobin (Chapter 36) and are richly supplied with blood by dense networks of capillaries. Red muscle is found throughout the human body and forms the "dark meat" in poultry drumsticks and in portions of fish such as tuna. (Chickens and turkeys are primarily ground-dwelling birds, and they normally run more often and for longer periods than they fly. For this reason, the legs of these birds contain mostly red muscle. Similarly, red muscles in tuna enable those fish to swim continuously.)

White muscle, also called *fast-twitch* muscle, contracts more rapidly and generates more force than red muscle, but it obtains its energy primarily from anaerobic glycolysis and hence fatigues rapidly. The breast muscles of chickens and turkeys, which operate the seldom-used wings of these ground-dwelling birds, are mostly white muscle.

Smooth Muscle

Smooth muscle, found primarily in and around internal organs, arteries, and the digestive tract, lacks striations and tends to form sheets, rather than the bundles that skeletal muscle forms. Smooth muscle contracts much

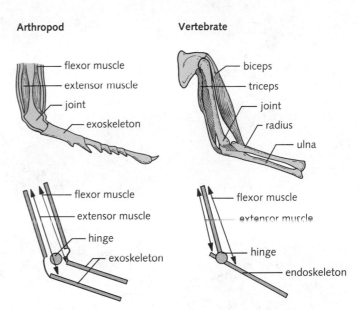

Figure 45.2 Arthropods wear their skeletons on the outside; muscles operate within its protective covering. Vertebrate bones are internal, so they are surrounded by the muscles that move them. The biceps and triceps muscles of the upper arm are antagonists; contraction of the biceps flexes the arm, whereas contraction of the triceps extends it.

more slowly than either type of skeletal muscle, but it can maintain contractions for a longer period of time. Both sympathetic and parasympathetic neurons supply smooth muscle, and because these two classes of neurons release different transmitters at their neuromuscular junctions, they have complementary effects. If sympathetic stimulation causes a particular smooth muscle to contract—as it does for most muscles of the circulatory system, for example—parasympathetic stimulation causes it to relax.

Cardiac Muscle

Cardiac muscle, discussed in detail in Chapter 35, has characteristics of both smooth and skeletal muscle. The tight electrical connections among cardiac muscle cells allow excitatory impulses to spread across the heart.

The Structure of Skeletal Muscle

We can progressively dissect a vertebrate skeletal muscle into smaller and smaller units to understand how it operates (Fig. 45.3). Each whole muscle is formed from numerous parallel bundles surrounded by connective tissue sheaths that attach the muscle to tendons at both ends. Each of those units, in turn, is composed of many still smaller parallel units called **muscle fibers.** Each mus-

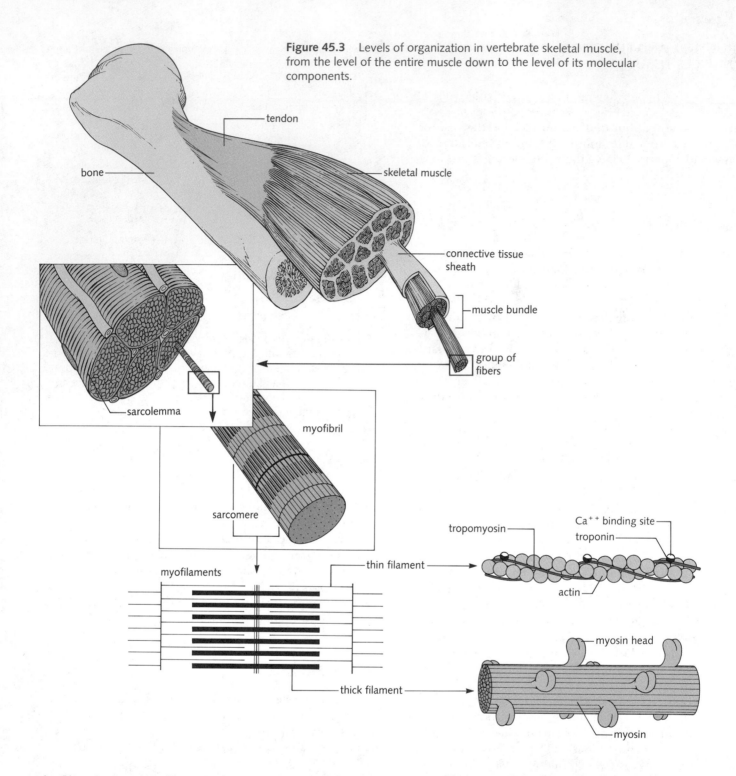

Figure 45.3 Levels of organization in vertebrate skeletal muscle, from the level of the entire muscle down to the level of its molecular components.

cle fiber is functionally a single cell, although numerous nuclei show that each was formed from the fusion of many embryonic cells. (Some of these fused-cell fibers are extremely long; single muscle cells in the long muscles of the leg may be half a meter in length.) Thus the terms "muscle cell" and "muscle fiber" are used interchangeably.

Inside the muscle cell membrane (called the **sarcolemma**) is a still smaller nested set of units called **myofibrils,** each of which is made up of two types of

myofilaments. *Thin filaments* are twisted strands of three types of proteins: **actin, troponin,** and **tropomyosin.** *Thick filaments* are made from a protein called **myosin.** Many myosin molecules, each of which looks somewhat like a matchstick with its head bent sideways, line up in a staggered parallel array to form each thick filament.

Thick and thin filaments are organized into units called **sarcomeres** ("muscle parts"), whose overlapping banded structures create the striations characteristic of skeletal muscle that are visible in micrographs (Fig. 45.4).

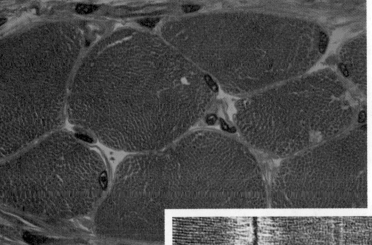

(a)

Figure 45.4 The organization of skeletal muscle, shown by light and electron micrographs and schematically. **(a)** This light micrograph shows a *cross section* of several muscle fibers, including the connective tissue that surrounds each bundle and the myofibrils that comprise them. (Magnification factor: 1000) **(b)** This electron micrograph shows a *longitudinal section* of striated muscle, revealing the banded structures that give this tissue its name. **(c)** This schematic illustration of a striated muscle in longitudinal section relates the principal molecular components of the sarcomere—actin and myosin—to the structures visible in the electron micrograph above it. **(d)** Schematic cross sections through (LEFT TO RIGHT) I band, A band, and M line, showing the relative positions of actin and myosin filaments at various positions in the schematic sarcomere.

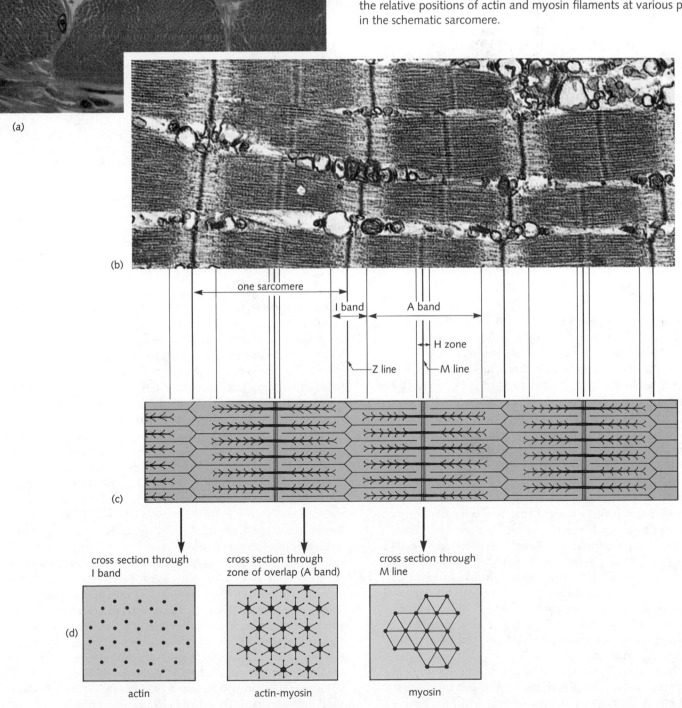

(b)

one sarcomere

I band   A band

H zone

Z line   M line

(c)

cross section through
I band

cross section through
zone of overlap (A band)

cross section through
M line

(d)

actin

actin-myosin

myosin

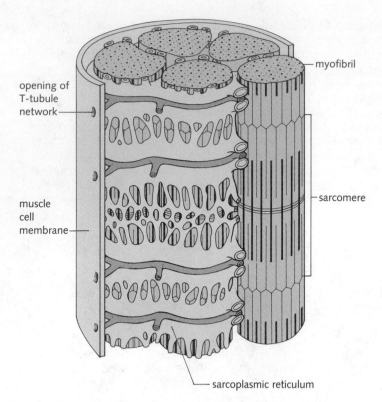

Figure 45.5 The three-dimensional structure of a muscle cell.

Sarcomeres are bounded on each end by fibrous structures called *Z lines* to which numerous thin filaments are attached. The Z line and the light area immediately on either side of it consist exclusively of actin filaments and form the *I band*. In the middle of the sarcomere, the myosin filaments that form the *A band* are linked to one another at the *M line*. The center of the A band, which contains only myosin filaments, is called the H band. Thick and thin filaments overlap extensively, the area of overlap becoming the darkest, most dense region of the sarcomere.

When the sarcomere is at rest, the heads of the myosin molecules project toward the thin actin filaments, but most do not touch them. Although there are *active sites* on the thin filaments to which the myosin heads would readily attach, most of those sites are normally blocked by troponin and tropomyosin molecules. In this position, the myosin heads are bound to ADP and phosphate derived from previously hydrolyzed ATP. The energy released in that process has "set" the myosin head like a loaded spring now poised to release that stored energy during contraction.

To understand how muscle contraction works, we must back up a bit and place the sarcomere in context within the muscle cell. As shown in Fig. 45.5, the sarco-

meres run down the length of the myofibrils within the muscle cell. All myofibrils are surrounded by two networks of tubules, the **sarcoplasmic reticulum** and the **T-tubule** network. The sarcoplasmic reticulum is a closed network of tubes and storage sacs within the cell that collects and stores calcium ions when the muscle is at rest. The T-tubules are extensions of the cell membrane that penetrate into the cell and run between the reservoirs of the sarcoplasmic reticulum. The cell membrane itself is an excitable membrane that generates and conducts action potentials just as nerve cells do.

Contraction in Individual Muscle Cells

The stimulus Muscle contraction begins when a nerve impulse reaches the neuromuscular junction. There, enough acetylcholine is released to depolarize the muscle cell membrane and fire off an action potential. That action potential races along the cell membrane and dives down into the cell along the T-tubule network, where it causes the sarcoplasmic reticulum to become "leaky" to calcium ions, which flood out into the myofibrils.

The force generators: Sliding filaments The process that occurs next, according to the **sliding filament model,** is shown in Fig. 45.6. The calcium ions released by the sarcoplasmic reticulum bind in large numbers to the troponin–tropomyosin complex on the thin filaments. This binding changes the shape of those molecules sufficiently to uncover the active sites on the actin molecules themselves. The heads of the myosin molecules bind rapidly to the closest open active sites, forming what are called **cross bridges** between each myosin filament and the actin filaments around it. As soon as they have formed, these cross bridges release their stored energy by bending towards the center of the sarcomere and pulling the actin molecules along with them. As each myosin head completes this motion, it releases the ADP and phosphate to which it had been attached but remains attached to the thin filament.

If there are any ATP molecules in the vicinity (and there usually are in healthy muscle), the head quickly binds one, hydrolyzes it, releases the active site, and resets to its original "spring-loaded" position. As long as enough calcium ions remain bound to the troponin–tropomyosin complex, this process repeats, as rapidly as five times each second. As the 350-odd heads on each myosin filament tug in this fashion, the actin filaments from both sides of the sarcomere are pulled towards the center, shortening the sarcomere. As all the sarcomeres in the entire fiber contract at once, the whole fiber shortens.

Contraction continues for as long as the muscle is stimulated, or until it fatigues (see below). When no further impulses stimulate the cell, the sarcoplasmic reticu-

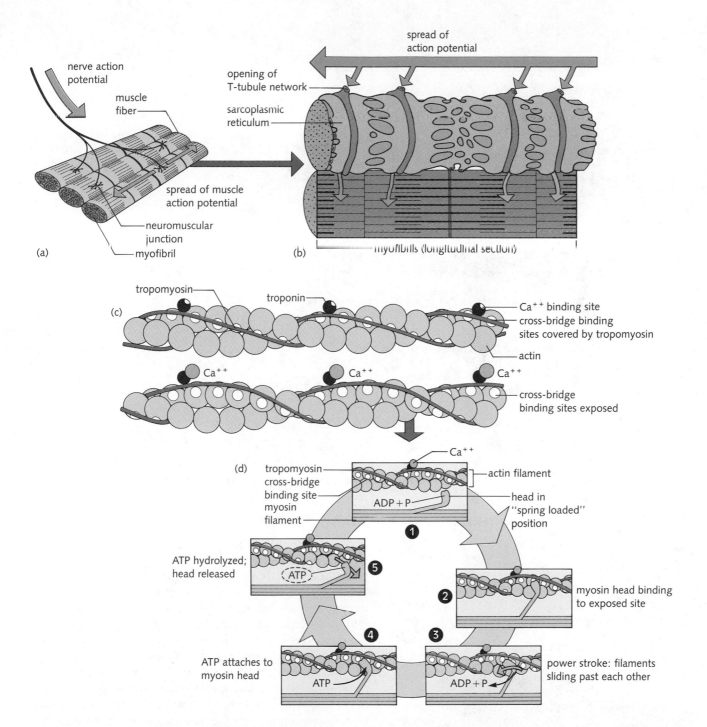

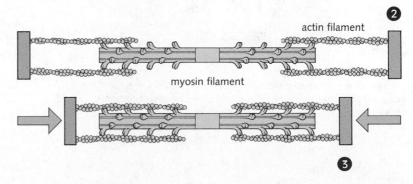

Figure 45.6 Events in muscle contraction. **(a)** An action potential arrives at the neuromuscular junction. **(b)** The muscle cell action potential enters the interior of the muscle fiber via the T-tubule network. **(c)** Ca^{++} ions attach to the troponin–tropomyosin complex, exposing active sites on the actin filaments. **(d)** The molecular events during muscle contraction according to the sliding filament model. (1) Calcium ions flooding into the sarcomere expose the myosin binding sites on the actin filaments. (2) Myosin heads attach to those binding sites and change shape, (3) pulling the ends of the sarcomere closer together. (4) ATP combines with the myosin heads and (5) causes them to release and reset.

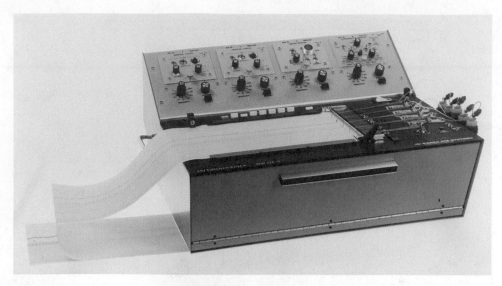

Figure 45.7 A physiograph converts the contraction of an isolated, intact muscle into movements of a magnetically driven pen that leaves a trace on a long chart of moving paper. An electrical stimulator is hooked up to both the muscle and to a second pen on the physiograph. Whenever the stimulator excites the muscle, this second pen leaves a trace on the drum at the exact moment the stimulus is applied.

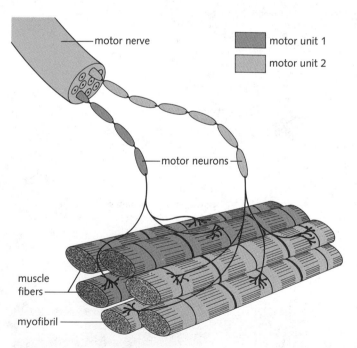

Figure 45.8 Muscle innervation and motor units. Each motor neuron, together with the several muscle fibers it innervates, can function independently, and is called a motor unit.

lum resorbs calcium ions by active transport, the troponin and tropomyosin settle back over the active sites on actin, and the sarcomere rests (lengthens) once again.

Contractions of whole muscles Each muscle cell responds in an all-or-nothing fashion to an action potential. When the fiber is stimulated, it contracts. There is no large or small contraction, because the muscle action potential is always the same. But entire muscles don't work that way; we can hammer nails and stroke a baby with the same muscles because we can control the force our muscles generate. The nervous system exercises that control by adjusting both the *rate* at which individual muscle cells are stimulated and the *number of cells* called upon to contract. Those adjustments are possible because of the way motor neurons are connected to muscle cells and because of the way those cells respond to repeated stimulation. To explain these phenomena, we use records of intact, isolated muscle activity produced by a device called a **physiograph** (Fig. 45.7).

Motor units Each motor neuron branches as it reaches its target muscle and forms synapses on numerous muscle fibers (cells). A single motor neuron and the collection of muscle fibers it serves form a unified, functioning element called a **motor unit** (Fig. 45.8). All the cells in a motor unit contract simultaneously whenever that unit is stimulated, but because action potentials do not spread between cells in skeletal muscle, each motor unit can be controlled individually.

Each motor nerve contains many motor neurons, so different patterns of activity among those neurons can stimulate the target muscle to contract to varying degrees. If only a few motor neurons fire, for example, only those muscle cells connected to them are stimulated, and the resulting contraction represents only a fraction of the muscle's total potential strength. As more neurons fire, more fibers are called into action, and the overall contraction of the muscle increases in intensity.

We can see this by conducting an experiment that directly stimulates a muscle with electrical pulses, as shown in Fig. 45.9. If the stimulus is small enough, nothing happens, but we can slowly increase the strength of the stimulus until we activate a few muscle fibers near the stimulating electrodes. When those few fibers fire, they contract and generate the first small contraction, or **muscle twitch,** shown in Fig. 45.9. This simulates what happens when only a few motor units are stimulated. If we increase the stimulus intensity still further, we fire more fibers, and the muscle contracts more strongly. The stronger we make the stimulus, the more fibers fire, and the stronger the contraction, up to a certain maximum.

Muscles that perform very different tasks in the body differ both in size and in the "wiring" of their motor units. In some of the body's smallest muscles—such as those that control the precision movements of the eye—each motor unit contains only a few muscle cells. This allows for extremely accurate control of muscle activity. In other, much larger and more powerful muscles—such as those of the thigh—each motor unit consists of many muscle cells. Though this wiring pattern results in coarser control, it is an efficient way to serve the hundreds of thousands of fibers in large muscles.

Single twitch, summation, and tetanus If we speed the movement of the paper chart on our apparatus, we can see clearly the components of a single twitch (Fig. 45.10). Each twitch has a *latency period,* a *contraction time,* and a *relaxation time.* Fast-twitch muscle and slow-twitch muscle respond differently to stimulation (Fig. 45.10).

So far in our demonstrations, we have always waited until the end of the muscle's relaxation time before delivering the next stimulus. Under these conditions, identical stimuli produce identical twitches (Fig. 45.11). If we deliver identical stimuli rapidly enough to overlap the relaxation period, however, the muscle responds by contracting more strongly, as shown in Fig. 45.11. This increase in response to closely spaced stimuli is called **summation.** By stimulating the muscle more and more rapidly (represented by more closely spaced stimuli in the figure), we can increase the amount of summation until all the responses fuse into the steady, powerful contraction called **tetanus.**

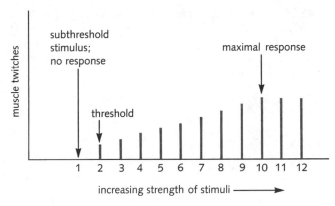

Figure 45.9 These traces show the response of an isolated, intact muscle to stimuli of progressively higher intensity. Note that the muscle twitch reaches a certain maximum, beyond which increasing stimulus intensity has no further effect.

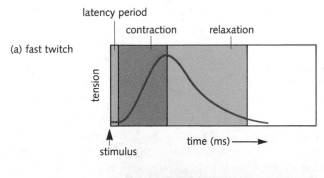

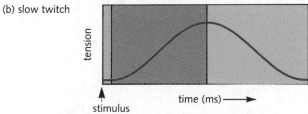

Figure 45.10 These traces of single twitches in fast **(a)** and slow **(b)** muscle show the differences in their response speed. The latency period is the time between stimulation and the beginning of the twitch. The relaxation time is the time it takes the muscle to return to its resting length after contraction.

Figure 45.11 Summation in muscle preparations. **(a)** Identical, widely spaced stimuli produce identical twitches. **(b)** When the same stimuli are spaced so that the second occurs during the relaxation period of the first twitch, the second twitch generates more force. **(c)** When many stimuli are closely spaced, the resulting twitches fuse and increase in intensity, producing a response called a tetanus.

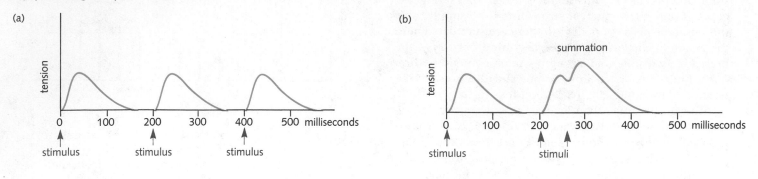

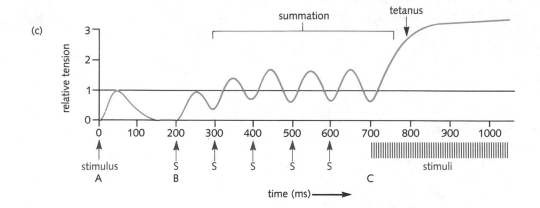

Thus the central nervous system can adjust the strength and extent of muscle contraction both by controlling the *number of motor units* firing at any given time and by controlling *the rate at which each motor unit is stimulated*. Even in muscles that are not actively involved in exercise, a certain proportion of motor units are always stimulated. This basal level of stimulation produces a resting tension called **muscle tone** that helps maintain body posture and muscular readiness for action.

Muscle Physiology: The Power Behind the Force

Powering muscle contractions over extended periods requires more energy than can be stored effectively in muscle in the form of ATP. Rather, muscles' ATP stores are replenished from several sources during and after contraction. Most of the quickly available energy in vertebrate muscle is stored in **creatine phosphate,** a compound that, like ATP, contains a high-energy phosphate bond. Both the energy and the phosphate can be rapidly transferred to ADP during muscle contraction, maintaining the concentration of ATP during brief periods of muscle activity.

If contractions continue for very long, however, the supply of creatine phosphate runs low, and the muscle tissue must obtain energy from other sources. Red muscle, as we noted earlier, relies mostly on oxidative phosphorylation of glucose and fatty acids. This is a very energy-efficient pathway, as you will recall from Chapter 18, but it requires substantial amounts of oxygen and is relatively slow. (It takes 2 to 3 minutes for a muscle to "gear up" to produce maximal power aerobically. ATP production from anaerobic glycolysis, on the other hand, can reach maximum power in less than 5 seconds.) Accordingly, if the muscle is worked beyond a certain point, it switches pathways and forms ATP without the use of oxygen through glycolysis.

Fatigue But muscle power has its limits. If we stimulate our muscle to tetanus for a long time, at some point the muscle response begins to fall off (Fig. 45.12). This failure to sustain contraction is called **muscle fatigue.** The full story of why muscles in intact animals fatigue is complex and is still not completely understood, despite decades of intensive research.

It was once believed that muscles fatigued when they ran out of ATP, but it is not so simple. When muscle *really* runs out of ATP, the myosin cross bridges cannot disconnect from actin, and the muscle is bound in a rigid state called **rigor.** The only time this usually happens is after death, when all the body's muscles stiffen in **rigor mortis.**

Most muscles, when overworked, fail to contract long before ATP is completely exhausted. This failure is at least partially due to a buildup of lactic acid, the end product of anaerobic glycolysis. Lactic acid buildup in muscle tissue lowers pH (increases the H^+ concentration) within the muscle tissue, producing the short-term muscle pain familiar to anyone who exercises strenuously. There is also evidence that the increase in H^+ ions inhibits two enzymes in the glycolytic pathway and interferes with the activation of thin filaments by calcium ions. At the same time, repeated or constant contractions may cause a buildup of potassium ions around the muscle cells, altering the membrane potential and decreasing the sensitivity of the neuromuscular junction.

This process explains these symptoms during and just after intense exercise, but it cannot account for fatigue experienced after brief workouts. Nor can it explain the weakness, stiffness, and muscle aches experienced long after unusually heavy or prolonged activity. It appears that muscular strength and endurance are related not only to structures and biochemical events within muscles themselves but also to the state of the blood supply to individual muscles, to the overall cardiovascular fitness of the individual, and to both long-term and short-term intake of energy through food in various forms. We will return to these issues when we discuss endurance and the effects of exercise.

THE SKELETAL SYSTEM: LEVERS AND HINGES

The skeletal systems of vertebrates are composed of bones and joints that form moveable and adaptable backbones, jaws, limbs, and even the grasping tails of some primates. In physical terms, bones act as levers that apply the force generated by skeletal muscles to other bones and to the environment. Our leg bones, for example, apply force to the ground, enabling us to stand, walk, or run, whereas the bones in birds' forelimbs apply the

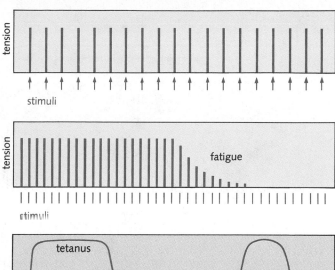
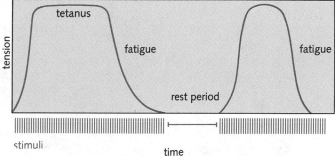

Figure 45.12 When muscles are stimulated to tetanus for prolonged periods, response falls off as muscle fatigue sets in.

power of their breast muscles to the air, enabling them to fly. So closely linked are skeletal elements, the muscles that power them, and the functions they perform that paleontologists can often reconstruct both the muscles of extinct animals and their habits by studying the areas of fossil bones to which muscles and tendons were once attached.

In vertebrates, the skeleton is divided into axial and appendicular portions.

Axial Skeleton

The **axial skeleton** (Fig. 45.13) is composed of the central supporting elements, including the skull, the vertebral column (backbone), the ribs, and the sternum (breastbone). Most of the bones that form the skull are joined by immobile joints called *sutures,* although a few, such as the bones of the lower jaw, do move. The skull rests on the upper two bones of the vertebral column, the *atlas* and the *axis,* which allow the head to nod up and down and to swivel from side to side. The bones of the vertebral

Exercise and Skeletal Muscles

Adult muscle cells cannot divide, but they can change in length and diameter under the influence of exercise. In some way not yet understood, repeated stimulation of muscles under heavy load close to tetanus alters—possibly through tearing—the fine structure of myofibrils. This process, which is thought to play a role in delayed muscle soreness, stimulates regenerating muscle fibers to increase in diameter, producing the muscle growth (hypertrophy) seen in athletes. Disuse of muscles, on the other hand, leads to a decrease in fiber diameter (atrophy).

Increased endurance under training involves both physical and chemical changes within muscle fibers, changes in the gross structure of muscles, and changes in the cardiovascular system. Endurance training not only improves the energy reserves of muscles but also minimizes fatigue by stimulating growth in the muscles' local blood supply, by improving the muscles' ability to extract oxygen from blood, and by increasing both heart volume and lung efficiency.

It is well known that weight lifters and sprinters rely primarily on fast-twitch fibers for quick power, and marathon runners depend primarily on slow-twitch fibers for endurance. Fiber proportions vary, however, even among different classes of runners. Olympic-class marathoners' leg muscles are 80–90 percent slow-twitch fibers; sprinters' leg muscles can be up to 70 percent fast-twitch.

Marathoners reap two benefits from their predominantly slow-twitch muscles. On the one hand, they gain the greater long-term muscle power of predominantly slow-twitch fibers. But they may also enjoy greater freedom from muscle pain, because it is the anaerobic, fast-twitch fibers that generally overload with lactic acid.

Genetic factors strongly influence the relative numbers of fast- and slow-twitch muscle fibers an individual develops early in life. It is not clear, however, whether exercises chosen to prepare for different sports can convert one existing fiber type into another or whether world-class athletes stumble (or are coached) into the events best suited to their genetically determined muscle composition.

In an effort to clarify this situation, researchers in Stockholm documented changes in the muscles of an extraordinary 46-year-old man who ran more than 3500 km in only 7 weeks. With the athlete's permission, researchers removed and examined small samples of muscle tissue before and after his feat. These samples confirmed that his muscles were composed mostly of the slow-twitch fibers at the beginning of the run. But during the 2-month run, his

muscles changed dramatically; fiber shapes were altered, and he developed even higher percentages of slow-twitch muscle.

These changes apparently did not occur because of the sort of fiber damage seen in weight lifting but because of insufficient blood supply during the run. However, the researchers could still not say for certain whether preexisting fast-twitch fibers had converted to slow-twitch, or whether those fast-twitch fibers had simply died from lack of oxygen, to be replaced by regenerating slow-twitch fibers.

Many athletes prepare nutritionally for marathons and other endurance events by spending several days preceding the event depleting their muscle carbohydrate stores and then following that with the intake of massive meals of high-carbohydrate foods. This strategy is known as *carbohydrate loading*. Although there are no clear-cut, quantitative data on the actual benefits athletes derive from this procedure, it is based on chemical events understood at the molecular level.

Recall that the energy for prolonged muscular work is largely provided by glycogen (a form of carbohydrate) stored in muscle tissue. On the average, human muscles store about 1.5 grams of glycogen per 100 grams of muscle. It has been shown, however, that muscle glycogen stores can be increased to as much as 4 or 5 grams per 100 grams of muscle through dietary control. Stored glycogen appears to be increased the most by eating only protein and fat for several days (depleting carbohydrate) and then gorging on carbohydrates for a day or two prior to the athletic event. Because creatine phosphate levels are not affected by this procedure, however, it probably does not improve performance in short-term activities such as sprinting.

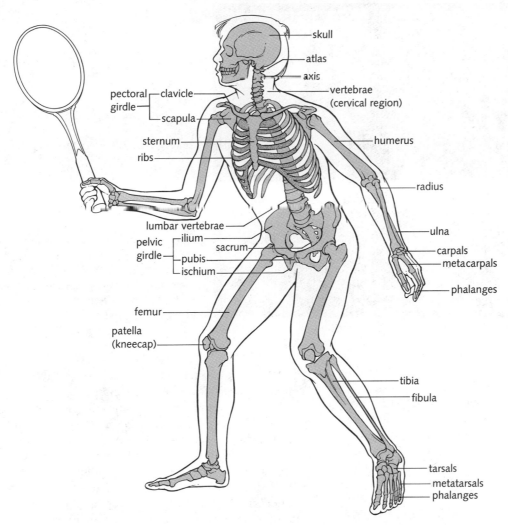

Figure 45.13 The human skeleton.

column are firmly but flexibly connected to each other by ligaments and cushioned by pads called intervertebral disks.

Appendicular Skeleton

The **appendicular skeleton** consists of the arm and leg bones, together with the **pectoral** and **pelvic** girdles that attach them to the axial skeleton. The pectoral girdle, composed of the paired **clavicles** ("collarbones") and **scapulae** (shoulder blades), are held together and to the head of the **humerus** (upper arm bone) by tendons and ligaments to form the shoulder joint. The clavicles are firmly attached to the breastbone in front; the scapulae are hung in back by ligaments, tendons, and muscles.

The bones of the forearm, the **radius** and **ulna,** are arranged such that the radius swivels at the elbow and the ulna swivels at the wrist. This allows us to turn our arms and position our hands. The wrist bones and hand bones (the **carpals** and **metacarpals**) and finger bones (**phalanges**) form the versatile human hand.

The pelvic girdle (hip bone) is created by the fusion of paired bones, the **ilium,** the **ischium,** and the **pubis.** This assembly, in turn, is tied firmly to the **sacrum** at the base of the vertebral column by a network of ligaments. The largest bone in the body, the **femur** (thigh bone), is connected to the **tibia** (shin bone) at the knee joint, one of the most complex and important joints. The much smaller **fibula** runs alongside the tibia. The ankle and foot bones (**tarsals** and **metatarsals**) are in turn attached to the small **phalanges** of the toes.

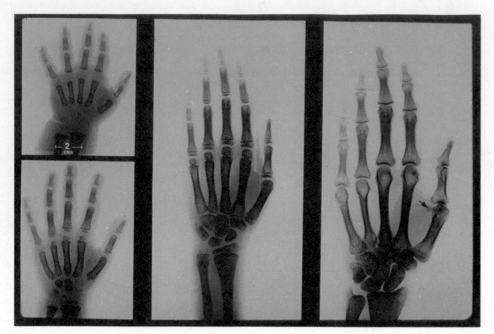

Figure 45.14 The first X-ray in this series shows just how much of the skeleton of a two-year-old boy is still cartilage. Note how the lighter cartilage is replaced by bone at each stage from 2 years, to 3 years, 14 years, and 60 years.

Cartilage

Cartilage is a dense, fibrous connective tissue (Chapter 34) that absorbs shocks and provides support for body parts that don't carry much weight. The skeleton of most vertebrates is predominantly cartilage at birth, but bone gradually replaces the cartilage as the animal grows (Fig. 45.14). During growth, cartilage becomes restricted to areas called *epiphysial plates* at either end of the bones. In adults, epiphysial plates disappear, and cartilage remains only in the nose, ears, larynx, trachea, where it serves as support, and around and inside joints, where it cushions the impact of movement and aids in lubricating bone surfaces as they slide past one another.

Bone

Bone is a complex tissue in which living cells and long, twisted collagen fibers are supported by crystals of *hydroxyapatite,* a mineral formed from calcium, phosphate, and water. Like steel bars imbedded in reinforced concrete, this conglomerate of flexible and rigid elements gives bone a remarkable combination of strength, rigidity, and resistance to impact that exceeds the characteristics of any of its components alone. Compact bone can take nearly as much stress as cast iron, though bone weighs only one-third as much. (The breaking stress of bone is 15.5 metric tons per square inch, compared with 18 metric tons per square inch for cast iron.) This strength is necessary because during many normal activities, bones and joints are subjected to much more stress than you might expect. Sprinting, for example, subjects leg bones to forces equal to nearly five times the runner's weight.

Most bones contain several distinctly different types of bone tissue, as shown in a longitudinal section of a young, long bone from the arm or leg (Fig. 45.15a). The outer bone shaft is composed of dense **compact bone,** and most of the interior is made up of less dense **spongy bone.** At either end, where the bone forms part of a moveable joint, a network of **trabeculae** is arranged like the supporting elements in a bridge. Trabeculae transmit stress applied to the bone ends down onto the compact bone along the shaft (Fig. 45.15b). In the center of long bones is a soft tissue called **bone marrow,** the source of the cells that ultimately give rise to both red blood cells and white blood cells of all types (Chapter 35).

The structure of compact bone Mature compact bone consists of many roughly cylindrical units called **Haversian systems.** At the center of each is a tube called the *Haversian canal,* which contains a bundle of blood vessels and nerves and is surrounded by concentric rings of bone tissue. Within those rings, spaces or *lacunae* house living bone cells (Fig. 45.15c).

Bone growth and remodeling Although we tend to think of bones as inert (like fingernails and hair), bone is a dynamic, living tissue that grows, remodels, and repairs itself. Active skeletal growth begins during embryonic life and continues through roughly age 25 years. This growth is under the control of three different factors: the action

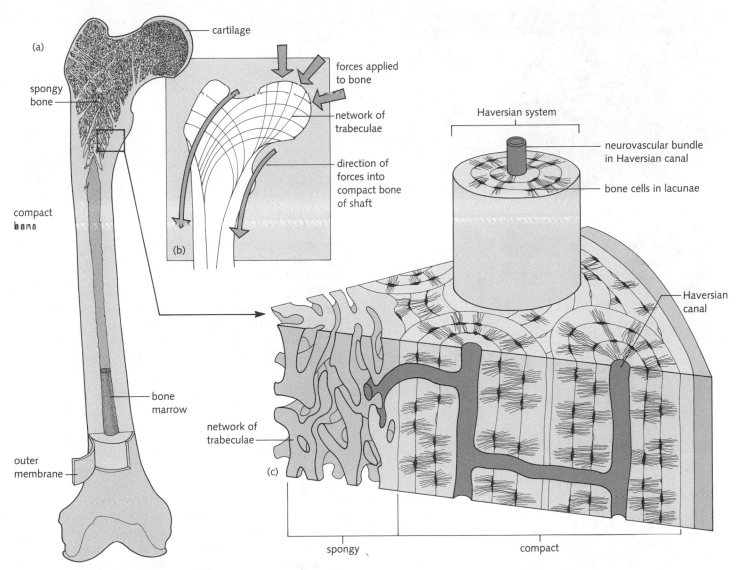

Figure 45.15 Anatomy of a human long bone. **(a)** In this cross section of long bone, notice the relative positions of compact bone, spongy bone, and marrow. **(b)** The trabeculae that comprise the spongy bone at the top of the femur channel the combination of vertical and sideways forces on the hip joint into mostly vertical force onto the shaft. This allows the compact bone to absorb forces that would snap it if applied laterally. **(c)** This schematic of bone fine structure shows the location of living bone cells and blood vessels within the Haversian systems of compact bone.

of growth hormone during early life, the combination of stresses that exercise and gravity apply to the skeleton on a daily basis, and the levels of circulating calcium in the bloodstream, which are determined by diet and general body physiology.

During embryological development, nearly the entire skeleton of most vertebrates is present in a sort of "scale model" made out of cartilage. Each part of that model grows, but as new cartilage is added, the older cartilage is degraded and replaced by bone (Fig. 45.16).

The rigid bone matrix is first laid down by bone cells called **osteoblasts** near the bone surface. Some of these cells become imbedded in the bone and mature into **osteocytes** that continue to live within the bone matrix. These cells together control the formation of compact and spongy bone that replaces the original cartilage

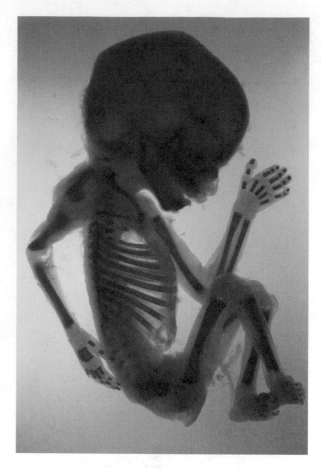

Figure 45.16 Early centers of bone growth in the cartilage of a human fetus about 3 months old.

model, beginning in the center of the bone shaft and continuing towards the ends.

Near each end of the bone, the cartilage is organized into the epiphyseal plates, which separate the growing bone shaft from the very end of the bone. In the epiphyseal plates, new cartilage is continually produced and steadily replaced by new bone tissue. At the same time, the addition of bone by cells around the outside of the bone enables it to grow in girth. As bone growth nears completion, cartilage production in the epiphyseal plates slows down and finally stops, and the cartilage is completely replaced by bone, except within the joints.

Bone growth, however, is not a one-way process. Cells called **osteoclasts,** derived from macrophage cells in the bloodstream, can literally tunnel through bone by dissolving hydroxyapatite and releasing calcium and phosphate into the blood. Osteoblasts, osteocytes, and osteo-clasts establish a dynamic balance of bone formation and destruction.

Day-to-day activity and the force of gravity produce physical stresses essential to maintaining normal bone structure, although the mechanisms that control this response are not understood. Bone, particularly in young children, can change markedly in response to changes in forces applied to them, so any unusual, long-term changes in stress on growing bones can dramatically change their final shape. (That's why you were told to sit up straight when you were a child; chronically bad posture can result in permanent deformation of the vertebral column.) When stress on bones is removed altogether, as it is during weightlessness in space travel, osteoclasts begin to dissolve bone minerals. Even in young astronauts at peak physical condition, prolonged weightlessness triggers bone resorption, though the balance is rapidly restored upon the individual's return to earth. The ability of bone cells to remodel the matrix around them also allows broken bones to heal.

Additionally, as we noted in Chapter 39, bone serves as a major body reservoir for calcium. Should calcium levels fall, parathyroid hormone stimulates osteoclasts to dissolve bone; when calcium levels rise, calcium is redeposited by osteoblasts via the action of the hormone calcitonin.

Changes in the balance between bone production and destruction generally cause bone mass to decline slowly after age 20 to 30. In some older individuals, particularly postmenopausal women, mineral loss can so weaken bones that they break under minimal stress. This syndrome, called **osteoporosis,** has several physiological causes, but it is also related to decreased activity level. In numerous clinics around the country, regular, supervised, moderate exercise in aging men and women has been shown to slow bone loss significantly, probably by applying stress to the skeleton. The most beneficial exercises seem to be those involving weight bearing against gravity, such as walking, and not those where body weight is supported, like swimming.

Joint Structure and Function

Bones meet at **joints,** whose structure determines the nature and extent of possible movement. Joints are just as important to movement as muscle and bone are; if you have ever injured one of your joints, you know that we usually take their remarkable structures and functions for granted. There are three major classes of joints: *fibrous* or "fixed" joints, *cartilaginous* or "slightly moveable" joints, and *synovial* or "freely moveable" joints.

The most important type of fibrous joint is the *suture,* an immobile joint found only among the bones of the skull. At birth, while the skull is still partly cartilage, the sutures between bones are slightly flexible. (This property enables the fetus's head to flex as it squeezes through the birth canal.) During the 18 months following birth, however, the skull solidifies and the edges of the bones become as irregular as pieces of a jigsaw puzzle, interlocking with one another, ensuring stability. After bone growth stops, the membrane lining the suture is replaced by bone.

Cartilaginous joints are common throughout the axial skeleton, and they range from the tight, scarcely moveable joints between ribs and sternum to the more flexible joints between the vertebrae in the spinal column. The ribs, though firmly attached to the vertebrae and sternum by ligaments and cartilage, can move slightly during breathing, and they can be pried apart by surgeons who must gain access to the chest cavity. The cartilaginous joints of the spinal column, designed for support and resilience rather than extensive movement, can be damaged if bent too far under pressure. Lifting heavy weights with the back bent, for example, can damage the cartilaginous intervertebral disks, causing a painful condition known as a herniated disk.

Synovial joints are remarkable both for their strength and for the freedom and efficiency of movement they allow. Synovial joints consist of a combination of tendons and ligaments that form a fibrous *joint capsule,* which helps hold the bones together. Inside the capsule, bone surfaces are lined with resilient cartilage and "oiled" by an extraordinarily effective natural lubricant called *synovial fluid.*

Synovial joints vary enormously in structure. The shoulder joint, whose capsule is reinforced by tendons of the many muscles that run across it, allows exceptional range and freedom of movement. The knee joint, one of the most complicated and critical joints in the body, not only allows a great range of motion but also absorbs powerful shocks during such exercises as jumping and running (Fig. 45.17).

Flattened sacs called *bursae,* also filled with synovial fluid, often occur to ease friction in places where muscle, tendons, or ligaments rub against each other or against bones.

Joints, particularly synovial joints, are subject to several kinds of injuries and diseases. **Bursitis,** an inflammation of bursae, and **arthritis,** a more general set of problems, are common, both in young athletes who push themselves too far in training and in elderly individuals who undertake too little exercise. In both of these cases, moderate exercise can often eliminate discomfort while maintaining joint flexibility. More serious types of arthritis result from genetic defects, acquired disease (such as untreated gonorrhea or Lyme disease), or malfunctions of the immune system.

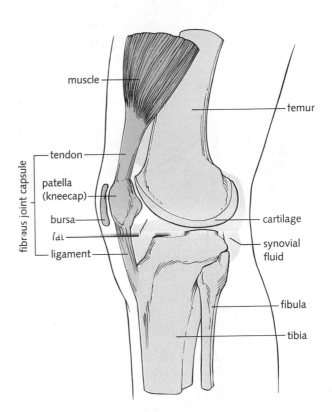

Figure 45.17　The knee joint, which absorbs forces several times body weight during running, is reinforced both internally and externally by ligaments and tendons, and is cushioned by a fluid-filled sac and pads of resilient fat.

MUSCLES AND BONES TOGETHER: A DYNAMIC SYSTEM

Neural control of movement is a complex phenomenon that is only partially understood. As you have already learned, conscious movement is initiated in the motor areas of the cerebral cortex, whereas coordination involves the cerebellum and relay areas in the brainstem (Fig. 45.18). Thus, although what we call *voluntary* movement is initiated consciously, the carrying out of that movement is accomplished only through a great deal of unconscious, or involuntary, central nervous system activity.

Once movements are started, feedback from stretch receptors in muscle, and from other receptors in tendons, joints, skin, and inner ears, provides constantly changing information about body position, acceleration, and balance. Some of this information we are conscious of; other parts of it are monitored subconsciously. When we learn any task involving movement, we are modifying neural pathways in a manner similar to the more abstract

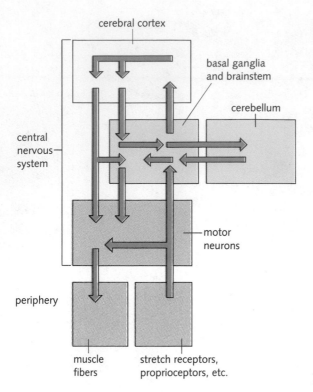

cerebral cortex

basal ganglia
and brainstem

cerebellum

central
nervous
system

motor
neurons

periphery

muscle
fibers

stretch receptors,
proprioceptors, etc.

Figure 45.18 Neural pathways involved in body movements. Note, even in this highly simplified schematic, that coordinated movement requires a constant exchange of information among muscles, cortex, cerebellum, and other relay centers within the brain.

processes of learning. Whether the task involved is walking, writing, driving a car, or skiing, what were once conscious activities become more and more automatic—and hence more efficient.

Coordination of Muscular Activity

The task of coordinating muscular activity is made somewhat easier by the fact that many complex motor patterns are "hard-wired" into the spinal cord and operate without any input from the brain. Try the following demonstration. Have a friend hold his or her arm at a right angle, with all the muscles of the upper arm tensed rigidly in place (Fig. 45.19a). Note that both biceps and triceps

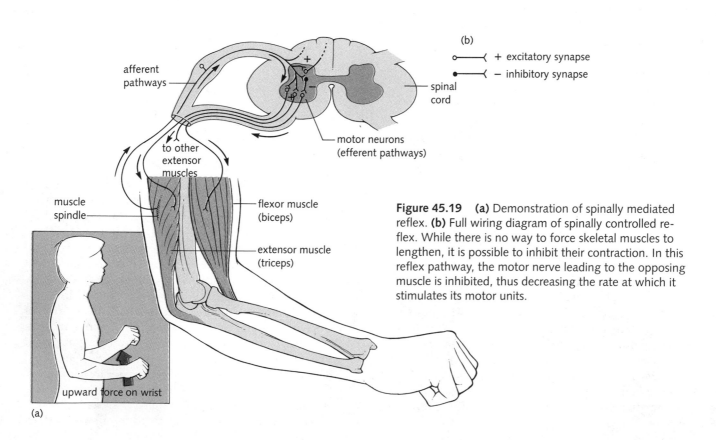

afferent
pathways

(b)

○—⊏ + excitatory synapse
●—⊏ − inhibitory synapse

spinal
cord

motor neurons
(efferent pathways)

to other
extensor
muscles

flexor muscle
(biceps)

extensor muscle
(triceps)

muscle
spindle

upward force on wrist

(a)

Figure 45.19 **(a)** Demonstration of spinally mediated reflex. **(b)** Full wiring diagram of spinally controlled reflex. While there is no way to force skeletal muscles to lengthen, it is possible to inhibit their contraction. In this reflex pathway, the motor nerve leading to the opposing muscle is inhibited, thus decreasing the rate at which it stimulates its motor units.

stand out and are rigid to the touch. Now place your hand at the point shown by the arrow in the figure and apply force upwards without warning. Note that your friend's biceps relaxes while the triceps contracts to hold the arm in place.

Reciprocal inhibition   The demonstration above is an example of **reciprocal inhibition,** in which stretching an extensor muscle (the triceps) inhibits the flexor muscle (the biceps). Reciprocal inhibition is controlled by the same sort of circuit responsible for the knee-jerk reflex we examined in Chapter 43. Recall that the sensory neurons that detect the stretch and stimulate motor neurons leading to the stretched muscle simultaneously inhibit the motor neuron leading to that muscle's antagonist (Fig. 45.19b).

This sort of circuit, which works at most of the body's joints, helps keep antagonistic muscles and muscle groups working with each other smoothly. When someone hands you an unexpectedly heavy object, for example, your arms respond to the extra load before your brain realizes what has happened. Similarly, if you trip or stub your toe on an unseen object while walking, reflex leg movements help you keep your balance.

Walking: A Complex "Simple" Activity

Watching a baby trying to toddle across a room suggests just how complicated the "simple" act of walking really is. Most of us don't remember, of course, but it took us a good year of regular practice to walk unaided on two limbs instead of four. This is because walking is controlled by scores of muscles organized into several complementary groups that stretch from our toes all the way up to the center of our bodies. The contractions of all these elements must follow one another like clockwork. Those movements eventually become second nature; we can walk—and even run—while listening to music, talking, and thinking of other things.

SUMMARY

The musculoskeletal system that enables us to move efficiently is composed of several important components. Muscles generate contractile power by hydrolyzing ATP. Bones provide rigid structural supports that apply the force generated by muscles to the environment, allowing the animal to run, swim, or fly. Joints between bones permit the skeleton to move. Tendons and ligaments control the application of muscular force within the body

and, together with the shapes of joint surfaces, determine the range of possible movements.

Muscles, like nerves, are composed of excitable cells that conduct action potentials. According to the sliding filament theory of muscle contraction, action potentials in the muscle cell membrane enter the cell interior through a network of T-tubules. There they stimulate the release of calcium from storage sites in the sarcoplasmic reticulum. Through a series of events, this calcium causes actin and myosin filaments to slide past one another, shortening the muscle. Both the structure and the physiology of muscles can be altered through exercise.

Bone, though largely composed of crystalline calcium salts, is an active and dynamic tissue. Living bone cells can lay down or resorb bone throughout most of an individual's life, reshaping the skeleton in response to applied stress. Bone also serves as an important reservoir of calcium for the body.

STUDY FOCUS

After studying this chapter, you should be able to:

- Explain the structure and function of muscle, and describe the process of skeletal muscle contraction.

- Describe the major components of the skeletal system, and explain how bone is a dynamic living tissue.

- Explain the interrelationship between muscles and bones.

TERMS AND CONCEPTS

endoskeleton *948*	tetanus *955*
sarcolemma *950*	creatine phosphate *956*
actin *950*	rigor mortis *957*
troponin *950*	axial skeleton *957*
tropomyosin *950*	appendicular
myosin *950*	skeleton *959*
sarcomere *950*	osteoblast *961*
sliding filament	osteoclast *962*
model *952*	osteoporosis *962*
motor unit *954*	arthritis *963*
summation *955*	reciprocal inhibition *965*

REVIEW

Objective Questions (Answers in Appendix)

1. The cells of skeletal muscle
   (a) contain actin and myosin.
   (b) are generally under involuntary control.

(c) are arranged in sheets.

(d) are found in internal organs.

2. Each muscle fiber is a
   (a) nucleus.   (c) cell.
   (b) sarcomere.   (d) sarcolemma.

3. As an action potential moves along a skeletal muscle cell,
   (a) phosphate ions are released by the T-tubules.
   (b) energy is stored in the creatine phosphate.
   (c) calcium ions are stored in the T-tubule network.
   (d) calcium ions are released by the sarcoplasmic reticulum.

4. The sliding filament theory refers to the movement of _____ by each other.
   (a) sarcolemmas
   (b) ATP molecules
   (c) calcium ion binding sites
   (d) actin and myosin

5. Which muscle region almost disappears when a muscle myofibril contracts?
   (a) A band   (c) H band
   (b) I band   (d) Z line

Discussion Questions

6. Why are both muscles and some sort of skeletal system required for effective movement? Can you imagine how a "spineless" person—if one existed—could move effectively on land?

7. Compare and contrast the advantages and disadvantages of endoskeletons and exoskeletons. Why do you think there will never be any spiders 10 m tall in terrestrial environments? Why can arthropods grow so much larger in aquatic habitats than on land?

8. Explain, in order, the events that are involved in muscle contraction, according to the sliding filament theory.

9. What is the phenomenon called fatigue as shown in laboratory muscle preparations? How is it different from the fatigue you feel hours after heavy exercise? What do we know about exercise-induced fatigue?

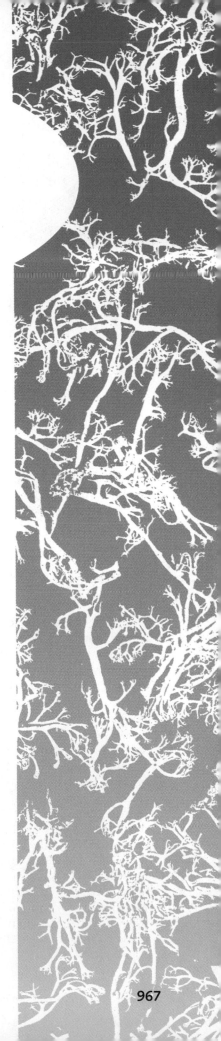

46

Animal Behavior

On the arid grasslands of Ethiopia, a male baboon guards a small group of females who have associated with him for months. The offspring he sired with those females play with one another or cling to their mothers' backs. Suddenly a foreign male appears, battles the resident male, and displaces him. The newcomer then methodically kills the nursing infants in the troop, over their mothers' vocal but ineffectual protests.

This is but one striking example of **animal behavior,** the response of animals to stimuli in their environments. As we have seen in earlier chapters, behavior is essential to animals in finding food, in selecting habitats, in mediating predator–prey and symbiotic interactions, and in creating the reproductive isolation necessary for speciation. In this chapter, we will discuss the events that underlie behavior and examine them from an evolutionary perspective.

THE BEHAVIORAL SCIENCES

Behavior ranges from simple responses of protozoans to light and chemicals to extremely complex interactions within chimpanzee and human societies. These diverse phenomena attract researchers ranging from biochemists to anthropologists—each of whom brings to the study of behavior different techniques and preconceptions.

This chapter examines behavior primarily from the perspective of the discipline called **ethology** (*ethos* means "manner or behavior"). Ethologists concentrate on the normal behavioral repertoire of animals, either in their natural environment or in laboratory settings that simulate natural conditions as closely as possible. Modern ethology was founded by Konrad Lorenz, Niko Tinbergen, and Karl von Frisch, who shared a Nobel Prize in 1973. But ethology's roots date back to Darwin, who observed that individual *variations* in behavior made behavioral evolution possible.

Terminology and Approach in Behavioral Studies

Behavioral scientists are often criticized for describing animal behavior in human terms. In most cases, these criticisms result either from a misunderstanding of the manner in which ethologists use such terms

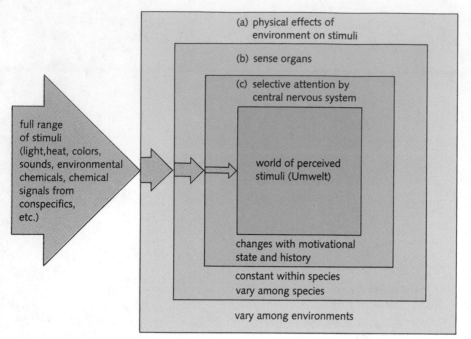

Figure 46.1 This schematic diagram illustrates the physical and physiological "filters" that allow some environmental stimuli to enter an animal's awareness while blocking the perception of others. **(a)** Certain potential stimuli pass through the environment in which the animal lives, while others cannot. Light cannot pass through soil, for example, and electricity does not travel through air. **(b)** The sense organs of each species detect certain stimuli and fail to detect others. Human eyes, for example, cannot detect ultraviolet light, but the eyes of honeybees can. **(c)** The central nervous system sifts through the reports of sense organs, paying attention to some stimuli and ignoring others, depending on the individual's motivational state.

or from a lack of appreciation of the complexities of animal behavior.

As an example of the first case, there are species of ants that invade the nests of related species and carry the helpless pupae back to their own nests. When those pupae emerge, they become workers in the colony of the invading species. Although the technical name for this behavior is "dulosis," it is usually referred to as "slave making."

Use of such language in scientific description does *not* ascribe human motives to the ant "captors." Neither does it justify slavery in humans by implying that slavery is a "natural" phenomenon. Such language is employed *strictly* as a convenient shorthand. To remind you of that point in this chapter, we will place such words in quotation marks. Similarly, when discussing the evolution of behavior, we might say "it is advantageous for the female to act this way" or "the best strategy for the female is to behave in this manner." This language does *not* mean that any individual female "knows" her behavior is advantageous; it is a convenient way of saying "females that behave in this manner will leave more offspring, so over time, this trait is favored by selection."

Most behavioral researchers scrupulously avoid ascribing human emotions to animals. This is easy when dealing with lower organisms, for no one would suggest that a *Paramecium* swims toward a light because it "knows" it will find food there. But animals such as chimpanzees often present problems. When observing a female chimpanzee whose infant has been killed, for example, some field researchers find the simplest explanation to be that the female "loved" the infant and "mourns" its death.

ELEMENTS OF BEHAVIOR

Sensory Worlds

As we noted in Chapter 44, animals are bombarded with information about their environment. Species differ in their ability to receive that information, because each has its own sense organs and information-processing abilities that enable it to attend to certain stimuli while ignoring others.

German ethologists, recognizing the importance of this phenomenon, coined the term *Umwelt* (*um* means "around"; *welt* means "world"; hence *Umwelt* means "the world around") to describe the sensory world perceived by an organism at any time. An animal's Umwelt may vary not only with its location but also with its age and *motivational state* or "mood" (whether, for example, it is hungry) (Fig. 46.1).

Researchers must consider carefully the Umwelt of their subjects when designing and interpreting experiments. Early in the study of honeybee behavior, for example, one researcher wondered whether bees had color vision. To test this hypothesis, he designed a wooden box with two circular holes. He left one hole open and blocked the other hole with a transparent piece of glass.

When he placed bees in the dark interior of the box, they immediately tried to escape by flying toward the light. Naturally, they could get out through the open hole but not through the glass.

To test whether or not the bees could discriminate between two different colors, the experimenter shone light of one color through the open hole and light of a different color through the hole blocked with glass. He hypothesized that if the bees could discriminate between the two colors, they would associate one color with freedom and learn to fly toward light of that color to make good their escape. The bees, however, never learned which color signaled the way out of the box. The investigator concluded that bees are colorblind.

Karl von Frisch, on the other hand, reasoned that bees *must* have color vision. Why else would bee-pollinated flowers be so brightly colored, and how else could bees single out from a distance individual types of flowers in fields containing many species? So he designed an experiment to test honeybee color vision in as natural a situation as possible (Fig. 46.2). Von Frisch trained bees to feed on sugar water in a glass dish placed on a colored square. He then cleaned the square thoroughly (to re-move any odor cues that might be left over) and placed it among other squares of different colors and shades of grey. Identical dishes on *all* squares held only distilled water, but the bees flew unerringly toward the color on which they had been fed earlier. Thus, when operating in "food-gathering mode" in a setting similar to that in which they normally use color vision, bees can distinguish colors. When trying to escape from a dark box, however, they simply ignore color information.

Simple Behaviors: Programmed Responses

In certain species, an individual's Umwelt is restricted to a simple sensory input at any point in time, and its responses are totally preprogrammed and inflexible. An extreme example occurs in the females of one species of tick, which require a meal of mammalian blood to lay their eggs. From the time these animals hatch until they lay their eggs and die, they pay attention to only three sensory stimuli of all those present in the world around them.

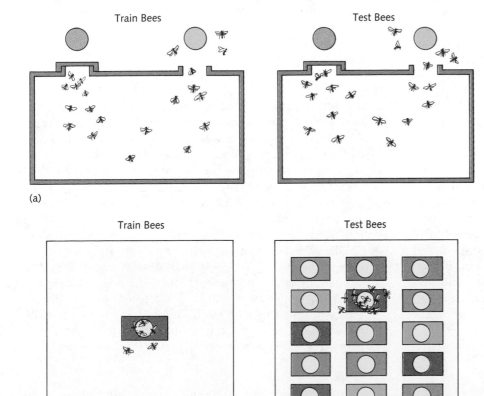

(a)

(b)

Figure 46.2 The importance of context in animal behavior experiments, shown by tests of color vision in bees. (a) One early experimenter placed bees in a dark box and attempted to teach them that following light of one color would lead to freedom, but flying towards light of another color would not. The bees failed to learn this lesson. (b) Karl von Frisch trained bees to find sugar water in a glass dish placed on a background of a particular color. He then attempted to "hide" an identical dish and colored square amidst a host of identical dishes placed on differently colored backgrounds. The bees, however, unerringly selected the color to which they had been trained.

When a female tick hatches, she responds positively to light by climbing upward on grass and shrubs. Once perched, she ignores all stimuli except one: the odor of butyric acid, a compound found in mammalian sweat. Time is apparently not relevant to the tick at this stage of her life cycle, for she may wait, immobile, up to 20 years for the scent indicating that a mammal is passing beneath her. When that stimulus comes, she lets go of the leaf and drops—if she is fortunate—onto the skin of the mammal below. She then responds blindly to heat (a sign of blood near the skin surface) and upon finding a warm, hairless spot, she burrows in. At this stage, experiments have shown that she will drink any warm liquid, regardless of taste, until she fills to capacity. She then drops to the ground, lays her eggs, and dies.

Each stimulus in the tick's sensory world (light, butyric acid, and heat) serves as what ethologists call a **sign stimulus** or **releaser;** each triggers a specific pattern of behavior. Although we usually associate the word *sign* with a visual sign, sign stimuli can occur through any sensory channel.

A neurally preprogrammed, invariant response to a sign stimulus is called a **fixed-action pattern.** Some fixed-action patterns are as simple as the responses of the female tick, while others may be quite involved. One of the best known among higher vertebrates is the egg-retrieving behavior of the greylag goose, first reported by Lorenz and Tinbergen. When a goose incubating her nest notices an egg nearby, she rises, stretches out her neck, rolls the egg neatly into the nest, and settles down upon it. Even if the egg is removed after she has begun to stretch out her neck, the animal completes the sequence, gingerly retrieving and incubating an egg that is no longer there! Once begun, this preprogrammed behavior pattern continues to completion automatically, even in the absence of further sensory cues.

Furthermore, this behavior can be triggered by objects (including beer cans and baseballs) that resemble eggs in some ways. The goose's nervous system apparently evolved to respond with retrieval behavior to convex objects with rounded edges. Because geese in nature never encounter beer cans or baseballs, there was no selective pressure to distinguish and exclude them as stimuli, and in the birds' Umwelt for this behavior, all these objects are identical. Interestingly, a separate behavioral program, which controls removal of empty eggshells and foreign objects from the nest, focuses on a different set of sensory cues. In that program, the carefully retrieved beer can is recognized as a "non-egg" and ejected from the nest.

Complex Behaviors: Inherited and Acquired

Some simple behaviors (such as those of the female ticks described above) are controlled by genetically prepro-grammed wiring in the nervous system and are therefore called **innate behaviors** or **instincts.** The existence of instinct enables animals to perform certain acts important to their survival without ever having had the opportunity to learn them.

For many years, most behavioral scientists believed a sharp distinction existed between such instinctive behaviors and acquired behaviors, which they classified as **learning.** To many, a behavior had to be *either* instinctive *or* learned. Ethologists, viewing behavior from an evolutionary perspective, emphasized the importance of instinctive behaviors to survival in nature. Behavioral psychologists, interested in mechanisms of learning under controlled conditions, believed that all complex behaviors are acquired by a nervous system that starts off at birth as a clean slate. To some, instinct seemed nearly irrelevant.

This philosophical conflict between the importance of genes (or "nature") and the significance of environmental factors and learning during development (or "nurture") raged fiercely for many years. We now know that both nature and nurture are important in shaping many complex behaviors. Genetic instructions guide the growth of the neurons, effectors, and synapses that make behavior possible, but individual experience can change both the structure and the function of cells and synapses. The behaviors that mature animals display, like many other phenotypic characteristics, depend on both genes and environment. To understand this still-controversial area, we must look closely at both instinctive behaviors and learned responses.

Instinctive behaviors   Many instinctive behaviors are relatively rigid and stereotyped in form. Even these responses, such as the fixed-action patterns of female ticks and the suckling response of newborn human infants (Fig. 46.3), may involve the action of numerous muscles over long periods of time.

Other instincts are far more complex. Both bird nests and spider webs, for example, can be surprisingly elaborate in design and construction, yet the animals begin building these structures the first time with no prior experience (Fig. 46.4). Somehow, local sensory cues guide the animals in combining many short, stereotyped actions into long behavioral routines that create a nest architecture well suited to a precise location. Over the course of construction, however, the animals do learn, and their performance becomes more skilled as time goes on.

Learning   The ability to modify behavior through experience, a process known as learning, is critical to the success of animals from honeybees to humans. Learning specialists often divide learning into several related phenomena.

Habituation (also called extinction), a simple form of learning, occurs when an animal decreases or stops its response to "insignificant" stimuli. A sea anemone presented with a small disk of paper on one of its tentacles, for example, bends that tentacle to place the disk in its mouth. The anemone finds the paper neither rewarding (it is indigestible) nor noxious (it contains no poisons or irritants). After several presentations, the anemone "learns" to ignore the paper as irrelevant.

Classical conditioning (also called associative learning) occurs when animals associate one sensory stimulus with the occurrence of another stimulus. The most famous example of classical conditioning was noted by the Russian physiologist Ivan Pavlov, who studied digestion in dogs. Pavlov first determined that dogs salivate when presented with meat or meat extract. He then discovered that if he repeatedly rang a bell just before he presented food or food odors, the dogs learned to salivate when they heard the bell alone. The dogs thus associated the sound of the bell with the presentation of food.

Operant conditioning (which is also called trial-and-error learning) occurs when an animal learns to perform a particular operation to receive a reward (**positive reinforcement**) or to avoid some painful experience such as an electric shock (**negative reinforcement**). The classic examples are the rats used in learning experiments performed by psychologist B. F. Skinner. Skinner trained rats to press a lever in their boxes, in response to sound

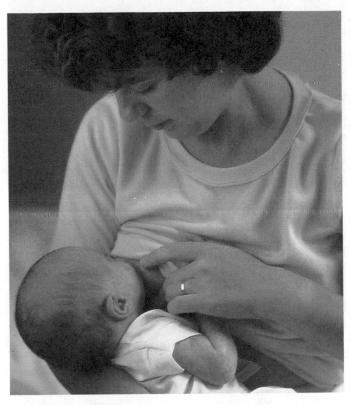

Figure 46.3   Presented with a breast, a newborn human infant will instinctively begin suckling.

Figure 46.4   The construction of a weaverbird nest involves a series of individual knots and weaving patterns determined by the shape of the specific nesting site. This complex instinctive behavior can be performed with no prior experience yet becomes more accomplished with practice.

Figure 46.5 A chimpanzee uses a plant stem as a tool to probe into a termite nest for food.

Figure 46.6 Geese, imprinted on Konrad Lorenz at birth, follow him across a field just as they would normally follow their mother.

or light stimuli, either to receive a food pellet or to avoid an electric shock. Pigeons can be similarly trained to peck at various objects or colored lights in order to receive food.

Insight learning occurs when an animal applies prior experience in a completely new situation without trial and error. Neither positive nor negative reinforcement is involved in the original learning process; rather, the animal "thinks" about what to do and then acts. Insight learning is most pronounced in our own species and has been documented in other primates, but it is rare in the rest of the animal kingdom. Chimpanzees and orangutans, for example, spontaneously use objects as tools in novel situations to obtain food or to play tricks on one another (Fig. 46.5). It is not unusual for chimpanzees to pile up objects to reach a food item hanging out of reach above their heads, even though they have never encountered that situation before.

Instinctively Guided Learning

It is now clear that instinct and learning, rather than operating independently, often work together, for learning is often shaped or guided by instinct. The fact that important components of complex behaviors can be inherited allows those behaviors to evolve over time. The flexibility of acquired behaviors, on the other hand, enables individual organisms to adjust to specific conditions they encounter.

Imprinting is learning in which an animal "customizes" an inherited behavior on the basis of environmental information it encounters, usually during a brief period early in life. This information is often learned without the benefit of any immediate reward or punishment. Imprinting was first described by Lorenz in experiments with newly hatched geese that begin to waddle behind their mothers a day or so after hatching. This act of following an adult indelibly imprints the image of the parent on the young birds' minds. Thereafter, they follow wherever their parents go.

Goslings will, in fact, trundle after any object that presents sign stimuli even remotely resembling the movement and sound of a goose. Lorenz discovered, for example, that by mimicking goose parenting calls, he could imprint on himself goslings that had been hatched away from their mothers. The result (in addition to numerous important ethological papers) was a series of endearing photographs of the ethologist followed by his "brood" (Fig. 46.6).

It has since been determined that many birds are most sensitive to imprinting within a fairly narrow period of time after hatching. Birds not imprinted by the end of that period may never imprint.

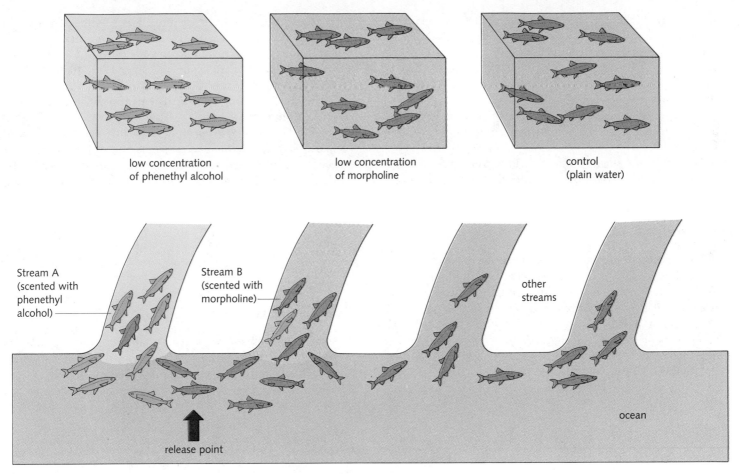

low concentration
of phenethyl alcohol

low concentration
of morpholine

control
(plain water)

Stream A
(scented with
phenethyl
alcohol)

Stream B
(scented with
morpholine)

other
streams

ocean

release point

Figure 46.7 Experiments performed by Arthur Hasler on salmon in Lake Michigan. Young fish were raised in holding tanks where 1/3 were exposed to very low concentrations of the chemical morpholine, 1/3 were exposed to phenethyl alcohol, and controls were not exposed to any artificial odor. Fish were released into the sea from a spot on the shore between two streams. At spawning season, one of these streams was "flavored" with morpholine, the other with the alcohol. The vast majority of fish "homed" to the streams flavored with the compound on which they had imprinted.

Though imprinting usually occurs early in life, the stored information is often used again much later. While goslings imprint on adults as objects to follow, for example, they are also learning to identify members of their species as future mates. When Lorenz's human-imprinted birds matured, they repeatedly attempted to breed with him. Though the birds had continuously associated with others of their species, when the time came to choose mates, their brains called up the imprinted image of Lorenz!

Migration and homing Imprinted information is also used by animals that migrate over long distances. Salmon, for example, hatch in freshwater streams but migrate hundreds of miles out to sea where they feed and grow

for a year or more. When the fish mature, they return to breed in the stream of their birth. Among the processes that facilitate this remarkable feat is an imprint of their home stream odor that they carry throughout their lives. By searching out this olfactory stimulus, each fish locates the precise stream in which it was born. An experiment demonstrating the role of olfaction in salmon homing is shown in Fig. 46.7.

Genetic Influences on Song Learning in Birds

Woods and forests are alive with bird song in early spring, as male birds of many species court females and stake out their nesting and feeding territories. The songs of each species are distinct and identifiable; white-crowned sparrows, chaffinches, red-winged blackbirds, and scores of other birds sing species-specific songs. But those songs vary subtly from place to place and from individual to individual within each species. Because many birds use their songs to identify themselves not only as members of their species but also as members of local populations and as individuals, each song must somehow encode that information.

But how can birds create individualized songs without losing the critical patterns that identify their species? Sophisticated experiments over the years have shown that in many birds there is an intricate interaction between innate mental processes and individual experience in song learning.

Normally, young birds hear the songs of their parents and other members of their species as hatchlings. Later, as the males mature and begin to sing, they produce what is called *subsong*—a rambling, babbling song that crystallizes over a period of weeks into a fully formed adult song. This adult song, though it is specific to the species, often displays both regional variations, or *dialects*, and individual variations from male to male.

Hearing adult song is important to song learning in many species; white-crowned sparrows reared in isolation sing when they mature, but they produce only stilted, simplified versions of their species' song (Fig. 46.8). If, however, experimenters play tapes of recorded adult songs to young birds during a critical learning period, the birds learn from those tapes.

But if songs are learned, how do birds in the wild learn only their own species' songs and not those of other birds they also hear while young? The answer seems to be that the birds' brains contain some sort of "song-learning program" that accepts input only from its own species' song. Given a choice, birds favor songs of their own species as models. Exposed only to the "wrong" song, they fail to learn at all.

Behavioral Genetics

Innate influences on learning and complex behaviors that appear in the absence of experience provide solid evidence of a strong genetic component to behavior. But

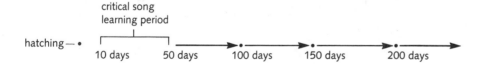

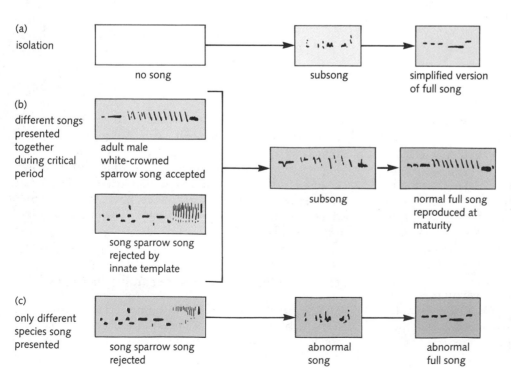

Figure 46.8 **(a)** Exposed to no song at all, birds produce subsong but develop only a rudimentary version of their species' normal song. **(b)** Exposed to tapes of both their own species' song and that of the related song sparrow, they produce more complex subsong and a fully developed song characteristic of their own species. **(c)** Exposed only to the other species' song, they fail to learn.

the only way to *prove* that behaviors are genetically influenced is to show that behaviors can be inherited in the same ways as other phenotypic characteristics, such as flower color in Mendel's experiments with peas. Ideally, one could observe the behavioral phenotype of hybrids and cross them back to parental types to see how the traits assort. Ultimately, these kinds of experiments could show not only *which* behaviors are shaped by genes but also *how* that influence is produced.

Genes and mating behavior The activities by which individuals mate and rear their young are of utmost importance to the survival of species and are therefore good candidates for genetic influence. We have already discussed one important example of such behavior: species-specific song, which is used to attract and identify mates.

In an effort to study the genetics behind mating songs in crickets, one group of researchers performed laboratory crosses of two cricket species that produce different calls. The resulting hybrids produced calls intermediate between those of their parents: strong evidence of genetic control (Fig. 46.9). Hoping that only a few genes were involved in controlling song production, the researchers made backcrosses to the parent species. Those F_2 hybrids, however, were intermediate between the F_1 hybrids and the parents. Further experiments strongly indicated that, unlike the trait of color in peas, this behavior is subject to the control of several genes.

Genes and learning The complexity of neural function and the number of interacting genes involved in even the simplest innate behaviors make it difficult to conceive of how individual genes, or even small groups of genes, can have powerful effects on complex behaviors. Genes, after all, code for structural proteins and enzymes that affect behavior, not for the behaviors themselves. Some experiments in behavioral genetics, however, remind us that certain gene products can indeed affect learning and memory.

Wild-type fruit flies, for example, can be trained to avoid unpleasant stimuli, but two single-locus (single-allele) mutants, called *dunce* and *amnesiac*, cannot learn so efficiently. Homozygous dunce mutants simply do not learn. Homozygous amnesiac mutants, although they appear to *learn* normally, *forget* much more quickly than wild-type flies. Molecular studies have shown that the dunce allele codes for an enzyme in the cyclic AMP pathway, which—as you may recall from the discussion of second messengers in Chapter 43—is involved in the sorts of changes in synaptic activity that accompany learning.

Extrapolating these results to mammalian behavior, however, is both difficult and dangerous, as shown by a series of experiments involving maze-running ability in rats. The experimenter ran many rats through a particular

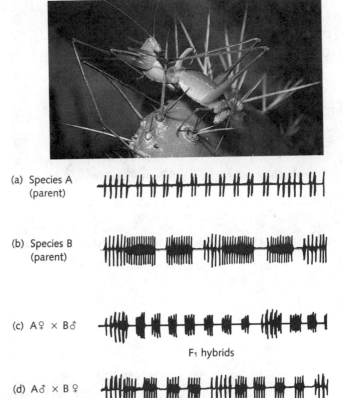

(a) Species A (parent)

(b) Species B (parent)

(c) A♀ × B♂

F_1 hybrids

(d) A♂ × B♀

Figure 46.9 The dynamics of cricket mating (TOP) involve complex mating songs. **(a)** Males of species A begin each mating song with a five-pulse chirp and follow with roughly nine, two-pulse chirps. **(b)** Males of species B begin their mating songs with a six-pulse chirp and follow with one or two chirps of about eleven pulses each. **(c,d)** Hybrids are intermediate in number of pulses/chirp and total chirp number.

type of maze and recorded the number of errors the animals made in learning the correct route. He then selected the animals with the very best scores and those with the very worst scores, inbred them within each group, and continued this artificial selection for several generations. He had no problem demonstrating that one group of progeny did far better in that maze than the other.

This researcher called his groups "bright" and "dull," implying that he had bred lines of smart and stupid rats. But when other researchers used the same rats in a different kind of maze, the scores of the two groups were indistinguishable. Though the original experiments had clearly manipulated genetic influences on behavior, they had manipulated only some influence on ability to migrate through a single sort of maze, not "intelligence" overall.

Figure 46.10 Elephants care for one another and protect their young, as shown in this defensive formation of adults and calves.

Figure 46.11 Deceit in animal communication. TOP: A leaf katydid conceals itself from predatory birds. BOTTOM: Visual camouflage is used adroitly by a scorpion fish to blend in with its surroundings.

SOCIAL BEHAVIOR

Animals that reproduce sexually cannot live alone. They are forced to link destinies for at least part of their lives with one or more of their fellows to form social units ranging from mated pairs to highly complex societies. Social life has significant costs, but as the existence of animal social systems proves, benefits often outweigh those costs. Mated pairs of animals, for example, can often gather food, defend territories, protect one another and their young from predators, and raise offspring far more successfully than a single animal working alone (Fig. 46.10). And the most complex animal societies, such as those of ants, bees, termites, and other types of *social insects,* can accomplish tasks that far exceed the abilities of any single individual.

Communication

The coordination of activity between individuals requires **communication,** the passing from one animal to another of information that influences subsequent behaviors of the recipient of the information. In *social communication* within a species, information exchange is often two-way and facilitates cooperation and coordination between breeding males and females or parents and offspring, for example, or among members of a larger society. Communication may also occur between animals of different species; in such cases, the purpose of communication is not always to transmit accurate information (Fig. 46.11).

Intraspecific communication Communication within species often revolves around a variety of **social releasers,** sign stimuli that trigger socially related behaviors in mates or members of the signaler's social group. As is the case with all behaviors, responses to social releasers depend heavily on the individual's motivational state at the time the message is received. Social releasers can take the form of signals presented to any of the senses; virtually all sensory modalities—from vision to electroreception—are used in the intraspecific communication systems of one species or another. In general, however, the most important social signals are presented to the senses of vision, hearing, and olfaction. In this general discussion, we will use examples taken from the world of chemical communication, but the concepts discussed apply to all types of social signals.

Olfactory communication is extremely important in many species and is often mediated by **pheromones,** chemicals released by one animal that influence the behavior of other individuals of the same species. Some animals produce only one pheromone, but others—particularly social insects such as ants, bees, and ter-

Figure 46.12 Each of these numerous glands produces a specific pheromone used in the intraspecific communication of this ant species.

Figure 46.13 A queen bee is constantly surrounded by workers who feed her, groom her, and provision the eggs she lays within the honeycomb. The queen, marked here by a researcher for purposes of observation, constantly excretes pheromones that are spread among the workers through mutual grooming.

mites—are practically walking pheromone factories (Fig. 46.12).

Releaser pheromones trigger immediate behavioral responses in the individuals that receive them. Sexually mature female moths of several species, for example, release a *sex pheromone* that attracts males from miles away. Ants, on the other hand, use many releaser pheromones to coordinate the activities of their colony. A worker ant that finds food outside the nest, for example, returns home dragging her abdomen and leaving a trace of *trail pheromone* behind her. Her nestmates follow that trail from the nest back to the food. Ants and bees also have *alarm pheromones* that, released in times of distress, signal threats to the colony.

Primer pheromones act like externally applied hormones by initiating long-term physiological and behavioral changes in the animal that detects them. Among honeybees, for example, the queen bee releases a primer pheromone called *queen substance* to control her workers (Fig. 46.13). Although the entire population of the hive is female for most of the year, only the queen can lay eggs because queen substance suppresses the worker bees' ovaries. Primer pheromones are common in mammals as well. Female mice raised in complete isolation from males, for example, often fail to enter estrus. But when the odor of male mouse urine is introduced into their cages, a pheromone it contains causes the female to begin reproductive cycling immediately.

Courtship behavior In many species, courtship involves an intricate exchange of successive signals called a **stimulus–response chain,** in which each behavior by one partner serves as a stimulus that releases the next behavior in the other partner. Often, stimuli received by the animals' nervous systems interact through the hypothalamic–pituitary axis (Chapter 40) to synchronize mating with hormonal cycles that control reproductive organs (Fig. 46.14).

Male–male competition and female choice In the half-light of dawn on an American prairie, scores of male sage grouse fight fiercely with one another on a communal mating ground, jockeying for the preferred position in the center of the arena. Physical competition ensures that the coveted center spot will be held by the largest, strongest, and most experienced male. This raucous display, though amusing to humans, is irresistible to female grouse, who wander through the melee eyeing the competitors. The females bypass the younger, weaker males at the periphery and head straight for the center, where one after another mates with the top-ranking male (Fig. 46.15).

This display is a prime example of two widespread behaviors that together create a phenomenon known as **sexual selection:** *male–male competition* for females and *female choice* in mating. The situation arises because reproductively, females of most species represent a "limiting resource" in the ecological sense.

Because sperm take so little energy to produce, males can produce millions of them easily. Females, on the other hand, expend far more energy to produce far fewer eggs. This difference in *parental investment* in offspring can have profound evolutionary consequences.

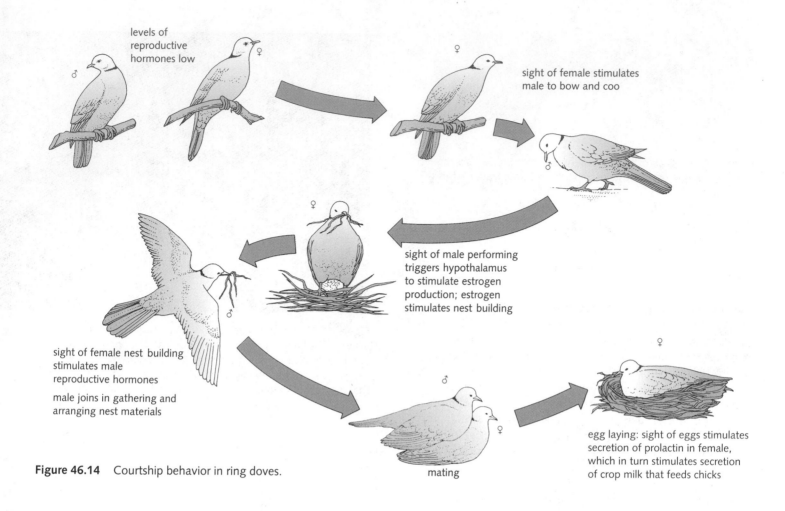

Figure 46.14 Courtship behavior in ring doves.

levels of reproductive hormones low

sight of female stimulates male to bow and coo

sight of male performing triggers hypothalamus to stimulate estrogen production; estrogen stimulates nest building

sight of female nest building stimulates male reproductive hormones

male joins in gathering and arranging nest materials

mating

egg laying: sight of eggs stimulates secretion of prolactin in female, which in turn stimulates secretion of crop milk that feeds chicks

Figure 46.15 Bird mating rituals can be both graceful, as shown by these courting Japanese red-crowned cranes (LEFT), and fierce, as displayed on this sage grouse lekking ground (RIGHT). These males compete with each other for the attention of females.

Figure 46.16 The results of sexual selection. LEFT: Male peacocks carry large, gaudy tails, used only in mating behavior. RIGHT: Red deer carry antlers used in fierce combat over feeding territory.

In situations where females by themselves can successfully raise offspring, males can maximize their reproductive success (and Darwinian fitness) by mating as frequently as possible. Females' reproductive success, on the other hand, is limited by the number of eggs they can produce and the number of offspring they can raise. For that reason, the evolutionarily most advantageous strategy for females is to select the "best" male available.

The evolutionary consequence of this basic situation (which has many variations depending on the characteristics of individual species) is selective pressure on males to advertise their sexual readiness, to compete with one another for mates, and to mate with any female who comes along. As a corollary of this competition, males often carry either sexual ornaments to attract the attention of females or weapons to battle competitors (Fig. 46.16).

Females, on the other hand, tend to be "choosy" about selecting their mates—hence the concept of *female choice*. Some, such as grouse females, allow competition among males in an arena to pinpoint the most powerful, aggressive individuals. Others, such as peacock females, select their mates on the basis of individual displays. Still others, such as the females of many grazing animals, do not choose the male himself but, rather, the territory he has "won" in competition with other males. Wandering about to feed, females spend the most time in the most fertile and hospitable territory and thus mate most frequently with the males guarding those territories.

Exceptions to this "promiscuous" mating behavior occur when neither male nor female alone can provide sufficient care to ensure the survival of offspring. Sexual selection still operates in such cases; males advertise and court, while females make the ultimate mating decision. But because no young survive unless both parents are involved in raising them, behavioral evolution has channeled these species into more permanent relationships. Among many bird species, for example, hatchlings must be fed prodigious amounts of food that only *both* parents can provide. In such cases, males and females mate—often for life—to form **monogamous** pairs.

Among many mammals, on the other hand, the young must be fed milk, which only females can provide, but they often require at least the part-time protection of males. In such cases, several females often choose to form long-term associations with individual males, a situation called **polygyny.**

It is important to realize that these relationships between caring for offspring and mating systems are powerful but not absolute; mating systems ranging from monogamy and polygamy to strictly casual reproductive encounters are found in many vertebrate groups from fishes all the way through primates. Some birds, such as red-winged blackbirds, are polygynous. By the same token, monogamy—though rare among mammals—is by no means unheard of. Lest the exceptions obscure the rule, however, it is worth noting that nearly 93 percent of bird species that have been studied are monogamous.

In contrast, only 10 percent of mammalian species are monogamous.

Language

Ever since Darwin, most biologists have assumed that animal communication is little more than a collection of simple social releasers and fixed-action patterns used to express simple mental states such as fear or sexual arousal. Philosophers and linguists alike built a wall between human-style communication and that of other species, asserting that *Homo sapiens* alone uses abstract symbols (words) to represent objects in communicating to its fellows.

Although there are important differences between human language and all other forms of animal communication (no other species seems to come close to ours in grammatical complexity and vocabulary), a host of experiments over the years has given ethologists great insight into and respect for the complexity and subtlety of animal communication.

The dance language of the bees   One of the most famous and best-studied examples of animal communica-tion is the "dance language" of the bees—a combination of tactile, olfactory, and visual communication elements that enables scout bees to direct their nestmates to food sources they have discovered. This information is encoded in a manner so complex and so unusual that von Frisch published his description of the behavior only after 20 years of experiments.

To decode bees' language, von Frisch first trained various scout bees to feed on dilute sugar-water solution at different locations around the hive. (When quality of food is low, scouts do not recruit other nestmates to it.) He then marked these bees with identification tags and enriched the food source sufficiently that the scouts began to recruit their nestmates.

Von Frisch determined that the recruitment process is simple when food is close to the hive. Returning scouts perform a maneuver called the *round dance*, vibrating and circling on the honeycomb within the hive (Fig. 46.17). This action, combined with the scent of food carried on the scouts' bodies, directs nestmates to fly out and search around for food nearby.

But for distant food sources (and single scouts can search areas over 25 square miles), encouraging nestmates to search around randomly would be very inefficient. When scout bees locate rich food sources farther

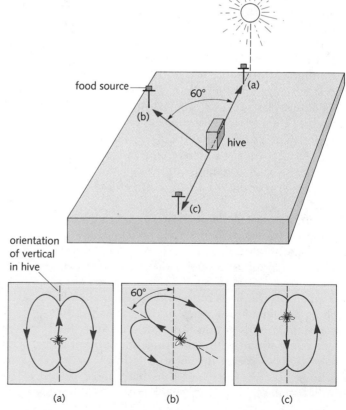

Figure 46.17   The dance language of honeybees. LEFT: The round dance. RIGHT: The waggle dance contains specific information on the location of the food source. In this imaginary depiction of von Frisch's experiments, different groups of scouts have been trained to feed at locations A, B, and C. The relationship between the food source and the sun for each location is translated into waggle dances on the honeycomb.

than about 80 m from their hive, therefore, they return and perform a different dance, the *waggle dance,* on the honeycomb (Fig. 46.17). This waggle dance conveys to the scout's nestmates both the approximate distance and the direction from the hive to the food source.

Distance from the hive is coded in the number of waggles in the dance and in the total length of time each dance circuit takes. The longer the dance, the longer the distance to the food source.

Direction from the hive to the food source is coded by the angle between the waggling part of the dance and an imaginary vertical line down the honeycomb. Apparently, the bee measures the angle between the sun and the food source as she leaves the hive and flies towards the food. She then duplicates that angle on the honeycomb with her dance. As the sun moves across the sky, the dancing bees change the angle they depict on the hive, thus keeping their information accurate.

Language in primates   Our closest relatives, the higher primates, use a wide variety of gestures, body postures, facial expressions, and vocalizations to communicate within and among their social groups. In the past few years, a series of elegant experiments with apes and monkeys in the wild has revealed great sophistication in these animals' use of vocalization.

Peter Marler of Rockefeller University knew that vervet monkeys (like many other animals) emit alarm calls when they spot predators, but he decided to determine whether the calls carried any more specific information (Fig. 46.18). To many people's surprise, Marler and his co-workers discovered different calls for each of the predators that threaten vervets: snakes, eagles, and leopards. Upon hearing the "snake alarm," the monkeys jump on their hind legs and retreat, searching in the grass where the snake might hide. Upon hearing the eagle alarm call, on the other hand, they look skyward, trying to spot the bird as they move towards cover.

Other researchers have studied the alarm calls of rhesus monkeys, which live in large troops. As is common in primate societies, rhesus troops develop a pecking order called a **dominance hierarchy** through aggressive interactions. That hierarchy is continually tested and reshaped by regular squabbles in which numerous individuals participate.

Whenever fights break out among troop members, the individual being attacked emits an alarm call that was long thought to be either a simple emotional outlet or a behavior intended to appease the attacker. In fact, a substantial vocabulary of alarm calls carries a great deal of specific information; the calls inform the members of the troop not only which animal is being attacked but also by whom it is being attacked. Other animals react to that information and, on the basis of their relationships to the combatants, decide whether or not to join the fray.

The Evolution of Social Behavior

Most behaviors we have discussed can be easily explained in evolutionary terms, for most of them directly enhance organisms' Darwinian fitness—their genetic contribution

Figure 46.18   Upon seeing the proverbial snake in the grass, vervet monkeys produce a call that clearly means "snake," and not just "danger" or "run." The call issued here has warned nearby individuals to stand up and search the ground for the python.

Figure 46.19 A killdeer distracts a potential predator by feigning a broken wing.

Figure 46.20 Lions, like elephants, are matriarchal. Lion prides, which may contain between 4 and 37 individuals, center on a core of several related adult females and their cubs, as shown here. Females cooperate and hunt in groups to bring down large prey. Males (after watching the kill) push in rudely to eat their fill first. Males do, however, assist in defending the group against foreign males and other intruders.

to the next generation. Certain behaviors, however, seem to fly in the face of evolutionary logic. Sounding a predator alarm, for example, benefits the social group but places the *caller* at greater risk. Alarm calls, sharing of food, and joining fights among other group members are all examples of **altruistic behaviors:** actions that are either apparently self-destructive or place an individual at risk to benefit others.

The subfield of behavior called **sociobiology,** which concerns itself with the special problems of social behavior, explains such behaviors in evolutionary terms through two related phenomena. The concept of **inclusive fitness** states that an individual can increase the amount of its genomes that survives not only by reproducing itself but also by helping *relatives* (who share some of that genome) to survive and reproduce. The theory of **kin selection** holds that an altruistic behavior can be adaptive if the extent to which it enhances an animal's inclusive fitness is greater than the loss of personal fitness the animal incurs.

Two siblings (brothers and/or sisters), for example, share on average half of their genetic makeup, whereas first cousins share one-eighth of their genome. The logic of inclusive fitness and kin selection argues that it is evolutionarily advantageous for an organism to sacrifice its life for a sibling if that sacrifice doubles the sibling's success in raising offspring. A similar sacrifice for a cousin would be advantageous for the altruist only if it allowed that cousin to rear eight times as many offspring.

There is no way to prove that kin selection is universal, but a host of observations on social behavior suggest that it exists in many species. The most extreme and widespread self-sacrificing behaviors involve parents and their offspring. When caring for their young in a nest, for example, many birds (such as nighthawks and wood ducks) decoy predators away from the nest by feigning near-fatal injury and hopping along the ground as though crippled with a broken wing. Once the predator has been led astray by the performance, the bird flies off and returns circuitously to the nest (Fig. 46.19).

Altruism involving more distant relatives is common only among animals (such as birds and mammals) whose complex behaviors and social systems enable siblings, parents, half-siblings, and cousins to keep track of one another. We now know, for example, that members of many bird flocks, lion prides, elephant herds, and monkey troops are composed of close relatives—usually parents and one or more generations of their offspring (Fig. 46.20). We also know that many animals have an uncanny ability to recognize relatives among many individuals that appear identical to the human eye.

There is a dark side to relatedness and behavior, however, for it also explains such behaviors as the killing of infants described at the beginning of this chapter. There is no evolutionary advantage in protecting the offspring of unrelated males. In such cases, a new male that gains control of a female group eliminates the offspring of his predecessor, inducing the females to enter estrus and bear *his* young. Among some primates, analysis of the situation at takeover time is remarkably sophisticated; males seem to "know" that infants born within a few weeks of their arrival do not belong to them—and they kill those infants as well.

Insect Societies

The most extreme cases of altruistic behavior occur among the social insects. For most of the year, colonies of ants, bees, and wasps are composed entirely of female workers, all of whom are daughters of a single queen and who are therefore referred to as "sisters." (The queen produces male offspring only seasonally, and these are not tolerated within the colony for long.) Workers devote their lives not to reproducing themselves but to helping their mother raise yet more sisters.

This unusual turn of events, according to sociobiological theory, has been brought about by the peculiar genetics of social insects and the effects of that genetics on inclusive fitness and kin selection. For in most hymenoptera, all fertilized eggs develop into diploid females, while unfertilized eggs develop into haploid males—a condition known as **haplodiploidy.**

Because workers in a colony share half the alleles they received from their mother but share *all* the alleles they received from their haploid father, the average number of alleles shared among sisters is 3/4 (Fig. 46.21). Haplodiploid reproduction thus creates a situation wherein a worker gains more in inclusive fitness by "helping" her mother raise sisters than by raising her own offspring. The results in terms of behavioral and social evolution are little short of extraordinary.

The insect colony If you looked into an ant or bee colony, you would see thousands of insects scurrying around, buzzing, clicking, and rubbing antennae. These apparently random meetings transmit information that connects individual insects, creating an entirely new level of biological organization—the colony. And that colony is more intelligent and adaptable than individual insects could ever be. For just as cells in multicellular organisms differentiate to serve varied functions, individuals in social insect colonies develop into strikingly divergent **castes** whose body forms and behaviors serve different purposes within the colony (Fig. 46.22).

Individual workers isolated from their sisters seem to be capable of little more than random activities. Join those workers into a colony, however, and complex, highly ordered group behaviors emerge. Tropical leaf-cutter ants, for example, don't eat the leaves they strip from

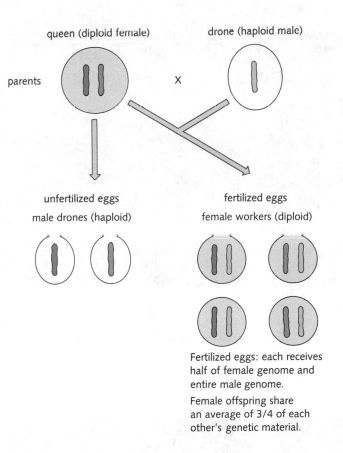

Figure 46.21 Fertilization, sex determination, and relatedness among social insects. In these insects, males (produced from unfertilized eggs) are haploid; females (from fertilized eggs) are diploid. Daughters of the same queen share approximately 3/4 of their genome with one another.

Figure 46.22 LEFT: These bloated members of a honey ant nest spend their lives as "living honey casks" that store liquid food for distribution to their sisters during lean times. RIGHT: The termite caste system includes soldiers, nymphs, and workers.

Figure 46.23 These tropical leaf-cutter ants are carrying leaves back to their nest, where they will be used as food for a fungus garden whose produce feeds the ant colony.

plants. Instead they take them back to their nest, chew them into a pulp, and use them as food to grow a colony of fungus on which they feed (Fig. 46.23). That fungus garden is tended, watered, "weeded," and harvested as meticulously as any human farmer could do it. In addition to growing fungus gardens, several species of termites build complex, highly ordered, three-dimensional nests with built-in "air conditioning" ducts whose convection currents keep the colony from overheating (Fig. 46.24).

Some researchers view the insect colony not merely as a social unit but also as a sort of "superorganism" because of its extreme degree of specialization and integration. Both ant and bee colonies, for example, appear to be able to "learn" and respond to environmental changes in ways that individual workers cannot. There are, in fact, striking parallels between the way individual insects join to form a colony and the way individual nerve cells connect to form the brains of higher animals.

Primate Societies

Some of the most fascinating studies in animal behavior—and among the most difficult to interpret—involve members of our own order, the primates (Fig. 46.25). Many of the most important observations on primates in the wild have been conducted by three primatologists sometimes nicknamed the "trimates": Jane Goodall, who works with chimpanzees in Tanzania; Birute Galdikas, who observes orangutans in Borneo; and Dian Fossey, who studied and protected the mountain gorillas of

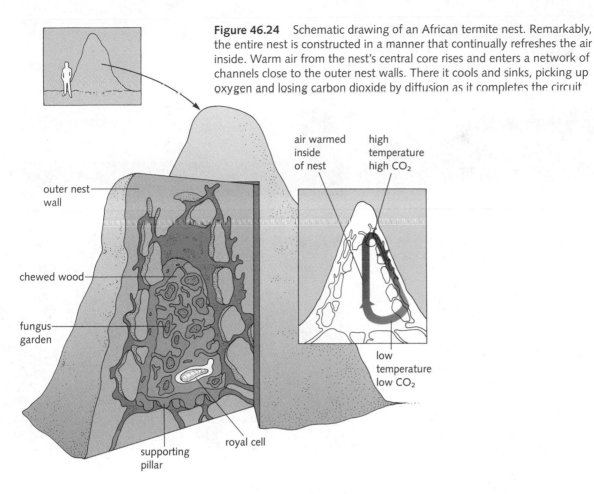

Figure 46.24 Schematic drawing of an African termite nest. Remarkably, the entire nest is constructed in a manner that continually refreshes the air inside. Warm air from the nest's central core rises and enters a network of channels close to the outer nest walls. There it cools and sinks, picking up oxygen and losing carbon dioxide by diffusion as it completes the circuit

air warmed inside of nest

high temperature high CO_2

outer nest wall

chewed wood

fungus garden

low temperature low CO_2

royal cell

supporting pillar

Figure 46.25 Research on primate societies, as primatologist Sarah Hrdy demonstrates here, requires countless hours of patient observation (often over several years) and great skill at recognizing differences among individual animals. The langurs under observation here in India exhibit a number of surprising behaviors, including infanticide.

Figure 46.26 LEFT: Two young snow monkeys at play. Many young animals, particularly mammals, engage in play for varying lengths of time. RIGHT: Grooming, seen here among yellow baboons, is a common behavior among many primates. In addition to removing bothersome parasites, grooming helps cement positive relationships among troop members.

Rwanda until her death, presumably at the hands of poachers.

These and other primate researchers have shown that many higher primates have personalities as varied and individual as those of humans. Like humans, many apes develop distinctive styles of interpersonal relations and varied strategies for survival within their troops. Some fight their way to top positions in the dominance hierarchy; others gain status and protection by making "friends" with dominant individuals or by forming coalitions within the troop. The animals cooperate in hunting, squabble among themselves, appear to exhibit such emotions as "jealousy," "affection," and "grief," and—in the case of chimpanzee troops—make war on one another (Fig. 46.26).

It is often difficult not to see primate behavior in human terms. Pygmy chimps, for example, seem to "play games." In the San Diego Zoo, they often amuse themselves by climbing down a chain that hangs into the moat surrounding their enclosure. Periodically, when the dominant male in the group is alone in the moat, an adolescent runs over to the chain and pulls it up, trapping the older male at the bottom. Several younger males then gather at the top and make play-faces (the chimpanzee equivalent of laughing) at the dominant below. This suggests to some that the pranksters can place themselves in the position of the older male sufficiently to realize that they're playing a trick on him.

HUMAN BEHAVIOR

> What a piece of work is man! how infinite in faculty! in form and moving how express and admirable! in action how like an angel! in apprehension how like a god!
> —Shakespeare, *Hamlet*

Homo sapiens is a behaviorally complex primate whose behavior concerns us a great deal. We approach the study of our own behavior from the perspectives of several disciplines: anthropology, sociology, psychology, and—of particular interest to us here—ethology and sociobiology.

We have seen that the spectrum of animal behaviors ranges from rigidly preprogrammed fixed-action patterns to flexible combinations of instinct and learning. From a biological perspective, researchers ask whether humans stand at some locatable point on this continuum or have leapt off the scale altogether. As individuals, how much of what we think, feel, and do today is programmed within our DNA?

Sociobiologists emphasize that our genes have shaped most of our evolutionary history, and they argue that despite our behavioral flexibility, it is unlikely that we have outgrown completely the influences of genotype on behavioral phenotype. Other researchers counter forcibly that the relatively recent innovations of language and culture have introduced unparalleled behavioral flexibility and variety—a "nurture" that overwhelms the influences of any genetically predetermined "human nature."

These fascinating and provocative positions have been debated fiercely for centuries. Unfortunately, the history of studies into genetic influences on the human mind includes a great deal of bad science, distorted by the sorts of prejudice we discussed in Chapters 1 and 2. For it is particularly difficult for anyone to investigate links between genes and brain activity in humans without falling prey to their own preconceptions.

Central to this issue is the question of how single genes or groups of genes could possibly influence behavior patterns as complex as those found in humans. It is one thing to find a single gene that affects learning in a fruit fly or genes that affect mating calls in crickets. It is entirely another to postulate (and extremely difficult—some would say impossible—to *prove*) that genes—which can code directly only for structural or regulatory proteins and not for behavior—control such difficult-to-define human traits as "intelligence," "aggressiveness," and "dominance."

For a short while during the 1980s, it seemed that identifiable single mutations could powerfully influence complex behaviors. A decades-long study of the Amish had yielded remarkably strong evidence for genetic predisposition to manic depression in that tightly inbred, socially cohesive group. Other studies published soon afterwards purported to have found either single genes (or closely linked genes) or genetic markers linked to behaviors as diverse as alcoholism, childhood depression, and schizophrenia.

For better and for worse, however, those studies began to unravel soon after they were published. By the beginning of January 1993, all the claims mentioned above had been retracted. Part of the problem seems to have arisen from oversimplified diagnosis of complex, multifaceted psychological conditions; several individuals in the original study were subsequently rediagnosed and placed in different categories. In other cases, when information on additional individuals was obtained from expanded family pedigrees, the new and expanded data set no longer supported the original hypothesis.

By late 1992, researchers in another study had linked two different forms of early-onset, familial Alzheimer's disease (FA) to two different regions of the genome. One form of FA, which runs in families and can appear in individuals as young as 45 years old, was linked to markers on chromosome 14; and the other was linked to markers on chromosome 21. Neither gene had yet been isolated and cloned. Hopefully, identification of those genes will ultimately provide clues to the more common form of Alzheimer's that affects many people after the age of 60.

Yet the question remains: What do normal genes affecting the brain have to do with the essence of being human? Molecular biology is now providing the tools to permit a new generation of experiments correlating, for example, the genetically influenced production of neurotransmitter molecules and receptors with brain activity and behavior. Still, the study of genes and human behavior will always be plagued by three fundamental obstacles. The first is that human behaviors are extremely difficult to categorize neatly. Manic depression and schizophrenia, for example, were once thought to be unitary conditions—meaning that each was caused by one or a very small number of underlying physiological defects. Today they are recognized as multifaceted and elusive, as evidenced by the wide variety of drugs used to treat or control them. This difficulty in precisely specifying phenotype—which is even more pronounced with nonpathological behavioral traits such as intelligence—makes correlation with genotype extremely difficult. It is possible that improved diagnostic tools may someday overcome this diagnostic uncertainty.

The other two obstacles plaguing human behavioral genetics, however, will always (and thankfully!) exist, due to ethical constraints on human experimentation. First, we obviously will not allow the controlled breeding of humans to facilitate genetic studies. Second, we cannot control the environment of experimental subjects rigorously enough to eliminate nongenetic influences on nervous system development and learning. The answer to those experiments we *can* perform may someday tell us more about who we are than some of us want to know. What we as a society choose to do with the information will tell us even more.

SUMMARY

The activities of the nervous and muscular systems in response to environmental stimuli are ultimately reflected in animal behaviors vital to survival and reproduction. One approach to the study of behavior is ethology, which emphasizes the importance of examining animals in a natural context.

Behaviors that are governed primarily by genetically programmed neural wiring and require little or no experience to develop are called innate or instinctive. Behaviors that depend on changes in the nervous system as a result of experience are called learned. Among invertebrates and lower vertebrates, strong genetic influences on many behaviors can be demonstrated through breeding experiments. In higher animals, most behaviors result from complex interactions between nature and nurture. The fact that many behaviors are influenced by genes means that behavior can respond to natural selection by evolving over time. The influence of genes on human behavior and the extent to which such behaviors as language and emotion are uniquely human are subjects of heated debate.

Social behavior, coordinated by intraspecific communication through visual, chemical, auditory, tactile, and even electric signals, includes courtship, mating, parental care of their young, and more intricate activities of complex animal societies. The sophisticated organization of social insect colonies enables those societies to build elaborate nests and to exchange information.

STUDY FOCUS

After studying this chapter, you should be able to:

- Discuss behavior as a critical part of evolutionary adaptation.
- Explain "innate" and "learned" behavior.
- Describe the evolution of social behavior.
- Explore the biological roots of human behavior.

TERMS AND CONCEPTS

sign stimulus/releaser   *970*
fixed-action pattern   *970*
innate behavior   *970*
instinct   *970*
learning   *970*
habituation   *971*
classical conditioning   *971*
operant conditioning   *971*
insight learning   *972*
imprinting   *972*
communication   *976*

social releaser   *976*
pheromone   *976*
stimulus–response
   chain   *977*
sexual selection   *977*
dominance hierarchy   *981*
altruistic behavior   *982*
inclusive fitness   *982*
kin selection   *982*
castes   *983*

REVIEW

Objective Questions (Answers in Appendix)

1. Any stimulus that activates a specific motor response is called a(n)
   (a) ritualized cue.   (c) releaser.
   (b) imprinting cue.   (d) pheromone.

2. The different sexes of song sparrows look alike. A pair mates only once in a season, and both help in nest building, food gathering, and defense. These behaviors are examples of maximizing reproductive success

   (a) through monogamy.
   (b) through polygamy.
   (c) by not competing with other mates for females.
   (d) by selecting the best female available.

3. Using the concept of inclusive fitness, an organism is "fit" if
   (a) its genome is passed to the offspring through its own efforts or that of its relatives.
   (b) it produces many healthy sex cells.
   (c) it survives beyond its statistical life expectancy.
   (d) it is nurtured for a long time by its parents.

4. Applying prior experience to a new situation is called
   (a) habituation.
   (b) classical conditioning.
   (c) operant conditioning.
   (d) insight learning.

5. Chemicals that are released by one animal and that influence the behavior of other individuals of the same species are known as
   (a) imprints.   (c) castes.
   (b) pheromones.   (d) insights.

Discussion Questions

6. What is an animal's Umwelt? How and why can Umwelt vary from species to species, from individual to individual, and from time to time?

7. What is "instinctive" behavior? Are any behaviors in any animals *totally* instinctive? If so, how can such behaviors appear in the absence of experience?

8. Describe the four basic kinds of learning and give examples of each.

9. Explain how many complex behaviors are shaped both by "nature" and by "nurture."

10. Explain the phenomenon of female choice in shaping courtship and mating behavior.

11. What is the biological definition of altruistic behavior? How is altruism in nonhuman animals explained in terms of the principles of inclusive fitness?

READINGS

"If you were a slug and could choose your own sex, which would you pick? Are you sure?" *Discover* (June 1992): 88.

"Mind and brain." *Scientific American* 267 (3) (Special issue, September 1992).

Robinson, Michael H. "An ancient arms race shows no sign of letting up." *Smithsonian* (April 1992): 75.

Premack, David. *Gavagai! or The Future History of the Animal Language Controversy.* Cambridge, MA: M.I.T. Press, 1986. An intensely personal and stimulating view of the similarities and differences between human language and the communication systems of a variety of animals.

Appendix

A N S W E R S

Chapter 1

1. c; 2. d; 3. c; 4. b

Chapter 2

1. b; 2. a

Chapter 3

1. a; 2. d; 3. a; 4. a; 5. c; 6. a; 7. a

Chapter 4

1. a; 2. a; 3. c; 4. c; 5. d

Chapter 5

1. b; 2. a; 3. c; 4. d; 5. c

Chapter 6

1. c; 2. a; 3. a

Chapter 7

1. c; 2. b; 3. c

Chapter 8

1. a; 2. d; 3. b; 4. d; 5. a

Chapter 9

1. a; 2. a; 3. d; 4. b; 5. a; 6. b

Chapter 10

1. d; 2. a; 3. b; 4. c; 5. c; 6. c

Chapter 10 In-Text Questions

Page 195   The ideal bird, of course, is the white one—the double-recessive one.

Page 212   Parakeets are not epistatic. In fact, the two loci are codominant because each is expressed.

Chapter 11

1. b; 2. a; 3. a; 4. c; 5. a

Chapter 12

1. d; 2. a; 3. a; 4. d; 5. c

Hardy–Weinberg Problems

Problem 1

Remember that $p^2 + 2pq + q^2 = 1$.

We assume, because of the nature of the disease as explained in the problem, that the population is in Hardy–Weinberg equilibrium. (Nothing in the data presented suggests any effects of the gene on the reproductive success of either homozygous or heterozygous individuals.)

Here the homozygous recessives in the population number 1 in 5000, so, expressed as a percentage of the total population, those homozygotes represent

$$q^2 = 1/5000 = 0.0002.$$

Therefore,

$$q = 0.014142.$$

Then, because $p + q = 1$,

$$p = 0.98585.$$

The number of heterozygotes in the population is

$$2pq = 0.027884$$

so the total number of heterozygotes in a population of 5000 is 0.027884(5000) or roughly 139 individuals.

To show that this result is not trivial, imagine a similar sample population that differs only in the number of $A'A'$ homozygotes as follows:

1627 1469 138 20
total AA AA' $A'A'$

In this case, the observed percentages of the total population for each genotype are

0.90289 0.08481 0.01229
AA AA' $A'A'$

so

$$p^2 = 0.90289 \quad \text{and} \quad p = 0.9502$$
$$q^2 = 0.01229 \quad \text{and} \quad q = 0.1109.$$

Thus the expected percentage of the heterozygotes in the population is

$$2pq = 0.2175$$

or more than twice the observed percentage of heterozygotes. If this were indeed the case, what hypotheses could you devise to explain this difference between expected distribution and observed distribution? What experiments could you design to test your hypotheses?

Problem 2

	Individuals collected	% of total population
Total population	1612	100
AA	1469	0.91129
AA'	138	0.08561
$A'A'$	5	0.003102

There are two ways to determine whether this population is close to H–W equilibrium.

Most simply, for the population sampled,

$$AA \;\; = p^2 = 0.91129, \quad \text{so } p = 0.9546$$
$$A'A' = q^2 = 0.00310, \quad \text{so } q = 0.0556$$

Then $AA' = 2pq = 0.1061$, the calculated percentage of heterozygotes, whereas the observed percentage was 0.08561.

This result indicates that the population conforms reasonably well to theoretical expectations, which, however, predict a slightly higher percentage of heterozygotes at equilibrium.

Chapter 13

1. b; 2. a; 3. b; 4. a; 5. c

Chapter 14

1. a; 2. d; 3. b; 4. a; 5. b; 6. c

Chapter 15

1. b; 2. d; 3. b; 4. a; 5. d; 6. b; 7. a; 8. b

Chapter 16

1. b; 2. b; 3. d; 4. b; 5. c; 6. d; 7. c

Chapter 17

1. a; 2. a; 3. b; 4. c; 5. d; 6. d

Chapter 18

1. c; 2. d; 3. b; 4. c; 5. a; 6. b

Chapter 19

1. b, 2. a, 3. c, 4. b, 5. c

Chapter 20

1. b; 2. c; 3. b; 4. c

Chapter 21

1. a; 2. a; 3. b; 4. d

Chapter 22

1. d; 2. d; 3. b; 4. c; 5. d

Chapter 23

1. a; 2. b; 3. b

Chapter 24

1. b

Chapter 25

1. a; 2. c; 3. d; 4. c; 5. b; 6. d

Chapter 26

1. a; 2. c; 3. d; 4. b

Chapter 27

1. c; 2. a; 3. a; 4. b;

Chapter 28

1. d; 2. a; 3. a; 4. c; 5. d

Chapter 29

1. a; 2. d; 3. c; 4. b; 5. c

Chapter 30

1. b; 2. a; 3. c; 4. d; 5. d; 6. d

Chapter 31

1. a; 2. d; 3. b; 4. d; 5. c

Chapter 32

1. c; 2. a; 3. d; 4. a; 5. b; 6. b

Chapter 33

1. c; 2. c; 3. b; 4. a; 5. a; 6. c; 7. c

Chapter 34

1. d; 2. b; 3. c; 4. b; 5. a

Chapter 35

1. a; 2. c; 3. a; 4. b; 5. a

Chapter 36

1. b; 2. a; 3. d; 4. a; 5. d

Chapter 37

1. a; 2. d; 3. c; 4. d; 5. d

Chapter 38

1. d; 2. d; 3. a; 4. b; 5. a; 6. c; 7. b

Chapter 39

1. c; 2. a; 3. a; 4. d; 5. b

Chapter 40

1. d; 2. c; 3. c; 4. c; 5. a; 6. a

Chapter 41

1. a; 2. d; 3. c; 4. b; 5. c

Chapter 42

1. c; 2. c; 3. b; 4. c; 5. b

Chapter 43

1. c; 2. a; 3. a; 4. a; 5. d

Chapter 44

1. c; 2. b; 3. a; 4. b

Chapter 45

1. a; 2. c; 3. d; 4. d; 5. c

Chapter 46

1. c; 2. a; 3. a; 4. d; 5. b

GENETICS PROBLEM SET
Mitosis, Meiosis, and Mendelian Genetics

Problem Set

1. Distinguish between two sister chromatids and two homologous chromosomes. Distinguish between a diploid nucleus and a haploid nucleus.

2. If we let X be the minimum amount of genetic material that carries all the information of a species, then in an organism with a diploid chromosome number of 2, how much DNA (X, $2X$, $4X$, etc.) is found in each of the following? an egg nucleus, a sister chromatid, a daughter nucleus following mitosis, a homologue following mitosis, a nucleus at the onset of mitotic coiling

3. A karyotype of a mitotic nucleus from a female cat shows 76 sister chromatids. What are the diploid and haploid chromosome numbers of the cat? How many homologous chromosome pairs are present?

4. Consider an organism with a haploid chromosome number of 7. How many sister chromatids are present in its mitotic metaphase nucleus? In its meiotic metaphase I nucleus? In its meiotic metaphase II nucleus?

5. Consider an organism with a haploid chromosome number of 3. Write out the eight possible combinations of maternal and paternal homologues in its gametes (for example, 1M2M3P). For each case, diagram the alignments of the nonhomologues with respect to the metaphase I plate.

6. Cite six human traits that you would classify as wild-type traits (that is, highly invariant) and six traits that commonly appear in a number of variant forms in the human population.

7. A pair of alleles is always observed to segregate at the first meiotic division. How might this observation be explained?

8. A cross was made between two *Drosophila* that exhibited the wild phenotype. Their progeny were found to have the same phenotype. A sample was taken of the progeny, each of which was crossed with a fly that had purple eye color. Half of the crosses gave only wild-type flies, and the other half gave 50 percent wild-type and 50 percent purple-eyed progeny. What were the genotypes of the original pair of wild-type flies?

9. Thalassemia is a type of human anemia that is rather common in Mediterranean populations but relatively rare in other peoples. The disease occurs in two forms, minor and major; the latter is much more severe. Persons with thalassemia major are homozygous for an aberrant recessive gene; mildly affected persons (with thalassemia minor) are heterozygous; persons normal in this regard are homozygous for the normal allele.
   (a) A man with thalassemia minor marries a normal woman. With respect to thalassemia, what types of children, and in what proportions, can they expect? (Let t = the allele for thalassemia minor and T = its normal allele.)
   (b) Both father and mother in a particular family have thalassemia minor. What is the chance that their baby will be severely affected? Mildly affected? Normal? Diagram the possible unions of germ cells in this family.
   (c) An infant has thalassemia major. According to the information given, what results would you expect if you checked the infant's parents for anemia?
   (d) Thalassemia major is almost always fatal in childhood. How does this fact modify your answer to part (c)?

10. In humans the most frequent type of albinism (itself quite rare) is inherited as a simple recessive characteristic. Standard symbols are C = normal pigmentation and c = albino. Assume that the genes for thalassemia and albinism assort independently.
   (a) A husband and wife, both normally pigmented and neither afflicted with severe anemia, have an albino child who dies in infancy of thalassemia major. What are the probable genotypes of the parents?
   (b) If these people have another child, what are its chances of being phenotypically normal with respect to pigmentation? Of having entirely normal (that is, nonthalassemic) blood? Of being phenotypically normal in both regards? Of being homozygous for the normal alleles of both genes?

11. A cross was made between two wild-type *Drosophila*. Their progeny were 187 males with raspberry eye color, 194 wild-type males, and 400 wild-type females. Is *ras* a sex-linked gene? Explain. What are the parental genotypes? What are the genotypes of the F_1 wild-type females, and what is their ratio?

12. A cross was made between a female heterozygous for the recessive genes *ct* (cut wings) and *se* (sepia eye color) and a sepia male. Among their female progeny, half the phenotypes were wild-type and half sepia. Among their

male progeny, the phenotypes were 1/4 wild-type, 1/4 cut, 1/4 sepia, and 1/4 cut and sepia. Is either of these genes sex-linked? What are the genotypes of the parents and their offspring?

13. If a woman with normal color vision has a colorblind father, what is the probability that her sons will be colorblind if she marries a man with normal color vision? What genotypes are possible among her male and female offspring? What is the probability of her having a colorblind child if she marries a colorblind man, and does this probability differ depending on whether the child is a male or a female?

14. Hemophilia is a sex-linked recessive trait. A son born to phenotypically normal parents has Klinefelter syndrome and hemophilia. In which meiotic division of which parent is nondisjunction of the X chromosome most likely to have occurred? Explain and state your assumptions.

15. Neither Tsar Nicholas II, his wife Empress Alexandra, nor their daughter Princess Anastasia had the disease hemophilia. However, their son, the Tsarevich Alexis, did have the disease. Can one automatically assume that Anastasia was a carrier? Why or why not?

16. In *Drosophila,* the gene for red eye is dominant to its white allele, and the gene for long wing is dominant to its vestigial allele. The $+/w$ locus is on the X chromosome; the $+/vg$ locus is not on a sex chromosome. Two red-eyed, long-winged flies are bred together and produce offspring in the following ratios:

Females 3/4 red long, 1/4 vestigial
Males   3/8 red long, 3/8 white long, 1/8 red vestigial, 1/8 white vestigial

What are the genotypes of the parents?

Solutions to Genetics Problem Set

1. Sister chromatids are the identical copies of a single chromosome produced by DNA replication. Homologous chromosomes are two chromosomes that contain genes determining the same traits. The genes on two

homologous chromosomes need not be identical! A diploid nucleus has a complete set of all chromosomes, including *both* copies of each pair of homologous chromosomes. A haploid nucleus has only *one* representative of each pair of homologous chromosomes.

2. It depends on what you mean by "minimum amount"! We'll take it to be the haploid amount, OK?

Egg nucleus	X
Sister chromatid	X
Daughter nucleus	$2X$
Homologue after mitosis	$2X$
Nucleus before mitosis	$4X$

3. Diploid number, 38. Haploid number, 19. There are 19 pairs of homologous chromosomes present at metaphase.

4. The haploid number is 7, so the diploid number is 14.

Metaphase of mitosis	28 chromatids
Meiosis I	28
Meiosis II	14

5. Here are the possible combinations, written as though they were aligned along a single metaphase plate running right to left:
MMM MMP MPM MPP PPP
PPM PMP PMM

6. This one you can work out for yourselves. Invariant traits (generally) include such traits as number of eyes, fingers, or arms. Highly variable traits in our species include skin color, height, and body structure.

7. Both alleles must be both located on the same pair of homologous chromosomes. Because they separate after the first meiotic division, the alleles would always be observed to segregate as well.

8. If we use P for purple eyes and W for wild-type eye color, we can answer this pretty easily. The half of the offspring that gave *only* wild-type flies when crossed with the purple ones must have been WW. The other half must have been WP (and the purple flies PP).

9. (a) Normal children (Tt), 50%; mild anemia (Tt), 50%.
(b) (Tt) × (Tt) = 25% severe (tt), 50%

mild (Tt), 25% normal (TT).
(c) The parents must both be either Tt or tt.
(d) They must *both* be Tt.

10. (a) The fact that the child was both albino and tt for thalassemia means that the child's genotype must have been $cctt$. Therefore, each parent must have at least one copy of each recessive gene. However, both parents were phenotypically normal. This means that they must both have been heterozygous for each trait: $CcTt$.
(b) Normal pigmentation: 3 chances in 4 (75% chance)
Normal blood: 1 chance in 4 (25% chance)
Phenotype normal in both: 3 chances in 16
Homozygous normal for both alleles: 1 in 16

11. You bet ras is a classic sex-linked gene! The fact that no females are ras and 1/2 of the males are ras is a classic sex-linked pattern. It suggests that 1/2 of the males have the ras gene on their X chromosomes. The phenotypes of the parents must have been XrX and XY. The wild-type females of the F_1 generation are 50% XX and 50% XrX (r represents the ras gene.)

12. This is tough. Let's look at each gene individually to get the answer. First, sepia. 1/2 of the offspring are sepia without regard to sex. Therefore, sepia must not be sex-linked. The parents must have been $sel+$ (female) and se/se (male). No problem so far.

Now, cut wings. Note that none of the females have cut wings. However, 1/2 of the males do have cut wings. Cut wings must be sex-linked. Here are the genotypes of parents and offspring:

PARENTS
Female $XctX$
      $sel+$
Male X
      se/se
OFFSPRING
Female XX $XXct$ XX $XXct$
      se/se se/se $sel+$ $sel+$
Male Xct Xct X X
      se/se $sel+$ se/se $sel+$

13. The "colorblind gene" is on the *X* chromosome. Because a woman gets one of her *X* chromosomes from her father, and he has only one *X* chromosome, this woman must have the colorblind gene on one of her *X* chromosomes.

Father with
normal vision: 50% of sons
                  color-blind

Possible
genotypes: Male *XY* and
                  XcbY
                  Female *aXX* and
                  XXcb

Colorblind
father: 50% of both sons
                  and daughters
                  colorblind

14. This one takes a little close reasoning. A person with Klinefelter syndrome has 47 chromosomes and the genotype 44*XXY*. This poor fellow also has hemophilia. He couldn't have gotten his hemophilia genes from his father, however (or Dad would have had hemophilia too). Mom must have been a carrier for the disease (*XXh*), but how did he get two *Xh* chromosomes? The homologous *X* and *Xh* chromosomes from Mom separated at the first meiotic division. Then at the second meiotic division, the two identical *Xh* chromosomes must have failed to separate when the haploid gametes were formed. Those two *Xh* chromosomes gave this poor boy both Klinefelter syndrome and hemophilia.

15. No. Because Alexis got his only *X* chromosome from his mother, it must have had the hemophilia gene on it, meaning that Alexandra must have been a carrier. However, Anastasia got one *X* from Nicholas that didn't have the gene, and she got the other *X* from her mother. Because Alexandra was *XXh*, she had a 50–50 chance of giving the good *X* to her daughter. (The events of October 1918 made all of this somewhat academic. Putting it briefly, this family left no offspring.)

16. This is a nice exercise in genetic analysis. We know that one gene is sex-linked and one is not. Let's look at the vestigial gene first. 1/4 of the offspring were vestigial. This number is part of the classic 3:1 ratio for crosses of heterozygotes. Thus the original flies, both with long wings, must both have been +/*vg* at the wing locus. So far so good. Now, we know that red eyes (+) are dominant over white eyes (*w*). The male in our cross must therefore have been *X, Y* (he had red eyes, remember). Similarly, the female must have been *X, Xw.*

PARENTAL GENOTYPES

Sex	*X, Y*	*X, Xw*
Somatic	+/*vg*	+/*vg*

THE METRIC SYSTEM

History

The original basis for the metric system was the diameter of the earth. The original length of the meter, the basic metric unit, was chosen such that the distance from one pole to the equator would be approximately 10,000,000 meters.

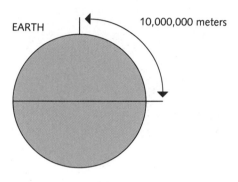

EARTH

10,000,000 meters

Commonly Used Metric Units

Metric	Prefixes	Mass	Volume	Distance
mega	10^6	—	—	—
kilo	10^3	kilogram	—	kilometer
hecto	10^2	—	—	—
deka	10	—	—	—
—	1	gram	liter	meter
deci	10^{-1}	—	—	—
centi	10^{-2}	—	—	centimeter
milli	10^{-3}	milligram	milligram	millimeter
micro	10^{-4}	—	—	—
nano	10^{-9}	—	—	nanometer
—	10^{-10}	—	—	angstrom
pico	10^{-12}	—	—	—

Relationship Between Volume, Distance and Mass

One *liter* is the volume enclosed by a cube 10 *centimeters* on a side.

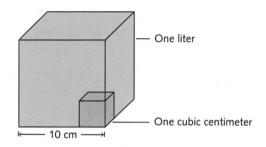

One liter

One cubic centimeter

10 cm

One liter (1000 ml) of water has a mass of one kilogram. One *cubic centimeter* of water, with a volume of 1 *milliliter* (1 ml) under standard conditions, has a mass of 1 gram.

Square Units

Square meter = 1 meter on each side
Hectare = 100 meters on each side
Square kilometer = 1000 meters on each side

Temperature

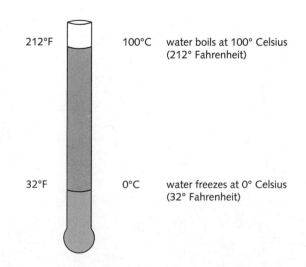

212°F   100°C   water boils at 100° Celsius (212° Fahrenheit)

32°F   0°C   water freezes at 0° Celsius (32° Fahrenheit)

THE PERIODIC TABLE OF THE ELEMENTS

Legend:
- 1 — atomic number
- H — symbol for element
- 1.008 — atomic weight

1	2		3	4	5	6	7	8	9	10	11	12	13	14	15	16	17	18
1 **H** 1.008																		2 **He** 4.003
3 **Li** 6.941	4 **Be** 9.012												5 **B** 10.81	6 **C** 12.01	7 **N** 14.01	8 **O** 16.00	9 **F** 19.00	10 **Ne** 20.18
11 **Na** 22.99	12 **Mg** 24.31												13 **Al** 26.98	14 **Si** 28.09	15 **P** 30.97	16 **S** 32.07	17 **Cl** 35.45	18 **Ar** 39.95
19 **K** 39.10	20 **Ca** 40.08		21 **Sc** 44.96	22 **Ti** 47.88	23 **V** 50.94	24 **Cr** 52.00	25 **Mn** 54.94	26 **Fe** 55.85	27 **Co** 58.93	28 **Ni** 58.69	29 **Cu** 63.55	30 **Zn** 65.38	31 **Ga** 69.72	32 **Ge** 72.59	33 **As** 74.92	34 **Se** 78.96	35 **Br** 79.90	36 **Kr** 83.80
37 **Rb** 85.47	38 **Sr** 87.62		39 **Y** 88.91	40 **Zr** 91.22	41 **Nb** 92.91	42 **Mo** 95.94	43 **Tc** (98)	44 **Ru** 101.1	45 **Rh** 102.9	46 **Pd** 106.4	47 **Ag** 107.9	48 **Cd** 112.4	49 **In** 114.8	50 **Sn** 118.7	51 **Sb** 121.8	52 **Te** 127.6	53 **I** 126.9	54 **Xe** 131.3
55 **Cs** 132.9	56 **Ba** 137.3		57 **La*** 138.9	72 **Hf** 178.5	73 **Ta** 180.9	74 **W** 183.9	75 **Re** 186.2	76 **Os** 190.2	77 **Ir** 192.2	78 **Pt** 195.1	79 **Au** 197.0	80 **Hg** 200.6	81 **Tl** 204.4	82 **Pb** 207.2	83 **Bi** 209.0	84 **Po** (209)	85 **At** (210)	86 **Rn** (222)
87 **Fr** (223)	88 **Ra** 226		89 **Ac** (227)	104 **Unq**	105 **Unp**	106 **Unh**	107 **Uns**	108 **Uno**	109 **Une**									

*Lanthanides

58 **Ce** 140.1	59 **Pr** 140.9	60 **Nd** 144.2	61 **Pm** (145)	62 **Sm** 150.4	63 **Eu** 152.0	64 **Gd** 157.3	65 **Tb** 158.9	66 **Dy** 162.5	67 **Ho** 164.9	68 **Er** 167.3	69 **Tm** 168.9	70 **Yb** 173.0	71 **Lu** 175.0

Actinides

90 **Th** 232.0	91 **Pa** (231)	92 **U** 238.0	93 **Np** (237)	94 **Pu** (244)	95 **Am** (243)	96 **Cm** (247)	97 **Bk** (247)	98 **Cf** (251)	99 **Es** (252)	100 **Fm** (257)	101 **Md** (258)	102 **No** (259)	103 **Lr** (260)

NOTE: 1. Group numbers 1—18 represent the system recommended by the International Union of Pure and Applied Chemistry.

2. Atomic weights in parentheses represent the weight of the longest—lived radioactive isotope.

CLASSIFICATION SCHEME

KINGDOM MONERA: Prokaryotes
(under constant revision; assigned many more divisions by some researchers concentrating on biochemical pathways and genetics)

Division Archaebacteria: "Ancient" bacteria
Division Eubacteria: "True" bacteria
Division Cyanobacteria: "Blue-Greens"

KINGDOM PROTISTA: Predominantly unicellular and colonial eukaryotes (A "catch-all" category for predominantly single-celled eukaryotes; extensively revised or even eliminated by some researchers)

PHYLUM Mastigophora: Flagellates
PHYLUM Sarcodina: Amoebae
PHYLUM Sporozoa: Sporozoans
PHYLUM Euglenophyta: Euglenas
PHYLUM Chrysophyta: Diatoms, golden-brown algae
PHYLUM Pyrrophyta: Dinoflagellates

KINGDOM FUNGI: Predominantly multicellular, heterotrophic eukaryotes with cell walls containing chitin

Division Zygomycota: Common molds
Division Ascomycota: Sac fungi and yeasts
Division Basidiomycota: Mushrooms
Division Oomycota: Water molds
Division Deuteromycota: Imperfect fungi

KINGDOM PLANTAE: Predominantly multicellular, photosynthetic, autotrophic eukaryotes; cell walls containing cellulose

Division Chlorophyta: Green algae
Division Rhodophyta: Red algae
Division Phaeophyta: Brown algae
Division Bryophyta: Mosses, liverworts, hornworts
Division Lycophyta: Club mosses
Division Sphenophyta: Horsetails
Division Pterophyta: Ferns
Division Cycadophyta: Cycads
Division Ginkgophyta: Ginkgoes

Division Coniferophyta: Conifers
Division Anthophyta: Flowering plants
  Class Monocotyledones: Monocots (lillies, grasses, etc.)
  Class Dicotyledones: Dicots (roses, deciduous trees, etc.)

KINGDOM ANIMALIA: Multicellular, heterotrophic eukaryotes

PHYLUM Porifera: Sponges
PHYLUM Cnidaria
  Class Hydrozoa: Hydras
  Class Scyphozoa: Jellyfishes
  Class Anthozoa: Sea anemones and corals
PHYLUM Platyhelminthes
  Class Turbellaria: Flatworms
  Class Trematoda: Flukes
  Class Cestoda: Tapeworms
PHYLUM Nematoda: Roundworms
PHYLUM Mollusca: Snails, clams, and squid
  Class Polyplacophora: Chitons
  Class Gastropoda: Snails and slugs
  Class Bivalvia: Clams, oysters, and scallops
  Class Cephalopoda: Octopi, squid, and nautilus
PHYLUM Annelida
  Class Polychaeta: Feather duster worms, etc.
  Class Oligochaeta: Earthworms
  Class Hirudinea: Leeches
PHYLUM Arthropoda
  Subphylum Trilobita: Trilobites (extinct)
  Subphylum Chelicerata: Scorpions, spiders, and mites
  Subphylum Crustacea: Shrimps, crabs, and lobsters
  Subphylum Uniramia: Centipedes, millipedes, and insects
    Class Chilopoda: Centipedes
    Class Diplopoda: Millipedes
    Class Insecta: Insects
PHYLUM Echinodermata
  Class Asteroidea: Starfish
  Class Echinoidea: Sea urchins
  Class Crinoidea: Sea lillies
  Class Holothuroidea: Sea cucumbers
  Class Ophiuroidea: Brittle stars

PHYLUM Chordata
  (major groups with at least 1 living member)
  Subphylum Urochordata: Tunicates
  Subphylum Cephalochordata: Lancelets
  Subphylum Vertebrata:
    Class Agnatha*: Jawless fishes
    (lampreys, hagfishes)
    Class Chondrichthyes: Cartilaginous fishes
    (sharks, skates, and rays)
    Class Osteichthyes: Bony fishes
      Subclass Actinopterygii: Ray-finned fishes
      (contains most modern fishes)
      Subclass Sarcopterygii: Fleshy-finned fishes
        Order Dipnoi: Lungfishes
        Order Crossopterygii: Coelacanths (and ancestors
        of land vertebrates)

Class Amphibia: Amphibians
  Order Anura: Frogs and toads
  Order Urodela: Salamanders and newts
  Order Apoda: Limbless amphibians
Class Reptilia: Reptiles
  Subclass Anapsida: Turtles
  Subclass Lepidosauria: Snakes, lizards
  Subclass Archosauria: Dinosaurs, crocodilians
Class Aves: Birds
Class Mammalia: Mammals
  Subclass Protheria: Egg-laying mammals
  Subclass Theria
    Infraclass Metatheria: Pouched mammals
    Infraclass Eutheria: Placental mammals (contains many
    orders, including our own, the Order Primates)

*"Agnatha," though now rarely used by ichthyologists as a formal class,
is still a convenient grouping under which to list these fishes.

ABO gene The hereditary unit containing the A, B, and O alleles that determine the configuration of antigens on the surface of red blood cells.

abscisic acid (*ab-SISS-ik*) A plant hormone that inhibits growth and plays a role in preparing seeds, buds, and woody tissues for dormant periods.

abscission layer (*ab-SIH-zhun*) A corky cell barrier that forms when hormonal action at the base of a leaf stalk shuts off the flow of water and nutrients to the leaf; this layer is the "break point" when the leaf falls off.

accessory pigments Compounds that absorb energy from wavelengths of light that chlorophyll either does not absorb or does not absorb as well.

accommodation Adjustments in the lens of the eye that focus images on the retina.

acetylcholine (*a-SEE-till-KOH-leen*) A neurotransmitter released and hydrolyzed at certain nerve endings, notably at neuromuscular junctions.

acid A compound that releases hydrogen ions (H⁺) into solution when it is dissolved in water, thus lowering the pH. An acidic solution has a pH lower than 7. Examples include lemon juice, vinegar, and battery acid.

acid rain Rain that, because of acid-forming industrial pollutants introduced into the atmosphere, has a pH lower than normal. Acid rain is currently causing serious environmental damage in the northern United States and Europe.

acoelomates (*ay-SEE-loh-MATES*) Bilaterally symmetrical animals with no body cavity (*coelom*).

acrosome (*A-krow-soam*) The vesicle at the forward end of a sperm containing enzymes that help break through protective layers surrounding the egg cell.

actin With myosin, one of the two major proteins making up the thin filaments of muscle; functions in muscle contraction.

action potential An impulse in a nerve cell; a sudden, brief change in the electric charge across the plasma membrane that is propagated along the membrane.

activation energy The energy necessary to initiate a chemical reaction.

active site The place on an enzyme where substrate molecules bind during catalysis.

active transport A process in which energy is used in the movement of ions and molecules across a cell membrane against a concentration gradient.

adaptation (*a-dap-TAY-shun*) 1. A structure, behavior, or other feature of an organism that enables it to survive or reproduce successfully under existing environmental conditions. 2. The process by which organisms are modified over many generations by natural selection to become better suited or "fitted" to their environment, or to retain fitness under changing environmental conditions.

adaptive radiation The evolution, from a common ancestor, of a number of species specialized for survival in diverse environments.

adenine A nitrogen-containing base found in nucleotides, belonging to the chemical group known as purines, which have two carbon-nitrogen rings in their basic structures. Found in ATP, DNA, and RNA.

adenylate cyclase (*a-DEN-a-late SIGH-klase*) An enzyme that, when activated, transforms ATP into cyclic AMP.

ADH (antidiuretic hormone) A hormone produced by the hypothalamus and released from the posterior pituitary that affects the kidneys and blood vessels, stimulating water retention, affecting blood vessel constriction, and concentrating the urine.

adhering junction Also called **desmosome** (*DEZ-mo-sowm*). An attachment between adjacent animal cells that strongly binds them together but permits materials to move in the spaces between the attachments; common in tissues subject to mechanical stress, such as skin.

adipose tissue (*AD-ih-pos*) A type of connective tissue that serves as a storage reservoir for fat; found throughout the body.

adrenal cortex (*a-DREE-nul KOR-tex*) The outer part of the adrenal glands, which are located on top of the kidneys. The adrenal cortex produces a wide variety of steroid hormones essential to normal body function, including glucocorticoids such as cortisol, mineralocorticoids such as aldosterone, and sex hormones such as testosterone and estrogen.

adrenalin (*a-DRE-na-lin*) A hormone produced by the adrenal medulla and released into the bloodstream when danger threatens; also called *epinephrine*.

adrenal medulla (*a-DREE-nul me-DULE-uh*) The inner part of the adrenal glands, which are located on top of the kidneys. The adrenal medulla produces a variety of hormones, including adrenalin (epinephrine) and noradrendalin (norepinephrine).

aerobe (*AIR-obe*) An organism that requires oxygen to carry on respiration.

aerobic (*air-OH-bik*) Requiring free oxygen.

aggregate fruit A fruit formed from multiple unfused ovaries within a single flower, as in a raspberry, strawberry, or pea.

agnathans (*ag-NAY-thuns*) A group of jawless fishes, fossils of which are the earliest evidence of vertebrate life-forms. Includes today's lampreys and hagfishes.

Agrobacterium tumefaciens A soil bacterium that infects several types of plants and produces large tumor-like growths called galls by transmitting a small piece of DNA called the *Ti* plasmid to plant cells. Plant molecular biologists use recombinant-DNA techniques to alter the *Ti* plasmid

and use it as a vehicle for inserting new genes into plants.

AIDS (acquired immune deficiency syndrome) A disorder of the immune system, caused by a retrovirus, the Human Immunodeficiency Virus (HIV), that renders victims vulnerable to a variety of infections. At the time of this writing, AIDS is incurable and fatal.

albinism (*al-BYE-nizm*) A recessive gene disorder in which the affected individual cannot synthesize the pigment melanin, which is responsible for most human skin and hair color.

alimentary canal (*al-im-EN-ta-ree*) The passageway that extends from the mouth through a variety of specialized digestive organs to the anus.

allantois (*a-LAN-twahz*) An extraembryonic sac that contributes to the formation of the umbilical cord and placenta; develops from the primitive gut.

alleles (*a-LEELZ*) Two or more forms of a single gene.

allergy A disorder in which the immune system overreacts to a foreign molecule in the environment, initiating an inflammatory response.

allopatric speciation (*al-lo-PAT-rick spee-shee-AY-shun*) The development of new species following the physical separation of two or more populations of an organism in such a way that gene flow ceases between these populations.

allosteric site (*AL-oh-STEER-ik*) A location on an enzyme that is not an active site but that affects the shape of the active site when a regulatory molecule binds to it.

alpha helix A coiled chemical configuration of the secondary structure of proteins stabilized by hydrogen bonds between every fourth amino acid.

alternation of generations A pattern in the life cycle of plants in which a diploid sporophyte stage alternates with a haploid gametophyte stage.

altruistic behavior (*AL-true-ISS-tik*) Behavior that benefits others of the same species but is destructive or potentially destructive to the individual that performs it.

alveoli (singular, **alveolus**) (*al-VEE-a-lie*) The small, thin-walled sacs in the lungs where gas exchange takes place.

amino acid (*a-MEE-no*) A molecule containing an amino group (–NH₂), a carboxyl group (–COOH), and an "R group" that varies from one amino acid to another. Amino acids are the building blocks of proteins.

ammonification The process by which decomposers such as bacteria and fungi break down organic molecules in dead organisms, releasing nitrogen in the form of ammonia (NH₃). Ammonification is part of the nitrogen cycle.

amniocentesis (*AM-nee-oh-SENT-ee-sis*) A medical procedure in which a needle is inserted into the uterus of a pregnant female to withdraw a small amount of amniotic fluid containing cells from the developing offspring. These cells are then grown in culture and examined for genetic disorders.

amniotic egg (*AM-nee-AWE-tik*) A watertight egg produced by internal fertilization and wrapped in three protective membranes.

amniotic sac (*AM-nee-AWE-tik*) An extraembryonic sac filled with fluid that protects the embryo from desiccation and physical shock.

amoeba (plural, **amoebae** or **amoebas**) (*a-MEE-bah*) Members of the phylum Sarcodina. Simple, animal-like protists that generally live in water or soil habitats or occur as parasites. Includes radiolarians and foraminiferans, important members of plankton communities.

amphibians Members of the class Amphibia, a group of four-legged ectothermic vertebrates, many of which exhibit an aquatic larval phase and a semiterrestrial or terrestrial adult phase.

anabolic pathway (*AN-a-BOL-ic*) A chemical pathway in which smaller, simpler molecules are assembled into larger, more complex molecules.

anaerobe (*AN-ur-obe*) An organism capable of living in an environment in which there is no oxygen.

analogous structures Organic structures of similar function (and sometimes similar appearance) that result from convergent evolution rather than common ancestry.

anaphase (*AN-a-faze*) The third stage of mitosis, during which the centromeres break and the chromosomes move to the poles of the nuclear spindle.

angiosperms (*AN-jee-oh-sperms*) More recent terminology, *anthophyta* (*an-THOFF-uh-tah*). The flowering plants: a group of vascular plants that produce seeds enclosed in ovaries.

Animalia The kingdom containing animals (various heterotrophs without cell walls).

anion (*AN-eye-un*) A negatively charged ion.

annelid (*AN-el-id*) A segmented worm. Member of the phylum Annelida, segmented worms characterized by bristle-like setae, body segments sealed off by septa, and a hydrostatic "skeleton." The earthworm is an annelid.

anterior Toward the head or front.

anther In flowering plants, the enlarged sacs where pollen is produced and released. Located at the tip of the male reproductive structure (*stamen*).

anthophytes (*AN-tho-FIGHTS*) Also called *angiosperms*. Members of the plant division Anthophyta. The flowering plants: a group of vascular plants that produce seeds enclosed in ovaries.

anthropoids (*AN-thra-poyds*) The more advanced branch of the order of primates; includes the New World and Old World monkeys, the great apes, and humans.

antibody (*AN-tee-bod-ee*) A Y-shaped protein capable of binding to an antigen and either inactivating it or tagging it for destruction by the body's defense systems.

antibody-mediated immune response Also called *humoral immune response*. The production of antibodies by B-lymphocytes of the immune system, triggered by foreign proteins (antigens).

anticodon (*AN-tee-CODE-on*) A string of three nucleotide bases on a tRNA molecule that is complementary to a particular mRNA codon.

antigen (*AN-tih-jun*) A molecule recognized as foreign by the body's immune system.

anus (*AY-nus*) The opening at the end of the alimentary canal through which waste material is expelled.

aorta (*ay-OR-tuh*) The thick artery that carries oxygenated blood from the left side of the heart to all parts of the body except the lungs.

apical dominance (*AY-pih-kul*) The tendency for plant growth to be most vigorous at the tip, or apex, of branches.

apical meristem (*AY-pih-kul MARE-ih-stem*) The undifferentiated embryonic tissue in root tips or stem buds where cells are produced, causing increases in length.

apomixis In a few flowering plants such as the common dandelion, an asexual reproductive process in which diploid egg cells develop into embryos and produce seeds that are genetically identical to the mother plant, often in a manner that mimics sexual reproduction.

appendicular skeleton (*AP-pen-DIK-you-lar*) The arm and leg bones and the pectoral and pelvic girdles that attach them to the axial skeleton.

aquaculture (*AH-kwuh-kul-chur*) The cultivation of aquatic organisms to harvest for human use.

aquifer (*AH-kwi-fer*) An underground rock stratum that holds water.

archaebacteria (*ARK-ee-bak-TEER-ee-a*) Three groups of bacteria—the methanogens, thermoacidophiles, and halophiles—that resemble the first prokaryotes to evolve on Earth.

archenteron (*ark-EN-ter-on*) The cavity formed during gastrulation that ultimately forms the embryo's digestive system.

area effect The relationship between the size of an island and the diversity of species inhabiting it. Generally, the number of species an island can hold doubles for every tenfold increase in the area of an island.

arteries In vertebrate circulatory systems, thick-walled, elastic blood vessels that carry blood away from the heart.

arterioles In vertebrate circulatory systems, small blood vessels that lead from the arteries to the capillaries.

arthritis Inflammation of the tissues of a moveable skeletal joint.

arthropod (*AR-throw-pod*) Member of the phylum Arthropoda, a group of animals characterized by jointed legs, a segmented body, and a hard exoskeleton that must be shed to permit new growth. The phylum includes the insects, crustaceans, spiders, millipedes, centipedes, and the extinct trilobites.

artificial selection Specialized breeding by humans of plants or animals possessing valued characteristics.

ascomycetes (*ASS-koh-MY-seets*) Members of the division Ascomycota. Also called *sac fungi*, a diverse group of unicellular or multicellular fungi that produce spores inside a sac, or ascus; includes the yeasts and most of the fungi that live symbiotically in lichens.

asexual reproduction Reproduction mediated not by the union of gametes but by budding, binary fission, fragmentation, or other nonsexual means.

atherosclerosis (*ATH-er-row-skle-ROE-sis*) The build-up of fatty tissue on the inner surface of arteries, which can slow down or stop blood flow and result in strokes or heart attacks.

atom A unit of matter, the smallest unit of an element, made up of three kinds of subatomic particles: negatively charged electrons, positively charged protons, and (in all but the simplest atom, hydrogen) electrically neutral neutrons.

atomic number The number of protons in an atomic nucleus.

atomic orbital The volumes around an atom's nucleus wherein an electron is most likely to be found.

ATP (adenosine triphosphate) (*a-DEN-a-seen try-FOSS-fate*) A major energy-carrying molecule in cells, this nucleotide can release the energy stored in its phosphate bonds; the energy is used in many metabolic activities.

ATP synthetase An enzyme bound to the inner mitochondrial membrane and the chloroplast thylakoid membrane that produces ATP from ADP and phosphate, capturing energy from a protein gradient formed across these membranes during electron transport.

atrioventricular (AV) node (*AY-tree-oh-ven-TRIK-you-lar*) The area of the heart where the fibers originate that conduct the impulse to contract from the atria to the ventricles.

atrioventricular valves (*AY-tree-oh-ven-TRIK-you-lar*) The heart valves that permit blood to flow from the atria to the ventricles but not in the other direction.

atrium (plural, *atria* or *atriums*) (*AY-tree-um*) A chamber of the heart that receives blood from a vein and pumps it into a ventricle. The human heart has two atria.

australopithecines (*AWE-strah-low-PITH-a-seens*) Members of a genus of early bipedal hominids thought by some to have been the common ancestor of all other hominids.

autoimmune disease (*AWE-toe-i-MUNE*) A disorder in which the immune system malfunctions and turns its destructive power against the body's own tissues.

autonomic nervous system The involuntary portion of the peripheral nervous system, connecting with glands, smooth muscles, and cardiac muscles.

autosomal chromosomes (*AWE-toe-SOHM-ul*) Generally, chromosomes that do not differ in males and females; in humans, the 22 pairs of homologous chromosomes.

autotroph (*AWE-toe-trowfe*) An organism, such as a plant, capable of producing its own food from inorganic materials and an environmental energy source.

auxin (*AWEK-sin*) Any of several plant hormones that can produce a variety of effects in different types of cells and in different concentrations, including promoting cell growth and, in some circumstances, inhibiting it.

axial skeleton (*AXE-ee-ul*) The central supporting skeleton, including the skull, backbone, ribs, and breastbone.

axon (*AXE-on*) The long, slender process of a neuron that conducts impulses away from a cell body.

bacteriophage (*back-TEER-ee-oh-fahj*, "*bacteria eaters*") A class of viruses that attack bacteria.

Barr body The condensed, inactive X chromosome found within the nuclei of female cells.

basal bodies Barrel-like structures in the cell cytoplasm to which cilia and flagella are attached.

basal firing rate The slow, steady pace at which a resting muscle spindle produces action potentials.

basal metabolic rate The minimal rate of energy use that is required to sustain vital functions in an organism at complete rest.

base A compound that removes hydrogen ions (H^+) from solution, raising the pH. A basic solution has a pH higher than 7. Examples include ammonia, hair remover, and oven cleaner.

base pairing The specific hydrogen bonding between complementary nucleotide bases, in which a specific purine pairs with a specific pyrimidine. In the two strands of a DNA molecule, adenine pairs with thymine, and guanine pairs with cytosine. During transcription, when an mRNA molecule forms along a single strand of a DNA molecule, adenine pairs with uracil instead of thymine.

basidiomycetes (*bass-ID-ee-oh-MY-seets*) Members of the division Basidiomycota. Fungi that produce spores in reproductive structures known as basidia; this group includes the mushrooms, puffballs, and shelf fungi.

Batesian mimicry (*BAIT-see-un*) A form of mimicry in which a harmless species resembles a poisonous or otherwise protected species; the harmless species thereby gains protection.

behavioral isolation A prezygotic isolating mechanism through which two species do not interbreed because of behavioral incompatibility.

benign tumor (*bih-NINE*) A type of tumor that does not metastasize and can therefore often be removed surgically.

beta carotene (*BAY-ta CARE-a-teen*) A reddish-orange accessory photosynthetic pigment that absorbs light in regions of the spectrum where chlorophyll absorbs less effectively.

beta particle (*BAY-ta*) A fast-moving electron such as those released during radioactive decay.

beta sheet (*BAY-ta*) A chemical configuration of the secondary structure of proteins in which hydrogen bonds link adjacent parallel or antiparallel polypeptide chains.

bicarbonate ion A soluble ion (HCO_3^-) that forms in the blood from a chemical reaction involving the carbon dioxide (CO_2) that diffuses into the bloodstream as a waste product from cells. In the capillaries of the lungs, bicarbonate ion in blood is converted back to carbon dioxide, which diffuses out of the blood and is exhaled. Bicarbonate ion is also important in maintaining the pH of the blood.

bilateral symmetry (*bye-LAT-er-al*) Symmetrical organization in which, when an organism is divided down a central longitudinal plane, the two halves are mirror images of one another. Worms and many higher organisms exhibit bilateral symmetry.

binary fission (*BYE-na-ree*) A process of prokaryotic cell division. DNA is replicated while attached to the cell membrane; the membrane between the DNA molecules grows, separating the DNA molecules; ultimately, the cell splits, each daughter cell receiving a copy of the DNA.

binomial nomenclature (*bye-NOME-ee-al NO-men-clay-chur*) The naming system in which each organism is designated by a two-part Latin name, designating genus and species.

biogeochemical cycles The cycling or passage of chemical elements that serve as nutrients for organisms through living organisms, geological processes, and geological features.

biogeochemical view The thesis that the physical and chemical characteristics of the earth and/or its atmosphere are governed by both physical and biological processes.

biogeography (*BYE-oh-jee-OG-ra-fee*) The study of the geographic distribution of species.

biological clock Internal time-keeping mechanisms that adjust the biochemical processes within an organism to the length of day and the season of year.

biological diversity Also called *biodiversity*. The variety of organisms in an area, encompassing diversity in such realms as genetics, species, and ecosystems, among others.

biological magnification The process by which substances such as toxic pollutants come to be found in increasing concentrations in the tissues of organisms at higher trophic levels.

biome (*BYE-ome*) A major ecological community inhabited by certain characteristic types of plants and animals. The terrestrial biomes of the world include tundra, taiga, temperate forest, desert, grassland, and tropical rain forest.

bioreactor A device that nurtures cultures of genetically altered bacteria or eukaryotic cells and collects the genetically engineered proteins that they produce.

biosphere (*BYE-oh-sfeer*) Collectively, the regions of the earth that support life, including the lower atmosphere, the earth's surface, and water environments.

biotic environment (*bye-AH-tik*) Also called *biological environment*. The living aspects of an organism's environment.

bivalve (*BYE-valve*) A clam, oyster, or scallop. A member of the class Bivalvia, a group of molluscs with two shells, or *valves*; most bivalves are sedentary filter feeders.

bladder In an excretory system, the membranous sac in which urine is stored before being excreted.

blastocyst (*BLASS-toe-sist*) In mammals, the equivalent of the blastula phase of embryonic development, in which an inner cell mass produces embryo tissues and an outer trophoblast produces the tissues that draw nourishment from the mother.

blastodisc (*BLASS-toe-disk*) In birds, the equivalent of the blastula phase of embryonic development, in which a double layer of cells on top of the yolk mass gradually separates into two layers with a hollow blastocoel between them.

blastula (*BLASS-chew-lah*) The hollow ball of cells that results from cleavage in early embryonic development.

blood In animals, the fluid connective tissue containing many specialized cells that moves through the circulatory system.

blood–brain barrier A highly selective membrane formed by the membranes of brain capillaries, which do not let pass as many substances as capillaries in other parts of the body. Many large proteins, including certain hormones and drugs, do not pass into the brain.

blood pressure The force exerted by the blood against the walls of blood vessels with each beat of the heart.

B-lymphocyte (*B-LIM-foe-site*) Also called *B-cell*. A type of white blood cell that produces antibodies.

bone A type of strong, rigid connective tissue made up of a matrix of collagen that has gradually been mineralized by calcium crystals. The skeleton consists of bones, which provide structural support and protect internal organs.

Bowman's capsule In the kidney, the cup-shaped structure that much of the blood plasma enters after being filtered by the glomerulus.

breathing Also called *ventilation*. The movement of air into and out of the lungs.

bronchioles (*BRONG-kee-oles*) The progressively finer tubes into which the bronchi split; bronchioles bear alveoli at their tips.

bryophytes (*BRI-oh-fights*) A division of the plant kingdom that includes the mosses, liverworts, and hornworts; a large, ancient group of small, nonvascular plants commonly found in moist terrestrial habitats.

buffer A substance that tends to stabilize pH by maintaining the relative concentrations of hydrogen and hydroxyl ions in solution. Buffers occur in living cells.

calorie (*KAL-a-ree*) The amount of heat energy needed to raise the temperature of one gram of water one degree Celsius. The "calorie" commonly used to express the amount of energy in food is actually a *kilocalorie* (*kcal*), or Calorie, which equals one thousand calories.

Calvin–Benson cycle Also called *C3 cycle*. A stage of the light-independent reactions of photosynthesis in which two three-carbon molecules (PGA) are produced by adding a carbon dioxide molecule to a five-carbon sugar; the reaction is catalyzed by the enzyme RuBP carboxylase. Energy for the reaction is provided by NADPH and ATP produced by the light reactions.

cambium (*KAM-bee-um*) The lateral meristem, the cells in vascular plants that are responsible for lateral growth. Vascular cambium produces secondary xylem and phloem; cork cambium produces cork that fills spaces where phloem tissue has been split by growth.

cAMP (cyclic adenosine monophosphate) (*SICK-lick ah-DEN-a-seen mon-oh-FOSS-fate*) A molecule that functions as a chemical messenger in slime molds; in vertebrate endocrine systems, cAMP functions as an intracellular second messenger.

cancer An invasive, uncontrolled growth of certain cells of an organism at the expense, and to the detriment, of other cells and the organism.

cannibalism The eating of members of one's own species.

capillaries In closed circulatory systems, small blood vessels with walls seldom more than one cell thick, permitting diffusion between body cells and blood.

capillary action (*KA-pill-air-ee*) The movement of a liquid along a tube caused by adhesion of the liquid's molecules to the tube walls and those molecules' cohesion to each other. Plants rely on capillary action to bring liquids to their upper branches.

carbohydrate (*KAR-bow-HIGH-drate*) An organic molecule in which carbon, hydrogen, and oxygen are present in a ratio of about 1:2:1. Sugars, starches, and cellulose are carbohydrates.

carbon cycle The cycle that carbon naturally follows through the biosphere by such processes as photosynthesis, cellular respiration, and the decomposition and combustion of inorganic compounds.

carbon monoxide (*KAR-bun mun-OX-ide*) A colorless, odorless gas (CO) produced when fuel is burned under conditions of limited oxygen availability; can supplant oxygen at the binding sites of hemoglobin and cause rapid death.

carcinogen (*kar-SIN-oh-jen*) A chemical capable of causing cancer.

cardiac cycle The two-stage contraction-and-relaxation (systole/diastole) pattern of the heartbeat that moves blood efficiently through the heart's chambers and produces the characteristic "lub-dub" sounds of a heartbeat.

carnivore (*KAR-nih-vore*) An organism (usually an animal but occasionally a fungus or plant) that subsists by trapping and digesting other animals.

carpel (*KAR-pel*) The female portion of a flower, which contains the ovary, style, and stigma.

carrying capacity The maximum population of a species that a given environment can sustain for an extended period of time.

cartilage (*KAR-tih-lidg*) A type of tough, resilient connective tissue composed of collagen and other fibers; in humans, cartilage occurs at bone joints and in flexible structures such as the ears and nose.

Casparian strip (*kass-PAR-ee-un*) A thin, waxy strip contained in endodermal cell walls that prevents water from moving between cells.

caste Among social insects, a class of individuals of similar body form and behavior that performs a particular type of activity within the colony.

catabolic pathway (*CAT-a-BALL-ik*) A chemical pathway in which larger, more complex molecules are broken down into simpler, smaller molecules.

catalyst (*CAT-a-list*) A chemical that changes—especially that increases—the rate of a reaction without being consumed in the process. In effect, catalysts lower the activation energy of the reaction.

catastrophism (*KAT-a-STROFF-ism*) The archaic view that accounted for the fossil record by contending that, after creation, life was destroyed in a series of catastrophes and repopulated by survivors or created anew.

cation (*CAT-eye-un*) A positively charged ion.

cell cycle The sequence of events during the life span of a growing eukaryotic cell, consisting of four phases: mitosis (M) or cell division, a "gap" (G$_1$) between mitosis and the next phase, which is DNA synthesis (S), followed by another "gap" (G$_2$) between DNA synthesis and mitosis.

cell division The splitting of a cell to form two cells.

cell junction Specific attachments between adjacent cells in multicellular organisms.

cell-mediated immune response The cell-to-cell process in which activated T-lymphocytes of the immune system recognize and destroy foreign cells or tissues.

cell membrane The outer membrane that separates a cell from its environment.

cell plate The double membrane that forms between the two halves of a dividing plant cell and produces the new cell wall.

cell theory The theory that all living organisms are composed of individual, living, self-reproducing structures known as cells.

cell wall Tough, rigid structure surrounding the cell membrane of many plant, algal, and bacterial cells.

cellular respiration Collectively, the pathways of glycolysis, the Krebs cycle, and electron transport, which produce ATP through the consumption of oxygen in reaction with an organic fuel.

cellular slime molds Fungus-like protists that spend part of their lives as independent, free-living amoeboid cells, but when food becomes scarce, they swarm together to produce a multicellular, slug-like *plasmodium* that eventually produces a tall fruiting body that releases spores.

central nervous system (CNS) The brain and spinal cord.

centrioles (*SEN-tree-oles*) Barrel-shaped structures near the cell nucleus that serve as foci of the poles of mitotic spindles in animal cells.

centromere (*SEN-troe-mere*) The specialized region of a chromosome to which two sister chromatids are joined during mitosis.

cephalization (*SEFF-a-lies-ZAY-shun*) The evolutionary concentration of sensory and nerve tissue at one end of an organism, forming a head (or head end).

cephalopod (*SEFF-a-low-pod*) An octopus, squid, or nautilus. A member of the class Cephalopoda, a group of predatory marine molluscs. Octopi and squid are shell-less, have good vision and advanced nervous systems, and are good swimmers.

cerebellum A region of the hindbrain that coordinates body movements through sophisticated integrating functions. It regulates balance, controls muscle tone, and coordinates opposing muscle groups.

cerebral cortex (*su-REE-brul KOR-tex*) The convoluted outer layer of gray tissue on the surface of the cerebrum that is responsible for most of the higher functioning of the nervous system.

cerebrum The largest part of the human brain and the one that has changed the most through evolution. Divided into two hemispheres.

C4 pathway A pathway of carbon fixation in which carbon dioxide is first incorporated into oxaloacetate (which contains four carbon atoms) before the Calvin cycle is initiated.

chelicerate (*ke-LISS-er-ate*) A member of the subphylum Chelicerata, a group of arthropods characterized by a two-part body consisting of a cephalothorax and an abdomen and by the presence of chelicera and pedipalps. This group includes ticks, mites, scorpions, spiders, and horseshoe crabs.

chemical energy A form of potential energy stored in molecular structure, such as the energy in gasoline.

chemiosmosis (*KEM-ee-oz-MOE-sis*) The production of ATP from ADP through the operation of an electrochemical gradient across a cell membrane. The gradient results from the buildup of protons pumped through the membrane by the electron transport chain.

chemosynthetic bacteria Bacteria that make their own organic molecules using certain inorganic molecules (such as hydrogen sulfide) as a source of energy.

chimera (*ki-MEER-a*) In genetic engineering, a combination of a DNA fragment and a plasmid; generally, an organism that has tissues from two or more genetically distinct parents.

chiton (*KYE-tin*) A member of the class Polyplacophora, a group of marine herbivorous molluscs with shells divided into eight connected plates.

chlamydia (*kla-MID-ee-ah*) An increasingly common sexually transmitted disease caused by the bacterium *Chlamydia trachomatis*, which can infect individuals for long periods without overt symptoms, leading to sterility in women.

chlorophyll (*KLOR-oh-fill*) Any of several green pigment molecules that absorb light during the process of photosynthesis.

chlorophytes (*KLOR-oh-FIGHTS*) Members of the plant division Chlorophyta. The green algae, a group of aquatic or semiterrestrial photosynthetic organisms that vary enormously in size and complexity, including single-cell organisms,

colonial forms, and multicellular species such as *ulva*, the common sea lettuce.

chloroplast (*KLOR-oh-plast*) A eukaryotic organelle that contains chlorophyll and other substances associated with photosynthesis.

chlorosis (*klor-OH-sis*) Yellowing of older leaves resulting from a magnesium deficiency that prevents the plant from replacing damaged chlorophyll quickly enough to maintain the normal green color.

cholesterol (*kol-ESS-tur-awl*) An essential steroid component of animal cell membranes and a precursor of other steroids, which, if present in consistently high concentrations in the bloodstream, can lead to atherosclerosis.

Chondrichthyes (*kon-DRIK-thee-eez*) Cartilaginous fishes, the class of fishes with flexible skeletons made of cartilage rather than bone; includes the sharks and rays.

chordates (*KOR-dates*) Members of the phylum Chordata, a group of animals with a hollow dorsal nerve cord and, at least in embryonic stages, a notocord, a series of pharyngeal gill slits, and a tail that continues past the end of the digestive tract.

chorion (*KOR-ee-on*) The outermost layer of tissue surrounding the embryos of mammals, birds, and reptiles; in mammals it is a major component of the placenta.

chorionic villi biopsy (*KORE-ee-ON-ic VIL-eye BYE-op-see*) A recent medical procedure in which a thin tube is inserted through the vagina into the tissue surrounding the placenta of a pregnant female to withdraw cells from the chorion, which are derived from the developing offspring. These cells can be examined immediately for genetic disorders.

chromatids (*KROW-ma-tids*) The two strands of a duplicated chromosome that are connected to the centromere before and during nuclear division.

chromatin (*KROW-ma-tin*) The complex of protein and DNA in the eukaryotic nucleus.

chromosome (*KROW-ma-sowm*) A structure in the nucleus of a eukaryotic cell, consisting of protein and DNA, that contains part or all of an organism's genetic inheritance.

chromosome mutations Mutations that occur at the level of the chromosome, as when large sections of DNA are deleted, inverted, or transposed from one region of the genome to another.

chrysophytes (*KRICE-oh-FIGHTS*) Members of the phylum Chrysophyta. Photosynthetic plant-like protists, including single-celled algae such as diatoms, golden-brown algae, and yellow-green algae. Many are flagellated; diatoms have silica shells.

chylomicrons (*KI-low-MY-krons*) Lipid droplets in the endoplasmic reticulum, formed from fatty acids that have been resynthesized into triglycerides. Chylomicrons are absorbed by the lacteals of the lymphatic system and eventually enter the general circulation.

chyme (*KIME*) The product of the digestive processes of the stomach, a nutrient-rich milky slurry that is sent to the small intestine.

cilia (*SILL-ee-uh*) (singular, **cilium**) Hair-like structures less than 20 micrometers long that protrude from the cell surface and are used in cell locomotion or in moving the air or fluids surrounding the cell.

ciliates (*SILL-ee-ates*) Members of the phylum Ciliata. Animal-like protists characterized by the presence of many rapidly beating hairs, or cilia; may be either free-living or sedentary. Includes the well-known genus *Paramecium*.

circadian rhythm (*sir-KAY-dee-an*) A 24-hour cycle of regular physiological activity.

circulating tissue The blood and lymph, including both cellular and fluid components.

circulatory system In multicellular animals, a group of organs and tissues that circulates a fluid substance (generally called blood) through the body, carrying respiratory gases, nutrients, chemical messages, waste materials, and living cells from one place to another.

cladistics A school of thought in systematics that groups organisms together on the basis of their shared evolutionary history. Organisms are grouped with their closest relatives rather than with organisms that happen to resemble them.

classical conditioning The association of an initially neutral stimulus with a stimulus that elicits a particular response such that the formerly neutral stimulus alone also comes to elicit the response.

cleavage (*KLEE-vej*) The rapid division of cells of the early embryo from the stage of the fertilized ovum to the blastula or blastocyst stage.

cleavage furrow (*KLEE-vej*) The progressive constriction in the cytoplasm between two nuclei that forms during the process of cytokinesis in animal cells.

climate The weather conditions, including temperature, rainfall, wind, hours of sunlight, and other factors, that *on average* or *typically* prevail in a particular area; compare *weather*.

climax community The ultimate stage in a succession sequence; during this stage, the ecological community remains stable as long as environmental conditions do not change.

clone (*KLOHN*) A group of genetically identical cells (or organisms) descended from a single ancestor cell.

closed circulatory system A circulatory system in which the circulatory fluid is confined to blood vessels.

clotting factor One of the proteins needed for blood clotting; absent in hemophiliacs.

cnidarians (*NID-air-ee-ans*) Members of the invertebrate phylum Cnidaria, a varied group of invertebrates with a radially symmetric body plan and specialized tissues including a nerve net. Includes the jellyfish, hydroids, sea anemones, and corals.

codominance (*koh-DOM-ih-nance*) In genetics the situation in which two alleles of a gene are both expressed in the heterozygote.

codon (*KOH-dawn*) A nucleotide triplet of three bases that codes for insertion of an amino acid or for chain termination at a particular location during protein synthesis.

coelacanth (*SEEL-a-kanth*) A primitive fleshy-finned fish long thought to be extinct but rediscovered in 1938.

coelom (*SEE-lum*) In many animals, a body cavity lined with mesoderm tissue, which differentiates to form structures called mesenteries from which internal organs are suspended.

coelomates (*SEE-loh-MATES*) Bilaterally symmetrical animals with a fluid-filled body cavity called a *coelom* lined with mesoderm tissue.

coenzyme (*KOH-en-zime*) A cofactor that is an organic molecule. Many vitamins that humans require in their diets are coenzymes that they cannot synthesize themselves.

coenzyme A (*KOH-en-zime*) A molecule of adenine, ribose, pantothenic acid, and sulfur that permits two carbon atoms at a time to enter the Krebs cycle by passing the acetyl group to a molecule of oxaloacetic acid.

coevolution (*KOH-ehv-eh-LEW-shun*) The evolution of two species in close association such that they reciprocally influence one another's adaptations.

cofactor (*KOH-fak-tore*) A molecule or ion required by some enzymes to function properly.

colon The large intestine.

columnar epithelium (*call-UM-nar ep-ih-THEE-lee-um*) A tissue composed of large, roughly columnar cells that function in secretion and absorption; found in such locations as the stomach and intestinal linings.

commensalism (*kum-MEN-sul-izm*) A type of symbiosis in which one organism benefits and the other is neither harmed nor benefited.

common descent Descent of diverse types of organisms from a shared ancestor.

communication The passage of information from one animal to another such that the subsequent behavior of the recipient is affected.

community A coherent assemblage of populations of organisms that inhabit the same environment and interact with each other.

competition Rivalry between organisms that require the same or similar resources over access to these resources when they are in limited supply.

competitive exclusion The principle that when two species live in the same environment at the same time and both require one or more of the same limiting resources, one species always drives the other to extinction locally.

competitive inhibitor A molecule that blocks the active site of an enzyme and prevents it from being occupied by a substrate.

complement proteins A group of circulating blood proteins that, when they encounter an antibody bound to a cell surface, attach to the cell membrane and puncture it, causing the cell to burst. Complement proteins destroy cells too large to be eaten by phagocytes.

compound A substance that is composed of two or more elements in fixed proportions and the characteristics of which generally differ from those of its constituent elements.

compound eye The type of eye found among arthropods, consisting of many individual "eyes" called ommatidia, each with a lens-like structure and a "retinula," or small retina, containing photoreceptor cells.

conception (*kon-SEP-shun*) Fertilization of an ovum.

condensation reaction A chemical reaction common in the formation of biological polymers in which monomers join through the removal of water molecules. One monomer loses a hydrogen ion and the other loses a hydroxyl group; the monomers form a new bond, and the hydrogen and oxygen can join to form a water molecule.

cones One of the two classes of photoreceptors in the neural retina. Cones function well only in relatively bright light and are responsible for the sharp, full-color visual images we receive in daylight.

conifers (*KON-ih-ferz*) A group of gymnosperms, the reproductive structures of which are contained in male and female cones; the conifers include the pines.

connective tissue The tissue that makes up the basic support structures of the body, including the bones, ligaments, and tendons.

contact inhibition The phenomenon commonly observed in cells growing in tissue culture that is marked by cessation of cell division when the cells come in contact with each other.

continental shelf A relatively shallow, downward-sloping, shelf-like extension of the continental shoreline into the sea, terminating abruptly at a steep, downward declination into the oceanic abyss.

continuous variation The existence in a population of a full range of phenotypes between two extremes.

contraceptive (*KON-trah-SEP-tiv*) A device or technique designed to prevent conception.

contractile vacuole (*kon-TRAK-tile VAK-you-ole*) An organelle found in some single-celled organisms that accumulates the cell's excess water and contracts, pumping it out of the cell.

control In an experiment, the subject or group to which the experimental variable is not applied.

controlled experiment A scientific procedure in which two parallel trials are performed, their conditions differing in only one introduced factor. Variations in results can then be reliably attributed to the introduced factor.

convergent evolution The process by which unrelated organisms exposed to similar environments and selective pressures come to resemble one another.

convoluted tubule (*KON-va-LOO-tid TUBE-yule*) The kidney tubule into which the primary filtrate empties; functions in reabsorption of salt and water molecules into the blood.

coral reef A large limestone structure constructed by the secretions of small animals in nutrient-poor tropical seas; home to a diverse community of plant and animal life.

coronary thrombus (*KOR-a-NAIR-ee THRAHM-bus*) A small blood clot that becomes lodged in one of the coronary arteries, blocking blood flow to the heart.

corpus callosum A narrow band of axons that connects the left and right hemispheres of the human brain.

corpus luteum (*KOR-pus LOO-tee-um*) A structure that develops from a ruptured ovarian follicle after that follicle produces an egg; secretes estrogen and progesterone.

countercurrent flow The flow of adjacent fluids in opposite directions. In gills, countercurrent flow maximizes rates of gas exchange; in kidneys, a countercurrent system functions to concentrate waste products in urine.

countercurrent multiplier system A system in which the configuration of the tubules in the loop of Henle in the mammalian kidney amplifies the abilities of the membranes to produce an osmotic gradient as fluids flow past one another in opposite directions. This system functions to concentrate waste products in urine.

coupling factor An enzyme that connects electron transport to the synthesis of ATP during photophosphorylation.

covalent bond (*koh-VAY-lent*) A chemical bond formed when two atoms share a pair of electrons.

Crassulacean Acid Metabolism (CAM) A form of carbon fixation used by members of the *Crassulaceae* family (such as many desert cacti and the "ice plants" along West Coast freeways).

creatine phosphate (*KREE-a-teen*) The compound in which most of the quickly available energy in vertebrate muscles is stored; contains a high-energy phosphate bond.

cristae (*KRISS-tee*) Intrusive convolutions of the inner membrane of mitochondria.

Cro-Magnon (*krow-MAG-nun*) An early form of *Homo sapiens* that appeared in Africa about 100,000 years ago and migrated throughout Europe and Asia.

crossing over Also called *genetic recombination*. In meiosis, the exchange of genetic material between homologous chromosomes during synapsis.

crustacean (*krust-AY-shun*) Member of the invertebrate subphylum Crustacea, a group of arthropods with two limb branches. Includes shrimp, crabs, lobsters, copepods, and barnacles.

C3 pathway In photosynthesis, a pathway of carbon fixation in which the first stable compound produced has three carbons; the compound is created through the Calvin–Benson cycle.

cuboidal epithelium (*kyew-BOY-dull ep-ith-EEL-ee-um*) A tissue composed of roughly cubical cells that generally function in secretion; found in such locations as the kidneys and a variety of glands.

cuticle (*CUE-tih-kul*) A protective outer layer, such as the waxy, waterproof material covering the leaf surfaces of land plants.

cyanobacteria (*sy-AN-oh-bak-TEER-ee-uh*) Formerly called *blue–green algae* or *blue–green bacteria*. Prokaryotic bacteria that carry on photosynthesis; some are capable of nitrogen fixation.

cycads (*SIGH-cads*) Members of the plant division Cycadophyta, a group of slow-growing, tropical and subtropical palm-like gymnosperms that diversified successfully during the Permian period but has only about 100 species today.

cyclic electron flow A pattern of electron flow involving only photosystem I during the light-dependent reactions of photosynthesis. High-energy electrons pass along a series of carrier molecules and return to the reaction center from which they originated, creating a powerful proton gradient across the photosynthetic membrane and driving the synthesis of ATP.

cystic fibrosis A heritable, progressive, chronic disease affecting primarily the lungs and pancreas and leading to death, usually in early childhood.

cytochrome c (*SIGH-toe-KROHM see*) A molecule in electron transport systems. Its structure varies slightly in a large number of organisms, ranging from yeasts and molds to sunflowers to humans, making it useful as the basis for constructing evolutionary trees.

cytokinesis (*SY-toe-kie-NEE-sis*) The biological process by which the cytoplasm of a cell divides following mitosis.

cytokinins (*SY-toe-KYE-nins*) Plant hormones that work with auxin and complement its effects, stimulating cell division and regulating plants' shape and growth.

cytoplasm (*SIGH-toe-plaz-im*) ("cell fluid") The major compartment in most living cells, consisting of everything bounded by the cell membrane but outside of the eukaryotic cell nucleus.

cytosine A nitrogen-containing base found in nucleotides, belonging to the chemical group known as pyrimidines, which have a single carbon-nitrogen ring in their basic structure. Found in DNA and RNA.

cytoskeleton (*SY-toe-SKEL-e-ton*) Found in the cell, a skeleton-like structure composed of microtubules, microfilaments, and intermediate filaments, that helps support the cell and aids in locomotion.

cytotoxic T-cell (*SY-toe-TOK-zik*) A type of T-cell that directly attacks an antigen-bearing foreign cell and destroys it by attacking the cell membrane.

day-neutral plant A plant the flowering of which is not affected by the duration of daylight.

DDT (*dichlorodiphenyltrichloroethane*) An insecticide that persists in the environment, accumulates in fatty tissues, and can have harmful environmental consequences; once widely used in the United States but now banned.

deductive reasoning Reasoning in which a conclusion by necessity follows from a given premise. (In the scientific method, the hypothesis derived by induction represents the premise used in deduction, and the conclusion that follows may represent a prediction).

deforestation The destruction of forests.

deletion An alteration in chromosome structure in which part of a chromosome is missing.

demographic transition A change in the pattern of human population growth that occurs as economic development takes place: first death rates drop due to improvements in health care, nutrition, and sanitation, and later birth rates drop. Today, the U.S., Europe, and Japan have completed the demographic transition and have slowly growing populations. Much of the rest of the world is undergoing the demographic transition, having lower death rates but high birth rates and the resulting higher population growth.

demography (*de-MAWE-gra-fee*) Study of the characteristics of a population, such as size, age distribution, fertility, mortality, and survivorship.

denature To disrupt the structure of a protein to such an extent that it no longer performs its functions. For example, very high temperatures may disrupt enzyme molecules near their active site, so they no longer catalyze reactions.

dendrite (*DEN-drite*) One of the many finely branched processes extending from the cell body of a neuron. Generally, one set of dendrites picks up information from the environment and carries it to the cell body; another set receives information from axons and distributes it to other neurons or to effector cells.

denitrification The process by which certain bacteria convert nitrogen from nitrogen-containing organic compounds into nitrogen gas (N_2) and release it into the atmosphere. Denitrification is part of the nitrogen cycle.

density-dependent factors Factors that influence population growth particularly strongly when population density is high. Competition, predation, parasitism, emigration, immigration, cannibalism, the accumulation of toxic metabolic wastes are density-dependent factors.

density-independent factors Factors that influence population growth to the same extent whatever the population density. Droughts, floods, cold spells, heat waves, forest fires, and volcanic eruptions are examples of density-independent factors.

dermal tissue (*DER-mul*) The outer layer of cells that protects a plant.

desert A biome characterized by very low annual rainfall and by plants and animals specially adapted to living in dry conditions.

desertification The conversion of forest, grassland, or other ecological communities to desert, often through habitat destruction or poor farming methods.

detritus (*deh-TRITE-us*) A combination of bits of decaying organisms and dung containing partially digested food.

deuteromycetes (*DEW-ter-oh-MY-seets*) Members of the fungal division Deuteromycota. Also called *imperfect fungi*. A varied group of fungi that lack a sexual stage in their life cycle. Includes disease-causing parasites of plants and animals, as well as several *Penicillium* species that produce antibiotics.

deuterostome (*DEW-ter-oh-stome*) A member of one of two evolutionary lines distinguished by differences in embryonic development. In deuterostomes, cell division produces a radial arrangement, the ultimate function of each cell being determined relatively late in development. The coelom is formed by buds that branch and separate from the embryonic gut. The embryonic gut opening becomes the anus, and a new opening forms the mouth.

diabetes mellitus (*die-a-BEE-tis MEL-ih-tus*) A disorder caused by failure of the pancreas to produce enough insulin, which prevents tissues from absorbing sugar from the blood at the proper rate.

diaphragm (*DI-a-fram*) The muscular partition that separates the thoracic and abdominal cavities; functions in expanding and contracting the thoracic cavity to carry out respiration.

diastole (*die-AS-ta-lee*) The state of the heart when the ventricles relax.

dicot (*DIE-cot*) Also called *dicotyledon (DIE-cot-uh-LEE-don)*. A member of a class of angiosperms (flowering plants) characterized by two seed leaves (cotyledons) and numerous other structural similarities, such as net-like veins, vascular tissue arranged in a ring, and floral parts in fours or fives.

differentiation The process through which cells produce specialized structures or come to perform specialized functions over the course of development.

diffusion The process through which a substance moves passively from areas where it is in higher concentration to areas of lower concentration.

dinoflagellates (*DINE-oh-FLAJ-a-lates*) Also called *pyrrophytes*. Members of the phylum *Pyrrophyta*. "Armored" flagellates, protists with bodies encased in tough cellulose plates. Most are capable of photosynthesis; they are an important component of phytoplankton.

diploid (*DIP-loid*) Having the full complement of two sets of chromosomes, one set from each parent.

directional selection A form of natural selection that favors individuals at one end of a range of phenotypic expression and thereby shifts the phenotype of the population as a whole.

discontinuous variation The existence in a population of a few well-separated categories of phenotypes.

disruptive selection A form of natural selection that reduces the numbers of individuals that exhibit the moderate range of a trait and favors expression at the extremes of the range.

distance effect The relationship between the distance of an island from a continent and the diversity of species inhabiting the island. Generally, an island close to a continent is likely to have more species than an equal-sized island farther away.

disulfide bond (*die-SUL-fide*) A sulfur-to-sulfur covalent bond between two cysteines that under certain conditions links two regions of a polypeptide chain. The two cysteine molecules may be remote from one another in the primary sequence of the molecule.

DNA (deoxyribonucleic acid) (*DEE-oxx-ee-RYE-bow-new-CLAY-ik*) A double-stranded helical nucleic acid molecule containing genetic information and capable of replicating itself, thereby passing on genetic instructions from one generation of cells to the next. The material of which genes are made.

DNA fingerprinting A technique for identifying individuals on the basis of certain DNA regions, which vary greatly from one person to the next. These regions, properly selected and sequenced, can theoretically be used to create a "DNA fingerprint" that is unique to each person, opening up new possibilities in criminology.

DNA polymerase (*pol-IM-er-aze*) The most important enzyme in DNA replication. DNA polymerase attaches the nucleotides to the new complementary strands of DNA that form along the existing strands. It also "proofreads" newly formed base pairs, removing and replacing any incorrect nucleotides.

dominance hierarchy A social ranking among members of a group, usually established and continually reshaped through aggressive interactions.

dominance, principle of The principle that when two or more forms of the same gene exist, one expresses itself rather than the other(s).

dominant gene A gene that expresses itself in the phenotype when paired with either a dominant or a recessive gene.

dorsal Toward the back.

double-blind study An experimental procedure for ensuring the objectivity of an observation by concealing from the observer, until the conclusion of the experiment, the identity of both the experimental sample and the control sample.

double fertilization In flowering plants, the fusion of a sperm nucleus with an egg nucleus to produce a diploid zygote, along with the fusion of a second sperm nucleus with the two polar nuclei of the nearby endosperm mother cell to form a triploid endosperm cell, which develops into nutritive triploid endosperm tissue.

Down syndrome Also called *trisomy 21*. A genetic disorder in humans caused by the presence of three copies of chromosome 21; results in mental retardation, reduced resistance to disease, and greatly lowered life expectancy.

duodenum (*doo-oh-DEE-num*) The upper portion of the small intestine that is connected to the stomach.

ecdysone (*EK-die-sone*) A hormone that is produced by the arthropod prothoracic gland and induces molting.

echinoderms (*e-KINE-oh-derms*) Members of the phylum Echinodermata, a group of marine invertebrates characterized by radial symmetry, often spiny skin, and a water-vascular system used in feeding, locomotion, and respiration. Includes starfish, sea urchins, sea lilies, and sea cucumbers.

echolocation (*EK-oh-low-KAY-shun*) A type of radar system through which bats, marine mammals, and certain other animals use echoes of their own cries to navigate and to locate prey.

ecological isolation (*eek-oh-LOJ-ih-KUL*) A prezygotic isolating mechanism that functions when two species live in close proximity but are adapted to different environments.

ecological pyramid (*eek-oh-LOJ-ih-KUL*) The pyramid-shaped representation of successive trophic levels, each of which is smaller than the one below because only about 10 percent of the energy stored in one level is ultimately contributed to the biomass of organisms at the next higher level.

ecology The science of the relationships among organisms and between organisms and their environment.

ecosystem (*oikos* means "house," *systema* means "that which is put together") Also called *ecological system*. A sizable interacting system composed both of living organisms and their physical environment.

ectoderm The outermost tissue or "germ" layer of triploblastic embryos.

ectotherm (*EK-toe-therm*) An animal that depends on external sources of energy to regulate its body heat.

effector (*ee-FEK-tore*) 1. A molecule that regulates the activity of an enzyme, either increasing or decreasing the rate of an enzyme-catalyzed reaction. 2. Any nonneural cell whose actions are under neural control. Effectors form effector organs, such as muscles and glands, which are under neural control.

egg cell The gamete, produced by the female gametophyte, with which a sperm nuclei fuses to form a diploid zygote.

electron A negatively charged subatomic particle in the orbitals surrounding the nucleus of an atom.

electron transport chain A sequence of enzymes on the inner mitochondrial membrane that is critical to oxidation-reduction reactions as electrons are carried from one enzyme to the next. The process releases energy used in ATP formation and in other reactions.

electrophoresis (*eh-LEK-troh-fer-EES-iss*) The migration of weakly charged molecules or colloidal particles through a fluid or gelatinous medium under the influence of an electric field imposed on that medium.

element A pure substance that cannot be chemically decomposed into simpler substances. Each element consists of one type of atom with its own atomic number (number of protons).

El Niño (*el NEEN-yo*) ("The Child") A periodic variation in the global climate that develops when warm water from the western equatorial Pacific Ocean spreads eastward in concert with shifting patterns of atmospheric pressure known as the Southern Oscillation. These interrelated changes in global wind and ocean current patterns cause warming that alters typical weather patterns around the world, bringing rains to some areas and droughts to others.

embolus (*EM-buh-luss*) A detached blood clot that drifts through the circulatory system until it lodges in an artery too small for it to pass.

embryo sac The seven-celled female gametophyte of an angiosperm, in which the embryo develops.

emigration The departure of individuals from a population or geographic area.

endemic species (*ǝn-DEM-ik*) A species that is native to a particular location and occurs nowhere else.

endocrine (*ENN-doh-krinn*) Pertaining to a ductless gland that secretes into the blood or tissue fluids, or to the secretion (a hormone) of such a gland.

endocytosis (*EN-doh-sigh-TOH-sis*) The process by which materials (such as molecules bound to proteins at the cell membrane surface) are brought into the cell enclosed in vesicles. The cell membrane turns inward, forming an *endocytic vesicle* that can be delivered to a destination within the cell.

endoderm The innermost tissue layer of triploblastic embryos.

endogonic reaction (*en-doh-GON-ic*) A chemical reaction that requires energy. In living organisms, endogonic reactions are often coupled with exogonic reactions, which release energy.

endorphins (*en-DOR-fins*) Polypepides secreted by the pituitary and brain that mimic opiates in alleviating pain.

endoskeleton (*EN-doe-skell-e-ton*) A system of internal bones and joints, such as that of vertebrates, that functions in support and locomotion.

endosperm (*EN-doe-sperm*) The triploid nutritive tissue, unique to flowering plants, that nourishes the embryo during early growth.

endotherm (*EN-doe-therm*) An animal capable of sustaining a constant body temperature by generating heat in its tissues.

entropy (*EN-troe-pee*) A measure of the degree of disorganization of a system. As the energy in a system is dispersed, the entropy increases.

environment The external biological and physical conditions that influence an organism.

enzyme (*EN-zime*) A substance that serves as a biological catalyst, allowing chemical reactions to take place at normal cell temperatures at rates that would otherwise be far too slow to support life.

epiglottis (*EP-ih-GLOT-iss*) A small flap that closes over the entrance of the trachea and prevents food from entering it during swallowing.

epistasis (*ep-uh-STAY-sis*) ("standing above") The condition when the genes at one locus control the expression of genes at another locus. The genes that control the expression of other genes are said to be *epistatic* to the others.

epithelial tissues (*EP-ih-THEE-lee-al*) The tissues that form surfaces on and within the body. These surfaces include the skin; the linings of the digestive system, blood vessels, and body cavities; and many secretory glands. Such tissues are composed of sheets of tightly packed cells.

equilibrium 1. A steady state or condition of balance, in which all ongoing changes balance or cancel out in such a way that there is no net change. 2. The stage in a chemical reaction at which the rates of the forward and reverse reactions are equal, such that the net amount of products or reactants is constant. 3. The sense of balance. In humans, the sense of equilibrium is mediated by hair cells in chambers in the inner ear. A jelly-like material presses down certain hair cells whenever the head changes position, causing a neural message to be sent to the brain.

equilibrium species Also called *K-selected species*. Organisms, such as many tropical birds and mammals, that collectively often make up a large proportion of species found in mature, stable ecosystems and have evolved a reproductive strategy that enables them to produce larger, stronger, and better-developed offspring with better chances of survival in a highly competitive environment. They typically mature and reproduce later in life, and produce fewer offspring, which they spend time and energy caring for.

erythrocytes (*ur-RITH-row-sites*) The red blood cells, which are produced by cells in the bone marrow and lack nuclei.

esophagus (*ee-SOFF-a-gus*) The muscular tube that transports food from the pharynx to the stomach.

essential amino acids Those amino acids that animals cannot synthesize and must acquire in their food (about half of the 22 amino acids commonly found in cellular proteins).

essential fatty acids Those unsaturated fatty acids that animals cannot synthesize and must acquire in their food. Essential fatty acids are needed to synthesize certain lipid components of cell membranes and the hormones known as prostaglandins.

essential mineral nutrients At least 18 inorganic nutrients required in very small amounts for normal human growth (includes calcium, phosphorus, sulfur, potassium, chlorine, sodium, magnesium, iron, fluorine, zinc, copper, manganese, iodine, cobalt).

essential nutrients Chemical elements that an organism requires for the construction of body tissues and cannot synthesize itself. For proper growth, most organisms require approximately 17 elements. Six of these (carbon, hydrogen, oxygen, nitrogen, phosphorus, and potassium) are required in large amounts, and 11 others (calcium, magnesium, sulfur, iron, manganese, sodium, boron, molybdenum, copper, zinc, and chlorine) are needed in smaller amounts.

estrogen (*ESS-troe-jen*) A female steroid sexual hormone that is produced in a follicle in the ovary and helps prepare the uterus lining for pregnancy. Estrogen is also active in the development of secondary female sexual characteristics.

estrus (*ESS-truss*) In nonprimate mammals, the period when a female is sexually receptive.

estuary (*ESS chew air ee*) A highly fertile coastal area where freshwater runoff from land mixes with seawater.

ethology (*ee-THOL-oh-gee*) The study of animal behavior in its natural environment.

ethylene (*ETH-a-leen*) A gaseous plant hormone that influences fruit formation and ripening. Its production is triggered by auxin.

eubacteria (*YOU-bak-TEER-ee-a*) A diverse group of prokaryotic bacteria, including many of importance to humans. Most of the bacteria that exist today are eubacteria.

euglenids (*you-GLEE-nids*) Flagellated photosynthetic protists, often common in fresh water. Members of the genus *Euglena* can function as autotrophs in sunlight but may exist as heterotrophs in darkness and under certain other conditions.

euglenoids (*you-GLEE-noyds*) Members of the phylum Euglenophyta. Plant-like protists, some of which have chloroplasts and carry on photosynthesis, and others of which absorb complex molecules from their environment to survive as heterotrophs or absorb simpler nutrients from their environment as fungi do. Contains the genus *Euglena*.

eukaryotic cell (*YOU-car-ee-AWE-tik*) A cell containing a true nucleus and other organelles that are individually bounded by membranes.

eutherians (*you-THEER-ee-uns*) The "true" or placental mammals, each female of which produces a placenta for nourishing a developing fetus.

eutrophic (*you-TROFE-ik*) Containing many nutrients. Because of heavy phytoplankton growth, eutrophic lakes are usually not clear.

evolution 1. The process by which populations of organisms change over time through such mechanisms as mutation, natural selection, and genetic drift. 2. In genetic terms, any change over successive generations in the relative frequencies of different genes in the gene pool of a population.

evolutionary adaptation A genetically controlled characteristic that increases an organism's evolutionary fitness.

evolutionary fitness The measure of an organism's capacity to pass on its genes to the next generation.

evolution through inheritance of acquired characteristics Lamarck's theory that changes in organisms over time resulted from the passing on to offspring of characteristics that individuals developed during their lives.

excretion The process by which metabolic wastes such as ammonia are eliminated from the body.

exergonic reaction (*ex-er-GON-ic*) A chemical reaction that releases energy. In living organisms, exergonic reactions are often coupled with endogonic reactions, which require energy.

exocrine (*EX-o-krinn*) Pertaining to a gland that secretes onto a free surface either directly or via a duct, or to the secretion (a hormone) of such a gland.

exocytosis (*EX-oh-sigh-TOH-sis*) The process by which materials (such as proteins synthesized within the cell) are released from the cell. The membrane surrounding a vesicle enclosing the material fuses with the cell membrane.

exon (*EX-on*) Regions of a eukaryotic gene (DNA) that actually code for the amino acid sequences of proteins.

exoskeleton A hard external skeleton.

experimental variable In an experiment, the factor introduced into the experimental group (but not into the control group) to assess the factor's influence.

exponential growth Population growth in which the growth rate increases as the size of the population increases. For example, a pair of rabbits might have 6 offspring, which, if unrestrained by environmental factors, might have 36 offspring, which could have 216 offspring, and so on. Also known as *logarithmic growth* and typically represented on a graph as a J-shaped curve.

exteroceptor (*ex-TARE-oh-sep-tor*) A receptor that monitors conditions impinging on the organism from the exterior, such as taste, temperature, light, touch, and certain kinds of pain.

extracellular fluid In multicellular organisms, the fluid outside the cells of the body, which bathes the cells and makes up their immediate environment. Its temperature, nutrient and oxygen content, and salt concentration must remain within narrow ranges for cellular activities to continue.

extracellular matrix A layer of fibers and a ground substance surrounding and produced by the cells of connective tissues; may be fluid or solid, loose or dense; responsible for many of the properties of the various connective tissues.

facilitated diffusion A form of diffusion in which molecules cross membranes by passing through special pores or by using transport molecules that facilitate their passage. Like simple diffusion, facilitated diffusion is driven by a concentration gradient.

fallopian tubes (*fah-LO-pee-un*) In the female reproductive system, the tubes that convey the mature ova from the ovaries to the uterus.

fatty acid A monomer consisting of a long hydrocarbon chain with a carboxyl group at one end. Fatty acids are a building block of most lipids.

feedback control A self-regulatory mechanism by which the effect of a given activity or function influences that very activity or function. In negative feedback, the activity is inhibited or depressed. In positive feedback, the activity is enhanced.

feedback inhibition A control process in which the accumulation of an end product inhibits the process that produces that end product. Many chemical reactions in the cell are regulated by feedback inhibition.

fermentation A pathway of carbohydrate metabolism that produces relatively small amounts of ATP from glucose without an electron transport chain. A substance such as ethanol or lactic acid is also produced.

fertility rate The rate at which organisms in a given population produce offspring, usually measured in number of births per female.

fertilization The union of the nuclei of two haploid gametes to form a diploid zygote.

fertilizers Essential nutrients applied by humans to improve the condition of soil for growing crops.

fibroblast (*FYE-bro-blast*) A type of cell, common in loose connective tissue, that produces the fiber-like proteins of the extracellular matrix.

fitness The physical and behavioral characteristics of an organism that enable it to survive and reproduce in a particular environment.

fixed action pattern A neurally preprogrammed, unvarying response to a sign stimulus.

flagellates (*FLAJ-e-lates*) Also called *mastigophora*. Members of the phylum Mastigophora. Simple animal-like protists, many of which are parasites, such as members of the genus *Trypanosoma* that cause African sleeping sickness.

flagellum (plural, **flagella**) (*fla-JEL-um*) A whip-like structure between 20 and 100 micrometers long; used in cell locomotion.

flame cell A component of the excretory system of flatworms that pumps out of the body the water and nitrogen wastes collected in tubules that extend throughout the organism.

flower The reproductive organ of an angiosperm, typically made up of sepals, petals, stamens, and carpels.

fluid mosaic model The most widely accepted theory of cell membrane structure, which posits that the membrane consists of a fluid lipid bilayer in which individual protein molecules drift.

food chain A sequence of consumption among organisms; for example, a plant may be eaten by a beetle, which is eaten by a bird, which is eaten by a cat.

food web The complex set of interrelationships among members of different trophic levels in an ecosystem.

fossil A remnant or trace of an organism of a past geological age, such as a skeleton, footprint, or leaf imprint, embedded in the earth's crust.

founder effect The influences on the genetic composition of an isolated population that result from the accidental—and occasionally idiosyncratic—genetic configuration of its restricted number of founders.

frameshift mutation A mutation caused by the insertion or deletion of a number of nucleotides that is not a multiple of 3. Frameshift mutations disrupt the reading of every condon "downstream" from the point of the alteration.

fruit The ripened ovary of an anthophyte, which serves as a protective covering for seeds and, in many plants, aids in seed dispersal.

functional group An atom or group of atoms that has similar chemical functions and properties in a variety of different compounds.

fundamental niche The broadest of all possible niches that an organism theoretically can occupy.

fungi (*FUN-jie*) Heterotrophic decomposers and parasites with cell walls. The fungi exhibit a broad range of adaptations, functioning as decomposers, symbionts, predators, and parasites. Includes the yeasts, molds, and mushrooms.

Fungi The kingdom containing heterotrophic decomposers and parasites with cell walls.

Gaia hypothesis A contested hypothesis (named after the earth goddess of the ancient Greeks) which holds that the biosphere functions and evolves in the manner of a global superorganism, Gaia, whose parts and processes have evolved in ways that maintain the delicate balance necessary for life.

gall bladder A sac that accumulates secretions of bile from the liver and passes them on to the duodenum.

gamete incompatibility (*GAM-eet*) A prezygotic isolating mechanism that functions when two species cannot interbreed because their sperm and eggs cannot successfully unite in fertilization.

gametes (*GAM-eets*) The haploid germ cells (sperm and egg) that unite during fertilization to form a zygote.

gametogenesis (*gam-ee-toh-JENN-eh-siss*) The process by which gametes, or sex cells, are produced.

gametophyte (*gam-EET-oh-fight*) The haploid, gamete-producing stage of the plant life cycle.

gap junction A small, pore-like protein channel that connects the cytoplasm of one cell to that of another and permits certain compounds to pass between them. A gap junction functioning as an electrical synapse permits the free flow of ions between cells and thereby avoids the transmission delay of several milliseconds that occurs in chemical synapses.

gastropod (*GAS-troe-pod*) A snail or slug. Member of the class Gastropoda, a group of molluscs with a large, fleshy foot. Many have a spiraled shell, and most are herbivorous and aquatic. Some carnivorous and many land-dwelling forms also exist.

gastrovascular cavity (*GAS-troe-VAS-kyou-lar*) A sac-like digestive cavity with only one opening to the exterior, which is used both to ingest food and to expel undigested material; found in simple organisms such as hydras and planarians.

gastrula (*GAS-true-la*) A stage in embryonic development in which cells migrate and differentiate to produce distinct layers, generally ectoderm, endoderm, and mesoderm.

gene A unit of hereditary information. Each gene consists of a linear sequence of nucleotides along the length of a DNA molecule and may specify the sequence of amino acids in a polypeptide chain.

gene imprinting The hypothesis that certain genes are chemically "imprinted" or altered by their presence in a male or female parent in way that affects their activity in an offspring. Because of imprinting, a gene inherited from a parent of one sex might not have the same effect as an identical gene inherited from the parent of the opposite sex.

gene linkage The tendency of certain genes to be inherited together because they are located on the same chromosome.

gene pool The totality of alleles of genes available for reproduction in a given population at a given time.

gene therapy Treating or curing genetic disorders by inserting functional DNA sequences to replace or supplement non-functional ones.

genetic code The language of protein synthesis, in which a triplet of three bases, or codon, stands for each of the 20 amino acids that make up proteins. During protein synthesis, the sequence of triplets along an mRNA molecule is translated codon by codon into the sequence of amino acids in a polypeptide chain.

genetic diversity The heritable variability among the individual members of a single species, which is essential to a species' long-term ability to adapt—and survive—as environmental conditions change.

genetic drift Changes in a population's gene pool that result from chance alone.

genetic engineering The use of recent technologies to "cut and paste" genes from one organism to another, introducing new genes and new characteristics into organisms.

genetic map A representation of the physical locations of genes on a chromosome.

genome All of the genetic content of a species.

genotype (*JEEN-oh-type*) An organism's hereditary makeup.

genus (*JEEN-us*) (plural, **genera**) The taxonomic group just one step more inclusive than the species level, designated by the first of the two parts of an organism's Latin name.

geochemical view The thesis that the physical and chemical characteristics of the earth and/or its atmosphere are governed by strictly physical processes.

geographic isolation A prezygotic isolating mechanism that functions when two populations of a species live in different locations separated by a physical barrier, such as a mountain range or body of water.

geotropism (*JEE-oh-TROE-pizm*) A plant's response to gravity. Roots are positively geotropic, growing in the direction from which the force of gravity is exerted; stems are negatively geotropic, growing in the direction opposite to the force of gravity.

germination The reactivation of growth in a plant embryo that leads to the plant's sprouting from its seed.

germ layers In animal embryos, the three layers of cells that give rise to all the cells and tissues of the body through the process of differentiation.

gibberellin (*JIB-er-ELL-in*) A plant hormone that is thought to regulate the elongation of the internode regions of plant stems; also functions in seed germination.

gills Respiratory organs specialized for gas exchange in water; they collect oxygen and release carbon dioxide.

ginkgoes (*GING-kohs*) Members of the plant division Ginkgophyta, a group of gymnosperms that diversified successfully in the Permian period but today has only a single living species, *Ginkgo biloba*, a tree that was believed extinct until it was rediscovered in China where it had been maintained in horticulture.

global warming The hypothesis that an increase in concentrations of carbon dioxide in the atmosphere caused by human activity might lead to a rise in global temperatures.

glomerulus (*glow-MARE-you-luss*) The net of thin capillaries through which blood plasma is filtered into the Bowman's capsule in the kidney.

glucagon (*GLEW-ka-gone*) A polypeptide hormone that is secreted by the pancreas and causes cells in the liver to release sugar into the bloodstream.

glucose (*GLEW-kose*) A six-carbon sugar that is a major source of nutrients for cells. Starch, cellulose, and glycogen are all polymers of glucose.

glycerol (*GLISS-er-all*) A molecule with three carbons, to each of which a hydroxyl group is covalently bonded. Glycerol is a component of many triglycerides and has a variety of commercial uses.

glycolysis (*gly-KOL-lih-sis*) The initial chemical pathway involved in the breakdown of glucose; common to all unicellular and multicellular plants and animals, glycolysis results in a relatively small net gain of ATP for the cell and produces pyruvate.

Golgi apparatus (*GOAL-jee*) A stack of flattened vesicles associated with the rough endoplasmic reticulum that function in modifying and storing secretions within the cell.

gonadotropins (*go-NAD-oh-TROE-pins*) Hormones, produced by the anterior pituitary, that control male and female sex organs. The major gonadotopins are luteinizing hormone (LH) and follicle-stimulating hormone (FSH).

gonads (*GO-nads*) In animals, the primary sexual organs that produce gametes and, in some organisms, sex hormones.

gonorrhea (*gone-or-EE-ah*) A serious sexually transmitted disease caused by the bacterium *Neisserea gonorrhoeae*, once readily treated with antibiotics but now spreading in several antibiotic-resistant strains.

gradualism The view that evolution proceeds through the gradual accumulation of small changes within species.

grassland Also called *savannah, prairie, pampas, veldt,* or *steppes.* A biome found in temperate and tropical areas with low to moderate rainfall. Grasses are the dominant plants.

greenhouse effect 1. Earth's atmosphere's functioning like the glass of a greenhouse to retain heat. Solar energy in the form of visible and ultraviolet light is absorbed by plants and other materials and reradiated as infrared energy, which is trapped by certain gases in the atmosphere and thus prevented from escaping into space. 2. The exacerbation of the above effect; global warming as a consequence of elevated concentrations of carbon dioxide, water vapor, and certain other gases in the atmosphere that trap the heat of solar radiation.

green revolution A dramatic change in agricultural methods that produced a marked increase in world agricultural output beginning in the 1950s.

ground tissue Plant tissue that provides structural support; also contains photosynthetic cells.

groundwater Water that exists beneath the earth's surface in porous rock formations and often supplies wells and springs.

guanine A nitrogen-containing base found in nucleotides, belonging to the chemical group known as purines, which have two carbon-nitrogen rings in their basic structures. Found in DNA and RNA.

gustation (*gus-TAY-shun*) The sense of taste.

gymnosperms (*JIM-no-sperms*) A group of vascular plants that produce "naked" seeds not enclosed in ovaries; includes the conifers.

habitat The location within an environment in which an organism actually lives or is commonly found.

habituation (*ha-BIT-you-AY-shun*) A simple form of learning in which an organism learns to ignore a stimulus that is of no importance to it.

half-life The time it takes for half of a radioactive substance to undergo decay.

halophiles (*HAL-oh-files*) Archaebacteria that inhabit extremely salty environments.

haplo-diploidy A system of sex determination, found in bees and ants, where there are no sex chromosomes. Instead, males develop from unfertilized eggs and are haploid, while females develop from fertilized eggs and are diploid.

haploid (*HAP-loyd*) Having only one set of unpaired chromosomes.

Hardy–Weinberg law The principle that sexual reproduction does not change the relative frequencies of genes in a population.

HCG (human chorionic gonadotropin) (*kor-ee-ON-ik go-NAD-oh-TROPE-in*) A hormone, produced by a developing embryo and the surrounding tissue, that prepares the body to deal with pregnancy and prevents the hypothalamus from inducing development of new ova.

heart A muscular organ that pumps blood throughout the body of an animal.

helper T-cell The kind of T-cell that secretes the interleukin that stimulates B-cells to differentiate and produce antibodies or to secrete several other substances that influence immune system response.

heme (*HEEM*) In the hemoglobin molecule, the ring-like, iron-containing component that binds with oxygen.

hemoglobin (*HEE-moh-glow-bin*) The reddish protein in red blood cells that functions in carrying oxygen.

hemophilia (*HEE-moh-FILL-ee-yah*) Disease characterized by failure of blood to clot. Hemophilia is caused by a defective gene on the human *X* chromosome.

herbicide (*ER-biss-ide*) A substance used to kill plants.

herbivore (*ER-biv-ore*) An animal that subsists solely on plants.

heritable (*HER-it-a-bull*) Capable of being passed on genetically by an organism to its offspring.

Herpes (*HERP-eez*) A family of viruses that cause a number of ailments in various locations in the body. *Herpes type I* produces cold sores, often on the lips. *Herpes type II* infects the genital area, producing a sexually transmitted disease characterized by periodic outbreaks of painful, small red blisters.

hertz (Hz) (*HURTS*) A unit of frequency equal to 1 cycle per second.

heterocysts (*HET-er-oh-sists*) Specialized cells in some cyanobacteria that carry out nitrogen fixation.

heterotroph (*HET-er-oh-trofe*) An organism incapable of producing its own food from inorganic materials; such organisms depend, directly or indirectly, on primary producers such as green plants to meet their food requirements.

heterozygous (*HET-er-oh-ZYE-gus*) Having two different alleles for a single trait.

histamine (*HISS-ta-meen*) A substance released in response to injury by most cells near the skin surface; causes blood vessels near the wound to expand and attracts and directs the actions of white blood cells.

histone (*HISS-tone*) A member of a group of proteins that contain large numbers of positively charged amino acids, which bind to DNA strands making up chromatin.

HIV (human immunodeficiency virus) (*im-MYUNE-oh-dee-FISH-en-see*) A retrovirus that causes AIDS by impairing the body's immune system.

homeostasis (*HOME-ee-oh-STAY-sis*) The maintenance of stable internal conditions by an organism despite variations in the external environment.

hominid (*HOM-ih-nid*) A member of the family Hominidae, which includes *Homo sapiens* (modern humans) and extinct ancestral human species in the genera *Australopithecus* and *Homo*. Hominids are distinguished from the great apes (the family Pongidae) by such characteristics as much larger brains and more erect posture permitting more efficient bipedal walking.

hominoids (*HOM-ih-noyds*) An advanced group of anthropoids that includes modern and extinct ancestral humans (the family Hominidae, also called the hominids) and the living great apes (the family Pongidae).

homologous chromosomes (*home-ALL-a-gus*) Chromosomes that possess genes for the same characteristics at the same loci. Sexually reproducing organisms generally inherit one such chromosome from the male parent and one from the female parent.

homologous structures (*home-ALL-a-gus*) Structures found in different species that have similar form or configuration because of the species' common ancestry. An example is the human arm and the bird wing.

homozygous (*HOME-oh-ZYE-gus*) Having identical alleles for a particular trait.

hormone A substance that is produced in one part of an organism and affects the physiology of another part.

hormone–receptor complex A unit consisting of a hormone bound to a receptor in a target cell. Such a complex can activate or deactivate specific genes or groups of genes in the cell's nucleus.

Human Genome Project An international research effort that is now underway to map the genes on all of the human chromosomes.

humoral immune response Also called *antibody-mediated immune response*. The production of antibodies by B-lymphocytes of the immune system, triggered by foreign proteins (antigens).

humus (*HYUME-us*) Decomposing organic materials, which serve as an important source of nutrients and water-retaining capacity in soil.

Huntington's disease A rare human genetic disorder, caused by a dominant allele on the fourth chromosome, that leads to degeneration of the nervous system and death.

hybrid The offspring produced by breeding plants or animals of different varieties or species.

hybrid infertility A postzygotic isolating mechanism that functions when two species mate and then produce offspring, but those offspring are sterile.

hybrid inviability (*in-VYE-a-BILL-a-tee*) A postzygotic isolating mechanism that functions when two species can interbreed and produce a fertilized egg, but the embryo or newborn organism cannot survive to reproduce.

hydrocarbon A chemical compound composed solely of hydrogen and carbon.

hydrogen bond A weak chemical bond in which an electronegative atom is attracted to a hydrogen atom that is already involved in a polar covalent bond.

hydrogen ion (H⁺) A hydrogen atom that has lost its electron, becoming simply a proton.

hydrological cycle (*HIGH-droe-LOJ-ih-kull*) The cycle that water follows through an ecosystem via evaporation, condensation, precipitation, and runoff.

hydrolysis (*high-DRAHL-ih-sis*) A chemical reaction in the breakdown of polymers in which bonds between molecules are broken through the consumption of water molecules. Hydrogen becomes bonded to one monomer, and a hydroxyl group joins the adjacent monomer.

hydrophilic molecules (*HIGH-droe-FILL-ik*) Polar molecules that interact strongly with water and dissolve freely.

hydrophobic molecules (*HIGH-droe-FOE-bik*) Nonpolar molecules that do not interact with water and tend to cluster together; an example is oil in water.

hydroponic culture (*HIGH-droe-PAWN-ik*) A technique for growing plants in liquids; used to assess a plant's needs for nutrients and also to grow certain commercial crops.

hypertension Excessively high blood pressure, a serious medical problem that increases the heart's work load.

hypertonic (*hyper* means "greater," *tonic* means "strength") Having a greater concentration of dissolved material than another fluid.

hyphae (singular, **hypha**) (*HIGH-fee*) The thin filaments that compose the mycelium of a mushroom or other fungi.

hypothalamus (*HIGH-poe-THAL-a-muss*) The part of the brain that lies below the thalamus and functions in the regulation of temperature and various other autonomic activities.

hypothesis (*high-POTH-ih-sis*) An assertion that can be tested through experimentation.

hypothyroidism (*HIGH-poe-THIGH-royd-ism*) A condition resulting from insufficient thyroxine levels; often caused by lack of iodine in the diet. Symptoms include lowered heart rate and blood pressure, sleepiness, energy loss, and weight gain.

hypotonic (*hypo* means "lesser," *tonic* means "strength") Having a lower concentration of dissolved material than another fluid.

ideal types In Greek philosophy, the idea that actual objects are imperfect manifestations of the mental images we create of their perfect forms.

imbibition (*IM-bie-BIH-shun*) A seed's absorption of water, which causes it to swell and initiates germination.

immigration Movement of organisms born elsewhere into a population of their own kind.

immune system The cells and tissues of the body that identify foreign substances and organisms and attack them.

imprinting A type of learning in which an animal "customizes" an inherited behavior on the basis of the environmental information it encounters, often during a brief, critical period early in life.

inclusive fitness The idea promulgated by sociobiologists that an individual can increase the number of its genes that survive not only by reproducing itself but also by helping its relatives (and thus their genes) survive and reproduce.

incomplete dominance The condition in which neither gene for a trait is dominant, and the offspring's trait blends characteristics of the trait found in both parents.

independent assortment, principle of The principle that each pair of alleles segregates independently during formation of gametes.

indeterminate growth The growth pattern of most plants, in which there are no absolute restrictions on the exact size or shape of the plant, but controls on growth rates and patterns help shape the plant to fit the conditions such as light, gravity, wind, and moisture in its environment.

indicator species A species whose welfare is used to monitor the overall health of an ecosystem because that species requires significant amounts of habitat in good condition.

induced fit A change in the shape of an enzyme as it binds substrate, producing an even tighter fit between the enzyme molecule and the substrate molecule.

inductive reasoning A process of reasoning whereby a general principle or hypothesis is derived from a series of specific observations.

industrial melanism The phenomenon in which several moth species in Britain changed color through natural selection during the Industrial Revolution. Moth populations became darker over many generations because tree trunks blackened by coal soot in certain industrial areas conferred an advantage upon darker individuals, which were harder for predatory birds to detect than lighter individuals.

infertile Incapable of producing functional gametes.

inflammatory response One of the body's nonspecific defenses against infection, often marked by swelling, redness, and heat at a wound site, and involving several types of white blood cells which attack foreign cells.

inhibitor A molecule that decreases the catalytic activity of an enzyme.

innate behavior Also called *instinct*. A behavior that is genetically preprogrammed such that, given the proper stimulus, an animal can perform it without having learned it or having been exposed previously to the stimulus that elicits it.

insectivorous (*IN-sek-TIV-er-us*) Insect-eating. Certain plants trap and digest insects as a source of nutrients.

insects Members of the class Insecta, a large and enormously diverse group of arthropods with six legs and a body divided into three major sections. Many have two pairs of wings.

insight learning Learning that occurs when an animal applies previous experience to a completely new situation without trial and error.

instinct Also called *innate behavior*. Behavior that is genetically preprogrammed such that, given the proper stimulus, an animal can perform it without having learned it or having been exposed previously to the stimulus that elicits it.

insulin (*IN-su-lin*) A hormone, produced by beta cells of the pancreas in the islets of Langerhans, that stimulates cells to remove excess sugar from the blood and store it.

integument The outer covering or skin of an organism.

interleukin A type of messenger molecule that activates or stimulates various cells of the immune system, often resulting in their proliferation.

intermediate filaments A class of cytoskeletal filaments about 10 nm in diameter, intermediate in size between microtubules and microfilaments, and composed of a range of related proteins that vary from one cell type to another.

interneurons Nerve cells (neurons) that collect and relay information gathered from sensory neurons.

interoceptors (*in-TARE-oh-sep-tors*) Receptors in the internal organs that monitor conditions within the body, reporting such sensations as cramps, hunger, thirst, and the need to defecate or urinate.

interphase (*in-ter-FAZE*) In eukaryotic cells, the period between cell divisions when the cell is rapidly growing, increasing in size, developing new structures, and synthesizing new molecules, including DNA and DNA-associated proteins.

interspecific competition Competition among members of two or more species.

interstitial fluid (*in-ter-STISH-ul*) The extracellular fluid in the spaces between cells and tissues.

intraspecific competition Competition occuring among members of a single species.

intron (*IN-tron*) A noncoding sequence of nucleotides located within a eukaryotic gene.

inversion An alteration in chromosome structure in which part of a chromosome is turned upside down with respect to the rest of the chromosome.

invertebrate An animal that lacks a backbone.

ion An atom or molecule that has become charged as a result of gaining or losing one or more electrons.

ionic bond (*eye-ON-ik*) A chemical bond formed between oppositely charged ions.

isotonic (*iso* means "the same," *tonic* means "strength") Having the same concentration of dissolved material as another fluid.

isotopes (*ICE-oh-topes*) Forms of the atoms of an element that have differing numbers of neutrons.

juvenile hormone A hormone, produced by the corpora allata of arthropods, that prevents the development of adult characteristics in larvae.

karyotyping (*CARE-ee-oh-type-ing*) A technique for examining human chromosomes that entails adding colchicine to cultured white blood cells; the cells become locked in metaphase and the chromosomes can be analyzed in photomicrographs. The technique is often used to diagnose genetic disorders.

kelp A type of giant cold-water alga that forms large coastal "forests" that support a distinctive food web.

keystone predator The top carnivore in a food chain; such predators often have a powerful influence on the structure of the communities in which they live.

kidneys In vertebrates, the two organs located in the abdominal cavity that remove nitrogenous waste, salts, and excess water from the blood and form urine.

kinetic energy (*ki-NET-ik*) Energy of motion.

kinetochore (*kin-EE-to-KORE*) The point at which each chromatid is attached to the mitotic spindle during mitosis.

kin selection theory The theory that altruistic behavior can be adaptive if its benefit to an animal's inclusive fitness is greater than the loss to the animal's personal fitness that it entails.

Klinefelter syndrome A genetic nondisjunction abnormality in which the individual has two *X* chromosomes and one *Y* chromosome (47*XXY*); afflicted persons are male, but they are sterile, often unusually tall, and frequently mentally retarded.

Krebs cycle A cyclic chemical pathway, occurring in mitochondria, in which acetyl coenzyme A from pyruvic acid is used in the production of two molecules of carbon dioxide and four pairs of hydrogen atoms.

*K***-selected species** Species that maximize their carrying capacity. Such species exhibit late maturity, infrequent mating, and the production of relatively few offspring, to the rearing of which they may devote a good deal of care and energy. Also known as *equilibrium species*.

lac **operon** (*LACK OP-er-on*) A set of protein-coding genes, regulated as a cluster, that affects the metabolism of lactose in *E. coli*.

larva (plural, **larvae**) An immature, free-living stage in the development of an animal. Example: the tadpole is the larval stage of a frog.

larynx (*LARE-inks*) The upper part of the windpipe, which contains the vocal cords.

lateral line A sensory organ of fishes and amphibians that permits them to detect minute movements in water.

lateral meristem (*MARE-uh-stem*) Also called *cambium*. In vascular plants, the regions of rapidly dividing unspecialized tissue containing cells that are responsible for lateral growth—the thickening of roots and stems.

learning The modification of behavior as a result of past experience.

leukocytes (*LEW-ko-sites*) The white blood cells, which function in the immune system.

lichen (*LYE-kun*) An organism formed by a symbiotic partnership between a fungus and an alga or cyanobacterium.

light-dependent reactions Formerly called *light reactions*. Chemical reactions of the first stage of photosynthesis, which trap the energy of sunlight and transform it into a chemical form, producing ATP and reducing NADP$^+$ to NADPH. These reactions are associated with the photosynthetic membranes, or thylakoids, of the chloroplasts.

light-independent reactions Formerly called *dark reactions*. Chemical reactions of the second stage of photosynthesis, which use the ATP and NADPH produced in the first stage to make sugars and other compounds. These reactions occur in the stroma of the chloroplasts and do not require sunlight directly.

light microscope An instrument that makes use of the magnifying properties of lenses to produce an enlarged image of minute objects.

limbic system (*LIM-bik*) The section of the mid-brain responsible for the control of emotions.

limiting factors, law of The principle that the growth of an organism is limited when any factor essential to its growth is lacking, regardless of the quantity available of other factors.

lipid bilayer (*LIP-id*) The structural foundation of a biological membrane, consisting of lipids arranged in two layers with their hydrophilic groups facing outward at the two surfaces of the bilayer and their hydrophobic groups gathered in the center of the bilayer.

lipids (*LIP-ids*) A diverse group of waxy or oily, generally hydrophobic, substances that are soluble in organic solvents; most are hydrocarbons.

loam Soil containing a mixture of sand, silt, and clay particles.

locus (plural, **loci**) The position or location of a gene on a chromosome.

logarithmic growth (*LOG-a-rith-mik*) Population growth in which the rate of growth increases as the size of the population increases. For example, a pair of rabbits might have 6 offspring, which if unrestrained by environmental factors might have 36 offspring, which could have 216 offspring, and so on. Also known as *exponential growth* and typically represented on a graph as a J-shaped curve.

logistic growth (*low-JIST-ik*) A pattern of population growth. Growth is initially slow because the number of reproducing individuals is low; the growth rate increases as the population grows, and it subsequently decreases as environmental limitations begin to have an effect. Growth ultimately levels off as the environment's carrying capacity is reached. Logistic growth is typically represented on a graph as an S-shaped curve.

long-day plant A plant that flowers only when daylight lasts longer than a certain minimum. Such plants flower in late spring and early summer.

loop of Henle A section of the nephron (kidney) tubule, consisting of an ascending and a descending branch, that functions in the concentration of dissolved waste products in urine.

lungs The organs that carry out respiration in terrestrial vertebrates and certain other organisms. Gas exchange is conducted as the blood in the lungs flows past the thousands of tiny air sacs into which the lungs have ramified.

lycophytes (*LIKE-oh-fights*) The club mosses; an ancient group of seedless vascular plants dominant during the Carboniferous period. About 1000 small species survive today.

lymph (*LIMF*) The fluid in the vessels of the lymphatic system, which returns accumulated interstitial fluids to circulation, absorbs fat from the digestive tract, and helps move white blood cells to aid in the immune response.

lymphatic system The interconnected system of spaces and vessels between tissues and organs by which lymph is circulated throughout the body, returning excess interstitial fluid to the blood, picking up fat from the digestive tract, and moving white blood cells to where they are needed for defense.

lymphocyte (*LIMF-oh-site*) A type of white blood cell (leukocyte) that produces special proteins known as antibodies that attach to and help to destroy foreign cells and foreign matter.

Lyonization (*LYE-on-eye-ZAY-shun*) The model proposed by Mary Lyon suggesting that in the cells of a female mammal, one of the two *X* chromosomes becomes inactivated during embryonic development.

lysosome (*LYE-so-soam*) A small, membrane-bounded organelle filled with enzymes that function in digestion, in the destruction of damaged organelles, and in the killing of cells in locations where space for development is needed.

macroevolution The accumulation of genotypic and phenotypic changes that are great enough to create new species, genera, and higher taxonomic categories. Speciation is an example of macroevolution.

macromolecule (*MAK-roe-MOL-a-kyool*) A giant molecule such as occurs in living cells, formed by the aggregation of smaller molecules. Examples include nucleic acids, proteins, carbohydrates, and lipids.

macronutrients (*MAK-roe-NEW-tree-ents*) The nine elements that must be present in relatively large amounts for a plant to thrive: oxygen, carbon, hydrogen, nitrogen, potassium, calcium, magnesium, phosphorus, and sulfur.

macrophage (*MAK-roe-fayj*) ("big eater") A phagocyte that engulfs and digests pathogens, thereby beginning a process that helps T-cells recognize antigens and that galvanizes the immune system into action.

malignant (*ma-LIGG-nant*) In pathology, a term descriptive of the life-threatening, uncontrolled growth and invasive character of certain tumors.

malignant tumor A cancerous tumor, the cells of which tend to break off from the original mass and spread throughout the body, forming new tumors.

malnourishment The condition of receiving an inadequate supply of one or more essential nutrients.

Malpighian tubules (*mal-PIG-ee-un TOOB-yules*) In insects, the system of ducts that removes nitrogen wastes, which are emptied into the digestive system and excreted through the anus.

mammals Members of the class Mammalia, a group of endothermic vertebrates that have body hair and produce milk for their offspring.

mandibles (*MAN-dih-bulls*) Jaw-like structures found on the second or third body segment of crustaceans and uniramians.

mangrove (*MANG-grove*) A type of salt-tolerant tree or shrub found growing in shallow waters along tropical and subtropical coasts. Mangrove forests host distinctive communities of marine and terrestrial life.

marsupials (*mar-SOO-pee-als*) An order of mammals among whom newborn offspring are reared in a maternal pouch.

mass extinction event An episode in the history of life when large numbers of species (and often higher taxa as well) become extinct.

mechanical isolation A prezygotic isolating mechanism that functions when two species cannot interbreed because their reproductive organs are structurally or functionally incompatible.

medulla oblongata A portion of the brain, located just above the spinal chord, that contains centers for controlling various vital functions, including a breathing center. It also controls blood pressure, heart rate, and coughing.

meiosis (*my-OH-sis*) The two-stage process by which the number of chromosomes in a cell nucleus is halved during the formation of gamete cells.

melanin (*MEL-a-nin*) The pigment molecule responsible for most human skin coloring.

memory cell Antigen-stimulated B- and T-lymphocytes that remain after the immune response has moderated, ready to attack again if the same antigen appears again. Memory cells make the immune system's second response to the same antigen swifter and stronger and form the basis of permanent immunity to certain diseases.

menopause (*MEN-oh-pawz*) The cessation of menstrual cycles, signifying that a woman's reproductive potential has ended.

menstruation (*MEN-strew-AY-shun*) The process by which the lining of the uterus is broken down and discharged if pregnancy has not occurred.

meristem (*MARE-uh-stem*) Regions of plant tissue containing rapidly dividing undifferentiated cells capable of developing into any of the plant's specialized tissues.

mesoderm (*MEZ-o-derm*) The middle tissue or cellular layer of triploblastic embryos.

messenger RNA (mRNA) A type of RNA molecule that contains the instructions for protein synthesis. Messenger RNA is synthesized (transcribed) from one strand of a DNA molecule, and then its sequence of bases is "read" (translated) to create the sequence of amino acids in a polypeptide.

metabolism (*meh-TAB-oh-lizm*) The sum of all the chemical reactions associated with life processes.

metaphase (*MET-a-faze*) The second stage of mitosis, during which the chromosomes come to be aligned along the mitotic spindle equator.

metastasize (*me-TAS-ta-size*) To spread from one site to another throughout the body, as malignant tumors may.

metatherians (*meh-ta-THEER-ee-uns*) The order of mammals also known as *marsupials*. Its members, such as kangaroos, rear newborn offspring in a maternal pouch.

methanogens (*meth-AN-a-jens*) A type of archaebacterium that lives in anaerobic mud and produces methane as a metabolic by-product.

metric system The decimal system of length and mass used by scientists. The meter (about 39 inches) is the standard unit of length, and the kilogram (about 2.2 pounds) is the standard unit of mass.

MHC markers Proteins that identify body cells as "self" because no two individuals (except identical twins) have exactly the same MHC markers. Produced by a region of the genome called the major histocompatibility complex (MHC complex).

microclimate (*MIKE-roe-kly-met*) The climate of a microhabitat, including such characteristics as temperature, humidity, and wind speed.

microenvironment (*MIKE-roe-en-VYE-run-ment*) A small or minute environment. Ecologists study microenvironments because all of the conditions relevant to a particular organism may exist in a very small area.

microevolution The accumulation over time of small changes within a species. Examples include changes in beak size in finches and the darkening of wing color in moths.

microfilaments (*MIKE-roe-FILL-a-ments*) Fibers in the cytoskeleton, composed of the protein actin, that function in stabilizing cell shape, in cell movement, and in cell growth.

micronutrients (*MIKE-roe-NEW-tree-ents*) Elements essential to plant growth that need be present only in minute amounts to meet the plant's requirements: boron, chlorine, copper, iron, manganese, molybdenum, and zinc.

microtubules (*MIKE-roe-TOOB-yules*) Tube-like structural components of the cytoskeleton, composed of the protein tubulin, that are involved in providing support for the cell surface in mitosis and in constructing the cell's motile structures.

microvilli Tiny projections of the cell membrane of the epithelial cells lining the small intestine. They dramatically increase surface area available for absorbing nutrients.

mimicry (*MIM-ik-ree*) The resemblance of one species to another; evolves through natural selection when such a resemblance has survival value for the mimic.

mineral nutrients In biology, the various inorganic nutrients necessary for an organism's health. Humans require at least 18 mineral nutrients.

minichromosome (*MIN-ee-KROME-oh-sowm*) A small chromosome. Also a structure used in the transformation of yeast cells; similar to the plasmids used for the transformation of prokaryotic cells.

mitochondrion (*my-toe-KON-dree-on*) (plural, **mitochondria**) The membrane-bound organelles in which the chemical energy from food molecules is used to produce ATP. These key energy-releasing organelles are self-replicating and contain their own DNA.

mitogen (*MY-toh-jen*) A substance that stimulates the initiation of cell division.

mitosis (*my-TOE-sis*) A process in eukaryotic cell division in which chromosomes in the cell nucleus are duplicated such that each daughter cell receives a complete set.

mitotic spindle (*my-TOT-ik*) A structure composed of microtubules that extends from one pole of a cell to another during mitosis and functions in chromosome distribution.

molecular biology The field of biology in which the structure and development of biological systems are analyzed in terms of the physics and chemistry of their molecular constituents, especially with reference to the sequence and function of their genetic material.

molecular clock The rate of neutral mutations in DNA sequences coding for similar proteins in different organisms, which researchers use to help determine the timing of evolutionary events.

molecule (*MOLL-ih-kule*) A bonded aggregation of atoms of one or more elements. A molecule is the smallest unit of a compound that displays the characteristics of that compound.

mollusc (*MOLL-usk*) Member of the phylum Mollusca, a diverse group of more than 100,000 soft-bodied animals, including slugs, snails, oysters, mussels, octopi, and squid. Molluscs are characterized by a visceral mass containing the internal organs; a muscular foot; a mantle that may secrete a shell; and, in many cases, a head.

Monera The kingdom containing all the prokaryotic organisms (bacteria).

monocot (*MON-oh-COT*) Also called *monocotyledon* (*MON-oh-COT-a-LEE-don*). A member of a class of angiosperms (flowering plants) characterized by a single seed leaf (cotyledon) and numerous other structural similarities, such as parallel veins, scattered vascular tissues, and floral parts in threes.

monoculture (*MON-oh-kull-chur*) A form of agriculture in which a genetically uniform variety of a single species of crop is grown over large areas.

monocyte (*MAHN-oh-site*) Small, roundish cells in the circulation that, when attracted to a wound, change into larger granulocytes that engulf bacteria.

monomer A small individual molecule (such as a simple sugar or a nucleotide) that can serve as a building block in a larger molecule (polymer).

monosaccharide Also called *simple sugars*. Individual sugar molecules, such as glucose, which can be linked together to form complex carbohydrates.

monozygotic twins Commonly called "identical twins." Twins that develop from a single cell that divides at an early stage to form a pair of genetically identical embryos.

morphology The size, shape, and structure of an organism.

mortality In a life table, the percentage of a population that can be expected to die at a certain age.

mosaic evolution (*moe-ZAY-ik*) The evolution of different sets of traits at different rates as organisms exploit new ecological opportunities.

motor neurons Nerve cells (neurons) that carry information to the muscles and glands, generating appropriate movements or responses.

motor unit A functional unit consisting of a single motor neuron and the collection of muscle fibers that it serves.

mouth The opening at the beginning of the alimentary canal through which an organism ingests food.

mucosa (*mew-KOE-sa*) The innermost layer of the gastrointestinal tract, consisting of a layer of epithelial cells that releases mucus, digestive enzymes, and other substances into the tract; a thin layer of connective tissue; and a thin layer of muscle.

Müllerian mimicry (*mew-LAIR-ee-un*) A form of mimicry in which two or more species that share a similar defense mechanism (such as unpalatability) resemble one another.

multiple fruits Fruits formed from several separate flowers clustered tightly together, such as a pineapple or fig.

muscle tissue Tissue composed of cells specialized to contract and cause internal and external movement. Muscles also produce heat and so are important in homeostasis. There are three types: smooth, skeletal, and cardiac.

mutagen (*MEW-ta-jen*) An agent capable of causing a mutation, such as radiation (including ultraviolet light) and certain chemicals.

mutation (*mew-TAY-shun*) A heritable change in an organism's DNA.

mutualism (*MEW-chew-al-izm*) A form of symbiosis in which both organisms benefit.

mycelium (*my-SEEL-ee-um*) The tangled, branched structure of hyphae that makes up the main body of many fungi.

mycoplasm (*MY-co-PLAZ-um*) A kind of bacteria that are the smallest living cells, measuring less than 0.2 μm and visible only with an electron microscope.

mycorrhizae (*MY-ko-RYE-za*) Fungi that live in a mutualistic relationship with the roots of many plants, providing the roots with moisture and mineral nutrients. The plant, in turn, provides the fungi with organic nutrients.

myelin sheath An insulating sheath around the axons of many vertebrate neurons that greatly speeds the conduction of action potentials along myelinated axons. Formed by Schwann cells that wrap around the axon, leaving small sections (nodes of Ranvier) exposed.

myocardial infarction (*my-oh-KARD-ee-ul in FARK-shun*) A heart attack, which occurs when a vessel that supplies blood to the heart is blocked and part of the heart muscle dies.

myoglobin (*MY-oh-GLOW-bin*) A molecule resembling hemoglobin that binds oxygen in muscle cells and releases it as needed during vigorous activity.

myosin (*MY-a-sin*) The protein that makes up the thick filaments of the myofibrils; myosin functions with actin to produce contractions.

NADH The reduced or electron-carrying form of the electron carrier molecule nicotinamide adenine dinucleotide (NAD⁺), which plays an important role in glycolysis, the Krebs cycle, and electron transport.

NADPH The reduced or electron-carrying form of the soluble electron carrier molecule nicotinamide adenine dinuceotide phosphate (NADP⁺), which plays an important role in photosynthetic electron transport. NADPH is a product of the light-dependent reactions of photosynthesis and is used in the light-independent reactions.

natural selection A mechanism first proposed by Darwin to explain the differential survival and reproduction of groups of organisms that differ from one another in one or more inheritable traits. In modern biology parlance, natural selection applies to a change in the relative frequencies of alleles in a population, or between two or more populations, that is caused by differential fitness of different genotypes. The differential mortality, or reproductive success, that leads to these changes cannot be due to chance alone.

Neanderthals (*nee-AN-der-thawls*) An extinct subspecies of *Homo sapiens* that originated in Africa about 500,000 years ago.

negative feedback system A mechanism that maintains homeostasis in which a shift in a physiological variable ultimately produces conditions that inhibit the process that led to the shift.

nematodes (*NEE-ma-toads*) Members of the invertebrate phylum Nematoda, also called *roundworms,* an ecologically diverse group of worm-like invertebrates that are the most numerous multicellular organisms on Earth today. Some are free-living and others are parasitic.

nephridia (*nef-RID-ee-a*) A pair of excretory organs, found in each segment of an earthworm's body, that collect metabolic wastes and excess water and send them to a bladder for excretion.

nephrons (*NEF-rons*) The structures in the kidney that remove excess water and nitrogen wastes from the blood and produce urine. Each human kidney possesses about 1 million nephrons.

nerve A bundle of neurons in a sheath of connective tissue; nerves function in communication between the central and peripheral nervous systems.

nervous tissue Specialized tissue that senses stimuli and rapidly transmits information between parts of the body.

neuromuscular junction (*NURE-oh-MUSS-kyou-ler*) The junction between motor neurons and muscle cells.

neuron (*NURE-on*) A nerve cell, the principal component of the nervous system, consisting of a cell body, dendrites, and a long, slender axon.

neurotransmitter (*NURE-oh-TRANZ-mit-er*) A chemical messenger that diffuses from a presynaptic neuron across the synaptic cleft to the membrane of the postsynaptic cell, which it binds to and stimulates.

neurulation (*noor-yu-LAY-shun*) The formation of a neural tube in chordate animals.

neutral mutations Mutations that do not change the functional characteristics of the protein that the genes specify.

neutron An electrically neutral subatomic particle in the nucleus of an atom.

niche (*NISH*) The range of physical and biological conditions within which an organism can exist, and the manner in which the organism makes use of these conditions.

nitrification The conversion of ammonia (NH_3) to nitrite (NO_2^-) by bacteria of the genus *Nitrosomonas*, followed by the conversion of the nitrite into nitrate (NO_3^-) by bacteria of the genus *Nitrobacter*. Taken together, these two conversions change ammonia into compounds that are more readily used by plants. Nitrification is important in the nitrogen cycle.

nitrogen cycle The cycle that nitrogen follows through organisms and the environment by such processes as nitrification, ammonification, nitrogen fixation, and denitrification.

nitrogen fixation The process by which certain bacteria convert nitrogen gas (N_2) from the atmosphere into reduced nitrogen compounds such as ammonia (NH_3), making the nitrogen accessible to other organisms. Important nitrogen-fixing bacteria include those living on the roots of peas, soybeans, and other legumes, as well as photosynthetic cyanobacteria ("blue–green algae") in the sea. Nitrogen fixation is part of the nitrogen cycle.

noncompetitive inhibitor A negative effector; a molecule that decreases the rate of an enzyme-catalyzed reaction.

noncyclic electron flow (*non-SIH-klick*) An pattern of electron flow that occurs during the light-dependent reactions of photosynthesis; involves both photosystem I and photosystem II and produces oxygen, ATP, and NADPH. Noncyclic electron flow is the major pathway by which the photosynthetic reaction centers trap the energy of sunlight.

nondisjunction (*NON-dis-JUNK-shun*) Failure of two chromosomes to separate properly during the first meiotic division; causes a variety of chromosomal abnormalities, including Turner, Klinefelter, and Down syndromes.

nonrenewable energy Energy derived from limited sources such as oil, coal, and natural gas.

nonteleological Without any ultimate goal, purpose, or overall design.

notochord (*NO-toe-kord*) A flexible supporting structure that runs the length of the body between the gut and the nerve cord. Present in all chordates at some stage in their development.

nuclear envelope (*NEW-klee-er*) The double membranes that surround the eukaryotic cell nucleus.

nuclear pores (*NEW-klee-er*) Openings in the nuclear envelope that permit materials to move in and out of the nucleus without passing directly through a membrane.

nucleic acid (*new-CLAY-ik*) A polymer composed of nucleotide monomers; the two main classes, ribonucleic acid and deoxyribonucleic acid, play an important role in the transmission of hereditary information.

nucleic acid hybridization A laboratory technique involving the creation of a DNA–RNA hybrid

molecule held together by hydrogen bonds. A specially engineered RNA "probe" having the base sequence representing a particular protein is allowed to bind to the complementary sequence in heat-treated DNA fragments, forming the DNA–RNA hybrid.

nucleolus (*new-KLEE-oh-lus*) The section of the nucleus in which ribosomes are assembled from ribosomal RNA and the appropriate proteins.

nucleosome (*NEW-klee-oh-soam*) A tightly organized, bead-like structure containing a double coil of DNA wound around eight protein molecules; thought to function in packing DNA and perhaps in controlling the expression of genes.

nucleotide A small organic molecule consisting of three main parts: a five-carbon sugar (pentose), a phosphate group, and a nitrogen-containing base. Nucleotides are the building blocks or monomers of nucleic acids, as well as serving other functions.

nucleus (*NEW-klee-us*) 1. In eukaryotic cells, the organelle that contains the DNA and controls various cell processes. 2. The positively charged central region of an atom, where most of the mass of the atom is concentrated, made up of positively charged protons and (in all atoms except the hydrogen atom) electrically neutral neutrons.

nutrient limitation The phenomenon in which the productivity of an entire ecosystem is limited or curtailed because a single nutrient is in short supply.

olfaction (*ole-FAK-shun*) The sense of smell.

oligotrophic (*ol-ih-go-TROE-fik*) Containing few nutrients. Oligotrophic lakes are clear and have little phytoplankton growth.

omnivore An organism that consumes both plants and animals as food.

oncogene (*ONG-koe-jeen*) A gene capable of transforming normal body cells into cancerous cells.

oomycetes (*oh-oh-MY-seets*) Members of the fungal division Oomycota, also called *water molds*, a group of fungi that grow mainly in water or in wet soil and include many parasites that cause mildew or blights.

open circulatory system A circulatory system in which the circulating fluid travels throughout the body and bathes nearly all of its cells. Most molluscs and arthropods have open circulatory systems.

operant conditioning (*OP-er-ent*) Conditioning in which the subject learns to perform a particular operation in order to receive a reward or to avoid a painful experience.

opportunistic species Also called *R-selected species*. Organisms, such as many bacteria, protozoans, and plant "weeds," that have evolved reproductive strategies that enable them to invade new or disturbed habitats quickly. They typically reproduce at a young age, have large numbers of offspring requiring little or no parental care, and may die after reproducing.

organ A structure made up of more than one tissue and specialized for performing a particular function or group of related functions; examples include the brain, liver, skin, pancreas, nerves, and lungs.

organelles (*or-gan-ELZ*) Specialized structures in the cytoplasm of eukaryotic cells that perform various functions.

organic chemistry The chemistry of carbon compounds.

organism A living thing.

organogenesis (*or-GAN-oh-JEN-ih-sis*) The origin and development of organs in an embryo.

organ system A group of organs that are physically or functionally related; the mouth, stomach, and large intestine, for example, form part of the digestive system.

orgasm (*OR-gazm*) The culmination of sexual excitement, accompanied by ejaculation in the male and by pleasurable sensations in both sexes.

oscillations Regular fluctuations in the populations of both predator and prey species, as they both alternately rise and fall in numbers.

osmosis (*oz-MOE-sis*) The movement of water molecules across a semipermeable membrane in response to a concentration gradient.

Osteichthyes (*OS-tee-IK-thee-eez*) The bony fishes, members of the class of fishes with skeletons of bone rather than cartilage; contains about 30,000 species.

osteoblast (*OS-tee-oh-blast*) A cell that secretes bone tissues.

osteoclasts (*OSS-tee-oh-CLASTS*) Bone-dissolving cells that play a vital role in the modeling of developing bone and as an adjunct to calcium metabolism.

osteoporosis (*OS-tee-oh-pore-OH-sis*) A condition found in older persons in which mineral loss results in weakened bones that break under minimal stress.

ostracoderms (*os-TRAK-a-DERMS*) An extinct type of jawless fish. The bodies of ostracoderms were encased in an armored covering of bony plates.

ovary (*OH-var-ee*) In animals, the primary female sexual organs that produce the female gametes and sex hormones; in plants, the part of the pistil that contains the ovules.

overexploitation (*OH-ver-EX-ploy-TAY-shun*) The harvesting of stock (such as fish) at a rate exceeding the stock's ability to replenish its numbers.

ovulation (*oh-vyou-LAY-shun*) The release of a secondary oocyte (an egg cell that has not yet completed the second meiotic division) from a mature follicle in the ovary.

oxidation (*OX-ih-DAY-shun*) Any chemical reaction in which electrons are removed from an atom or a compound.

oxygen-evolving apparatus In plants, a set of enzymes in the photosynthetic membranes that are capable of removing electrons from water; these electrons are used to replace those lost by the chlorophyll in photosystem II.

oxytocin (*OX-ee-TOE-sin*) A hormone released from the anterior pituitary that stimulates uterine contractions during labor.

ozone (*OH-zone*) O^3, a gas each molecule of which consists of three atoms of oxygen. The ozone layer in the earth's atmosphere, which protects the planet's surface from potentially dangerous ultraviolet radiation, is currently threatened by compounds, such as some found in aerosol sprays, released into the atmosphere.

pacemaker The sinoatrial node, the region of the right atrium that initiates the contraction of the heart.

palate (*PAL-et*) The roof of the mouth. Near the front of the oral cavity, the palate is hard and bony; at the rear, it is soft and fleshy. The soft palate functions in separating breathing and swallowing.

palindrome (*PAL-in-drome*) A sequence of nucleotides in DNA that reads the same in one direction as in the other.

pancreas (*PAN-kree-us*) A large "double" gland located between the stomach and the duodenum. Its exocrine portion secretes sodium bicarbonate into the small intestine, as well as digestive enzymes that break down proteins, fats, and carbohydrates. Its endocrine islets of Langerhans produce insulin and glycogen, two important hormones that regulate blood sugar.

parapatric speciation (*PARE-a-PAT-rik*) The development of new species that occurs when a small population on the fringe of a large population diverges despite a modest degree of gene flow between the two populations.

parasite (*PARE-a-site*) An organism that lives and feeds on or in a host organism for at least part of its life cycle; a parasite may or may not kill its host.

parasympathetic division of the autonomic nervous system The portion of the autonomic nervous system that slows the body down to a relaxed state conducive to bodily functions such as digestion.

parathyroid glands (*PARE-a-THIGH-royd*) In the endocrine system, four tiny glands embedded in the thyroid gland. They produce parathyroid hormone (PTH), which increases calcium levels in the blood.

parathyroid hormone (PTH) (*PARE-a-THIGH-royd*) A hormone, secreted by the parathyroid, that promotes release of calcium into the bloodstream.

partial pressure The pressure that can be directly attributed to one gas in a mixture of gases.

pathogen (*PATH-oh-jen*) An organism or agent that infects the body and causes disease, either by causing direct physical damage to tissues, by releasing toxic chemicals, or by taking over cells' genetic machinery.

PCBs (polychlorinated biphenyls) (*POL-ee-KLOR-in-ate-ed buy-FEN-uls*) Toxic compounds used in electronics parts manufacturing that have persistent and damaging environmental consequences.

pedigree (*PED-ih-gree*) Lineage, ancestry, or genealogical chart or table.

peptide bond The carbon-to-nitrogen covalent bond that links one amino acid to another.

pericycle (*PARE-ih-sigh-kul*) A core of cells within the endodermis of a root. During rapid growth, the pericycle is the source of new tissue as the root sends out lateral branches.

peripheral nervous system (PNS) The nerves leading to and from the spinal cord and brain, and the sense organs.

peristalsis (*PARE-ih-STALL-sis*) The series of muscular contractions that move food through the esophagus.

petals The often brightly colored segments of flowers that in many plants function in attracting pollinators.

PGA (phosphoglycerate) An intermediate compound in glycolysis and in the Calvin–Benson cycle.

PGAL (phosphoglyceraldehyde) An intermediate compound in glycolysis and in the Calvin–Benson cycle.

pH A measure of the degree of acidity or alkalinity of a solution. On the pH scale, 7 is neutral, an acid solution has a pH of less than 7, and a basic solution has a pH of greater than 7.

phaeophytes (*FAY-oh-FIGHTS*) Members of the plant division Phaeophyta. The brown algae, a group of photosynthetic organisms that includes most seaweeds, such as rockweeds, giant kelps, and sargassum weeds.

phagocyte (*FAG-oh-site*) ("eating cells") A type of white blood cell (leukocyte) that is capable of engulfing and destroying foreign cells and foreign matter. Includes granulocytes and macrophages.

phagocytosis (*FAG-oh-sigh-TOE-sis*) ("cell eating") Endocytosis that brings a very large particle into the cell. An example is a white blood cell engulfing potentially harmful bacteria.

pharyngeal gill slits (*fa-RINJ-ee-al*) Openings or pouches in the lining of the upper digestive tract of chordates.

pharynx (*FARE-inks*) The throat, a muscular passage that leads from the nasal cavities to the esophagus and trachea; used in moving both food and air.

phenetics A school of thought in systematics that groups organisms together on the basis of a weighted analysis of their anatomical and physiological similarities.

phenotype (*FEE-no-type*) The observable form of an organism; phenotype reflects both genetic inheritance and environmental factors.

pheromone (*FER-o-mone*) A hormone-like, volatile chemical substance that is secreted by one individual and elicits a physiological or behavioral response from another individual of the same species.

phloem (*FLOW-em*) The vascular tissue that circulates nutritive sap throughout a plant.

phospholipid (*FOSS-foe-LI-pid*) A polar lipid consisting of a glycerol molecule covalently bonded to two fatty acids and a phosphate group to which another molecule may be attached. Phospholipids are an important component of cell membranes.

phosphorylation (*foss-FOUR-a-LAY-shun*) The attachment of a high-energy phosphate group to another molecule.

photic zone (*FOE-tik*) The relatively shallow region near the surface of a body of water where enough sunlight penetrates to allow photosynthesis.

photolysis (*foe-TAHL-a-sis*) The splitting of water molecules, fueled by radiant energy, during the first stage of noncyclic photophosphorylation. Photolysis yields electrons for reducing photosystem II, yields oxygen, and releases a pair of protons into the thylakoid sac.

photomorphogenesis (*FOE-toe-MOR-foe-JEN-ih-sis*) The process through which exposure to light influences the shape and growth pattern of a plant.

photoperiodism (*FOE-toe-PEER-ee-ud-izm*) A plant's physiological response to variations in duration of daylight.

photorespiration A metabolic pathway used by plants in bright, arid circumstances when oxygen concentrations in cells exceed carbon dioxide concentrations. The process reduces the efficiency of photosynthesis, releases carbon dioxide, consumes oxygen, and does not produce ATP.

photosynthesis (*foe-toe-SIN-the-sis*) The process by which plants use the energy of sunlight to construct organic molecules, primarily from water and carbon dioxide.

photosynthetic electron transport (*FOE-toe-sin-THET-ik*) The process by which electrons ejected from a photosystem pass through a sequence of molecules in a series of oxidation–reduction reactions, releasing energy that is used to attach an inorganic phosphate to ADP, thereby forming ATP.

photosystem I The photosynthetic reaction center in the thylakoid membrane that absorbs light with a maximum wavelength of 700 nm. Photosystem I is associated with the reduction of NADP$^+$ to NADPH in noncyclic electron flow.

photosystem II The photosynthetic reaction center in the thylakoid membrane that absorbs light with a maximum wavelength of 680 nm. Photosystem II is associated with the splitting of water to remove electrons and produce oxygen.

phototropism (*FOE-toe-TROE-piz-em*) A plant's differential growth in response to light.

physical environment Also called *abiotic environment*. The nonliving components of an organism's surroundings.

phytochrome (*FY-toe-krome*) A plant pigment molecule that is activated and deactivated by certain wavelengths of light and functions in seed germination, flowering, leaf expansion, and other processes of plant growth.

phytoplankton (*FIE-toe-PLANK-ton*) (*phyto* means "plant," *plankton* means "drifter") An assemblage of floating or weakly swimming photosynthetic organisms that make up an important part of fresh- and saltwater ecosystems.

pigment A molecule that absorbs most wavelengths of light and reflects light of particular wavelengths, which gives it a distinctive color. Chlorophyll and hemoglobin are pigments.

pineal gland (*PINE-ee-al*) In the endocrine system, a tiny gland located in the midbrain region of the skull that produces the hormone melatonin, which influences light/dark cycles and body rhythms.

pituitary gland (*pih-TOO-it-AIR-ee*) In the endocrine system, a small gland about twice the size of a pea, located within the human skull beneath the hypothalamus and closely linked to the hypothalamus both anatomically and functionally. The posterior lobe of the pituitary stores and releases oxytocin and antidiuretic hormone (ADH). The pituitary's anterior lobe produces seven of the most important hormones, including several that affect other endocrine glands: melanocyte-stimulating hormone (MSH), follicle-stimulating hormone (FSH), luteinizing hormone (LH), prolactin (PRL), thryotropic hormone (TSH), adenocorticotropic hormone (ACTH), and growth hormone (GH).

placebo (*plah-SEE-bow*) An inert substance containing no active ingredient; used as a control in an experiment.

placenta (*pla-SEN-ta*) An internal organ possessed by female mammals that nourishes a developing fetus.

Plantae The kingdom containing plants (predominantly photosynthetic primary producers with cell walls).

plasma (*PLAZ-ma*) The fluid in which blood cells are suspended. Plasma consists of water in which plasma proteins, gases, salts, sugars, and other substances are dissolved.

plasma cell A descendant of an activated B-cell that synthesizes and releases antibodies targeted at the antigen that originally activated the cell.

plasmid (*PLAZ-mid*) A type of "minichromosome" found in some bacteria that replicates independently of the bacteria's single DNA molecule.

platelets (*PLATE-lets*) Small fragments of cells with sticky surfaces that function in blood clotting; also called *thrombocytes*.

plate tectonics (*tek-TAWN-iks*) The drifting of large slabs that make up the earth's crust, producing a variety of geological effects.

platyhelminthes (*PLAT-ee-HEL-min-theez*) Members of the invertebrate phylum Platyhelminthes, also called *flatworms*, a group of simple worm-like invertebrates that includes both parasitic and free-living organisms.

pleiotropy (*PLEE-oh-troe-pee*) The capacity of a single gene to have several effects on an organism's phenotype.

pleural sac (*PLUR-al*) The two sacs, made up of epithelial membrane, that contain the lungs and reduce friction against the walls of the thoracic cavity.

point mutation A mutation involving a change in only one or two bases in a DNA molecule, such as a base substitution or the insertion or deletion of bases.

polar molecule A molecule in which different parts have different electrical charges produced by an uneven distribution of electrons.

pollen The immature male gametophytes produced by anthophytes and gymnosperms.

pollen sacs The chambers in the anther of a flower that contain the microspore mother cells.

polygenic (*polly-JEN-ik*) ("many genes") Controlled by many genes. Many human characteristics, including height and skin color, are polygenic.

polymerase chain reaction (PCR) An automated technique for creating large numbers of copies of a segment of a DNA molecule. The strands of DNA are repeatedly separated and copied by DNA polymerase, doubling the amount of DNA with each round until enormous quantities have been produced.

polymerization (*PAWL-ee-mer-eyes-AY-shun*) The formation of large molecules (polymers) by the bonding of many small molecules (monomers). The four major classes of biological macromolecules produced through polymerization are carbohydrates, proteins, lipids, and nucleic acids.

polymorphism (*PAWL-ee-MORF-izm*) The condition of two or more distinctly different manifestations of form.

polypeptide (*PAWL-ee-PEP-tide*) A chain of amino acids linked by peptide bonds between the nitrogen and carbon atoms of adjacent amino acids.

polysaccaride (*PAWL-ee-SAK-ah-ride*) A group of three or more bonded monosaccharides.

polytene chromosome (*PAWL-ee-teen*) A thickened, banded chromosome resulting from repeated chromosome replication without cell division or separation of sister chromatids. The copies remain packed together in parallel rows, creating distinctive bands easily seen with a microscope.

polyunsaturated The condition of a hydrocarbon (such as a fatty acid) in which carbon atoms are not bonded to the maximum number of hydrogen atoms, so the hydrocarbon contains at least two carbon-to-carbon double bonds.

population A group of individuals of a single species that interact and interbreed in a particular area.

population bottleneck The drastic reduction in size of a population, such that subsequent generations of offspring descend from relatively few individuals and a greatly reduced gene pool.

Population bottlenecks tend to reduce genetic diversity.

porifera (*pour-IF-er-ah*) Members of the invertebrate phylum Porifera, also called *sponges*, a group of invertebrates with several different cell types but no organs or well-differentiated tissues.

positive feedback system In biology, a mechanism that operates such that a shift in a physiological variable produces conditions that intensify the process that led to the shift.

posterior Toward the tail or back.

posterior pituitary (*pih-TOO-ih-tair-ee*) The posterior lobe of the pituitary gland, which produces two neurohormones, ADH and oxytocin. ADH increases water retention in the kidneys; oxytocin stimulates uterine contractions during labor.

postzygotic isolating mechanism (*post-zye-GOT-ik*) A mechanism that promotes reproductive isolation by preventing hybrid zygotes from maturing prenatally, from surviving after birth, or from reproducing if they do survive to maturity.

potential energy Stored energy capable of being released under the proper conditions.

power of 10 The number 10 multiplied by itself a certain number of times; indicated by the use of an exponent, such as 10^{-1} (0.1) or 10^3 (1000).

predator An organism that feeds upon other living organisms.

pressure-flow hypothesis A hypothesis that explains the movement of material through a vascular plant's phloem system as a combination of active transport and osmotic pressure. Sugar and water flow steadily between tissues that produce sugar and tissues that consume sugar, as sugars are actively transported out of and into those tissues, creating osmotic pressure that makes water follow the sugars.

prey An organism that serves as food for another living organism.

prezygotic isolating mechanism A mechanism that causes reproductive isolation by preventing eggs and sperm from uniting to form a zygote.

primary consumer An organism that eats plants; an herbivore.

primary growth Plant growth that occurs in the apical meristems.

primary producer An organism that can produce its own food from inorganic materials and an environmental energy source; an autotroph.

primary root The major "trunk" root that a plant sends into the soil.

primary structure The amino acid sequence of a polypeptide.

primary succession Ecological succession that occurs for the first time in newly formed habitats such as lava flows, deep bogs or lakes, or sand dunes.

primates An order of mammals characterized by highly dexterous digits, sophisticated binocular vision, a complex central nervous system, few offspring per pregnancy, and social systems.

progesterone (*pro-JES-te-rone*) A female sexual hormone, produced by the corpus luteum, that helps prepare the uterus for pregnancy.

prokaryotic cell (*pro-KAR-ree-OT-ik*) A type of cell that lacks membrane-bounded organelles and a distinct nucleus.

prophase (*PRO-faze*) The first stage of mitosis, during which the chromosomes gather near the center of the cell.

proprioceptors (*pro-PREE-oh-sep-tors*) Interoceptors embedded in muscles, tendons, joints, skin, and connective tissue that report on the movement and relative positions of body parts.

prosthetic group A non–amino acid portion of a protein molecule, such as the heme group that is responsible for binding oxygen in hemoglobin and myoglobin.

prosimians (*pro-SIH-mee-uns*) The more primitive branch of the order of primates; includes the tarsiers and lemurs.

prostaglandins (*PRAHS-tuh-GLANN-dinz*) A somewhat heterogeneous group of modified fatty acids secreted by a variety of tissues in response to cell injury or other hormones and having a variety of hormonal-like effects upon local tissues near their site of release.

protein (*PRO-teen*) A diverse class of macromolecules formed by the polymerization of amino acids. Proteins perform a variety of important functions in living cells.

protein kinase (*KYE-nase*) A protein that is an inactive enzyme until cAMP binds with it, at which time the kinase is activated. The activated kinase catalyzes chemical reactions that activate other enzymes.

Protista The kingdom containing single-celled eukaryotes and some multicellular algae.

proton A positively charged subatomic particle in the nucleus of an atom.

protoplast Plant cells that lack cell walls, produced under certain growing conditions and used in techniques for manipulating plant genes. For example, protoplasts from different plants or different species may be fused to form a hybrid cell with a new combination of genes, which then grows into a complete organism.

protostome (*PRO-toe-stome*) A member of one of two evolutionary lines distinguished by differences in embryonic development. In protostomes, cell division produces a spiral arrangement of cells, the ultimate function of each cell being determined early in development. The coelom is formed by the splitting of solid masses of mesoderm tissue, and the embryonic gut opening becomes the animal's mouth.

prototherians (*PRO-toe-THEER-ee-yuns*) Members of the mammalian subclass Prototheria, a group of egg-laying mammals. Includes the platypus and the spiny anteater.

pseudocoelomates (*SOO-doh-SEE-loh-MATES*) Bilaterally symmetrical animals with a simple body cavity lined with tissue other than mesoderm.

pterophytes (*TAIR-oh-fights*) The ferns; an ancient group of vascular plants that have compound leaves (fronds) and reproduce with spores. About 12,000 species exist today.

pulmonary circulation A circulatory pathway leading from the heart to the lungs.

punctuated equilibrium The proposition that evolution proceeds through long periods during which species change very little and brief periods during which change is relatively rapid.

Punnett square A powerful predictive technique in genetics, consisting of a diagram that represents all of the possible combinations of alleles that might occur in the offspring of two parents.

purine (*PURE-een*) Nitrogenous bases that occur in nucleotides and have two carbon-nitrogen rings in their structure; examples include adenine and guanine.

pyrimidine (*pih-RIM-ih-deen*) Nitrogenous bases that occur in nucleotides and have one carbon-nitrogen ring in their structure; examples include thymine and cytosine.

pyrrophytes (*PYE-roe-fights*) Also called *dinoflagellates*, or *"armored" flagellates*. Members of the phylum Pyrrophyta. Plant-like protists with bodies encased in tough cellulose plates. Most are capable of photosynthesis, and they are an important component of phytoplankton.

pyruvate (*pie-ROO-vate*) The three-carbon compound produced by the breakdown of glucose in glycolysis. Pyruvate may then be broken down further in another chemical pathway, such as the Krebs cycle, alcoholic fermentation, or lactate fermentation, depending on the organism.

quarternary structure The spatial arrangement of polypeptide chains in a protein that contains more than one polypeptide subunit.

radial symmetry (*RAY-dee-ul*) Symmetrical organization around a central axis, like the spokes of a bicycle wheel; sponges, starfish, and sea anemones are radially symmetrical.

radioisotope (*RAY-dee-oh-EYE-soh-tohp*) The unstable form of an element that emits radioactive radiation.

realized niche The portion of its fundamental niche that an organism actually occupies. An organism's realized niche is typically more limited than its fundamental niche because of interactions with other organisms, such as competition and predation.

receptor In cells, a molecule to which a hormone binds, forming a receptor–hormone complex that affects cellular metabolism.

receptor potential A brief change in voltage across the plasma membrane of a receptor cell. Receptor potentials are graded potentials; that is, the size of the potential varies with the intensity of the stimulus.

recessive gene A gene that is expressed in the phenotype when paired with another recessive gene but not when paired with a dominant gene.

reciprocal inhibition The characteristic of antagonistic muscle pairs whereby stretching an extensor muscle inhibits the flexor muscle; mediated by the spinal cord.

recombinant DNA (*re-KOM-bin-ent*) DNA created through the recombination of genes from different sources.

redox reactions (*reduction–oxidation*) Oxidation and reduction reactions that occur together as one compound loses electrons (oxidation) and another compound gains electrons (reduction).

reduction A chemical reaction in which electrons are added to an atom or compound.

reflex arc A sequence of several neurons that regulates a simple body movement without requiring conscious thought. A reflex arc is responsible for the knee-jerk reflex in humans.

releasing hormones Hormones that bind to receptors on cell membranes in the anterior pituitary, where they regulate the release of other hormones.

renal artery (*REE-nul*) The artery that brings blood to a kidney, where wastes and water are removed from the blood.

renal vein (*REE-nul*) The vein that returns blood to the circulatory system after wastes have been removed by the kidneys.

renewable energy Energy derived from replenishable sources such as sunlight, wind, and plants.

replication (*REP-li-KAY-shun*) The process by which DNA produces an exact duplicate of itself.

reproductive isolating mechanisms Characteristics that prevent physically similar organisms from interbreeding.

reproductive isolation The absence of interbreeding between members of different populations of the same species.

reptiles Members of the class Reptilia, a group of ectothermic, egg-laying, usually terrestrial vertebrates that have a scaly skin and breathe by means of lungs. Includes lizards, snakes, and turtles.

resolution limit The ultimate limit of minuteness at which a microscope can distinguish two or more structures a certain distance apart.

respiration The exchange of gases between an animal's body and the environment: the release of carbon dioxide and the uptake of oxygen.

respiratory surface The place where gases enter and leave an organism during respiration.

respiratory system A group of cells, tissues, and organs that exchanges gases with the environment and helps to distribute those gases within the organism.

resting potential The electrical potential across nerve cell membrane: the characteristic difference in electrical charge between the more negative inside of a nerve cell at rest and the more positive outside, equal to about − 70 millivolts.

restriction enzyme (*EN-zime*) An enzyme, produced by certain bacteria, that cuts DNA at specific sequences. The name is derived from the observation that such enzymes limit the types of DNA sequences that can survive in the bacteria and thereby help protect the bacteria from invading viruses.

reticular formation (*re-TIK-you-lar*) The net-like structure of nerve cells in the midbrain that determines which stimuli the cortex attends to and which it ignores.

retina (*RET-in-uh*) The light-receptive membrane at the back of the eye. The vertebrate retina consists of two layers, the neural retina and the pigment epithelium.

retrovirus (*RE-troe-vye-rus*) An RNA virus that constructs a DNA molecule from its RNA by using the enzyme reverse transcriptase. AIDS and many types of tumors are caused by retroviruses.

reverse transcriptase (*tran-SKRIP-tase*) An enzyme encoded by certain RNA viruses that can synthesize DNA from an RNA template.

RFLP Acronym for *r*estriction *f*ragment *l*ength *p*olymorphism. A change in the length of a restriction fragment in DNA that is due to inactivation of an adjacent restriction site by mutation.

Rh factor A blood antigen the gene for which has two alleles, positive and negative. Rh-factor incompatibility between mother and fetus can cause the mother's immune system to attack the fetus's blood if the proper medical treatment is not instituted.

rhodophytes (*RODE-oh-FIGHTS*) Members of the plant division Rhodophyta. The red algae, a diverse group of primarily marine photosynthetic organisms. Various accessory pigments give them red, violet, yellowish, or brownish colors and efficiently absorb the blue and green wavelengths of light that penetrate deep into seawater.

ribosomal RNA (rRNA) A type of RNA molecule that makes up ribosomes, together with a large number of ribosomal proteins.

ribosome (*RYE-boe-soam*) An organelle consisting of two subunits, each composed of proteins and RNA, that is the site of protein synthesis in the cell.

rigor mortis (*RIG-or MOR-tiss*) A condition of muscle rigidity that occurs after death when muscles run out of ATP. When this occurs, myosin cross-bridges cannot disconnect from actin.

RNA (ribonucleic acid) (*RYE-boe-new-CLAY-ik*) A single-stranded nucleic acid, containing ribose as its pentose sugar, that functions in protein synthesis.

rods One of the two classes of photoreceptors in the neural retina; responsible for "black and white" vision in very dim light.

root hairs Hair-like projections from the surface of a root that greatly increase the surface area in direct contact with the soil.

root nodules Growths in the roots of legumes such as peas and soybeans in which nitrogen-fixing bacteria live, getting protection and nutrients from the plants and converting nitrogen into a form that the plants can use to make proteins.

rough endoplasmic reticulum (*en-doe-PLAZ-mik ri-TIK-you-lum*) The portions of the endoplasmic reticulum that are associated with ribosomes.

r-selected species Also called *opportunistic species*. Species that maximize their intrinsic rate of reproduction. Such species reproduce frequently and at a young age, produce many offspring, devote little care to their offspring, and are often found in unpredictable environments.

RBP (ribulose bisphosphate) A five-carbon sugar involved in the first, key carbon-fixing step of the Calvin–Benson cycle in photosynthesis.

salinization (*SAL-in-eye-ZAY-shun*) The buildup of salts in irrigated soil in arid regions; salinization can result in dramatic reductions in crop yields.

saltatory conduction (*SAWL-ta-tor-ee*) Conduction down a myelated axon in which the action potential "jumps" from one node of Ranvier to the next.

salt marsh A saltwater shoreline ecosystem, characterized in temperate areas by the growth of marsh grass, in which the soil is alternately inundated and drained by the tides.

saprophyte (*SAP-roh-fite*) A decomposer: an organism that obtains energy by breaking down complex molecules in tissues of dead organisms. Many fungi and bacteria are saprophytes.

saprophytic (*SAP-roh-FIT-ic*) Living on dead or decaying organic matter.

sarcolemma (*SAR-koh-LEM-a*) The plasma membrane that surrounds a muscle fiber.

sarcomere (*SAR-koh-MERE*) A single structured unit in muscle; it is composed of thick and thin filaments and bounded on both ends by Z lines.

saturated The condition of a hydrocarbon (such as a fatty acid) in which all carbon atoms are bonded to the maximum number of hydrogen atoms.

scanning electron microscope A microscope that produces an image by scanning an electron beam across an object's surface and dislodging electrons, which are measured by a detector and the image displayed on a video screen.

Schwann cell (*SHWAHN*) A type of glial cell that forms an insulating myelin sheath around the axons of many vertebrate neurons; functions in increasing the speed at which action potentials are conducted along the axon.

science A way of interpreting and predicting events in the natural world based on an ongoing process of observing events in a systematic way, forming hypotheses that explain natural phenomena, testing those hypotheses, and revising conclusions.

scientific method The systematic approach to investigation characteristic of scientific research. It typically entails making observations (or recognizing a question), developing a hypothesis, testing the hypothesis through observation and experimentation, and drawing and reporting conclusions. In reality, scientific methodology exists in many forms.

scientific notation The system used by scientists to denote numbers efficiently. The system is based on powers of 10; 4,800,000 is expressed as 4.8×10^6, for example.

secondary consumer A carnivore that eats herbivores (primary consumers).

secondary growth Lateral plant growth, produced by the cambium, that results in the thickening of plant stems or roots.

secondary roots The smaller roots that branch off a primary root.

secondary structure The localized folding of a polypeptide chain, as typified by the alpha helix and beta sheet patterns.

secondary succession Ecological succession that occurs in places where human activity or natural events have destroyed or disturbed ecosystems, such as a plowed field or an area damaged by fire, storm, or disease.

second messenger A molecule such as cAMP that conveys a signal from a primary messenger (a hormone) across a cell membrane into the cell interior.

secretory vesicle (*sih-KREE-tore-ee VES-ik-al*) A vesicle involved in releasing proteins from the Golgi bodies to the exterior of the cell.

seed A plant reproductive structure consisting of an embryo and a food supply encased in a protective coat.

segregation, principle of The principle that even though an adult organism has two genes for each trait, the two genes separate when reproductive cells are formed.

semen (*SEE-men*) The fluid in which sperm are suspended. Semen is produced by glands in the reproductive tract and expelled through the penis during sexual intercourse.

semiconservative replication The process by which the DNA molecule reproduces itself. The two strands of the DNA molecule separate and a new complementary strand forms along each of the existing strands. Two identical DNA molecules result, each composed of one "old" strand and one "new" strand.

seminiferous tubules (*SEM-in-IF-er-us TOOB-yules*) Tightly coiled tubes in the testes that produce sperm cells in their cell walls.

senescence (*se-NESS-ense*) Generally, a state of agedness. Leaf senescence is the process by which plant leaves begin to cease metabolic activity and drop from branches in the fall.

sensory adaptation The process by which the eye adjusts its sensitivity to varying degrees of light and darkness. It involves pupil expansion and contraction, bleaching and reactivation of visual pigment, and adjustments in the sensitivity of photoreceptors and interneurons.

sensory neurons Nerve cells (neurons) that respond to light, heat, pressure, or chemicals in the environment and turn that information into electrochemical signals.

sepals (*SEE-puls*) Leaf-like structures that enclose the flower bud during early development and remain green throughout the life of the flower.

sex chromosomes Generally, chromosomes that differ in the two sexes and that, by their presence or absence, determine the sex of a zygote; in humans, the *XX* chromosomes in females and the *XY* chromosomes in males. Note that in some species—notably fish and reptiles—nongenetic factors determine the sex of an individual.

sex-linked genes Genes carried on either the *X* or the *Y* chromosome.

sex-linked inheritance A pattern of inheritance in which a phenotype is preferentially expressed in one sex or the other, due to the fact that the genes controlling it are located on the sex chromosomes.

sexual reproduction The process by which two parents generate offspring through fusion of gametes (eggs and sperm).

sexual selection A natural selection process in which traits that give an animal an advantage in the competition for mates tend to be passed on to offspring; often manifested in male-male competition and female choice in mating.

short-day plant A plant that flowers only when daylight is briefer than a certain maximum. Such plants flower in late summer and early fall.

sickle-cell anemia An inherited disorder of the hemoglobin in red blood cells that can cause fatal internal bleeding and blood clots in the lungs in persons who receive the sickle-cell allele from both parents.

sieve tubes The elongated, interconnected cells in phloem that convey phloem sap.

sign stimulus An external stimulus that elicits a particular pattern of behavior; a releaser.

simple dominance Also called *complete dominance*. The condition in which one gene for a trait is dominant and another is recessive.

simple fruit A fruit produced from a single carpel or from several fused carpels, such as a tomato or grape.

sinoatrial (SA) node (*SIGH-no-AY-tree-ul*) The region in the right atrium that initiates the contraction of the heart.

sliding filament model A model of muscle activity suggesting that muscles contract (shorten) when thin actin filaments slide over thick myosin filaments, pulling the Z lines together.

smog A collection of airborne dust, smoke, particles produced by internal combustion engines, and a variety of gases, such as ozone.

smooth endoplasmic reticulum (*EN-doe-PLAZ-mik re-TIK-you-lum*) The portions of the endoplasmic reticulum that do not have ribosomes attached.

social releasers Sign signals that trigger socially related behaviors in mates or members of the signaler's social group.

sodium-potassium pump A protein in nerve cell plasma membranes that uses energy from ATP to pump sodium ions out of and cell and potassium ions into the cell.

soil The medium, composed of pulverized rock and organic substances, in which plants grow.

somatic nervous system The voluntary portion of the peripheral nervous system, including most sensory neurons from the skin and skeletal muscles and most motor neurons leading to skeletal muscles.

somites (*SOE-mites*) Distinct blocks of mesodermal tissue in the embryo that develop into bone, muscle, and internal organs.

Southern blotting A powerful laboratory technique for finding a particular gene in the DNA from a cell, using nucleic acid hybridization and gel electrophoresis. A radioactive RNA "probe" having the base sequence representing a particular protein is created to bind to the complementary sequence in heat-treated DNA fragments, forming a DNA–RNA hybrid that can be detected on photographic film after gel electrophoresis.

specialization The process through which undifferentiated cells differentiate to perform a variety of functions.

speciation (*spee-shee-AY-shun*) The process by which populations of organisms diversify genetically, evolve intrinsic isolating mechanisms, and form new species.

species (*SPEE-shees*) A natural population or group of interbreeding natural populations that produces fertile offspring and is reproductively isolated from other groups.

spermatogonia (*sper-MAT-oh-GOE-nee-a*) Cells that are found near the periphery of the seminiferous tubules and divide by mitosis to produce the primary spermatocytes.

sphenophytes (*SFEEN-oh-fights*) The horsetails; an ancient group of seedless vascular plants. The sphenophytes were a diverse group during the Carboniferous; only about 20 species survive today.

spiracle (*SPEER-ih-kul*) An opening in the exoskeleton of an insect through which air enters and gases are exchanged.

sporophyte (*SPORE-oh-fight*) The diploid, spore-producing stage of the plant life cycle.

sporozoa (*SPORE-oh-ZOE-a*) Members of the phylum Sporozoa. Parasitic animal-like protists, many of which have complex life cycles involving two hosts. Includes the genus *Plasmodium* which causes malaria in humans.

squamous epithelium (*SKWAY-muss EP-ih-THEEL-ee-um*) A tissue made up of flat, scale-like cells that functions in diffusion and filtration; found in such locations as the air sacs of lungs and the linings of blood vessels.

stabilizing selection A form of natural selection that reduces the prevalence of extreme expressions of a trait and favors a more moderate manifestation of the trait.

stamen (*STAY-men*) The male portions of a flower, which produce pollen in the anthers.

statocyst (*STAT-oh-sist*) The spherical organ found in many invertebrates that is used to convey information about the body's orientation relative to the pull of gravity. The interior of the statocyst is typically lined with hair cells, which are stimulated by small particles that gravitate to the lowest part of the statocyst's interior.

steroid (*STEER-oid*) Member of a class of lipids characterized by four carbon ring structures with various side groups and including a number of hormones, such as sex hormones and hormones of the adrenal cortex.

steroid hormone A hormone with the chemical composition of a steroid (a kind of lipid) and the potential to alter the expression of genes. Steroid hormones act by crossing the lipid bilayers of cell and nuclear membranes and binding to the appropriate receptor to form a hormone-receptor complex, which then binds to chromatin and either activates or deactivates genes.

stimulus-response chain A sequence of signals, characteristic of the courtship of many animals,

stimulus-response chain (*continued*) in which one partner's behavior serves as a stimulus that releases the next behavior in the other partner, and so on.

stomata (singular, **stoma**) (*sto-MAH-tah*) Openings in leaf epidermis that allow evaporation of water and intake of carbon dioxide. Two guard cells around the opening regulate the stoma's closing and opening.

stop codon A triplet of bases along an mRNA molecule that signifies the end of a polypeptide chain, instead of standing for a particular amino acid.

stroke The death of brain cells from lack of oxygen, often due to a blood clot blocking the flow of blood in the blood vessel serving those cells.

stroma (*STROH-mah*) The semifluid matrix within the chloroplast in which the light-independent reactions of photosynthesis take place.

stromatolite (*stroh-MAT-oh-lite*) Fossilized domes consisting of layered sediment that were produced by early prokaryotic organisms; the fossils of these organisms are the earliest known evidence of life on Earth.

subatomic particle (*SUB-a-TOM-ik*) The particles of which atoms are composed: electrons, protons, and neutrons.

substrate (*SUB-strait*) A reactant in an enzyme-catalyzed reaction.

substrate phosphorylation (*SUB-strait foss-FOUR-a-LAY-shun*) The formation of ATP by the direct transfer of a phosphate group from a substrate molecule to ADP.

succession In an ecosystem, the gradual process of species replacement in response to changes in the environment.

summation A characteristic of the transmission of action potentials between neurons whereby the membrane potential of the postsynaptic cell is affected by the total activity of all excitatory and inhibitory potentials of presynaptic neurons.

suppressor T-cell A type of T-cell that moderates the immune response by slowing down the rate of cell division and limiting the production of antibodies once an infection has been brought under control.

surface tension The resistance to rupture of the surface of a liquid caused by the mutual attraction of molecules. Water has high surface tension.

survivorship In a life table, the percentage of a population that can expect to live to a given age.

sustainable agriculture An agricultural system that preserves a functional balance and resiliency in the local ecosystem, is economically efficient and productive, and relies on technology and practices that are suited to the local culture and economy.

symbiosis (*sim-bee-OH-sis*) A protracted association of two organisms in an intimate relationship: may range from mutualism to parasitism.

sympathic division of the autonomic nervous system The portion of the autonomic nervous system that prepares the body for intense physical exertion and stimulates the "flight-or-fight" response.

sympatric speciation (*sim-PAT-rik spee-shee-AY-shun*) The development of a new species in the midst of its parent populations, following the emergence of genetic or other reproductive isolating mechanisms.

synapse (*SIN-naps*) The point of junction between the axon of one neuron and the cell body or dentrite of another.

synapsis (*sin-AP-sis*) The pairing of homologous chromosomes during the first stage of meiosis.

syphilis (*SIF-uh-liss*) A serious sexually transmitted disease caused by the bacterium *Treponema pallidum*, which at first produces a hard sore called a chancre and then gradually infects the circulatory and nervous systems. Syphilis can be treated with massive doses of antibiotics in its early stages before permanent physical damage has occurred.

systemic circulation A circulatory pathway leading from the lungs to the heart, on to the rest of the body, and back to the heart.

systematics The classification of organisms on the basis of degree of relationship and evolutionary history.

systole (*SIS-ta-lee*) The state of the heart when the ventricles contract and push blood into the aorta and major arteries.

taiga (*TYE-guh*) A biome of predominantly evergreen forest characterized by harsh winters, warmer and longer summers than tundra areas experience, and a diverse, complex array of plants and animals.

target cell A cell that contains receptor sites for a particular hormone molecule.

taste bud The site of the taste receptor cells in grooves in the surface of the tongue. The actual sensation of taste occurs when a particular molecular shape binds to a receptor molecule.

taxa (singular, **taxon**) (*TAX-a*) The categories used in naming and classifying organisms, including phylum, class, order, family, genus, and species.

taxonomy (*tax-ON-a-mee*) The science of naming and classifying organisms.

Tay–Sachs disease (*TAY-SAKS*) A relatively rare, heritable, debilitating, lethal disease leading progressively to blindness, emaciation, and early death.

TCE (trichloroethylene) A toxic, colorless liquid and a suspected carcinogen discovered in the groundwater supplies of several states.

teleology (*TEEL-ee-AWL-a-jee*) The study of the ultimate purpose or goal of natural processes. Lamarck's theory of evolution was teleological in that he thought evolution was purposeful, motivated by organisms' striving for "perfection" as they attempted to rise through the "chain of being."

telophase (*TEE-loe-faze*) The final stage of mitosis, the stage during which the chromosomes organize into two cell nuclei.

temperate forest A biome found in the middle latitudes, characterized by predominantly broad-leaved forests, relatively rich soil, ample rainfall, and a great diversity of plant and animal life.

template (*TEM-plit*) A pattern for making accurate reproductions. In DNA replication, each strand of the double helix serves as a template for synthesis of a new strand.

temporal isolation A prezygotic isolating mechanism that functions when two species cannot interbreed because they breed at different times.

ten, rule of The assumption that roughly 10 percent of the energy produced by organisms at one trophic level is ultimately stored by organisms at the next higher level.

tertiary consumer (*TUR-shee-AIR-ee*) A carnivore that eats other carnivores.

tertiary structure The complete, large-scale, three-dimensional folding of a polypeptide chain.

test cross In genetics, the mating of individuals of uncertain genotype with individuals that are homozygous recessive for the genes in question. The phenotypes of the offspring reveal what the uncertain genotype must have been.

testes (singular, **testis**) (*TES-teeze*) The primary male sexual organs, which produce male gametes (sperm) and sex hormones.

testosterone (*tes-TOSS-ter-own*) The male sexual hormone, which acts on spermatogonia to induce sperm development and, at puberty, induces the development of secondary sexual characteristics.

tetanus (*TET-nus*) The powerful, sustained muscle contraction that results from a steady, rapid succession of action potentials caused by constant stimulation.

tetrads Also called *bivalents*. In meiosis, the bundles consisting of four chromatids each, formed when the duplicated homologous chromosomes pair up in synapsis during the first meiotic division.

theory A systematically organized set of related hypotheses that explain and accurately predict a broad, complex range of natural phenonoma and rests on a long-established body of observations and experimental tests.

therapsids (*thir-AP-sidz*) A group of terrestrial animals intermediate between reptiles and mammals; flourished during the Permian.

thermal stratification In bodies of water, the condition in which surface layers are separated from deeper layers by sharp differences in temperature (thermoclines), creating distinctly different environments.

thermoacidophiles (*THUR-moe-ass-ID-oh-files*) Archaebacteria that inhabit hot, acid environments such as sulfur springs and volcanic vents.

thermocline (*THUR-moe-kline*) An area of rapid temperature change separating surface and deeper water masses.

thermodynamics, first law of (*THUR-moe-dye-NAM-iks*) The principle of conservation of energy, which states that the total amount of energy available in the universe does not change.

thermodynamics, second law of (*THUR-moe-dye-NAM-iks*) The principle that, in the universe as a whole and in any isolated system not at equilibrium, the energy becomes more dispersed and less organized such that, once it has been used to do work, it is less available to do additional work.

thermoregulation (*THUR-moe-REG-you-LAY-shun*) The process by which an organism controls its body temperature.

thigmotropism (*THIG-moe-TROE-pizm*) A plant's growth or movement in response to contact with a solid object.

thoracic cavity (*thoe-RASS-ik*) The sealed cavity in the chest where the lungs are located.

thylakoids (*THIGH-la-koyds*) Also called *photosynthetic membranes*. The sac-like membranes within the chloroplast, where the light-dependent reactions of photosynthesis take place. In some parts of the chloroplast, the thylakoids are stacked to form *grana*.

thymine A nitrogen-containing base found in nucleotides, belonging to the chemical group known as pyrimidines, which have a single carbon-nitrogen ring in their basic structure. Found in DNA but not in RNA.

thymine dimer (*THIGH-mean DYE-mer*) A molecule of two covalently bonded thymidine molecules that cannot be copied properly during DNA replication, resulting in a pair of adjacent base substitutions.

thymus gland (*THIGH-mus*) In the endocrine and immune systems, a gland near the center of the chest that is important in the maturation of white blood cells and produces the hormone thymosin, which stimulates the development of certain types of lymphocytes. According to current theory, the thymus functions in the development of the immune system by eliminating T-cells that would otherwise attack the host's tissues.

thyroid gland (*THIGH-royd*) In the endocrine system, a large gland in the neck that regulates growth and metabolic rate by producing thyroxine and reduces calcium levels in the blood by producing calcitonin.

tight junction An attachment between adjacent cells that seals off the space between the cells and thereby prevents materials from leaking between them; important in the lining of fluid filled cavities.

tissues Groups of similar cells that perform specialized functions throughout the body in multicellular organisms.

T-lymphocyte (*LIMF-oh-sight*) Also called *T-cell*. A type of lymphocyte capable of mounting a cell-to-cell attack on a foreign organism. There are three kinds of T-cells (helper, killer, and suppressor) that differentiate and mature only after passing through the thymus gland.

tolerance, law of The ecological principle that the growth of a species is limited by the environmental factor for which organisms of the species have the narrowest range of tolerance.

trace fossil Petrified tracks, borrows, or other evidence left in sediment by organisms that were not preserved.

trachea (*TRAY-key-a*) The windpipe, a tube leading from the larynx to the two bronchi that enter the lungs.

tracheids (*TRAY-key-idz*) Specialized elongated cells in vascular plants that function in the conduction of fluids and in providing structural support.

tracheoles (*TRAY-key-ohlz*) In arthropods, minute branches of the trachea, which make direct contact with many cells and function in gas exchange.

transcription The synthesis of RNA as the complement of one strand of a double-stranded DNA molecule.

transfer RNA (tRNA) A specialized class of RNA molecules that recognize the codons on an mRNA molecule held by a ribosome and bring the appropriate amino acids into position to bond to the newly forming polypeptide chain.

transformation The process by which a cell acquires genetic material from an external source.

transgenic organism (*tranz-JEEN-ik*) In genetics, an organism that incorporates functional DNA derived from another organism, usually one of a different species.

translation The conversion of the information from an RNA molecule into a particular series of amino acids in a polypeptide or protein.

translocation An alteration in chromosome structure in which part of a chromosome is broken off and attached to another chromosome.

transmission electron microscope A microscope that produces an image by using magnetic lenses to project an electron beam through a specimen and ultimately onto a screen of photosensitive film.

transpiration (*TRANS-purr-AY-shun*) The loss of water from a plant as a result of evaporation.

transposable elements (*tranz-POHZ-uh-bul*) Also called *transposons*. Relatively short segments of DNA that are replicated at one part of a chromosome and inserted in another part.

triglyceride (*try-GLISS-er-ide*) A lipid consisting of a single glycerol molecule covalently bonded to three fatty acids.

trilobite (*TRY-loe-bite*) One of an extinct group of marine arthropods that were common in early Cambrian seas.

trisomy-21 (*TRY-so-mee*) See *Down syndrome*.

trophic hormones (*TROH-fik*) Hormones that stimulate the secretion of other hormones.

trophic level (*TROE-fik*) A stage in a food chain. Primary producers represent one trophic level, herbivores a second, carnivores who eat herbivores a third, and so on.

tropical rain forest A biome found in tropical areas and characterized by substantial rainfall, lush growth, and an extraordinary diversity of plant and animal life.

tropism (*TROE-pizm*) A plant growth response to an environmental factor such as gravity, light, a chemical concentration, or physical contact with an object. The response typically entails differential rates of cell elongation, causing the plant organ as it grows to move toward or away from the source of the stimulus.

tropomyosin (*TROE-poe-MY-oh-sin*) One of three proteins that compose the thin filaments in muscle cells. With troponin, tropomyosin helps to regulate contraction.

troponin (*TROE-poe-nin*) One of three proteins that compose the thin filaments in muscle cells. A calcium-binding protein that helps to regulate contraction.

tubulin Proteins (alpha-tubulin and beta-tubulin) that make up microtubules.

tumor Any abnormal mass of cellular tissue, usually benign, that may or may not be circumscribed and may or may not become malignant.

tumor suppressor A gene that normally prevents the uncontrolled growth of cells and the formation of tumors. Its absence or inactivation of mutation allows a cancer to develop.

tundra (*TON-drah*) A biome found in polar regions and at high altitudes and characterized by permanently frozen subsoil, a brief summer, and low-growing plants.

Turner syndrome A genetic nondisjunction abnormality in which the individual has only one *X* chromosome. Afflicted persons are female, but they are infertile, do not reach sexual maturity, and have a variety of other abnormal characteristics.

undernourishment The condition of having less than the amount or kind of food required for normal health or development.

uniformitarianism (*YOUNE-if-for-mih-TARE-ee-un-izm*) The theory proposed by Lyell that natural laws remain constant and operate gradually and that, as a result, past events can be explained in terms of processes occurring today.

uniramia (*you-nih-RAY-me-ah*) Members of the invertebrate subphylum Uniramia, a group of arthropods with one limb branch. Includes centipedes, millipedes, and insects, and may contain more living species than all other groups of organisms put together.

upwelling The movement of large quantities of nutrient-laden deep water up toward the surface due to particular combinations of winds and currents.

uracil A nitrogen-containing base found in nucleotides, belonging to the chemical group known as pyrimidines, which have a single carbon-nitrogen ring in their basic structure: found in RNA but not DNA, and can base-pair with adenine.

urea (*yur-EE-a*) In mammals and some amphibians, the soluble nitrogenous waste released by the liver into the bloodstream, from which it is removed by the kidneys and excreted in urine.

ureters (*yur-IH-ters*) The tubes that transport urine from the kidneys to the bladder.

urethra (*yur-EE-thra*) The tube (channel) that transports urine from the bladder out of the body.

uric acid (*YUR-ik*) In birds, insects, and reptiles, the insoluble nitrogenous waste into which ammonia is processed for excretion.

vaccination (*VAK-sin-AY-shun*) Inoculation with a weakened or dead form of a pathogen, which causes the body to develop antibodies for the pathogen and thereby produces immunity to it.

vacuole Organelle consisting of a single large membranous sac enclosing such things as food particles, waste products, or pigments. Vacuoles are especially prominent in plant cells.

variations Differences in characteristics among individuals in a species.

vascular bundles Parallel bundles of xylem and phloem tissue running lengthwise in the stems of tracheophytes.

vascular cylinder (*VAS-kew-ler*) The cylinder of water-transporting endodermis in the central region of a root or stem.

vascular plants Plants with specialized tissues for transporting fluids.

vascular tissue (*VAS-kew-ler*) Tissue specialized for carrying fluids.

vector pollination The transfer of pollen from the anthers to the stigmata of flowers by another organism, such as a bee.

vegetative reproduction (*VEJ-ih-TAY-tiv*) Asexual reproduction in plants through such means as adventitious roots, runners, cuttings, and rhizomes.

veins In vertebrate circulatory systems, large blood vessels that carry blood back toward the heart. Many veins contain valves that prevent blood from flowing in a reverse direction.

venereal disease A sexually transmitted disease (STD), caused by a bacteria (as in the case of syphilis, gonorrhea, and chlamydia) or a virus (as in the case of *Herpes type II* infections, genital warts, hepatitis type B, and AIDS).

ventral Toward the belly.

ventricle (*VEN-tre-kul*) A muscular chamber in the heart that pumps blood out of the heart. The human heart has two ventricles.

venules In vertebrate circulatory systems, small blood vessels that channel blood from the capillaries into larger veins leading back to the heart.

vertebrates Animals with a segmented backbone.

vertical zonation The pattern of horizontal bands in which plants and animals are distributed along shorelines between the upper and lower reaches of the tide.

vestigial structure (*veh-STIJ-ee-ul*) A degenerate or rudimentary form of an organic structure that

vestigial structure (*continued*)
functioned more fully in ancestors that occurred in earlier stages of the organism's evolution.

virus (*VYE-russ*) A disease-causing subcellular particle, consisting of a core of nucleic acids surrounded by a coat of protein, that cannot carry on its own metabolism but can direct the metabolic processes of host cells to its own use and can reproduce only within the cells of a host.

visual pigments Photosensitive compounds concentrated in the outer segment of the receptor cells of the eye, the molecules of which change shape when they absorb light.

vitamins Various complex organic compounds needed in small amounts in the diet, most of which function as cofactors in catalyzing chemical reactions.

weather The atmospheric conditions (rainfall, cloudiness, humidity, sunlight, wind speed, and other factors) that exist in an area at a given time; compare *climate*.

xylem (*ZYE-lem*) The vascular tissue that conducts water and minerals up from a plant's roots.

zooplankton (*zoo* means "animals," *plankton* means "drifters") An assemblage of floating or weakly swimming heterotrophs that are mostly microscopic and make up an important part of fresh- and saltwater ecosystems.

zygomycetes (*ZYE-go-MY-seets*) Members of the division Zygomycota. Also called *common molds*. A group of terrestrial or aquatic saprophytic or parasitic fungi. An example is *Rhizopus*, the common black bread mold.

zygote (*ZYE-goat*) ("fertilized egg") The diploid cell formed by the fusion of gametes.

The following abbreviations are used for sources that are referred to frequently: Animals Animals–AA; Bruce Coleman, Inc.–BC; Biological Photo Service–BPS; Earth Scenes–ES; Grant Heilman–GH; Horvath & Cuthbertson–H&C; Peter Arnold, Inc.–PA; Phototake–PHT; PhotoNats–PN; Photo Researchers–PR; Tom Stack & Associates, Inc.–TSA; Visuals Unlimited–VU.

Preface

p. v William Munoz; *p. vi* Maryland CartoGraphics, Inc.; *p. vii* William H. Mullins.

Part 1 David Muench

Chapter 1

Fig. 1.1 John Harrington / Black Star; *Fig. 1.2* NASA; *Fig. 1.3 (left to right)* Dennis Kunkel / PHT, David M. Phillips / VU, Arthur Siegelman / VU; *Fig. 1.4* Patrice Rossi; *Fig. 1.5* Richard Feldman; *Box p. 6* Dick Canby / DRK; *Fig. 1.6* Stephen Jay Gould, *The Mismeasure of Man,* New York: W. W. Norton, 1981, p. 93; *Fig. 1.7* The Bettmann Archive; *Fig. 1.8* and *Fig. 1.9* P. Rossi; *Fig. 1.10* Dr. Porter Kier / Smithsonian Museum; *Fig. 1.14* P. Rossi.

Chapter 2

Fig. 2.1 Roger Ressmeyer / Starlight; *Fig. 2.2* Wellcome Institute Library, London; *Fig. 2.3* Lynn Johnson / Black Star; *Fig. 2.4* Michel de la Sabliere / Hillstrom Stock Photo, Inc.; *Box p. 25* Randall Hyman; *Fig. 2.5* David J. Cross; *Fig. 2.6* Doug Wechsler / ES; *Fig. 2.7* Time.

Part 2 Steve C. Wilson / Entheos

Chapter 3

Fig. 3.1a Michael Fogden / DRK Photo; *Fig. 3.1b* North Wind Picture Archives; *Fig. 3.1c* Jim Foster / The Stock Market; *Fig. 3.1d* Claus C. Meyer / Black Star; *Fig. 3.2 (top)* Jeff Rotman; *Fig. 3.2 (bottom)* J. Lotter / TSA; *Fig. 3.3* Patti Murray / ES; *Fig. 3.4a* Patrice Rossi; *Fig. 3.4b* Jeff Rotman; *Fig. 3.5* and *Fig. 3.6* P. Rossi; *Fig. 3.7* Sanderson Associates; *Fig. 3.8* and *Fig. 3.9* P. Rossi; *Box p. 41* Richard Legekis; *Fig. 3.10* Sanderson Associates; *Fig. 3.11* P. Rossi; *Fig. 3.12 (top)* Gene Ahrens / BC; *Fig. 3.12 (bottom left)* Breck P. Kent / ES; *Fig. 3.12 (bottom right)* Arthur Gloor / ES; *Fig. 3.12 (center)* Sanderson Associates; *Fig. 3.13* through *Fig. 3.16* P. Rossi; *Fig. 3.17* Andrew Robinson / Virgil Pomfret Agency; *Fig. 3.18* Sanderson Associates; *Fig. 3.19* Stephen J. Krasemann / PA; *Fig. 3.20* Peter Arnold / PA; *Box pp. 50–51* P. Rossi; *Fig. 3.21 (left)* Richard Thom / TSA; *Fig. 3.21 (right)* Scott Blackman / TSA; *Fig. 3.22* Jeff Foott / TSA; *Fig. 3.23* Jocelyn Burt / BC; *Fig. 3.24* Stephen J. Krasemann / PA; *Fig. 3.25* UPI / Bettmann Newsphotos; *Fig. 3.26 (left)* Lynette Cook; *Fig. 3.26 (right)* Mickey Gibson / TSA; *Fig. 3.27a* David L. Brown / TSA; *Fig. 3.27b* E. R. Degginger; *Fig. 3.27c* Michael P. Gadomsi / PR; *Fig. 3.28* Marlene DenHouter; *Fig. 3.29* Alan D. Briere / TSA; *Fig. 3.30* M. Timothy O'Keefe / TSA; *Fig. 3.31* Anne Wertheim; *Fig. 3.32* Jeff Rotman; *Fig. 3.33* Jeff Foott / BC; *Fig. 3.34 (both)* Jeff Rotman; *Fig. 3.35* P. Rossi.

Chapter 4

Fig. 4.1 Barbara Von Hoffmann / TSA; *Fig. 4.2* William H. Amos / BC; *Fig. 4.3* Patrice Rossi; *Fig. 4.6 (top)* Bruce Davidson / AA; *Fig. 4.6 (center)* Jeff Foott / TSA; *Fig. 4.6 (bottom)* Kevin Schafer / TSA; *Fig. 4.7* P. Rossi; *Fig. 4.8 (top to bottom)* Walter E. Harvey / PR, Stephen Dalton / PR, Inc., Stephen J. Krasemann / PR; *Fig. 4.9* Ed Reschke / PA; *Fig. 4.10* David T. Overcash / BC; *Fig. 4.11* Dave Woodward / TSA; *Fig. 4.12* Jeff Rotman; *Fig. 4.13* and *Fig. 4.14* Marlene DenHouter; *Fig. 4.15* and *Fig. 4.16* P. Rossi; *Fig. 4.17* and *Fig. 4.18* Michael Woods / Linden Artists Ltd.; *Fig. 4.19* P. Rossi; *Fig. 4.20* and *Fig. 4.21* Michael Woods / Linden Artists Ltd.; *Fig. 4.22* Lyrl Ahern; *Fig. 4.23* P. Rossi; *Box p. 82* P. Rossi.

Chapter 5

Fig. 5.1 Ray Pfortner / PA; *Fig. 5.2* through *Fig. 5.6* Patrice Rossi; *Fig. 5.7* through *Fig. 5.11* P. Rossi; *Box p. 97* P. Rossi; *Fig. 5.12 (both)* Division of Entomology / CSIRO.

Chapter 6

Fig. 6.1 The Bettmann Archive; *Fig. 6.2* through *Fig. 6.4* Patrice Rossi; *Fig. 6.5* Jack D. Swenson / TSA; *Fig. 6.6* and *Fig. 6.7* P. Rossi; *Fig. 6.8* Anne Wertheim; *Fig. 6.9* and *Fig. 6.10* P. Rossi; *Box p. 109* P. Rossi; *Fig. 6.11 (left)* Jeff Rotman; *Fig. 6.11 (right)* E. R. Degginger / ColorPic, Inc.; *Fig. 6.12* Zig Leszczynski / AA; *Fig. 6.13 (left)* Zig Leszczynski / ES; *Fig. 6.13 (right)* P. Rossi; *Fig. 6.14* Darlyne Murawski; *Fig. 6.15* P. Rossi; *Fig. 6.16a* William J. Weber / VU; *Fig. 6.16b* P. Rossi; *Fig. 6.17* through *Fig. 6.19* P. Rossi.

Chapter 7

Fig. 7.1 Photri; *Fig. 7.2* Scott Camazine / PR, Inc.; *Fig. 7.3* Tom Sobolik / Black Star; *Fig. 7.4* Patrice Rossi; *Fig. 7.5* © 1975 Aileen and W. Eugene Smith / Black Star; *Fig. 7.6* P. Rossi; *Fig. 7.7* Sanderson Associates; *Fig. 7.8* P. Rossi; *Fig. 7.9 (both)* K. H. Mills, from D. W. Schindler et al., *Science,* 288 (21 June 1985): 1395–1401, First Light, Toronto; *Fig. 7.10* NASA; *Fig. 7.11* and *Fig. 7.12* Maryland CartoGraphics, Inc.; *Fig. 7.13* Frans Lanting / Minden Pictures; *Fig. 7.14* GH; *Fig. 7.15* Marlene DenHouter; *Fig. 7.16* Catherine M. Pringle / BPS; *Fig. 7.17a* Maryland CartoGraphics, Inc.; *Fig. 7.17b* David C. Turnley / Black Star; *Fig. 7.18* Sanderson Associates; *Fig. 7.19a* Bruce Coleman / BC; *Fig. 7.19b* M. DenHouter; *Fig. 7.19c* P. Rossi; *Fig. 7.20* through *Fig. 7.21a* P. Rossi; *Fig. 7.21b* Jeff Rotman; *Fig. 7.22* Mark Moffet / Minden Pictures; *Fig. 7.23* Maryland CartoGraphics, Inc.; *Fig.*

7.24a–b James Porter; *Fig. 7.25* and *Fig. 7.26* Maryland CartoGraphics, Inc.

Part 3 David M. Stone / PN

Chapter 8
Fig. 8.1 Royal College of Surgeons / Charles Darwin House; *Fig. 8.2* Patrice Rossi; *Fig. 8.3 (both)* James R. Hill, 3rd / New York State Museum; *Fig. 8.4 (left)* P. Rossi; *Fig. 8.4 (right)* The British Museum of Natural History; *Box p. 153* National Portrait Gallery, London; *Fig. 8.5* Sanderson Associates; *Fig. 8.7 (left)* Hans and Judy Beste / AA; *Fig. 8.7 (right)* Barrie E. Watts / OSF / AA; *Fig. 8.8 (left)* L. L. Rue, 3rd / BC; *Fig. 8.8 (right)* W. H. Hodge / PA; *Fig. 8.9* The Royal Geographical Society / Bridgeman Art Library; *Fig. 8.10 (right top to bottom)* David M. Cavagnaro; *Fig. 8.11* Enid Kitschnig / *Scientific American*; *Fig. 8.12 (top)* Marlene DenHouter; *Fig. 8.12 (bottom)* The *Illustrated London News*; *Fig. 8.13* Andrew Robinson / Virgil Pomfret Agency; *Fig. 8.14* The Bettmann Archive / BBC Hulton; *Fig. 8.16* M. DenHouter; *Fig. 8.17* and *Fig. 8.18* Stephen Dalton / AA; *Fig. 8.19* and *Fig. 8.20* M. DenHouter; *Fig. 8.21 (left)* Patti Murray / ES; *Fig. 8.21 (right)* John M. Trager / VU; *Fig. 8.22 (left)* Jack W. Dykinga / BC; *Fig. 8.22 (right)* P. Rossi; *Fig. 8.23* P. Rossi; *Fig. 8.24 (top)* Charles Marden Fitch; *Fig. 8.24 (bottom)* Catherine Pringle / BPS; *Fig. 8.25* Lawrence E. Gilbert; *Fig. 8.26 (clockwise)* E. R. Degginger / ColorPic, Inc., L. West / BC, Breck P. Kent / AA; *Fig. 8.27* Kenneth W. Fink / PR; *Fig. 8.28 (both)* Lincoln Brower / Univ. of Florida, Gainesville; *Fig. 8.29* P. Rossi.

Chapter 9
Fig. 9.1 (clockwise) David M. Phillips / VU, L. L. Sims / VU, © Paul W. Johnson 1987 / BPS, Peter Parks / OSF / AA; *Fig. 9.2 (left)* Arleen Frasca; *Fig. 9.2 (photos)* Ken Miller; *Fig. 9.3 (top)* Dr. Dennis Kunkel / PHT; *Fig. 9.3 (bottom)* Richard H. Gross / Biological Photography; *Fig. 9.4* K.G. Murti / VU; *Fig. 9.5* Patrice Rossi; *Fig. 9.6* H&C; *Fig. 9.7 (left)* Hans Ris, Univ. of Wisconsin; *Fig. 9.7 (right)* H&C; *Fig. 9.8* H&C; *Fig. 9.9* Keith Porter; *Fig. 9.10* P. Rossi; *Fig. 9.11* Andrew S. Bajer, Univ. of Oregon; *Box p. 182* Peter Hepler; *Fig. 9.12 (left)* H&C; *Fig. 9.12 (right)* Conly L. Rieder; *Fig. 9.13 (top)* Andrew S. Bajer, Univ. of Oregon; *Fig. 9.13 (bottom)* H&C; *Fig. 9.14* Andrew S. Bajer, Univ. of Oregon; *Fig. 9.15 (left)* Andrew S. Bajer, Univ. of Oregon; *Fig. 9.15 (right)* Jeremy Pickett-Heaps; *Fig. 9.16* and *Fig. 9.17 (both)* Andrew S. Bajer, Univ. of Oregon;

Fig. 9.18 David Phillips / VU; *Fig. 9.19 (left)* P. Rossi; *Fig. 9.19 (right)* Peter Hepler; *Fig. 9.20* P. Rossi.

Chapter 10
Fig. 10.1 Hans Reinhard / BC; *Fig. 10.2* Moravske Muzeum, Brne Czechoslovakia; *Fig. 10.3* through *Fig. 10.12* Patrice Rossi; *Box p. 199* Dr. Tokuyasu, Univ. of California, San Diego; *Fig. 10.13* through *Fig. 10.17* P. Rossi; *Fig. 10.18* Marsha Goldberg; *Fig. 10.19* M. L. Pardue, M.I.T.; *Fig. 10.20* P. Rossi; *Box p. 207* Carnegie Institute of Washington; *Fig. 10.21* P. Rossi; *Fig. 10.22 (left)* P. Rossi; *Fig. 10.22 (right)* Cabisco / VU; *Box p. 210* P. Rossi; *Fig. 10.24 (photos)* John Elseley; *Fig. 10.25 (top left)* J. D. Cunningham / VU; *Fig. 10.25 (top right)* Richard Kolar / AA; *Fig. 10.25 (bottom left and right)* Hans Reinhard / BC; *Fig. 10.26 (photos top to bottom)* Henry Ausloos / AA, Robert Pearcy / AA, Jean-Claude Carton / BC, Ralph A. Reinhold / AA; *Fig. 10.27* and *Fig. 10.28* P. Rossi; *Fig. 10.29* A. Novak / Photri.

Chapter 11
Fig. 11.1 H&C; *Fig. 11.2* Dr. Anne Richardson; *Fig. 11.3 (above)* Universitetsbiblioteket, Oslo; *Fig. 11.4 (bottom)* Richard Dranitzke / PR; *Fig. 11.4 (right)* H&C; *Fig. 11.5* Ken Miller; *Fig. 11.6* Lester V. Bergman; *Fig. 11.7* H&C; *Fig. 11.8* David Falconer / BC; *Fig. 11.10* K. Miller; *Fig. 11.11* Patricia Barry Levy / Profiles West; *Fig. 11.12* Patrice Rossi; *Fig. 11.14* and *Fig. 11.15* H&C; *Fig. 11.17* Sally Myers; *Fig. 11.18* and *Fig. 11.19* H&C; *Fig. 11.20 (left)* Michael and Elvan Habicht / ES; *Fig. 11.20 (right)* Dr. Ram S. Verma, Interfaith Medical Center, Div. of Cytogenetics / PHT; *Box p. 233* Mikki Senkarik; *Fig. 11.22* K. Miller; *Fig. 11.23* P. Rossi; *Fig. 11.24 (both)* The Children's Hospital of Philadelphia; *Fig. 11.25* Dr. David Ward; *Box p. 236* P. Rossi; *Fig. 11.26* P. Rossi.

Chapter 12
Fig. 12.1 H&C; *Fig. 12.2* Patrice Rossi; *Fig. 12.3* F. Blakeslee, *Journal of Heredity*, 1914; *Fig. 12.4* P. Rossi; *Fig. 12.5* H&C; *Fig. 12.6 (clockwise)* M. A. Chappell / AA, David Cavagnaro, Phyllis Greenberg / Comstock; *Fig. 12.7* through *Fig. 12.19* P. Rossi; *Fig. 12.20 (both)* Breck P. Kent / AA.

Chapter 13
Fig. 13.1 © 1987 Laura Riley / BC; *Fig. 13.2 (top)* Frank T. Awbrey / VU; *Fig. 13.2 (bottom)* Dan Suzio; *Fig. 13.3* Patrice Rossi; *Box p. 263* Robert Noonan; *Fig. 13.4* and *Fig. 13.5* P. Rossi; *Fig. 13.6* H&C; *Fig. 13.7* through *Fig. 13.12* P. Rossi.

Part 4 © Lennart Nilsson, *The Incredible Machine*

Chapter 14
Fig. 14.1 Biblioteque Royal Albert, Brussells; *Fig. 14.2* and *Fig. 14.3* Illustrious, Inc.; *Fig. 14.4* H&C; *Fig. 14.5a* Illustrious, Inc.; *Fig. 14.5b* SIU School of Medicine / PA; *Fig. 14.6* H&C; *Fig. 14.7* Illustrious, Inc.; *Fig. 14.9* H&C; *Fig. 14.10 (left)* NASA / ColorPic, Inc.; *Fig. 14.10 (below)* Illustrious, Inc.; *Fig. 14.11* Illustrious, Inc.; *Fig. 14.12* H&C; *Fig. 14.13* Illustrious, Inc.; *Fig. 14.14 (left)* Patrice Rossi; *Fig. 14.14 (right)* Herman Eisenbeiss / PR; *Fig. 14.15* Illustrious, Inc.

Chapter 15
Fig. 15.1 through *Fig. 15.4* Peggy Jefferson; *Fig. 15.5* through *Fig. 15.16* Illustrious, Inc.; *Table 15.1* Illustrious, Inc.; *Fig. 15.17* M. DenHouter; *Box p. 309* Illustrious, Inc.

Chapter 16
Fig. 16.1 Patrice Rossi; *Fig. 16.2* Lans Taylor; *Fig. 16.3* P. Rossi; *Fig. 16.4 (top)* Fritz Goro; *Fig. 16.4 (bottom)* P. Rossi; *Fig. 16.5* Ken Miller; *Fig. 16.6* Charles Boyter; *Fig. 16.7 (photo)* Daniel S. Friend, M.D., Univ. of California, SF; *Fig. 16.7 (art and details)* P. Rossi; *Fig. 16.8 (photo)* Dr. Daniel S. Friend, M.D., Univ. of California, SF; *Fig. 16.8 (art and details)* P. Rossi; *Fig. 16.9a* Helen Padykula / Shirwin Pockwins; *Fig. 16.9b* Nigel Unwin, Cambridge, England; *Fig. 16.9c* Marlene DenHouter; *Fig. 16.9d* Jenny Hinshaw; *Fig. 16.10 (photo)* Dr. James Lake; *Fig. 16.10 (art)* P. Rossi; *Fig. 16.11* George E. Palade, Yale Univ.; *Fig. 16.12* Daniel S. Friend, M.D., Univ. of California, SF; *Fig. 16.13* Mikki Senkarik; *Fig. 16.14* G. T. Cole / BPS; *Fig. 16.15 (top)* Dr. Don Fawcett/S. Ito & A. Like / PR; *Fig. 16.15 (bottom)* Biophoto Associates / PR; *Fig. 16.16 (top)* P. M. Novikoff; *Fig. 16.16 (bottom)* M. DenHouter; *Fig. 16.17 (art)* P. Rossi; *Fig. 16.17 (photo)* K. Miller; *Fig. 16.18 (art)* P. Rossi; *Fig. 16.18 (photo)* K. Miller; *Fig. 16.19* P. Rossi; *Fig. 16.20a* M. Schliwa / VU; *Fig. 16.20b* P. Rossi; *Fig. 16.20c* Lewis Tilney; *Fig. 16.20d* K. G. Murti / VU; *Fig. 16.21a* Michael Abbey / PR; *Fig. 16.21b* Biophoto Associates / Science Source / PR; *Fig. 16.21c* Zosia Rybkowski; *Fig. 16.21d* P. Rossi; *Fig. 16.22* and *Fig. 16.23* David M. Phillips / VU; *Fig. 16.24a* Roger Craig; *Fig. 16.24b–c* P. Rossi; *Fig. 16.25* Lan BoChen; *Fig. 16.26* M. DenHouter; *Fig. 16.27a–c* M. DenHouter; *Fig. 16.27d* K. Miller; *Fig. 16.28* M. DenHouter; *Fig. 16.29* through *Fig. 16.32* P. Rossi; *Fig. 16.33 (all)* © Dennis Kunkel / PHT; *Box p. 338 (photo)* Leonard Rome;

Box p. 338 (art) Nancy Kedersha; *Fig. 16.34* P. Rossi; *Fig. 16.35a* K. Miller; *Fig. 16.35b* M. DenHouter; *Box p. 340* Lans Taylor.

Chapter 17
Fig. 17.1 through *Fig. 17.15* Patrice Rossi; *Box p. 353* P. Rossi; *Box p. 354* P. Rossi.

Chapter 18
Fig. 18.1 through *Fig. 18.5* Patrice Rossi; *Fig. 18.6 and details* P. Rossi; *Fig. 18.7* through *18.11* P. Rossi; *Fig. 18.12 (top)* P. Rossi; *Fig. 18.12 (bottom)* Ken Miller; *Fig. 18.13* and *Fig. 18.14* P. Rossi; *Fig. 18.15 (art)* P. Rossi; *Fig. 18.15 (photo)* R. Btatnasar / VU; *Fig. 18.16* through *Fig. 18.18* P. Rossi; *Fig. 18.19 (art)* P. Rossi; *Fig. 18.19 (photo)* Peter Menzel; *Fig. 18.20 (photo)* Chris Cole / Allsport; *Fig. 18.20 (art)* P. Rossi; *Box p. 376* Nancy J. Newman, M.D., Emory Univ.; *Box p. 377 (art)* P. Rossi; *Box p. 377 (photo)* Lloyd Matsumoto, Rhode Island College.

Chapter 19
Fig. 19.1 and *Fig. 19.2* Patrice Rossi; *In-text art p. 380* P. Rossi; *Fig. 19.3* through *Fig. 19.6* P. Rossi; *Fig. 19.7 and details* P. Rossi; *Fig. 19.8* P. Rossi; *Fig. 19.9a* Marlene DenHouter; *Fig. 19.9b* Ken Miller; *Fig. 19.10* P. Rossi; *Fig. 19.11 (left)* Peter A. Simon / PHT; *Fig. 19.11 (right)* Illustrious, Inc.; *Fig. 19.12a–c* P. Rossi; *Fig. 19.12d* Peter Goro; *Fig. 19.13* Larry Ulrich / DRK; *Fig. 19.14* M. DenHouter; *Fig. 19.15* through *Fig. 19.18* P. Rossi; *Fig. 19.19 (top)* W. H. Hodge / PA; *Fig. 19.19 (bottom)* Liz Ball / PN; *Fig. 19.20* P. Rossi; *Box p. 394* John Bova / PR.

Chapter 20
Fig. 20.1 H&C; *Fig. 20.2* Patrice Rossi; *Fig. 20.3* Illustrious, Inc.; *Fig. 20.4 (art)* H&C; *Fig. 20.4 (photo)* Laboratory of Molecular Biology, Medical Research Council, Cambridge, England; *Fig. 20.5* Illustrious, Inc.; *Fig. 20.6* Dr. James Watson, Cold Spring Harbour Laboratory; *Fig. 20.7 (photo)* Laboratory of Molecular Biology, Medical Research Council, Cambridge, England; *Fig. 20.7 (art)* Illustrious, Inc.; *Fig. 20.8 (art)* Illustrious, Inc.; *Fig. 20.8 (photo)* Nelson Max / PA; *Box p. 405* Alberts, *Molecular Biology of the Cell*, 2nd ed., Garland Publishing Co.; *Fig. 20.9* Illustrious, Inc.; *Box p. 407* P. Rossi; *Fig. 20.10* Illustrious, Inc.; *Fig. 20.11* and *Fig. 20.12* P. Rossi; *Fig. 20.13* Illustrious, Inc.; *Fig. 20.14* P. Rossi; *Fig. 20.15 (top)* Dr. James Lake, UCLA, reprinted with permission from James A. Lake, *Journal of Molecular Biology*, 105 (1976): 131–50; *Fig. 20.15 (bottom)* P. Rossi; *Fig. 20.16* P. Rossi;

Fig. 20.17 P. Rossi; *Box p. 415* P. Rossi; *Fig. 20.18 (left)* Jonathan King / M.I.T.; *Fig. 20.18 (right)* Dr. Timothy S. Baker / Purdue Univ.; *Fig. 20.19* through *Fig. 20.21* H&C; *Fig. 20.22* From H. Delius and A. Worcel, *Journal of Molecular Biology*, 82 (1974): 108. Reprinted with permission; *Fig. 20.23a* Ada L. Olins and Donald E. Olins; *Fig. 20.23b* H&C; *Box p. 421 (photo)* Oscar Miller; *Box p. 421 (art)* P. Rossi; *Fig. 20.24* and *Fig. 20.25* P. Rossi; *Box p. 424* Thomas Steitz, Yale Univ.; *Fig. 20.26* Alberts, *Molecular Biology of the Cell*, 2nd ed., Garland Publishing Co.; *Fig. 20.27* Illustrious, Inc.; *Fig. 20.28* and *Fig. 20.29* Lyrl Ahern.

Chapter 21
Fig. 21.1 through *Fig. 21.3* Illustrious, Inc.; *Fig. 21.4* Patrice Rossi; *Fig. 21.5* Lyrl Ahern; *Fig. 21.5 (photo)* Susan Dibartolomeis; *Fig. 21.6* TechDocPub, Inc.; *Box p. 436* Illustrious, Inc.; *Fig. 21.7* TechDocPub, Inc.; *Box pp. 439–440* Illustrious, Inc.; *Fig. 21.8 (left)* Photo Courtesy of Vetex Pharmaceuticals Incorporated, Cambridge, Mass.; *Fig. 21.8 (right)* Genentech; *Fig. 21.9 (top)* "Genetically Engineering Plants for Crop Improvement," Gasser, C. S. and Robert T. Fraley, *Science* 244 (16 June 1989): 1281–1288, Fig. I. © 1989 by AAAS; *Fig. 21.9 (bottom)* Photo Courtesy of Monsanto; *Fig. 21.10 (photo)* Vitale / Mass. Biotech Council; *Fig. 21.10 (art)* TecDocPub, Inc.; *Fig. 21.11* Dr. Philip Leder, Harvard Univ.; *Fig. 21.12* Thomas Hovland / GH; *Fig. 21.13* TecDocPub, Inc.; *Box p. 447 (photo)* DNA Plant Technology, Inc.; *Box p. 447 (art)* Illustrious, Inc.; *Box p. 448* Discover; *Fig. 21.14* Courtesy of *Cell Magazine*; *Box p. 451 (art)* Illustrious, Inc.; *Box p. 451 (both photos)* Dan Sears, Univ. of N. Carolina; *Fig. 21.15* Gary Crespo.

Chapter 22
Fig. 22.1 Douglas Waugh / PA; *Fig. 22.2* David M. Phillips / VU; *Fig. 22.3* Lucien Neumeyer; *Fig. 22.4* H&C; *Fig. 22.5 (photo)* del Regato, et al., reprinted with permission, Ackerman and del Regato's *Cancer: Diagnosis, Treatment and Prognosis*; *Fig. 22.6a–b* TecDocPub, Inc.; *Fig. 22.6c* Nina Fedorof / Carnegie Institute; *Fig. 22.7* and *Fig. 22.8* Gary Crespo; *Fig. 22.9* H&C; *Box pp. 462–463 (left)* Donald W. Kyhos, *(others)* Bruce Baldwin, Botany Dept., Duke Univ.; *Fig. 22.10* and *Fig. 22.11* G. Crespo; *Fig. 22.12* TecDocPub, Inc.; *Fig. 22.13* and *Fig. 22.14* Mikki Senkarik; *Fig. 22.15* H&C; *Fig. 22.16* through *Fig. 22.19* G. Crespo; *Fig. 22.20* Mark J. Rauzon; *Fig. 22.21* M. P. Kahl / DRK Photo; *Box p. 473* G. Crespo; *Fig. 22.22 (clockwise)* M. P. Kahl

/ PR, S. R. Maglione / PR, Paolo Koch / PR, Bas Von Beek / Leo de Wys, Inc., Fritz Henee / PR, Charles Preitner / VU, George Holton / PR.

Part 5 Sunni / FPC

Chapter 23
Fig. 23.1 Giraudon / Art Resource; *Fig. 23.2 (left)* Tom Branch / PR; *Fig. 23.2 (center)* Nell Bolen / PR; *Fig. 23.2 (right)* Bruce Cushing / VU; *Fig. 23.3* Patrice Rossi; *Box p. 485 (left to right)* Tom & Pat Leeson / PR, Gary Milburn / TSA, Tom Brakefield / The Stock Market; *Fig. 23.4* P. Rossi; *Fig. 23.5* Gary Crespo.

Chapter 24
Fig. 24.1 Steven Brook Studios; *Fig. 24.2 (top)* Richard Weiss / PA; *Fig. 24.2 (bottom)* William Ferguson; *Box p. 493* Gary Crespo; *Box p. 494 (art)* P. Rossi; *Box p. 495* WHOI / D. Foster / VU; *Fig. 24.3 (left)* P. W. Johnson & J. McN. Sieburth, Univ. of Rhode Island / BPS; *Fig. 24.3 (right)* Sinclair Stammers / Science Photo Library / PR; *Fig. 24.4* Dr. Paul Strother, Boston Univ.; *Fig. 24.5 (clockwise) (top left & top center)* Dr. G. Kite, J. Robert Waaland, Univ. of Washington / BPS, Linda Thomashow, Washington State Univ. / BPS; *Fig. 24.6* H&C; *Fig. 24.7* P. Rossi.

Chapter 25
Fig. 25.1a Patrice Rossi; *Fig. 25.1b* Sanderson Associates; *Fig. 25.2* P. Rossi; *Fig. 25.3 (photo)* S. M. Awramik, Univ. of California / BPS; *Fig. 25.3 (art)* Arleen Frasca; *Fig. 25.4 (photo)* Field Museum of Natural History; *Fig. 25.4 (art)* A. Frasca; *Fig. 25.5* T. A. Wiewandt / DRK; *Fig. 25.6* Field Museum of Natural History; *Fig. 25.7 (photo)* Field Museum of Natural History; *Fig. 25.7 (art)* A. Frasca; *Fig. 25.8* Mikki Senkarik; *Fig. 25.9* A. Frasca; *Fig. 25.10 (left)* Greg Retallack, Oregon Univ.; *Fig 25.10 (top right)* Courtesy of the Royal Museum of Scotland; *Fig 25.10 (bottom right)* Beverly Benner; *Fig. 25.11* A. Frasca; *Fig. 25.12* H&C; *Fig. 25.13 (art)* A. Frasca; *Fig. 25.13 (photo)* Field Museum of Natural History; *Fig. 25.14* P. Rossi; *Fig. 25.15a–b* Marlene DenHouter; *Fig. 25.15c* Smithsonian Institution / Div. of Paleontology; *Fig. 25.16* William Shear; *Fig. 25.17* Field Museum of Natural History; *Fig. 25.18 (photo left)* G. R. Roberts; *Fig. 25.18 (photo right)* Tom McHugh / PR; *Fig. 25.19* P. Rossi; *Fig. 25.20* M. DenHouter; *Fig. 25.21 (photo)* Douglas Henderson; *Fig. 25.22* Douglas Henderson; *Fig. 25.23* Mark Hallett, *Ranger Rick's Dinosaur Book*, © 1984

National Wildlife Federation; *Box p. 523* M. DenHouter; *Fig. 25.24* NMNH / PR.

Chapter 26

Fig. 26.1 Arleen Frasca; *Fig. 26.2* H&C; *Fig. 26.3* David M. Phillips / VU; *Fig. 26.4* Mikki Senkarik; *Fig. 26.5* Paul W. Johnson / BPS; *Fig. 26.6* S. Sharnoff / VU; *Fig. 26.7* Paul W. Johnson / BPS; *Fig. 26.8* William Patterson / TSA; *Fig. 26.9 (top)* C. Rosebush Visions Corp. / PHT; *Fig. 26.9 (bottom left to right)* K. G. Murti / VU, Science / VU, G. Musil / VU.

Chapter 27

Fig. 27.1 (art) Mikki Senkarik; *Fig. 27.1 (left photo)* Charlie Ott / PR; *Fig. 27.1 (center)* Stephen Krasemann / DRK; *Fig. 27.1 (right)* D. Wilder / TSA; *Fig. 27.2 (top)* William Ferguson; *Fig. 27.2 (bottom)* R. L. Peterson, Univ. of Guelph / BPS; *Fig. 27.3* M. Senkarik; *Fig. 27.4 (top)* D. P. Wilson / Eric David Hosking / PR; *Fig. 27.4 (center)* Robert Evans / PA; *Fig. 27.4 (bottom)* William Ferguson; *Fig. 27.5 (art)* M. Senkarik; *Fig. 27.5 (top photo)* Brian Parker / TSA; *Fig. 27.5 (bottom left)* John Cunningham / VU; *Fig. 27.5 (bottom right)* John Forsythe / VU; *Fig. 27.6 (art)* M. Senkarik; *Fig. 27.6 (photos)* and *Fig. 27.7* William Ferguson; *Fig. 27.8 (art)* M. Senkarik; *Fig. 27.8 (left photo)* Pat O'Hara / DRK Photo; *Fig. 27.8 (right photo)* C. R. Arndt / VU; *Fig. 27.9* Marlene DenHouter; *Fig. 27.10 (art)* M. Senkarik; *Fig. 27.10 (photo)* Michael Ederegger / DRK; *Fig. 27.11 (art)* M. Senkarik; *Fig. 27.11 (left photo)* Townsend Dickinson / PR; *Fig. 27.11 (right)* J. Anderson / Taurus; *Fig. 27.12 (top)* S. J. Krasemann / PA; *Fig. 27.12 (bottom)* John Shaw / TSA.

Chapter 28

Fig. 28.1 (photo) Charles Seaborn / Odyssey Productions; *Fig. 28.1 (art)* Mikki Senkarik; *Fig. 28.2* M. Senkarik; *Fig. 28.3 (both)* Charles Seaborn / Odyssey Productions; *Table 28.1* Arleen Frasca; *Fig. 28.4 (art)* M. Senkarik; *Fig. 28.4 (photo)* Denise Tackett / TSA; *Fig. 28.5 (left)* M. Senkarik; *Fig. 28.5 (right)* H&C; *Fig. 28.6 (art)* M. Senkarik; *Fig. 28.6 (top photo)* John Cunningham / VU; *Fig. 28.6 (bottom photo)* Charles Seaborn / Odyssey Productions; *Fig. 28.7a* Ed Robinson / TSA; *Fig. 28.7b* Gregory Dimijian / PR; *Fig. 28.8 (art)* M. Senkarik; *Fig. 28.8 (left photo)* Runk / Schoenberger / GH; *Fig. 28.8 (right photo)* Stan Elems / VU; *Fig. 28.9 (art)* M. Senkarik; *Fig. 28.9 (top photo)* Peter Bryant / BPS; *Fig. 28.9 (bottom photo)* C. Allan Morgan / PA; *Fig. 28.10* M. Senkarik; *Fig. 28.11* A. Frasca; *Fig. 28.12 (left)* M. Senkarik; *Fig. 28.12 (right)* Marlene DenHouter; *Fig. 28.13 (art)* M.

Senkarik; *Fig. 28.13 (left photo)* Brian Parker / TSA; *Fig. 28.13 (center & right photo)* Charles Seaborn / Odyssey Productions; *Fig. 28.14* Ed Robinson / TSA; *Fig. 28.15 (photo)* Charles Seaborn / Odyssey Productions; *Fig. 28.15 (art)* M. Senkarik; *Fig. 28.16* A. Frasca; *Box p. 569* Michael Fogden / OSF / AA.

Chapter 29

Fig. 29.1 (art) Mikki Senkarik; *Fig. 29.1 (photo)* Patrice / VU; *Fig. 29.2 (art)* M. Senkarik; *Fig. 29.2 (top photo)* Carl Roessler / BC; *Fig. 29.2 (bottom photo)* M. P. Kahl / BC; *Fig. 29.3 (art)* M. Senkarik; *Fig. 29.3 (photo)* Charles Seaborn / Odyssey Productions; *Fig. 29.4* Patrice / VU; *Fig. 29.5* Peter Scoones / Planet Earth Pictures; *Fig. 29.6 (left art)* Arleen Frasca; *Fig. 29.6 (right art)* M. Senkarik; *Fig. 29.6 (photo)* John MacGreggor / PA; *Fig. 29.7 (left photo)* John Burnley / PR; *Fig. 29.7 (right photo)* Jane Burton / BC; *Fig. 29.8 (art)* M. Senkarik; *Fig. 29.8 (left photo)* William Ferguson; *Fig. 29.8 (top right photo)* G. R. Roberts; *Fig. 29.8 (bottom right photo)* David G. Barker / TSA; *Fig. 29.9 (art)* M. Senkarik; *Fig. 29.9 (photos left to right)* Stephen J. Krasemann / DRK Photo, Kennan Ward / DRK Photo, Roy Morsch / The Stock Market; *Fig. 29.10* H&C; *Fig. 29.11* Hans and Judy Beste / AA; *Fig. 29.12* David C. Fitts / AA; *Fig. 29.13* Milton Tierney, Jr. / VU; *Fig. 29.14* Patrice Rossi; *Fig. 29.15* R. Van Nostrand / PR; *Fig. 29.16 (photos)* David Brill / © National Geographic Society; *Fig. 29.16 (art)* Marlene DenHouter; *Fig. 29.17* David Brill / © National Geographic Society; *Fig. 29.18* M. DenHouter; *Fig. 29.19* H&C; *Fig. 29.20* David Brill / © National Geographic Society; *Fig. 29.21* David Brill, National Museum of Kenya; *Fig. 29.23* Denver Museum of National History; *Box pp. 590–591* Gary Crespo.

Part 6 David Muench

Chapter 30

Fig. 30.1 Gregory Dimijian / PR; *Fig. 30.2* Patrice Rossi; *Fig. 30.3 (top)* Wayne Lynch / DRK; *Fig. 30.3 (center)* William Ferguson; Fig. 30.3 (bottom) Zig Leszczynski / ES; *Fig. 30.4* Dr. Ursula Goodenough, Washington Univ.; *Fig. 30.5 (top)* Peter Parks / Oxford Scientific Films / AA; *Fig. 30.5 (bottom)* D. P. Wilson / Eric David Hosking / PR; *Fig. 30.6 (both)* and *Fig. 30.7* William Ferguson; *Box p. 603* Hans Reinhard / BC; *Fig. 30.8* through *Fig. 30.10* P. Rossi; *Fig. 30.11 (left)* William Ferguson; *Fig. 30.11 (right)* P. Dayanadan / PR ; *Fig. 30.11 (art)* P. Rossi; *Fig. 30.12* and *Fig. 30.13* P. Rossi; *Fig. 30.14 (left)* R. Moore / VU; *Fig. 30.14 (art)* P. Rossi; *Fig. 30.15* P. Rossi; *Box p. 610*

Raymond Mendez / AA; *Fig. 30.16* and *Fig. 30.17* P. Rossi; *Fig. 30.18 (art)* L. Cook; *Fig. 30.18 (top photo)* Ed Reschke / PA; Fig 30.18 *(bottom photo)* Ken Wagner / PHT; *Fig. 30.19 (top)* P. Rossi; *Fig. 30.19 (bottom)* Richard Weiss / PA; *Fig. 30.19 (photo)* David Stone / PN; *Fig. 30.20 (photo)* David Stone / PN; *Fig. 30.20 (art)* Marcia Smith; *Fig. 30.21 (top)* Bruce Iverson; *Fig. 30.21 (bottom)* M. Fogden / BC; *Fig. 30.22 (art)* M. Smith; *Fig. 30.22 (photos)* Dr. Jeremy Burgess / Science Photo Library / PR.

Chapter 31

Fig. 31.1 (left) William Ferguson; *Fig. 31.1 (right)* Runk / Schoenberger / GH; *Fig. 31.2* and *Fig. 31.3* Patrice Rossi; *Fig. 31.4 (flower)* Marcia Smith; *Fig. 31.4 (details)* P. Rossi; *Fig. 31.5* Dr. Jeremy Burgess / Science Photo Library / PR; *Fig. 31.6* Lynette Cook; *Fig. 31.7* P. Rossi; *Fig. 31.8 (top left)* J. Barden / PN; *Fig. 31.8 (center)* Valorie Hodgson / PN; *Fig. 31.8 (bottom right)* L.L.T. Rhodes / ES; *Fig. 31.8 (art)* P. Rossi; *Fig. 31.9 (left)* William Ferguson; *Fig. 31.9 (center)* James Carmichael / BC; *Fig. 31.9 (right)* Gay Bumgarner / PN; *Fig. 31.10 (both)* N.A.S. / M.W.F. Tweedie / PR; *Fig. 31.11* Wayne van Kinen / DRK; *Fig. 31.12* Merlin Tuttle / PR; *Box p. 633* Phil Gates / BPS; *Fig. 31.13* Leonard Lee Rue III / BC; *Box p. 635* Ken Miller; *Fig. 31.14 (left)* Gay Bumgarner / PN; *Fig. 31.14 (right)* G. R. Roberts.

Chapter 32

Fig. 32.1 Lynette Cook; *Fig. 32.2* Barry Runk / GH; *Fig. 32.3* Patrice Rossi; *Fig. 32.4* Holt Studios / ES; *Fig. 32.5* and *Fig. 32.6* P. Rossi; *Fig. 32.7 (both)* E. R. Degginger / ColorPic, Inc.; *Box p. 646* W. B. Rudman / Australian National Museum; *Fig. 32.8 (left)* Dana Richter / VU; *Fig. 32.8 (right)* Runk / Schoenberger / GH; *Fig. 32.9 (top)* Hugh Spencer / PR; *Fig. 32.9 (bottom)* Science Photo Library / PR; *Fig. 32.10 (left)* L. L. T. Rhodes / AA; *Fig. 32.10 (right)* Dr. A. A. Szalay, Plant Molecular Genetics Laboratories, Univ. of Alberta; *Fig. 32.11 (left)* Abe Blank / BC; *Fig. 32.11 (center)* William Ferguson; *Fig. 32.11 (right)* Denny Barret / PN.

Chapter 33

Fig. 33.1 Patrice Rossi; *Fig. 33.2* E. R. Degginger / ColorPic, Inc.; *Fig. 33.3* Runk / Schoenberger / GH; *Fig. 33.4* William Ferguson; *Fig. 33.5* through *Fig. 33.7* P. Rossi; *Box p. 655* Ronald Davis, Stanford Univ.; *Fig. 33.8* P. Rossi; *Fig. 33.9* Runk / Schoenberger / GH; *Fig. 33.10* P. Rossi; *Fig. 33.11* Alfred Owczarzak / Taurus Photos; *Fig. 33.12* Wide World Photos; *Fig. 33.13* E. R. Degginger / ColorPic, Inc.; *Fig. 33.14* Steve Raye / Taurus Photos; *Fig.*

33.15 Breck P. Kent / ES; *Fig. 33.16* and *Fig. 33.17* P. Rossi; *Fig. 33.18 (both)* William Ferguson; *Fig. 33.19* P. Rossi; *Fig. 33.20* D. Cavagnaro / DRK; *Fig. 33.21* Dr. Jeremy Burgess / Science Photo Library / PR; *Fig. 33.22 (top)* C. Raymond / PR; *Fig. 33.22 (bottom)* P. Rossi; *Fig. 33.23* David Ow / Science; *Fig. 33.24 (photos)* Courtesy of Monsanto; *Fig. 33.24 (art)* P. Rossi.

Part 7 © Boehringer Ingelheim International GmbH, Photo: Lennart Nilsson, *The Incredible Machine*

Chapter 34
Fig. 34.1 Biblioteque de Geneve, *Fig. 34.2 (art)* Mikki Senkarik; *Fig. 34.2 (top left)* Robert Knauft / BioMedia / PR; *Fig. 34.2 (top right)* Ed Reschke / PA, *Fig. 34.2 (center left)* Manfred Kage / PA; *Fig. 34.2 (center right)* Ed Reschke / PA; *Fig. 34.2 (bottom left)* Ed Reschke / PA; *Fig. 34.2 (bottom right)* John Cunningham / VU; *Fig. 34.3 (top)* Arleen Frasca; *Fig. 34.3 (photos)* L. A. Stacehelin and B. E. Hull / Univ. of Boulder, CO; *Fig. 34.4 (left)* A. Frasca; *Fig. 34.4 (right)* Manfred Kage / PA; *Fig. 34.5* Ed Reschke / PA; *Fig. 34.6 (top)* M. Senkarik; *Fig. 34.6 (bottom)* Ed Reschke / PA; *Fig. 34.7 (art)* M. Senkarik; *Fig. 34.7 (photo)* Manfred Kage / PA; *Fig. 34.8 (photo)* Ed Reschke / PA; *Fig. 34.8 (art)* M. Senkarik; *Fig. 34.9 (photos: top)* M. I. Walker / PR; *Fig. 34.9 (center)* Dwight Kuhn; *Fig. 34.9 (bottom)* Eric V. Grave / PR; *Fig. 34.9 (art)* M. Senkarik; *Fig. 34.10* Patrice Rossi; *Table 34.1* and *Fig. 34.11* M. Senkarik; *Fig. 34.12* through *Fig. 34.14* P. Rossi; *Fig. 34.15* David Madison; *Fig. 34.16* P. Rossi; *Fig. 34.17* A. Frasca.

Chapter 35
Fig. 35.1 Arleen Frasca; *Fig. 35.2* Mikki Senkarik; *Fig. 35.3* A. Frasca; *Fig. 35.4* and *Fig. 35.5* M. Senkarik; *Fig. 35.6* through *Fig. 35.8* Charles Boyter; *Fig. 35.9* Dr. Don W. Fawcett; *Fig. 35.10* A. Frasca; *Fig. 35.11* through *Fig. 35.13* M. Senkarik; *Fig. 35.14* Patrice Rossi; *Fig. 35.15* Sloop-Ober / VU; *Box p. 701 (art)* P. Rossi; *Box p. 701 (photo)* Mary Ann Fittipaldi; *Fig. 35.16* M. Senkarik; *Box p. 704* David Binder; *Fig. 35.17* P. Rossi; *Fig. 35.18 (left)* Martin M. Rotker / Taurus Photos; *Fig. 35.18 (right)* Bill Longcore / Science Source / PR; *Fig. 35.19 (all)* Alfred Owczarzak / Taurus Photos; *Fig. 35.20 (top)* P. Rossi; *Fig. 35.20 (bottom)* David M. Phillips / VU.

Chapter 36
Fig. 36.1 and *Fig. 36.2* Patrice Rossi; *Fig. 36.3 (left)* Marlene DenHouter; *Fig. 36.3 (top)* R. Calentine / VU; *Fig. 36.4* and *Fig. 36.5* M. DenHouter; *Fig. 36.6* Mikki Senkarik; *Box p. 715* David G. Campbell;

Fig. 36.7 and *Fig. 36.8* P. Rossi; *Fig. 36.9* M. DenHouter; *Fig. 36.10 (art)* M. Senkarik; *Fig. 36.10 (left photo)* Cabisco / VU; *Fig. 36.10 (right photo)* R. Dirksen / VU; *Fig. 36.11* and *Fig. 36.12* P. Rossi; *Fig. 36.13* through *Fig. 36.16* P. Rossi; *Box p. 724* P. Rossi; *Fig. 36.17 (both)* Oscar C. Williams.

Chapter 37
Fig. 37.1 and *Fig. 37.2* Patrice Rossi; *Fig. 37.3* Wide World Photos; *Box p. 732* P. Rossi; *Fig. 37.4* TecDocPub, Inc.; *Fig. 37.5* P. Rossi; *Fig. 37.6* Douglas Kirkland / Contact Press Images; *Fig. 37.7 (clockwise)* Steven C. Kaufman / PA, Hans Pfletschinger / PA, Fred Bavendam / PA, John Cancalosi / PA; *Fig. 37.8* P. Rossi; *Fig. 37.9 (left)* P. Rossi; *Fig. 37.9 (right)* Michael Abbey / PR, *Fig. 37.10* P. Rossi; *Fig. 37.11* and *Fig. 37.12* Mikki Senkarik; *Fig. 37.13 (left)* David M. Stone / PN; *Fig. 37.13 (right)* Tom McHugh / PR; *Fig. 37.14* and *Fig. 37.15* P. Rossi; *Fig. 37.16* and *Fig. 37.17* M. Senkarik; *Fig. 37.18 (art)* M. Senkarik; *Fig. 37.18 (photo)* Dr. Barbara Hull / Wayne State Univ.; *Fig. 37.19* P. Rossi; *Fig. 37.20* M. Senkarik; *Fig. 37.21* P. Rossi.

Chapter 38
Fig. 38.1 Patrice Rossi; *Fig. 38.2 (left)* P. Rossi; *Fig. 38.2 (right)* M. P. Kahl / PR; *Fig. 38.3 (both)* Cabisco / VU; *Fig. 38.4* P. Rossi; *Fig. 38.5* Mikki Senkarik; *Fig. 38.6 (left)* M. Senkarik; *Fig. 38.6 (top right)* Biophoto Associates / PR; *Fig. 38.6 (bottom right)* David M. Phillips / VU; *Box p. 765* M. Senkarik; *Fig. 38.7* and *Fig. 38.8* M. Senkarik; *Fig. 38.9* and *Fig. 38.10* P. Rossi; *Fig. 38.11* M. Senkarik; *Fig. 38.12* and *Fig. 38.13* P. Rossi; *Fig. 38.14* Stephen Dalton / PR; *Fig. 38.15* Alford W. Cooper / PR; *Fig. 38.16* P. Rossi; *Fig. 38.17 (left)* M. Warren Williams / Taurus; *Fig. 38.17 (center)* Joe McDonald / VU; *Fig. 38.17 (right)* Leonard Lee Rue / PR; *Fig. 38.18* P. Rossi; *Fig. 38.19* Custom Medical Stock Photo.

Chapter 39
Fig. 39.1 Richard Vogel / Gamma Liaison; *Fig. 39.2 (left)* Mikki Senkarik; *Fig. 39.2 (right both)* Dr. Peter Newell / Oxford Univ.; *Fig. 39.3* and *Fig. 39.4* M. Senkarik; *Box p. 785* Greg Crisci / PN; *Fig. 39.6* through *Fig. 39.10* Patrice Rossi; *Fig. 39.11* and *Fig. 39.12* M. Senkarik; *Fig. 39.13 (left)* The Bettmann Archive; *Fig. 39.13 (right)* Carol Publishing Group; *Fig. 39.14* and *Fig. 39.15* M. Senkarik; *Fig. 39.16 (left)* M. Senkarik; *Fig. 39.16 (right)* Arthur M. Siegelman; *Fig. 39.17* P. Rossi; *Fig. 39.18* M. Senkarik; *Fig. 39.19* Mitch Reardon / PR.

Chapter 40
Fig. 40.1 Patrice Rossi; *Fig. 40.2* through

Fig. 40.8 Mikki Senkarik; *Fig. 40.9a* M. Senkarik; *Fig. 40.9b–c* P. Rossi; *Fig. 40.10* and *Fig. 40.11* P. Rossi; *Fig. 40.12* M. Senkarik; *Fig. 40.13* P. Rossi; *Fig. 40.14* M. Senkarik.

Chapter 41
Fig. 41.1 Patrice Rossi; *Fig. 41.2 (both)* Carol Vater and Robert Jackson / Dartmouth; *Fig. 41.3* and *Fig. 41.4* P. Rossi; *Fig. 41.5* and *Fig. 41.6* Mikki Senkarik; *Fig. 41.7 (photos)* Dr. Richard G. Kessel and Dr. C. Y. Shin; *Fig. 41.7 (art)* M. Senkarik; *Fig. 41.8* M. Senkarik; *Fig. 41.9 (photos)* Carolina Biological Supply Company; *Fig. 41.9 (art)* M. Senkarik; *Fig. 41.10* through *Fig. 41.16* M. Senkarik; *Fig. 41.17 (top and middle rows)* Lennart Nilsson, Bonna Fakta from *A Child Is Born*, Dell Publishing Company; *Fig. 41.17 (bottom)* Lennart Nilsson, Nestle / PR; *Fig. 41.18* M. Senkarik; *Fig. 41.19* P. Rossi; *Fig. 41.20* Stanley Schoenberger / GH; *Fig. 41.21* M. Senkarik; *Fig. 41.23* Lori Grinker / Contact; *Fig. 41.24 (both)* F. R. Turner, Indiana University / BPS; *Fig. 41.25* P. Rossi; *Fig. 41.26* and *Fig. 41.27* M. Senkarik; *Box p. 853* Dr. Edward Kollar / Univ. of Connecticut Health Center; *Fig. 41.28* and *Fig. 41.29* M. Senkarik; *Box p. 856* Camera M.D. Studios; *Fig. 41.30 (top)* K. G. Murti / Boston Univ.; *Fig. 41.30 (bottom)* P. Rossi; *Fig. 41.31* P. Rossi; *Box p. 858* © 88 J. Gilman / Custom Medical Stock; *Box p. 860* Greg Vaughn / TSA.

Chapter 42
Fig. 42.1 (left) Alan Reiniger / Contact Press Images; *Fig. 42.1 (top)* Mary Ann Fittipaldi; *Fig. 42.1 (bottom)* © Mary Ann Fittipaldi; *Fig. 42.1 (right)* Georg Gerster / Comstock; *Fig. 42.2* Beinecke Library, Yale Univ.; *Fig. 42.3* Grapes / Michaud / PR; *Fig. 42.4* and *Fig. 42.5* Mikki Senkarik; *Fig. 42.6 (photo)* Dr. Amit and Dr. Poljack, Pasteur Institute Paris; *Fig. 42.6 (art)* Illustrious, Inc.; *Fig. 42.7* M. Senkarik; *Fig. 42.8* and *Fig. 42.9* Illustrious, Inc.; *Box pp. 874–875* P. Rossi; *Fig. 42.10* P. Rossi; *Fig. 42.11* Illustrious, Inc.; *Fig. 42.12* © Boehringer Ingelheim International, GMBH, Lennart Nilsson, photographer; *Box p. 884* P. Rossi; *Fig. 42.13 (all)* Robert Gallo; *Fig. 42.14 (left)* George Kelvin, *Scientific American*; *Fig. 42.14 (right)* Lyrl Ahern; *Fig. 42.15* and *Fig. 42.16* P. Rossi; *Box p. 890* Billy F. Barnes / Photo Edit.

Chapter 43
Fig. 43.1 through *Fig. 43.3* Mikki Senkarik; *Fig. 43.4* through *Fig. 43.8* H&C; *Fig. 43.9* through *Fig. 43.11* Patrice Rossi; *Fig. 43.12* and *Fig. 43.13* H&C; *Box p. 910* Ciba-Geigy Corporation; *Fig. 43.14* and *Fig. 43.15* P. Rossi; *Fig. 43.16* Marsha Goldberg; *Fig.*

43.17 and *Fig. 43.18* M. Senkarik; *Fig. 43.19* and *Fig. 43.20* P. Rossi; *Fig. 43.22* H&C; *Fig. 43.23* Ohio–Nuclear Corporation, Science Photo Library / PR; *Fig. 43.24 (left)* © Roger Ressmeyer 1988; *Fig. 43.24 (right)* UCLA Medical School.

Chapter 44
Fig. 44.1 (photo) A. J. Deane / BC; *Fig. 44.1 (art)* H&C; *Fig. 44.2* Mikki Senkarik; *Fig. 44.3* Patrice Rossi; *Fig. 44.4* Reproduced from Barlow, H. B. and Mollon, J. D. (1982) *The Senses,* Cambridge Univ. Press; *Fig. 44.5* P. Rossi; *Fig. 44.6 (top)* Richard H. Gross / Biological Photography; *Fig. 44.6 (center)* Lynn Funkhouser / PA; *Fig. 44.6 (bottom)* © 1979 Darwin Dale / PR; *Fig. 44.6 (art)* Marlene DenHouter; *Fig. 44.7* Illustrious, Inc.; *Fig. 44.8* M. Senkarik; *Fig. 44.10* H&C; *Fig. 44.12* M. DenHouter; *Fig. 44.13* H&C; *Fig. 44.15* H&C; *Fig. 44.16* through *Fig. 44.18* M. Senkarik; *Fig. 44.19* H&C; *Fig. 44.20* M. DenHouter; *Fig. 44.21* M. Senkarik; *Fig. 44.22* Frithfoto / BC; *Fig. 44.23* through *Fig. 44.25* M. Senkarik; *Fig.*

44.26 M. DenHouter; *Fig. 44.27* J. Cancalosi / PA; *Fig. 44.28* P. Rossi.

Chapter 45
Fig. 45.1 and *Fig. 45.2* Marlene DenHouter; *Fig. 45.3* Mikki Senkarik; *Fig. 45.4 (top photo)* Ed Reschke / PA; *Fig. 45.4 (bottom photo)* Biology Media / PR; *Fig. 45.4 (art)* Patrice Rossi; *Fig. 45.5* and *Fig. 45.6* M. Senkarik; *Fig. 45.7* Narco Bio-Systems; *Fig. 45.8* M. Senkarik; *Fig. 45.9* through *Fig. 45.12* P. Rossi; *Box p. 958* R. Leatherwood / FPG; *Fig. 45.13* M. Senkarik; *Fig. 45.14* Biophoto Associates / PR; *Fig. 45.15* M. Senkarik; *Fig. 45.16* © 1986 Scott Camazine / PR; *Fig. 45.17* M. Senkarik; *Fig. 45.18* P. Rossi; *Fig. 45.19* M. Senkarik.

Chapter 46
Fig. 46.1 and *Fig. 46.2* Patrice Rossi; *Fig. 46.3* David Austin / Stock Boston; *Fig. 46.4* P. Rossi; *Fig. 46.5* Hugo van Lawick / Nature Photographers; *Fig. 46.6* Nina Leen © 1964 Time, Inc.; *Fig. 46.7* P. Rossi;

Fig. 46.8 (left) P. Rossi; *Fig. 46.8 (right)* Jeff March / PN; *Fig. 46.9 (top)* K. G. Preston-Mafham / AA; *Fig. 46.9 (bottom)* P. Rossi; *Fig. 46.10* Stephen G. Mora / PN; *Fig. 46.11 (top)* Michael Fogden / AA; *Fig. 46.11 (bottom)* Carl Roessler / BC; *Fig. 46.12* P. Rossi; *Fig. 46.13* OFS / AA; *Fig. 46.14* P. Rossi; *Fig. 46.15 (left)* Steven C. Kaufman / PA; *Fig. 46.15 (right)* Charles G. Summer, Jr. / TSA; *Fig. 46.16 (left)* Dwight Kuhn; *Fig. 46.16 (right)* Y. Arthus-Bertrand / PA; *Fig. 46.17* P. Rossi; *Fig. 46.18* Wrangham / AnthroPhoto; *Fig. 46.19* Edward Hodgson / PN; *Fig. 46.20* Priscilla Connell / PN; *Fig. 46.21* P. Rossi; *Fig. 46.22 (left)* Mervin w. Larson / BC; *Fig. 46.22 (right)* Donald Specker / AA; *Fig. 46.23* David M. Dennis / TSA; *Fig. 46.24* P. Rossi; *Fig. 46.25* AnthroPhoto; *Fig. 46.26 (left)* David C. Fritts / AA; *Fig. 46.26 (right)* Duncan Anderson & Rachel Wilder / AA.

Appendix
p. A9 Illustrious, Inc.

INDEX

NOTE: Boldfaced page numbers indicate terms defined in the text, and italicized page numbers indicate material in tables, figures, and illustrations.